MECHANICAL DESIGN OF
MACHINE COMPONENTS

SECOND EDITION

MECHANICAL DESIGN of Machine Components

SECOND EDITION

ANSEL C. UGURAL

with Contributors

Youngjin Chung

Errol A. Ugural

CRC Press
Taylor & Francis Group
Boca Raton London New York

CRC Press is an imprint of the
Taylor & Francis Group, an **informa** business

CRC Press
Taylor & Francis Group
6000 Broken Sound Parkway NW, Suite 300
Boca Raton, FL 33487-2742

Printed on acid-free paper
Version Date: 20150325

International Standard Book Number-13: 978-1-4398-8780-6 (Hardback)

To my father,

without whose encouragement this would not have been possible.

Contents

Section II Failure Prevention

Section III Applications

Preface

Introduction

This book developed from classroom notes prepared in connection with junior–senior undergraduate and first-year graduate courses in mechanical design, machine design, mechanical engineering design, and engineering design and analysis. The scope of this book is wider than any other book on the subject. In addition to its applicability to *mechanical engineering,* and to some extent, aerospace, agricultural, and nuclear engineering and applied engineering mechanics curricula, I have endeavored to make the book useful to *practicing engineers* as well. The book offers a simple, comprehensive, and methodical presentation of the fundamental concepts and principles in the design and analysis of machine components and basic structural members. The coverage presumes knowledge of the mechanics of materials and material properties. However, topics that are particularly significant to understanding the subject are reviewed as they are taken up. Special effort has been made to present a book that is as self-explanatory as possible, thereby reducing the work of the instructor.

The presentation of the material in this book strikes a balance between the theory necessary to gain insight into mechanics and the design methods. I, therefore, attempt to stress those aspects of theory and application that prepare a student for more advanced study or professional practice in design. Above all, I have made an effort to provide a visual interpretation of equations and present the material in a form useful to a diverse audience. The analysis presented should facilitate the use of computers and programmable calculators. The commonality of the analytical methods needed to design a wide variety of elements and the use of computer-aided engineering as an approach to design are emphasized.

Mechanical Design of Machine Components provides unlimited opportunities for the use of computer graphics. Computer solutions are usually preferred because the evaluation of design changes and "what-if" analyses require only a few keystrokes. Hence, many examples, case studies, and problems in the book are discussed with the aid of a computer. Generally, solid modeling serves as a design tool that can be used to create finite element (*FE*) models for analysis and dynamic simulation. Instructors may use a simple PC-based *FE* program to give students exposure to the method applied to stress concentration and axisymmetrically loaded and plane stress problems. The website for the book (see Optional Media Supplements) allows the user to treat problems more realistically and demonstrates the elements of good computational practice. The book is *independent* of any software package.

Traditional analysis in design, based on the methods of mechanics of materials, is given full treatment. In some instances, the methods of the applied theory of elasticity are employed. The role of the theory of elasticity in this book is threefold: It places limitations on the application of the mechanics of materials theory, it is used as the basis of FE formulation, and it provides exact solutions when configurations of loading and component shape are simple. Plates, shells, and structural members are discussed to enable the reader to solve real-life problems and understand interactive case studies. Website addresses of component and equipment manufacturers and open-ended web problems are given in many chapters to provide the reader access to additional information on those topics. Also presented is finite element analysis (FEA) in computer-aided design. The foregoing unified methods of analysis give the reader opportunity to expand his or her ability to perform the design process in a more realistic setting. The book attempts to fill what I believe to be a void in the world of textbooks on machine design.

The book is divided into three sections. The basics of loading, stress, strain, materials, deflection, stiffness, and stability are treated first. Then fracture mechanics, failure criteria, fatigue phenomena, and surface damage of components are dealt with. These are followed by applications to machine and miscellaneous mechanical and structural components. All the sections attempt to provide an integrated approach that links together a variety of topics by means of case studies. Some chapters and sections in the book are also carefully integrated through cross-referencing. Throughout the book, most case studies provide numerous component projects. They present different aspects of the same design or analysis problem in successive chapters. Case studies in the preliminary design of two machines are taken up in the last chapter.

Attention is given to the presentation of the fundamentals and necessary empirical information required to formulate design problems. Important principles and applications are illustrated with numerical examples, and a broad range of practical problems are provided to be solved by students. This book offers numerous worked-out examples and case studies, aspects of which are presented in several sections of the book; many problem sets, most of which are drawn from engineering practice; and a multitude of formulas and tabulations from which design calculations can be made. Most problems can be readily modified for in-class tests. Answers to selected problems and References (identified in *brackets*) are given at the end of the book.

A sign convention consistent with vector mechanics is used throughout for loads, internal forces (with the exception of the shear in beams), and stresses. This convention has been carefully chosen to conform to that used in most classical mechanics of materials, elasticity, and engineering design texts as well as to that most often employed in the numerical analysis of complex machines and structures. Both the international system of units (SI) and the US customary system of units are used, but since in practice the former is replacing the latter, this book places a greater emphasis on SI units.

Text Arrangement

A glance at the table of contents shows the topics covered and the way in which they are organized. Because of the extensive subdivision into a variety of topics and the use of alternative design and analysis methods, the book should provide flexibility in the choice of assignments to cover courses of varying length and content. A discussion of the design process and an overview of the material included in the book are given in Sections 1.1 through 1.4. Most chapters are substantially self-contained. Hence, the order of presentation can be smoothly altered to meet an instructor's preference. It is suggested, however, that Chapters 1 and 2 be studied first. The sections and chapters marked with an asterisk (*) deal with special or advanced topics. These are optional for a basic course in design and can be skipped without disturbing the continuity of the book.

This book attempts to provide synthesis and analysis that cut through the clutter and save the reader's time. Every effort has been made to eliminate errors. I hope I have maintained a clarity of presentation, as much simplicity as the subject permits, unpretentious depth, an effort to encourage intuitive understanding, and a shunning of the irrelevant. In this context, emphasis is placed on the use of fundamentals to build students' understanding and ability to solve more complex problems throughout.

Features

The following overview highlights key features of this innovative machine design book.

Large Variety of Interesting and Engaging Worked Examples and Homework Problems

Providing fresh, practically based problem content, the text offers *731 homework problems, 185 worked examples, and 11 case studies.*

Consistent Problem-Solving Approach

To provide students a consistent framework for organizing their work, worked examples and case studies use a standard problem-solving format:

1. Problem statement (given)
2. Find
3. Assumptions
4. Solution
5. Comments

Unique Case Studies

Eleven text cases provide additional applications of the use of design processes. Two major case studies—the *crane with winch study* and the *high-speed cutting machine study*—concern system design allowing students to see how the stress and displacement of any one member may be invariably affected by the related parts. These also add to the skill sets they need as practicing engineers. The cases are interesting and relevant with special emphasis on industry uses, material selection, safety considerations, and cost factors.

Three Aspects of Solid Mechanics Emphasized

Equilibrium, material behavior, and geometry of deformation. The book reinforces the importance of these *basic principles of analysis.*

Strong Visual Approach

The book includes about 600 figures and 35 photographs, many with multiple parts, to aid students' comprehension of the concepts. All regular figures include explanatory captions.

Introduction

The author provides solid pedagogical tools and objectives for each chapter, including an excellent summary at the beginning.

Additional Features

Free-body diagrams, review of key stress analysis concepts, material properties and applications, rational design procedure, role of analysis in design, FEA in design.

This Edition Promise

Text Accuracy

The author, a proof reader, and a production editor checked all final pages for accuracy.

Solution Accuracy

Fully worked-out solutions written and class-tested by the author. An accuracy checker independently checked all final solutions.

Reliability

Over the last three decades, Ansel Ugural has written best-selling books on advanced mechanics of materials, elasticity, mechanics of materials, beams, plates and shells, and mechanical design.

Time-Saving Support Material

Available on the companion site at http://www.physicalpropertiesofmaterials.com/book/?isbn=9781439866511.

Meeting ABET Criteria

This book addresses the following ABET criteria:

1. An ability to apply knowledge of mathematics, science, and engineering
2. An ability to design and conduct experiments, as well as to analyze and interpret data
3. An ability to design a system, a component, or a process to meet desired needs within realistic constraints such as economic, environmental, social, political, ethical, health and safety, manufacturability, and sustainability
4. An ability to identify, formulate, and solve engineering problems
5. An understanding of professional and ethical responsibilities
6. An ability to use the techniques, skills, and modern engineering tools necessary for engineering practice

Optional Media Supplements

Supplements: A Solutions Manual is available upon request to instructors in printed form. Written and class-tested by the author, it features complete solutions to problems in the text. Additional material is available from the CRC Web site: http://www.crcpress.com/product/isbn/9781439887806. This includes solutions using MATLAB® for a variety of examples and case studies of practical importance presented in the text.

MATLAB® is a registered trademark of The MathWorks, Inc. For product information, please contact:

The MathWorks, Inc.
3 Apple Hill Drive
Natick, MA 01760-2098 USA
Tel: 508-647-7000
Fax: 508-647-7001
E-mail: info@mathworks.com
Web: www.mathworks.com

Acknowledgments

To acknowledge everyone who contributed to this book in some manner is clearly impossible. However, the author is especially thankful to the following reviewers, who offered constructive suggestions and made detailed comments: D. Beale, Auburn University; D. M. McStravick, Rice University; T. R. Grimm, Michigan Technological University; R. E. Dippery, Kettering University; Yuen-Cjen Yong, California Polytechnic University–San Luis Obispo; A. Shih, North Carolina State University; J. D. Gibson, Rose-Hulman Institute of Technology; R. Paasch, Oregon State University; J. P. H. Steele, Colorado School of Mines; C. Nuckolls, the University of Central Florida; D. Logan, the University of Wisconsin–Platteville; E. Conley, New Mexico State University; L. Dabaghian, California State University–Sacramento; E. R. Mijares, California State University–Long Beach; T. Kozik, Texas A&M University; C. Crane, the University of Florida; B. Bahram, North Carolina State University; A. Mishra, Auburn University; S. Yurgartis, Clarkson University; M. Corley, Louisiana Tech. University; R. Rowlands, the University of Wisconsin; B. Hyman, the University of Washington; McDonald, G.H., the University of Tennessee; J.D. Leland, the University of Nevada; O. Safadi, the University of Southern California; Zhu, Y., North Carolina State University; Sepahpour, B., Trenton College of New Jersey; S.G. Hall, Louisiana State University; A. Almqvist, Lulea University of Technology, Sweden; J. Svenninggaard, VIA University College, Horshens, Denmark; G.S. Tarrant, University of Montana; J.A. Hanks, University of Nebraska, Lincoln; T.D. Coburn, California State Polytechnic University, Pomona; A. Bazar, CSU Fullerton; and G. R. Pennock, Purdue University. P. Brackin, Rose-Hulman Institute of Technology, checked the accuracy of the text. R. Sodhi, New Jersey Institute of Technology, offered valuable perspectives on some case studies based on student design projects. I am pleased to express my gratitude to all of these colleagues for their invaluable advice.

Executive editor Jonathan Plant made extraordinary efforts in developing this edition of the book. Also, production was managed and handled efficiently by the staff of CRC Press and SPi Global–Content Solutions. I thank them for their professional help. Accuracy checking of the problems, typing of solutions manual, proof reading, and solutions of MATLAB® problems on the website were done by my former student, Dr. Youngjin Chung. In addition, contributing considerably to this volume with computer work, typing new inserts, assisting with some figures, and cover design was Errol A. Ugural. Their work is much appreciated. Last, I deeply appreciate the encouragement of my wife, Nora, daughter, Aileen, and son, Errol, during the preparation of the text.

Ansel C. Ugural
Holmdel, New Jersey

Author

Ansel C. Ugural is a visiting and research professor of mechanical engineering at the New Jersey Institute of Technology, Newark, New Jersey. He was a National Science Foundation fellow and has taught at the University of Wisconsin–Madison, Madison, Wisconsin. Dr. Ugural held positions at Fairleigh Dickinson University, where he served for two decades as a professor and chairman of the mechanical engineering department. He has considerable and diverse industrial experience in both full-time and consulting capacities as a design, development, and research engineer.

Dr. Ugural earned his MS in mechanical engineering and PhD in engineering mechanics from the University of Wisconsin–Madison. He has been a member of the American Society of Mechanical Engineers and the American Society of Engineering Education. He is also listed in *Who's Who in Engineering.*

Dr. Ugural is the author of several books, including *Mechanical Design: An Integrated Approach* (McGraw-Hill, 2004); *Stresses in Plates and Shells* (McGraw-Hill, 1999); *Stresses in Beams, Plates, and Shells* (CRC Press, 3rd ed., 2010); *Mechanics of Materials* (McGraw-Hill, 1990); and *Mechanics of Materials* (Wiley, 2008). These books have been translated into Korean, Chinese, and Portuguese. Dr. Ugural is also the coauthor (with S.K. Fenster) of *Advanced Mechanics of Materials and Applied Elasticity* (Prentice Hall, 5th ed., 2012). In addition, he has published numerous articles in trade and professional journals.

Symbols

See Sections 11.2, 11.4, 11.9, 11.11, 12.3, 12.5, 12.6, 12.8, and 12.9 for some gearing symbols.

Roman Letters

A	Amplitude ratio, area, coefficient, cross-sectional area
A_e	Effective area of clamped parts, projected area
A_f	Final cross-sectional area
A_o	Original cross-sectional area
A_t	Tensile stress area, tensile stress area of the thread
a	Acceleration, crack depth, distance, radius, radius of the contact area of two spheres
B	Coefficient
b	Distance, width of beam, band, or belt; radius
C	Basic dynamic load rating, bolted-joint constant, centroid, constant, heat transfer coefficient, specific heat, spring index
C_c	Limiting value of column slenderness ratio
C_f	Surface finish factor
C_r	Reliability factor, contact ratio
C_s	Basic static load rating, size factor
c	Distance from neutral axis to the extreme fiber, radial clearance, center distance
D	Diameter, mean coil diameter, plate flexural rigidity $[Et^3/12(1-\nu^2)]$
d	Diameter, distance, pitch diameter, wire diameter
d_{avg}	Average diameter
d_c	Collar (or bearing) diameter
d_m	Mean diameter
d_p	Pitch diameter
d_r	Root diameter
E	Modulus of elasticity
E_b	Modulus of elasticity for the bolt
E_k	Kinetic energy
E_p	Modulus of elasticity for clamped parts, potential energy
e	Dilatation, distance, eccentricity, efficiency
F	Force, tension
F_a	Axial force, actuating force
F_b	Bolt axial force
F_c	Centrifugal force
F_d	Dynamic load
F_i	Initial tensile force or preload
F_n	Normal force
F_p	Clamping force for the parts, proof load
F_r	Radial force
F_t	Tangential force
F_u	Ultimate force
f	Coefficient of friction, frequency
f_c	Collar (or bearing) coefficient of friction

f_n Natural frequency
G Modulus of rigidity
g Acceleration due to gravity
H Time rate of heat dissipation, power
H_B Brinell hardness number (Bhn)
H_V Vickers hardness number
h Cone height, distance, section depth, height of fall, weld size, film thickness
h_f Final length, free length
h_0 Minimum film thickness
h_s Solid height
I Moment of inertia
I_e Equivalent moment of inertia of the spring coil
J Polar moment of inertia, factor
K Bulk modulus of elasticity, constant, impact factor, stress intensity factor, system stiffness
K_c Fracture toughness
K_f Fatigue stress concentration factor
K_r Life adjustment factor
K_s Service factor, shock factor, direct shear factor for the helical spring
K_t Theoretical or geometric stress concentration factor
K_w Wahl factor
k Buckling load factor for the plate, constant, element stiffness, spring index or stiffness
k_b Stiffness for the bolt
k_p Stiffness for the clamped parts
L Grip, length, lead
L_e Equivalent length of the column
L_f Final length
L_0 Original length
L_5 Rating life for reliability greater than 90%
L_{10} Rating life
l Direction cosine, length
M Moment
M_a Alternating moment
M_f Moment of friction forces
M_m Mean moment
M_n Moment of normal forces
m Direction cosine, mass, module, mass
N Normal force, number of friction planes, number of teeth, fatigue life or cycles to failure
N_a Number of active spring coils
N_{cr} Critical load of the plate
N_t Total number of spring coils
N_θ Hoop force
N_ϕ Meridional force
n Constant, direction cosine, factor of safety, modular ratio, number, number of threads, rotational speed
n_{cr} Critical rotational speed
P Force, concentrated load, axial load, equivalent radial load for a roller bearing, radial load per unit projected area

P_a	Alternating load
P_{all}	Allowable load
P_{cr}	Critical load of the column or helical spring
P_m	Mean load
p	Pitch, pressure, probability
p_{all}	Allowable pressure
p_i	Internal pressure
p_{max}	Maximum pressure
p_{min}	Minimum pressure
p_o	Outside or external pressure
p_0	Maximum contact pressure
$p(x)$	Probability or frequency function
Q	First moment of area, imaginary force, volume, flow rate
Q_S	Side leakage rate
q	Notch sensitivity factor, shear flow
R	Radius, reaction force, reliability, stress ratio
R_B	Rockwell hardness in B scale
R_C	Rockwell hardness in C scale
r	Aspect ratio of the plate, radial distance, radius, radius of gyration
r_{avg}	Average radius
r_i	Inner radius
r_o	Outer radius
S	Section modulus, Saybolt viscometer measurement in seconds, Sommerfeld number, strength
S_e	Endurance limit of mechanical part
S_e'	Endurance limit of specimen
S_{es}	Endurance limit in shear
S_f	Fracture strength
S_n	Endurance strength of mechanical part
S_n'	Endurance strength of specimen
S_p	Proof strength, proportional limit strength
S_u	Ultimate strength in tension
S_{uc}	Ultimate strength in compression
S_{us}	Ultimate strength in shear
S_y	Yield strength in tension
S_{ys}	Yield strength in shear
s	Distance, sample standard deviation
T	Temperature, tension, torque
T_a	Alternating torque
T_d	Torque to lower the load
T_f	Friction torque
T_m	Mean torque
T_o	Torque of overhauling
T_t	Transition temperature
T_u	Torque to lift the load
t	Temperature, distance, thickness, time
t_a	Temperature of surrounding air
t_o	Average oil film temperature
U	Strain energy, journal surface velocity

U_o Strain energy density
U_{od} Distortional strain energy density
U_{ov} Dilatational strain energy density
U_r Modulus of resilience
U_t Modulus of toughness
U^* Complementary energy
U_o^* Complementary energy density
u Radial displacement, fluid flow velocity
V Linear velocity, a rotational factor, shear force, volume
V_s Sliding velocity
υ Displacement, linear velocity
W Work, load, weight
w Distance, unit load, deflection, displacement
X A radial factor
Y Lewis form factor based on diametral pitch or module, a thrust factor
y Distance from the neutral axis, Lewis form factor based on circular pitch, quantity
$\bar{y}$ Distance locating the neutral axis
z Number of standard deviations

Greek Letters

α Angle, angular acceleration, coefficient, coefficient of thermal expansion, cone angle, form factor for shear, thread angle
α_n Thread angle measured in the normal plane
β Angle, coefficient, half-included angle of the V belt
γ Included angle of the disk clutch or brake, pitch angle of the sprocket, shear strain, weight per unit volume; γ_{xy}, γ_{yz}, and γ_{xz} are shear strains in the xy, yz, and xz planes
γ_{max} Maximum shear strain
Δ Gap, material parameter in computing contact stress
δ Deflection, displacement, elongation, radial interference or shrinking allowance, a virtual infinitesimally small quantity
δ_{max} Maximum or dynamic deflection
δ_s Solid deflection
δ_{st} Static deflection
δ_w Working deflection
ϵ Eccentricity ratio
ε Normal strain; ε_x, ε_y, and ε_z are normal strains in the x, y, and z directions
ε_f Normal strain at fracture
ε_t True normal strain
ε_u Ultimate strain
η Absolute viscosity or viscosity
θ Angle, angular displacement, slope
θ_p Angle to a principal plane or to a principal axis
θ_s Angle to a plane of maximum shear
λ Lead angle, helix angle, material constant
μ Population mean
ν Kinematic viscosity, Poisson's ratio
ρ Mass density

σ Normal stress; σ_x, σ_y, and σ_z are normal stresses in the *x*, *y*, and *z* planes, standard deviation
σ_a Alternating stress
σ_{all} Allowable stress
σ_{cr} Critical stress
σ_e Equivalent stress
σ_{ea} Equivalent alternating stress
σ_{em} Equivalent mean stress
$\sigma_{\max}$ Maximum normal stress
$\sigma_{\min}$ Minimum normal stress
σ_{nom} Nominal stress
σ_{oct} Octahedral normal stress
σ_{res} Residual stress
τ Shear stress; τ_{xy}, τ_{yz}, and τ_{xz} are shear stresses perpendicular to the *x*, *y*, and *z* axes and parallel to the *y*, *z*, and *x* axes; direct shear stress; torsional shear stress
τ_{avg} Average shear stress
τ_{all} Allowable shear stress
τ_{oct} Octahedral shear stress
$\tau_{\max}$ Maximum shear stress
$\tau_{\min}$ Minimum shear stress
τ_{nom} Nominal shear stress
ϕ Angle, angle giving the position of minimum film thickness, pressure angle, angle of twist, angle of wrap
$\phi_{\max}$ Position of maximum film pressure
ψ Helix angle, spiral angle
ω Angular velocity, angular frequency ($\omega = 2\pi f$)
ω_n Natural angular frequency

Abbreviations

all	Allowable
avg	Average
Bhn	Brinell hardness number
CCW	Counterclockwise
CD	Cold drawn
cr	Critical
CW	Clockwise
fpm	Foot per minute
ft	Foot, feet
h	Hour
HD	Hard drawn
hp	Horsepower
HT	Heat treated
Hz	Hertz (cycles per second)
ID	Inside diameter
in.	Inch, inches
ipm	Inch per minute
ips	Inch per second
J	Joule
kg	Kilogram(s)
kip	Kilopound (1000 lb)
kips	Kilopounds
ksi	Kips per square inch (10^3 psi)
kW	Kilowatt
lb	Pound(s)
ln	Naperian natural logarithm
log	Common logarithm (base 10)
m	Meter
max	Maximum
min	Minimum
mph	Miles per hour
m/s	Meter per second
N	Newton
NA	Neutral axis
OD	Outside diameter
OQ&T	Oil quenched and tempered
OT	Oil tempered
Pa	Pascal
psi	Pounds per square inch
Q&T	Quenched and tempered
R_C	Rockwell hardness, C scale
rad	Radian
req	Required
res	Residual

rpm	Revolutions per minute
rps	Revolutions per second
s	Second
SI	System of international units
st	Static
SUS	Saybolt universal seconds
SUV	Saybolt universal viscosity
VI	Viscosity index
W	Watt
WQ&T	Water quenched and tempered

Section I

Basics

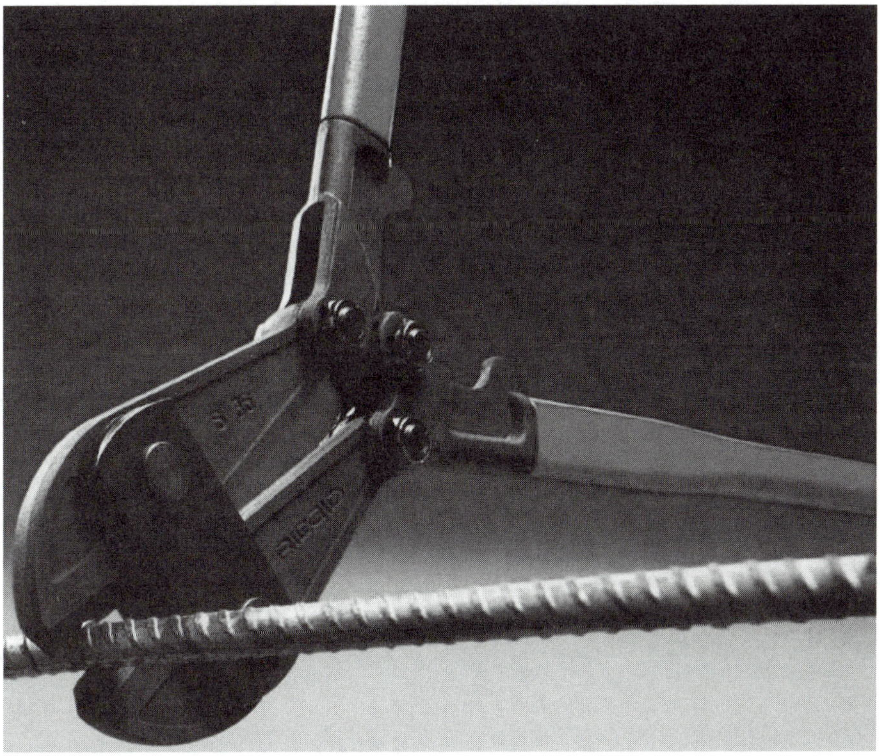

A bolt cutter suited for professional users (www.ridgit.com). We will examine such a tool in Case Studies 1.1, 3.1 and 4.1. Section I is devoted to the analysis of load, material properties, stress, strain, deflection, and elastic stability of variously loaded machine and structural components.

1

Introduction

1.1 Scope of the Book

As an applied science, engineering uses scientific knowledge to achieve a specific objective. The mechanism by which a requirement is converted to a meaningful and functional plan is called a *design*. The design is an innovative, iterative, and decision-making process. This book deals with the analysis and design of *machine elements or components* and basic *structural members* that compose the system or assembly. Typical truss, frame, plate, and shell-like structures also are considered. The purpose and scope of this text may be summarized as follows: it presents a body of knowledge that will be useful in component design for performance, strength, and durability; provides treatments of *design to meet strength requirements* of members and other aspects of design involving prediction of the displacements and buckling of a given component under prescribed loading; presents classical and numerical methods amenable to electronic digital computers for the analysis and design of members and structural assemblies; and presents many examples, case studies, and problems of various types to provide an opportunity for the reader to develop competence and confidence in applying the available design formulas and deriving new equations as required.

The text consists of three sections. Section I focuses on fundamental principles and methods, a synthesis of stress analysis, and materials engineering, which forms the cornerstone of the subject and has to be studied carefully. We begin with a discussion of basic concepts in design and analysis and definitions relating to properties of a variety of engineering materials. Detailed equilibrium and energy methods of analysis for determining stresses and deformations in variously loaded members, design of bars and beams, buckling, failure criteria, and reliability are presented in Section II. A thorough grasp of these topics will prove of great value in attacking new and complex problems. Section III is devoted mostly to machine component design. The fundamentals are applied to specific elements such as shafts, bearings, gears, belts, chains, clutches, brakes, and springs and typical design situations that arise in the selection and application of these members and others. Power screws; threaded fasteners; bolted, riveted, and welded connections; adhesive bonding; and axisymmetrically loaded components are also considered in some detail. In conclusion, introductory finite element analysis (FEA) and case studies in design are covered.

The full understanding of both terminology in statics and principles of mechanics is an essential prerequisite to the analysis and design of machines and structures. Design methods for members are founded on the methods of mechanics of materials; and the theory of applied elasticity is used or referred to in design of certain elements. The objective of this chapter is to provide the reader the basic definitions and process of the design, load analysis, and the concepts of solid mechanics in a condensed form. Selected references provide readily available sources where additional analysis and design information can be obtained.

1.2 Mechanical Engineering Design

Design is the formulation of a plan to satisfy a particular need, real or imaginary. Fundamentally, design represents the process of problem solving. *Engineering design* can be defined as the process of applying science and engineering methods to prescribe a component or a system in sufficient detail to permit its realization. A system constitutes several different elements arranged to work together as a whole. Design is thus the essence, art, and intent of engineering. *Design function* refers to the process in which mathematics, computers, and graphics are used to produce a plan. Engineers with more scientific insight are able to devise better solutions to practical problems. Interestingly, there is a similarity between the engineer and the physician. Although they are not scientists, both use scientific evidence complimented by empirical data and professional judgment—in dealing with demanding problems.

Mechanical design means the design of components and systems of a mechanical nature—machines, structures, devices, and instruments. For the most part, mechanical design utilizes the stress analysis methods and materials engineering and energy concepts. That is, it applies to design of mechanical systems or components where structures, motion, and energy or heat transfer can be involved. A *machine* is an apparatus consisting of interrelated elements or a device that modifies force motion or energy (see Section 1.9). *Machine design* is the art of planning or devising new or improved machines to accomplish a specific purpose. The field of machine design is a subset of mechanical design in which focus is on the structures and motion *only*.

Mechanical engineering design deals with the conception, design, development, and application of machines and mechanical apparatus of all types. It involves *all* the disciplines of mechanical engineering. Although *structural design* is most directly associated with civil engineering, it interacts with any engineering field that requires a structural system or member. As noted earlier, the topic of machine design is the main focus of this text.

The ultimate goal in a mechanical design process is to size and shape the elements and choose appropriate materials and manufacturing processes so that the resulting system can be expected to perform its intended function without failure. An *optimum design* is the best solution to a design problem within prescribed constraints. Of course, such a design depends on a seemingly limitless number of variables. When faced with many possible choices, a designer may make various design decisions based on experience, reducing the problem to that with one or few variables.

Generally, it is assumed that a good design meets performance, safety, reliability, aesthetics, and cost goals. Another attribute of a good design is robustness, a resistance to quality loss, or deviation from desired performance. Knowledge from the entire engineering curricula goes into formulating a good design. Communications is as significant as technology. Basically, the means of communication are in written, oral, and graphical forms. The first fundamental canon in the *Code of Ethics for Engineers* [1] states that "Engineers shall hold paramount the safety, health, and welfare of the public in the performance of their professional duties." Therefore, engineers must design products that are safe during their intended use for the life of the products. Product safety implies that the product will protect humans from injury, prevent property damage, and prevent harm to the environment.

A plan for satisfying a need often includes preparation of individual preliminary design. A *preliminary design*, or sometimes also referred to as conceptual design, mainly concerns with analysis, synthesis, evaluation, and comparison of proposed machine components or machines. Each preliminary design involves a thorough consideration of the loads and actions that the structure or machine has to support. For each case, a mechanical analysis is necessary. *Design decisions*, or choosing reasonable values of the factors, is important in the design process. As a designer gains more experience, decisions are reached more readily. Both individual talent and creativeness are needed in engineering design.

1.2.1 ABET Definition of Design

The Accreditation Board for Engineering and Technology (ABET) defines engineering design as the process of devising a system, component, or process to meet desired needs. It is a decision-making process (often iterative), in which the basic science and mathematics and engineering sciences are applied to convert resources optimally to meet a stated objective. Among the fundamental elements of the design process are the establishment of objectives and criteria, synthesis, analysis, construction, testing, and evaluation.

The engineering design component of a curriculum must include most of the following features: development of student creativity, use of open-ended problems, development and use of modern design theory and methodology, formulation of design problem statements and specification, consideration of alternative solutions, feasibility considerations, production processes, concurrent engineering design, and detailed system description. Further, it is essential to include a variety of realistic constraints, such as economic factors, safety, reliability, aesthetics, ethics, and social impact. The ABET criteria (see page xxiv) for accreditation emphasizes the use of teams in solving problems and performing designs.

1.3 Design Process

The *process* of *design* is basically an exercise in creativity. The complete process may be outlined by design flow diagrams with feedback loops [2–6]. Figure 1.1 shows some aspects of such a diagram. In this section, we discuss the *phases of design* common to all disciplines in the field of engineering design. Most engineering designs involve safety, ecological, and

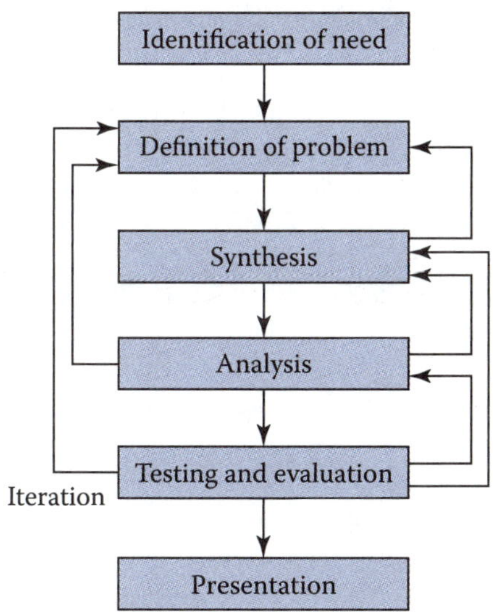

FIGURE 1.1
Design process.

societal considerations. It is a challenge to the engineer to recognize all of these in proper proportion. Fundamental actions proposed for the design process are establishing a need as a design problem to be solved, understanding the problem, generating and evaluating possible solutions, and deciding on the best solution.

1.3.1 Phases of Design

The design process is independent of the product and is based on the concept of product life cycle. The content of each engineering design problem is unique, but the methodology for solving these problems is universal and can be described in a specific way. To understand fully all that must be considered in the process of design, here, we explain the characteristics of each phase of Figure 1.1. The process is neither exhaustive nor rigid and will probably be modified to suit individual problems. Number of authorities on the methodology of design has presented similar descriptions of the process.

1.3.1.1 Identification of Need

The design process begins with a recognition of a *need*, real or imagined, and a *decision* to do something about it. For example, present equipment may require improving durability, efficiency, weight, speed, or cost. New equipment may be needed to perform an automated function, such as computation, assembly, or servicing. The identification aspect of design can have origin in any number of sources. Customer reports on the product function and quality may force a redesign. Business and industrial competition constantly force the need for new or improved apparatus, processes, and machinery designs. Numerous other sources of needs give rise to contemporary design problems.

1.3.1.2 Definition of the Problem

This phase in design conceives the mechanisms and arrangements that will perform the needed function. For this, a broad knowledge of members is desirable, because new equipment ordinarily consists of new members, perhaps with changes in size and material. *Specification* is a form of input and output quantities. A number of decisions must be made to establish the *specification set*, which is a collection of drawings, text, bills of materials, and detailed directions. All specifications must be carefully spelled out. Often, this area is also labeled *design and performance requirements*. The specifications also include the definitions of the member to be manufactured, the cost, the range of the operating temperature, expected life, and the reliability.

A *standard* is a set of specifications for parts, materials, or processes intended to achieve uniformity, efficiency, and a specified quality. A *code* is a set of specifications for the analysis, design, manufacture, and construction of something. The purpose of a code is to achieve a specified degree of safety, efficiency, and performance or quality. All organizations and technical societies (listed in Section 1.6) have established specifications for standards and safety or design codes.

Once the specifications have been prepared, relevant design information is collected to make a *feasibility study*. The purpose of this study is to verify the possible success or failure of a proposal both from the technical and economic standpoints. Frequently, as a result of this study, changes are made in the specifications and requirements of the project. The designer often considers the engineering feasibility of various alternative proposals. When some idea as to the amount of space needed or available for a project has been determined, to-scale layout drawings may be started.

1.3.1.3 Synthesis

The synthesis (putting together) of the solution represents perhaps the most challenging and interesting part of the design. Frequently termed the *ideation and invention phase,* it is where the largest possible number of creative solutions is originated. The philosophy, functionality, and uniqueness of the product are determined during synthesis. In this step, the designer combines separate parts to form a complex whole of various new and old ideas and concepts to produce an overall new idea or concept.

1.3.1.4 Analysis

Synthesis and analysis are the main stages that constitute the design process. Analysis has as its objective satisfactory performance as well as durability with minimum weight and competitive cost. Synthesis cannot take place without both analysis or resolution and optimization, because the product under design must be analyzed to determine whether the performance complies with the specifications. If the design fails, the synthesis procedure must begin again. After synthesizing several components of a system, we analyze what effect this has on the remaining parts of the system. It is now necessary to draw the layouts, providing details, and make the supporting calculations that will ultimately result in a prototype design. The designer must specify the dimensions, select the components and materials, and consider the manufacturing, cost, reliability, serviceability, and safety.

1.3.1.5 Testing and Evaluation

At this juncture, the working design is first fabricated as a *prototype.* Product evaluation is the final proof of a successful design and usually involves testing a prototype in a laboratory or on a computer that provides the analysis database. More often, computer prototypes are utilized because they are less expensive and faster to generate. By evaluation, we discover whether the design really satisfies the need and other desirable features. Subsequent to many *iterations* (i.e., repetitions or returns to a previous state), the process ends with the vital step of communicating the design to others.

1.3.1.6 Presentation

The designer must be able to understand the need and describe a design graphically, verbally, and in writing. This is the presentation of the plans for satisfying the need. A successful presentation is of utmost importance as the final step in the design process. Drawings are utilized to produce blueprints to be passed to the manufacturing process. A number of references are available on the design process for those seeking a more-thorough discussion [2,3].

It is interesting to note that individual parts should be designed to be easily fabricated, assembled, and constructed. The goal of the *manufacturing process* is to construct the designed component or system. Manufacturability plays an important role in the success of commercial products. Individual parts should be designed to be easily fabricated, assembled, and constructed. The process planning attempts to determine the most effective sequence to produce the component. The produced parts are inspected and must pass certain quality control or assurance requirements. Components surviving inspection are assembled, packaged, labeled, and shipped to customers.

The features of a product that attract consumers and how the product is presented to the marketplace are significant functions in the success of a product. Marketing is a crucial last stage of the manufacturing process. Market feedback is very important in enhancing products. These feedback loops are usually incorporated into the first stage of a design process [7]. Many disciplines are involved in product development. Therefore, design engineers need to be familiar with other disciplines, at least from a communications standpoint, to integrate them into the design process.

1.3.2 Design Considerations

Usually engineering designs involve quite a number of considerations that must be properly recognized by the engineer. *Traditional considerations* for a mechanical component, or perhaps the entire system, include strength, deflection, weight, size and shape, material properties, operating conditions, processing, cost, availability, usability, utility, and life. Examples of *modern consideration* are safety, quality of life, and ecology. *Miscellaneous considerations* include reliability, maintainability, ergonomics, and aesthetics.

We shall consider some of the foregoing factors throughout this text. Frequently, fundamentals will be applied to resolve a problem based on the design decisions. A final point to be noted is that often a variety of design considerations may be incompatible until the engineer puts together a sufficiently imaginative and ingenious solution. The design of the winch crane (see Figure 18.1) provides a simple example. Here, achieving a desired aesthetic appearance is almost incompatible with cost limitations.

In concluding this section, we note that a degree of caution is necessary when employing formulas for which there is uncertainty in applicability and restriction of use. The relatively simple form of many formulas usually results from idealizations made in their derivations. These assumptions include simplified boundary conditions and loading on a member and approximation of shape or material properties. Designers and stress analysts must be aware of such constraints.

1.4 Design Analysis

The objective of the design analysis is, of course, to attempt to predict the stress or deformation in the component so that it may safely carry the loads that will be imposed on it. The analysis begins with an attempt to put the conceptual design in the context of the abstracted engineering sciences to evaluate the performance of the expected product. This constitutes design modeling and simulation.

1.4.1 Engineering Modeling

Geometric modeling is the method of choice for obtaining the data necessary for failure analysis early in design process. Creating a useful engineering model of a design is probably the most difficult, challenging part of the whole process. It is the responsibility of the designer to ensure the adequacy of a chosen geometric model to a particular design. If the structure is simple enough, theoretical solutions for basic configurations may be adequate for obtaining the stresses involved. For more complicated structures, finite element models can not only estimate the stresses but also utilize them to evaluate the failure criteria for each element in a member.

We note that the geometric model chosen and subsequent calculations made merely approximate reality. Assumptions and limitations, such as linearity and material homogeneity, are used in developing the model. The choice of a geometric model depends directly on the kind of analysis to be performed. Design testing and evaluation may require changing the geometric model before finalizing it. When the final design is achieved, the drafting and detailing of the models start, followed by documentation and production of final drawings.

1.4.2 Rational Design Procedure

The rational design procedure to meet the *strength requirements* of a load-carrying member attempts to take the results of fundamental tests, such as tension, compression, and fatigue, and apply them to all complicated and involved situations encountered in present-day structures and machines. However, not all topics in design have a firm analytical base from which to work. In those cases, we must depend on a semi-rational or empirical approach to solving a problem or selecting a design component.

In addition, details related to actual service loads and various factors, discussed in Section 7.7, have a marked influence on the strength and useful life of a component. The static design of axially loaded members, beams, and torsion bars are treated by the rational procedure in Chapters 3 and 9. Suffice it to say that complete design solutions are not unique and often trial and error is required to find the best solution.

1.4.3 Methods of Analysis

Design methods are based on the mechanics of materials theory generally used in this text. Axisymmetrically loaded mechanical components are analyzed by methods of the elasticity theory in Chapter 16. The former approach employs assumptions based on experimental evidence along with engineering experience to make a reasonable solution of the practical problem possible. The latter approach concerns itself largely with more mathematical analysis of the *exact* stress distribution on a loaded body [8,9]. The difference between the two methods of analysis is further discussed at the end of Section 3.17.

Note that solutions based on the mechanics of materials give average stresses at a section. Since, at concentrated forces and abrupt changes in cross section, irregular local stresses (and strains) arise, only at distance about equal to the depth of the member from such disturbances are the stresses in agreement with the mechanics of materials. This is due to **Saint-Venant's Principle**: the stress of a member at points away from points of load application may be obtained on the basis of a statically equivalent loading system; that is, the manner of force application on stresses is significant only in the vicinity of the region where the force is applied. This is also valid for the disturbances caused by the changes in the cross section. The mechanics of materials approach is therefore best suited for relatively slender members.

The complete analysis of a given component subjected to prescribed loads by the method of equilibrium requires consideration of three conditions. These *basic principles of analysis* can be summarized as follows:

1. *Statics*. The equations of equilibrium must be satisfied.
2. *Deformations*. Stress–strain or force deformation relations (e.g., Hooke's law) must apply to the behavior of the material.
3. *Geometry*. The conditions of compatibility of deformations must be satisfied; that is, each deformed part of the member must fit together with adjacent parts.

Solutions based on these requirements must satisfy the boundary conditions. Note that it is not always necessary to execute the analysis in this exact order. Applications of the foregoing procedure are illustrated in the problems involving mechanical components as the subject unfolds. Alternatively, stress and deformation can also be analyzed using the energy methods. The roles of both methods are twofold. They can provide solutions of acceptable accuracy, where configurations of loading and member are regular, and they can be employed as a basis of the numerical methods, for more complex problems.

1.5 Problem Formulation and Computation

The discussion of Section 1.3 shows that synthesis and analysis are the *two faces* of the design. They are opposites but symbiotic. These are the phases of the mechanical design process addressed in this book. Most examples, case studies, and problems are set up so the identification of need, specifications, and feasibility phases already have been defined. As noted previously, this text is concerned with the fundamentals involved and mostly with the application to specific mechanical components. The machine and structural members chosen are widely used and will be somewhat familiar to the reader. The emphasis in treating these components is on the methods and procedures used.

1.5.1 Solving Mechanical Component Problems

Ever-increasing industrial demand for more sophisticated machines and structures calls for a good grasp of the concepts of analysis and design and a notable degree of ingenuity. Fundamentally, design is the process of problem solving. It is very important to formulate a mechanical element problem and its solution accurately. This requires consideration of physical and its related mathematical situations. The reader may find the following format helpful in problem formulation and solution:

1. **Given**: Define the problem and known quantities.
2. **Find**: State consistently what is to be determined.
3. **Assumptions**: List simplifying idealizations to be made.
4. **Solution**: Apply the appropriate equations to determine the unknowns.
5. **Comments**: Discuss the results briefly.

We illustrate most of these steps in the solution of the sample problems throughout the text.

Assumptions expand on the given information to further constrain the problem. For example, one might take the effects of friction to be negligible or the weight of the member can be ignored in a particular case. The student needs to understand what assumptions are made in solving a problem. Comments present the key aspects of the solution and discuss how better results might be obtained by making different analysis decisions, relaxing the assumptions, and so on.

This book provides the student the ideas and information necessary for understanding the mechanical analysis and design and encourages the creative process based on that understanding. It is significant that the reader visualize the nature of the quantities being computed. Complete, carefully drawn, free-body diagrams (FBDs) facilitate visualizations, and we provide these, knowing that the subject matter can be mastered best by solving practical problems. It should also be pointed out that the relatively simple form of many equations usually results from simplifying assumptions made with respect to the deformation and load patterns in their derivation. Designers and analysts must be aware of such restrictions.

1.5.1.1 Significant Digits

In practical engineering problems, the data are seldom known with an accuracy of greater than 0.2%; answers to such problems should not exceed this accuracy. Note that, calculations when performed by electronic calculators and computers (usually carrying eight or nine digits); the possibility exists that numerical result will be reported to an accuracy that has no physical meaning. Consistently throughout this text, we generally shall follow a common engineering rule to report the final results of calculations:

- Numbers beginning with "1" are recorded to four significant digits.
- All other numbers (that begin with "2" through "9") are recorded to three significant digits.

Hence, a force of 15 N, for example, should read 15.00 N, and a force of 32 N should read 32.0 N. Intermediate results, if recorded for further calculations, are recorded to several additional digits to preserve the numerical accuracy. We note that the values of π and trigonometric functions are calculated to many significant digits (10 or more) within the calculator or computer.

1.5.2 Computational Tools for Design Problems

A wide variety of computational tools can be used to perform design calculations with success. A high-quality scientific calculator may be the best tool for solving most of the problems in this book. General purpose analysis tools such as spreadsheets and equation solvers have particular merit for certain computational tasks. These mathematical software packages include MATLAB®, TK Solver, and MathCAD. The tools have the advantage of allowing the user to document and save completed work in a detailed form. Computer-aided design (CAD) software may be used throughout the design process [10], but it supports the analysis stages of the design more than conceptual phases.

In addition, there are proprietary software developed by a number of organizations to implement the preliminary design and proposal presentation stage. This is particularly true, for cases in which existing product lines need to be revised to meet new specifications or codes.

The computer-aided drafting software packages can produce realistic 3D representations of a member or solid models. The CAD software allows the designer to visualize without costly models, iterations, or prototypes. Most CAD systems provide an interface to one or more FEA or boundary element analysis (BEA) programs. They permit direct transfer of the model's geometry to an FEA or BEA package for analysis of stress and vibration as well as fluid and thermal analysis. However, usually, these analyses of design problems require the use of special purpose programs. The FEA techniques are briefly discussed in Chapter 17.

TABLE 1.1

Optimum Time to Do Everything

Time	Activity
6:00 a.m.	Wake
6:00–6:30 a.m.	Unsuitable to concentrate
6:30–8:30 a.m.	Suitable for creativity
8:30 a.m.–12:00 noon	*Suitable for problem solving*
12:00–2:30 p.m.	Unsuitable to concentrate
2:30–4:30 p.m.	*Suitable for problem solving*
4:30–8:00 p.m.	Rejuvenation
8:00–10:00 p.m.	Unsuitable for problem solving
10:00 (or 11:00) p.m.–6:00 a.m.	Sleep

As noted earlier, the website available with the text contains MATLAB simulations for mechanical design. The computer-based software may be used as a tool to assist students with design projects and lengthy homework assignments. However, computer output providing analysis results must not be accepted on faith alone; the designer must always check computer solutions. It is necessary that fundamentals of analysis and design be thoroughly understood.

1.5.3 Best Time to Solve Problems

Daily planning can help us make the best of our time. A tentative schedule [11] for the *morning person* who prefers to wake up early and go to sleep early is presented in Table 1.1. It is interesting to note that the so-called evening person works late and wakes up late. Most people may shift times from one to another, and others combine some characteristics of both.

We point out that creativity refers to the state or quality of being creative and serves well for open-ended thinking. Rejuvenation is a phenomenon of vitality and freshness being restored and achieved by renewing the mind with activities like reading, artwork, and puzzle solving. During times suitable for problem solving, concentration is the highest for analysis. To concentrate is unsuitable when body's biological clock changes.

1.6 Factor of Safety and Design Codes

It is sometimes difficult to determine accurately the various factors involved in the phases of design of machines and structures. An important area of uncertainty is related to the assumptions made in the stress and deformation analysis. An equally significant item is the nature of failure. If failure is caused by ductile yielding, the consequences are likely to be less severe than if caused by brittle fracture. In addition, a design must take into account such matters as the following: types of service loads, variations in the properties of the material, whether failure is gradual or sudden, the consequences of failure (minor damage or catastrophe), and human safety and economics.

1.6.1 Definitions

Engineers employ a safety factor to ensure against the foregoing unknown uncertainties involving strength and loading. This factor is used to provide assurance that the load applied to a member does not exceed the largest load it can carry. The factor of safety, n, is the ratio of the maximum load that produces failure of the member to the load allowed under service conditions:

$$n = \frac{\text{Failure load}}{\text{Allowable load}} \tag{1.1}$$

The allowable load is also referred to as the *service load* or *working load*. The preceding represents the basic definition of the factor of safety. This ratio must always be greater than unity, $n > 1$. Since the allowable service load is a known quantity, the usual design procedure is to multiply this by the safety factor to obtain the failure load. Then, the member is designed so that it can just sustain the maximum load at failure.

A common method of design is to use a safety factor with respect to the strength of the member. In most situations, a linear relationship exists between the load and the stress produced by the load. Then, the factor of safety may also be defined as

$$n = \frac{\text{Material strength}}{\text{Allowable stress}} \tag{1.2}$$

In this equation, the materials strength represents either static or dynamic properties. Obviously, if loading is static, the material strength is either the yield strength or the ultimate strength. For fatigue loading, the material strength is based on the endurance limit, discussed in Chapter 7. The allowable stress is also called the *applied stress*, *working stress*, or *design stress*. It represents the required strength.

The foregoing definitions of the factor of safety are used for all types of member and loading conditions (e.g., axial, bending, shear). Inasmuch as there may be more than one potential mode of failure for any component, we can have more than one value for the factor of safety. The smallest value of n for any member is of the greatest concern, because this predicts the most likely mode of failure.

1.6.2 Selection of a Factor of Safety

Modern engineering design gives a rational accounting for all factors possible, leaving relatively few items of uncertainty to be covered by a factor of safety. The following numerical values of factor of safety are presented as a guide. They are abstracted from a list by J.P. Vidosic [12]. These safety factors are based on the yield strength S_y or endurance limit S_e of a *ductile material*. When they are used with a *brittle material* and the ultimate strength S_u, the factors must be approximately doubled:

1. n = 1.25–1.5 is for exceptionally reliable materials used under controllable conditions and subjected to loads and stresses that can be determined with certainty. It is used almost invariably where low weight is a particularly important consideration.
2. n = 1.5–2 is for well-known materials under reasonably constant environmental conditions, subjected to loads and stresses that can be determined readily.

3. n = 2–2.5 is for average materials operated in ordinary environments and subjected to loads and stresses that can be determined.
4. n = 2.5–4 is for less-tried (or 3–4 for untried) materials under average conditions of environment, load, and stress.
5. n = 3–4 is also for better-known materials used in uncertain environments or subjected to uncertain stresses.

Where higher factors of safety might appear desirable, a more-thorough analysis of the problem should be undertaken before deciding on their use.

In the field of aeronautical engineering, in which it is necessary to reduce the weight of the structures as much as possible, the term factor of safety is replaced by the *margin of safety*:

$$\frac{\text{Ultimate load}}{\text{Design load}} - 1 \tag{a}$$

In the nuclear reactor industries, the safety factor is of prime importance in the face of many unknown effects, and hence, the factor of safety may be as high as five. The value of factor of safety is selected by the designer on the basis of experience and judgment.

The simplicity of Equations 1.1 and 1.2 sometimes mask their importance. A large number of problems requiring their use occur in practice. The employment of a factor of safety in a design is a reliable, time-proven approach. When properly applied, sound and safe designs are obtained. We note that the factor of safety method to safe design is based on rules of thumb, experience, and testing. In this approach, the strengths used are always the *minimum* expected values.

A concept closely related to safety factor is termed *reliability*. It is the statistical measure of the probability that a member will not fail in use. In the reliability method of design, the goal is to achieve a reasonable likelihood of survival under the loading conditions during the intended design life. For this purpose, mean strength and load distributions are determined, and then, these two are related to achieve an acceptable safety margin. Reliability is discussed in Chapter 6.

1.6.3 Design and Safety Codes

Numerous engineering societies and organizations publish standards and codes for specific areas of engineering design. Most are merely recommendations, but some have the force of law. For the majority of applications, relevant factors of safety are found in various construction and manufacturing codes, for instance, the American Society of Mechanical Engineers (ASME) Pressure Vessel Codes. Factors of safety are usually embodied into computer programs for the design of specific members. Building codes are legislated throughout this country and often deal with publicly accessible structures (e.g., elevators and escalators). Underwriters Laboratories (UL) has developed its standards for testing consumer products. When a product passes their tests, it may be labeled *listed UL*. States and local towns have codes as well, relating mostly to fire prevention and building standards.

It is clear that, where human safety is involved, high values of safety factor are justified. However, members should not be overdesigned to the point of making them unnecessarily costly, heavy, bulky, or wasteful of resources. The designer and stress analyst must be apprehensive of the codes and standards, lest their work lead to inadequacies.

The following is a partial list of societies and organizations* that have established specifications for standards and safety or design codes.

AA	Aluminum Association
AFBMA	Anti-Friction Bearing Manufacturing Association
AGMA	American Gear Manufacturing Association
AIAA	American Institute of Aeronautics and Astronautics
AISC	American Institute of Steel Construction
AISI	American Iron and Steel Institute
ANSI	American National Standards Institute
API	American Petroleum Institute
ASCE	American Society of Civil Engineers
ASLE	American Society of Lubrication Engineers
ASM	American Society of Metals
ASME	American Society of Mechanical Engineers
ASTM	American Society for Testing and Materials
AWS	American Welding Society
NASA	National Aeronautics and Space Administration
NIST	National Institute for Standards and Technology
IFI	Industrial Fasteners Institute
ISO	International Standards Organization
SAE	Society of Automotive Engineers
SEM	Society for Experimental Mechanics
SESA	Society for Experimental Stress Analysis
SPE	Society of Plastic Engineers

1.7 Units and Conversion

The units of the physical quantities employed in engineering calculations are of major significance. The most recent universal system is the International System of Units (SI). The U.S. customary units have long been used by engineers in this country. Both systems of units, reviewed briefly here, are used in this text. However, greater emphasis is placed on the SI units, in line with international conventions. Some of the fundamental quantities in SI and the U.S. customary systems of units are listed in Table 1.2. For further details, see, for example, Reference 13.

We observe from the table that, in SI, force F is a derived quantity (obtained by multiplying the mass m by the acceleration a, in accordance with Newton's second law, $F = ma$). However, in the U.S. customary system, the situation is reversed, with mass being the derived quantity. It is found from Newton's second law, as $\text{lb} \cdot \text{s}^2/\text{ft}$, sometimes called the *slug*.

Temperature is expressed in SI by a unit termed kelvin (K), but for common purposes, the degree Celsius (°C) is used (as shown in the table). The relationship between the two units: temperature in Celsius = temperature in kelvins −273.15. The temperature is expressed in

* The address and data on their publications can be obtained in any technical library or from a designated website; for example, for specific titles of ANSI standards, see www.ansi.org.

TABLE 1.2

Basic Units

	SI Unit		U.S. Unit	
Quantity	**Name**	**Symbol**	**Name**	**Symbol**
Length	Meter	m	Foot	ft
Force[a]	Newton	N[a]	Pound force	lb
Time	Second	s	Second	s
Mass	Kilogram	kg	Slug	$lb \cdot s^2/ft$
Temperature	Degree Celsius	°C	Degree Fahrenheit	°F

[a] Derived unit ($kg \cdot m/s^2$).

U.S. units by the degree Fahrenheit (°F). Conversion formulas between the temperature scales is given by

$$t_c = \frac{5}{9}(t_f - 32) \tag{1.3}$$

and

$$t_k = (t_f - 32) + 273.15 \tag{1.4}$$

where t is the temperature. Subscripts c, f, and k denote the Celsius, Fahrenheit, and kelvin, respectively.

It is sufficiently accurate to assume that the acceleration of gravity, denoted by g, near earth's surface equals

$$g = 9.81\ \mathrm{m/s^2} \quad (\text{or } 32.2\ \mathrm{ft/s^2})$$

From Newton's second law, it follows that, in SI, the weight W of a body of mass 1 kg is $W = mg = (1\ \mathrm{kg})(9.81\ \mathrm{m/s^2}) = 9.81\ \mathrm{N}$. In the U.S. customary system, the weight is expressed in pounds (lb). The unit of force is of particular importance in engineering analysis and design, because it is involved in calculations of the force, moment, torque, stress (or pressure), work (or energy), power, and elastic modulus. Interestingly, in SI units, a newton is approximately the weight of (or earth's gravitational force on) an average apple.

Tables A.1 and A.2 furnish conversion factors and SI prefixes in common usage. The use of prefixes avoids unusually large or small numbers. Note that a dot is to be used to separate units that are multiplied together. Thus, for instance, a newton meter is written $N \cdot m$ and must not be confused with mN, which stands for millinewtons. The reader is cautioned always to check the units in any equation written for a problem solution. If properly written, an equation should cancel all units across equal sign.

1.8 Loading Classes and Equilibrium

External forces, or loads acting on a structure or member, may be classified as surface forces and body forces. A surface force acts at a point or is distributed over a finite area. Body forces are distributed throughout the volume of a member. All forces acting on a body, including the reactive forces caused by supports and the body forces, are considered as external forces. Internal forces are the forces holding together the particles forming the member.

Line loads and *concentrated forces* are considered to act along a line and at a single *point*, respectively. Both of these forces are thus idealizations. Nevertheless, they permit accurate analysis of a loaded member except in the immediate vicinity of the loads. Loads and internal forces can be further classified with respect to location and method of application: normal, shear, bending, and torsion loads and combined loadings. There are only few types of loading that may commonly occur on machine or structural members.

A *static load* is applied slowly, gradually increasing from zero to its maximum value and thereafter remaining constant. Thus, a static load can be a stationary (i.e., unchanging in magnitude, point of application, and direction) force, torque, moment, or a combination of these acting on a member. In contrast, *dynamic loads* may be applied very suddenly, causing vibration of structure, or they may change in magnitude with time. Note that, unless otherwise stated, we assume in this book that the *weight* of the body can be *neglected* and that the load is static. As observed earlier, in SI, force is expressed in newtons (N). But, because the newton is a small quantity, the kilonewton (kN) is often used in practice. The unit of force in the U.S. customary system is pounds (lb) or kilopounds (kips).

1.8.1 Conditions of Equilibrium

When a system of forces acting on a body has zero resultant, the body is said to be in equilibrium. Consider the equilibrium of a body in space. The conditions of equilibrium require that the following *equations of statics* need be satisfied

$$\begin{aligned} \sum F_x &= 0 \qquad \sum F_y = 0 \qquad \sum F_z = 0 \\ \sum M_x &= 0 \qquad \sum M_y = 0 \qquad \sum M_z = 0 \end{aligned} \tag{1.5}$$

If the forces act on a body in equilibrium in a single (xy) plane, a planar problem, the most common forms of the static equilibrium equations are

$$\sum F_x = 0 \qquad \sum F_y = 0 \qquad \sum M_z = 0 \tag{1.6}$$

By replacing either or both force summations by equivalent moment summations in Equation 1.6, two *alternate* sets of equations can be obtained [9].

When bodies are accelerated, that is, the magnitude or direction of their velocity changes, it is necessary to use Newton's second law to relate the motion of the body with the forces acting on it. The *plane motion* of a body, symmetrical with respect to a plane (xy) and rotating about an axis (z), is defined by

$$\sum F_x = ma_x \qquad \sum F_y = ma_y \qquad \sum M_z = I\alpha \tag{1.7}$$

in which

m represents the mass

I is the principal centroidal mass moment of inertia about the z axis.

The quantities a_x, a_y, and α represent the linear and angular accelerations of the mass center about the principal x, y, and z axes, respectively. The preceding relationships express that the system of external forces is equivalent to the system consisting of the inertia forces (ma_x and ma_y) attached at the mass center and the couple moment $I\alpha$. Equation 1.7 can be written for all the connected members in a 2D system and an entire set solved simultaneously for forces and moments.

A structure or system is said to be *statically determinate* if all forces on its members can be obtained by using only the equilibrium conditions; otherwise, the structure is referred to as *statically indeterminate*. The degree of static indeterminacy is equal to the difference between the number of unknown forces and the number of pertinent equilibrium equations. Since any reaction in excess of those that can be found by statics alone is called *redundant*, the number of redundants is the same as the degree of indeterminacy. To effectively study a structure, it is usually necessary to make simplifying idealizations of the structure or the nature of the loads acting on the structure. These permit the construction of an *FBD*, a sketch of the isolated body and all external forces acting on it. When internal forces are of concern, an imaginary cut through the body at the section of interest is displayed, as illustrated in the next section.

1.8.2 Internal Load Resultants

Distributed forces within a member can be represented by statically equivalent *internal forces*, so-called *stress-resultants* or load resultants. Usually, they are exposed by an imaginary cutting plane containing the centroid C through the member and resolved into components normal and tangential to the cut section. This process of dividing the body into two parts is called the **method of sections**. Figure 1.2a shows only the isolated left part of a slender member. A bar whose least dimension is less than about 1/10 its length may usually be considered a *slender* member. Note that the sense of moments follows the right-hand screw rule and, for convenience, is often represented by double-headed vectors. In 3D problems, the four modes of load transmission are axial force P (also denoted F or N), shear forces V_y and V_z, torque or twisting moment T, and bending moments M_y and M_z.

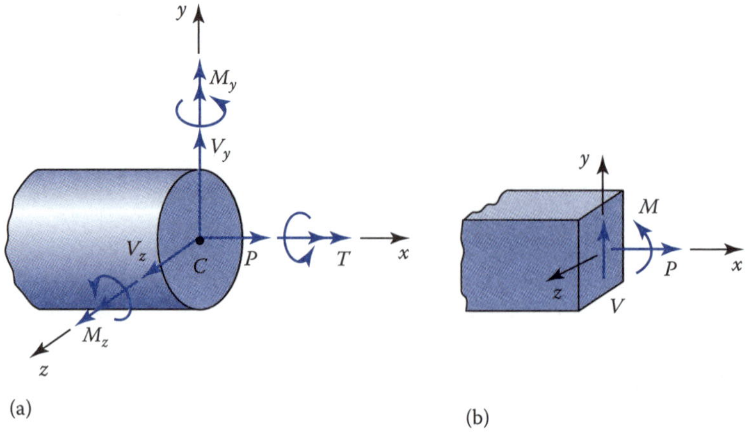

FIGURE 1.2
Internal forces and moments by the method of sections: (a) the general or three-dimensional (3D) case and (b) the two-dimensional (2D) case.

In planar problems, we find only three components acting across a section: the axial force P, the shear force V, and the bending moment M (Figure 1.2b). The cross-sectional face, or *plane*, is defined as positive when its outward normal points in a positive coordinate direction and as negative when its *outward* normal points in the negative coordinate direction. According to Newton's third law, the forces and moments acting on the faces at a cut section are equal and opposite. The location in a plane where the largest internal force resultants develop and failure is most likely to occur is called the *critical section*.

1.8.3 Sign Convention

When both the outer normal and the internal force or moment vector component point in a positive (or negative) coordinate direction, the force or moment is defined as positive. Therefore, Figure 1.2 depicts positive internal force and moment components. However, it is common practice for the direction shown in the figure to represent a negative internal shear force. In this text, we use a sign convention for *shear force* in a beam that is contrary to the definition given in Figure 1.2 (see Section 3.6). Note also that the sense of the *reaction* at a support of a structure is arbitrarily assumed; the positive (negative) sign of the answer obtained by the equations of statics will indicate that the assumption is correct (incorrect).

1.9 Free-Body Diagrams and Load Analysis

Application of equilibrium conditions requires a complete specification of all loads and reactions that act on a structure or machine. So, the first step in the solution of an equilibrium problem should consist of drawing an *free-body diagram* (FBD) of the body under consideration. An FBD is simply a sketch of a body, with all of the appropriate forces, both known and unknown, acting on it. This may be of an entire structure or a substructure of a larger structure. The general procedure in drawing a complete FBD includes the following steps:

1. Select the free body to be used.
2. Detach this body from its supports and separate from any other bodies. (If internal force resultants are to be determined, use the method of sections.)
3. Show on the sketch all of the external forces acting on the body. Location, magnitude, and direction of each force should be marked on the sketch.
4. Label significant points and include dimensions. Any other detail, however, should be omitted.

Clearly, the prudent selection of the free body to be used is (item 1) of primary significance. The reader is strongly urged to adopt the habit to draw clear and complete FBDs in the solution of problems concerning equilibrium. Examples 1.1 and 1.2 and Case Study 1.1 will illustrate the construction of the FBDs and the use of equations of statics.

A *structure* is a unit composed of interconnected members supported in a manner capable of resisting applied forces in static equilibrium. The constituents of such units or systems are bars, beams, plates, and shells, or their combinations. An extensive variety of structures are used in many fields of engineering. Structures can be considered in four broad categories: frames, trusses, machines, and thin-walled structures. Adoption of thin-walled structure behavior allows certain simplifying assumptions to be made in the structural analysis (see Section 4.10). The American Society of Civil Engineers (ASCE) lists design loads for buildings and other common structures [14].

Here, we consider load analysis dealing with the assemblies or structures made of several connected members. A *frame* is a structure that always contains at least one multiforce member, that is, a member acted on by three or more forces, which generally are not directed along the member. A *truss* is a special case of a frame, in which all forces are directed along the axis of a member. Machines are similar to frames in that of the elements that may be multiforce members. However, as noted earlier, a *machine* is designed to transmit and modify forces (or energy) and always contains moving parts.

Usually, the whole machine requires a *base* (a frame, housing) into or upon which all subassemblies are mounted. For this purpose, a variety of structural types may be used. A *baseplate* represents the simplest kind of machine frame. A machine room floor consists of a number of spaced cross beams forming a grid pattern. Basically, components of machines and their bases are designed on the similar principles. In both cases, recognition must be given to growing necessity for integration of manufacturing, assembly, and inspection requirements into the design process at an early stage (Section 1.3).

The approach used in the load analysis of a pin-jointed structure may be summarized as follows. First, consider the entire structure as a free body, and write the equations of static equilibrium. Then, dismember the structure, and identify the various members as either two-force (axially loaded) members or multiforce members. Pins are taken to form an integral part of one of the members they connect. Draw the FBD of each member. Clearly, when two-force members are connected to the same member, they are acted on by that member with equal and opposite forces of unknown magnitude but known direction. Finally, the equilibrium equations obtained from the FBDs of the members may be solved to yield various internal forces.

Example 1.1: Member Forces in a Pin-Connected Frame

The assembly shown in Figure 1.3a, which carries a load of 30 kN, consists of two beams *ABCD* and *CEF*, and one bar *BE*, connected by pins; it is supported by a pin at *A* and a cable *DG*. The dimensions are in meters.

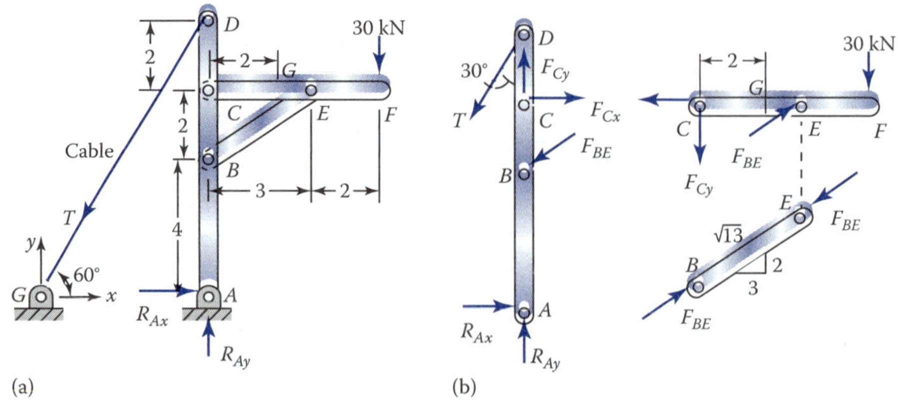

FIGURE 1.3

Example 1.1. (a) Structural assembly and (b) dismembered structure: FBDs of beams *ABCD* and *CEF*.

Find

a. The components of the forces acting on each member.
b. The axial force, shear force, and moment acting on the cross section at point G.

Assumptions

Friction forces in the pin joints will be omitted. All forces are coplanar and 2D.

Solution

There are two components R_{Ax} and R_{Ay} of the reaction at A and the force T exerted by the cable at D. Therefore, we can compute the reactions by considering the free body of the entire frame (Figure 1.3a):

$$\Sigma M_A = 30(5) - T\sin 30°(8) = 0, \quad T = 37.5\text{ kN}$$

$$\Sigma F_x = R_{Ax} - 37.5\sin 30° = 0, \quad R_{Ax} = 18.75\text{ kN}$$

$$\Sigma F_y = R_{Ay} - 37.5\cos 30° - 30 = 0, \quad R_{Ay} = 62.48\text{ kN}$$

a. The frame is now dismembered, since only two members are connected at each joint; equal and opposite components or resultants are shown on each member at each joint (Figure 1.3b). We note that BE is a two-force member, with *relative dimensions* are shown by a small triangle. Observe that the slope of the force F_{BE} is 2/3 and we can write the proportionalities:

$$\frac{F_{Bx}}{3} = \frac{F_{By}}{2} = \frac{F_{BE}}{\sqrt{13}}$$

Hence, for computational convenience, the force F_{BE} may be resolved into the x and y components:

$$F_{Bx} = \frac{3}{\sqrt{13}}F_{BE}, \qquad F_{By} = \frac{2}{\sqrt{13}}F_{BE} \tag{a}$$

Now, we write the following equilibrium conditions for the member CEF:

$$\Sigma F_y = 30(5) - \frac{2}{\sqrt{13}}F_{BE}(3) = 0, \quad F_{BE} = 90.14\text{ kN}$$

$$\Sigma M_E = 30(2) - F_{Cy}(3) = 0, \quad F_{Cy} = 20\text{ kN}$$

$$\Sigma F_x = \frac{3}{\sqrt{13}}(90.14) - F_{Cx} = 0, \quad F_{Cx} = 75\text{ kN}$$

Member ABCD: All internal forces have been found. To check the results, using the equations of statics, we verify that the beam $ABCD$ is in equilibrium.

Comment: The positive values obtained means that the directions shown for the force components are correct.

b. Cut member *CEF* at point *G*. Choosing the free body of segment *CG*, we have

$$M_G = F_{Cy}(2) = 20(2) = 40 \text{ kN} \cdot \text{m}, \quad F_G = F_{Cx} = 75 \text{ kN}, \quad V_G = F_{Cy} = 20 \text{ kN}.$$

The internal forces at *G* are equivalent to a couple, an axial force, and a shear.

Comment: MATLAB solution of this sample problem and many others are on the website (see Appendix E).

Example 1.2: Load Resultants at a Section of a Piping

An L-shaped pipe assembly of two perpendicular parts *AB* and *BC* is connected by an elbow at *B* and bolted to a rigid frame at *C*. The assembly carries a vertical load P_a, a torque T_a at *A*, as well as its own weight (Figure 1.4a). Each pipe is made of steel of unit weight *w* and nominal diameter *d*.

Find

What are the axial force, shear forces, and moments acting on the cross section at point *O*?

Given

$a = 0.6$ m, $b = 0.48$ m, $d = 63.5$ mm (2.5 in.), $P_a = 100$ N, $T_a = 25$ N·m, $w = 5.79$ lb/ft (see Table A.4).

Assumption

The weight of the pipe assembly is uniformly distributed over its entire length.

Solution

See Figure 1.4 and Equation 1.5.

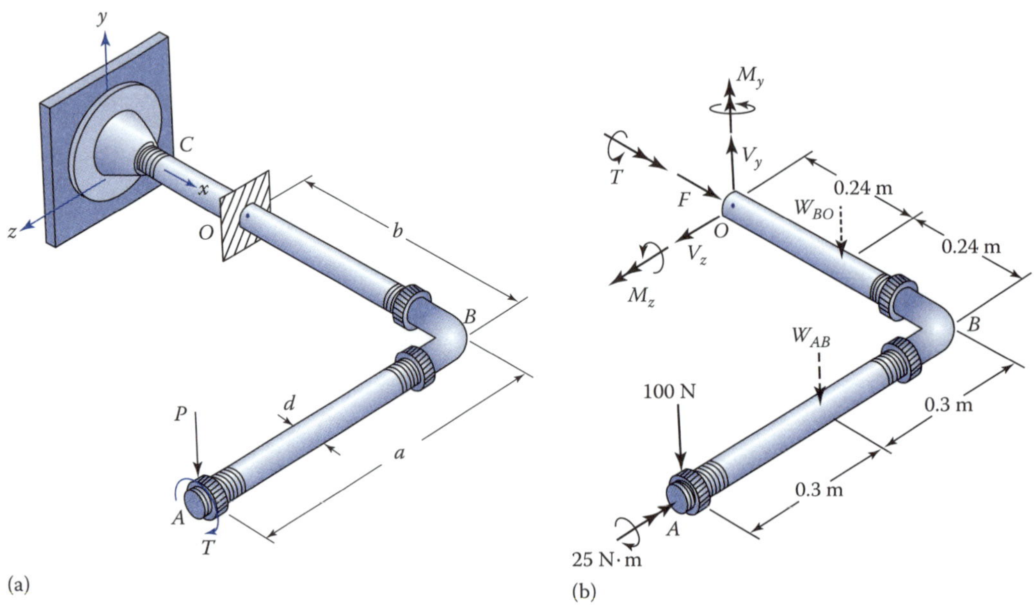

FIGURE 1.4
Example 1.2. (a) Pipe assembly and (b) FBD of part *ABO*.

Using conversion factor from Table A.1, w = 5.79 (N/m)/(0.0685) = 84.53 N/m. Thus, the weight of the pipes AB and BO are equal to

$$W_{AB} = (84.53)(0.6) = 50.72\text{ N},\quad W_{BO} = (84.53)(0.48) = 40.57\text{ N}$$

Free-body: Part ABO. We have six equations of equilibrium for the 3D force system of six unknowns (Figure 1.4b). The first three of Equations 1.5 results in the internal forces on the pipe at point O as follows:

$$\Sigma F_x = 0:\quad F = 0$$

$$\Sigma F_y = 0:\quad V_y - 50.72 - 40.57 - 100 = 0:\quad V_y = 191.3\text{ N}$$

$$\Sigma F_z = 0:\quad V_z = 0$$

Applying the last three of Equations 1.5, the moments about point O are found to be

$$\Sigma M_x = 0:\quad T + (50.72)(0.3) + 100(0.6) = 0,\quad T = -75.2\text{ N}\cdot\text{m}$$

$$\Sigma M_y = 0:\quad M_y = 0$$

$$\Sigma M_z = 0:\quad M_z - 25 - 100(0.48) - (50.72)(0.48) - (40.57)(0.24) = 0,\quad M_z = 107.1\text{ N}\cdot\text{m}$$

Comment: The negative value calculated for T means that the torque vector is directed opposite to that indicated in the figure.

1.10 Case Studies in Engineering

An engineering case is an account of an engineering activity, event, or problem. Good case studies are taken from real-life situations and include sufficient data for the reader to treat problem. They may come in the following varieties: the history of an engineering activity, illustration of some form of engineering process, an exercise (such as stress and deformation analysis), a proposal of problems to be solved, or a preliminary design project. Design analysis has its objective satisfactory performance as well as durability with minimum weight and competitive cost. Through case studies, we can create a bridge between systems theory and actual design plans.

The basic geometry and loading on a member must be given to the engineer before any analysis can be done. The stress that would result, for example, in a bar subjected to a load would depend on whether the loading gives rise to tension, transverse shear, direct shear, torsion, bending, or contact stresses. In this case, *uniform stress* patterns may be more efficient at carrying the load than others. Therefore, making a careful study of the types of loads and stress patterns that can arise in structures or machines, considerable insight can be gained into improved shapes and orientations of components. This type of study allows the designer and analyst in choosing the shape or volume (weight) of members that will optimize the use of the material provided under the conditions of applied loads.

Case studies presented in select chapters of this text involve situations found in engineering practice. Among these are various preliminary design projects: the assemblies containing a variety of elements such as links under combined axial and bending loads, ductile–brittle transition of steel, shafts subjected to bending and torsion simultaneously, gear sets and bearings subject to steady and fluctuating loads, and compression springs, connections, a floor crane with electric winch, and a high-speed cutting machine. This book offers a number of case studies of which aspects are discussed in selective chapters. Next, Case Study 1.1 involving a bolt cutter demonstrates the simplest form of force determination.

CASE STUDY 1.1 Bolt Cutter Loading Analysis

Many components, such as bicycle levers, automotive scissors jacks, bolt cutting tools, various types of pliers, and pin-connected symmetrical assemblies, may be treated by applying Equation 1.5, similar to that will be illustrated here. We note that a mechanical linkage system is designed to transform a given input force and movement into a desired output force and movement. In this case, accelerations on moving bars require that a dynamic analysis be done through the use of Equation 1.7. *Bolt cutters* can be used for cutting rods (see Page 1), wire mesh, and bolts. Often, a bolt cutter's slim cutting head permits cutting close to surfaces and incorporates one-step internal cam mechanism to maintain precise *jaw* or *blade* alignment. *Handle* design and handle grips lend to controlled cutting action. Jaws are manufactured from heat-treated, hardened alloy steel.

Figure 1.5 depicts schematic drawing of a bolt cutter, a pin-connected tool in the closed position in the process of gripping its jaws into a bolt. The user provides the input loads between the handles, indicated as the reaction pairs *P*. Determine the force exerted on the bolt and the pins at joints *A*, *B*, and *C*.

Given

The geometry is known. The data are

$$P = 2\text{ lb},\quad a = 1\text{ in.},\quad b = 3\text{ in.},\quad c = \frac{1}{2}\text{ in.},\quad d = 8\text{ in.},\quad e = 1\text{ in.}$$

Assumptions

Friction forces in the pin joints are omitted. All forces are coplanar, 2D, and static. The weights of members are neglected as being insignificant compared to the applied forces.

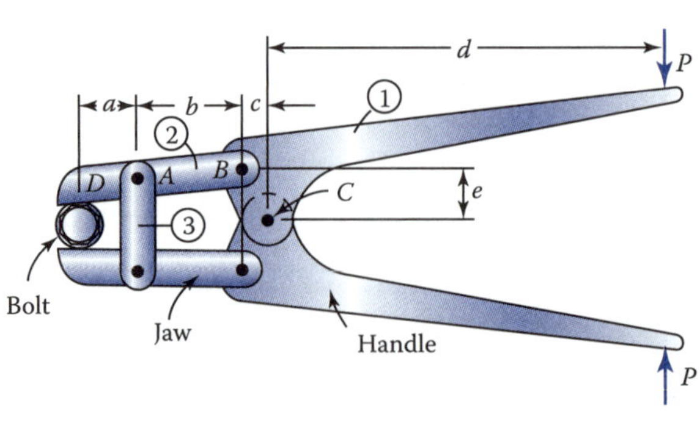

FIGURE 1.5
Sketch of a bolt cutter.

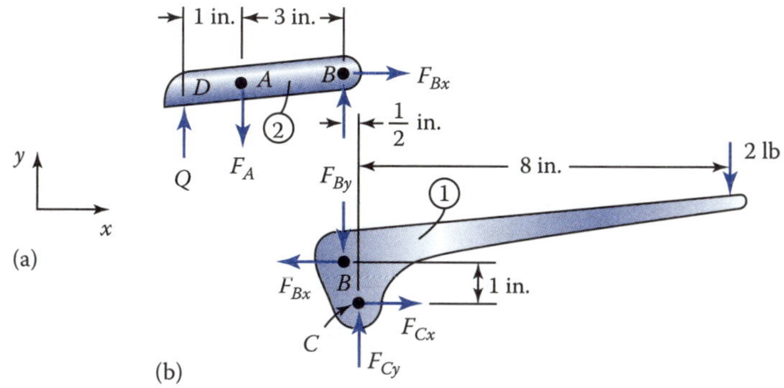

FIGURE 1.6
FBDs of bolt cutter shown in Figure 1.5, (a) jaw and (b) handle.

Solution

The equilibrium conditions are fulfilled by the entire cutter. Let the force between the bolt and the jaw be Q, whose direction is taken to be normal to the surface at contact (point D). Due to the symmetry, only two FBDs shown in Figure 1.6 need to be considered. Inasmuch as link 3 is a two-force member, the orientation of force F_A is known. Note also that the force components on the two elements at joint B must be equal and opposite, as shown on the diagrams.

Conditions of equilibrium are applied to Figure 1.6a to give $F_{Bx} = 0$ and

$$\sum F_y = Q - F_A + F_{By} = 0 \qquad F_A = Q + F_{By}$$

$$\sum M_B = Q(4) - F_A(3) = 0 \qquad F_A = \frac{4Q}{3}$$

from which $Q = 3F_{By}$. In a like manner, referring to Figure 1.6b, we obtain

$$\sum F_y = -F_{By} + F_{Cy} - 2 = 0 \qquad F_{Cy} = \frac{Q}{3} + 2$$

$$\sum M_C = F_{Bx}(1) + F_{By}(0.5) - 2(8) = 0 \qquad F_{By} = 32 \text{ lb}$$

and $F_{Cx} = 0$. Solving $Q = 3(32) = 96$ lb. The shear forces on the pins at the joints A, B, and C are

$$F_A = 128 \text{ lb}, \quad F_B = F_{By} = 32 \text{ lb}, \quad F_C = F_{Cy} = 34 \text{ lb}$$

Comments

Observe that the high mechanical advantage of the tool transforms the applied load to a large force exerted on the bolt at point D. The handles and jaws are under combined bending and shear forces. Stresses and deflections of the members are taken up in Case Studies 3.1 and 4.1 in Chapters 3 and 4, respectively. MATLAB solution of this case study and some others are on the website (see Appendix E).

1.11 Work, Energy, and Power

This section provides a brief introduction to the method of work and energy, which is particularly useful in solving problems dealing with buckling design and components subjected to combined loading. All machines or mechanisms consisting of several connected members involve loads and motion that, in combination, represent work and energy. The concept of *work* in mechanics is presented as the product of the magnitudes of the force and displacement vectors and the cosine of the angle between them. The work W done by a constant force F moving through a displacement s in the direction of force can be expressed as

$$W = Fs \tag{1.8}$$

Similarly, the work of a couple of forces or torque T during a rotation θ of the member, such as the wheel, is given by

$$W = T\theta \tag{1.9}$$

The work done by a force, torque, or moment can be regarded as a transfer of energy to the member. In general, the work is stored in a member as potential energy, kinetic energy, internal energy, or any combination of these or dissipated as heat energy. The magnitude of the energy a given component can store is sometimes a significant consideration in mechanical design. Members, when subjected to impact loads, are often chosen based on their capacity to absorb energy. *Kinetic energy* E_k of a member represents the capacity to do work associated with the speed of the member. The kinetic energy of a component in rotational motion may be written as

$$E_k = \frac{1}{2} I\omega^2 \tag{1.10}$$

The quantity I is the mass moment of inertia and ω represents the angular velocity or speed. Table A.5 lists mass moments of inertia of common shapes. The work of the force is equal to the change in kinetic energy of the member. This is known as the *principle of work and energy*. Under the action of conservative forces, the sum of the kinetic energy of the member remains constant.

The units of work and energy in SI is the newton meter (N · m), called the joule (*J*). In the U.S. customary system, work is expressed in foot pounds (ft · lb) and British thermal units (Btu). The unit of energy is the same as that of work. The quantities given in either unit system can be converted quickly to the other system by means of the conversion factors listed in Table A.1. Specific facets are associated with work, energy, and power, as will be illustrated in the analysis and design of various components in the chapters to follow.

Example 1.3: Camshaft Torque Requirement

A rotating camshaft (Figure 1.7) of an intermittent motion mechanism moves the follower in a direction at right angles to the cam axis. For the position shown, the follower is being moved upward by the lobe of the cam with a force F. A rotation of θ corresponds to a follower motion of s. Determine the average torque T required to turn the camshaft during this interval.

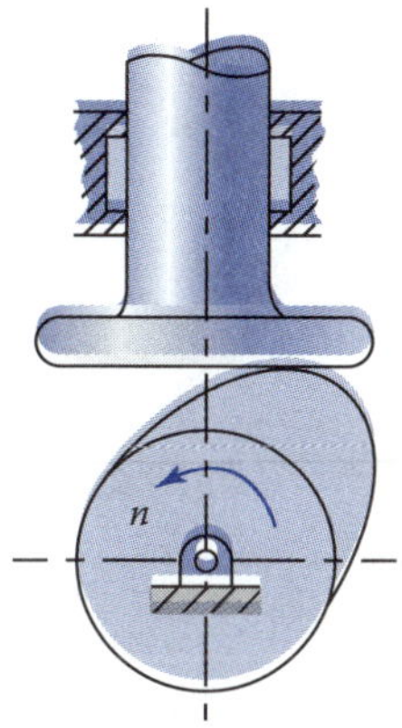

FIGURE 1.7
Example 1.3. Camshaft and follower.

Given

$F = 0.2$ lb, $\theta = 8° = 0.14$ rad, $s = 0.05$ in.

Assumptions

The torque can be considered to be constant during the rotation. The friction forces can be omitted.

Solution

The work done on the camshaft equals to the work done by the follower. Therefore, by Equations 1.8 and 1.9, we write

$$T\theta = Fs \tag{a}$$

Substituting the given numerical values,

$$T(0.14) = 0.2(0.05) = 0.01 \text{ lb} \cdot \text{in.}$$

The foregoing gives $T = 0.071$ lb · in.

Comments

Using conversion factor (Table A.1), in SI units, the answer is

$$T = 0.071 \frac{10^3 \text{N} \cdot \text{mm}}{(0.7376)12} = 8.02 \text{ N} \cdot \text{m}$$

The stress and deflection caused by force F at the contact surface between the cam and follower are considered in Chapter 8.

Example 1.4: Automobile Traveling at a Curved Road

A car of mass m is going through a curve of radius r at speed of V. Calculate the centrifugal force F_c.

Given

$m = 2$ tons $= 2000$ kg, $r = 120$ m, $V = 153$ km/h $= 153(1000)/3600 = 42.5$ m/s

Assumption

The speed is constant.

Solution

The *centrifugal force* is expressed in the form

$$F_c = \frac{mV^2}{r} \tag{1.11}$$

Introducing the given data,

$$F_c = \frac{2000(42.5)^2}{120} = 30{,}104\ \text{kg}\cdot\text{m/s}^2 = 30.1\ \text{kN}$$

Comment: Since the automobile moves at constant speed along its path, the tangential component of inertia force is zero. Centrifugal force (normal component) represents the tendency of the car to leave its curved path.

Power is defined as the time rate at which work is done. Note that, in selecting a motor or engine, power is a much more significant criterion than the actual amount of work to be performed. When work involves a force, the rate of energy transfer is the product of the force F and the velocity V at the point of application of the force. The power is therefore defined

$$\textit{Power}\ \ P = FV \tag{1.12}$$

In the case of a member, such as a shaft rotating with an angular velocity or speed ω in radians per unit time and acted on by a torque T, we have

$$\textit{Power}\ \ P = T\omega \tag{1.13}$$

The mechanical *efficiency*, designated by e, of a machine may be defined as follows:

$$e = \frac{\text{Power output}}{\text{Power input}} \tag{1.14}$$

Because of energy losses due to friction, the power output is always smaller than the power input. Therefore, machine efficiency is always less than 1. Inasmuch as power is defined as the time rate of doing work, it can be expressed in units of energy and time. Hence, the unit of power in SI is the watt (W), defined as the joule per second (J/s). If U.S. customary units are used, the power should be measured in ft · lb/s or in horsepower (hp).

1.11.1 Transmission of Power by Rotating Shafts and Wheels

The power transmitted by a rotating machine component such as a shaft, flywheel, gear, pulley, or clutch is of keen interest in the study of machines. Consider a circular shaft or disk of radius r subjected to a constant tangential force F. Then, the torque is expressed as $T = Fr$. The velocity at the point of application of the force is V. A relationship between the power, speed, and the torque acting through the shaft is readily found, from first principles, as follows.

In SI units, the power transmitted by a shaft is measured by kilowatt (kW), where 1 kW equals 1000 W. One watt does the work of 1 N·m/s. The speed n is expressed in revolutions per minute; then, the angle through which the shaft rotates equals $2\pi n$ rad/min. Thus, the work done per unit time is $2\pi nT$. This is equal to the power delivered: $2\pi nT/60 = 2\pi nFr/60 = \text{kW}(1000)$. Since $V = 2\pi rn/60$, the foregoing may be written as $FV = \text{kW}(1000)$. For convenience, power transmitted may be expressed in two forms:

$$\text{kW} = \frac{FV}{1000} = \frac{Tn}{9549} \tag{1.15}$$

where

T = the torque (N·m)
n = the shaft speed (rpm)
F = the tangential force (N)
V = the velocity (m/s)

We have one horsepower (hp) equals 0.7457 kW, and the preceding equation may be written as

$$\text{hp} = \frac{FV}{745.7} = \frac{Tn}{7121} \tag{1.16}$$

In U.S. customary units, horsepower is defined as a work rate of 550 × 60 = 33,000 ft·lb/m. An equation similar to that preceding can be obtained:

$$\text{hp} = \frac{FV}{33{,}000} = \frac{Tn}{63{,}000} \tag{1.17}$$

Here, we have

T = the torque in lb·in.
n = the shaft speed in rpm
F = the tangential force in lb
V = the velocity in fpm

Example 1.5: Power Capacity of Punch Press Flywheel

A high-strength steel flywheel of outer and inner rim diameters d_o and d_i, and length in axial direction of l, rotates at a speed of n (Figure 1.8). It is to be used to punch metal during two-thirds of a revolution of the flywheel. What is the average power available?

Given

$d_o = 0.5$ m, $n = 1000$ rpm, $\rho = 7860$ kg/m^3, (Table B.1)

Assumptions

1. Friction losses are negligible.
2. Flywheel proportions are $d_i = 0.75d_o$ and $l = 0.18d_o$.
3. The inertia contributed by the hub and spokes is omitted: the flywheel is considered as a rotating ring free to expand.

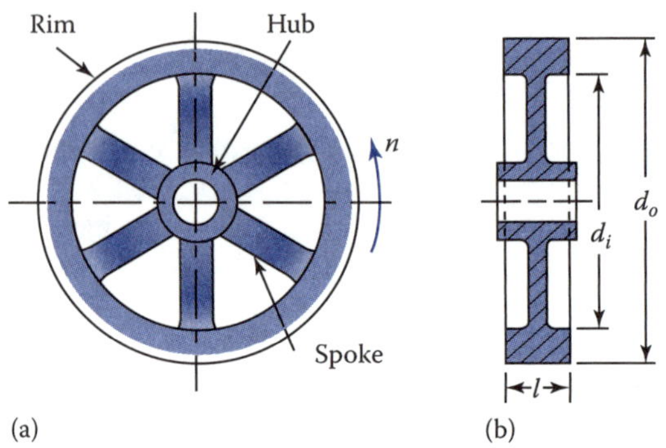

FIGURE 1.8
Example 1.5. (a) Punch press flywheel and (b) its cross section.

Solution

Through the use of Equations 1.9 and 1.10, we obtain

$$T\theta = \frac{1}{2} I\omega^2 \tag{1.18}$$

where

$$\theta = \frac{2}{3}(2\pi) = 4\pi/3 \text{ rad}$$

$$\omega = 1000(2\pi/60) = 104.7 \text{ rad/s}$$

$$I = \frac{\pi}{32}\left(d_o^4 - d_i^4\right)l\rho \quad \text{(Case 5, Table A.5)}$$

$$= \frac{\pi}{32}\left[(0.5)^4 - (0.375)^4\right](0.09)(7860) = 2.967 \text{ kg}\cdot\text{m}^2$$

Introducing the given data into Equation 1.18 and solving $T = 3882$ N · m. Equation 1.15 is therefore

$$\text{kW} = \frac{Tn}{9549} = \frac{3882(1000)}{9549}$$

$$= 406.5$$

Comment: The braking torque required to stop a similar disk in two-third revolution would have an average value of 3.88 kN · m (see Section 16.5).

1.12 Stress Components

Stress is a concept of paramount importance to a comprehension of solid mechanics. It permits the mechanical behavior of load-carrying members to be described in terms essential to the analyst and designer. Applications of this concept to typical members are discussed in Chapter 3. Consider a member in equilibrium, subject to external forces.

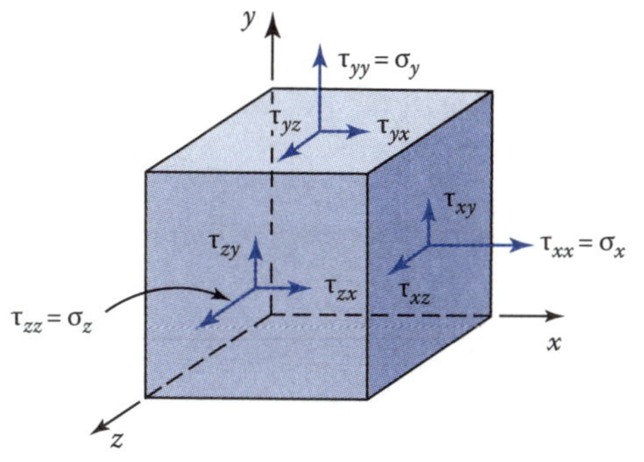

FIGURE 1.9
Element in three-dimensional (3D) stress. (Only stresses acting on the positive faces are shown.)

Under the action of these forces, internal forces and hence stresses are developed between the parts of the body [9]. In SI units, the stress is measured in newtons per square meter (N/m^2) or pascals. Since the pascal is very small quantity, the megapascal (MPa) is commonly used. Typical prefixes of the SI units are given in Table A.2. When the U.S. customary system is used, stress is expressed in pounds per square inch (psi) or kips per square inch (ksi).

The 3D state of stress at a point, using three mutually perpendicular planes of a cubic element isolated from a member, can be described by nine *stress components* (Figure 1.9). Note that only three (positive) faces of the cube are actually visible in the figure and that oppositely directed stresses act on the hidden (negative) faces. Here, the stresses are considered to be identical on the mutually parallel faces and uniformly distributed on each face. The general state of stress at a point can be assembled in the form

$$\begin{bmatrix} \tau_{xx} & \tau_{xy} & \tau_{xz} \\ \tau_{yx} & \tau_{yy} & \tau_{yz} \\ \tau_{zx} & \tau_{zy} & \tau_{zz} \end{bmatrix} = \begin{bmatrix} \sigma_x & \tau_{xy} & \tau_{xz} \\ \tau_{yx} & \sigma_y & \tau_{yz} \\ \tau_{zx} & \tau_{zy} & \sigma_z \end{bmatrix} \tag{1.19}$$

This is a matrix presentation of the *stress tensor*. It is a second-rank tensor requiring two indices to identify its elements. (A vector is a tensor of first rank; a scalar is of zero rank.) The double-subscript notation is explained as follows: the first subscript denotes the direction of a normal to the face on which the stress component acts; the second designates the direction of the stress. Repetitive subscripts are avoided in this text. Therefore, the *normal stresses* are designated σ_x, σ_y, and σ_z, as shown in Equation 1.19. In Section 3.16, it is demonstrated rigorously for the *shear stresses* that $\tau_{xy} = \tau_{yx}$, $\tau_{yz} = \tau_{zy}$, and $\tau_{xz} = \tau_{zx}$.

1.12.1 Sign Convention

When a stress component acts on a positive plane (Figure 1.9) in a positive coordinate direction, the stress component is *positive*. Also, a stress component is considered positive when it acts on a negative face in the negative coordinate direction. A stress component is considered negative when it acts on a positive face in a negative coordinate direction (or vice versa). Hence, tensile stresses are always positive and compressive stresses are always negative. The sign convention can also be stated as follows: a stress component is positive if both the outward normal of the plane on which it acts and its direction are in coordinate directions of the same sign; otherwise, it is negative. Figure 1.9 depicts a system of positive normal and shearing stresses. This sign convention for stress, which agrees with that adopted for internal forces and moments, is used throughout the text.

1.12.2 Special Cases of State of Stress

The general state of stress reduces to simpler states of stress commonly encountered in practice. An element subjected to normal stresses σ_1, σ_2, and σ_3, acting in mutually perpendicular directions alone with respect to a particular set of coordinates, is said to be in a state of *triaxial stress*. Such a stress can be represented as

$$\begin{bmatrix} \sigma_1 & 0 & 0 \\ 0 & \sigma_2 & 0 \\ 0 & 0 & \sigma_3 \end{bmatrix}$$

The absence of shearing stresses indicates that these stresses are the principal stresses for the element (Section 3.15).

In the case of two-dimensional (2D) or *plane stress*, only the x and y faces of the element are subjected to stresses (σ_x, σ_y, τ_{xy}), and all the stresses act parallel to the x and y axes, as shown in Figure 1.10a. Although the 3D aspect of the stress element should not be forgotten, for the sake of convenience, we usually draw only a 2D view of the plane stress element (Figure 1.10b). A thin plate loaded uniformly over the thickness, parallel to the plane of the plate, exemplifies the case of plane stress. When only two normal stresses are present, the state of stress is called *biaxial*.

In *pure shear*, the element is subjected to plane shear stresses acting on the four side faces only, for example, $\sigma_x = \sigma_y = 0$ and τ_{xy} (Figure 1.10b). Typical pure shear occurs over the cross sections and on longitudinal planes of a circular shaft subjected to torsion. Examples include axles and drive shafts in machinery, propeller shafts (Chapter 9), drill rods, torsional pendulums, screwdrivers, steering rods, and torsion bars (Chapter 14). If only one normal stress exists, the one-dimensional (1D) stress (Figure 1.10c) is referred to as a *uniaxial* tensile or compressive stress.

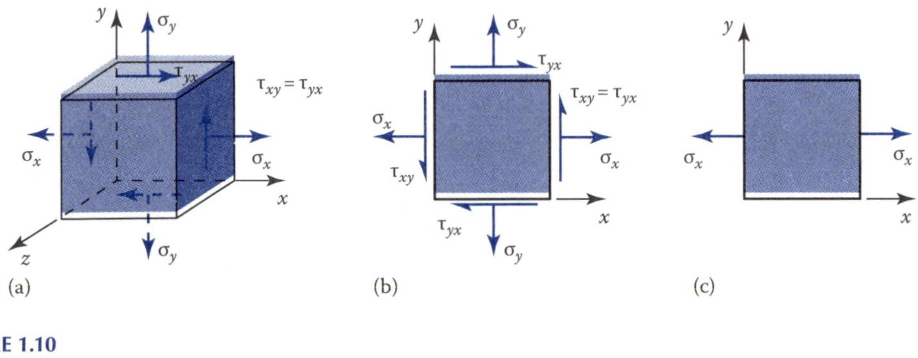

FIGURE 1.10
(a) Element in plane stress and (b and c) 2D and 1D presentations of plane stress.

1.13 Normal and Shear Strains

In the preceding section, our concern was with the stress within a loaded member. We now turn to deformation caused by the loading, the analysis of which is as important as that of stress. The analysis of deformation requires the description of the concept of strain, that is, the intensity of deformation. As a result of deformation, extension, contraction, or change of shape of a member may occur. To obtain the actual stress distribution within a member, it is necessary to understand the type of deformation occurring in that member. Only small displacements, commonly found in engineering structures, are considered in this text.

The strains resulting from small deformations are small compared with unity, and their products (higher-order terms) are neglected. The preceding assumption leads to one of the fundamentals of solid mechanics, the *principle of superposition* that applies whenever the quantity (deformation or stress) to be obtained is directly proportional to the applied loads. It permits a complex loading to be replaced by two or more simpler loads and thus renders a problem more amenable to solution, as will be observed repeatedly in the text.

The fundamental concept of *normal strain* is illustrated by considering the deformation of the homogenous prismatic bar shown in Figure 1.11a. A prismatic bar is a straight bar having constant cross-sectional area throughout its length. The initial length of the member is L. Subsequent to application of the load, the total deformation is δ. Defining the normal strain ε as the unit change in length, we obtain

$$\varepsilon = \frac{\delta}{L} \tag{1.20}$$

A positive sign designates elongation; a negative sign, contraction. The foregoing state of strain is called *uniaxial strain*. When an unconstrained member undergoes a temperature change ΔT, its dimensions change and a normal strain develops. The uniform thermal strain for a homogeneous and isotropic material is expressed as

$$\varepsilon_t = \alpha \Delta T \tag{1.21}$$

The coefficient of expansion α is approximately constant over a moderate temperature change. It represents a quantity per degree Celsius (1/°C) when ΔT is measured in °C.

Shear strain is the tangent of the total change in angle taking place between two perpendicular lines in a member during deformation. Inasmuch as the displacements considered are

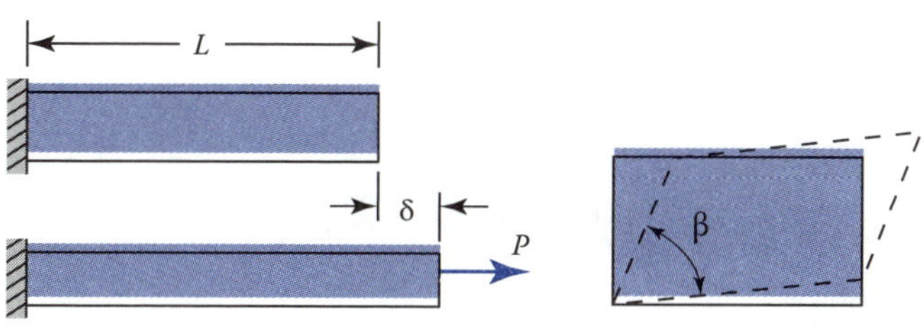

FIGURE 1.11
(a) Deformation of a bar and (b) distortion of a rectangular plate.

small, we can set the tangent of the angle of distortion equal to the angle. Thus, for a rectangular plate of unit thickness (Figure 1.11b), the shear strain γ measured in radians is defined as

$$\gamma = \frac{\pi}{2} - \beta \tag{1.22}$$

Here, β is the angle between the two rotated edges. The shear strain is positive if the right angle between the reference lines decreases, as shown in the figure; otherwise, the shearing strain is negative. Because normal strain s is the ratio of the two lengths, it is a dimensionless quantity. The same conclusion applies to shear strain. Strains are also often measured in terms of units mm/mm, in./in., and radians or microradians. For most engineering materials, strains rarely exceed values of 0.002 or 2000 μ in the elastic range. We read this as 2000 μ.

Example 1.6: Strains in a Plate

Given

A thin, triangular plate *ABC* is uniformly deformed into a shape *ABC′*, as depicted by the dashed lines in Figure 1.12.

Find

a. The normal strain along the centerline *OC*.
b. The normal strain along the edge *AC*.
c. The shear strain between the edges *AC* and *BC*.

Assumptions

The edge *AB* is built in to a rigid frame. The deformed edges $AC' = BC'$ are straight lines.

Solution

We have $L_{OC} = a$ and $L_{AC} = L_{BC} = a\sqrt{2} = 1.41421a$ (Figure 1.12).

a. *Normal strain along OC*. Since the contraction in length *OC* is $\Delta a = -0.0015a$, Equation 1.20 gives

$$\varepsilon_{OC} = -\frac{0.0015a}{a} = -0.0015 = -1500\ \mu$$

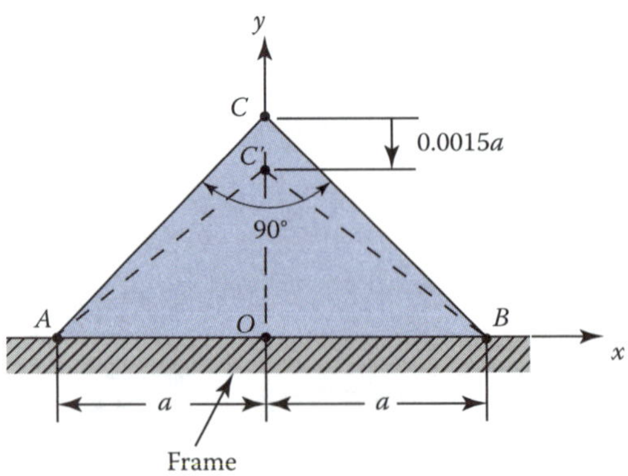

FIGURE 1.12
Example 1.6. Deformation of a triangular plate with one edge fixed.

b. *Normal Strain along AC and BC*. The lengths of the deformed edges are equal to $L_{AC'} = L_{BC'} = [a^2 + (a - 0.0015)^2]^{1/2} = 1.41315a$. It follows that

$$\varepsilon_{AC} = \varepsilon_{BC} = -\frac{1.41315a - 1.41421a}{1.41421a} = -750\ \mu$$

c. *Shear Strain between AC and BC*. After deformation, angle *ACB* is therefore

$$AC'B = 2\tan^{-1}\left[\frac{a}{a - 0.0015a}\right] = 90.086°$$

So, the change in the right angle is 90 – 90.086 = – 0.086°. The associated shear strain (in radians) equals $\gamma = -0.086\left(\frac{\pi}{180}\right) = -1501\ \mu$.

Comment

Inasmuch as the angle *ACB* is increased, the shear strain is negative.

Problems

Sections 1.1 through 1.9

1.1 A right angle bracket *ABC* of a control mechanism is subjected to loads *F*, *P*, and *T*, as shown in Figure P1.1. Draw FBD of the member and find

a. The value of the force *F*

b. The magnitude and direction of reaction at support *B*

1.2 A frame consists of three pin-connected members *ABC* of length 3*a*, and *ADE* and *BD* carry a vertical load *W* at point *E* as shown in Figure P1.2.

Find

a. The reactions at supports *A* and *C*

b. The internal forces and moments acting on the cross section at point *O*

1.3 A beam *CAB* with simple supports at *A* and *B* and an overhang *AC* carries loads as shown in Figure P1.3. All forces are coplanar and 2D. Determine the shear force and moment acting on the cross sections at the points *D* and *E*.

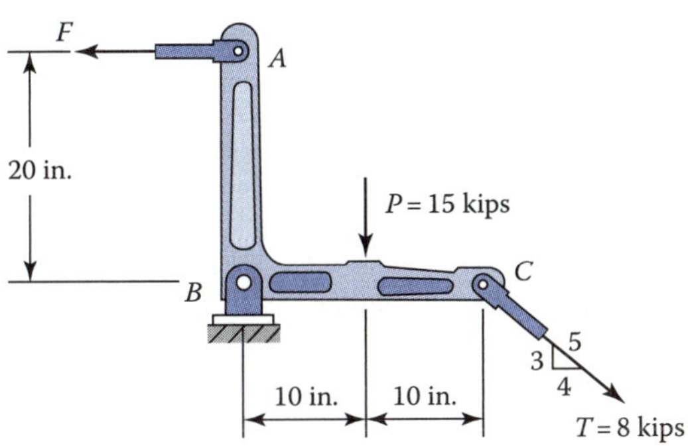

FIGURE P1.1

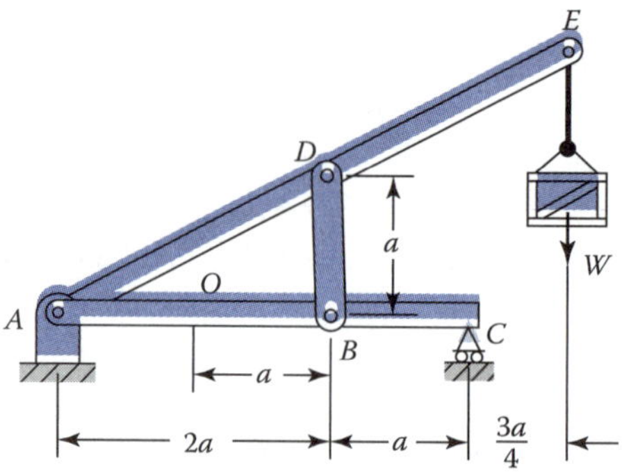

FIGURE P1.2

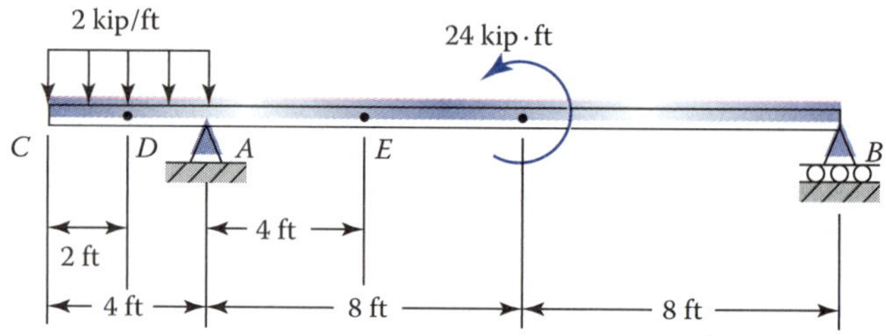

FIGURE P1.3

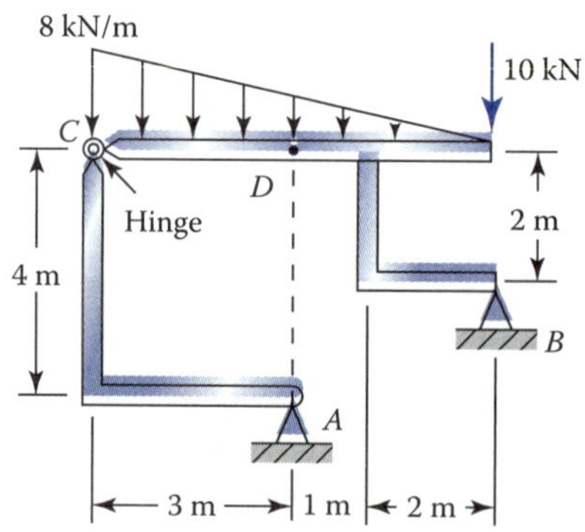

FIGURE P1.4

1.4 and **1.5** Two planar pin-connected frames are supported and loaded as shown in Figures P1.4 and P1.5. For each structure, determine

a. The components of reactions at B and C

b. The axial force, shear force, and moment acting on the cross section at point D

1.6 The piston, connecting rod, and crank of an engine system are shown in Figure P1.6. Calculate

a. The torque T required to hold the system in equilibrium

b. The normal or axial force in the rod AB

Given: A total gas force $P = 4$ kips acts on piston as indicated in the figure.

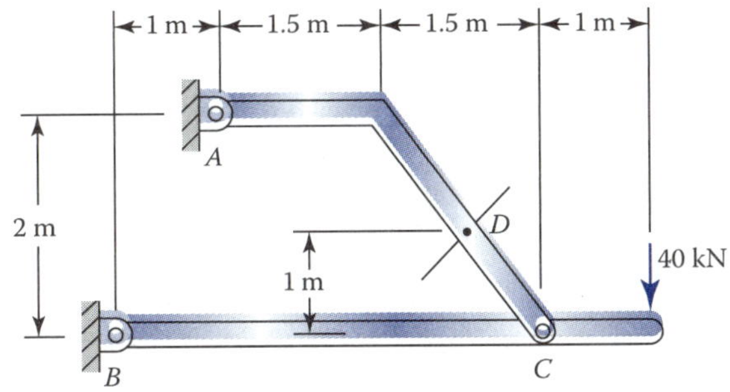

FIGURE P1.5

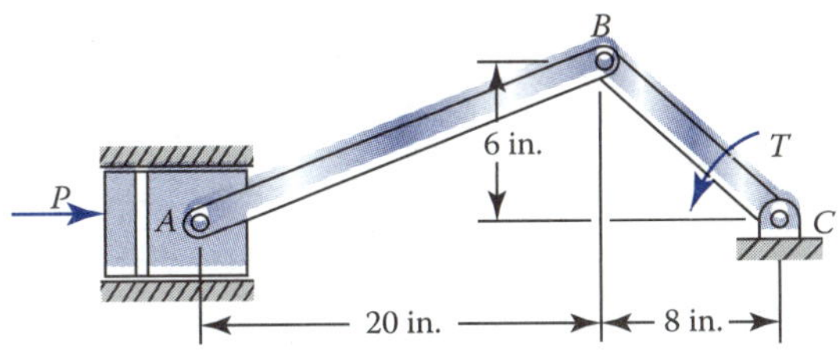

FIGURE P1.6

1.7 A crankshaft supported by bearings at A and B is subjected to a horizontal force $P = 4$ kN at point C, and a torque T at its right end is in static equilibrium (Figure P1.7).

Find:

a. The value of the torque T and the reactions at supports

b. The shear force, moment, and torque acting on the cross section at D

Given: $a = 120$ mm, $b = 50$ mm, $d = 70$ mm, $P = 4$ kN.

1.8 A structure, constructed by joining a beam AB with bar CD by a hinge, is under a weight $W = 30$ kN and a horizontal force $P = 60$ kN as depicted in Figure P1.8. Draw FBD of the beam AB and compute the reactions at support A.

1.9 A planar frame is supported and loaded as shown in Figure P1.9. Determine the reaction at hinge B.

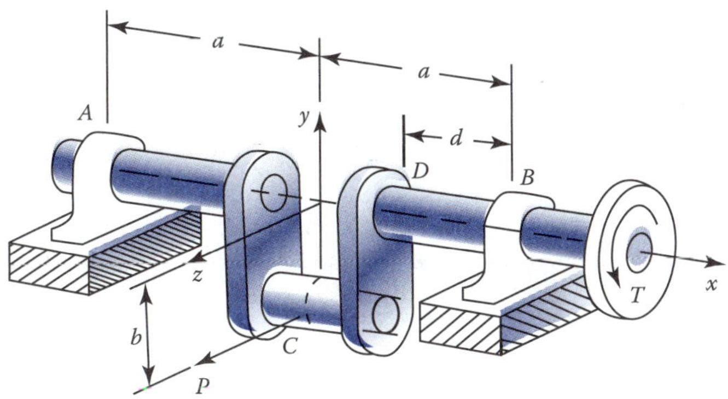

FIGURE P1.7

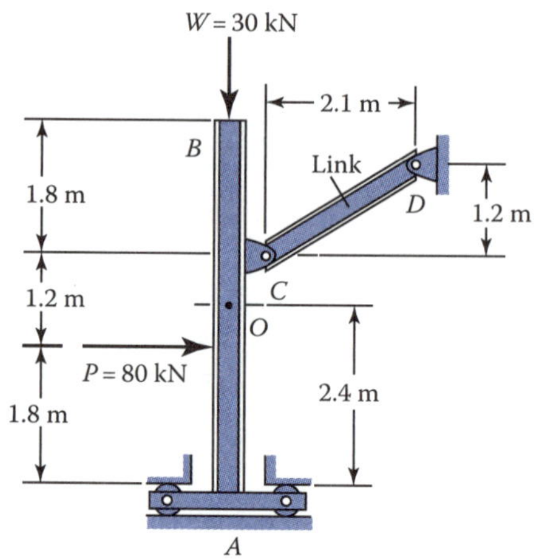

FIGURE P1.8

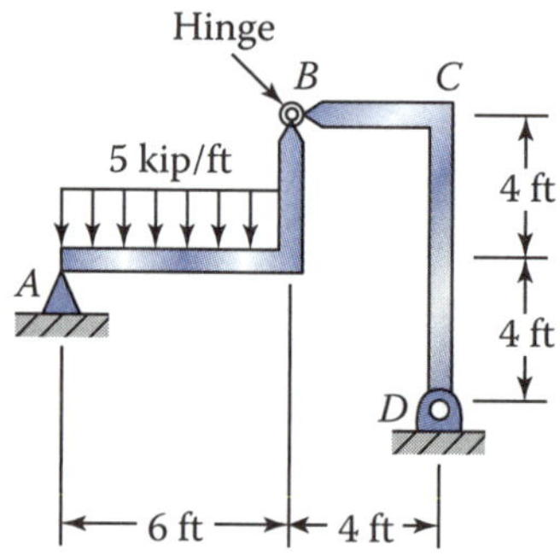

FIGURE P1.9

1.10 A hollow transmission shaft *AB* is supported at *A* and *E* by bearings and loaded as depicted in Figure P1.10. Calculate

a. The torque *T* required for equilibrium

b. The reactions at the bearings

Given: $F_1 = 4$ kN, $F_2 = 3$ kN, $F_3 = 5$ kN, $F_4 = 2$ kN.

1.11 A crank is built in at left end *A* and subjected to a vertical force $P = 2$ kN at *D*, as shown in Figure P1.11.

a. Sketch FBDs of the shaft *AB* and the arm *BC*.

b. Find the values and directions of the forces, moments, and torque at *C*, at end *B* of arm *BC*, at end *B* of shaft *AB*, and at *A*.

1.12 A pipe formed by three perpendicular arms *AB*, *BC*, and *CD* lying in the *x*, *y*, and *z* directions, respectively, is fixed at left end *A* (Figure P1.12). The force $P = 200$ N acts at point *E* by a wrench. Draw the FBD of the entire pipe and determine the reactions at *A*.

1.13 Resolve Problem 1.12 for the case in which the entire piping is constructed of a 75 mm (3 in.) nominal diameter standard steel pipe.

Assumption: The weight of the pipe (see Table A.4) will be taken into account.

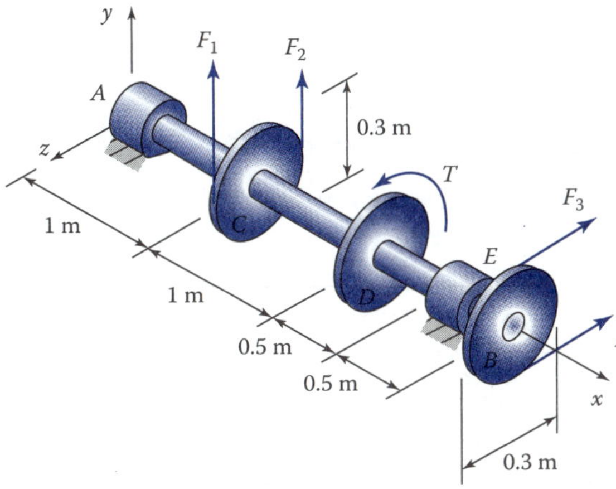

FIGURE P1.10

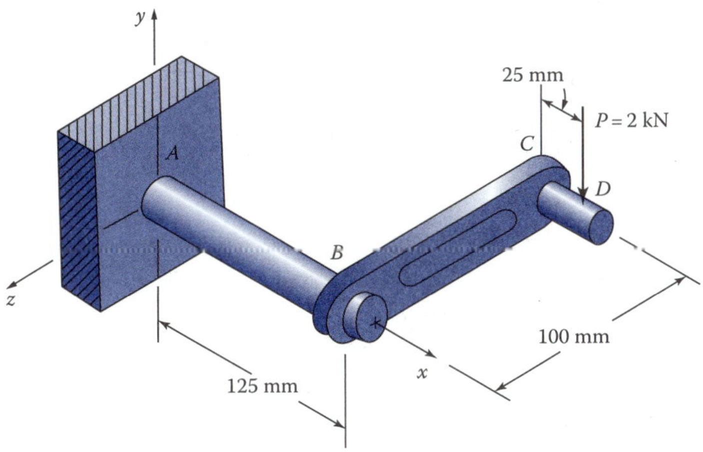

FIGURE P1.11

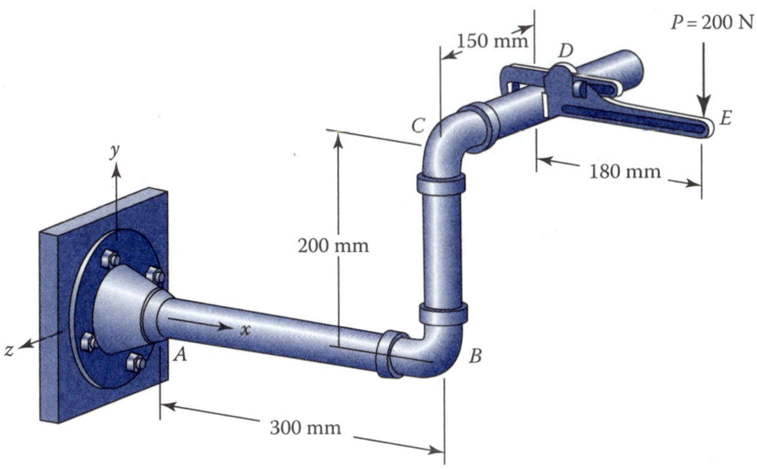

FIGURE P1.12

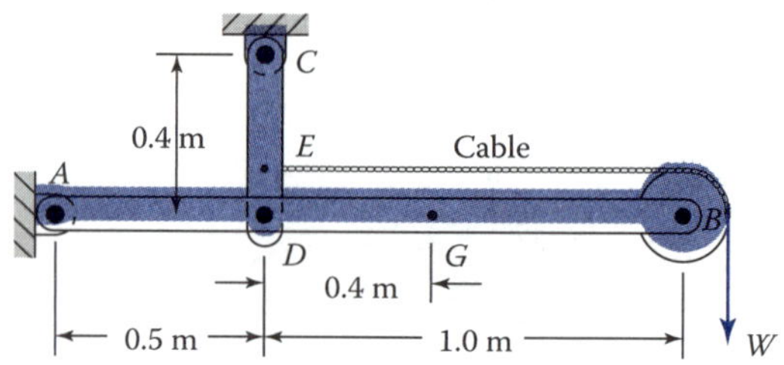

FIGURE P1.14

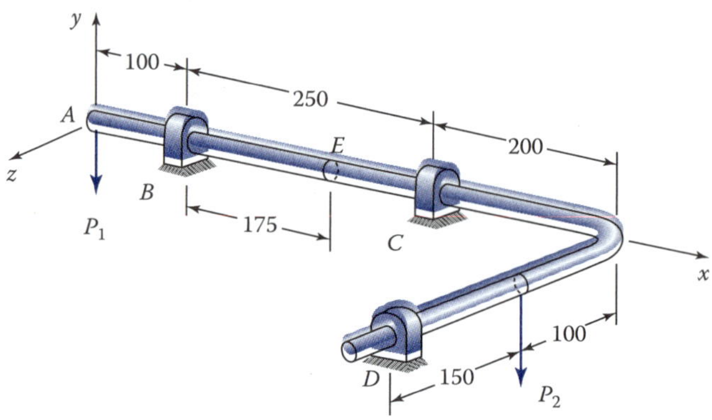

FIGURE P1.15

1.14 Pin-connected members *ADB* and *CD* carry a load *W* applied by a cable–pulley arrangement, as shown in Figure P1.14. Determine

a. The components of the reactions at *A* and *C*

b. The axial force, shear force, and moment acting on the cross section at point *G*

Given: The pulley at *B* has a radius of 150 mm. Load $W = 1.6$ kN.

1.15 A bent rod is supported in the *xz* plane by bearings at *B*, *C*, and *D* and loaded as shown in Figure P1.15. Dimensions are in millimeters. Calculate the moment and shear force in the rod on the cross section at point *E*, for $P_1 = 200$ N and $P_2 = 300$ N.

1.16 Redo Problem 1.15, for the case in which $P_1 = 0$ and $P_2 = 400$ N.

1.17 A gear train is used to transmit a torque $T = 150$ N·m from an electric motor to a driven machine (Figure P1.17). Determine the torque acting on driven machine shaft, T_d, required for equilibrium.

1.18 A planar frame formed by joining a bar with a beam with a hinge is loaded as shown in Figure P1.18. Calculate the axial force in the bar *BC*.

1.19 A frame *AB* and a simple beam *CD* are supported as shown in Figure P1.19. A roller fits snugly between the two members at *E*. Determine the reactions at *A* and *C* in terms of load *P*.

Section 1.10

1.20 Consider a conventional air compressor, like a small internal combustion engine, which has a crankshaft, a connecting rod and piston, a cylinder, and a valve head. The crankshaft is driven by either an electric motor or a gas engine. Note that the

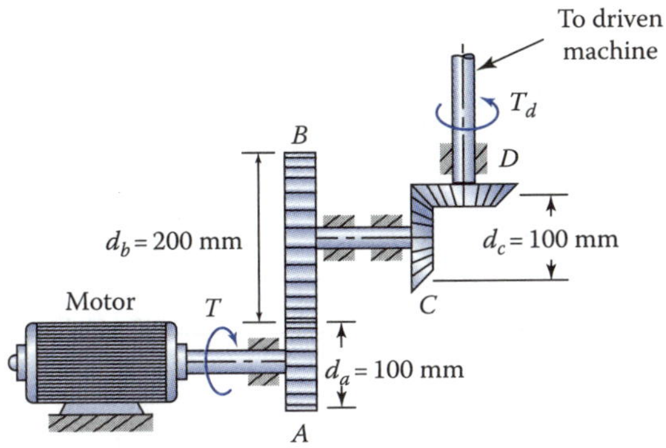

FIGURE P1.17

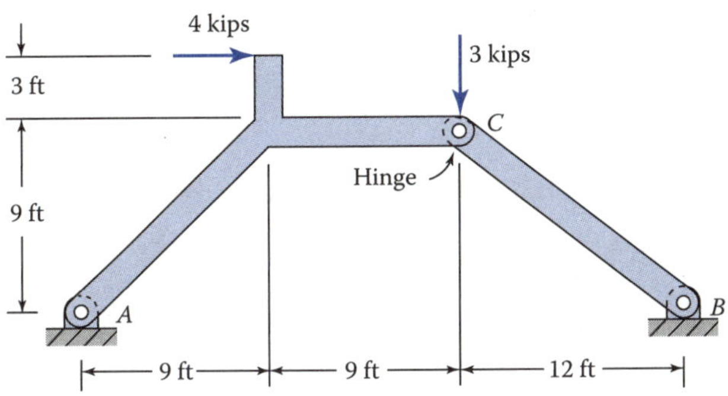

FIGURE P1.18

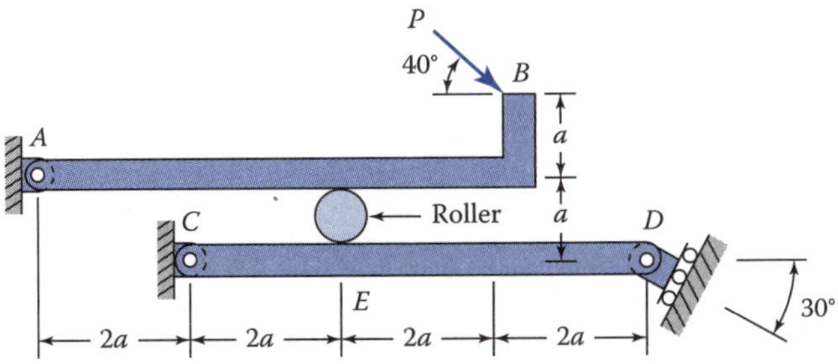

FIGURE P1.19

compressor has an air tank to hold a quantity of air within a preset pressure range that drives the air tools.

Given: The compressor's crankshaft (such as in Figure P1.7) is rotating at a constant speed n. Mean air pressure exerted on the piston during the compression period equals p. The piston area, piston stroke, and compressor efficiency are A, L, and e, respectively. *Data*:

$A = 2100\text{ mm}^2$, $L = 60\text{ mm}$, $e = 90\%$, $n = 1500\text{ rpm}$, $p = 1.2\text{ MPa}$.

Find

a. Motor power (in kW) required to drive the crankshaft.

b. Torque transmitted through the crankshaft.

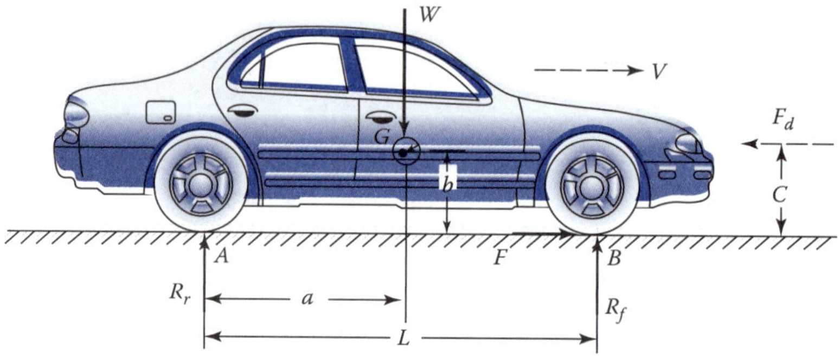

FIGURE P1.21

1.21 A car of weight with its center of gravity located at G is shown in Figure P1.21. Find the reactions between the tires and road

a. When the car travels at a constant speed V with an aerodynamic drag of 18 hp.

b. If the car is at rest

Given: $a = 60$ in., $b = 22$ in., $c = 25$ in., $L = 110$ in., $V = 65$ mph, $W = 3.2$ kips.

Assumptions: The car has front wheel drive. Vertical aerodynamic forces are omitted. Drag force F_d may be approximated by Equation 1.17.

1.22 Redo Problem 1.21, for the case in which the car has rear-wheel drive and its load acting at G is increased about 1.2 kips.

1.23 A shaft ABC is driven by an electric motor, which rotates at a speed of n and delivers 35 kW through the gears to a machine attached to the shaft DE (Figure P1.23). Draw the FBD of the gears and find

a. Tangential force F between the gears

b. Torque in the shaft DE

Given: $r_A = 125$ mm, $r_D = 75$ mm, $n = 500$ rpm.

1.24 The input shaft to a gearbox operates at speed of n_1 and transmits a power of 30 hp. The output power is 27 hp at a speed of n_2. What is the torque of each shaft (in kip·in.) and the efficiency of the gearbox?

Given: $n_1 = 1800$ rpm, $n_2 = 425$ rpm.

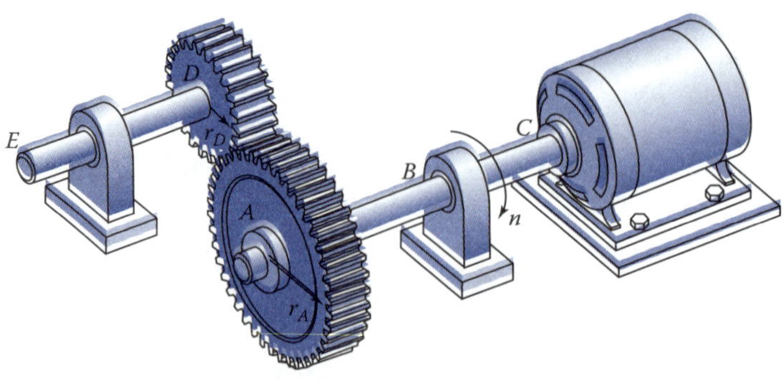

FIGURE P1.23

1.25 A punch press with a flywheel produces N punching strokes per minute. Each stroke provides an average force of F over a stroke of s. The press is driven through a gear reducer by a shaft. Overall efficiency is e. Determine

a. The power output

b. The power transmitted through the shaft

Given: $N = 150$, $F = 500$ lb, $s = 2.5$ in., $e = 88\%$.

1.26 A rotating ASTM A-48 cast iron flywheel has outer rim diameter d_o, inner rim diameter d_i, and length in the axial direction of l (Figure 1.8). Calculate the braking energy required in slowing the flywheel from 1200 to 1100 rpm.

Assumption: The hub and spokes add 5% to the inertia of the rim.

Given: $d_o = 400$ mm, $d_i = 0.75d_o$, $l = 0.25d_o$, $\rho = 7200$ kg/m^3 (see Table B.1).

Sections 1.11 and 1.12

1.27 A pin-connected frame $ABCD$ consists of three bars and a wire (Figure P1.27). Following the application a horizontal force F at joint B, joint C moves 0.4 in. to the right, as depicted by the dashed lines in the figure. Compute the normal strain in the wire.

Assumptions: The bars will be taken as rigid and weightless. Inasmuch as the angle of rotation of bar DC is very small, the vertical coordinate of C' can be taken equal to its length: $L_{DC} \approx L_{DC'} \cos \alpha$. Similarly, $L_{AB} \approx L_{AB'} \cos \alpha$.

1.28 A hollow cylinder is under an internal pressure that increases its 300 mm inner diameter and 500 mm outer diameter by 0.6 and 0.4 mm, respectively. Calculate

a. The maximum normal strain in the circumferential direction

b. The average normal strain in the radial direction

1.29 A thin triangular plate ABC is uniformly deformed into a shape $AB'C$, as shown by the dashed lines in Figure P1.29. Determine

a. The normal strain in the direction of the line OB

b. The normal strain for the line AB

c. The shear strain between the lines AB and AC

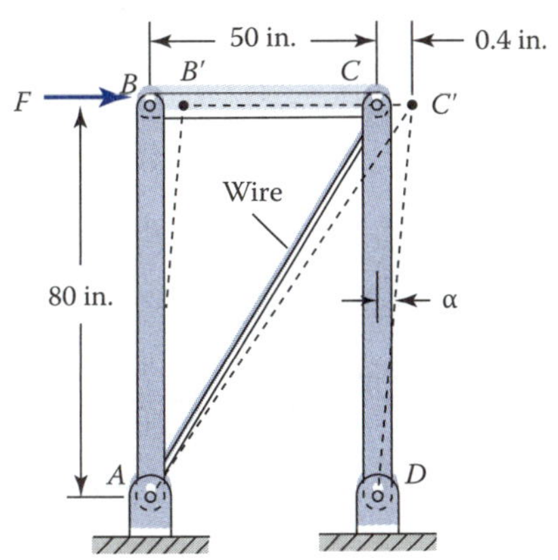

FIGURE P1.27

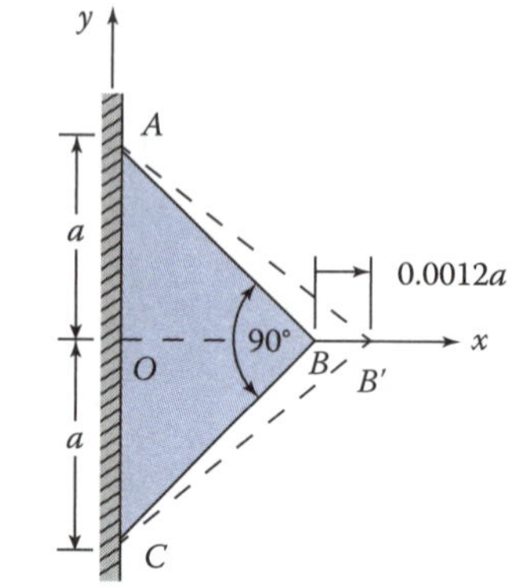

FIGURE P1.29

1.30 A 200 mm × 250 mm rectangle *ABCD* is drawn on a thin plate prior to loading. After loading, the rectangle has the dimensions (in millimeters) shown by the dashed lines in Figure P1.30. Calculate, at corner point *A*,

a. The normal strains ε_x and ε_y
b. The final length of side *AD*

1.31 A thin rectangular plate, $a = 200$ mm and $b = 150$ mm (Figure P1.31), is acted on by a biaxial tensile loading, resulting in the uniform strains $\varepsilon_x = 1000\ \mu$ and $\varepsilon_y = 800\ \mu$. Determine the change in length of diagonal *BD*.

1.32 When loaded, the plate of Figure P1.32 deforms into a shape in which diagonal *AC* elongates 0.2 mm and diagonal *BD* contracts 0.5 mm while they remain perpendicular. Calculate the average strain components ε_x, ε_y, and γ_{xy}.

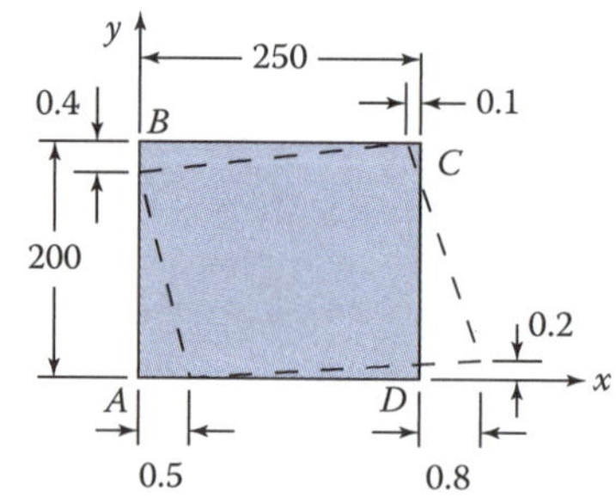

FIGURE P1.30

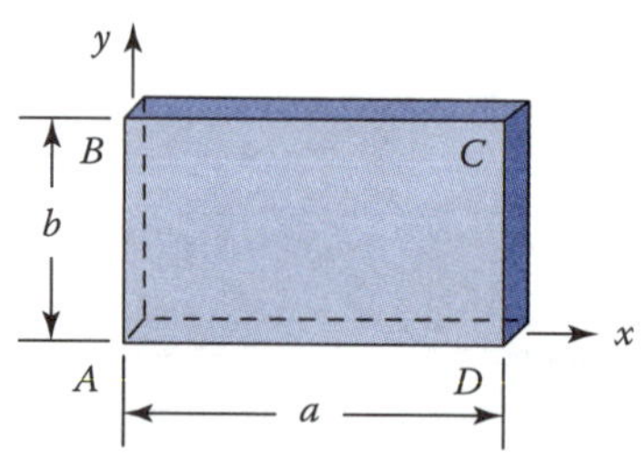

FIGURE P1.31

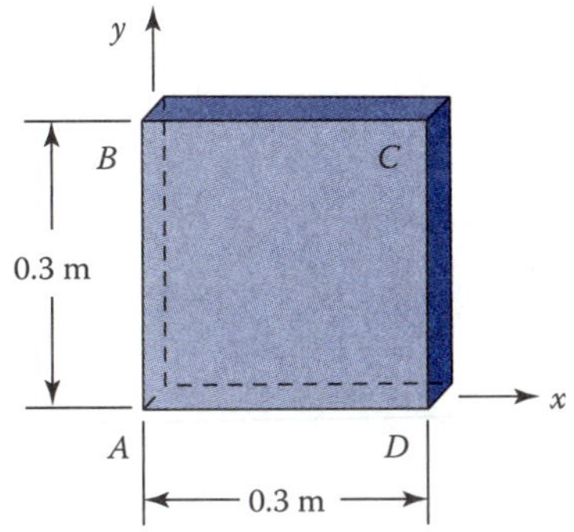

FIGURE P1.32

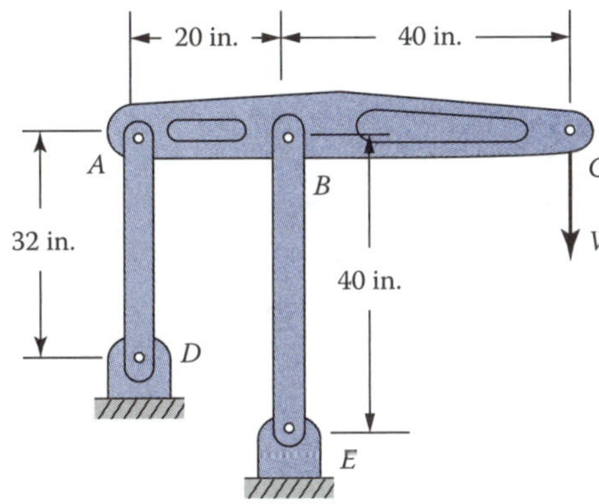

FIGURE P1.33

1.33 A rigid bar *ABC* is attached to the links *AD* and *BE* as illustrated in Figure P1.33. After the load *W* is applied, point *C* moves 0.2 in. downward, and the axial strain in the bar *AD* equals 800 μ. What is the axial strain in the bar *BE*?

1.34 As a result of loading, the thin rectangular plate (Figure P1.31) deforms into a parallelogram in which sides *AB* and *CD* shorten 0.004 mm and rotate 1000 μ rad counterclockwise, while sides *AD* and *BC* elongate 0.006 mm and rotate 200 μ rad clockwise. Determine, at corner point *A*,

a. The normal strains ε_x and ε_y and the shear strain γ_{xy}

b. The final lengths of sides *AB* and *AD*

Given: $a = 50$ mm, $b = 25$ mm.

2

Materials

2.1 Introduction

A great variety of materials has been produced, and more are being produced in seemingly endless diversification. Material may be crystalline or noncrystalline. A crystalline material is made up of a number of small units called *crystals* or *grains*. Most materials must be processed before they are usable. Table 2.1 gives a general classification of engineering materials. This book is concerned with the macroscopic structural behavior: properties are based on experiments using samples of materials of appreciable size. It is clear that a macroscopic structure includes a number of elementary particles forming a continuous and homogeneous structure held together by internal forces. The website at www.matweb.com offers extensive information on materials.

In this chapter, the mechanical behavior, characteristics, treatment, and manufacturing processes of some common materials are briefly discussed. A review of the subject matter presented emphasizes how a viable as well as an economic design can be achieved. Later chapters explore typical material failure modes in more detail. The average properties of selected materials are listed in Table B.1 [1–4]. Unless specified otherwise, we assume in this text that the material is homogeneous and isotropic. With the exception of Sections 2.10 and 5.10, our considerations are limited to the behavior of elastic materials. Note that the design of plate and shell-like members, for example, as components of a missile or space vehicle, involves materials having characteristics dependent on environmental conditions. We refer to the ordinary properties of engineering materials in this volume. It is assumed that the reader has had a course in material science.

2.2 Material Property Definitions

The mechanical properties are those that indicate how the material is expected to behave when subjected to varying conditions of load and environment. These characteristics are determined by standardized destructive and nondestructive test methods outlined by the American Society for Testing and Materials (ASTM). A thorough understanding of material properties permits the designer to determine the size, shape, and method of manufacturing mechanical components.

Durability denotes the ability of a material to resist destruction over long periods of time. The destructive conditions may be chemical, electrical, thermal, or mechanical in nature or combinations of these conditions. The relative ease with which a material may

TABLE 2.1

Some Commonly Used Engineering Materials

Metallic Materials	
Ferrous metals	Nonferrous metals
Cast iron	Aluminum
Malleable iron	Chromium
Wrought iron	Copper
Cast steel	Lead
Plain carbon steel	Magnesium
Steel alloys	Nickel
Stainless steel	Platinum
Tool steel	Silver
Special steels	Tin
Structural steel	Zinc
Nonmetallic Materials	
Carbon and graphite	Plastics
Ceramics	Brick
Cork	Stone
Felt	Elastomer
Glass	Silicon
Concrete	Wood

be machined, or cut with sharp-edged tools, is termed its *machinability*. *Workability* represents the ability of a material to be formed into required shape. Usually, *malleability* is considered a property that represents the capacity of a material to withstand plastic deformation in compression without fracture. We see in Section 2.10 that *hardness* may represent the ability of a material to resist scratching, abrasion, cutting, or penetration.

Frequently, the limitations imposed by the materials are the controlling factors in design. Strength and stiffness are main factors considered in the selection of a material. However, for a particular design, durability, malleability, workability, cost, and hardness of the materials may be equally significant. In considering the cost, attention focuses on not only the initial cost but also the maintenance and replacement costs of the part. Therefore, selecting a material from both its functional and economic standpoints is vitally important.

An elastic material returns to its original dimensions on removal of applied loads. This elastic property is called *elasticity*. Usually, the elastic range includes a region throughout which stress and strain have a linear relationship. The elastic portion ends at a point called the *proportional limit*. Such materials are linearly elastic. In a viscoelastic solid, the state of stress is function of not only the strain but the time rates of change of stress and strain as well. A *plastically* deformed member does not return to its initial size and shape when the load is removed.

A homogenous solid displays identical properties throughout. If properties are the same in all directions at a point, the material is *isotropic*. A *composite* material is made up of two or more distinct constituents. A nonisotropic, or *anisotropic*, solid has direction-dependent properties. Simplest among them is that the material properties differ in three mutually perpendicular directions. A material so described is *orthotropic*. Some wood material may be modeled by orthotropic properties. Many manufactured materials are approximated as orthotropic, such as corrugated and rolled metal sheet, plywood, and fiber-reinforced concrete.

The capacity of a material to undergo large strains with no significant increase in stress is called *ductility*. Thus, a ductile material is capable of substantial elongation prior to failure. Such materials include mild steel, nickel, brass, copper, magnesium, lead, and teflon. The converse applies to a *brittle material*. A brittle material exhibits little deformation before rupture, for example, concrete, stone, cast iron, glass, ceramic materials, and many metallic alloys. A member that ruptures is said to *fracture*. Metals with strains at rupture in *excess* of 0.05 in./in. in the tensile test are sometimes considered to be ductile [5]. Note that, generally, ductile materials fail in *shear*, while brittle materials fail in *tension*. Further details on material property definitions are found in Sections 2.12 through 2.14, where description of metal alloys, the numbering system of steels, typical nonmetallic materials and material selection are included.

2.3 Static Strength

In analysis and design, the mechanical behavior of materials under load is of primary importance. Experiments, mainly in tension or compression tests, provide basic information about overall response of specimens to the applied loads in the form of stress–strain diagrams. These curves are used to explain a number of mechanical properties of materials. Data for a stress–strain diagram are usually obtained from a *tensile test*. In such a test, a specimen of the material, usually in the form of a round bar, is mounted in the grips of a testing machine and subjected to tensile loading, applied slowly and steadily or statically at room temperature (Figure 2.1). The ASTM specifies precisely the dimensions and construction of standard tension specimens.

The tensile test procedure consists of applying successive increments of load while taking corresponding electronic extensometer readings of the elongation between the two gage marks (gage length) on the specimen. During an experiment, the change in gage length is noted as a function of the applied load. The specimen is loaded until it finally ruptures. The force necessary to cause rupture is called the *ultimate load*. Figure 2.2 illustrates a steel specimen that has fractured under load and the extensometer attached at the right by two arms to it. Based on the test data, the stress in the specimen is found by dividing the force by the cross-sectional area, and the strain is found by dividing the elongation by the gage length. In this manner, a complete stress–strain diagram, a plot of strain as abscissa and stress as the ordinate, can be obtained for the material. The stress–strain diagrams differ widely for different materials.

2.3.1 Stress–Strain Diagrams for Ductile Materials

A typical stress–strain plot for a ductile material such as structural or mild steel in tension is shown in Figure 2.3a. Curve *OABCDE* is a conventional or *engineering* stress–strain diagram. The other curve, *OABCF*, represents the true stress–strain. The *true stress* refers to the load divided by the actual instantaneous cross-sectional area of the bar; the *true strain* is the sum of the elongation increments divided by the corresponding momentary length. For most practical purposes, the conventional stress–strain diagram provides satisfactory information for use in design.

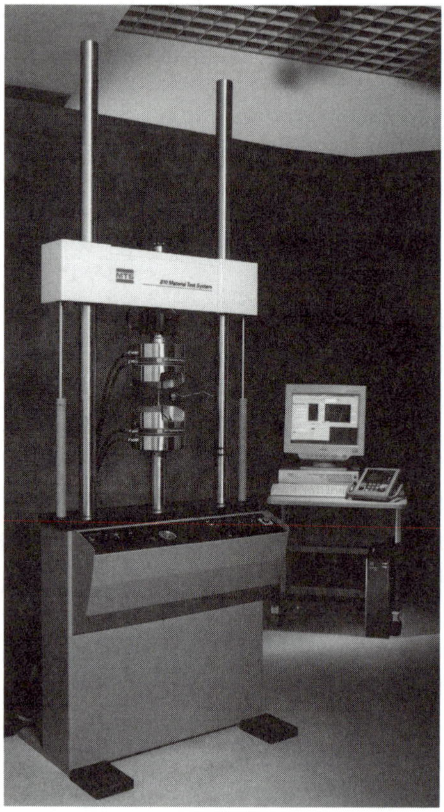

FIGURE 2.1
Tensile loading machine with automatic data-processing system. (Courtesy of MTS Systems Corp.)

We note that engineering stress (σ) is defined as load per unit area, and for the tensile, specimen is calculated from

$$\sigma = \frac{P}{A} \tag{a}$$

where
P is the applied load at any instant
A represents the original cross-sectional area of the specimen

The stress is assumed to be uniformly distributed across the cross section. The engineering strain (ε) is given by Equation 1.20. A detailed analysis of stress and strain will be taken up in the next chapter.

2.3.1.1 Yield Strength

The portion *OA* of the diagram is the elastic range. The linear variation of stress–strain ends at the *proportional limit*, S_p, point *A*. The lowest stress (point *B*) at which there is

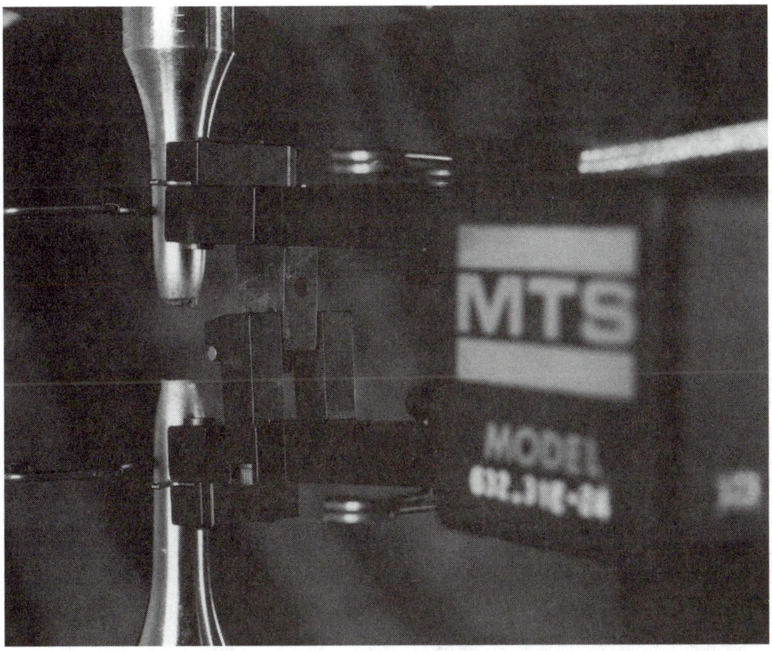

FIGURE 2.2
A tensile test specimen with extensometer attached; the specimen has fractured. (Courtesy of MTS Systems Corp.)

a marked increase in strain without corresponding increase in stress is referred to as the *yield point* or *yield strength* S_y. For most cases, in practice, the proportional limit and yield point are assumed to be one: $S_p \approx S_y$. In the region between *B* and *C*, the material becomes *perfectly plastic*, meaning that it can deform without an increase in the applied load.

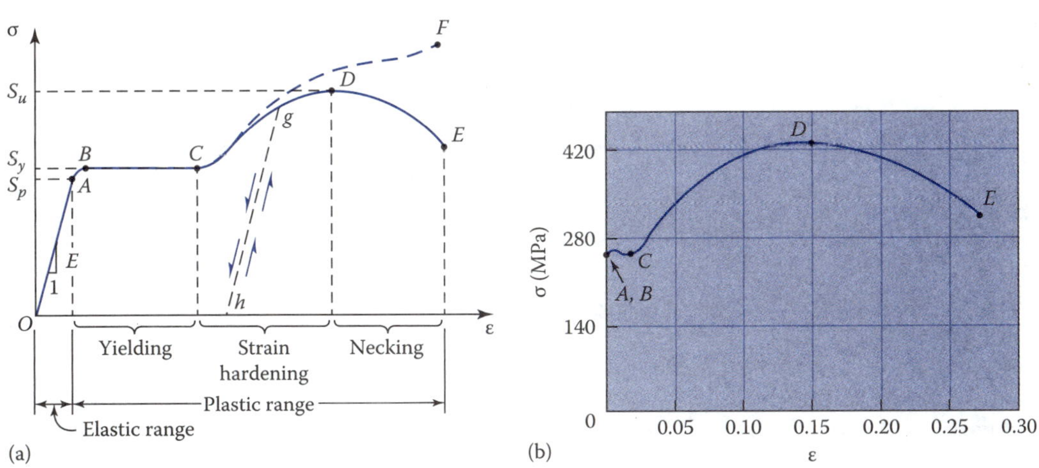

FIGURE 2.3
Stress–strain diagram for a typical structural steel in tension: (a) drawn not to scale and (b) drawn to scale.

2.3.1.2 Strain Hardening: Cold Working

The elongation of a mild steel specimen in the yield (or perfect plasticity) region BC is typically 10–20 times the elongation that occurs between the onset of loading and the proportional limit. The portion of the stress–strain curve extending from A to the point of fracture (E) is the plastic range. In the range CD, an increase in stress is required for a continued increase in strain. This is called the *strain hardening* or *cold working*. If the load is removed at a point g in region CD, the material returns to no stress at a point h along a new line parallel to the line OA: *a permanent set Oh* is introduced. If the load is reapplied, the new stress–strain curve is $hgDE$. Note that there is now new yield point (g) that is higher than before (point B) but reduced ductility. This process can be repeated until the material becomes brittle and fractures.

2.3.1.3 Ultimate Tensile Strength

The engineering stress diagram for the material when strained beyond C displays a typical ultimate stress (point D), referred to as the *ultimate* or *tensile strength* S_u. Additional elongation is actually accompanied by a reduction in the stress, corresponding to fracture strength S_f (point E) in the figure. Failure at E occurs, by separation of the bar into two parts (Figure 2.2), along the cone-shaped surface forming an angle of approximately 45° with its axis that corresponds to the planes of maximum shear stress. In the vicinity of the ultimate stress, the reduction of the cross-sectional area or the lateral contraction becomes clearly visible and a pronounced *necking* of the bar occurs in the range DE. An examination of the ruptured cross-sectional surface depicts a fibrous structure produced by stretching of the grains of the material.

Interestingly, the *standard measures of ductility* of a material are defined on the basis of the geometric change of the specimen, as follows:

$$\text{Percent elongation} = \frac{L_f - L_o}{L_o}(100) \tag{2.1}$$

$$\text{Percent reduction in area} = \frac{A_o - A_f}{A_o}(100) \tag{2.2}$$

Here, A_o and L_o denote, respectively, the original cross-sectional area and gage length of the specimen. Clearly, the ruptured bar must be pieced together to measure the final gage length L_f. Similarly, the final area A_f is measured at the fracture site where the cross section is minimal. Note that the elongation is not uniform over the length of the specimen but concentrated in the region of necking. Therefore, percent elongation depends on the gage length.

The diagram in Figure 2.3a depicts the general characteristics of the stress–strain diagram for mild steel, but its proportions are *not* realistic. As already noted, the strain between B and C may be about 15 times the strain between O and A. Likewise, the strains from C to E are many times greater than those from O to A. Figure 2.3b shows a stress–strain curve for mild steel drawn to scale. Clearly, the strains from O to A are so small that the initial part of the curve appears to be a vertical line.

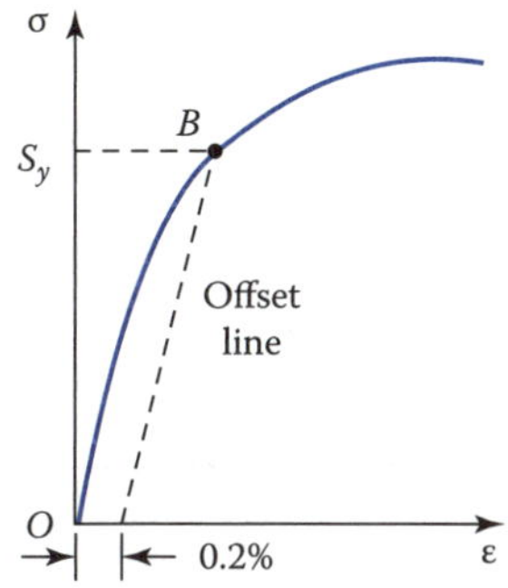

FIGURE 2.4
Determination of yield strength by the offset method.

2.3.1.4 Offset Yield Strength

Certain materials, such as heat-treated steels, magnesium, aluminum, and copper, do not show a distinctive yield point, and it is usual to use a yield strength S_y at an arbitrary strain. According to the so-called 0.2% offset method, a line is drawn through a strain of 0.002 (that is 0.2%), parallel to the initial slope at point O of the curve, as shown in Figure 2.4. The intersection of this line with the stress–strain curve defines the *offset yield strength* (point B). For the materials mentioned in the preceding discussion, the offset yield strength is slightly above the proportional limit.

2.3.2 Stress–Strain Diagram for Brittle Materials

The tensile behavior of gray cast iron, a typical brittle material, is shown in Figure 2.5a. We observe from the diagram that rupture occurs with no noticeable prior change in the rate of elongation. Therefore, for brittle materials, there is no difference between the ultimate strength and the fracture strength. Also, the strain at the rupture is much smaller for brittle materials than ductile materials. The stress–strain diagrams for brittle materials are characterized by having no well-defined linear region. The fracture of these materials is associated with the tensile stresses. Therefore, a brittle material breaks normal to the axis of the specimen (Figure 2.5b), because this is the plane of maximum tensile stress.

2.3.3 Stress–Strain Diagrams in Compression

Compression stress–strain curves, analogous to those in tension, may also be obtained for a variety of materials. Most ductile materials behave approximately the same in tension and compression over the elastic range. For these materials, the yield strength is about the same in tension and compression: $S_y \approx S_{yc}$, where the subscript c denotes compression. But, in the plastic range, the behavior is quite different. Since compression specimens expand instead of necking down, the compressive stress–strain curve continues to rise instead of reaching a maximum and dropping off.

A material having basically equal tensile and compressive strengths is termed an *even material*. For brittle materials, the entire compression stress–strain diagram has a shape similar to the shape of the tensile diagram. However, brittle materials usually

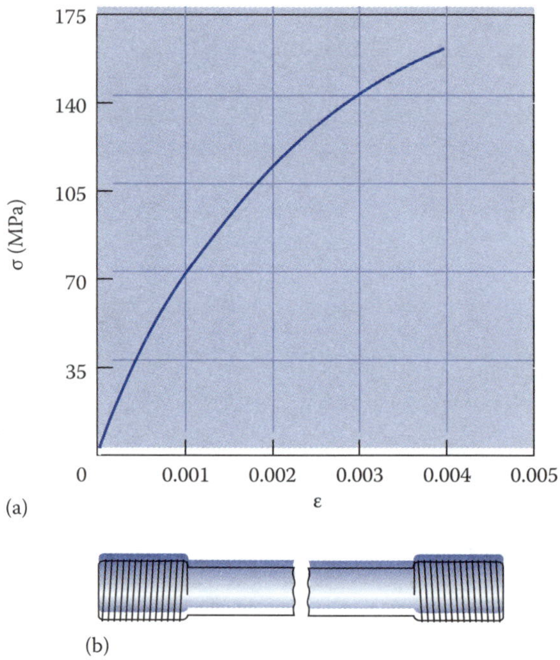

FIGURE 2.5
Gray cast iron in tension: (a) stress–strain diagram and (b) fractured specimen.

have characteristic stresses in compression that are much greater than in tension. A material that has different tensile and compressive strengths is referred to as an *uneven material*.

It is interesting to note that the strength of a machine component depends on its geometry and material as well as the type of loading it will experience. The strength of most *metals* is directly associated with the yield strength S_y of the material. When dealing with *polymers* or *ceramics*, the strength of interest is the *ultimate* strength S_u at the break or fracture, respectively, rather than the yield strength as for metals (see Example 2.5). Properties of a variety of nonmetals will be discussed in later sections.

2.4 Hooke's Law and Modulus of Elasticity

Most engineering materials have an initial region on the stress–strain curve where the material behaves both elastically and linearly. The linear elasticity is a highly important property of materials. For the straight-line portion of the diagram (Figure 2.3), the stress is directly proportional to the strain. Therefore,

$$\sigma = E\varepsilon \tag{2.3}$$

This relationship between stress and strain for a bar in tension or compression is known as *Hooke's law*. The constant E is called the *modulus of elasticity*, *elastic modulus*, or *Young's modulus*.

Inasmuch as ε is a dimensionless quantity, E has units of σ. In SI units, the elastic modulus is measured in newtons per square meter (or pascals), and in the U.S. customary system of units, it is measured in pounds per square inch (psi).

Equation 2.3 is highly significant in most of the subsequent treatment: the derived formulas are based on this law. We emphasize that Hooke's law is valid only up to the proportional limit of the material. The modulus of elasticity is seen to be the *slope* of the stress–strain curve in the linearly elastic range and is different for various materials. The E represents the stiffness of material in tension or compression. It is obvious that a material has a high elastic modulus value when its deformation in the elastic range is small.

Similarly, linear elasticity can be measured in a member subjected to pure shear loading. Referring to Equation 2.3, we now have

$$\tau = G\gamma \tag{2.4}$$

This is the Hooke's law for shear stress τ and shear strain γ. The constant G is called the *shear modulus of elasticity* or modulus of rigidity of the material, expressed in the same units as E: pascals or psi. The values of E and G for common materials are included in Table B.1.

We note that the slope of the stress–strain curve above the proportional limit is the *tangent modulus* E_t. That is, $E_t = d\sigma/d\varepsilon$. Likewise, the slope of a line from the origin to the point on the stress–strain curve above the proportional limit is known as the *Secant modulus* E_s. Therefore, $E_s = \sigma/\varepsilon$. Below the proportional limit, both E_t and E_s equal E.

In the elastic range, the ratio of the lateral strain to the axial strain is constant and known as *Poisson's ratio*:

$$\nu = \frac{\text{Lateral strain}}{\text{Axial strain}} \tag{2.5}$$

Here, the minus sign means that the lateral or transverse strain is of sense opposite to that of the axial strain.* Figure 2.6 depicts the lateral contraction of a rectangular parallelepiped element of side lengths a, b, and c in tension. Observe that the faces of the element at the origin

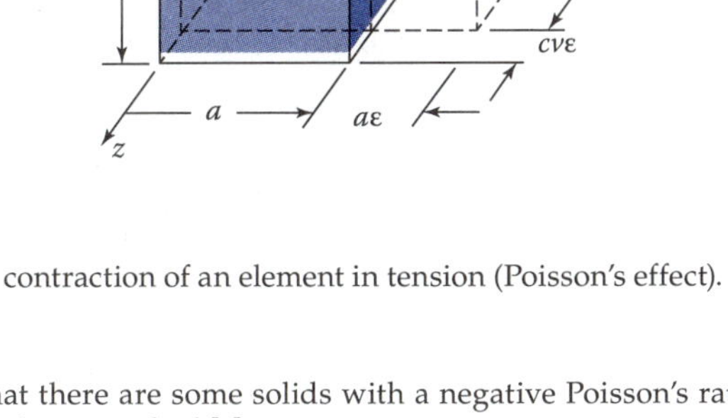

FIGURE 2.6
Axial elongation and lateral contraction of an element in tension (Poisson's effect).

* It should be mentioned that there are some solids with a negative Poisson's ratio. These materials become fatter in the cross section when stretched [6].

are assumed fixed in position. The deformations are greatly exaggerated, and the final shape of the element is shown by the dashed lines in the figure. The preceding definition is valid only for a uniaxial state of stress. Experiments show that, in most common materials, the values of ν are in the range 0.25–0.35. For steels, Poisson's ratio is usually assumed to be 0.3. Extreme cases include $\nu = 0.1$ for some concretes and $\nu = 0.5$ for rubber (Table B.1).

Example 2.1: Deformation and Stress in a Tension Bar

A tensile test is performed on an aluminum specimen of diameter d_o and gage length of L_o (see Figure 2.7). When the applied load reaches a value of P, the distance between the gage marks has increased by ΔL while the diameter of the bar has decreased by Δd.

Given: $d_o = 50$ mm, $\Delta d = 0.01375$ mm, $L_o = 250$ mm, $\Delta L = 0.2075$ mm, $P = 114$ kN

Find:

a. Axial and lateral strains
b. Poisson's ratio
c. Normal stress and modulus of elasticity

Solution

a. Lateral or transverse strain is equal to

$$\varepsilon_l = \frac{\Delta d}{d_o} = -\frac{0.01375}{50} = -0.000275 = -275\mu$$

where the transverse strain is negative, since the diameter of the bar decreases by Δd.
Axial strain, from Equation 1.20, is

$$\varepsilon_d = \frac{\Delta L}{L_o} = \frac{0.2075}{250} = 0.00083 = 830\mu$$

b. *Poisson's ratio,* using Equation 2.5 is

$$\nu = -\frac{\varepsilon_l}{\varepsilon_d} = -\frac{(-275)}{830} = 0.33$$

c. We have (see Section 2.3) the normal stress:

$$\sigma_d = \frac{P}{A} = \frac{114(10^3)}{\frac{\pi}{4}(0.05)^2} = 58.06 \text{ MPa}$$

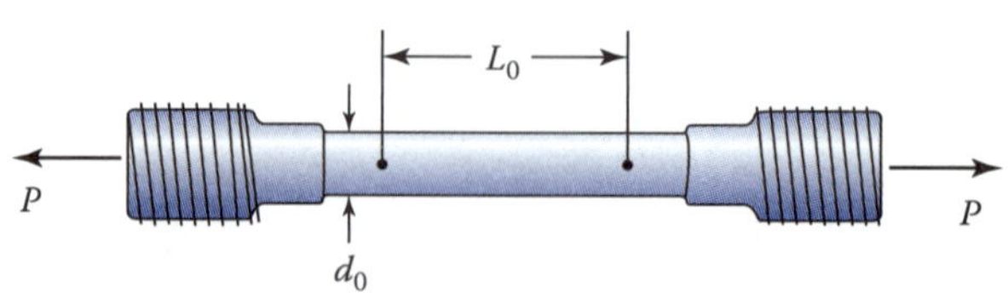

FIGURE 2.7
Example 2.1. A tensile specimen.

Modulus of elasticity, by Equation 2.3, is then

$$E = \frac{\sigma_a}{\varepsilon_a} = \frac{58.06(10^6)}{830(10^{-6})} = 70 \text{ GPa}$$

Comments: The stress obtained (58.06 MPa) is well within the yield strength of the material (260 MPa, from Table B.1). We note that, practically, when properties such as Poisson's ratio and modulus of elasticity are studied, it is best to work with the corresponding stress–strain diagram, assuring that these quantities are associated with the elastic range of the material behavior.

2.5 Generalized Hooke's Law

For a 2D or 3D state of stress, each of the stress components is taken to be a linear function of the components of strain within the linear elastic range. This assumption usually predicts the behavior of engineering materials with good accuracy. In addition, the principle of superposition applies under multiaxial loading, since strain components are small quantities. In the following development, we rely on certain experimental evidence to derive the stress–strain relations for linearly elastic isotropic materials: a normal stress creates no shear strain whatsoever, and shear stress produces only shear strain.

Consider now an element of unit thickness subjected to a biaxial state of stress (Figure 2.8). Under the action of the stress σ_x, not only would the direct strain σ_x/E occur, but a y contraction as well, $-\nu\sigma_x/E$. Likewise, were σ_y to act only, an x contraction $-\nu\sigma_y/E$ and a y strain σ_y/E would result. Therefore, simultaneous action of both stresses σ_x and σ_y results in the following strains in the x and y directions:

$$\varepsilon_x = \frac{\sigma_x}{E} - \nu\frac{\sigma_y}{E} \tag{2.6a}$$

$$\varepsilon_y = \frac{\sigma_y}{E} - \nu\frac{\sigma_x}{E} \tag{2.6b}$$

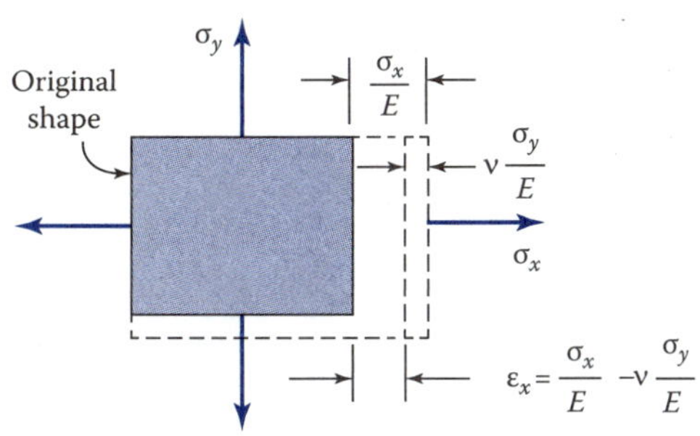

FIGURE 2.8
Element deformations caused by biaxial stress.

The elastic stress–strain relation, Equation 2.4, for the state of 2D pure shear, is given by

$$\gamma_{xy} = \frac{\tau_{xy}}{G} \tag{2.6c}$$

Inversion of Equations 2.6 results in the stress–strain relationships of the form

$$\begin{aligned}
\sigma_x &= \frac{E}{1-\nu^2}\left(\varepsilon_x + \nu\varepsilon_y\right) \\
\sigma_y &= \frac{E}{1-\nu^2}\left(\varepsilon_y + \nu\varepsilon_x\right) \\
\tau_{xy} &= G\gamma_{xy}
\end{aligned} \tag{2.7}$$

Equations 2.6 and 2.7 represent *Hooke's law for 2D stress.*

The foregoing procedure is easily extended to a 3D stress state (Figure 1.9). Then, the strain–stress relations, known as the *generalized Hooke's law,* consist of the following expressions:

$$\begin{aligned}
\varepsilon_x &= \frac{1}{E}\left[\sigma_x - \nu\left(\sigma_y + \sigma_z\right)\right] \\
\varepsilon_y &= \frac{1}{E}\left[\sigma_y - \nu\left(\sigma_x + \sigma_z\right)\right] \\
\varepsilon_z &= \frac{1}{E}\left[\sigma_z - \nu\left(\sigma_x + \sigma_y\right)\right] \\
\gamma_{xy} &= \frac{\tau_{xy}}{G}, \quad \gamma_{yz} = \frac{\tau_{yz}}{G}, \quad \gamma_{xz} = \frac{\tau_{xz}}{G}
\end{aligned} \tag{2.8}$$

The shear modulus of elasticity G is related to the modulus of elasticity E and Poisson's ratio ν. It can be shown that

$$G = \frac{E}{2(1+\nu)} \tag{2.9}$$

So, for an isotropic material, there are only two independent elastic constants. The values of E and G are determined experimentally for a given material, and ν can be found from the preceding basic relationship. Since the value of Poisson's ratio for ordinary materials is between 0 and 1/2, we observe from Equation 2.9 that G must be between $(1/3)E$ and $(1/2)E$.

2.5.1 Volume Change

The unit change in volume e, the change in volume ΔV per original volume V_o, in elastic materials subjected to stress is defined by

$$e = \frac{\Delta V}{V_o} = \varepsilon_x + \varepsilon_y + \varepsilon_z \tag{2.10}$$

The shear strains cause no change in volume. The quantity e is also referred to as *dilatation*. Equation 2.10 can be used to calculate the increase or decrease in volume of a member under loading, provided that the strains are known.

Based on the generalized Hooke's law, the dilatation can be found in terms of stresses and material constants. Using Equation 2.8, the stress–strain relationships may be expressed as follows:

$$
\begin{aligned}
\sigma_x &= 2G\varepsilon_x + \lambda e \qquad & \tau_{xy} &= G\gamma_{xy} \\
\sigma_y &= 2G\varepsilon_y + \lambda e \qquad & \tau_{yz} &= G\gamma_{yz} \\
\sigma_z &= 2G\varepsilon_z + \lambda e \qquad & \tau_{xz} &= G\gamma_{xz}
\end{aligned}
\tag{2.11}
$$

In the preceding, we have

$$
e = \varepsilon_x + \varepsilon_y + \varepsilon_z = \frac{1-2\nu}{E}\left(\sigma_x + \sigma_y + \sigma_z\right) \tag{2.12}
$$

$$
\lambda = \frac{\nu E}{(1+\nu)(1-2\nu)} \tag{2.13}
$$

where λ is an elastic constant.

When an elastic member is subjected to a hydrostatic pressure p, the stresses are $\sigma_x = \sigma_y = \sigma_z = -p$ and $\tau_{xy} = \tau_{yz} = \tau_{xz} = 0$. Then, Equation 2.12 becomes $e = -3(1 - 2\nu)p/E$. This may be written in the following form:

$$
K = -\frac{p}{e} = \frac{E}{3(1-2\nu)} \tag{2.14}
$$

The quantity K represents the modulus of volumetric expansion or the so-called *bulk modulus of elasticity*. Equation 2.14 shows that, for incompressible materials ($e = 0$), $\nu = 1/2$. For most materials, however, $\nu < 1/2$ (Table B.1). Note that, in the perfectly plastic region behavior of a material, no volume change occurs, and hence, Poisson's ratio may be taken as 1/2.

Example 2.2: Determination of Displacements of a Plate

A steel panel of a device is approximated by a plate of thickness t, width b, and length a, subjected to stresses σ_x and σ_y, as shown in Figure 2.9. Calculate

a. The value of σ_x for which length a remains unchanged
b. The final thickness t' and width b'
c. The normal strain for the diagonal AC
d. The change in volume of the plate

Given: $a = 400$ mm, $b = 300$ mm, $t = 6$ mm, $E = 200$ GPa, $\nu = 0.3$, and $\sigma_y = 220$ MPa.

Assumption: The plate is in plane state of stress.

Solution

Inasmuch as the length does not change, we have $\varepsilon_x = 0$. In addition, plane stress $\sigma_z = 0$. Then, Equation 2.8 becomes

$$
\sigma_x = \nu\sigma_y \tag{2.15a}
$$

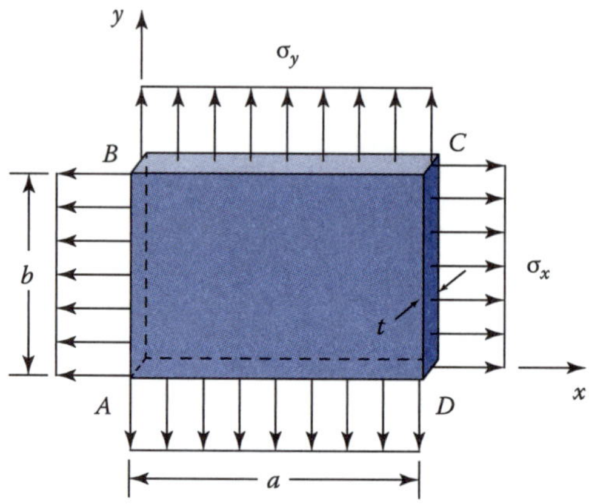

FIGURE 2.9
Example 2.2. Plate in biaxial stress.

$$\varepsilon_y = \frac{1}{E}(\sigma_y - \nu\sigma_x)$$
$$\varepsilon_z = -\frac{\nu}{E}(\sigma_x + \sigma_y) \tag{2.15b}$$

a. The given data are carried into Equation 2.15a to yield

$$\sigma_x = 0.3(220 \times 10^6) = 66 \text{ MPa}$$

b. Through the use of Equation 2.15b, we obtain

$$\varepsilon_y = \frac{10^6}{200(10^9)}[220 - 0.3(66)] = 1001\ \mu$$

$$\varepsilon_z = -\frac{0.3}{200(10^3)}(66 + 220) = -429\ \mu$$

In the foregoing, a minus sign means a decrease in the thickness. Therefore,

$$t' = t(1 + \varepsilon_z) = 6(0.9996) = 5.998 \text{ mm}$$

$$b' = b(1 + \varepsilon_y) = 300(1.0010) = 300.300 \text{ mm}$$

c. The original and final lengths of the diagonal are, respectively,

$$AC = (300^2 + 400^2)^{1/2} = 500 \text{ mm}$$

$$A'C' = (300.300^2 + 400^2)^{1/2} = 500.180 \text{ mm}$$

Note that $A'C'$ is not shown in Figure 2.9. The normal strain for the diagonal is

$$\varepsilon_{AC} = \frac{500.180 - 500}{500} = 360\ \mu$$

Comment: Alternatively, this result may readily be found by using the strain transformation equations, to be discussed in Section 3.11.

d. Change in volume, applying Equation 2.12, is

$$e = \frac{1-2\nu}{E} = \left(\sigma_x + \sigma_y + \sigma_z\right)$$

$$= \frac{1-2(0.3)}{200\left(10^9\right)}\left[66 + 220 + 0\right]10^6 = 0.57 \times 10^{-3}$$

Equation 2.10 is therefore

$$\Delta V = eV_o = \left(0.57 \times 10^{-3}\right)\left(400 \times 300 \times 6\right) = 410 \text{ mm}^3$$

Comment: The positive sign indicates an increase in the volume of the plate.

Example 2.3: Volume Change of a Cylinder under Biaxial Loads

A solid brass cylinder of diameter *d* and length *L* (Figure 2.10) is under axial and radial pressures 30 and 12 ksi, respectively.

Find: The change in

a. The length ΔL and diameter Δd
b. The volume of the cylinder ΔV

Given: $d = 5$ in., $L = 8$ in., and $E = 15 \times 10^6$ psi, $\nu = 0.34$ (from Table B.1).

Assumption: Cylinder deforms uniformly.

Solution

Axial stress $\sigma_x = -30$ ksi and along any diameter radial stresses $\sigma_y = \sigma_z = \sigma = -12$ ksi. Applying Equation 2.8, associated strains are found as follows:

$$\varepsilon_x = -\frac{1}{E}\left[\sigma_x - \nu(\sigma + \sigma)\right]$$

$$= -\frac{10^3}{15 \times 10^6}\left[30 - (0.34)(12 + 12)\right] = -1456\ \mu$$

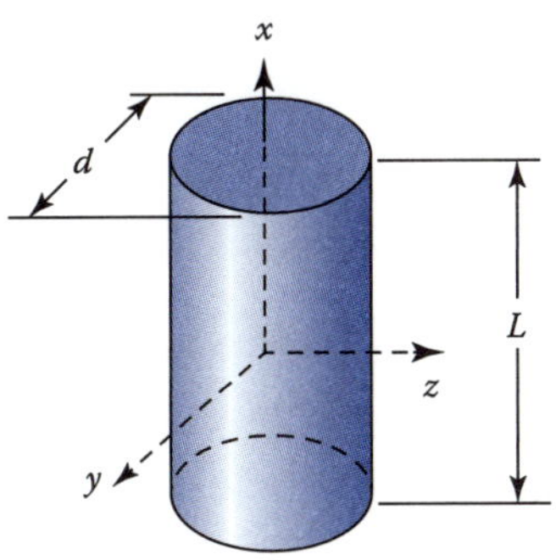

FIGURE 2.10
Example 2.3. A solid cylinder.

and

$$\varepsilon_y = \varepsilon_z = \varepsilon = -\frac{1}{E}\left[\sigma - \nu\left(\sigma + \sigma_x\right)\right]$$

$$= -\frac{10^3}{15\times 10^6}\left[12-(0.34)(12+30)\right] = 152\ \mu$$

a. *Changes in length and diameter.* Decrease in length and increase in diameter are, respectively,

$$\Delta L = \varepsilon_x L = \left(-1456\times 10^{-6}\right)(8) = -11.65\left(10^{-3}\right)\text{ in.}$$

$$\Delta d = \varepsilon d = \left(152\times 10^{-6}\right)(5) = 0.76\left(10^{-3}\right)\text{ in.}$$

b. Volume change. Using Equation 2.12, we have

$$e = \varepsilon_x + 2\varepsilon$$

$$= (-1456 + 2\times 152)10^{-6} = -1152\times 10^{-6}$$

It follows, from Equation 2.10, that

$$\Delta V = eV_o$$

$$= \left(-1152\times 10^{-6}\right)\left[\pi(2.5)^2(8)\right] = -181\left(10^{-3}\right)\text{in}^3.$$

Comment: A negative sign means a decrease in the volume of the cylinder.

2.6 Thermal Stress–Strain Relations

When displacements of a heated isotropic member are prevented, thermal stresses occur. The effects of such stresses can be severe, particularly since the most adverse thermal environments are frequently associated with design requirements dealing with unusually stringent constraints as to weight and volume. The foregoing is especially true in aerospace and machine design (e.g., engine, power plant, and industrial process) applications.

The total strains are obtained by adding thermal strains of the type described by Equation 1.21 and the strains owing to the stress resulting from mechanical loads. In doing so, for instance, referring to Equation 2.6 for 2D stress,

$$\varepsilon_x = \frac{1}{E}\left(\sigma_x - \nu\sigma_y\right) + \alpha T$$

$$\varepsilon_y = \frac{1}{E}\left(\sigma_y - \nu\sigma_x\right) + \alpha T \qquad (2.16)$$

$$\gamma_{xy} = \frac{\tau_{xy}}{G}$$

From these equations, we obtain the stress–strain relations as

$$\begin{aligned}\sigma_x &= \frac{E}{1-\nu^2}\left(\varepsilon_x+\nu\varepsilon_y\right)-\frac{E\alpha T}{1-\nu}\\ \sigma_y &= \frac{E}{1-\nu^2}\left(\varepsilon_y+\nu\varepsilon_x\right)-\frac{E\alpha T}{1-\nu}\\ \tau_{xy} &= G\gamma_{xy}\end{aligned} \tag{2.17}$$

The quantities T and α represent the temperature change and the coefficient of expansion, respectively. Equations for 3D stress may be readily expressed in a like manner.

Note that because free thermal expansion causes no distortion in an isotropic material, the shear strain is unaffected, as shown in the preceding expressions. The differential equations of equilibrium are based on purely mechanical considerations and unchanged for thermoelasticity. The same is true of the strain–displacement relations and hence the conditions of compatibility, which are geometrical in character (see Section 3.17). Thermoelasticity and ordinary elasticity therefore differ only to the extent of Hooke's law. Solutions for the problems in the former are usually harder to obtain than solutions for the problems in the latter.

In statically determinate structures, a uniform temperature change will not cause any stresses, as thermal deformations are permitted to occur freely. On the other hand, a temperature change in a structure supported in a statically indeterminate manner induces stresses in the members. Detailed discussions and illustrations of thermal loads and stresses in components and assemblies are given in Chapters 4 and 16.

2.7 Temperature and Stress–Strain Properties

A large deviation in temperature may cause a change in the properties of a material. In this section, temperature effects on stress–strain properties of materials are considered. Effects of temperature on impact and fatigue strengths are treated in Sections 2.9 and 7.7, respectively. Another important thermal effect results because most materials expand with an increase in temperature.

2.7.1 Short-Time Effects of Elevated and Low Temperatures

For the static short-time testing of metals at elevated temperatures, it is generally found that the ultimate strength, yield strength, and modulus of elasticity are lowered with increasing temperature, whereas the ductility increases with temperature. Here, *elevated temperatures* refer to the absolute temperatures in excess of about one-third of the melting point absolute temperature of the material. On the contrary, at low temperatures, there is an increase in yield strength, ultimate strength, modulus of elasticity, and hardness and a decrease in ductility for metals. Therefore, when the operating temperatures are lower than the transition temperature, defined in Section 2.9, the possibility arises that a component could fail due to a brittle fracture.

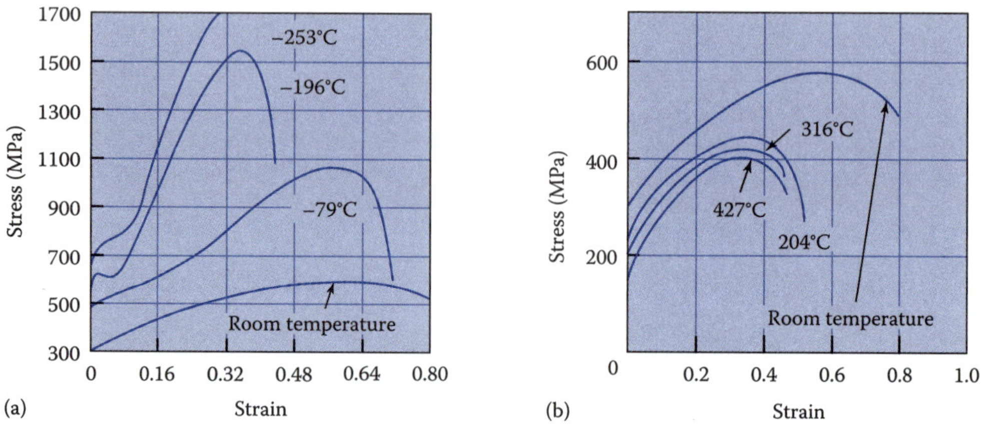

FIGURE 2.11
Stress–strain diagrams for AISI type 304 stainless steel in tension: (a) at low temperatures and (b) at elevated temperatures.

The problem of designing for extreme temperatures is a special one, in that information concerning material properties is not overly abundant. Figure 2.11 depicts the effect of low and high temperatures on the strength of a type 304 stainless steel [7]. The considerable property variations illustrated by these curves are caused by metallurgical changes that take place as the temperature increases or decreases.

2.7.2 Long-Time Effects of Elevated Temperatures: Creep

Most metals under a constant load at elevated temperatures over a long period develop additional strains. This phenomenon is called *creep*. Creep is time dependent because deformation increases with time until a rupture occurs. For some nonferrous metals and a number of nonmetallic materials such as plastics, wood, and concrete, creep may also be produced at low stresses and normal (room) temperatures.

A typical creep curve, for a mild steel specimen in tension at elevated temperatures, consists of three regions or stages (Figure 2.12). In the first region, the material is becoming stronger because of strain hardening, and the strain rate or creep rate ($d\varepsilon/dt$) decreases continuously. This stage is important if the load duration is short. The second region begins at a minimum strain rate and remains constant because of the balancing effects of strain hardening and annealing. *Annealing* refers to a process involving softening of a metal by heating and slowly cooling, discussed in Section 2.11. The secondary stage is usually the dominant interval of a creep curve. In the third region, the annealing effect predominates, and the deformation occurs at an accelerated creep rate until a rupture results.

When a component is subjected to a steady loading at elevated temperature and for a long period, the creep-rupture strength of the material determines its failure. However, failure at elevated temperatures due to dynamic loading will most likely occur early in the life of the material. Interest in the phenomenon of creep is not confined to possible failure by rupture but includes failure by large deformations that can make equipment inoperative. Therefore, in many designs, creep deformation must be maintained small. However, for some applications and within certain temperatures, stress, and time limits, creep effects need not be considered, and the stress–strain properties determined from static short-time testing are adequate.

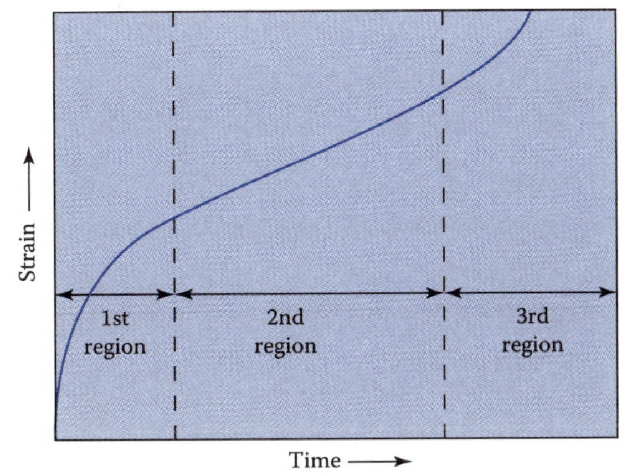

FIGURE 2.12
Creep curve for structural steel in tension at high temperatures.

2.8 Moduli of Resilience and Toughness

Some machine and structural elements must be designed more on the basis of absorbing energy than withstanding loads. Inasmuch as energy involves both loads and deflections, stress–strain curves are particularly relevant. A detailed discussion of strain energy and its application is found in Chapter 5. Here, we limit ourselves to the case of a member in tension to illustrate how the energy-absorbing capacity of a material is determined.

2.8.1 Modulus of Resilience

Resilience is the capacity of a material to absorb energy within the elastic range. The modulus of resilience U_r represents the energy absorbed per unit volume of material, or the *strain energy density*, when stressed to the proportional limit. This is equal to the area under the straight-line portion of the stress–strain diagram (Figure 2.13a), where proportional

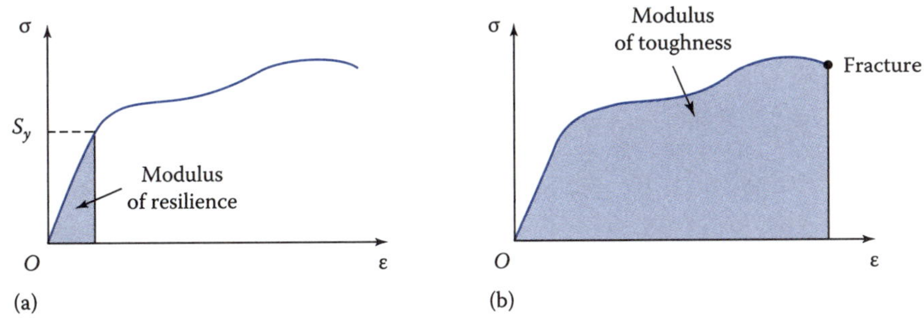

FIGURE 2.13
Stress–strain diagram: (a) modulus of resilience and (b) modulus of toughness.

limit S_p and yield strength S_y are taken approximately the same. The value of modulus of resilience, setting $\sigma_x = S_y$ into Equation 5.1b, has the following form:

$$U_r = \frac{S_y^2}{2E} \tag{2.18}$$

where E is the modulus of elasticity. Therefore, resilient materials are those having high strength and low moduli of elasticity.

2.8.2 Modulus of Toughness

Toughness is the capacity of a material to absorb energy without fracture. The modulus of toughness U_t represents the energy absorbed per unit volume of material up to the point of fracture. It is thus equal to the entire area under stress–strain diagram (Figure 2.13b). Expressed mathematically, the modulus of toughness is

$$U_t = \int_0^{\varepsilon_f} \sigma d\varepsilon \tag{2.19}$$

The quantity ε_f is the *strain at fracture*. Clearly, the toughness of a material is related to its ductility as well as to its ultimate strength. It is often convenient to perform the foregoing integration graphically. A planimeter can be used to determine this area.

Sometimes, the modulus of toughness is approximated by representing the area under the stress–strain curve of ductile materials as the average of the yield strength S_y and ultimate strength S_u times the fracture strain. Therefore,

$$U_t = \frac{S_y + S_u}{2} \varepsilon_f \tag{2.20}$$

For brittle materials (e.g., cast iron), the approximation of the area under the stress–strain curve (Figure 2.5), as given by Equation 2.20, would be considerably in error. In such cases, the modulus of toughness is occasionally estimated by assuming that the strain–stress curve is a parabola. Then, using Equation 2.19 with $\varepsilon_f = \varepsilon_u$, the modulus of toughness is

$$U_t = \frac{2}{3} S_u \varepsilon_u \tag{2.21}$$

in which ε_u is the strain at the ultimate strength.

Toughness is usually associated with the capacity of a material to withstand an impact or shock load. Two common tests, the Charpy and Izod tests, discussed in the next section, determine the impact strength of materials at various temperatures. We observe that toughness obtained from these tests is as dependent on the geometry of the specimen as on the load rate. The units of both the modulus of toughness and modulus of resilience are expressed in joules (N · m) per cubic meter (J/m^3) in SI and in in. · lb per cubic inch in the U.S. customary system. These are the same units of stress, so we can also use pascals or psi as the units for U_r, and U_t. As an example, consider a structural steel having $S_y = 250$ MPa, $S_u = 400$ MPa, $\varepsilon_f = 0.3$, and $E = 200$ GPa (Table B.1). For this material, by Equations 2.18 and 2.20, we have $U_r = 156.25$ kPa and $U_t = 97.5$ MPa, respectively.

Note that fracture toughness is another material property that defines its ability to resist further crack propagation at the tip of a crack. When stress intensity reaches the fracture toughness, a fracture takes place with no warning. The study of this phenomenon is taken up in Section 6.3.

Example 2.4: Material Resilience on an Axially Loaded Rod

During the manufacturing process, a prismatic round steel rod must acquire an elastic strain energy of $U_{app} = 200$ in. · lb (Figure 2.14). Determine the required yield strength S_y for a factor of safety of $n = 2.5$ with respect to permanent deformation.

Given: $E = 30 \times 10^6$ psi, diameter $d = 7/8$ in., length $L = 4$ ft.

Solution

The volume of the member is

$$V = AL = \frac{\pi}{4}\left(\frac{7}{8}\right)^2 (4 \times 12) = 28.9 \text{in.}^3$$

The rod should be designed for a strain energy:

$$U = nU_{app} = 2.5(200) = 500 \text{ in.} \cdot \text{lb}$$

The strain energy density is therefore 500/28.9 = 17.3 lb/in.^{3}.

Through the use of Equation 2.18, we have

$$U_r = \frac{S_y^2}{2E}; \quad 17.3 = \frac{S_y^2}{2(30 \times 10^6)}$$

Solving,

$$S_y = 32.2 \text{ ksi}$$

Comment: Observe that the factor of safety is applied to the energy load and not to the stress.

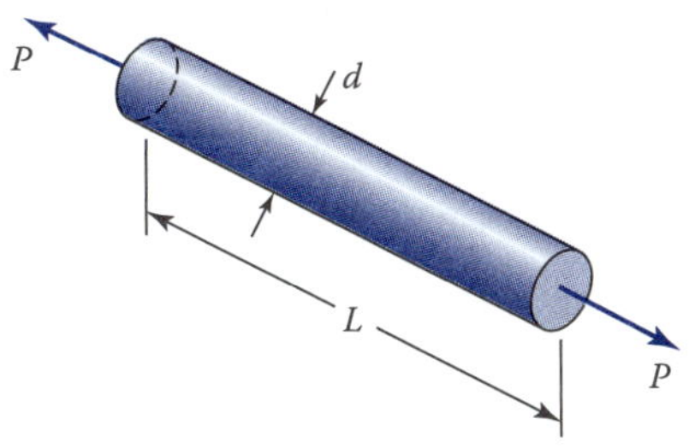

FIGURE 2.14
Example 2.4. Prismatic bar in tension.

Example 2.5: Most Efficient Rubber Bearing for Impact

A vibration absorption unit (such as in Figure P2.10) is to be designed using natural rubber (NR) or a synthetic rubber (SR), polyurethane rubber, and bearing material.

Find: Which is the most efficient choice?

Assumptions: Resilience is an important factor to be considered in this evaluation. As noted in Section 2.3, ultimate strength at the break will be substituted for the yield strength in Equation 2.18.

Solution

NR. Approximate mean values of the yield strength and modulus of elasticity, by Table B.10, are

$$S_u = 28 \text{ MPa} \quad \text{and} \quad E = 4.6 \text{ MPa} = 0.0046 \text{ GPa}$$

Through the use of Equation 2.18, the modulus of resilience is

$$(U_r)_{NR} = \frac{S_u^2}{2E}$$

$$= \frac{(28)^2\left(10^{12}\right)}{2(4.6)\left(10^6\right)} = 85.217\left(10^6\right)\text{ Pa} = 85.2 \text{ MPa}$$

Polyurethane rubber. Material properties from Table B.10 are

$$S_u = 30 \text{ MPa} \quad \text{and} \quad E = 17 \text{ MPa} = 0.017 \text{ GPa}$$

Equation 2.18 is thus

$$(U_r)_{SR} = \frac{S_u^2}{2E}$$

$$= \frac{(30)^2\left(10^{12}\right)}{2(17)\left(10^6\right)} = 26.471\left(10^6\right)\text{ Pa} = 26.5 \text{ MPa}$$

Comments: Results show that NR is about 3.2 times resilient than synthetic polyurethane rubber. NR is elastically more stretchy and flexible; it should be the choice for this application.

2.9 Dynamic and Thermal Effects

A *dynamic load* applied to a structure or machine is called the *impact load*, also referred to as the *shock load*, if the time of application is less than one-third of the lowest natural period of the structure. Otherwise, it is termed the *static load*. Examples of a shock load include rapidly moving loads, such as those caused by a railroad train passing over a bridge, or direct impact loads, such as a result from a drop hammer. In machine operation, impact loads are due to gradually increasing clearances that develop between mating parts with progressive wear, for example, steering gears and axle journals of automobiles; sudden application of loads, as occurs during explosion stroke of a combustion engine; and inertia loads, as introduced by high acceleration, such as in a flywheel.

2.9.1 Strain Rate

Strains and stresses in dynamic loading are much greater than those found in static loading, and hence, effects of impact loading are significant. Physical properties of materials depend on loading and speed. When a body is subjected to dynamic loading, strain rate $d\varepsilon/dt$ and its strengths increase. Here, ε and t represent normal strain and time, respectively. That is, the more rapid the loading, the higher both the yield and ultimate strengths of the material as illustrated in Figure 2.15. However, the curves indicate little change in elongation: ductility remains about the same. Observe that for strain rates from 10^{-1} to 10^3 s^{-1}, the yield strength increases significantly.

2.9.2 Ductile–Brittle Transition

We now discuss the conditions under which metals may manifest a change from ductile to brittle or from brittle to ductile behavior. The matter of *ductile–brittle transition* has important applications where the operating environment includes a wide variation in temperature or when the rate of dynamic loading changes. The stress raisers such as grooves and notches also have a significant effect on the transition from brittle to ductile failure. The *transition temperature* represents roughly the temperature a material's behavior changes from ductile to brittle. While most ferrous metals have well-defined transition temperature, some nonferrous metals do not. Therefore, the width of the temperature range over which the transition from brittle to ductile failure occurs is material dependent.

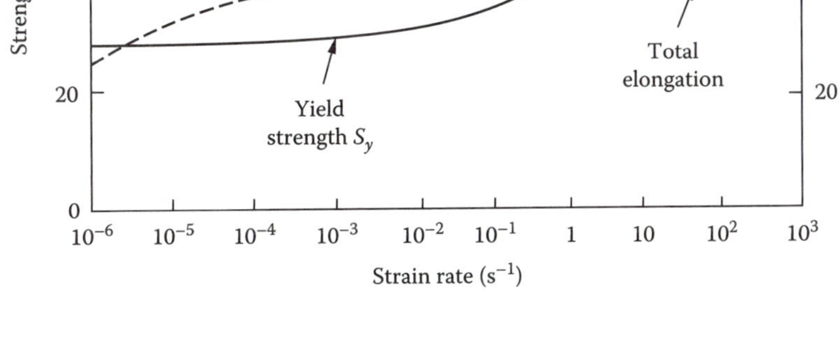

FIGURE 2.15
Influence of strain rate on tensile properties of a mild steel at room temperature.

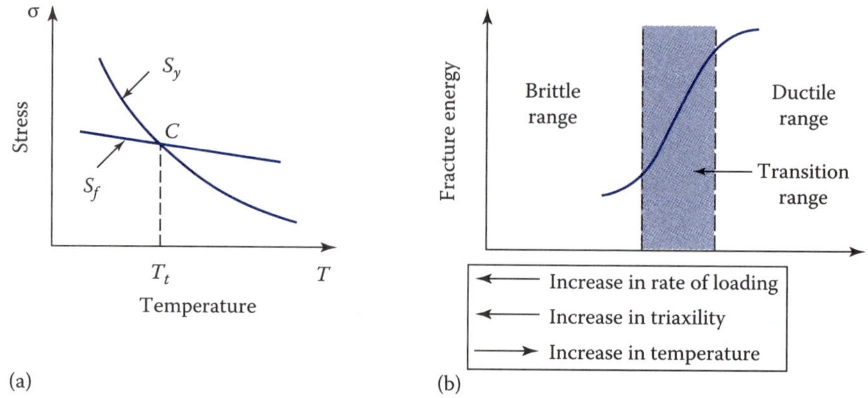

FIGURE 2.16
Typical transition curves for metals: (a) variation of yield strength S_y and fracture strength S_f with temperature and (b) effects of loading rate, stress around a notch, and temperature on impact toughness.

Let us, to begin with, refer to Figure 2.16a, where yield strength S_y and fracture strength S_f in tension are shown as functions of the temperature of a metal. Note that S_f exhibits only a small decrease with increasing temperature. The point of intersection (C) of the two strength curves in the figure defines the critical or transition temperature, T_t. If, at a given temperature above T_t, the stress is progressively increased, failure will occur by yielding, and the fracture curve will never be encountered. Likewise, for a test conducted at $T < T_t$, the yield curve is not intercepted, since failure occurs by fracture. At temperatures close to T_t, the material generally exhibits some yielding prior to a partially brittle fracture.

A transition phenomenon is more commonly examined from the viewpoint of the energy required to fracture a notched or unnotched specimen, the impact toughness rather than the stress (Figure 2.16b). The transition temperature is then defined as the temperature at which there is a sudden decrease in impact toughness. The *Charpy* and *Izod* method notched-bar impact bending *tests* made at various temperatures utilize specimens to determine the impact toughness. In both tests, the specimen is struck by a pendulum released from a fixed height, and the energy absorbed is computed from the height of swing (or indicated on a dial) after fracture. We note that the state of stress around the notch is triaxial and nonuniformly distributed throughout the specimen. Notches (and grooves) reduce the energy required to fracture and shift the transition temperature, normally very low, to the range of normal temperatures. This is the reason why most experiments are performed on notched specimens.

The foregoing discussion shows that, under certain conditions, a material said to be ductile will behave in a brittle fashion and vice versa. The principal factors governing whether failure occurs by fracture or yielding are summarized as follows (Figure 2.16):

1. *Temperature*: If the temperature increases (exceeds T_t), the resistance to yielding is less than the resistance to fracture ($S_y < S_f$), and the specimen yields. On the contrary, if the temperature decreases (less than T_t), the specimen fractures without yielding.
2. *Loading rate*: Increasing the rate at which the load is applied increases a metal's ability to resist yielding.
3. *Triaxiality*: The effect on the transition of a 3D stress condition around the notch, the so-called triaxiality, is similar to that of the loading rate.

In addition, other factors may also affect a ductile material to undergo a fracture similar to that of a brittle material. Some of these are fatigue, cyclic loading at normal temperatures (see Section 8.3); creep, long-time static loading at elevated temperatures; severe quenching, in heat treatment, if not followed by tempering; and work hardening by sufficient amount of yielding. Internal cavities or voids in casting or forging may have an identical effect.

CASE STUDY 2.1 Rupture of Titanic's Hull

Titanic was designed by expert engineers, employing the most advanced technologies and extensive features of the day, and was called *unsinkable*. The world's largest at the time, this passenger steamship was on her maiden voyage from Southampton, England, to New York City when it struck an iceberg in the North Atlantic. Titanic sank on April 15, 1912, resulting in the deaths of 1517 out of 2223 people on board in one of the deadliest peacetime maritime disaster in history. The huge loss of life, including noteworthy victims, changes in maritime law, and later, discovery of the famous underwater wreck has all driven a continuing interest in this complex case.
General characteristics of the Titanic include the following:

Weight (tonnage): 46,328 gross register tons (GRT)
Length: 269.1 m
Height: 53.3 m (keel to top of funnels)
Depth: 19.7 m
Propulsion: Two bronze triple-blade wing propellers and one bronze quadruple-blade center propeller
Installed power: 46,000 hp (total)
29 marine boilers feeding two four-cylinder steam engines, each producing 15,000 hp for the wing propellers
A low-pressure turbine producing 16,000 hp for the center propeller
Speed: 21 knots (39 km/h)
Capacity: 3547 passenger and crew (fully loaded)

Among the many possible reasons for the sinking (Figure 2.17), Titanic's construction has often been cited. Particular focus has been given to the quality of the ship's hull. Initially, historians thought the iceberg had simply cut a gash into the hull. Because the part of the ship that the iceberg damaged is now buried, investigators used sonar to examine the area in question. They discovered that the massive iceberg had actually caused the hull to buckle, allowing water to enter Titanic between her 25–38 mm thick steel plates. Metallurgical analysis of small pieces of this hull plating revealed that the steel had a high *ductile–brittle transition* temperature, making it dangerous for icy water and leaving the hull vulnerable to dent-induced ruptures.

Although probably the best plain carbon steel available at the time, the hull fragments were found to have very high contents of *phosphorus* and *sulfur* and a low content of *manganese*, compared with modern steels. Excessive amount of phosphorus initiates fractures, sulfur forms grains of iron sulfide that facilitates propagation of cracks (particularly at punched rivet holes), and lack of manganese makes steel less ductile. Charpy V-notch tests on the recovered samples showed them undergoing ductile–brittle transition at around 32°C for longitudinal samples and 56°C for transverse samples, compared with a transition temperature of about −17°C common for modern steel [8].

FIGURE 2.17
Depiction of Titanic sinking. (Courtesy of google.com.)

Since Titanic was sailing in −2°C ocean water, the ship's hull was extremely brittle. Therefore, Titanic's steel was unsuitable for use at such low temperatures and contributed significantly to its sinking.

Comments: A number of other fatal static tank failures by brittle fracture also occurred in the early 1900s. The ductile–brittle transition temperature of parts under various environmental conditions is an important factor in design.

2.10 Hardness

Selection of a material that has good resistance to wear and erosion very much depends on the hardness and the surface condition. Hardness is the ability of a material to resist indentation and scratching. The kind of hardness considered depends on the service requirements to be met. For example, gears, cams, rails, and axles must have a high resistance to indentation. In mineralogy and ceramics, the ability to resist scratching is used as a measure of hardness. The *indentation hardness*, generally used in engineering, is briefly discussed in this section.

Hardness testing is one of the principal methods for ascertaining the suitability of a material for its intended purpose. It is also a valuable inspection tool for maintaining uniformity of quality in heat-treated parts. Indentation hardness tests most often involve one of the three methods: Brinell, Rockwell, or Vickers [5]. The shore scleroscope hardness testing is sometimes employed as well. These nondestructive tests yield a relative numerical measure or scale of hardness, showing how well a material resists indentation.

2.10.1 Brinell Hardness

The Brinell hardness test uses a spherical ball in contact with a flat specimen of the material and subjected to a selected compressive load. Subsequent to removal of the load, the diameter of the indentation is measured with an optical micrometer. The hardness is then defined as the *Brinell hardness number* (Bhn), H_B, which is equal to the applied load (in kg) divided by the area of the surface of indentation (in mm^2). Therefore, the units of H_B (and other hardness numbers) are the same as those of stress. However, they are seldom stated.

Tables of hardness values are given in the standards of the ASTM. The Brinell test is used mainly for materials whose thickness is 6.25 mm or greater. As a rule, the case-hardened steels are unsuitable for Brinell testing. The test is used to determine the hardness of a wide variety of materials. The harder is the material, the smaller the indentation and the higher the Brinell number.

2.10.2 Rockwell Hardness

The Rockwell test uses an indenter (steel ball or diamond cone called *brale*) pressed into the material. The relationships of the total test force to the depth of indentation provide a measure of Rockwell hardness, which is indicated on a dial gage. Depending on the size of the indenter, the load used, and the material being tested, the Rockwell test furnishes hardness data on various scales. Two common scales, R_B and R_C (i.e., Rockwell *B* and *C*), are frequently used for soft metals (such as mild steel or copper alloys) and hard metals (such as hardened steel or heat-treated alloy steel), respectively. In standard tests, the thickness should be at least 10 times the indentation diameter.

The Rockwell test is simple to perform and the most widely employed method for determining hardness of metals and alloys, ranging from the softest bearing materials to the hardest steels. It can also be used for certain plastics, such as acrylics, acetates, and phenolics. Optical measurements are not required; all readings are direct. Routine testing is usually performed with bench-type Rockwell machines.

2.10.3 Vickers Hardness

The Vickers hardness test is similar to the Brinell test. However, it uses a four-sided inverted diamond pyramid with an apex angle of 136°. The Vickers hardness number (H_V) is the ratio of the impressed load to the square indented area. The Vickers hardness test is of particular value for hard, thin materials where hardness at a spot is required.

2.10.4 Shore Scleroscope

The shore scleroscope uses a small diamond-tipped pointer or hammer that is allowed to fall from a fixed height onto the specimen. Hardness is measured by the height of the rebound. The method is easy and rapid to apply. However, the results obtained are the least reliable of all machine methods. The hardness of soft plastics and wood felts is measured by this scleroscope.

2.10.5 Relationships among Hardness and Ultimate Strength in Tension

Figure 2.18 shows the conversion plot between Brinell, Rockwell (*B* and *C*), and the tensile strength of steel. Note that the curves for R_B and R_C are nonlinear and the related values are only approximate. However, the results of the Brinell hardness test have been found to correlate linearly with the tensile strength S_u of most *steels* as follows:

$$S_u = 500H_B \text{ psi} \tag{2.22}$$

This is indicated by a nearly straight line in the figure. In addition, for stress-relieved (not cold drawn) steels, the tensile yield strength S_y is given by

$$S_y = 1.05S_u - 30,000 \text{ psi} \tag{2.23}$$

Substituting Equation 2.22,

$$S_y = 525H_B - 30,000 \text{ psi} \tag{2.24}$$

Formulas (2.22) through (2.24) are estimates and should be used only when definite strain hardening data are lacking.

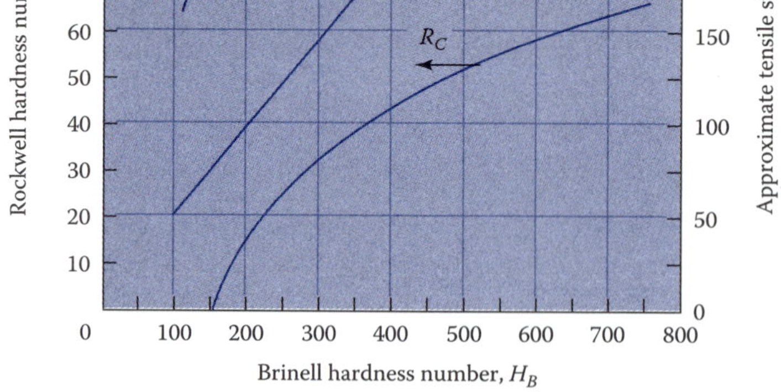

FIGURE 2.18
Hardness conversion to ultimate strength in tension of steel.

Example 2.6: Finding the Strength of Steel from Hardness

An American Iron and Steel Institute (AISI) 4140 steel component is heat treated to 217 Bhn (Brinell hardness number). Determine the corresponding values of ultimate tensile strength S_u and yield strength in tension S_y.

Assumption: Relationships among hardness and ultimate strength are sufficiently accurate.

Solution

Through the use of Equation 2.22, the ultimate strength is equal to

$$S_U = 500H_B \text{ psi}$$

$$= 500\ (217) = 108{,}500 \text{ psi} = 108.5 \text{ ksi}$$

In a like manner, Equation 2.24 gives the following yield strength:

$$S_y = 525\ (217) - 30{,}000 \text{ psi} = 83{,}925 \text{ psi} = 83.9 \text{ ksi}$$

Comment: Experimental data obtained for the material under consideration would serve to refine the preceding formulas.

2.11 Processes to Improve Hardness and the Strength of Metals

A material with metallic properties consisting of two or more elements, one of which is a basic metal, is called an *alloy*. An alloying element is deliberately added to a metal to alter its physical or mechanical properties. For example, the addition of alloying elements to iron results in cast iron and steel. By general usage, the term *metal* is used in a generic sense, often referring to both a simple metal and metallic alloys. Unless specified otherwise, we adhere to this practice.

There are a number of ways to increase the hardness and strength of metals. These include suitably varying the composition or alloying, mechanical treatment, and heat treatment. Various alloys are considered in the next section. Numerous coatings and surface treatments are also available for materials. Several of these have the main purpose of preventing corrosion while the others are aimed to improve surface hardness and wear. In this section, we shall discuss only a few treatments and coating types.

2.11.1 Mechanical Treatment

Mechanical forming and hardening consists of hot-working and cold-working processes. A metal can be shaped and formed when it is above a certain temperature, known as the *recrystallization temperature*. Below this temperature, the effects of mechanical working are cold worked. On the other hand, in hot working, the material is worked mechanically above its recrystallization temperature. Note that hot working gives a finer, more uniform grain structure and improves the soundness of the material. However, in general, cold working leaves the part with residual stress on the surface. Thus, resulting mechanical properties in the foregoing processes are quite different.

2.11.1.1 Cold Working

Cold working, also called strain hardening, is a process of forming the metal usually at a room temperature (see Section 2.3). This results in an increase in hardness and yield strength, with a loss in toughness and ductility (that can be recovered by a heat treatment process termed annealing). Cold working is used to gain hardness on low-carbon steels, which cannot be heat treated. Typical of cold-working operations include cold rolling, drawing, spinning, stamping, and forming. As noted previously, the relative ease with which a given material may be machined, or cut with sharp-edged tools, is called its machinability.

The most common and versatile of the cold-working treatments is shot peening. It is widely used with springs, gears, shafts, connecting rods, and many other components. In *shot peening*, the surface is bombarded with high-velocity iron or steel shot (small, spherical pellets) discharged from a rotating wheel or pneumatic nozzle. The process leaves the surface in compression and alters its smoothness. Since fatigue cracks are not known to initiate or propagate in a compression region (see Section 7.1), shot peening has proven very successful in raising the fatigue life of most members. Machine parts made of very high-strength steels (about 1400 MPa), such as springs, are particularly benefited. Shot peening has also been used to reduce the probability of stress corrosion cracking in a turbine rotor and blades.

2.11.1.2 Hot Working

Hot working reduces the strain hardening of a material but avoids the ductility and toughness loss attributed to cold working. However, hot-rolled metals tend to have greater ductility, lower strength, and a poorer surface finish than cold-worked metals of the identical alloy. Examples of hot-working processes are rolling, forging, hot extrusion, and hot pressing, where the metal is heated sufficiently to make it plastic and easily worked. Forging is an automation of blacksmithing. It uses a series of hammer dies shaped to gradually form the hot metal into the final configuration. Practically, any metal can be forged. Extrusion is used mainly for nonferrous metals and it typically uses steel dies.

2.11.2 Heat Treatment

The heat treatment process refers to the controlled heating and subsequent cooling of a metal. It is a complicated process, employed to obtain properties desirable and appropriate for a particular application. For instance, an intended heat treatment may be to strengthen and harden a metal, relieving its internal stresses, hardening its surface only, softening a cold-worked piece, or improving its machinability. The heating is done in the furnace, and the maximum temperature must be maintained long enough to refine the grain structure. Cooling is also done in the furnace or an insulated contained. The definitions that follow are concerned with some common heat-treating terms [1,2].

Quenching: The rapid cooling of a metal from an elevated temperature by injecting or spraying the metal with a suitable cooling medium, such as oil or water, to increase hardness. The stress–strain curve as a result of quenching a mild steel is depicted in Figure 2.19.

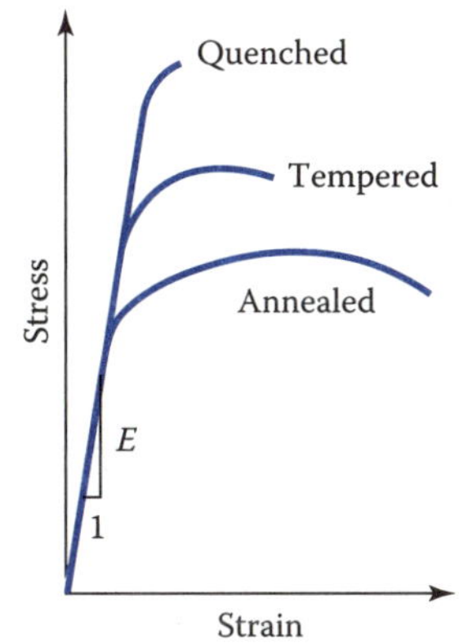

FIGURE 2.19
Stress–strain diagrams for annealed, quenched, and tempered steel.

Tempering or drawing: A process of stress relieving and softening by heating then quenching. Figure 2.19 shows a stress–strain curve for a mild steel after tempering.

Annealing: A process involving heating and slowly cooling, applied usually to induce softening and ductility. The quenching and tempering process is reversible by annealing; that is, annealing effectively returns a part to original stress–strain curve (Figure 2.19).

Normalizing: A process that includes annealing, except that the material is heated to a slightly higher temperature than annealing. The result is a somewhat stronger, harder metal than a fully annealed one.

Case hardening or carburizing: A process where the surface layer (or case) is made substantially harder than the metal's interior core. This is done by increasing the carbon content at the surface. Surface hardening by any appropriate method is a desirable hardening treatment for various applications. Some of the more useful case-hardening processes are carburizing, nitriding, cyaniding, induction hardening, and flame hardening. In the induction-hardening process, a metal is quickly heated by an induction coil followed by quenching in oil.

Through hardening: With a sufficiently high-carbon content, 0.35%–0.50%, the material is quenched and drawn at suitable temperatures to obtain the desired physical properties. Alloy steels will harden and retain their shape better than plain carbon steels when heat treated. The greater strength and hardness of surface and core for heat-treated material is accompanied by loss of ductility.

2.11.3 Coatings

Coating is a covering that is applied to the surface of a part or often referred to as the *substrate*. A variety of metallic and nonmetallic surface coatings are used for metals. As noted before, they mainly improve surface hardness, wear resistance, and scratch resistance. In many situations, coatings are also applied to alter dimensions slightly and physical properties (i.e., appearance, color, reflectance, and resistance). Some examples of practical importance are fasteners that are plated to reduce corrosion resistance, selective parts of automobiles chrome plated for appearance and corrosion resistance, piston rings that are chrome plated to increase wear resistance. Here, *plating* refers to a thin coating of metal.

The coating is an essential part of the finished products involving printing processes and semiconductor device fabrication. Printing and coating processes deal with the application of a thin film of material to a substrate, such as paper, fabric, and foil. It is to be noted that most coatings are usually porous that promote undesirable crack growth and reduce fatigue strength of metals. So, they should not be employed on components that are fatigue loaded. Coatings also, applied with high temperatures, may thermally induce tensile residual stresses to the surface of parts. Frequently, coatings are applied as solids, liquids, or gases. Numerous protective *chemical coatings* and *paints* are in widespread usage.

2.11.3.1 Galvanization

The process of galvanization involves applying a protective metallic coating to the surface of another dissimilar metal to prevent rusting. A *galvanic action* may be created electrochemically. Combination of metals, like steel and cast iron, are considerably safe from galvanic action. On the other hand, combinations like aluminum and copper will experience severe corrosion in moist surrounding. Zinc coatings represent the common practice of galvanizing ferrous materials to prevent them against corrosion.

2.11.3.2 Electroplating

The process of electroplating is the creation of a galvanic action, where the part to be plated is the *cathode* (negative electrode) and the plating material is the *anode* (positive electrode). These two metals are immersed in a solution, called *electrolyte* bath, and a direct current is applied from anode to cathode. *Ions* or atoms of the plating are driven through the solution and cover the part with a thin coating. Electroplating thus is mainly used for depositing a layer of material to create a desired property, such as abrasion and wear resistance, corrosion protection, and aesthetic qualities, to a surface. Another application of the electroplating is to build up thickness of undersized parts. As the name suggests, an *electroless* plating puts a coating of metal on a part without any electric current. For this purpose, zinc is the most commonly used metal.

2.11.3.3 Anodizing

The process of anodizing is used to increase the thickness of the oxide layer on the surface of the metal parts. It is a relatively inexpensive treatment with good corrosion resistance and wear. This process provides better adhesion for paint primers and glues than does bare metal. Thin anodic coats or *anodic films* are customarily applied to protect aluminum alloys. Aluminum oxide is naturally very hard and abrasion resistant. Titanium, zinc, and magnesium can also be anodized. However, the process is not a useful treatment for iron or carbon steel. Anodic films are usually much stronger and more adherent than most paint kinds and metal plating.

2.12 General Properties of Metals

There are various nonferrous and ferrous engineering metals (Table 2.1). This section attempts to provide some general information for the readers to help identify the types of a few selected metal alloys. Appendix B lists common mechanical properties of the foregoing materials.

2.12.1 Iron and Steel

Iron is a metal that, in its pure form, has almost no commercial use. The addition of other elements to iron essentially changes its characteristics, resulting in a variety of cast and wrought irons and steel. Cast iron and cast steel result from pouring the metal into molds of the proper form. To make wrought iron and varieties of wrought steel, the metal is cast into a suitable size and shape (e.g., slabs) then hot rolled to form bars, tubing, plate, structural shapes, pipe, nails, wires, and so on. Wrought iron is tough and welds easily. Cast steel can also be readily welded. Steel is difficult to cast because it shrinks considerably. Castings are ordinarily inferior to corresponding wrought metals in impact resistance.

2.12.2 Cast Irons

Cast iron is an iron alloy containing over 2% carbon. Cast irons constitute a whole family of materials. Having such a high-carbon content, cast iron is brittle, has a low ductility, and hence cannot be cold worked. While relatively weak in tensile strength, it is very strong in compression. Bronze welding rods are widely used in cast iron that is not easily welded. The common composition of cast iron is furnished in Table 2.2.

The characteristics of cast iron can be altered extensively with the addition of alloying metals (such as copper, silicon, manganese, phosphorus, and sulfur) and proper heat treatment. Cast iron alloys are widely used as crankshafts, camshafts, and cylinder blocks in engines, gearing, dies, railroad brake shoes, rolling mill rolls, and so on. Cast iron is inexpensive, easily cast, and readily machined. It has superior vibration characteristics and resistance to wear.

Since the physical properties of a cast iron casting are fully affected by its cooling rate during solidification, there are various cast iron types. Gray cast iron is the most widely used form of cast iron. It is common to refer gray cast iron just as *cast iron*. Other basic types of cast iron include malleable cast irons and nodular or ductile cast irons. Nodular cast iron shrinks more than gray cast iron, but its melting temperature is lower than for cast steel. A particular type of cast iron, called *Meehanite iron*, is made using a patented process by the addition of a calcium–silicon alloy. In practice, cast irons are classified with respect to ultimate strength. In the *ASTM numbering system* for cast irons, the class number corresponds to the minimum ultimate strength. Thus, an ASTM No. 30 cast iron has a minimum tensile strength of 30 ksi (210 MPa). Some average properties and minimum ultimate strengths of cast irons are shown in Tables B.1 and B.2, respectively.

2.12.3 Steels

Steel is an alloy of iron containing less than 2% of carbon. Additional alloying elements ease the hardening of steel. Nevertheless, carbon content, almost alone, induces the maximum

TABLE 2.2

Composition of Cast Iron

Carbon	2.00%–4.00%
Silicon	0.50%–3.00%
Manganese	0.20%–1.00%
Phosphorus	0.05%–0.80%
Sulfur	0.04%–0.15%
Iron	Remainder

TABLE 2.3

Groups and Typical Uses of Plain Carbon Steels

Type	Carbon Content (%)	Area of Use
Low carbon	0.03–0.25	Plate, sheet, structural parts
Medium carbon	0.30–0.55	Machine parts, crane hooks
High carbon	0.60–1.40	Spring, tools, cutlery

hardness that can be developed in steel. Steels are used extensively in machine construction. They can be classified as plain carbon steels, alloy steels, high-strength steels, cast steels, stainless steel, tool steel, and special purpose steel.

2.12.3.1 Plain Carbon Steels

These steels contain only carbon, usually less than 1%, as a significant alloying element. Carbon is a potent alloying element, and wide changes in strength and hardness can be obtained by changing the amount of this element. Carbon steel owes its distinct properties chiefly to the carbon it contains. The range of desired characteristics can be further gained by heat treatment. Plain carbon steel is the least expensive steel, manufactured in larger quantities than any other. Table 2.3 summarizes general uses for steels having various levels of carbon content.

Low-carbon or mild steels, also referred to as the *structural steels*, are ductile and thus readily formable. If welded, they do not become brittle. Where a wear-resistant surface is needed, this steel can be case hardened. A minimum 0.30% carbon is necessary to make a heat-treatable steel. Therefore, medium- and high-carbon steels can be heat treated to achieve the desired characteristics.

2.12.3.2 Alloy Steels

There are many effects of any alloy addition to a basic carbon steel. The primary reason is to improve the ease with which steel can be hardened. That is, potential hardness and strength, controlled by the carbon content, can be accomplished with less drastic heat treatment by alloying. When a proper alloy is present in a carbon steel, the metallurgical changes take place during quenching at a faster rate, the cooling effects penetrate deeper, and a large portion of the part is strengthened.

Ordinary alloying elements (in addition to carbon) include, singly or in various combinations, manganese, molybdenum, chromium, vanadium, and nickel. Added to the steel, nickel and chromium also bring significant impact resistance and provide considerable wear as well as corrosion resistance, respectively. A variety of carbon and alloy steels are employed for the construction of machinery and structures.

2.12.3.3 Stainless Steels

The so-called stainless steels are in widespread use for resisting corrosion (see Section 8.2) and heat-resisting applications. They contain (in addition to carbon) at least 12% chromium as the basic alloying element. Stainless steel is of three types: austenitic (18% chromium, 8% nickel), ferritic (17% chromium), and martensitic (12% chromium). Particularly, the austenitic kind of stainless steel polishes to a luster and finish. All the chromium–nickel steels have greater corrosion-resistant properties than the plain chromium steels and may be welded. The mechanical characteristics of various wrought steels are given in Table B.5.

2.12.3.4 Steel Numbering Systems

Various numbering systems of steels are used. The Society of Automotive Engineers (SAE), the AISI, and the ASTM have devised codes to define the alloying elements and the carbon content in steels. These designations can provide a simple means by which any particular steel can be specified. A brief description of the common systems follows.

The *AISI/SAE numbering system* generally uses a number composed of four digits. The first two digits indicate the principal alloying element. The last two digits give the approximate carbon content, expressed in hundredths of percent. The AISI number for steel is similar to the SAE number, but a letter prefix is included to indicate the process of manufacture (such as A and C for the steels). For instance,

AISI C1020 (SAE 1020) steel

represents a plain carbon steel denoted by the basic number 10, containing 0.20% carbon. In a like manner,

AISI A3140 (SAE 3140) steel

is an alloy steel (with nickel and chromium designated by 31) with 0.40% carbon. We see from the foregoing examples that the first two digits are not so systematic.

The *ASTM numbering system* for steels is based on the ultimate strength [2]. Mostly used specifications are ASTM-A27 mild- to medium-strength carbon-steel castings for general application and A148 high-strength steel castings for structural purposes [1]. Tables B.1 and 14.2 include average properties of a few ASTM steels. Samples of the minimum yield strengths S_y and ultimate strengths S_u for certain ASTM steels are shown in Table 2.4, where Q and T denote quenched and tempered (Q&T), respectively. Note that the ASTM-A36 is the all purpose carbon grade steel extensively used in building and bridge construction. ASTM-A572 is a high-strength low alloy steel, A588 represents atmospheric corrosion-resistant high-strength low alloy steel, and A514 is an alloy Q&T steel. For complete information on each steel, reference should be made to the appropriate ASTM specification.

The ASTM, the AISI, and the SAE developed the *unified numbering system* (UNS) for metals and alloys. This system also contains cross-reference numbers for other material specifications. The UNS uses a letter prefix to indicate the material (e.g., *G* for the carbon and

TABLE 2.4

Structural Steel Strengths in the ASTM Numbering System

		S_y		S_u		Max. Thickness	
Steel Type	**ASTM No.**	**MPa**	**(ksi)**	**MPa**	**(ksi)**	**mm**	**(in.)**
Carbon	A36	248	(36)	400	(58)	200	(8)
Low alloy	A572	290	(42)	414	(60)	150	(6)
Stainless	A588	345	(50)	480	(65)	100	(4)
Alloy Q&T	A514	690	(100)	758	(110)	62.5	(2.5)

alloy steels). The mechanical properties of selected carbon and alloy steels are furnished in Tables B.3 and B.4. The AISI, the SAE, the ASTM, and UNS lists are continuously being revised, and it is necessary to confer the latest edition of a material handbook. We mostly use the AISI/SAE designations for steels.

2.12.4 Aluminum and Copper Alloys

Aluminum alloys are very versatile materials, having good electrical and thermal conductivity as well as light reflectivity. They possess a high strength-to-weight ratio, which can be a very important consideration in the design of, for example, aircraft, missiles, and trains. Aluminum has a high resistance to most corrosive atmospheres because it readily forms a passive oxide surface coating. Lightweight aluminum alloys have extensive applications in manufactured products. Aluminum is readily formed, drawn, stamped, spun, machined, welded, or brazed. The high-strength aluminum alloys have practically the same strength as mild steel.

Numerous aluminum alloys are available in both wrought and cast form. The aluminum *casting alloys* are indicated by three-digit numbers. The *wrought alloys,* shaped by rolling or extruding, use four-digit numbers. Silicon alloys are preferred for casting. Typical wrought aluminum alloys include copper and silicon–magnesium. The comparison and mechanical properties of some typical aluminum alloys are given in Table B.6. The temper of an aluminum alloy is a main factor governing its strength, hardness, and ductility. Temper designation is customarily specified by cold work such as rolling, drawing, or stretching. Other alloys are heat treatable, and their properties can be enhanced considerably by appropriate thermal processing.

Copper alloys are very ductile materials. Copper may be spun, stamped, rolled into a sheet, or drawn into wire and tubing. Owing to its high electrical and thermal conductivity, resistance to corrosion, but relatively low ratio of strength to weight, copper is used extensively in the electrical, telephone, petroleum, and power industries. The most notable copper-base alloys are brass and bronze. *Brass* is a copper–zinc alloy, and *bronze* is composed mainly of copper and tin. Brass and bronze are used in both cast and wrought form. The strength of brass increases with the zinc content. Brass is about equal to copper in corrosion resistance, but bronze is superior to both.

Die castings generally are made from zinc, aluminum, magnesium, and, to a lesser extent, brass. They are formed by forcing a molten alloy into metal molds or dies under high pressure. The die cast process is applicable for parts containing very thin sections of intricate forms. Copper and most of its alloys can be fabricated by soldering, welding, or brazing. They can be worked strengthened by cold working but cannot be heat treated. The machinability of the brass and bronze is satisfactory. The properties of copper alloys can often be significantly improved by adding small amounts of additional alloying elements (Table B.7).

2.13 General Properties of Nonmetals

The three common categories of nonmetals are of engineering interest: plastics, ceramics, and composites (see Table 2.1). Plastics represent a vast and growing field of synthetics. Sometimes, optimum properties can be obtained by combination of dissimilar materials or composites. Ceramics are hard, heat-resistant, brittle materials. Here, we briefly discuss the general properties of these materials.

2.13.1 Plastics

Plastics are synthetic materials known as *polymers*. They are employed increasingly for structural purposes, and thousands of different types are available. Table 2.5 presents several common plastics. The mechanical properties of these materials vary tremendously, with some plastics being brittle and others ductile. When designing with plastics, it is significant to remember that their properties are greatly affected by both change in temperature and the passage of time. Observe from the table that polymers are of two principal classes: thermoplastics and thermosets.

Thermoplastic materials repeatedly soften when heated and harden when cooled. There are also highly elastic flexible materials known as thermoplastic *elastomers*. *Thermosets* or thermosetting plastics sustain structural change during processing to become permanently insoluble and infusible. Thermoplastic materials may be formed into a variety of shapes by the simple application of heat and pressure, while thermoset plastics can be formed only by cutting or machining.

Rubber is a common elastomer. Elastomer's industrial applications include belts, hoses, gaskets, seals, machinery mounts, and vibration dampers. *NR* was originally derived from latex, a milky colloid produced by some plants. The purified form of NR is the chemical polyisoprene, which can also be produced synthetically. NR has long fatigue life and high strength even without reinforcing fillers. It has good creep and stress relaxation resistance and is low cost, but its main disadvantage is its poor oil resistance. SRs are artificially produced materials with properties similar to NR. A wide

TABLE 2.5
Selected Plastics

Chemical Classification	Trade Name
Thermoplastic materials	
Acetal	Delrin, Celcon
Acrylic	Lucite, Plexiglas
Cellulose acetate	Fibestos, Plastacele
Cellulose nitrate	Celluloid, Nitron
Ethyl cellulose	Gering, Ethocel
Polyamide	Nylon, Zytel
Polycarbonate	Lexan, Merlon
Polyethylene	Polythene, Alathon
Polystyrene	Cerex, Lustrex
Polytetrafluoroethylene	Teflon
Polyvinyl acetate	Gelva, Elvacet
Polyvinyl alcohol	Elvanol, Resistoflex
Polyvinyl chloride	PVC, Boltaron
Polyvinylidene chloride	Saran
Thermosetting materials	
Epoxy	Araldite, Oxiron
Phenol-formaldehyde	Bakelite, Catalin
Phenol-furfural	Durite
Polyester	Beckosol, Glyptal
Urea-formald	Beetle, Plaskon

range of different SRs have been produced with chemical and mechanical properties for a variety of applications (see Table B.9).

An examination of tensile stress–strain diagrams at various low temperatures, for instance, a cellulose nitrate and similar ones for other plastics, indicates that the ultimate strength, yield strength, and modulus of elasticity, but not ductility, increase with the decrease in temperature. These modifications of the stress–strain properties are similar to those found for metals (see Section 2.7). Experiments also show an appreciable decrease in impact toughness with a decrease in temperature for some plastics but not all. A loaded plastic may stretch gradually over a time until it no longer is serviceable. Interestingly, because of its light weight, the strength-to-weight ratio for nylon is about the same as for structural steel.

The principal advantage of plastics is their ability to be readily processed, as previously noted; most plastics can be easily molded into complicated shapes. Large elastic deflections allow the design of polymer components that snap together, making assembly fast and inexpensive. Furthermore, many plastics are low-cost materials and display exceptional resistance to wear and corrosive attacks by chemicals.

Fiber reinforcement increases the stiffness, hardness, strength, and resistance to environmental factors and reduces the shrinking of plastics. A glass-reinforced plastic has improved strength by a factor of about two or more. Further improvement is gained by carbon reinforcement. The foregoing relatively new materials (with 10%–40% carbon) have tensile strengths as high as 280 MPa. Reinforced plastics increasingly are being employed for machine and structural components requiring light weight or high strength-to-weight ratios. Tables B.1 and B.8 show that the ranges of properties that can be obtained with plastics that are very large.

2.13.2 Ceramics and Glasses

Ceramics are basically compounds of nonmetallic as well as metallic elements, mostly oxides, nitrides, and carbides. Generally, silica and graphite ceramics dominate the industry. However, newer ceramics, often called *technical ceramics*, play a major role in many applications. Glasses also consist of metallic and nonmetallic elements; however, they have crystal structure. Glass ceramics are in widespread usage as electrical, electronic, and laboratory ware.

Ceramics have high hardness and brittleness and high compressive but low tensile strengths. Both ceramics and glasses exhibit behavior and are typically 15 times stronger in compression than in tension. High temperature and chemical resistance, high dielectric strength, and low weight characterize many of these materials. Therefore, attempts are being made to replace customary metals with ceramics in some machine and structural members.

2.13.3 Composites

As mentioned earlier, a composite material is made up of two or more unique elements. Composites usually consist of high-strength reinforcement material embedded in a surrounding material. They have a relatively large strength-to-weight ratio compared to a homogeneous material and additional other desirable characteristics [10]. Furthermore, there are many situations where different materials are used in combination so that the maximum advantage is gained from each component part. For instance, graphite-reinforced epoxy gets strength from the graphite fibers while the epoxy protects the graphite from oxidation and provides toughness. In this text, the discussions concern isotropic composites like reinforced-concrete beams and multilayer members and filament-wound anisotropic composite cylinders.

2.13.3.1 Fiber-Reinforced Composite Materials

It is recalled from Section 2.2 that a material whose characteristics rely on direction is termed anisotropic. Here, we briefly discuss an important class of widely used anisotropic materials known as fiber-reinforced composites. Typical examples are thick-walled vessels under high pressure, marine and aircraft wind shields, portions of space vehicles, and components of many other machines and structures. Publications associated with the theory and applications of composites contain extensive practical information (see [10]).

A *fiber-reinforced composite* is made by fixing firmly fibers of a strong, stiff material into a weaker reinforcing material or *matrix*. Familiar materials used for fibers include carbon, glass, polymers, graphite, and some metals, while various resins are employed as a matrix (such as in glass filament/epoxy rocket motor cases). Figure 2.20 illustrates the cross sections of two typical fiber-reinforced composite materials. Fiber length is an important parameter for strengthening and stiffening of fiber-reinforced composites. For example, for a number of glass-and-carbon-fiber-reinforced composites, the fiber length is about 1 mm, or 20–150 times its diameter.

A layer, also called *lamina*, of a composite material consists of a large number of parallel fibers embedded in a matrix (Figure 2.20a). So, a *laminate* is composed of arbitrarily oriented variety of bonded layers or laminas. Each layer may have a different thickness, orientation of fiber directions, and anisotropic properties. Some layers are positioned so their fibers are oriented usually at 30, 45, or 60 relative to one another. This increases the resistance of the laminate to the applied loads. A special case of an anisotropic material is an orthotropic material. When the fibers of all layers are positioned in the same orientation, the laminate represents an *orthotropic* material. Usually, a composite is composed of bonded three-layer orthotropic materials.

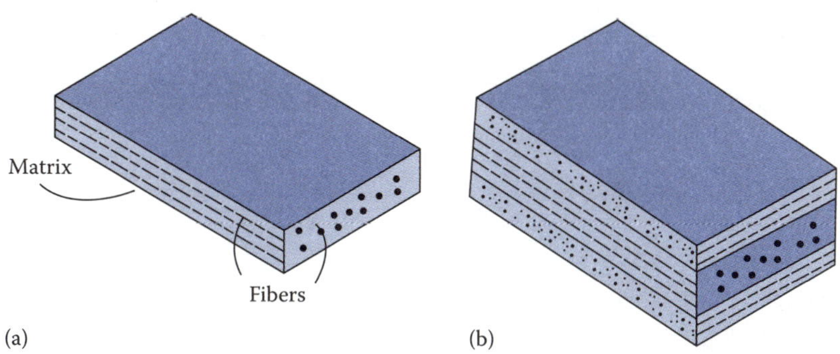

FIGURE 2.20
Fiber-reinforced materials: (a) single layer and (b) three-cross layer.

Figure 2.20b depicts a *cross-ply* laminate, in which the fibers of the midlayer and two outer layers are positioned along axial and lateral directions, respectively.

2.14 Selecting Materials

Material selection plays a very significant role in machine design. Each material should be chosen carefully according to the specific requirements imposed on the components, since these members operate in various environments. Preference of a specific material for the members relies on the purpose and kind of operation as well as the expected failure mode of these elements. Strength and stiffness present essential factors taken into account in the choice of a material. But selecting a material from both its functional and economical viewpoint is utmost important.

2.14.1 Strength Density Chart

The common properties of materials are not sufficient for selecting a material for a particular application. Rather, one or several combinations of properties are required. Some important property combinations include stiffness versus density (E–ρ), strength versus density (S–ρ), strength versus temperature (S–T), and stiffness versus strength (E–S). Various other combinations might be useful in material selection; however, the foregoing are primary considerations in designing machine elements.

In this section, we shall discuss briefly only one of Ashby's material selection charts [3]. These graphs are very useful reference for the practicing engineers. It should be mentioned that the information contained in Ashby's charts are for rough calculations and not for final design analysis. Ordinary properties of a material chosen should be employed in the final design followed by experimental verification and testing. Table B.9 in Appendix B furnishes the types and abbreviations for the material selection charts.

Figure 2.21 portrays S strength ρ -to-density (weight per unit volume) relationships for a number of materials. The sketched values for the strength are (1) *yield* strength for metals and polymers, (2) *compressive* strength for ceramics and glasses, (3) *tensile* strength for composites, and (4) *tear* strength for elastomers. It is seen that the *brittle* materials are enclosed by *dashed* envelopes. The *guide lines* of constants

$$\frac{S}{\rho} = C, \quad \frac{S^{2/3}}{\rho} = C, \quad \frac{S^{1/2}}{\rho} = C$$

are used, respectively, in minimum weight design of tension members, beams and shafts, plates, as shown in the figure.

Materials placed at the greatest distance from a selected guide line (up and left) are superior. In other words, materials with the greatest strength-to-weight ratios are placed in the upper left corner. Observe from the graph that the strength-to-weight ratios of some woods are as good as high-strength steel and better than most other metals. For the preceding reason, wood is a favorite material in building construction.

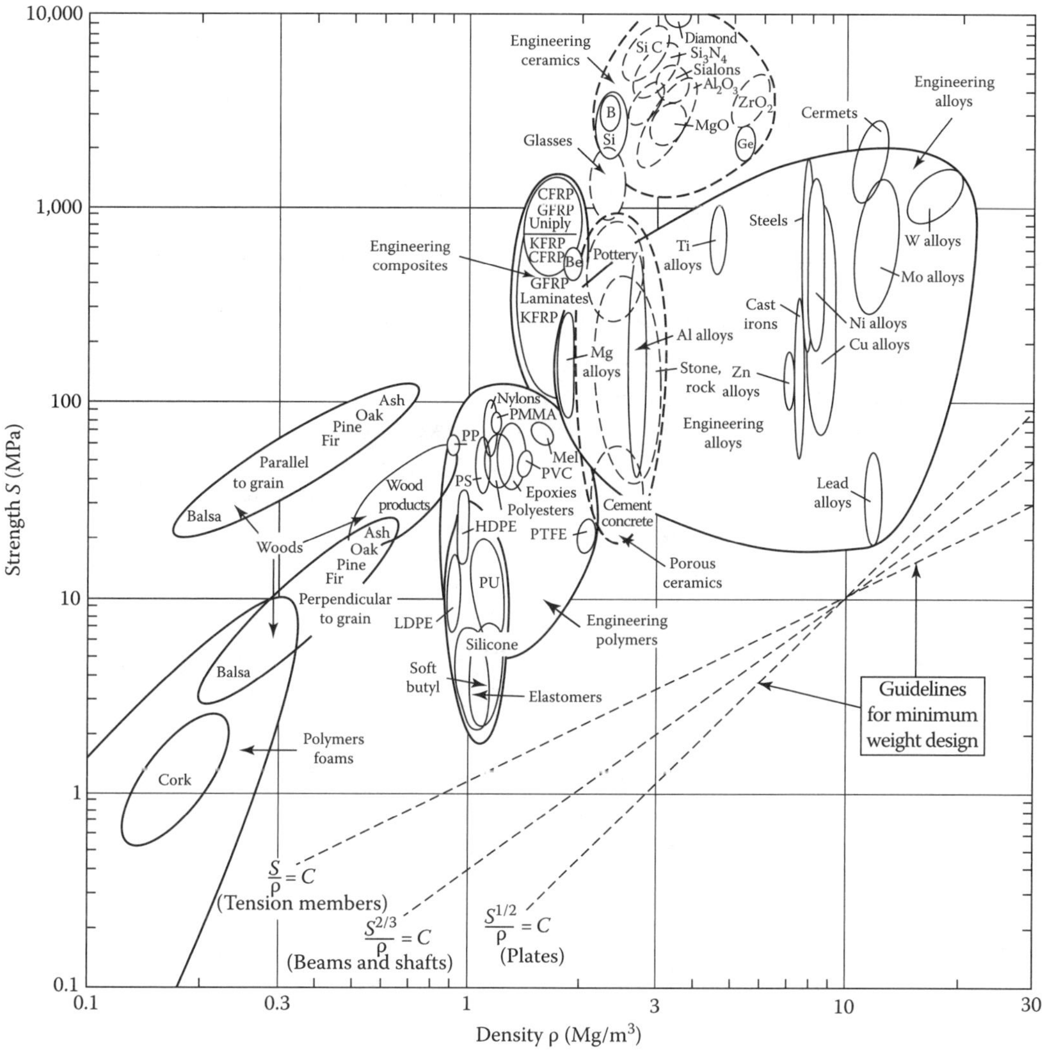

FIGURE 2.21
Strength versus density for engineering materials. The envelopes enclose data for a prescribed class of material. (From Ashby, M.J., *Material Selection in Mechanical Design*, 4th ed., Butterworth Heinemann, Oxford, U.K., 2011.

Example 2.7: Selecting Fishing Rod Material

A fishing rod is to be made from a tapered tube. Determine the material that makes the rod as strong as possible for a given weight.

Assumption: The material is tentatively selected for preliminary design purposes.

Solution

Figure 2.21 shows that the strongest materials for a given density are diamond, silicon carbide, and other ceramics.

Comment: It is not practical and expensive to use such materials for fishing rods. We thus choose a carbon-fiber-reinforced plastic or glass-fiber-reinforced plastic with 800–1000 MPa strength for density of 1.5 mg/m^3.

Problems

Sections 2.1 through 2.7

2.1 A bar of diameter d, gage length L, is loaded to the proportional limit in a tensile testing machine. A strain gage is placed on the surface of the bar to measure normal strains in the longitudinal direction. Under an axial load P, the bar is elongated $12(10^{-3})$ in. and its diameter reduced $0.24(10^{-3})$ in. Calculate the proportional limit, modulus of elasticity, Poisson's ratio, reduction in area, and percentage elongation.

Given: $d = 0.5$ in., $L = 8$ in., $P = 4$ kips.

2.2 A ⅛ in. diameter and 18 ½ ft long steel wire of yield strength $S_y = 50$ ksi stretches by 0.3 in. when subjected to a 500 lb tensile load. Compute the modulus of elasticity E.

2.3 A tensile test is performed on a flat-bar ASTM-A243 high-strength steel *specimen* (Figure P2.3). At certain instant, the applied load is P, while the distance between the gage marks is increased by ΔL, and the width w_o of the bar is decreased by Δw. Find

a. The axial strain and axial stress

b. The modulus of elasticity

c. Decrease in original width Δw and the original thickness Δt

Given: $L_o = 2.5$ in., $\Delta L = 0.00331$ in., $t_o = 0.24$ in., $w_o = 0.5$ in., $P = 4.8$ kips, $\nu = 0.3$.

2.4 When a 5 mm diameter brass bar is stretched by an axial force P, its diameter decreases by 1.5 mm. Find the magnitude of the load P.

Given: $E = 105$ GPa, $S_y = 250$ MPa, and $\nu = 0.34$ (by Table B.1).

2.5 A 15 mm × 15 mm square $ABCD$ is drawn on a member prior to loading. After loading, the square becomes the rhombus shown in Figure P2.5. Determine

a. The modulus of elasticity

b. Poisson's ratio

c. The shear modulus of elasticity

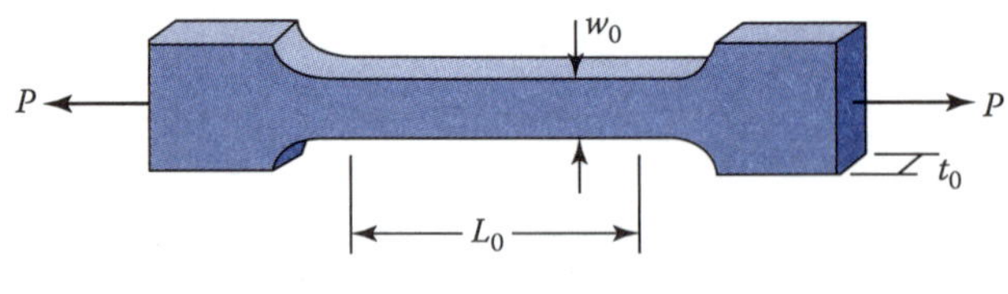

FIGURE P2.3

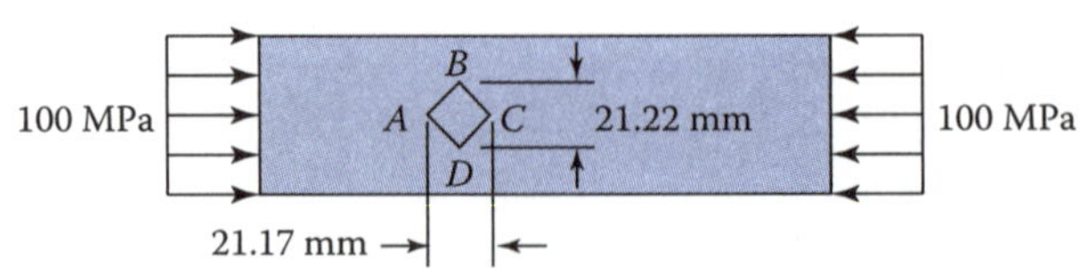

FIGURE P2.5

2.6 A bar of any given material is subjected to uniform triaxial stresses. Determine the maximum value of Poisson's ratio.

2.7 A rectangular block of width a, depth b, and length L is subjected to an axial tensile load P, as shown in Figure P2.7. Subsequent to the loading, dimensions b and L are changed to 1.999 in. and 10.02 in., respectively. Calculate

a. Poisson's ratio

b. The modulus of elasticity

c. The final value of the dimension a

d. The shear modulus of elasticity

Given: $a = 3$ in., $b = 2$ in., $L = 10$ in., $P = 100$ kips.

2.W Search the website at www.matweb.com. Review the material property *database* and select

a. Four metals with a tensile strength $S_y < 50$ ksi (345 MPa), modulus of elasticity $E > 26 \times 10^6$ psi (179 GPa), and Brinell hardness number $H_B < 200$.

b. Three metals having elongation greater than 15%, $E > 28 \times 10^6$ psi (193 GPa) and Poisson's ratio $\nu < 0.32$

c. One metallic alloy that has $E > 30 \times 10^6$ psi (207 GPa) and ultimate strength in compression $S_{uc} = 200$ ksi (1378 MPa)

2.8 The block shown in Figure P2.7 is subjected to an axial load P. Calculate the axial strain.

Given: $P = 25$ kN, $a = 20$ mm, $b = 10$ mm, $L = 100$ mm, $E = 70$ GPa, and $\nu = 0.3$. Assumption: The block is constrained against y- and z-directed contractions.

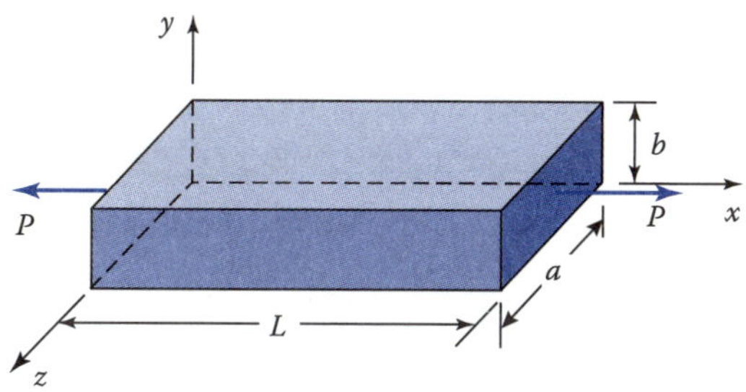

FIGURE P2.7

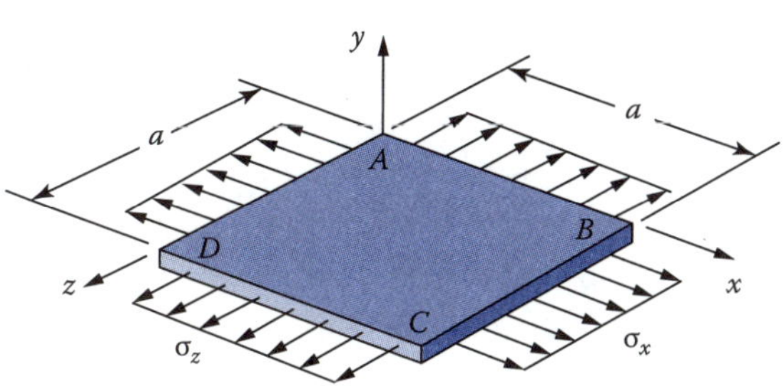

FIGURE P2.9

2.9 An aluminum alloy 2014-T6 square plate of sides a and thickness t is subjected to normal stresses σ_x and σ_z as shown in Figure P2.9. Find the change in

a. The length AB

b. The thickness of the plate

c. Volume of the plate

Given: $a = 320$ mm, $t = 15$ mm, $\sigma_x = 80$ MPa, $\sigma_z = 140$ MPa, $E = 72$ GPa, $\nu = 0.3$ (Table B.1).

2.10 A vibration damper unit is composed of a rubber cylinder of diameter d compressed inside of a wrought iron cylinder by a force F applied to the steel rod (Figure P2.10). Develop, in terms of d, F, and Poisson's ratio ν for the rubber, the following expression for the lateral pressure p between the rubber and the wrought iron cylinder:

$$p = \frac{4\nu F}{\pi d^2(1-\nu)} \tag{P2.10}$$

Compute the value of p for the following data: $d = 2.5$ in., $\nu = 0.5$, and $F = 2$ kips.

Assumptions: Friction between the rubber and cylinder as well as between the rod and cylinder can be negligible. The cylinder and rod are taken to be rigid.

2.11 Consider Figure P2.11 with a bronze block ($E = 100$ GPa, $\nu = 1/3$) subjected to uniform stresses σ_x, σ_y, and σ_z. Calculate the new dimensions after the loading.

Given: $L = 100$ mm, $a = 50$ mm, and $b = 10$ mm prior to the loading

$$\sigma_x = 150 \text{ MPa}, \quad \sigma_y = -90 \text{ MPa}, \quad \text{and} \quad \sigma_z = 0.$$

2.12 Resolve Problem 2.11 assuming that the block is under a uniform pressure of only $p = 120$ MPa on all its faces.

2.13 An ASTM A-48 gray cast iron solid sphere of radius r is under a uniform pressure p (Figure P2.13). Find

a. The decrease in circumference of the sphere

b. Decrease in the volume ΔV of the sphere

Given: $r = 5$ in., $p = 24$ ksi, $E = 10 \times 10^6$ psi, $\nu = 0.25$ (from Table B.1). *Note*: The volume of the sphere is $V_o = 4\pi r^3/3$.

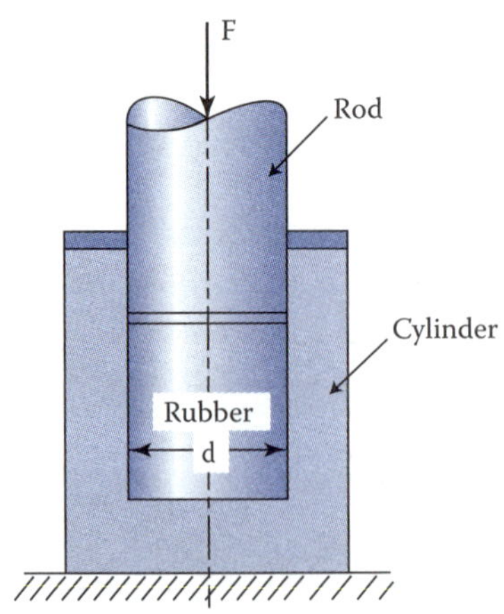

FIGURE P2.10

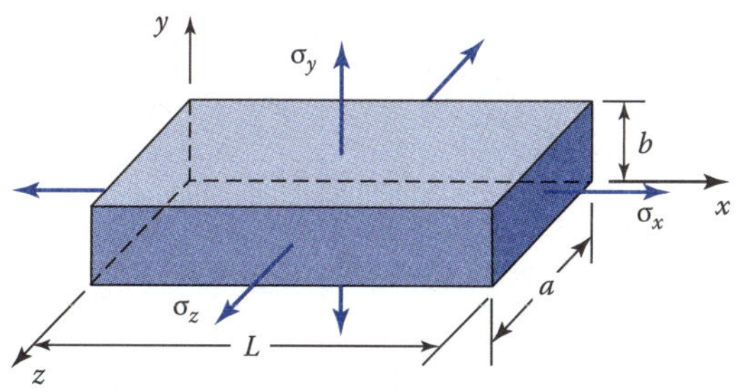

FIGURE P2.11

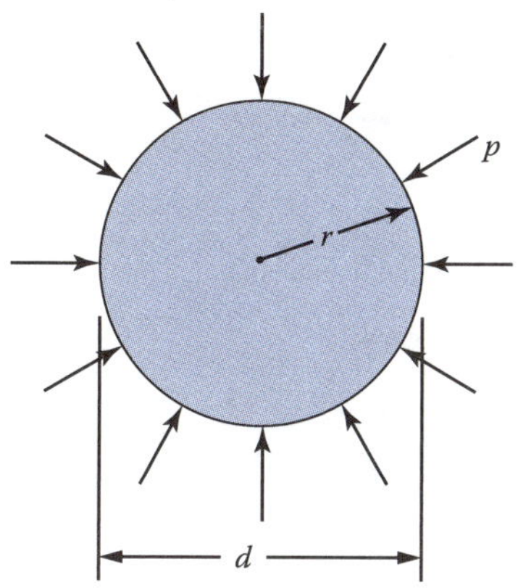

FIGURE P2.13

Sections 2.8 through 2.14

2.14 Determine the approximate value of the modulus of toughness for a structural steel bar having the stress–strain diagram of Figure 2.3b. What is the permanent elongation of the bar for a 50 mm gage length?

2.15 A strain energy of 9J must be acquired by a 6061-T6 aluminum alloy rod of diameter d and length L, as an axial load is applied (Figure P2.15). Determine the factor of safety n of the rod with respect to permanent deformation.

Given: $d = 5$ mm and $L = 3$ m.

2.16 Compute the modulus of resilience for two grades of steel (see Table B.1):

a. ASTM-A242

b. Cold-rolled, stainless steel (302)

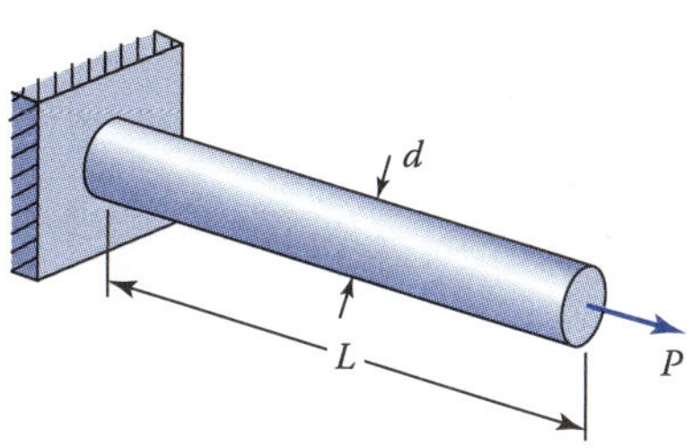

FIGURE P2.15

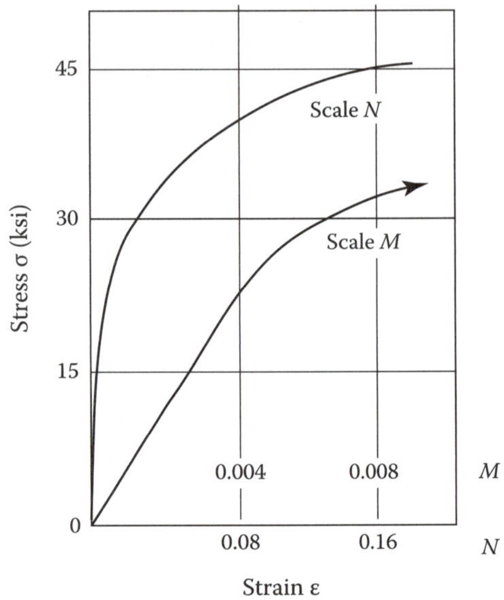

FIGURE P2.18

2.17 Compute the modulus of resilience for the following two materials (see Table B.1):

a. Aluminum alloy 2014-T6

b. Annealed yellow brass

2.18 A bar is made from a magnesium alloy, stress–strain diagram shown in Figure P2.18. Estimate the values of

a. The modulus of resilience

b. The modulus of toughness

2.19 A square steel machine component of sides a by a and length L is to resist an axial energy load of 400 N · m. Determine

a. The required yield strength of the steel

b. The corresponding modulus of resilience for the steel

Given: a = 50 mm, L = 1.5 m, factor of safety with respect to yielding n = 1.5, and G = 200 GPa.

2.20 A strain energy of U_{app} = 150 in. · lb must be acquired by an ASTM-A36 steel rod of diameter d and length L when the axial load is applied (Figure P2.15). Calculate the diameter d of the rod with a factor of safety n with respect to permanent deformation.

Given: L = 8 ft and n = 5.

2.21 The stress–strain diagrams of a structural steel bar are shown in Figure P2.21. Find

a. The modulus of resilience

b. The approximate modulus of toughness

2.22 A 50 mm square steel rod with modulus of elasticity E = 210 GPa and length L = 1.2 m is to resist axial energy load of 150 N · m. On the basis of a safety factor n = 1.8, find

a. The required proportional limit of steel

b. The corresponding modulus of resilience for the steel

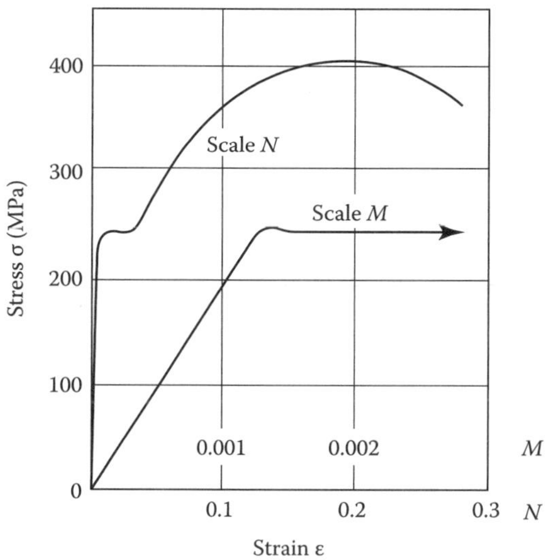

FIGURE P2.21

2.23 An AISI 1030 steel machine component is normalized to 149 Bhn. Using relationships of Section 2.10, determine the values of S_u and S_y for this component.

2.24 An AISI 1060 steel part is annealed to 179 Bhn. Using relationships given in Section 2.10, calculate the values of S_u and S_y for this part.

2.25 An AISI 4130 steel machine element is annealed to 156 Bhn. Using relationships given in Section 2.10, estimate the values of S_u and S_y for this element.

2.26 An AISI 1095 steel component is annealed to 293 Bhn. Using relationships given in Section 2.10, compute the values of S_u and S_y for this component.

3

Stress and Strain

3.1 Introduction

This chapter provides a review and insight into the stress and strain analyses. Expressions for both stresses and deflections in mechanical elements are developed throughout the text as the subject unfolds, after examining their function and general geometric behavior. With the exception of Sections 3.13 through 3.17, we employ mechanics of materials approach, simplifying the assumptions related to the deformation pattern so that strain distributions for a cross section of a member can be determined. A fundamental assumption is that *plane sections remain plane*. This hypothesis can be shown to be exact for axially loaded elastic prismatic bars and circular torsion members and for slender beams, plates, and shells subjected to pure bending. The assumption is approximate for other stress analysis problems. Note, however, that there are many cases where applications of the *basic formulas of mechanics of materials*, so-called elementary formulas for stress and displacement, lead to useful results for slender members under any type of loading.

Our coverage presumes a knowledge of mechanics of materials procedures for determining stresses and strains in a homogeneous and an isotropic bar, shaft, and beam. In Sections 3.2 through 3.9, we introduce the basic formulas, the main emphasis being on the underlying assumptions used in their derivations. Next to be treated are the transformation of stress and strain at a point and measurement of normal strains on the free surface of a member. Then attention focuses on stresses arising from various combinations of fundamental loads applied to members and the stress concentrations. The chapter concludes with discussions on the states of stress and strain.

In the treatment presented here, the study of complex stress patterns at the supports or locations of concentrated load is not included. According to Saint-Venant's Principle (Section 1.4), the actual stress distribution closely approximates that given by the formulas of the mechanics of materials, except near the restraints and geometric discontinuities in the members. For further details, see texts on solid mechanics and theory of elasticity, for example, References 1–3.

3.2 Stresses in Axially Loaded Members

Axially loaded members are structural and machine elements having straight longitudinal axes and supporting only axial forces (tensile or compressive). Figure 3.1a shows a homogeneous prismatic bar loaded by tensile forces P at the ends. To determine the

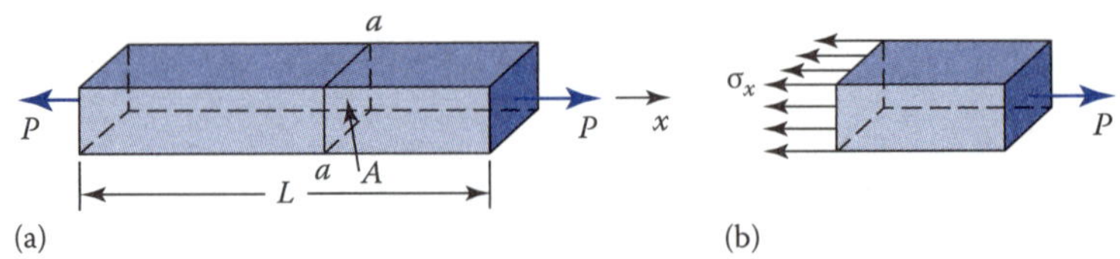

FIGURE 3.1
(a) Prismatic bar in tension and (b) free-body diagram of an isolated portion.

normal stress, we make an imaginary cut (section A–A) through the member at right angles to its axis (x). A free-body diagram of the isolated part is shown in Figure 3.1b. Here, the stress is substituted on the cut section as a replacement for the effect of the removed part.

Assuming that the stress has a uniform distribution over the cross section, the equilibrium of the axial forces, the first of Equation 1.5, yields $P = \int \sigma_x dA$ or $P = A\sigma_x$. The normal stress is therefore

$$\sigma_x = \frac{P}{A} \tag{3.1}$$

where A is the cross-sectional area of the bar. The remaining conditions of Equations 1.5 are also satisfied by the stress distribution pattern shown in Figure 3.1b. When the member is being stretched as depicted in the figure, the resulting stress is a uniaxial tensile stress; if the direction of the forces is reversed, the bar is in compression, and uniaxial compressive stress occurs. Equation 3.1 is applicable to tension members and chunky, short compression bars. For slender members, the approaches discussed in Chapter 6 must be used.

Stress due to the restriction of thermal expansion or contraction of a body is called *thermal stress*, σ_t. Using Hooke's law and Equation 1.21, we have

$$\sigma_t = \alpha(\Delta T)E \tag{3.2}$$

The quantity ΔT represents a temperature change. We observe that a high modulus of elasticity E and high coefficient of expansion α for the material increase the stress.

3.2.1 Design of Tension Members

Tension members are found in bridges, roof trusses, bracing systems, and mechanisms. They are used as tie rods, cables, angles, channels, or combinations of these. Of special concern is the design of prismatic tension members for strength under static loading. In this case, a rational design procedure (see Section 1.6) may be briefly described as follows:

1. *Evaluate the mode of possible failure.* Usually the normal stress is taken to be the quantity most closely associated with failure. This assumption applies regardless of the type of failure that may actually occur on a plane of the bar.
2. *Determine the relationships between load and stress.* This important value of the normal stress is defined by $\sigma = P/A$.

3. *Determine the maximum usable value of stress.* The maximum usable value of σ without failure, σ_{max}, is the yield strength S_y or the ultimate strength S_u. Use this value in connection with equation found in step 2, if needed, in any expression of failure criteria, discussed in Chapter 6.
4. *Select the factor of safety.* A safety factor n is applied to σ_{max} to determine the allowable stress $\sigma_{all} = \sigma_{max}/n$. The required cross-sectional area of the member is therefore

$$A = \frac{P}{\sigma_{all}} \tag{3.3}$$

If the bar contains an abrupt change of cross-sectional area, the foregoing procedure is repeated, using a stress-concentration factor to find the normal stress (step 2).

Example 3.1: Design of a Hoist

A pin-connected two-bar assembly or hoist is supported and loaded as shown in Figure 3.2a. Determine the cross-sectional area of the round aluminum eyebar *AC* and the square wood post *BC*.

Given: The required load is $P = 50$ kN. The maximum usable stresses in aluminum and wood are 480 and 60 MPa, respectively.

Assumptions: The load acts in the plane of the hoist. Weights of members are insignificant compared to the applied load and omitted. Friction in pin joints and the possibility of member *BC* buckling are ignored.

Design Decision: Use a factor of safety of $n = 2.4$.

Solution

Members *AC* and *BC* carry axial loading with relative dimensions shown by small triangles in Figure 3.2b. We see that the slopes of the forces F_A and F_B are 5/12 and 1/1, respectively. It follows that $F_{Ay}/5 = F_A/13$ and $F_{By} = F_B/\sqrt{2}$ and that the vertical force components are $F_{Ay} = (5/13)F_A$ and $F_{By} = \left(1/\sqrt{2}\right)F_B$. Hence, applying equations of statics to the free-body diagram of Figure 3.2b, we have

$$\Sigma M_B = -40(2.5) - 30(2.5) + \frac{5}{13}F_A(3.5) = 0 \qquad F_A = 130 \text{ kN}$$

$$\Sigma M_A = -40(2.5) - 30(6) + \frac{1}{\sqrt{2}}F_B(3.5) = 0 \qquad F_B = 113.1 \text{ kN}$$

Note, as a check, that $\Sigma F_x = 0$.

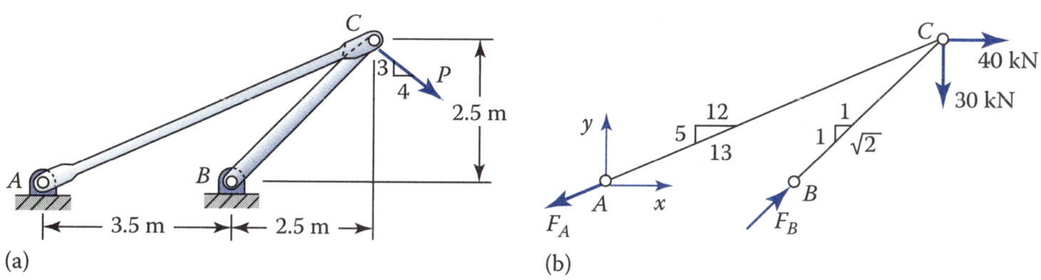

FIGURE 3.2
Example 3.1. (a) a loaded hoist and (b) its free-body diagram.

The allowable stress, from design procedure steps 3 and 4, is

$$(\sigma_{all})_{AC} = \frac{480}{2.4} = 200 \text{ MPa}, \quad (\sigma_{all})_{BC} = \frac{60}{2.4} = 25 \text{ MPa}$$

By Equation 3.3, the required cross-sectional areas of the bars are

$$A_{AC} = \frac{130(10^3)}{200} = 650 \text{ mm}^2, \quad A_{BC} = \frac{113.1(10^3)}{25} = 4524 \text{ mm}^2$$

Comment: A 29 mm diameter aluminum eyebar and a 68 mm × 68 mm wood post should be used.

3.3 Direct Shear Stress and Bearing Stress

A *shear stress* is produced whenever the applied forces cause one section of a body to tend to slide past its adjacent section. As an example, consider the connection shown in Figure 3.3a. This joint consists of a blade or bracket, a clevis, and a pin that passes through holes in the bracket and clevis. A *force-flow path* through the connection is depicted by the dashed lines in Figure 3.3b. Observe that in this symmetrical design, the load P is equally divided between the two prongs of the clevis. The pin resists the shear across the two cross-sectional areas at b–b and c–c; hence, it is said to be in *double shear.* At each cut section,

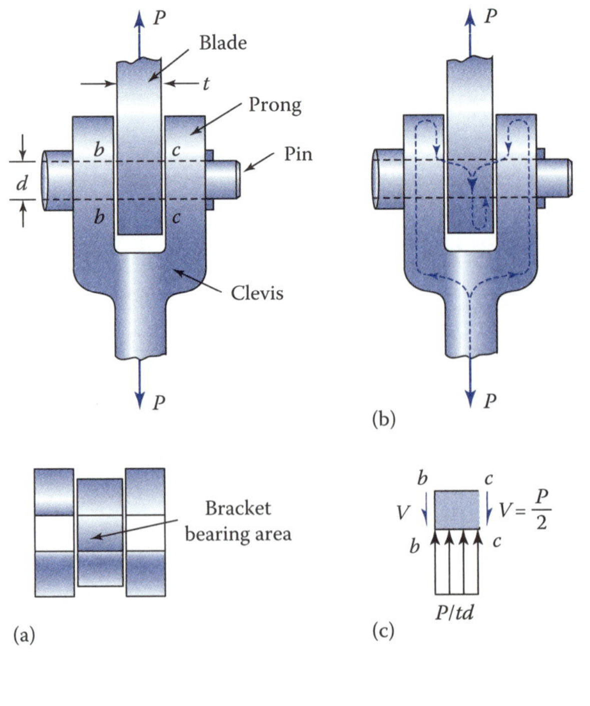

FIGURE 3.3
(a) A clevis–pin connection, with the bracket bearing area depicted, (b) force-flow lines, and (c) portion of pin subjected to direct shear stresses and bearing stress.

a shear force V equivalent to $P/2$ (Figure 3.3c) must be developed. Thus, the shear occurs over an area parallel to the applied load. This condition is termed *direct shear.*

The distribution of shear stress τ across a section cannot be taken as uniform. Dividing the total shear force V by the cross-sectional area A over which it acts, we can obtain the average shear stress in the section:

$$\tau_{\text{avg}} = \frac{V}{A} \tag{3.4}$$

The average shear stress in the pin of the connection shown in the figure is therefore $\tau_{\text{avg}} = (P/2)/(\pi d^2/4) = 2P/\pi d^2$. Direct shear arises in the design of bolts, rivets, welds, glued joints, as well as in pins (Sections 15.13 through 15.18). In each case, the shear stress is created by a direct action of the forces in trying to cut through the material. Shear stress also arises in an indirect manner when members are subjected to tension, torsion, and bending, as discussed in the following sections.

Note that under the action of the applied force, the bracket and the clevis press against the pin in bearing and a nonuniform pressure develops against the pin (Figure 3.3b). The average value of this pressure is determined by dividing the force P transmitted by the *projected area* A_p of the pin into the bracket (or clevis). This is called the *bearing stress*:

$$\sigma_b = \frac{P}{A_p} \tag{3.5}$$

Therefore, bearing stress in the bracket against the pin is $\sigma_b = P/td$, where t and d represent the thickness of bracket and diameter of the pin, respectively. Similarly, the bearing stress in the clevis against the pin may be obtained. In the preceding, it is assumed that the diameter of the pin and the hole (in bracket and clevis) are about the same.

Example 3.2: Design of a Monoplane Wing Rod

The wing of a monoplane is approximated by a pin-connected structure of beam AD and bar BC, as depicted in Figure 3.4a. Determine

a. The shear stress in the pin at hinge C
b. The diameter of the rod BC

Given: The pin at C has a diameter of 15 mm and is in double shear.

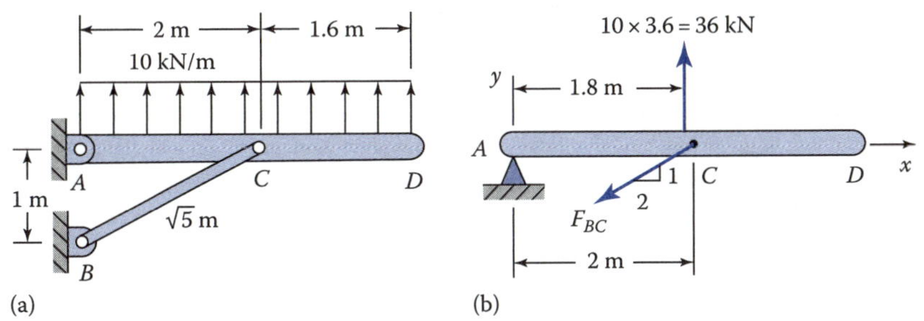

FIGURE 3.4
Example 3.2. (a) A uniformly loaded wing and (b) its free-body diagram.

Assumptions: Friction in pin joints is omitted. The air load is distributed uniformly along the span of the wing. Only rod *BC* is under tension. A round 2014-T6 aluminum alloy bar (see Table B.1) is used for rod *BC* with an allowable axial stress of 210 MPa.

Solution

Referring to the free-body diagram of the wing *ACD* (Figure 3.4b),

$$\Sigma M_A = 36(1.8) - F_{BC}\frac{1}{\sqrt{5}}(2) = 0 \quad F_{BC} = 72.45 \text{ kN}$$

where F_{BC} $(1/\sqrt{5})$ is the vertical component of the axial force in member *BC*.

a. Through the use of Equation 3.4,

$$\tau_{avg} = \frac{F_{BC}}{2A} = \frac{72,450}{2\left[\pi(0.0075)^2\right]} = 205 \text{ MPa}$$

b. Applying Equation 3.1, we have

$$\sigma_{BC} = \frac{F_{BC}}{A_{BC}}, \quad 210(10^6) = \frac{72,450}{A_{BC}}$$

Solving

$$A_{BC} = 3.45(10^{-4})\text{m}^2 = 345 \text{ mm}^2$$

Hence,

$$345 = \frac{\pi d^2}{4}, \quad d = 20.96 \text{ mm}$$

Comments: A 21 mm diameter rod should be used. Note that for steady inverted flight, the rod *BC* would be a compression member.

3.4 Thin-Walled Pressure Vessels

Pressure vessels are closed structures that contain liquids or gases under pressure. Common examples include tanks for compressed air, steam boilers, and pressurized water storage tanks. Although pressure vessels exist in a variety of different shapes (see Sections 16.10 through 16.13), only thin-walled cylindrical and spherical vessels are considered here. A vessel having a wall thickness less than about 1/10 of inner radius is called *thin walled.* For this case, we can take $r_i \approx r_o \approx r$, where r_i, r_o, and r refer to inner, outer, and mean radii, respectively. The contents of the pressure vessel exert internal pressure, which produces small stretching deformations in the membrane-like walls of an inflated balloon. In some cases, external pressures cause contractions of a vessel wall. With either internal or external pressure, stresses termed *membrane stresses* arise in the vessel walls.

Section 16.11 shows that application of the equilibrium conditions to an appropriate portion of a thin-walled tank suffices to determine membrane stresses. Consider a thin-walled *cylindrical vessel* with closed ends and internal pressure p (Figure 3.5a). The longitudinal or

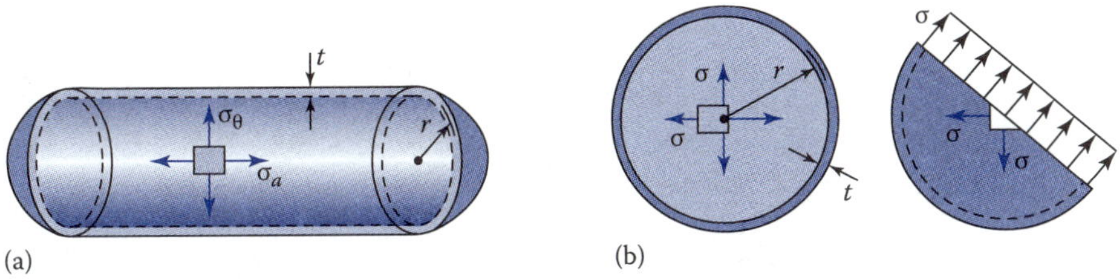

FIGURE 3.5
Thin-walled pressure vessels: (a) cylindrical and (b) spherical.

axial stress σ_a and circumferential or *tangential stress* σ_θ acting on the side faces of a stress element shown in the figure are principal stresses from Equations 16.69:

$$\sigma_a = \frac{pr}{2t} \tag{3.6a}$$

$$\sigma_\theta = \frac{pr}{t} \tag{3.6b}$$

The circumferential strain as a function of the change in radius δ_c is $\varepsilon_\theta = \left[2\pi(r+\delta_c) - 2\pi r\right]/2\pi r = \delta_c/r$. Using Hooke's law, we have $\varepsilon_\theta = (\sigma_\theta - \nu\sigma_a)/E$, where ν and E represent Poisson's ratio and modulus of elasticity, respectively. The extension of the radius of the cylinder, $\delta_c = \varepsilon_\theta r$, under the action of the stresses given by Equations 3.6 is therefore

$$\delta_c = \frac{pr^2}{2Et}(2-\nu) \tag{3.7}$$

The tangential stresses σ act in the plane of the wall of a *spherical vessel* and are the same in any section that passes through the center under internal pressure p (Figure 3.5b). *Sphere stress* is given by Equation 16.66:

$$\sigma = \frac{pr}{2t} \tag{3.8}$$

They are half the magnitude of the tangential stresses of the cylinder. Thus, sphere is an optimum shape for an internally pressurized closed vessel. The radial extension of the sphere, $\delta_s = \varepsilon r$, applying Hooke's law $\varepsilon = (\sigma - \nu\sigma)/E$ is then

$$\delta_s = \frac{pr^2}{2Et}(1-\nu) \tag{3.9}$$

Note that the stress acting in the radial direction on the wall of a cylinder or sphere varies from $-p$ at the inner surface of the vessel to 0 at the outer surface. For thin-walled vessels, radial stress σ_r is much smaller than the membrane stresses and is usually omitted. The state of stress in the wall of a vessel is therefore considered biaxial. To conclude, we mention that a pressure vessel design is essentially governed by ASME Pressure Vessel Design Codes, discussed in Section 16.12.

Thick-walled cylinders are often used as vessels or pipe lines. Some applications involve air or hydraulic cylinders, gun barrels, and various mechanical components. Equations for exact elastic and plastic stresses and displacements for these members are developed in Chapter 16.* Composite thick-walled cylinders under pressure, thermal, and dynamic loading are discussed in detail. Numerous illustrative examples also are given.

Example 3.3: Design of Spherical Pressure Vessel

A spherical vessel of radius r is subjected to an internal pressure p. Determine the critical wall thickness t and the corresponding diametral extension.

Assumption: A safety factor n against bursting is used.

Given: $r = 2.5$ ft, $p = 1.5$ ksi, $S_u = 60$ ksi, $E = 30 \times 10^6$ psi, $\nu = 0.3$, $n = 3$

Solution

We have $r = 2.5 \times 12 = 30$ in. and $\sigma = S_u/n$. Applying Equation 3.8,

$$t = \frac{pr}{2S_u/n} = \frac{1.5(30)}{2(60/3)} = 1.125 \text{ in.}$$

Then Equation 3.9 results in

$$\delta_s = \frac{pr^2(1-\nu)}{2Et} = \frac{1500(30)^2(0.7)}{2(30\times 10^6)(1.125)} = 0.014 \text{ in.}$$

The diametral extension is therefore $2\delta_s = 0.028$ in.

3.5 Stress in Members in Torsion

In this section, attention is directed toward stress in prismatic bars subject to equal and opposite end torques. These members are assumed free of end constraints. Both circular and rectangular bars are treated. *Torsion* refers to twisting a structural member when it is loaded by couples that cause rotation about its longitudinal axis. Recall from Section 1.8 that, for convenience, we often show the moment of a couple or torque by a vector in the form of a double-headed arrow.

3.5.1 Circular Cross Sections

Torsion of circular bars or shafts produced by a torque T results in a shear stress τ and an angle of twist or angular deformation ϕ, as shown in Figure 3.6a. The *basic assumptions* of the formulations on the torsional loading of a circular prismatic bar are as follows:

1. A plane section perpendicular to the axis of the bar remains plane and undisturbed after the torques are applied.
2. Shear strain γ varies linearly from 0 at the center to a maximum on the outer surface.
3. The material is homogeneous and obeys Hooke's law; hence, the magnitude of the maximum shear angle γ_{max} must be less than the yield angle.

* Within this chapter, some readers may prefer to study Section 16.3.

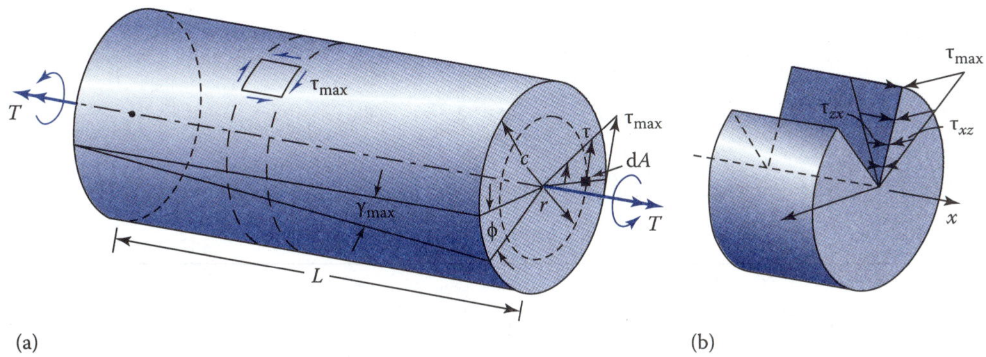

FIGURE 3.6
(a) Circular bar in pure torsion and (b) shear stresses on transverse (xz) and axial (zx) planes in a circular shaft segment in torsion.

The maximum shear stress occurs at the points most remote from the center of the bar and is designated τ_{max}. For a linear stress variation, at any point at a distance r from center, the shear stress is $\tau = (r/c)\tau_{max}$, where c represents the radius of the bar. On a cross section of the shaft, the resisting torque caused by the stress distribution must be equal to the applied torque T. Hence,

$$T = \int r\left(\frac{r}{c}\tau_{max}\right)dA$$

The preceding relationship may be written in the form

$$T = \frac{\tau_{max}}{c}\int r^2 dA$$

By definition, the polar moment of inertia J of the cross-sectional area is

$$J = \int r^2 dA \qquad \text{(a)}$$

For a solid shaft, $J = \pi c^4/2$. In the case of a circular tube of inner radius b and outer radius c, $J = \pi(c^4 - b^4)/2$.

Shear stress varies with the radius and is largest at the points most remote from the shaft center. This stress distribution leaves the external cylindrical surface of the bar free of stress distribution, as it should. Note that the representation shown in Figure 3.6a is purely schematic. The maximum shear stress on a cross section of a circular shaft, either solid or hollow, is given by the *torsion formula:*

$$\tau_{max} = \frac{Tc}{J} \qquad (3.10)$$

The shear stress at distance r from the center of a section is

$$\tau = \frac{Tr}{J} \qquad (3.11)$$

The *transverse* shear stress found in Equation 3.10 or 3.11 is accompanied by an axial shear stress of equal value, that is, $\tau = \tau_{xz} = \tau_{zx}$ (Figure 3.6b), to satisfy the conditions of static equilibrium of an element. Since the shear stress in a solid circular bar is maximum at the outer boundary of the cross section and 0 at the center, most of the material in a solid shaft is stressed significantly below the maximum shear stress level. When weight reduction and savings of material are important, it is advisable to use hollow shafts (see also Example 3.4).

3.5.2 Noncircular Cross Sections

In treating torsion of noncircular prismatic bars, cross sections initially plane experience out-of-plane deformation or *warping*, and the first two assumptions stated previously are no longer appropriate. Figure 3.7 depicts the nature of distortion occurring in a rectangular section. The mathematical solution of the problem is complicated. For cases that cannot be conveniently solved by applying the theory of elasticity, the governing equations are used in conjunction with the experimental techniques. The finite element analysis is also very efficient for this purpose. Torsional stress (and displacement) equations for a number of noncircular sections are summarized in references such as [3,4]. Table 3.1 lists the *exact* solutions of the maximum shear stress and the angle of twist ϕ for a few common cross sections. Note that the values of coefficients α and β depend on the ratio of the side lengths a and b of a rectangular section. For thin sections ($a \gg b$), the values of α and β approach 1/3.

The following approximate formula for the maximum shear stress in a *rectangular member* is of interest:

$$\tau_{max} = \frac{T}{ab^2}\left(3 + 1.8\frac{b}{a}\right) \tag{3.12}$$

As in Table 3.1, a and b represent the lengths of the long and short sides of a rectangular cross section, respectively. The stress occurs along the centerline of the wider face of the bar. For a *thin section*, where a is much greater than b, the second term may be neglected. Equation 3.12 is also valid for equal-leg angles; these can be considered as two rectangles, each of which is capable of carrying half the torque.

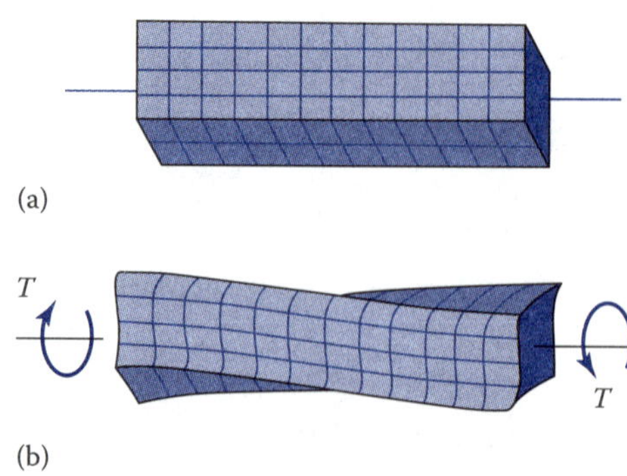

FIGURE 3.7
Rectangular bar (a) before and (b) after a torque is applied.

TABLE 3.1

Expressions for Stress and Deformation in Some Cross-Sectional Shapes in Torsion

Cross Section	Maximum Shearing Stress	Angle of Twist
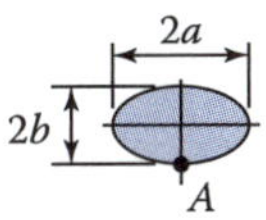 Ellipse for circle: $a = b$	$\tau_A = \dfrac{2T}{\pi ab^2}$	$\phi = \dfrac{(a^2 + b^2)T}{\pi a^3 b^3 G}$
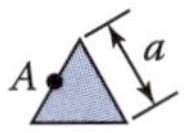 Equilateral triangle	$\tau_A = \dfrac{20T}{a^3}$	$\phi = \dfrac{46.2T}{a^4 G}$
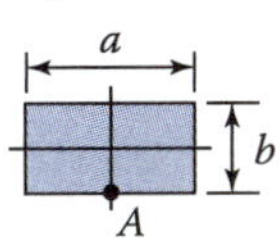 Rectangle	$\tau_A = \dfrac{T}{\alpha ab^2}$	$\phi = \dfrac{T}{\beta ab^3 G}$

a/b	α	β
1.0	0.208	0.141
1.5	0.231	0.196
2.0	0.246	0.229
2.5	0.256	0.249
3.0	0.267	0.263
4.0	0.282	0.281
5.0	0.292	0.291
10.0	0.312	0.312
∞	0.333	0.333

Cross Section	Maximum Shearing Stress	Angle of Twist
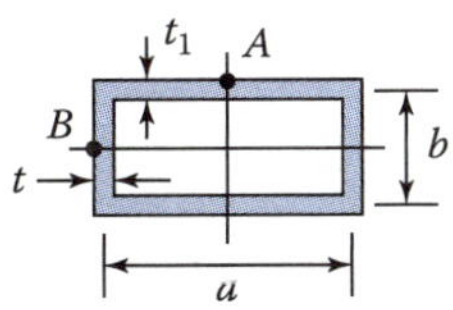 Hollow rectangle	$\tau_A = \dfrac{T}{2abt_1}$ $\tau_B = \dfrac{T}{2abt}$	$\phi = \dfrac{(at + bt_1)T}{2tt_1 a^2 b^2 G}$
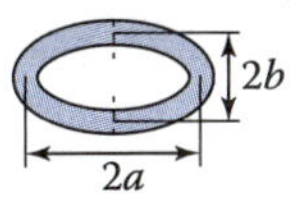 Hollow ellipse for hollow circle: $a = b$	$\tau_A = \dfrac{T}{2\pi abt}$	$\phi = \dfrac{\sqrt{2(a^2 + b^2)}T}{4\pi a^2 b^2 tG}$
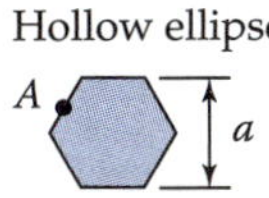 Hexagon	$\tau_A = \dfrac{5.7T}{a^3}$	$\phi = \dfrac{8.8T}{a^4 G}$

Example 3.4: Torque Transmission Efficiency of Hollow and Solid Shafts

A hollow shaft and a solid shaft (Figure 3.8) are twisted about their longitudinal axes with torques T_h and T_s, respectively. Determine the ratio of the largest torques that can be applied to the shafts.

Given: $c = 1.15b$

Assumptions: Both shafts are made of the same material with allowable stress, and both have the same cross-sectional area.

Solution

The maximum shear stress τ_{max} equals τ_{all}. Since the cross-sectional areas of both shafts are identical, $\pi(c^2 - b^2) = \pi a^2$:

$$a^2 = c^2 - b^2$$

For the hollow shaft, using Equation 3.10,

$$T_h = \frac{\pi}{2c}\left(c^4 - b^4\right)\tau_{all}$$

Likewise, for the solid shaft,

$$T_s = \frac{\pi}{2}a^3\tau_{all}$$

We therefore have

$$\frac{T_h}{T_s} = \frac{c^4 - b^4}{ca^3} = \frac{c^4 - b^4}{c\left(c^2 - b^2\right)^{3/2}} \tag{3.13}$$

Substituting $c = 1.15b$, this quotient gives

$$\frac{T_h}{T_s} = 3.56$$

Comments: The result shows that hollow shafts are more efficient in transmitting torque than solid shafts. Interestingly, thin shafts are also useful for creating an essentially uniform shear (i.e., $\tau_{min} \approx \tau_{max}$). However, to avoid buckling (see Section 5.9), the wall thickness cannot be excessively thin.

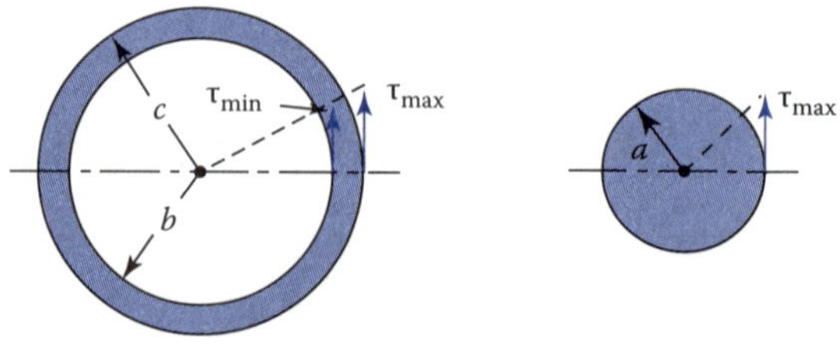

FIGURE 3.8
Example 3.4. Hollow and solid shaft cross sections.

3.6 Shear and Moment in Beams

In beams loaded by transverse loads in their planes, only two components of stress resultants occur: the shear force and bending moment. These loading effects are sometimes referred to as *shear* and *moment in beams.* To determine the magnitude and sense of shearing force and bending moment at any section of a beam, the method of sections is applied. The sign conventions adopted for internal forces and moments (see Section 1.8) are associated with the deformations of a member. To illustrate this, consider the positive and negative shear forces V and bending moments M acting on segments of a beam cut out between two cross sections (Figure 3.9). We see that a positive shear force tends to raise the left-hand face relative to the right-hand face of the segment and a positive bending moment tends to bend the segment concave upward, so it *retains water.* Likewise, a positive moment compresses the upper part of the segment and elongates the lower part.

3.6.1 Load, Shear, and Moment Relationships

Consider the free-body diagram of an element of length dx, cut from a loaded beam (Figure 3.10a). Note that the distributed load w per unit length, the shears, and the bending moments are shown as positive (Figure 3.10b). The changes in V and M from position x to x + dx are denoted by dV and dM, respectively. In addition, the resultant of the distributed load (w dx) is indicated by the dashed line in the figure. Although w is not uniform, this is permissible substitution for a very small distance dx.

Equilibrium of the vertical forces acting on the element of Figure 3.10b, $\sum F_x = 0$, results in $V + w\,dx = V + dV$. Therefore,

$$\frac{dV}{dx} = w \tag{3.14a}$$

FIGURE 3.9
Sign convention for beams: definitions of positive and negative shear and moment.

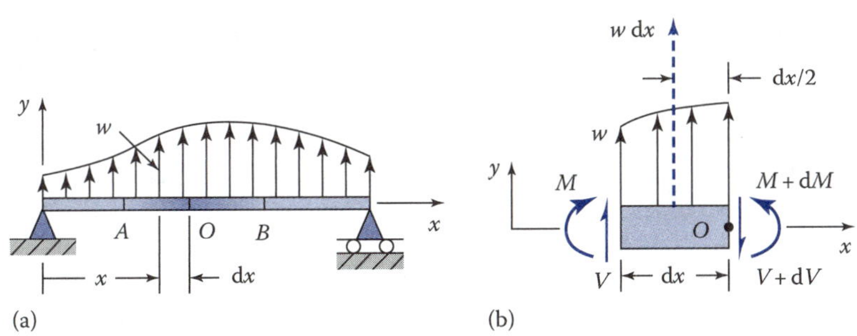

FIGURE 3.10
(a) Beam and (b) an element isolated from it.

This states that at any section of the beam, the slope of the shear curve is equal to w. Integration of Equation 3.14a between points A and B on the beam axis gives

$$V_B - V_A = \int_A^B w \, dx = \text{Area of load diagram between } A \text{ and } B \tag{3.14b}$$

Clearly, Equation 3.14a is not valid at the point of application of a concentrated load. Similarly, Equation 3.14b cannot be used when concentrated loads are applied between A and B. For equilibrium, the sum of moments about O must also be $0: \sum M_0 = 0$ or $M + dM - (V + dV)dx - M = 0$. If second-order differentials are considered as negligible compared with differentials, this yields

$$\frac{dM}{dx} = V \tag{3.15a}$$

The foregoing relationship indicates that the slope of the moment curve is equal to V. Therefore, the shear force is inseparably linked with a change in the bending moment along the length of the beam. Note that the maximum value of the moment occurs at the point where V (and hence dM/dx) is 0. Integrating Equation 3.15a between A and B, we have

$$M_B - M_A = \int_A^B V dx = \text{Area of shear diagram between } A \text{ and } B \tag{3.15b}$$

The differential equations of equilibrium, Equations 3.14a and 3.15a, show that the shear and moment curves, respectively, always are 1° and 2° higher than the load curve. We note that Equation 3.15a is not valid at the point of application of a concentrated load. Equation 3.15b can be used even when concentrated loads act between A and B, but the relation is not valid if a couple is applied at a point between A and B.

3.6.2 Shear and Moment Diagrams

When designing a beam, it is useful to have a graphical visualization of the shear force and moment variations along the length of a beam. A shear diagram is a graph where the shearing force is plotted against the horizontal distance (x) along a beam. Similarly, a graph showing the bending moment plotted against the x axis is the bending moment diagram. The signs for shear V and moment M follow the general convention defined in Figure 3.9. It is convenient to place the shear and bending moment diagrams directly below the free-body, or load, diagram of the beam. The maximum and other significant values are generally marked on the diagrams.

We use the so-called summation method of constructing shear and moment diagrams. The procedure of this semigraphical approach is as follows:

1. Determine the reactions from free-body diagram of the entire beam.
2. Determine the value of the shear, successively summing from the *left end* of the beam to the vertical external forces or using Equation 3.14b. Draw the shear diagram obtaining the shape from Equation 3.14a. Plot a positive V upward and a negative V downward.

3. Determine the values of moment, either continuously summing the external moments from the left end of the beam or using Equation 3.15b, whichever is more appropriate. Draw the moment diagram. The shape of the diagram is obtained from Equation 3.15a.

A check on the accuracy of the shear and moment diagrams can be made by noting whether or not they close. Closure of these diagrams demonstrates that the sum of the shear forces and moments acting on the beam is 0, as they must be for equilibrium. When any diagram fails to close, you know that there is a construction error or an error in calculation of the reactions. The following example illustrates the procedure.

Example 3.5: Shear and Moment Diagrams for a Simply Supported Beam by Summation Method

Draw the shear and moment diagrams for the beam loaded as shown in Figure 3.11a.

Assumptions: All forces are coplanar and 2D.

Solution

Applying the equations of statics to the free-body diagram of the entire beam, we have (Figure 3.11b)

$$R_A = 6.4 \text{ kN}, \quad R_B = 9.6 \text{ kN}$$

In the *shear diagram* (Figure 3.11c), the shear at end *C* is $V_C = 0$. Equation 3.14b yields

$$V_A - V_C = \frac{1}{2}w(1.5) = \frac{1}{2}(-4)(1.5) = -3, \quad V_A = -3 \text{ kN}$$

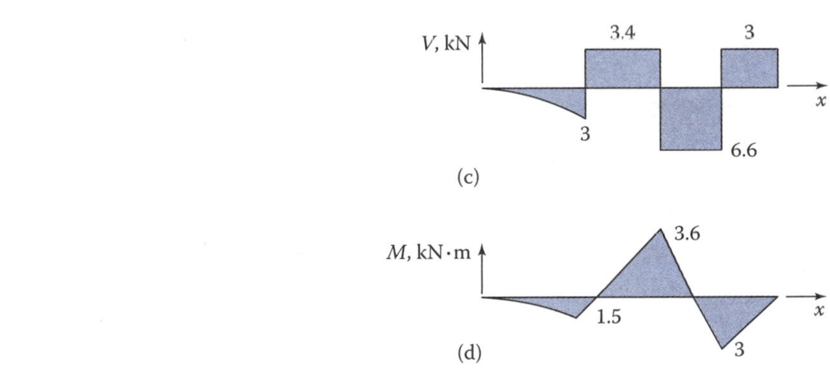

FIGURE 3.11
Example 3.5. (a) an overhanging beam, (b) free-body or load diagram, (c) shear diagram, and (d) moment diagram.

the upward force near to the left of A. From C to A, the load increases linearly; hence, the shear curve is parabolic, which has a negative and increasing slope. In the regions AD, DB, and BE, the slope of the shear curve is 0 or the shear is constant. At A, the 6.4 kN upward reaction force increases the shear to 3.4 kN. The shear remains constant up to D where it decreases by a 10 kN downward force to −6.6 kN. Likewise, the value of the shear rises to 3 kN at B. No change in the shear occurs until point E, where the downward 3 kN force closes the diagram. The maximum shear $V_{max} = -6.6$ kN occurs in region BD.

In the *moment diagram* (Figure 3.11d), the moment at end C is $M_C = 0$. Equation 3.15b gives

$$M_A - M_C = -\int_0^{1.5}\left(\frac{1}{2}\frac{4x}{1.5}x\right)dx \qquad M_A = -1.5\ \text{N}\cdot\text{m}$$

$$M_D - M_A = 3.4(1.5) \qquad M_D = 3.6\ \text{kN}\cdot\text{m}$$

$$M_B - M_D = -6.6(1) \qquad M_B = -3\ \text{kN}\cdot\text{m}$$

$$M_E - M_B = 3(1) \qquad M_E = 0$$

Since M_E is known to be 0, a check on the calculations is provided. We find that from C to A, the diagram takes the shape of a cubic curve concave downward with 0 slope at C. This is in accordance with $dM/dx = V$. Here, V, prescribing the slope of the moment diagram, is negative and increases to the right. In the regions AD, DB, and BE, the diagram forms straight lines. The maximum moment, $M_{max} = 3.6$ kN · m, occurs at D.

A procedure identical to the preceding one applies to axially loaded bars and twisted shafts. The applied axial forces and torques are positive if their vectors are in the direction of a positive coordinate axis. When a bar is subjected to loads at several points along its length, the internal axial forces and twisting moments vary from section to section. A graph showing the variation of the axial force along the bar axis is called an *axial force diagram*. A similar graph for the torque is referred to as a *torque diagram*. We note that the axial force and torque diagrams are *not* used as commonly as shear and moment diagrams.

3.7 Stresses in Beams

A *beam* is a bar supporting loads applied laterally or transversely to its (longitudinal) axis. This flexure member is commonly used in structures and machines. Examples include the main members supporting floors of buildings, automobile axles, and leaf springs. We see in Sections 4.10 and 4.11 that the following formulas for stresses and deflections of beams can readily be reduced from those of rectangular plates.

3.7.1 Assumptions of Beam Theory

The basic assumptions of the technical or engineering *theory for slender beams* are based on geometry of deformation. They can be summarized as follows [2]:

1. The deflection of the beam axis is *small* compared with the depth and span of the beam.
2. The slope of the deflection curve is very small, and its square is negligible in comparison with unity.

3. Plane sections through a beam taken normal to its axis remain plane after the beam is subjected to bending. This is the fundamental hypothesis of the flexure theory.
4. The effect of shear stress τ_{xy} on the distribution of bending stress σ_x is omitted. The stress normal to the neutral surface, σ_y, may be disregarded.

A generalization of the preceding presuppositions forms the basis for the theories of plates and shells [5]. In deep, *short beams* (where $L/h < 5$), shear stresses are important. Such beams are treated by means of the theory of elasticity because assumptions 3 and 4 are no longer appropriate.

It is interesting to note that in practice, the span/depth ratio is approximately 10 or more for *metal* beams of compact section, 15 or more for beams with relatively *thin webs*, and 24 or more for rectangular *timber* beams [4]. In addition, the slope of the deflection curve of the beam is almost always less than 5° or 0.087 rad, and hence, $(0.087)^2 = 0.0076 \ll 1$. Therefore, the equations developed in this book generally give results of good accuracy for beams of customary proportions.

When treating the bending problem of beams, it is frequently necessary to distinguish between pure bending and nonuniform bending. The former is the flexure of a beam subjected to a constant bending moment; the latter refers to flexure in the presence of shear forces. We discuss the stresses in beams in both cases of bending.

3.7.2 Normal Stress

Consider a linearly elastic beam having the y axis as a vertical axis of symmetry (Figure 3.12a). Based on assumptions 3 and 4, the normal stress σ_x over the cross section (such as *A*-*B*, Figure 3.12b) varies linearly with y, and the remaining stress components are 0:

$$\sigma_x = ky \qquad \sigma_y = \tau_{xy} = 0 \tag{a}$$

Here,

k is a constant

$y = 0$ contains the neutral surface

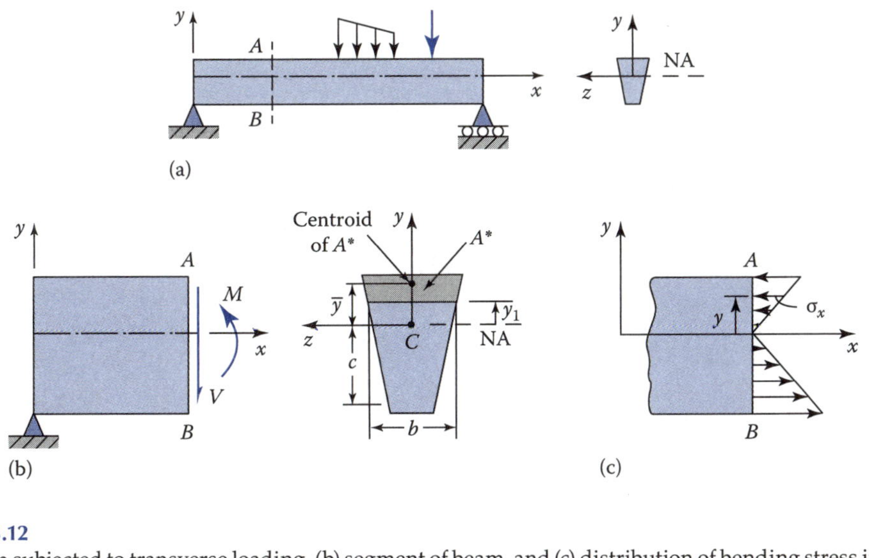

FIGURE 3.12
(a) A beam subjected to transverse loading, (b) segment of beam, and (c) distribution of bending stress in a beam.

The intersection of the neutral surface and the cross section locates the *neutral axis* (NA). Figure 3.12c depicts the linear stress field in section *A-B*.

Conditions of equilibrium require that the resultant normal force produced by the stresses σ_x be 0 and the moments of the stresses about the axis be equal to the bending moment acting on the section. Hence,

$$\int_A \sigma_x dA = 0, \quad -\int_A (\sigma_x dA)y = M \tag{b}$$

in which A represents the cross-sectional area. The negative sign in the second expression indicates that a positive moment M is one that produces compressive (negative) stress at points of positive y. Carrying Equation (a) into Equation (b),

$$k\int_A y dA = 0 \tag{c}$$

$$-k\int_A y^2 dA = M \tag{d}$$

Since $k = 0$, Equation (c) shows that the first moment of cross-sectional area about the NA is 0. This requires that the neutral and centroidal axes of the cross section coincide. It should be mentioned that the symmetry of the cross section about the y axis means that the y and z axes are principal centroidal axes. The integral in Equation (d) defines the moment of inertia, $I = \int y^2 dA$, of the cross section about the z axis of the beam cross section. It follows that

$$k = -\frac{M}{I} \tag{e}$$

An expression for the normal stress, known as the elastic *flexure formula* applicable to initially straight beams, can now be written by combining Equations (a) and (e):

$$\sigma_x = -\frac{My}{I} \tag{3.16}$$

Here, y represents the distance from the neutral axis (NA) to the point at which the stress is calculated. It is common practice to recast the flexure formula to yield the maximum normal stress σ_{max} and denote the value of $|y_{max}|$ by c, where c represents the distance from the NA to the outermost fiber of the beam. On this basis, the flexure formula becomes

$$\sigma_{max} = \frac{Mc}{I} = \frac{M}{S} \tag{3.17}$$

The quantity $S = I/c$ is known as the section modulus of the cross-sectional area. Note that the flexure formula also applies to a beam of *unsymmetrical* cross-sectional area, provided I is a principal moment of inertia and M is a moment around a principal axis [2].

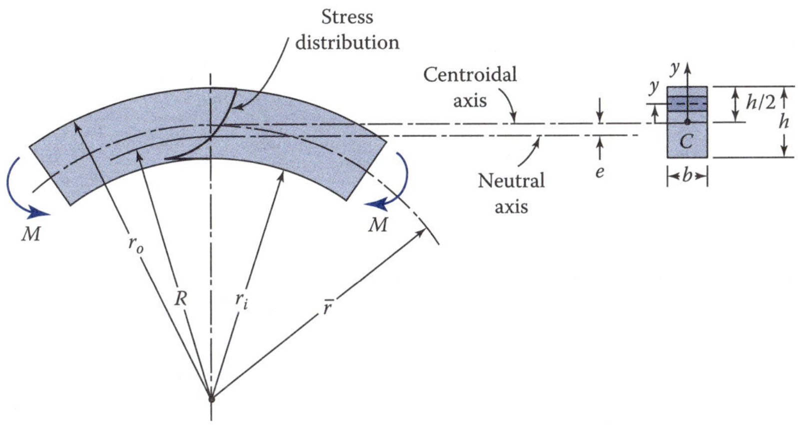

FIGURE 3.13
Curved bar of rectangular cross section in pure bending.

3.7.2.1 Curved Beam of a Rectangular Cross Section

Many machine and structural components loaded as beams, however, are not straight. When beams with initial curvature are subjected to bending moments, the stress distribution is not linear on either side of the NA but increases more rapidly on the inner side. The flexure and displacement formulas for these axisymmetrically loaded members are developed in the later chapters, using energy, elasticity, or exact, approximate technical theories.*

Here, the general equation for stress in curved members is adapted to the rectangular cross section shown in Figure 3.13. Therefore, for pure bending loads, the normal stresses σ_i and σ_o at the inner and outer surfaces of a curved rectangular beam, respectively, from Equation 16.55 are

$$\sigma_i = \frac{M(R - r_i)}{Aer_i} \qquad \sigma_o = \frac{M(R - r_o)}{Aer_o} \tag{3.18}$$

The quantities R and e by Table 16.1 and Figure 3.13 are

$$R = \frac{h}{\ln(r_o/r_i)} \qquad e = \bar{r} - R \tag{3.19}$$

In the foregoing expressions, we have

A = the cross-sectional area
h = the depth of beam
R = the radius of curvature to the NA
M = the bending moment, *positive* when directed toward the *concave* side, as shown in the figure
r_i, r_o = radii of the curvature of the inner and outer surfaces, respectively
$\bar{r}$ = the radius of curvature of the centroidal axis
e = the distance between the centroid and the NA

Accordingly, a positive value obtained from Equation 3.18 means tensile stress.

* Some readers may prefer to study Section 16.9.

The NA shifts toward the center of curvature by distance e from the centroidal axis ($y = 0$), as shown in Figure 3.13. Expression for R and e for many common cross-sectional shapes can be found referring to Table 16.1. Combined stresses in curved beams are presented in Chapter 16. A detailed comparison of the results obtained by various methods is illustrated in Example 16.6. Deflections of curved members due to bending, shear, and normal loads are discussed in Section 5.5.

3.7.3 Shear Stress

We now consider the distribution of shear stress τ in a beam associated with the shear force V. The vertical shear stress τ_{xy} at any point on the cross section is numerically equal to the horizontal shear stress at the same point (see Section 1.11). Shear stresses as well as the normal stresses are taken to be uniform across the width of the beam. The shear stress $\tau_{xy} = \tau_{yx}$ at any point of a cross section (Figure 3.12b) is given by the *shear formula*:

$$\tau_{xy} = \frac{VQ}{Ib} \tag{3.20}$$

Here,

V = the shearing force at the section

b = the width of the section measured at the point in question

By definition, Q is the first moment with respect to the NA of the area A^* beyond the point at which the shear stress is required. We thus have

$$Q = \int_{A^*} y\mathrm{d}A = A^* \bar{y} \tag{3.21}$$

The quantity A^* represents the area of the part of the section beyond the point in question, and $\bar{y}$ is the distance from the NA to the centroid of A^*. Clearly, if $\bar{y}$ is measured above the NA, Q equals the *first moment of the area* above the level where the shear stress is to be found, as shown in Figure 3.12b. It is obvious that, shear stress varies in accordance with the shape of the cross section.

3.7.3.1 Rectangular Cross Section

To ascertain how the shear stress varies, we must examine how Q varies, because V, I, and b are constants for a rectangular cross section. In so doing, we find that the distribution of the shear stress on a cross section of a rectangular beam is parabolic. The stress is 0 at the top and bottom of the section ($y_1 = \pm h/2$) and has its maximum value at the NA ($x_1 = 0$) as shown in Figure 3.14. Therefore,

$$\tau_{\max} = \frac{V}{Ib} A^* \bar{y} = \frac{V}{\left(bh^3/12\right)b} \frac{bh}{2} \frac{h}{4} = \frac{3}{2} \frac{V}{A} \tag{3.22}$$

where $A = bh$ is the cross-sectional area of a beam having depth h and width b. For *narrow beams* with sides parallel to the y axis, Equation 3.20 gives solutions in good agreement

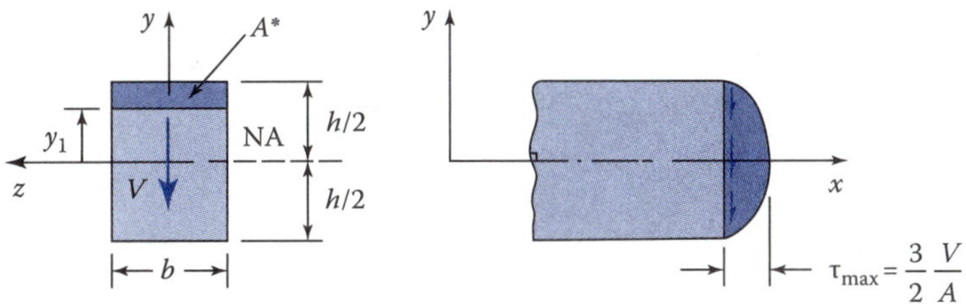

FIGURE 3.14
Shear stresses in a beam of rectangular cross section.

with the *exact* stress distribution obtained by the methods of the theory of elasticity. Equation 3.22 is particularly useful, since beams of rectangular section form are often employed in practice. Stresses in a wide beam and plate are discussed in Section 4.10 after derivation of the strain–curvature relations.

The shear force acting across the width of the beam per unit length along the beam axis may be found by multiplying τ_{xy} in Equation 3.22 by b (Figure 3.12b). This quantity is denoted by q, known as the *shear flow*,

$$q = \frac{VQ}{I} \tag{3.23}$$

This equation is valid for any beam having a cross section that is symmetrical about the y axis. It is very useful in the analysis of *built-up beams*. A beam of this type is fabricated by joining two or more pieces of material. Built-up beams are generally designed on the basis of the assumption that the parts are adequately connected so that the beam acts as a single member. Structural connections are taken up in Chapter 15.

3.7.3.2 Various Cross Sections

It should be noted that the shear formula, also called shear stress formula, for beams is derived on the basis of the flexure formula. Hence, the limitations of the bending formula apply. A variety of cross sections are treated upon following procedures similar to that for rectangular section discussed earlier and for circular section (described in Example 3.13). Table 3.2 lists some common cases. Observe that shear stress can always be expressed as a constant times the average shear stress (P/A), in which the constant is a function of the cross-sectional shape.

Example 3.6: Maximum Stresses in a Simply Supported Beam

A simple beam of *T*-shaped cross section is loaded as shown in Figure 3.15a. Determine

a. The maximum shear stress
b. The shear flow q_j and the shear stress τ_j in the joint between the flange and the web
c. The maximum bending stress

Given: $P = 4$ kN and $L = 3$ m

TABLE 3.2

Maximum Shearing Stress for Some Typical Beam Cross-Sectional Forms

Cross Section		Maximum Shearing Stress	Location
Rectangle	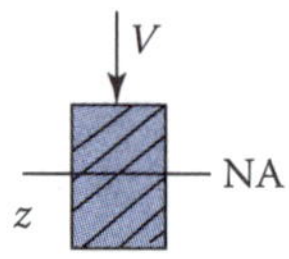	$\tau_{max} = \frac{3}{2}\frac{V}{A}$	NA
Circle	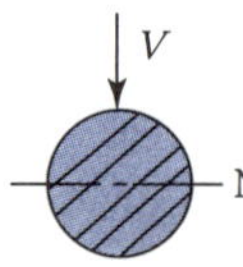	$\tau_{max} = \frac{4}{3}\frac{V}{A}$	NA
Hollow circle	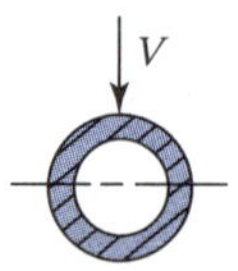	$\tau_{max} = 2\frac{V}{A}$	NA
Triangle	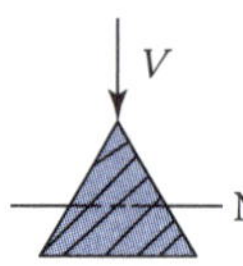	$\tau_{max} = \frac{3}{2}\frac{V}{A}$	Halfway between top and bottom
Diamond	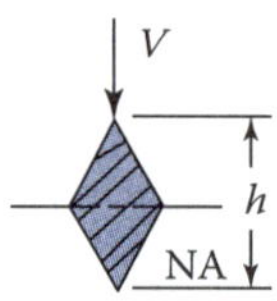	$\tau_{max} = \frac{9}{8}\frac{V}{A}$	At $h/8$ above and below the NA

Notes: A, cross-sectional area; V, transverse shear force; NA, the neutral axis.

Solution

The distance $\bar{y}$ from Z axis to the centroid is determined as follows (Figure 3.15b):

$$\bar{y} = \frac{A_1\bar{y}_1 + A_2\bar{y}_2}{A_1 + A_2} = \frac{20(60)70 + 60(20)30}{20(60) + 60(20)} = 50 \text{ mm}$$

The moment of inertia I about the NA is found using the parallel axis theorem:

$$I = \frac{1}{12}(60)(20)^3 + 20(60)(20)^2 + \frac{1}{12}(20)(60)^3 + 20(60)(20)^2 = 136 \times 10^4 \text{ mm}^4$$

The shear and moment diagrams (Figure 3.15c,d) are drawn using the method of sections.

a. The maximum shearing stress in the beam occurs at the NA on the cross section supporting the largest shear force V. Hence,

$$Q_{NA} = 50(20)25 = 25 \times 10^3 \text{ mm}^3$$

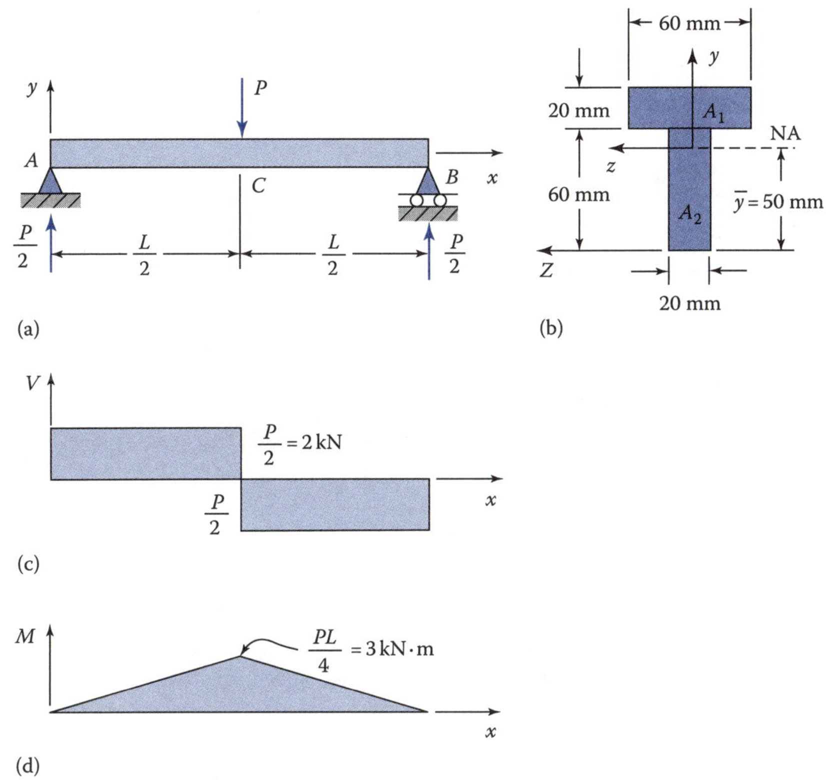

FIGURE 3.15
Example 3.6. (a) loading diagram, (b) beam cross section, (c) shear diagram, and (d) moment diagram.

Since the shear force equals 2 kN on all cross sections of the beam (Figure 3.12c), we have

$$\tau_{\max} = \frac{V_{\max} Q_{NA}}{Ib} = \frac{2\times 10^3 (25\times 10^{-6})}{136\times 10^{-8}(0.02)} = 1.84 \text{ MPa}$$

b. The first moment of the area of the flange about the NA is

$$Q_f = 20(60)20 = 24\times 10^3 \text{ mm}^3$$

Applying Equations 3.23 and 3.20,

$$q_j = \frac{VQ_f}{I} = \frac{2\times 10^3 (24\times 10^{-6})}{136\times 10^{-8}} = 35.3 \text{ kN/m}$$

$$\tau_j = \frac{VQ_f}{Ib} = \frac{35.3(10^3)}{0.02} = 1.765 \text{ MPa}$$

c. The largest moment occurs at midspan, as shown in Figure 3.15d. Therefore, from Equation 3.17, we obtain

$$\sigma_{max} = \frac{Mc}{I} = \frac{3\times10^3(0.05)}{136\times10^{-8}} = 110.3 \text{ MPa}$$

3.8 Design of Beams

We are here concerned with the elastic design of beams for strength. Beams made of single and two different materials are discussed. We note that some beams must be selected based on allowable deflections. This topic is taken up in Chapters 4 and 5. Occasionally, beam design relies on plastic moment capacity, the so-called limit design [2].

3.8.1 Prismatic Beams

We select the dimensions of a beam section so that it supports safely applied loads without exceeding the allowable stresses in both flexure and shear. Therefore, the design of the member is controlled by the largest normal and shear stresses developed at the critical section, where the maximum value of the bending moment and shear force occur. Shear and bending moment diagrams are very helpful for locating these critical sections. In heavily loaded short beams, the design is usually governed by shear stress, while in slender beams, the flexure stress generally predominates. Shearing is more important in wood than steel beams, as wood has relatively low shear strength parallel to the grain.

Application of the rational procedure in design, outlined in Section 3.2, to a beam of ordinary proportions often includes the following steps:

1. It is assumed that failure results from yielding or fracture and flexure stress is considered to be most closely associated with structural damage.
2. The significant value of bending stress is $\sigma = M_{max}/S$.
3. The maximum usable value of σ without failure, σ_{max}, is the yield strength S_y or the ultimate strength S_u.
4. A factor of safety n is applied to σ_{max} to obtain the allowable stress: $\sigma_{all} = \sigma_{max}/n$. The *required section modulus* of a beam is then

$$S = \frac{M_{max}}{\sigma_{all}} \tag{3.24}$$

There are generally several different beam sizes with the required value of S. We select the one with the lightest weight per unit length or the smallest sectional area from tables of beam properties. When the allowable stress is the same in tension and compression, a doubly symmetric section (i.e., section symmetric about the y and z axes) should be chosen. If σ_{all} is different in tension and compression, a singly symmetric section (e.g., a T beam) should be selected so that the distances to the extreme fibers in tension and compression are in a ratio nearly the same as the respective σ_{all} ratios.

We now check the *shear-resistance requirement* of the beam tentatively selected. After substituting the suitable data for Q, I, b, and V_{max} into Equations 3.20, we determine the maximum shear stress in the beam from the formula

$$\tau_{max} = \frac{V_{max}Q}{Ib} \tag{3.25}$$

When the value obtained for τ_{max} is smaller than the allowable shearing stress τ_{all}, the beam is acceptable; otherwise, a stronger beam should be chosen and the process repeated.

Example 3.7: Design of a Beam of Doubly Symmetric Section

Select a wide-flange steel beam to support the loads shown in Figure 3.16a.

Given: The allowable bending and shear stresses are 160 and 90 MPa, respectively.

Solution

Shear and bending moment diagrams (Figure 3.16b) show that M_{max} = 110 kN·m and V_{max} = 40 kN. Therefore, Equation 3.24 gives

$$S = \frac{110 \times 10^3}{160(10^6)} = 688 \times 10^3 \text{ mm}^3$$

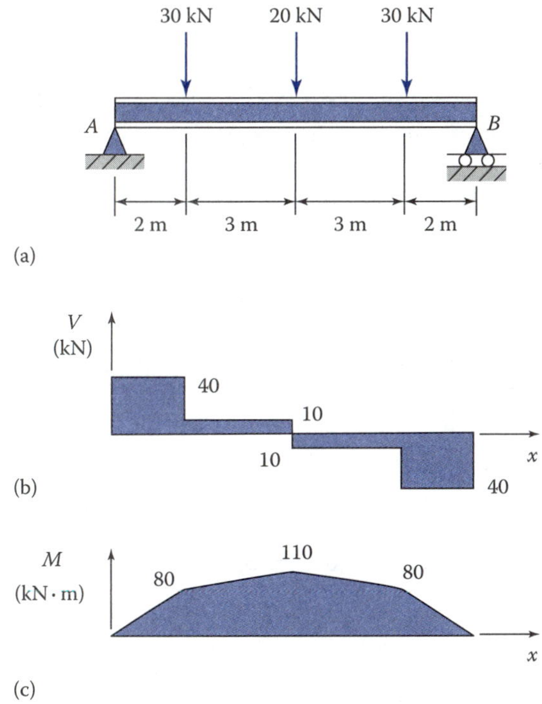

FIGURE 3.16
Example 3.7. (a) load diagram, (b) shear diagram, and (c) moment diagram.

Using Table A.6, we select the lightest member that has a section modulus larger than this value of S: a 200 mm W beam weighing 71 kg/m ($S = 709 \times 10^3$ mm^3). Since the weight of the beam ($71 \times 9.81 \times 10 = 6.97$ kN) is small compared with the applied load (80 kN), it is neglected.

The approximate or average maximum shear stress in beams with flanges may be obtained by dividing the shear force V by the web area:

$$\tau_{avg} = \frac{V}{A_{web}} = \frac{V}{ht} \tag{3.26}$$

In this relationship, h and t represent the beam depth and web thickness, respectively. From Table A.6, the area of the web of a W 200 × 71 section is $216 \times 10.2 = 2.203(10^3)$ mm^2. We therefore have

$$\tau_{avg} = \frac{40 \times 10^3}{2.203\left(10^{-3}\right)} = 18.16 \text{ MPa}$$

Comment: Inasmuch as this stress is well within the allowable limit of 90 MPa, the beam is acceptable.

3.8.2 Beams of Constant Strength

When a beam is stressed to a uniform allowable stress, σ_{all} throughout, then it is clear that the beam material is used to its greatest capacity. For a prescribed material, such a design is of minimum weight. At any cross section, the required section modulus S is given by

$$S = \frac{M}{\sigma_{all}} \tag{3.27}$$

where M presents the bending moment on an arbitrary section. Tapered beams designed in this manner are called *beams of constant strength*. Note that shear stress at those beam locations where the moment is small controls the design.

Beams of uniform strength are exemplified by leaf springs and certain forged or cast machine components (see Section 14.10). For a structural member, fabrication and design constraints make it impractical to produce a beam of constant stress. So welded cover plates are often used for parts of prismatic beams where the moment is large, for instance, in a bridge girder. If the angle between the sides of a tapered beam is small, the flexure formula allows little error. On the other hand, the results obtained by using the shear stress formula may not be sufficiently accurate for nonprismatic beams. Usually, a modified form of this formula is used for design purposes. The exact distribution in a rectangular wedge is obtained by the theory of elasticity [3].

Example 3.8: Design of a Constant Strength Beam

A cantilever beam of uniform strength and rectangular cross section is to support a concentrated load P at the free end (Figure 3.17a). Determine the required cross-sectional area for two cases: (1) the width b is constant; (2) the height h is constant.

Solution

a. At a distance x from A, $M = Px$ and $S = bh^2/6$. Through the use of Equation 3.27, we write

$$\frac{bh^2}{6} = \frac{Px}{\sigma_{all}} \tag{a}$$

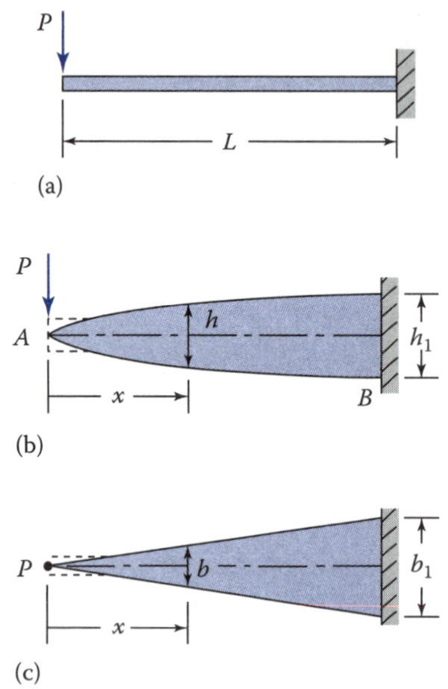

FIGURE 3.17
Example 3.8. (a) uniform strength cantilever, (b) side view, and (c) top view.

Similarly, at a fixed end ($x = L$ and $h = h_1$),

$$\frac{bh_1^2}{6} = \frac{PL}{\sigma_{all}}$$

Dividing Equation (a) by the preceding relationship results in

$$h = h_1\sqrt{\frac{x}{L}} \tag{b}$$

Therefore, the depth of the beam varies parabolically from the free end (Figure 3.17b).

b. Equation (a) now yields

$$b = \left(\frac{6P}{h^2\sigma_{all}}\right)x = \frac{b_1}{L}x \tag{c}$$

Comments: In Equation (c), the expression in parentheses represents a constant and is set equal to b_1/L so that when $x = L$, the width is b_1 (Figure 3.17c). In both cases, obviously the cross section of the beam near end A must be designed to resist the shear force, as shown by the dashed lines in the figure.

Example 3.9: Design of Traffic Light Support Beam

A three-phase traffic light of weight W carries a steel beam (Figure 3.18a) of yield strength S_y. The beam may be modeled as a prismatic member having constant cross-sectional area and length L, as illustrated in Figure 3.18b.

Given: $L = 4.5$ m, $S_y = 250$ MPa, $W = 200$ N

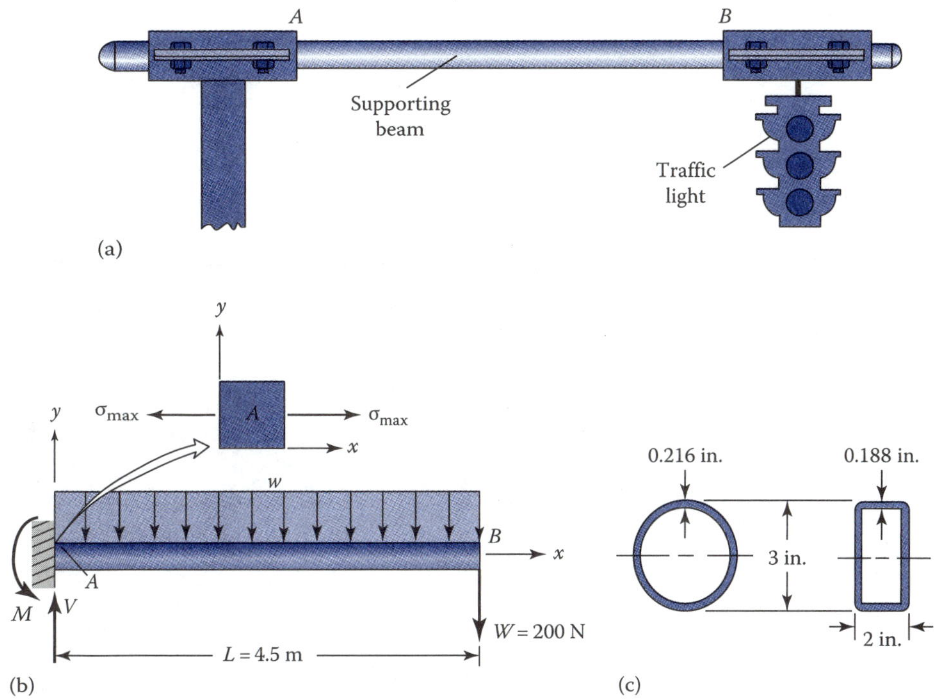

FIGURE 3.18
Example 3.9. (a) Schematic model, part of structure, (b) free-body diagram of the beam *AB*, and (c) two standard beam cross sections.

Find: The safety factor *n* of the beam associated with yielding for the two choices of the *same nominal depth* beam geometries, shown in Figure 3.18c,

a. The circular tube
b. The rectangular tube

Assumptions:

1. The weight of the beam *w* per unit length and the weight *W* of the traffic light will be taken into account. The wind loading is disregarded.
2. The left end of the beam is taken as built-in to a rigid tapered pipe.
3. According to loading and support conditions, the beam develops an internal shear force *V* and bending moment *M* at each section along its length. Inasmuch as the ratio of the length to nominal depth of the beam equals 4500/75 = 60, the effect of shear on bending stress can be disregarded (see Section 3.7).

Solution

See Figure 3.18, Table A.4, and Equation 3.24.

We observe from Figure 3.18b that the largest bending stress takes place at point *A* on the top surface of the left end of the beam. Hence, *A* represents the critical point, where the weight of the traffic light *W* and the beam weight *w* per unit length produce a bending moment *M* (and a shear force $V = wL$) in the vertical plane. At the left end of the beam, moment is expressed as

$$M = WL + \frac{1}{2}wL^2 \tag{3.28}$$

a. *Circular pipe* (3 in. nominal diameter, Table A.4)

$$w = 7.58\ \text{lb/ft} = 110.7\ \text{N/m} \quad S = 1.72\ \text{in}^3. = 28.2 \times 10^3\ \text{mm}^3$$

Equation 3.28 yields

$$M = 200(4.5) + \frac{1}{2}(110.7)(4.5)^2 = 2.021\ \text{kN} \cdot \text{m}$$

The flexure formula Equation 3.27 results in

$$\sigma_{\text{all}} = \frac{M}{S}; \quad \frac{250}{n} = \frac{2021}{28.2}, \quad n = 3.49$$

b. Rectangular tube (3 × 2 in. nominal size and 3/16 in. thick, Table A.4)

$$w = 5.59\ \text{lb/ft} = 81.6\ \text{N/m} \quad S = 0.977\ \text{in}^3. = 16 \times 10^3\ \text{mm}^3$$

Through the use of Equation 3.28, we find

$$M = 200(4.5) + \frac{1}{2}(81.6)(4.5)^2 = 1.726\ \text{kN} \cdot \text{m}$$

It follows that

$$\sigma_{\text{all}} = \frac{M}{S}; \quad \frac{250}{n} = \frac{1726}{16}, \quad n = 2.32$$

Comments: The results indicate that a circular pipe is very efficient for bending loads, and rectangular tubing seems to be *the weaker* beam.

3.8.3 Composite Beams

Beams fabricated of two or more materials having different moduli of elasticity are called *composite beams*. The advantage of this type of construction is that large quantities of low-modulus material can be used in regions of low stress and small quantities of high-modulus materials can be used in regions of high stress. Two common examples are wooden beams whose bending strength is bolstered by metal strips, either along its sides or along its top or bottom, and reinforced concrete beams. The assumptions of the technical theory of homogenous beams, discussed in Section 3.7, are valid for a beam of more than one material. We use the common *transformed-section method* to ascertain the stresses in a composite beam. In this approach, the cross section of several materials is transformed into an equivalent cross section of one material in that the resisting forces are the same as on the original section. The flexure formula is then applied to the transformed section.

To demonstrate the method, a typical beam with symmetrical cross section built of two different materials is considered (Figure 3.19a). The moduli of elasticity of materials are denoted by E_1 and E_2. We define the modular ratio, n, as follows:

$$n = \frac{E_2}{E_1} \tag{d}$$

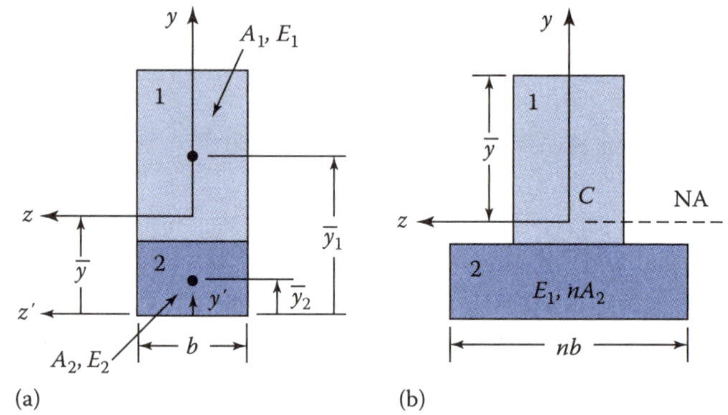

FIGURE 3.19
Beam of two materials: (a) cross section and (b) equivalent section.

Although $n > 1$ in Equation (d), the choice is arbitrary: the technique applies well for $n < 1$. The transformed section is composed of only material 1 (Figure 3.19b). The moment of inertia of the entire transformed area about the NA is then denoted by I_t. It can be shown [1] that the flexure formulas for a composite beam are in the forms

$$\sigma_{x1} = -\frac{My}{I_t}, \quad \sigma_{x2} = -\frac{nMy}{I_t} \tag{3.29}$$

where σ_{x1} and σ_{x2} are the stresses in materials 1 and 2, respectively. Obviously, when $E_1 = E_2 = E$, this equation reduces to the flexure formula for a beam of homogeneous material, as expected. The following sample problems illustrate the use of Equations 3.29.

Example 3.10: Stresses in a Composite Beam of Two Materials

A composite beam is made of wood and steel having the cross-sectional dimensions shown in Figure 3.20a. The beam is subjected to a bending moment $M_z = 25$ kN·m. Calculate the maximum stresses in each material.

Given: The modulus of elasticity of wood and steel are $E_w = 10$ GPa and $E_s = 210$ GPa, respectively.

Solution

The modular ratio $n = E_s/E_w = 21$. We use a transformed section of wood (Figure 3.20b). The centroid and the moment of inertia about the NA of this section are

$$\bar{y} = \frac{150(200)(112) + 3150(12)(6)}{150(200) + 3150(12)} = 52.9 \text{ mm}$$

$$I_t = \frac{1}{12}(150)(200)^3 + 150(200)(59.1)^2 + \frac{1}{12}(3150)(12)^3 + 3150(12)(46.9)^2$$

$$= 288 \times 10^6 \text{ mm}^4$$

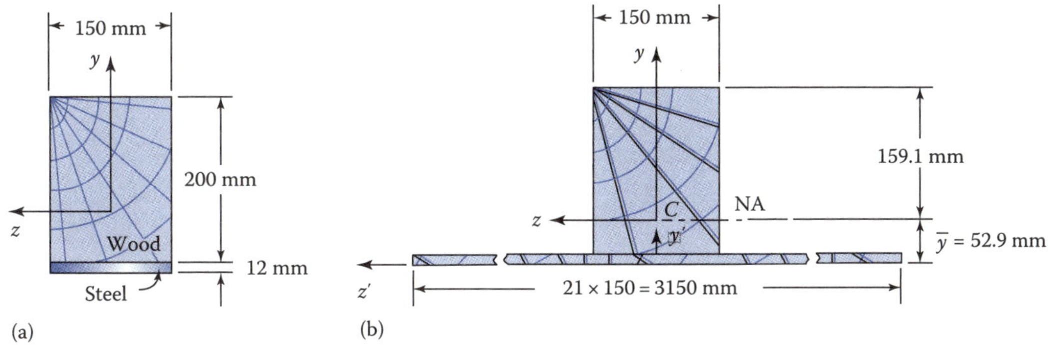

FIGURE 3.20
Example 3.10. (a) composite beam and (b) equivalent section.

The maximum stress in the wood and steel portions are therefore

$$\sigma_{w,\max} = \frac{Mc}{I_t} = \frac{25(10^3)(0.1591)}{288(10^{-6})} = 13.81 \text{ MPa}$$

$$\sigma_{s,\max} = \frac{nMc}{I_t} = \frac{21(25 \times 10^3)(0.0529)}{288(10^{-6})} = 96.43 \text{ MPa}$$

At the juncture of the two parts, we have

$$\sigma_{w,\min} = \frac{Mc}{I_t} = \frac{25(10^3)(0.0409)}{288(10^{-6})} = 3.55 \text{ MPa}$$

$$\sigma_{s,\min} = n(\sigma_{w,\min}) = 21(3.55) = 74.55 \text{ MPa}$$

Stress at any other location may be determined likewise.

3.9 Plane Stress

The stresses and strains treated thus far have been found on sections perpendicular to the coordinates used to describe a member. This section deals with the states of stress at points located on inclined planes. In other words, we wish to obtain the stresses acting on the sides of a stress element oriented in any desired direction. This process is termed a *stress transformation*. The discussion that follows is limited to 2D, or plane, stress. A 2D state of stress exists when the stresses are independent of one of the coordinate axes, here taken as z. The plane stress is therefore specified $\sigma_z = \tau_{yz} = \tau_{xz} = 0$, where σ_x, σ_y, and τ_{xy} have nonzero values. Examples include the stresses arising on inclined sections of an axially loaded bar, a shaft in torsion, a beam with transversely applied force, and a member subjected to more than one load simultaneously.

Consider the stress components σ_x, σ_y, τ_{xy} at a point in a body represented by a 2D stress element (Figure 3.21a). To portray the stresses acting on an inclined section, an infinitesimal wedge is isolated from this element and depicted in Figure 3.21b. The angle θ, locating

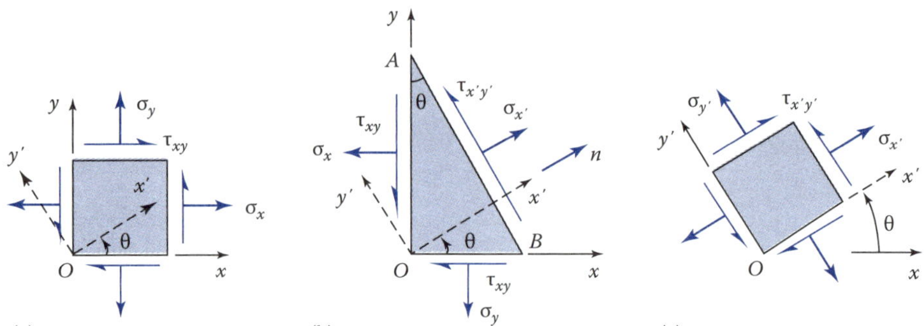

FIGURE 3.21
Elements in plane stress.

the x' axis or the unit normal n to the plane AB, is assumed positive when measured from the x axis in a counterclockwise direction. Note that according to the sign convention (see Section 1.11), the stresses are indicated as positive values. It can be shown that equilibrium of the forces caused by stresses acting on the wedge-shaped element gives the following transformation equations for plane stress [1]:

$$\sigma_{x'} = \sigma_x \cos^2\theta + \sigma_y \sin^2\theta + 2\tau_{xy}\sin\theta\cos\theta \tag{3.30a}$$

$$\tau_{x'y'} = \tau_{xy}\left(\cos^2\theta - \sin^2\theta\right) + \left(\sigma_y - \sigma_x\right)\sin\theta\cos\theta \tag{3.30b}$$

The stress $\sigma_{y'}$ may readily be obtained by replacing θ in Equation 3.30a by $\theta + \pi/2$ (Figure 3.21c). This gives

$$\sigma_{y'} = \sigma_x \sin^2\theta + \sigma_y \cos^2\theta - 2\tau_{xy}\sin\theta\cos\theta \tag{3.30c}$$

Using the double-angle relationships, the foregoing equations can be expressed in the following useful alternative form:

$$\sigma_{x'} = \frac{1}{2}\left(\sigma_x + \sigma_y\right) + \frac{1}{2}\left(\sigma_x - \sigma_y\right)\cos 2\theta + \tau_{xy}\sin 2\theta \tag{3.31a}$$

$$\tau_{x'y'} = -\frac{1}{2}\left(\sigma_x - \sigma_y\right)\sin 2\theta + \tau_{xy}\cos 2\theta \tag{3.31b}$$

$$\sigma_{y'} = \frac{1}{2}\left(\sigma_x + \sigma_y\right) - \frac{1}{2}\left(\sigma_x - \sigma_y\right)\cos 2\theta - \tau_{xy}\sin 2\theta \tag{3.31c}$$

For design purposes, the largest stresses are usually needed. The two perpendicular directions $\left(\theta_p' \text{ and } \theta_p''\right)$ of planes on which the shear stress vanishes and the normal stress has extreme values can be found from

$$\tan 2\theta_p = \frac{2\tau_{xy}}{\sigma_x - \sigma_y} \tag{3.32}$$

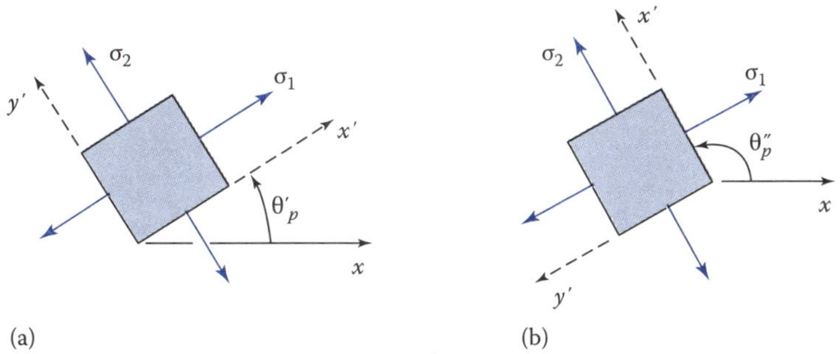

FIGURE 3.22
Planes of principal stresses.

The angle θ_p defines the orientation of the principal planes (Figure 3.22). The in-plane principal stresses can be obtained by substituting each of the two values of θ_p from Equation 3.32 into Equations 3.31a and c as follows:

$$\sigma_{\max,\min} = \sigma_{1,2} = \frac{\sigma_x + \sigma_y}{2} \pm \sqrt{\left(\frac{\sigma_x - \sigma_y}{2}\right)^2 + \tau_{xy}^2} \tag{3.33}$$

The plus sign gives the algebraically larger maximum principal stress σ_1. The minus sign results in the minimum principal stress $\sigma_{2.}$ It is necessary to substitute θ_p into Equation 3.31a to learn which of the two corresponds to σ_1.

Example 3.11: Stresses in a Cylindrical Pressure Vessel Welded along a Helical Seam

Figure 3.23a depicts a cylindrical pressure vessel constructed with a helical weld that makes an angle ψ with the longitudinal axis. Determine

a. The maximum internal pressure p
b. The shear stress in the weld

Given: r = 10 in., t = ¼ in., and ψ = 55°. Allowable tensile strength of the weld is 14.5 ksi.

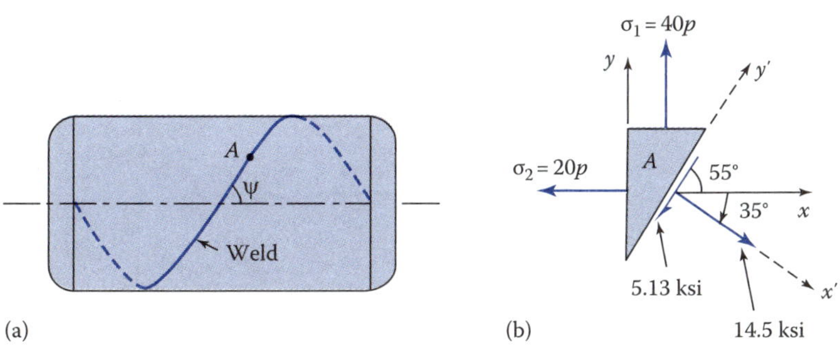

FIGURE 3.23
Example 3.11.

Assumptions: Stresses are at point A on the wall away from the ends. Vessel is a thin-walled cylinder.

Solution

The principal stresses in axial and tangential directions are, respectively,

$$\sigma_a = \frac{pr}{2t} = \frac{p(10)}{2(1/4)} = 20p = \sigma_2, \quad \sigma_\theta = 2\sigma_a = 40p = \sigma_1$$

The state of stress is shown on the element of Figure 3.23b. We take the x' axis perpendicular to the plane of the weld. This axis is rotated $\theta = 35°$ clockwise with respect to the x axis.

a. Through the use of Equation 3.31a, the tensile stress in the weld is

$$\begin{aligned}\sigma_{x'} &= \frac{\sigma_2 + \sigma_1}{2} + \frac{\sigma_2 - \sigma_1}{2}\cos 2(-35°) \\ &= 30p - 10p\cos(-70°) \le 14{,}500\end{aligned}$$

from which $p_{\max} = 546$ psi.

b. Applying Equation 3.31b, the shear stress in the weld corresponding to the foregoing value of pressure is

$$\begin{aligned}\tau_{x'y'} &= -\frac{\sigma_2 - \sigma_1}{2}\sin 2(-35°) \\ &= 10p\sin(-70°) = -5.13 \text{ ksi}\end{aligned}$$

The answer is presented in Figure 3.23b.

3.9.1 Mohr's Circle for Stress

Transformation equations for plane stress, Equations 3.31, can be represented with σ and τ as coordinate axes in a graphical form known as *Mohr's circle* (Figure 3.24b). This representation is very useful in visualizing the relationships between normal and shear stresses acting on various inclined planes at a point in a stressed member. Also, with the aid of this graphical construction, a quicker solution of stress-transformation problem can be facilitated.

The coordinates for point A on the circle correspond to the stresses on the x face or plane of the element shown in Figure 3.24a. Similarly, the coordinates of a point A' on Mohr's circle are to be interpreted, representing the stress components $\sigma_{x'}$ and $\tau_{x'y'}$ that act on x' plane. The center is at $(\sigma', 0)$ and the circle radius r equals the length CA. In Mohr's circle representation, the normal stresses obey the *sign convention* of Section 1.13. However, for the purposes of *only constructing and reading values* of stress from a Mohr's circle, the shear stresses on the y planes of the element are taken to be positive (as before), but those on the x faces are now negative (Figure 3.24c).

The magnitude of the maximum shear stress is equal to the radius R of the circle. From the geometry of Figure 3.24b, we obtain

$$\tau_{\max} = \sqrt{\left(\frac{\sigma_x - \sigma_y}{2}\right)^2 + \tau_{xy}^2} \tag{3.34}$$

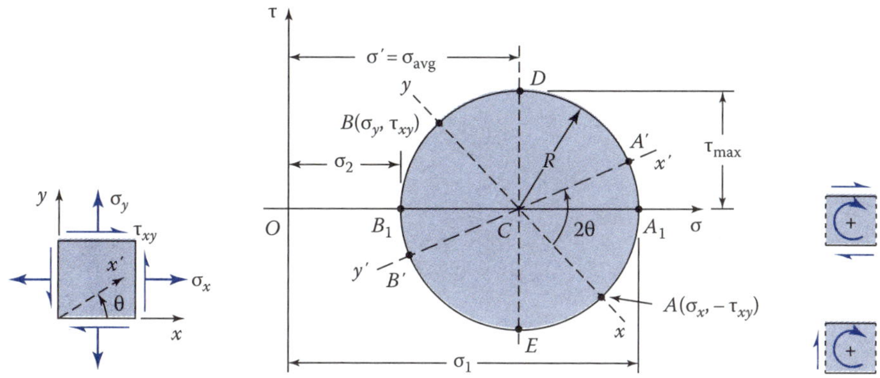

FIGURE 3.24
(a) Stress element, (b) Mohr's circle of stress, and (c) interpretation of positive shear stress.

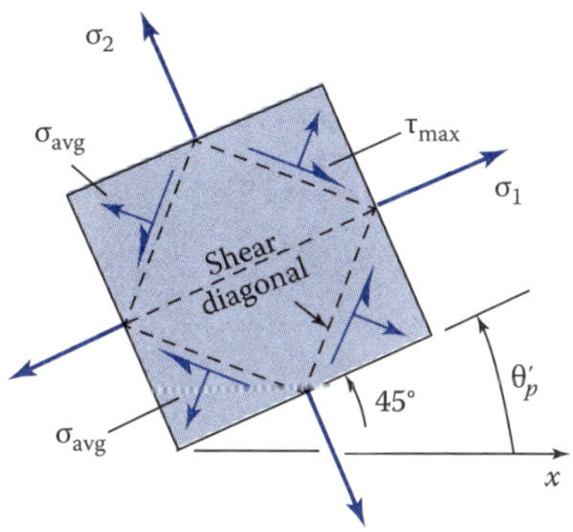

FIGURE 3.25
Planes of principal and maximum shear stresses.

Mohr's circle shows the planes of maximum shear are always oriented at 45° from planes of principal stress (Figure 3.25). Note that a diagonal of a stress element along which the algebraically larger principal stress acts is called the *shear diagonal*. The maximum shear stress acts toward the shear diagonal. The normal stress occurring on planes of maximum shear stress is

$$\sigma' = \sigma_{avg} = \frac{1}{2}\left(\sigma_x + \sigma_y\right). \tag{3.35}$$

It can readily be verified using Equations 3.30 or Mohr's circle that on any mutually perpendicular planes,

$$I_1 = \sigma_x + \sigma_y = \sigma_{x'} + \sigma_{y'} \quad I_2 = \sigma_x\sigma_y - \tau_{xy}^2 = \sigma_{x'}\sigma_{y'} - \tau_{x'y'}^2 \tag{3.36}$$

The quantities I_1 and I_2 are known as 2D *stress invariants*, because they do not change in value when the axes are rotated positions. This assertion is also valid in the case of a 3D stress. Interestingly, in mathematical terms, the stress which components transform in the foregoing way upon rotation is termed *tensor*. Equations 3.36 are particularly useful in checking numerical results of stress transformation.

Note that in the case of triaxial stresses σ_1, σ_2, and σ_3, Mohr's circle is drawn corresponding to each projection of a 3D element. The three-circle cluster represents Mohr's circle for triaxial stress (see Figure 3.28). The general state of stress at a point is discussed in some detail in the later sections of this chapter. Mohr's circle construction is of fundamental importance because it applies to all (second-rank) tensor quantities: that is, Mohr's circle may be used to determine strains, moments of inertia, and natural frequencies of vibration [6]. It is customary to draw for Mohr's circle only a rough sketch: distances and angles are determined with the help of trigonometry. Mohr's circle provides a convenient means of obtaining the results for the stresses under the following two common loadings.

3.9.1.1 Axial Loading

In this case, we have $\sigma_x = \sigma_1 = P/A$, $\sigma_y = 0$, and $\tau_{xy} = 0$, where A is the cross-sectional area of the bar. The corresponding points A and B define a circle of radius $R = P/2A$ that passes through the origin of coordinates (Figure 3.26b). Points D and E yield the orientation of the planes of the maximum shear stress (Figure 3.26a), as well as the values of τ_{max} and the corresponding normal stress σ':

$$\tau_{max} = \sigma' = R = \frac{P}{2A} \tag{a}$$

Observe that the normal stress is either maximum or minimum on planes for which shearing stress is 0.

3.9.1.2 Torsion

Now we have $\sigma_x = \sigma_y = 0$ and $\tau_{xy} = \tau_{max} = Tc/J$, where J is the polar moment of inertia of cross-sectional area of the bar. Points D and E are located on the τ axis, and Mohr's circle is a circle of radius $R = Tc/J$ centered at the origin (Figure 3.27b). Points A_1 and B_1 define the principal stresses:

$$\sigma_{1,2} = \pm R = \pm\frac{Tc}{J} \tag{b}$$

So, it becomes evident that, for a material such as cast iron that is weaker in tension than in shear, failure occurs in tension along a helix indicated by the dashed lines in Figure 3.27a.

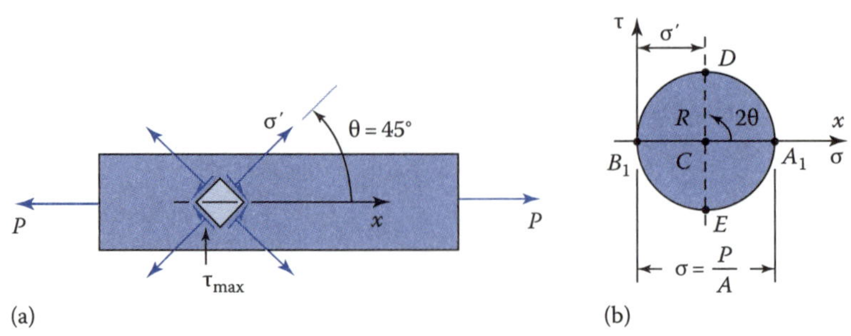

FIGURE 3.26
(a) Maximum shear stress acting on an element of an axially loaded bar and (b) Mohr's circle for uniaxial loading.

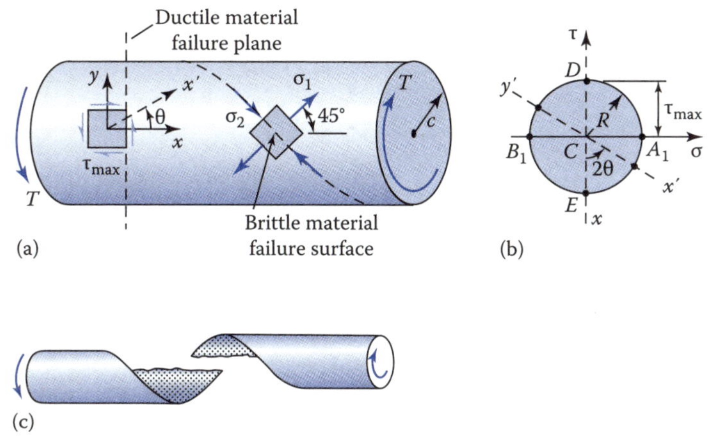

FIGURE 3.27
(a) Stress acting on a surface element of a twisted shaft, (b) Mohr's circle for torsional loading, and (c) brittle material fractured in torsion.

Fracture of a bar that behaves in a brittle manner in torsion is depicted in Figure 3.27c: ordinary chalk behaves this way. Shafts made of materials weak in shear strength (e.g., structural steel) break along a line perpendicular to the axis. Experiments show that a very thin-walled hollow shaft buckles or wrinkles in the direction of maximum compression, while in the direction of maximum tension, tearing occurs.

Example 3.12: Stress Analysis of Cylindrical Pressure Vessel Using Mohr's Circle

Redo Example 3.11 using Mohr's circle. Also determine maximum in-plane and absolute shear stresses at a point on the wall of the vessel.

Solution

Mohr's circle (Figure 3.28), constructed referring to Figure 3.23 and Example 3.11, describes the state of stress. The x' axis is rotated $2\theta = 70°$ on the circle with respect to x axis.

a. From the geometry of Figure 3.28, we have $\sigma_{x'} = 30p - 10p\cos 70° \leq 14{,}500$. This results in $p_{max} = 546$ psi.
b. For the preceding value of pressure, the shear stress in the weld is

$$\tau_{x'y'} = \pm 10(546)\sin 70° = \pm 5.13 \text{ ksi}$$

The largest in-plane shear stresses are given by points D and E on the circle. Hence,

$$\tau = \pm\frac{1}{2}(40p - 20p) = \pm 10(546) = \pm 5.46 \text{ ksi}$$

The third principal stress in the radial direction is 0, $\sigma_3 = 0$. The three principal stress circles are shown in the figure. The absolute maximum shear stresses are associated with points D' and E' on the major principal circle. Therefore,

$$\tau_{max} = \pm\frac{1}{2}(40p - 0) = \pm 20(546) = \pm 10.92 \text{ ksi}$$

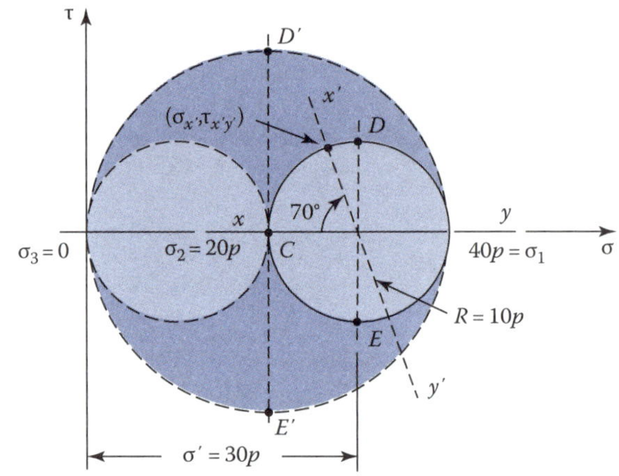

FIGURE 3.28
Example 3.12.

3.10 Combined Stresses

Basic formulas of mechanics of materials for determining the state of stress in elastic members are developed in Sections 3.2 through 3.7. Often these formulas give either a normal stress or a shear stress caused by a single load component being axially, centric, or lying in one plane. Note that each formula leads to stress as directly proportional to the magnitude of the applied load. When a component is acted on simultaneously by two or more loads, causing various internal-force resultants on a section, it is assumed that each load produces the stress as if it were the only load acting on the member. The final or combined stress is then found by superposition of several states of stress. As we see throughout the text, under combined loading, the critical points may not be readily located. Therefore, it may be necessary to examine the stress distribution in some detail.

Consider, for example, a solid circular cantilevered bar subjected to a transverse force P, a torque T, and a centric load F at its free end (Figure 3.29a). Every section experiences an

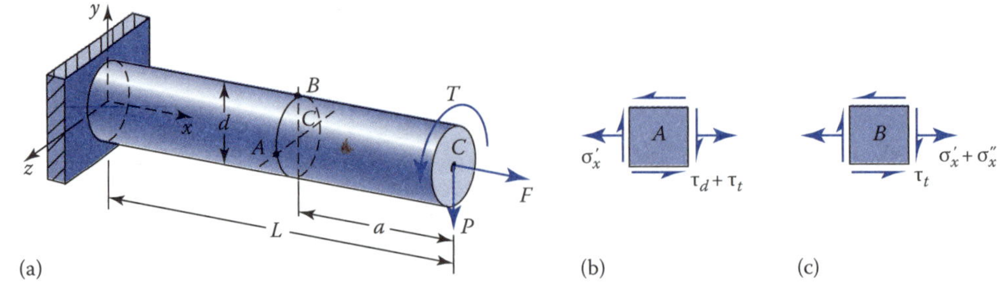

FIGURE 3.29
(a) Combined stresses owing to torsion, tension, and direct shear; (b,c) stress elements at points A and B.

axial force F, torque T, a bending moment M, and a shear force $P = V$. The corresponding stresses may be obtained using the applicable relationships:

$$\sigma'_x = \frac{F}{A}, \quad \tau_t = -\frac{Tc}{J}, \quad \sigma''_x = -\frac{Mc}{I}, \quad \tau_d = -\frac{VQ}{Ib}$$

Here, τ_t and τ_d are the torsional and direct shear stresses, respectively. In Figure 3.29b, the stresses shown are those acting on an element B at the top of the bar and on an element A on the side of the bar at the NA. Clearly, B (when located at the support) and A represent the critical points at which most severe stresses occur. The principal stresses and maximum shearing stress at a critical point can now be ascertained as discussed in the preceding section.

The following examples illustrate the general approach to problems involving combined loadings. Any number of critical locations in the components can be analyzed. These either confirm the adequacy of the design or, if the stresses are too large (or too small), indicate the design changes required. This is used in a seemingly endless variety of practical situations, so it is often not worthwhile to develop specific formulas for most design use. We develop design formulas under combined loading of common mechanical components, such as shafts, shrink or press fits, flywheels, and pressure vessels in Chapters 9 and 16.

Example 3.13: Determining the Allowable Combined Loading in a Cantilever Bar

A round cantilever bar is loaded as shown in Figure 3.29a. Determine the largest value of the load P.

Given: Diameter $d = 60$ mm, $T = 0.1P$ N · m, and $F = 10P$ N

Assumptions: Allowable stresses are 100 MPa in tension and 60 MPa in shear on a section at $a = 120$ mm from the free end.

Solution

The normal stress at all points of the bar is

$$\sigma'_x = \frac{F}{A} = \frac{10P}{\pi(0.03)^2} = 3536.8P \tag{a}$$

The torsional stress at the outer fibers of the bar is

$$\tau_t = -\frac{Tc}{J} = -\frac{0.1P(0.03)}{\pi(0.03)^4/2} = -2357.9P \tag{b}$$

The largest tensile bending stress occurs at point B of the section considered. Therefore, for $a = 120$ mm, we obtain

$$\sigma''_x = \frac{Mc}{I} = \frac{0.12P(0.03)}{\pi(0.03)^4/4} = 5658.8P$$

Since $Q = A\bar{y} = (\pi c^2/2)(4c/3\pi) = 2c^3/3$ and $b = 2c$, the *largest* direct shearing stress at point A is

$$\tau_d = -\frac{VQ}{Ib} = -\frac{4V}{3A} = -\frac{4P}{3\pi(0.03)^2} = -471.57P \tag{c}$$

The maximum principal stress and the maximum shearing stress at point A (Figure 3.29b), applying Equations 3.33 and 3.34 with $\sigma_y = 0$ and Equations (a), (b), and (c), are

$$(\sigma_1)_A = \frac{\sigma'_x}{2} + \left[\left(\frac{\sigma'_x}{2}\right)^2 + (\tau_d + \tau_t)^2\right]^{1/2}$$

$$= \frac{3536.8P}{2} + \left[\left(\frac{3536.8P}{2}\right)^2 + (-2829.5P)^2\right]^{1/2}$$

$$= 1768.4P + 3336.7P = 5105.1P$$

$$(\tau_{\max})_A = 3336.7P$$

Likewise, at point B (Figure 3.29c),

$$(\sigma_1)_B = \frac{\sigma'_x + \sigma''_x}{2} + \left[\left(\frac{\sigma'_x + \sigma''_x}{2}\right)^2 + \tau_t^2\right]^{1/2}$$

$$= \frac{9195.6P}{2} + \left[\left(\frac{9195.6P}{2}\right)^2 + (-2357.9P)^2\right]^{1/2}$$

$$= 4597.8P + 5167.2P = 9765P$$

$$(\tau_{\max})_B = 5167.2P$$

It is observed that the stresses at B are more severe than those at A. Inserting the given data into the foregoing, we obtain

$$100(10^6) = 9765P \quad \text{or} \quad P = 10.24\ \text{kN}$$

$$60(10^6) = 5167.2P \quad \text{or} \quad P = 11.61\ \text{kN}$$

Comment: The magnitudes of the largest allowable transverse, axial, and torsional loads that can be carried by the bar are $P = 10.24$ kN, $F = 102.4$ kN, and $T = 1.024$ kN·m, respectively.

Example 3.14: Pressure Capacity of a Hydraulic Cylinder

Pressurized hydraulic fluid (liquid or air) produces stresses and deformation of a cylinder. Hydraulic systems are widespread usage in brakes, control mechanisms, and actuators in positioning devices. A hydraulic cylinder of a loader truck is shown in Figure P18.1 of Problem 18.1. Design of a pressurized duplex conduit is illustrated in Example 16.2.

Given: A hydraulic cylinder of radius r and thickness t subjected to internal pressure p is simultaneously compressed by an axial load P through the piston of diameter $d \approx 2r$ as shown in Figure 3.30a. Note that, the vessel is inadvertently subjected to torque T at its mounting. Data: $r = 60$ mm, $t = 5$ mm, and $T = 300$ N·m. Allowable in-plane shear stress in the cylinder wall will be 75 MPa.

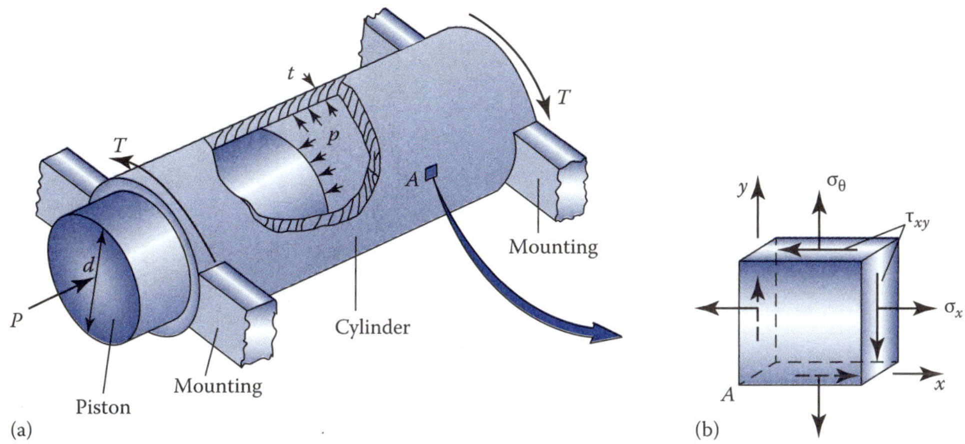

FIGURE 3.30
Example 3.14. (a) Schematic hydraulic cylinder and (b) element in plane stress.

Find: The largest value of p that can be applied to the cylinder.

Assumptions: The critical stress is at point A on cylinder remote from the ends. The effect of bending of the cylinder on stresses is disregarded.

Solution

Combined stresses act at a critical point on an element in the wall of the pipe (Figure 3.30b). We have

$$\tau_{xy} = \frac{Tr}{J} = \frac{T}{2\pi r^2 t}$$

$$= \frac{300}{2\pi(0.06)^2(0.005)} = 2.65\,\text{MPa}$$

$$\sigma_x = \frac{pr}{2t} = \frac{p(60)}{2(5)} = 6p$$

$$\sigma_\theta = \frac{pr}{t} = 12p$$

Applying Equation 3.34,

$$\tau_{\max} = \sqrt{\left(\frac{6p - 12p}{2}\right) + (2.65)^2}$$

$$= \sqrt{9p^2 + 7.023} \le 75$$

from which

$$p_{\max} = 25\text{ MPa}$$

Comment: The largest allowable axial load that can be applied to piston is about $P_{max} = 25(\pi \times 60^2) = 283$ kN.

CASE STUDY 3.1 Bolt Cutter Stress Analysis

A bolt cutting tool is shown in Figure 1.5. Determine the stresses in the members.

Given: The geometry and forces are known from Case Study 1.1. Material of all parts is AISI 1080 HR steel. Dimensions are in inches. We have

$$S_y = \frac{420}{6.895} = 60.9 \text{ ksi (Table B.3)}, \quad S_{yx} = 0.5S_y = 30.45 \text{ ksi}, \quad E = 30 \times 10^6 \text{ psi}$$

Assumptions:

1. The loading is taken to be static. The material is ductile, and stress-concentration factors can be disregarded under steady loading.
2. The most likely failure points are in link 3, the hole where pins are inserted, the connecting pins in shear, and jaw 2 in bending.
3. Member 2 can be approximated as a simple beam with an overhang.

Solution

See Figures 1.5 and 3.31.

The largest force on any pin in the assembly is at joint *A*.

Member 3 is a pin-ended tensile link. The force on a pin is 128 lb, as shown in Figure 3.31a. The normal stress is therefore

$$\sigma = \frac{F_A}{(w_3 - d)t_3} = \frac{128}{\left(\frac{3}{8} - \frac{1}{8}\right)\left(\frac{1}{8}\right)} = 4.096 \text{ ksi}$$

For the bearing stress in joint *A*, using Equation 3.5, we have

$$\sigma_b = \frac{F_A}{dt_3} = \frac{128}{\left(\frac{1}{8}\right)\left(\frac{1}{8}\right)} = 8.192 \text{ ksi}$$

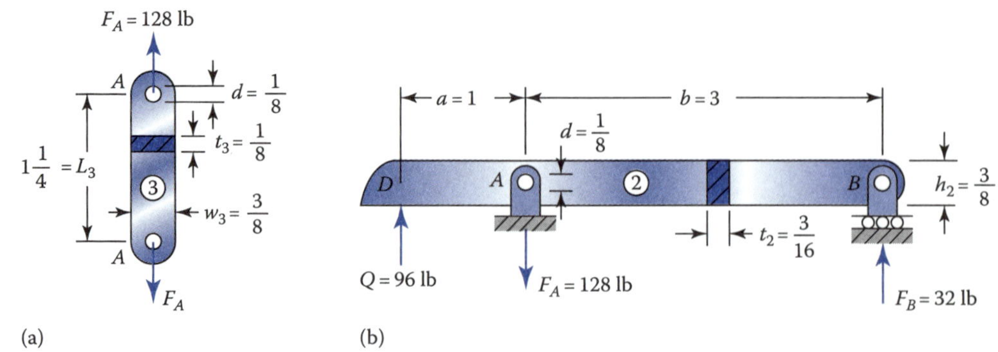

FIGURE 3.31
Some free-body diagrams of bolt cutter shown in Figure 1.5: (a) link 3 and (b) jaw 2.

The link and other members have ample material around holes to prevent tearout. The ⅛ in. diameter pins are in single shear. The worst-case direct shear stress, from Equation 3.4, is

$$\tau = \frac{4F_A}{\pi d^2} = \frac{4(128)}{\pi\left(\frac{1}{8}\right)^2} = 10.43 \text{ ksi}$$

Member 2, the jaw, is supported and loaded as shown in Figure 3.31b. The moment of inertia of the cross-sectional area is

$$I = \frac{t_2}{12}\left(h_2^3 - d^3\right)$$

$$= \frac{3/16}{12}\left[\left(\frac{3}{8}\right)^3 - \left(\frac{1}{8}\right)^3\right] = 0.793\left(10^{-3}\right) \text{in.}^4$$

The maximum moment that occurs at point A of the jaw equals $M = F_B b = 32(3) = 96$ lb · in. The bending stress is then

$$\sigma_c = \frac{Mc}{I} = \frac{96\left(\frac{3}{16}\right)}{0.793 \times 10^{-3}} = 22.7 \text{ ksi}$$

It can readily be shown that the shear stress is negligibly small in the jaw.

Member 1, the handle, has an irregular geometry and is relatively massive compared to the other components of the assembly. Accurate values of stresses as well as deflections in the handle may be obtained by the finite element analysis.

Comment: The results show that the maximum stresses in members are well under the yield strength of the material.

3.11 Plane Strain

In the case of two-dimensional (2D) or plane strain, all points in the body before and after the application of the load remain in the same plane. Therefore, in the xy plane, the strain components ε_x, ε_y, and γ_{xy} may have nonzero values. The normal and shear strains at a point in a member vary with direction in a way analogous to that for stress. We briefly discuss expressions that give the strains in the inclined directions. These in-plane strain transformation equations are particularly significant in experimental investigations, where strains are measured by means of strain gages. The site at www.measurementgroup.com includes general information on strain gages as well as instrumentation.

Mathematically, in every respect, the transformation of strain is the same as the stress transformation. It can be shown that [3] *transformation* expressions of stress are converted into strain relationships by substitution:

$$\sigma \rightarrow \varepsilon \quad \text{and} \quad \tau \rightarrow \frac{\gamma}{2} \tag{a}$$

These replacements can be made in all the analogous 2D and 3D transformation relations. Therefore, the principal strain directions are obtained from Equation 3.32 in the form, for example,

$$\tan 2\theta_p = \frac{\gamma_{xy}}{\varepsilon_x - \varepsilon_y} \tag{3.37}$$

Using Equation 3.33, the magnitudes of the in-plane principal strains are

$$\varepsilon_{1,2} = \frac{\varepsilon_x + \varepsilon_y}{2} \pm \sqrt{\left(\frac{\varepsilon_x - \varepsilon_y}{2}\right)^2 + \left(\frac{\gamma_{xy}}{2}\right)^2} \tag{3.38}$$

In a like manner, the in-plane transformation of strain in an arbitrary direction proceeds from Equations 3.31:

$$\varepsilon_{x'} = \frac{1}{2}(\varepsilon_x + \varepsilon_y) + \frac{1}{2}(\varepsilon_x - \varepsilon_y)\cos 2\theta + \frac{\gamma_{xy}}{2}\sin 2\theta \tag{3.39a}$$

$$\gamma_{x'y'} = -(\varepsilon_x - \varepsilon_y)\sin 2\theta + \gamma_{xy}\cos 2\theta \tag{3.39b}$$

$$\varepsilon_{y'} = \frac{1}{2}(\varepsilon_x + \varepsilon_y) - \frac{1}{2}(\varepsilon_x - \varepsilon_y)\cos 2\theta - \frac{\gamma_{xy}}{2}\sin 2\theta \tag{3.39c}$$

An expression for the maximum shear strain may also be found from Equation 3.34. Similarly, the transformation equations of 3D strain may be deduced from the corresponding stress relations, given in Section 3.17.

3.11.1 Mohr's Circle for Strain

In Mohr's circle for strain, the normal strain ε is plotted on the horizontal axis, positive to the right. The vertical axis is measured in terms of $\gamma/2$. The abscissa of the center C and the radius R of the circle, respectively, are

$$\varepsilon_{avg} = \varepsilon' = \frac{\varepsilon_x + \varepsilon_y}{2}, \quad R = \sqrt{\left(\frac{\varepsilon_x - \varepsilon_y}{2}\right)^2 + \left(\frac{\gamma_{xy}}{2}\right)^2} \tag{b}$$

When the shear strain is *positive*, the point representing the x axis strain is plotted a distance $\gamma/2$ *below* the axis and vice versa when shear strain is negative. Note that this convention for shearing strain, used *only* in constructing and reading values from Mohr's circle, agrees with the convention used for stress in Section 3.9.

Example 3.15: Determination of Principal Strains Using Mohr's Circle

It is observed that an element of a structural component elongates 450 μ along the x axis, contracts 120 μ in the y direction, and distorts through an angle of −360 μ (see Section 1.12). Calculate

a. The principal strains
b. The maximum shear strains

Given: $\varepsilon_x = 450\ \mu$, $\varepsilon_y = -120\ \mu$, $\gamma_{xy} = -360\ \mu$

Assumption: Element is in a state of plane strain.

Solution

A sketch of Mohr's circle is shown in Figure 3.32, constructed by finding the position of point C at $\varepsilon' = (\varepsilon_x + \varepsilon_y)/2 = 165\ \mu$ on the horizontal axis and of point A at $(\varepsilon_x, -\gamma_{xy}/2) =$ (450 μ, 180 μ) from the origin O.

a. The in-plane principal strains are represented by points A and B. Hence,

$$\varepsilon_{1,2} = 165\ \mu \pm \left[\left(\frac{450+120}{2}\right)^2 + (-180)^2\right]^{1/2}$$

$$\varepsilon_1 = 502\ \mu \quad \varepsilon_2 = -172\ \mu$$

Note, as a check, that $\varepsilon_x + \varepsilon_y = \varepsilon_1 + \varepsilon_2 = 330\ \mu$. From geometry,

$$\theta_p' = \frac{1}{2}\tan^{-1}\frac{180}{285} = 16.14°$$

It is seen from the circle that θ_p' locates the ε_1 direction.

b. The maximum shear strains are given by points D and E. Hence,

$$\gamma_{\max} = \pm(\varepsilon_1 - \varepsilon_2) = \pm 674\ \mu$$

Comments: Mohr's circle depicts that the axes of maximum shear strain make an angle of 45° with respect to principal axes. In the directions of maximum shear strain, the normal strains are equal to $\varepsilon' = 165\ \mu$.

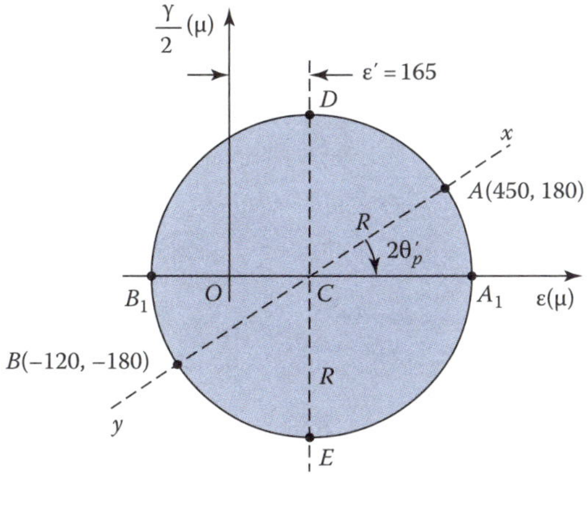

FIGURE 3.32
Example 3.15.

3.12 Measurement of Strain; Strain Rosette

The equations of transformation for strain and Mohr's circle facilitate the interpretation of strain gage measurements. But, literature on the subject should be consulted to gain further insight into experimental stress analysis. A variety of mechanical, electrical, and optical systems have been manufactured for measuring the *normal strain* on the *free surface* of a member where a state of plane stress occurs [7]. A widely employed, convenient, and accurate method uses electrical strain gages. We shall now briefly discuss a typical bonded strain gage and its special combinations.

Strain gage. A typical *strain gage* composed of a grid of fine wire or foil filament cemented between two sheets of treated paper foil or plastic backing as depicted in Figure 3.33. The backing serves to insulate the grid from the metal surface on which it is to be bonded. Usually, 0.03 mm diameter wire or 0.003 mm foil filament is used. The gages are manufactured of various gage lengths, changing from 4 to 150 mm, and are designed for different environmental conditions. As the surface is strained, the grid is lengthened or shortened, which changes the electrical resistance of the gage. A bridge circuit, connected to the gage by means of wires, is then used to translate variations in electrical resistance into strains. An instrument employed for this purpose is the *Wheatstone bridge*.

Strain rosette. At least three strain measurements in three different directions at a point on the surface of a member are needed to find the average state of strain at that point. So three gages are often clustered to form a *strain rosette*, which may be cemented to the free surface of a member. Two customary kinds of rosettes are the *rectangular rosette* with three gages spaced at 45° angles and the *delta rosette* with three gages spaced at 60° angles (Figure 3.34).

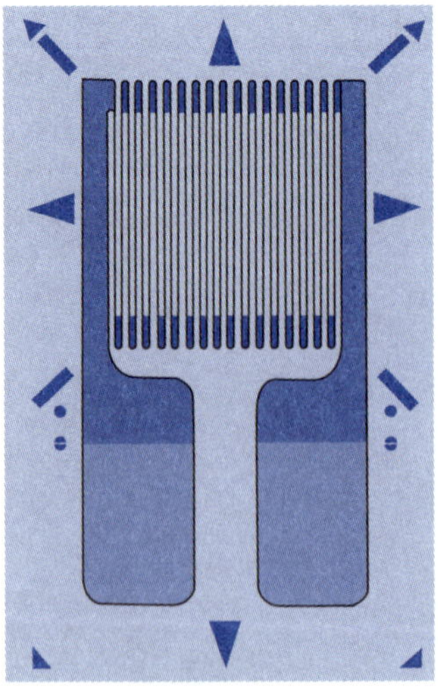

FIGURE 3.33
Strain gage. (Courtesy: Micro Measurements Division, Vishay Intertechnology, Inc., Malvern, PA.)

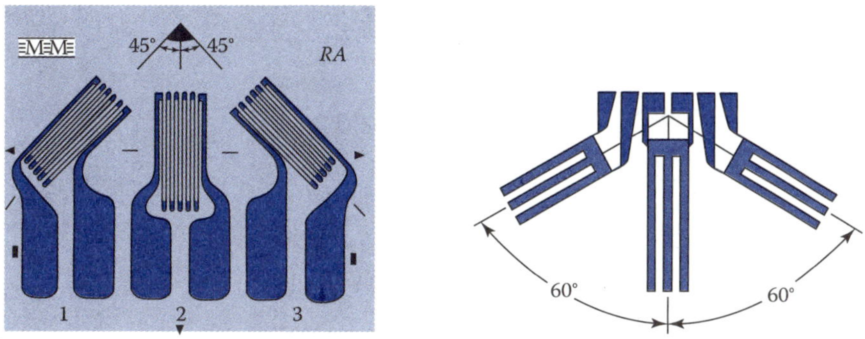

(a) (b)

FIGURE 3.34
Rosette strain gages: (a) rectangular rosette and (b) delta rosette. (Courtesy: Micro-Measurements Division; Vishay Intertechnology, Inc., Malvern, PA.)

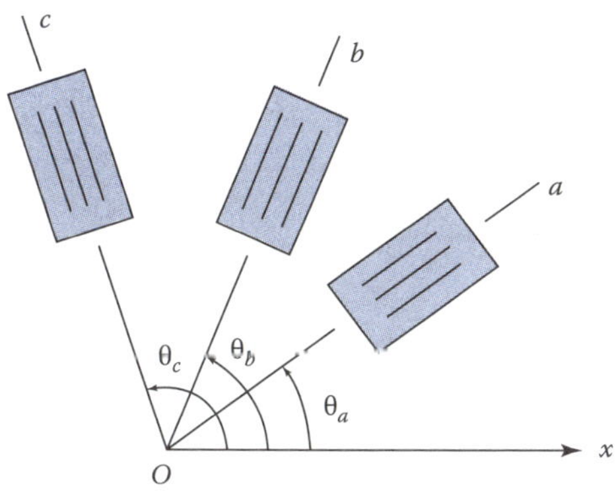

FIGURE 3.35
A schematic strain rosette.

Let us consider an arbitrary arrangement of strain gages with angles θ_a, θ_b, and θ_c about to the reference x axis, as shown in Figure 3.35. Then a-, b-, and c-directed normal strains, referring to Equations 3.30 and 3.39, are expressed as follows:

$$
\begin{aligned}
\varepsilon_a &= \varepsilon_x \cos^2\theta_a + \varepsilon_y \sin^2\theta_a + \gamma_{xy}\sin\theta_a\cos\theta_a \\
\varepsilon_b &= \varepsilon_x \cos^2\theta_b + \varepsilon_y \sin^2\theta_b + \gamma_{xy}\sin\theta_b\cos\theta_b \\
\varepsilon_c &= \varepsilon_x \cos^2\theta_c + \varepsilon_y \sin^2\theta_c + \gamma_{xy}\sin\theta_c\cos\theta_c
\end{aligned} \tag{3.40}
$$

If the values of ε_a, ε_b, and ε_c are measured for given θ_a, θ_b, and θ_c, the values of ε_x, ε_y, and γ_{xy} can be obtained by simultaneous solution of Equation 3.40. Usually, one of the axes is taken to be aligned with one arm of the rosette, say, the arm a. Hence, $\varepsilon_x = \varepsilon_a$, the strain in the direction a. The components ε_y, γ_{xy} may then be written in terms of the measured strains ε_a, ε_b, and ε_c in the directions of the three rosette arms a, b, c, respectively.

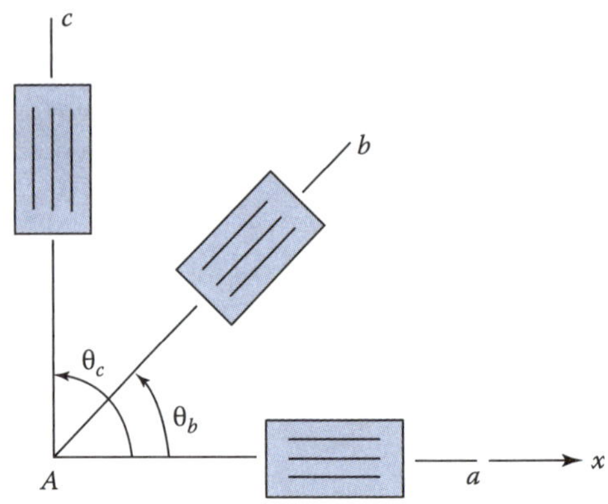

FIGURE 3.36
Rectangular strain rosette.

Example 3.16: Measured Strains in a Frame Component

Given: At point *A* on the free surface of a frame during a static testing, the 45° rosette readings show the normal strains:

$$\varepsilon_a = 800\ \mu, \quad \varepsilon_b = 600\ \mu, \quad \varepsilon_c = -150\ \mu$$

at $\theta_a = 0°$, $\theta_b = 45°$, and $\varepsilon_c = 90°$, respectively (Figure 3.36).

Find: The plane strain components.

Solution

For the 45° rosette arrangement, Equations 3.40 reduce to

$$\varepsilon_a = \varepsilon_x, \quad \varepsilon_c = \varepsilon_y, \quad \varepsilon_b = \frac{1}{2}\left(\varepsilon_x + \varepsilon_y + \gamma_{xy}\right)$$

from which

$$\varepsilon_x = \varepsilon_a, \quad \varepsilon_y = \varepsilon_c, \quad \gamma_{xy} = 2\varepsilon_b - \left(\varepsilon_a + \varepsilon_c\right) \tag{3.41}$$

Substituting the given data, we obtain $\varepsilon_x = 800\ \mu$, $\varepsilon_y = -150\ \mu$, and $\gamma_{xy} = 550\ \mu$.

Comment: The principal strains and the maximum shear strains for these numerical values may then readily be calculated.

3.13 Stress-Concentration Factors

The condition where highly localized stresses are produced as a result of an abrupt change in geometry is called the stress concentration. The abrupt change in form or discontinuity occurs in such frequently encountered stress raisers as holes, notches, keyways, threads, grooves, and fillets. Note that the stress concentration is a primary cause of fatigue failure and static failure in brittle materials, discussed in the next section. The formulas of mechanics of materials apply as long as the material remains linearly elastic and shape variations are gradual. In some cases, the stress and accompanying deformation near a discontinuity can be analyzed by applying the theory of elasticity. In those instances that do not yield to analytical methods, it is more usual to rely on experimental techniques or the finite element method (see Case Study 17.3). In fact, much research centers on determining stress-concentration effects for combined stress.

A geometric or theoretical *stress-concentration factor* K_t is used to relate the maximum stress at the discontinuity to the nominal stress. The factor is defined by

$$K_t = \frac{\sigma_{\text{max}}}{\sigma_{\text{nom}}} \quad \text{or} \quad K_t = \frac{\tau_{\text{max}}}{\tau_{\text{nom}}} \tag{3.42}$$

Here, the nominal stresses are stresses that would occur if the abrupt change in the cross section did not exist or had no influence on stress distribution. It is important to note that a stress-concentration factor is applied to the stress computed for the net or reduced cross section. Stress-concentration factors for several types of configuration and loading are available in technical literature [8,9].

The stress-concentration factors for a variety of geometries, provided in Appendix C, are useful in the design of machine parts. Curves in the Appendix C figures are plotted on the basis of dimensionless ratios: the shape, but not the size, of the member is involved. Observe that all these graphs indicate the advisability of streamlining junctures and transitions of portions that make up a member: that is, stress concentration can be reduced in intensity by properly proportioning the parts. Large fillet radii help at reentrant corners.

The values shown in Figures C.1, C.2, and C.7 through C.9 are for fillets of radius r that join a part of depth (or diameter) d to the one of larger depth (or diameter) D at a step or shoulder in a member (see Figure 3.37). A **full fillet** is a 90° arc with radius $r = (D - d_f)/2$. The stress-concentration factor decreases with increases in r/d or d/D. Also, results for the axial tension pertain equally to cases of axial compression. However, the stresses obtained are valid only if the loading is not significant relative to that which would cause failure by buckling.

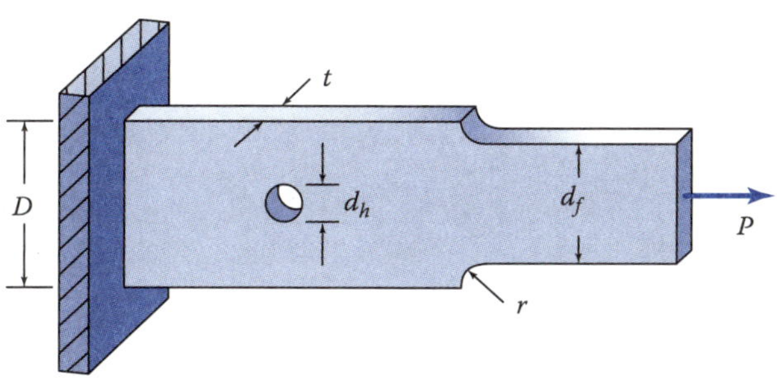

FIGURE 3.37
A flat bar with fillets and a centric hole under axial loading.

Example 3.17: Design of Axially Loaded Thick Plate with a Hole and Fillets

A filleted plate of thickness t supports an axial load P (Figure 3.37). Determine the radius r of the fillets so that the same stress occurs at the hole and the fillets.

Given: $P = 50$ kN, $D = 100$ mm, $d_f = 66$ mm, $d_h = 20$ mm, $t = 10$ mm

Design Decisions: The plate will be made of a relatively brittle metallic alloy: we must consider stress concentration.

Solution

For the *circular hole,*

$$\frac{d_h}{D} = \frac{20}{100} = 0.2, \quad A = (D - d_h)t = (100 - 20)10 = 800 \text{ mm}^2$$

Using the lower curve in Figure C.5, we find that $K_t = 2.44$ corresponding to $d_h/D = 0.2$. Hence,

$$\sigma_{\max} = K_t \frac{P}{A} = 2.44 \frac{50 \times 10^3}{800(10^{-6})} = 152.5 \text{ MPa}$$

For *fillets,*

$$\sigma_{\max} = K_t \frac{P}{A} = K_t \frac{50 \times 10^3}{660(10^{-6})} = 75.8K_t \text{ MPa}$$

The requirement that the maximum stress for the hole and fillets be identical is satisfied by

$$152.5 = 75.8K_t \quad \text{or} \quad K_t = 2.01$$

From the curve in Figure C.1, for $D/d_f = 100/66 = 1.52$, we find that $r/d_f = 0.12$ corresponding to $K_t = 2.01$. The necessary fillet radius is therefore

$$r = 0.12 \times 66 = 7.9 \text{ mm}$$

3.14 Importance of Stress-Concentration Factors in Design

Under certain conditions, a normally ductile material behaves in a brittle manner and vice versa. So, for a specific application, the distinction between ductile and brittle materials must be inferred from the discussion of Section 2.9. Also remember that the determination of stress-concentration factors is based on the use of Hooke's law.

3.14.1 Fatigue Loading

Most engineering materials may fail as a result of propagation of cracks originating at the point of high dynamic stress. The presence of stress concentration in the case of fluctuating (and impact) loading, as found in some machine elements, must be considered, regardless of whether the material response is brittle or ductile. In machine design, then, fatigue stress concentrations are of paramount importance. However, its effect on the nominal stress is not as large, as indicated by the theoretical factors (see Section 7.7).

3.14.2 Static Loading

For static loading, stress concentration is important only for *brittle* material. However, for some brittle materials having internal irregularities, such as cast iron, stress raisers usually have little effect, regardless of the nature of loading. Hence, the use of a stress-concentration factor appears to be unnecessary for cast iron. Customarily, stress concentration is ignored in static loading of *ductile* materials. The explanation for this restriction is quite simple. For ductile materials slowly and steadily loaded beyond the yield point, the stress-concentration factors decrease to a value approaching unity because of the redistribution of stress around a discontinuity.

To illustrate the foregoing inelastic action, consider the behavior of a mild-steel flat bar that contains a hole and is subjected to a gradually increasing load P (Figure 3.38). When σ_{max} reaches the yield strength S_y, stress distribution in the material is of the form curve *mn*, and yielding impends at *A*. Some fibers are stressed in the plastic range, but enough others remain elastic, and the member can carry additional load. We observe that the area under stress distribution curve is equal to the load P. This area increases as overload P increases, and a contained plastic flow occurs in the material. Therefore, with the increase in the value of P, the stress distribution curve assumes forms such as those shown by line *mp* and finally *mq*. That is, the effect of an abrupt change in geometry is nullified, and $\sigma_{max} = \sigma_{nom}$, or $K_t = 1$; prior to necking, a nearly *uniform* stress distribution across the net section occurs. Hence, for most practical purposes, the bar containing a hole carries the same static load as the bar with no hole.

The effect of ductility on the strength of the shafts and beams with stress raisers is similar to that of axially loaded bars. That is, localized inelastic deformations enable these members to support high stress concentrations. Interestingly, material ductility introduces a certain element of forgiveness in analysis while producing acceptable design results; for example, rivets can carry equal loads in a riveted connection (see Section 15.13).

When a member is yielded nonuniformly throughout a cross section, *residual stresses* remain in this cross section after the load is removed. An overload produces residual stresses favorable to future loads in the same direction and unfavorable to future loads in the opposite direction. Based on the idealized stress–strain curve, the increase in load capacity in one direction is the same as the decrease in load capacity in the opposite direction. Note that coil springs in compression are good candidates for favorable residual stresses caused by yielding.

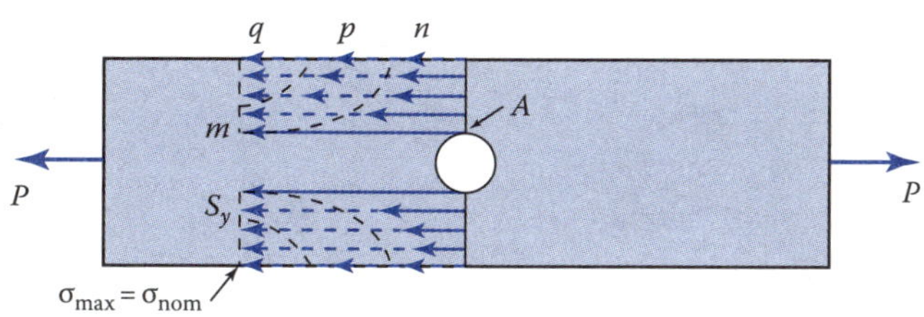

FIGURE 3.38
Redistribution of stress in a flat bar of mild steel.

Example 3.18: Load Capacity of a Stepped Steel Shaft in Tension

A round stepped ASTM-A36 structural steel shaft of diameters d and D with shoulder fillet radius r is loaded by an axial tensile load P as shown in Figure 3.39. Compute

a. The value of P that may be applied to the bar without causing the steel to yield
b. The maximum value of P that the bar can carry

Given: $d = 25$ mm, $D = 50$ mm, $r = 3.75$ mm, $S_y = 250$ MPa (Table B.1)

Assumption: The mild steel is idealized to be elastic–plastic material.

Solution

a. *Yield load.* Material behaves elastically. From the given dimensions, the geometric proportions of the bar are

$$\frac{D}{d} = \frac{50}{25} = 2, \quad \frac{r}{d} = \frac{3.75}{25} = 0.15$$

The corresponding stress-concentration factor from Figure C.7 is found as $K_t = 1.8$.

The largest load without causing yielding takes place when $\sigma_{max} = S_y$. Equation 3.42 becomes then

$$S_y = K_t \frac{P_y}{A} = K_t \frac{P_y}{\pi d^2/4}$$

Introducing the given data leads to

$$250\left(10^6\right) = 1.8 \frac{4P_y}{\pi(0.025)^2}$$

from which

$$P_y = 68.2 \text{ kN}$$

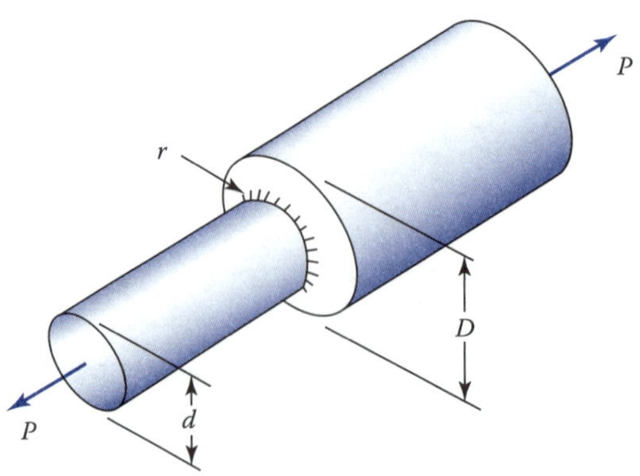

FIGURE 3.39
Example 3.18. A filleted shaft in an axial tensile load.

b. *Ultimate load.* The maximum load supported by the shaft causes all the material at the smallest cross section to yield uniformly. Consequently, $\sigma_{max} = \sigma_{nom}$ or $K_t = 1$. We thus have

$$S_u = \frac{P_u}{A} = \frac{P_u}{\pi d^2/4}$$

Inserting the given numerical values, we have

$$250(10^6) = \frac{4P_u}{\pi(0.025)^2}$$

Solving,

$$P_u = 122.7 \text{ kN}$$

Comment: Inasmuch as $P_u > P_y$, the elastic design is conservative.

*3.15 Three-Dimensional Stress

In the most general case of 3D stress, an element is subjected to stresses on the orthogonal x, y, and z planes, as shown in Figure 1.9. Consider a tetrahedron, isolated from this element and represented in Figure 3.40. Components of stress on the perpendicular planes (intersecting at the origin O) can be related to the normal and shear stresses on the oblique plane ABC, by using an approach identical to that employed for the 2D state of stress.

Orientation of plane ABC may be defined in terms of the *direction cosines*, associated with the angles between a unit normal n to the plane and the x, y, z coordinate axes:

$$\cos(\mathbf{n}, x) = l, \quad \cos(\mathbf{n}, y) = m, \quad \cos(\mathbf{n}, z) = n \tag{3.43}$$

The sum of the squares of these quantities is unity:

$$l^2 + m^2 + n^2 = 1 \tag{3.44}$$

Consider now a new coordinate system x', y', z', where x' coincides with n and y', z' lie on an oblique plane. It can readily be shown that [3] the normal stress acting on the oblique x' plane shown in Figure 3.40 is expressed in the form

$$\sigma_{x'} = \sigma_x l^2 + \sigma_y m^2 + \sigma_z n^2 + 2(\tau_{xy} lm + \tau_{yz} mn + \tau_{xz} ln) \tag{3.45}$$

where l, m, and n are direction cosines of angles between x' and the x, y, z axes, respectively. The shear stresses $\tau_{x'y'}$ and $\tau_{x'z'}$ may be written similarly. The stresses on the three mutually perpendicular planes are required to specify the stress at a point. One of these planes is the oblique (x') plane in question. The other stress components $\sigma_{y'}$, $\sigma_{z'}$, and $\tau_{y'z'}$ are obtained by considering those (y' and z') planes perpendicular to the oblique plane. In so doing, the resulting six expressions represent *transformation equations* for 3D stress.

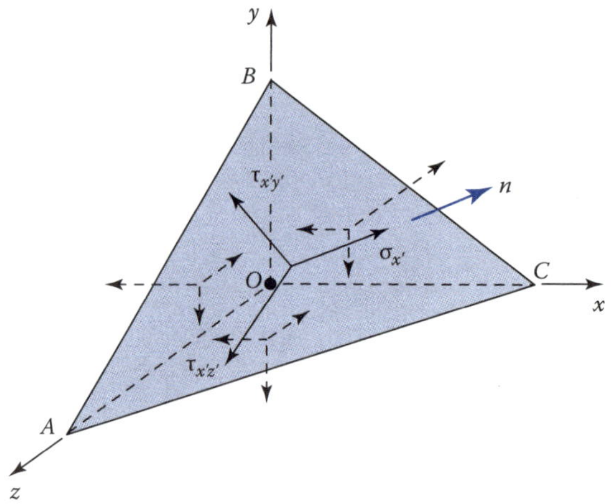

FIGURE 3.40
Components of stress on a tetrahedron.

3.15.1 Principal Stresses in Three Dimensions

For the 3D case, three mutually perpendicular planes of zero shear exist, and on these planes, the normal stresses have maximum or minimum values. The foregoing normal stresses are called *principal stresses* σ_1, σ_2, and σ_3. The algebraically largest stress is represented by σ_1 and the smallest by σ_3. Of particular importance are the direction cosines of the plane on which $\sigma_{x'}$ has a maximum value, determined from the following equations:

$$\begin{bmatrix} \sigma_x - \sigma_i & \tau_{xy} & \tau_{xz} \\ \tau_{xy} & \sigma_y - \sigma_i & \tau_{yz} \\ \tau_{xz} & \tau_{yz} & \sigma_z - \sigma_i \end{bmatrix} \begin{Bmatrix} l_i \\ m_i \\ n_i \end{Bmatrix} = 0, \quad (i = 1, 2, 3) \tag{3.46}$$

A nontrivial solution for the direction cosines requires that the characteristic determinant vanishes. Thus,

$$\begin{vmatrix} \sigma_x - \sigma_i & \tau_{xy} & \tau_{xz} \\ \tau_{xy} & \sigma_y - \sigma_i & \tau_{yz} \\ \tau_{xz} & \tau_{yz} & \sigma_z - \sigma_i \end{vmatrix} = 0 \tag{3.47}$$

Expanding Equation 3.47, we obtain the following stress cubic equation:

$$\sigma_i^3 - I_1\sigma_i^2 + I_2\sigma_i - I_3 = 0 \tag{3.48}$$

where

$$\begin{aligned} I_1 &= \sigma_x + \sigma_y + \sigma_z \\ I_2 &= \sigma_x\sigma_y + \sigma_x\sigma_z + \sigma_y\sigma_z - \tau_{xy}^2 - \tau_{yz}^2 - \tau_{xz}^2 \\ I_3 &= \sigma_x\sigma_y\sigma_z + 2\tau_{xy}\tau_{yz}\tau_{xz} - \sigma_x\tau_{yz}^2 - \sigma_y\tau_{xz}^2 - \sigma_z\tau_{xy}^2 \end{aligned} \tag{3.49}$$

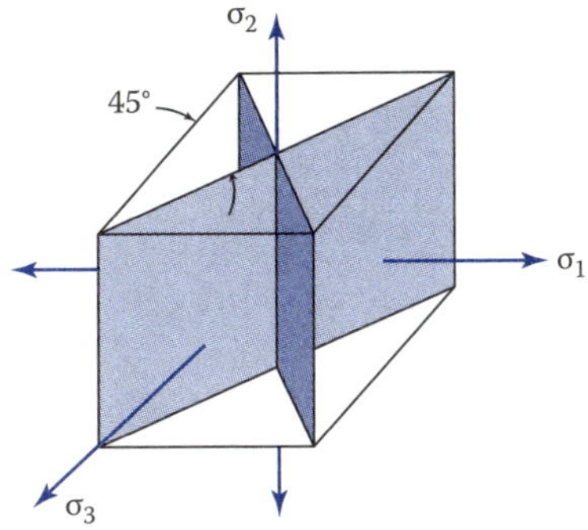

FIGURE 3.41
Planes of maximum 3D shear stress.

The quantities I_1, I_2, and I_3 represent *invariants* of the 3D stress. For a given state of stress, Equation 3.48 may be solved for its three roots, σ_1, σ_2 and σ_3. Introducing each of these principal stresses into Equation 3.46 and using $l_i^2 + m_i^2 + n_i^2 = 1$, we can obtain three sets of direction cosines for three principal planes. Note that the direction cosines of the principal stresses are occasionally required to predict the behavior of members. A convenient way of determining the roots of the stress cubic equation and solving for the direction cosines is given in Appendix D.

After obtaining the 3D principal stresses, we can readily determine the maximum shear stresses. Since no shear stress acts on the principal planes, it follows that an element oriented parallel to the principal directions is in a state of triaxial stress (Figure 3.41). Therefore,

$$\tau_{max} = \frac{1}{2}(\sigma_1 - \sigma_3) \tag{3.50}$$

The **maximum shear stress** acts on the planes that bisect the planes of the maximum and minimum principal stresses as shown in the figure.

Example 3.19: Three-Dimensional State of Stress in a Member

At a critical point in a loaded machine component, the stresses relative to x, y, z coordinate system are given by

$$\begin{bmatrix} \sigma_x & \tau_{xy} & \tau_{xz} \\ \tau_{xy} & \sigma_y & \tau_{yz} \\ \tau_{xz} & \tau_{yz} & \sigma_z \end{bmatrix} = \begin{bmatrix} 60 & 20 & 20 \\ 20 & 0 & 40 \\ 20 & 40 & 0 \end{bmatrix} \text{MPa} \tag{a}$$

Determine the principal stresses σ_1, σ_2, σ_3 and the orientation of σ_1 with respect to the original coordinate axes.

Solution

Substitution of Equation (a) into Equation 3.48 gives

$$\sigma_i^3 - 60\sigma_i^2 - 2{,}400\sigma_i + 64{,}000 = 0, \quad (i = 1, 2, 3)$$

The three principal stresses representing the roots of this equation are

$$\sigma_1 = 80\text{ MPa},\quad \sigma_2 = 20\text{ MPa},\quad \sigma_3 = -40\text{ MPa}$$

Introducing σ_1 into Equation 3.46, we have

$$\begin{bmatrix} 60-80 & 20 & 20 \\ 20 & 0-80 & 40 \\ 20 & 40 & 0-80 \end{bmatrix} \begin{Bmatrix} l_1 \\ m_1 \\ n_1 \end{Bmatrix} = 0 \tag{b}$$

Here l_1, m_1, and n_1 represent the direction cosines for the orientation of the plane on which σ_1 acts.

It can be shown that only two of Equation (b) are independent. From these expressions, together with $l_1^2 + m_1^2 + n_1^2 = 1$, we obtain

$$l_1 = \frac{2}{\sqrt{6}} = 0.8165,\quad m_1 = \frac{1}{\sqrt{6}} = 0.4082,\quad n_1 = \frac{1}{\sqrt{6}} = 0.4082$$

Comment: The direction cosines for σ_2 and σ_3 are ascertained in a like manner. The foregoing computations may readily be performed by using the formulas given in Appendix D.

3.15.2 Simplified Transformation for Three-Dimensional Stress

Often we need the normal and shear stresses acting on an arbitrary oblique plane of a tetrahedron in terms of the principal stresses acting on perpendicular planes (Figure 3.42). In this case, the *x*, *y*, and *z* coordinate axes are parallel to the principal axes: $\sigma_{x'} = \sigma$, $\sigma_x = \sigma_1$, $\tau_{xy} = \tau_{xz} = 0$, and so on, as depicted in the figure. Let *l*, *m*, and *n* denote the direction cosines of oblique plane *ABC*. The normal stress σ on the oblique plane, from Equation 3.45, is

$$\sigma = \sigma_1 l^2 + \sigma_2 m^2 + \sigma_3 n^2 \tag{3.51a}$$

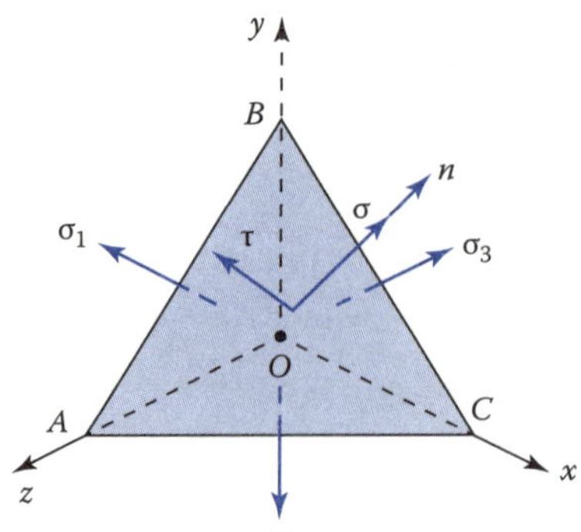

FIGURE 3.42
Triaxial stress on a tetrahedron.

It can be verified that the shear stress τ on this plane may be expressed in the following convenient form:

$$\tau = \left[(\sigma_1-\sigma_2)^2 l^2m^2 + (\sigma_2-\sigma_3)^2 m^2n^2 + (\sigma_3-\sigma_1)^2 n^2l^2\right]^{1/2} \tag{3.51b}$$

The preceding expressions are the simplified transformation equations for 3D state of stress.

3.15.3 Octahedral Stresses

Let us consider an oblique plane that forms equal angles with each of the principal stresses, represented by face *ABC* in Figure 3.42 with *OA* = *OB* = *OC*. Thus, the normal *n* to this plane has equal direction cosines relative to the principal axes. Inasmuch as $l^2 + m^2 + n^2 = 1$, we have

$$l = m = n = \frac{1}{\sqrt{3}} \tag{c}$$

There are eight such plane or *octahedral planes*, all of which have the same intensity of normal and shear stresses at a point *O* (Figure 3.43). Interestingly, the same normal stresses acting on all eight planes serve to enlarge (or contract) the octahedral plane but not to distort it. On the contrary, identical shear stresses occurring on all eight planes serve to distort the octahedron without altering its volume.

Substitution of the preceding equation into Equations 3.51 results in the magnitudes of the *octahedral normal stress* and *octahedral shear stress*, in the following forms:

$$\sigma_{oct} = \frac{1}{3}(\sigma_1+\sigma_2+\sigma_3) \tag{3.52a}$$

$$\tau_{oct} = \frac{1}{3}\left[(\sigma_1-\sigma_2)^2 + (\sigma_2-\sigma_3)^2 + (\sigma_3-\sigma_1)^2\right]^{1/2} \tag{3.52b}$$

Equation 3.52a indicates that the normal stress acting on an octahedral plane is the mean of the principal stresses. It should be noted that equality of stresses in eight octahedral planes is a powerful factor in failure of ductile materials. That is, the octahedral stresses play an important role in certain failure criteria, discussed in Sections 5.2 and 6.8.

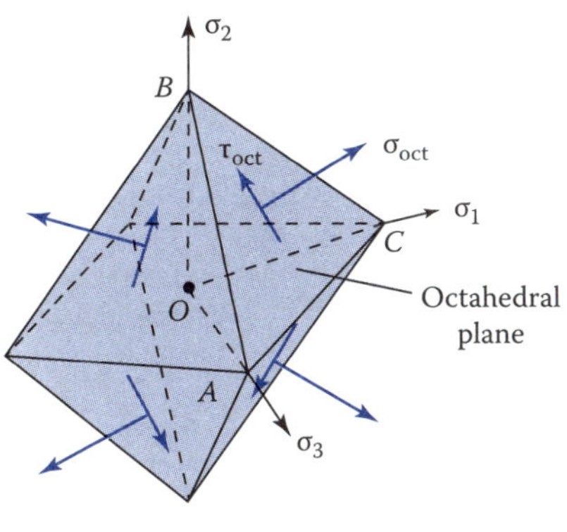

FIGURE 3.43
Stresses on an octahedron.

Example 3.20: Principal Stresses Using Mohr's Circle

Figure 3.44a depicts a point in a loaded machine base subjected to the 3D stresses. Determine at the point

a. The principal planes and principal stresses
b. The maximum shear stress
c. The octahedral stresses

Solution

We construct Mohr's circle for the transformation of stress in the xy plane as indicated by the solid lines in Figure 3.44b. The radius of the circle is $R = (12.5^2 + 30^2)^{1/2} =$ 32.5 MPa.

a. The principal stresses in the plane are represented by points A and B:

$$\sigma_1 = 47.5 + 32.5 = 80 \text{ MPa}$$

$$\sigma_2 = 47.5 - 32.5 = 15 \text{ MPa}$$

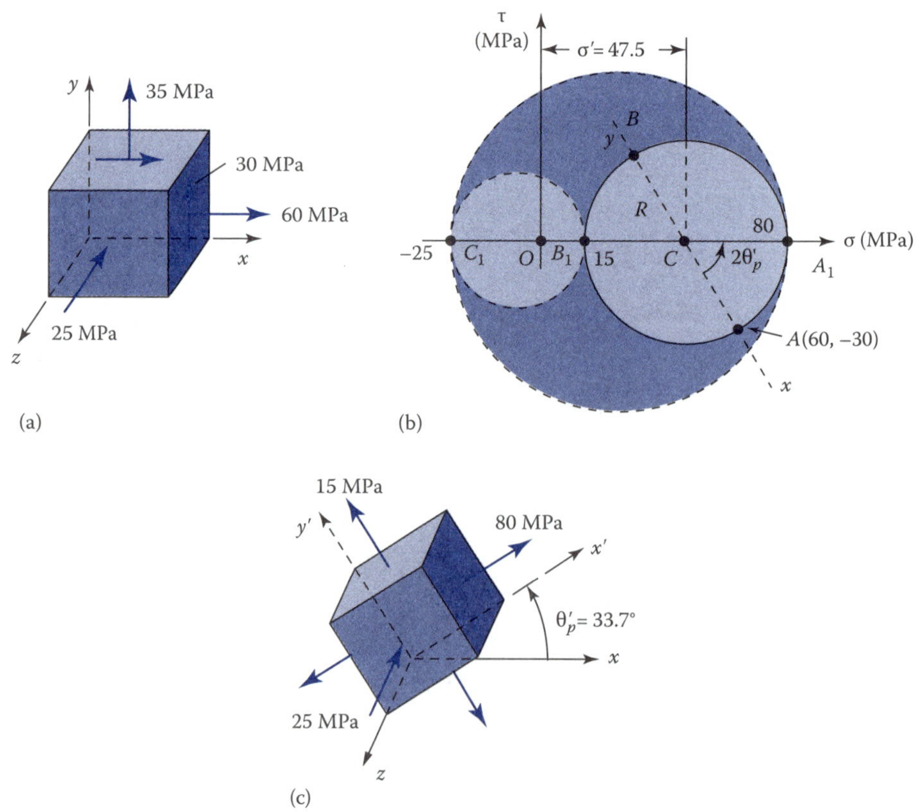

FIGURE 3.44
Example 3.20. (a) stress element for $\theta = 0°$, (b) Mohr's circle for strain, and (c) stress element for $\theta = 33.7°$.

The z faces of the element define one of the principal stresses: $\sigma_3 = -25$ MPa. The planes of the maximum principal stress are defined by θ'_p, the angle through which the element should rotate about the z axis:

$$\theta'_p = \frac{1}{2}\tan^{-1}\frac{30}{12.5} = 33.7°$$

The result is shown on a sketch of the rotated element (Figure 3.44c).

b. We now draw circles of diameters C_1B_1 and C_1A_1, which correspond, respectively, to the projections in the $y'z'$ and $x'z'$ planes of the element (Figure 3.44b). The maximum shearing stress, the radius of the circle of diameter C_1A_1, is therefore

$$\tau_{max} = \frac{1}{2}(80 + 25) = 52.5 \text{ MPa}$$

Planes of the maximum shear stress are inclined at 45° with respect to the x' and z faces of the element of Figure 3.44c.

c. Through the use of Equations 3.52, we have

$$\sigma_{oct} = \frac{1}{3}(80 + 15 - 25) = 23.3 \text{ MPa}$$

$$\tau_{oct} = \frac{1}{3}\left[(80-15)^2 + (15+25)^2 + (-25-80)^2\right]^{1/2} = 43.3 \text{ MPa}$$

*3.16 Equations of Equilibrium for Stress

As noted earlier, the components of stress generally vary from point to point in a loaded member. Such variations of stress, accounted for by the theory of elasticity, are governed by the equations of statics. Satisfying these conditions, *the differential equations of equilibrium* are obtained. To be physically possible, a stress field must satisfy these equations at every point in a load-carrying component.

For the 2D case, the stresses acting on an element of sides dx and dy and of unit thickness are depicted in Figure 3.45. The body forces per unit volume acting on the element, F_x and F_y, are independent of z, and the component of the body force $F_z = 0$. In general, stresses are functions of the coordinates (x, y). For example, from the lower-left corner to the upper-right corner of the element, one stress component, say, σ_x, changes in value: $\sigma_x + (\partial\sigma_x/\partial x)\mathrm{d}x$. The components σ_y and τ_{xy} change in a like manner. The stress element must satisfy the equilibrium condition $\sum M_z = 0$. Hence,

$$\left(\frac{\partial\sigma_y}{\partial y}\mathrm{d}x\mathrm{d}y\right)\frac{\mathrm{d}x}{2} - \left(\frac{\partial\sigma_x}{\partial x}\mathrm{d}x\mathrm{d}y\right)\frac{\mathrm{d}y}{2} + \left(\tau_{xy} + \frac{\partial\tau_{xy}}{\partial x}\mathrm{d}x\right)\mathrm{d}x\mathrm{d}y$$

$$-\left(\tau_{yx} + \frac{\partial\tau_{yx}}{\partial y}\mathrm{d}y\right)\mathrm{d}x\mathrm{d}y + F_y\,\mathrm{d}x\mathrm{d}y\frac{\mathrm{d}x}{2} - F_x\mathrm{d}x\mathrm{d}y\frac{\mathrm{d}y}{2} = 0$$

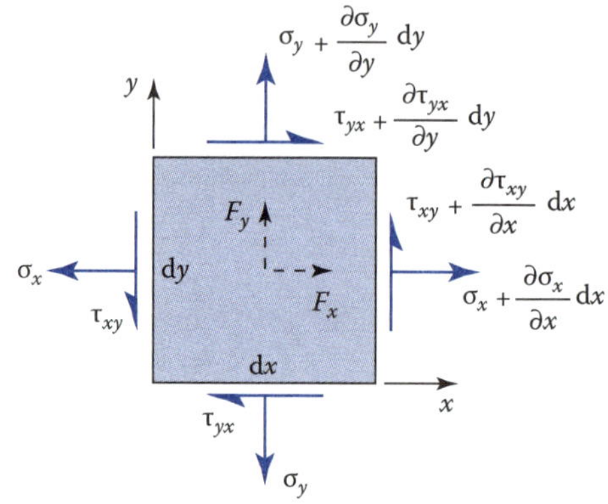

FIGURE 3.45
Stresses and body forces on an element.

After neglecting the triple products involving dx and dy, this equation results in $\tau_{xy} = \tau_{yx}$. Similarly, for a general state of stress, it can be shown that $\tau_{yz} = \tau_{zy}$ and $\tau_{xz} = \tau_{zx}$. Hence, the shear stresses in mutually perpendicular planes of the element are equal.

The equilibrium condition of x-directed forces must sum to 0, $\sum F_x = 0$. Therefore, referring to Figure 3.45, we write

$$\left(\sigma_x + \frac{\partial \sigma_x}{\partial x} \mathrm{d}x\right) \mathrm{d}y - \sigma_x \mathrm{d}y + \left(\tau_{xy} + \frac{\partial \tau_{xy}}{\partial y} \mathrm{d}y\right) \mathrm{d}x - \tau_{xy} \mathrm{d}x + F_x \mathrm{d}x \mathrm{d}y = 0$$

Summation of the forces in the y direction yields an analogous result. After reduction, we obtain the differential equations of equilibrium for a *2D stress* in the form [3]

$$\begin{aligned} \frac{\partial \sigma_x}{\partial x} + \frac{\partial \tau_{xy}}{\partial y} + F_x = 0 \\ \frac{\partial \sigma_y}{\partial y} + \frac{\partial \tau_{xy}}{\partial x} + F_y = 0 \end{aligned} \tag{3.53}$$

In the general case of an element under *3D stresses*, it can be shown that the differential equations of equilibrium are determined similarly.

We observe that two relations of Equations 3.53 involve the three unknown (σ_x, σ_y, τ_{xy}) stress components. Therefore, problems in stress analysis are *internally* statically indeterminate. In the mechanics of materials method, this indeterminacy is eliminated by introducing simplifying assumptions regarding the stresses and considering the equilibrium of the finite segments of a load-carrying component.

*3.17 Strain–Displacement Relations: Exact Solutions

If deformation is distributed uniformly over the original length, the normal strain may be written $\varepsilon_x = \delta/L$, where L and δ are the original length and the change in length of the member, respectively (see Figure 1.11a). However, the strains generally vary from point to point in a member. Hence, the expression for strain must relate to a line of length dx, which elongates by an amount du under the axial load. The definition of normal strain is therefore

$$\varepsilon_x = \frac{du}{dx} \tag{3.54}$$

This represents the strain at a point.

As noted earlier, in the case of 2D or *plane strain*, all points in the body, before and after application of load, remain in the same plane. Therefore, the deformation of an element of dimensions dx and dy and of unit thickness can contain normal strain (Figure 3.46a) and a shear strain (Figure 3.46b). Note that the partial derivative notation is used, since the displacement u or υ is function of x and y. Recalling the basis of Equations 3.54 and 1.22, an examination of Figure 3.46 yields

$$\varepsilon_x = \frac{\partial u}{\partial x}, \quad \varepsilon_y = \frac{\partial \upsilon}{\partial y}, \quad \gamma_{xy} = \frac{\partial \upsilon}{\partial x} + \frac{\partial u}{\partial y} \tag{3.55}$$

Obviously, γ_{xy} is the shear strain between the x and y axes (or y and x axes); hence, $\gamma_{xy} = \gamma_{yx}$. A long prismatic member subjected to a lateral load (e.g., a cylinder under pressure) exemplifies the state of plane strain. In an analogous manner, the strains at a point in a rectangular prismatic element of sides dx, dy, and dz are found in terms of the displacements u, υ, and w.

3.17.1 Problems in Applied Elasticity

In many problems of practical importance, the stress or strain condition is one of plane stress or plane strain. These 2D problems in elasticity are *simpler* than those involving three dimensions. A finite element solution of 2D problems is taken up in Chapter 17. In examining Equations 3.55, we see that the three strain components depend linearly on the derivatives of the two displacement components. Therefore, the strains cannot be independent of one another. An equation, referred to as the *condition of compatibility*, can be developed showing the relationships among ε_x, ε_y, and γ_{xy} [3]. The condition of compatibility asserts that the displacements are continuous. Physically, this means that the body must be pieced together.

To conclude, *exact solution* by the theory of elasticity is based on the following requirements: strain compatibility, stress equilibrium, general relationships between the stresses and strains, and boundary conditions for a given problem. In Chapter 16, we discuss various

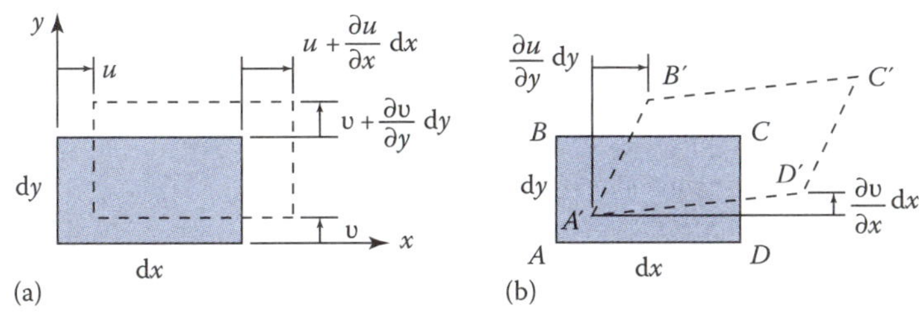

FIGURE 3.46
Deformations of a 2D element: (a) normal strain and (b) shear strain.

axisymmetric problems using the applied elasticity approaches. In the method of mechanics of materials, simplifying assumptions are made with regard to the distribution of strains in the body as a whole or the finite portion of the member. Thus, the difficult task of solving the condition of compatibility and the differential equations of equilibrium are avoided.

Problems

Sections 3.1 through 3.8

3.1 Two plates are fastened by a bolt and nut as shown in Figure P3.1. Calculate

a. The normal stress in the bolt shank

b. The average shear stress in the head of the bolt

c. The shear stress in the threads

d. The bearing stress between the head of the bolt and the plate

Assumption: The nut is tightened to produce a tensile load in the shank of the bolt of 10 kips.

3.2 A short steel pipe of yield strength S_y is to support an axial compressive load P with factor of safety of n against yielding. Determine the minimum required inside radius a.

Given: $S_y = 280$ MPa, P = 1.2 MN, and $n = 2.2$

Assumption: The thickness t of the pipe is to be one-fourth of its inside radius a.

3.3 The landing gear of an aircraft is depicted in Figure P3.3. What are the required pin diameters at A and B?

Given: Maximum stress of 28 ksi in shear. A factor of safety $n = 2$ will be used.

Assumption: Pins act in double shear.

3.4 The frame of Figure P3.4 supports a concentrated load P. Calculate

a. The normal stress in the member BD if it has a cross-sectional area A_{BD}

b. The shearing stress in the pin at A if it has a diameter of 25 mm and is in double shear

Given: $P = 5$ kN, $A_{BD} = 8 \times 10^3$ mm²

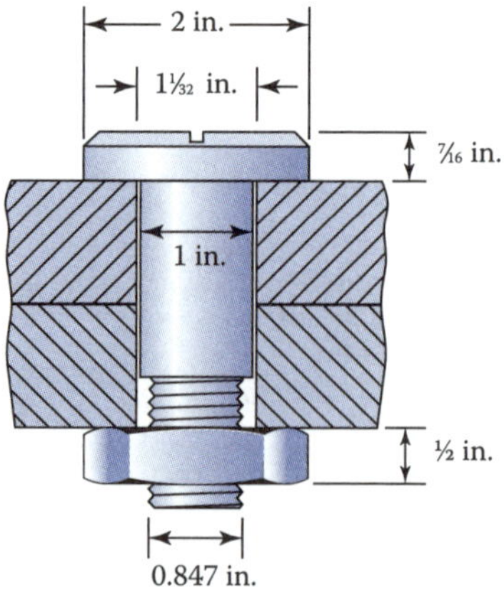

FIGURE P3.1

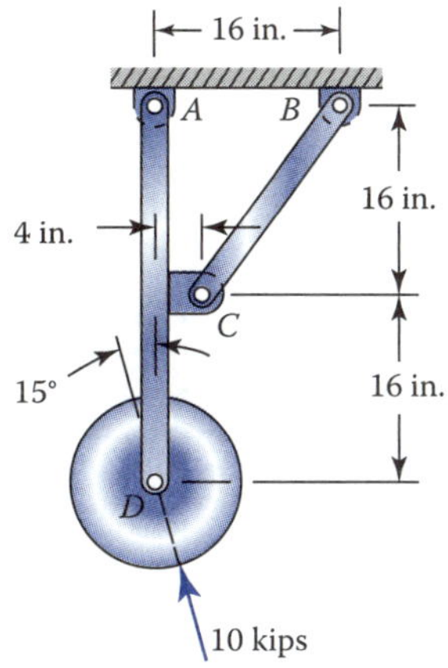

FIGURE P3.3

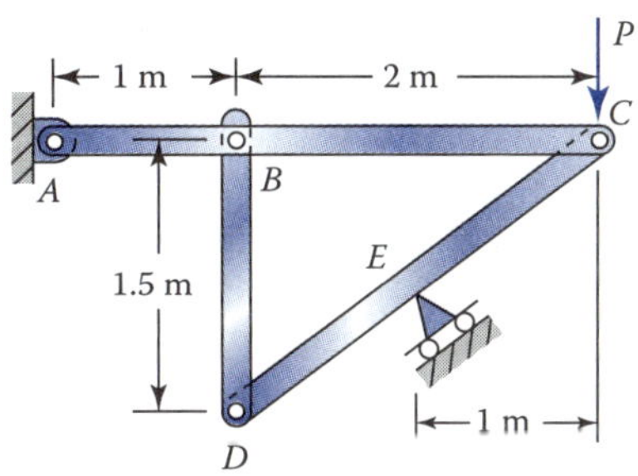

FIGURE P3.4

3.5 Two bars AC and BC are connected by pins to form a structure for supporting a vertical load P at C (Figure P3.5). Determine the angle α if the structure is to be of minimum weight.

Assumption: The normal stresses in both bars are to be the same.

3.6 A uniform thickness steel tube of length L and rectangular cross section having mean width a by mean depth b is built in at one end and carries a torque T at the free end (Figure P3.6). What is the minimum wall thickness t if the shearing stress and the angle of twist are limited to τ_{all} and θ_{all}, respectively?

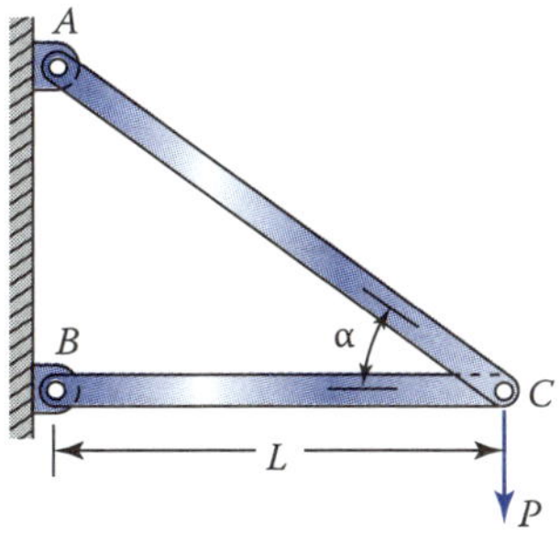

FIGURE P3.5

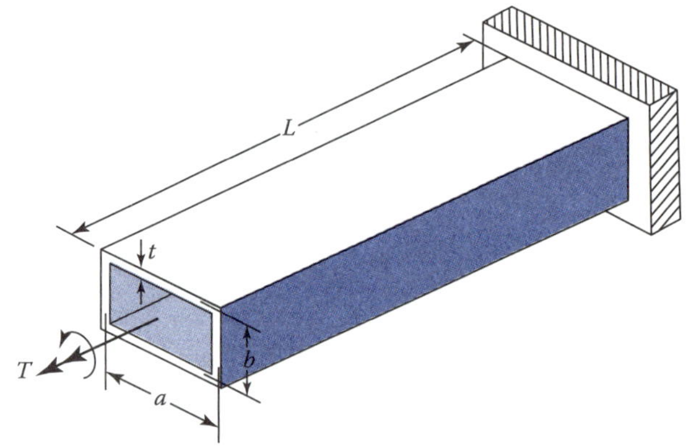

FIGURE P3.6

Given: $a = 2b = 90$ mm, $L = 0.8$ m, $T = 1.2$ kN m, $G = 28$ GPa, $\tau_{all} = 30$ MPa, $\theta_{all} = 1.5°$

Assumption: The effect of stress concentration at the corners is neglected.

3.7 Redo Problem 3.6 for the case in which the cross section of the tube is square ($a = b = 70$ mm) box of uniform thickness t.

3.8 An aluminum alloy 2014-T6 tube of modulus of rigidity G, length L, and $a \times a$ square cross section of uniform thickness t is under a torque as illustrated in Figure P3.6. The allowable yield strength in shear and angle of twist are τ_{all} and θ_{all}, respectively. What is the largest value of the torque that can be applied to the tube?

Given: $a = 1\frac{1}{4}$ in., $t = \frac{3}{16}$ in., $L = 8$ ft, $G = 4.1 \times 10^6$ psi (Table B.1), $\tau_{all} = 15$ ksi, $\theta_{all} = 12°$

3.9 A circular cylindrical tube having an outside radius of r_o and inside radius of r_i is twisted at its ends by a torque T. Compare the shear stresses in the tube obtained by Equation 3.11 with that estimated by $\tau = T/2\pi abt$ in Table 3.1.

Given: $r_o = 12.4$ mm and $r_i = 11$ mm

3.10 For a pin-and-clevis joint (see Figure 3.3a), it is found that forces act on the pin as depicted in the free-body diagram of Figure P3.10. Draw the shear and bending moment diagrams. Design the pin (find diameter d) on the basis of

1. Bending strength, $\sigma_{all} = 250$ MPa
2. Shear strength, $\tau_{all} = 150$ MPa

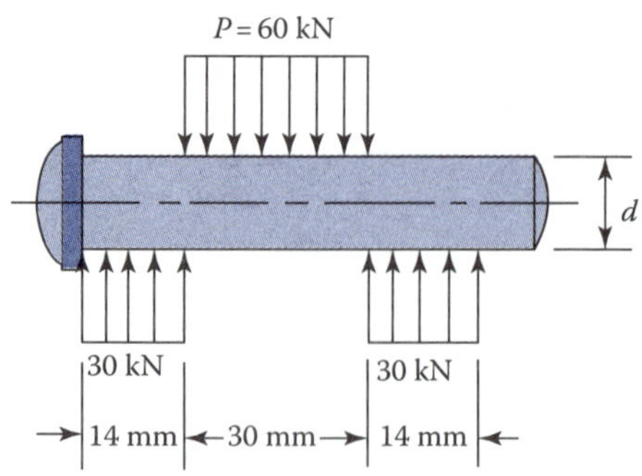

FIGURE P3.10

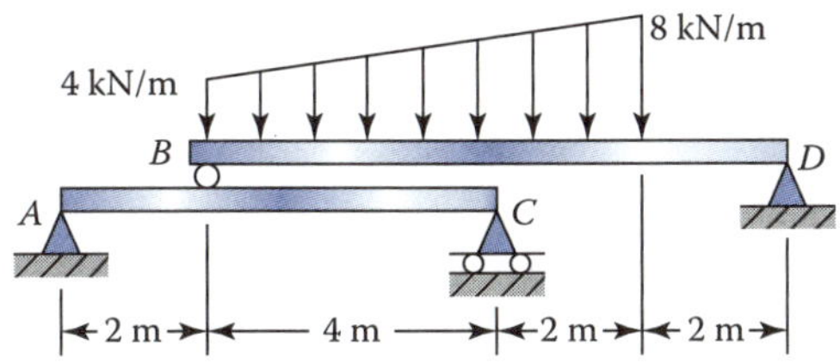

FIGURE P3.11

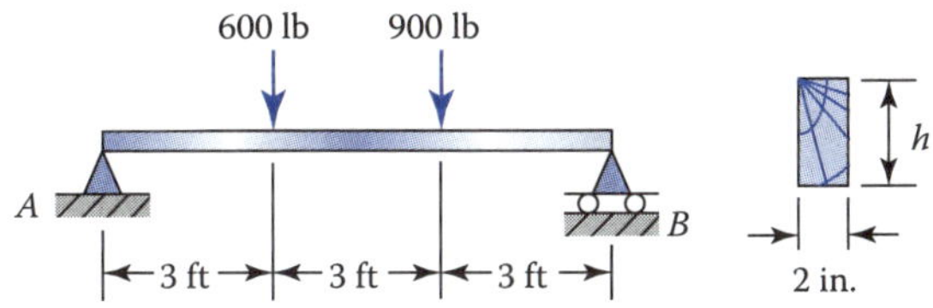

FIGURE P3.12

3.11 Two beams *AC* and *BD* are supported as shown in Figure P3.11. A roller fits snugly between the two beams at point *B*. Draw the shear and moment diagrams of the lower beam *AC*.

3.12 Design the cross section (determine *h*) of the simply supported beam loaded at two locations as shown in Figure P3.12.

Assumption: The beam will be made of timber of $\sigma_{all} = 1.8$ ksi and $\tau_{all} = 100$ psi.

3.13 A rectangular beam is to be cut from a circular bar of diameter *d* (Figure P3.13). Determine the dimensions *b* and *h* so that the beam will resist the largest bending moment.

3.14 The T beam, whose cross section is shown in Figure P3.14, is subjected to a shear force *V*. Calculate the maximum shear stress in the web of the beam.

Given: $b = 200$ mm, $t = 15$ mm, $h_1 = 175$ mm, $h_2 = 150$ mm, $V = 22$ kN

3.15 A box beam is made of four 50 mm × 200 mm planks, nailed together as shown in Figure P3.15. Determine the maximum allowable shear force *V*.

Given: The longitudinal spacing of the nails, $s = 100$ mm; the allowable load per nail, $F = 15$ kN.

3.16 For the beam and loading shown in Figure P3.16, design the cross section of the beam for $\sigma_{all} = 12$ MPa and $\tau_{all} = 810$ kPa.

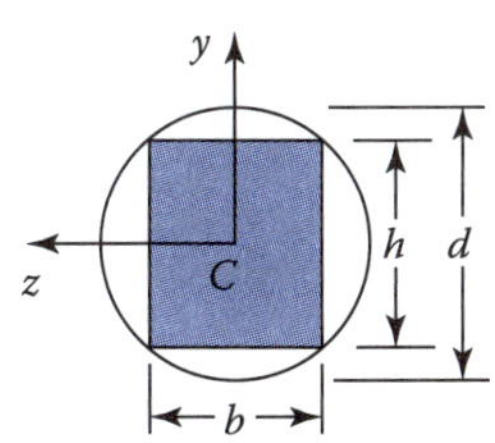

FIGURE P3.13

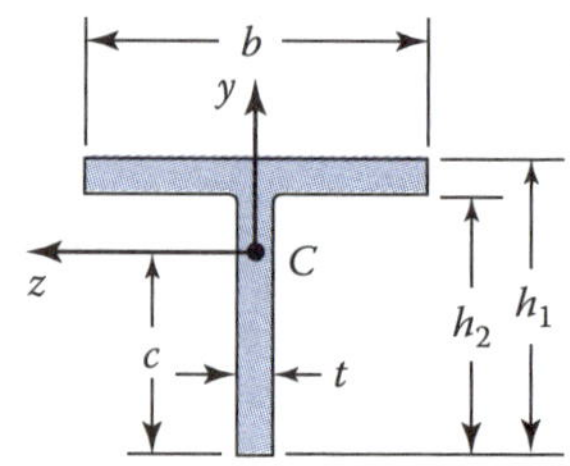

FIGURE P3.14

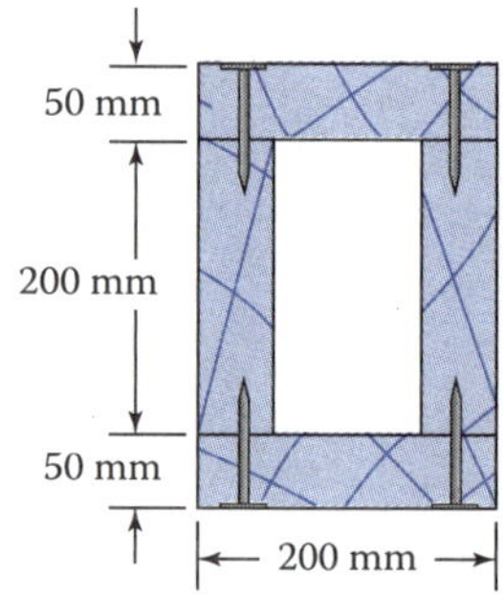

FIGURE P3.15

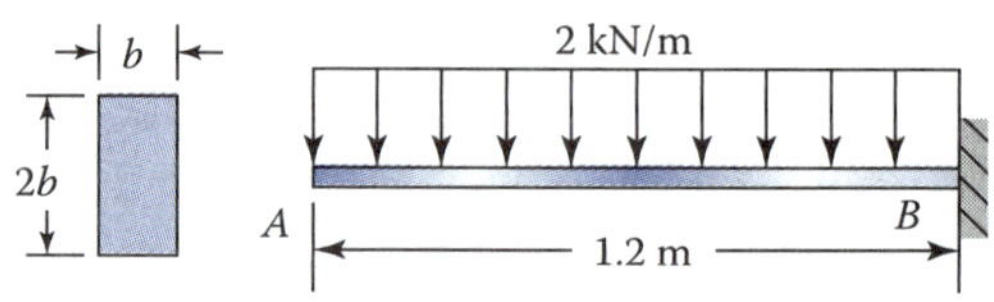

FIGURE P3.16

3.17 Select the S shape of a simply supported 6 m long beam subjected a uniform load of intensity 50 kN/m, for σ_{all} = 170 MPa and τ_{all} = 100 MPa.

3.18 and **3.19** The beam *AB* has the rectangular cross section of constant width *b* and variable depth *h* (Figures P3.18 and P3.19). Derive an expression for *h* in terms of *x*, *L*, and h_1, as required.

Assumption: The beam is to be of constant strength.

3.20 A wooden beam 8 in. wide × 12 in. deep is reinforced on both top and bottom by steel plates 0.5 in. thick (Figure P3.20). Calculate the maximum bending moment *M* about the *z* axis.

Design Assumptions: The allowable bending stresses in the wood and steel are 1.05 and 18 ksi, respectively. Use $n = E_x/E_w = 20$.

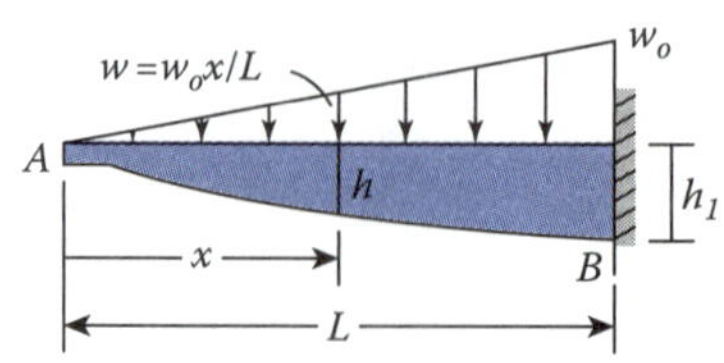

FIGURE P3.18

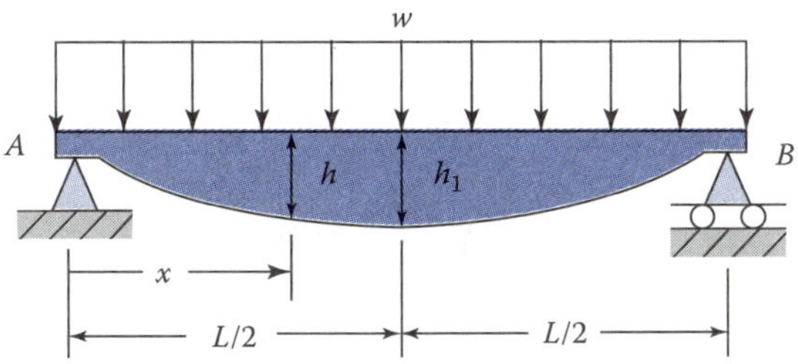

FIGURE P3.19

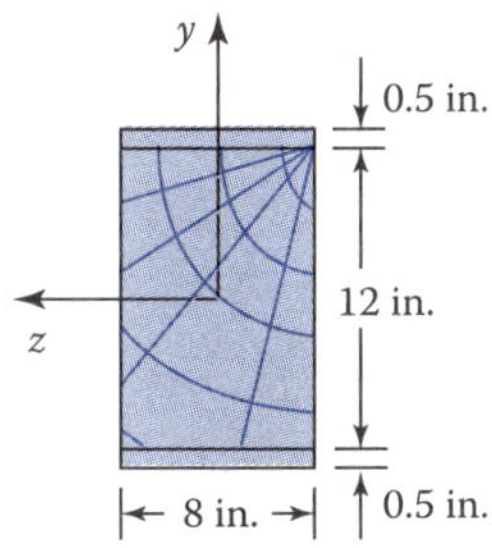

FIGURE P3.20

3.21 A simply supported beam of span length 8 ft carries a uniformly distributed load of 2.5 kip/ft. Determine the required thickness t of the steel plates.

Given: The cross section of the beam is a hollow box with wood flanges ($E_w = 1.5 \times 10^6$ psi) and steel ($E = 30 \times 10^6$ psi), as shown in Figure P3.21.

Assumptions: The allowable stresses are 19 ksi for the steel and 1.1 ksi for the wood.

3.22 and **3.23** For the composite beam with cross section as shown (Figures P3.22 and P3.23), determine the maximum permissible value of the bending moment M about the z axis.

Given: $(\sigma_b)_{\text{all}} = 120 \text{ MPa}, \quad (\sigma_s)_{\text{all}} = 140 \text{ MPa}$

$$E_b = 100 \text{ GPa}, \quad E_x = 200 \text{ GPa}$$

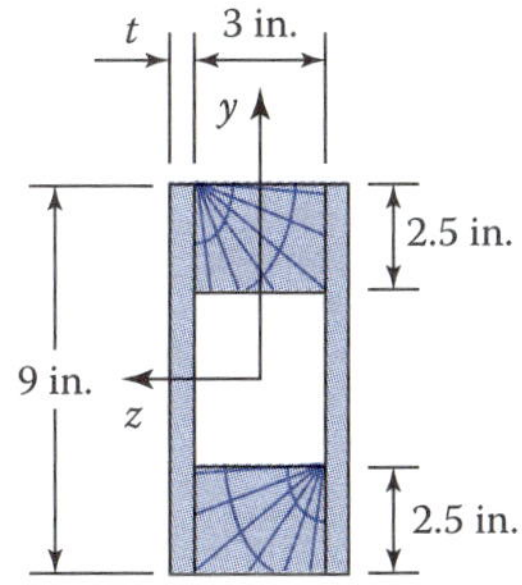

FIGURE P3.21

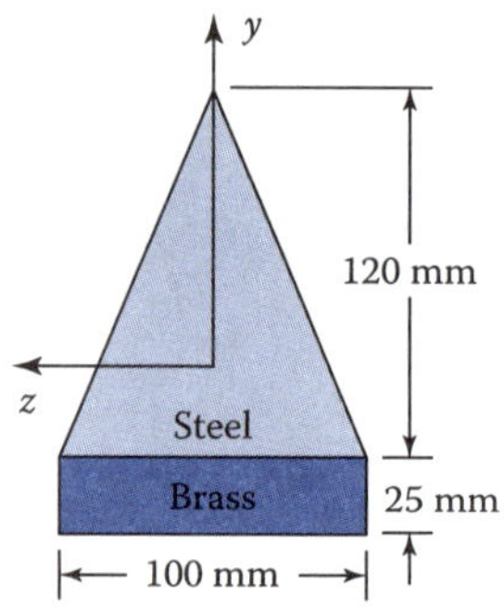

FIGURE P3.22

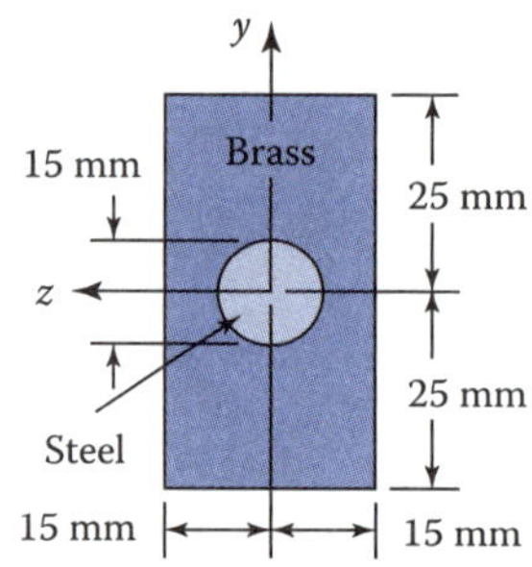

FIGURE P3.23

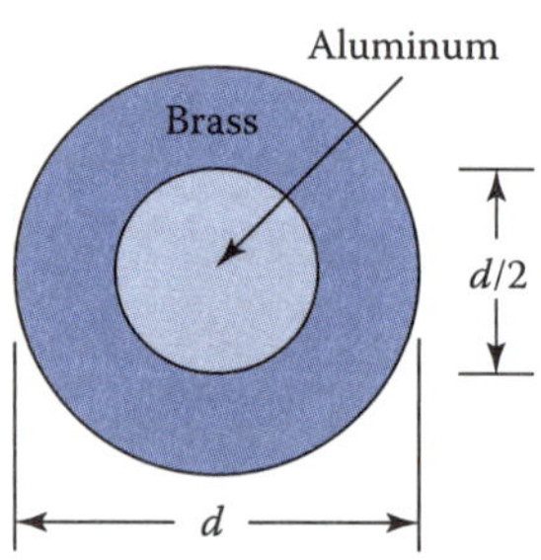

FIGURE P3.24

3.24 A round brass tube of outside diameter d and an aluminum core of diameter $d/2$ are bonded together to form a composite beam (Figure P3.24). Determine the maximum bending moment M that can be carried by the beam, in terms of E_b, E_s, σ_b, and d, as required. What is the value of M for $E_b = 15 \times 10^6$ psi, $E_a = 10 \times 10^6$ psi, $\sigma_b = 50$ ksi, and $d = 2$ in.?

Design Requirement: The allowable stress in the brass is σ_b.

Sections 3.9 and 3.10

3.25 The state of stress at a point in a loaded machine component is represented in Figure P3.25. Determine

a. The normal and shear stresses acting on the indicated inclined plane *a-a*

b. The principal stresses

Sketch results on properly oriented elements.

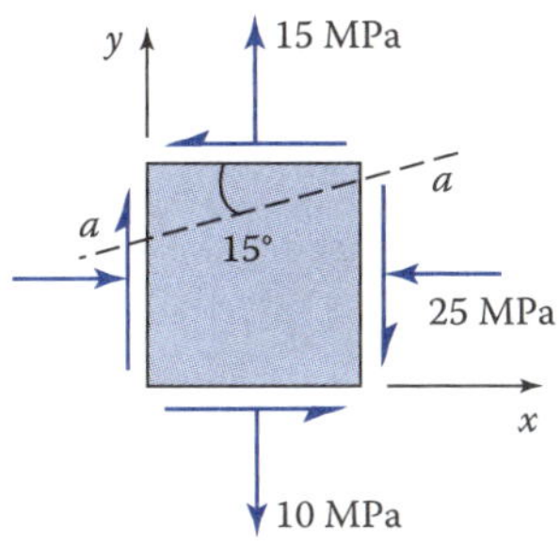

FIGURE P3.25

3.26 At point A on the upstream face of a dam (Figure P3.26), the water pressure is –70 kPa, and the measured tensile stress parallel to this surface is 30 kPa. Calculate

a. The stress components σ_x, σ_y, and τ_{xy}

b. The maximum shear stress

Sketch the results on a properly oriented element.

3.27 The stress acting uniformly over the sides of a skewed plate is shown in Figure P3.27. Determine

a. The stress components on a plane parallel to a-a

b. The magnitude and orientation of principal stresses

Sketch the results on properly oriented elements.

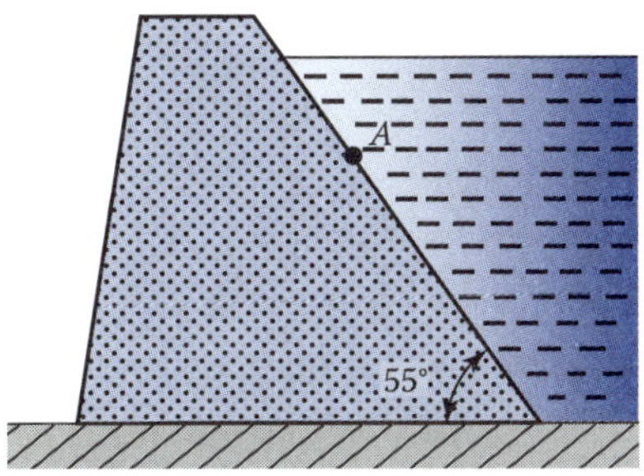

FIGURE P3.26

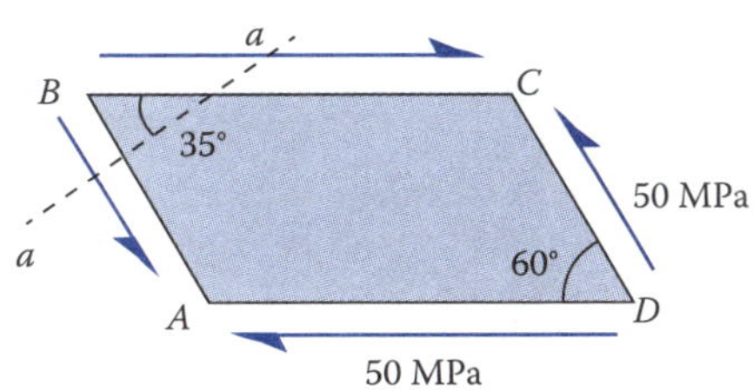

FIGURE P3.27

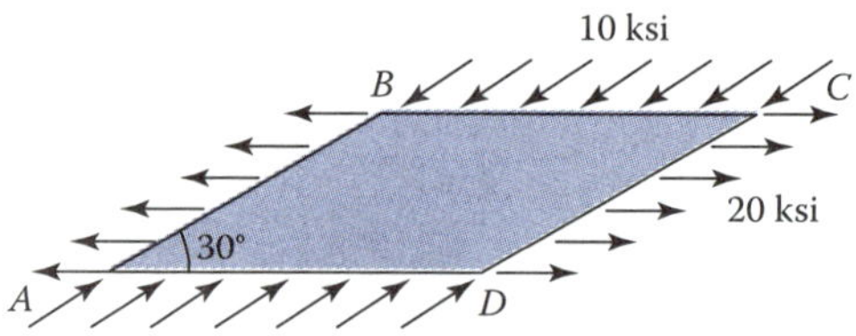

FIGURE P3.29

3.28 A thin skewed plate is depicted in Figure P3.27. Calculate the change in length of

a. The edge *AB*

b. The diagonal *AC*

Given: $E = 200$ GPa, $\nu = 0.3$, $AB = 40$ mm, and $BC = 60$ mm

3.29 The stresses acting uniformly at the edges of a thin skewed plate are shown in Figure P3.29. Determine

a. The stress components σ_x, σ_y, and τ_{xy}

b. The maximum principal stresses and their orientations

Sketch the results on properly oriented elements.

3.30 For the thin skewed plate shown in Figure P3.29, determine the change in length of the diagonal *BD*.

Given: $E = 30 \times 10^6$ psi, $\nu = \frac{1}{4}$, $AB = 2$ in., and $BC = 3$ in.

3.31 The stresses acting uniformly at the edges of a wall panel of a flight structure are depicted in Figure P3.31. Calculate the stress components on planes parallel and perpendicular to *a-a*. Sketch the results on a properly oriented element.

3.32 A rectangular plate is subjected to uniformly distributed stresses acting along its edges (Figure P3.32). Determine

The normal and shear stresses on planes parallel and perpendicular to *a-a*

a. The maximum shear stress

Sketch the results on properly oriented elements.

3.33 For the plate shown in Figure P3.32, calculate the change in the diagonals *AC* and *BD*.

Given: $E = 210$ GPa, $\nu = 0.3$, $AB = 50$ mm, and $BC = 75$ mm

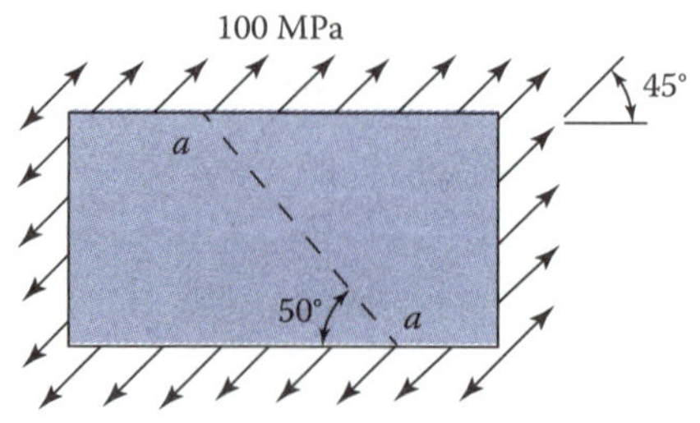

FIGURE P3.31

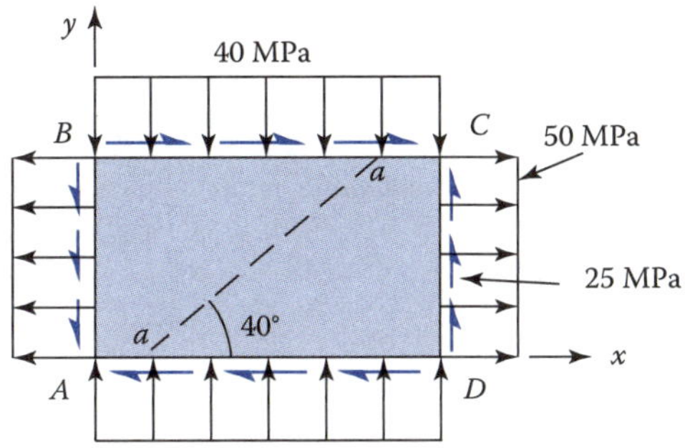

FIGURE P3.32

3.34 A cylindrical pressure vessel of diameter $d = 3$ ft and wall thickness $t = 1/8$ in. is simply supported by two cradles as depicted in Figure P3.34. Calculate, at points A and C on the surface of the vessel,

a. The principal stresses

b. The maximum shear stress

Given: The vessel and its contents weigh 84 lb/ft of length, and the contents exert a uniform internal pressure of $p = 6$ psi on the vessel.

3.35 Redo Problem 3.34, considering point B on the surface of the vessel.

3.36 Calculate and sketch the normal stress acting perpendicular and shear stress acting parallel to the helical weld of the hollow cylinder loaded as depicted in Figure P3.36.

3.37 A 40 mm wide × 120 mm deep bracket supports a load of $P = 30$ kN (Figure P3.37). Determine the principal stresses and maximum shear stress at point A. Show the results on a properly oriented element.

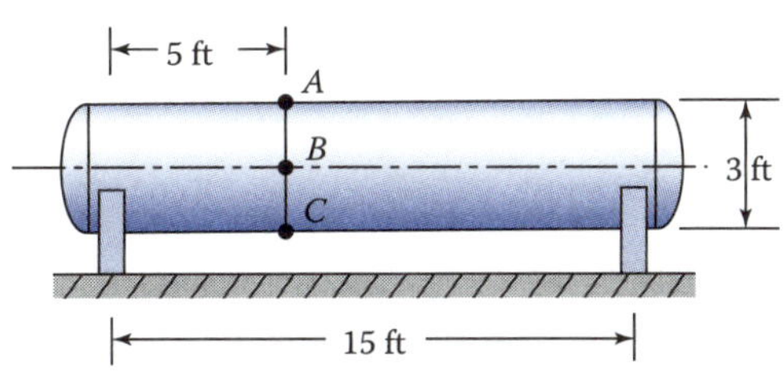

FIGURE P3.34

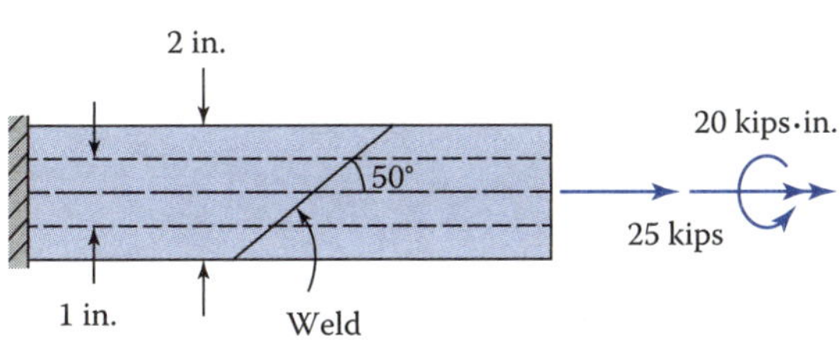

FIGURE P3.36

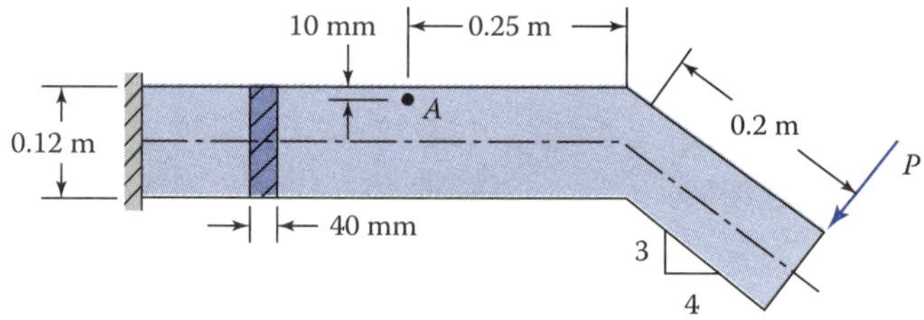

FIGURE P3.37

3.38 A link having a T section is subjected to an eccentric load P as illustrated in Figure P3.38. Compute at section A-B the maximum normal stress.

3.39 Figure P3.39 shows an eccentrically loaded bracket of $b \times h$ rectangular cross section. Find the maximum normal stress.

Given: $b = 25$ mm, $h = 100$ mm, $P = 50$ kN

3.40 What is the largest load P that the bracket of Figure P3.39 can support?

Given: $h = 6b = 150$ mm, $\sigma_{all} = 120$ MPa

3.41 A pipe of 120 mm outside diameter and 10 mm thickness is constructed with a helical weld making an angle of 45° with the longitudinal axis, as shown in Figure P3.41. What is the largest torque T that may be applied to the pipe?

Given: Allowable tensile stress in the weld, $\sigma_{all} = 80$ MPa

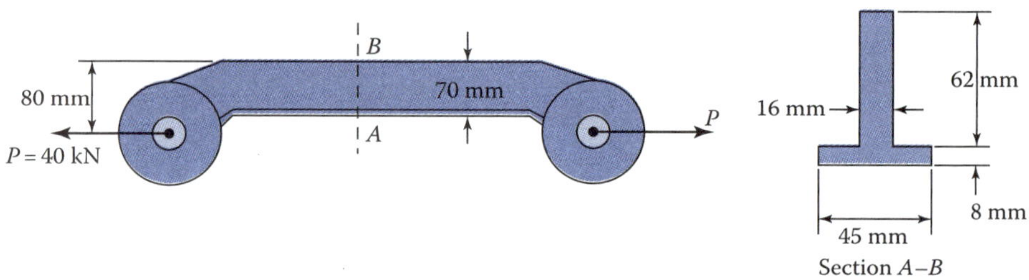

FIGURE P3.38

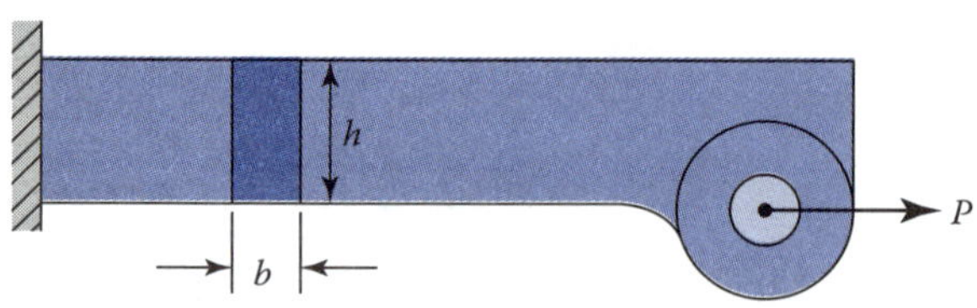

FIGURE P3.39

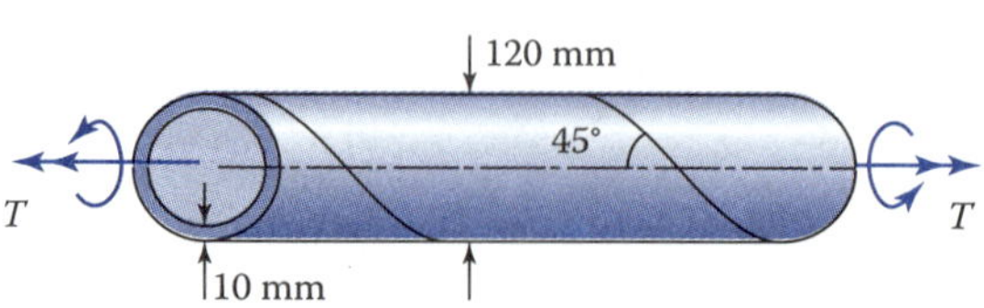

FIGURE P3.41

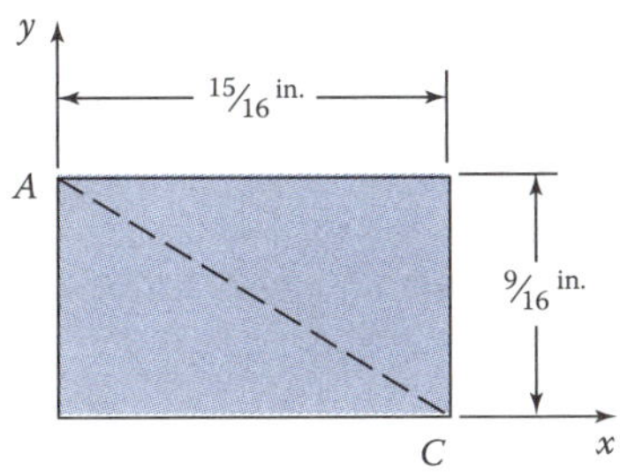

FIGURE P3.43

3.42 The strain at a point on a loaded shell has components $\varepsilon_x = 500\ \mu$, $\varepsilon_y = 800\ \mu$, $\varepsilon_z = 0$, and $\gamma_{xy} = 350\ \mu$. Determine

a. The principal strains

b. The maximum shear stress at the point

Given: $E = 70$ GPa and $\nu = 0.3$

3.43 A thin rectangular steel plate shown in Figure P3.43 is acted on by a stress distribution, resulting in the uniform strains $\varepsilon_x = 200\ \mu$ and $\gamma_{xy} = 400\ \mu$. Calculate

a. The maximum shear strain

b. The change in length of diagonal *AC*

3.44 The strain at a point in a loaded bracket has components $\varepsilon_x = 50\ \mu$, $\varepsilon_y = 250\ \mu$, and $\gamma_{xy} = -150\ \mu$. Determine the principal stresses.

Assumptions: The bracket is made of a steel of $E = 210$ GPa and $\nu = 0.3$.

3.W Review the website at www.measurementsgroup.com. Search and identify

a. Websites of three strain gage manufacturers

b. Three grid configurations of typical foil electrical resistance strain gages

3.45 A thin-walled cylindrical tank of 500 mm radius and 10 mm wall thickness has a welded seam making an angle of 40° with respect to the axial axis (Figure P3.45). What is the allowable value of p?

Given: The tank carries an internal pressure of p and an axial compressive load of $P = 20\pi$ kN applied through the rigid end plates.

Assumption: The normal and shear stresses acting simultaneously in the plane of welding are not to exceed 50 and 20 MPa, respectively.

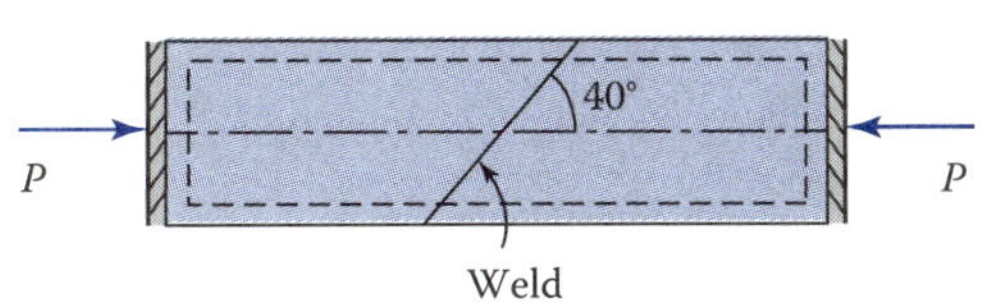

FIGURE P3.45

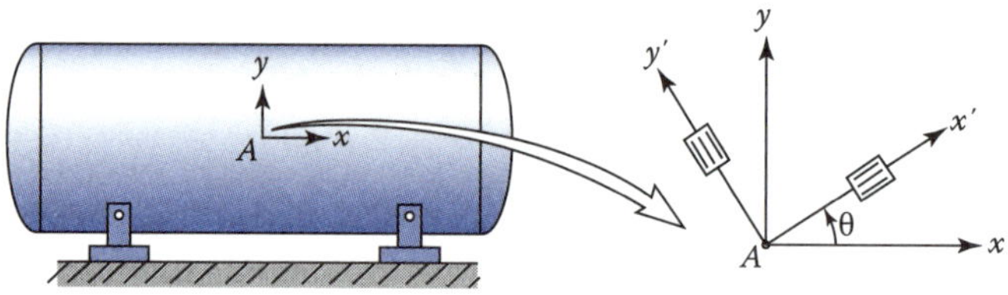

FIGURE P3.46

Sections 3.11 through 3.17

3.46 At point A on the surface of a steel vessel, a strain gage measures $\varepsilon_{x'}$ and $\varepsilon_{y'}$ in the x' and y' directions at an angle θ to the x and y axes, respectively (Figure P3.46). Find

a. Strain components ε_x, ε_y, and $\gamma_{x'y'}$

b. Poisson's ratio ν for the vessel

Given: $\varepsilon_{x'} = 240$, $\varepsilon_{y'} = 410$, $\gamma_{xy} = 0$, $\theta = 34°$

3.47 The strain measurements from a 60° rosette mounted at point A on a loaded C-clamp, a portion depicted in Figure P3.47, are

$$\varepsilon_a = 880\ \mu, \quad \varepsilon_b = 320\ \mu, \quad \varepsilon_c = -60\ \mu$$

Find the magnitudes and directions of principal strains.

3.48 An ASTM-A242 high-strength steel shaft of radius c is subjected to a torque T (Figure P3.48). A strain gage placed at point A measures the strain ε_ϕ at an angle ϕ to the axis of the shaft. Compute the value of torque T.

Given: $c = 1\frac{3}{4}$ in., $G = 11.5 \times 10^6$ psi, $\varepsilon_\phi = 600\ \mu$, $\phi = 25°$

3.49 During a static test, the strain readings from a 45° rosette (Figure P3.49) mounted at point A on an aircraft panel are as follows:

$$\varepsilon_a = -300\ \mu, \quad \varepsilon_b = -375\ \mu, \quad \varepsilon_c = 150\ \mu$$

Determine the magnitudes and directions of principal strains.

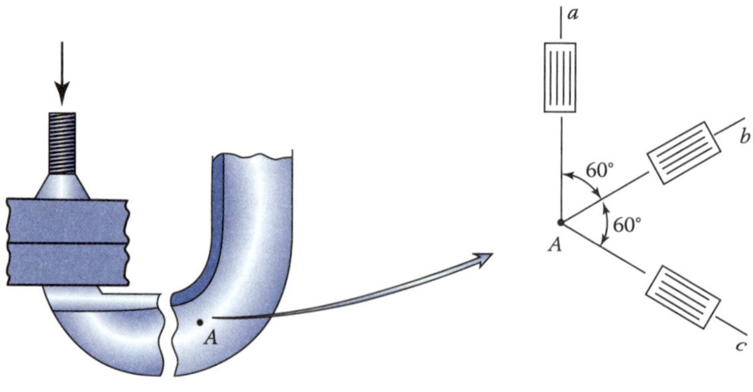

FIGURE P3.47

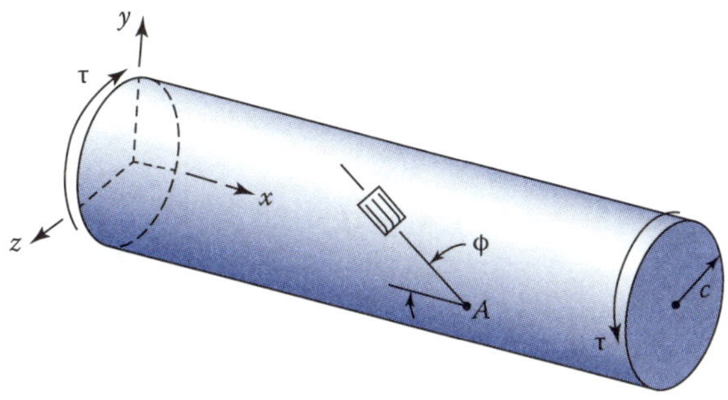

FIGURE P3.48

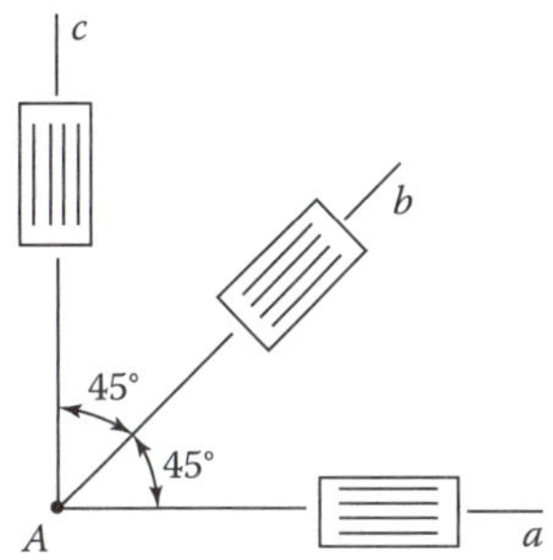

FIGURE P3.49

3.50 The 15 mm thick metal bar is to support an axial tensile load of 25 kN as shown in Figure P3.50 with a factor of safety of n = 1.9 (see Appendix C). Design the bar for minimum allowable width h.

Assumption: The bar is made of a relatively brittle metal having S_y = 150 MPa.

3.51 Calculate the largest load P that may be carried by a relatively brittle flat bar consisting of two portions, both 12 mm thick and, respectively, 30 and 45 mm wide, connected by fillets of radius r = 6 mm (see Figure C.1).

Given: S_y = 210 MPa and a factor of safety of n = 1.5

3.52 A steel symmetrically filleted plate with a central hole and uniform thickness t is under an axial load P (Figure P3.52). Compute the value of the maximum stress at both the hole and the fillet.

Given: d = 15 mm, D = 90 mm, r = 7.5 mm, t = 10 mm, P = 12 kN

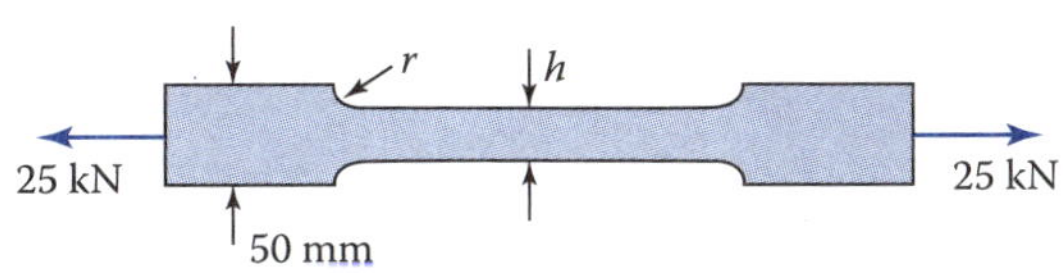

FIGURE P3.50

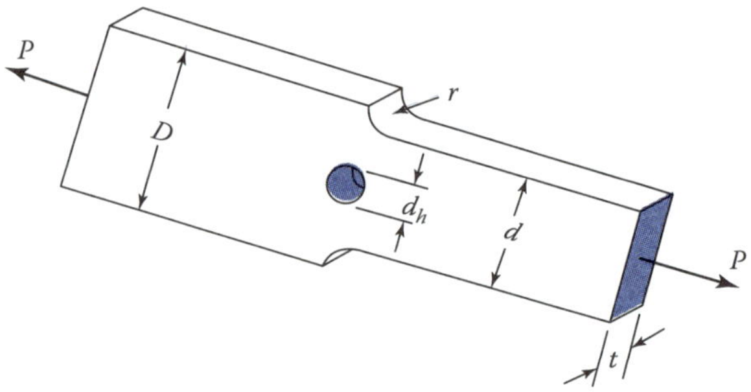

FIGURE P3.52

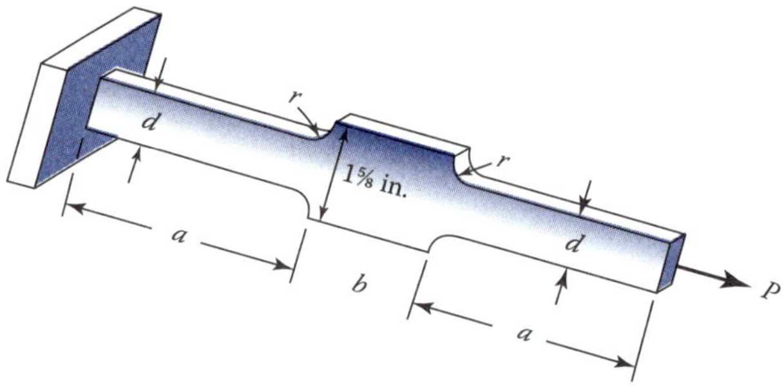

FIGURE P3.53

3.53 What are the *full-fillet* radius r and width d of the steel plate under tension shown in Figure P3.53? Use the maximum permissible stress of σ_{max} and permissible nominal stress in the reduced section of σ_{nom}.

Given: $D/d = 1.5$, $\sigma_{max} = 26$ ksi, $\sigma_{nom} = 16$ ksi

3.54 For the ½ in. thick *full-fillet* ASTM-A242 high-strength steel bar of Figure P3.53, what is the value of maximum axial load P that can be applied without causing permanent deformation?

Given: $r/d = 0.2$ and $S_y = 50$ ksi (from Table B.1)

3.55 Consider a point in a loaded machine component subjected to the 3D state of stress represented in Figure P3.55. Find, using the Mohr's circle,

a. The principal stresses

b. The maximum shear stress

Given: $\sigma_x = 24$ ksi, $\sigma_y = 12$ ksi, $\tau_{xy} = 6$ ksi, $\sigma_z = -3$ ksi

3.56 Rework Problem 3.55 for the case in which the state of stress is as follows:

Given: $\sigma_x = 50$ MPa, $\sigma_y = 0$, $\tau_{xy} = 25$ MPa, $\sigma_z = -60$ MPa

3.57 Redo Problem 3.55 knowing that the state of stress is represented by

Given: $\sigma_x = 10$ ksi, $\sigma_y = 2$ ksi, $\tau_{xy} = -8$ ksi,, $\sigma_z = 5$ ksi

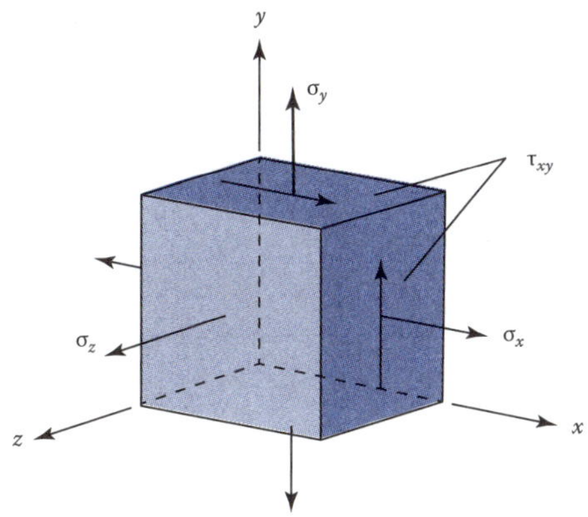

FIGURE P3.55

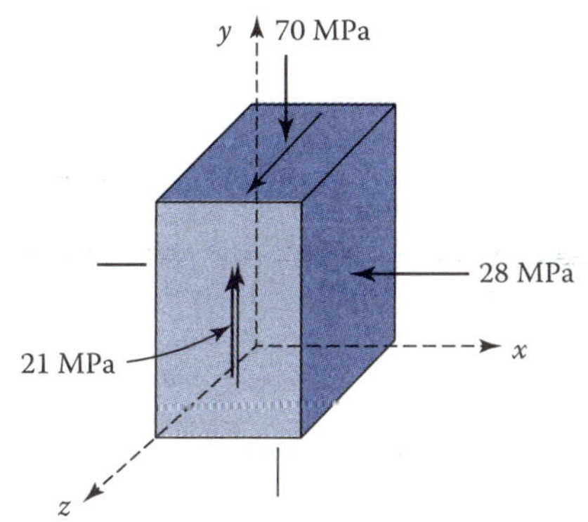

FIGURE P3.58

3.58 The 3D state of stress at a point in a loaded frame is represented in Figure P3.58. Determine

a. The principal stresses, using Mohr's circle
b. The octahedral shearing stresses and maximum shearing stress

3.59 At a point in a structural member, stresses with respect to an x, y, z coordinate system are

$$\begin{bmatrix} -10 & 0 & -8 \\ 0 & 2 & 0 \\ -8 & 0 & 2 \end{bmatrix} \text{ksi}$$

Calculate

a. The magnitude and direction of the maximum principal stress
b. The maximum shear stress
c. The octahedral stresses

3.60 The state of stress at a point in a member relative to an *x*, *y*, *z* coordinate system is

$$\begin{bmatrix} 9 & 0 & 0 \\ 0 & 12 & 0 \\ 0 & 0 & -18 \end{bmatrix} \text{ksi}$$

Determine

a. The maximum shear stress

b. The octahedral stresses

3.61 At a critical point in a loaded component, the stresses with respect to an *x*, *y*, *z* coordinate system are

$$\begin{bmatrix} 42.5 & 0 & 0 \\ 0 & 5.26 & 0 \\ 0 & 0 & -7.82 \end{bmatrix} \text{ksi}$$

Determine the normal stress σ and the shear stress τ on a plane whose outer normal is oriented at angles of 40°, 60°, and 66.2° relative to the *x*, *y*, and *z* axes, respectively.

4

Deflection and Impact

4.1 Introduction

Strength and stiffness are considerations of basic importance to the engineer. The stress level is frequently used as a measure of strength. Stress in members under various loads was discussed in Chapter 3. We now turn to deflection, the analysis of which is as important as that of stress. Moreover, deflections must be considered in the design of statically indeterminate systems, although we are interested only in the forces or stresses.

Stiffness relates to the ability of a part to resist deflection or deformation. Elastic deflection or stiffness, rather than stress, is frequently the controlling factor in the design of a member. The deflection, for example, may have to be kept within limits so that certain clearances between components are maintained. Structures such as machine frames must be extremely rigid to maintain manufacturing accuracy. Most components may require great stiffness to eliminate vibration problems. We begin by developing basic expressions relative to deflection and stiffness of variously loaded members using the equilibrium approaches. The integration, superposition, and moment-area methods are discussed. Then, the *impact* or shock loading and bending of plates are treated. The theorems based upon work–energy concepts, classic methods, and finite element analysis (FEA) for determining the displacement on members are considered in the chapters to follow.

4.1.1 Comparison of Various Deflection Methods

When one approach is preferred over another, the advantages of each technique may be briefly summarized as follows. The governing differential equations for beams on integration give the solution for deflection in a problem. However, it is best to limit their application to prismatic beams: otherwise, considerable complexities arise. In practice, the deflection of members subjected to several or complicated loading conditions is often synthesized from simpler loads, using the principle of superposition.

The dual concepts of strain energy and complementary energy provide the basis for some extremely powerful methods of analysis, such as Castigliano's theorem and its various forms. These approaches may be employed very effectively for finding deflection due to applied forces and are not limited at all to linearly elastic structures. Similar problems are treated by the principles of virtual work and minimum potential energy for obtaining deflections or forces caused by any kind of deformation. They are of great importance in the matrix analysis of structures and in finite elements. The moment-area method, a specialized procedure, is particularly convenient if deflection of only a few points on a beam or frame is desired. It can be used to advantage in the solution of statically indeterminate problems as a check. An excellent insight into the kinematics is obtained by applying this technique. The FEA is perfectly general and can be used for the analysis of statically indeterminate as well as determinate, both linear and nonlinear, problems.

4.2 Deflection of Axially Loaded Members

Here, we are concerned with the elongation or contraction of slender members under axial loading. The axial stress in these cases is assumed not to exceed the proportional limit of the linearly elastic range of the material. The definitions of normal stress and normal strain and the relationship between the two, given by Hooke's law, are used.

Consider the deformation of a prismatic bar having a cross-sectional area A, length L, and modulus of elasticity E, subjected to an axial load P (see Figure 3.1a). The magnitudes of the axial stress and axial strain at a cross section are found from $\sigma_x = P/A$ and $\varepsilon_x = \sigma_x/E$, respectively. These results are combined with $\varepsilon_x = \delta/L$ and integrated over the length L of the bar to give the following equation for the *deformation* δ of the bar:

$$\delta = \frac{PL}{AE} \tag{4.1}$$

The product AE is known as the *axial rigidity* of the bar. The positive sign indicates elongation. A negative sign would represent contraction. The deformation δ has units of length L. Note that, for tapered bars, the foregoing equation gives results of acceptable accuracy provided the angle between the sides of the rod is no larger than 20° [1].

Most of the force–displacement problems encountered in this book are linear, as in the preceding relationship. The *spring rate*, also known as spring constant or stiffness, of an axially loaded bar is then

$$k = \frac{P}{\delta} = \frac{AE}{L} \tag{4.2}$$

The units of k are often kilonewtons per meter or pounds per inch. Spring rate, a deformation characteristic, plays a significant role in the design of members.

A change in temperature of $\Delta T°$ causes a strain $\varepsilon_t = \alpha\Delta T$, defined by Equation 1.21, where α represents the coefficient thermal expansion. In an elastic body, thermal axial deformation caused by a uniform temperature is therefore

$$\delta_t = \alpha(\Delta T)L \tag{4.3}$$

The thermal strain and deformation usually are positive if the temperature increases and negative if it decreases.

Example 4.1: Analysis of a Duplex Structure

A steel rod of cross-sectional area A_s and modulus of elasticity E_s has been placed inside a copper tube of cross-sectional area A_c and modulus of elasticity E_c (Figure 4.1a). Determine the axial shortening of this system of two members, sometimes called an isotropic duplex structure, when a force P is exerted on the end plate as shown.

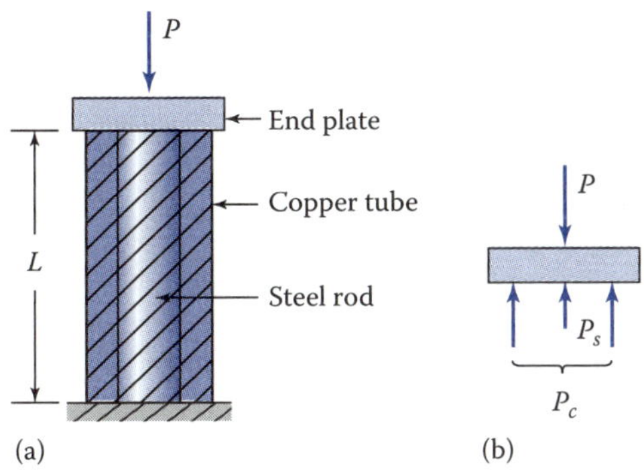

FIGURE 4.1
Example 4.1.

Assumptions: Members have the same length L. The end plate is rigid.

Solution

The forces produced in the rod and in the tube are designated by P_s and P_c, respectively.

Statics: The equilibrium condition is applied to the free body of the end plate (Figure 4.1b):

$$P_c + P_s = P \tag{a}$$

This is the only equilibrium equation available, and since it contains two unknowns (P_c and P_s), the structure is statically indeterminate to the first degree (see Section 1.8).

Deformations: Through the use of Equation 4.1, the shortening of the members are

$$\delta_c = \frac{P_c L}{A_c E_c}, \quad \delta_s = \frac{P_s L}{A_s E_s}$$

Geometry: Axial deformation of the copper tube is equal to that of the steel rod:

$$\frac{P_c L}{A_c E_c} = \frac{P_s L}{A_s E_s} \tag{b}$$

Solution of Equations (a) and (b) gives

$$P_c = \frac{(A_c E_c)P}{A_c E_c + A_s E_s}, \quad P_s = \frac{(A_s E_s)P}{A_c E_c + A_s E_s} \tag{4.4}$$

The foregoing equation show that the forces in the members are proportional to the axial rigidities.

Compressive stresses σ_c in copper and σ_s in steel are found by dividing P_c and P_s by A_c and A_s, respectively. Then, applying Hooke's law together with Equation 4.4, we obtain the compressive strain

$$\varepsilon = \frac{P}{A_c E_c + A_s E_s} \tag{4.5}$$

The shortening of the assembly is therefore $\delta = \varepsilon L$.

Comments: Equation 4.5 indicates that the strain equals the applied load divided by the sum of the axial rigidities of the members. Composite duplex structures are treated in Chapter 16.

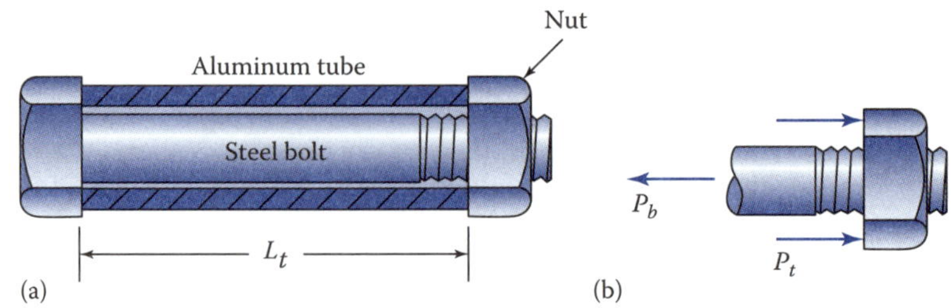

FIGURE 4.2
Examples 4.2 and 4.3.

Example 4.2: Analysis of Bolt–Tube Assembly

In the assembly of the aluminum tube (cross-sectional area A_t, modulus of elasticity E_t, length L_t) and steel bolt (cross-sectional area A_b, modulus of elasticity E_b) shown in Figure 4.2a, the bolt is single threaded, with a 2 mm pitch. If the nut is tightened one-half turn after it has been fitted snugly, calculate the axial forces in the bolt and tubular sleeve.

Given: $A_t = 300\text{ mm}^2$, $E_t = 70\text{ GPa}$, $L_t = 0.6\text{ m}$, $A_b = 600\text{ mm}^2$, and $E_b = 200\text{ GPa}$

Solution

The forces in the bolt and in the sleeve are denoted by P_b and P_t, respectively.

Statics: The only equilibrium condition available for the free body of Figure 4.2b gives

$$P_b = P_t$$

That is, the compressive force in the sleeve is equal to the tensile force in the bolt. The problem is therefore statically indeterminate to the first degree.

Deformations: Using Equation 4.1, we write

$$\delta_b = \frac{P_b L_b}{A_b E_b}, \quad \delta_t = \frac{P_t L_t}{A_t E_t} \tag{c}$$

where

δ_b is the axial extension of the bolt
δ_t represents the axial contraction of the tube

Geometry: The deformations of the bolt and tube must be equal to $\Delta = 0.002/2 = 0.001$ m, and the movement of the nut on the bolt must be

$$\delta_b + \delta_t = \Delta$$
$$\frac{P_b L_b}{A_b + E_b} + \frac{P_t L_t}{A_t E_t} = \Delta \tag{4.6}$$

Setting $P_b = P_t$ and $L_b = L_t$, the preceding equation becomes

$$P_b\left(\frac{1}{A_b E_b} + \frac{1}{A_t E_t}\right) = \frac{\Delta}{L_t} \tag{4.7}$$

Introducing the given data, we have

$$P_b\left[\frac{1}{600(200)10^3}+\frac{1}{300(70)10^3}\right]=\frac{0.001}{0.6}$$

Solving, $P_b = 29.8$ kN.

Example 4.3: Thermal Stresses in a Bolt–Tube Assembly

Determine the axial forces in the assembly of bolt and tube (Figure 4.2a), after a temperature rise of ΔT.

Given: $\Delta T = 100°C$, $\alpha_b = 11.7 \times 10^{-6}/°C$, and $\alpha_t = 23.2 \times 10^{-6}/°C$

Assumptions: The data presented in the preceding example remain the same.

Solution

Only force–deformation relations, Equation (c), change from Example 4.2. Now the expressions for the *extension* of the bolt and the *contraction* of the sleeve are

$$\delta_b = \frac{P_b L_b}{A_b E_b} + \alpha_b(\Delta T)L_b$$
$$\delta_t = \frac{P_t L_t}{A_t E_t} - \alpha_t(\Delta T)L_t \qquad \text{(d)}$$

Note that, in the foregoing, the minus sign indicates a decrease in tube contraction due to the temperature rise.

We have $L_b = L_t$ and $P_b = P_t$. These, carried into $\delta_b + \delta_t = \Delta$, give

$$P_b\left(\frac{1}{A_b E_b}+\frac{1}{A_t E_t}\right)+(\alpha_b-\alpha_t)\Delta T=\frac{\Delta}{L_t} \qquad (4.8)$$

where, as before, Δ is the movement of the nut on the bolt. Substituting the numerical values into Equation 4.8, we obtain

$$P_b\left[\frac{1}{600(200)10^3}+\frac{1}{300(70)10^3}\right]+(11.7-23.2)10^{-6}(100)=\frac{0.001}{0.6}$$

This yields $P_b = 50.3$ kN.

Comment: The final elongation of the bolt and the contraction of the tube can be calculated by substituting the axial force of 50.3 kN into Equation (d). Interestingly, when the bolt and tube are made of the same material ($\alpha_b = \alpha_t$), the temperature change does not affect the assembly. That is, the forces obtained in Example 4.2 still hold.

Example 4.4: Deflections of a Three-Bar Device

The rigid member *BC* is attached by the 12 mm diameter rod *AB* and the 10 mm diameter rod *CD* (Figure 4.3a). Each rod is made of cold-rolled yellow brass with yield strength S_y and elastic modulus of elasticity *E*. What is the displacement of point *E* of the bar caused by a vertical load *P* applied at this point?

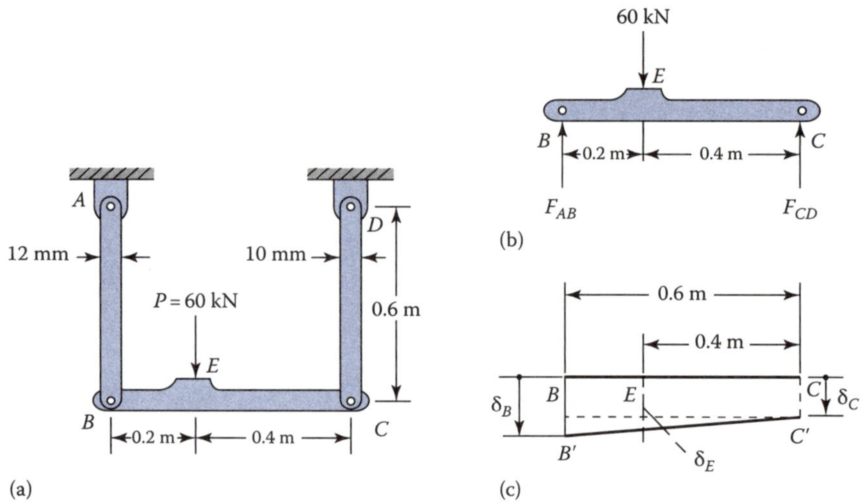

FIGURE 4.3
Example 4.4. (a) A three-bar assembly, (b) free-body diagram of bar *BC*, and (c) displacement diagram of centerline of bar *BC*.

Given: S_y = 435 MPa, P = 60 kN, E = 105 GPa (by Table B.1)

Assumptions: Failure will not occur at the pin-connected joints. Both rods deform uniformly.

Solution

The cross-sectional areas of the bars equal

$$A_{AB} = \frac{\pi}{4}(12)^2 = 113.1\ \text{mm}^2 \quad A_{CD} = \frac{\pi}{4}(10)^2 = 78.5\ \text{mm}^2.$$

Free body of bar BC. Applying equations of equilibrium to Figure 4.3b, the tensile forces in each bar are

$$\sum M_C = 0: \quad -F_{AB}(0.6) + 60(0.4) = 0 \quad F_{AB} = 40\ \text{kN}$$

$$\sum F_y = 0: \quad -60 + 40 + F_{CD} = 0 \quad F_{CD} = 20\ \text{kN}$$

Displacements of B and C. The deflections of the bottom of the links, respectively, are

$$\delta_B = \frac{F_{AB}L_{AB}}{A_{AB}E} = \frac{40(10^3)(0.6)}{(113.1)(105)(10^3)} = 2.02\ \text{mm}$$

$$\delta_C = \frac{F_{CD}L_{CD}}{A_{CD}E} = \frac{20(10^3)(0.6)}{(78.5)(105)(10^3)} = 1.46\ \text{mm}$$

Displacement of E. A diagram showing the centerline displacements of points *B*, *C*, and *E* on the rigid bar is represented in Figure 4.3b. It follows that, by proportion of the shaded triangle, the displacement of point *E* is found as

$$\delta_E = \delta_C + (\delta_B - \delta_C)\left(\frac{0.4}{0.6}\right) = \left[1.46 + (0.56)\left(\frac{2}{3}\right)\right]$$

or

$$\delta_E = 1.83 \text{ mm}$$

Comments: A positive sign means downward displacement. The largest axial stress is in rod *AB*, $\sigma_{AB} = 40(10^3)/113.1 \times 10^{-6} = 354$ MPa < 435 MPa. Therefore, the bar will not deform permanently.

4.3 Angle of Twist of Shafts

In Section 3.5, the concern was with torsion stress. We now treat angular displacement of twisted prismatic bars or shafts. We assume that the entire bar remains elastic. For most structural materials, the amount of twist is small and hence the member behaves as before. But in a material such as rubber, where twisting is large, the basic assumptions must be reexamined.

4.3.1 Circular Sections

Consider a circular prismatic shaft of radius *c*, length *L*, and modulus of elasticity in shear *G* (Figure 3.6). The maximum shear stress τ_{max} and maximum shear strain γ_{max} are related by Hooke's law: $\gamma_{max} = \tau_{max}/G$. Moreover, by the torsion formula, $\tau_{max} - Tc/J$, where *J* is the polar moment of inertia. Substitution of the latter expression into the former results in $\gamma/_{max} = Tc/GJ$. For small deformations, by taking $\tan \gamma_{max} = \gamma_{max}$, we also write $\gamma_{max} = c\phi/L$. These expressions lead to the *angle of twist*, representing the angle through which one end of a cross section of a circular shaft rotates with respect to another:

$$\phi = \frac{TL}{GJ} \tag{4.9}$$

Angle ϕ is measured in radians. The product *GJ* is called the *torsional rigidity* of the shaft. Equation 4.9 can be used for either solid or hollow bars having circular cross sections. We observe that the *spring rate* of a circular torsion bar is given by

$$k = \frac{T}{\phi} = \frac{GJ}{L} \tag{4.10}$$

Typical units of the *k* are kilonewton-meters per radian or pound-inches per radian.

Examining Equation 4.9 implies a method for obtaining the modulus of elasticity in shear G for a given material. A circular prismatic specimen of the material, of known diameter and length, is placed in a torque-testing machine. As the specimen is twisted, increasing the value of the applied torque T, the corresponding values of the angle of twist ϕ between the two ends of the specimen are recorded as a torque-twist diagram. The slope of this curve (T/ϕ) in the linearly elastic region is the quantity GJ/L. From this, the magnitude of G can be calculated.

4.3.2 Noncircular Sections

As pointed out in Section 3.5, determination of stresses and displacements in noncircular members is a difficult problem and beyond the scope of this book. However, the following angle of twist formula for *rectangular bars* is introduced here for convenience:

$$\phi = \frac{TL}{CG} \tag{4.11}$$

where

$$C = \frac{ab^3}{16}\left[\frac{16}{3} - 3.36\frac{b}{a}\left(1 - \frac{b^4}{12a^4}\right)\right] \tag{4.12}$$

In Equation 4.12, a and b denote the wider and narrower sides of the rectangular cross section, respectively. Table 3.1 gives the exact solutions of the angle of twist for a number of commonly encountered cross sections [1,2].

Example 4.5: Determination of Angle of Twist of a Rod with Fixed Ends

A circular brass rod (Figure 4.4a) is fixed at each end and loaded by a torque T at point D. Find the maximum angle of twist.

Given: $a = 20$ in., $b = 40$ in., $d = 1$ in., $T = 500$ lb · in., and $G = 5.6 \times 10^6$ psi

Solution

The reactions at the end are designated by T_A and T_B.

Statics: The only available equation of equilibrium for the free-body diagram of Figure 4.4b yields

$$T_A + T_B = T$$

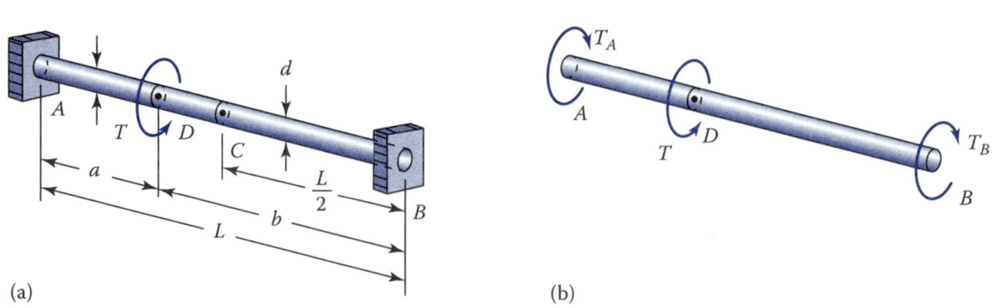

FIGURE 4.4
Example 4.5.

Therefore, the problem is statically indeterminate to the first degree.

Deformations: The angle of twist at section D for the left and right segments of the bar are

$$\phi_{AD} = \frac{T_A a}{GJ}, \quad \phi_{BD} = \frac{T_B b}{GJ} \tag{a}$$

Geometry: The continuity of the bar at section D requires that

$$\phi_{AD} = \phi_{BD} \quad \text{or} \quad T_A a = T_B b \tag{b}$$

Equations (a) and (b) can be solved simultaneously to obtain

$$T_A = \frac{Tb}{L}, \quad T_B = \frac{Ta}{L} \tag{4.13}$$

The maximum angle of rotation occurs at section D. Therefore,

$$\phi_{max} = \frac{T_A a}{GJ} = \frac{Tab}{GJL}$$

Substituting the given numerical values into this equation, we have

$$\phi_{max} = \frac{500(20)40}{5.6\left(10^6\right)\frac{\pi}{32}(1)^4\left(60\right)} = 0.012 \text{ rad} = 0.7°$$

4.4 Deflection of Beams by Integration

Beam deflections due to bending are determined from deformations taking place along a span. Analysis of the deflection of beams is based on the assumptions of the beam theory outlined in Section 3.7. As we see in Section 5.4, for slender members, the contribution of shear to deflection is regarded as negligible, since for static bending problems, the shear deflection represents no more than a few percent of the total deflection. Direct integration and superposition methods for determining elastic beam deflection are discussed in the sections to follow.

Governing the differential equations relating the deflection υ to the internal bending moment M in a linearly elastic beam whose cross section is symmetrical about the plane (xy) of loading is given by [3]

$$\frac{d^2\upsilon}{dx^2} = \frac{M}{EI} \tag{4.14}$$

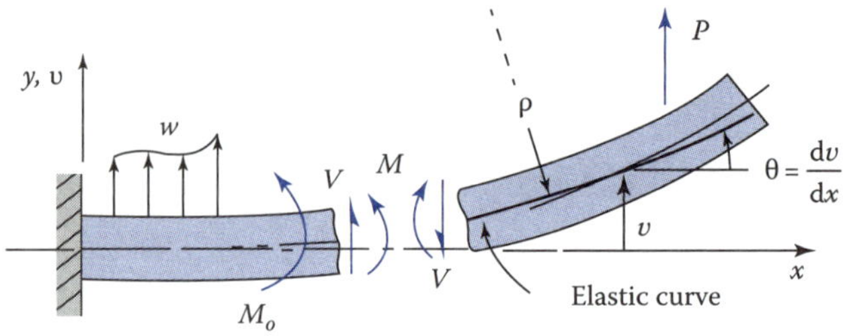

FIGURE 4.5
Positive loads and internal forces.

The quantity EI is called the *flexural rigidity*. The sign convention for applied loading and the internal forces, according to that defined in Section 1.8, is shown in Figure 4.5. The deflection and slope θ (in radians) of the deflection curve are related by the equation

$$\theta = \frac{d\upsilon}{dx} = \upsilon' \tag{4.15}$$

Positive (and negative) θ, like moments, follow the right-hand rule, as depicted in the figure.

As shown in Section 3.6, internal shear force V, bending moment M, and the load intensity w are connected by Equations 3.14 and 3.15. These, combined with Equation 4.14, give the useful sequence of relationships, for constant EI, in the following form:

$$\text{Moment} = M = EI\frac{d^2\upsilon}{dx^2} = EI\upsilon'' \tag{4.16a}$$

$$\text{Shear} = V = EI\frac{d^3\upsilon}{dx^3} = EI\upsilon''' \tag{4.16b}$$

$$\text{Load} = w = EI\frac{d^4\upsilon}{dx^4} = EI\upsilon'''' \tag{4.16c}$$

The deflection υ of a beam can be found by solving any one of the foregoing equations by successive integrations. The choice of equation depends on the ease with which an expression of load, shear, or moment can be formulated and individual preference. The approach to solving the deflection problem beginning with Equations 4.16c or 4.16b is known as the *multiple-integration method*. When Equation 4.16a is used, because two integrations are required to obtain the υ, this is called the *double-integration method*.

The constants of the integration are evaluated using the specified conditions on the ends of the beam, that is, the boundary conditions. Frequently encountered conditions that may apply at the ends ($x = a$) of a beam are shown in Figure 4.6. We see from the figure that the force (static) variables M, V and the geometric (kinematic) variables υ, θ are 0 for common situations.

If the beam has a cross-sectional width b that is large compared to the depth h (i.e., $b \gg h$), the beam is stiffen and the deflection is less than that determined by Equation 4.16 for narrow beams. The large cross-sectional width prevents the lateral expansion and contraction of the material, and the deflection is thereby reduced, as shown in Section 4.9. An improved value for the deflection υ of *wide beams* is obtained by multiplying the result given by the equation for a narrow beam by $(1 - \nu^2)$, where ν is Poisson's ratio.

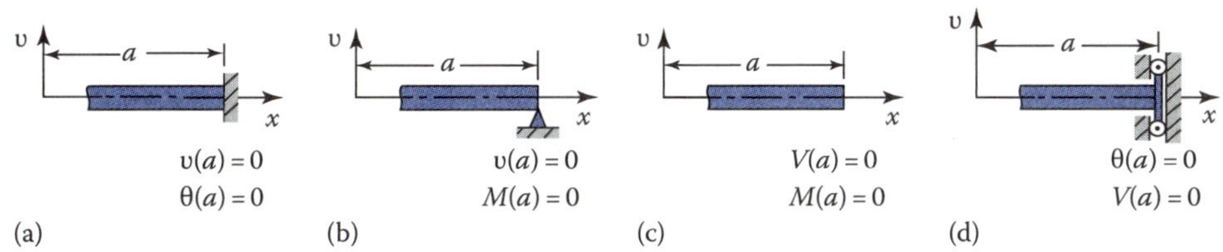

FIGURE 4.6
Boundary conditions: (a) fixed end, (b) simply supported end, (c) free end, and (d) guided or sliding support.

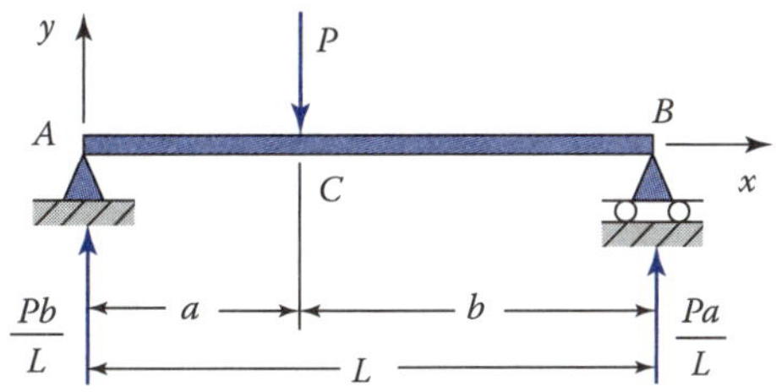

FIGURE 4.7
Example 4.6.

Example 4.6: Simple Beam Deflections by the Double-Integration Method

A simply supported beam is subjected to a concentrated load at a distance a from the left end as shown in Figure 4.7. Develop

a. The expressions for the elastic curve
b. The deflection at point C for the case in which $a = b = L/2$

Solution

The reactions are noted in the figure.

a. The moments for the segments AC and CB of the beam are expressed as

$$M_1 = \frac{Pb}{L}x \quad (0 \le x \le a)$$

$$M_2 = \frac{Pb}{L}x - P(x-a) \quad (a \le x \le L)$$

Double integrations of these equations give the results

For segment AC	For segment CB
$EI\upsilon_1'' = \frac{Pb}{L}x$	$EI\upsilon_2'' = \frac{Pb}{L}x - P(x-a)$
$EI\upsilon_1' = \frac{Pb}{2L}x^2 + c_1$	$EI\upsilon_2' = \frac{Pb}{2L}x^2 - \frac{P}{2}(x-a)^2 + c_3$
$EI\upsilon_1 = \frac{Pb}{6L}x^3 + c_1x + c_2$	$EI\upsilon_2 = \frac{Pb}{6L}x^3 - \frac{P}{6}(x-a)^3 + c_3x + c_4$

From the boundary and the continuity conditions, we easily obtain

$$\upsilon_1'(a) = \upsilon_2'(a): \quad c_1 = c_3$$

$$\upsilon_1(a) = \upsilon_2(a): \quad c_2 = c_4$$

$$\upsilon_1(0) = 0: \quad 0 = c_2, \quad \upsilon_2(L) = 0: \quad c_3 = -\frac{Pb}{6L}\left(L^2 - b^2\right)$$

The elastic curves for the left- and right-hand segments are therefore

$$\begin{aligned} \upsilon_1 &= -\frac{Pbx}{6EIL}\left(L^2 - b^2 - x^2\right) \quad (0 \le x \le a) \\ \upsilon_2 &= -\frac{Pbx}{6EIL}\left(L^2 - b^2 - x^2\right) - \frac{P(x-a)^3}{6EI} \quad (a \le x \le L) \end{aligned} \tag{4.17}$$

Then, through the use of Equation 4.15, the slopes for the two parts of the beam can readily be found.

b. Force P acts at the middle of the beam span ($a = b = L/2$) and hence Equation 4.17 results in

$$\upsilon_{max} = \upsilon_c = -\frac{PL^3}{48EI} \tag{4.18}$$

Comments: The minus sign means that the deflection is downward. In this case, the elastic curve is symmetric about the center of the beam.

4.5 Beam Deflections by Superposition

The elastic deflections (and slopes) of beams subjected to simple loads have been solved and are readily available (see Tables A.9 and A.10). In practice, for combined load configurations, the method of superposition may be applied to simplify the analysis and design. The method is valid whenever displacements are linearly proportional to the applied loads. This is the case if Hooke's law holds for the material and deflections are small.

To demonstrate the method, consider the beam of Figure 4.8a, replaced by the beams depicted in Figure 4.8b and c. At point C, the beam undergoes deflections $(\upsilon)_P$ and $(\upsilon)_M$, due to P and M, respectively. Hence, the deflection υ_C due to combined loading is $\upsilon_C = (\upsilon_C)_P + (\upsilon_C)_M$. From the cases 1 and 2 of Table A.9, we have

$$\upsilon_C = \frac{5PL^3}{48EI} + \frac{ML^2}{8EI} \tag{4.19}$$

Similarly, the deflection and the angle of rotation at any point of the beam can be found by the foregoing procedure.

The method of superposition can be effectively applied to obtain deflections or reactions for *statically indeterminate* beams. In these problems, the redundant reactions are considered unknown loads and the corresponding supports are removed or modified accordingly. Next, superposition is employed: The load diagrams are drawn and expressions are written for the deflections produced by the individual loads (both known and unknown);

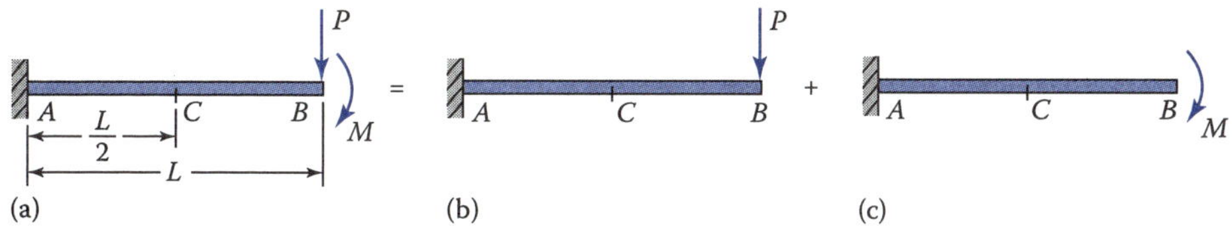

FIGURE 4.8
Deflections of a cantilevered beam.

the redundant reactions are computed by satisfying the geometric boundary conditions. Following this, all other reactions can be found from equations of static equilibrium.

The steps described in the preceding paragraph can be made clearer though the illustration of a beam statically indeterminate to the first degree (Figure 4.9a). Reaction R_B is selected as redundant and treated as unknown load by eliminating the support at B. Decomposition of the loads is shown in Figure 4.9b and c. Deflections due to R_B and the redundant R_B are (see cases 5 and 8 of Table A.9)

$$(\upsilon_B)_w = -\frac{5wL^4}{24EI}, \quad (\upsilon_B)_R = \frac{R_B L^3}{6EI}$$

From geometry of the original beam,

$$\upsilon_B = -\frac{5wL^4}{24EI} + \frac{R_B L^3}{6EI} = 0$$

or

$$R_B = \frac{5}{4} wL \tag{4.20}$$

The remaining reactions are $R_A = R_C = 3wL/8$, as determined by applying the equations of equilibrium. Having the reactions available, deflection can be obtained using the method discussed in the preceding section.

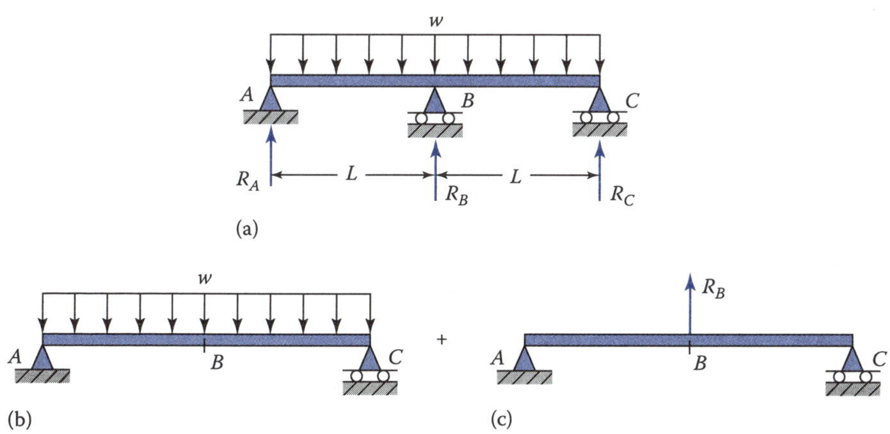

FIGURE 4.9
Deflections of a two-span continuous beam.

Example 4.7: Interaction Force between Two Beams Placed One on Top of the Other

Two simply supported beams are situated one on top of the other as shown in Figure 4.10. The top beam is subjected to a uniformly distributed load of intensity w.

Find:

a. The interaction force R at midspan C acting upward on the beam AB and acting downward on beam DE
b. The maximum moment and deflection in beam AB

Assumption: The beams are supported in such a way that they are to deflect by the same amount at the junction C.

Solution

Both beams are statically indeterminate to the first degree. We select R as redundant and treat it as an unknown load. Considering the two beams in turn and using the data in Table A.9, the deflections at the center are as follows. For *beam AB*, due to load W and owing to R,

$$\upsilon_w = \frac{5w(2a)^4}{384EI_a}, \quad \upsilon_R = -\frac{R(2a)^3}{48EI_a}$$

The total downward deflection is therefore

$$\upsilon_a = \frac{5wa^4}{24EI_a} - \frac{Ra^3}{6EI_a}$$

For *beam DE*, due to R, the downward deflection is

$$\upsilon_b = \frac{Rb^3}{6EI_b}$$

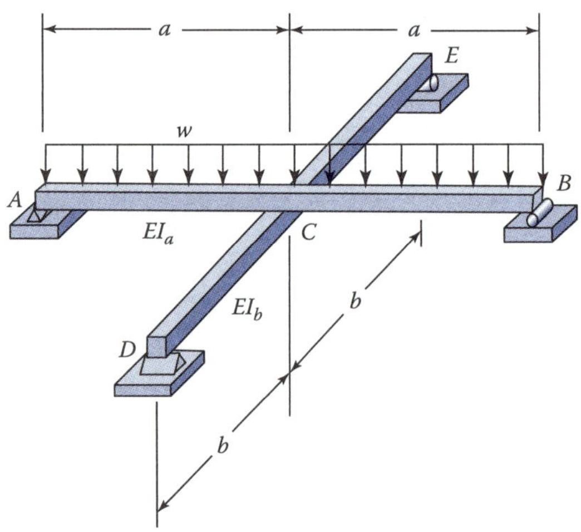

FIGURE 4.10
Example 4.7.

a. Equating the two expressions for deflections $\upsilon_a = \upsilon_b$ and solving for the interaction force, we have

$$R = \frac{1.25wa}{\alpha} \tag{4.21}$$

where

$$\alpha = 1 + \left(\frac{b}{a}\right)^3 \frac{I_a}{I_b}$$

Comment: If beam DE is rigid (i.e., $I_b \to \infty$), then $\alpha = 1$ and $R = 1.25wa$, which is equal to the central reaction for the beam resting on three simple supports.

b. Maximum bending moment and deflection in beam AB occurring at the center are, respectively,

$$M_C = \frac{1}{2}wa^2 - \frac{1}{2}Ra = \frac{1}{2}wa^2\left(1 - \frac{1.25}{\alpha}\right) \tag{4.22a}$$

$$\upsilon_C = \frac{5w(2a)^4}{384EI_a}\left(1 - \frac{1}{\alpha}\right) = \frac{5wa^4}{24EI_a}\left(1 - \frac{1}{\alpha}\right) \tag{4.22b}$$

Comments: We see from the preceding results that, as beam DE is made stiffer by either reducing its span $2b$ or increasing its moment of inertia I_b, the value of α decreases and hence the value of R increases. This decreases the deflection and also reduces the bending moment in the loaded beam AB.

CASE STUDY 4.1 Bolt Cutter Deflection Analysis

Members 2 and 3 of the bolt cutter shown in Figure 3.31 are critically stressed. Determine the deflections employing the superposition method.

Given: The dimensions (in inches) and loading are known from Case Study 3.1. The parts are made of AISI 1080 HR steel having $E = 30 \times 10^6$ psi.

Assumptions: The loading is static. The member 2 can be approximated as a simple beam with an overhang.

Solution

See Figures 3.31 (page 136) and 4.11 and Table A.9.

Member 3. The elongation of this tensile link (Figure 3.31a) is obtained from Equation 4.1. So, due to symmetry in the assembly, the displacement of each end point A is

$$\delta_A = \frac{1}{2}\left(\frac{PL}{AE}\right) = \frac{F_A L_3}{2AE}$$

$$= \frac{128(1.25)}{2\left(\frac{3}{8}\right)\left(\frac{1}{8}\right)\left(30 \times 10^6\right)} = 56.9\left(10^{-6}\right) \text{ in.}$$

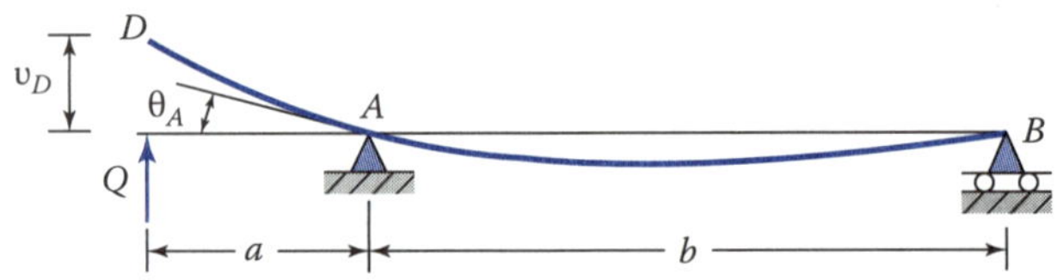

FIGURE 4.11
Deflection of simple beam with an overhang.

Member 2. This jaw is loaded as shown in Figure 3.31b. The deflection of point *D* is made up of two parts: a displacement υ_1 owing to bending of part *DA* acting as a cantilever beam and a displacement υ_2 caused by the rotation of the beam axis at A (Figure 4.11).

The deflection υ_1 at *D* (by case 1 of Table A.9) is

$$\upsilon_1 = \frac{Qa^3}{3EI}$$

The angle θ_A at the support *A* (from case 7 of Table A.9) is

$$\theta_A = \frac{Mb}{3EI}$$

where $M = Qa$. The displacement υ_2 of point *D*, due to only the rotation at *A*, is equal to $\theta_A\, a$, or

$$\upsilon_2 = \frac{Qba^2}{3EI}$$

The total deflection of point *D*, shown in Figure 4.11, $\upsilon_1 + \upsilon_2$, is then

$$\upsilon_D = \frac{Qba^2}{3EI}(a+b)$$

In the foregoing, we have

$$I = \frac{1}{12}t_2h_2^3$$

$$= \frac{1}{12}\left(\frac{3}{16}\right)\left(\frac{3}{8}\right)^3 = 0.824\left(10^{-3}\right)\text{ in.}^4$$

Substitution of the given data results in

$$\upsilon_D = \frac{96\left(1^2\right)(1+3)}{3\left(30\times10^6\right)\left(0.824\times10^{-3}\right)} = 5.18\times10^{-3}\text{ in.}$$

Comment: Only very small deflections are allowed in members 2 and 3 to guarantee the proper cutting stroke, and the values found are acceptable.

4.6 Beam Deflection by the Moment-Area Method

In this section, we consider a semigraphical technique called the *moment-area method* for determining deflections of beams. The approach uses the relationship between the derivatives of the deflection v and the properties of the area of the bending moment diagram. Usually, it gives more rapid solution than integration methods when the deflection and slope at only one point of the beam are required. The moment-area method is particularly effective in the analysis of beams of variable cross sections with uniform or concentrated loading [4].

4.6.1 Moment-Area Theorems

Two theorems form the basis of the moment-area approach. These principles are developed by considering a segment AB of the deflection curve of a beam under an arbitrary loading. The sketches of the M/EI diagram and greatly exaggerated deflection curve are shown in Figure 4.12a. Here, M is the bending moment in the beam and EI represents the flexural rigidity. The changes in the angle $d\theta$ of the tangents at the ends of an element of length dx and the bending moment are connected through Equations 4.14 and 4.15:

$$d\theta = \frac{M}{EI}dx \tag{a}$$

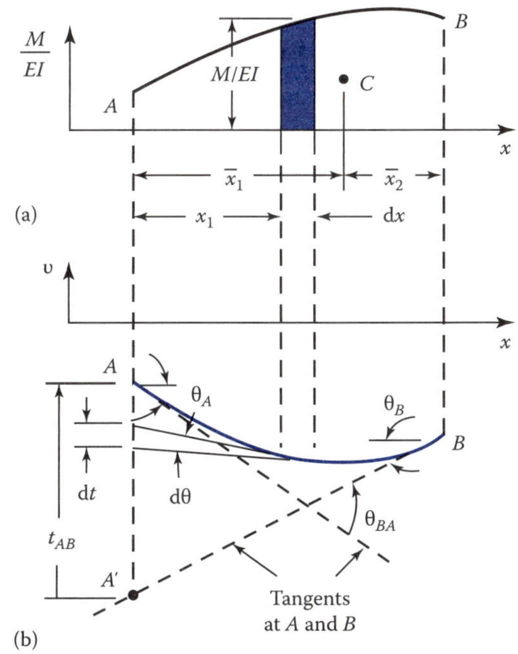

FIGURE 4.12
Moment-area method: (a) M/EI diagram and (b) elastic cure.

The difference in slope between any two points A and B for the beam (Figure 4.12) can be expressed as follows:

$$\theta_{BA} = \theta_B - \theta_A = \int_A^B \frac{Mdx}{EI} = \left[\text{area of } \frac{M}{EI} \text{ diagram between } A \text{ and } B\right] \tag{4.23}$$

This is called *the first moment-area theorem:* The change in angle θ_{BA} between the tangents to the elastic curve at two points A and B equals the area of the M/EI diagram between those points. Note that the angle θ_{BA} and the area of the M/EI diagrams have the same sign. That means a positive (negative) area corresponds to a counterclockwise (clockwise) rotation of the tangent to the elastic curve as we proceed in the x direction. Hence, θ_{BA} shown in Figure 4.12b is positive.

Inasmuch as the deflection of a beam is taken to be small, we see from Figure 4.12b that the vertical distance dt due to the effect of curvature of an element of length dx equals x dθ, where dθ is defined by Equation (a). Therefore, vertical distance AA', the *tangential deviation* t_{AB} of point A from the tangent at B, is

$$t_{AB} = \int_A^B x_1 \frac{Mdx}{EI} = \left[\text{area of } \frac{M}{EI} \text{ diagram between } A \text{ and } B\right]\bar{x}_1 \tag{4.24}$$

in which x is the horizontal distance to the centroid C of the area from A. This is the *second moment-area theorem:* The tangential deviation t_{AB} of point A with respect to the tangent at B equals the moment with respect to A of the area of the M/EI diagram between A and B.

Likewise, we have

$$t_{BA} = \left[\text{area of } \frac{M}{EI} \text{ diagram between } A \text{ and } B\right]\bar{x}_2 \tag{4.25}$$

The quantity $\bar{x}_2$ represents the horizontal distance from point B to the centroid C of the area (Figure 4.12a). Note that $t_{AB} \neq t_{BA}$ generally. Also observe from Equations 4.24 and 4.25 that the *signs* of t_{AB} and t_{BA} depend on the sign of the bending moments. In many beams, it is obvious whether the beam deflects upward or downward and whether the slope is clockwise or counterclockwise. When this is the case, it is not necessary to follow the sign conventions described for the moment-area method: We calculate the absolute values and find the directions by inspection.

4.6.2 Application of the Moment-Area Method

Determination of beam deflections by moment-area theorems is fairly routine. They are equally applicable for rigid frames. In continuous beams, the two sides of a joint are 180° to one another, whereas in rigid frames, the sides of a joint often are at 90° to one another. Our discussion is limited to beam problems. A correctly constructed M/EI diagram and a sketch of the elastic curve are always necessary. Table A.3 may be used to obtain the areas and centroidal distances of common shapes. The slopes of points on the beam with respect to one another can be found from Equation 4.23, and the deflection, using Equation 4.24 or 4.25.

The moment-area procedure is readily used for beams in which the direction of the tangent to the elastic curve at one or more points is known (e.g., cantilevered beams). For computational simplicity, often M/EI diagrams are drawn and the formulations made in terms of the quantity EI: that is, numerical values of EI may be substituted in the final step of the solution.

For a statically determinate beam with various loads or an indeterminate beam, the displacements determined by the moment-area method are usually best found by superposition. This requires a series of diagrams indicating the moment due to each load or reaction drawn on a separate sketch. In this manner, calculations can be simplified, because the areas of the separate M/EI diagrams may be simple geometric forms. When treating statically indeterminate problems, each additional compatibility condition is expressed by a moment-area equation to supplement the equations of statics.

Example 4.8: Shaft Deflection by the Moment-Area Method

A simple shaft carries its own weight of intensity w, as depicted in Figure 4.13a. Determine the slopes at the ends and center deflection.

Assumption: Bearings act as simple supports.

Solution

Inasmuch as the flexural rigidity EI is constant, the M/EI diagram has the same parabolic shape as the bending moment diagram (Figure 4.13b), where the area properties are taken from Table A.3. The elastic curve is depicted in Figure 4.13c, with the tangent drawn at A.

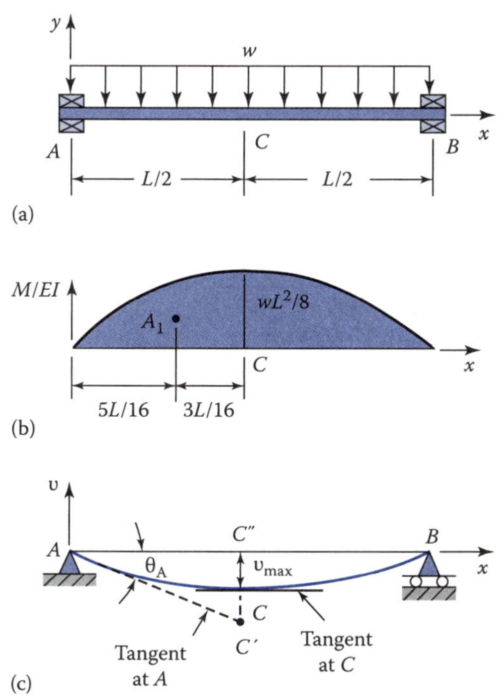

FIGURE 4.13
Example 4.8.

The shaft and loading are symmetric about the center C; hence, the tangent to the elastic curve at C is horizontal: $\theta_c = 0$. Therefore, $\theta_{CA} = 0 - \theta_A$ or $\theta_A = -\theta_{CA}$ and

$$A_1 = \frac{2}{3}\left(\frac{L}{2}\right)\left(\frac{wL^2}{8EI}\right) = \frac{wL^3}{24EI}$$

By the first moment-area theorem, $\theta_{CA} = A_1$,

$$\theta_A = -\frac{wL^3}{24EI} = -\theta_B \tag{4.26}$$

The minus sign means that the end A of the beam rotates clockwise, as shown in the figure.

Through the use of the second moment-area theorem, Equation 4.25,

$$t_{CA} = A_1\left(\frac{3L}{16}\right) = \frac{wL^4}{128EI}$$

in which $t_{CA} = CC'$ and $\theta_A E/2 = C'C''$ (Figure 4.13c). The maximum deflection, $\upsilon_{max} = CC''$, is

$$\upsilon_{max} = -\frac{wL^3}{24EI}\left(\frac{L}{2}\right) + \frac{wL^3}{24EI} = -\frac{5wL^4}{384EI} \tag{4.27}$$

The minus sign indicates that the deflection is downward.

Comment: Alternatively, the moment of area A_1 about point A, Equation 4.24, readily gives the numerical value of υ_{max}.

Example 4.9: Displacements of a Stepped Cantilevered Beam by the Moment-Area Method

A nonprismatic cantilevered beam with two different moments of inertia carries a concentrated load P at its free end (Figure 4.14a). Find the slope at B and deflection at C.

Solution

The M/EI diagram is divided conveniently into its component parts, as shown in Figure 4.14b:

$$A_1 = -\frac{PL^2}{8EI}, \quad A_2 = -\frac{PL^2}{16EI}, \quad A_3 = -\frac{PL^2}{8EI}$$

The elastic curve is in Figure 4.14c. Inasmuch as $\theta_A = 0$ and $v_A = 0$, we have $\theta_C = \theta_{CA}$, $\theta_B = \theta_{BA}$, $\upsilon_C = t_{CA}$, and $\upsilon_B = t_{BA}$.

Applying the first moment-area theorem,

$$\theta_B = A_1 + A_2 + A_3 = -\frac{5PL^2}{16EI} \tag{4.28a}$$

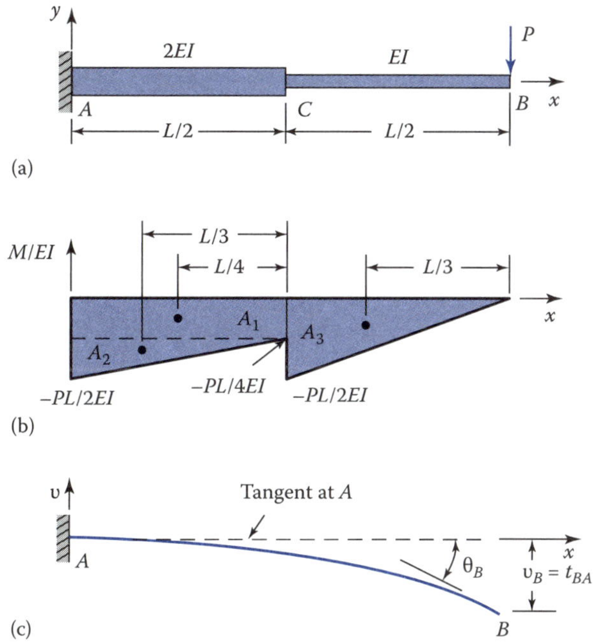

FIGURE 4.14
Example 4.9.

The minus sign means that the rotations are clockwise. From the second moment-area theorem,

$$\upsilon_C = A_1\left(\frac{L}{4}\right) + A_2\left(\frac{L}{3}\right) = -\frac{5PL^3}{96EI} \tag{4.28b}$$

The minus sign shows that the deflection is downward.

Example 4.10: Reactions of a Propped Up Cantilever by the Moment-Area Method

A propped cantilevered beam is loaded by a concentrated force P acting at the position shown in Figure 4.15a. Determine the reactional forces and moments at the ends of the beam.

Solution

The reactions indicated in Figure 4.15a show that the beam is statically indeterminate to the first degree. We select R_B as a redundant (or unknown) load and remove support B (Figure 4.15b). The corresponding M/EI diagram is in Figure 4.15c, with the component areas

$$A_1 = \frac{R_B L^2}{2EI}, \quad A_2 = -\frac{Pa^2}{2EI}$$

One displacement compatibility condition is required to find the redundant load. Observe that the slope at the fixed end and the deflection at the supported end are 0;

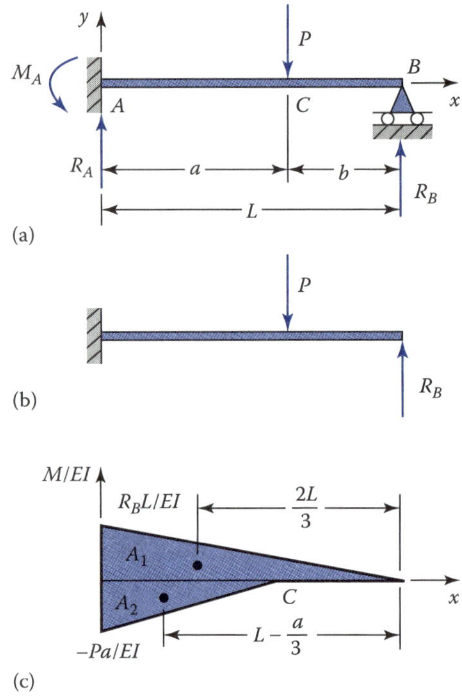

FIGURE 4.15
Example 4.10.

the tangent to the elastic curve at A passes through B, or $t_{BA} = 0$. Therefore, by the second moment-area theorem,

$$\frac{R_B L^2}{2EI}\left(\frac{2L}{3}\right) - \frac{Pa^2}{2EI}\left(L - \frac{a}{3}\right) = 0$$

Solving,

$$R_B = \frac{Pa^2}{2L^3}(3L - a) \tag{4.29}$$

Comments: The remaining reactions are obtained from equations of statics. Then, the slope and deflection are found as needed by employing the usual moment-area procedure.

4.7 Impact Loading

A moving body striking a structure delivers a suddenly applied dynamic force that is called an *impact* or *shock load*. Details concerning the material behavior under dynamic loading are presented in Section 2.9 and Chapter 7. Although the impact load causes elastic members to vibrate until equilibrium is reestablished, our concern here is with only the influence of impact or shock force on the maximum stress and deformation within the member.

Note that the design of engineering structures subject to suddenly applied loads is complicated by a number of factors, and theoretical considerations generally serve only qualitatively to guide the design [5,6]. In Sections 4.8 and 4.9, typical impact problems are analyzed using the *energy method* of the mechanics of materials theory together with the following common *assumptions*:

1. The displacement is proportional to the loads.
2. The material behaves elastically, and a static stress–strain diagram is also valid under impact.
3. The inertia of the member resisting impact may be neglected.
4. No energy is dissipated because of local deformation at the point of impact or at the supports.

Obviously, the energy approach leads to an approximate value for impact loading. It presupposes that the stresses throughout the impacted member reach peak values at the same time. In a more exact method, the stress at any position is treated as a function of time, and waves of stress are found to sweep through the elastic material at a propagation rate. This wave method gives higher stresses than the energy method. However, the former is more complicated than the latter and not discussed in this text. The reader is directed to references for further information [7,8].

4.8 Longitudinal and Bending Impact

Here, we determine the stress and deflection caused by linear or longitudinal and bending impact loads. In machinery, the longitudinal impact may take place in linkages, hammer-type power tools, coupling-connected cars, hoisting rope, and helical springs. Examples of bending impact are found in shafts and structural members, such as beams, plates, shells, and vessels.

4.8.1 Freely Falling Weight

Consider the free-standing spring of Figure 4.16a, on which is dropped a body of mass m from a height h. Inasmuch as the velocity is 0 initially and 0 again at the instant of

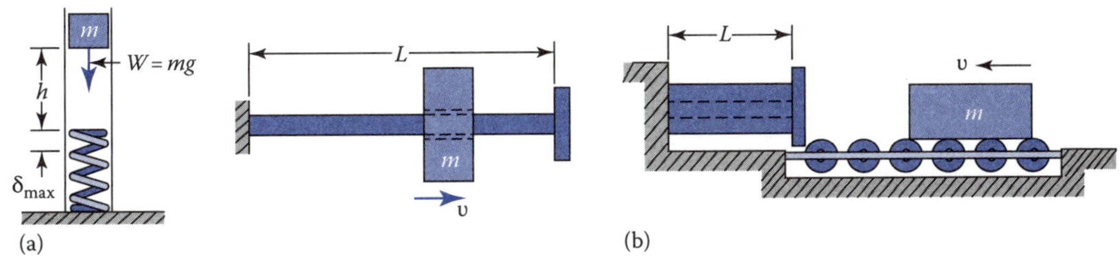

FIGURE 4.16
(a) Freely falling body and (b) horizontal moving body.

maximum deflection of the spring (δ_{max}), the change in kinetic energy of the system is 0. Therefore, the work done by gravity on the body as it falls is equal to the resisting work done by the spring:

$$W\left(h+\delta_{max}\right)=\frac{1}{2}k\delta_{max}^{2} \tag{4.30}$$

in which k is the spring constant.

The deflection corresponding to a static force equal to the weight of the body is simply W/k. This is called the *static* deflection, δ_{st}. The general expression of *maximum dynamic deflection* is, using Equation 4.30,

$$\delta_{max}=\delta_{st}+\sqrt{(\delta_{st})^{2}+2\delta_{st}h} \tag{4.31a}$$

This may be written in the form

$$\delta_{max}=\left(1+\sqrt{1+\frac{2h}{\delta_{st}}}\right)\delta_{st} \tag{4.31b}$$

The term in the parenthesis in this equation, termed the *impact factor*, will be designated by

$$K=1+\sqrt{1+\frac{2h}{\delta_{st}}} \tag{4.32}$$

Multiplying the K by W gives an equivalent static, or *maximum dynamic load*:

$$P_{max}=KW \tag{4.33}$$

To compute the maximum stress and deflection resulting from impact loading, P may be used in the formulas for static loading. The *maximum stress* and *maximum deflection* resulting from the impact loading may be obtained by using P_{max} in expressions for static loading. Thus,

$$\sigma_{max}=K\sigma_{st} \tag{4.34}$$

and

$$\delta_{max}=K\delta_{st} \tag{4.35}$$

Special Cases

Two extreme situations are clearly of particular interest. When $h \gg \delta_{max}$, the work term $W\delta_{max}$ in Equation 4.30 may be neglected, reducing the expression to

$$\delta_{max}=\sqrt{2\delta_{st}h} \tag{a}$$

On the other hand, when $h = 0$, the load is suddenly applied, and Equation 4.30 reduces to

$$\delta_{max}=2\delta_{st} \tag{b}$$

4.8.2 Horizontally Moving Weight

An analysis similar to the preceding one may be used to develop expressions for the case of a mass ($m = W/g$) in horizontal motion with a velocity υ, stopped by an elastic body. In this case, kinetic energy $E_k = m\upsilon^2/2$ replaces $W(h + \delta_{max})$, the work done by W, in Equation 4.30. By so doing, the maximum dynamic deflection and load are

$$\delta_{max} = \delta_{st}\sqrt{\frac{\upsilon^2}{g\delta_{st}}} = \sqrt{\frac{2E_k}{k}} \tag{4.36a}$$

$$P_{max} = m\sqrt{\frac{\upsilon^2 g}{\delta_{st}}} = \sqrt{2E_k k} \tag{4.36b}$$

The quantity δ_{st} is the static deflection caused by a horizontal force W. Note that m is measured in kg in SI or lb · s²/in. in U.S. units. Likewise expressed are υ (in m/s or in./s), the gravitational acceleration g (in m/s² or in./s²), and E_k (in N · m or in. · lb).

When the body hits the end of a *prismatic bar* of length L and axial rigidity AE (Figure 4.16b), we have $k = AE/L$ and hence $\delta_{st} = mgL/AE$. Equations 4.36 are therefore

$$\delta_{max} = \sqrt{\frac{m\upsilon^2 L}{AE}} \tag{4.37a}$$

$$P_{max} = \sqrt{\frac{m\upsilon^2 AE}{L}} \tag{4.37b}$$

The corresponding maximum dynamic compressive stress, taken to be uniform through the bar, is

$$\sigma_{max} = \sqrt{\frac{m\upsilon^2 E}{AL}} \tag{4.38}$$

The foregoing shows that the stress can be reduced by increasing the volume AL or decreasing the kinetic energy E_k and the modulus of elasticity E of the member. We note that the stress concentration in the middle of a notched bar would reduce its capacity and tend to promote brittle fracture. This point has been treated in Section 2.9.

Example 4.11: Impact Loading on a Rod

The prismatic rod depicted in Figure 4.17 has length L, diameter d, and modulus of elasticity E. A rubber compression washer of stiffness k and thickness t is installed at the end of the rod.

a. Calculate the maximum stress in the rod caused by a sliding collar of weight W that drops from a height h onto the washer.
b. Redo part a, with the washer removed.

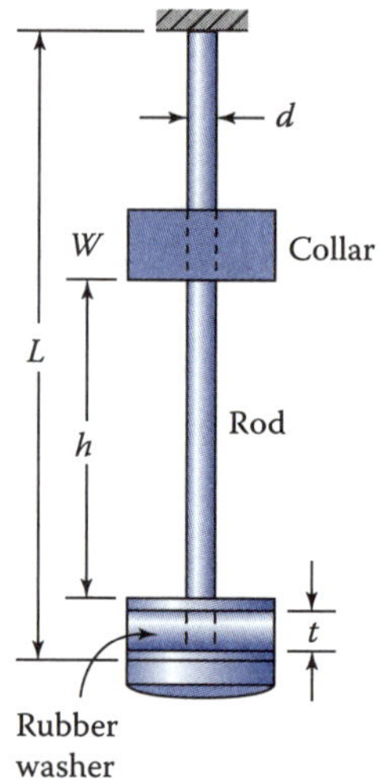

FIGURE 4.17
Example 4.11.

Given: $L = 5\text{ ft},\quad h = 3\text{ ft},\quad d = \frac{1}{2}\text{ in.}\quad t = \frac{1}{4}\text{ in.}$

$$E = 30 \times 10^6 \text{ psi},\quad k = 25 \text{ lb/in.},\quad W = 8 \text{ lb}$$

Solution

The cross-sectional area of the rod $A = \pi(1/2)^2/4 = \pi/16\ \text{in.}^2$

a. For the rod *with* the washer, the static deflection is

$$\delta_{st} = \frac{WL}{AE} + \frac{W}{k}$$

$$= \frac{8(16)(5 \times 12)}{\pi(30 \times 10^6)} + \frac{8}{25} = 0.081 \times 10^{-3} + 0.32 \text{ in.}$$

The maximum dynamic stress, from Equations 4.33 and 4.32, is

$$\sigma_{\max} = \frac{WK}{A}$$

$$= \frac{8 \times 16}{\pi}\left[1 + \sqrt{1 + \frac{2(3 \times 12)}{0.32}}\right] = 653 \text{ psi}$$

b. In the absence of the washer, this equation results in

$$\sigma_{\max} = \frac{8 \times 16}{\pi}\left[1 + \sqrt{1 + \frac{2(3 \times 12 + 0.25)}{0.081 \times 10^{-3}}}\right] = 38.6 \text{ ksi}$$

Comments: The difference in stress for the preceding two solutions is large. This suggests the need for flexible systems for withstanding impact loads. Interestingly, bolts subjected to dynamic loads, such as those used to attach the ends to the tube in pneumatic cylinders, are often designed with long grips (see Section 15.9) to take advantage of the more favorable stress conditions.

Example 4.12: Impact Loading on a Beam

A weight W is dropped from a height h, striking at midspan a simply supported steel beam of length L. The beam is of rectangular cross section of width b and depth d (Figure 4.18). Calculate the maximum deflection and maximum stress for these two cases:

a. The beam is rigidly supported at each end.
b. The beam is supported at each end by springs.

Given: $W = 100$ N, $h = 150$ mm, $L = 2$ m, $b = 30$ mm, and $d = 60$ mm

Assumptions: Modulus of elasticity $E = 200$ GPa and spring rate $k = 200$ kN/m.

Solution

We have $M_{max} = WL/4$ at point C and $I = bd^3/12$. The maximum deflection, due to a static load, is (from case 5 of Table A.9)

$$\delta_{st} = \frac{WL^3}{48EI} = \frac{100(2)^3(12)}{48(200\times 10^9)(0.03)(0.06)^3} = 0.154 \text{ mm}$$

The maximum static stress equals

$$\sigma_{st} = \frac{M_{max}C}{I} = \frac{100(2)(0.03)(12)}{4(0.03)(0.06)^3} = 2.778 \text{ MPa}$$

a. The impact factor, using Equation 4.32, is

$$K = 1 + \sqrt{1 + \frac{2(0.15)}{0.154(10^{-3})}} = 45.15$$

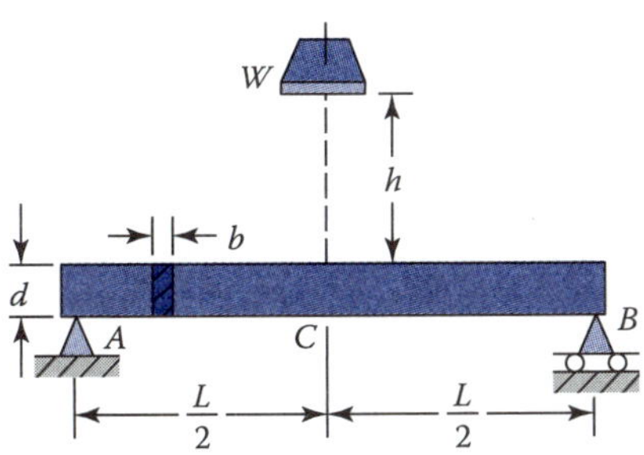

FIGURE 4.18
Example 4.12.

Therefore,

$$\delta_{max} = 45.15(0.154) = 6.95 \text{ mm}$$

$$\sigma_{max} = 45.15(2.778) = 125 \text{ MPa}$$

b. The static deflection of the beam due to its own bending and the deformation of the springs is

$$\delta_{st} = 0.154 + \frac{50}{200} = 0.404 \text{ mm}$$

The impact factor is then

$$K = 1 + \sqrt{1 + \frac{2(0.15)}{0.404(10^{-3})}} = 28.27$$

Hence,

$$\delta_{max} = 28.27(0.404) = 11.42 \text{ mm}$$

$$\sigma_{max} = 28.27(2.778) = 78.53 \text{ MPa}$$

Comments: Comparing the results, we observe that dynamic loading considerably increases deflection and stress in a beam. Also noted is a reduction in stress with increased flexibility, owing to the spring added to the supports. However, the values calculated are probably somewhat higher than the actual quantities, because of our simplifying assumptions 3 and 4.

Example 4.13: Impact Analysis of a Diving Board

A *diving board*, also referred to as *springboard*, is a flexible board from which a dive may be executed, secured at one end and projecting over water at the other. Spring constant of a diving board is customarily adjusted by way of a fulcrum (roller support). *Commercial fiberglass diving boards* are made of molded fiberglass and a laminated Douglas fir wood core, with additional fiberglass that reinforces the tip and fulcrum area. *Residential diving boards* are usually fabricated of Douglas fir wood core with acrylic coating. Some springboards are made out of aluminum and there is frequently textured gripping material, such as crashed glass or sand mixed with the paint, to provide grip to persons not wearing shoes. It was shown in Example 2.3 using Ashby's chart that glass-reinforced plastic is one of the low-cost materials that make a beam (like diving board) as strong as possible for a given weight.

Given: A diver of weight W is springing on a rectangular cross-sectional diving board ABC (Figure 4.19). On a particular bounce, the diver reaches a height h above the end C of the board. Data are as follows:

$$a = 3 \text{ m} \quad L = 4 \text{ m} \quad b = 400 \text{ mm} \quad d = 65 \text{ mm}$$

$$h = 500 \text{ mm} \quad E = 12.6 \text{ GPa} \quad W = 600 \text{ N}$$

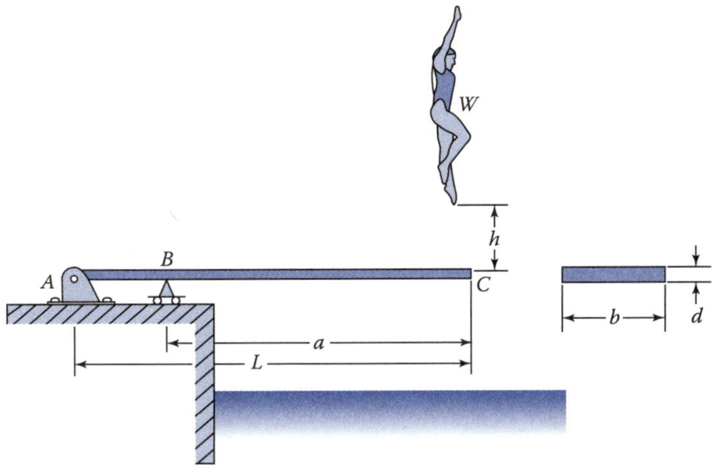

FIGURE 4.19
Example 4.13. Model of diving board.

Find:

a. The maximum static deflection and stress in the board
b. The maximum dynamic deflection at end C
c. The maximum dynamic stress in the board

Assumptions:

1. The diver remains as a rigid mass when strikes the very end C of the board.
2. The flexural rigidity EI of the beam is constant.
3. The weight of the beam (e.g., fiberglass board) is much smaller than that of the diver and is neglected.
4. The deformation of the board due to the shear force is disregarded.

Solution

The beam cross-sectional area properties are

$$I = \frac{1}{12}bd^3 = \frac{1}{12}(400)(65)^3 = 9.154(10^6)\ \text{mm}^4$$

$$c = \frac{1}{2}(65) = 32.5\ \text{mm}$$

a. *Static loading*. The deflection at point C, from Case 10 of Table A.9, is

$$\upsilon_{st} = \frac{Wa^2L}{3EI} \tag{c}$$

Introducing the given numerical values, we have

$$\upsilon_{st} = \frac{600(3)^2(4)}{3(12.6\times 10^9)(9.154)(10^{-6})} = 62.4\ \text{mm}$$

The magnitude of the maximum moment $M = Wa$ takes place at point B of the beam. The flexure formula results in therefore

$$\sigma_{st} = \frac{Mc}{I} = \frac{(600 \times 3)(32.5 \times 10^{-3})}{9.154(10^{-6})} = 6.39 \text{ MPa}$$

b. *Maximum deflection at C*. The impact factor, using Equation 4.32, is found to be

$$K = 1 + \sqrt{1 + \frac{2h}{\upsilon_{st}}} = 1 + \sqrt{1 + \frac{2(500)}{62.4}} = 5.126$$

Hence,

$$\upsilon_{max} = K\upsilon_{st} = 5.126(62.4) = 319.9 \text{ mm}$$

c. *Maximum stress in the board*. Through the use of Equation 4.34, we have

$$\sigma_{max} = K\sigma_{st} = 5.126(6.39) = 32.8 \text{ MPa}$$

which takes place at point B of the beam.

Comments: The results show that the dynamic deflection and dynamic stress in the diving board are much greater than the corresponding static quantities. But the actual values could be less than the calculated because of simplifying assumptions 3 and 4 of Section 4.7. A safety factor should be used to ensure against uncertainties related to strength of the board and the loading applied by the diver.

4.9 Torsional Impact

In machinery, torsional impact occurs in the rotating shafts of punches and shears; in geared drives; at clutches, brakes, and torsional suspension bars; and numerous other components. Here, we discuss the stress and deflection in members subjected to impact torsion. The problem is analyzed by the approximate energy method used in the preceding section. Advantage will be taken of the analogy between linear and torsional systems to readily write the final relationships.

Consider a circular prismatic shaft of flexural rigidity GJ and length L, fixed at one end and subjected to a suddenly applied torque T at the other end (Figure 4.20). The shaft stiffness, from Equation 4.10, is $k = GJ/L$, where $J = \pi d^4/32$ and d is the diameter. The maximum dynamic angle of twist (in rad), from Equation 4.36a, is

$$\phi_{max} = \sqrt{\frac{2E_k L}{GJ}} \tag{4.39a}$$

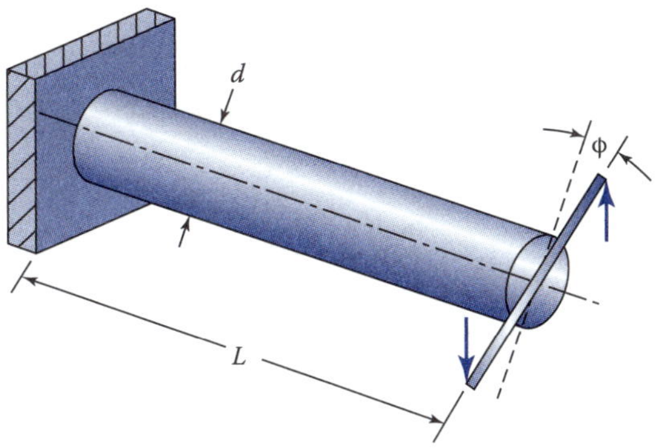

FIGURE 4.20
Bar subjected to impact torsion.

in which E_k is the kinetic energy. Similarly, the maximum dynamic torque, referring to Equation 4.36b, is

$$T_{\max} = \sqrt{\frac{2E_k GJ}{L}} \tag{4.39b}$$

The maximum dynamic shear stress, $\tau_{\max} = 16T_{\max}/\pi d^3$, is therefore

$$\tau_{\max} = 2\sqrt{\frac{E_k G}{AL}} \tag{4.40}$$

Here, A represents the cross-sectional area of the shaft.

Recall from Section 1.10 that, for a rotating wheel of constant thickness, the kinetic energy is expressed in the form

$$E_k = \frac{1}{2} I\omega^2 \tag{4.41}$$

with

$$I = \frac{1}{2} mb^2 \tag{4.42a}$$

$$m = \pi b^2 t\rho \tag{4.42b}$$

In the foregoing, we have

I = the mass moment of inertia (N·s²·m or lb·s²·in.)
ω = the angular velocity (rad/s)
m = the mass (kg or lb·s²/in.)
b = the radius
t = the thickness
ρ = the mass density (kg/m³ or lb·s²/in.⁴)

As before, W and g are the weight and acceleration of gravity, respectively. A detailed treatment of stress and displacement in disk flywheels is given in Section 16.5.

Note that in the case of wheel of variable thickness, the mass moment of inertia may conveniently be obtained from the expression

$$I = mr^2 \tag{4.43}$$

The quantity r is called the *radius of gyration* for the mass. It is a hypothetical distance from the wheel center at which the entire mass could be concentrated and still have the same moment of inertia as the original mass.

Example 4.14: Impact Loading on a Shaft

A shaft of diameter d and length L has a flywheel (radius of gyration r, weight W, modulus of rigidity G, yield strength in shear S_{ys}) at one end and runs at a speed of n. If the shaft is instantly stopped at the other end, determine

a. The maximum shaft angle of twist
b. The maximum shear stress

Given: d = 3 in., L = 2.5 ft, W = 120 lb, r = 10 in., n = 150 rpm

Assumption: The shaft is made of ASTM-A242 steel. So, by Table B.1, $G = 11.5 \times 10^6$ psi and S_{ys} = 30 ksi.

Solution

The area properties of the shaft are

$$A = \frac{\pi(3)^2}{4} = 7.069 \text{ in.}^2, \quad J = \frac{\pi(3)^4}{32} = 7.952 \text{ in.}^4$$

The angular velocity equals

$$\omega = n\left(\frac{2\pi}{60}\right) = 150\left(\frac{2\pi}{60}\right) = 5\pi \text{ rad/s}$$

a. The kinetic energy of the flywheel must be absorbed by the shaft. So, substituting Equation 4.43 into Equation 4.41, we have

$$E_k = \frac{W\omega^2 r^2}{2g} \tag{a}$$

$$= \frac{120(5\pi)^2(10)^2}{2(386)} = 3835 \text{ in} \cdot \text{lb}$$

From Equation 4.39a,

$$\phi_{max} = \sqrt{\frac{2E_k L}{GJ}} = \left[\frac{2(3835)(2.5 \times 12)}{(11.5 \times 10^6)(7.952)}\right]^{1/2}$$

$$= 0.05 \text{ rad} = 2.87°$$

b. Through the use of Equation 4.40,

$$\tau_{\max} = 2\sqrt{\frac{E_k G}{AL}} = 2\left[\frac{(3835)(11.5\times10^6)}{(7.069)(2.5\times12)}\right]^{1/2}$$

$$= 28.84 \text{ ksi}$$

Comment: The stress is within the elastic range, 28.84 < 30, and hence assumption 2 of Section 4.7 is satisfied.

*4.10 Bending of Thin Plates

A plate is an initially flat structural member with smaller thickness compared with remaining dimensions. It is usual to divide the plate thickness t into equal halves by a plane parallel to the faces. This plane is called the *midsurface* of the plate. The plate thickness is measured in a direction normal to the midsurface at each point under consideration. Plates of technical importance are usually defined as thin when the ratio of the thickness to the smaller span length is less than 1/20. Here, we discuss briefly the bending of thin plates. For a detailed treatment of the subject, see [9].

4.10.1 Basic Assumptions

Consider a plate before deformation, depicted in Figure 4.21a, where the xy plane coincides with the midsurface and hence the z deflection is 0. When deformation occurs due to external loading, the midsurface at any point x_A, y_A undergoes a deflection w. Referring to the

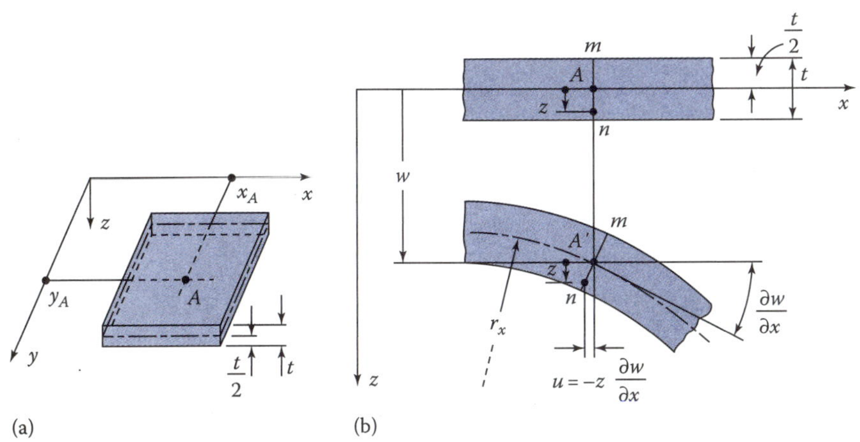

FIGURE 4.21
Plate in bending: (a) before deformation and (b) rotation of a plane section due to deflection.

coordinate system shown, the basic *assumptions* of the small deflection theory of bending for isotropic, homogenous, thin plates may be summarized as follows:

1. The deflection w of the midsurface is small in comparison with the thickness t of the plate; hence, the slope of the deflected surface is much less than unity.
2. Straight lines (such as mn) initially normal to the midsurface remain straight and normal to that plane after bending.
3. No midsurface straining occurs due to bending. This is equivalent to stating that strains γ_{yz}, γ_{xz}, and ε_z are negligible.
4. The component of stress normal to the midsurface, σ_z, may be neglected.

These presuppositions are analogous to those associated with the simple bending theory of beams.

4.10.2 Strain–Displacement Relations

On the basis of the foregoing assumptions, the strain–displacement relations are expressed by Equation 3.55. Therefore,

$$\varepsilon_x = \frac{\partial u}{\partial x}, \quad \varepsilon_x = \frac{\partial \upsilon}{\partial y}, \quad \gamma_{xy} = \frac{\partial \upsilon}{\partial x} + \frac{\partial u}{\partial y} \tag{a}$$

It can be shown that [9]

$$u = -z\frac{\partial w}{\partial x}, \quad \upsilon = -z\frac{\partial w}{\partial y} \tag{b}$$

Combining Equations (a) and (b), we have

$$\varepsilon_x = -z\frac{\partial^2 w}{\partial x^2}, \quad \varepsilon_y = -z\frac{\partial^2 w}{\partial y^2}, \quad \gamma_{xy} = -2z\frac{\partial^2 w}{\partial x \partial y} \tag{4.44}$$

Because, in small deflection theory, the square for a slope may be considered negligible, the partial derivatives of these relations represent the curvatures of the plate. The *curvatures* at the midsurface in planes parallel to the zx (Figure 4.21b), yz, and xy planes are, respectively,

$$\frac{1}{r_x} = \frac{\partial^2 w}{\partial x^2}, \quad \frac{1}{r_y} = \frac{\partial^2 w}{\partial y^2}, \quad \frac{1}{r_{xy}} = \frac{\partial^2 w}{\partial x \partial y} \tag{4.45}$$

The quantity r represents the radius of curvature. Clearly, the curvatures are the rates at which the slopes vary over the plate.

Examining the preceding relationships, we are left to conclude that a circle of curvature can be constructed similarly to Mohr's circle of strain. The curvatures hence transform in the same manner as strains. It can be verified by using Mohr's circle that $(1/r_x) + (1/r_y) = \nabla^2 w$. The sum of the curvatures in perpendicular directions, called the *average curvature*, is invariant with respect to rotation of the coordinate axis. This assertion is valid at any location on the midsurface.

4.10.3 Plate Stress, Curvature, and Moment Relations

For a thin plate, substituting Equation 4.44 into Hooke's law, we obtain

$$\sigma_x = -\frac{Ez}{1-\nu^2}\left(\frac{\partial^2 w}{\partial x^2} + \nu\frac{\partial^2 w}{\partial y^2}\right)$$

$$\sigma_y = -\frac{Ez}{1-\nu^2}\left(\frac{\partial^2 w}{\partial y^2} + \nu\frac{\partial^2 w}{\partial x^2}\right) \tag{4.46}$$

$$\tau_{xy} = -\frac{Ez}{1+\nu}\frac{\partial^2 w}{\partial x \partial y}$$

where $\tau_{xy} = \tau_{yx}$. We see from these relations that the stresses are 0 at the midsurface and vary linearly over the thickness of the plate.

The stresses distributed over the side surfaces of the plate, while producing no net force, result in bending and twisting moments. These moment resultants per unit length (in SI units N · m/m, or simply N) are designated M_x, M_y, and M_{xy}. With reference to Figure 4.22a,

$$\int_{-t/2}^{t/2} z\sigma_x \, dy \, dz = dy \int_{-t/2}^{t/2} z\sigma_x \, dz = M_x \, dy$$

$$M_x = \int_{-t/2}^{t/2} z\sigma_x \, dz \tag{4.47}$$

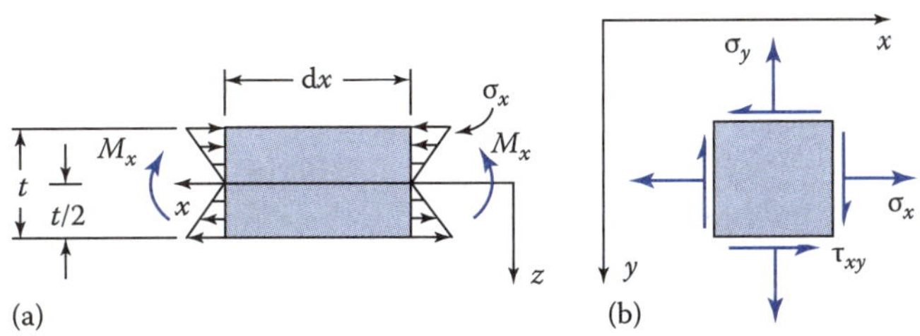

FIGURE 4.22
(a) Plate segment in pure bending and (b) positive stresses on an element in the bottom half of a plate.

Here, t *is* the thickness of the plate. Substituting into this expression the stress σ_x given by Equations 4.46, we obtain M_x in terms of curvatures. Expressions involving M_y and $M_{xy} = M_{yx}$ are derived in a like manner. In so doing, we have

$$M_x = -D\left(\frac{\partial^2 w}{\partial x^2} + \nu \frac{\partial^2 w}{\partial y^2}\right)$$

$$M_y = -D\left(\frac{\partial^2 w}{\partial y^2} + \nu \frac{\partial^2 w}{\partial x^2}\right) \tag{4.48}$$

$$M_{xy} = -D(1-\nu)\frac{\partial^2 w}{\partial x \partial y}$$

The quantity D represents the *flexural rigidity*:

$$D = \frac{Et^3}{12\left(1-\nu^2\right)} \tag{4.49}$$

Interestingly, if a plate element of unit width were free to expand sidewise under the given loading, the flexural rigidity would be $Et^3/12$. The remainder of the plate does not allow this action, however. Because of this, a plate shows *greater stiffness* than a narrow *beam* by a factor $1/(1 - \nu^2)$ or about 10% for $\nu = 0.3$.

According to the *sign convention*, a positive moment is one that results in positive stresses in the positive (bottom) half of the plate (see Section 1.11), as depicted in Figure 4.22b. The maximum stresses occurring on the surface of the plate are obtained by substituting $z = t/2$ into Equation 4.46, together with the use of Equations 4.48, as

$$\sigma_{x,\max} = \frac{6M_x}{t^2}, \quad \sigma_{y,\max} = \frac{6M_y}{t^2}, \quad \tau_{xy,\max} = \frac{6M_{xy}}{t^2} \tag{4.50}$$

Since there is a direct correspondence between the moments and stresses, the equation for transforming the stresses should be identical with that used for the moments. Mohr's circle therefore may be applied to moments as well as to stresses.

4.11 Deflection of Plates by Integration

Variations in the moment and force resultants are governed by the conditions of equilibrium. Application of the equations of statics to an element $dxdy$ of the plate subject to a uniformly distributed load per unit area p leads to a single differential equation in terms of the moments. This, when combined with Equations 4.48, results in [9]

$$\frac{\partial^4 w}{\partial x^4} + 2\frac{\partial^4 w}{\partial x^2 \partial y^2} + \frac{\partial^4 w}{\partial y^4} = \frac{p}{D} \tag{4.51a}$$

or, in concise form,

$$\nabla^4 w = \frac{p}{D} \tag{4.51b}$$

The preceding expression, first derived by Lagrange in 1811, is the governing differential equation for deflection of thin plates.

4.11.1 Boundary Conditions

Two common boundary conditions that apply along the edge at $x = a$ of the rectangular plate with edges parallel to the x and y axes (Figure 4.23) may be expressed as follows.

In the *clamped edge* (Figure 4.23a), both the slope and the deflection must vanish:

$$w = 0, \quad \frac{\partial w}{\partial x} = 0, \quad (x = a) \tag{4.52}$$

In the *simply supported* edge (Figure 4.23b), both the deflection and bending moment are 0. The latter implies that, at edge $x = a$, $\partial w/\partial y = 0$ and $\partial^2 w/\partial y^2 = 0$. Therefore,

$$w = 0, \quad \frac{\partial^2 w}{\partial x^2} = 0, \quad (x = a) \tag{4.53}$$

Other typical conditions at the boundaries may be expressed similarly.

To determine the deflection w, we must integrate Equation 4.51 with the constants of integration dependent on the appropriate boundary conditions. Having the deflection available, the stress (as well as strain and curvature) components are obtained using the formulas derived in the preceding section.

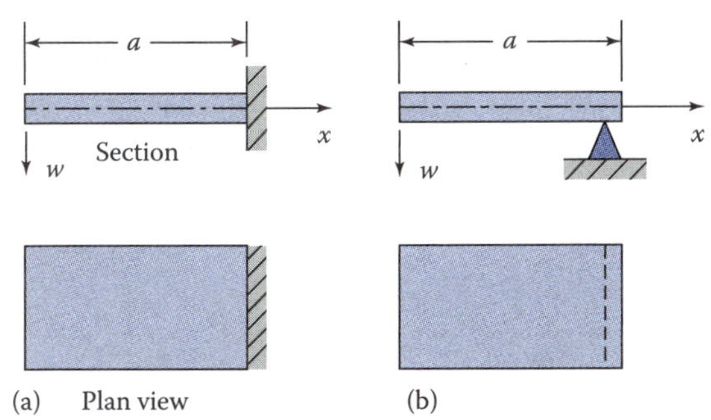

FIGURE 4.23
Two common boundary conditions: (a) fixed edge and (b) simply supported edge.

Discussions here are limited to *long and narrow rectangular plates bent into a cylindrical surface* with its generating line parallel to the x axis. For this case, $w = w(y)$, and the *governing equation for deflection*, Equation 4.51, becomes

$$\frac{d^4 w}{dy^4} = \frac{p}{D} \tag{4.54}$$

The foregoing equation is the same as the wide beam equation, and we conclude that the solution proceeds as in the *case of a beam*. This is illustrated in the numerical example to follow.

Example 4.15: Determination of Deflection and Stress in a Plate

A long, narrow plate of width b and thickness t, a so-called plate strip, is simply supported at edges $y = 0$ and $y = b$, as depicted in Figure 4.24. The plate carries the loading of the form

$$p(y) = p_o \sin\left(\frac{\pi y}{b}\right) \tag{a}$$

The quantity p_o represents the load intensity along the x' axis. Let $\nu = 1/3$. Determine

a. The equation of the deflection surface and maximum stresses
b. The maximum deflection and stresses for $p_o = 15$ kPa, $b = 0.5$ m, $t = 12.5$ mm, and $G = 210$ GPa

Solution

Due to symmetry in the loading and end restraints about the x axis, the plate deforms into a cylindrical surface with its generating line parallel to the x' axis. Since, for this situation, $\partial w/\partial x = 0$ and $\partial^2 w/\partial x \partial y = 0$, Equations 4.48 reduce to

$$M_x = -\nu D \frac{d^2 w}{dy^2}, \quad M_y = -D \frac{d^2 w}{dy^2} \tag{4.55}$$

a. Introducing Equation (a) into Equation 4.54 and integrating and satisfying the boundary conditions

$$w = 0, \quad \frac{d^2 w}{dy^2} = 0, \quad (y = 0, y = b)$$

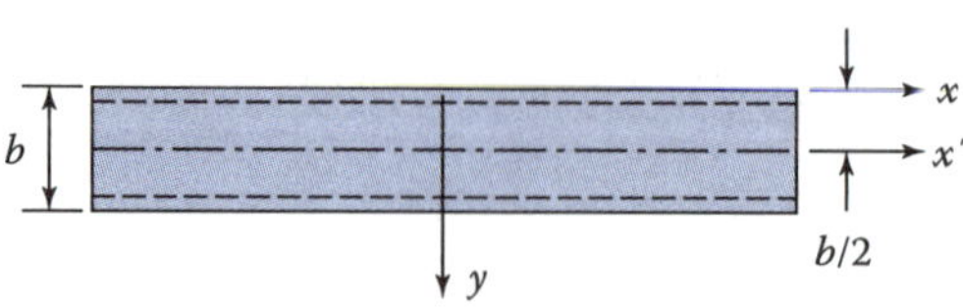

FIGURE 4.24
Example 4.15.

we obtain the deflection

$$w_{\max} = \left(\frac{b}{\pi}\right)^4 \frac{p_o}{D} \sin\left(\frac{\pi y}{b}\right) \tag{4.56}$$

The largest deflection of the plate occurs at $y = b/2$. Therefore,

$$w_{\max} = \left(\frac{b}{\pi}\right)^4 \frac{p_o}{D} \tag{4.57}$$

The moments are now readily determined carrying Equation 4.56 into Equation 4.55. Then, the maximum stresses, occurring at $y = b/2$, are found applying Equation 4.50 as

$$\sigma_{y,\max} = 0.6p_o\left(\frac{b}{t}\right)^2, \quad \sigma_{x,\max} = \nu\sigma_{y,\max} = 0.2p_o\left(\frac{b}{t}\right)^2, \quad \tau_{xy} = 0 \tag{4.58}$$

b. Substituting the given data into Equation 4.49, we have

$$D = \frac{210(10^9)(0.0125)^3}{12\left(1-\frac{1}{9}\right)} = 38{,}452\ \text{N}\cdot\text{m}$$

Similarly, Equations 4.57 and 4.58 lead to

$$w_{\max} = \left(\frac{0.5}{\pi}\right)^4 \frac{15(10^3)}{38{,}452} = 0.25\ \text{mm}$$

$$\sigma_{y,\max} = 0.6\left(15\times10^3\right)\left(\frac{500}{12.5}\right)^2 = 14.4\ \text{MPa}$$

$$\sigma_{x,\max} = \frac{1}{3}(14.4) = 4.8\ \text{MPa}$$

Comment: The result, $w_{\max}/t = 0.02$, shows that the deflection surface is extremely flat, as is often the case for small deflections.

Problems

Sections 4.1 through 4.6

4.1 A high-strength steel rod of length L, used in control mechanism, must carry a tensile load of P without exceeding its yield strength S_y, with a factor of safety n, nor stretching more than δ.

a. What is the required diameter of the rod?

b. Calculate the spring rate for the rod.

Given: $P = 10$ kN, $E = 200$ GPa. $S_y = 250$ MPa, $L = 6$ m, $\delta = 5$ mm

Design Decision: The rod will be made of ASTM-A242 steel. Take $n = 1.2$.

4.2 In Figure 4.3a of Example 4.4, let the bar AB be 12 mm wide and 8 mm thick uniform rectangular cross-sectional bar. Compute the stress, the deformation, and the stiffness of the bar.

4.3 A hollow aluminum alloy 2014-T6 bar of length L must support an axial tensile load of P at a normal stress of σ_{max}. The outside and inside diameters of the bar are D and d, respectively. Calculate the outside diameter, axial deformation, and spring rate of the bar.

Given: $E = 10.6 \times 10^6$ psi (by Table B.1), $d = 0.6D$, $L = 15$ in., $P = 1.5$ kips, $\sigma_{max} = 20$ ksi

4.4 Before loading, there is a gap Δ between the wall and the right end of the copper rod of diameter d (Figure P4.4). Calculate the reactions at A and B, after the rod is subjected to an axial load of P.

Given: $\Delta = 0.014$ in., $d = ½$ in., $P = 8$ kips, $E = 17 \times 10^6$ psi

4.5 At room temperature (20°C), a gap Δ exits between the wall and the right end of the bars shown in Figure P4.5. Determine

a. The compressive axial force in the bars after temperature reaches 140°C

b. The corresponding change in length of the aluminum bar

Given:

$$A_a = 1000 \text{ mm}^2, \quad E_a = 70 \text{ GPa}, \quad \alpha_a = 23 \times 10^{-6}/°\text{C}, \quad \Delta = 1 \text{ mm}$$

$$A_s = 500 \text{ mm}^2, \quad E_s = 210 \text{ GPa}, \quad \alpha_s = 12 \times 10^{-6}/°\text{C}.$$

4.6 Redo Problem 4.5 for the case in which $\Delta = 0$.

4.7 A rigid beam AB is hinged at the left end A and kept horizontally by a vertical steel pipe at point C (Figure P4.7). The pipe has an outer diameter D, inside diameter d, and length L. Find the vertical deflection δ_B of the right end B of the beam caused by the applied load P.

Given: $a = 52$ in., $b = 14$ in., $D = 4.2$ in., $d = 3.8$ in., $L = 25$ in., $P = 2.5$ kips, $E = 29 \times 10^6$ psi

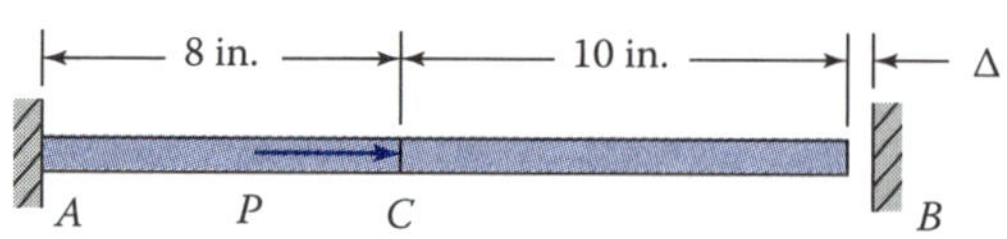

FIGURE P4.4

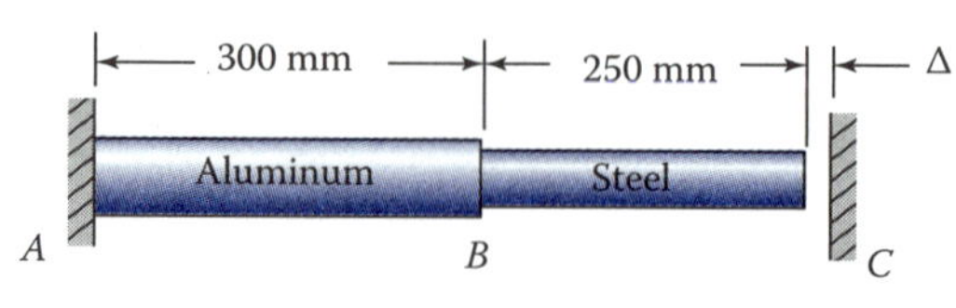

FIGURE P4.5

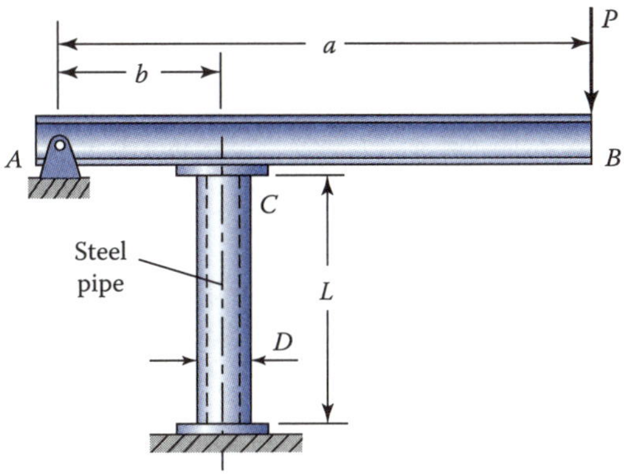

FIGURE P4.7

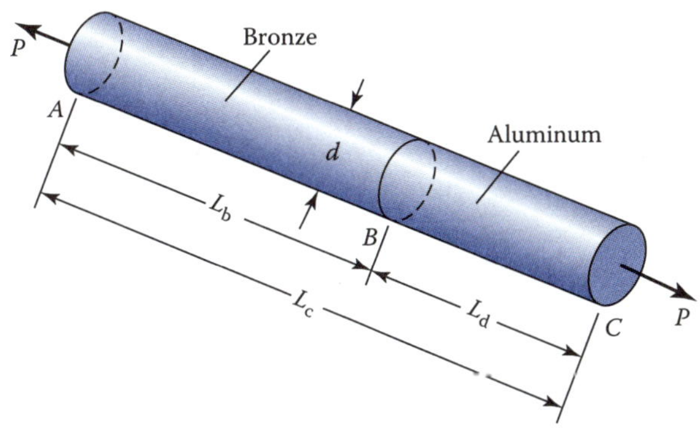

FIGURE P4.8

4.8 A rod *ABC* is composed of two materials joined and has a diameter *d* and total length *L* (Figure P4.8). Part *AB* is cold-rolled 510 bronze and part *BC* is aluminum alloy 6061-T6. The rod is subjected to an axial tensile load *P*. Find

a. The lengths L_a and L_b in order that both parts have the same elongation

b. The total elongation of the rod

Given: $d = 50$ mm, $L = 0.6$ m, $P = 120$ kN, $E_a = 70$ GPa, $E_b = 110$ GPa (by Table B.1)

4.9 Figure P4.9 shows an aluminum bar *CE* of cross-sectional area *A* hinged at upper end *C* and pin connected at point *D* to a rigid beam *AD*. Find the vertical displacement produced by loads *P* and *R*

a. At the end *A* of beam *AD*

b. At the end *E* of bar *CE*

Given: $E = 70$ GPa, $A_a = 130$ mm^2, $P = 12$ kN, $R = 4$ kN, $a = 2b$, $L_1 = 2L_2 = 0.3$ m

4.10 Redo Problem 4.9, for the case in which force *R* is directed upward at end *E* of bar *CE*.

4.11 Two steel shafts are connected by gears and subjected to a torque *T*, as shown in Figure P4.11. Calculate

a. The angle of rotation in degrees at *D*

b. The maximum shear stress in shaft *AB*

Given: $G = 79$ GPa, $T = 500$ N·m, $d_1 = 45$ mm, $d_2 = 35$ mm

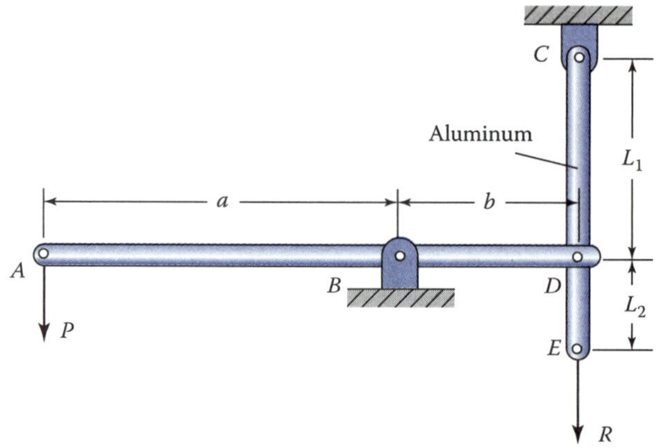

FIGURE P4.9

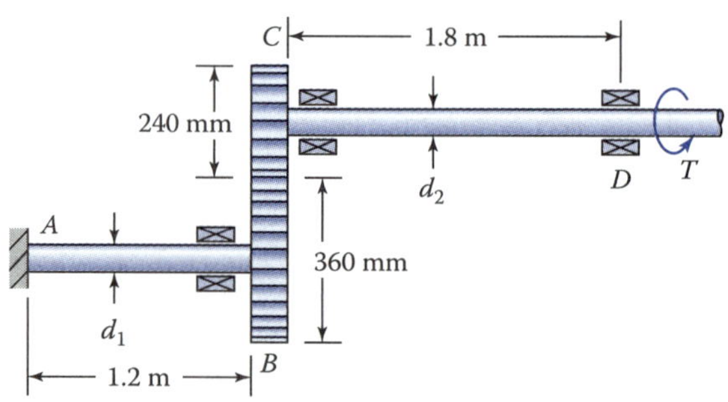

FIGURE P4.11

4.12 Determine the diameter d_1 of shaft *AB* shown in Figure P4.11, for the case in which the maximum shear stress in each shaft is limited to 150 MPa.

Design Decisions: $d_2 = 65$ mm. The factor of safety against shear is $n = 1.2$.

4.13 A hollow high-strength ASTM-A242 steel shaft is subjected to a torque *T* at a maximum shear stress of τ_{max}. The outside radius, inside radius, and length of the shaft are *c*, *b*, and *L*, respectively.

Find:

a. The outside diameter

b. The angle of twist

c. The spring rate

Given: $c = 2b$, $L = 10$ in., $G = 11.5 \times 10^6$ psi (from Table B.1), $\tau_{max} = 20$ ksi, $T = 40$ kips·in.

4.14 Three pulleys are fastened to a solid stepped steel shaft and transmit torques as illustrated in Figure P4.14.

Find:

a. The angle of twist ϕ_{BC} between *B* and *C*

b. The angle of twist ϕ_{BD} between *B* and *D*

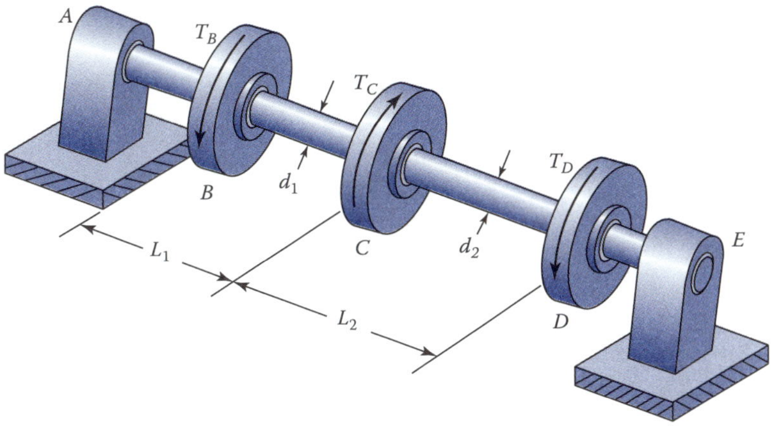

FIGURE P4.14

Given: $d_1 = 1\frac{3}{8}$ in., $d_2 = 1$ in., $L_1 = 25$ in., $L_2 = 30$ in., $G = 11.5 \times 10^6$ psi, $T_B = 5$ kips·in., $T_C = 12$ kips·in., $T_D = 4$ kips·in.

4.15 A high-strength ASTM-A242 steel shaft *AE* of outer diameter *D* and inside diameter *d* is supported by bearing at *B* and carries torques T_1, T_2, and T_3 at *A*, *C*, and *D* as seen in Figure P4.15. The shaft is connected to a gear box at *E*.

Determine:

a. The angle of twist ϕ_A at end *A*

b. The safety factor *n* on the basis of the yield strength

Given: $d = 1.4$ in., $D = 2$ in., $L_1 = 18$ in., $L_2 = 15$ in., $L_3 = 25$ in., $T_1 = 10$ kips·in., $T_2 = 25$ kips·in., $T_3 = 6$ kips·in., $G = 11.2 \times 10^6$ psi, $\tau_y = 30$ ksi (from Table B.1)

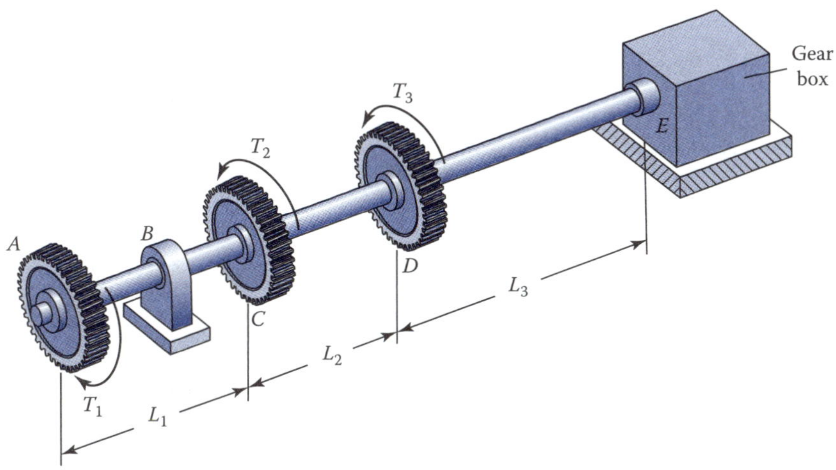

FIGURE P4.15

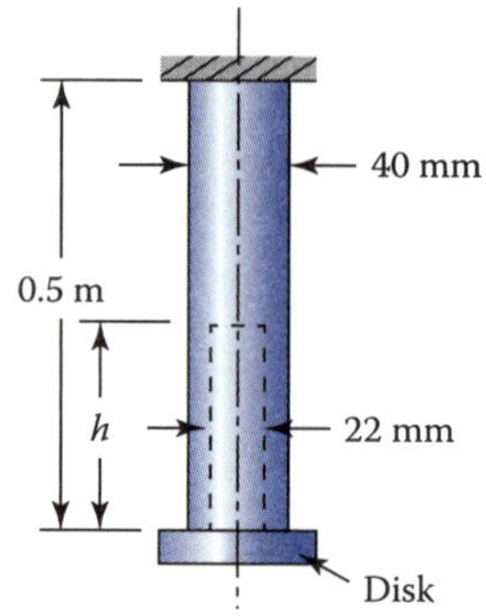

FIGURE P4.16

4.16 A disk is attached to a 40 mm diameter, 0.5 m long steel shaft ($G = 79$ GPa) as depicted in Figure P4.16.

Design Requirement: To achieve the desired natural frequency of torsional vibrations, the stiffness of the system is specified such that the disk will rotate 1.5° under a torque of kN · m. How deep (h) must a 22 mm diameter hole be drilled to satisfy this requirement?

4.17 A solid round shaft with fixed ends is under a distributed torque of intensity $T(x) = T_1$, lb · in./in., as shown in Figure P4.17. Determine the reactions at the walls.

4.18 A structural steel beam AB supported at the ends as illustrated in Figure P4.18 is subjected to a concentrated load P at the midspan C. Find the vertical deflection of the beam at end B.

4.19 A simply supported beam AB carries a triangularly distributed load of maximum intensity w_o (Figure P4.19).

a. Employ the fourth-order differential equation of the deflection to derive the expression for the elastic curve.

b. Determine the maximum deflection υ_{max}.

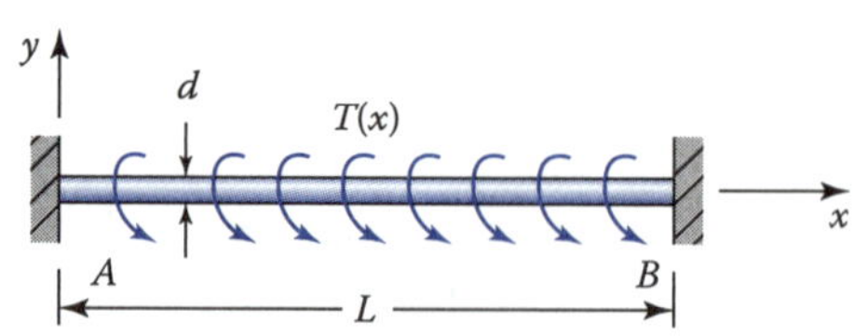

FIGURE P4.17

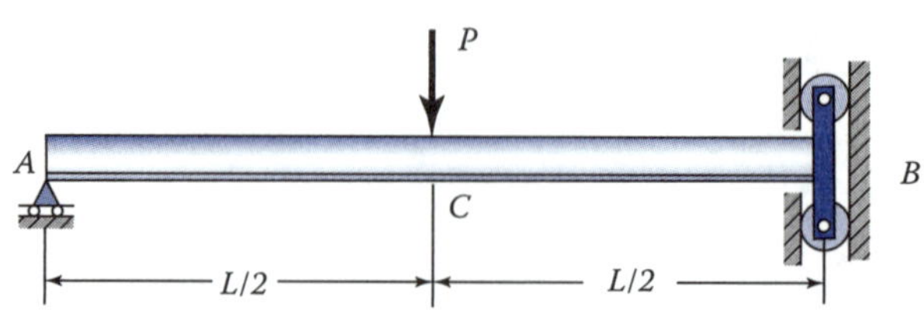

FIGURE P4.18

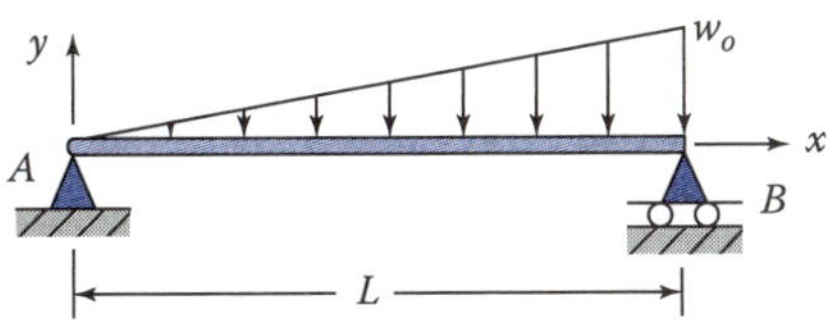

FIGURE P4.19

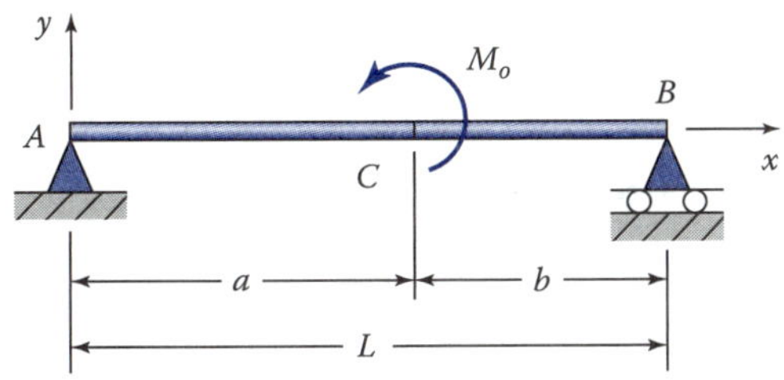

FIGURE P4.20

4.20 A simply supported beam is loaded with a concentrated moment M_o, as shown in Figure P4.20. Derive the equation of the elastic curve for the segment AC of the beam.

4.21 A simple beam of wide-flange cross section carries a uniformly distributed load of intensity w (Figure P4.21). Determine the span length L.

Given. $h = 12.5$ in., $E = 30 \times 10^6$ psi

Design Requirements: $\sigma_{max} = 10$ ksi, $\upsilon_{max} = 1/8$ in.

4.22 The overhanging beam ABC supports a concentrated load P at the free end, as shown in Figure P4.22, for the segment BC of the beam.

a. Derive the equation of the elastic curve.

b. What is the maximum deflection υ_{max}?

c. Calculate the value of the υ_{max} for the following data:

$$I = 5.12 \times 10^6 \text{ mm}^4, \quad E = 200 \text{ GPa}, \quad P = 25 \text{ kN}$$

$$L = 2 \text{ m}, \quad a = 0.5 \text{ m}$$

4.23 A shaft-pulley assembly with an overhang is shown in Figure P4.23. Observe that the pulley rotates freely and delivers no torque but tension load to its shaft.

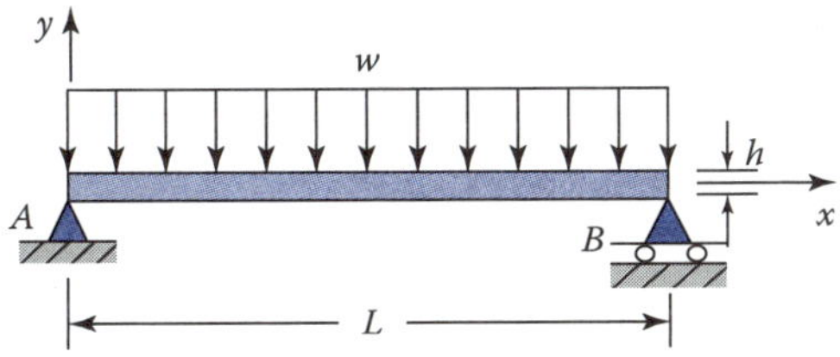

FIGURE P4.21

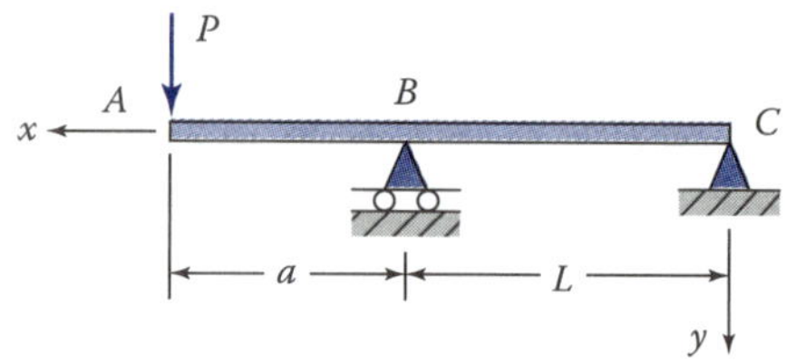

FIGURE P4.22

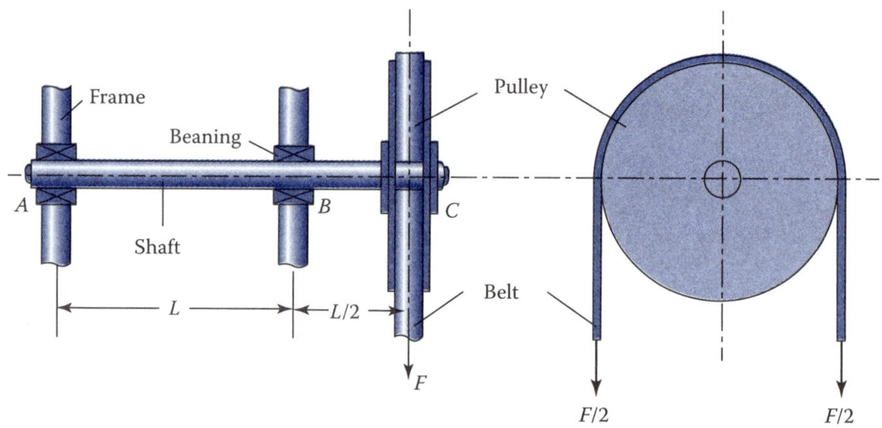

FIGURE P4.23

Determine:

a. Equations of the elastic curve using Equation 4.16a
b. The deflection at point C

Assumption: The bearings at A and B act as simple supports.

4.24 Two cantilever beams AB and CD are supported and loaded as shown in Figure P4.24. What is the interaction force R transmitted through the roller that fits snugly between the two beams at point C? Use the method of superposition and the deflection formulas of the beams from Table A.9.

4.25 Figure P4.25 shows a compound beam with a hinge at point B. It is composed of two parts: a beam BC simply supported at C and a cantilevered beam AB fixed at A. Apply the superposition method using Table A.9 to determine the deflection υ_B at the hinge.

4.26 A steel cantilever beam AB built-in at end A and reinforced at location C by a steel rod CD is to carry a load W at the free end B (Figure P4.26). After the loading, the beam

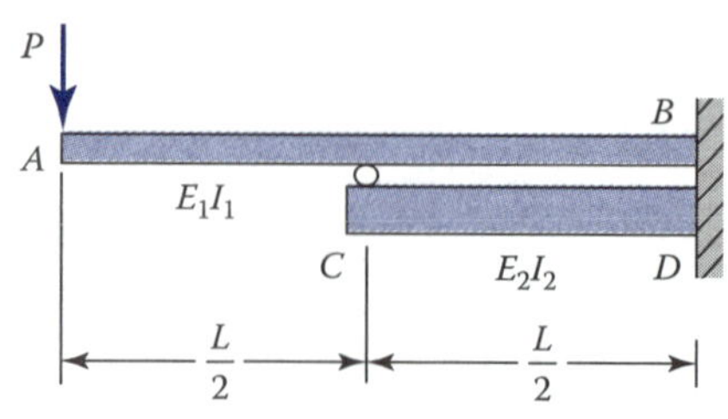

FIGURE P4.24

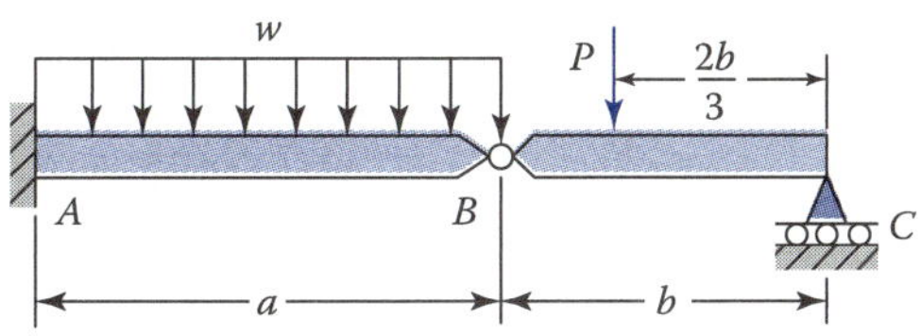

FIGURE P4.25

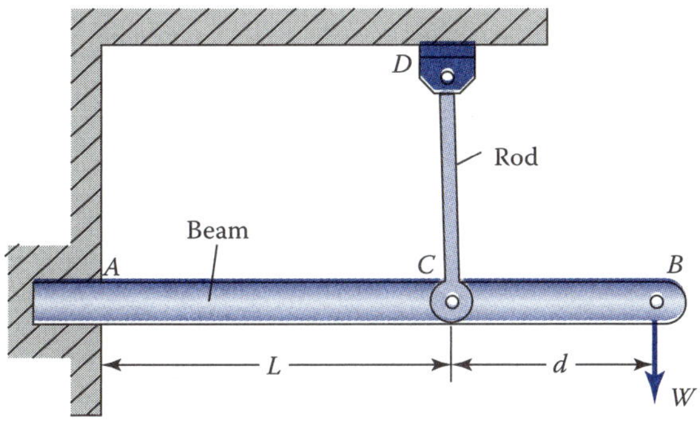

FIGURE P4.26

deflects downward at C and develops a reactive tensile force F in the rod. Verify that equation for this force using the method of superposition is expressed as follows:

$$F = \frac{WL^2k(2L + 3a)}{2(kL^3 + 3EI)} \tag{P4.26}$$

In this expression, we have EI as the flexural rigidity of the beam, and $k = F/\delta_C$ represents the spring rate of the rod with δ_C its elongation.

4.27 A propped cantilevered beam carries a uniform load of intensity w (Figure P4.27). Determine the reactions at the supports, using the second-order differential equation of the beam deflection.

4.28 A fixed-ended beam AB is under a symmetric triangular load of maximum intensity as shown in Figure P4.28. Determine all reactions, the equation of the elastic curve, and the maximum deflection.

Requirement: Use the second-order differential equation of the deflection.

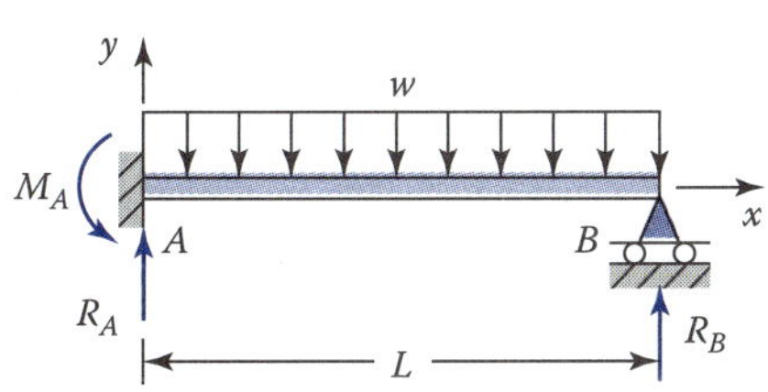

FIGURE P4.27

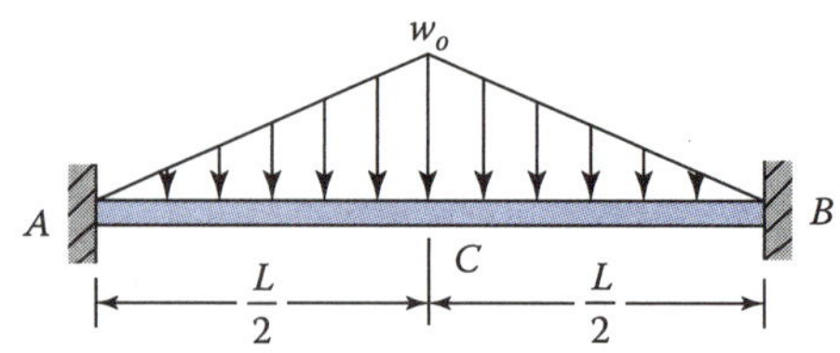

FIGURE P4.28

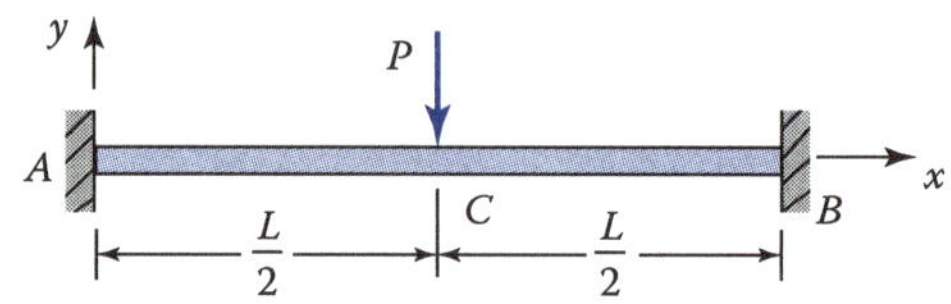

FIGURE P4.29

4.29 A fixed-ended beam supports a concentrated load P at its midspan (Figure P4.29). Determine all reactions and the equation of the elastic curve.

4.30 Redo Problem 4.27, using the method of superposition together with Table A.9.

4.31 A cantilever beam is subjected to a partial loading of intensity w, as shown in Figure P4.31. Use the area moments to determine

a. The slope at the free end

b. The deflection at the free end

4.32 A simple beam with two different moments of inertia is under a center load P, as shown in Figure P4.32. Apply the area moments to find

a. The slope at point B

b. The maximum deflection

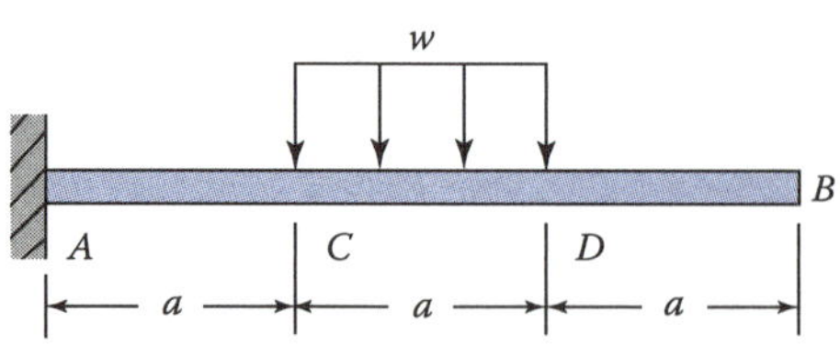

FIGURE P4.31

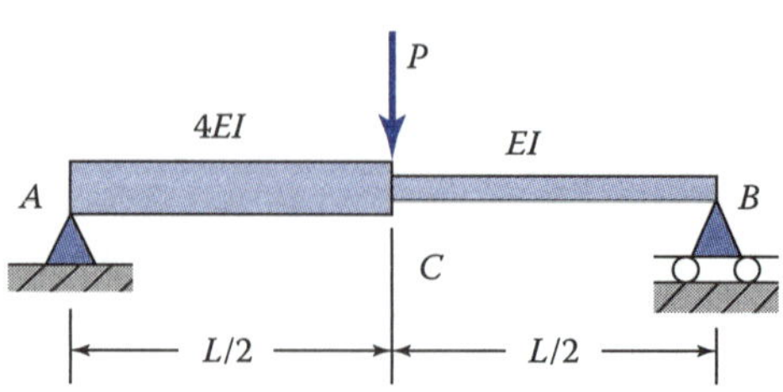

FIGURE P4.32

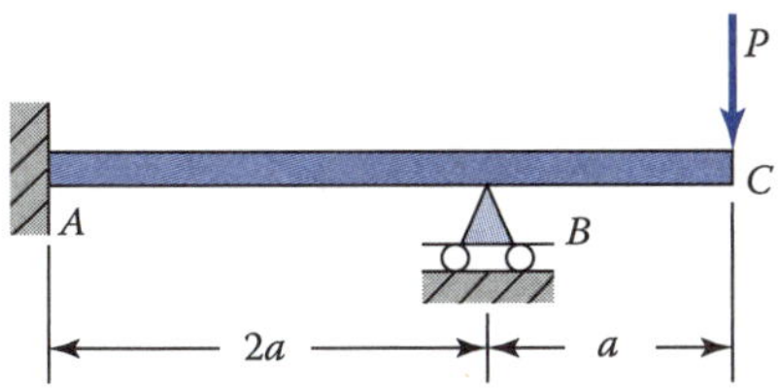

FIGURE P4.33

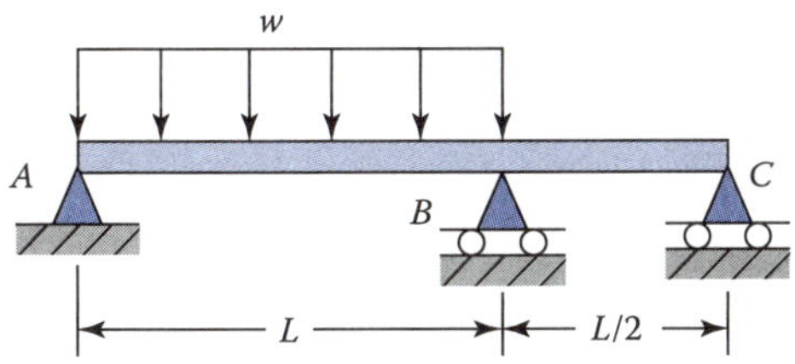

FIGURE P4.34

4.33 and **4.34** A simple beam with an overhang and a continuous beam are supported and loaded, as shown in Figures P4.33 and P4.34, respectively. Use the area moments to determine the support reactions.

Sections 4.7 through 4.11

4.35 The uniform rod *AB* is made of steel. Collar *D* moves along the rod and has a speed of υ = 3.5 m/s as it strikes a small plate attached to end *A* of the rod (Figure P4.35). Determine the largest allowable weight of the collar.

Given: S_y = 250 MPa, E = 210 GPa

Design Requirement: A factor of safety of n = 3 is used against failure by yielding.

4.36 The 20 kg block *D* is dropped from a height *h* onto the steel beam *AB* (Figure P4.36). Determine

a. The maximum deflection of the beam

b. The maximum stress in the beam

Given: h = 0.5 m, E = 210 GPa

4.37 Collar of weight *W*, depicted in Figure P4.37, is dropped from a height *h onto* a flange at the end *B* of the round rod. Determine the *W*.

Given: h = 3.5 ft, d = 1 in., L = 15 ft, $E = 30 \times 10^6$ psi

Requirement: The maximum stress in the rod is limited to 35 ksi.

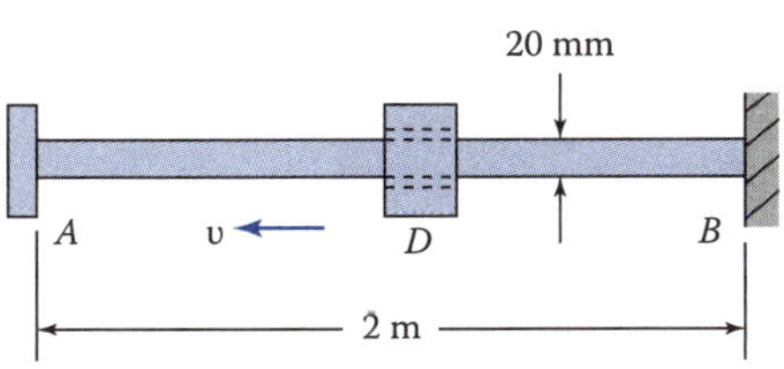

FIGURE P4.35

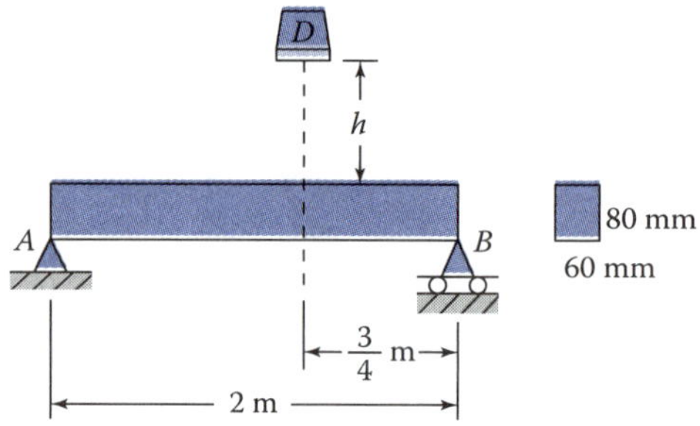

FIGURE P4.36

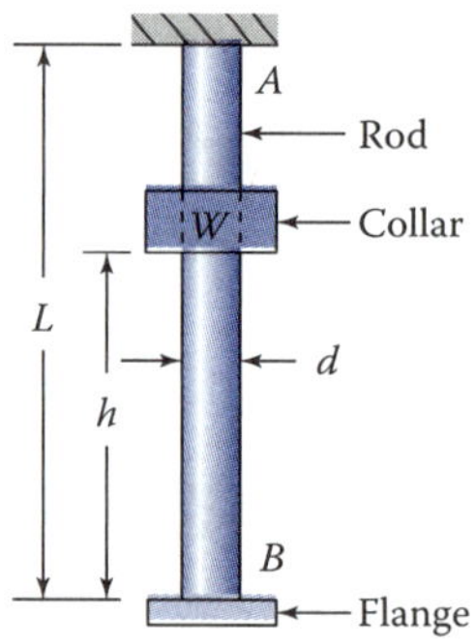

FIGURE P4.37

4.38 The collar of weight W falls a distance h when it comes into contact with end B of the round steel rod (Figure P4.37). Determine diameter d of the rod.

Given: $W = 20$ lb, $h = 4$ ft, $L = 5$ ft, $E = 30 \times 10^6$ psi

Design Requirement: The maximum stress in the rod is not to exceed 18 ksi.

4.39 The collar of weight W falls onto a flange at the bottom of a slender rod (Figure P4.37). Calculate the height h through which the weight W should drop to produce a maximum stress in the rod.

Given: $W = 500$ N, $L = 3$ m, $d = 20$ mm, $E = 170$ GPa

Design Requirement: Maximum stress in the rod is limited to $\sigma_{max} = 350$ MPa.

4.40 A block of weight W falls from a height h onto the midspan C of a simply supported beam. The beam is also reinforced at C by a spring of stiffness k as shown in Figure P4.40.

Find:

a. The maximum deflection

b. The maximum stress

Given: $h = 50$ mm, $k = 180$ kN/m, $W = 24$ kg, $E = 70$ GPa

4.41 Design the shaft (determine the minimum required length L_{min}), described in Example 4.14, so that yielding does not occur. Based on this length and the given impact load, what is the angle of twist?

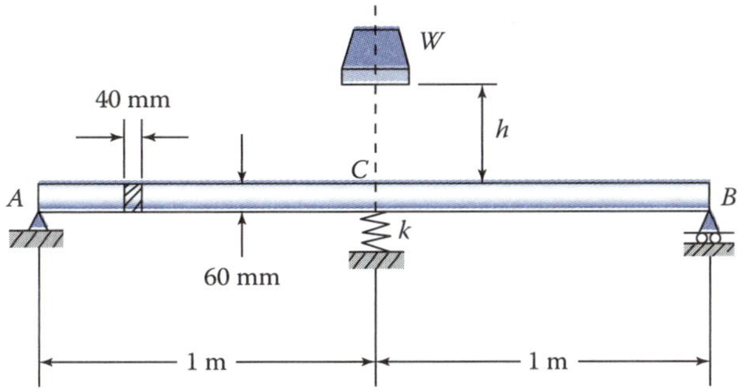

FIGURE P4.40

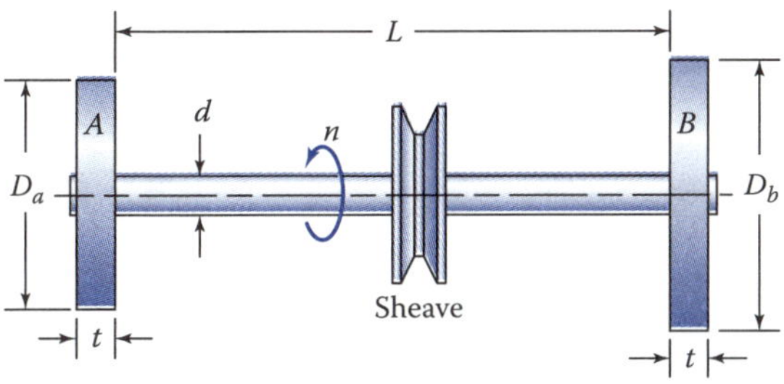

FIGURE P4.42

4.42 The steel shaft and abrasive wheels A and B at the ends of a belt-drive sheave rotates at n rpm (Figure P4.42). If the shaft is suddenly stopped at the wheel A because of jamming, determine

a. The maximum angle of twist of the shaft

b. The maximum shear stress in the shaft

Given: $D_a = 125$ mm, $D_h = 150$ mm, $d = t = 25$ mm, $L = 0.3$ m, $n = 1500$ rpm, $G = 19$ GPa, $S_{ys} = 250$ MPa, density of wheels $\rho = 1800$ kg/m^3

Assumption: Abrasive wheels are considered solid disks.

4.43 Redo Problem 4.42, for the case in which the shaft runs at $n = 1200$ rpm and the wheel end B is suddenly stopped because of jamming.

4.44 A rectangular sheet plate of thickness t is bent into circular cylinder of radius r. Determine the diameter D of the cylinder and the maximum moment M_{max} developed in the plate.

Given: $t = 1/8$ in., $E = 30 \times 10^6$ psi, $\nu = 0.3$

Design Assumption: The maximum stress in the plate is limited to 18 ksi.

4.45 A long, narrow rectangular plate is under a nonuniform loading

$$p = p_o \sin \frac{\pi y}{b}$$

and clamped at edges $y = 0$ and $y = b$ (Figure P4.45). Determine

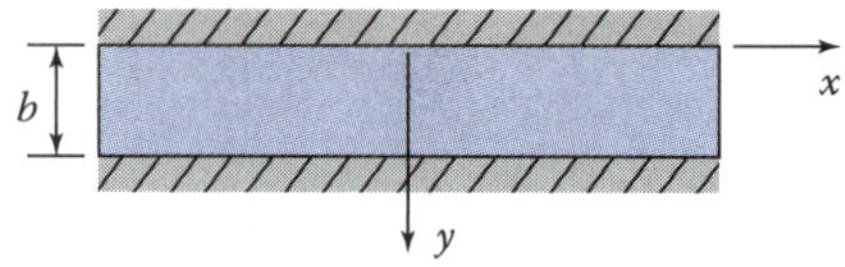

FIGURE P4.45

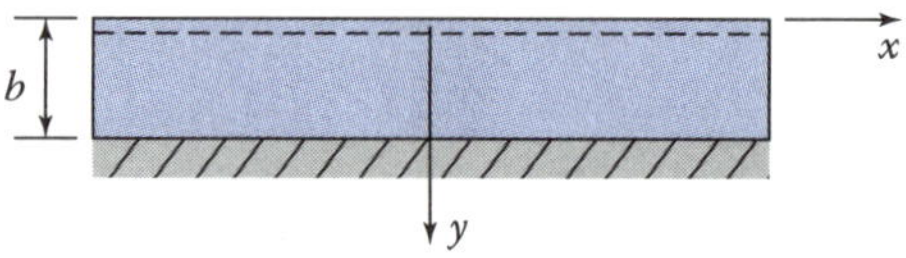

FIGURE P4.46

a. An expression for the deflection surface w
b. The maximum bending stress
c. The values of maximum deflection and stress for the data (Figure P4.45)

$$b = 20 \text{ in.}, \quad t = 0.4 \text{ in.}, \quad E = 10 \times 10^6 \text{ psi}, \quad \nu = 0.3, \quad p_o = 5 \text{ psi}$$

4.46 Figure P4.46 depicts a long, narrow rectangular plate with edge $y = 0$ simply supported and edge $y = b$ clamped. The plate is under a uniform load of intensity p_o. Determine

a. An expression for the deflection surface w
b. The maximum bending stress at $y = b$
c. The values of maximum deflection at $y = b/2$ and maximum bending stress at $y = b$. based on the data given in Problem 4.45

5

Energy Methods and Stability

5.1 Introduction

As pointed out in Section 1.4, instead of the equilibrium methods, displacements and forces can be ascertained through the use of energy methods. The latter are based on the concept of strain energy, which is of fundamental importance in analysis and design. Application of energy techniques is effective in cases involving members of variable cross sections and problems dealing with elastic stability, trusses, and frames. In particular, strain energy approaches can greatly ease the chore of obtaining the displacement of members under combined loading.

In this chapter, we explore two principal energy methods and illustrate their use with a variety of examples. The first deals with the *finite* deformation experienced by load-carrying components (Sections 5.2 through 5.6). The second, the variational methods, based on a *virtual* variation in stress or displacement, is discussed in Sections 5.7 and 5.8. Literature related to the energy approaches is extensive [1–6].

Elastic stability relates to the ability of a member or structure to support a given load without experiencing a sudden change in configuration. A buckling response leads to instability and collapse of the member. Some designs may thus be governed by the possible instability *of* a system that commonly arises in buckling of components. Here, we are concerned primarily with the column buckling, which presents but one case of structural stability [7–13]. Critical stresses in rectangular plates are discussed briefly in Section 5.16. The problem of buckling in springs is examined in Section 14.6. Buckling of thin-walled cylinders under axial loading and pressure vessels are taken up in the last section of Chapter 16, after discussing the bending of shells.

Both equilibrium and energy methods are applied in determining the critical load. The choice depends on the particulars of the problem considered. Although the equilibrium approach gives exact solutions, the results obtained by the energy approach (sometimes approximate) usually are preferred due to the physical insight that may be more readily gained. A vast number of other situations involve structural stability, such as the buckling of pressure vessels under combined loading; twist–bend buckling of shafts in torsion; lateral buckling of deep, narrow beams; and buckling of thin plates in the form of an angle or channel in compression. Analysis of such problems is mathematically complex and beyond the scope of this text.

5.2 Strain Energy

Internal work stored in an elastic body is the internal energy of deformation or the elastic strain energy. It is often convenient to use the quantity, called *strain energy per unit volume* or *strain energy density*. The area under the stress–strain diagram represents the strain energy density, designated U_o, of a tensile specimen (Figure 5.1). Therefore,

$$U_o = \int \sigma_x d\varepsilon_x \tag{5.1a}$$

The area above the stress–strain curve is termed the *complementary energy density*:

$$U_o^* = \int \varepsilon_x d\varepsilon_x \tag{5.2}$$

Observe from Figure 5.1b that, for a nonlinearly elastic material, these energy densities have different values.

In the case of a linearly elastic material, from origin up to the proportional limit, substituting σ_x/E for ε_x, we have

$$U_o = \frac{1}{2}\sigma_x\varepsilon_x = \frac{1}{2E}\sigma_x^2 \tag{5.1b}$$

and the two areas are equal $U_o = U_o^*$, as shown in Figure 5.1a. In SI units, the strain energy density is measured in joules per cubic meter (J/m^3) or in pascals; in U.S. customary units, it is expressed in inch-pounds per cubic inch (in. · lb/in.^{3}) or psi. Similarly, strain energy density for shear stress is given by

$$U_o = \frac{1}{2}\tau_{xy}\gamma_{xy} = \frac{1}{2G}\tau_{xy}^2 \tag{5.3}$$

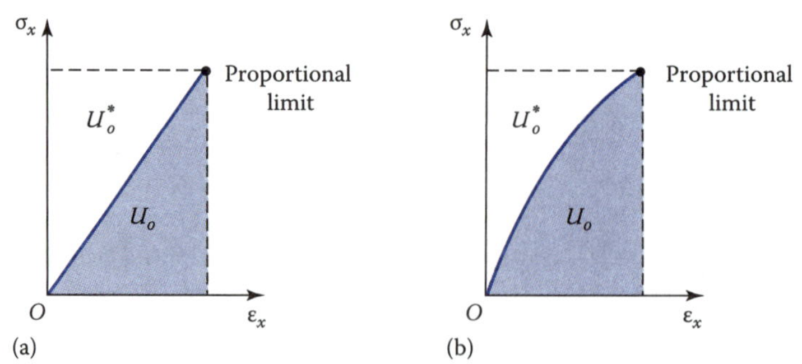

FIGURE 5.1
Work done by uniaxial stress: (a) linearly elastic material and (b) nonlinearly elastic material.

When a body is subjected to a general state of stress, the total strain energy density equals simply the sum of the expressions identical to the preceding equations. We have then

$$U_o = \frac{1}{2}\left(\sigma_x\varepsilon_x + \sigma_y\varepsilon_y + \sigma_z\varepsilon_z + \tau_{xy}\gamma_{xy} + \tau_{yz}\gamma_{yz} + \tau_{xz}\gamma_{xz}\right) \tag{5.4}$$

Substitution of generalized Hooke's law into this expression gives the following equation, involving only stresses and elastic constants:

$$U_o = \frac{1}{2E}\left[\sigma_x^2 + \sigma_y^2 + \sigma_z^2 - 2\nu(\sigma_x\sigma_y + \sigma_y\sigma_z + \sigma_x\sigma_z) + \frac{1}{2G}\left(\tau_{xy}^2 + \tau_{yz}^2 + \tau_{xz}^2\right)\right] \tag{5.5}$$

When the principal axes are used as coordinate axes, the shear stresses are 0. The preceding equation then becomes

$$U_o = \frac{1}{2E}\left[\sigma_1^2 + \sigma_2^2 + \sigma_3^2 - 2\nu\left(\sigma_1\sigma_2 + \sigma_2\sigma_3 + \sigma_1\sigma_3\right)\right] \tag{5.6}$$

in which σ_1, σ_2, and σ_3 are the principal stresses.

The elastic strain energy U stored within an elastic body can be obtained by integrating the strain energy density over the volume V. Thus,

$$U = \int_V U_o dV = \iiint U_o dxdydz \tag{5.7}$$

This equation is convenient in evaluating the strain energy for a number of commonly encountered shapes and loading. It is important to note that the strain energy is a nonlinear (quadratic) function of load or deformation. The principle of superposition therefore is not valid for the strain energy.

5.2.1 Components of Strain Energy

The 3D state of stress at a point (Figure 5.2a) may be separated into two parts. The state of stress in Figure 5.2b is associated with the volume changes, so-called dilatations. In the figure, σ_m represents the mean stress or the octahedral stress σ_{oct}, defined in Section 3.15. On the other hand, the shape changes, or distortions, are caused by the set of stresses shown in Figure 5.2c.

The dilatational strain energy density can be obtained through the use of Equation 5.6 by letting $\sigma_1 = \sigma_2 = \sigma_3 = \sigma_m$. In so doing, we have

$$U_{ov} = \frac{3(1-2\nu)}{2E}\left(\sigma_m\right)^2 = \frac{1-2\nu}{6E}\left(\sigma_1 + \sigma_2 + \sigma_3\right)^2 \tag{5.8}$$

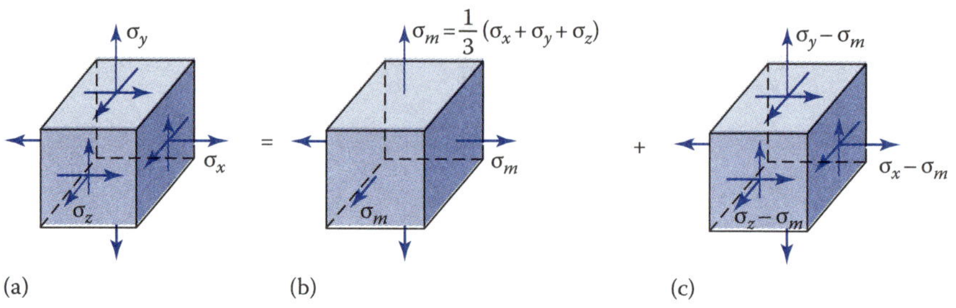

FIGURE 5.2
(a) Principal stresses, resolved into (b) dilatational stresses and (c) distortional stresses.

The *distortional strain energy density* is readily found by subtracting the foregoing from Equation 5.6:

$$U_{od} = \frac{1}{12G}\left[(\sigma_1 - \sigma_2)^2 + (\sigma_2 - \sigma_3)^2 + (\sigma_3 - \sigma_1)^2\right] = \frac{3}{4G}\tau_{oct}^2 \tag{5.9}$$

The quantities G and E are related by Equation 2.9. The octahedral planes where σ_{oct} and τ_{oct} act are shown in Figure 3.43.

Test results indicate that the dilatational strain energy is ineffective in causing failure by yielding of ductile materials. The energy of distortion is assumed to be completely responsible for material failure by inelastic action. This is discussed further in Chapter 7. Stresses and strains associated with both components of the strain energy are also very useful in describing the plastic deformation [4,5].

Example 5.1: Components of Strain Energy in a Prismatic Bar

A structural steel bar having uniform cross-sectional area A carries an axial tensile load P. Find the strain energy density and its components.

Solution

The state of stress is uniaxial tension, $\sigma_x = \sigma = P/A$, and the remaining stress components are 0 (Figure 5.2a). We therefore have the stresses causing volume change $\sigma_m = \sigma/3$ and shape change $\sigma_x - \sigma_m = 2\sigma/3$, $\sigma_y - \sigma_m = \sigma_z - \sigma_m = \sigma/3$ (Figure 5.2b and c). The strain energy densities for the stresses in Figure 5.2, from Equations 5.5, 5.8, and 5.9, are

$$U_o = \frac{\sigma^2}{2E}$$

$$U_{ov} = \frac{(1-2\nu)\sigma^2}{6E} = \frac{\sigma^2}{12E}$$

$$U_{od} = \frac{(1+\nu)\sigma^2}{3E} = \frac{5\sigma^2}{12E}$$

Comments: We observe that $U_o = U_{ov} + U_{od}$ and that $5U_{ov} = U_{od}$. That is, to change the shape of a unit volume element subjected to simple tension, five times more energy is absorbed than to change the volume.

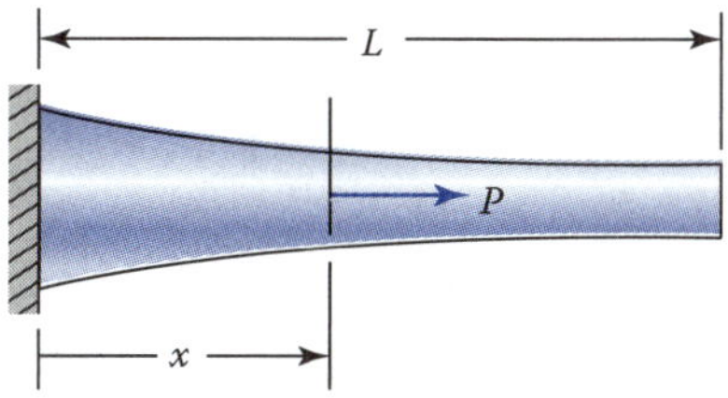

FIGURE 5.3
Nonprismatic bar with varying axial loading.

5.3 Strain Energy in Common Members

Recall from Section 5.2 that the method of superposition is not applicable to strain energy; that is, the effects of several forces or moments on strain energy are not simply additive, as illustrated in Example 5.2. In this section, the following types of loads are considered for the various members of a structure: axial loading, torsion, bending, and shear. Note that the equations derived are restricted to linear material behavior.

5.3.1 Axially Loaded Bars

The normal stress at any transverse section through a bar subjected to an axial load P is $\sigma_x = P/A$, where A represents the cross-sectional area and x is the axial axis (Figure 5.3).

Substitution of this and Equation 5.1 into Equation 5.7 and setting $dV = Adx$, we obtain

$$U_a = \frac{1}{2}\int_0^L \frac{P^2 dx}{AE} \tag{5.10}$$

For a prismatic bar, subjected to end forces of magnitude P, Equation 5.10 becomes

$$U_a = \frac{P^2 L}{2AE} \tag{5.11}$$

The quantity E represents the modulus of elasticity and L is the length of the member.

Example 5.2: Strain Energy Stored in a Bar due to Combined Loading

A prismatic bar suspended from one end carries, in addition to its own weight, an axial load F (Figure 5.4). Find the strain energy stored in the member.

Solution

The axial load, in an element at a distance x from the fixed end, is expressed as

$$P = \gamma A(L - x) + F$$

Here,
γ represents the weight of the material
A is the cross-sectional area

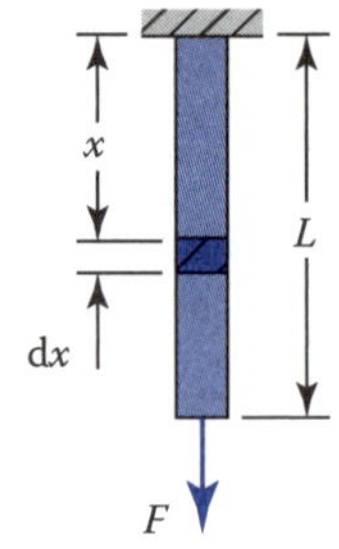

FIGURE 5.4
Example 5.2.

Introducing this equation into Equation 5.10,

$$U_a = \int_0^L \frac{\left[\gamma A(L-x)+F\right]^2}{2AE}\,dx = \frac{\gamma^2 AL^3}{6E} + \frac{\gamma FL^2}{2E} + \frac{F^2L}{2AE} \tag{5.12}$$

Comments: The first and third terms on the right side of this equation are the strain energy of the bar due to its own weight and the strain energy of the bar supporting only axial force *F*. The presence of the middle term shows that the strain energy produced by the two loads acting simultaneously is not equal to the sum of the strain energies associated with the loads acting separately.

Example 5.3: Energy Absorbed by a Bolt Fastener

A stainless (302) cold-rolled steel bolt is under a tension force *P* when used as a fastener as shown in Figure 5.5. The shank and thread diameters of the bolt are d_s and d_t, respectively. Detailed discussion of threaded fasteners will be taken up in Chapter 15.

Given: Prescribed numerical values are $d_s = 0.75$ in., $d_t = 0.63$ in., $L_s = 3$ in., $L_t = 0.115$ in., $E = 28 \times 10^6$ psi, $S_y = 75$ ksi (by Table B.1)

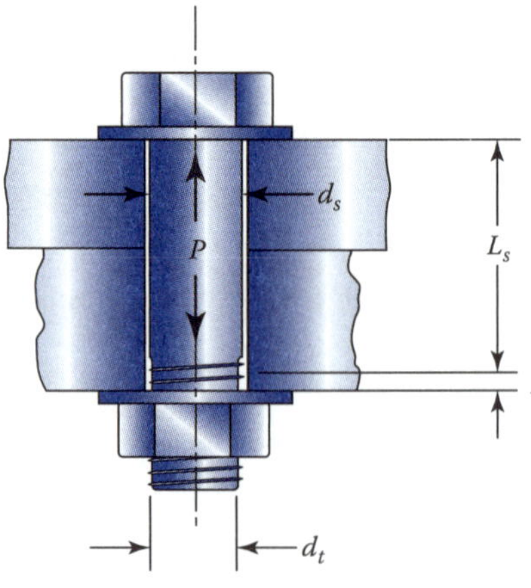

FIGURE 5.5
Example 5.3. A bolted connection.

Find: (a) The maximum tension force that bolt can carry without yielding and (b) the maximum elastic strain energy that bolt can absorb.

Assumption: The extra material that makes up the threads can be disregarded.

Solution

The major (shank) and minor (thread root) cross-sectional areas of the bolt, respectively, are

$$A_s = \frac{\pi}{4}d_s^2 = \frac{\pi}{4}(0.75)^2 = 0.442 \text{ in.}^2$$

$$A_t = \frac{\pi}{4}d_t^2 = \frac{\pi}{4}(0.63)^2 = 0.312 \text{ in.}^2$$

a. *Allowable load*. The largest stress of 75 ksi takes place within region $L_t = 0.115$ in. It follows that

$$P_{\text{all}} = S_y A_t = 75(0.312) = 23.4 \text{ kips}$$

b. *Strain energy capacity of bolt*. Applying Equation 5.11 within each portion, we have

$$U = \Sigma \frac{P_{\text{all}}^2 L}{2AE} = \frac{P_{\text{all}}^2}{2E}\left(\frac{L_s}{A_s} + \frac{L_t}{A_t}\right) \tag{5.13}$$

Introducing the data leads to

$$U = \frac{(23.4)^2}{2(28)}\left(\frac{3}{0.442} + \frac{0.115}{0.312}\right)$$
$$= 9.778(6.787 + 0.369) = 70 \text{ in.} \cdot \text{lb}$$

Comments: It is interesting to note that, if the bolt has a uniform diameter of $d_t = 0.63$ in. throughout its 3.115 in. length, we obtain

$$U = 9.778\left(\frac{3.115}{0.312}\right) = 97.6 \text{ in.} \cdot \text{lb}$$

It would then absorb about 28% *more* elastic energy, although it has a *smaller* cross section along its shank.

5.3.2 Circular Torsion Bars

In the case of pure torsion of a bar, Equation 3.11 for an arbitrary distance r from the centroid of the cross section gives $\tau = Tr/J$. The strain energy density, Equation 5.3, becomes then $U_o = T^2r^2/2J^2G$. Inserting this into Equation 5.7, the strain energy owing to torsion is

$$U_t = \int_0^L \frac{T^2}{2GJ^2}\left(\int r^2 \mathrm{d}A\right)\mathrm{d}x \tag{5.14}$$

We have $dV = dA\ dx$; dA is an element of the cross-sectional area. By definition, the term in parentheses is the polar moment of inertia J of the cross-sectional area. Hence,

$$U_t = \frac{1}{2}\int_0^L \frac{T^2 dx}{GJ} \tag{5.15}$$

For a prismatic bar subjected to end torques T (Figure 3.6), Equation 5.15 appears as

$$U_t = \frac{T^2 L}{2GJ} \tag{5.16}$$

in which L is the length of the bar.

5.3.3 Beams

Consider a beam in pure bending. The flexure formula gives the axial normal stress $\sigma_x = My/I$. Using Equation 5.1, the strain energy density is $U_o = M^2y^2/2EI^2$. After carrying this into Equation 5.7 and noting that $M^2/2EI^2$ is a function of x alone, we obtain

$$U_b = \int_0^L \frac{M^2}{2EI^2}\left(\int y^2 dA\right) dx \tag{5.17}$$

Since the integral in parentheses defines the moment of inertia I of the cross-sectional area about the neutral axis, the strain energy due to *bending* is

$$U_b = \frac{1}{2}\int_0^L \frac{M^2 dx}{EI} \tag{5.18}$$

This, integrating along beam length L, gives the required quantity.

For a beam of constant flexural rigidity EI, Equation 5.18 may be written in terms of deflection by using Equation 4.14 as follows:

$$U_b = \frac{EI}{2}\int_0^L \left(\frac{d^2 \upsilon}{dx^2}\right)^2 dx \tag{5.19}$$

The transverse shear force V produces shear stress τ_{xy} at every point in the beam. The strain energy density, inserting Equation 3.20 into Equation 5.3, is $U_o = V^2Q^2/2GI^2b^2$. Integration of this, over the volume of the beam of cross-sectional area A, results in the strain energy for beams in *shear*:

$$U_s = \int_0^L \frac{\alpha V^2 dx}{2AG} \tag{5.20}$$

TABLE 5.1

Form Factor for Shear for Various Beam Cross Sections

Cross Section	Form Factor α
Rectangle	6/5
I section, box section, or channels[a]	A/A_{web}
Circle	10/9
Thin-walled circular	2

[a] *A*, area of the entire section; A_{web}, area of the web *ht*, where *h* is the beam depth and *t* is the web thickness.

In Equation 5.20, the *form factor for shear* is

$$\alpha = \frac{A}{I^2}\int_A \frac{Q^2}{b^2}\mathrm{d}A \tag{5.21}$$

This represents a dimensionless quantity specific to a given cross-sectional geometry.

Example 5.3 illustrates the determination of the form factor for shear for a rectangular cross section. Other cross sections can be treated similarly. Table 5.1 furnishes several cases [4]. Subsequent to finding α, the strain energy due to shear is obtained using Equation 5.20.

Example 5.4: Total Strain Energy Stored in a Beam

A cantilevered beam with a rectangular cross section supports a concentrated load *P* as depicted in Figure 5.6. Find the total strain energy and compare the values of the bending and shear contributions.

Solution

The first moment of the area, by Equation 3.21, is $Q = (b/2)\left[(h/2)^2 - y_1^2\right]$. Inasmuch as $A/I^2 = 144/bh^5$, Equation 5.21 gives

$$\alpha = \frac{144}{bh^5}\int_{-h/2}^{h/2} \frac{1}{4}\left(\frac{h^2}{4} - y_1^2\right)^2 b\mathrm{d}y_1 = \frac{6}{5}$$

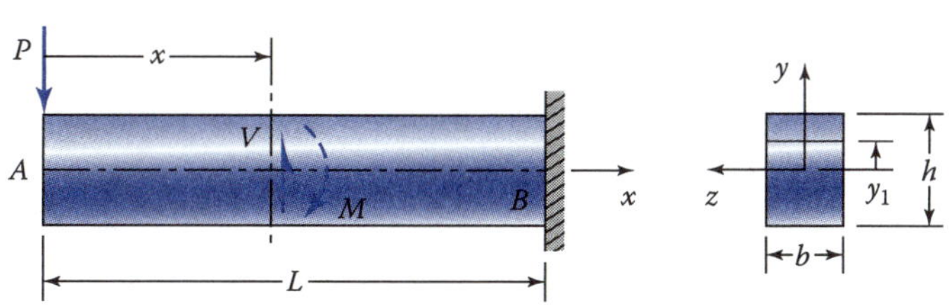

FIGURE 5.6
Examples 5.4 and 5.5.

From the equilibrium requirements, the bending moment $M = -Px$ and the shear force $V = P$ at x (Figure 5.6). Carrying these and $\alpha = 6/5$ into Equations 5.18 and 5.20 then integrating, we obtain

$$U_b = \int_0^L \frac{P^2x^2}{2EI}\,dx = \frac{P^2L^3}{6EI} \tag{5.22}$$

$$U_s = \int_0^L \frac{6}{5}\frac{V^2}{2AG}\,dx = \frac{3P^2L}{5AG} \tag{5.23}$$

Note that $I/A = h^2/12$. The total strain energy stored in the cantilever beam is

$$U = U_b + U_s = \frac{P^2L^3}{6EI}\left[1 + \frac{3E}{10G}\left(\frac{h}{L}\right)^2\right] \tag{5.24}$$

Through the use of Equations 5.22 and 5.23, we find the ratio of the shear strain energy to the bending strain energy in the beam as follows:

$$\frac{U_s}{U_b} = \frac{3E}{10G}\left(\frac{h}{L}\right)^2 = \frac{3}{5}(1+\nu)\left(\frac{h}{L}\right)^2 \tag{5.25}$$

Comments: When, for example, $L = 10h$ and $\nu = 1/3$, this quotient is only 1/125: the strain energy owing to the shear is less than 1%. For a slender beam, $h \ll L$, it is observed that the energy is due mainly to bending. Therefore, it is usual to neglect the shear in evaluating the strain energy in beams of ordinary proportions. Unless stated otherwise, we adhere to this practice.

5.4 Work–Energy Method

The strain energy of a structure subjected to a set of forces and moments may be expressed in terms of the forces and resulting displacements. Suppose that all forces are applied gradually and the final values of the force and displacement are denoted by P_k ($k = 1, 2, \ldots, n$) and δ_k, the total work $W, 1/2\sum P_k\delta_k$ is equal to the strain energy gained by the structure, provided no energy is dissipated. That is,

$$U = W = \frac{1}{2}\sum_{k=1}^{n} P_k\delta_k \tag{5.26}$$

In other words, the work done by the loads acting on the structure manifests as elastic strain energy.

Consider a member or structure subjected to a single concentrated load P. Equation 5.26 then becomes

$$U = \frac{1}{2}P\delta \tag{5.27}$$

The quantity δ is the displacement through which the force P moves. In a like manner, it can be shown that

$$U = \frac{1}{2} M\theta \tag{5.28}$$

$$U = \frac{1}{2} T\phi \tag{5.29}$$

Note that M (or T) and θ (or ϕ) are, respectively, the moment (or torque) and the associated slope (or angle of twist) at a point of a structure.

The foregoing relationships provide a convenient approach for finding the displacement. This is known as the *method of work–energy*. In the next section, we present a more general approach that may be used to obtain the displacement at a given structure even when the structure carries combined loading.

Example 5.5: Beam Deflection by the Work–Energy Method

A cantilevered beam with a rectangular cross section is loaded as shown in Figure 5.6. Find the deflection υ_A at the free end by considering the effects of both the internal bending moments and shear force.

Solution

The total strain energy U of the beam, given by Equation 5.24, is equated to the work, $W = P\upsilon_A/2$. Hence,

$$\upsilon_A = \frac{PL^3}{3EI}\left[1 + \frac{3E}{10G}\left(\frac{h}{L}\right)^2\right] \tag{5.30}$$

Comment: If the effect of shear is disregarded, note that the relative error is identical to that found in the previous example. As already shown, this is less than 1% for a beam with ratio $L/h = 10$.

5.5 Castigliano's Theorem

Castigliano's theorems are in widespread use in the analysis of structural displacements and forces. They apply with ease to a variety of statically determinate as well as indeterminate problems. Two theorems were proposed in 1879 by A. Castigliano (1847–1884). The first theorem relies on a virtual (imaginary) variation in deformation and is discussed in Section 5.7. The second concerns the finite deformation experienced by a member under load. Both theorems are limited to small deformations of structures. The first theorem is pertinent to structures that behave nonlinearly as well as linearly. We deal mainly with Castigliano's second theorem, which is restricted to structures composed of linearly elastic materials. Unless specified otherwise, we refer in this text to the second theorem as *Castigliano's theorem*.

Consider a linearly elastic structure subjected to a set of gradually applied external forces P_k ($k = 1, 2, \ldots, n$). Strain energy U of the structure is equal to the work done W by the applied forces, as given by Equation 5.26. Let us permit a single load, say, P_i, to be increased at a small amount dP_i, while the other applied forces P_k remain unchanged. The increase in strain energy is then $dU = (\partial U/\partial P_i)\, dP_i$ where $\partial U/\partial P_i$ represents the rate of change of the strain energy with respect to P_i. The total energy is

$$U' = U + \left(\frac{\partial U}{\partial P_i}\right) dP_i$$

Alternatively, an expression for U' may be written by reversing the order of loading. Suppose that dP_i is applied first, followed by the force P_k. Now the application of dP_i causes a small displacement $d\delta_i$. The work, $dP_i \cdot d\delta_i/2$, corresponding to this load increment, can be omitted because it is of the second order. The work done during the application of the forces P_k is unaffected by the presence of dP_i. But the latter force dP_i performs work in moving an amount δ_i. Here, δ_i is the displacement caused by the application of P_k. The total strain energy due to the work done by this sequence of loads is therefore

$$U' = U + dP_i \cdot \delta_i$$

Equating the preceding equations, we have the Castigliano's theorem:

$$\delta_i = \frac{\partial U}{\partial P_i} \tag{5.31}$$

The foregoing states that, for a linear structure, *the partial derivative of the strain energy with respect to an applied force is equal to component of displacement at the point of application and in the direction of that force.*

Castigliano's theorem can similarly be shown to be valid for applied moments M (or torques T) and the resulting slope θ (or angle of twist ϕ) of the structure. Therefore,

$$\theta_i = \frac{\partial U}{\partial M_i} \tag{5.32}$$

$$\phi = \frac{\partial U}{\partial T_i} \tag{5.33}$$

In using Castigliano's theorem, we must express the strain energy in terms of the external forces or moments. In the case of a prismatic beam, we have $U = \int M^2 dx/2EI$. To obtain the deflection υ_i corresponding to load P_i, it is often much simpler to differentiate under the integral sign. In so doing, we have

$$\upsilon_i = \frac{\partial U}{\partial P_i} = \frac{1}{EI}\int M \frac{\partial M}{\partial P_i} dx \tag{5.34}$$

Similarly, the slope may be expressed as

$$\theta_i = \frac{\partial U}{\partial M_i} = \frac{1}{EI}\int M \frac{\partial M}{\partial M_i} \mathrm{d}x \tag{5.35}$$

Generally, the total strain energy in a straight or curved member subjected to a number of common loads (axial force *F*, bending moment *M*, shear force *V*, and torque *T*) equals the sum of the strain energies given by Equations 5.11, 5.16, 5.18, and 5.20. So applying Equation 5.31, the displacement at any point in the member is obtained in the following convenient form:

$$\delta_i = \frac{1}{AE}\int F \frac{\partial F}{\partial P_i} \mathrm{d}x + \frac{1}{EI}\int M \frac{\partial M}{\partial P_i} \mathrm{d}x + \frac{1}{AG}\int \alpha V \frac{\partial V}{\partial P_i} \mathrm{d}x + \frac{1}{GJ}\int T \frac{\partial T}{\partial P_i} \mathrm{d}x \tag{5.36}$$

Clearly, the last term of this equation applies only to circular bars. An expression may be written for the angle of rotation in a like manner. If it is necessary to obtain the displacement at a point where no corresponding load acts, the problem is treated as follows. We place a fictitious force *Q* (or couple C) at the point in question in the direction of the desired displacement δ (or θ). We then apply Castigliano's theorem and set the fictitious load $Q = 0$ (or $C = 0$) to obtain the desired displacements.

Example 5.6: Deflection of a Simple Beam by Castigliano's Theorem

A simple beam is subjected to a uniform load of intensity *w*, as shown in Figure 5.7a. What is the deflection υ_C at an arbitrary distance *a* from the left support?

Solution

As the deflection is sought, a fictitious force *Q* is introduced at point *C* (Figure 5.7b). From the conditions of equilibrium, the reactions are found to be

$$R_A = \frac{wL}{2} + \frac{Qb}{L}, \quad R_B = \frac{wL}{2} + \frac{Qa}{L}$$

The appropriate moment equations are then

$$M_{AC} = R_A x - \frac{wx^2}{2}, \quad M_{BC} = R_B x' - \frac{wx'^2}{2}$$

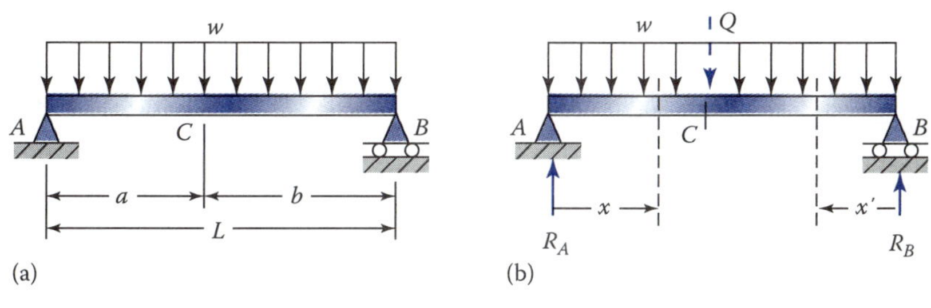

FIGURE 5.7
Example 5.6.

Therefore,

$$\frac{\partial M_{AC}}{\partial Q} = \frac{bx}{L}, \quad \frac{\partial M_{BC}}{\partial Q} = \frac{ax'}{L}$$

Introducing the foregoing into Equation 5.36, we have

$$\upsilon_C = \frac{1}{EI}\left\{\int_0^a \left[\left(\frac{wL}{2} + \frac{Qb}{L}\right)x - \frac{wx^2}{2}\right]\left(\frac{bx}{L}\right)\mathrm{d}x + \int_0^b \left[\left(\frac{wL}{2} + \frac{Qa}{L}\right)x' - \frac{wx'^2}{2}\right]\left(\frac{ax'}{L}\right)\mathrm{d}x'\right\}$$

Setting $Q = 0$ leads to

$$\upsilon_C = \frac{w}{2EIL}\left[\int_0^a (Lx - x^2)(bx)\mathrm{d}x + \int_0^b (Lx' - x'^2)(ax')\mathrm{d}x'\right]$$

Integration results in

$$\upsilon_C = \frac{wab}{24EIL}\left[4L(a^2 + b^2) - 3(a^3 + b^3)\right] \tag{a}$$

Note, as a check, that, for $a = b$, the preceding equation reduces to $5wL^4/384EI$ (see case 8 of Table A.9).

Example 5.7: Deflection of a Curved Frame Using Castigliano's Theorem

A load of P is applied to a steel curved frame, as depicted in Figure 5.8a. Develop an expression for the vertical deflection δ_v of the free end by considering the effects of the internal normal and shear forces in addition to the bending moment. Calculate the value of δ_v for the following data:

$$a = 60 \text{ mm}, \quad h = 30 \text{ mm}, \quad b = 15 \text{ mm},$$
$$P = 10 \text{ kN}, \quad E = 210 \text{ GPa}, \quad G = 80 \text{ GPa}.$$

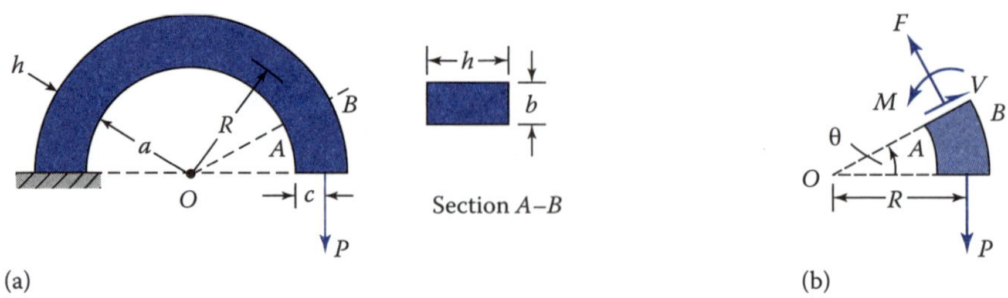

FIGURE 5.8
Example 5.7. (a) steel curved frame and (b) free-body diagram.

Solution

A free-body diagram of the part of the bar defined by angle θ is depicted in Figure 5.8b, where the internal forces (*F* and *V*) and moment (*M*) are positive as shown. Referring to the figure, we write

$$M = PR(1-\cos\theta), \quad V = P\sin\theta, \quad F = P\cos\theta \tag{5.37}$$

Therefore,

$$\frac{\partial M}{\partial P} = R(1-\cos\theta), \quad \frac{\partial V}{\partial P} = \sin\theta, \quad \frac{\partial F}{\partial P} = \cos\theta$$

The form factor for shear for the rectangular section is $\alpha = 6/5$ (Table 5.1). Substitution of the preceding expressions into Equation 5.36 with $dx = R\, d\theta$ results in

$$\delta_v = \frac{PR^3}{EI}\int_0^{\pi}(1-\cos\theta)^2 d\theta + \frac{6PR}{5AG}\int_0^{\pi}\sin^2\theta d\theta + \frac{PR}{AE}\int_0^{\pi}\cos^2\theta d\theta \tag{b}$$

Using the trigonometric identities $\cos^2\theta = (1 + \cos2\theta)/2$ and $\sin^2\theta = (1 - \cos2\theta)/2$, we obtain, after integration,

$$\delta_v = \frac{3\pi PR^3}{2EI} + \frac{3\pi PR}{5AG} + \frac{\pi PR}{2AE} \tag{c}$$

The geometric properties of the cross section of the bar are

$$A = 0.015(0.03) = 4.5\times10^{-4}\ \text{m}^2, \quad R = 0.075\ \text{m},$$

$$I = \frac{1}{12}\left[0.015(0.03)^3\right] = 337.5\times10^{-10}\ \text{m}^4$$

Carrying these values into Equation (c) gives

$$\delta_v = (2.81 + 0.04 + 0.01)\times10^{-3} = 2.86\ \text{mm}$$

Comments: If the effects of the normal and shear forces are neglected, we have $\delta_v = 2.81$ mm. Then, the error in deflection is approximately 1.7%. For this curved bar, where $R/c = 5$, the contribution of *V* and *F* to the displacement can therefore be disregarded. It is common practice to neglect the first and the third terms in Equation 5.36 when $R/c > 4$.

Example 5.8: Displacements of a Split Ring by Castigliano's Theorem

A slender, cross-sectional, circular ring of radius *R* is cut open at θ = 0, as shown in Figure 5.9a. The ring is fixed at one end and loaded at the free end by a *z*-directed force *P*. Find, at the free end *A*,

a. The *z*-directed displacement
b. The rotation about the *y* axis

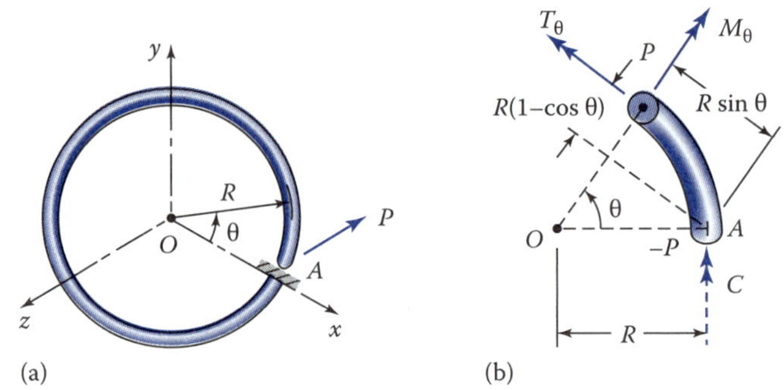

FIGURE 5.9
Example 5.8.

Solution

Since the rotation is sought, a fictitious couple C is applied at point A. The bending and twisting moments at any section are (Figure 5.9b)

$$M_\theta = -PR\sin\theta - C\sin\theta, \quad T_\theta = PR(1-\cos\theta) - C\cos\theta \tag{5.38}$$

a. Substitution of these quantities with $C = 0$ and $dx = Rd\theta$ into Equation 5.36 gives

$$\delta_z = \frac{\partial U}{\partial P} = \frac{1}{EI}\int_0^{2\pi}(-PR\sin\theta)(-R\sin\theta)Rd\theta + \frac{1}{GJ}\int_0^{2\pi} PR(1-\cos\theta)R(1-\cos\theta)Rd\theta$$

$$= \frac{\pi PR^3}{EI} + \frac{3\pi PR^3}{GJ}$$

b. Introducing Equations 5.38 into 5.35 with $dx = Rd\theta$ and setting $C = 0$, we have

$$\theta_y = \frac{\partial U}{\partial C} = \frac{1}{EI}\int_0^{2\pi}(-PR\sin\theta)(-\sin\theta)Rd\theta + \frac{1}{GJ}\int_0^{2\pi} PR(1-\cos\theta)(-\cos\theta)Rd\theta$$

$$= \frac{\pi PR^2}{EI} + \frac{\pi PR^2}{GJ}$$

5.5.1 Application to Trusses

We now apply Castigliano's theorem to plane trusses. As pointed out in Section 1.9, it is assumed that the connection between the members is pinned and the only force in the member is an axial force, either tensile or compressive. Note that, in practice, members of a plane truss are usually riveted, welded, or bolted together by means of so-called gusset plates. However, due to the slenderness of the members, the internal forces can often be computed on the basis of frictionless joints that prevent translation in two directions, corresponding to reactions of two unknown force components. The methods of joints and sections are commonly used for the analysis of trusses. The *method of joints* consists of analyzing the truss, joint by joint, to determine the forces in the members by applying the conditions of equilibrium to the free-body diagram for each joint. This approach is most effective when the forces in all members of a truss are to be determined. If, however, the force in only a few members of a truss is desired, the method of sections applied to a portion (containing two or more joints) of the truss isolated by imagining that the members in which the forces are to be ascertained is cut.

The strain energy U for a truss is equal to the sum of the strain energies of its members. In the case of a truss consisting m members of length L_j, axial rigidity $A_j E_j$, and internal axial force F_j, the strain energy can be found from Equation 5.11 as

$$U = \sum_{j=1}^{m} \frac{F_j^2 L_j}{2A_j E_j} \tag{5.39}$$

The displacement δ_i of the joint of application of load P_i can be obtained by substituting Equation 5.39 into Castigliano's theorem, Equation 5.31. Therefore,

$$\delta_i = \frac{\partial U}{\partial P_i} = \sum_{j=1}^{m} \frac{F_j L_j}{A_j E_j} \frac{\partial F_j}{\partial P_i} \tag{5.40}$$

The preceding discussion applies to statically determinate and indeterminate linearly elastic trusses.

Example 5.9: Displacements of a Crane Boom by Castigliano's Theorem

A planar truss with pin and roller supports at A and B, respectively, is subjected to a vertical load P at joint E, as shown in Figure 5.10. Determine

a. The vertical displacement of point E
b. The horizontal displacement of point E

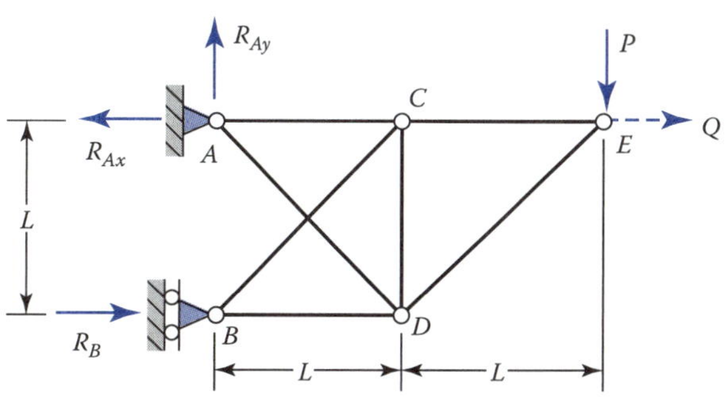

FIGURE 5.10
Example 5.9.

Assumption: All members are of equal axial rigidity AE.

Solution

a. The equilibrium conditions for the entire truss (Figure 5.10) result in $R_{Ax} = 2P$, $R_{Ay} = P$, and $R_B = 2P$. Applying the method of joints at A, E, C, and D, we obtain

$$F_{AD} = \sqrt{2}P, \quad F_{AC} = F_{CE} = P, \quad F_{DE} = -\sqrt{2}P,$$
$$F_{BC} = F_{CD} = 0, \quad F_{BD} = -2P \tag{d}$$

Through the use of Equation 5.40,

$$\begin{aligned}\delta_v &= \frac{1}{AE}\Sigma FL\frac{\partial F}{\partial P} \\ &= \frac{1}{AE}\left[2(PL)(1)+2\left(\sqrt{2}P\right)\left(\sqrt{2}L\right)\left(\sqrt{2}\right)+(-2P)(L)(-2)\right] \\ &= 11.657\frac{PL}{AE}\end{aligned}$$

The positive sign for δ_v indicates that the displacement has the same sign as that assumed for P; it is downward.

b. As the horizontal displacement is sought, a fictitious load Q is applied at point E (Figure 5.10). Now $F_{AC} = F_{CE} = P + Q$; all other forces remain the same as given by Equation (d). From Equation 5.40, we have

$$\delta_h = \frac{1}{AE}\Sigma FL\frac{\partial F}{\partial Q} = \frac{2L}{AE}(P+Q)$$

However, $Q = 0$, and the preceding reduces to

$$\delta_h = 2\frac{PL}{AE}$$

5.6 Statically Indeterminate Problems

Castigliano's theorem or the unit load method may be applied as a supplement to the equations of statics in the solutions of support reactions of a statically indeterminate structure. Consider, for instance, a structure indeterminate to the first degree. In this case, we select one of the reactions as the redundant (or unknown) load, say, R_1, by removing the corresponding support. All external forces, including both loads and redundant reactions, must generate displacements compatible with the original supports. We first express the strain energy in terms of R_1 and the given loads. Equation 5.34 may be applied at the removed support and equated to the given displacement:

$$\frac{\partial U}{\partial R_1} = \delta_1 = 0 \tag{5.41}$$

This expression is solved for the redundant reaction R_1. Then, we can find from static equilibrium the other reactions.

In an analogous manner, the case of statically indeterminate structure with n redundant reactions, assuming no support movement, can be expressed in the following form:

$$\frac{\partial U}{\partial R_1} = 0, \ldots, \frac{\partial U}{\partial R_n} = 0 \tag{5.42}$$

By solving these equations simultaneously, the magnitude of the redundant reactions is obtained. The remaining reactions are found from equations of equilibrium. Note that Equation 5.42, Castigliano's second theorem, is also referred to as the *principle of least work* in some literature.

Analytical techniques are illustrated in solution of the following sample problems.

Example 5.10: Reactions of a Propped Cantilevered Beam Using Castigliano's Theorem

A propped cantilevered beam carries a concentrated load P at its midspan (Figure 5.11a). Find the support reactions.

Solution

The free-body diagram of Figure 5.11b shows that the problem is statically indeterminate to the first degree. We select R_A as redundant; hence, the expressions for the moments are

$$M_{AD} = R_A x, \quad M_{DB} = R_A x - P\left(x - \frac{L}{2}\right)$$

It is important to note that the remaining unknowns, R_B and M_B, do not appear in preceding equations.

The deflection υ_A at A must be 0. We have $\partial M_{AD}/\partial R_A = x$ and $\partial M_{DB}/\partial R_A = x$. Equation 5.34 is thus

$$\upsilon_A = \frac{1}{EI}\left\{\int_0^{L/2} R_A x(x)\mathrm{d}x + \int_{L/2}^{L}\left[R_A x - P\left(x - \frac{L}{2}\right)\right] x\mathrm{d}x\right\} = 0$$

The preceding, after integrating, results in

$$R_A = \frac{5}{16}P$$

The other two reactions are $R_B = 11P/16$ and $M_B = 3PL/16$, as determined from the equations of statics.

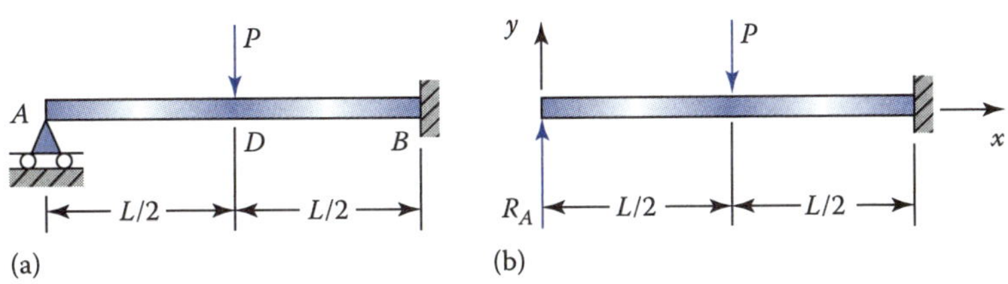

FIGURE 5.11
Example 5.10.

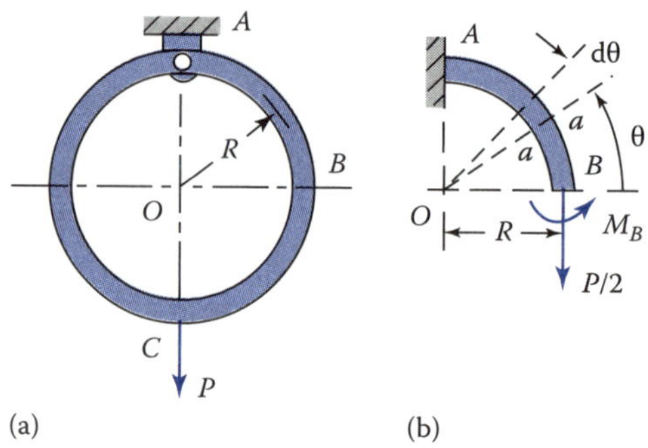

FIGURE 5.12
Example 5.11. (a) ring and (b) quarter segment.

Example 5.11: Deflection of a Ring by Castigliano's Theorem

A ring of radius R is hinged and subjected to force P as shown in Figure 5.12a. Taking into account only the strain energy due to bending, determine the vertical displacement of point C.

Solution

Owing to symmetry, it is necessary to analyze only a quarter segment (Figure 5.12b). Inasmuch as M_A and M_B are unknowns, the problem is statically indeterminate. The moment at any section A-A is

$$M_\theta = M_B - \frac{1}{2}PR(1-\cos\theta)$$

Since the slope is 0 at B, substituting this expression into Equation 5.35 with $dx = Rd\theta$, we have

$$\theta_B = \frac{1}{EI}\int_0^{\pi/2}\left[M_B - \frac{1}{2}PR(1-\cos\theta)\right](1)Rd\theta = 0$$

from which $M_B = PR[1 - (2/\pi)]/2$. The first equation then becomes

$$M_\theta = \frac{PR}{2}\left(\cos\theta - \frac{2}{\pi}\right)$$

The displacement of point C, by Equation 5.34, is

$$\delta_C = \frac{4}{EI}\int_0^{\pi/2} M_\theta \frac{\partial M_\theta}{\partial P} Rd\theta$$

$$= \frac{4}{EI}\int_0^{\pi/2} P\left[\frac{R}{2}\left(\cos\theta - \frac{2}{\pi}\right)\right]^2 Rd\theta$$

Integration of the foregoing results in

$$\delta_C = 0.15\frac{PR^3}{EI} \tag{a}$$

The positive sign of δ_C means that the displacement has the same sense as the force P, as shown.

Example 5.12: Deflection and Reaction of a Frame Using Castigliano's Theorem

A frame of constant flexural rigidity EI supports a downward load P, as depicted in Figure 5.13a. Find

a. The deflection at E
b. The horizontal reaction at E, if the point E is a fixed pin (Figure 5.13b)

Solution

a. Inasmuch as a displacement is sought, a fictitious force Q is introduced at point E (Figure 5.13a). Because of symmetry about a vertical axis through point C, we need to write expressions for moment associated with segments ED and DC, respectively:

$$M_1 = Qx, \quad M_2 = Qa + \frac{Px}{2}$$

The horizontal displacement at E is found by substituting these equations into Equation 5.36:

$$\delta_E = \frac{1}{EI}\int_0^a (Qx)x\,dx + \frac{2}{EI}\int_0^{b/2}\left(Qa + \frac{Px}{2}\right)a\,dx$$

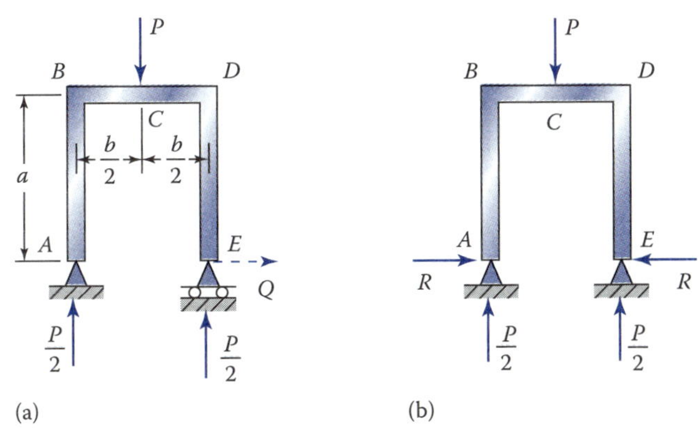

FIGURE 5.13
Example 5.12.

Setting $Q = 0$ and integrating,

$$\delta_E = \frac{Pab^2}{8EI} \tag{b}$$

b. The problem is now statically indeterminate to the first degree; there is one unknown reaction after satisfying three equations of statics (Figure 5.13b). Since $\delta_E = 0$, setting $Q = -R$, we equate the deflection given by Equation (a) to zero. In so doing, we have

$$\frac{1}{EI}\int_0^a (-Rx)x\mathrm{d}x + \frac{2}{EI}\int_0^{b/2}\left(-Ra + \frac{Px}{2}\right)a\mathrm{d}x = 0$$

The preceding gives

$$R = \frac{3}{8}\frac{Pb^2}{a^2 + 3ab} \tag{c}$$

5.7 Virtual Work Principle

In connection with virtual work, we use the symbol δ to denote a virtual infinitesimally small quantity. An arbitrary or imaginary incremental displacement is termed a *virtual displacement*. A virtual displacement results in no geometric alterations of a member. It also must not violate the boundary or support conditions for a structure. In the brief development to follow, p_x, p_y, and p_z represent the x, y, and z components of the surface forces per unit area and the body forces are taken to be negligible.

The virtual work, δW, done by surface forces on a member under a virtual displacement is given by

$$\delta W = \int \left(p_x\delta u + p_y\delta \upsilon + p_z\delta w\right)\mathrm{d}A \tag{5.43}$$

The quantity A is the boundary surface area, and δu, $\delta \upsilon$, and δw represent the x-, y-, and z-directed components of a virtual displacement. In a like manner, the virtual strain energy, δU, acquired of a member of volume V caused by a virtual straining is expressed as follows:

$$\delta U = \int_V \left(\sigma_x\delta\varepsilon_x + \sigma_y\delta\varepsilon_y + \sigma_z\delta\varepsilon_z + \tau_{xy}\delta\gamma_{xy} + \tau_{yz}\delta\gamma_{yz} + \tau_{xz}\delta\gamma_{xz}\right)\mathrm{d}V \tag{5.44}$$

It can be shown that [1,4] the total work done during the virtual displacement is 0: $\delta W - \delta U = 0$. Therefore,

$$\delta W = \delta U \tag{5.45}$$

This is called the *principle of virtual work*.

It is essential to note that during a virtual displacement, the magnitudes and directions of applied forces do not change. Application of the virtual work principle to find deflections of typical beams is shown in the next section.

5.7.1 Castigliano's First Theorem

Consider now a structure subjected to a set of external forces P_k ($k = 1.2, \ldots, n$). Suppose that the structure experiences a continuous virtual displacement in such a manner that it vanishes at all points of loading except under a single load, say, P_i. The virtual displacement in the direction of this force is denoted by $\delta(\delta_i)$. From Equation 5.45, we have $\delta U = P_i \cdot \delta(\delta_i)$. In the limit, the principle of virtual work results in

$$P_i = \frac{\partial U}{\partial \delta_i} \tag{5.46}$$

This is known as *Castigliano's first theorem:* for a linear or nonlinear structure, *the partial derivative of the strain energy with respect to an applied virtual displacement is equal to the load acting at the point in the direction of that displacement*. Similarly, it can be demonstrated that

$$M_i = \frac{\partial U}{\partial \theta_i} \tag{5.47}$$

in which θ_i is the angular rotation and M_i represents the resulting moment.

Note that Castigliano's first theorem is also known as the theorem of virtual work. It is the basis for the derivation of the finite element stiffness equations. In applying the Castigliano's first theorem, the strain energy must be expressed in terms of the displacements.

*5.8 Use of Trigonometric Series in Energy Methods

Certain problems in the structural analysis and design are amenable to solutions by the use of trigonometric series. This technique offers a significant advantage because a single expression may apply to the entire length or surface of the member. A disadvantage in the trigonometric series is that arbitrary support conditions can make it impossible to write a series that is simple. The solution by trigonometric series is applied for variously loaded members in this and the sections to follow [1,4].

The method is now illustrated for the case of a simple beam loaded as depicted in Figure 5.14. The deflection curve can be represented by the following Fourier sine series:

$$\upsilon = \sum_{m=1}^{\infty} a_m \sin \frac{m\pi x}{L} \tag{5.48}$$

that satisfies the boundary conditions ($\upsilon = 0$, $\upsilon'' = 0$ at $x = 0$ and $x = L$). The Fourier coefficients a_m are the maximum coordinates of the sine curves, and the values of m show the

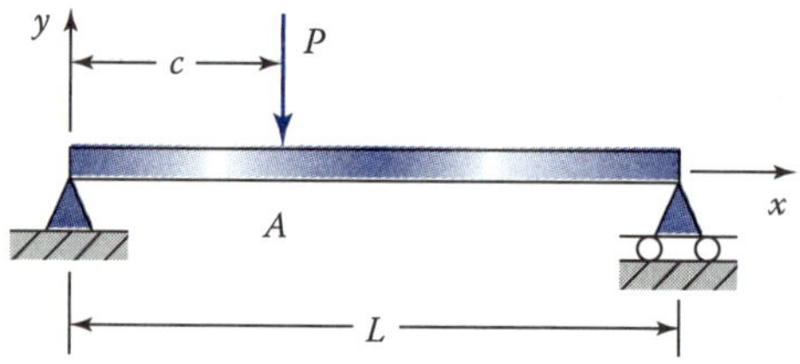

FIGURE 5.14
Simply supported beam under a force P at an arbitrary distance c from the left support.

number of half-waves in the sine curves. The accuracy can be improved by increasing the number of terms in the series. We apply the principle of virtual work to determine the coefficients.

The strain energy of the beam, substituting Equation 5.48 into Equation 5.19, is expressed in the following form:

$$U = \frac{EI}{2}\int_0^L \left(\frac{d^2 \upsilon}{dx^2}\right)^2 dx = \frac{EI}{2}\int_0^L \left[\sum_{m=1}^{\infty} a_m \left(\frac{m\pi}{L}\right)^2 \sin\frac{m\pi x}{L}\right]^2 dx \tag{a}$$

The term in brackets, after expanding, can be expressed as

$$U = \sum_{m=1}^{\infty}\sum_{n=1}^{\infty} a_m a_n \left(\frac{m\pi}{L}\right)^2 \left(\frac{n\pi}{L}\right)^2 \sin\frac{m\pi x}{L}\sin\frac{n\pi x}{L}$$

For the orthogonal functions $\sin(m\pi x/L)$ and $\sin(nx/L)$, by direct integration, it can be verified that

$$\int_0^L \sin\frac{m\pi x}{L}\sin\frac{n\pi x}{L}dx = \begin{cases} 0, & (\text{for } m \neq n) \\ L/2, & (\text{for } m = n) \end{cases} \tag{5.49a}$$

The strain energy given by Equation (a) is therefore

$$U = \frac{\pi^4 EI}{4L^3}\sum_{m=1}^{\infty} m^4 a_m^2 \tag{5.50}$$

The virtual work done by a force P acting through a virtual displacement at A increases the strain energy of the beam by δU. Applying Equation 5.45,

$$-P \cdot \delta\upsilon_A = \delta U \tag{b}$$

The minus sign means that P and $\delta\upsilon_A$ are oppositely directed. So by Equations 5.50 and (b),

$$-P\sum_{m=1}^{\infty}\sin\frac{m\pi c}{L}\delta a_m = \frac{\pi^4 EI}{4L^3}\sum_{m=1}^{\infty} m^4 \delta a_m^2$$

The foregoing gives

$$a_m = -\frac{2PL^3}{\pi^4 EI}\frac{1}{m^4}\sin\frac{m\pi c}{L}$$

Carrying this equation into Equation 5.48, we have the equation of the deflection curve:

$$\upsilon = -\frac{2PL^3}{\pi^4 EI}\sum_{m=1}^{\infty}\frac{1}{m^4}\sin\frac{m\pi c}{L}\sin\frac{m\pi x}{L} \tag{5.51}$$

Using this infinite series, the deflection for any prescribed value of x can be readily obtained.

Example 5.13: Deflection of a Cantilevered Beam by the Principle of Virtual Work

A cantilevered beam is subjected to a concentrated load P at its free end, as shown in Figure 5.15. Derive the equation of the deflection curve.

Assumptions: The origin of the coordinates is placed at the fixed end. Deflection is taken in the following form:

$$\upsilon = \sum_{m=1,3,5,\ldots}^{\infty} a_m\left(1-\cos\frac{m\pi x}{2L}\right) \tag{c}$$

Solution

The boundary conditions, $\upsilon(0) = 0$ and $\upsilon'(0) = 0$, are satisfied by the preceding equation. Substitution of this series into Equation 5.19 gives

$$U = \frac{EI}{2}\int_0^L\left[\sum_{m=1,3,5,\ldots,}^{\infty} a_m\left(\frac{m\pi}{2L}\right)^2\cos\frac{m\pi x}{2L}\right]^2 dx$$

Squaring the term in brackets and noting the orthogonality relation,

$$\int_0^L \cos\frac{m\pi x}{2L}\cos\frac{m\pi x}{2L}dx = \begin{cases} 0, & (\text{for } m \neq n) \\ L/2, & (\text{for } m = n) \end{cases} \tag{5.49b}$$

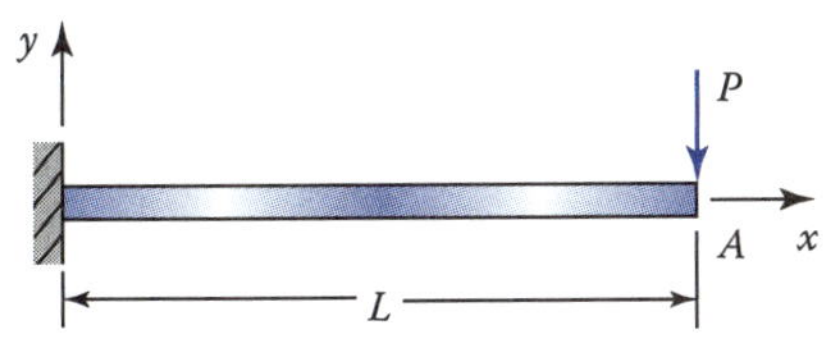

FIGURE 5.15
Example 5.13.

the strain energy becomes

$$U = \frac{\pi^4 EI}{64L^3} \sum_{m=1,3,5,\ldots}^{\infty} m^4 a_m^2 \tag{5.52}$$

By the principle of virtual work, $-P \cdot \delta\upsilon_A = \delta U$, we have

$$-P \sum_{m=1,3,5,\ldots}^{\infty} \left(1 - \cos\frac{m\pi}{2}\right)\delta a_m = \frac{\pi^4 EI}{64L^3} \sum_{m=1,3,5,\ldots}^{\infty} m^4 \delta a_m^2$$

This results in $a_m = -32\,PL^3/m^4\pi^4\,EI$. The beam deflection is found by inserting the value of a_m obtained into Equation (c).

Comments: At the free end ($x = L$), retaining only the first three terms of the solution, we have the value of the maximum deflection $\upsilon_{max} = PL^3/3.001EI$. The exact solution owing to bending is $PL^3/3EI$.

5.9 Buckling of Columns

A prismatic bar loaded in compression is called *column*. Such bars are commonly used in trusses and the framework of buildings. They are also encountered in machine linkages, jack screws, coil springs in compression, and a wide variety of other elements. The buckling of a column is its sudden, large, lateral deflection due to a small increase in existing compressive load. A wooden household yardstick with a compressive load applied at its ends illustrates the basic buckling phenomenon. Failure from the viewpoint of instability may take place for a load that is 1% of the compressive load alone that would cause failure based on a strength criterion. That is, consideration of material strength (stress level) alone is insufficient to predict the behavior of such a member. Railroad rails, if subjected to an axial compression because of temperature rise, could fail similarly.

5.9.1 Pin-Ended Columns

Consider a slender pin-ended column centrically loaded by compressive forces P at each end (Figure 5.16a). In Figure 5.16b, load P has been increased sufficiently to cause a small lateral deflection. This is a condition between stability and instability or neutral equilibrium. The bending moment at any section is $M = -P\upsilon$. So Equation 4.14 becomes

$$EI\frac{d^2\upsilon}{dx^2} = -P\upsilon \tag{5.53}$$

or

$$\frac{d^2\upsilon}{dx^2} + k^2\upsilon = 0 \tag{5.54}$$

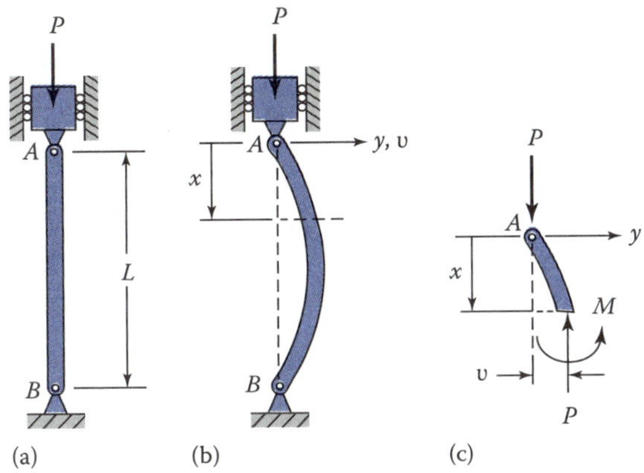

FIGURE 5.16
Column with pinned ends: (a) initially straight, (b) buckled form for number of half-wave n = 1, and (c) free-body diagram of a segment.

For simplification, the following notation is used:

$$k^2 = \frac{P}{EI} \tag{5.55}$$

The solution of Equation 5.54 is

$$\upsilon = A\sin kx + B\cos kx \tag{a}$$

The constants A and B are obtained from the end conditions:

$$\upsilon(0) = 0 \quad \text{and} \quad \upsilon(L) = 0$$

The first requirement gives $B = 0$ and the second leads to for $A = 0$ or $\sin kL = 0$. If $A = 0$, the solution of Equation 5.54 is called *trivial*. The other possibility, $\sin kL$, is satisfied by

$$\sqrt{\frac{P}{EI}}L = n\pi \qquad (n = 1,2,\ldots) \tag{b}$$

The quantity n represents the number of half-waves in the buckled column shape. Note that n = 2,... are usually of no practical interest. The only way to obtain higher modes of buckling is to provide lateral support of the column at the points of 0 moments on the elastic curve, the so-called *inflection points*.

When n = 1, solution of Equation (b) results in the value of the smallest critical load, Euler's buckling load:

$$P_{cr} = \frac{\pi^2 EI}{L^2} \tag{5.56}$$

This is also called *Euler's column formula*. The quantities I, L, and E are moment of inertia of the cross-sectional area, original length of the column, and modulus of elasticity, respectively. Note that the strength is not a factor in the buckling load. Introducing the foregoing results back in Equation (a), we obtain the buckled shape of the column as

$$\upsilon = A\sin\frac{\pi x}{L}$$

The value of the maximum deflection, $\upsilon_{max} = A$, is undefined. Therefore, the critical load sustains only a small or no lateral deflection [4,7].

It is clear that EI represents the flexural rigidity for bending in the plane of buckling. If the column is free to deflect in any direction, it tends to bend about the axis having the smallest principal moment of inertia I. By definition, $I = Ar^2$, where A is the cross-sectional area and r is the *radius of gyration* about the axis of bending. We may consider the r of an area to be the distance from the axes at that entire area that could be concentrated and yet has the same value for I. Substitution of the preceding relationship into Equation 5.56 gives

$$P_{cr} = \frac{\pi^2 EA}{(L/r)^2} \tag{5.57}$$

We seek the minimum value of P_{cr}: hence, the smallest radius of gyration should be used in this equation. The quotient L/r, called the *slenderness ratio*, is an important parameter in the classification of columns.

5.9.2 Columns with Other End Conditions

For columns with various combinations of fixed, free, and pinned supports, Euler's formula can be written in the following form:

$$P_{cr} = \frac{\pi^2 EI}{L_e^2} \tag{5.58}$$

in which L_e is called the *effective length*. As shown in Figure 5.17, it develops that the effective length is the distance between the inflection points on the elastic curves. In a like manner, Equation 5.57 can be expressed as

$$P_{cr} = \frac{\pi^2 EA}{(L_e/r)^2} \tag{5.59}$$

The quantity L_e/r is referred to as the *effective slenderness ratio* of the column. In the actual design of a column, the designer endeavors to configure the ends, using bolts, welds, or pins, to achieve the required ideal end condition.

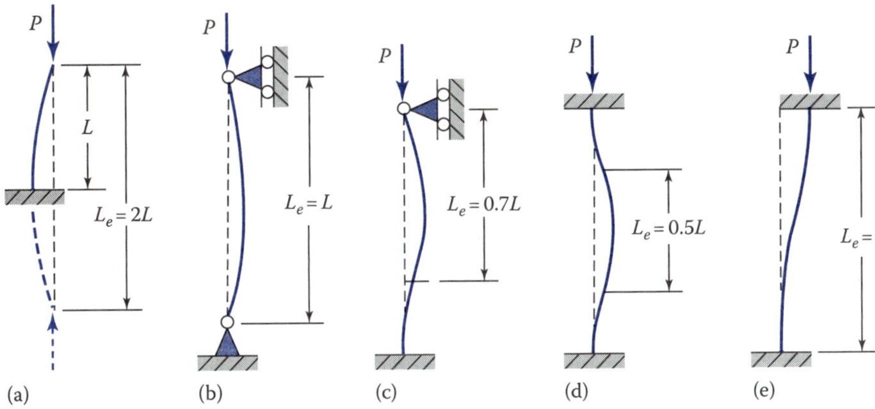

FIGURE 5.17
Effective lengths of columns for various end conditions: (a) fixed–free, (b) pinned–pinned, (c) fixed–pinned, (d) fixed–fixed, and (e) fixed–nonrotating.

Minimum AISI recommended *actual* effective lengths for *steel* columns [10] are as follows:

$$\begin{aligned} L_e &= 2.1L \quad \text{(fixed–free)} \\ L_e &= L \quad \text{(pinned–pinned)} \\ L_e &= 0.80L \quad \text{(fixed–pinned)} \\ L_e &= 0.65L \quad \text{(fixed–fixed)} \\ L_e &= 1.2L \quad \text{(fixed–nonrotating)} \end{aligned} \tag{5.60}$$

Note that only a steel column with pinned–pinned ends has the same actual length and the theoretical value noted in Figure 5.17b. Also observe that steel columns with one or two fixed ends always have actual lengths longer than the theoretical values. The foregoing apply to end construction, where ideal conditions are approximated. The distinction between the theoretical analyses and empirical approaches necessary in design is discussed in Section 5.13.

5.10 Critical Stress in a Column

As previously pointed out, a column failure is always sudden, total, and unexpected. There is no advance warning. The behavior of an ideal column is often represented on a plot of average critical stress P_{cr}/A versus the slenderness ratio L_e/r (Figure 5.18). Such a representation offers a clear rationale for the classification of compression bars. The range of L_e/r is a function of the material under consideration.

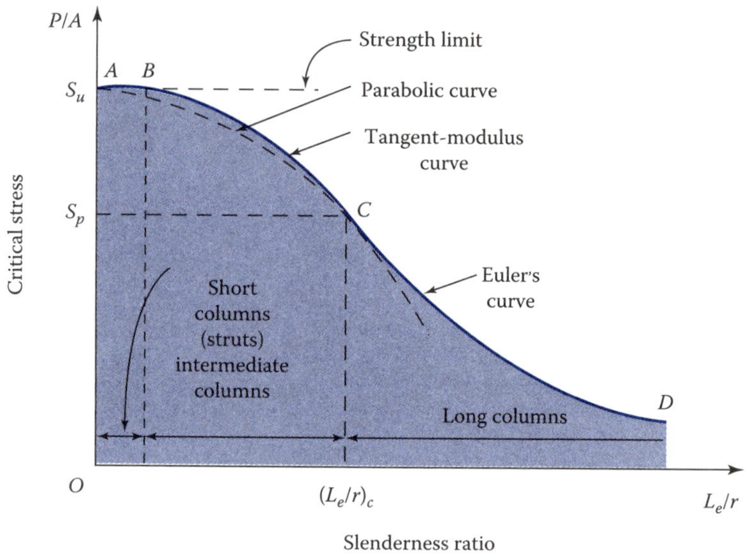

FIGURE 5.18
Average stress in columns versus the slenderness ratio.

5.10.1 Long Columns

For a long column, that is, a member with a sufficiently large slenderness ratio, buckling occurs elastically at stress that does not exceed the proportional limit of the material. Hence, Euler's load of Equation 5.59 is appropriate in this case, and the critical stress is

$$\sigma_{cr} = \frac{P_{cr}}{A} = \frac{\pi^2 E}{(L_e/r)^2} \tag{5.61}$$

The corresponding portion *CD* of the curve (Figure 5.18) is labeled as *Euler's curve*. The smallest value of the slenderness ratio for which Euler's formula applies is found by equating σ_{cr} to the proportional limit or yield strength ($S_p \approx S_y$) of the specific material:

$$\left(\frac{L_e}{r}\right)_c = \pi\sqrt{\frac{E}{S_y}} \tag{5.62}$$

For instance, in the case of a structural steel with $E = 210$ GPa and $S_y = 250$ MPa, this equation gives $(L_e/r)_c = 91$.

We see from Figure 5.18 that very slender columns buckle at low levels of stress; they are much less stable than short columns. Equation 5.62 shows that the critical stress is increased by using a material of higher modulus of elasticity E or by increasing the radius of gyration r. A tubular column, for example, has a much larger value of r than a solid column of the same cross-sectional area. However, there is a limit beyond which the buckling strength cannot be increased. The wall thickness eventually becomes so thin as to cause the member to crumble due to a change in the shape of a cross section.

5.10.2 Short Columns or Struts

Compression members having low slenderness ratios (for instance, steel rods with $L_e/r < 30$) show essentially no instability and are called *short columns*. For these bars, failure occurs by yielding or crushing, without buckling, at stresses above the proportional limit of the material. Therefore, the maximum stress

$$\sigma_{max} = \frac{P}{A} \tag{5.63}$$

represents the strength limit of such a column, shown by horizontal line *AB* in Figure 5.18. This is equal to the yield strength or ultimate strength in compression.

5.10.3 Intermediate Columns

Most structural columns lie in a region between the short and long classifications, represented by part *BC* in Figure 5.18. Such intermediate columns fail by inelastic buckling at stress levels above the proportional limit. Substitution of the tangent modulus E_t, slope of the stress–strain curve beyond the proportional or yield point, for the elastic modulus E, is the only modification necessary to make Equation 5.59 applicable in the inelastic range. Hence, the critical stress may be expressed by the generalized Euler's buckling formula, *tangent modulus formula*:

$$\sigma_{cr} = \frac{\pi^2 E_t}{(L_e/r)^2} \tag{5.64}$$

If the tangent moduli corresponding to the given stresses can be read from the compression stress–strain diagram, the value of L_e/r at which a column buckles can readily be calculated by applying Equation 5.64. When, however, L_e/r is known and σ_{cr} is to be obtained, a trial-and-error approach is necessary.

Over the years, many other formulas have been proposed and employed for intermediate columns. A number of these formulas are based on the use of linear or parabolic relationships between the slenderness ratio and the critical stress. The parabolic J.B. *Johnson formula* has the following form:

$$\sigma_{cr} = S_y - \frac{1}{E}\left(\frac{S_y}{2\pi}\frac{L_e}{r}\right)^2 \tag{5.65a}$$

or, in terms of the critical load,

$$P_{cr} = \sigma_{cr}A = S_yA\left[1 - \frac{S_y(L_e/r)^2}{4\pi^2 E}\right] \tag{5.65b}$$

where A represents the cross-sectional area of the column. The relation (5.65) seems to be the preferred one among designers in the machine, aircraft, and structural steel construction fields. Despite much scatter in the test results, the Johnson formula has been found to agree reasonably well with experimental data. However, the dimensionless form of tangent modulus curves has very distinct advantage when structures of new materials are analyzed.

Note that Equations 5.61, 5.62, and 5.64 or 5.65 determine the ultimate stresses, not the working stresses. It is therefore necessary to divide the right side of each formula by an appropriate factor of safety, often 2–3, depending on the material, to determine the allowable values. Some typical relationships for allowable stress are introduced in Section 5.13.

Example 5.14: The Most Efficient Design of a Rectangular Column

A steel column of length L and an $a \times b$ rectangular cross section is fixed at the base and supported at the top, as shown in Figure 5.19. The column must resist a load P with a factor of safety n with respect to buckling.

a. What is the ratio of a/b for the most efficient design against buckling?
b. Design the most efficient cross section for the column, using L = 400 mm, E = 200 GPa, P = 15 kN, and n = 2.

Assumption: Support restrains end A from moving in the yz plane but allows it to move in the xz plane.

Solution

The radius of gyration r of the cross section is

$$r_z^2 = \frac{I_z}{A} = \frac{ba^3/12}{ab} = \frac{a^2}{12} \quad \text{or} \quad r_z = a/\sqrt{12}$$

Similarly, we obtain $r_y = b/\sqrt{12}$. The effective lengths of the column shown in Figure 5.19 with respect to buckling in the xy and xz planes, from Figure 5.17a and c, are $L_e = 0.7L$ and $L_e = 2L$, respectively. Thus,

$$\frac{L_e}{r_z} = \frac{0.7L}{a/\sqrt{12}}$$
$$\frac{L_e}{r_y} = \frac{2L}{b/\sqrt{12}} \qquad \text{(a)}$$

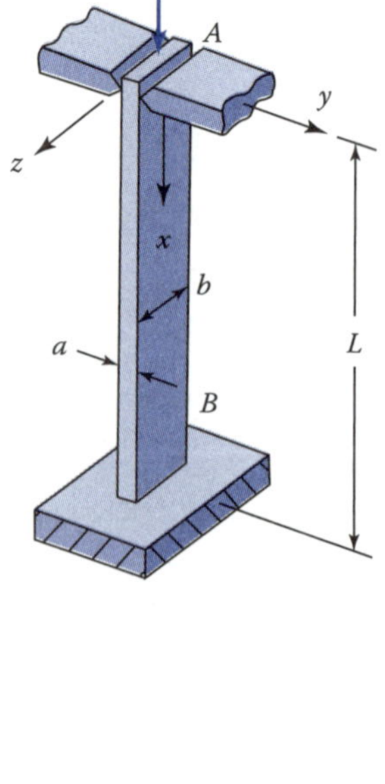

FIGURE 5.19
Example 5.14.

a. For the most effective design, the critical stresses corresponding to the two possible modes of buckling are to be identical. Referring to Equation 5.61, it is concluded therefore that

$$\frac{0.7L}{a/\sqrt{12}} = \frac{2L}{b/\sqrt{12}}$$

Solving,

$$\frac{a}{b} = 0.35 \tag{b}$$

b. Designing for the given data, based on a safety factor of $n = 2$, we have $P_{cr} = 2(15) = 30$ kN and

$$\sigma_{cr} = \frac{P_{cr}}{A} = \frac{30(10^3)}{0.35b^2}$$

The second of Equation (a) gives $L_e/r = 2(0.4)\sqrt{12}/b = 2.771/b$. Through the use of Equation 5.61, we write

$$\frac{30(10^3)}{0.35b^2} = \frac{\pi^2(200\times 10^9)}{(2.771/b)^2}$$

from which $b = 24$ mm and hence $a = 8.4$ mm.

Example 5.15: Development of Specific Johnson's Formula

Derive specific Johnson formula for the intermediate sizes of columns having

a. Round cross sections
b. Rectangular cross sections

Solution

a. For a solid circular section, $A = \pi d^2/4$, $I = \pi d^4/64$, and

$$r = \sqrt{\frac{I}{A}} = \frac{d}{4} \tag{c}$$

Using Equation 5.65a,

$$\frac{4P_{cr}}{\pi d^2} = S_y - \frac{1}{E}\left(\frac{S_y}{2\pi}\frac{4L_e}{d}\right)^2$$

Solving,

$$d = 2\left(\frac{P_{cr}}{\pi S_y} + \frac{S_y^2 L_e^2}{\pi^2 E}\right)^{1/2} \tag{5.66}$$

b. In the case of a rectangular section of height h and width b with $h \le b$: $A = bh$, $I = bh^3/12$; hence, $r^2 = h^2/12$. Introducing these into Equation 5.65a,

$$\frac{P_{cr}}{bh} = S_y - \frac{1}{E}\left(\frac{S_y}{2\pi}\frac{\sqrt{12}L_e}{h}\right)^2$$

or

$$b = \frac{P_{cr}}{hS_y\left(1 - \dfrac{3L_e^2 S_y}{\pi^2 E h^2}\right)} \tag{5.67}$$

Example 5.16: Load-Carrying Capacity of a Pin-Ended Braced Column

A steel column braced at midpoint C as shown in Figure 5.20. Determine the allowable load P_{all} on the basis of a factor of safety n.

Given: $L = 15$ in., $a = \frac{1}{4}$ in., $b = 1$ in., $S_y = 36$ ksi, $E = 30 \times 10^6$ psi

Assumptions: Bracing acts as simple support in the xy plane. Use $n = 3$.

Solution

Referring to Example 5.14, the cross-sectional area properties are

$$r_z = \frac{a}{\sqrt{12}} = \frac{0.25}{\sqrt{12}} = 0.072 \text{ in.}$$

$$r_y = \frac{b}{\sqrt{12}} = \frac{1}{\sqrt{12}} = 0.289 \text{ in.}$$

and $A = ab = (0.25)(1) = 0.25$ in.^{2}

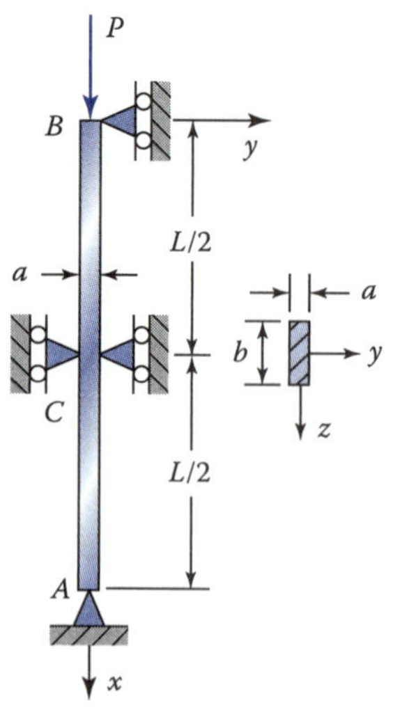

FIGURE 5.20
Example 5.16.

Buckling in the xz plane (unrestrained by the brace). The slenderness ratio is L/r_y = 15/0.289 = 51.9 < 91 and the Johnson equation is valid per Equation 5.62. Relation 5.65b results in

$$P_{cr} = (36{,}000)(0.25)\left[1 - \frac{36{,}000(51.9)^2}{4\pi^2(30\times10^6)}\right]$$

$$= 8.263 \text{ kips}$$

Buckling in the xy plane (braced). We have L_e/r = 7.5/0.072 = 104.2. Hence, applying Euler's equation,

$$P_{cr} = \frac{\pi^2 EA}{(L_e/r)^2} = \frac{\pi^2(30\times10^6)(0.25)}{(104.2)^2}$$

$$= 6.818 \text{ kips}$$

Comment: The working load P_{all}, therefore, must be based on buckling in the xy plane:

$$P_{all} = \frac{P_{cr}}{n} = \frac{6.818}{3} = 2.273 \text{ kips}$$

Example 5.17: Critical Stress in a Simple Truss

A plane truss, consisting of members *AB*, *BC*, and *AC*, is subjected to vertical and horizontal loads P_1 and P_2 at point *B* and supported at *A* and *C*, as shown in Figure 5.21a. Although the members are actually joined together by riveted or welded connections, it is usual to consider them pinned together (Figure 5.21b). Further information on trusses may be found in Section 5.5.1. Determine the stresses in each member.

Given: The geometry of the assembly is known. Each member is 6061-T6 aluminum alloy pipe of outer diameter *D* and inner diameter *d*.

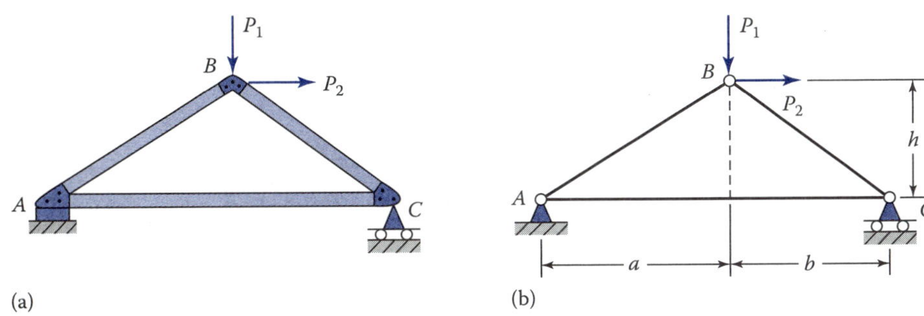

FIGURE 5.21
Example 5.17. (a) Basic truss and (b) schematic representation of basic truss.

Data

$$D = 80\text{ mm},\quad d = 65\text{ mm},\quad a = 4\text{ m},$$
$$b = 2.25\text{ m},\quad h = 3\text{ m},$$
$$P_1 = 50\text{ kN},\quad P_2 = 25\text{ kN},$$
$$E = 70\text{ GPa},\quad S_y = 260\text{ MPa},\quad (\text{from Table B.1})$$

Assumptions:

1. The weight of members is small compared to the applied loading and is neglected.
2. The loading is static.
3. Each member is treated as a two-force member.
4. Friction in the pin joints is ignored.

Solution

See Figure 5.21b.

The axial forces in members, applying the method of joints, are found as

$$F_{AB} = 10\text{ kN}(C),\quad F_{BC} = 55\text{ kN}(C),$$
$$F_{AC} = 33\text{ kN}(T)$$

For a tube cross section, we have the following properties:

$$A = \frac{\pi}{4}\left(D^2 - d\right)^2,\quad I = Ar^2 = \frac{\pi}{64}\left(D^4 - d^4\right)$$

The radius of gyration is therefore

$$r = \frac{1}{4}\sqrt{D^2 + d^2} \tag{5.68}$$

Substitution of the given numerical values results in

$$A = \frac{\pi}{4}\left(80^2 - 65^2\right) = 1708\text{ mm}^2$$
$$r = \frac{1}{4}\sqrt{80^2 + 65^2} = 25.77\text{ mm}$$

Hence, $L/r = 5000/25.77 = 194$ for member AB and $L/r = 3750/25.77 = 145.5$ for member BC. Using Equation 3.1, the tensile stress in the member AC is

$$\sigma_{AC} = \frac{F_{AC}}{A} = \frac{33\left(10^3\right)}{1708\left(10^{-6}\right)} = 19.32\text{ MPa}$$

The critical stresses from Equation 5.61, for members *AB* and *BC*, respectively, are

$$(\sigma_{cr})_{AB} = \frac{\pi^2(70\times 10^9)}{(194)^2} = 18.36 \text{ MPa}$$

$$(\sigma_{cr})_{BC} = \frac{\pi^2(70\times 10^9)}{(145.5)^2} = 32.63 \text{ MPa}$$

Comments: Interestingly, 18.36 and 32.63 MPa, compared with the yield strength of 260 MPa, demonstrate the significance of buckling analysis in predicting a safe working load. A somewhat detailed finite element analysis (FEA) of the member forces, displacements, and design of a basic truss are discussed in Case Study 17.1.

5.11 Initially Curved Columns

In an actual structure, it is not always possible for a column to be perfectly straight. As might be expected, the load-carrying capacity and deflection under load of a column are significantly affected by even a small initial curvature. To determine the extent of this influence, consider a pin-ended column with the unloaded form described by a half sine wave:

$$\upsilon_0 = a_0 \sin\frac{\pi x}{L} \tag{a}$$

This is shown by the dashed lines in Figure 5.22, where a_0 represents the maximum initial displacement.

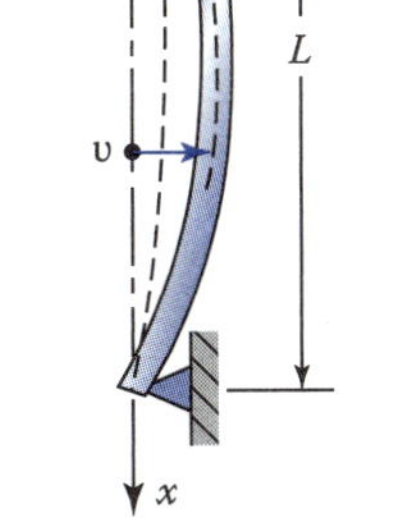

FIGURE 5.22
Initially curved column with pinned ends.

5.11.1 Total Deflection

An additional displacement υ_1 accompanies a subsequently applied load P. Therefore,

$$\upsilon = \upsilon_0 + \upsilon_1$$

The differential equation of the column, using Equation 4.14, is

$$EI\frac{d^2\upsilon_1}{dx^2} = P(\upsilon_0 + \upsilon_1) = -P\upsilon \tag{b}$$

Introducing Equation (a) and setting $k^2 = P/EI$, we have

$$\frac{d^2\upsilon_1}{dx^2} + k^2\upsilon_1 = -\frac{P}{EI}a_0 \sin\frac{\pi x}{L}$$

For simplicity, let b designate the ratio of the axial load to its critical value:

$$b = \frac{P}{P_{cr}} = \frac{PL^2}{\pi^2 EI} \tag{5.69}$$

The trial particular solution of this equation, $\upsilon_{1p} = B\sin(\pi x/L)$, when inserted into Equation (b) gives

$$B = \frac{Pa_0}{(\pi^2 EI/L^2) - P} = \frac{a_0}{(1/b) - 1}$$

The general solution of Equation (b) is

$$\upsilon_1 = c_1 \sin kx + c_2 \cos kx + B\sin\frac{\pi x}{L}$$

The constants c_1 and c_2 are evaluated, from the end conditions $\upsilon_1(0) = \upsilon_1(L) = 0$, as $c_1 = c_2 = 0$. The column deflection is then

$$\upsilon = a_0 \sin\frac{\pi x}{L} + B\sin\frac{\pi x}{L} = \frac{a_0}{1-b}\sin\frac{\pi x}{L} \tag{5.70}$$

This equation indicates that the axial force P causes the initial deflection of the column to increase by the factor $1/(1 - b)$. Since $b < 1$, this factor is always greater than unity. Clearly, if $b = 1$, deflection becomes infinitely large. Note that an initially curved column deflects with any applied load P in contrast to a perfectly straight column that does not bend until P_{cr} is reached.

5.11.2 Critical Stress

We begin with the formula applicable to combined axial loading and bending: $\sigma_x = (P/A) + (My/I)$, in which A and I are the cross-sectional area and the moment of inertia.

Substituting $M = -P\upsilon$ together with Equation 5.70 into this expression, the maximum compressive stress at midspan is found as

$$\sigma_{max} = \frac{P}{A}\left(1 + \frac{a_o A}{S}\frac{1}{1-b}\right) \tag{5.71}$$

The quantity S represents the section modulus I/c, in which c is the distance measured in the y direction from the centroid of the cross section to the outermost fibers.

By imposing the yield strength in compression (and tension) S_y as σ_{max}, we write Equation 5.71 in the following form:

$$S_y = \frac{P_y}{A}\left(1 + \frac{a_o A}{S}\frac{1}{1-b}\right) \tag{5.72}$$

In the foregoing, P_y is the limit load that results in impending yielding and subsequent failure. Given S_y, a_o, E, and the column dimensions, Equation 5.72 may be solved exactly by solving a quadratic or by trial and error for P. The allowable load P_{all} can then be obtained by dividing P_y by an appropriate factor of safety n.

5.12 Eccentric Loads and the Secant Formula

In the preceding sections, we deal with the buckling of columns for which the load acts at the centroid of a cross section. We here treat columns under an eccentric load. This situation is obviously of great practical importance because, frequently, problems occur in which load eccentricities are unavoidable.

Let us consider a pinned-end column under compressive forces applied with a small eccentricity e from the column axis (Figure 5.23a). We assume the member is initially straight and that the material is linearly elastic. By increasing the load, the column deflects as depicted by the dashed lines in the figure. The bending moment at distance x from the midspan equals $M = -P(\upsilon + e)$. Then, the differential equation for the elastic curve appears in the following form:

$$EI\frac{d^2\upsilon}{dx^2} + P(\upsilon + e) = 0$$

The boundary conditions are $\upsilon(L/2) = \upsilon(-L/2) = 0$. This equation is solved by following a procedure in a manner similar to that of Sections 5.9 and 5.10. In so doing, expressed in the terms of the critical load $P_{cr} = \pi^2 EI/L^2$, the midspan ($x = 0$) deflection is found as

$$\upsilon_{max} = e\left[\sec\frac{\pi}{2}\sqrt{\frac{P}{P_{cr}}} - 1\right] \tag{5.73}$$

We observe from the foregoing expression that, as P approaches P_{cr}, the maximum deflection goes to infinity. It is therefore concluded that P should not be allowed to reach the critical value found in Section 5.9 for a column under a centric load.

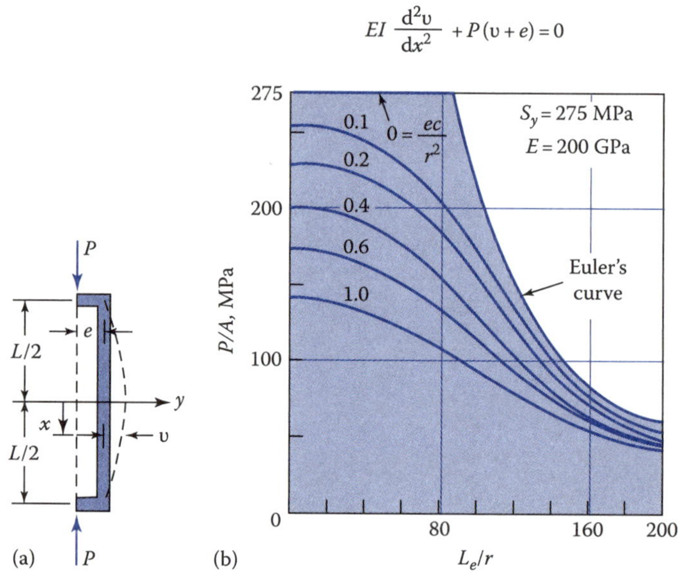

FIGURE 5.23
(a) Eccentrically loaded pinned–pinned column and (b) graph of the secant formula.

The maximum compressive stress σ_{max} takes place at $x = 0$ on the concave side of the column. Hence,

$$\sigma_{max} = \frac{P}{A} + \frac{M_{max}c}{I}$$

The quantity r is the radius of gyration and c is the distance from the centroid of the cross section to the outermost fibers, both in the direction of eccentricity. Carrying $M_{max} = -P(\upsilon_{max} + e)$ and Equation 5.73 into the foregoing expression,

$$\sigma_{max} = \frac{P}{A}\left[1 + \frac{ec}{r^2}\sec\left(\frac{\pi}{2}\sqrt{\frac{P}{P_{cr}}}\right)\right] \tag{5.74a}$$

Alternatively, we have

$$\upsilon_{max} = \frac{P}{A}\left[1 + \frac{ec}{r^2}\sec\left(\frac{L}{2r}\sqrt{\frac{P}{AE}}\right)\right] \tag{5.74b}$$

This expression is referred to as the *secant formula*. The term ec/r^2 is called the *eccentricity ratio*.

We now impose the yield strength S_y as σ_{max}. Then, Equation 5.74b can be written as

$$\frac{P}{A} = \frac{S_y}{1 + \left(ec/r^2\right)\sec\left[\left(L_e/2r\right)\sqrt{P/AE}\right]} \tag{5.75}$$

Here, the effective length is used: hence, the formula applies to columns with different end conditions. Figure 5.23b is a plot of the preceding expression for steel having a yield strength of 275 MPa and modulus of elasticity of 200 GPa. Note how the P/A contours asymptotically approach Euler's curve as L_e/r increases. Equation 5.75 may be solved for the load P by trial and error or a root-finding technique using numerical methods. Design charts in the fashion of Figure 5.23b can be used to good advantage (see Problem 5.59). Obviously, column design by the secant formula might best be programmed on a computer.

Formula 5.74 is an excellent description of column behavior, provided the eccentricity e of the load is known with a degree of accuracy. We point out that the foregoing derivation of the secant formula is on the assumption that buckling takes place in the xy plane. It may also be necessary to analyze buckling in the plane; Equation 5.74 does not apply in this plane. This possibility correlates especially to narrow columns.

Example 5.18: Analysis of Buckling of a Column Using the Secant Formula

A 20 ft long pin-ended ASTM-A36 steel column of $S8 \times 23$ section (Figure 5.24a) is subjected to a centric load P_1 and an eccentrically applied load P_2, as shown in Figure 5.24b. Determine

a. The maximum deflection
b. The factor of safety n against yielding

Given: The geometry of the column and applied loading are known.

Data: $S_y = 36$ ksi, $E = 29 \times 10^6$ psi, (from Table B.1).

Solution

See Figure 5.24 and Table A.7.

The loading may be replaced by a statically equivalent load $P = 100$ kips acting with an eccentricity $e = 1.2$ in. (Figure 5.24c). Using the properties of an $S8 \times 23$ section given in Table A.7, we obtain

$$\frac{P}{A} = \frac{100}{6.77} = 14.77 \text{ ksi}, \quad \frac{L_e}{r} = \frac{20 \times 12}{3.10} = 77.42$$

$$\frac{ec}{r^2} = \frac{eA}{S} = \frac{1.2(6.77)}{16.2} = 0.501$$

$$P_{cr} = \frac{\pi^2 EI}{L^2} = \frac{\pi^2 (29 \times 10^6)(64.9)}{(20 \times 12)^2} = 322.5 \text{ kips}$$

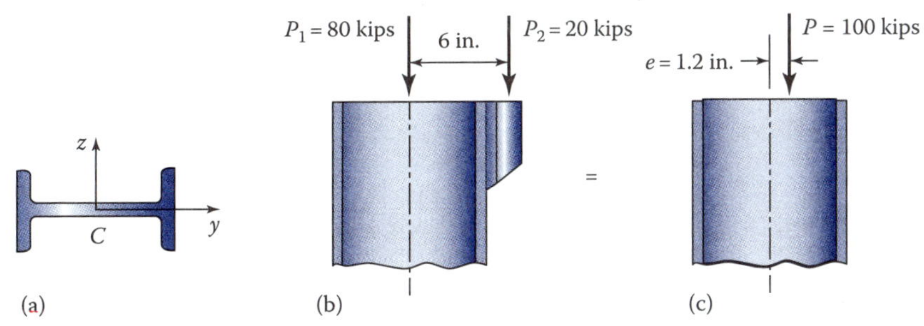

FIGURE 5.24
Example 5.18.

a. Carrying $P/P_{cr} = 1/3.225$ and $e = 1.2$ in. into Equation 5.73 leads to the value of midspan deflection of the column as

$$\upsilon_{max} = e\left[\sec\left(\frac{\pi}{2}\sqrt{\frac{P}{P_{cr}}}\right) - 1\right] = 1.2(1.559 - 1) = 0.671 \text{ in.}$$

b. In a like manner, through the use of Equation 5.74a, we have

$$\sigma_{max} = \frac{P}{A}\left[1 + \frac{ec}{r^2}\sec\left(\frac{\pi}{2}\sqrt{\frac{P}{P_{cr}}}\right)\right] = 26.31 \text{ ksi}$$

Hence,

$$n = \frac{S_y}{\sigma_{max}} = \frac{36}{26.31} = 1.37$$

Comments: Since the maximum stress and the slenderness ratio are within the elastic limit of 36 ksi and the slenderness ratio of about 200, the secant formula is applicable.

5.12.1 Short Columns

A strut or short column under a centric compression load P will shorten prior to the stress approaches the elastic limit of the material. As previously noted, at this point, yielding occurs and usefulness of a component may be at an end. For the case in which there is eccentricity in loading, the elastic strength is reached at small loads (Figure 5.23a).

The secant formula for short columns returns to usual form for small values of the slenderness ratio L/r. In this situation, the secant is nearly equal to 1, and consequently, Equation 5.74b reduces to

$$\sigma_{max} = \frac{P}{A}\left(1 + \frac{ec}{r^2}\right) \tag{5.76}$$

The foregoing is a relationship often employed for short columns. Clearly, unlike Equations 5.74, Equation 5.76 does not incorporate the bending deflection, which has the effect of increasing the eccentricity e and thus stress at the midspan.

Example 5.19: Stress in a Strut of a Clamping Assembly

A piece of work in the process of manufacture is attached to a cutting machine table by a bolt tightened to a tension of T (Figure 5.25a). The clamp contact is offset from a centroidal axis of the strut AB by a distance e as shown in Figure 5.25b. The strut is made of a structural ASTM 36 steel of diameter d and length L.

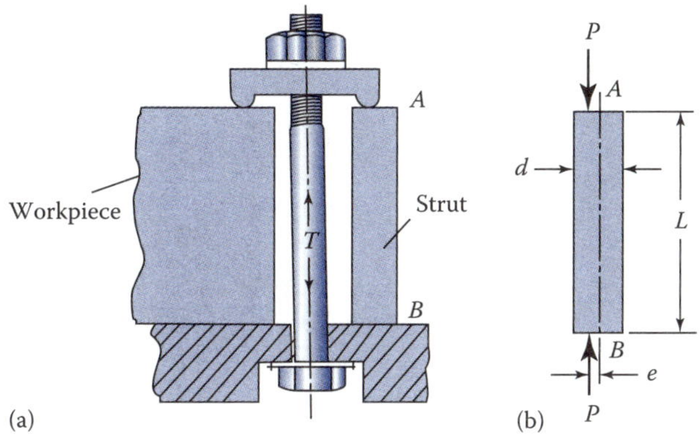

FIGURE 5.25
Example 5.19. (a) Fastening assembly and (b) strut with eccentric compressive load.

Given: The numerical values are

$$d = 30 \text{ mm} \quad e = 3.5 \text{ mm} \quad L = 125 \text{ mm}$$

$$T = 2P = 7 \text{ kN} \quad E = 200 \text{ GPa (by Table B.1)}$$

Find: The largest stress in the strut, applying

a. Equation 5.76 for a short compression bar
b. The secant formula

Assumption: The strut is taken as a pin connected at both ends, and hence $L_e = L$.

Solution

The *cross-sectional area properties* of the strut equal to

$$c = \frac{d}{2} = \frac{1}{2}(30) = 15 \text{ mm}$$

$$A = \frac{\pi}{4}d^2 = \frac{\pi}{4}(30)^2 = 706.9 \text{ mm}^2$$

$$I = \frac{\pi}{64}d^4 = \frac{\pi}{64}(30)^4 = 39.76 \times 10^3 \text{ mm}^4$$

$$r = \sqrt{I/A} = 7.5 \text{ mm}$$

$$\frac{L_e}{r} = \frac{12.5}{7.5} = 16.6667$$

Substitution of the given data leads to

$$\frac{P}{A} = \frac{3500}{706.9} = 4.951 \text{ kPa}$$

$$\frac{ec}{r^2} = \frac{(3.5)(15)}{(7.5)^2} = 0.9333$$

a. *Short-column formula.* Insertion of the foregoing values into Equation 5.76 results in

$$\sigma_{max} = \frac{P}{A}\left(1 + \frac{ec}{r^2}\right) = 9572 \text{ kPa}$$

b. *Secant formula.* Introducing the data into Equation 5.74, we obtain

$$\sigma_{max} = \frac{P}{A}\left[1 + \frac{ec}{r^2}\sec\left(\frac{L_e}{2r}\sqrt{\frac{P}{AE}}\right)\right] = 9572 \text{ kPa}$$

Comment: The results indicate that for this short column, the effect of the lateral deflection on stress can be omitted.

5.13 Design Formulas for Columns

In Sections 5.9 and 5.10, we obtain the critical load in a column by applying Euler's formula. This is followed by the investigation of the deformations and stresses in initially curved columns and eccentrically loaded columns by using the combined axial load and bending formula and the secant formula, respectively. In each case, we assume that all stresses remain below the proportional point or yield limit and that the column is a homogeneous prism.

The foregoing idealizations are important in understanding column behavior. However, the design of actual columns must be based on empirical formulas that consider the data obtained by laboratory tests. Care must be used in applying such *special purpose formulas.* Specialized references should be consulted prior to design of a column for a particular application. Typical *design formulas for centrically loaded columns* made of three different materials follow. These represent specifications recommended by the American Institute of Steel Construction (AISC), the Aluminum Association, and the National Forest Products Association (NFPA). Various computer programs are readily available for the analysis and design of columns with any cross section (including variable) and any boundary conditions.

Column formulas for structural steel [10]:

$$\sigma_{all} = \frac{S_y}{n}\left[1 - \frac{1}{2}\left(\frac{L_e/r}{C_e}\right)^2\right] \qquad \left(\frac{L_e}{r} < C_c\right) \tag{5.77a}$$

$$\sigma_{all} = \frac{\pi^2 E}{1.92(L_e/r)^2} \qquad \left(C_c \le \frac{L_e}{r} \le 200\right) \tag{5.77b}$$

where

$$C_c = \sqrt{2\pi^2 E/S_y} \tag{5.78a}$$

$$n = \frac{5}{3} + \frac{3}{8}\left(\frac{L_e/r}{C_c}\right) - \frac{1}{8}\left(\frac{L_e/r}{C_c}\right)^3 \tag{5.78b}$$

Column formulas for aluminum 6061-T6 alloy [11]:

$$\sigma_{\text{all}} = 19 \text{ ksi} = 130 \text{ MPa} \qquad \left(\frac{L_e}{r} \le 9.5\right) \tag{5.79a}$$

$$\begin{aligned}\sigma_{\text{all}} &= \left[20.2 - 0.126\left(\frac{L_e}{r}\right)\right] \text{ksi} \\ &= \left[140 - 0.87\left(\frac{L_e}{r}\right)\right] \text{MPa} \qquad \left(9.5 < \frac{L_e}{r} < 66\right)\end{aligned} \tag{5.79b}$$

$$\begin{aligned}\sigma_{\text{all}} &= \frac{51{,}000 \text{ ksi}}{(L_e/r)^2} \\ &= \frac{350 \times 10^3}{(L_e/r)^2} \text{MPa} \qquad \left(\frac{L_e}{r} \ge 66\right)\end{aligned} \tag{5.79c}$$

Column formulas for timber of a rectangular cross section [12]:

$$\sigma_{\text{all}} = 1.2 \text{ ksi} = 8.27 \text{ MPa} \qquad \left(\frac{L_e}{d} \le 11\right) \tag{5.80a}$$

$$\sigma_{\text{all}} = 1.2\left[1 - \frac{1}{3}\left(\frac{L_e/d}{26}\right)^2\right] \text{ksi} \qquad \left(11 < \frac{L_e}{d} \le 26\right) \tag{5.80b}$$

$$\sigma_{\text{all}} = \frac{0.3E}{(L_e/d)^2} \qquad \left(26 < \frac{L_e}{d} \le 50\right) \tag{5.80c}$$

where d is the *smallest* dimension of the member. The allowable stress is not to exceed the value of stress (1.2 ksi) for compression parallel to the grain of the timber used. Equation 5.80c may be used with either SI or U.S. customary units.

Note that, for the structural steel columns, in Equations 5.77 and 5.78, C_c defines the limiting value of the slenderness ratio between intermediate and long bars. This is taken to correspond to one-half the yield strength S_y of the steel. By Equation 5.61, we therefore have

$$C_c = \frac{L_e}{r} = \sqrt{\frac{2\pi^2 E}{S_y}} \tag{a}$$

Clearly, by applying a variable factor of safety, Equation 5.78b renders a consistent formula for intermediate and short columns. Also observe in Equation 5.79 that for short and intermediate aluminum columns. σ_{all} is constant and linearly related to L_e/r. For long columns, a Euler-type formula is applied in both steel and aluminum columns. Equation 5.80c for timber columns is also Euler's formula, adjusted by a suitable factor of safety.

Example 5.20: Design of a Wide-Flange Steel Column

Select the lightest wide-flange steel section to support an axial load of P on an effective length of L_e.

Given: $S_y = 250$ MPa, $E = 200$ GPa, (from Table B.1), $P = 408$ kN, $L_e = 4$ m

Solution

See Table A.6.

A suitable size for a prescribed shape may conveniently be obtained using tables in the AISC manual. However, we use a trial-and-error procedure here. Substituting the given data, Equation 5.78a results in

$$C_c = \sqrt{\frac{2\pi^2 E}{S_y}} = \sqrt{\frac{2\pi^2\left(200\times 10^3\right)}{250}} = 126$$

as the slenderness ratio.

First Try: Let $L_e/r = 0$. Equation 5.78b yields $n = 5/3$, and by Equation 5.77a, we have $\sigma_{\text{all}} = 250/n = 150$ MPa. The required area is then

$$A = \frac{P}{\sigma_{\text{all}}} = \frac{408\times 10^3}{150\times 10^6} = 2720\ \text{mm}^2$$

Using Table A.6, we select a $W150 \times 24$ section having an area $A = 3060$ mm^2 (greater than 2720 mm^2) and with a minimum r of 24.6 mm. The value of $L_e/r = 4/0.0246 = 163$ is greater than $C_c = 126$, which was obtained in the foregoing. Therefore, applying Equation 5.77b,

$$\sigma_{\text{all}} = \frac{\pi^2\left(200\times 10^9\right)}{1.92(163)^2} = 38.7\ \text{MPa}$$

Hence, the permissible load, $38.7 \times 10^6(3.06 \times 10^{-3}) = 118$ kN, is less than the design load of 408 kN, and a column with a larger A, a larger r, or both must be selected.

Second Try: Consider a $W150 \times 37$ section (see Table A.6) with $A = 4740$ mm^2 and minimum $r = 38.6$ mm. For this case, $L_e/r = 4/0.0386 = 104$ is less than 126. From Equation 5.77a, we have

$$n = \frac{5}{3} + \frac{3}{8}\left(\frac{104}{126}\right) - \frac{1}{8}\left(\frac{104}{126}\right)^3 = 1.91$$

$$\sigma_{\text{all}} = \frac{250}{1.91}\left[1 - \frac{1}{2}\left(\frac{104}{126}\right)^2\right] = 86.3\ \text{MPa}$$

The permissible load for this section, $86.3 \times 4740 = 409$ kN, is slightly larger than the design load.

Comment: A $W150 \times 37$ steel section is acceptable.

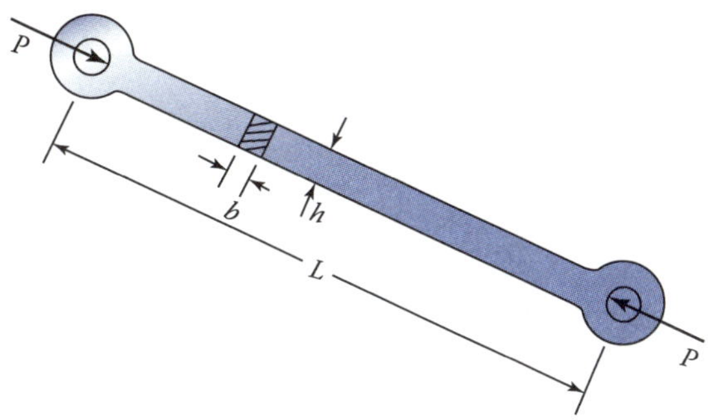

FIGURE 5.26
Example 5.21. Free-body diagram of the connecting rod.

Example 5.21: Steel Connecting Rod Buckling Analysis

A pin-ended steel rod with a rectangular cross-sectional area of $b \times h$ is subjected to a centric compression load P as illustrated in Figure 5.26.

Given: $b = 30$ mm, $h = 50$ mm, $E = 200$ GPa, $P = 60$ kN, $S_y = 250$ MPa

Find: Through the use of AISC formulas, compute

a. The permissible stress for the rod length of $L = 1.2$ m
b. The maximum length L_{max} for which the rod can safely support the loading

Assumptions: The pinned ends are designed to create an effective length of $L = L_e$. Friction in the joints is disregarded.

Solution

The cross-sectional area properties of the rod are

$$A = bh = 30 \times 50 = 1.5 \times 10^3 \text{ mm}^2$$

$$I_{\min} = \frac{1}{12}hb^3 = \frac{1}{12}(50)(30^3) = 112.5 \times 10^3 \text{ mm}^4$$

$$r = r_{\min} = \sqrt{\frac{I_{\min}}{A}} = 8.66 \text{ mm}$$

The limiting value of the slenderness ratio C_c, from Equation 5.78a, is equal to

$$C_c = \sqrt{\frac{2\pi^2 E}{S_y}} = \sqrt{\frac{2\pi^2(200 \times 10^3)}{250}} = 125.7$$

a. *Allowable stress.* For the 1.2 m rod column, $L_e/r = 1200/8.66 = 138.6 > C_c$, and Equation 5.77b applies. Therefore,

$$\sigma_{\text{all}} = \frac{\pi^2 E}{1.92(L_e/r)^2} = \frac{\pi^2(200 \times 10^9)}{1.92(138.6)^2} = 53.5 \text{ MPa}$$

Comment: The foregoing stress is much lower than specified material strength; rod will not yield.

b. *Largest column length.* When the 60 kN load is to be safely carried, the required value of the allowable stress equals

$$\sigma_{all} = \frac{P}{A} = \frac{60(10^3)}{1.5(10^{-3})} = 40 \text{ MPa}$$

Assuming $L_e/r > C_c$, Equation 5.77b leads to

$$\sigma_{all} = \frac{\pi^2(200 \times 10^9)}{1.92(L_{max}/r)^2}$$

Equating the preceding equations results in

$$\frac{L_{max}}{r} = 160.3$$

Inasmuch as $L_{max}/r > C_c$, our assumption was correct. It follows that

$$\frac{L_{max}}{r} = \frac{L_{max}}{8.66} = 160.3, \quad L_{max} = 1.388 \text{ m}$$

Comment: Should the length of this connecting rod be more than 1.388 m, it would buckle.

*5.14 Beam–Columns

Beams subjected simultaneously to axial compression and lateral loads are called *beam–columns.* Deflections in these members are not proportional to the magnitude of the axial load, like the previously discussed columns with initial curvature and eccentrically loaded columns. In this and the next sections, beam–columns of symmetrical cross section and with some common conditions of support and loading are analyzed.

Consider a beam subjected to an axial force P and a distributed lateral load w, as shown in Figure 5.27. The relationships between axial load P, shear force V, and bending moment M are obtained from the equilibrium of an isolated element of length dx between two cross

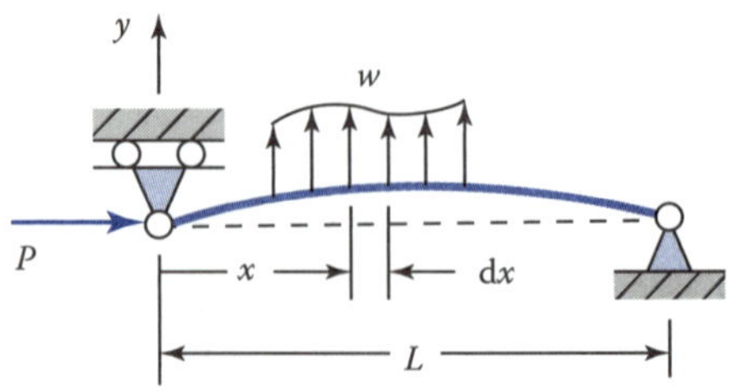

FIGURE 5.27
Beam–column with force P and lateral load w. Initial beam axis is shown by the dashed line.

sections taken normal to the original axis of the beam. On following a procedure similar to that used in Section 3.6, it can be shown that [7]

$$w = \frac{dV}{dx} \tag{5.81a}$$

$$V = \frac{dM}{dx} + P\frac{d\upsilon}{dx} \tag{5.81b}$$

We observe from Equation 5.81b that, for beam–columns, shear force V, in addition to dM/dx as in beams, now also depends on the magnitude of axial force P and slope of the deflection curve.

For the analysis of beam–columns, it is sufficiently accurate to use the usual differential equation for deflection curve of beams. That is,

$$\frac{d^2\upsilon}{dx^2} = \frac{M}{EI} \tag{4.14}$$

where quantity EI is the flexural rigidity of the beam in the xy plane. However, in applying the foregoing expression, the bending moment due to the lateral loads as well as the axial forces must be written for the deflected member. Combining Equation 4.14 with Equations 5.81a and b, we can express the two alternative governing differential equations for beam–columns:

$$\frac{d^2M}{dx^2} + k^2M = w \tag{5.82a}$$

$$\frac{d^4\upsilon}{dx^4} + k^2\frac{d^2\upsilon}{dx^2} = \frac{w}{EI} \tag{5.82b}$$

In the preceding, as before, $k^2 = P/EI$. Clearly, if $P = 0$, the preceding equations reduce to the usual expressions for bending by lateral loads only.

Example 5.22: Analysis of a Beam–Column by the Equilibrium Method

A pin-ended beam–column of length L is subjected to a concentrated transverse load F at its midspan as shown in Figure 5.28a. Determine

a. An expression for the elastic curve
b. Maximum deflection and moment
c. The critical axial load

Assumption: The member has constant flexural rigidity EI.

Solution

a. The bending moment in the left segment of the beam (Figure 5.28b) is

$$M = -P\upsilon - \frac{1}{2}Fx, \qquad \left(0 < x \le \frac{L}{2}\right) \tag{a}$$

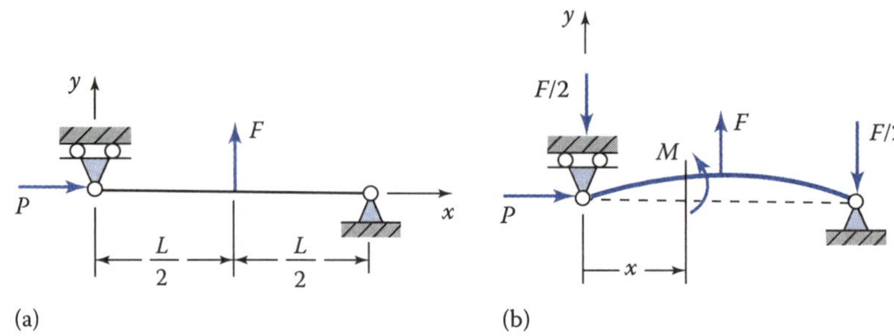

FIGURE 5.28
Example 5.22. (a) a beam–column under lateral and axial forces and (b) free-body diagram for the deflected beam axis.

Equation 4.14 then becomes

$$EI\upsilon'' + P\upsilon = -\frac{1}{2}Fx$$

The governing equation, setting $k^2 = P/EI$, is written as

$$\frac{d^2\upsilon}{dx^2} + k^2\upsilon = -\frac{k^2F}{2P}x, \quad \left(0 < x \leq \frac{L}{2}\right) \tag{b}$$

The general solution is

$$\upsilon = A\sin kx + B\cos kx - \frac{F}{2P}x \tag{c}$$

The boundary condition $\upsilon(0) = 0$ and the condition of symmetry $\upsilon'(L/2) = 0$ are applied to Equation (c) to yield, respectively,

$$\upsilon(0) = 0: \quad B = 0$$

$$\upsilon'\left(\frac{L}{2}\right) = 0: \quad A = \frac{F}{2Fk\cos(kL/2)}$$

Substituting these constants into Equation (c), we have

$$\upsilon = \frac{F}{2Pk}\frac{\sin kx}{\cos(kL/2)} - \frac{F}{2P}x \tag{5.83}$$

By differentiating this equation, we obtain the expressions for the slope ($d\upsilon/dx$) of the deflection curve and the bending moment ($EI\, d^2\upsilon/dx^2$) at any section of the beam–column.

b. The maximum deflection takes place at the center. Substituting $x = L/2$ into Equation 5.83, after some simplifications, we obtain

$$\upsilon_{max} = \frac{F}{2Pk}\left[\tan\frac{kL}{2} - \frac{kL}{2}\right] \tag{5.84}$$

The absolute maximum bending moment occurring at the midspan is readily obtained from Equations (a) and 5.84:

$$M_{\max} = \frac{FL}{4} + P\upsilon_{\max} = \frac{F}{2k}\tan\frac{kL}{2} \tag{5.85}$$

c. Equations 5.83 through 5.85 become infinite if $kL/2$ approaches $\pi/2$, since then $\cos(kL/2) = 0$ and $\tan(kL/2) = \infty$. When $kL/2 = \pi/2$, we obtain

$$\frac{kL}{2} = \sqrt{\frac{P}{EI}}\frac{L}{2} = \frac{\pi}{2}$$

The critical load is therefore

$$P_{cr} = \frac{\pi^2 EI}{L^2} \tag{d}$$

Clearly, as $P \to P_{cr}$, even the smallest lateral load F produces considerable deflection and moment.

Comment: An alternate solution of this problem, obtained by the energy method (see Example 5.23), sheds further light on the deflection produced by the lateral load.

*5.15 Energy Methods Applied to Buckling

Energy approaches often more conveniently yield solution than equilibrium techniques in the analysis of elastic stability and buckling. The energy methods always result in buckling loads *higher* than the exact values if the assumed deflection of a slender member subject to compression differs from the true elastic curve. An efficient application of these approaches may be realized by selecting a series approximation for the deflection. Since a series involves a number of parameters, the approximation can be improved by increasing the number of terms in the series.

Reconsider the column hinged at both ends as depicted in Figure 5.16a. The configuration of this column in the first buckling mode is illustrated in Figure 5.16b. It can be shown [4] that the displacement of the column in the direction of load P is given by $\delta u \approx (1/2)\int(\mathrm{d}\upsilon/\mathrm{d}x)^2\mathrm{d}x$. Inasmuch as the load remains constant, the work done is

$$\delta W = \frac{1}{2}P\int_0^L \left(\frac{\mathrm{d}\upsilon}{\mathrm{d}x}\right)^2 \mathrm{d}x \tag{5.86}$$

The strain energy associated with column bending is given by Equation 5.19 in the following form:

$$U_1 = \int_0^L \frac{M^2}{EI} dx = \int_0^L \frac{EI}{2}\left(\frac{d^2\upsilon}{dx^2}\right)^2 dx$$

Likewise, the strain energy owing to a uniform compressive load P is, from Equation 5.11,

$$U_2 = \frac{P_2 L}{2AE}$$

Because U_2 is constant, it does not enter to the analysis. Since the initial strain energy equals 0, the change in strain energy as the column proceeds from its initial to its buckled configuration is

$$\delta U = \int_0^L \frac{EI}{2}\left(\frac{d^2\upsilon}{dx^2}\right)^2 dx \tag{5.87}$$

From the principle of virtual work, $\delta W = \delta U$, it follows that

$$\frac{1}{2}\int_0^L P\left(\frac{d\upsilon}{dx}\right)^2 dx = \frac{1}{2}\int_0^L EI\left(\frac{d^2\upsilon}{dx^2}\right)^2 dx \tag{5.88a}$$

This results in

$$P = \frac{\int_0^L EI(\upsilon'')^2}{\int_0^L (\upsilon')^2} \tag{5.88b}$$

The end conditions are fulfilled by a deflection curve of the following form:

$$\upsilon = a \sin\frac{\pi x}{L}$$

in which a represents a constant. Carrying this deflection into Equation 5.88b and integrating, the critical load is found as

$$P_{cr} = \frac{\pi^2 EI}{L^2}$$

We observe from Equation 5.88a that, for $P > P_{cr}$, the work done by P exceeds the strain energy stored in the column: that is, a straight column is unstable if $P > P_{cr}$. This point,

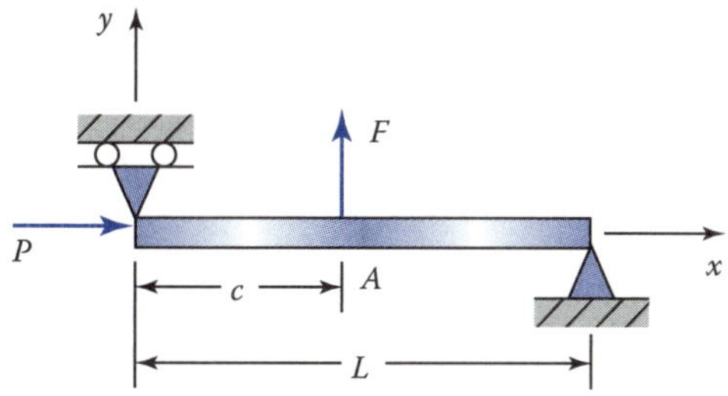

FIGURE 5.29
Example 5.23.

with regard to stability, corresponds to $A = 0$ in Section 5.9 and could not be found as readily by the equilibrium method. When $P = P_{cr}$, the column is in neutral equilibrium. If $P < P_{cr}$, a column is in stable equilibrium.

Example 5.23: Buckling Load and Deflection Analysis of a Beam–Column by the Principle of Virtual Work

A simply supported beam–column is under a lateral force F at point A and axial loading P, as shown in Figure 5.29. Develop the equation of the elastic curve.

Solution

The total work done is obtained by addition of the work owing to the force F to Equation 5.86. This problem has already been solved for $P = 0$ by employing the following series for deflection:

$$\upsilon = \sum_{m=1}^{\infty} a_m \sin\frac{m\pi x}{L} \tag{5.48}$$

On following a procedure similar to that used in Section 5.8, the condition $\delta U = \delta W$ now becomes

$$\frac{\pi^4 EI}{4L^3}\sum_{m=1}^{\infty} m^4\delta\left(a_m^2\right) = \frac{\pi^2 P}{4L}\sum_{m=1}^{\infty} m^2\delta\left(a_m^2\right) + F\sum_{m=1}^{\infty}\sin\frac{m\pi c}{L}\delta\left(a_m\right)$$

Solving,

$$a_m = \frac{2FL^3}{\pi^4 EI}\frac{1}{m^2}\frac{\sin(m\pi c/L)}{m^2 - b}$$

The quantity $b = P/P_{cr}$ is defined by Equation 5.69. Carrying this equation into Equation (a) results in

$$\upsilon - \frac{2FL^3}{\pi^4 EI}\sum_{m=1,3,\ldots}^{\infty}\frac{\sin\left(m\pi c/L\right)}{m^2\left(m^2 - b\right)}\sin\frac{m\pi x}{L}, \quad \left(0 \le x \le L\right) \tag{a}$$

Comments: When P approaches its critical value we have $b \to 1$ and the first term in Equation (a),

$$\upsilon = \frac{2FL^3}{\pi^4 EI}\frac{1}{1-b}\sin\frac{\pi c}{L}\sin\frac{\pi x}{L} \tag{b}$$

shows that the deflection becomes infinite, as expected. Comparing the Equation (b) with Equation 5.51, we see that the axial force increases the deflection produced by the lateral force by a factor of $1/(1 - b)$.

*5.16 Buckling of Rectangular Plates

The strength and deflection theory of plates, discussed in Sections 4.10 and 4.11, applies when such plates are in static equilibrium. Should a plate be compressed in its midplane, it becomes unstable and begins to buckle at a certain critical value of the in-plane force. The buckling of plates is qualitatively analogous to column buckling. However, a buckling analysis of the former case is not performed as readily as the latter. This is especially true in plates having other than simply supported edges. Often, in these cases, the energy method is used to good advantage to obtain the approximate buckling loads [7]. Thin plates or sheets, although quite capable of carrying tensile loadings, are poor in resisting compression. Usually, buckling or wrinkling phenomena observed in compressed plates (and shells) occur rather suddenly and are very dangerous. Fortunately, there is close correlation between theory and experimental data concerned with buckling of plates under a variety of loads and edge conditions.

In this section, we consider only simply supported rectangular plates subjected to uniaxial in-plane compressive forces per unit length N (Figure 5.30). It can be shown that the minimum value of N occurs when $n = 1$. That is, when the simply supported plate buckles, the buckling mode can be only one-half sine wave across the span, while several half sine waves can occur in the direction of compression.

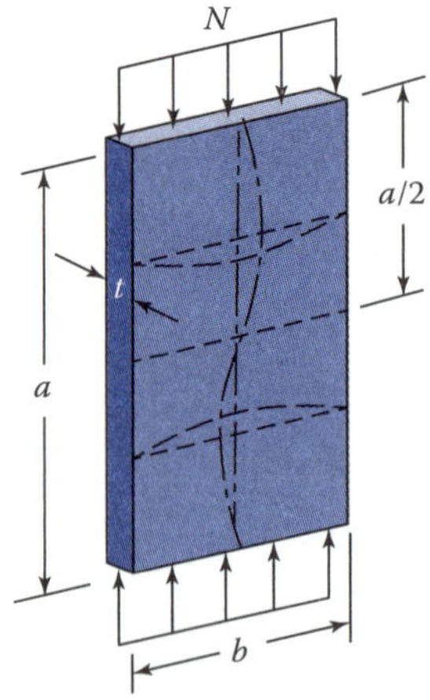

FIGURE 5.30
Compression of plate, simply supported on all four edges.

The critical load N_{cr} per unit length of the plate is expressed as follows [13]:

$$N_{cr} = \frac{\pi^2 D}{b^2}\left(\frac{m}{r} + \frac{r}{n}\right)^2 = k^2 \frac{\pi^2 D}{b^2} \tag{5.89}$$

Here, the buckling load factor, aspect ratio, and flexural rigidity are, respectively,

$$k = \frac{m}{r} + \frac{r}{n}, \quad r = \frac{a}{b}, \quad D = \frac{Et^3}{12(1-\nu^2)}$$

The quantity E represents the modulus of elasticity and ν is Poisson's ratio. The ratio $m/r = 1$ provides the following minimum value of the critical load

$$N_{cr} = \frac{4\pi^2 D}{b^2} \tag{5.90}$$

The corresponding critical stress, N_{cr}/t, is given by

$$\sigma_{cr} = \frac{\pi^2 E}{3(1-\nu^2)}\left(\frac{t}{b}\right)^2 \tag{5.91}$$

where t is the thickness of the plate.

Variations in the buckling load factor k as functions of the aspect ratio r for $m = 1, 2, 3, 4$ are sketched in Figure 5.31. Obviously, for a specific m, the magnitude of k depends on r only. With reference to the figure, the magnitude of N_{cr} and the number of half-waves m for any value of the aspect ratio r can easily be found. When $r = 1.5$, for example, by Figure 5.30, $k = 4.34$ and $m = 2$. The corresponding critical load is equal to $N_{cr} = 4.34\pi^2 D/b^2$,

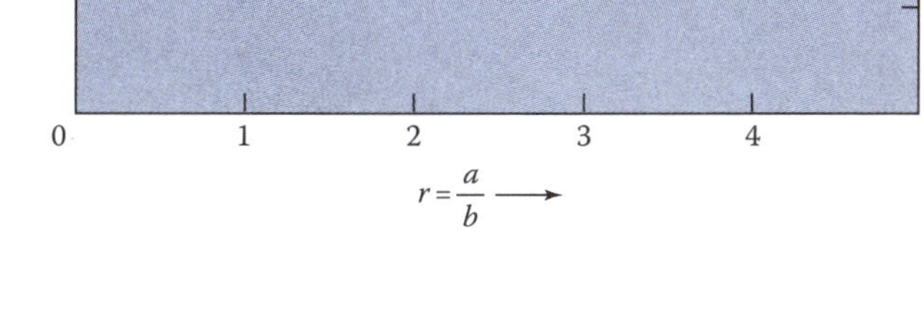

FIGURE 5.31
Variation of buckling load factor k with aspect ratio r for various numbers of half sine waves in the direction of compression.

under which the plate buckles into two half sine waves in the direction of the loading, as depicted in Figure 5.30. We also see from Figure 5.31 that a plate m times as long as it is wide buckles in m half sine waves. Therefore, a long plate ($b \ll a$) with simply supported edges under a uniaxial compression tends to buckle into a number of square cells of side dimensions b, and its critical load for all practical purposes is defined by Equation 5.90.

Problems

The beams, frames, and trusses described in the following problems have constant flexural rigidity EI, axial rigidity AE, and shear rigidity JG.

Sections 5.1 through 5.8

5.1 The stepped bar having a square cross section with sides a and a circular cross section with diameter d is under an axial tensile force P as shown in Figure P5.1. What is the ratio of d/a in order that the strain energy in both parts is the same?

5.2 For the axially loaded stepped-bar illustrated in Figure P5.1, determine

a. The strain energy

b. The deflection at the free end, using the work–energy method

5.3 The stepped-shaft ABC (G = 40 GPa) carries torques as shown in Figure P5.3. For the case in which $T_B = 4$ kN·m and $T_C = 1.5$ kN·m, find the strain energy of the shaft.

5.4 Redo Problem 5.3 knowing that the part AB of the shaft is hollow, the inside diameter is 20 mm, and $T_B = 2$ kN·m and $T_C = 5$ kN·m.

5.5 Compute the strain energy of the aluminum alloy shaft (G = 28 GPa) seen in Figure P5.3 if $T_B = 0$ and the angle of twist ϕ is 0.06 rad.

5.6 The shaft ABC (G = 80 GPa) of Figure P5.3 is subjected to torques $T_B = 0$ and T_A = 1.4 kN·m. Find the angle of twist ϕ using the work–energy approach.

5.7 and **5.8** A beam of rectangular cross-sectional area A is supported and loaded as shown in Figures P5.7 and P5.8. Determine the strain energy of the beam caused by the shear deformation.

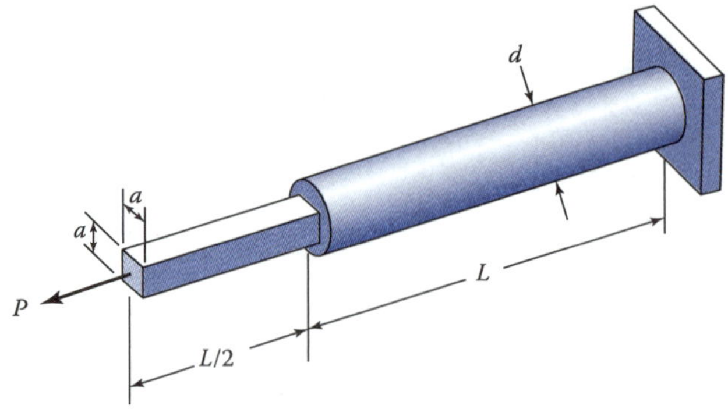

FIGURE P5.1

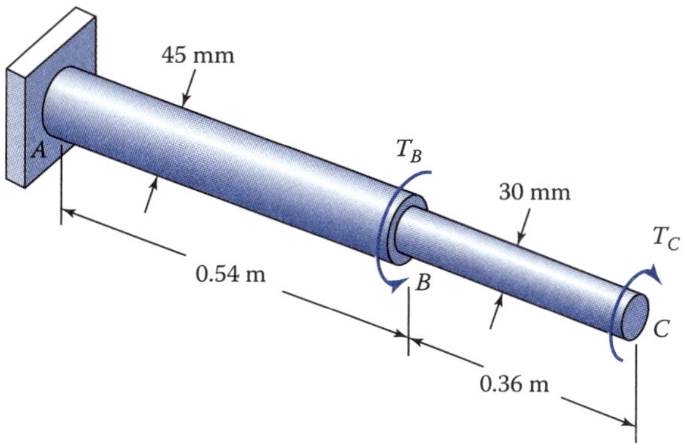

FIGURE P5.3

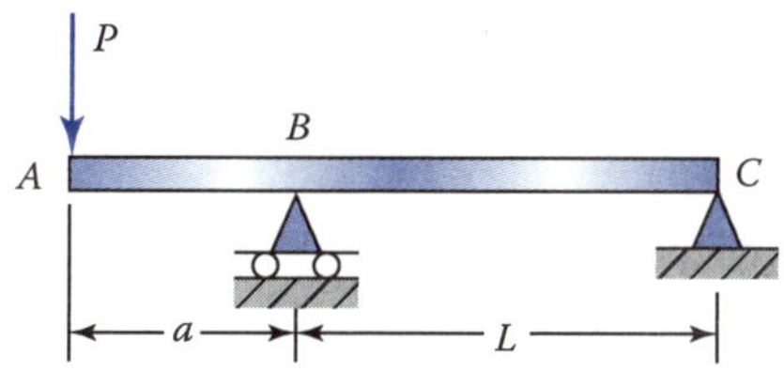

FIGURE P5.7

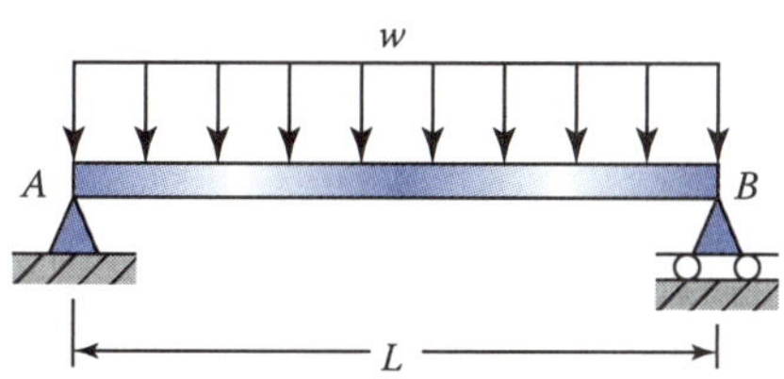

FIGURE P5.8

5.9 An overhanging beam *ABC* of modulus of elasticity *E* is loaded as shown in Figure P5.9. Taking into account only the effect of normal stresses, determine

a. The total strain energy

b. The maximum strain energy density

Assumption: The beam has a rectangular cross section of width *b* and depth *h*.

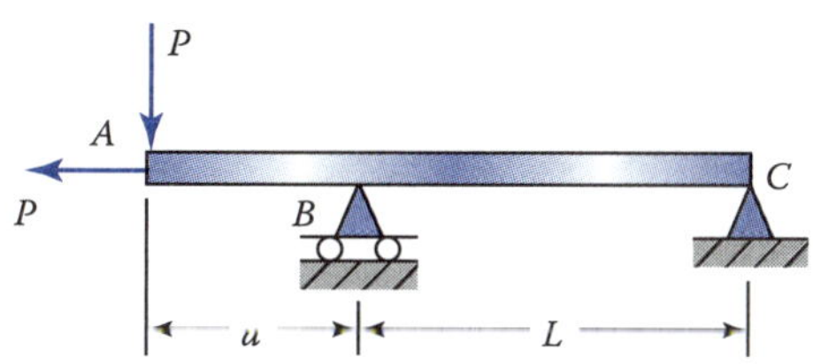

FIGURE P5.9

5.10 A simply supported rectangular beam of depth h, width b, and length L is under a uniform load w (Figure P5.8). Show that the maximum strain energy density due to bending is given by $U_{0,\max} = 45U/8V$. The quantities U and V represent the strain energy and volume of the beam, respectively.

5.11 An overhang beam is loaded as shown in Figure P5.7. Determine the vertical deflection υ_A at the free end A due to the effects of bending and shear. Apply the work–energy method.

5.12 and **5.13** A beam is supported and loaded as illustrated in Figures P5.12 and P5.13. Find the strain energy in the beam due to bending.

5.14 and **5.15** A beam of rectangular cross section is supported and loaded as seen in Figures P5.12 and P5.13. What is the strain energy of the beam owing to shear deformation?

5.16 A beam is supported and loaded as illustrated in Figure P5.13. Applying the work–energy approach, find the deflection at point C (or D) owing to bending and shear.

5.17 A workpiece is clamped to a milling machine by a steel bolt of diameter d tightened to a tension of T (Figure P5.17). The steel link AB has a rectangular cross section of width b and depth h. Find

a. The strain energy and deflection of the bolt caused by the axial load P

b. The strain energy in the link due to bending

Given: $b = 12$ mm, $h = 8$ mm, $d = 6$ mm, $T = 630$ N, $E = 200$ GPa

5.18 A cantilever carries a concentrated load P as shown in Figure P5.18. Using Castigliano's theorem, determine the vertical deflection υ_A at the free end A.

5.19 A cantilevered spring of constant flexural rigidity EI is loaded as depicted in Figure P5.19. Applying Castigliano's theorem, determine the vertical deflection at point B.

Assumption: The strain energy is attributable to bending alone.

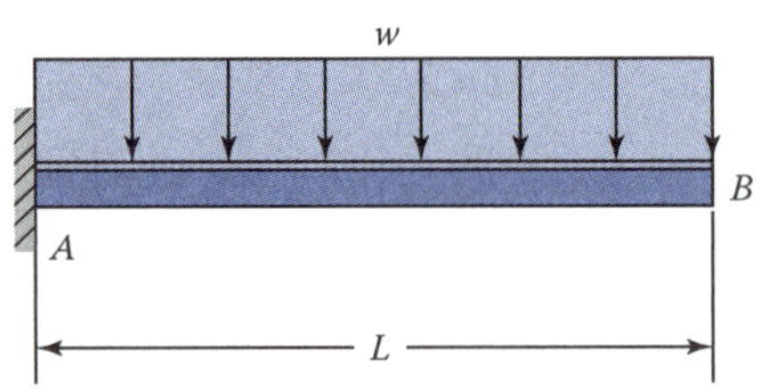

FIGURE P5.12

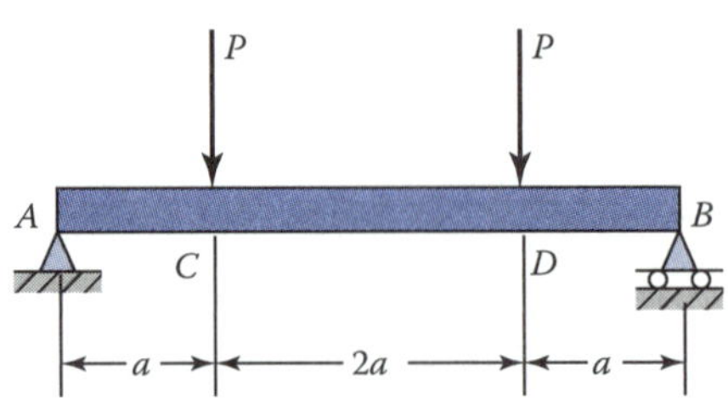

FIGURE P5.13

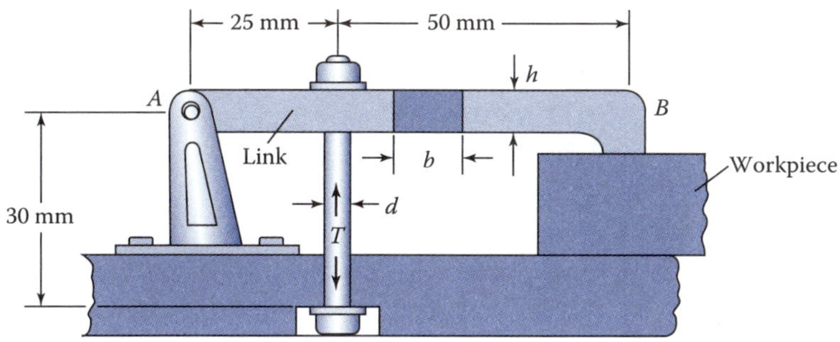

FIGURE P5.17

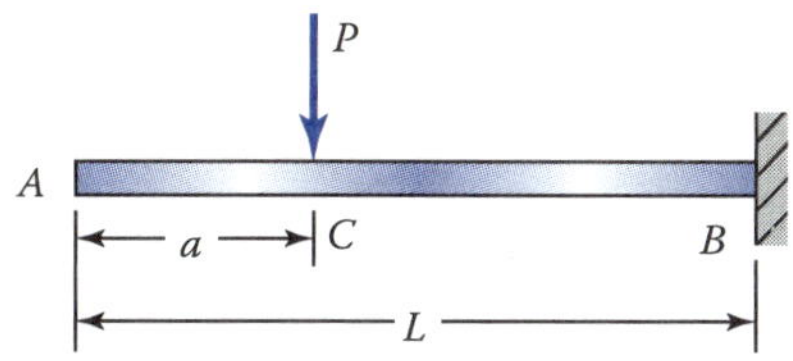

FIGURE P5.18

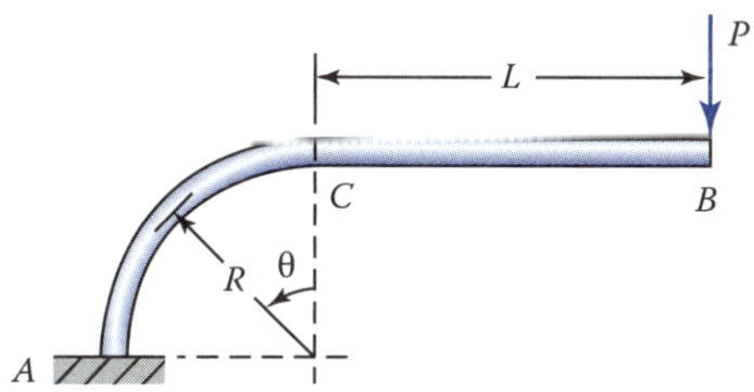

FIGURE P5.19

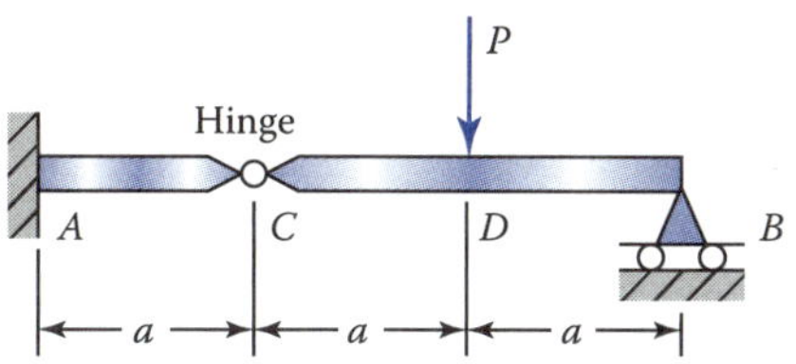

FIGURE P5.20

5.20 Figure P5.20 shows a compound beam with a hinge at C. It is composed of two portions: a beam BC, simply supported at B, and a cantilever AC, fixed at A. Employing Castigliano's theorem, determine the deflection υ_D at the point of application of the load P.

5.21 A continuous beam is subjected to a bending moment M_o at support C (Figure P5.21). Applying Castigliano's theorem, find the reaction at each support.

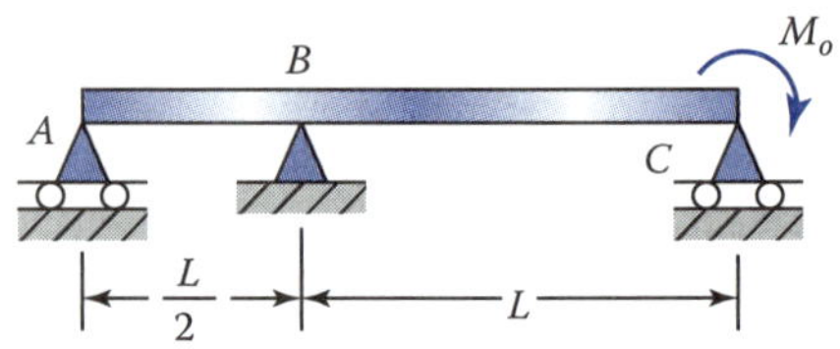

FIGURE P5.21

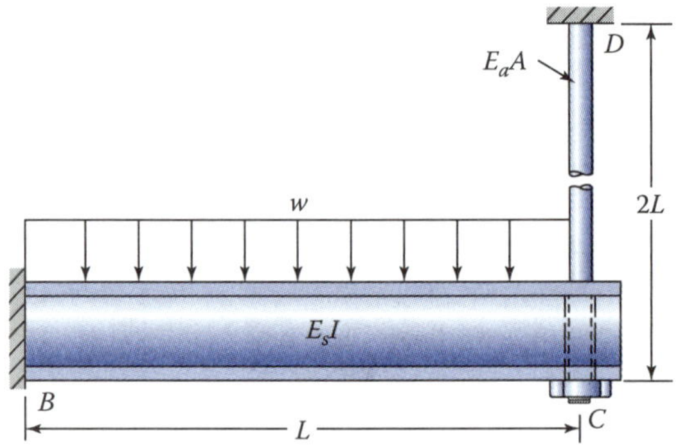

FIGURE P5.22

5.22 A steel I-beam is fixed at B and supported at C by an aluminum alloy tie rod CD of cross-sectional area A (Figure P5.22). Using Castigliano's theorem, determine the tension P in the rod caused by the distributed load depicted, in terms of w, L, A, E_a, E_s, and I, as needed.

5.23 and **5.24** A bent frame is supported and loaded as shown in Figures P5.23 and P5.24. Employing Castigliano's theorem, determine the horizontal deflection δ_A for point A.

Assumption: The effect of bending moment is considered only.

5.25 A semicircular arch is supported and loaded as shown in Figure P5.25. Using Castigliano's theorem, determine the horizontal displacement of the end B.

Assumption: The effect of bending moment is taken into account alone.

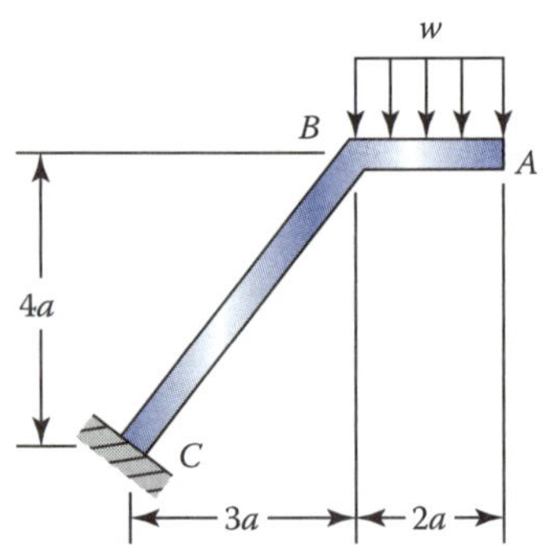

FIGURE P5.23

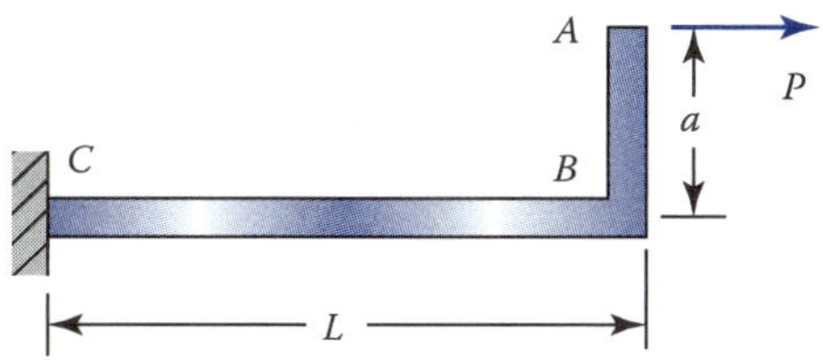

FIGURE P5.24

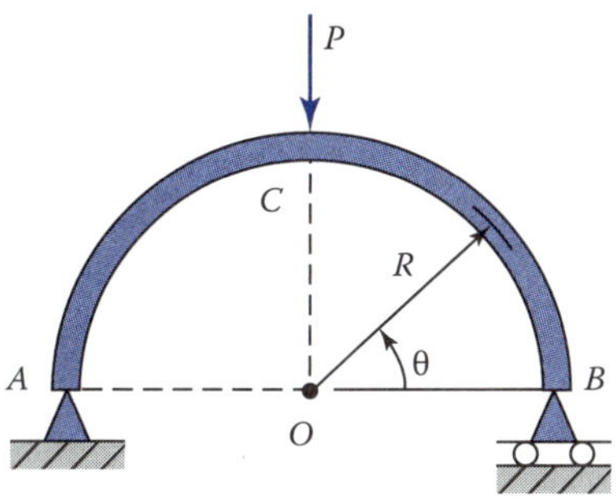

FIGURE P5.25

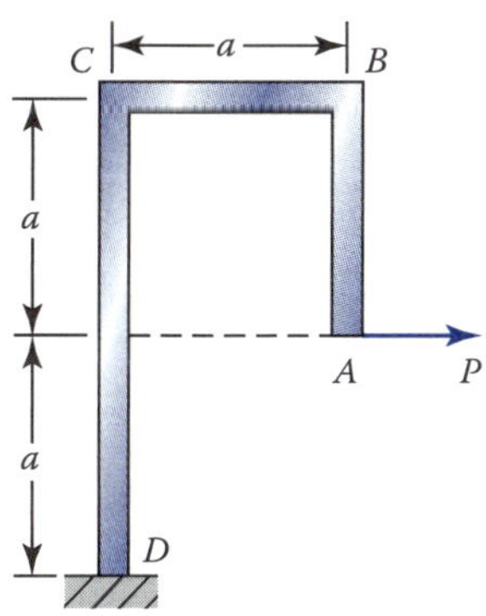

FIGURE P5.26

5.26 A frame is fixed at one end and loaded at the other end as depicted in Figure P5.26. Apply Castigliano's theorem to determine

a. The horizontal deflection δ_A at point A

b. The slope θ_A at point A

Assumption: The effects of axial force as well as shear are omitted.

5.27 A frame is fixed at one end and loaded as shown in Figure P5.27. Employing Castigliano's theorem, determine

a. The vertical deflection δ_A at point A

b. The angle of twist at point B

Assumption: The effect of bending moment is considered only.

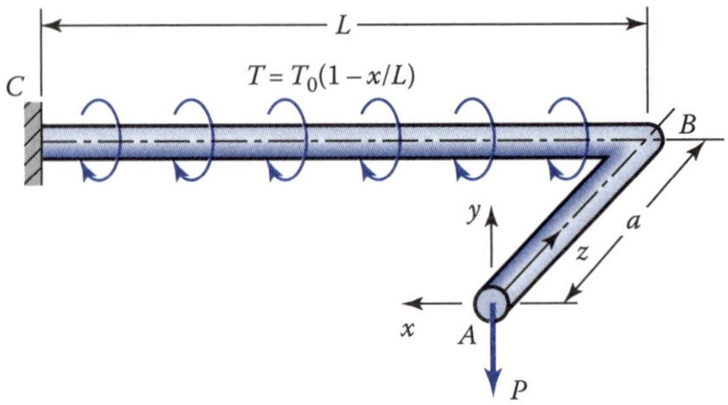

FIGURE P5.27

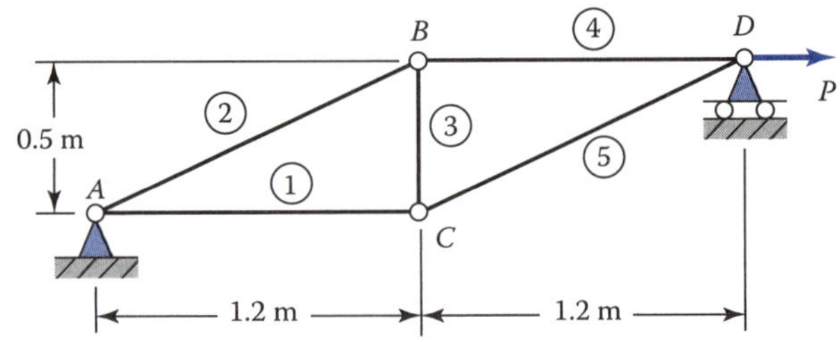

FIGURE P5.28

5.28 A simple truss carries a concentrated force P as indicated in Figure P5.28. Applying the work–energy method, determine the horizontal displacement of the joint D.

Given: $P = 60$ kN, $A = 250$ mm^2, $E = 210$ GPa

5.29 The basic truss shown in Figure P5.29 carries a vertical load $2P$ and a horizontal load P at joint B. Apply Castigliano's theorem, to obtain horizontal displacement δ_C of point C.

Assumption: Each member has an axial rigidity AE.

5.30 A pin-connected truss supports the loads, as shown in Figure P5.30. Using Castigliano's theorem, find the vertical and horizontal displacements of the joint C.

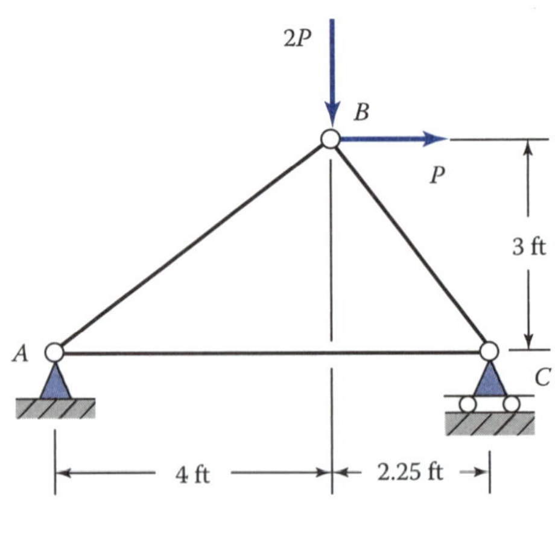

FIGURE P5.29

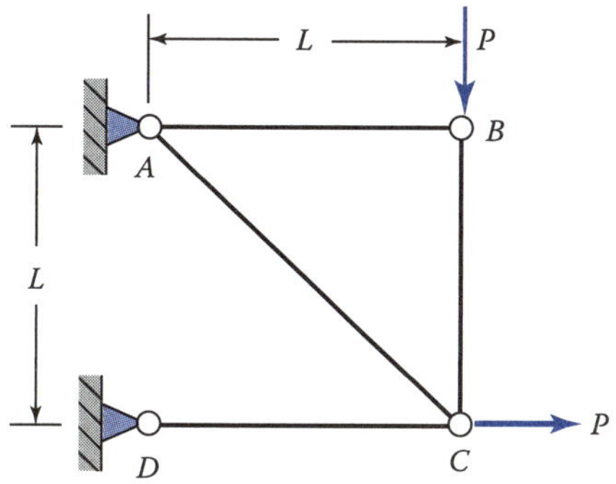

FIGURE P5.30

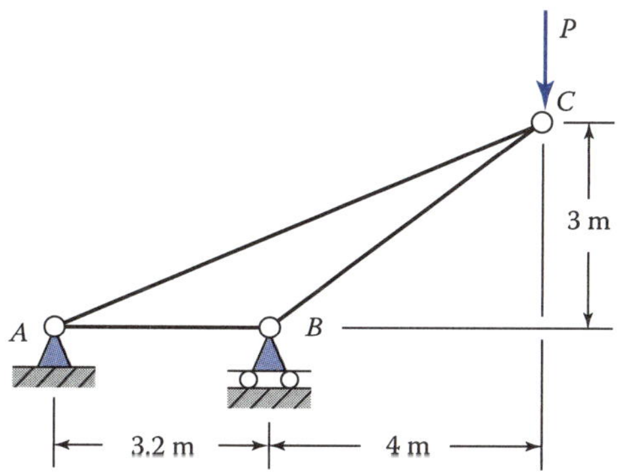

FIGURE P5.31

5.31 The truss *ABC* shown in Figure P5.31 is subjected to a vertical load *P*. Employing Castigliano's theorem, determine the horizontal and vertical displacements of joint *C*.

5.32 A planar truss supports a horizontal load *P*, as shown in Figure P5.32. Apply Castigliano's theorem to obtain the vertical displacements of joint *D*.

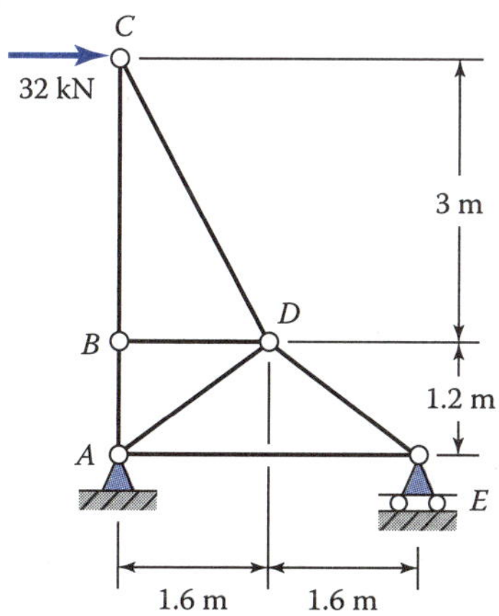

FIGURE P5.32

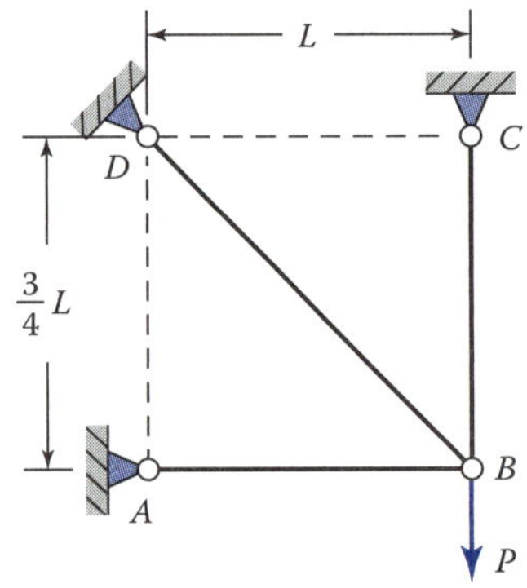

FIGURE P5.33

5.33 A three-member truss carries load P, as shown in Figure P5.33. Applying Castigliano's theorem, determine the force in each member.

5.34 A curved frame of a structure is fixed at one end and simply supported at another, where a horizontal load P applies (Figure P5.34). Determine the roller reaction F at the end B, using Castigliano's theorem.

Assumption: The effect of bending moment is considered only.

5.35 Two-hinged frame ACB carries a concentrated load P at C, as shown in Figure P5.35. Determine, using Castigliano's theorem,

a. The horizontal displacement δ_B at B

b. The horizontal reaction R at B, if the support B is a fixed pin

Assumption: The strain energy is attributable to bending alone.

5.36 Figure P5.36 shows a structure that consists of a cantilever AB, fixed at A, and bars BC and CD, pin connected at both ends. Find the vertical deflection of joint C, considering the effects of normal force and bending moment. Employ Castigliano's theorem.

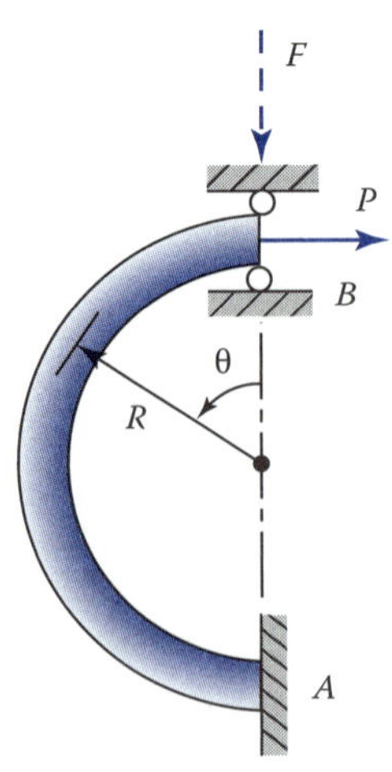

FIGURE P5.34

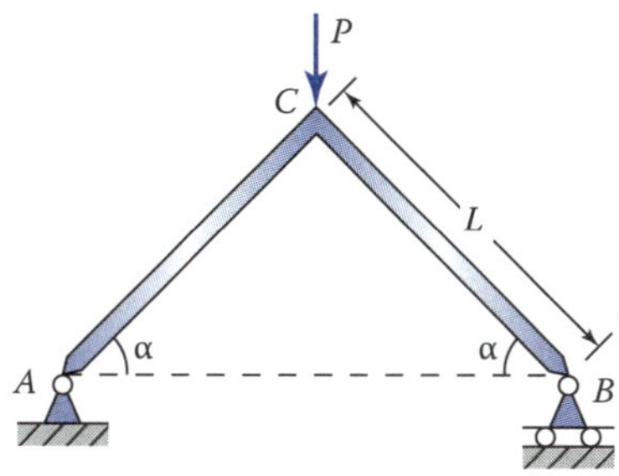

FIGURE P5.35

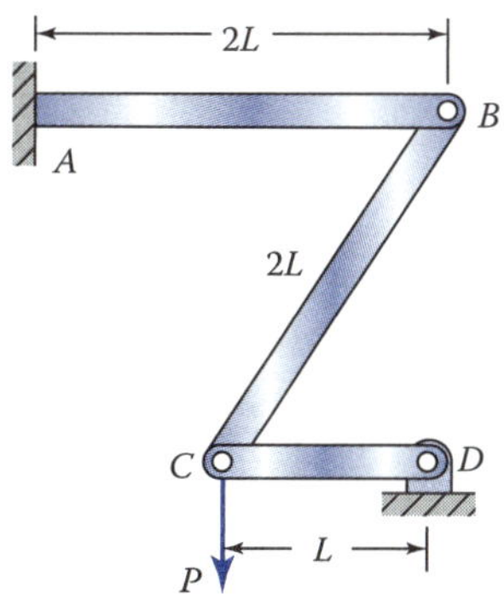

FIGURE P5.36

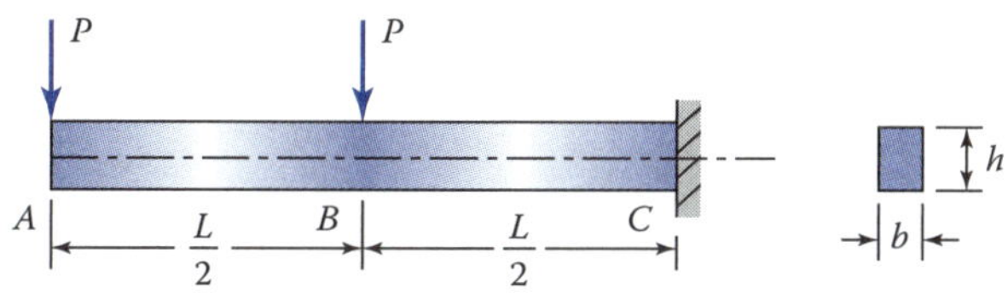

FIGURE P5.37

5.37 A cantilevered beam with a rectangular cross section carries concentrated loads P at free end and at the center as shown in Figure P5.37. Determine, using Castigliano's theorem,

a. The deflection of the free end, considering the effects of both the bending and shear

b. The error, if the effect of shear is neglected, for the case in which $L = 5h$ and the beam is made of ASTM-A36 structural steel

5.38 A curved frame ABC is fixed at one end, hinged at another, and subjected to a concentrated load P, as shown in Figure P5.38. What are the horizontal H and vertical F reactions? Use Castigliano's theorem.

Assumption: The strain energy is attributable to bending only.

5.39 A pin-connected structure of three bars supports a load W at joint D (Figure P5.39). Apply Castigliano's theorem to determine the force in each bar.

Given: $a = 0.6L$, $h = 0.8L$

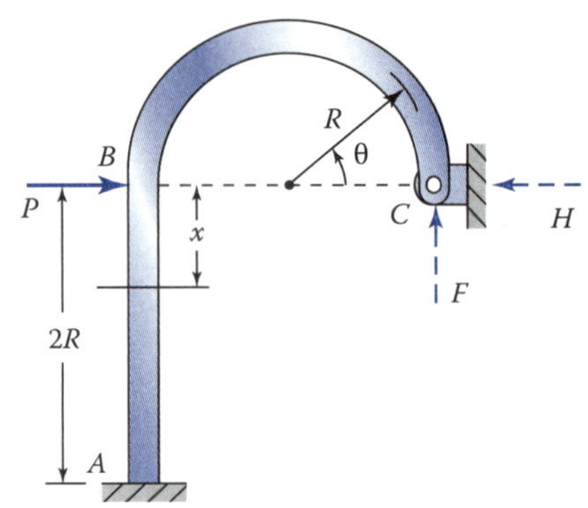

FIGURE P5.38

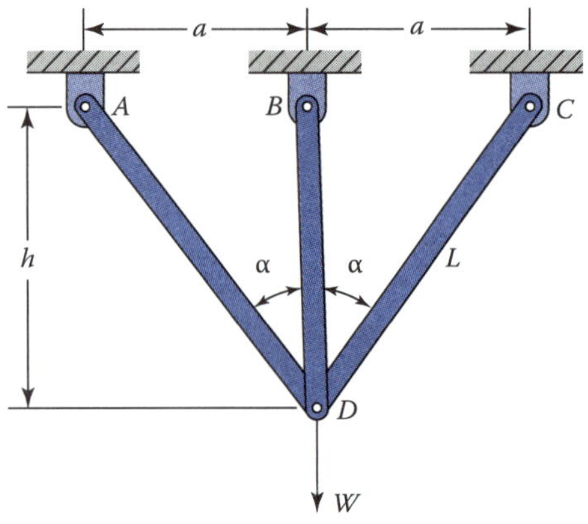

FIGURE P5.39

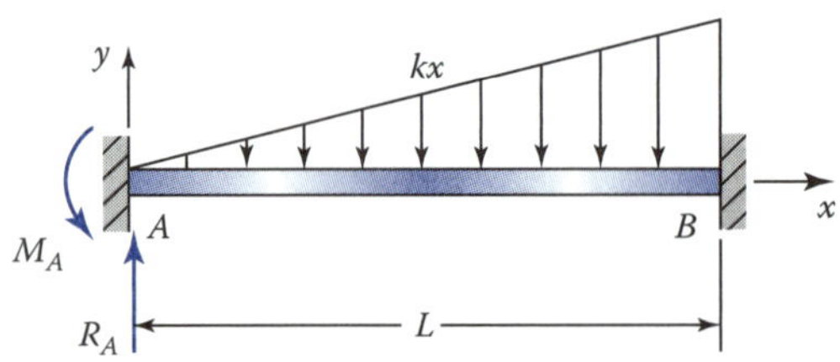

FIGURE P5.40

5.40 A fixed-ended beam supports a uniformly increasing load $w = kx$ (Figure P5.40), where k is a constant. Determine the reactions R_A and M_A at end A, using Castigliano's theorem.

5.41 A simply supported beam carries a distributed load of intensity $w = w_0 \sin \pi x/L$ as shown in Figure P5.41. Using the principle of virtual work, determine the expression for the deflection curve υ.

Assumption: The deflection curve has the form $\upsilon = a \sin \pi x/L$, where a is to be found.

5.42 Applying Castigliano's first theorem, find the load W required to produce a vertical displacement of ¼ in., at joint D in the pin-connected structure depicted in Figure P5.39.

Given: $h = 10$ ft, $\alpha = 30°$, $E = 30 \times 10^6$ psi, $A = 1$ in.^{2}

Assumption: Each member has the same cross-sectional area A.

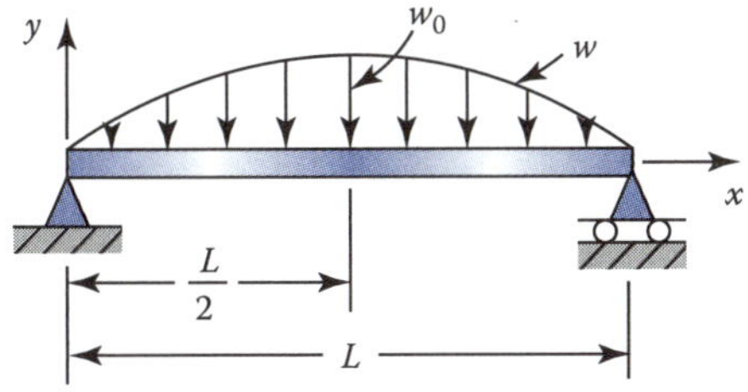

FIGURE P5.41

5.43 A cantilevered beam is subjected to a concentrated load P at its free end (Figure 5.15). Apply the principle of virtual work to determine

a. An expression for the deflection curve υ

b. The maximum deflection and the maximum slope

Assumption: Deflection curve of the beam has the form $\upsilon = ax^2\,(3L - x)/2L^3$, where a is a constant.

5.44 A simply supported beam is loaded as shown in Figure 5.14. Using the virtual work principle, determine the deflection at point A.

Assumption: The deflection curve of the beam is of the form $\upsilon = ax(L - x)$, in which a is a constant.

Sections 5.9 through 5.12

5.45 An industrial machine requires a solid, round steel piston connecting rod of length L that carries a maximum compressive load P. Determine the required diameter.

Given: $L = 1.2$ m, $P = 50$ kN, $E = 210$ GPa, $S_y = 600$ MPa

Assumption: The ends are taken to be pinned.

5.46 Redo Problem 5.45 for a cold-rolled (510) bronze connecting rod.

Given: $L = 250$ mm, $P = 220$ kN, $E = 110$ GPa, $S_y = 520$ MPa (Table B.1)

5.47 Repeat Example 5.15 for a long column using Euler's formula.

5.48 Figure P5.48 shows the cross sections of two aluminum alloy 2014-T6 bars used as compression members, each having an effective length L_e. Find

a. The wall thickness t of the hollow square bar in order that they have the same cross-sectional area

b. The critical load of each bar

Given: $D = 50$ mm, $d = 35$ mm, $a_0 = 50$ mm, $L_e = 2.2$ m, $E = 72$ GPa (by Table B.1)

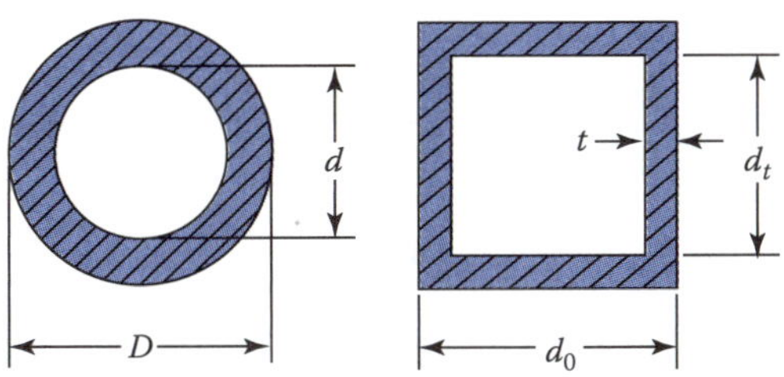

FIGURE P5.48

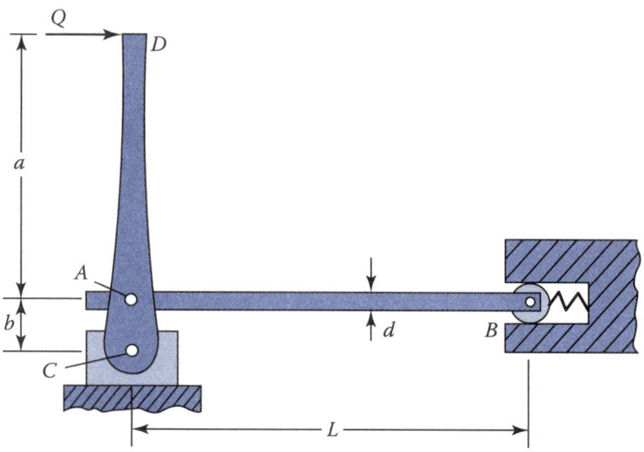

FIGURE P5.49

5.49 A control linkage is composed of a structural ASTM-A36 steel rod *AB* of diameter *d* and a pivot arm *CD* as depicted in Figure P5.49. The load is transmitted to the rod through pin *A*. Compute the maximum value of *Q* that can be applied based on the buckling strength of the rod and a safety factor of *n*.

Given: $a = 150$ mm, $b = 30$ mm, $L = 0.4$ m, $d = 8$ mm, $n = 1.4$, $E = 200$ GPa, $S_y = 250$ MPa (see Table B.1)

5.50 A round steel column with length *L* is built in at its base and pinned at its top and carries a maximum allowable load *P* as shown in Figure P5.50. What is the diameter *d* of the member on the basis of factor of safety *n* with respect to buckling?

Given: $L = 40$ in., $P = 5$ kips, $n = 2.6$, $E = 29 \times 10^6$ psi, $S_y = 36$ ksi (from Table B.1)

5.51 Resolve Problem 5.50 knowing that the length of the column is $L = 25$ in. and the largest allowable load is $P = 25$ kips.

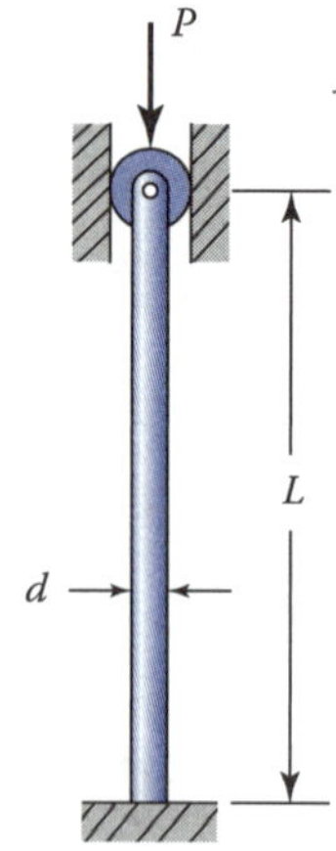

FIGURE P5.50

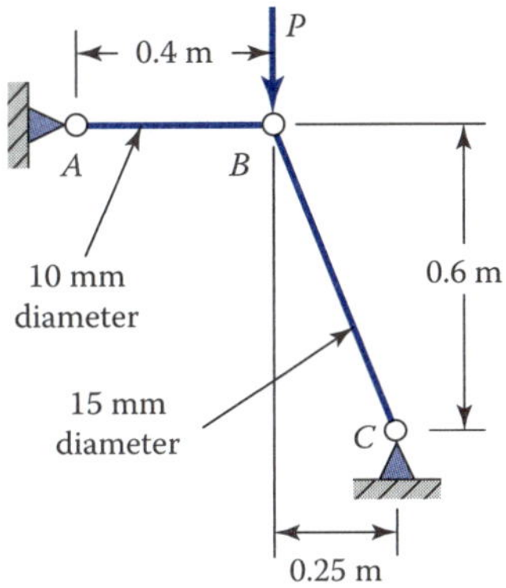

FIGURE P5.52

5.52 A two-member pin-connected structure supports a concentrated load P at joint B as shown in Figure P5.52. Calculate the largest load P that may be applied with a factor of safety n.

Given: $n = 2.5$, $E = 210$ GPa

Assumption: Buckling occurs in the plane of the structure.

5.53 A simple truss is loaded as shown in Figure P5.53. Calculate the diameter d necessary for

a. The bar AB

b. The bar BC

Given: $E = 210$ GPa, $S_y = 250$ MPa

Assumption: Failure occurs by yielding. Buckling occurs in the plane of the truss. Euler's formula applies.

5.54 A structure that consists of the beam AB and the column CD is supported and loaded as shown in Figure P5.54. What is the largest load F that may be applied?

Given: The member CD is a round steel bar with $E = 200$ GPa

Design Assumption: Failure is due to buckling only. Use Euler's formula with a factor of safety $n = 1.5$.

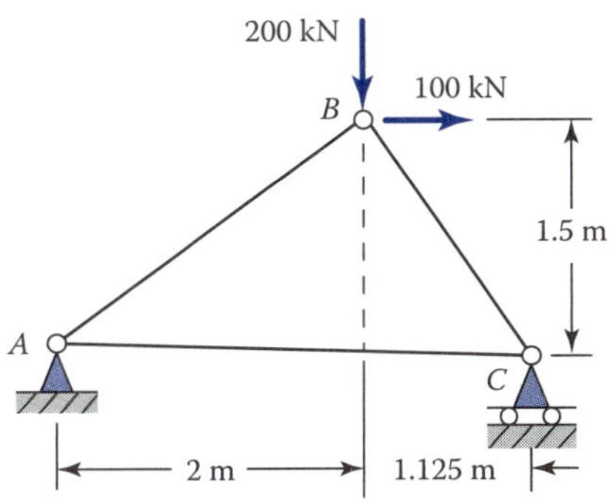

FIGURE P5.53

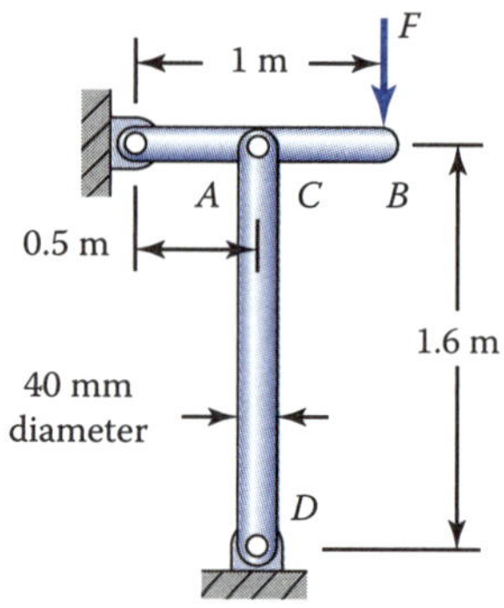

FIGURE P5.54

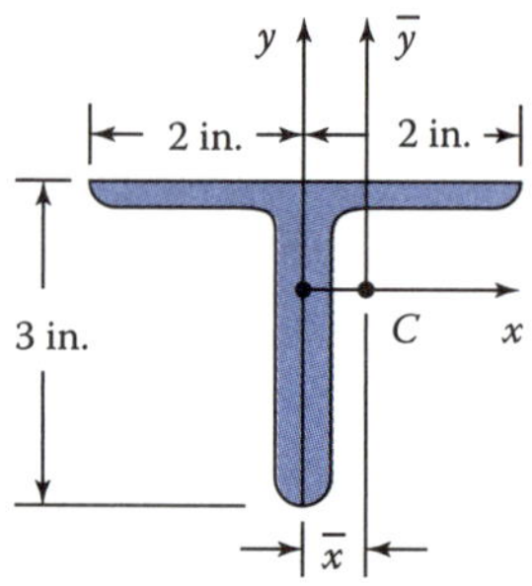

FIGURE P5.55

5.55 A compression member of 9 ft effective length is made by welding together two $3 \times 2 \times \frac{1}{4}$ in. steel angles ($E = 30 \times 10^6$ psi), as shown in Figure P5.55. Determine the allowable centric load for the member if a factor of safety of 2.5 is required.

Given: For one angle, the area properties $\bar{x} = 0.413$ in., $A = 1.19$ in.^{2}, $I_{\bar{x}} = 1.09$ in.^{4}, $I_{\bar{y}} = 0.392$ in.^{4}

5.56 A solid circular steel column of length L and diameter d is hinged at both ends. Calculate the load capacity for a safety factor of n.

Given: $S_y = 350$ MPa, $E = 210$ GPa, $L = 1$ m, $d = 60$ mm, $n = 3$

Assumption: The initial crookedness is 2 mm.

5.57 A rectangular aluminum alloy 2014-T6 tube of uniform thickness t (Figure P5.57) is used as a column of length L fixed at both ends. What is the critical stress in the column?

Given: $b = 160$ mm, $h = 80$ mm, $L = 5.5$ m, $t = 15$ mm, $E = 72$ GPa (by Table B.1)

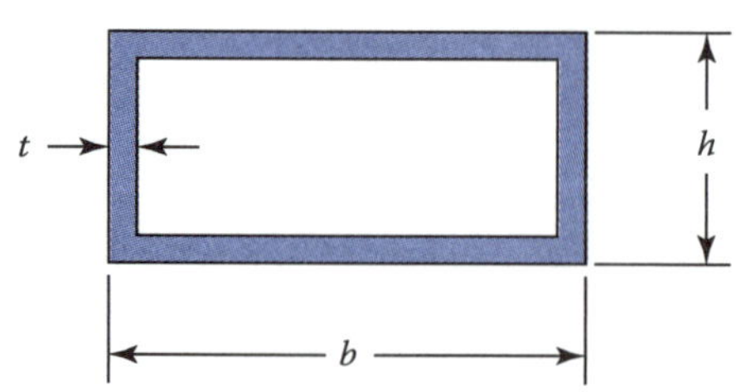

FIGURE P5.57

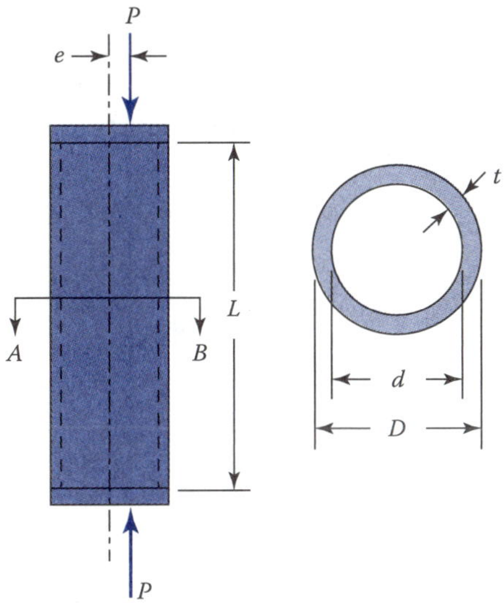

FIGURE P5.59

5.58 Redo Problem 5.57, for the case in which the column is pinned at one end and fixed at the other.

5.59 A steel pipe of outer diameter D and inner diameter d is employed as a 2 m column (Figure P5.59). Use Figure 5.23 to determine the largest allowable load P_y when the eccentricity e is

a. 12 mm

b. 9 mm

Given: $D = 100$ mm, $d = 88$ mm, $L = 2$ m, $S_y = 275$ MPa, $E = 200$ GPa

5.60 A pin-ended wrought iron pipe of outer diameter D, inside diameter d, and length L carries an eccentric load P with a safety factor of n as shown in Figure P5.59. What is the largest allowable eccentricity e?

Given: $D = 8$ in., $d = 7$ in., $L = 12$ ft, $P = 100$ kips, $n = 1.5$, $E = 27 \times 10$ psi, $S_y = 30$ ksi

5.61 A pin-ended steel tube is subjected to an eccentrically applied load P (Figure P5.59). For the case in which the maximum deflection at the mid length is υ_{max}, find

a. The eccentricity e

b. The maximum stress in the rod

Given: $D = 8$ in., $d = 7$ in., $L = 15$ ft, $P = 10$ kips, $\upsilon_{max} = 0.05$ in., $E = 29 \times 10^6$ psi

5.62 Redo Problem P5.61, knowing that the bottom and top ends of the column are fixed and free, respectively.

5.63 A steel hollow box column of length L is fixed at its base and free at its top as illustrated in Figure P5.63. For the case in which a concentrated load P acts at the middle of side AB (i.e., $e = 37.5$ mm) of the free end, find the maximum stress in the column.

Given: $b = 150$ mm, $h = 75$ mm, $t = 15$ mm, $L = 1.8$ m, $E = 200$ GPa, $P = 160$ kN

5.64 Resolve Problem 5.63 knowing that the load acts at the middle of side AC (i.e., $e = 75$ mm).

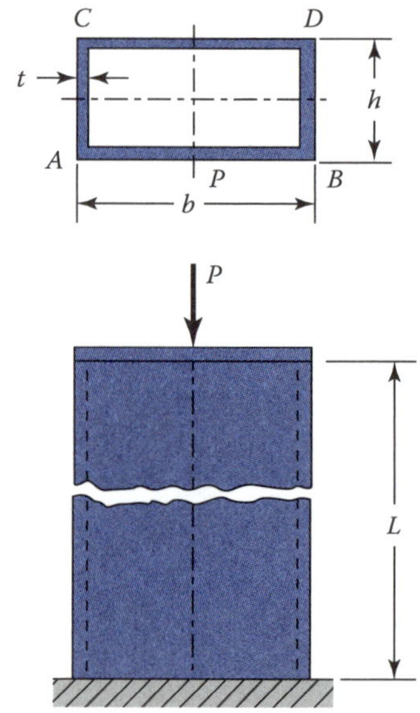

FIGURE P5.63

5.65 An aluminum alloy 6061-T6 hollow box column of square cross section and length L is built in at its base and free at its top (Figure P5.65). The column is under a compressive load P acting with an eccentricity of e. Find

a. The horizontal deflection of the top of the column

b. The maximum stress in the column

Given: a_o = 120 mm, a_i = 100 mm, e = 55 mm, L = 2 m, P = 200 kN, E = 70 GPa, S_y = 260 MPa (from Table B.1)

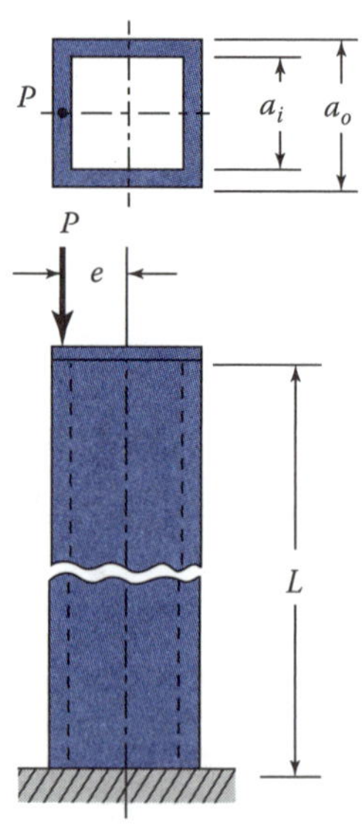

FIGURE P5.65

5.66 Figure P5.65 shows a steel tubular column subjected to a concentrated load P with an eccentricity e. When the horizontal deflection at the top caused by the loading is υ_{max}, find

a. The eccentricity e

b. The largest stress in the column

Given: $a_o = 120$ mm, $a_i = 100$ mm, $L = 1.9$ m, $P = 300$ kN, $\upsilon_{max} = 15$ mm, $E = 200$ GPa, $S_y = 250$ MPa (from Table B.1)

5.67 A brass column ($E = 105$ GPa) with of one end fixed and the other free is under axial compression. Compute the critical buckling load for the following two forms (Figure P5.48):

a. A cylindrical tube with $D = 51$ mm outer diameter

b. A square tube with an $a_o = 40$ mm outer dimension

Requirement: The column cross-sectional area and length are $A = 500$ mm^2 and $L = 2$ m, respectively.

5.68 Consider a $b \times h$ rectangular cross-sectional structural steel column with one end fixed and one end pinned. Determine the buckling load for the two lengths:

a. $L = 180$ mm

b. $L = 500$ mm

Given: $b = 35$ mm, $h = 10$ mm, $S_y = 250$ MPa, $E = 200$ GPa (by Table B.1)

5.69 A 30 in. long hollow aluminum alloy 6061-T6 tube having pinned ends has a $D =$ 2.5 in. outer diameter and inside diameter $d = 2.4$ in. Determine

a. The critical load for a concentric loading

b. The maximum stress under an eccentric loading (Figure P5.59) with $P =$ 3.5 kips and $e = 0.12$ in.

Given: $E = 10 \times 10^6$ psi, $S_y = 38$ ksi

Sections 5.13 and 5.16

5.70 Resolve Problem 5.55 using the AISC formulas.

Given: $S_y = 35$ ksi

5.71 A steel rod of length L is required to support a concentric load P. Compute the minimum diameter required.

Given: $S_y = 350$ MPa, $E = 200$ GPa, $L = 3$ m, $P = 50$ kN

Assumption: Both ends of the rod are fixed.

5.72 A 4 m long fixed-ended timber column must safely carry a 100 kN centric load. Design the column using a square cross section.

Given: $E = 12.4$ GPa and $\sigma_{all} = 10$ MPa for compression parallel to the grain

5.73 An aluminum alloy 6061-T6 pipe has an outer diameter D, inner diameter d, and length L. Determine the allowable axial load P if the pipe is used as a column fixed at one end and pinned at the other.

Given: $D = 14$ in., $d = 12$ in., $L = 20$ ft

5.74 A pin-ended steel column is subjected to a vertical load P. Determine the allowable stress.

Given: The cross section of the column is 80×120 mm and length is 3.5 m.

$$p = 600 \text{ kN}, \quad E = 210 \text{ GPa}, \quad S_y = 280 \text{ MPa}$$

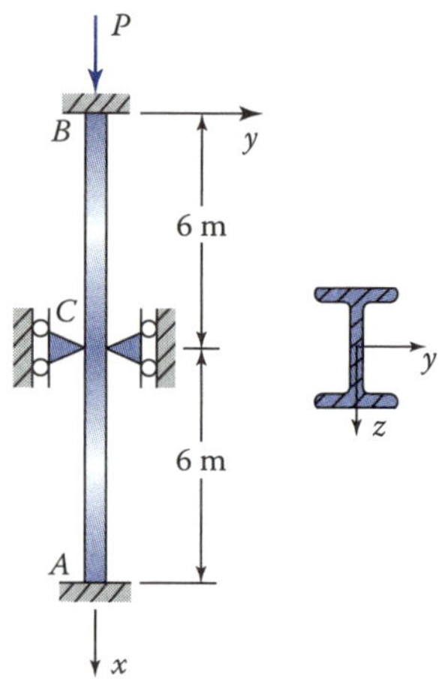

FIGURE P5.75

5.75 A W 360 × 216 rolled-steel column (see Table A.6) with built-in ends is braced at midpoint C, as depicted in Figure P5.75. Calculate the allowable axial load P.

Given: $E = 200$ GPa, $S_y = 280$ MPa

Assumption: Bracing acts as a simple support in the xy plane.

5.76 An $a \times a$ square cross-sectional aluminum alloy 6061-T6 column of length $L = 0.5$ m is pinned at both ends. Compute the minimum allowable width a when the member is to support an axial load of $P = 200$ kN.

5.77 Figure P5.77 shows the rectangular cross section of an aluminum alloy 6061-T6 (see Table B.1) tube that is to be used as a column of an effective length L_e. What is the largest allowable axial centric load P?

Given: $L_e = 14$ ft, $E = 10 \times 10^6$ psi (Table B.1)

5.78 A 3.5 in. by 2.5 in. rectangular cross-sectional timber column of length L is fixed at its base and free at its top. Compute the maximum allowable axial load P that it can support.

Given: $L = 4$ ft, $E = 1.7 \times 10^6$ psi (Table B.1)

5.79 Redo Problem 5.78 for the case in which the length of the column is $L = 30$ in.

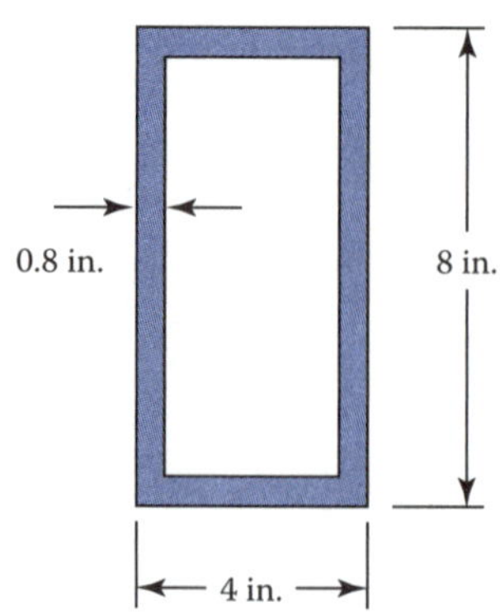

FIGURE P5.77

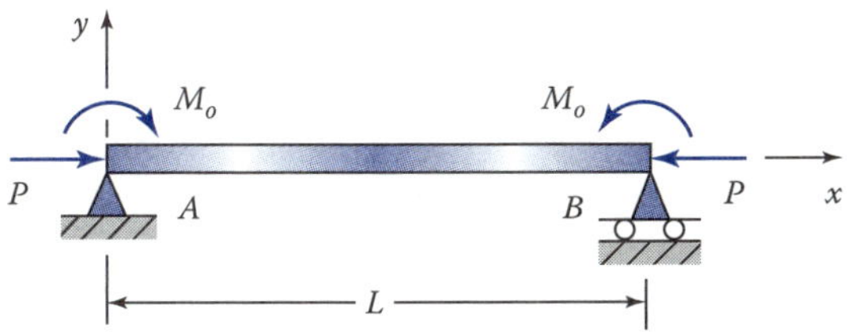

FIGURE P5.80

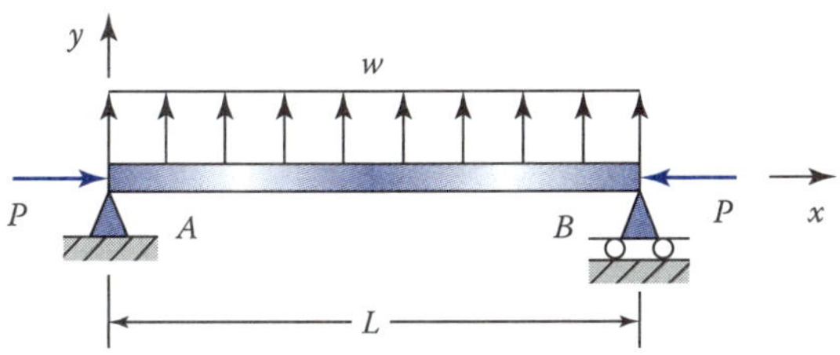

FIGURE P5.81

5.80 A simply supported beam–column is subjected to end moments M_o and axial compression forces P, as depicted in Figure P5.80. Employ the equilibrium method to determine an expression for the deflection υ.

5.81 A simply supported beam–column carries a uniformly distributed lateral load w and axial compression forces P (Figure P5.81). Using the equilibrium approach, determine an expression for the deflection υ.

5.82 Redo Problem 5.81, employing the principle of virtual work. Let the deflection curve be expressed by Equation 5.48.

Section II

Failure Prevention

An automobile crankshaft split into pieces because of fatigue loading (www.google.com). Prevention or reduction of such fatigue failures as well as surface damage due to corrosion and wear is one of the greatest challenges to modern engineering. Many surface treatments and coatings tailored to suit different materials and environments can be used to enhance the life and performance of products. Section II is concerned with the formulations of static and dynamic failure criteria, the reliability method in design, and an empirical approach to surface wear. Practical applications to variously loaded components are included.

6

Static Failure Criteria and Reliability

6.1 Introduction

A proper design includes the prediction of circumstances under which failure is likely to occur. In the most general terms, *failure* refers to any action that causes the member of the structure or machine to cease to function satisfactorily. The strength, stiffness, and stability of various load-carrying members are possible types or modes of failure. Failure may also be associated with poor appearance, poor adaptability to new demands, or other considerations not directly related to the ability of the structure to carry a load. Important variables associated with the failure include the type of material, the configuration and rate of loading, the shape and surface peculiarities, and the operational environment.

This chapter is devoted to the study of *static failure criteria* and the *reliability method* in design. We are concerned mainly with the failure of homogeneous and isotropic materials by yielding and fracture. The mechanical behavior of materials associated with failure also is discussed. In addition to possible failure by yielding or fracture, a member can fail at much lower stresses by crack propagation, should a crack of sufficient size be present. The *fracture mechanics theory* provides a means to predict a sudden failure on the basis of a computed stress-intensity factor compared to a tested toughness criterion for the material (Sections 6.2 through 6.4). Other modes of failure include excessive elastic deflection of some element, rendering the machine or structure useless, or failure of a component by buckling. Various types of failure are considered in the problems presented as the subject unfolds [1–4].

Unless we are content to overdesign members, it is necessary to predict the most probable modes of failure and product reliability. Of necessity, the strength theories of failure are used in the majority of machine and structural designs. The actual failure mechanism in an element may be quite complicated; each failure theory is only an attempt to model the mechanism of failure for a given class of material. In each case, a factor of safety is employed to provide the required safety and reliability. Clearly, composite materials that do not exhibit uniform properties require more complex failure criteria [3].

6.2 Introduction to Fracture Mechanics

Fracture is defined as the separation or fragmentation of a member into two or more pieces. It normally constitutes a *pulling apart* associated with the tensile stress. A relatively brittle material fractures without yielding occurring throughout the fractured cross section.

Thus, a brittle fracture occurs with little or no deformation or reduction in area and hence very little energy absorption. This type of failure usually takes place in some materials in an instant. The mechanisms of brittle fracture are the concern of fracture mechanics. It is based on a stress analysis in the vicinity of a crack, flaw, inclusion, or defect of unknown small radius in a part. A crack is a microscopic flaw that may exist under normal conditions on the surface or within the material.

As pointed out in Section 3.13, the stress-concentration factors are limited to elastic structures for which all dimensions are precisely known, particularly the radius of the curvature in regions of high stress concentration. When there exists a crack, the stress-concentration factor approaches infinity as the root radius approaches 0, thus rendering the concept of stress-concentration factor useless. Furthermore, even if the radius of curvature of the flaw tip is known, the high local stresses there lead to some local plastic deformation. Elastic stress-concentration factors become meaningless for this situation. On the basis of the foregoing, it may be concluded that analysis from the point of view of stress-concentration factors is inadequate when cracks are present.

In 1920, Griffith postulated that an existing crack rapidly propagates (leading to rupture in an instant) when the strain energy released from the stressed body equals or exceeds that required to create the surfaces of the crack. On this basis, it has been possible to calculate the average stress (if no crack were present) that causes crack growth in a part. For relatively brittle materials, the crack provides a mechanism by which energy is supplied continuously as the crack propagates.

Since major catastrophic failures of ships, bridges, and pressure vessels in the 1940s and 1950s, increasing attention has been given by design engineers to the conditions of the growth of a crack. Griffith's concept has been considerably expanded by Irwin [5]. Although Griffith's experiments dealt primarily with glass, his criterion has been widely applied to other materials, such as hard steels, strong aluminum alloys, and even low-carbon steel below the ductile–brittle transition temperatures. Inasmuch as failure does not occur then in an entirely brittle manner, application to materials that lie between relatively brittle and ductile requires modification of the theory and remains an active area of contemporary design and research in solid mechanics.

Adequate treatment of the subject of fracture mechanics is beyond the scope of this text. However, the basic principles and some important results simply are stated. Briefly, the fracture mechanics approach starts with an assumed initial minute crack (or cracks), for which the size, shape, and location can be defined. If brittle failure occurs, it is because the conditions of loading and environment are such that they cause an almost sudden propagation to failure of the original crack. Under fatigue loading, the initial crack may grow slowly until it reaches a critical size at which the rapid fracture occurs.

6.3 Stress–Intensity Factors

In the fracture mechanics approach, a stress-intensity factor, K, is evaluated, as contrasted to stress-concentration factors. This can be thought of as a measure of the effective local stress at the crack root. The three modes of crack deformation of a plate are depicted in Figure 6.1. The most currently available values of K are for tensile loading normal to the crack, which is called *mode I* (Figure 6.1a). Accordingly, it is designated as K_I. Other types,

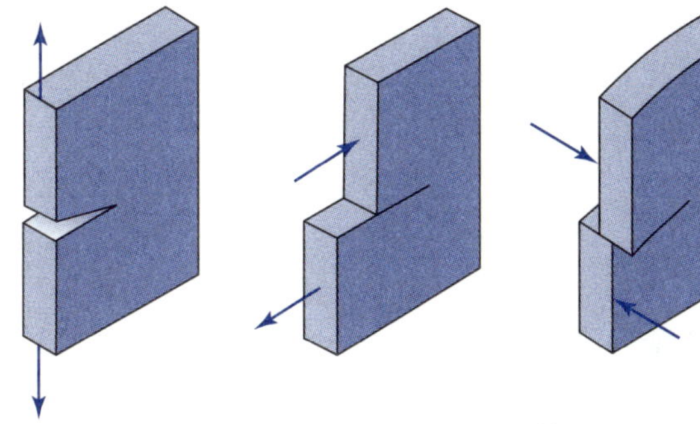

FIGURE 6.1
Crack deformation types: (a) mode I, opening; (b) mode II, sliding; and (c) mode III, tearing.

modes II and III, essentially pertain to the in-plane and out-of-plane shear loads, respectively (Figures 6.1b and c). The treatment here concerns only mode I, and we eliminate the subscript and let $K = K_I$.

Acceptable solutions for many configurations, specific initial crack shapes, and orientations have been developed analytically and by computational techniques, including finite element analysis (FEA). For plates and beams, the *stress-intensity factor* is defined in the form [6]

$$K = \lambda\sigma\sqrt{\pi a} \tag{6.1}$$

where

σ = the normal stress
λ = the geometry factor, depending on a/w, listed in Table 6.1
a = the crack length (or half crack length)
w = the member width (or half width of the member)

We observe from Equation 6.1 and Table 6.1 that the stress-intensity factor depends on the applied load and geometry of the specimen as well as the size and shape of the crack. Clearly, the K may be increased by increasing either the stress or the crack size. The units of the stress-intensity factors are commonly $\text{MPa}\sqrt{\text{m}}$ in SI and $\text{ksi}\sqrt{\text{in.}}$ in U.S. customary system.

Most cracks are not as simple as depicted in Table 6.1. They may be at an angle, embedded in a body, or sunken into a surface. A shallow surface crack in a member may be considered semielliptical. A circular or elliptical shape has proven to be adequate for various studies. Books on fracture mechanics provide methods of analysis, applications, and voluminous references [7].

Note that crack propagation occurring after an increase in load may be interrupted if a small inelastic zone forms ahead of the crack. But, stress intensity has risen with the increase in crack length and, in time, the crack may advance again a short amount. If stress continues to increase due to the reduced load-carrying area or otherwise, the crack may grow, leading to failure. The use of the stress-intensity factors to predict the rate of growth of a fatigue crack is discussed in Section 7.14.

TABLE 6.1

Geometry Factors λ for Some Initial Crack Shapes

Case A. Tension of a long plate with a central crack

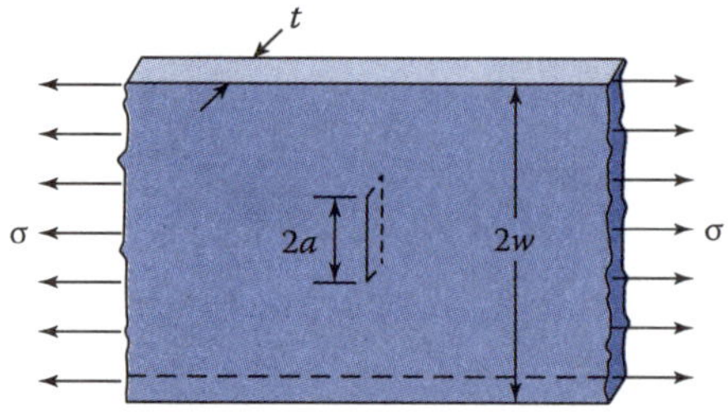

a/w	λ
0.1	1.01
0.2	1.03
0.3	1.06
0.4	1.11
0.5	1.19
0.6	1.30

Case B. Tension of a long plate with an edge crack

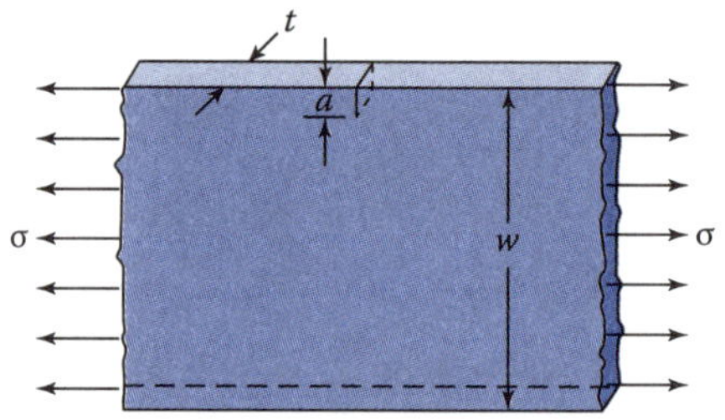

a/w	λ
0 ($w \to \infty$)	1.12
0.2	1.37
0.4	2.11
0.5	2.83

Case C. Tension of a long plate with double-edge cracks

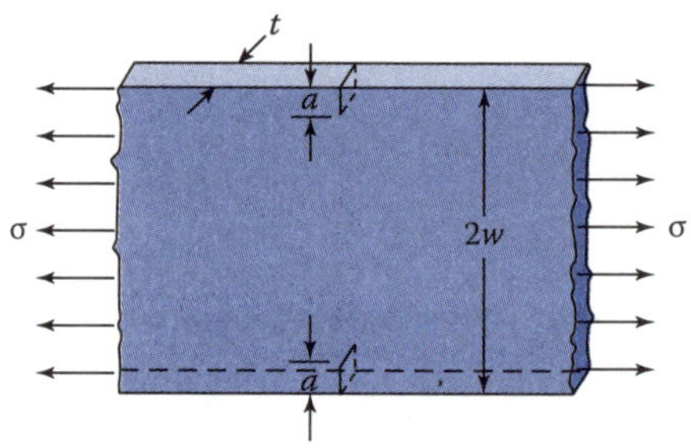

a/w	λ
0 ($w \to \infty$)	1.12
0.2	1.12
0.4	1.14
0.5	1.15
0.6	1.22

Case D. Pure bending of a beam with an edge crack

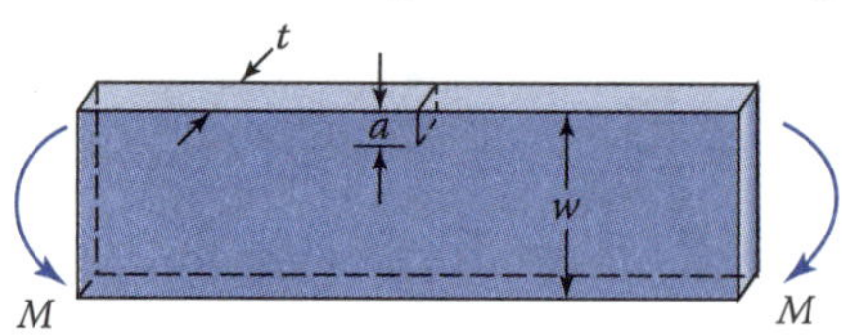

a/w	λ
0.1	1.02
0.2	1.06
0.3	1.16
0.4	1.32
0.5	1.62
0.6	2.10

6.4 Fracture Toughness

In a toughness test of a given material, the stress-intensity factor at which a crack propagates is measured. This is the critical stress-intensity factor, known as *the fracture toughness* and designated by the symbol K_c. Usually, testing is done on an ASTM standard specimen, either a beam or a tension member with an edge crack at the root of a notch. Loading is applied slowly and a record is made of load versus notch opening. The data are interpreted for the value of fracture toughness [8].

For a known applied stress acting on a part of known or assumed crack length, when the magnitude of stress-intensity factor K reaches fracture toughness K_c, the crack propagates leading to rupture in an instant. The *factor of safety for fracture mechanics*, strength–stress ratio, is therefore

$$n = \frac{K_c}{K} \tag{6.2}$$

Substituting the stress-intensity factor from Equation 6.1, the foregoing becomes

$$n = \frac{K_c}{\lambda \sigma \sqrt{\pi a}} \tag{6.3}$$

Table 6.2 presents the values of the yield strength and fracture toughness for some metal alloys, measured at room temperature in a single edge-notch test specimen [9]. For consistency of results, the ASTM specifications require a crack length a or member thickness t given by

$$a,\, t \geq 2.5\left(\frac{K_c}{S_y}\right)^2 \tag{6.4}$$

This ensures plane strain and flat crack surfaces. The values of a and t obtained by Equation 6.4 are also included in the table.

Application of the preceding equations is illustrated in the solution of various numerical problems to follow.

TABLE 6.2

Yield Strength S_y and Fracture Toughness K_c for Some Engineering Materials

	S_y		K_c		Minimum Values of a and t	
Metals	**MPa**	**(ksi)**	**MPa$\sqrt{\text{m}}$**	**(ksi$\sqrt{\text{in.}}$)**	**mm**	**(in.)**
Steel						
AISI 4340	1503	(218)	59	(53.7)	3.9	(0.15)
Stainless steel						
AISI 403	690	(100)	77	(70.1)	31.1	(1.22)
Aluminum						
2024-T851	444	(64.4)	23	(20.9)	6.7	(0.26)
7075-T7351	392	(56.9)	31	(28.2)	15.6	(0.61)
Titanium						
Ti-6A1–4V	798	(116)	111	(101)	48.4	(1.91)
Ti-6a1–6V	1149	(167)	66	(60.1)	8.2	(0.32)

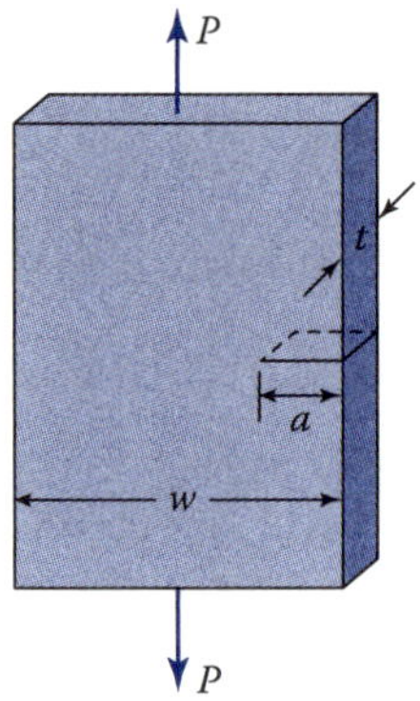

FIGURE 6.2
Example 6.1.

Example 6.1: Edge Crack on Aircraft Panel

An aircraft panel of width w and thickness t is loaded in tension, as shown in Figure 6.2. Estimate the maximum load P that can be applied without causing sudden fracture when an edge crack grows to length of a.

Given: w = 100 mm, t = 16 mm, a = 20 mm.

Design Decision: The plate will be made of 7075-T7351 aluminum alloy. This decision should result in low weight.

Solution

From Table 6.2,

$$K_c = 31\text{ MPa}\sqrt{\text{m}} = 31\sqrt{1000}\text{ MPa}\sqrt{\text{mm}}, \quad S_y = 392\text{ MPa}$$

for the aluminum alloy. Note that values of the length a and thickness t satisfy Table 6.2. Referring to Case B of Table 6.1, we have

$$\frac{a}{w} = \frac{20}{100} = 0.2, \quad \lambda = 1.37$$

Equation 6.3 with $n = 1$:

$$\sigma_{\text{all}} = \frac{K_c}{\lambda\sqrt{\pi a}} = \frac{31\sqrt{1000}}{1.37\sqrt{\pi(20)}} = 90.27\text{ MPa}$$

Hence,

$$P = \sigma_{\text{all}}(wt) = (90.27)(100 \times 16) = 144.4\text{ kN}$$

Comment: The nominal stress at the fracture, $P/(16)(100 - 20) = 112.8$ MPa, is well below the yield strength of the material.

Example 6.2: Design of a Wide Plate with a Central Crack

A large plate of width $2w$ carries a uniformly distributed tensile force P in a longitudinal direction with a safety factor of n. The plate has a central transverse crack that is $2a$ long. Calculate the thickness t required

1. To resist yielding
2. To prevent sudden fracture

Given: $w = 60$ mm, $P = 160$ kN, $n = 2.5$, $a = 9$ mm.

Design Decision: The plate is made of Ti-6A1-6V alloy.

Solution

By Table 6.2,

$$K_c = 66\sqrt{1000}\ \text{MPa}\sqrt{\text{mm}}, \quad S_y = 1149\ \text{MPa}$$

for the titanium alloy.

a. The allowable tensile stress based on the net area is

$$\sigma_{\text{all}} = \frac{S_y}{n} = \frac{P}{2(w-a)t}$$

Hence,

$$t = \frac{Pn}{2(w-a)S_y} = \frac{160(10^3)(2.5)}{2(60-9)1149} = 3.4\ \text{mm}$$

b. By Case A of Table 6.1,

$$\frac{a}{w} = \frac{9}{60} = 0.15, \quad \lambda = 1.02$$

Through the use of Equation 6.3 with $n = 2.5$, the stress at fracture is

$$\sigma = \frac{K_c}{\lambda n\sqrt{\pi a}} = \frac{66\sqrt{1000}}{1.02(2.5)\sqrt{\pi(9)}} = 153.9\ \text{MPa}$$

Since this stress is smaller than the yield strength, the *fracture governs* the design; $\sigma_{\text{all}} = 153.9$ MPa. Therefore,

$$t_{\text{red}} = \frac{P}{2w\sigma_{\text{all}}} = \frac{160(10^3)}{2(60)(153.9)} = 8.66\ \text{mm}$$

Comment: Use a thickness of 8.7 mm. Both values of a and t fulfill Table 6.2.

Example 6.3: Load Capacity of a Bracket with an Edge Crack

A bracket having an edge crack and uniform thickness t carries a concentrated load, as shown in Figure 6.3a. What is the magnitude of the fracture load P with a safety factor of n for crack length of a?

Given: $a = 0.16$ in., $w = 2$ in., $d = 4$ in., $t = 1$ in., $n = 1.2$.

Assumptions: The bracket is made of AISI 4340 steel. A linear elastic stress analysis is acceptable.

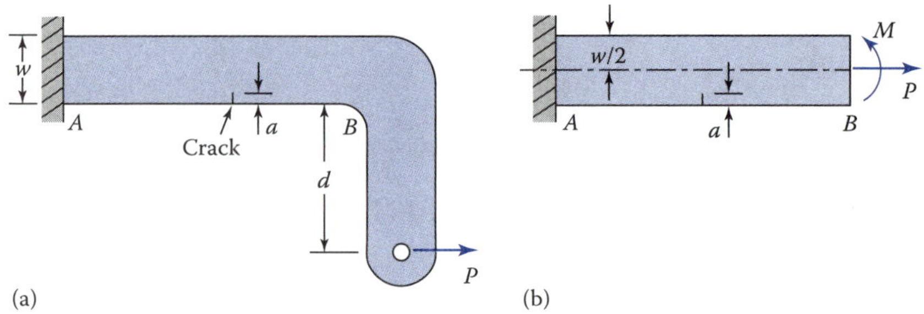

FIGURE 6.3
Example 6.3. (a) bracket with edge crack and (b) free-body diagram of segment *AB*.

Solution

Referring to Table 6.2,

$$K_c = 53.7 \text{ ksi}\sqrt{\text{in.}}, \quad S_y = 218 \text{ ksi}$$

Observe that values of a and t both satisfy the table. At the section through the point B (Figure 6.3b), the bending moment is

$$M = P\left(d + \frac{w}{2}\right) = P(4+1) = 5P$$

Nominal stress for the *combined loading*, by superposition of two states of stress for axial force P and moment M, is expressed as

$$\begin{aligned} \lambda\sigma &= \lambda_a\sigma_a + \lambda_b\sigma_b \\ &= \lambda_a \frac{P}{wt} + \lambda_b \frac{6M}{tw^2} \end{aligned} \tag{6.5}$$

Here, w and t represent the width and thickness of the member, respectively.

The ratio of crack length to bracket width is $a/w = 0.08$. For Cases B and D of Table 6.1, we have $\lambda_a = 1.12$ and $\lambda_b = 1.02$, respectively. Equation 6.5 leads to

$$\begin{aligned} \lambda\sigma &= 1.12\frac{P}{2(1)} + 1.02\frac{6(5P)}{1(2)^2} \\ &= 0.56P + 7.65P = 8.21P \end{aligned}$$

Then, through the use of Equation 6.3,

$$\lambda\sigma = \frac{K_c}{n\sqrt{\pi a}}; \quad 8.21P = \frac{53.7(10^3)}{1.2\sqrt{\pi(0.16)}}$$

Solving, $P = 7.69$ kips.

Comments: The normal stress at fracture, 7.69/[1(2 − 0.16)] = 4.18 ksi, is well below the yield strength and our assumption is valid.

6.5 Yield and Fracture Criteria

In many cases, a member fails when the material begins to yield or deform permanently. For metals, the shear stress plays an important role in yielding. Occasionally, a small dimensional change may be tolerated and a static load that exceeds the yield point permitted. The fracture at the ultimate strength of the material would then constitute failure. The fracture, or separation of the material under stress into two or more parts, usually is associated with a tensile stress. Particularly in a 1D stress field, compression stress is generally considered less damaging than tensile stress. The brittle or ductile character of a material is relevant to the mechanism of failure. The distinction between ductile and brittle material itself is not simple, however. The nature of the stress, the temperature, and the material itself all play a role, as is discussed in Section 2.9, by defining the boundary between ductility and brittleness.

Let us consider an element subjected to a triaxial state of stress, where $\sigma_1 > \sigma_2 > \sigma_3$. Recall that subscripts 1, 2, and 3 denote the principal directions. The state of stress in a uniaxial loading is defined by σ_1, equal to the normal force divided by the cross-sectional area, and $\sigma_1 = \sigma_3 = 0$. Corresponding to the onset of yielding and fracture in a simple tension test, the stresses and strain energy shown in the second column of Table 6.3 are determined as follows. When specimen starts to yield, we have $\sigma_1 = S_y$. Therefore, the maximum shear stress is $\tau_{max} = \sigma_1/2 = S_y/2$ by Equation 3.34, the maximum distortion energy density absorbed by the material is $U_{od} = S_y^2/6G$ using Equation 5.9 with $G = E/2(1 + \nu)$, and the maximum octahedral shear stress is $\tau_{oct} = (\sqrt{2}/3)S_y$ from Equation 3.52b. On the other hand, at an impending fracture, the maximum principal stress is $\sigma_1 = S_u$. Note that the forgoing quantities obtained in simple tension have special significance in predicting failure involving combined stress.

In the torsion test, the state of stress is specified by $\tau = \sigma_1 = -\sigma_2$ and $\sigma_3 = 0$. Here, the shear stress is calculated using the torsion formula. Corresponding to this case of pure shear, at the start of yielding and fracture, are the stresses and strain energy shown in the third column of the table. These quantities are readily obtained by a procedure similar to that described in the preceding for tension test.

The mechanical behavior of materials subjected to uniaxial normal stresses or pure shearing stresses is readily presented on stress–strain diagrams. The onset of failure by yielding or fracture in these cases is considerably more apparent than in situations involving combined stress. From the viewpoint of mechanical design, it is imperative that some practical guidelines be available to predict yielding or fracture under various

TABLE 6.3

Utilizable Values of a Material for States of Stress in Tension and Torsion Tests

Quantity	Tension Test	Torsion Test
Maximum shear stress	$S_{ys} = S_y/2$	S_{ys}
Maximum energy of distortion	$U_{od} = [(1+\nu)/3E]S_y^2$	$U_{od} = S_{ys}^2(1+\nu)/E$
Maximum octahedral shear stress	$\tau_{oct} = (\sqrt{2}/3)S_y$	$\tau_{oct} = (\sqrt{2}/\sqrt{3})S_{ys}$
Maximum principal stress	S_u	S_{us}

Notes: S_y, yield strength in tension; S_{ys}, yield strength in shear; S_u, ultimate strength in tension; S_{us}, ultimate strength in shear.

conditions of stress, as they are likely to exist in service. To meet this need, a number of failure criteria or theories consistent with the behavior and strength of material have been developed.

These strength theories are structured to apply to particular classes of materials. We discuss the two most widely accepted theories to predict the onset of inelastic behavior for ductile materials under combined stress in Sections 6.6 through 6.9. The three fracture theories pertaining to brittle materials under combined stress are presented in Sections 6.10 through 6.12. As we observe, the *theory* behind most static failure criteria is that *whatever is responsible for failure in the simple tensile test also is responsible for failure under combined loading.* It is important to note that yielding of ductile materials should be further qualified to yielding of ductile metals; many polymers are ductile but do not follow the standard yield theories.

6.6 Maximum Shear Stress Theory

The maximum shear stress theory is developed on the basis of the experimental observation that a ductile material yields as a result of slip or shear along crystalline planes. Proposed by Coulomb (1736–1806), it is also known as the *Tresca yield criterion* in recognition of the contribution by Tresca (1814–1885) to its application. This theory states that yielding begins whenever the maximum shear stress at any point in the body becomes equal to the maximum shear stress at yielding in a simple tension test. Hence, according to Equation 3.50 and Table 6.3,

$$\tau_{max} = \frac{1}{2}|\sigma_1 - \sigma_3| = S_{ys} = \frac{1}{2}S_y$$

The maximum shear stress theory is therefore given by

$$|\sigma_1 - \sigma_3| = \frac{S_y}{n} \tag{6.6}$$

for a factor of safety n.

In the case *of plane stress*, $\sigma_3 = 0$, two combinations of stresses are to be considered. When σ_1 and σ_2 have *opposite signs*, that is, one tensile and the other compressive, the maximum shear stress is $(\sigma_1 - \sigma_2)/2$. The yield condition then becomes

$$|\sigma_1 - \sigma_2| = \frac{S_y}{n} \tag{6.7}$$

The foregoing may be restated in the form, for $n = 1$:

$$\frac{\sigma_1}{S_y} - \frac{\sigma_2}{S_y} = \pm 1 \tag{6.8}$$

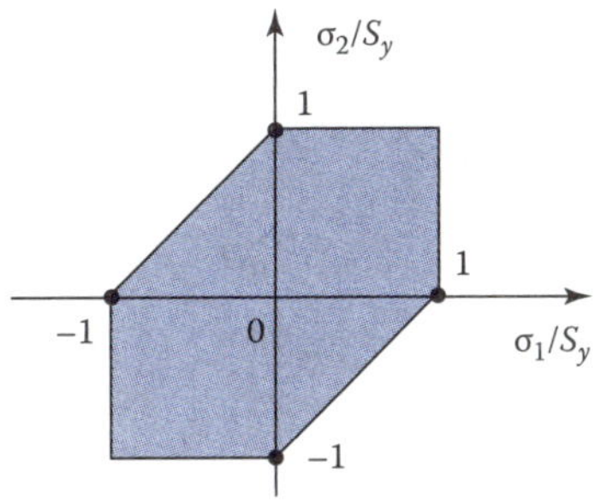

FIGURE 6.4
Yield criterion based on maximum shear stress.

When σ_1 and σ_2 carry the *same sign*, the maximum shear stress is $(\sigma_1 - \sigma_3)/2 = \sigma_1/2$. Then, for $|\sigma_1|>|\sigma_2|$ and $|\sigma_2|>|\sigma_1|$, we have the yield conditions, respectively,

$$|\sigma_1| = \frac{S_y}{n} \quad \text{and} \quad |\sigma_2| = \frac{S_y}{n} \tag{6.9}$$

Equations 6.8 and 6.9 for $n = 1$ are represented graphically in Figure 6.4. Note that Equation 6.8 applies to the second and fourth quadrants, while Equation 6.9 applies to the first and third quadrants. The boundary of the hexagon thus marks the onset of yielding, with points outside the shaded region depicting a yielded state. We demonstrate in Section 6.9 that the maximum shear stress theory has reasonably good agreement with the experiment for ductile materials. The theory offers an additional advantage in its ease of application. However, the maximum distortion criterion, discussed next, is recommended because it correlates better with the actual test data for yielding of ductile materials.

6.6.1 Typical Case of Combined Loading

In a common case of combined plane bending, torsion, axial, and transverse shear loadings, such as in Figure 3.29, we have $\sigma_y = \sigma_z = \tau_{xz} = \tau_{xz} = 0$. Hence, the principal stresses reduce to

$$\sigma_{1,2} = \frac{\sigma_x}{2} \pm \tau_{max} \tag{6.10}$$

where

$$\tau_{max} = \left[\left(\frac{\sigma_x}{2}\right)^2 + \tau_{xy}^2\right]^{1/2}$$

Substituting these into Equation 6.7, the maximum shear stress criterion becomes

$$\frac{S_y}{n} = \left(\sigma_x^2 + 4\tau_{xy}^2\right)^{1/2} \tag{6.11}$$

for the preceding special case.

Example 6.4: Failure of a Rod under Combined Torsion and Axial Loading

A circular rod, constructed of a ductile material of tensile yield strength S_y, is subjected to a torque T. Determine the axial tensile force P that can be applied simultaneously to the rod (Figure 6.5).

Given: $T = 500\pi\ \text{N}\cdot\text{m}$, $D = 50\,\text{mm}$, factor of safety $n = 1.2$.

Design Decisions: The rod is made of steel of $S_y = 300$ MPa. Use the maximum shear stress failure criterion.

Solution

For the situation described, the critical stresses occur on the elements at the surface of the shaft. Based on the maximum shear stress theory, from Equation 6.11,

$$\sigma_x = \left[\left(\frac{S_y}{n}\right)^2 - 4\tau_{xy}^2\right]^{1/2} \tag{a}$$

where

$$\sigma_x = \frac{P}{A} = \frac{4P}{\pi D^2}, \quad \tau_{xy} = \frac{Tr}{J} = \frac{16T}{\pi D^3} \tag{b}$$

Substituting the given numerical values, Equation (a) gives

$$\sigma_x = \left[\left(\frac{300\times10^6}{1.2}\right)^2 - 4\left(\frac{16\times500\pi}{\pi\times0.05^3}\right)^2\right]^{1/2} = 214.75\ \text{MPa}$$

The first of Equation (b) is therefore

$$P = \frac{\pi(0.05)^2(214.75\times10^6)}{4} = 421.7\ \text{kN}$$

Comment: This is the maximum force that can be applied without causing permanent deformation.

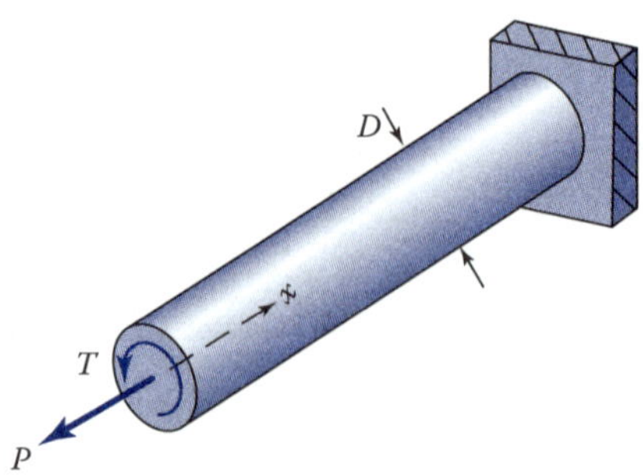

FIGURE 6.5
Example 6.4.

6.7 Maximum Distortion Energy Theory

The maximum distortion energy theory or criterion was originally proposed by Maxwell in 1856, and additional contributions were made in 1904 by Hueber, in 1913 by von Mises, and in 1925 by Hencky. Today, it is mostly referred to as the *von Mises–Hencky theory* or simply *von Mises theory*. This theory predicts that failure by yielding occurs when, at any point in the body, the distortion energy per unit volume in a state of combined stress becomes equal to that associated with yielding in a simple tension test. Hence, in accordance with Equation 5.9 and Table 6.3,

$$\frac{1}{12G}[(\sigma_1-\sigma_2)^2+(\sigma_2-\sigma_3)^2+(\sigma_3-\sigma_1)^2]=\frac{1+\nu}{3E}S_y^2$$

where $G = E/2(1 + \nu)$. The maximum energy of distortion criterion for yielding is therefore

$$\frac{\sqrt{2}}{2}\left[(\sigma_1-\sigma_2)^2+(\sigma_2-\sigma_3)^2+(\sigma_3-\sigma_1)^2\right]^{1/2}=\frac{S_y}{n} \tag{6.12}$$

for a safety factor n.

It is often convenient to replace S_y/n by an *equivalent stress* σ_e in the preceding equation. In so doing, we have

$$\frac{\sqrt{2}}{2}\left[(\sigma_1-\sigma_2)^2+(\sigma_2-\sigma_3)^2+(\sigma_3-\sigma_1)^2\right]^{1/2}=\sigma_e \tag{6.13}$$

Commonly used names for the equivalent stress are the effective stress and the *von Mises stress*. Observe from Equations 6.12 and 6.13 that only the differences of the principal stresses are involved. Consequently, the addition of an equal amount to each stress does not affect the conclusion with respect to whether or not yielding occurs. In other words, inelastic action does not depend on hydrostatic tensile or compressive stress.

For *plane stress* $\sigma_3 = 0$, the maximum energy of distortion theory becomes

$$\left(\sigma_1^2-\sigma_1\sigma_2+\sigma_2^2\right)^{1/2}=\frac{S_y}{n} \tag{6.14}$$

or

$$(\sigma_1^2-\sigma_1\sigma_2+\sigma_2^2)^{1/2}=\sigma_e \tag{6.15}$$

Equation 6.14 may alternatively be represented in the following form for $n = 1$:

$$\left(\frac{\sigma_1}{S_y}\right)^2-\left(\frac{\sigma_1}{S_y}\right)\left(\frac{\sigma_2}{S_y}\right)+\left(\frac{\sigma_2}{S_y}\right)^2=1$$

This expression defines the ellipse shown in Figure 6.6. As in the case of the maximum shear stress theory, points within the shaded area represent nonyielding states. The boundary of the ellipse indicates the onset of yielding, with the points outside the shaded area representing a yielded state. The maximum energy of distortion theory of failure agrees

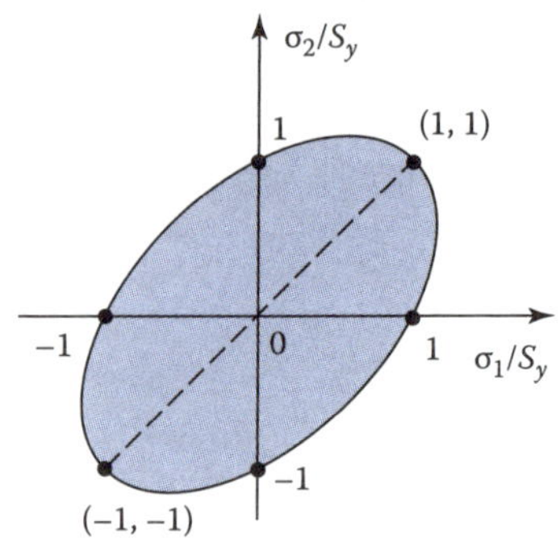

FIGURE 6.6
Yield criterion based on distortion energy.

quite well with test data for yielding of ductile materials and plane stress. It is commonly used in design and gives the same result as the octahedral shear stress theory, discussed in the next section.

6.7.1 Yield Surfaces for Triaxial State of Stress

We observe from Equation 6.12 that only the *differences* of the stresses are involved. Hence, addition of an equal amount to each stress does not affect the conclusion with respect to whether or not yielding will take place. That is, yielding does not depend on hydrostatic tensile or compressive stresses. Figure 6.7a depicts a state of stress defined $P(\sigma_1, \sigma_2, \sigma_3)$ in a principal stress coordinate system. Clearly, a hydrostatic alteration of the stress at point P requires shifting of this point P direction parallel to direction n, making equal angles with coordinate axes. This is because changes in hydrostatic stress involve changes of the normal stresses by equal amounts.

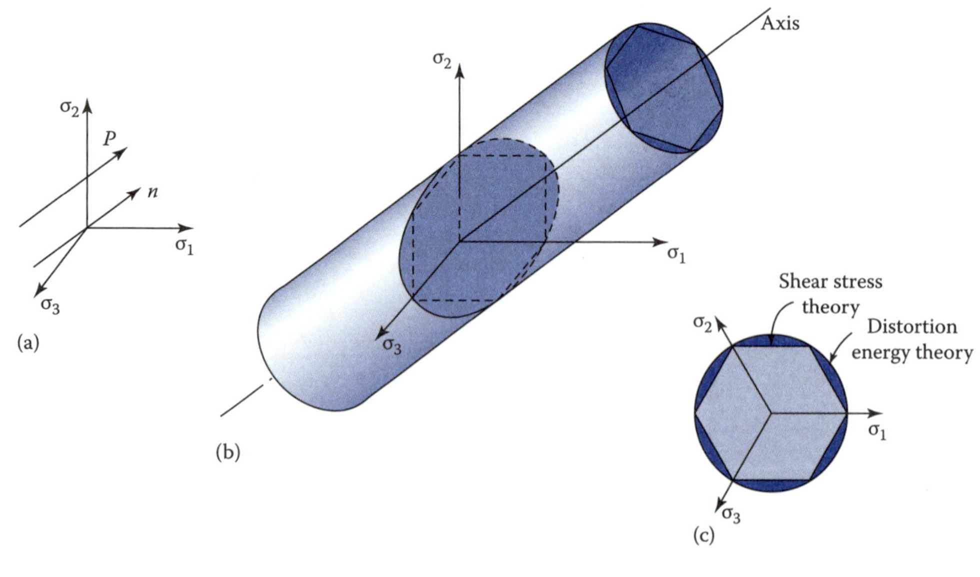

FIGURE 6.7
Yield criteria based on distortion energy and shear stress: (a) stress state defined by position, (b) 3D yield loci, and (c) view along (n), axis of cylinder and hexagon.

We are led to conclude therefore that the yield criterion *distortion energy theory* is properly described by the *cylinder* shown in Figure 6.7b and that the surface of the cylinder is the yield surface or also called yield locus. Points within the surface represent states of nonyielding. Thus, any calculations that predict a stress state outside the yield locus predict failure. The ellipse of Figure 6.6 is defined by the intersection of the cylinder with the σ_1, σ_2 plane. Note that the yield surface or yield locus appropriate to the *maximum-stress theory* (indicated by the dashed lines for plane stress) is described by a *hexagonal* surface placed within the cylinder.

6.7.2 Typical Case of Combined Loading

Reconsider the particular case of combined loading, where $\sigma_y = \sigma_z = \tau_{yz} = \tau_{xz} = 0$ (see Section 6.6). Substitution of Equation 6.10 into 6.14 leads to the expression

$$\frac{S_y}{n} = \left(\sigma_x^2 + 3\tau_{xy}^2\right)^{1/2} \tag{6.16}$$

Clearly, Equation 6.16 is based on the maximum energy of distortion criterion for the foregoing special case.

6.8 Octahedral Shear Stress Theory

The octahedral shear stress theory, also known as the *Mises–Hencky criterion* or simply the *Mises criterion*, predicts failure by yielding whenever the octahedral shear stress for any state of stress equals the octahedral shear stress for the simple tensile test. Accordingly, through the use of Equation 3.52b and Table 6.3, the octahedral shear stress theory is

$$\tau_{\text{oct}} = \frac{\sqrt{2}}{3} S_y \tag{6.17}$$

which gives Equation 6.12. Eichinger (in 1926) and Nadai (in 1937) independently developed this theory.

The octahedral shear stress criterion may also be considered in terms of distortion energy. In a general state of stress, from Equation 5.9, we have

$$U_{od} = \frac{3}{2}\frac{1+\nu}{E}\tau_{\text{oct}}^2$$

When the forgoing reaches the value given in Table 6.3, Equation 6.17 again is found.

We see that the octahedral shear stress criterion is equivalent to the distortion energy theory; that is, the former criterion enables us to apply the latter theory while dealing with stress rather than energy. We use the procedure of the energy of distortion criterion in this text.

Example 6.5: Design of a Torsion Bar

A cold-drawn AISI-1050 steel torsion bar *CB* of diameter *d* is fastened to a rigid arm at *A*, supported by a bearing at *C*, and fixed at *B*, as depicted in Figure 6.8. At the right end of the arm, the tire wheel is mounted on which the vertical force *P* acts from the ground.

Given:

$$a = 120 \text{ mm} \quad d = 34 \text{ mm} \quad L = 360 \text{ mm}$$

$$P = 2.6 \text{ kN} \quad S_y = 580 \text{ MPa (from Table B.3)}$$

Assumptions: Bearing *C* acts as a simple support. Effect of transverse shear due to *P* is negligible.

Find: Factor of safety n with respect to inelastic deformation of the torsion bar, using the maximum shear stress and the maximum distortion energy criteria

Solution

The critical stresses occur on the surface of the torsion bar. We have the torque $T = PL = 2600(0.36) = 936$ N·m and moment $M = Pa = 2600(0.12) = 312$ N·m act uniformly along this member. Thus, with reference to Figure 6.8, Equations 3.10 and 3.16 result in the maximum shear and bending stresses:

$$\tau_{xy} = \frac{16T}{\pi d^3} = \frac{16(936)}{\pi(0.034)^3} = 121.3 \text{ MPa}$$

and

$$\sigma_x = \frac{32M}{\pi d^3} = \frac{32(312)}{\pi(0.034)^3} = 80.86 \text{ MPa}$$

Maximum shear stress theory. Through the use of Equation 6.11,

$$\left[(80.86)^2 + 4(121.3)^2\right]^{1/2} = \frac{580}{n}$$

from which $n = 2.27$.

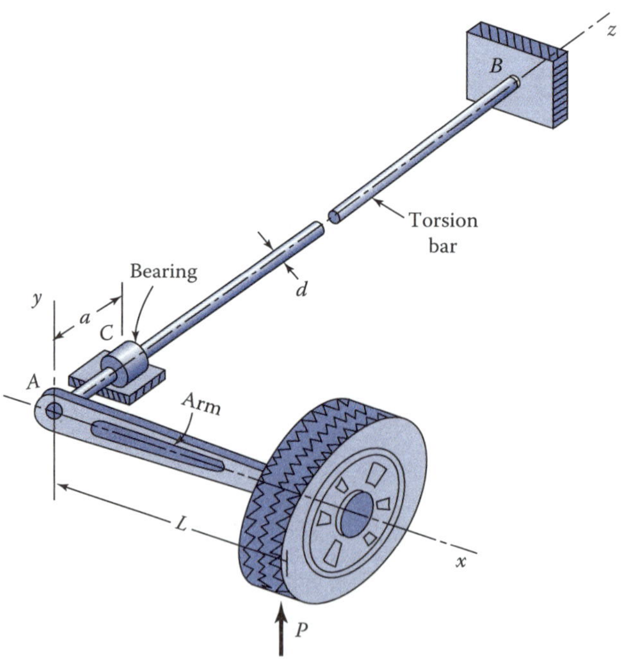

FIGURE 6.8
Example 6.5. Schematic of rear wheel suspension used in some autos.

Maximum energy of distortion theory. Applying Equation 6.14,

$$\left[(80.86)^2 + 3(121.3)^2\right]^{1/2} = \frac{580}{n}$$

Solving, we obtain $n = 2.58$.

Comments: Inasmuch as the maximum distortion energy criterion is more accurate, it makes sense for a higher factor of safety to be obtained by this theory.

Example 6.6: Failure Analysis of a Conical Liquid Storage Tank

A thin-walled conical vessel, or tank, is supported on its edge and filled with a liquid, as depicted in Figure 6.9. Expressions for stress in this shell are developed in Example 16.10. Determine the vessel wall thickness on the basis of the maximum shear stress and the energy of distortion failure theories.

Given: The geometry and loading of the tank are known.

Assumptions:

1. The vessel is made of structural steel of yield strength S_y.
2. The factor of safety against yielding is n.
3. The vessel is taken to be simply supported from the top.

Solution

The tangential stress $\sigma_\theta = \sigma_1$ and meridional stress $\sigma_s = \sigma_2$ in the tank are expressed by Equations 16.71a and 16.72a as follows:

$$\sigma_1 = \gamma(h-y)y\frac{\tan\alpha}{t\cos\alpha}$$
$$\sigma_2 = \gamma\left(h-\frac{2}{3}y\right)y\frac{\tan\alpha}{2t\cos\alpha} \qquad \text{(a)}$$

where
- h = the liquid height
- t = the vessel wall thickness
- α = the half angle at the apex of cone
- γ = the specific weight of liquid

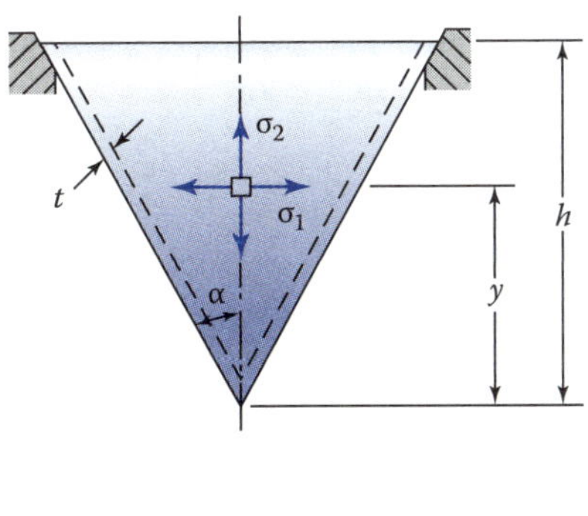

FIGURE 6.9
Example 6.6. Conical tank filled with a liquid.

The largest magnitudes of these principal stresses are given by

$$\begin{aligned} \sigma_{1,\max} &= \frac{\gamma h^2}{4t}\frac{\tan\alpha}{\cos\alpha}, \quad \sigma_2 = \frac{\gamma h^2}{12t}\frac{\tan\alpha}{\cos\alpha}, \quad \left(\text{at} \quad y = \frac{h}{2}\right) \\ \sigma_{2,\max} &= \frac{3\gamma h^2}{16t}\frac{\tan\alpha}{\cos\alpha}, \quad \sigma_1 = \frac{\gamma h^2}{4t}\frac{\tan\alpha}{\cos\alpha}, \quad \left(\text{at} \quad y = \frac{3h}{4}\right) \end{aligned} \tag{b}$$

Comment: Note that the maximum stresses occur at different locations.

Maximum shear stress criterion. Since σ_1 and σ_2 are of the same sign and $|\sigma_1| > |\sigma_2|$, the first of Equation 6.9 together with (b) results in

$$\frac{S_y}{n} = \frac{\gamma h^2}{4t}\frac{\tan\alpha}{\cos\alpha} \tag{6.18}$$

The thickness of the vessel is obtained from the preceding equation in the form

$$t = 0.25\frac{\gamma h^2 n}{S_y}\frac{\tan\alpha}{\cos\alpha} \tag{6.19a}$$

Maximum energy of distortion criterion. Inasmuch as the maximum magnitudes of σ_1 and σ_2 occur at different locations, we must first determine the section at which combined stresses are at a critical value. For this purpose, we substitute Equation (a) into Equation 6.14:

$$\begin{aligned} \frac{S_y^2}{n^2} = &\left[\gamma(h-y)y\frac{\tan\alpha}{t\cos\alpha}\right]^2 + \left[\gamma\left(h-\frac{2}{3}y\right)y\frac{\tan\alpha}{2t\cos\alpha}\right]^2 \\ &-\left[\gamma(h-y)y\frac{\tan\alpha}{t\cos\alpha}\right]\left[\gamma\left(h-\frac{2}{3}y\right)y\frac{\tan\alpha}{2t\cos\alpha}\right] \end{aligned} \tag{c}$$

Differentiation of Equation (c) with respect to the variable y and equating the result to 0 gives [10]

$$y = 0.52h$$

Introducing this value of y back into Equation (c), the thickness of the vessel is found to be

$$t = 0.225\frac{\gamma h^2 n}{S_y}\frac{\tan\alpha}{\cos\alpha} \tag{6.19b}$$

Comment: The thickness according to the maximum shear stress criterion is therefore 10% larger than that based on the maximum distortion energy criterion.

6.9 Comparison of the Yielding Theories

Two approaches may be used to compare the theories of yielding heretofore discussed. The first comparison equates, for each theory, the critical values corresponding to uniaxial loading and torsion. Referring to Table 6.3,

$$\text{Maximum shearing stress theory:} \quad S_{ys} = 0.50S_y$$

$$\text{Energy of distortion theory, or its equivalent,} \tag{6.20}$$

$$\text{the octahedral shear stress theory:} \quad S_{ys} = 0.577S_y$$

We observe that the difference in strength predicted by these criteria is not substantial. A second comparison may be made by means of superposition of Figures 6.4 and 6.6. This is left as an exercise for the reader.

Experiment shows that, for ductile materials, the yield stress obtained from a torsion test is 0.5–0.6 times than that determined from simple tension test. We conclude, therefore, that the energy of distortion criterion or octahedral shearing stress criterion is most suitable for ductile materials. However, the shear stress theory, which results in $S_{ys} = 0.50S_y$, is simple to apply and offers a conservative result in design.

As a third comparison, consider a solid shaft of diameter D and tensile yield strength S_y subjected to combined loading consisting of tension P and torque T. The yield criteria based on the maximum shear stress and energy of distortion theories, for $n = 1$, are given by Equations 6.11 and 6.16:

$$S_y = \left(\sigma_x^2 + 4\tau_{xy}^2\right)^{1/2}, \quad S_y = \left(\sigma_x^2 + 3\tau_{xy}^2\right)^{1/2} \tag{a}$$

In the preceding, σ_x and τ_{xy} represent axial tension and torsional stresses, respectively. Therefore,

$$\sigma_x = \frac{4P}{\pi D^2}, \quad \tau_{xy} = \frac{16T}{\pi D^3}$$

A dimensionless plot of Equation (a) and some experimental results are shown in Figure 6.10. We note again particularly good agreement between the maximum energy of distortion criterion and experimental data for ductile materials. The difference in results is not very great, however, and both theories are widely used in design of members.

6.10 Maximum Principal Stress Theory

In accordance with the maximum principal stress theory, credited to Rankine (1820–1872), a material fails by fracturing when the maximum principal stress reaches the ultimate strength S_u in a simple tension test. Thus, at the beginning of the fracture,

$$|\sigma_1| = \frac{S_u}{n} \quad \text{or} \quad |\sigma_3| = \frac{S_u}{n} \tag{6.21}$$

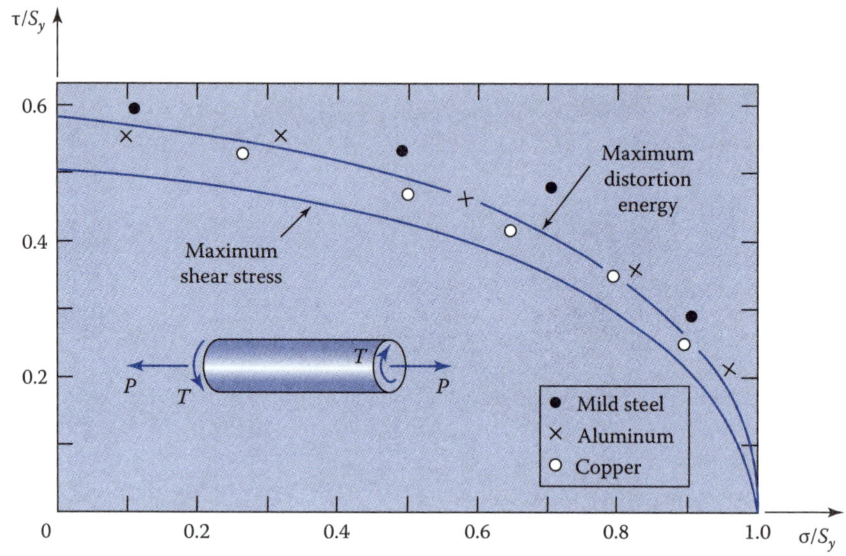

FIGURE 6.10
Yield curves for torsion–tension shaft. The points shown in this figure are based on experimental data.

for safety factor n. That is, a crack starts at the most highly stressed point in a brittle material when the maximum principal stress at the point reaches S_u. This criterion is suggested by the observation that fracture surfaces in brittle materials under tension are planes that carry the maximum principal stress. Clearly, the maximum principal stress theory is based on the assumption that the ultimate strength of the material is the same in tension and compression: $S_u = |S_{uc}|$.

For the case of *plane stress* ($\sigma_3 = 0$), Equation 6.21, the fracture condition, is given by

$$|\sigma_1| = \frac{S_u}{n} \quad \text{or} \quad |\sigma_2| = \frac{S_u}{n} \tag{6.22}$$

This may be restated in the, for $n = 1$,

$$\frac{\sigma_1}{S_u} = \pm 1 \quad \text{or} \quad \frac{\sigma_2}{S_u} = \pm 1 \tag{6.23}$$

Figure 6.11 is a plot of Equation 6.23. Note that points a, b, and c, d in the figure indicate the tensile and compressive principal stresses, respectively. As in other criteria, the boundary of the square indicates the onset of failure by fracture. The area within the boundary is therefore a region of no failure [11].

Note that, while a material may be weak in simple compression, it may nevertheless sustain very high hydrostatic pressure without fracturing. Furthermore, most brittle materials are much stronger in compression than in tension; that is, $S_{uc} \gg S_u$. These are inconsistent with the theory. Moreover, the theory makes no allowance for influences on the failure mechanism other than those of normal stresses. However, for brittle materials in all stress ranges, the maximum principal stress theory has good experimental verification, provided there is a tensile principal stress.

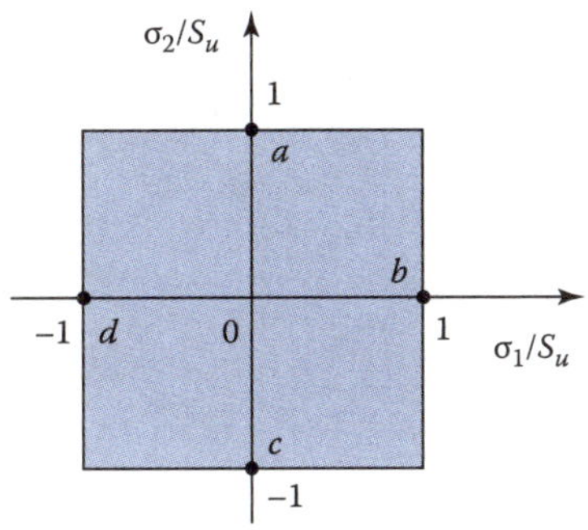

FIGURE 6.11
Fracture criterion based on maximum principal stress.

Example 6.7: Failure of a Pipe of Brittle Material under Static Torsion Loading

A cast pipe of outer diameter D and inner diameter d is made of an aluminum alloy having ultimate strengths in tension and compression S_u and S_{uc}, respectively. Determine the maximum torque that can be applied without causing rupture.

Given: $D = 100$ mm, $d = 60$ mm, $S_u = 200$ MPa, $S_{uc} = 600$ MPa

Design Decision: Use the maximum principal stress theory and a safety factor of $n = 2$.

Solution

The torque and the maximum shear stress are related by the torsion formula:

$$T = \frac{J\tau}{c} = \frac{\pi\left(0.05^4 - 0.03^4\right)\tau}{2(0.05)} = 170.9 \times 10^{-6}\tau \tag{a}$$

The state of stress is described by

$$\sigma_1 = -\sigma_2 = \tau, \quad \sigma_3 = 0$$

From Equation 6.22 and the preceding, we have $\tau = S_u/n$. Then, Equation (a) results in

$$T = 170.9 \times 10^{-6}\left(\frac{200 \times 10^6}{2}\right) = 17.09 \text{ kN} \cdot \text{m}$$

Comment: According to the maximum principal stress theory, the torque is limited to 17.09 kN m to avoid failure by fracture (Figure 6.11).

6.11 Mohr's Theory

Mohr's theory of failure is employed to predict the fracture of a material with different properties in tension and compression when the results of a variety of tests are available for that material. This criterion uses Mohr's circles of stress. Using the extreme values of principal stress enables one to apply Mohr's approach to either 2D or 3D cases.

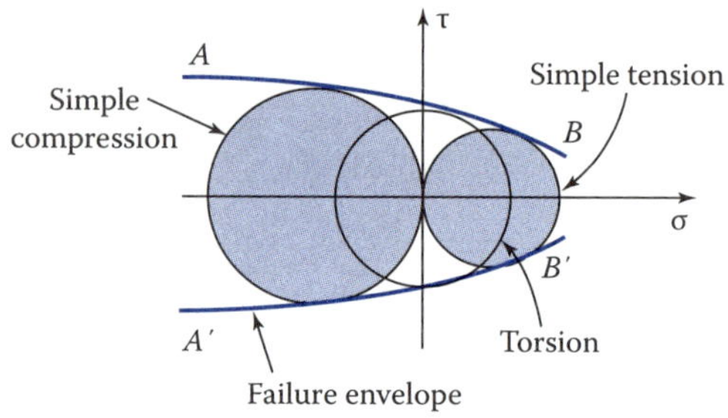

FIGURE 6.12
Mohr's fracture criterion.

Experiments are performed on a given material to determine the states of stress that result in failure. Each such state of stress defines Mohr's circle. When the data describing states limiting stress are derived from only simple tension, compression, and torsion tests, the three resulting circles are sufficient to construct the envelope, labeled by lines *AB* and *A′B′* in Figure 6.12.

Note that Mohr's envelope represents the locus of all possible failure states. Many solids, particularly those that are brittle, show greater resistance to compression than to tension. As a result, higher limiting shear stresses, for these materials, are found to the left of the origin as depicted in the figure.

6.12 Coulomb–Mohr Theory

The Coulomb–Mohr theory, like Mohr's criterion, may be employed to predict the effect of a given state of stress on a brittle material having different properties in tension and in compression. Mohr's circles for the uniaxial tension and compression tests are used to predict failure by the Coulomb–Mohr theory as shown in Figure 6.13a. The points of contact of the straight-line envelopes (*AB* and *A′B′*) with the stress circles define the state of stress at a fracture. For example, if such points are *C* and *C′*, the stresses and the planes on which they act can be obtained using the established procedure for Mohr's circle of stress.

In the case of *plane stress*, we have $\sigma_3 = 0$. When σ_1 and σ_2 have *opposite signs* (i.e., one is tensile and the other is compressive), it can be verified that [12] the onset of fracture is expressed by

$$\frac{\sigma_1}{S_u} - \frac{\sigma_2}{|S_{uc}|} = \frac{1}{n} \tag{6.24}$$

for safety factor n. Here, S_u and S_{uc} represent the ultimate strengths of the material in tension and compression, respectively. This equation may be rearranged into the form

$$n = \frac{S_u}{\sigma_1 - \sigma_2\,|S_u/S_{\text{uc}}|} \tag{6.25}$$

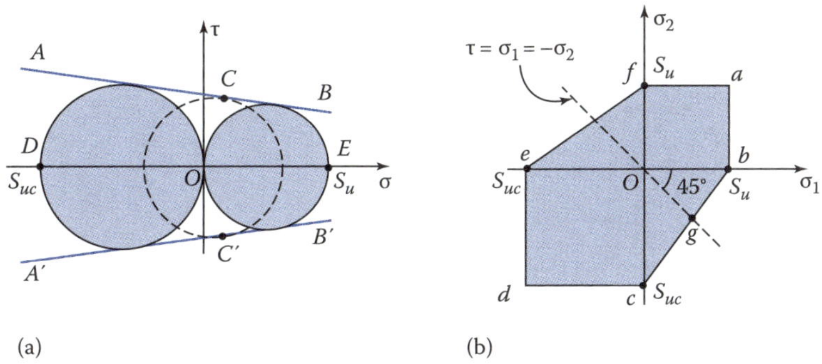

FIGURE 6.13
(a) Straight-line Mohr's envelopes and (b) the Coulomb–Mohr fracture criterion.

Relationships for the case where the principal stresses have the same sign may be deduced from Figure 6.13a. In the case of *biaxial tension*, the corresponding circle is represented by diameter *OE*. Hence, fracture occurs if either of the two tensile stresses achieves the value S_u; that is,

$$\sigma_1 = \frac{S_u}{n} \quad \text{or} \quad \sigma_2 = \frac{S_u}{n} \tag{6.26}$$

For *biaxial compression*, Mohr's circle of diameter *OD* is obtained. Failure by fracture occurs if either of the compressive stresses attains the value S_{uc}; therefore,

$$\sigma_2 = -\frac{S_{uc}}{n} \quad \text{or} \quad \sigma_1 = -\frac{S_{uc}}{n} \tag{6.27}$$

The foregoing expressions are depicted in Figure 6.13b, for the case in which $n = 1$. Lines *ab* and *af* represent Equation 6.26, and lines *dc* and *de*, Equation 6.27. The boundary *bc* is obtained by applying Equation 6.24. Line *ef* completes the hexagon in a way similar to Figure 6.4. Points within the shaded area represent states of nonfailure according to the Coulomb–Mohr theory. The boundary of the figure depicts the onset of failure due to fracture.

In the case of *pure shear*, the corresponding limiting point *g* represents the ultimate shear strength S_{us}. At point *g*, $\sigma_1 = S_{us}$ and $\sigma_2 = -\sigma_1 = -S_{us}$. Substituting for σ_1, σ_2, and $n = 1$ into Equation 6.24, we have

$$S_{us} = \frac{S_u}{1 + |S_u/S_{uc}|} \tag{6.28}$$

When $|S_{uc}| = S_u$, $S_{uc} = 0.5S_u$. If $|S_{uc}| = 4S_u$, typical of ordinary gray cast iron, then $S_{us} = 0.8S_u$.

Example 6.8: Rework Example 6.7, Employing the Coulomb–Mohr Theory

Solution

We have the following results, from Example 6.7:

$$T = 170.9 \times 10^{-6} \tau$$

and $\sigma_1 = -\sigma_2 = \tau$. So, applying Equation 6.24 with $n = 2$,

$$\frac{\tau}{200 \times 10^6} - \frac{-\tau}{600 \times 10^6} = \frac{1}{2}$$

Solving, $\tau = 75$ MPa. The first equation then gives $T = 12.82$ kN·m.

Comments: On the basis of the maximum principal stress theory, the torque that can be applied to the pipe 17.09 kN·m obtained in Example 6.7 is thus 25% larger than on the basis of the Coulomb–Mohr theory.

Example 6.9: Largest Load Supported by the Frame of a Punch Press

Figure 6.14 depicts a punch press frame made of ASTM A-48 gray cast iron having ultimate strengths in tension and compression S_u and S_{uc}, respectively. Calculate the allowable load P.

Given: $S_u = 170$ MPa, $S_{uc} = 650$ MPa

Design Decisions: Use the Coulomb–Mohr theory and a factor of safety of $n = 2.5$.

Solution

The centroid, total section area, and moment of inertia about the neutral axis (Figure 6.14) are

$$\bar{z} = \frac{(180 \times 80)(210) + (120 \times 240)60}{180 \times 80 + 120 \times 240} = 110 \text{ mm}$$

$$A = 180 \times 80 + 120 \times 240 = 43.2 \times 10^3 \text{ mm}^2$$

$$I = \frac{1}{12}(80)(180)^3 + (80 \times 180)(100)^2 + \frac{1}{12}(240)(120)^3 + (120 \times 240)(50)^2$$

$$= 289.44 \times 10^6 \text{ mm}^4$$

The internal force resultants in section A–B are equivalent to a centric force P and a bending moment $M = 0.51P$.

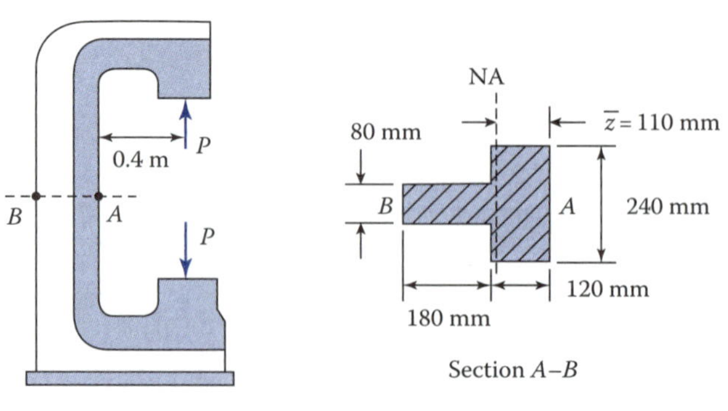

FIGURE 6.14
Example 6.9.

Note that cast iron has nonlinear stress–strain relationship. Therefore, bending stresses are not exactly given by the flexure formula, $\sigma = Mc/I$. For simplicity, however, it is generally used in the design of cast iron machine elements. Hence, the stress distribution across the section is taken to be linear.

The distances from the neutral axis to the extreme fibers are $c_A = 110$ mm and $c_B = 190$ mm. The greatest tensile and compressive stresses occur at points A and B, respectively:

$$\sigma_A = \frac{P}{A} + \frac{Mc_A}{I} = 23.148P + 193.823P$$

$$\sigma_B = \frac{P}{A} - \frac{Mc_B}{I} = 23.148P - 334.784P$$

We therefore have, on the tension and compression sides, respectively,

$$\begin{aligned} &\sigma_1 = 216.971P, \quad \sigma_2 = 0 \\ &\sigma_2 = -311.636P, \quad \sigma_1 = 0 \end{aligned} \tag{a}$$

The maximum allowable load P is the smaller of the two loads calculated from Equations (a), 6.26, and 6.27:

$$|216.971P| = \frac{170(10^6)}{2.5} \quad \text{or} \quad P = 313.4 \text{ kN}$$

$$|-311.636P| = \frac{650(10^6)}{2.5} \quad \text{or} \quad P - 834 \text{ kN}$$

Comment: The tensile stress governs the allowable load $P = 313.4$ kN that the member can carry.

6.13 Reliability

The concept of reliability is closely related to the factor of safety. Reliability is the probability that a member or structure will perform without failure a specific function under given conditions for a given period of time. It is very important for the designer and the manufacturer to know the reliability of the product. The reliability R can be expressed by a number that has the range

$$0 \le R < 1$$

For instance, reliability of $R = 0.98$ means that there is 98% chance, under certain operating conditions, that the part will perform its proper function without failure; that is, if 100 parts are put into service and an average of 2 parts fail, then the parts proved to be 98% reliable.

Recall from Section 1.6 that, in the conventional design of members, the possibility of failure is reduced to acceptable levels by factor of safety based on the judgment derived from past performances. In contrast, in the reliability method, the variability of material properties, fabrication-size tolerances, as well as uncertainties in loading and even design approximations can be appraised on a statistical basis. As far as possible, the proposed criteria are calibrated against well-established cases. The reliability method has the advantage of consistency in the safety factor, not only for individual members but also for complex structures and machines. Important risk analyses of complete engineering systems are based on the same premises. Clearly, the usefulness of the reliability approach depends on adequate information on the statistical distribution of loading applied to parts in service, from which can be calculated the significant stress and significant strength of production runs of manufactured parts.

Note that the reliability method to design is relatively new and analyses leading to an assessment of reliability address uncertainties. Of course, this approach is more expensive and time consuming than the factor of safety method of design, because a larger quantity of data must be obtained by testing. However, in certain industries, designing to a designated reliability is necessary. We use the factor of safety for most of the problems in this text. For an interactive statistics program from the engineering software database, see the website at www.mecheng.asme.org/database/STAT/MASTER.HTML.

6.14 Normal Distributions

To obtain quantitative estimates of the percentages of anticipated failures from a study, we must look into the nature of the distribution curves for significant quantities involved. We consider only the case involving the normal or Gauss distribution, credited to Gauss (1777–1855). This is the most widely used model for approximating the dispersion of the observed data in applied probability [12,13].

Several other distributions might prove useful in situations where random variables have only positive values or asymmetrical distributions. A formula, introduced by Weibull, is often used in mechanical design. This formula does not arise from classical statistics and is flexible to apply. The Weibull distribution is used in work dealing with experimental data, particularly reliability. The distribution of bearing failures at a constant load can be best approximated by the Weibull distribution (see Section 10.15).

In analytical form, the Gaussian, or normal, distribution is given as follows:

$$p(x)=\frac{1}{\sqrt{2\pi\sigma}}e^{-(x-\mu)/2\sigma^2}\quad(-\infty<x<\infty)\tag{6.29}$$

where

$p(x)$ = the probability or frequency function
σ = the standard deviation*
μ = the mean value
x = the quantity

The standard deviation is widely used and regarded as the usual index of dispersion or scatter of the particular quantity. The mean value and standard deviations are defined by

$$\mu=\frac{1}{n}\sum_{i=1}^{n}x_i\tag{6.30a}$$

* The symbol σ used for standard deviation here should not be confused with the symbol of stress, although often the units are the same.

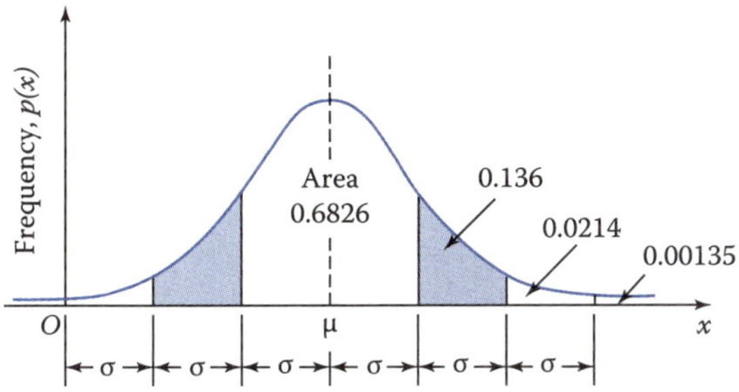

FIGURE 6.15
The standard normal distribution.

$$\sigma = \left[\frac{1}{n-1} \sum_{i=1}^{n} (x_i - \mu)^2 \right]^{1/2} \tag{6.30b}$$

where n is the total number of elements, called the population.

A plot of Equation 6.29, the standard normal distribution, is shown in Figure 6.15. This *bell-shaped curve* is symmetrical about the mean value μ. Since the probability that any value of x will fall between plus and minus infinity is 1, the area under the curve in the figure is unity. Note that about 68% of the population represented fall within the band $\mu + 1\sigma$, 95% fall within the band $\mu + 2\sigma$, and so on.

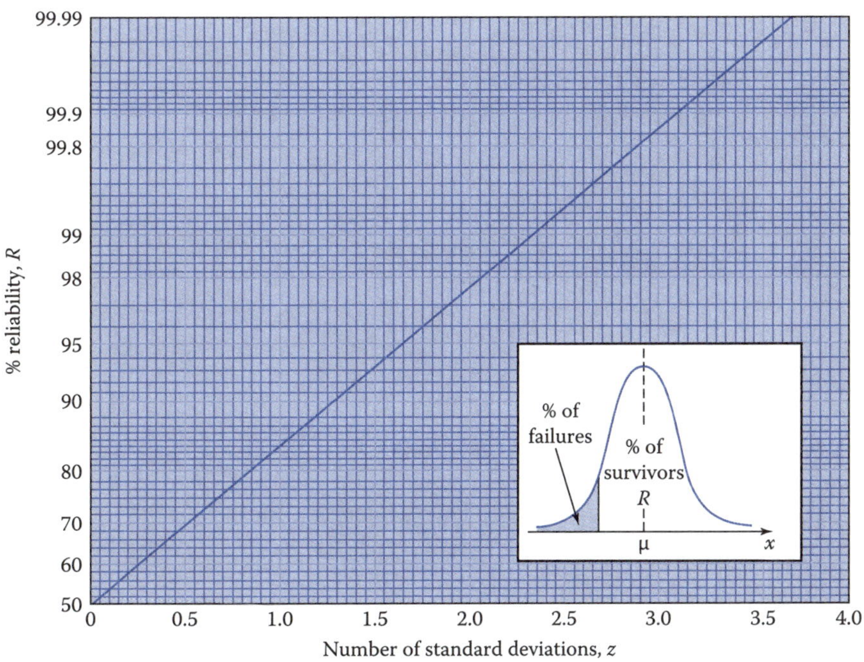

FIGURE 6.16
Reliability chart: generalized normal distribution curve plotted on special probability paper.

The reliability, or rate of survival, R is a function of the *number of standard deviations z*, also referred to as the safety index:

$$z = \left| \frac{x - \mu}{\sigma} \right| \tag{6.31}$$

The straight line of Figure 6.16, drawn on probability-chart paper, plots percentages of reliability R as an increasing function of number of standard deviations z. A larger z results in fewer failures, hence a more conservative design. We note that the percentage of the population corresponding to any portion of the standard normal distribution (Figure 6.15) can be read from the reliability chart.

6.15 Reliability Method and Margin of Safety

In the reliability method of design, for a given member, the distribution of loads and strengths are determined, and then these two are related to achieve an acceptable success rate. The designer's task is to make a judicious selection of materials, processes, and geometry (size) to achieve a reliability goal. The approach of reliability finds considerably more application with members subjected to wear and fatigue loading. We here introduce it in the simpler context of static loading. Section 7.6 discusses the reliability factor for fatigue endurance strength of materials.

Consider the distribution curves for the two main random variables, load L and strength S, which is also called capacity or resistance (Figure 6.17). For a given member, the frequency functions $p(l)$ and $p(s)$ define the behavior of critical parameters load and strength. The mean value of strength is denoted by μ_s and the mean value of the load by μ_l. So, based on mean values, there would be a *margin of safety*

$$m = S - L \tag{6.32}$$

However, the *interference* or shaded area of overlap in the figure indicates some possibility of a weak part in which failure could occur. The preceding margin of safety must not be confused with that used in aerospace industry (see Section 1.8).

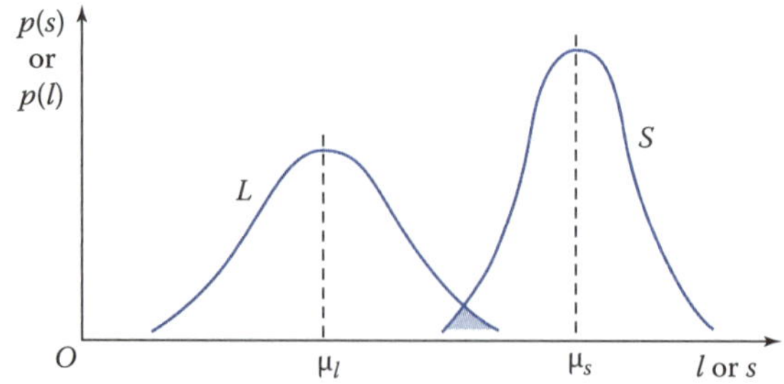

FIGURE 6.17
Normal distribution curves of load L and strength S.

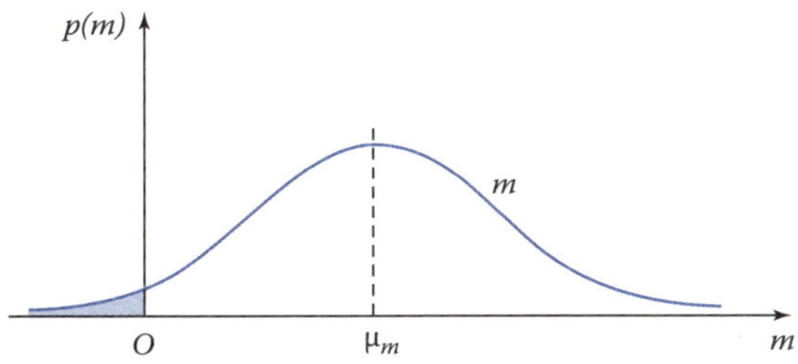

FIGURE 6.18
Normal distribution curve of margin of safety $m = S - L$.

Figure 6.18 shows a corresponding plot of the distribution of margin of safety. In this diagram, the probability of failure is given by the (shaded) area under the tail of the curve to the left of the origin. The member would survive in all instances to the right of the origin. By statistical theory, the difference between two variables with normal distributions has itself a normal distribution. Therefore, if the strength S and the load L are normally distributed, then the margin of safety m also has a normal distribution, as shown in Figures 6.17 and 6.18.

The margin of safety has a *mean value* μ_m and *standard deviation* σ_m expressed as follows [14]:

$$\mu_m = \mu_s - \mu_l \tag{6.33a}$$

$$\sigma_m = \sqrt{\sigma_s^2 + \sigma_l^2} \tag{6.33b}$$

Here, σ_s and σ_l are, respectively, the standard deviations for the strength S and the load L. When $S > L$, *in* is positive. The designer is interested in the probability that $m > 0$; that is, the area to the right of 0 in Figure 6.18. At $x = m = 0$, Equation 6.31 becomes $z = \mu/\sigma$. The number of standard deviations, on introducing Equation 6.32, may now be written in the following form:

$$z = \frac{\mu_m}{\sigma_m} = \left| \frac{\mu_s - \mu_l}{\sqrt{\sigma_s^2 + \sigma_l^2}} \right| \tag{6.34}$$

For the prescribed mean and deviation values of the strength and load, Equation 6.34 is solved to yield the number of standard deviations z. Then, the probability that a margin of safety exists may be read as the reliability R from the chart of Figure 6.16. Equation 6.34 is therefore called the coupling equation, because it relates the reliability, through z, to the statistical parameters of the normally distributed strength and load. For example, when the mean values of S and L are equal (i.e., $\mu_s = \mu_l$), it follows that $z = 0$ and the reliability of a part is 50%. Reliability of an assembly or system of parts may be found from their individual reliability values [14].

Application of the reliability theory is illustrated in the solution of the following numerical problems.

Example 6.10: Shipment of Control Rods

In a shipment of 600 control rods, the mean tensile strength is found to be 35 ksi and the standard deviation 5 ksi. How many rods can be expected to have

a. A strength of less than 29.5 ksi
b. A strength of between 29.5 and 48.5 ksi

Given: $\mu_s = 29.5$ and 48.5 ksi, $\mu_l = 35$ ksi, $\sigma_m = 5$ ksi.

Assumption: Both loading and strength have normal distributions.

Solution

a. Substituting given numerical values into Equation 6.34 results in the number of standard

$$z = \left| \frac{29.5 - 35}{5} \right| = 1.10$$

The corresponding reliability, obtained from Figure 6.16, is 86.5%. Note that $1 - 0.865 = 0.135$ represents the proportion of the total rods having a strength less than 29.5 ksi. Hence, the number of rods with a strength less than 29.5 ksi is $600(0.135) = 81$.

b. In this case, applying Equation 6.31,

$$z = \left| \frac{48.5 - 35}{5} \right| = 2.7$$

From Figure 6.16, we then have $R = 99.65\%$. The number of rods expected to have strength between 29.5 and 48.5 ksi is therefore $600(0.9965 - 0.135) = 517$.

Example 6.11: Machine Part in Service

At the critical point of a machine part in service, the load-induced mean stress and standard deviation are 30 and 5 ksi, respectively. If the material has a yield strength of 50 ksi with a standard deviation of 4 ksi, determine the reliability against yielding. What percentage of failure is expected in service?

Given: $\mu_s = 50$ ksi, $\mu_l = 30$ ksi, $\sigma_s = 4$ ksi, $\sigma_l = 5$ ksi.

Assumption: Both loading and strength have normal distribution.

Solution

Through the use of Equation 6.33, we have

$$\mu_m = 50 - 30 = 20 \text{ ksi}$$

$$\sigma_m = \sqrt{4^2 + 5^2} = 6.403 \text{ ksi}$$

Equation 6.34 then gives $z = 20/6.403 = 3.124$. Figure 6.16 shows that this corresponds to 99.91% reliability. So, the failure percentage expected in service would be $100 - 99.91 = 0.09\%$.

Example 6.12: Twisting-Off Strength of Bolts

Bolts, each of which has a mean twisting-off strength of 25 N·m with a standard deviation of 1.5 N·m, are tightened with automatic wrenches on a production line (see Section 15.7). If the automatic wrenches have a standard deviation of 2 N·m, calculate mean value of wrench torque setting that results in an estimated 1 bolt in 400 twisting off during assembly.

Given: $\mu_s = 25$ N·m, $\sigma_s = 1.54$ N·m, $\sigma_l = 2$ N·m.

Assumption: Both the wrench twist-off torque and the bolt twist-off strength have normal distributions.

Solution

Substitution of $\sigma_s = 1.5$ N·m and $\sigma_l = 2$ N·m in Equation 6.33b gives $\sigma_m = 2.5$ N·m. Figure 6.16 shows that a reliability of 399/400 = 0.9975, or 99.75%, corresponds to 2.8 standard deviation. The mean value is then $\mu_m = z\sigma_m = 2.8(2.5) = 7$ N·m. Since $\mu_s = 25$ N·m, we have, from Equation 6.33a, $\mu_l = 18$ N·m. This is the required value of wrench setting.

Problems

Sections 6.1 through 6.4

6.1 An A1SI-4340 steel ship deck of thickness t and width $2w$ is in tension. If a central transverse crack of length $2a$ is present (Case A of Table 6.1), estimate the maximum tensile load P that can be applied without causing sudden fracture. What is the nominal stress at fracture?

Given: $t = 25$ mm, $w = 250$ mm, $a = 25$ mm.

6.2 Estimate the maximum load P that the plate shown in Case B of Table 6.1 can carry. What is the mode of failure?

Given: $S_y = 650$ MPa, $K_c = 100\ \text{MPa}\sqrt{\text{m}}$, $w = 350$ mm, $a = 25$ mm, $t = 15$ mm, factor of safety $n = 1.2$.

6.3 A 2024-T851 aluminum alloy plate of width w and thickness t is subjected to a tensile loading. It contains a transverse crack of length a on one edge (Figure 6.2). There is concern that the plate will undergo sudden fracture. Calculate the maximum allow able axial load P. What is the nominal stress at fracture?

Given: $w = 125$ mm, $t = 25$ mm, $a = 20$ mm.

6.4 A thin, long AISI 4359 steel instrument panel of width $2w$ is under uniform longitudinal tensile stress σ. When a $2a$ long central transverse crack is present (Case A, Table 6.1), based on a safety factor n against yielding, compute the factor of safety for fracture.

Given: $a = 80 \times 10^{-3}$ in., $n = 2.5$, $a/w = 0.2$.

6.5 An 7073-T3351 aluminum alloy long plate of width w with and edge crack is subjected to tension (Case B, Table 6.1). The required factor of safety against yielding and the crack length are n and a, respectively. Find the safety factor on the basis of fracture.

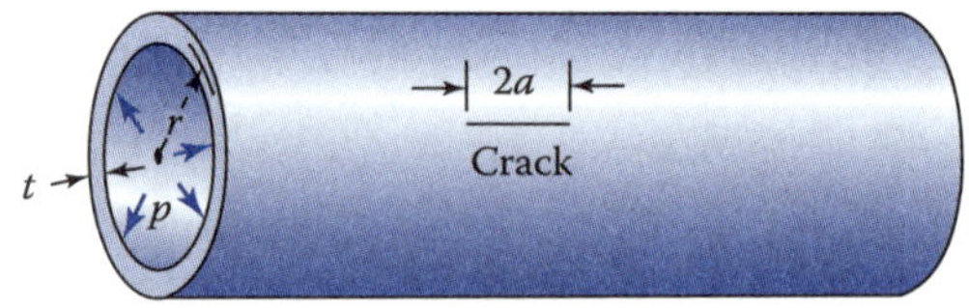

FIGURE P6.7

6.6 A long Ti-6A1-4V titanium panel of width $2w$ and thickness t carries a uniform tension. For the case in which a central transverse crack of length $2a$ exits (Case A, Table 6.1), determine

a. The safety factor for yielding and fracture

b. The tensile stress when fracture occurs

Given: $a = 50$ mm, $t = 25$ mm, $\sigma = 150$ MPa.

6.7 An AISI-4340 steel pipe of diameter d and wall thickness t contains a crack of length $2a$. Estimate the pressure p that will cause fracture when

a. The crack is longitudinal as in Figure P6.7

b. The crack is circumferential

Given: $d = 50$ mm, $t = 4$ mm, $a = 5$ mm.

Assumption: A factor of safety $n = 1.5$ and geometry factor $\lambda = 1.01$ are used (Table 6.1).

6.8 A 7075-T7351 aluminum alloy beam containing an edge crack of length a is in pure bending as shown in Case D of Table 6.1. Determine the maximum moment M that can be applied without causing sudden fracture.

Given: $a = 40$ mm, $w = 100$ mm, $t = 25$ mm.

6.9 An AISI-4340 steel plate of width $w = 5$ in. and thickness $t = 1$ in. is under uniaxial tension. A crack of length a is present on the edge of the plate, as shown in Figure 6.2. Determine

a. The axial load possible P_{all} for the case in which $a = ½$ in

b. The critical crack length a, if the plate is made of Ti-6A1-6V titanium alloy and subjected to the P_{all} calculated in part a

6.10 Rework Example 6.3, for the case in which the bracket is made of AISI-403 stainless steel and $a = 0.6$ in., $d = 6.25$ in., $w = 1.5$ in., $t = ½$ in., and $n = 2$.

6.11 An AISI-4340 steel ship deck panel of width w and thickness t is under tension. Calculate the maximum load P that can be applied without causing fracture when double-edge cracks grow to a length of a (Case C of Table 6.1).

Given: $t = 1$ in., $w = 2$ in., $a = 0.2$ in., $n = 1.4$.

Sections 6.5 through 6.12

6.12 A solid steel shaft having yield strength S_y and diameter D carries end loads P, M, and T (Figure P6.12). Calculate the factor of safety n, assuming that failure occurs according to the following criteria:

a. Maximum energy of distortion

b. Maximum shear stress

Given: $D = 100$ mm, $S_y = 260$ MPa, $P = 50$ kN, $M = 5$ kN·m, $T = 8$ kN·m.

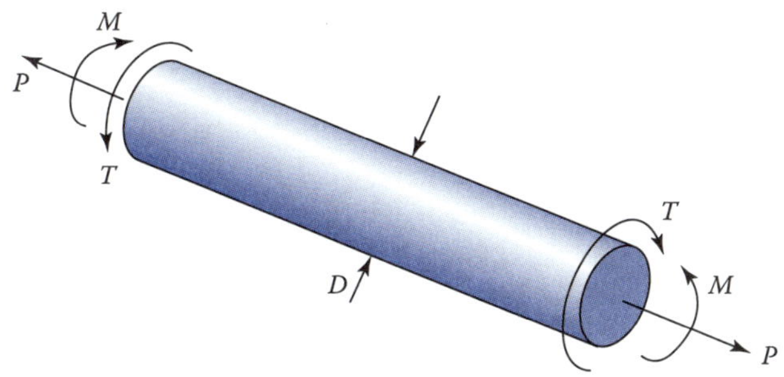

FIGURE P6.12

6.13 Redo Problem 6.9, applying the Coulomb–Mohr failure criterion and knowing that the shaft is constructed from ASTM A-48 gray cast iron.

6.14 A steel bar *AB* of diameter *D* and yield strength S_y supports an axial load *P* and vertical load *F* acting at the end of the arm *BC* (Figure P6.14). Determine the largest value of *F* according to the maximum energy of distortion theory of failure.

Given: $D = 40\,\text{mm}$, $S_y = 250$ MPa, $P = 20F$.

Assumptions: The effect of the direct shear is negligible and the factor of safety $n = 1.4$.

6.15 Resolve Problem 6.14 through the use of the maximum shear stress theory of failure.

6.16 A steel rod of diameter *D*, yield strength in tension S_y, and yield strength in shear S_{sy} is under loads *F* and $P = 0$ (Figure P6.14). Based on a safety factor of *n*, find the maximum allowable value of *F* applying

a. The maximum shear stress failure criterion

b. The maximum distortion energy failure criterion

Given: $D = 60$ mm, $n = 1.6$, $S_y = 260$ MPa, $S_{sy} = 140$ MPa.

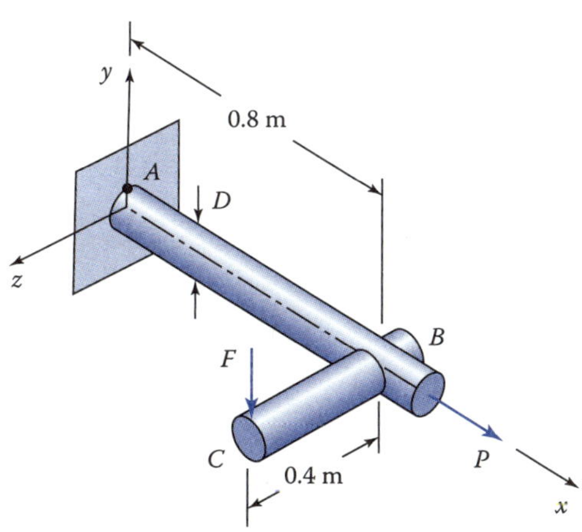

FIGURE P6.14

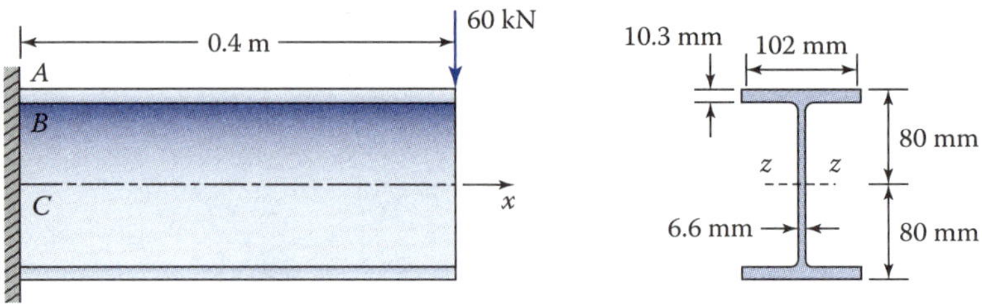

FIGURE P6.17

6.17 A cantilever *WF* aluminum alloy beam of yield strength S_y is loaded as shown in Figure P6.17. Using a factor of safety of n, determine whether failure occurs according to the maximum shear stress criterion.

Given: $S_y = 320$ MPa, $n = 2$, $I_z = 13.4 \times 10^6$ mm^4.

6.18 Resolve Problem 6.17 applying the maximum energy of distortion theory.

6.19 A thin-walled cylindrical pressure vessel of diameter d and constructed of structural steel with yield strength S_y must withstand an internal pressure p. Calculate the wall thickness t required.

Given: $S_y = 36$ ksi, $d = 20$ in., $p = 500$ psi, $n = 1.5$.

Design Decision: Use the following criteria:

a. Maximum shear stress

b. Maximum energy of distortion

6.20 Redo Problem 6.19, if the vessel is made of a material having $S_u = 50$ ksi and $S_{uc} = 90$ ksi.

Design Decision: Apply the following theories:

a. Maximum principal stress

b. Coulomb–Mohr

6.21 A cantilever *WF* cast iron beam of ultimate tensile strength S_u and ultimate compression strength S_{uc} is subjected to a concentrated load at its free end (Figure P6.17). What is the factor of safety n?

Given: $S_u = 280$ MPa, $S_{uc} = 620$ MPa.

Assumption: Failure occurs in accordance with the following theories:

a. Maximum principal stress

b. Coulomb–Mohr

6.22 Design the cross a $b \times 2b$ rectangular overhang beam (i.e., find the dimension b), loaded as illustrated in Figure P6.22, for $\sigma_{all} = 20$ ksi. Apply the maximum principal failure criterion.

6.23 Rework Example 6.9, if the cross section *A–B* of the punch press shown in Figure 6.14 is a 120 mm deep by 300 mm wide rectangle.

6.24 and **6.25** The state of stress shown (Figures P6.24 and P6.25) occurs at a critical point in an ASTM A-48 gray cast iron (Table B.1) component of a lawn mower. Calculate the factor of safety n with respect to fracture.

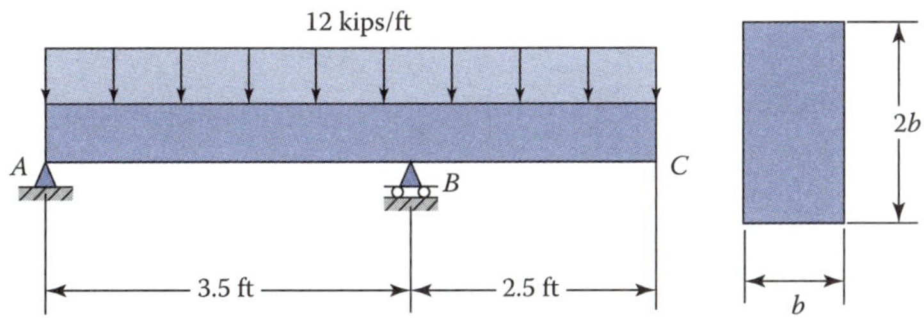

FIGURE P6.22

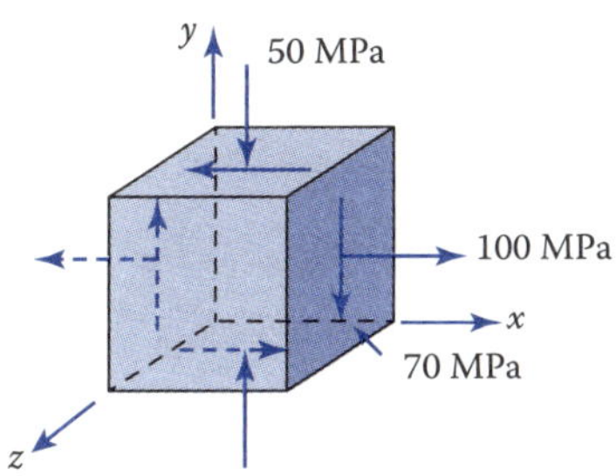

FIGURE P6.24

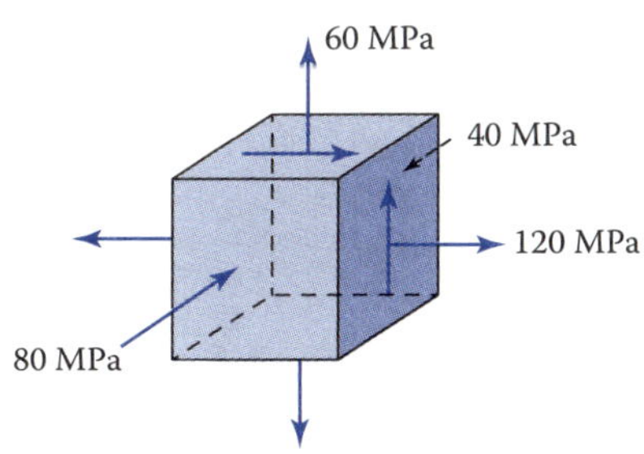

FIGURE P6.25

Design Decision: Apply the following criteria:

a. Maximum principal stress

b. Coulomb–Mohr

6.26 A closed-ended cylinder of radius r and wall thickness t is constructed of ASTM-A36 structural steel having tensile strength S_y rests on cradles as depicted in Figure P6.26.

Determine: The allowable pressure the shell can carry on the basis of a factor of safety n. Apply the two yield failure criteria:

a. The maximum shear

b. The maximum energy of distortion

Given: $r = 300$ mm, $t = 10$ mm, $n = 1.4$, $S_y = 250$ MPa (by Table B.1).

Assumption: The largest stresses take place on the elements outside of the cylinder wall, away from the supports and ends.

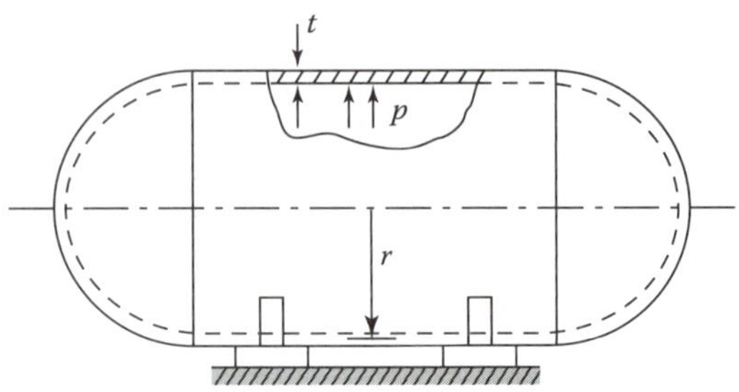

FIGURE P6.26

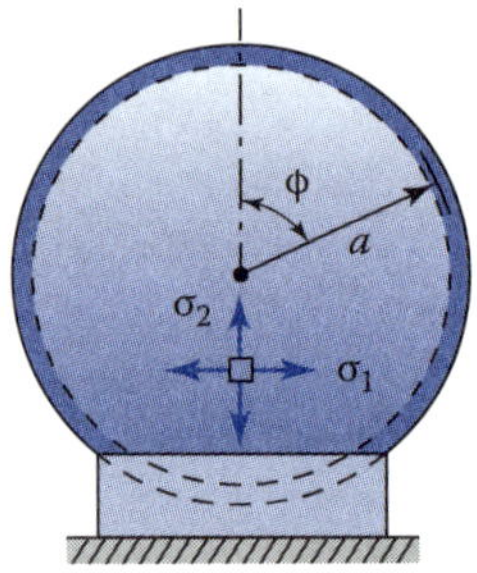

FIGURE P6.27

6.27 A thin-walled steel spherical storage tank is filled with liquid of density γ and supported on a ring, as shown in Figure P6.27. According to the maximum shear stress and maximum energy of distortion criteria, determine the factor of safety n, in terms of thickness t, mean radius a, yield strength S_y, and γ, as required.

Given: Tangential stress $\sigma_\theta = \sigma_1$ and meridian stress $\sigma_\phi = \sigma_2$ at $\phi = 150°$ are calculated from Equation P16.39 as follows:

$$\sigma_1 = 3.56\frac{\gamma a^2}{t}, \quad \sigma_2 = -1.70\frac{\gamma a^2}{t}$$

where the negative sign indicates compression.

6.28 Resolve Problem 6.27 for the condition that the tank is made of cast iron having ultimate compression strength S_{uc} five times ultimate tensile strength S_u.

Design Decision: The maximum principal stress and Coulomb–Mohr criteria are used.

6.29 A cylindrical rod of diameter D is made of ASTM-A36 steel (Table B.1). Use the maximum shear stress criterion to determine the maximum end torque T that can be applied to the rod simultaneously with an axial load of $P = 10$ kips (Figure 6.5).

Given: $D = 2$ in.

Assumption: $n = 1.5$.

6.30 Redo Problem 6.29, applying the maximum energy of distortion criterion.

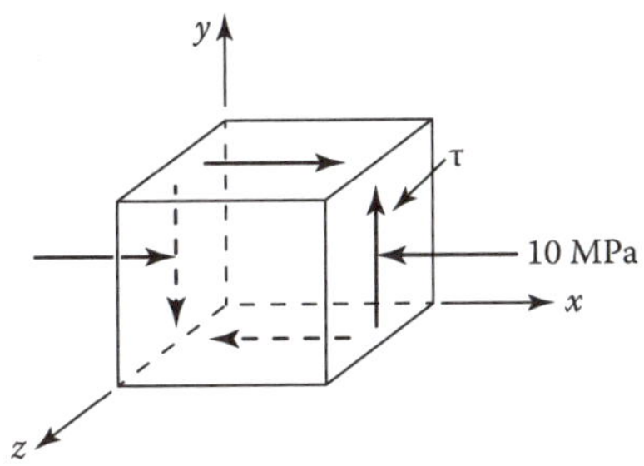

FIGURE P6.32

6.31 An ASTM-20 gray cast iron rod (Table B.2) is under pure torsion. Determine, with a factor of safety $n = 1.4$, the maximum shear stress τ that may be expected at impending rupture using

a. The Coulomb–Mohr criterion

b. The principal stress criterion

6.32 The state of stress shown in Figure P6.32 occurs at a critical point in a machine component made of ASTM-A47 malleable cast iron (Table B.1). Apply the Coulomb–Mohr theory to calculate the maximum value of the shear stress τ for a safety factor of $n = 2$.

6.33 Resolve Problem 6.32 for the condition that the machine component is made of an ASTM-A242 high-strength steel (Table B.1). Use

a. The maximum energy of distortion criterion

b. The maximum shear stress criterion

6.34 At a critical point in a cast metal (S_u = 8 ksi, S_{uc} = 22.5 ksi) machine frame, the state of stress is as depicted in Figure P6.34. Find whether failure occurs at the point in accordance with

a. The maximum principal stress theory

b. The Coulomb–Mohr theory

6.35 A cast iron ASTM grade A-47 round shaft is simultaneously subjected to torque T and load P, as shown in Figure P6.35. Find the diameter D, through the use of

a. The maximum normal stress criterion

b. The Coulomb–Mohr failure criterion

Given: $L = 12$ in., $T = 6$ kip·in., $P = 150$ kips, $n = 2.5$, $S_u = 50$ ksi, $S_{uc} = 90$ ksi (by Table B.1).

Assumption: The effect of transverse shear will be disregarded.

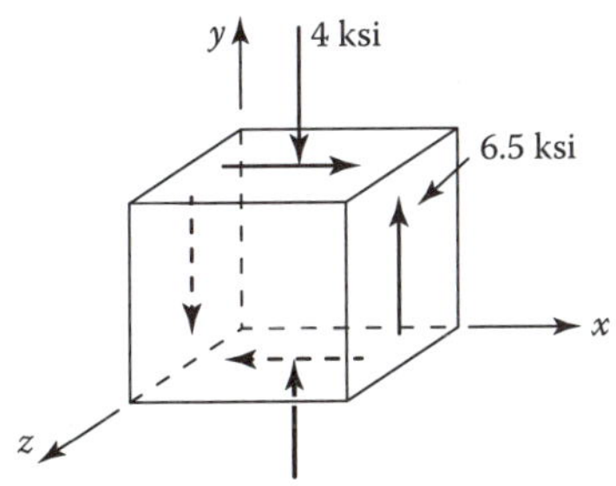

FIGURE P6.34

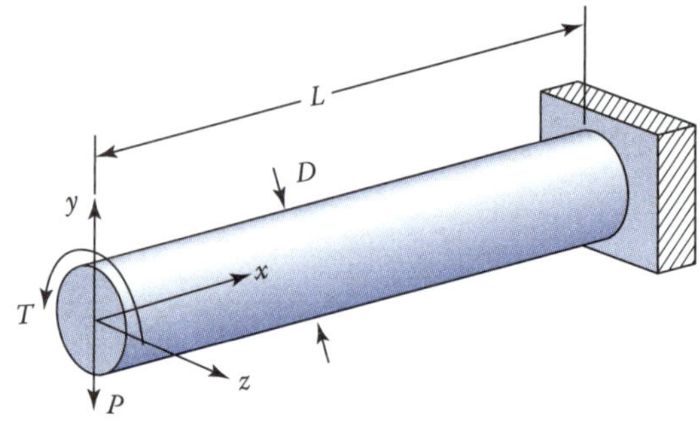

FIGURE P6.35

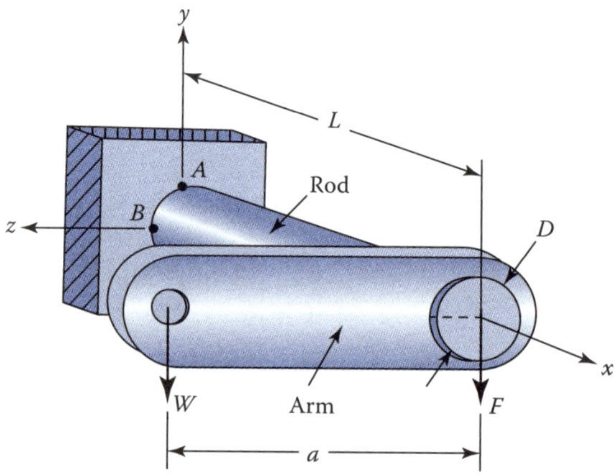

FIGURE P6.36

6.36 Figure P6.36 shows that a bracket arm of length a is acted on by a vertical loads W kips and P at its free ends. The ASTM-A242 high-strength steel rod has diameter D, length L, and shear yield strength S_y. Find the factor of safety n for the rod, using the maximum shear stress theory of failure.

Given: $D = 2$ in., $L = 10$ in., $a = 12$ in., $W = 1.8$ kips, $F = 400$ lb, $S_y = 30$ ksi.

6.37 An ASTM-A36 steel shaft of length L carries a torque T and its own weight per unit length w (see Table B.1), as depicted in Figure P6.37. Determine the required shaft diameter D, using the maximum energy of distortion criterion with a safety factor of $n = 2.1$.

Given: $L = 6$ m, $T = 400$ N · m.

Assumption: The bearings at the ends act as simple supports.

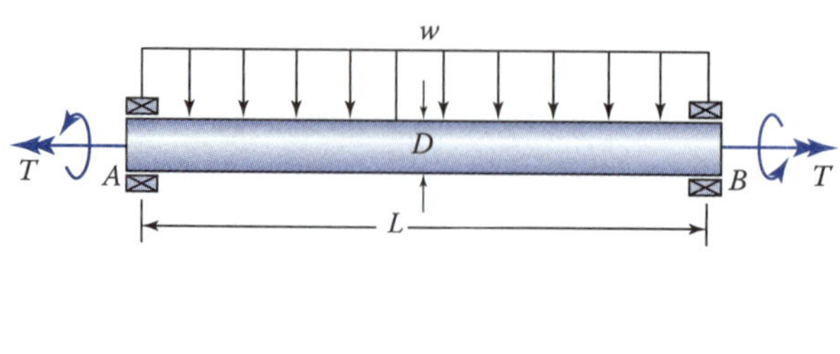

FIGURE P6.37

Sections 6.13 through 6.15

6.38 At a critical location in a component in tension, the load-induced stresses are μ_l = 250 MPa and σ_l = 35 MPa. What is the reliability R against yielding?

Given: The material yield strengths are μ_s = 400 MPa and σ_s = 30 MPa.

6.39 Calculate the diameter d of a bar subjected to an axial tensile load P for a desired reliability of R = 99.7%.

Given: The material yield strengths of μ_s = 50 ksi and σ_s = 5 ksi. The loads of μ_l = 40 kips and σ_l = 6 kips.

6.40 Determine the mean μ and the standard deviation σ for the grades of a sample of 12 students shown in the accompanying table.

n	1	2	3	4	5	6	7	8	9	10	11	12
x	77	85	48	94	80	60	65	96	70	86	69	82
y	78.2	82.1	60.3	91.5	84.6	70.8	68.4	90.8	75.0	92.5	61.8	80.1

Notes: n, number of students; x, final examination grade; y, course grade.

6.W Search and download the statistics shareware program on the website at www.mecheng.asme.org/database/STAT/MASTER.HTML for computing the mean and standard deviations for a normal distribution. Resolve Problem 6.40 using this program.

6.41 A total of 68 cold-drawn steel bars have been tested to obtain the 0.2% offset yield strength S_y in ksi. The results are as follows:

S_y	74	66	62	78	81	82	85	86	89	94
n	7	2	5	5	10	18	8	3	6	4

Based on normal distribution, determine

a. The mean μ and standard deviation σ of the population

b. The reliability for a yield strength of S_y = 75 ksi

6.42 A bar under a maximum load of 5 kips was designed to carry a load of 6 kips. The maximum load is applied with standard deviation of 600 lb and shaft strength standard deviation of 400 lb, both are normally distributed. Calculate the expected reliability.

6.43 A ½ in. diameter ASTM-A242 high-strength steel rod carries an axial nominal load of P_{nom}. Experimental data show that the yield strength is normally distributed with a mean value of $S_{y,nom}$ and standard deviation of σ_s. Owing to the variety of operational conditions, the load has been found to be actually normally distributed random variables with standard deviation of 500 lb. Estimate, on the basis of yielding failure,

a. The factor of n safety

b. The reliability R of the rod

Given: P_{nom} = 8 kips, σ_s = 4 ksi, $S_{y,nom} = \mu_s$ = 50 ksi (by Table B.1).

6.44 A structural member is subjected to a maximum load of 20 kN. Assume that the load and strength have normal distributions with standard variations of 3 and 2.5 kN, respectively. If the member is designed to withstand a load of 25 kN, determine failure percentage that would be expected.

7

Fatigue Failure Criteria

7.1 Introduction

A member may fail at stress levels substantially below the yield strength of the material if it is subjected to time-varying loads rather than static loading. The phenomenon of progressive fracture due to repeated loading is called *fatigue.* Its occurrence is a function of the magnitude of stress and a number of repetitions, so it is called *fatigue failure.* Photographs (Figure 7.1) represent two components failed by fatigue. We observe throughout this chapter that the fatigue strength of a component is significantly affected by a variety of factors. A fatigue crack most often is initiated at a point of high stress concentration, such as at the edge of a notch, or by minute flows in the material. Fatigue failure is of a brittle nature even for materials that normally behave in a ductile manner. The usual fracture occurs under tensile stress and with no warning. For combined fluctuating loading conditions, it is common practice to modify the static failure theories and material strength for the purposes of design.

The fatigue failure phenomenon was first recognized in the 1800s when railroad axles fractured after only a limited time in service. Until about the middle of the nineteenth century, repeated and static loadings were treated alike, with the exception of the use of safety factors. Poncelet's book in 1839 used the term *fatigue* owing to the fluctuating stress. At the present time, the development of modern high-speed transportation and machinery has increased the importance of the fatigue properties of materials. In spite of periodic inspection of parts for cracks and other flaws, numerous major railroad and aircraft accidents have been caused by fatigue failures. Fatigue is the single largest, estimated to be 90%, of failure in *metals.* Particularly, structural fatigue failures are catastrophic and dangerous, taking place suddenly and usually without any warning. The basic mechanism associated with fatigue failure is now reasonably well understood, although research continues on its many details [1–19]. The complexity of the problem is such that rational design procedures for fatigue are difficult to develop. The great variation in properties makes it necessary to apply statistical methods in the evaluation of the fatigue strength.

Essentially, fatigue is crack propagation, initially on a microscopic scale and then very rapid as the fatigue crack reaches a critical length. Experiments have shown that fatigue cracks often begin at a surface and propagate through the rest of the body, unless large subsurface flaws and stress raisers exist in the material. The fatigue life of a component presents the time it takes a crack to start plus the time it needs to propagate through the section. Design life can be extended by minimizing initial surface flows through processes like grinding or polishing, relieving tensile residual stresses on surface through manufacturing processes or by various surface treatments, maximizing propagation time using a material that does not have elongated grains in the direction of fatigue crack growth,

FIGURE 7.1
(a) Fatigue failure of a road bike hub, causing four spokes to break off its flange and (b) breaking apart of a typical crankshaft due to repeated dynamic loading.

and using material properties that permit for larger internal flaws. We shall here present methods to design for cyclic loading. It is important to note that in applying any of these techniques, generous factors of safety should always be used.

7.2 Nature of Fatigue Failures

The type of fracture produced in ductile metals subjected to fatigue loading differs greatly from that of fracture under static loading, considered in Section 2.3. In fatigue fractures, two regions of failure can be detected: the beachmarks (so termed because they resemble ripples left on sand by retracting waves) zone produced by the gradual development of the crack and the sudden-fracture zone. As the name suggests, the latter region is the portion that fails suddenly when a crack reaches its size limit. Figure 7.2 depicts fatigue fracture

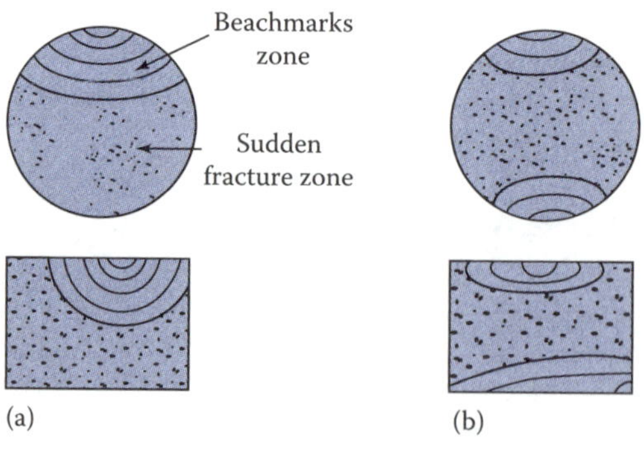

FIGURE 7.2
Schematic representation of fatigue fracture surfaces of circular and rectangular cross sections subjected to (a) tension–tension or tension–compression and (b) reversed bending. (Based on American Society of Metals, Failure analysis and prevention, *Metals Handbook*, 9th ed., Metals Park, OH, ASM International, 1986.)

surfaces of two common cross sections under high nominal stress conditions [1]. Note that the curvature of the beachmarks serves to indicate where the failure originates. The beachmarked area, also referred to as the *fatigue zone*, has a smooth, velvety texture. This contrasts with the sudden-fracture region, which is relatively dull and rough, looking like a static *brittle* type.

Microscopic examinations of ductile metal specimen subjected to fatigue stressing reveal that little if any distortion occurs, whereas failure due to static overload causes excessive distortion. The appearance of the surfaces of fracture greatly aids in identifying the cause of crack initiation to be corrected in redesign. For this purpose, numerous photographs and schematic representations of failed surfaces have been published in technical literature [1,5]. Figure 7.3 is a simplified sketch of the effect of state of stress on the origin, appearance, and location of fatigue fracture for variously loaded sections. For all axial and bending stress conditions, as well as the high-torsion smooth stress condition, the crack growth in the beachmarks region is indicated by curved vectors, starting from the point of crack initiation.

Note that the sudden-fracture region can be a small portion of the original cross section, particularly under bending and torsion fatigue stressing. Unless interrupted by notches, the fatigue crack under bending is normal to the tensile stresses, that is, perpendicular to the axis of a shaft. It follows that crack growth is owing to tensile stress and the crack grows along planes normal to the largest tensile stress. Clearly, fatigue stresses that are always compressive will not generate crack growth, since they tend to close the crack. In torsional fatigue failures, the crack is at a 45° angle to the axis of a notch-free shaft (or spring wire) under high nominal stress conditions (Figure 7.3). Finally, note that, if cracks initiate at several circumferential points, the sudden-fracture zone is more centered.

7.3 Fatigue Tests

To determine the strength of materials under the action of fatigue loads, four types of tests are performed: tension, torsion, bending, and combinations of these. In each test, specimens are subjected to repeated forces at specified magnitudes while the cycles or stress reversals to rupture are counted. A widely used fatigue testing device is the R.R. Moore high-speed rotating-beam machine (Figure 7.4a). To perform a test, the specimen is loaded with a selected weight *W*. Note that turning on the motor rotates the specimen, however, not the weight. There are various other types of fatigue testing machines [3]. A typical rotating-beam fatigue testing machine has an adjustable-speed spindle, operating at speeds in ranges of 500–10,000 rpm. The device can apply a moment up to 200 lb · in., to the specimen.

7.3.1 Reversed Bending Test

In the rotating-beam test, the machine applies a pure bending moment to the highly polished, so-called mirror finish, specimen of circular cross section (Figure 7.4b). As the specimen rotates at a point on its outer surface, the bending stress varies continuously from maximum tension to maximum compression. This fully or completely reversed bending stress can be represented on the stress *S*-cycles *N* axes by the curves of Figure 7.4c. It is obvious that the highest level of stress is at the center, where the smallest diameter

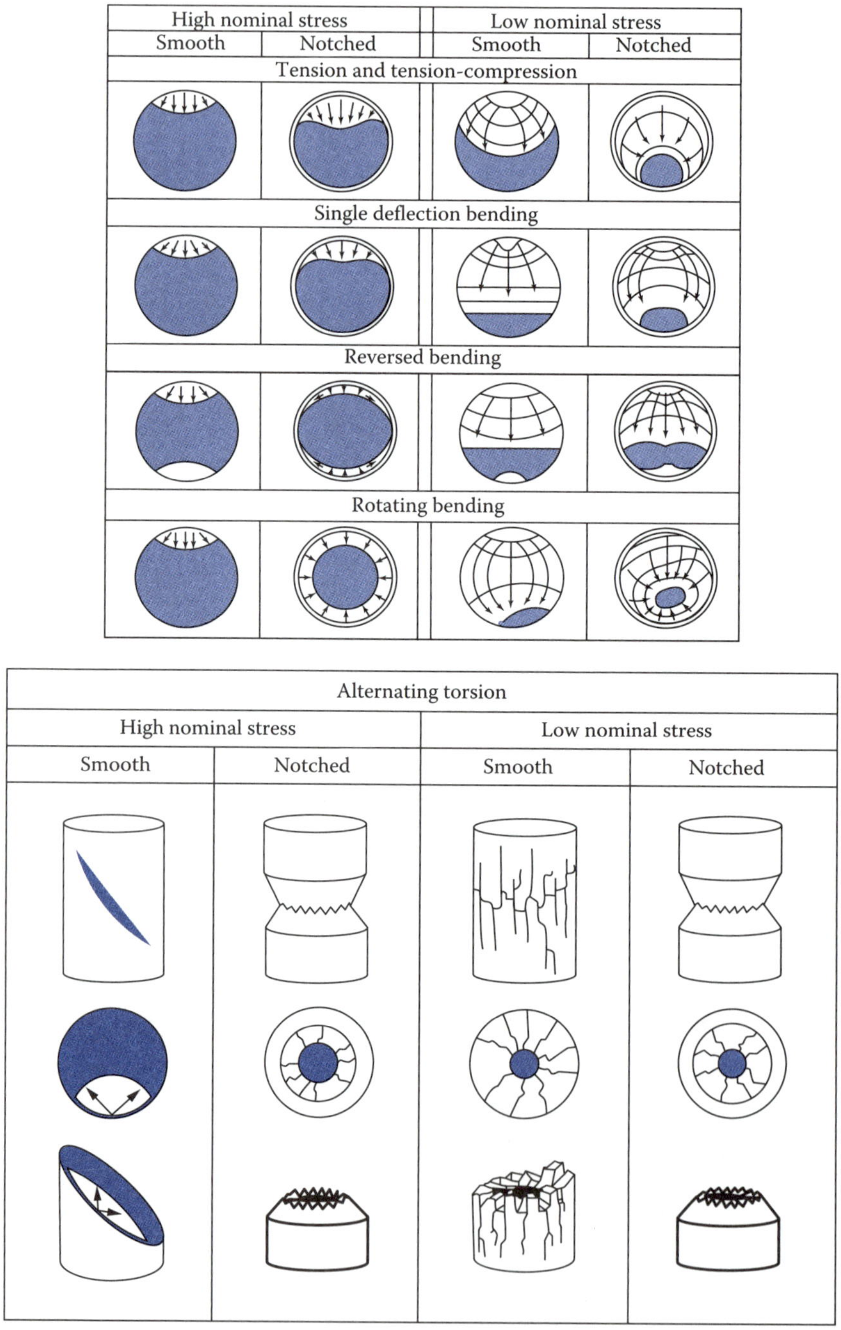

FIGURE 7.3
Effect of state of stress on fatigue fracture of circular smooth and notched cross sections under various loading conditions. (From Engel, L. and Klingele, H., *An Atlas of Metal Damage*, Hanser Verlag, Munich, Germany, 1981.)

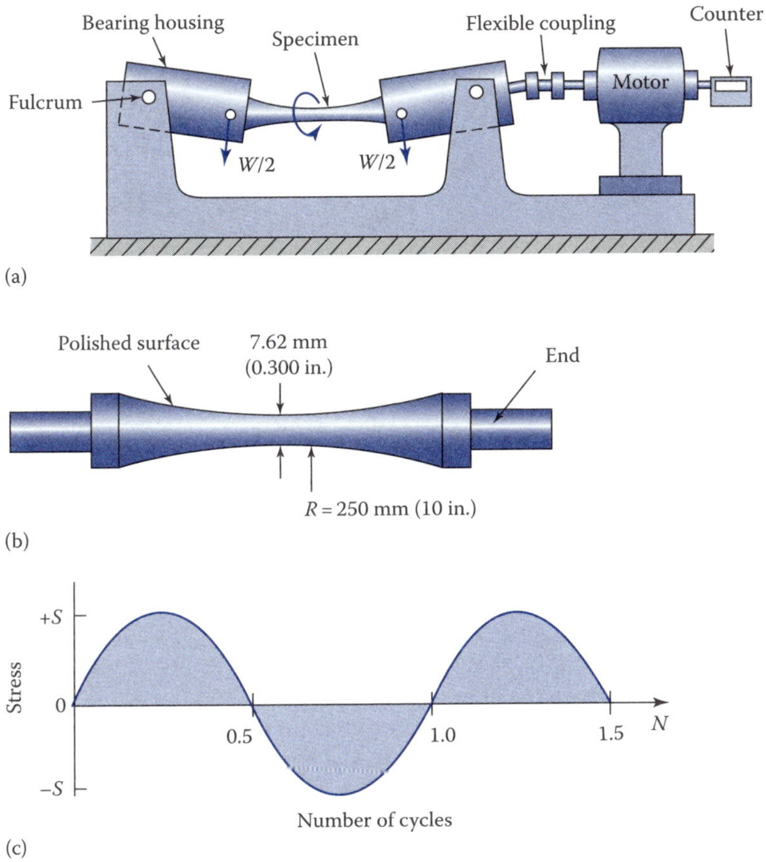

FIGURE 7.4
Bending fatigue: (a) schematic of the R.R. Moore rotating-beam fatigue testing machine, (b) standard round specimen, and (c) completely reversed (sinusoidal) stress.

is about 0.3 in. The large radius of curvature avoids stress concentration. Various standard types of fatigue specimens are used, including those for axial, torsion, and bending stresses described in the ASTM manual on fatigue testing.

In some fatigue testing machines, constant-speed (usually 1750 rpm) motors are used, which give the sinusoidal type or fully reversed cyclic stress variation shown in the figure. It takes about one-half of a day to reach 10^6 cycles and about 40 days to reach 10^8 cycles on one specimen. A series of tests performed with various weights and using multiple specimens, carefully made to be nearly the same as possible, gives results or the fatigue data.

7.4 S–N Diagrams

Fatigue test data are frequently represented in the form of a plot of *fatigue strength* S or completely reversed stress versus the number of cycles to failure or *fatigue life* N with a semilogarithmic scale, that is, S against $\log N$. Sometimes data are represented by plotting S versus

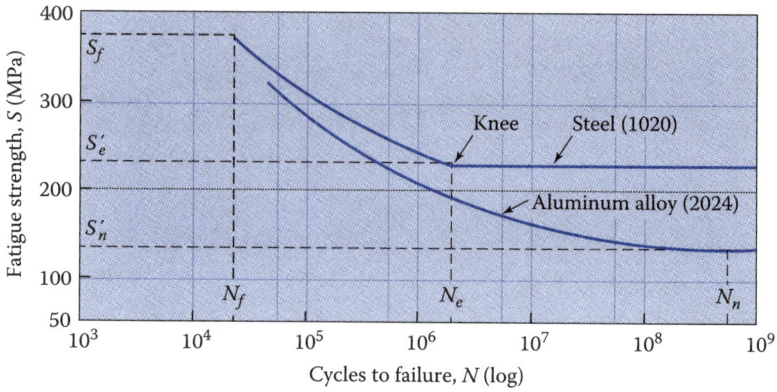

FIGURE 7.5
Fully reversed rotating-beam S–N curves for two typical materials.

N or log S versus log N. Inasmuch as fatigue failures originate at local points of relative weakness, usually the data contain a large amount of scatter. In any case, an average curve, tending to conform to certain generalized pattern, is drawn to represent the test results.

Figure 7.5 shows two typical S–N diagrams corresponding to rotating-beam tests on a series of identical round steel and aluminum specimens subjected to reversed flexural loads of different magnitude. As may be seen from the figure, when the applied maximum stress is high, a relatively small number of cycles cause fracture. Note that, most often, fatigue data represent the mean values based on a 50% survival rate (50% reliability) of specimens.

7.4.1 Endurance Limit and Fatigue Strength

The endurance limit and fatigue strength are two important cyclic properties of the materials. The *fatigue strength* (S'_n), sometimes also termed *endurance strength*, is the completely reversed stress under which a material fails after a specified number of cycles. Therefore, when a value for the fatigue strength of a material is stated, it must be accompanied by the number of stress cycles. The *endurance limit* (S'_e) or fatigue limit is usually defined as the maximum completely reversed stress a material can withstand *indefinitely* without fracture. The endurance limit is therefore stated with no associated number of cycles to failure.

7.4.1.1 Bending Fatigue Strength

For *ferrous materials*, such as steels, the stress where the curve levels off is the endurance limit S'_e (Figure 7.5). Note that the curve for steel displays a decided break or *knee* occurring before or near 1×10^6 cycles. This value is often used as the basis of the endurance limit for steel. Beyond the point $\left(S'_e N_e\right)$, failure does not occur, even for an infinitely large number of loading cycles. At $N = N_f$ cycles, rupture occurs at approximately static fracture stress $S_f \approx 0.9S_u$, where S_u is the ultimate strength in tension.

On the other hand, for *nonferrous metals*, notably aluminum alloys, the typical S–N curve indicates that the stress at failure continues to decrease as the number of cycles increases (Figure 7.5). That is, nonferrous materials do *not* show a break in their S–N curves, and as a result, a distinct endurance limit cannot truly be specified. For such materials, the stress corresponding to some arbitrary number of $5(10^8)$ cycles is commonly assigned as the endurance strength, S'_n.

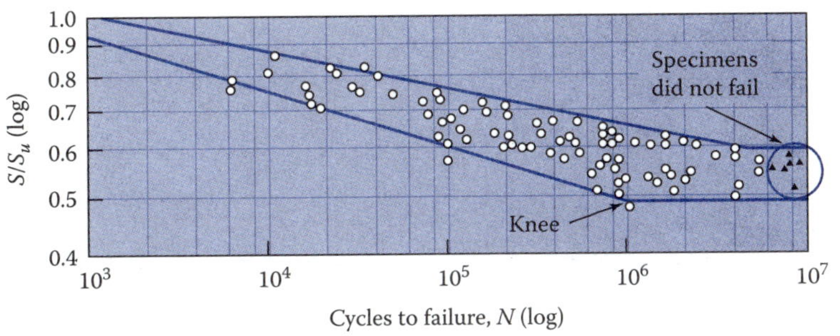

FIGURE 7.6
Fully reversed rotating-beam S–N curve for wrought steels of $S_s < 200$ ksi with superimposed data points. (From Juvinall, R.C., *Engineering Consideration of Stress, Strain, and Strength*, New York, McGraw-Hill, 1967.)

It is necessary to make the assumption that most ferrous materials must not be stressed above the endurance limit S'_e if about 10^6 or more cycles to failure is required. This is illustrated in Figure 7.6, presenting test results for wrought steels having ultimate strength $S_u < 200$ ksi. Note the large scatter in fatigue life N corresponding to a given stress level and the small scatter in fatigue stress corresponding to a prescribed life. The preceding is typical of fatigue strength tests. The figure also depicts that samples run at higher reversed stress levels break after fewer cycles, and some (labeled in the dotted circle) do not fail at all prior to their tests being stopped (here at 10^7 cycles). The data are bracketed by solid lines. Interestingly, at the lower bound of the scatter band, the endurance limit can be conservatively estimated as $0.5S_u$ for design purposes. We mention that, for most wrought steels, the endurance limit varies between 0.45 and 0.60 of the ultimate strength.

7.4.1.2 Axial Fatigue Strength

Various types of fatigue servohydraulic testing machines have been developed for applying fluctuating axial compression. A specimen similar to that used in static tensile tests (see Section 2.3) is used. The most common types apply an axial reversed sinusoidal stress as shown in Figure 7.4c. A comparison of the strengths obtained for uniaxial fatigue stresses and bending fatigue stresses indicate that, in some cases, the former strengths are about 10%–30% lower than the latter strengths for the same material [8,9]. Data for completely reversed axial loading test on AISI steel ($S_u = 125$ ksi) are shown in Figure 7.7. Observe the slope at around 10^3 cycles and the change to basically no slope at about 10^6 cycles corresponding to the endurance limit S'_e.

7.4.1.3 Torsional Fatigue Strength

A limited number of investigations have been made to determine the torsional fatigue strengths of materials using circular or cylindrical specimens subjected to complete stress reversal. For ductile metals and alloys, it was found that the torsional fatigue strength (or torsional endurance limit) for complete stress reversal is about equal to 0.577 times the fatigue strength (or endurance limit) for complete bending stress reversal [12].

For brittle materials, the ratio of the fatigue strength in reversed bending to reversed torsion is higher and may approach the value of 1. The failure points for reversed bending and reversed torsion in biaxial-stress tests are similar to that for static loading failure (e.g., Figure 6.10). We conclude therefore that the relationship between torsional strength and bending strength in cyclic loading is the same as in the static loading case.

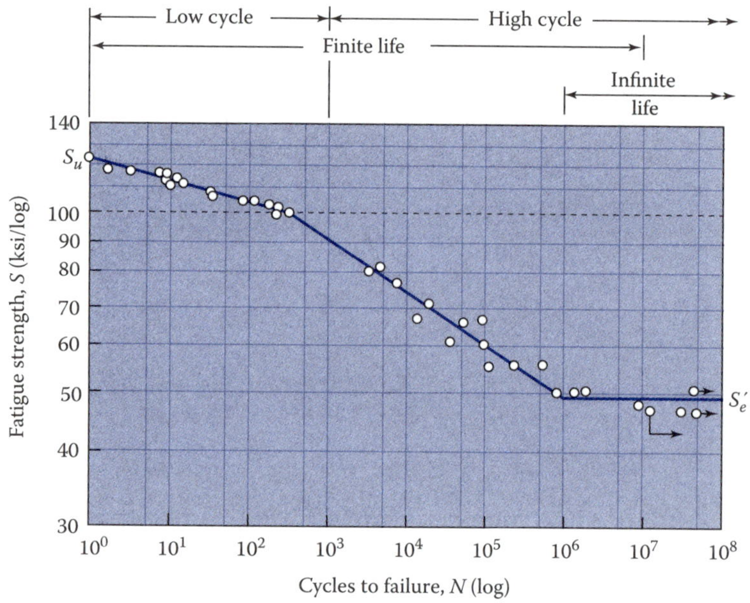

FIGURE 7.7
Fully reversed axial *S–N* curve for AISI 4310 steel, showing breaks at about the low-cycle/high-cycle transition and an endurance limit. (From Budynas, R. and Nisbett, K., *Shigley's Mechanical Engineering Design*, 9th ed., McGraw-Hill, New York, 2010.)

7.4.2 Fatigue Regimes

The stress–life (*S–N*) diagrams indicate different types of behavior as the cycles to failure increase. Two essential regimes are the low-cycle fatigue ($1 < N < 10^3$) and the high-cycle fatigue ($10^3 \leq N$). Note that there is no sharp dividing line between the two regions. In this text, we assume high-cycle fatigue starting at around $N = 10^3$ cycles. The infinite life begins at about 10^6 cycles, where failures may occur with only negligibly small plastic strains. The finite life portion of the curve is below about 10^7 cycles. The boundary between the infinite life and finite life lies somewhere between 10^6 and 10^7 cycles for steels, as shown in Figure 7.7. For low-cycle fatigue, the stresses are high enough to cause local yielding.

A fracture mechanics approach is applied to finite life problems in Section 7.14. We use the stress–life data in treating the high-cycle fatigue of components under any type of loading. For further details, see [5–7], which also provide discussion of the strain-life approach to fatigue analysis.

7.5 Estimating the Endurance Limit and Fatigue Strength

Many criteria have been suggested for interpreting fatigue data. No correlation exists between the endurance limit and such mechanical properties as yield strength and ductility. However, experiments show that the endurance limit, endurance strength, endurance limit in shear, and ultimate strength in shear can be related to the ultimate strength in tension.

Experimental values of ultimate strengths in tension S_u and shear should be used if they are available (see Appendix B). Recall from Section 2.10 that S_u can be estimated from a nondestructive hardness test. For reference purposes, Table 7.1 presents the relationships among the preceding quantities for a number of commonly encountered loadings and materials [9,11]. Steel product manufacturers customarily present stress data of this kind in terms of the ultimate tensile strength, because S_u is the easiest to obtain and most reliable experimental measure of part strength.

If necessary, the ultimate strength of a steel can be estimated by Equation 2.22 as $S_u = 3500H_B$ in MPa or $S_u = 500H_B$ in ksi. Here H_B denotes the Brinell hardness number (Bhn). It should be noted, however, that Equation 7.1 can be relied on only up to Bhn values of about 400. Test data show that the endurance limit S'_e may or may not continue to increase for greater hardness, contingent on the composition of the steel [11].

In the absence of test data, the values given in the table can be used for preliminary design calculations. The relations are based on testing a polished laboratory specimen of a fixed size and geometric shape and on a 50% survival rate. Therefore, these data must be modified by those factors adversely affecting results determined under laboratory conditions, discussed in the next section.

TABLE 7.1

Approximate Fatigue Strength of the Specimens for Fully Reversed Loads

Reversed bending			
Steels	$S'_e = 0.5S_u$	$[S_u < 1400 \text{ MPa } (200 \text{ ksi})]$	(7.1)
	$S'_e = 700 \text{ MPa } (100 \text{ ksi})$	$[S_u \geq 1400 \text{ MPa } (200 \text{ ksi})]$	
Irons	$S'_e = 0.4S_u$	$[S_u < 400 \text{ Mpa } (60 \text{ ksi})]$	(7.2a)
	$S'_e = 160 \text{ MPa } (24 \text{ ksi})$	$[S_u \geq 400 \text{ MPa } (60 \text{ ksi})]$	
Aluminums	$S'_n = 0.4S_u$	$[S_u < 330 \text{ MPa } (48 \text{ ksi})]$	(7.2b)
	$S'_n = 130 \text{ MPa } (19 \text{ ksi})$	$[S_u \geq 330 \text{ MPa } (48 \text{ ksi})]$	
Copper alloys	$S'_n = 0.4S_u$	$[S_u < 280 \text{ MPa } (40 \text{ ksi})]$	(7.2c)
	$S'_n = 100 \text{ MPa } (14 \text{ ksi})$	$[S_u \geq 280 \text{ MPa } (40 \text{ ksi})]$	
Axial loading			
Steels	$S'_e = 0.45S_u$		(7.3)
Torsional loading			
Steels	$S'_{es} = 0.29S_u$		
Irons	$S'_{es} = 0.32S_u$		(7.4)
Copper alloys	$S'_{es} = 0.22S_u$		
Also			
Steels	$S_{us} = 0.67S_u$		(7.5a)
	$S_{ys} = 0.577S_y$		(7.5b)

Notes: S'_e, endurance limit; S'_{es}, endurance limit in shear; S_u, ultimate tensile strength; S_y, yield strength in tension; S_{ys}, yield strength in shear; S_{us}, ultimate strength in shear; S'_n, endurance strength.

7.6 Modified Endurance Limit

The specimen used in the laboratory to determine the endurance limit is prepared very carefully and tested under closely controlled conditions. However, it is unrealistic to expect the endurance limit of a machine or structural member to match the values obtained in the laboratory. Material, manufacturing, environmental, and design conditions influence fatigue. Typical effects include the size, shape, and composition of the material; heat treatment and mechanical treatment; stress concentration; residual stresses, corrosion, and temperature; speed and type of stress; and life of the member [3].

To account for the most important of these effects, various endurance limit modifying factors are used. These empirical factors, when applied to steel parts, lead to results of good accuracy, because most of the data on which they are based are obtained from testing steel specimens. The corrected or modified endurance limit, also referred to as the *endurance limit*, representing the endurance limit of the mechanical element, is defined as follows:

$$S_e = C_f C_r C_s C_t \left(1/K_f\right) S'_e \tag{7.6}$$

where

S_e = the modified endurance limit
S'_e = the endurance limit of the test specimen
C_f = the surface finish factor
C_r = the reliability factor
C_s = the size factor
C_t = the temperature factor
K_f = the fatigue stress-concentration factor

This working equation for the endurance limit is extremely important in fatigue problems. It should be used when actual fatigue test data that pertain closely to the particular application are not available. Equation 7.6 can be applied with great confidence to steel components, since the data on which correction factors rely usually come from testing steel specimens.

Recall from Section 7.4 that nonferrous materials show no break in their *S*–*N* curves: hence, a definite endurance limit of test specimen cannot be specified. For these materials, the fatigue strength S'_n replaces S'_e in Equation 7.6. Likewise, for the case of reversed torsion loading, the modified endurance limit in shear S_{es} and endurance limit in shearing test specimen S'_{es} supersede S_e and S'_e, respectively, in the equation.

7.7 Endurance Limit Reduction Factors

The endurance limit modifying or reduction factors must be used in design application with great care, since the available information is related to specific specimens and tests. Only limited data are available for material strength in severe environments. Manufacturing processes can have significant effects on fatigue life characteristics. Most of the miscellaneous factors affecting the endurance limit, such as heat treatment, corrosion, mechanical surface treatment, metal spraying, residual stresses, and welding, have no quantitative values. Random variations occur in these factors, which are experimentally determined. The values assigned depend on the designer's experience and judgment.

Corrosion fatigue is a complex action, not yet entirely understood, but some experiments show its severity. Corrosion from water and acids may reduce the endurance limit to a very low value. The small pits that form on the surface as stress raisers. Based on empirical data, for carbon and low-carbon steels in freshwater, Equation 7.1 should be changed to

$$S_e' = 100 \text{ MPa} \quad (15 \text{ ksi}) \tag{a}$$

The only chromium and stainless steels retain considerable strength in water, since that alloying elements provide some corrosion protection. Most other material operating environments also have lowered fatigue strength.

The endurance limit modifying factors are nearly 1.0 for bending loads under 10^3 cycles. They increase progressively in some manner with the increase in the number of cycles. In the following brief discussion, some representative or approximate values for reduction factors are presented. These values are abstracted from [9,10,12].

7.7.1 Surface Finish Factor

Fatigue strength is sensitive to the condition of the surface, because the maximum stresses occur here in bending and torsion. As already noted, the rotating-beam specimen is polished to a mirror finish to preclude surface imperfections serving as stress raisers; rougher finishes lower the fatigue strength.

The surface finish factor C_f, which depends on the quality of the finish and tensile strength, may be expressed in the form

$$C_f = AS_u^b \tag{7.7}$$

where the ultimate strength S_u is in either MPa or ksi. Table 7.2 presents the values of the factor A and exponent b for a variety of finishes applied to steels. We observe from Equation 7.7 that (since $b < 0$) the values of C_f decrease with increases in tensile strength S_u. The surface factor for *mirror-polish* finish steels equals approximately 1, $C_f = 1$.

Equation 7.7 has the advantage of being computer programmable and eliminating the need to refer to charts such as Figure 7.8. Note that the surface conditions in this figure are poorly defined (e.g., the machined surface texture or degree of roughness). Interestingly, Figure 7.8 shows that corrosive environments drastically reduce the endurance limit.

TABLE 7.2
Surface Finish Factors C_f

	A		
Surface Finish	**MPa**	**ksi**	***b***
Ground	1.58	1.34	−0.085
Machined or cold drawn	4.51	2.7	−0.265
Hot rolled	57.7	14.4	−0.718
Forged	272.0	39.9	−0.995

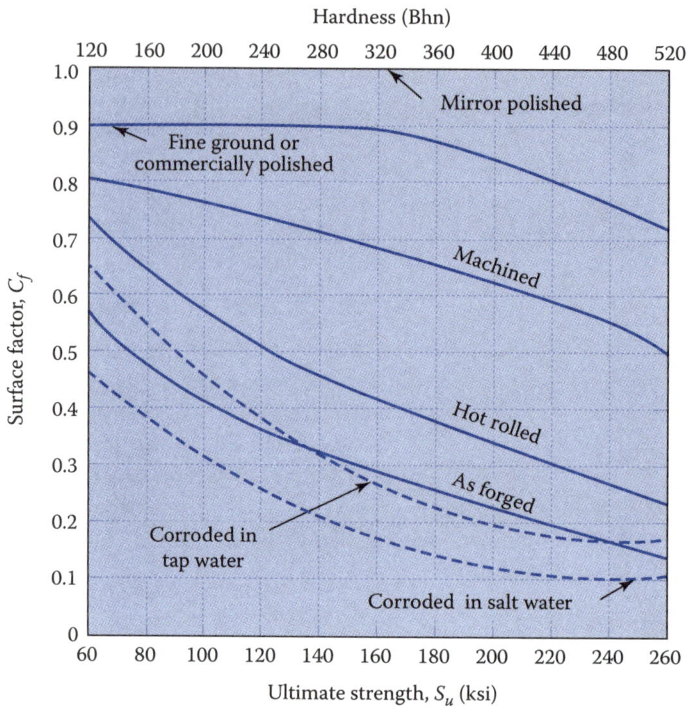

FIGURE 7.8
Surface factors for various finishes on steel. (From Juvinall, R.C., *Engineering Consideration of Stress, Strain, and Strength*, McGraw-Hill, New York, 1967.)

Table 7.2 may also be applied to aluminum alloys and other ductile metals with caution. It is important to mention that testing of actual parts under service loading conditions must be done in critical applications. The surface factor for ordinary cast irons is also taken as approximately 1, $C_f = 1$, since their internal discontinuities dwarf the effects of a rough surface.

7.7.2 Reliability Factor

The factor of reliability C_r accounts for material variation in fatigue data and depends on survival rate. It is defined by the following commonly used formula:

$$C_r = 1 - 0.08z \tag{7.8}$$

The quantity z is the number of standard deviations, discussed in Section 6.14.

For a required survival rate or percent reliability, Figure 6.16 gives the corresponding z, and then using Equation 7.8, we calculate a reliability factor. Table 7.3 presents a number of values of the C_r. Observe that a 50% reliability has a factor of 1 and the factor reduces with increasing survival rate.

7.7.3 Size Factor

The influence of size on fatigue strength can be a significant factor. The endurance limit decreases with increasing member size. This is owing to the probability that a larger part is more likely to have a weaker metallurgical defect at which a fatigue crack will start. There is notable scatter in reported values of the size factor C_s. Various researchers have suggested different formulas for estimating it.

TABLE 7.3
Reliability Factors

Survival Rate (%)	C_r
50	1.00
90	0.89
95	0.87
98	0.84
99	0.81
99.9	0.75
99.99	0.70

The approximate results for *bending* and *torsion* of a part of diameter D may be stated as follows:

$$C_s = \begin{cases} 0.85 & (13 \text{ mm} < D \le 50 \text{ mm}) \quad \left(\frac{1}{2} < D \le 2 \text{ in.}\right) \\ 0.70 & (D > 50 \text{ mm}) \quad (D > 2 \text{ in.}) \end{cases} \tag{7.9}$$

This applies to cylindrical parts, and for members of other shapes, few consistent data are available. For a rotating part of a rectangular cross section of width b and depth h, use the following equation [10]:

$$D - 0.8(bh)^{1/2} \tag{7.10}$$

Prudent design would suggest employing a factor $C_s = 0.7$, lacking other information. Note that, for *axial loading*, there is no size effect: $C_s = 1$.

7.7.4 Temperature Factor

Temperature effects vary with the material in most cases, and values of ultimate strength should be modified before determining the endurance limit S'_e in Equation 7.6. Alternatively, for steels, a temperature factor C_t can be approximated at moderately high temperatures by the formula [9]

$$C_t = \begin{cases} 1 \quad T \le 450°\text{C} \quad (840°\text{F}) \\ 1 - 0.0058(T - 450) \quad 450°\text{C} < T \le 550°\text{C} \\ 1 - 0.0032(T - 840) \quad 840°\text{F} < T \le 1020°\text{F} \end{cases} \tag{7.11}$$

A more accurate estimation of C_t is presented in [10,11]. Unless otherwise specified, we assume throughout the text the operating temperature is normal or room temperature and take $C_t = 1$.

7.7.5 Fatigue Stress-Concentration Factor

As pointed out earlier, the stress concentration is a very significant factor in failure by fatigue. For dynamic loading, the theoretical stress-concentration factor K_t (see Section 3.13) needs to be modified on the basis of the notch sensitivity of the material. *Notch* is a generic term in this context and can be a hole, a groove, or a fillet. The fatigue stress-concentration factor may be defined as

$$K_f = \frac{\text{Endurance limit of notch-free specimen}}{\text{Endurance limit of notched specimen}} \tag{7.12}$$

The tests show that K_f is often equal to or less than the K_t, owing to internal irregularities in the material structure. Therefore, even unnotched samples may suffer from these internal notches. An extreme case in point is gray cast iron.

The foregoing situation is dealt with by using a notch factor. The two stress-concentration factors are related by the ratio of the *notch sensitivity* q:

$$q = \frac{K_f - 1}{K_t - 1} \tag{7.13a}$$

This expression can be written in the form

$$K_f = 1 + q(K_t - 1) \tag{7.13b}$$

We observe from Equation 7.13 that q varies between 0 (giving $K_f = 1$) and 1 (giving $K_f = K_t$). Generally, the more ductile the material response, the less notch sensitive it is. Materials showing brittle behavior are more notch sensitive. Obviously, notch sensitivity also depends on the notch radius. Contrary to K_t, as notch radii approach 0, the q decreases and approaches 0.

Figure 7.9a and b provides approximate data for steels and 2024 aluminum alloys subjected to reversed bending, reversed axial loads, and reversed torsion [5,12]. Note that the actual test data from which the curves were plotted exhibit a large amount of scatter. So, it is always safe to use $K_f = K_t$ when there is doubt about the true value of q. The curves show that q is not far from unity for large notch radii. For larger notch radii, use the values of q corresponding to $r = 0.16$ in. (4 mm). In concluding this discussion, we note that the notch sensitivity of a cast iron is very low, $0 < q \leq 0.20$, depending on tensile strength. If one is uncertain, it would be conservative to use a value $q = 0.20$ for all grades of cast iron.

Example 7.1: Endurance Limit of a Torsion Bar

A round torsion bar machined from steel is under reversed torsional loading. Because of the design of the ends, a fatigue stress-concentration factor K_f exists. Estimate the modified endurance limit.

Given: The diameter of the bar is d = 1⅝ in. and $K_f = 1.2$. The operating temperature is 500°C maximum.

Assumption: Reliability is 98%.

Design Decision: The bar is made of AISI 1050 cold-drawn steel.

Solution

From Table B.3, we find the ultimate strength in tension as $S_u = 100$ ksi. Then, applying Equation 7.4, the endurance limit of the test specimen is

$$S'_{es} = 0.29S_u = 0.29(100) = 29 \text{ ksi}$$

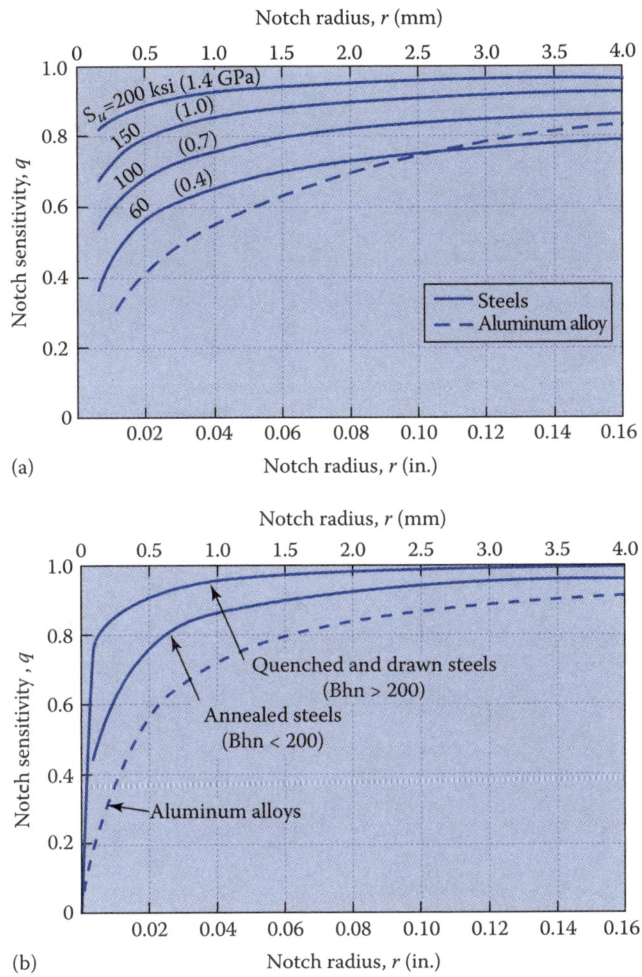

FIGURE 7.9
Fatigue notch sensitivity curves for (a) bending and axial loads and (b) torsion. (From Sines, G. and Waisman, J.L., eds., Metal Fatigue, McGraw-Hill, New York, 1959, pp. 296–298.)

By Equation 7.7 and Table 7.2, the surface finish factor is

$$C_f = AS_u^b = 2.7(100)^{-0.265} = 0.80$$

The reliability factor corresponding to 98% is $C_r = 0.84$ (Table 7.3). Using Equation 7.9, the size factor $C_s = 0.85$. Applying Equation 7.11,

$$C_t = 1 - 0.0058(500 - 450) = 0.71$$

Hence, the endurance limit for design is found to be

$$S_{es} = C_f C_r C_s C_t (1/K_f) S'_{es} \tag{b}$$

$$= (0.80)(0.84)(0.85)(0.71)(1/1.2)(29)$$

$$= 9.8 \text{ ksi}$$

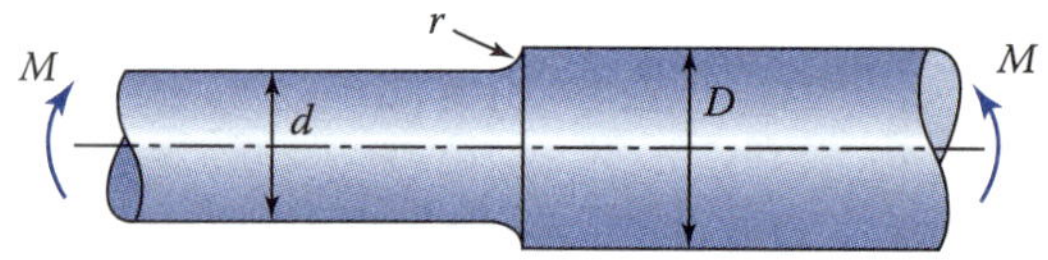

FIGURE 7.10
Example 7.2.

Example 7.2: Endurance Limit for a Stepped Shaft in Reversed Bending

Rework Example 7.1 for the condition that the critical point on the shaft is at a diameter change from d to D with a full fillet where there is reversed bending and no torsion, as shown in Figure 7.10.

Given: $d = 1\frac{5}{8}$ in., $D = 1\frac{7}{8}$ in., $S_u = 100$ ks.

Solution

We now have, by Equation 7.1, $S_e' = 0.5(100) = 50$ ksi. From the given dimensions, the full fillet radius is $r = (1\frac{7}{8} - 1\frac{5}{8})/2 = 0.125$ in. Therefore,

$$\frac{r}{d} = \frac{0.125}{1.625} = 0.08, \quad \frac{D}{d} = \frac{1.875}{1.625} = 1.15$$

Referring to Figure C.9, $K_t = 1.7$. For $r = 0.125$ in. and $S_u = 100$ ksi, by Figure 7.9a, $q = 0.82$. Hence, through the use of Equation 7.13b,

$$\begin{aligned} K_f &= 1 + q(K_t - 1) \\ &= 1 + 0.82(1.7 - 1) = 1.57 \end{aligned}$$

The endurance limit, given by Equation (b) of Example 7.1, becomes

$$\begin{aligned} S_e &= C_f C_r C_s C_t (1/K_f) S_e' \\ &= (0.80)(0.84)(0.85)(0.71)(1/1.57)(50) \\ &= 12.92 \text{ ksi} \end{aligned}$$

7.8 Fluctuating Stresses

Any loads varying with time can actually cause fatigue failure. The type of these loads may vary amply from one application to another. Hence, it is necessary to determine the fatigue resistance of parts corresponding to stress situations other than complete reversals discussed so far.

A common fluctuating stress pattern consists of an alternating (usually sinusoidal) stress superimposed on a uniform mean stress (Figure 7.11a). This loading is typical for an engine valve spring that is preloaded at installation and then further compressed when the valve is opened. The case in which the mean stress is 0 is called *fully* or *completely reversed stress* (Figure 7.11b), discussed in Section 7.3. Figure 7.11c shows repeated stress, where the minimum value equals 0. Figure 7.11d shows pulsating stress, varying between

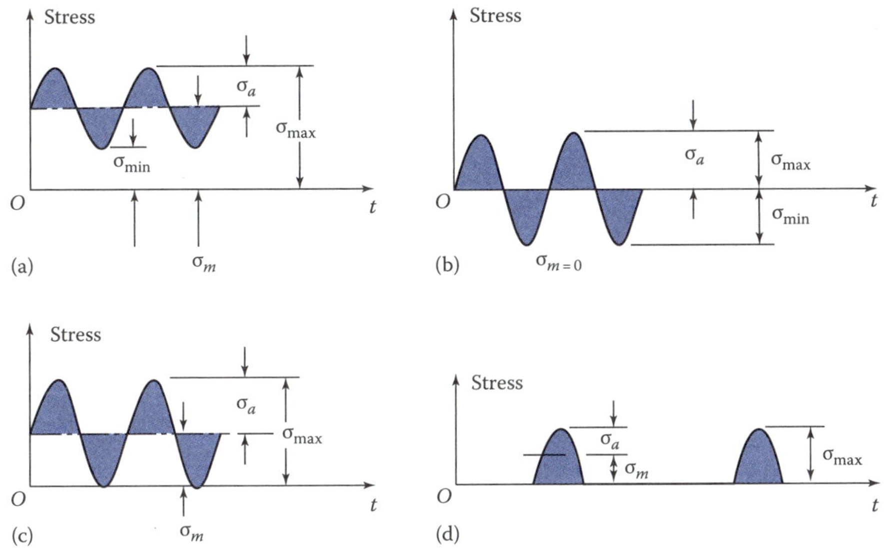

FIGURE 7.11
Some cyclic stress–time relations: (a) fluctuating, (b) completely reversed, (c) repeated, and (d) pulsating.

0 and the maximum value with each application of load, as on the teeth of gears. Note that the shape of the wave of the stress–time relation has no important effect on the fatigue failure, so usually the relation is schematically depicted as a sinusoidal or sawtooth wave.

Irrespective of the form of the stress–time relation, the stress varies from a maximum stress σ_{max} to a minimum stress σ_{min}. Therefore, the definitions of *mean stress* and range or *alternating stress* are

$$\sigma_m = \frac{1}{2}(\sigma_{max} + \sigma_{min})$$
$$\sigma_a = \frac{1}{2}(\sigma_{max} - \sigma_{min}) \tag{7.14}$$

Clearly, these components of the fluctuating stress are also independent of the shape of the stress–time curve. Two ratios can be formed:

$$R = \frac{\sigma_{min}}{\sigma_{max}}, \quad A = \frac{\sigma_a}{\sigma_m}$$

Here
- R is the stress ratio
- A represents the amplitude ratio

When the stress is fully reversed ($\sigma_m = 0$), we have $R = 1$ and $A = \infty$.

The mean stress is analogous to a static stress, which may have any value between σ_{max} and σ_{min}. We see that the presence of a mean stress component can have a significant effect on the fatigue life. The alternating stress represents the amplitude of the fluctuating stress. We have occasion to apply the subscripts (*a* and *m*) of these components to shear stresses as well as normal stresses.

7.9 Theories of Fatigue Failure

To predict whether the state of stress at a critical point in an element would result in failure, a fatigue criterion based on the mean and alternating stresses is used. Such a theory utilizes static and cyclic material characteristics. Table 7.4 presents frequently employed *fatigue failure theories* or *criteria,* also called the mean stress-alternating stress relations [13,14]. Equations as written apply only to materials with an endurance limit; either S_e, or S_e' can be used in these relationships. For a finite life (a given number of cycles), the corresponding fatigue strength S_n may be substituted for the endurance limit.

The criteria in the table together with given material properties form the basis for practical fatigue calculations for members subjected to a simple fluctuating loading. Note that the modified Goodman criterion is algebraically more involved, as there are two inequalities to check rather than one in the other relations. In the case of combined fluctuating loading, the static failure theories are modified according to a mean stress-alternating stress relation listed in Table 7.4, as will be shown in Section 7.12.

Figure 7.12 shows the foregoing relationships, plotted on mean stress (σ_m) versus alternating stress (σ_a) axes. *A fatigue failure diagram* of this type is usually constructed for analysis and design purposes; it is easy to use, and the results can be scaled off directly. For each criterion, points on or inside the respective line guard against failure. The mean stress axes of the diagrams have the fracture strength S_f, ultimate strength S_u, and yield strength S_y. Clearly, the yield strength plotted on the ordinate as well as endurance limit in Figure 7.12b indicates that yielding rather than fatigue might be the criterion of failure. The yield line connecting S_y on both axes shown in the figure serves as a limit on the first cycle of stress.

TABLE 7.4

Failure Criteria for Fatigue

Fracture Theory	**Goodman**	**Gerber**	**SAE**
Equation	$\frac{\sigma_a}{S_e}+\frac{\sigma_m}{S_u}=1$	$\frac{\sigma_a}{S_e}+\left(\frac{\sigma_m}{S_u}\right)^2=1$	$\frac{\sigma_a}{S_e}+\frac{\sigma_m}{S_f}=1$
Yield Theory	**Soderberg**	**Modified Goodman**	
Equation	$\frac{\sigma_a}{S_e}+\frac{\sigma_m}{S_y}=1$	$\frac{\sigma_a}{S_e}+\frac{\sigma_m}{S_u}=1$	$\left(\text{for } \frac{\sigma_a}{\sigma_m}\geq\beta\right)$
		$\frac{\sigma_a+\sigma_m}{S_y}=1$	$\left(\text{for } \frac{\sigma_a}{\sigma_m}\leq\beta\right)$

Notes: a, alternating; *y*, static tensile yield; *m*, mean; *u*, static tensile ultimate; *e*, modified endurance limit; *f*, fracture. Material constant, $\beta = S_e(S_u - S_y)/S_u(S_y - S_e)$.

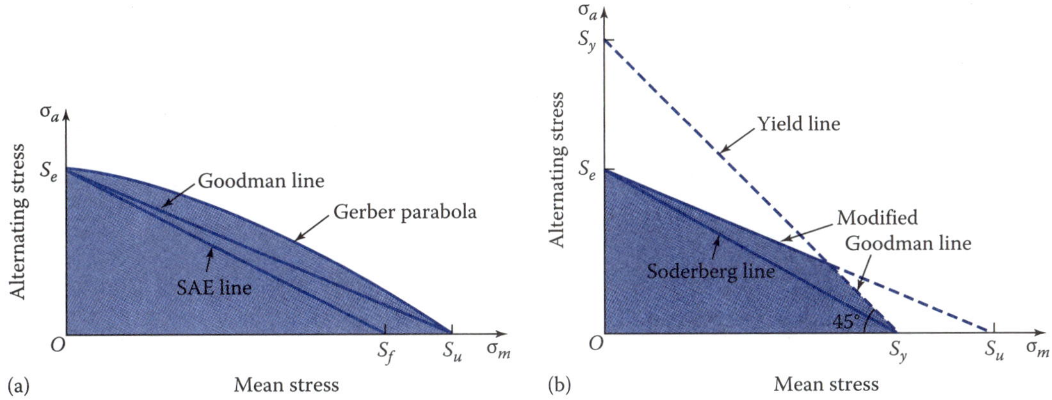

FIGURE 7.12
Fatigue diagrams showing various theories of failure: (a) fracture criteria and (b) yield criteria.

7.10 Comparison of the Fatigue Criteria

A comparison of the failure theories for fatigue may be made referring to Figure 7.12. We see from Figure 7.12a that the Gerber criterion leads to least conservative results for fracture. The Gerber (parabolic) line is a good fit to experimental data, making it useful for the analysis of failed parts. The Goodman criterion is more conservative than the SAE criterion. For hard steels, both theories give identical solutions, since for brittle materials $S_u = S_f$.

Figure 7.12b shows that the modified Goodman criterion resembles the Soderberg criterion, except that the former is slightly less conservative. Two line segments form the modified Goodman failure line, as shown in the figure. The 45° line segment implies failure when the maximum mean stress exceeds the yield strength. The modified Goodman line is particularly better for highly localized yielding occurring in many machine parts. Note that the Soderberg theory eliminates the need to involve the yield line.

Recall from Section 7.1 that fatigue failures appear brittle, even if the material shows some ductility in a static tension test, as high stresses and yielding are localized near the crack. Thus, the *Goodman criterion*, which gives reasonably good results for brittle materials while conservative values for ductile materials, is a realistic scheme for *most materials*. For most metals, the Soderberg relation also leads to conservative estimates. Both theories are in widespread use for mild steel. In this text, the Goodman criterion is used to derive readily the basic equations for the design and analysis of common components. The easy and quick graphical approach is applied for the modified Goodman criterion. The Soderberg line is employed less often.

Example 7.3: Allowable Fully Reversed Load of an Actuating Rod

A round rod is subjected to an axial tensile force F_m. Calculate the limiting value of the completely reversed load F_a that can be applied with a ½ in. eccentricity without causing fatigue failure at 10^6 cycles.

Given: The diameter of the rod $D = 2$ in. and $F_m = 40$ kips.

Assumptions: The rod is made of ASTM-A36 steel having $S_y = 36$ ksi, $S_u = 58$ ksi (Table B.1), and $S_n = 30$ ksi at 10^6 cycles.

Design Decisions: Apply the Soderberg and Goodman criteria.

Solution

The mean and maximum alternating stresses are

$$\sigma_m = \frac{F_m}{A} = \frac{40}{\pi(1)^2} = 12.73 \text{ ksi}$$

$$\sigma_a = \frac{M_a c}{I} = \frac{0.5F_a(1)}{\pi(1)^4/4} = 0.637F_a$$

Soderberg criterion. Substituting these and the given numerical values into Table 7.4,

$$\frac{0.637F_a}{30\times 10^3} + \frac{12.73\times 10^3}{36\times 10^3} = 1 \quad \text{or} \quad F_a = 30.4 \text{ kips}$$

Goodman criterion. Similar to the preceding, we now have

$$\frac{0.637F_a}{30\times 10^3} + \frac{12.73\times 10^3}{58\times 10^3} = 1 \quad \text{or} \quad F_a = 36.8 \text{ kips}$$

Comments: According to the Goodman theory, the eccentric load that can be carried by the rod is thus about 17% larger than on the basis of the Soderberg theory.

7.11 Design for Simple Fluctuating Loads

When tensile stress at a point occurs by an alternating stress σ_a and a mean stress σ_m as shown in Figure 7.11a, both these components contribute to failure. The failure line (Figure 7.12) is an approximate depiction of this effect. Usually, a fatigue theory of failure is not applied to problems where mean stress is negative.

As noted previously, the *Goodman criterion* may be used safely with almost any material for which the endurance limit S_e and ultimate strength S_u are known. For design purposes, these quantities are replaced by S_e/n and S_u/n, respectively, where n represents the factor of safety. In so doing, the Goodman criterion, given in Table 7.4, becomes

$$\frac{\sigma_a}{S_e} + \frac{\sigma_m}{S_u} = \frac{1}{n} \tag{7.15}$$

where σ_a and σ_m are defined by Equation 7.14. We note that, if a stress concentration exists at the cross section for which the stresses are computed, for ductile materials, it is commonly neglected as far as the mean stress is concerned (see Section 3.14). However, stress concentration must be taken into account for calculating the modified endurance limit S_e from Equation 7.6.

The Goodman criterion, Equation 7.15, may be rearranged in the following convenient form:

$$\frac{S_u}{n} = \sigma_m + \frac{S_u}{S_e}\sigma_a \tag{7.16}$$

The right-hand side of this equation can then be considered the static equivalent of the fluctuating state of stress. Hence, we define the *equivalent normal stress* as

$$\sigma_e = \sigma_m + \frac{S_u}{S_e}\sigma_a \tag{7.17}$$

Although Equation 7.17 refers to normal stress, the development could have been made equally well for shear stress by replacing σ by τ [8]. The equation for *equivalent shear stress* is then

$$\tau_e = \tau_m + \frac{S_u}{S_e}\tau_a \tag{7.18}$$

In this expression, it is assumed that

$$\frac{S_{us}}{S_{es}} \approx \frac{S_u}{S_e} \tag{7.19}$$

because data for the ultimate strength in shear S_{us} and modified endurance limit in shear S_{es} are ordinarily not available. However, recall from Sections 7.4 through 7.6 that there are methods for estimating these quantities.

In some situations, a member is to withstand a given ratio of the alternating load to mean load. Then, a solution may readily be obtained if the ratio of alternating stress to mean stress can be determined from some known stress–load relationship. In such cases, it is suitable to recast Equation 7.16 into the form

$$\sigma_m = \frac{S_u/n}{\dfrac{\sigma_a}{\sigma_m}\dfrac{S_u}{S_e} + 1} \tag{7.20}$$

Once mean stress σ_m is obtained, the stress–load relationship is used to determine the required dimension of the element. For shear stress, the foregoing equation may be expressed as

$$\tau_m = \frac{S_{us}/n}{\dfrac{\tau_a}{\tau_m}\dfrac{S_u}{S_e} + 1} \tag{7.21}$$

A reasonable design procedure ensures a significant safety factor against fatigue failure in the material. For a fluctuating stress, by inversion of Equation 7.16, we express the factor of safety n as follows:

$$n = \frac{S_u}{\sigma_m + \dfrac{S_u}{S_e}\sigma_a} \tag{7.22}$$

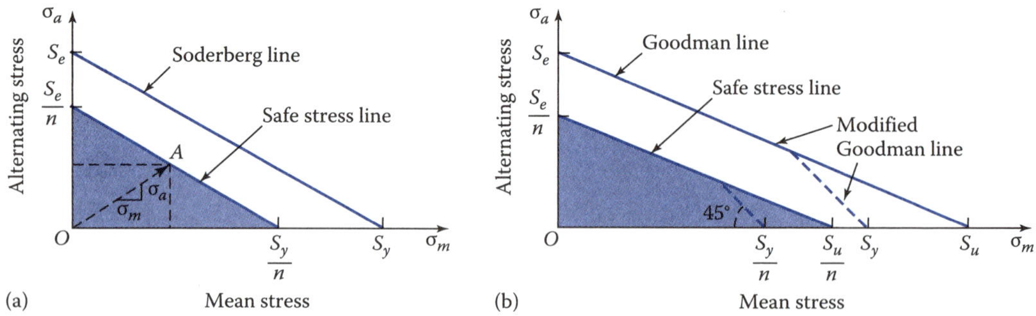

FIGURE 7.13
(a) Soderberg diagram and (b) Goodman diagrams.

The equations for the safety factor become, with simple steady stress,

$$n = \frac{S_u}{\sigma_m} \tag{7.23}$$

and with simple alternating stress,

$$n = \frac{S_e}{\sigma_a} \tag{7.24}$$

It should be pointed out that Equations 7.15 through 7.23 could also be written on the basis of the Soderberg criterion by substituting S_y for S_u.

7.11.1 Design Graphs of Failure Criteria

The graphical representations of the Soderberg, Goodman, and modified Goodman theories are shown in Figure 7.13. Note that other criteria listed in Table 7.4 may be plotted similarly. The Soderberg failure line is drawn between the yield point and the endurance limit on mean stress–range stress coordinates (Figure 7.13a). It is an approximate representation of the fatigue.

The *safe stress* line through any point $A(\sigma_m, \sigma_a)$ is constructed parallel to the Soderberg line. This line is the locus of all sets of σ_m and σ_a stresses having a factor of safety n. Any point on or below the safe line represents safe loading. The Goodman criteria are interpreted in a like manner (Figure 7.13b). The graphical approach permits rapid solution of the mean or range stress and provides an overview of the failure theory. Graphical solutions serve as a check to analytically obtained results.

The examples that follow illustrate the application of the Goodman criteria to the design of members under a simple fluctuating loading.

Example 7.4: Design of a Cylindrical Pressure Vessel for Fluctuating Loading

A thin-walled cylindrical pressure vessel of diameter d is subjected to an internal pressure p varying continuously from p_{min} to p_{max}. Determine the thickness t for an ultimate strength S_u, modified endurance limit S_e, and a safety factor of n.

Given: $D = 1.5$ m, $p_{min} = 0.8$ MPa, $p_{max} = 4$ MPa

$S_y = 300$ MPa, $S_u = 400$ MPa, $S_e = 150$ MPa.

Design Decision: The Goodman theories, based on maximum normal stress and a safety factor of $n = 2$, are used.

Solution

The state of stress on the cylinder wall is considered to be biaxial (see Section 3.4). Maximum principal stress, that is, tangential stress, in the cylinder has the mean and range values

$$\sigma_m = \frac{p_m r}{t}, \quad \sigma_a = \frac{p_a r}{t} \tag{a}$$

where

$$p_m = \frac{1}{2}(p_{max} + p_{min}) = \frac{1}{2}(4 + 0.8) = 2.4 \text{ MPa}$$

$$p_a = \frac{1}{2}(p_{max} - p_{min}) = \frac{1}{2}(4 - 0.8) = 1.6 \text{ MPa}$$

Since stresses are proportional to pressures, we have

$$\frac{\sigma_a}{\sigma_m} = \frac{p_a}{p_m} = \frac{1.6}{2.4} = \frac{2}{3}$$

Substitution of the given data into Equation 7.20 gives 400/2

$$\sigma_m = \frac{400/2}{\frac{2}{3}\frac{400}{150} + 1} = 72 \text{ MPa}$$

Using Equation (a), we have

$$t = \frac{p_m r}{\sigma_m} = \frac{2.4(750)}{72} = 25 \text{ mm}$$

This is the minimum safe thickness for the pressure vessel.

Alternatively, a graphical solution of σ_m by the modified Goodman criterion is obtained by plotting the given data to scale, as shown in Figure 7.14. We observe from the figure that the locus of points representing σ_m and σ_a for any thickness is a line through the origin with a slope of 2/3. Its intersection with the safe stress line gives the state of stress σ_m σ_a for the minimum safe value of the thickness t. The corresponding value of the mean stresses is $\sigma_m = 72$ MPa.

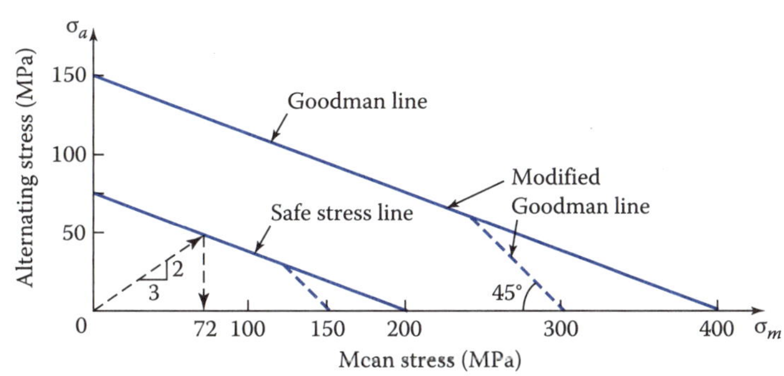

FIGURE 7.14
Example 8.4. Goodman criteria applied to design of pressure vessel.

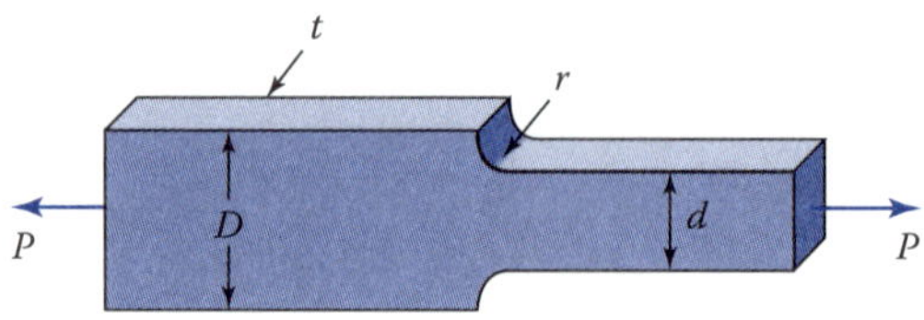

FIGURE 7.15
Example 7.5.

Example 7.5. Safety Factor against Fatigue Failure of a Tensile Link

A tensile link of thickness t with two fillets is subjected to a load fluctuating between P_{min} and P_{max} (Figure 7.15). Calculate the factor of safety n for unlimited life based on the Goodman criterion.

Given: $D = 120$ mm, $d = 80$ mm, $r = 16$ mm, $t = 15$ mm,

$$P_{\text{min}} = 90 \text{ kN}, P_{\text{max}} = 210 \text{ kN}.$$

Design Decisions: The link is made of steel with $S_u = 700$ MPa. The fillets and adjacent surfaces are ground. A reliability of 99.9% is desired.

Solution

The minimum cross-sectional area equals $A = 80 \times 15 = 1200$ mm^2. The maximum mean and range stresses are

$$\sigma_m = \frac{(210+90)(10^3)}{2(1200\times 10^{-6})} = 125 \text{ MPa}$$

$$\sigma_a = \frac{(210-90)(10^3)}{2(1200\times 10^{-6})} = 50 \text{ MPa}$$

Referring to Figure C.1, for $D/d = 1.5$ and $r/d = 0.2$, we obtain $K_t = 1.72$. Inasmuch as the given 16 mm fillet radius is large, we use the value of the notch sensitivity for the steel having $S_u = 700$ MPa, corresponding to $r = 4$ mm in Figure 7.9a; that is, $q = 0.85$. Hence, $K_f = 1 + 0.85(1.72 - 1) = 1.61$.

For axial loading, there is no size factor, $C_s = 1.0$. Corresponding to a round surface finish, from Equation 7.7 and Table 7.2,

$$C_f = A(S_u)^b$$
$$= 1.58(700)^{-0.085} = 0.91$$

By Table 7.3, for 99.9% material reliability, $C_r = 0.75$. The temperature is not elevated, $C_t = 1.0$. The modified endurance limit, using Equations 7.3 and 7.6, is

$$S_e = C_f C_r C_s C_t (1/K_f)(0.45S_u)$$
$$= (0.91)(0.75)(1.0)(1.0)(1/1.61)(0.45\times 700)$$
$$= 133.5 \text{ MPa}$$

The factor of safety, applying Equation 7.22, is therefore

$$n = \frac{700}{125 + \frac{700}{133.5}(50)} = 1.81$$

Comments: If the load is well controlled and there is no impact, this factor guards the link against the fatigue failure.

CASE STUDY 7.1 Camshaft Fatigue Design of Intermittent-Motion Mechanism

Figure 7.16 illustrates a rotating camshaft of an intermittent-motion mechanism in its peak lift position. The cam exerts a force P on the follower, because of a stop mechanism (not shown), only during less than half a shaft revolution. Calculate the factor of safety for the camshaft according to the Goodman criterion.

Given: The geometry is known and the shaft supports a pulsating force with P_{max} and P_{min}. The material of all parts is AISI 1095 steel, carburized on the cam surface and oil quenched and tempered (OQ&T) at 650°C. The fillet and adjacent surfaces are fine ground.

Data:

$$P_{max} = 1.6 \text{ kips}, \quad P_{min} = 0$$

$$S_u = 130 \text{ ksi}, \quad S_y = 80 \text{ ksi} \quad (\text{from Table B.4})$$

$$L_1 = 2.8 \text{ in.}, \quad L_2 = 3.2 \text{ in.},$$

$$L_3 = 0.5 \text{ in.}, \quad L_4 = 1.5 \text{ in.},$$

$$L_5 = L_1 + \frac{1}{2}(L_3 + L_4) = 3.8 \text{ in.},$$

$$L_6 = L_2 + \frac{1}{2}(L_3 + L_4) = 4.2 \text{ in.},$$

$$D_s = 1 \text{ in.}, \quad D_c = 1.6 \text{ in.},$$

$$r_c = 1.5 \text{ in.}, \quad r = 0.1 \text{ in.},$$

$$I/c = \pi D_s^3/32 = 98.175(10^{-3}) \text{ in.}^3$$

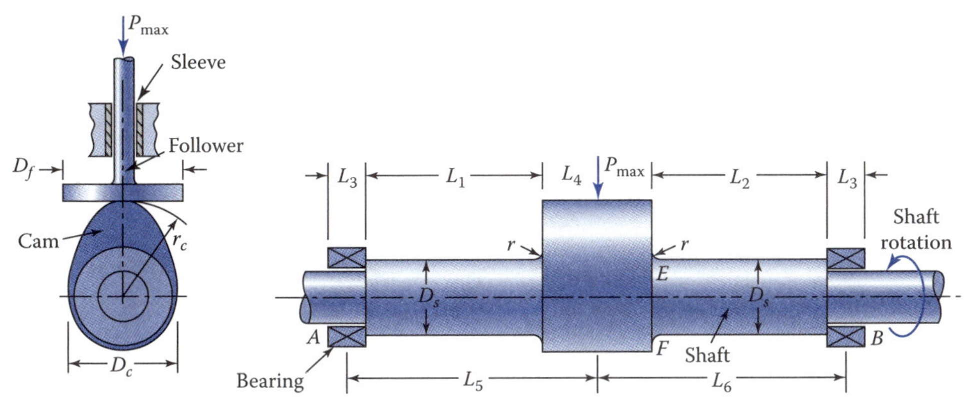

FIGURE 7.16
Layout of camshaft and follower of an intermittent-motion mechanism.

Assumptions:

1. Bearings act as simple supports.
2. The operating temperature is normal.
3. The torque can be regarded negligible.
4. A material reliability of 99.99% is required.

Solution

See Figures 7.16 and 7.17.

Alternating and mean stresses. The reactions at the supports A and B are determined by the conditions of equilibrium as

$$R_A = \frac{L_6}{L_5 + L_6} P_{\max}$$

$$= \frac{4.2}{8}(1600) = 840 \text{ lb}$$

$$R_B = P_{\max} - R_A = 760 \text{ lb}$$

and noted in Figure 7.17a.

The plot of the moment diagram, from a *maximum* moment of 760 × 4.2 = 3192 lb · in., is shown in Figure 7.17b. We observe that the moment on the right side

$$M = R_B\left(L_6 - \frac{1}{2}L_4\right)$$

$$= 760(3.45) = 2622 \text{ lb} \cdot \text{in.}$$

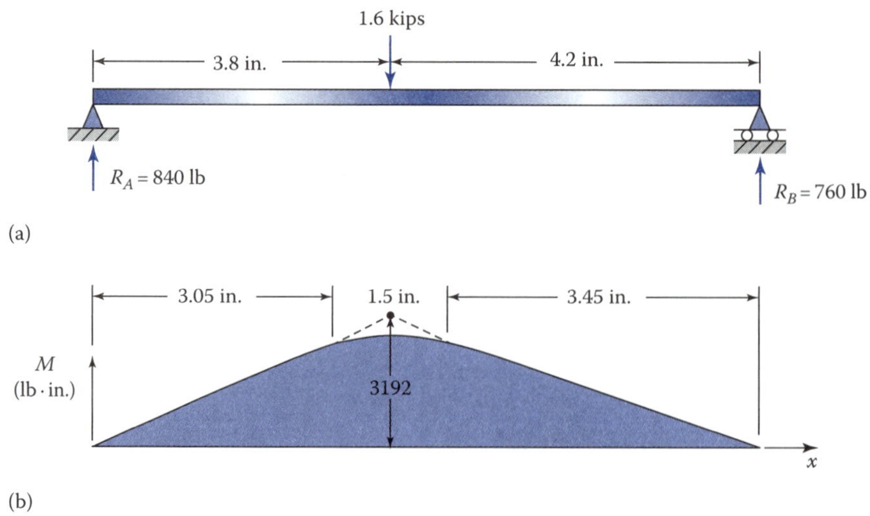

FIGURE 7.17
Diagrams of camshaft shown in Figure 7.16: (a) load and (b) bending moment.

is larger than (2562 lb · in.) at the left side. We have

$$\sigma_{\max} = \frac{M}{I/c} = \frac{2622}{98.175(10^{-3})} = 26.71 \text{ ksi}$$

$$\sigma_{\min} = 0$$

Equation 7.14 results in

$$\sigma_a = \sigma_m = \frac{26.71}{2} = 13.36 \text{ ksi}$$

Stress-concentration factors. The step in the shaft is asymmetrical. Stress at point E is influenced by the radius r = 1.5 in. (equivalent to a diameter of 3.0 in.) and at point F by the 0.8 in. cam radius (equivalent to D_c = 1.6 in. diameter). Hence, we obtain the following values:

At point E,

$$\frac{r}{d} = \frac{0.1}{1.0} = 0.1, \quad \frac{D}{d} = \frac{3.0}{1.0} = 3.0,$$

K_t = 1.8 (from Figure C.9)

At point F,

$$\frac{r}{d} = \frac{0.1}{1.0} = 0.1, \quad \frac{D}{d} = \frac{1.6}{1.0} = 1.6,$$

$$K_t = 1.7 \text{ (from Figure C.9)}$$

For r = 0.10 in. and S_u = 130 ksi, by Figure 7.9a, q = 0.86. It follows, from Equation 7.13b, that

$$(K_f)_E = 1 + 0.86(1.8 - 1) = 1.69$$

$$(K_f)_F = 1 + 0.86(1.7 - 1) = 1.60$$

Comments: Note that the maximum stress in the shaft is well under the material yield strength. The stress concentration at E is only 5% larger than that at F. Therefore, fatigue failure is expected to begin at point F, where the stress pulses are tensile and compressive at E.

Modified endurance limit. Through the use of Equation 7.6, we have

$$S_e = C_f C_r C_s C_t \left(\frac{1}{K_f}\right) S'_e$$

where

$$C_f = 1.34(130)^{-0.085} = 0.886 \quad (\text{from Table 7.2})$$

$$C_r = 0.75 \quad (\text{by Table 7.3})$$

$$C_s = 0.85 \quad (\text{from Equation 7.9})$$

$$C_t = 1.0 \quad (\text{room temperature})$$

$$K_f = 1.6$$

$$S'_e = 0.5(130) = 65 \text{ ksi} \quad (\text{by Equation 7.1})$$

Hence,

$$S_e = (0.886)(0.75)(0.85)(1.0)\left(\frac{1}{1.6}\right)(65)$$

$$= 22.95 \text{ ksi}$$

Factor of safety. The safety factor guarding against fatigue failure at point F is determined using Equation 7.22:

$$n = \frac{S_u}{\sigma_m + \dfrac{S_u}{S_e}\sigma_a} = \frac{130}{13.36 + \dfrac{130}{22.95}(13.36)} = 1.46$$

Comments: If the load is properly controlled so that there is no impact, the foregoing factor seems well sufficient. Inasmuch as lift motion is involved, the deflection needs to be checked accurately by FEA. Case Study 8.1 analyses contact stresses between cam and follower.

7.12 Design for Combined Fluctuating Loads

In numerous practical situations, structural and machine components are subject to combined fluctuating bending, torsion, and axial loading, for example, propeller shafts, crankshafts, and airplane wings. Often, under conditions of a general cyclic state of stress, static failure theories are modified for analysis and design. In this section, we consider the maximum shear stress and maximum distortion energy theories associated with the Goodman criterion. Note that the expressions to follow can also be written based on the Soderberg criterion, substituting the yield strength S_y for the ultimate strength S_u, as required.

For combined stresses, we treat the fatigue effect first by defining equivalent values of each principal stress. We designate the mean component of σ_{1a} owing to a steady loading by σ_{1m} and the alternating component due to the reversed load by σ_{1a}. Based on the Goodman relation, the equivalent principal stresses are then

$$\sigma_{ie} = \sigma_{im} + \frac{S_u}{S_e}\sigma_{ia} \quad (i = 1, 2, 3) \tag{7.25}$$

These equivalent values are then used in the expressions of static failure criteria applied to fatigue loading.

Equation 6.6, with S_u replacing the quantity S_y used thus far, and Equation 7.25 lead to the maximum shear stress theory applied to fatigue loading. Therefore,

$$\frac{S_u}{n} = (\sigma_1 - \sigma_3)_e = \sigma_{1m} - \sigma_{3m} + \frac{S_u}{S_e}(\sigma_{1a} - \sigma_{3a}) \tag{7.26}$$

Clearly, it is assumed that this modified static yield failure theory applies to brittle behavior as well. For the special case, where $\sigma_y = \sigma_z = \tau_{yz} = \tau_{xz} = 0$, Equation 6.11 results in

$$\frac{S_u}{n} = \left(\sigma_x^2 + 4\tau_{xy}^2\right)_e^{1/2} \tag{7.27}$$

Introducing the equivalent stresses σ_e and τ_e from Equations 7.17 and 7.18 for σ_x and τ_{xy}, respectively, into Equation 7.27, we have the maximum shear stress theory incorporated with the Goodman criterion:

$$\frac{S_u}{n} = \left[\left(\sigma_{xm} + \frac{S_u}{S_e}\sigma_{xa}\right)^2 + 4\left(\tau_{xym} + \frac{S_u}{S_e}\tau_{xya}\right)^2\right]^{1/2} \tag{7.28}$$

Similarly, Equations 7.13a and 7.25 give the maximum distortion energy theory applied to fatigue loading as

$$\frac{S_u}{n} = \left\{\frac{1}{2}\left[(\sigma_1 - \sigma_2)^2 + (\sigma_2 - \sigma_3)^2 + (\sigma_3 - \sigma_1)^2\right]\right\}_e^{1/2} \tag{7.29}$$

The 2D equivalent ($\sigma_3 = 0$) is

$$\frac{S_u}{n} = \left(\sigma_1^2 - \sigma_1\sigma_2 + \sigma_2^2\right)_e^{1/2} \tag{7.30}$$

For the special case in which $\sigma_y = \sigma_z = \tau_{yz} = \tau_{xz} = 0$ using Equation 6.16,

$$\frac{S_u}{n} = \left(\sigma_x^2 + 3\tau_{xy}^2\right)_e^{1/2} \tag{7.31}$$

Substitution of the equivalent stresses σ_e for σ_x and τ_e for τ_{xy} into this expression gives the *maximum energy of distortion theory* combined with the *Goodman criterion*:

$$\frac{S_u}{n} = \left[\left(\sigma_{xm} + \frac{S_u}{S_e}\sigma_{xa}\right)^2 + 3\left(\tau_{xym} + \frac{S_u}{S_e}\tau_{xya}\right)^2\right]^{1/2} \tag{7.32}$$

7.12.1 Alternative Derivation

Equivalent alternating stress/equivalent mean stress fatigue criteria are represented in Table 7.4, replacing σ_a and σ_m with σ_{ea} and σ_{em}. In so doing, the Goodman criterion, for example, becomes

$$\frac{1}{n} = \frac{\sigma_{ea}}{S_e} + \frac{\sigma_{em}}{S_u} \tag{7.33}$$

The static failure theories may also be modified, substituting a and m in the expressions given in Sections 6.6 through 6.12. So, the von Mises stresses for the alternating and mean components for a triaxial and biaxial states of stress are obtained by applying Equations 6.13 and 6.15, respectively. In a like manner, relations for the special case in which $\sigma_y = \sigma_z = \tau_{yz} = \tau_{xy} = 0$, through the use of Equation 6.16, may be written as

$$\sigma_{ea} = \left(\sigma_{xa}^2 + 3\tau_{xya}^2\right)^{1/2}, \quad \sigma_{em} = \left(\sigma_{xm}^2 + 3\tau_{xym}\right)^{1/2} \tag{7.34}$$

Carrying Equation 7.34 into Equation 7.33 yields the energy of distortion theory associated with the Goodman relation in the following alternate form:

$$\frac{1}{n} = \frac{1}{S_e}\left(\sigma_{xa}^2 + 3\tau_{xya}^2\right)^{1/2} + \frac{1}{S_u}\left(\sigma_{xm}^2 + 3\tau_{xym}\right)^{1/2} \tag{7.35}$$

in which n represents the factor of safety.

In conclusion, we note that Equations 7.28 and 7.32 or 7.35 can be employed to develop a series of design formulas. Their application to the design of transmission shafts is illustrated in the next chapter. Obviously, fatigue analysis should be considered wherever a simple or combined fluctuating load is present. Springs, for example, frequently fail in fatigue. Chapter 14 treats spring design by using the Soderberg and Goodman criteria. We discuss the preloaded threaded fasteners in fatigue in Section 15.12.

7.13 Prediction of Cumulative Fatigue Damage

Machine and structural members are not always subjected to the constant stress cycles, as shown in Figure 7.11. Many parts may be under different severe levels of reversed stress cycles or randomly varying stress levels. Examples include automotive suspension and aircraft structural components operating at stress levels between the fracture strength S_f and endurance limit S_e', say, S (Figure 7.5). If the reversed stress is higher than the endurance limit, S replaces S_e' in Equation 7.6 and the design may again be based on the formulas developed in the preceding section. However, when a machine part is to operate for a finite time at higher stress, the cumulative damage must be examined.

It is important to note that, predicting the cumulative damage of parts stressed previously, the endurance limit is at best a rough procedure. This point is demonstrated by the typical scatter band depicted in Figure 7.6 for completely reversed loads. Clearly, for parts subjected to randomly varying loads, the damage prognosis is further complicated.

7.13.1 Miner's Cumulative Rule

The simplest, most widely accepted criterion used to explain cumulative fatigue damage is called the *Miner's rule.* The procedure, also known as the linear cumulative damage rule, is expressed in the form

$$\frac{n_1}{N_1}+\frac{n_2}{N_2}+\cdots=1 \tag{7.36}$$

where

n represents the number of cycles of higher stress S applied to the specimen
N is the life (in cycles) corresponding to S, as taken from the appropriate S–N curve

Miner's equation assumes that the damage to the material is directly proportional to the number of cycles at a given stress. The rule also presupposes that the stress sequence does not matter and the rate of damage accumulation at a stress level is independent of the stress history. These have not been completely verified in tests. Sometimes specifications are used in which the right side of Equation 7.36 is taken to be between 0.7 and 2.2.

A typical set of plots of S versus N, for different types of surfaces, is shown in Figure 7.18 [5,13]. The values of N_1, N_2, and so on may be obtained from such curves. Employing these values, Equation 7.36 becomes the design criterion. The use of Miner's rule is illustrated in the solution of the following numerical problem.

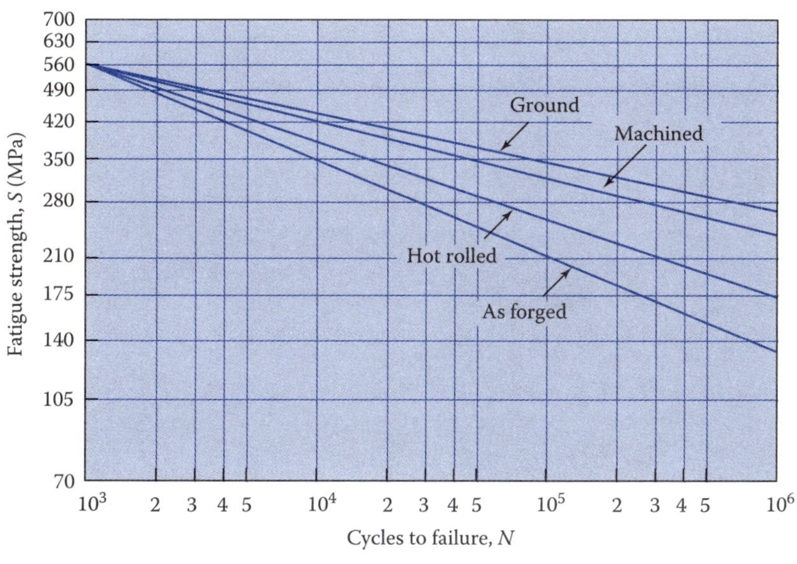

FIGURE 7.18
Allowable stress-cycle diagram for steel parts with 187–207 Bhn.

Example 7.6: Cumulative Fatigue Damage of a Machine Bracket

A steel bracket of a machine is subjected to a reversed bending stress of S_1 for n_1 cycles, S_2 for n_2 cycles, and S_3 for n_3 cycles. Determine whether failure will occur.

Given: $S_1 = 420$ MPa, $S_2 = 350$ MPa, $S_3 = 280$ MPa

$$n_1 = 5,000 \text{ cycles}, n_2 = 20,000 \text{ cycles}, n_3 = 30,000 \text{ cycles}.$$

Design Decisions: The bracket has a machined surface and Bhn = 200. Miner's cumulative damage rule is used.

Solution

The appropriate limiting number of cycles corresponding, respectively, to the preceding stress values is, from Figure 7.18,

$$N_1 = 12,000 \text{ cycles}, \quad N_2 = 50,000 \text{ cycles}, \quad N_3 = 280,000 \text{ cycles}$$

Applying Equation 7.36,

$$\frac{5,000}{12,000} + \frac{20,000}{50,000} + \frac{30,000}{280,000} = 0.924$$

Comment: Since 0.924 < 1, the member is safe.

7.14 Fracture Mechanics Approach to Fatigue

As discussed previously, the fatigue failures are progressive, starting with a very small crack at or near a surface, followed by their gradual increase in width and depth, and then sudden fracture through the remaining zone. We now present a procedure for estimating the life remaining in a part after the discovery of a crack. The method, known as the *fracture mechanics approach to fatigue*, applies to elastic isotropic materials.

To develop fatigue strength data in terms of a fracture mechanics approach, numerous specimens of the same material are tested to failure at certain levels of cyclical stress range $\Delta\sigma$. The test is usually done in an axial fatigue testing machine. The crack growth rate da/dN is continuously measured as the applied stress varies from σ_{min} to σ_{max} during the test. Here, a represents the initial crack length and N is the number of stress cycles.

For each loading cycle, the stress intensity range factor ΔK is defined as

$$\Delta K = K_{max} - K_{min} \tag{7.37a}$$

where K_{max} and K_{min} are the maximum and minimum stress intensity factors, respectively, around a crack. From Equation 6.1, we have

$$K_{max} = \lambda\sqrt{\pi a}\sigma_{max}, \quad K_{min} = \lambda\sqrt{\pi a}\sigma_{min} \tag{7.38}$$

in which λ is a geometry factor. Substitution of this into Equation 7.37a gives

$$\Delta K = \lambda\sqrt{\pi a}\left(\sigma_{max} - \sigma_{min}\right) \tag{7.37b}$$

The quantities σ_{max} and σ_{min} are the maximum and minimum nominal stresses, respectively. The critical or final crack length a_f at fracture, from Equation 6.3 taking factor of safety $n = 1$, may be expressed as

$$a_f = \frac{1}{\pi}\left(\frac{K_c}{\lambda\sigma_{max}}\right)^2 \tag{7.39}$$

where K_c is the fracture toughness.

Crack growth rate $\mathrm{d}a/\mathrm{d}N$ is often plotted on log–log paper against the stress intensity range factor ΔK. The major central portion of the curve plots as a straight line and is of interest in predicting fatigue life. The relationship in this region is defined in the form [16]

$$\frac{\mathrm{d}a}{\mathrm{d}N} = A(\Delta K)^n \tag{7.40}$$

This is known as the *Paris equation* after P.C. Paris. The empirical values of the factor A and exponent n for a number of steels are listed in Table 7.5. Details of the experimental basis for this expression are given in the technical literature [18].

Equation 7.40 is integrated to give the number of cycles N to increase the crack length from an initial value a to the critical length a_f at fracture, the remaining fatigue life, based on a particular load, geometry, and material parameters for a particular application. Considering λ independent of the initial crack length, it can be shown that [8,19]

$$N = \frac{a_f^{1-n/2} - a^{1-n/2}}{A\left(1-\dfrac{n}{2}\right)(1.77\lambda\Delta\sigma)^n} \tag{7.41}$$

where

N = the fatigue life cycles
a = the initial crack length
a_f = the crack length at fracture
λ = the geometry factor (see Table 6.1)
$\Delta\sigma$ = the stress range (in MPa or ksi)

The application of Equation 7.41, the fatigue life determination procedure, is illustrated in the simple example that follows.

TABLE 7.5

Paris Equation Parameters for Various Steels

	A		
Steel	**SI Units**	**U.S. Units**	n
Ferritic–pearlitic	6.90×10^{-12}	3.60×10^{-10}	3.00
Martensitic	1.35×10^{-10}	6.60×10^{-9}	2.25
Austenitic stainless	5.60×10^{-12}	3.00×10^{-10}	3.25

Source: Based on Barsom, J.M. and Rolfe, S.T., *Fracture and Fatigue Control in Structures*, 3rd ed., Oxford, U.K., Butterworth Heinemann, 1987.

Example 7.7: Fatigue Life of Instrument Panel with a Crack

A long plate of an instrument is of width $2w$ and thickness t. The panel is subjected to an axial tensile load that varies from P_{min} to P_{max} with a complete cycle every 15 s. Before loading, on inspection, a central transverse crack of length $2a$ is detected on the plate. Estimate the expected life.

Given: $a = 0.3$ in., $t = 0.8$ in., $w = 2$ in., $P_{max} = 2P_{min} = 144$ kips.

Assumption: The plate is made of an AISI 4340 tempered steel.

Solution

See Tables 6.1, 6.2, and 7.5.

The material and geometric properties of the panel are

$$A = 3.6 \times 10^{-10}, \quad n = 3, \quad K_c = 53.7 \text{ ksi}\sqrt{\text{in.}}, \quad S_y = 218 \text{ ksi},$$

$$\lambda = 1.02, \quad \text{for } a/w = 0.15 \quad (\text{Case A of Table 6.1})$$

Note that the values of a and t satisfy Table 6.2. The largest and smallest normal stresses are

$$\sigma_{max} = \frac{P_{max}}{2wt} = \frac{144}{2(2)(0.8)} = 45 \text{ ksi}, \quad \sigma_{min} = 22.5 \text{ ksi}$$

The cyclical stress range is then $\Delta\sigma = 45 - 22.5 = 22.5$ ksi.

The final crack length at fracture, from Equation 7.39, is found to be

$$a_f = \frac{1}{\pi}\left(\frac{K_c}{\lambda\sigma_{max}}\right)^2 = \frac{1}{\pi}\left(\frac{53.7}{1.02 \times 45}\right)^2 = 0.436 \text{ in.}$$

Substituting the numerical values, Equation 7.41 results in

$$N = \frac{0.436^{-0.5} - 0.3^{-0.5}}{3.6\left(10^{-10}\right)(-0.5)\left[(1.77)(1.02)(22.5)\right]^3} = 25{,}800 \text{ cycles}$$

With a period of 15 s, approximate fatigue life L is

$$L = \frac{25{,}800(15)}{60(60)} = 107.5 \text{ h}$$

Problems

Sections 7.1 through 7.7

7.1 A round bar made of 1020 steel having fatigue properties illustrated in Figure 7.7 is subjected to a completely reversed bending of $M = 4$ kN·m. Find the maximum diameter D of the bar, for

a. The infinite life

b. Least 10^5 cycles to failure

7.2 A circular aluminum alloy (2024) rod with the fatigue properties shown in Figure 7.7 is under a completely reversed bending moment $M = 1.5$ kN·m. Estimate the largest diameter D of the rod, at

a. The fatigue strength

b. Least 10^7 cycles to failure

7.3 A structural steel bar of thickness t with full fillets is loaded by reversed axial force P (Figure P7.3). Calculate

a. The maximum stress

b. The maximum fatigue stress-concentration factor

Given: $P = 15$ kN, $t = 10$ mm.

7.4 A bar with full fillets is forged from a structural steel (Figure P7.3). Determine the value of the endurance limit S_e.

Assumptions: A survival rate of 95% is used. The operating temperature is 475°C maximum.

7.5 A machined and full-filleted AISI 4140 annealed steel bar carries a fluctuating axial loading, as shown in Figure P7.5. What is the value of endurance limit S_e?

Given: $b = 20$ mm, $D = 30$ mm, $r = 2$ mm.

Assumptions: A reliability of 90% is used.

7.6 A stepped cantilever beam of diameters d and D, machined from an AISI 1060 annealed steel bar, is subjected to a fluctuating moment M, as depicted in Figure P7.6. Determine the modified endurance limit S_e.

Given: $d = 25$ mm, $D = 35$ mm, $r = 4$ mm.

Design Assumption: Reliability is 90%.

7.7 A notched beam, machined from AISI 1030 hot-rolled steel, is subjected to reversed bending. Determine the endurance limit S_e.

Assumptions: A survival rate of 98% and $C_s = 0.7$ are used. The fatigue stress-concentration factor is $K_f = 2.5$.

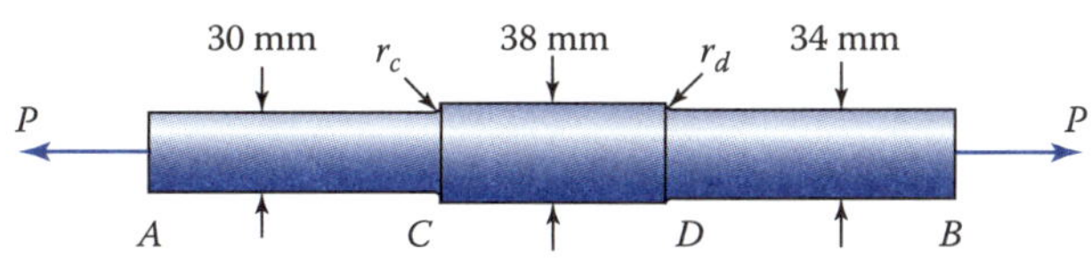

FIGURE P7.3

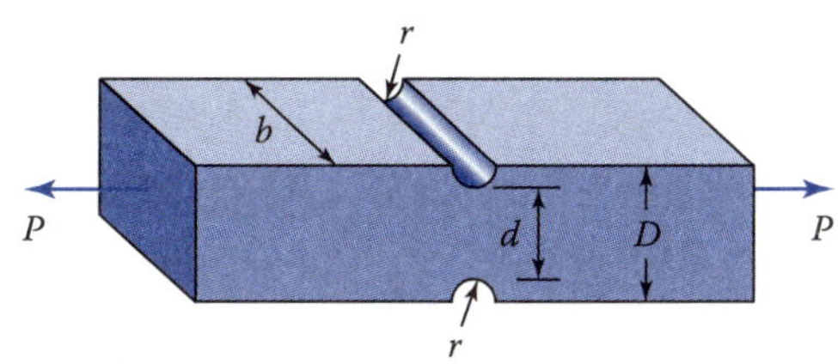

FIGURE P7.5

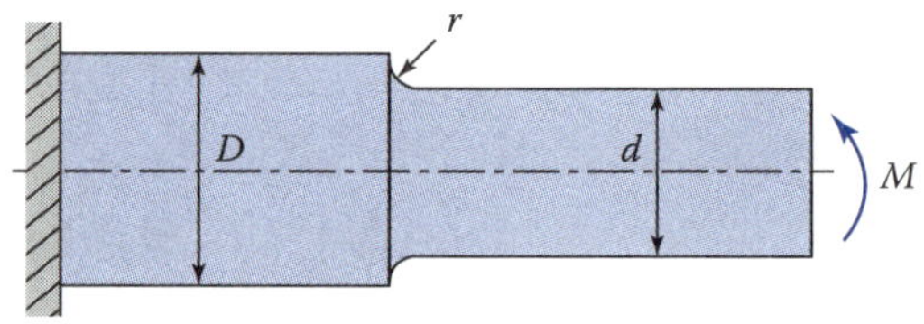

FIGURE P7.6

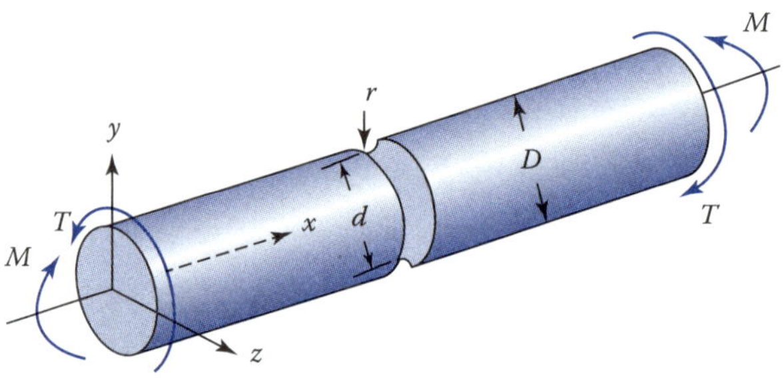

FIGURE P7.8

7.8 A circular shaft of diameter D, groove diameter d, and groove radius r is subjected to a moment M and a torque $T = 0$ (Figure P7.8). Find the endurance limit S_e, if the shaft is made from AISI 1095 annealed steel.

Given: $D = 30$ mm, $d = 25$ mm, $r = 2$ mm.

Design Assumption: A survival rate of 95% will be used. The operating temperature is 525°F maximum.

7.9 Resolve Problem 7.8, for the case in which the grooved shaft is subjected to a torque T and $M = 0$ (Figure P7.8).

Sections 7.8 through 7.14

7.10 At a critical point of a thin panel, the bending stress fluctuates. Compute the mean stress, range stress, stress ratio, and amplitude ratio for the three common cases:

a. Completely reversed ($\sigma_{max} = -\sigma_{min} = 12$ ksi)

b. Nonzero mean ($\sigma_{max} = 12$ ksi, $\sigma_{min} = -2$)

c. Released tension ($\sigma_{max} = 12$ ksi, $\sigma_{min} = 0$)

7.11 A stepped cantilevered beam, machined from steel having ultimate tensile strength S_u, is under reversed bending (Figure P7.6). Determine the maximum value of the bending moment M, using the Goodman criterion.

Given: $d = 1$ in., $D = 1.5$ in., $r = 0.05$ in., $S_u = 100$ ksi.

Design Assumptions: A survival rate of 95% is used. The factor of safety $n = 1.5$.

7.12 A cold-drawn AISI 1020 CD steel link is subjected to axial loading (which fluctuates from 0 to F) by pins that go through holes (Figure P7.12). What is the maximum value of F with a factor of safety of n, according to the Goodman criterion?

Given: $R = 10$ mm, $r = 4$ mm, $t = 2.5$ mm, $n = 1.4$.

Assumption: A reliability of 99.99% is used.

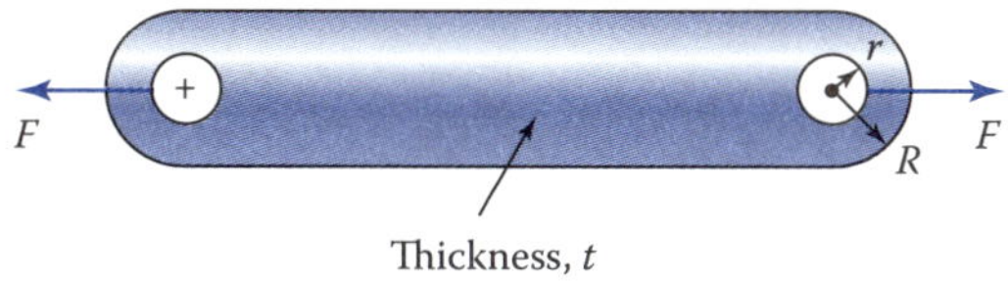

FIGURE P7.12

7.13 Consider the steel link described in Problem 7.11 operating at a temperature 1010°F maximum with a reliability factor of 95%. Find the largest value the axial tensile force F, using the SAE criterion.

Assumption: Fracture strength of the member will be $S_f = 415$ MPa with a factor of safety of $n = 1.2$.

7.14 What is the maximum value of the axial load F applied to the steel link of Problem 7.12, employing the Soderberg criterion?

Assumptions: Reliability factor is 90%. Factor of safety will be $n = 2.2$. The largest operating temperature equals 540°C.

7.15 A cold-drawn AISI 1050 steel plate with a central hole is under a tension load P that varies from 5 to 25 kN (Figure P7.15). Based on the Goodman criterion, determine the factor of safety n

a. Against yielding

b. Against fatigue failure

Given: $D = 25$ mm, $d = 5$ mm, $t = 10$ mm.

Assumption: A reliability of 98% and $C_r = 0.7$ are used.

7.16 Resolve Problem 7.15 for the condition that the load varies from −5 to 25 kN.

Assumption: Buckling does not occur.

7.17 A machined AISI 4130 normalized steel bar of diameter D carries an axial load P, as shown in Figure P7.17. Calculate the value of

a. The static force P to produce fracture

b. The completely reversed force P to produce fatigue failure

Given: $D = 2\frac{1}{8}$ in.

Assumptions: The survival rate is 95%. The operating temperature is 900°F maximum.

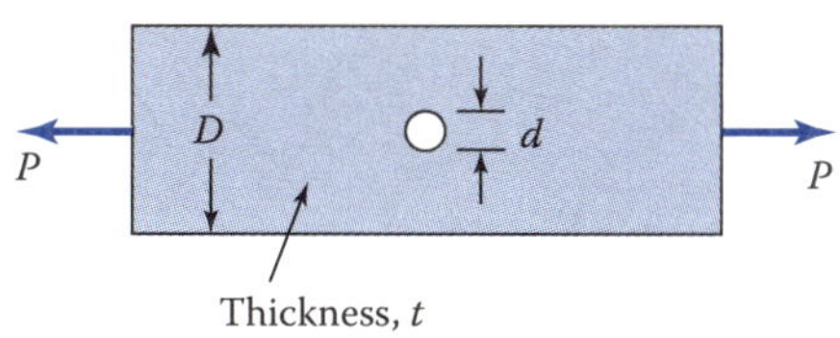

FIGURE P7.15

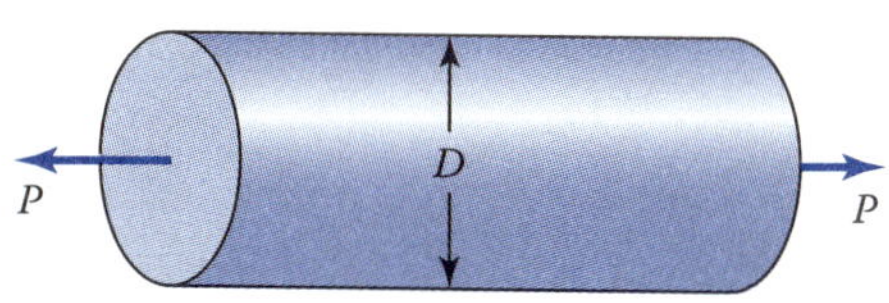

FIGURE P7.17

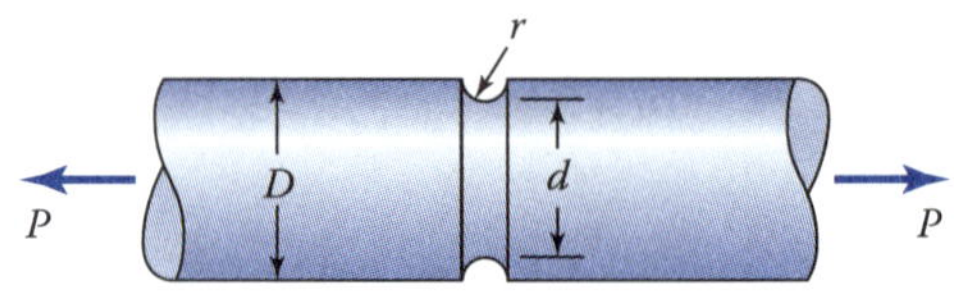

FIGURE P7.18

7.18 Redo Problem 7.17 for the case of a grooved shaft shown in Figure P7.18.

Given: $D = 2\frac{1}{8}$ in., $d = 2$ in., $r = 0.05$ in.

7.19 A stepped shaft ground from AISI 1040 annealed steel is subjected to torsion, as shown in Figure P7.19. Determine the value of

a. The torque T to produce static yielding

b. The torque T to produce fatigue failure

Given: $D = 50$ mm, $d = 25$ mm, $r = 1.25$ mm.

Assumption: Reliability is 98%.

7.20 Repeat Problem 7.19 for the condition that the shaft is subjected to axial loading and no torsion.

7.21 Redo Problem 7.19 for the case in which the shaft is machined from an AISI 1095 hot-rolled steel.

7.22 A shaft with a transverse hole ground from AISI 1095 annealed steel is under bending moment M that varies from 0.5 to 1.4 kips · in. (Figure P7.22). Determine the factor of safety n against fatigue failure, using the Goodman criterion.

Given: $D = 1$ in., $d = \frac{1}{8}$ in.

Assumption: A reliability of 99% is used.

7.23 Resolve Problem 7.22 for the condition that shaft is ground from AISI 1060 HR steel and is under axial loading varying from 5 to 15 kips.

7.24 A rotating AISI 1030-CD steel beam having a machined surface carries an axial load P and a moment M as shown in Figure P7.24. Compute the factor of safety based upon the Goodman criterion.

Given: $D = 25$ mm, $M = 150$ N · m, $P = 15$ kN.

Assumptions: Reliability factor will be 99.9%. Take $C_s = C_t = 1$.

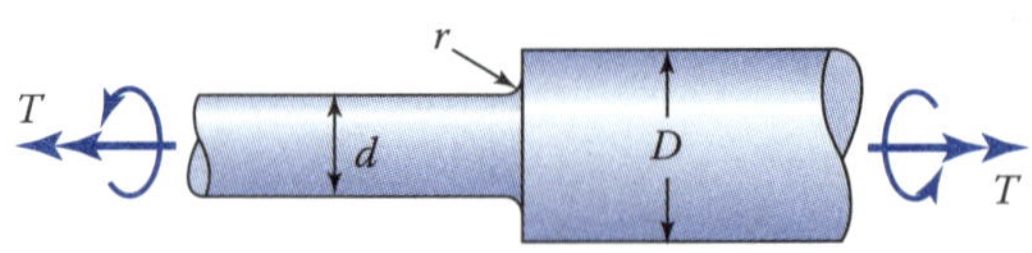

FIGURE P7.19

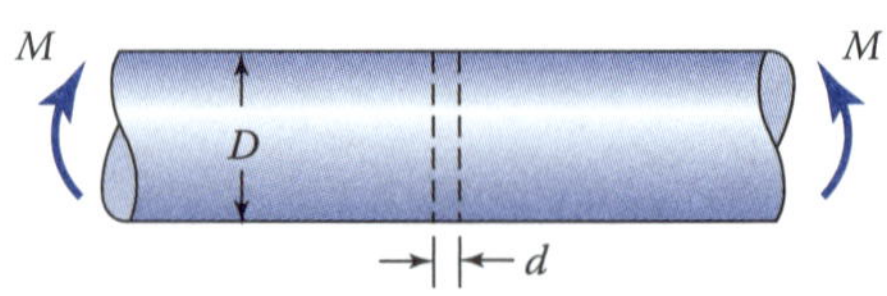

FIGURE P7.22

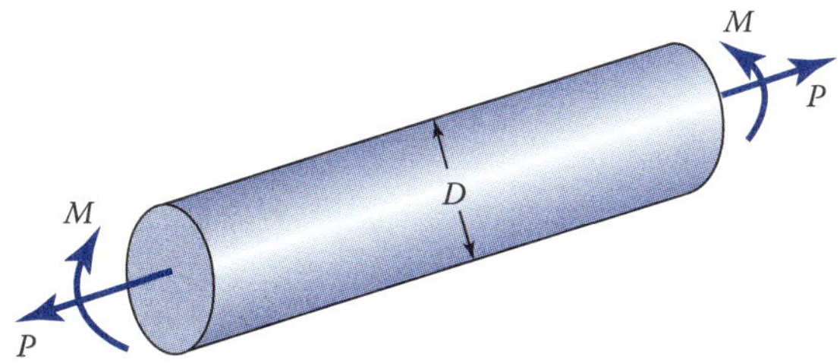

FIGURE P7.24

7.25 Consider the shaft described in Problem 7.22 operating at a temperature 850°F maximum with a reliability factor of 90%. Find the factor of safety n against failure by yielding using the Soderberg criterion.

7.26 A thin-walled cylindrical vessel of diameter d is subjected to an internal pressure varying from 60 to 300 psi continuously. Apply the maximum energy of distortion theory incorporated with the Soderberg relation to design the vessel.

Given: d = 80 in., S_y = 40 ksi, S_e = 30 ksi, n = 2.5.

7.27 A thin-walled cylindrical vessel of diameter d and thickness t is under internal pressure varying from 0.6 to 2.8 MPa continuously.

Given: d = 1.5 m, t = 25 mm, S_y = 250 MPa, S_u = 350 MPa, S_e = 150 MPa.

Design Decision: Use the maximum energy of distortion theory incorporated with the Goodman relation. Determine the factor of safety n.

7.28 A small leaf spring, b = 10 mm wide, 100 mm long, and h mm deep, is subjected to a concentrated center load P varying continuously from 0 to 20 N. The spring may be approximated to be a simply supported beam (Figure P7.28). Calculate the required depth for a factor of safety of 4.

Given: S_u = 980 MPa, S_e = 400 MPa.

Design Decision: Apply the Goodman theory, based on the maximum normal stress.

7.29 Redo Problem 7.28 using the Soderberg criterion and yield strength of S_y = 620 MPa.

7.30 An electrical contact includes a flat spring in the form of a cantilever, ⅛ in. wide × 1.5 in. long and h in. deep, is subjected at its free end to a load P that varies continuously from 0 to 0.5 lb (Figure P7.30). Compute the value of h for a factor of safety n = 1.2.

Given: S_u = 150 ksi, S_e = 72 ksi.

Design Decision: Employ the Goodman criterion, based on the maximum normal stress.

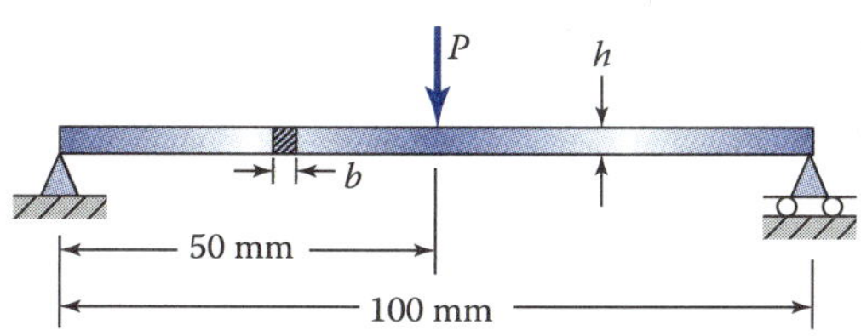

FIGURE P7.28

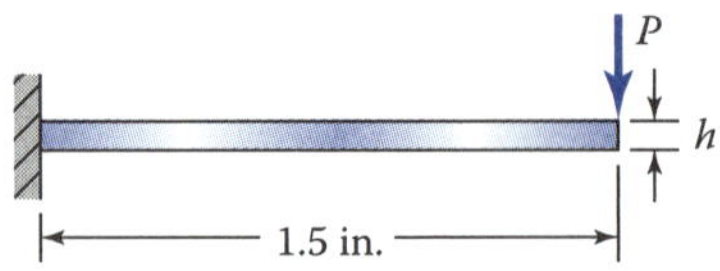

FIGURE P7.30

7.31 Consider the cantilever and loading described in Problem 7.30 operating at a temperature 880°F maximum. What is the value of the depth h according to the SAE criterion based on the maximum normal stress?

Given: Fracture strength $S_f = 98$ ksi, factor of safety $n = 1.5$.

7.32 A cantilever spring is subjected to a concentrated load P varying continuously from 0 to P_o (Figure P7.32). What is the greatest allowable load P_o for $n = 4$?

Given: $S_y = 850$ MPa, $S_e = 175$ MPa, $b = 5$ mm, $h = 10$ mm, $K_f = 2$.

Assumption: Failure occurs due to bending stress at the fillet.

Design Decision: Use the Soderberg criterion.

7.33 Resolve Problem 7.32 for the load varying from $P_o/2$ upward to P_o downward, $n = 2$.

7.34 A 24 mm wide, 4 mm thick, and 3 mm long leaf spring, made of AISI 1050CD steel, is straight and unstressed when the cam and shaft are removed (Figure P7.34). Use the Goodman theory to calculate the factor of safety n for the spring.

Given: $S_e = 250$ MPa, $E = 200$ GPa, $\nu = 0.3$.

Assumption: The cam rotates continuously. Leaf spring is considered as a *wide* cantilever beam.

7.35 Repeat Problem 7.34 for the case in which the cantilevered spring is made of normalized AISI 1095 steel and employing the Soderberg criterion.

7.36 Consider the long leaf spring of Problem 7.34 operating at a temperature 490°C maximum with a 99.9% reliability rate. What is the factor of safety n for the spring on the basis of the Gerber criterion?

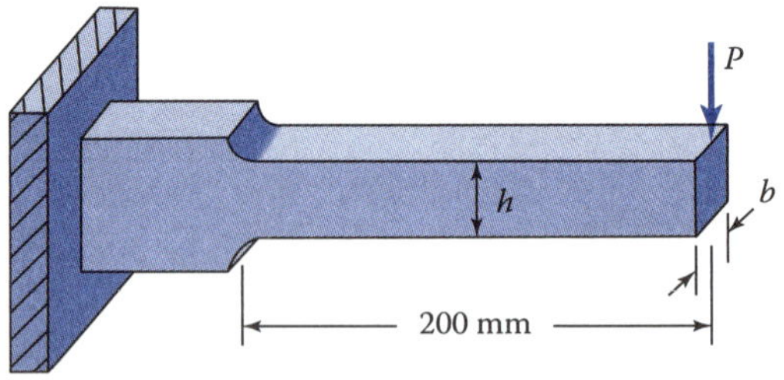

FIGURE P7.32

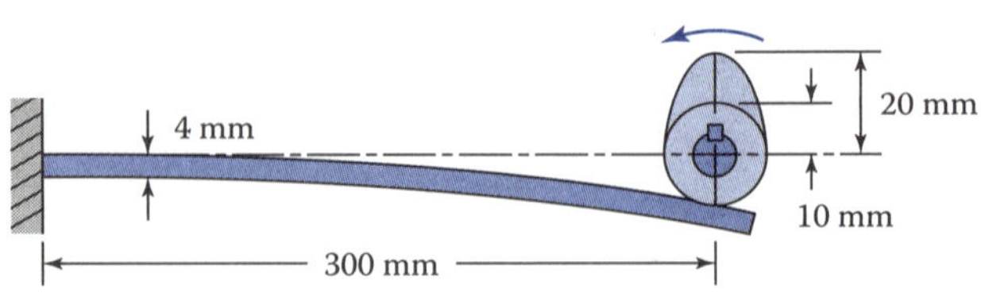

FIGURE P7.34

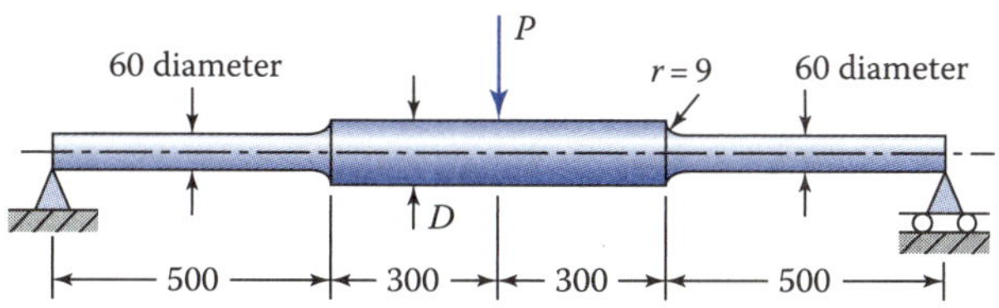

FIGURE P7.37

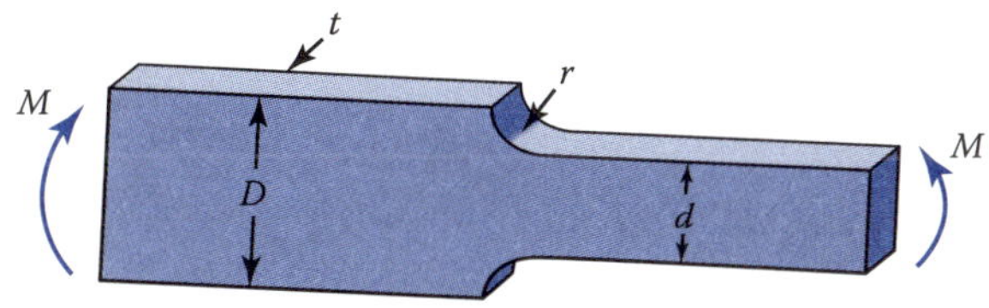

FIGURE P7.38

7.37 Figure P7.37 shows a circular aluminum bar having two shoulder fillets supporting a concentrated load P at its midspan. Determine the allowable value for diameter D if stress conditions at the fillets are to be satisfactory for conditions of operation. Dimensions shown are in millimeters.

Given: $S_u = 600$ MPa, $S_y = 280$ MPa, $n = 2.5$, $K_f S_e = 150$ MPa.

Assumptions: The load P varies from 2 to 6 kN. The Soderberg relation is employed.

7.38 The filleted flat bar shown in Figure P7.38 is made from 1040 steel OQ&T at 650°C. What is the factor of safety n, if the bending moment M varies from 0.6 to 3 kN · m?

Given: $K_f S_e = 400$ MPa, $D = 120$ mm, $d = 60$ mm, $r = 4$ mm, $t = 20$ mm.

Design Assumption: The Goodman criterion of fatigue failure is applied.

7.39 Redo Problem 7.38 for the case in which the moment M varies from 200 to 2200 N · m, through the use of the Gerber criterion.

7.40 A filleted bar in fluctuating bending, as described in Problem 7.38, is to operate at a temperature 475°C maximum with a reliability rate of 95%. Find the factor of safety n employing the SAE criterion.

Assumption: The moment M varies from 1 to 2 kN · m.

7.41 A long AISI 403 stainless steel equipment plate of width $2w$ and thickness t having a double-edge crack of length a is subjected to an axial load varying from P_{min} to P_{max} with a complete cycle every 20 s (Case C, Table 6.1). What is the expected life?

Given: $a = 32$ mm, $t = 34$ mm, $w = 60$ mm, $P_{max} = 2.2$, $P_{min} = 950$ kN.

8

Surface Failure

8.1 Introduction

So far, we have dealt with the modes of failure within the components by yielding, fracture, and fatigue. A variety of types of failure or damage can also occur to the surfaces of elements generally called *wear.* A surface failure or damage, a gradual process, may often render the part unfit for use. The surface may also corrode in a corrosive surrounding such as salt or water. A corrosive environment may reduce the fatigue strength of a metal (Section 7.7). Note that the combination of stress and corrosive surrounding increases the material corrosion more rapidly than if without stress. This chapter represents a brief discussion to the extensive topic of surface damage.

When two solid parts are pressed together, high contact stresses are caused that need special consideration. Two geometric cases are of practical significance: sphere on sphere and cylinder on cylinder (Sections 8.6 through 8.8). It will be observed that the former will have a circular contact patch and the latter will create a rectangular patch. Under repeated loading, contact stresses lead to surface-fatigue failure. Often two machine elements, such as cam and follower and the teeth of a pair of gears, *mate* with one another by rolling, sliding, or a combination of *rolling* and *sliding contact* (Section 8.9).

The surface strength of materials is of utmost importance to design machines having a long and satisfactory life. Surface engineering, a multidisciplinary activity, tailors the properties of the surfaces and near-surface regions of a material to allow the surface to perform functions that are distinct from those functions demanded from the bulk of the material [1,2]. Thus, it improves the function and serviceability and increase the working life of the machine and structural components.

Surface damage prevention is an important scientific and engineering challenge (Section 8.10). Introduction of a lubricant to a sliding interface helps to reduce the friction. The role of lubrication in controlling wear for various machine elements and some other considerations with respect to material failure will be further discussed for specific applications in Section III.

8.2 Corrosion

Corrosion is the deterioration or destruction of a material because of a chemical reaction with its environment. It is the wearing away of metals owing to chemical reaction. Usually, this means electrochemical oxidation of metals in reaction with an oxidant such as oxygen. *Rusting* is a term commonly used for the oxidation of iron and steel. It represents formation of an oxide of iron due to oxidation of the iron atoms in solid solution. This kind of damage often produces oxide(s) and/or salt(s) of the original metal. The main culprits in corroding metals are hydrogen and oxygen. Pure metals are customarily more resistive to corrosion than those containing impurities or small amount of other elements.

Corrosion can also allude to other materials than metals, like polymers; however in this context, the term *degradation* is more proper. Ceramic materials are almost entirely immune to corrosion. Usually, corrosion can be concentrated locally to form a pit or crack, or it can extend across a wide area. *Galvanic corrosion* occurs when two different materials contact one another and are immersed in any substance that is capable of conducting an electric current. It is a major interest to marine industry and also anywhere where water contacts metal structures, such as pipes. Frequently, it is possible to chemically remove the corrosion to produce a clean surface. For instance, phosphoric acid is often applied to ferrous surfaces of tools to remove rust. Materials (typically metals) also chemically deteriorate when subjected to a high-temperature atmosphere containing oxidizing compounds.

Corrosion is a complex phenomenon and still to be fully understood. It is usually studied in the specialized field of corrosion engineering. Atmospheric corrosion is greatest at high temperatures and high humidity, such as that in tropical climates. For further details, see texts on corrosion and [1–4]. Figure 8.1 is a broad guidance only for showing comparative rankings of the resistance of a variety of materials to corrosive attack by six surroundings. Observe that comparative rankings range from A (excellent) to D (bad). Table B.9 furnishes the classes and abbreviations for Figure 8.1.

8.2.1 Corrosion and Stress Combined

When a member is stressed in the presence of a corrosive surrounding, the corrosion is accelerated and failure takes place at a more rapid rate than would be anticipated from either the stress only or the corrosion process only. Most structural alloys corrode only from exposure to moisture in the air, but the process can be strongly affected by exposure to certain substances. Corrosion is occasionally accelerated by relative movement between the metal and corrosive fluid, which prevents the formation of a passive surface film on the metal.

Since corrosion process takes place on exposed surfaces, methods to reduce the activity of the exposed surface, such as passivation and chromate conversion, can increase a material's corrosion resistance. However, some corrosion mechanisms are less visible and less predictable. Failure of mechanical equipment due to corrosion may be hazardous to the operating personnel. Most machine components fail (such as corroded exhaust systems and suspension joints of automobiles), by surface deterioration than by breakage. Rust is one of the most common causes of bridge accidents. Corrosion and wear damage to materials, both directly and indirectly, cost industrial economies hundreds of billions of dollars.

8.2.1.1 Stress Corrosion

Stress corrosion refers to combined condition of static tensile stress and corrosion. Static stresses hasten the corrosion process. *Stress corrosion cracking* (SCC) may take place from the simultaneous presence of a static tensile stress and a specific corrosive environment. The metal as a whole is usually unaffected; however, a network of fine cracks spreads over its surface. The action may proceed along the grain boundaries or may occur across the crystals. Tensile stress-concentration cracking represents the sum of the residual and

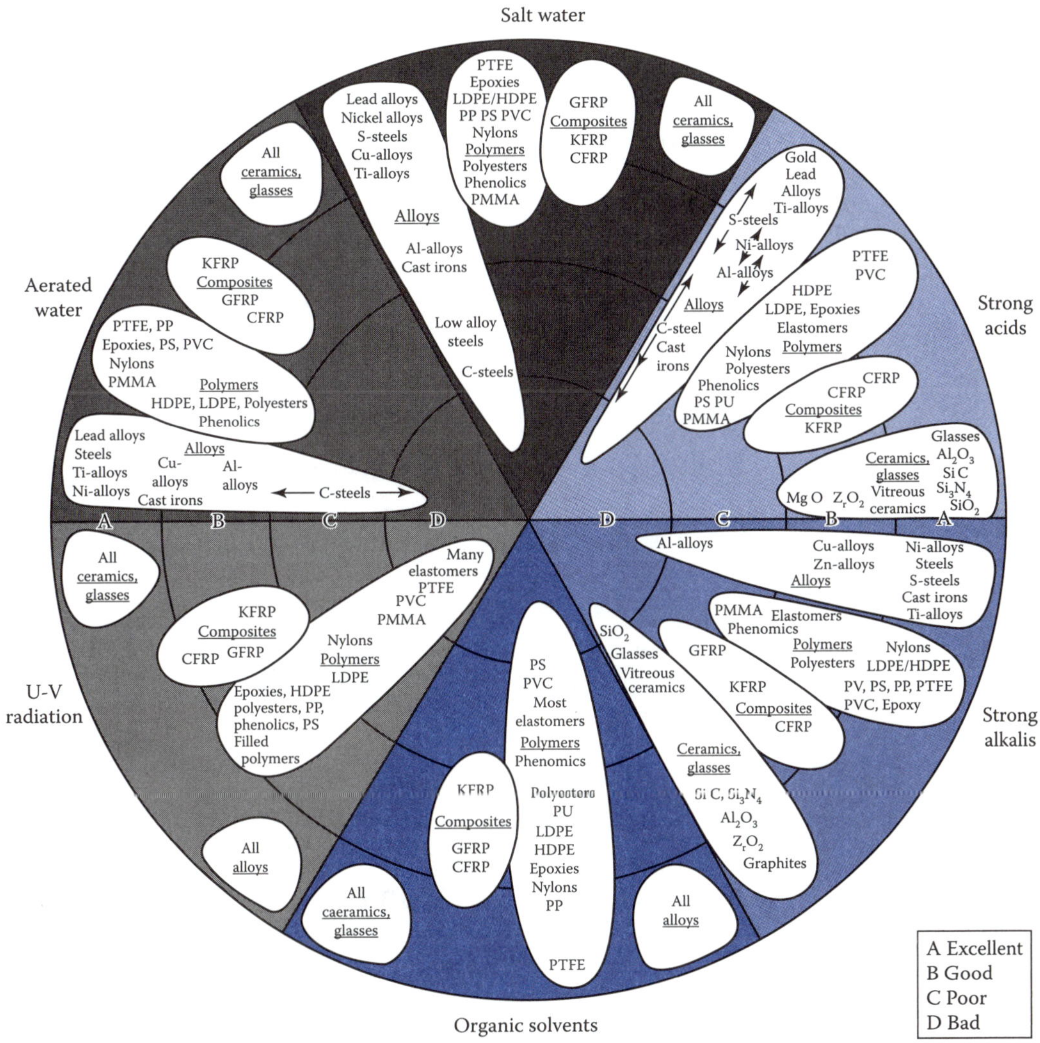

FIGURE 8.1
Comparative ranking ability of materials to resist corrosive attack from a variety of environments. (From Ashby, M.F., *Materials Selection in Mechanical Design*, 4th ed., Butterworth Heinemann, Oxford, U.K., 2011.)

operating stresses present at the local site where the cracks initiate. Examples of this kind of failure include steel boiler tubes operating with corrosive fluid, stainless steel aircraft parts, and bridge cables exposed to the salt and chemicals in the local atmosphere [4].

If a stress exists in a structure exposed to corrosive environment, the rate of corrosion can increase and is extremely localized. Furthermore, some specific chemicals are so aggressive that corrosion will occur at relatively low stress levels, such as those created during manufacturing. The residual stress in a component can then be enough to trigger crack growth and failure. Sometimes corrosion pits were found inside of the broken parts (Figure 8.2a), but often on the surface (Figure 8.2b and c), depending upon the applications.

For SCC to occur, it requires certain conditions, including susceptible material (stainless steel 304 is susceptible), tensile force, undesirable environment (such as with high chloride

FIGURE 8.2
(a) SCC of a bolt, (b) ship's propeller assembly corroded by seawater, and (c) corroded wing control shaft of an airplane.

concentration), higher temperatures, and tiny surface scratches created by machining that act as stress concentrations. In order to remove the surface scratches, bolts should be rolled instead of turned. One of the main causes of a structural component failure is corrosion as it threatens the strength and integrity of the member. Thus, the parts have to be periodically inspected for corrosion.

8.2.1.2 Corrosion Fatigue

For the case in which the member is cyclically stressed in a corrosive surrounding, the crack will grow more rapidly than from either factor only. The preceding is termed corrosion fatigue. It takes place most markedly with metals having little corrosion resistance. Corrosion can be highly concentrated locally to form a pit or crack, or it can extend across a wide area more or less uniformly corroding the surface. Intergranular corrosion, or the failure of grain boundaries, may occur particularly when impurities are present. Corrosion fatigue failures depict discoloration of the crack propagation surfaces.

8.2.2 Corrosion Wear

Corrosion wear designates a failure due to chemical reaction on the surface of a part. It adds to the corrosive environment a mechanical disruption of the surface layer owing to sliding or rolling contact of two bodies. Corrosion wear takes place if a corrosive atmosphere like oxygen is present on the surface of the material in combination with sliding that breaks the oxides free from the surfaces. This action exposes new material to the corrosive elements. Corrosion fatigue is the contamination of a corrosive surrounding with cyclic stresses, which greatly shortens the fatigue life of materials.

Surface corrosion combined with stresses produces a more destructive action than would be anticipated due to the corrosion and stress separately. Many processes are in widespread usage for applying a corrosion-resistant coating to the surface of steel and iron products. Application of corrosion-resistant coatings is one of the most widely employed means of protecting metals, such as steels, for outdoor applications.

A number of coatings are available to choose from. Proper selection is on the basis of the component size, the corrosive environment, the anticipated temperatures, the coating thickness, and the costs. Painting is probably the most widely used engineering coating employed to protect steel from corrosion. Corrosion-resistant plating like chromium are also frequently used. There are a variety of natural material combinations where corrosion can be reduced to a very low value. Often, such combinations will give the highest amount of corrosion protection at the lowest cost.

8.2.2.1 Fretting

Fretting corrosion occurs in tight joints (such as press fits and bolted or riveted connections) where practically no motion is present. Fluctuating loads that produce a slight relative movement are adequate to set up a corrosive wear termed *fretting*. This action can remove significant volume of material over time. It occurs typically in bearings, although most bearings have their surfaces hardened to resist the problem. Another situation occurs when cracks in either surfaces are created. The roughness and pitting produced by fretting reduce the fatigue strength. *Pitting* is among the most common and damaging forms of corrosion.

Resistance to fretting action varies considerably in different materials. Cobalt-base hardfacing and similar alloys are among the best. Usually, steel-on-steel and cast iron-on-cast iron are good. Unprotected bearings on large structures like bridges can suffer serious degradation in behavior, especially when salt is applied during winter to deice the highways on the bridges. Low-viscosity, highly adhesive lubricants help to reduce the intensity of fretting by keeping oxygen away from the active surface.

8.2.2.2 Cavitation Damage

High relative velocities between solid parts and liquid particles can produce cavitation of the liquid, which may destroy the surface of the part. Bubbles are produced if the liquid pressure drops lower than its vapor pressure. Cavitation ordinarily takes place on ship propellers, turbine blades, and centrifugal pump. Damage caused by cavitation to metal surfaces is mechanical. But in corrosive surroundings, cavitation can often damage protective oxide films on the surface that appears to be roughened with closely spaced pits. Enough material is taken off in severe situations, so that the surface has a spongy texture. Cast stainless steel, cast magnesium bronze, cast steel, bronze, cast iron, and aluminum are frequently used to reduce the cavitation [5,6]. The most effective way to deal with cavitation damage is generally to increase the surface hardness.

8.3 Friction

Friction is the force resisting relative movement between surfaces in contact. When a force is applied to a body, the resistive force of friction acts in the opposite direction, parallel to *mating surfaces.* The fundamental kinds of friction are *sliding* and *rolling.* The fundamental equation for determining the resistive force of friction when trying to slide two solid bodies together states that the force of friction equals the coefficient of friction times the normal compressive force pushing the two bodies together. Therefore,

$$F = fP \tag{8.1}$$

where

F = the friction force
f = the coefficient of friction
P = the normal force or perpendicular force pushing the two bodies together

The foregoing equation is valid for both *static* and *kinetic sliding* friction. The former is the friction before a body starts to move and the latter is the friction when the body is sliding. Static and sliding frictions have different friction coefficient or constant values [7] as shown in Table 8.1. In this text, f denotes the *sliding friction coefficient.* When a part rolls on another without sliding, the so-called rolling friction constant f_r is much smaller than that of sliding friction, $f_r \ll f$. In the case of sliding friction of hard surfaces, Equation 8.1 shows that friction is independent of the area of the mating surfaces. However, when it applies to soft surfaces, rotating friction and fluid friction, the coefficient of friction may depend upon area, shape, and viscosity factors. Surface roughness has influence on both sliding and rolling frictions.

Introduction of a lubricant between mating surfaces reduces the coefficient of friction considerably. Lubricants also serve to remove heat from the interface. They may be liquid or solid, which shares the properties of low shear strength and high compressive strength. Lower temperatures reduce surface interactions and wear. A somewhat detailed discussion of lubricants and lubrication phenomenon will be taken up in Chapter 10. A final point is to be noted that, in many situations, such as turbine and generator bearings, low friction is desirable. However, in brakes and clutches (Chapter 13), controlled high friction is needed.

TABLE 8.1

Coefficients of Friction for Various Material Combinations

		Static		Kinematic or Sliding	
Material 1	**Material 2**	**Dry**	**Lubricated**	**Dry**	**Lubricated**
Mild steel	Mild steel	0.74	—	0.57	0.09
Mild steel	Cast iron	—	0.183	0.23	0.133
Mild steel	Aluminum	0.61	—	0.47	—
Mild steel	Brass	0.51	—	0.44	—
Hard steel	Hard steel	0.78	0.11–0.23	0.42	0.03–0.19
Hard steel	Babbitt	0.42–0.70	0.08–0.25	0.34	0.06–0.16
Teflon	Teflon	0.04	—	—	0.04
Steel	Teflon	0.04	—	—	0.04
Cast iron	Cast iron	1.10	—	0.15	0.07
Cast iron	Bronze	—	—	0.22	0.077
Aluminum	Aluminum	1.05	—	1.4	—

8.4 Wear

As pointed out previously, *wear* is a broad term that encompasses numerous types of failures on the surface of the member. It is one of the most important and harmful processes in machine design. Failure from wear customarily involves the loss of some material from the mating surfaces of the parts in contact. When the parts are in *sliding contact*, various types of wear of deterioration occur that can be classed under the general heading wear. In this case, the severity of wear can be reduced by using a *lubricant* (i.e., oil, grease, or solid film) between the mating surfaces.

The study of the process of wear is part of the discipline of tribology [8]. Wear is usually divided, by the physical nature and underlying process, into *three common classes*. These are adhesive wear, abrasive wear, and corrosive wear (Section 8.2). The surface fatigue, an important surface deterioration, is sometimes also classed as *wear*, to be discussed in Section 8.9. All kinds of wear are greatly influenced by the presence of a lubricant. Wear ordinarily requires some relative motion to exist between two surfaces. Stresses introduced in two materials contacting at a rolling interface highly depend on the geometry of the surfaces in contact, on the loading, and on material properties.

8.4.1 Adhesive Wear

On a microscopic scale, sliding metal surfaces are never smooth, and inevitable peaks—usually termed *asperities*—and valleys take place, as depicted in Figure 8.3. At the locations indicated by arrows in the figure, local temperatures and pressures are very high, which cause adhesions. *Adhesive wear* is one of the most common kinds of wear and least preventable. It occurs if the asperities of two mating surfaces stick to one another and break during sliding, transferring material from one part to another or out of the system. In the former situation, the so-called scoring or scuffing results. In both cases,

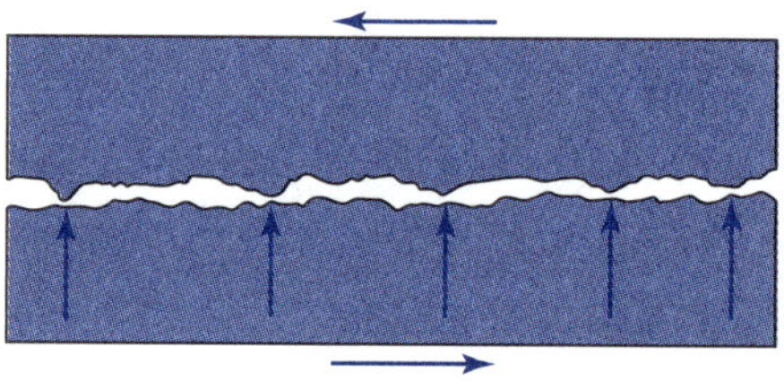

FIGURE 8.3
Adhesive wear simulation: schematic of greatly enlarged view of two nominally *smooth rubbing surfaces.*

surface failure occurs in the absence of adequate lubricant. Severe adhesive wear is termed *seizure* or *galling*.

Usually, the harder the surface, the greater the resistance to adhesive wear. Most solids will *adhere on contact* to some extent. But lubricants and contaminants usually suppress adhesion. Thus, adhesive wear is commonly encountered in conjunction with lubricant failure and often referred to as welding wear or *galling*. Metallurgically identical metals, called *compatible*, should not run together in unlubricated sliding contact. Metallurgically incompatible metals can slide on one another with relatively little scoring. Therefore, *incompatible pairs* can be run together and expected to *resist adhesive wear best*. Also, *partially incompatible* pairs are better in sliding contact than *partially compatible* pairs.

8.4.2 Abrasive Wear

The *abrasive wear* occurs when a hard surface slides across softer surface. The ASTM defines abrasive wear as the loss of material due to hard particles that are forced against and move along solid surface. Abrasion takes place in *two modes*, known as the two-body and three-body abrasive wear when two interacting surfaces are in *direct* physical *contact* and one is significantly harder than the other. Two-body wear occurs when the hard particles remove material from the opposite surface. Examples include soft babbitt bearings used with hard automotive crankshafts and wearing down of wood or soft metal with sandpaper.

Three-body abrasion arises when small and hard *particles are introduced* between the sliding surfaces, at least one of which is softer than the particles. That is, this type of abrasion is caused by the presence of foreign materials between the rubbing surfaces. Therefore, in the design of machinery, it is very important to use pertinent oil filters, dust covers, air filters, shaft seals, etc., to keep irrelevant particles away from the rubbing metal surfaces. In both modes, the harder the surface, the more resistant it is to abrasive wear.

8.5 Wear Equation

For two rubbing surfaces, the volume of material removed by wear is directly proportional to the sliding distance and applied normal force while inversely proportional to the surface hardness. However, the volume of the wear is independent of the velocity of sliding. No single predictive wear formula could be found for general and practical use.

A classic, commonly employed wear equation based on the theory of asperity contact is of the form

$$V = K\frac{PL}{H} \tag{8.2}$$

where
- V = the volume of material worn away
- K = the wear coefficient or constant (dimensionless)
- P = the compressive force between the surfaces
- L = the length of sliding
- H = the surface hardness,* MPa or ksi

It is interesting to note that an alternative form of Equation 8.2 may be written as

$$V = \frac{PL}{H/K} \tag{a}$$

Thus, we observe that for a given load P and length L, the material volume is a minimum when H/K is a maximum.

Often the depth of wear δ may be of interest in applications than the volume. Then, Equation 8.2 may be written as

$$\delta = K\frac{PL}{HA_a} \tag{8.3}$$

Here, the quantity A_a is the *apparent* area of contact of the interface. Application of this equation to a journal bearing is illustrated in Section 10.4. Clearly, both adhesive wear and abrasive wear obey to the foregoing relationships given by Equations 8.2 and 8.3.

Coefficient K represents a measure of the severity of wear. Typically for mild wear, $K \approx 10^{-8}$, whereas for severe wear, $K = 10^{-2}$. Figure 8.4 shows ranges of wear coefficient values determined with a variety of combinations of material compatibility and lubrication for three wear modes. We note that the values of wear coefficients belong to the *softer* of the two *rubbing materials*. Table 8.2 presents examples of approximate range of values of K for a few materials in contact.

The adhesive wear constants for some *metallic sliding* can be readily estimated from Table 8.3. The values listed depend on the tendency of the sliding metal to adhere, on the basis of metallurgical compatibility, and on the lubrication of the sliding surfaces. From a compatibility chart [9], it can be found that, for example, iron is compatible with aluminum, gold titanium, lead, and zinc. Also iron is partially compatible with copper, partially incompatible with tin and magnesium, and incompatible with lead and silver. Observe from Table 8.3 that partially compatible and partially incompatible pairs are placed in the same category, because the wear constants for sliding metals differ only slightly.

* Hardness may be expressed as Brinell, Vickers, and Rockwell in units kg/mm^2 (see Section 2.10). To convert to MPa or ksi, multiply by 9.81 or 1.424, respectively.

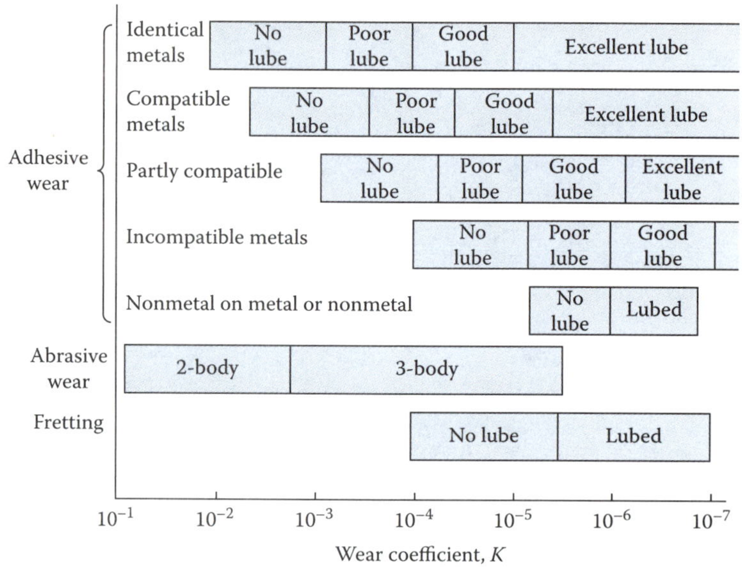

FIGURE 8.4
Some selected wear constants for a variety of general sliding situations. (Based on Peterson, M.B. and Winer, W.O. eds., *Wear Control Handbook*, The American Society of Mechanical Engineers, New York, 1980; Robinowicz, E., *Wear Coefficients-Metals*, American Society of Mechanical Engineers, New York, 1980, pp. 475–506.

TABLE 8.2
Coefficients of Adhesive Wear for Various Rubbing Materials

Material 1	Material 2	Adhesive Wear Coefficient, K
Copper	Copper	10^{-2}
Mild steel	Mild steel	10^{-2} to 10^{-1}
Brass	Hard steel	10^{-3}
Lead	Steel	2×10^{-3}
Polytetrafluoroethylene	Steel	2×10^{-5}
Stainless steel	Hard steel	2×10^{-5}
Tungsten carbide	Tungsten carbide	10^{-6}
Polyethylene	Steel	10^{-8} to 10^{-7}

TABLE 8.3
Adhesive Wear Coefficients K for Typical Metallic Sliding Surfaces

	Identical	Compatible	Partially Compatible or Partially Incompatible	Incompatible
Unlubricated	15×10^{-4}	5×10^{-4}	10^{-4}	15×10^{-6}
Poor lubrication	3×10^{-4}	10^{-4}	2×10^{-5}	3×10^{-6}
Good lubricant	3×10^{-5}	10^{-5}	2×10^{-6}	3×10^{-7}
Excellent lubrication	10^{-6}	3×10^{-7}	10^{-7}	3×10^{-8}

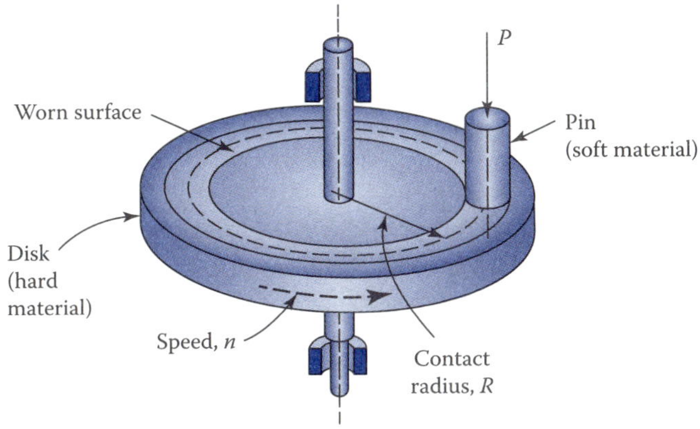

FIGURE 8.5
Schematic of pin-on-disk testing apparatus.

Note that *unlubricated surfaces* are those operated in air without the presence of a lubricant. Water, alcohol, and kerosene are among the *poor lubricant* category. *Good lubricants* include petroleum-based liquids and organic synthetic lubricants. *Excellent lubrication* is very difficult to attain for sliding between like and compatible metals; however, it is less difficult for partially compatible or incompatible metal surfaces [10]. Lubricants and lubrication will be studied in somewhat more details in Part A of Chapter 10.

Presently, there exist a few standard methods for different types of wear to obtain the amount of material removal during a specified time period under well-defined conditions. The ASTM International Committee attempts to update wear testing for specific application. The Society of Tribology and Lubrication Engineers (STLE) list a number of frictional wear and lubrications tests. The literature contains values of K for numerous combinations of metals that have been obtained under laboratory conditions. The results must be evaluated in service.

Test data to define wear constants K show considerable scatter and cannot be obtained with high precision. A common test used to estimate the wear volume is the pin-on-disk method. In this procedure, under controlled loading and lubrication conditions, a cylindrical round end pin is pressed against the surface of a rotary disk on the testing apparatus (Figure 8.5). Weight (and hence equivalent volume) losses of pin and disk can be measured for a specified test duration. Finally, Equation 8.2 is used to determine the wear coefficient, as illustrated in the solution of the following sample problem.

Example 8.1: Finding Wear Coefficients from Experimental Data

A component of a breaking system consists of the unlubricated rounded end of a 2011-T3 wrought aluminum alloy pin being pushed with a force P against the flat surface of a rotating AISI-1095 HR steel disk (see Figure 8.5). The rubbing contact is at a radius R and the disk rotates at a speed n. Consequent to a t minutes test duration, the disk and pin are weighed. It is found that adhesive wear produced weight losses equivalent to wear volumes of V_a and V_s for the aluminum and steel, respectively.

Find: Compute the wear coefficients.

Data: The given numerical values are as follows:
Steel disk: 248 Brinell hardness (Table B.3) wear volume $V_s = 0.98\ \text{mm}^3$
Aluminum pin: 95 Brinell hardness (Table B.6) wear volume $V_a = 4.1\ \text{mm}^3$
Contact force: $P = 30$ N at a radius $R = 24$ mm
Test duration: $t = 180$ min at a sliding speed $n = 120$ rpm

Solution

See Figure 8.5 and Equation 8.2.

Total length of sliding is expressed as

$$L = 2\pi Rnt \tag{b}$$

$$= 2\pi(24)(120)(180) = 3.26 \times 10^6\ \text{mm}$$

The values of hardness of pin and disk are

$$H_a = 9.81(95) = 932\ \text{MPa}$$
$$H_s = 9.81(248) = 2433\ \text{MPa}$$

Through the use of Equation 8.2, we have $K = VH/PL$. Therefore, introducing the numerical values, the wear coefficients for aluminum pin and steel disk are, respectively,

$$K_a = \frac{4.1(932)}{30(3.26 \times 10^6)} = 3.91 \times 10^{-5}$$

$$K_s = \frac{0.98(2433)}{30(3.26 \times 10^6)} = 2.44 \times 10^{-5}$$

Comment: Observe that the wear coefficient of the pin is about 1.6 times that of the disk wear coefficient. Interestingly, if the worn pin surface remains *flat*, for a given pin diameter, we approximately have

$$V_a = \frac{\pi d^2 h_p}{4} \qquad V_s = 2\pi R d h_d \tag{c}$$

Since wear volumes (as well as d and R) are known, then the linear pin wear depth h_p and the wear depth of h_d in the disk may readily be computed.

8.6 Contact-Stress Distributions

The application of a load over a small area of contact results in unusually high stresses. Situations of this nature are found on a microscopic scale whenever force is transmitted through bodies in contact. The original analysis of elastic contact stresses, by H. Hertz, was published in 1881. In his honor, the stresses at the mating surfaces of curved bodies in compression are called *Hertz contact stresses*. The Hertz problem relates to the stresses owing to the contact surface of a sphere on a plane, a sphere on a sphere, a cylinder on a cylinder, and the like. In addition to rolling bearings, the problem is of importance to cams, push-rod mechanisms, locomotive wheels, valve tappets, gear teeth, and pin joints in linkages.

Consider the contact without deflection of two bodies having curved surfaces of different radii (r_1 and r_2), in the vicinity of contact. If a collinear pair of forces (F) presses the bodies together, deflection occurs and the point of contact is replaced by a small area of contact. The first steps taken toward the solution of this problem are the determination of the size and shape of the contact area as well as the distribution of normal pressure acting on the area. The deflections and subsurface stresses resulting from the contact pressure are then evaluated. The following *basic assumptions* are generally made in the solution of the Hertz problem:

1. The contacting bodies are isotropic, homogeneous, and elastic.
2. The contact areas are essentially flat and small relative to the radii of curvature of the undeflected bodies in the vicinity of the interface.
3. The contacting bodies are perfectly smooth; therefore, friction forces need not be taken into account.

The foregoing set of presuppositions enables elastic analysis by theory of elasticity. Without going into the rather complex derivations, in this section, we introduce some of the results for both cylinders and spheres. The next section concerns the contact of two bodies of any general curvature. Contact problems of rolling bearings and gear teeth are discussed in the later chapters.*

8.7 Spherical and Cylindrical Surfaces in Contact

Figure 8.6 illustrates the contact area and corresponding stress distribution between two spheres, loaded with force F. Similarly, two parallel cylindrical rollers compressed by forces F are shown in Figure 8.7. We observe from the figures that, in each case, the maximum contact pressure exists on the load axis. The area of contact is defined by dimension a for

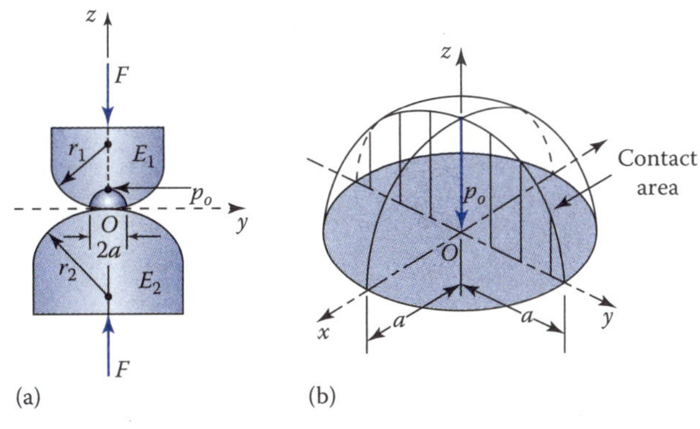

FIGURE 8.6
(a) Spherical surfaces of two members held in contact by force F and (b) contact-stress distribution. *Note*: The contact area is a circle of radius a.

* A summary and complete list of references dealing with contact-stress problems are given by References 10–15.

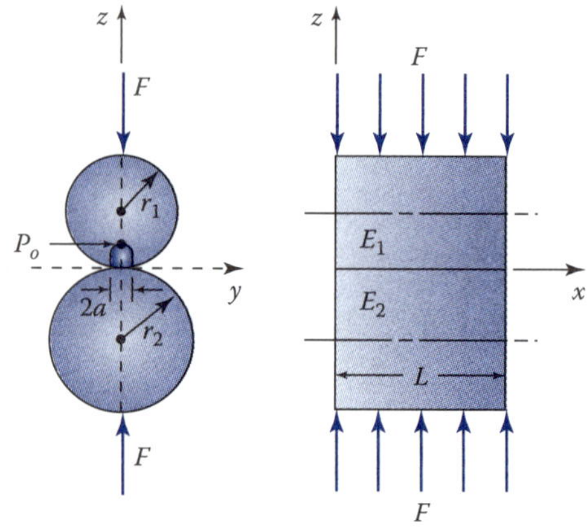

FIGURE 8.7
Two cylinders held in contact by force F uniformly distributed along cylinder length L. *Note*: The contact area is a narrow rectangle of $2a \times L$.

the spheres and a and L for the cylinders. The relationships between the force of contact F, *maximum pressure* p_o, and the *deflection* δ at the point of contact are given in Table 8.4. Obviously, the δ represents the relative displacement of the centers of the two bodies, owing to local deformation. The contact pressure within each sphere or cylinder has a semielliptical distribution; it varies from 0 at the side of the contact area to a maximum value p_o at its center, as shown in the figures. For spheres, a is the radius of the circular contact area (πa^2). But, for cylinders, a represents the half-width of the rectangular contact area ($2aL$), where L is the length of the cylinder. *Poisson's ratio* ν in the formulas is taken as 0.3.

The material along the axis compressed in the z direction tends to expand in the x and y directions. However, the surrounding material does not permit this expansion; hence, the compressive stresses are produced in the x and y directions. The maximum stresses occur along the load axis z, and they are principal stresses (Figure 8.8). These and the resulting maximum shear stresses are given in terms of the maximum contact pressure p_o by the equations to follow [11,14].

8.7.1 Two Spheres in Contact (Figure 8.6)

$$\sigma_x = \sigma_y = -p_o \left\{ \left(1 - \frac{z}{a} \tan^{-1} \frac{1}{z/a} \right) (1 + \nu) - \frac{1}{2\left[1 + (z/a)^2\right]} \right\} \tag{8.4a}$$

$$\sigma_z = -\frac{p_o}{1 + (z/a)^2} \tag{8.4b}$$

Therefore, we have $\tau_{xy} = 0$ and

$$\tau_{\max} = \tau_{yz} = \tau_{xz} = \frac{1}{2}(\sigma_x - \sigma_z) \tag{8.4c}$$

A plot of these equations is shown in Figure 8.9a.

TABLE 8.4

Maximum Pressure p_o and Deflection δ of Two Bodies at the Point of Contact

Configuration	Spheres: $p_o = 1.5\dfrac{F}{\pi a^2}$	Cylinders: $p_o = \dfrac{2}{\pi}\dfrac{F}{aL}$
A 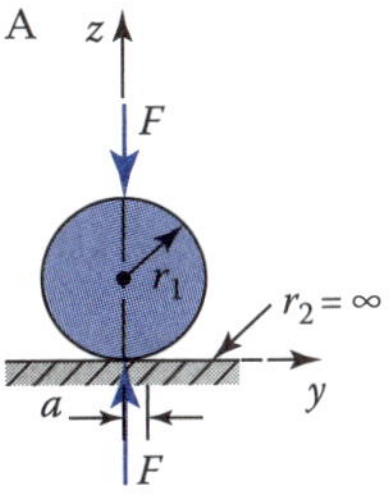	Sphere on a flat surface $a = 0.880\sqrt[3]{Fr_1\Delta}$ $\delta = 0.775\sqrt[3]{F^2\dfrac{\Delta^2}{r_1}}$	Cylinder on a flat surface $a = 1.076\sqrt{\dfrac{F}{L}r_1\Delta}$ For $E_1 = E_2 = E$ $\delta = \dfrac{0.579F}{EL}\left(\dfrac{1}{3} + \ln\dfrac{2r_1}{a}\right)$
B 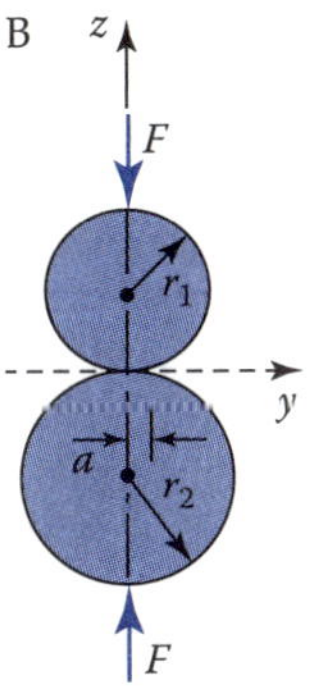	Two spherical balls $a = 0.880\sqrt[3]{F\dfrac{\Delta}{m}}$ $\delta = 0.775\sqrt[3]{F^2\Delta^2 m}$	Two cylindrical rollers $a = 1.076\sqrt{\dfrac{F\Delta}{Lm}}$
C 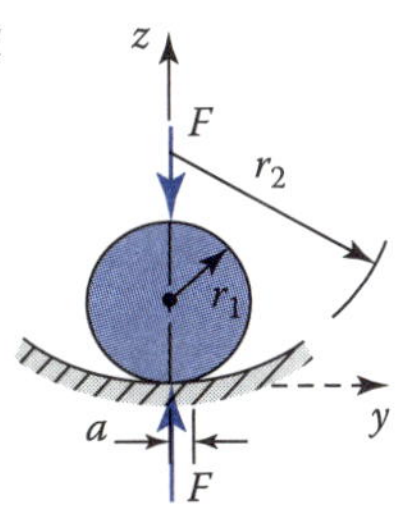	Sphere on a spherical seat $a = 0.880\sqrt[3]{F\dfrac{\Delta}{n}}$ $\delta = 0.775\sqrt[3]{F^2\Delta^2 n}$	Cylinder on a cylindrical seat $a = 1.076\sqrt{\dfrac{F\Delta}{Ln}}$

Source: Ugural, A.C. and Fenster, S.K., *Advanced Mechanics of Materials and Applied Elasticity*, 5th ed., Prentice Hall, Upper Saddle River, NJ, 2012.

Notes: $\Delta = \dfrac{1}{E_1} + \dfrac{1}{E_2}$, $m = \dfrac{1}{r_1} + \dfrac{1}{r_2}$, $n = \dfrac{1}{r_1} - \dfrac{1}{r_2}$, where the modulus of elasticity (E) and radius (r) are for the contacting members, 1 and 2. The L represents the length of the cylinder (Figure 8.7). The total force pressing two spheres or cylinder is F. Poisson's ratio ν in the formulas is taken as 0.3.

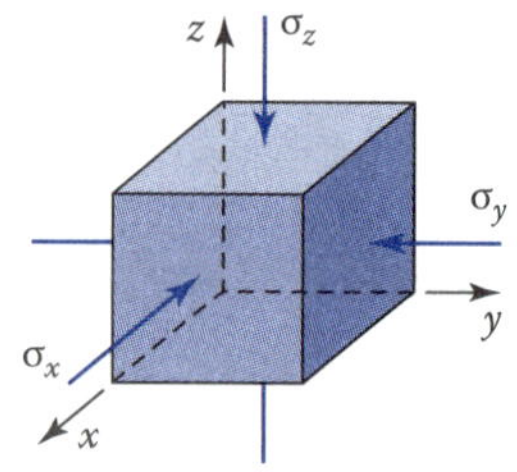

FIGURE 8.8
Principal stress below the surface along the load axis z.

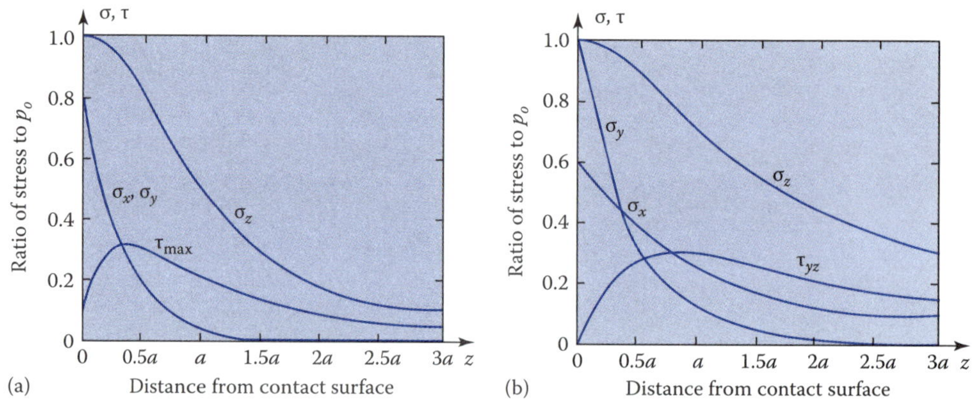

FIGURE 8.9
Stresses below the surface along the load axis (for $\nu = 0.3$): (a) two spheres and (b) two parallel cylinders. *Note*: All normal stresses are compressive stresses.

8.7.2 Two Cylinders in Contact (Figure 8.7)

$$\sigma_x = -2\nu p_o\left[\sqrt{1+\left(\frac{z}{a}\right)^2} - \frac{z}{a}\right] \tag{8.5a}$$

$$\sigma_y = -p_o\left\{\left[2-\frac{1}{1+(z/a)^2}\right]\sqrt{1+\left(\frac{z}{a}\right)^2} - 2\frac{z}{a}\right\} \tag{8.5b}$$

$$\sigma_z = -\frac{p_o}{\sqrt{1+(z/a)^2}} \tag{8.5c}$$

$$\tau_{xy} = \frac{1}{2}(\sigma_x - \sigma_y), \quad \tau_{yz} = \frac{1}{2}(\sigma_y - \sigma_z), \quad \tau_{xz} = \frac{1}{2}(\sigma_x - \sigma_z) \tag{8.5d}$$

Equations 8.5a through 8.5c and the second of Equation 8.5d are plotted in Figure 8.9b. For each case, Figure 8.9 illustrates how principal stresses diminish below the surface. It also shows how the *shear stress* reaches a *maximum* value slightly below the surface (at about $z = 0.75a$) and diminishes. The maximum shear stresses act on the planes bisecting the planes of maximum and minimum principal stresses.

The *subsurface* shear stresses are believed to be responsible for the surface-fatigue failure of contacting bodies (see Section 8.9). The explanation is that minute cracks originate at the point of maximum shear stress below the surface and propagate to the surface to permit small bits of material to separate from the surface. As already noted, all stresses considered in this section exist along the load axis z. The states of stress off the z axis are not required for design purposes, because the maxima occur on the z axis.

Example 8.2: Maximum Contact Pressure between a Cylindrical Rod and a Beam

A concentrated load F at the center of a narrow, deep beam is applied through a rod of diameter d laid across the beam width of width b. Determine

a. The contact area between rod and beam surface
b. The maximum contact stress
c. The maximum value of the subsurface shear stress

Given: $F = 4$ kN, $d = 12$ mm, $L = 125$ mm.

Assumptions: Both the beam and the rod are made of steel having $E = 200$ GPa and $\nu = 0.3$.

Solution

We use the equations on the third column of case A in Table 8.4.

a. Since $E_1 = E_2 = E$ or $\Delta = 2/E$, the half-width of the contact area is

$$a = 1.076\sqrt{\frac{F}{L} r_1 \Delta}$$

$$= 1.076\sqrt{\frac{4(10^3)}{0.125} \frac{(0.006)2}{200(10^9)}} = 0.0471 \text{ mm}$$

The rectangular contact area equals

$$2aL = 2(0.0471)\,(125) = 11.775 \text{ mm}^2$$

b. The maximum contact pressure is therefore

$$p_o = \frac{2}{\pi}\frac{F}{aL} = \frac{2}{\pi}\frac{4(10^3)}{5.888(10^{-6})} = 432.5 \text{ MPa}$$

c. Observe from Figure 8.9b that the largest value of the shear stress is at approximately $z = 0.75a$ for which

$$\frac{\tau_{yz,\max}}{p_o} = 0.3 \quad \text{or} \quad \tau_{yz,\max} = 0.3(432.5) = 129.8 \text{ MPa}$$

This stress occurs at a depth $z = 0.75(0.0471) = 0.0353$ mm below the surface.

CASE STUDY 8.1 Cam and Follower Stress Analysis of an Intermittent-Motion Mechanism

Figure 7.16 (see p. 367) shows a camshaft and follower of an intermittent-motion mechanism. For the position indicated, the cam exerts a force F_{max} on the follower. What are the maximum stress at the contact line between the cam and follower and the deflection?

Given: The shapes of the contacting surfaces are known. The material of all parts is AISI 1095 steel carburized on the surfaces, oil quenched, and tempered (Q&T) at 650°C.

Data:

$$F_{max} = P_{max} = 1.6 \text{ kips}, \quad r_c = 1.5 \text{ in.}, \quad D_f = L_4 = 1.5 \text{ in.}$$

$$E = 30 \times 10^6 \text{ psi}, \quad S_y = 80 \text{ ksi}$$

Assumptions: Frictional forces can be neglected. The rotational speed is slow so that the loading is considered static.

Solution

See Figure 7.16, Tables 8.4, B.1, and B.4.

Equations on the second column of case A of Table 8.4 apply. We first determine the half-width a of the contact patch. Since $E_1 = E_2 = E$ and $\Delta = 2/E$, we have

$$a = 1.076\sqrt{\frac{F_{max}}{L_4} r_c \Delta}$$

Substitution of the given data yields

$$a = 1.076\left[\frac{1600}{1.5}(1.5)\left(\frac{2}{30 \times 10^6}\right)\right]^{1/2}$$

$$= 11.113\left(10^{-3}\right) \text{ in.}$$

The rectangular patch area is

$$2aL_4 = 2(11.113 \times 10^{-3})(1.5) = 33.34(10^{-3}) \text{ in.}^2$$

Maximum contact pressure is then

$$p_o = \frac{2}{\pi}\frac{F_{max}}{aL_4}$$

$$= \frac{2}{\pi}\frac{1600}{\left(11.113 \times 10^{-3}\right)(1.5)} = 61.11 \text{ ksi}$$

The deflection δ of the cam and follower at the line of contact is obtained as follows:

$$\delta = \frac{0.579F_{\max}}{EL_4}\left(\frac{1}{3} + \ln\frac{2r_c}{a}\right)$$

Introducing the numerical values,

$$\delta = \frac{0.579(1600)}{30\times10^6(1.5)}\left(\frac{1}{3} + \ln\frac{2\times1.5}{11.113\times10^{-3}}\right)$$

$$= 0.122\left(10^{-3}\right)\text{ in.}$$

Comments: The contact stress is determined to be less than the yield strength and the design is satisfactory. The calculated deflection between the cam and the follower is very small and does not affect the system performance.

*8.8 Maximum Stress in General Contact

In this section, we introduce some formulas for the determination of the maximum contact stress or pressure p_o between the two contacting bodies that have any general curvature [10,15]. Since the radius of curvature of each member in contact is different in every direction, the equations for the stress given here are more complex than those presented in the preceding section. A brief discussion on factors affecting the contact pressure is given in Section 8.9.

Consider two rigid bodies of equal elastic modulus E, compressed by F, as shown in Figure 8.10. The load lies along the axis passing through the centers of the bodies and

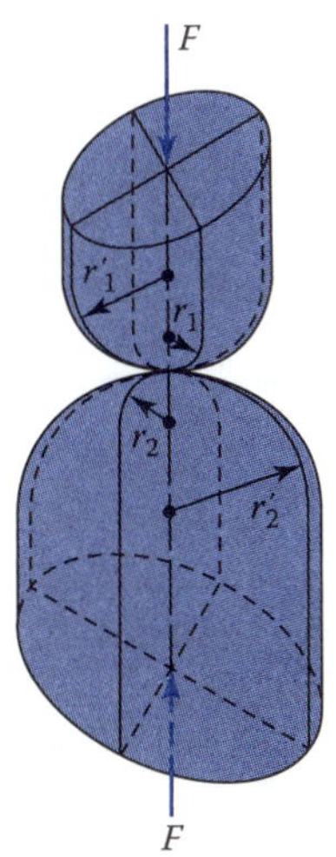

FIGURE 8.10
Curved surfaces of different radii of two bodies compressed by forces F.

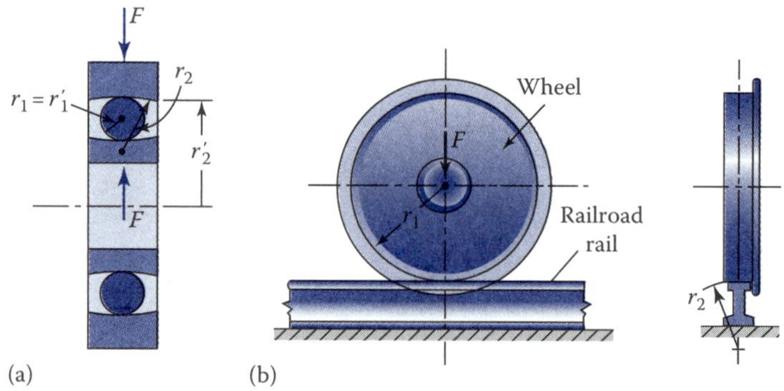

FIGURE 8.11
Contact loads in a (a) single-row ball bearing and (b) wheel and rail.

through the point of contact and is perpendicular to the plane tangent to both bodies at the point of contact. The minimum and maximum radii of curvature of the surface of the upper body are r_1 and r_1'; those of the lower body are r_2 and r_2' at the point of contact. Therefore, $1/r_1$, $1/r_1'$, $1/r_2$, and $1/r_2'$ are the principal curvatures. The *sign convention* of the *curvature* is such that it is positive if the corresponding center of curvature is inside the body; if the center of the curvature is outside the body, the curvature is negative. (For instance, in Figure 8.11, r_1, r_1' are positive, while r_2, r_2' are negative.)

Let θ be the angle between the normal planes in which radii r_1 and r_2 lie (Figure 8.10). Subsequent to the loading, the area of contact will be an ellipse with semiaxes a and b. The *maximum contact pressure is*

$$p_o = 1.5\frac{F}{\pi ab} \tag{8.6}$$

where

$$a = c_a\sqrt[3]{\frac{Fm}{n}} \qquad b = c_b\sqrt[3]{\frac{Fm}{n}} \tag{8.7}$$

In these formulas, we have

$$m = \frac{4}{\frac{1}{r_1}+\frac{1}{r_1'}+\frac{1}{r_2}+\frac{1}{r_2'}} \qquad n = \frac{4E}{3\left(1-\nu^2\right)} \tag{8.8}$$

The constants c_a and c_b are given in Table 8.5 corresponding to the value of α calculated from the formula

$$\cos\alpha = \frac{B}{A} \tag{8.9}$$

TABLE 8.5

Factors for Use in Equation 8.7

α (°)	c_a	c_b	α (°)	c_a	c_b
20	3.778	0.408	60	1.486	**0.717**
30	2.731	0.493	65	1.378	**0.759**
35	2.397	0.530	70	1.284	**0.802**
40	2.136	0.567	75	1.202	**0.846**
45	1.926	0.604	80	1.128	**0.893**
50	1.754	0.641	85	1.061	**0.944**
55	1.611	0.678	90	1.000	**1.000**

Here,

$$A = \frac{2}{m}, \quad B = \pm\frac{1}{2}\left[\left(\frac{1}{r_1} - \frac{1}{r_1'}\right)^2 + \left(\frac{1}{r_2} - \frac{1}{r_2'}\right)^2 + 2\left(\frac{1}{r_1} - \frac{1}{r_1'}\right)\left(\frac{1}{r_2} - \frac{1}{r_2'}\right)\cos 2\theta\right]^{1/2} \tag{8.10}$$

The proper sign in B must be chosen so that its values are positive.

Using Equation 8.6, many problems of practical importance may be solved. These include contact stresses in rolling bearings (Figure 8.11a) contact stresses in cam and push-rod mechanisms (see Problem P8.10), and contact stresses between a cylindrical wheel and rail (Figure 8.11b).

Example 8.3: Ball Bearing Load Capacity

A single-row ball bearing supports a radial load F as shown in Figure 8.11a. Calculate

a. The maximum pressure at the contact point between the outer race and a ball
b. The factor of safety, if the ultimate strength is the maximum usable stress

Given: $F = 1.2$ kN, $E = 200$ GPa, $\nu = 0.3$, and $S_u = 1900$ MPa. Ball diameter is 12 mm; the radius of the groove, 6.2 mm; and the diameter of the outer race, 80 mm.

Assumptions: The basic assumptions listed in Section 8.6 apply. The loading is static.

Solution

See Figure 8.11a and Table 8.5.
For the situation described, $r_1 = r_2' = 0.006$ m, $r_2 = -0.0062$ m, and $r_2' = -0.04$ m.

a. Substituting the given data into Equations 8.8 and 8.10, we have

$$m = \frac{4}{\dfrac{2}{0.006} - \dfrac{1}{0.0062} - \dfrac{1}{0.04}} = 0.0272, \quad n = \frac{4(200\times 10^9)}{3(0.91)} = 293.0403\times 10^9$$

$$A = \frac{2}{0.0272} = 73.5294, \quad B = \frac{1}{2}\left[(0)^2 + (-136.2903)^2 + 2(0)^2\right]^{1/2} = 68.1452$$

From Equation 8.9,

$$\cos\alpha = \frac{68.1452}{73.5294} = 0.9268, \quad \alpha = 22.06°$$

Corresponding to this value of α, interpolating in Table 8.5, we obtain c_a = 3.5623 and c_b = 0.4255. The semiaxes of the ellipsoidal contact area are found by using Equation 8.7:

$$a = 3.5623\left[\frac{1200 \times 0.0272}{293.0403 \times 10^9}\right]^{1/3} = 1.7140 \text{ mm}$$

$$b = 0.4255\left[\frac{1200 \times 0.0272}{293.0403 \times 10^9}\right]^{1/3} = 0.2047 \text{ mm}$$

The maximum contact pressure is then

$$p_o = 1.5\frac{1200}{\pi(1.7140 \times 0.2047)} = 1633 \text{ MPa}$$

b. Since contact stresses are not linearly related to load F, the safety factor is defined by Equation 1.1:

$$n = \frac{F_u}{F} \quad \text{(a)}$$

in which F_u is the ultimate loading. The maximum principal stress theory of failure gives

$$S_u = \frac{1.5F_u}{\pi ab} = \frac{1.5F_u}{\pi c_a c_b \sqrt[3]{(F_u m/n)}^2}$$

This may be written as

$$S_u = \frac{1.5\sqrt[3]{F_u}}{\pi c_a c_b (m/n)^{2/3}} \quad (8.11)$$

Introducing the numerical values into the preceding expression, we have

$$1900(10^6) = \frac{1.5\sqrt[3]{F_u}}{\pi(3.5623 \times 0.4255)\left(\dfrac{0.0272}{293.0403 \times 10^9}\right)^{2/3}}$$

Solving, F_u = 1891 N. Equation (a) gives then

$$n = \frac{1891}{1200} = 1.58$$

Comments: In this example, the magnitude of the contact stress obtained is quite large in comparison with the values of the stress usually found in direct tension, bending, and torsion. In all contact problems, 3D compressive stresses occur at the point, and hence a material is capable of resisting higher stress levels.

8.9 Surface-Fatigue Failure

Surface fatigue is a process by which the surface of a material is weakened by repeated loading. Fatigue damage is produced when the particles are detached by repeated crack growth of microcracks on the surface. These microcracks are either superficial cracks or subsurface cracks. The discussion of Section 8.6 shows that, when two solid members are pressed together, *contact stresses* are produced. *Pitting* is a *surface-fatigue failure*, also often referred to as *fatigue wear*, due to many repetitions of high contact stress: small pieces of material are lost from the surface, leaving behind pits. Pits grow into larger areas of flaked-off surface material, which is then termed *spalling*. An audible warning is often noticeable when the pitting process commences. In machine components such as rolling-element bearings, gears, friction drives, cams, and tappets, a prevalent form of failure is fatigue wear. In these situations, the removal of material results from a cyclic load variation.

Surface fatigue occurs in pure-rolling or roll-sliding contact, owing to many thousands of cycles of repeated contact stress. A typical stress–life diagram on the basis of computed maximum contact pressure p_o (see Sections 8.7 and 8.8) is shown in Figure 8.12. Note that the degree of sliding usually increases from the parallel rollers (top line) to spur gear teeth (bottom line). Other types of gear have essentially pure sliding at their interfaces.

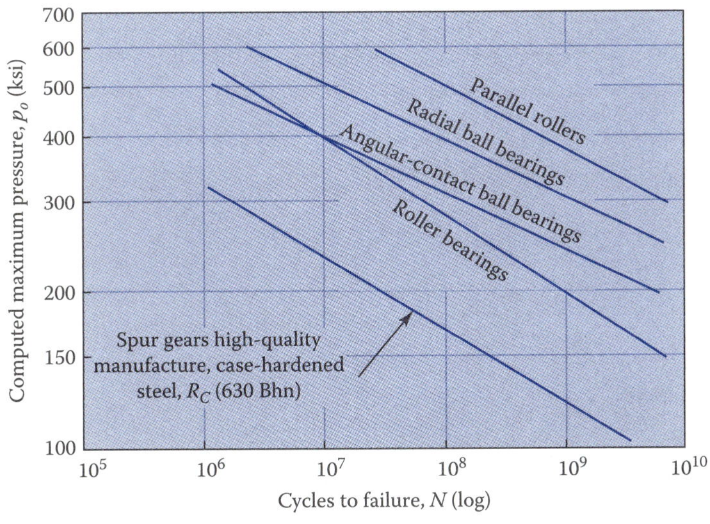

FIGURE 8.12
Average S–N curves for contact stresses, 10% failure probability. (From Juvinall, R.C. and Marshek, K.M., *Fundamentals of Machine Component Design*, 5th ed., Wiley, Hoboken, NJ, 2011.)

Observe from the figure that the tendency of surface-fatigue failure can be reduced by decreasing the sliding and decreasing loads.

High-strength smooth materials are required in contact-stress applications. No material has an endurance limit against surface fatigue. Therefore, a contact stress or surface-fatigue strength value for only a particular number of cycles is given for the materials. Usually, increased surface *hardness* increases resistance to surface fatigue. Also *compressive residual stresses* in the contacting surfaces increase resistance to surface-fatigue failure. These contact stresses can be introduced by methods such as surface treatments, thermal treatments, and mechanical treatments. Thermal stressing occurs whenever a part is heated and cooled, as in heat treatment. The most common methods for introducing surface compressive stresses are shot peening and cold forming (see Section 2.11). Mechanical prestressing refers to the prearranged overloading of the part in the same direction as its service loading, before its being placed in service.

8.9.1 Stresses Affecting Surface Fatigue

When two surfaces are in *pure-rolling* contact, shear stress τ (existing at any point below the surface and a distance from the load axis) reverses while going through the contact zone from A to A', as shown in Figure 8.12. This fully reversed shear stress, as well as the subsurface maximum shear stress occurring along the load axis and maximum contact pressure p_o, may be the cause of pits that begin at the subsurface.

If some *sliding accompanies rolling*, as shown in Figure 8.13, both fully reversed tangential surface shear and normal stresses are produced as any point on the surface rolls through the contact region: pitting begins at the surface. The resulting surface tensile stress leads to the propagation of surface-fatigue cracks. Figure 8.14 depicts the stresses produced below the surface that deform and weaken the metal. As cyclic loading continues, faults or cracks form below the surface (Figure 8.14a). Consequently, the faults merge near (Figure 8.14b) or on the surface. Material at the surface of the element is then readily broken away. In addition, significant factors that influence stresses in contact zone include highly localized heating and thermal expansion produced by sliding friction and the increase in viscosity of the oil due to high pressure in elastohydrodynamic lubrication (see Section 10.16).

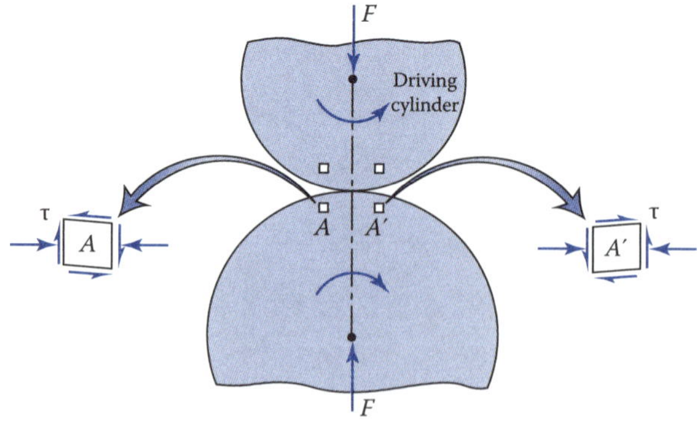

FIGURE 8.13
Two rotating cylinders compressed by force F. Note the subsurface shear stress that reverses when rolling through the contact zone.

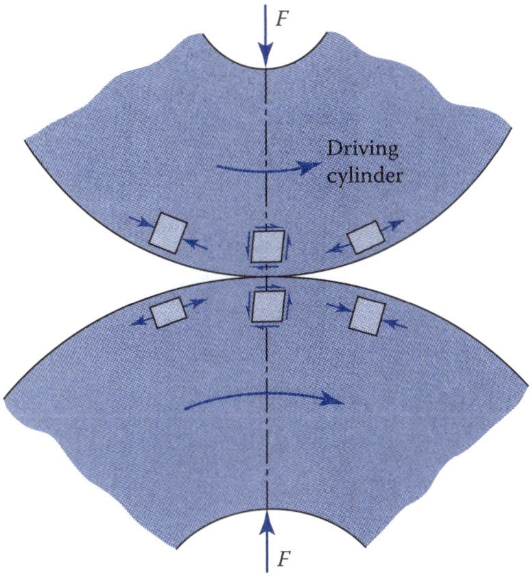

FIGURE 8.14
Two rotating and sliding thick-walled cylinders compressed by force *F*. Tangential normal and shear stresses due to the sliding friction between the members have the largest values at the surface in the locations shown.

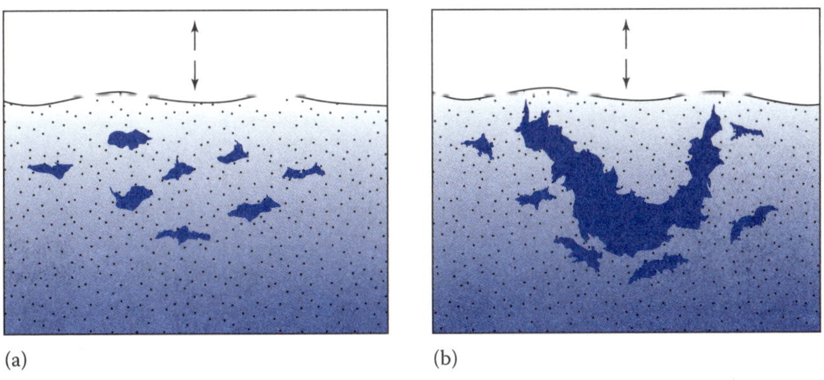

FIGURE 8.15
A simulation of surface-fatigue failure: (a) below the surface and (b) near the surface.

8.10 Prevention of Surface Damage

Corrosion and wear represent enormous ecological and economic burden. Prevention or reduction of surface failure is one of the greatest challenges to modern engineering. Machines should be designed to reduce surface failure as much as feasible and provide for easy replacement of worn-out components. *Smoothness* and *hardness* of a surface (see Sections 7.7 and 8.4) improve fatigue strength and provide resistance to wear. *Compressive residual* stresses in contacting surfaces increase fatigue strength, resistance to SCC, corrosion fatigue, and surface fatigue, as well as decrease damage from fretting corrosion.

In contemporary design of most machine elements, it is important to *choose different materials for the interior and for the surface*, that is, avoid making parts from a single material. When the material most suitable for the bulk of the component does not meet the surface requirements, a second material can usually be used on the surface. Steel parts, for instance, can be coated with chromium, zinc, nickel, or other metals to provide needed resistance. Soft metal components, including plastic parts, can be coated with hard bright surface metal to enhance abrasion resistance and appearance.

For low-friction and wear applications, coatings embodying plastics like Teflon, plating, and application of enamel are often used. Some other plastic coatings are employed for applications where *high* coefficient of friction is required, such as brakes, belts, and clutches. Ceramic material coatings may be used for components with surfaces under extreme heat. Frequently, the desired coating can be mixed with a paint-type material. Ecological and health concerns must be paid attention when choosing a coating material and coating process. Waste disposal without polluting the environment and the development of safe and economical processes for various coatings thus become very significant.

Numerous precautions may be taken into account by a designer to reduce a wear damage. Proper *choice of materials* and *lubricants*, *cleanliness of the surfaces*, and *avoiding stress concentrations* are among the most common remedies. Proper surface finish and hardness, strength to reduce abrasion, and increased surface life are necessary. Corrosive surroundings need special materials and thus coatings should be considered in some applications. Aluminum alloys often undergo a surface treatment. As discussed in Section 2.11, *anodizing*, a process whereby a surface is oxidized, increases corrosion resistance and wear resistance. It is usually applied to protect aluminum alloys.

Material homogeneity in contact-stress applications is useful. Often, higher surface hardness reduces wear and surface-fatigue damage. Hydrodynamic and hydrostatic lubrication are desirable where possible. More will be said about these in Section 10.3. Seals to protect bearings and other joints should be provided. A less-stiff material should be used to increase the contact-patch area and reduce stresses in surface-fatigue situations. Careful attention is required when any type of fatigue loading combined with a corrosive surrounding. Finally, attention is necessary considering the possibility of *fretting* failure when vibration is present in press or shrink fits or tight joints.

Problems

Sections 8.1 through 8.10

8.1 A bearing made of ASTM-A36 structural steel is used in a slow moving gate. In order to increase the bearing life only of the rubbing surfaces, it will be changed with lead of 3 Bhn, brass of 8 Bhn, or polyethylene of 7 Bhn. Which one of these materials will give the longest life?

8.2 A bronze part of 60 Bhn rubs back and forth over a distance of 80 mm in the slot of a 1010 CD steel link of 105 Hbn (Table B.3). Find the volumes of bronze and steel that will wear away during an average of 1500 times per 6 months for a compressive load between the surfaces of $P = 40$ N.

Assumption: Sliding surfaces are unlubricated and metals are partially compatible.

8.3 A steel follower stem of 450 Bhn moves up and down over a distance of 1.5 in. in the sleeve of a 160 Bhn cam follower systems. See, for example, Figure 7.16. The follower arm is to operate an average of 6000 times per month. Find the volumes of metal that will wear away from the stem and follower during a year.

Assumption: Metallic sliding surfaces are identical and have good lubrication. Stem exerts an average y compressive load of $P = 10$ lb on sleeve.

8.4 Reconsider Example 8.1, for the case in which the rotating disk is manufactured of wrought copper alloy with 110 Vickers hardness and a contact force of $P = 25$ kN.

8.5 Redo Example 8.1, knowing that the pin is made of wrought copper alloy with 85 Vickers hardness and a contact force of $P = 35$ kN.

8.6 Two identical 300 mm diameter balls of a rolling mill are pressed together with a force of $F = 500$ N. Determine

a. The width of contact
b. The maximum contact pressure
c. The maximum principal stresses and shear stress in the center of the contact area
d. The largest value of the subsurface shear stress

Assumption: Both balls are made of steel of $E = 210$ GPa and $\nu = 0.3$.

8.7 A spherical ball of radius r_1 fitting in a spherical bearing seat of radius r_2 supports a radial load F as depicted in Figure P8.7. Both ball and seat are made of AISI4130 normalized steel of $\nu = 0.3$, $E = 200$ GPa, and $S_y = 436$ MPa (Table B.4). Compute

a. The pressure at the contact point between the ball and seat
b. The deflection of the ball and seat at the point of contact
c. The maximum value of subsurface shear stress

Given: $r_1 = 6$ mm, $r_2 = 6.05$ mm, $F = 2.2$ kN.

8.8 In a machine, a cylindrical roller of radius r_1 is preloaded with a force F against a parallel cylindrical roller of radius r_2, as shown in Figure 8.7. The rollers have a length L and are made of AIS1 1010 HR steel (see Table B.3). Find

a. The width of contact and the maximum contact pressure
b. The largest value of the subsurface shear stress

Given: $r_1 = 25$ mm, $r_2 = 75$ mm, $L = 25$ mm, $F = 220$ N, $E = 200$ GPa, $S_y = 180$ MPa.

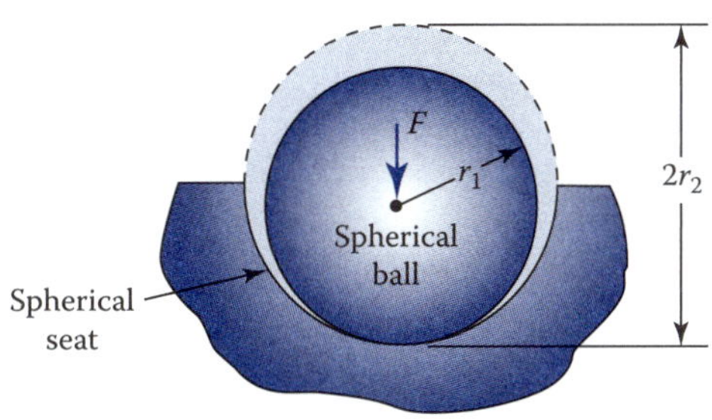

FIGURE P8.7
Ball and socket joint.

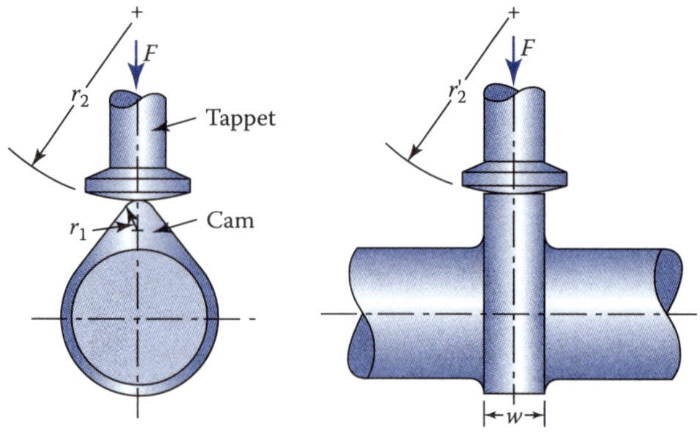

FIGURE P8.10
Valve tappet and cam shaft.

8.9 A 14 mm diameter cylindrical roller runs on the inside of a ring of inner diameter 90 mm (see Figure 10.23a). Calculate

a. The half-width a of the contact area
b. The value of the maximum contact pressure p_o

Given: The roller load is $F = 200$ kN/m of axial length.

Assumption: Both roller and ring are made of steel having $E = 210$ GPa and $\nu = 0.3$.

8.10 A spherical-faced (mushroom) follower or valve tappet is operated by a cylindrical cam (Figure P8.10). Determine the maximum contact pressure p_o.

Given: $r_2 = r_2' = 10$ in., $r_1 = 3/8$ in., and contact force $F = 500$ lb.

Assumptions: Both members are made of steel of $E = 30 \times 10^6$ psi and $\nu = 0.3$.

8.11 Resolve Problem 8.10, for the case in which the follower is flat faced.

Given: $w = 1/4$ in.

8.12 A hardened steel spherical ball of radius r_1 exerts a force F against a flat seat. Find

a. The largest contact stress that results from the loading
b. The deflection of the ball and seat at the point of contact

Given: $r_1 = 2\,1/4$ in., $F = 80$ lb, $S_y = 60$ ksi, $E = 30 \times 10^6$ psi, $\nu = 0.3$.

8.13 A cylindrical roller of radius r_1 and length L is subjected to a load F as it slowly runs inside of semicircular parallel groove with radius r_2 of a block (Figure P8.13). Both roller and block are made of AISI 1030 annealed steel. Determine

a. The width of contact and the largest contact pressure
b. The maximum principal stresses and shear stresses in the center of contact area
c. The largest value of the maximum subsurface shearing stress

Given: $r_1 = 0.6$ in., $r_2 = 0.65$ in., $L = 1.5$ in., $F = 3$ kips, $E = 29 \times 10^6$ psi, $\nu = 0.3$, $S_y = 46$ ksi (from Table B.4).

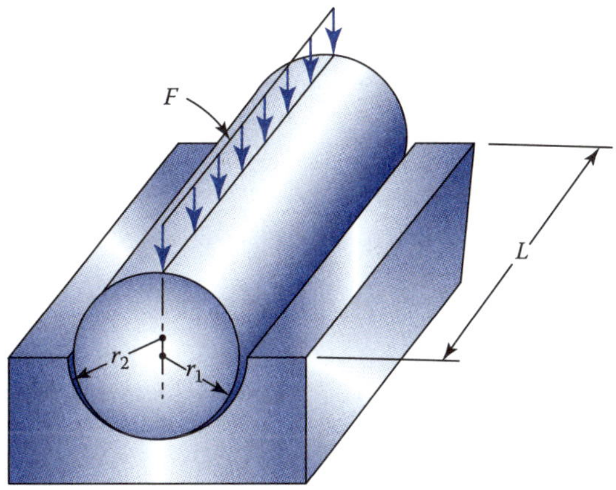

FIGURE P8.13

8.14 A ball of radius r_1 is pressed into a spherical seat of radius r_2 by a force F. Both ball and seat are made of steel. Find

a. The maximum radius of contact area

b. The largest contact pressure

c. The relative displacement of centers of the ball and seat

d. The maximum value of the subsurface shear stress

Given: $r_1 = 2$ in., $r_2 = 2.2$ in., $F = 125$ lb, $E = 30 \times 10^6$ psi, $\nu = 0.3$.

8.15 Determine the largest contact pressure in Problem 8.14 for the cases in which the 2 in. radius ball is pressed against

a. A flat surface

b. An identical ball of 2.5 in. radius

8.16 Consider a concentrated load of $F = 400$ lb at the center of a deep steel beam is applied through a steel rod of radius $r_1 = \frac{1}{2}$ in. laid across the 4 in. beam width. Both members are made of hardened steel with $E = 30 \times 10^6$ psi and $\nu = 0.3$. What are the width of the contact and the deflection between rod and beam surface?

8.17 What are the size of contact area and the largest pressure between two identical circular cylinders with mutually perpendicular axes?

Given: $r_1 = r_2 = 220$ mm, $F = 2$ kN, $E = 206$ GPa, $\nu = 0.25$.

8.18 A train wheel of radius r_1 runs slowly over a steel rail of crown radius r_2 (Figure 8.11b). What is the maximum contact pressure?

Given: $r_1 = 500$ mm, $r_2 = 300$ mm, $F = 5$ kN.

Assumption: Both wheel and rail are made of steel of $E = 206$ GPa and $\nu = 0.3$.

8.19 Redo Example 8.3 for a double-row ball bearing having $r_1 = r_1' = 5$ mm, $r_2 = -5.2$ mm, $r_2' = -30$ mm, $F = 600$ N, and $S_y = 1500$ MPa.

Assumptions: The remaining data are unchanged. The factor of safety is based on the yield strength.

8.20 Redo Problem 8.18, for the case in which the rail is 25 mm wide and flat. Assume the remaining data to be the same.

8.21 A steel cylindrical roller of radius r_1 runs on the inside of a steel ring of inner radius r_2. Compute

a. The width a of the contact area

b. The largest contact pressure

c. The largest value of the subsurface shear stress

Given: $r_1 = 10$ mm, $r_2 = 62.5$ mm, the roller load $F = 300$ kN per meter of axial length, $E = 200$ GPa, $\nu = 0.3$.

8.22 What is the maximum pressure at the contact point between the outer race and a ball in the single-row steel ball bearing assembly illustrated in Figure 8.11a? The ball radius is 20 mm; the radius of the grooves, 25 mm; the radius if the outer race, 200 mm; and the maximum compressive load on the ball $F = 1.2$ kN.

Given: $E = 210$ GPa and $\nu = 0.3$.

8.23 Reconsider Problem 8.22, but use a ball radius of 36 mm and a groove radius 20 mm, with the highest compressive load $F = 900$ N. Assume the remaining data to be unchanged.

Section III

Applications

Partial view of a motorcycle engine (www.google.com). Section III discusses the design of many components, some of which may be contained in this machine, such as shafts, bearings, gears, belts, chains, springs, clutches, brakes, and others. The main function of these elements is, of course, to serve as parts of a system. In the final chapter, we will present case studies in preliminary design of two complete machines

9

Shafts and Associated Parts

9.1 Introduction

Shafts are used in a variety of ways in all types of mechanical equipment. A *shaft*, usually a slender member of round cross section, rotates and transmits power or motion. However, a shaft can have a noncircular cross section and need not be rotating. An *axle*, a nonrotating member that carries no torque, is used to support rotating members. A *spindle* designates a short shaft. *A flexible shaft* transmits motion between two points (e.g., motor and machine), where the rotational axes are at an angle with respect to one another. The customary shaft types are straight shafts of constant or stepped cross section and crankshafts (Figure 9.1). The former two carry rotating members such as gears, pulleys, grooved pulleys (sheaves), or other wheels. The latter are used to convert reciprocating motion into rotary motion or vice versa.

Most shafts are under fluctuating loads of combined bending and torsion with various degrees of stress concentration. Many shafts are not subjected to shock or impact loading: however, some applications arise where such load takes place (Section 9.5). Thus, the associated considerations of static strength, fatigue strength, and reliability play a significant role in shaft design. A shaft designed from the preceding viewpoint satisfies strength requirements. Usually, the shaft geometry is such that the *diameter* will be the *variable* used to satisfy the design. Of equal importance in design is the consideration of shaft deflection and rigidity requirements. Excessive lateral shaft deflection can cause bearing wear or failure and objectionable noise. The operating speed of a shaft should not be close to a critical speed (Section 9.7), or large vibrations are likely to develop.

In addition to the shaft itself, the design usually must include calculation of the necessary keys and couplings. Keys, pins, snap rings, and clamp collars are used on shafts to secure rotating elements. The use of shaft shoulder is an excellent means of axially positioning the shaft elements. Figure 9.2 shows a stepped shaft supporting a gear, a crowned pulley, and a sheave. The mounting parts, discussed in Section 9.8, as well as shaft shoulders, are a source of stress raisers, and they must be properly selected and located to minimize the resulting stress concentrations. Press and shrink fits (Section 9.6) are also used for mounting. Shafts are earned in bearings, in a simply supported form, cantilevered or overhang, depending on the machine configuration. Couplings connect a shaft to a shaft of power source or load. Parameters that must be considered in the selection of a coupling to connect two shafts include the angle between the shafts, transmitted power, vibrations, and shock loads. The websites www.pddnet.com, www.powertransmission.com, and www.grainger.com present general information on shaft couplings.

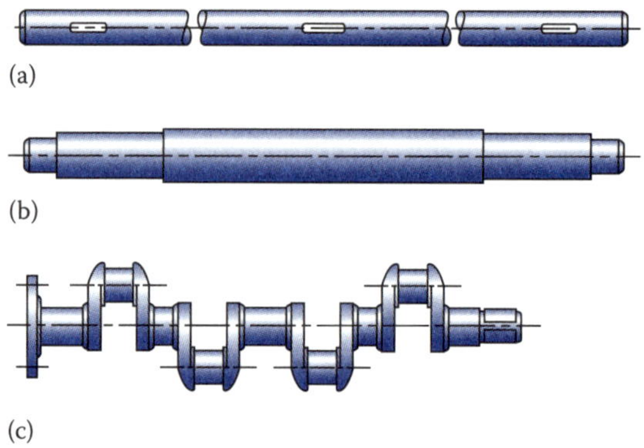

FIGURE 9.1
Common shaft types: (a) constant diameter, (b) stepped, and (c) crankshaft.

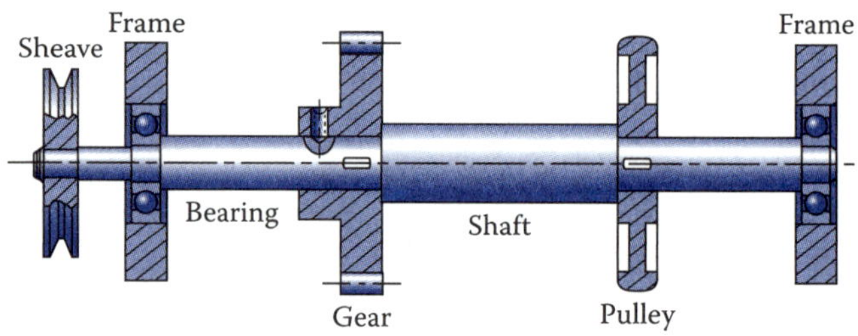

FIGURE 9.2
A stepped shaft with various elements attached.

9.2 Materials Used for Shafting

To minimize deflections, *shaft materials* are generally cold drawn or machined from hot-rolled, plain carbon steel. The shaft ends should be made with chamfers to facilitate forcing on the mounted parts and to avoid denting the surfaces. Cold drawing improves the physical properties. It raises considerably the values of ultimate tensile and yield strengths of steel. Where toughness, shock resistance, and greater strength are needed, alloy steels are used. The foregoing materials can be heat treated to produce the desired properties. If the service requirements demand resistance to wear rather than extreme strength, it is customary to harden only the surface of the shaft, and a carburizing grade steel can be used. Note that the hardening treatment is applied to those surfaces requiring it; the remainder of the shaft is left in its original condition.

Thick-walled seamless tubing is available for simpler, smaller shafts. Large-diameter members (> about 75 mm diameter), such as railroad axles and press cranks, are usually forged and machined to the required size. In addition to steels, high-strength nodular cast iron is used to make shaped shafts, for example, automotive engine crankshafts. Bronze or stainless steel is sometimes used for marine or other corrosive environments. Because *keys* and *pins* are loaded in shear, they are made of ductile materials. Soft, low-carbon steel is in widespread usage. Most keys and pins are usually made from cold-rolled bar stock, cut to length, and tapered if needed.

9.3 Design of Shafts in Steady Torsion

In the design of circular slender shafts that transmit power at a specified speed, the material and the dimensions of the cross section are selected to not exceed the allowable shearing stress or a limiting angle of twist when rotating. Therefore, a designer needs to know the torque acting on the power-transmitting shaft (see Section 1.11). Equations 1.15 through 1.17 may be used to convert the power supplied to the shaft into a constant torque exerted on it during rotation. After having determined the torque to be transmitted, the design of circular shafts to meet strength requirements can be accomplished by using the process outlined in Section 3.2:

1. Assume that, as is often the case, shear stress is closely associated with failure. Note, however, that in some materials, the maximum tensile and compression stresses occurring on planes at 45° (see Figure 3.27) to the shaft axis may cause the failure.
2. An important value of the shear stress is defined by $\tau_{max} = Tc/J$.
3. The maximum usable value of τ_{max} without failure is the yield shear strength S_{ys} or ultimate shear strength S_{us}.
4. A factor of safety n is applied to τ_{max} to determine the allowable stress $\tau_{all} = S_{ys}/n$ or $\tau_{all} = S_{us}/n$. The required parameter J/c of the shaft based on the strength specification of is

$$\frac{J}{c} = \frac{T}{\tau_{all}} \tag{9.1}$$

For a given allowable stress, Equation 9.1 can be used to design both solid and hollow circular shafts carrying torque only.

Example 9.1: Design of a Shaft for Steady Torsion Loading

A solid circular shaft is to transmit 500 kW at $n = 1200$ rpm without exceeding the yield strength in shear of S_{ys} or a twisting through more than 4° in a length of 2 m. Calculate the required diameter of the shaft.

Design Decisions: The shaft is made of steel having $S_{ys} = 300$ MPa and $G = 80$ GPa. A safety factor of 1.5 is used.

Solution

The torque, applying Equation 1.15,

$$T = \frac{9549 \text{ kW}}{n} = \frac{9549(500)}{1200} = 3979 \text{ N} \cdot \text{m}$$

Strength specification. Through the use of Equation 9.1, we have

$$\frac{\pi}{2} c^3 = \frac{3979(1.5)}{300(10^6)}$$

The foregoing gives $c = 23.3$ mm.

Distortion specification. The size of the shaft is now obtained from Equation 4.9:

$$\frac{\phi_{\text{all}}}{L} = \frac{T}{GJ} \tag{9.2}$$

Substituting the given numerical values,

$$\frac{4°(\pi/180)}{2} = \frac{3979}{\left(80 \times 10^9\right)\pi c^4/2}$$

This yields $c = 30.9$ mm.

Comments: The minimum allowable diameter of the shaft must be 61.8 mm. A 62 mm shaft should be used.

9.4 Combined Static Loadings on Shafts

The shaft design process is far simpler when only static loads are present than when the loading fluctuates. However, even with the fatigue loading, a preliminary estimate of shaft diameter may be needed many times, as is observed in the next section. Hence, the results of the rational design procedure of Section 3.2, presented here, is useful in getting the first estimate of shaft diameter for any type of combined static loading conditions.

9.4.1 Bending, Torsion, and Axial Loads

Consider a solid circular shaft of diameter D, acted on by bending moment M, torque T, and axial load P. To begin with, we determine the maximum normal and shear stresses occurring in the outer fibers at a critical section:

$$\begin{aligned} \sigma_x &= \frac{32M}{\pi D^3} + \frac{4P}{\pi D^2} \\ \tau_{xy} &= \frac{16T}{\pi D^3} \end{aligned} \tag{9.3}$$

in which the axial component of σ_x may be either additive or subtractive. The foregoing equations are used with a selected design criterion. Note that, for a hollow shaft, the preceding expressions become

$$\begin{aligned} \sigma_x &= \frac{32M}{\pi D^3\left[1-(d/D)^4\right]} + \frac{4P}{\pi\left(D^2-d^2\right)} \\ \tau_{xy} &= \frac{16T}{\pi D^3\left[1-(d/D)^4\right]} \end{aligned} \tag{9.4}$$

The quantities D and d represent the outer and inner diameters of the shaft, respectively.

Substituting Equation 9.3 into Equation 6.11, a shaft design formula based on the maximum shear theory of failure is

$$\frac{S_y}{n} = \frac{4}{\pi D^3}\left[(8M + PD)^2 + (8T)^2\right]^{1/2} \tag{9.5}$$

Similarly, carrying Equation 9.3 into Equation 6.16, the maximum energy of distortion theory of failure results in

$$\frac{S_y}{n} = \frac{4}{\pi D^3}\left[(8M + PD)^2 + 48T^2\right]^{1/2} \tag{9.6}$$

where S_y represents the yield strength in tension.

9.4.2 Bending and Torsion

Under many conditions, the axial force P in the preceding expressions is either 0 or so small that it can be neglected. Substituting $P = 0$ into Equations 9.5 and 9.6, we have the following shaft design equations based on the maximum shear stress theory of failure:

$$\frac{S_y}{n} = \frac{32}{\pi D^3}\left[M^2 + T^2\right]^{1/2} \tag{9.7}$$

And the *maximum energy of distortion theory* of failure is

$$\frac{S_y}{n} = \frac{32}{\pi D^3}\left[M^2 + \frac{3}{4}T^2\right]^{1/2} \tag{9.8}$$

Likewise, another expression based on the *maximum principal stress theory of failure* may be written as

$$\frac{S_u}{n} = \frac{16}{\pi D^3}\left[M + \sqrt{M^2 + T^2}\right] \tag{9.9}$$

in which S_u is the ultimate tensile strength.

Equations 9.5 through 9.9 can be used to determine the factor of safety n if the diameter D is given or to find the diameter if a safety factor is selected.

Example 9.2: Shaft Design for Combined Bending and Torsion

The gear A is attached to the AISI 1010 CD steel shaft AB of yield strength S_y that carries a vertical load of 360 lb (Figure 9.3a). The shaft is fitted with gear D that forms a set with gear E.

Find: (a) The value of the torque T_E applied on the gear E to support the loading and reactions at the bearings and (b) the required shaft diameter D, applying the maximum shear stress failure criterion.

Given: $S_y = 300/6.895 = 43.5$ ksi (from Table B.3).

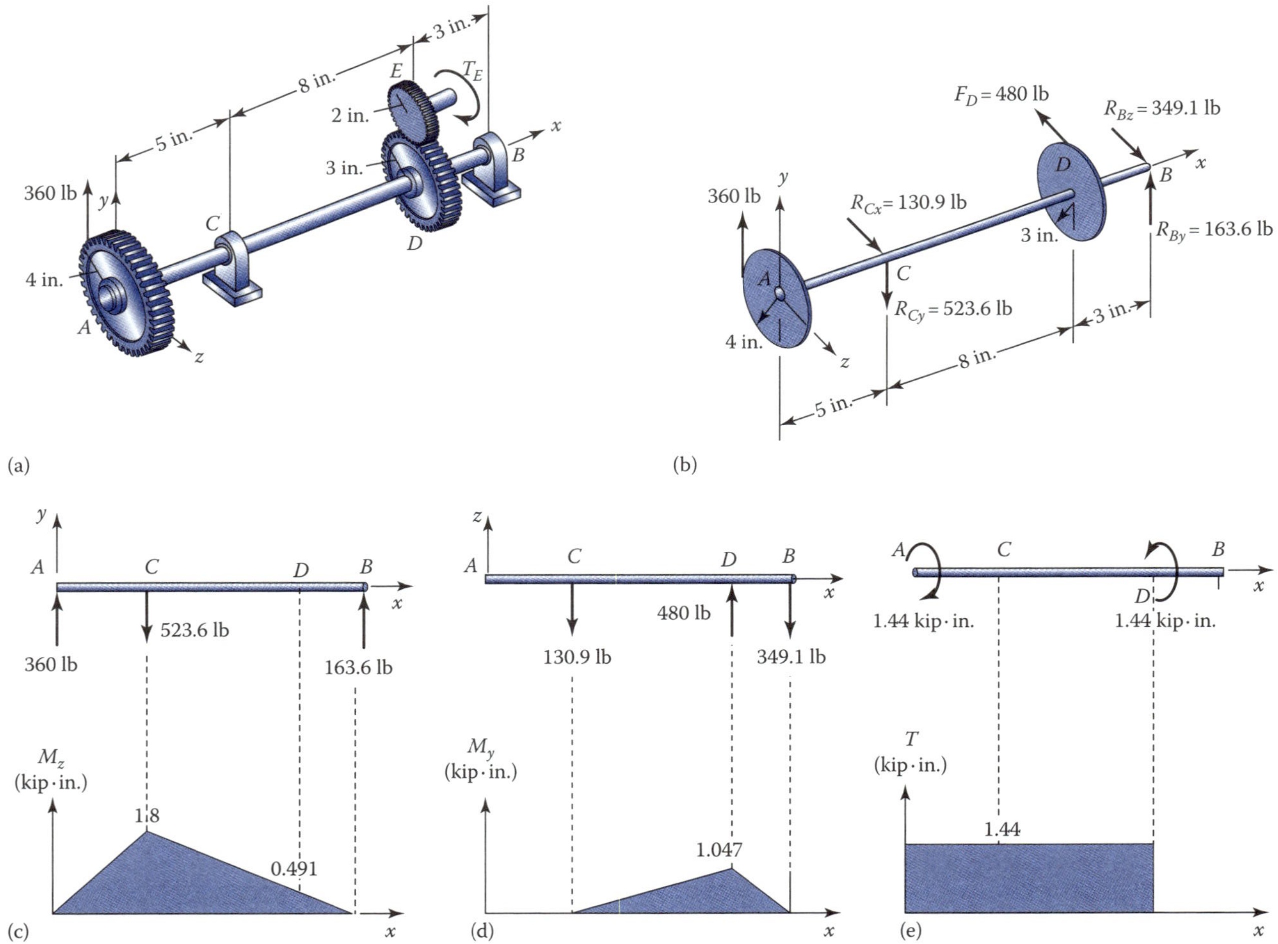

FIGURE 9.3
Example 9.2. (a) Assembly drawing, (b) free-body diagram, (c) moment diagram in *xy* plane, (d) moment diagram in *xz* plane, and (e) torque diagram.

Assumptions: The bearings at B and C are taken as simple supports. A safety factor of $n = 1.6$ is to be used with respect to yielding.

Solution

a. Conditions of equilibrium are applied to Figure 9.3b to find tangential force F_D acting on gear D. Then support reactions are determined using equilibrium conditions and marked on the figure. Referring to Figure 9.3a, we thus have $T_E = F_D(2) = 480(2) = 960 \text{ lb} \cdot \text{in.}$
b. Observe from Figure 9.3c through e that, since $M_C > M_D$, *critical section* where largest value of the stress is expected to occur is at C. Through the use of Equation 9.7, we have

$$D = \left[\frac{32n}{\pi S_y} \sqrt{M_C^2 + T_C^2} \right]^{1/3} \tag{a}$$

Substituting the numerical values results in

$$D = \left[\frac{32(1.6)}{\pi(43.5)} \sqrt{(1.8)^2 + (1.44)^2} \right]^{1/3}$$

$$= 0.952 \text{ in.}$$

Comment: It is interesting to note that, similar to the distortion energy criterion, Equation 9.8 gives $D = 0.936$ in. Thus, a standard diameter of 1.0 in. shaft can be safely used.

CASE STUDY 9.1 Motor-Belt-Drive Shaft Design for Steady Loading

A motor transmits the power P at the speed of n by a belt drive to a machine (Figure 9.4a). The maximum tensions in the belt are designated by F_1 and F_2 with $F_1 > F_2$. The shaft will be made of cold-drawn AISI 1020 steel of yield strength S_y. Note that design of main and drive shafts of a gear box will be considered in *Case Study 18.5*. Belt drives are discussed in details in Chapter 13.

Find: Determine the diameter D of the motor shaft according to the energy of distortion theory of failure, based on a factor of safety n with respect to yielding.

Given: Prescribed numerical values are

$$L = 230 \text{ mm} \quad a = 70 \text{ mm} \quad r = 51 \text{ mm} \quad P = 55 \text{ kW}$$

$$n_0 = 4500 \text{ rpm} \quad S_y = 390 \text{ MPa (from Table B.3)} \quad n = 3.5$$

Assumptions: Friction at the bearings is omitted; bearings act as simple supports. At maximum load $F_1 = 5F_2$.

Solution

Reactions at bearings. From Equation 1.15, the torque applied by the pulley to the motor shaft equals

$$T_{AC} = \frac{9549P}{n_0} = \frac{9549(55)}{4500} = 116.7 \text{ N} \cdot \text{m}$$

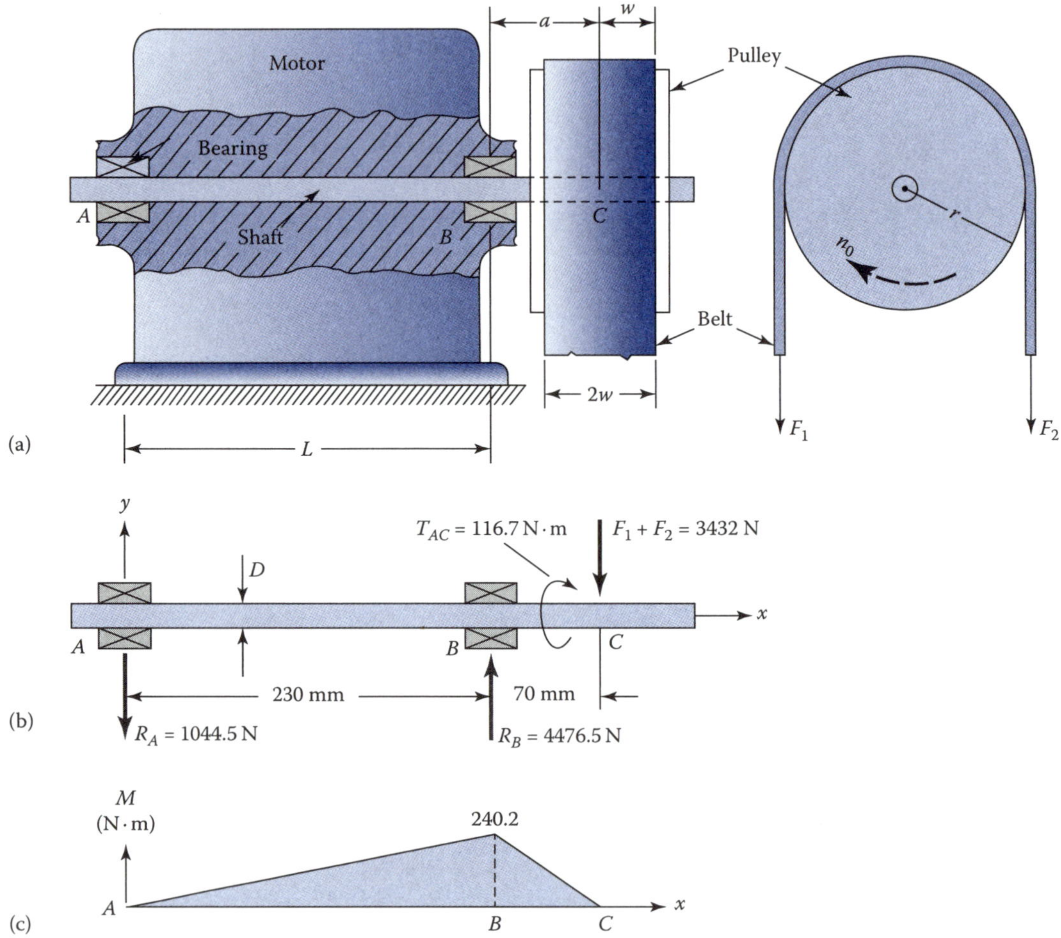

FIGURE 9.4
Motor belt drive: (a) assembly, (b) load diagram of the shaft, and (c) moment diagram of the shaft.

The *forces transmitted* through the belt is therefore

$$F_2 - \frac{F_1}{5} = \frac{T_{AC}}{r} = \frac{116.7}{0.051} = 2288 \text{ N}$$

or

$$F_1 = 2860 \text{ N} \quad \text{and} \quad F_2 = 572 \text{ N}$$

Applying the equilibrium equations to the free-body diagram of the shaft (Figure 9.4b), we have

$$\sum M_A = 3432(0.3) - R_B(0.23) = 0, \quad R_B = 4476.5 \text{ N}$$

$$\sum F_y = -R_A + R_B - 3432 = 0, \quad R_A = 1044.5 \text{ N}$$

The results indicate that R_A and R_B act in the directions shown in the figure.

Principal stresses. The largest moment takes place at support B (Figure 9.4c) and has a value of

$$M_B = 3432\ (0.07) = 240.2\ \text{N}\cdot\text{m}$$

Inasmuch as the torque is constant along the shaft, the critical sections are at B. It follows that

$$\tau = \frac{16T}{\pi D^3} = \frac{16(116.7)}{\pi D^3} = \frac{1867.2}{\pi D^3}$$

$$\sigma_x = \frac{32M}{\pi D^3} = \frac{32(240.2)}{\pi D^3} = \frac{7686.4}{\pi D^3}$$

and $\sigma_y = 0$. For the case under consideration, Equation 3.33 reduce to

$$\sigma_{1,2} = \frac{\sigma_x}{2} \pm \sqrt{\left(\frac{\sigma_x}{2}\right)^2 + \tau^2}$$

$$= \frac{3843.2}{\pi D^3} \pm \frac{1}{\pi D^3}\sqrt{\frac{(7686.4)^2}{4} + (1867.2)^2}$$

$$-\frac{1}{\pi D^3}(3843.2 \pm 4272.8)$$

from which

$$\sigma_1 = \frac{8116}{\pi D^3} \qquad \sigma_2 = -\frac{429.6}{\pi D^3} \tag{b}$$

Energy of distortion theory of failure. Through the use of Equation 6.14,

$$\left[\sigma_1^2 - \sigma_1\sigma_2 + \sigma_2^2\right]^{1/2} = \frac{S_y}{n}$$

This, after introducing Equation (b), leads to

$$\frac{1}{\pi D^3}\left[(8116)^2 - (8116)(-429.6) + (-429.6)^2\right]^{1/2} = \frac{390\left(10^6\right)}{3.5}$$

Solving,

$$D = 0.0288\ \text{m} = 28.8\ \text{mm}$$

Comment: A commercially available shaft diameter of 30 mm should be selected.

9.5 Design of Shafts for Fluctuating and Shock Loads

Shafts are used in a wide variety of machine applications. The design process for circular torsion members is described in Section 9.3: we are now concerned with the members carrying fluctuating and shock loads of combined bending and torsion, which is the case for most transmission shafts [1–5]. Referring to Section 7.8, the definitions of the mean and alternating moments and torques are

$$\begin{aligned} M_m &= \tfrac{1}{2}(M_{\max} + M_{\min}) \\ T_m &= \tfrac{1}{2}(T_{\max} + T_{\min}) \end{aligned} \quad \text{and} \quad \begin{aligned} M_a &= \tfrac{1}{2}(M_{\max} - M_{\min}) \\ T_a &= \tfrac{1}{2}(T_{\max} - T_{\min}) \end{aligned} \tag{9.10}$$

Although in practice design usually includes considerations for associated keys and couplings, these are neglected in the ensuing procedure. We note that all the shaft design formulas to be presented assume an infinite life design of a material with an endurance limit.

For a solid round shaft of diameter D subjected to bending moment M and torsion T, we have on an outermost element

$$\sigma_x = \frac{32M}{\pi D^3} \quad \text{and} \quad \tau_{xy} = \frac{16T}{\pi D^3}$$

We can replace σ_{xm}, σ_{xa}, τ_{xym}, and σ_{xya} by these formulas (using the appropriate subscripts on σ, τ, M, and T) to express the equations developed in Section 7.12 in terms of the bending moment and torque.

The maximum shear stress theory combined with the Goodman fatigue criterion, applying Equation 7.28, is thus obtained as

$$\frac{S_u}{n} = \frac{32}{\pi D^3}\left[\left(M_m + \frac{S_u}{S_e}M_a\right)^2 + \left(T_m + \frac{S_u}{S_e}T_a\right)^2\right]^{1/2} \tag{9.11}$$

In a similar manner, the *maximum energy of distortion theory* incorporated with the Goodman fatigue criterion, from Equation 7.32, is

$$\frac{S_u}{n} = \frac{32}{\pi D^3}\left[\left(M_m + \frac{S_u}{S_e}M_a\right)^2 + \frac{3}{4}\left(T_m + \frac{S_u}{S_e}T_a\right)^2\right]^{1/2} \tag{9.12}$$

The quantities S_u and S_e represent the ultimate strength and endurance limit, respectively.

Note that an alternate form of the maximum energy of distortion theory associated with the Goodman fatigue relation, through the use of Equation 7.35, may be expressed in the following form:

$$\frac{S_u}{n} = \frac{32}{\pi D^3}\left\{\left[\left(\frac{S_u}{S_e}M_a\right)^2 + \frac{3}{4}\left(\frac{S_u}{S_e}T_a\right)^2\right]^{1/2} + \left[(M_m)^2 + \frac{3}{4}(T_m)^2\right]^{1/2}\right\} \tag{9.12'}$$

Alternatively, the maximum shear stress criterion, Equation 9.11, may also be readily written.

9.5.1 Shock Factors

Effect of shock load on shaft has been neglected in the preceding derivation. For some equipment, where operation is jerky, this condition requires special consideration. To account for shock condition, multiplying coefficients (i.e., correction factors K_{sb} in bending and K_{st} in torsion) may be used in the foregoing equations. Thus, the maximum shear stress theory associated with Goodman fatigue relation. Equation 9.11 becomes

$$\frac{S_u}{n} = \frac{32}{\pi D^3}\left[K_{sb}\left(M_m + \frac{S_u}{S_e}M_a\right)^2 + K_{st}\left(T_m + \frac{S_u}{S_e}T_a\right)^2\right]^{1/2} \tag{9.13}$$

Likewise, the *maximum energy of distortion theory* combined with the *Goodman criterion* through the use of Equation 9.12 is

$$\frac{S_u}{n} = \frac{32}{\pi D^3}\left[K_{sb}\left(M_m + \frac{S_u}{S_e}M_a\right)^2 + \frac{3}{4}K_{st}\left(T_m + \frac{S_u}{S_e}T_a\right)^2\right]^{1/2} \tag{9.14}$$

The values for K_{sb} and K_{st} are listed in Table 9.1 [1]; Equations 9.13 and 9.14 represent the general form of design formulas of solid transmission shafts. As shown previously, for hollow shafts of outer diameter D and inner diameter d, D^3 is replaced by $D^3[1 - (d/D)^4]$ in these equations.

9.5.2 Steady-State Operation

Operation of shafts under steady loads involves a completely reversed alternating bending stress (σ_a) and an approximate torsional mean stress (τ_m). This is the case of a rotating shaft with constant moment $M = M_{max} = -M_{min}$ and torque $T = T_{max} = T_{min}$. Therefore,

Equation 9.10 gives

$$M_m = \frac{1}{2}[M + (-M)] = 0 \qquad M_a = \frac{1}{2}[M - (-M)] = M$$
$$T_m = \frac{1}{2}(T + T) = T \qquad T_a = \frac{1}{2}(T - T) = 0 \tag{9.15}$$

TABLE 9.1

Shock Factors in Bending and Torsion

Nature of Loading	K_{sb}, K_{st}
Gradually applied or steady	1.0
Minor shocks	1.5
Heavy shocks	2.0

and Equations 9.11 through 9.14 are simplified considerably. Then, Equation 9.12 results in

$$D^3 = \frac{32n}{\pi S_u}\left[\left(\frac{S_u}{S_e}M_a\right)^2 + \frac{3}{4}T_m^2\right]^{1/2} \tag{9.16a}$$

The preceding expressions could also be written on the basis of the *Soderberg criterion*, replacing the quantity S_u by the yield strength S_y, as needed. In so doing, for instance, Equation 9.16a becomes

$$D^3 = \frac{32n}{\pi S_y}\left[\left(\frac{S_y}{S_e}M_a\right)^2 + \frac{3}{4}T_m^2\right]^{1/2} \tag{9.16b}$$

This is essentially the *ASME shaft design equation* [2]. Note that, for a shaft with varying diameters or other causes of stress concentration, the section of worst combination of moment and torque may not be obvious. It might therefore be necessary to apply design equations at several locations. Clearly, unsteady operation produces fluctuations on the shaft torque (Example 9.3); hence, $T_a \neq 0$ in Equation 9.15.

The foregoing discussion shows that the design of shafts subjected to fluctuating and shock loads cannot be carried out in a routine manner, as in the case of static loads. Usually, the diameter of a shaft must be assumed and a complete analysis performed at a critical section where the maximum stress occurs. A design of this type may require several revisions. The FEA is in widespread use for such cases for final design. Alternatively, experimental methods are used, since formulas of solid mechanics may not be sufficiently accurate.

9.5.3 Displacements

Shaft deflections frequently can be a critical factor, since excessive displacements cause rapid wear of shaft bearings, misalignments of gears driven from the shaft, and shaft vibrations (see Section 9.6). Deflection calculations require that the entire shaft geometry be defined. Hence, a shaft typically is first designed for strength; then the displacements are calculated once the geometry is completely prescribed. Both transverse and twisting displacements must be analyzed. Approaches used in obtaining the deflections of a shaft include the methods of Chapters 4, 5, and 17.

Example 9.3: Shaft Design for Repeated Torsion and Bending

Power is transmitted from a motor through a gear at E to pulleys at D and C of a revolving solid shaft AB with ground surface. Figure 9.5a shows the corresponding load diagram of the shaft. The shaft is mounted on bearings at the ends A and B. Determine the required diameter of the shaft by employing the maximum energy of distortion theory of failure incorporating the Soderberg fatigue relation.

Given: The shaft is made of steel with an ultimate strength of 810 MPa and a yield strength of 605 MPa. Torque fluctuates 10% each way from the mean value. The fatigue stress-concentration factor for bending and torsion is equal to 1.4. The operating temperature is 500°C maximum.

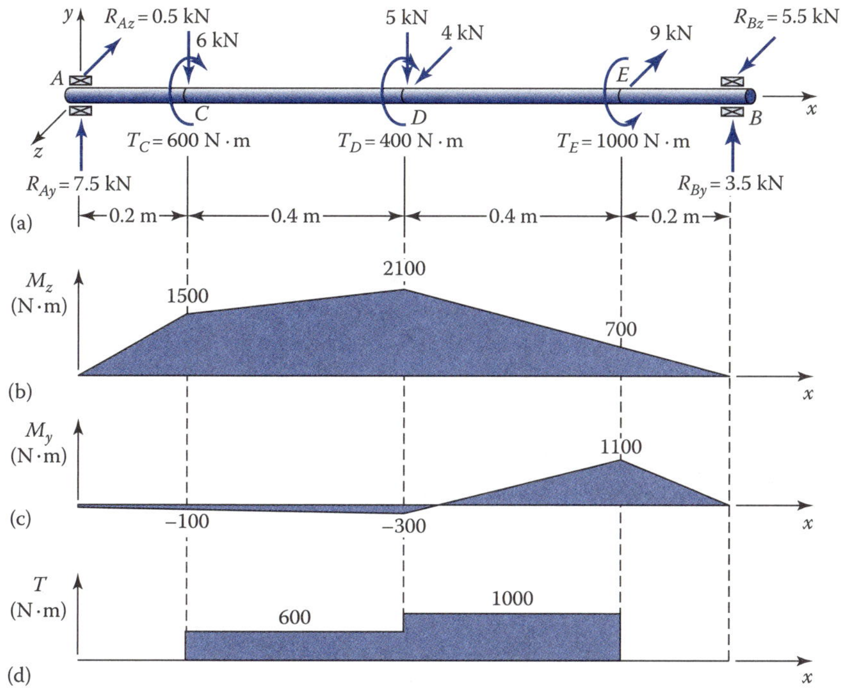

FIGURE 9.5
Example 9.3. (a) Free-body diagram, (b) moment diagram in *xy* plane, (c) moment diagram in *xz* plane, and (d) torque diagram.

Design Assumptions: Bearings act as simple supports. A factor of safety of $n = 2$ is used. The survival rate is taken to be 50%.

Solution

The reactions at *A* and *B*, as obtained from the equations of statics, are noted in Figure 9.5a. The determination of the resultant bending moment of $(M_y^2 + M_z^2)^{1/2}$ is facilitated by using the moment diagrams (Figure 9.5b and c). At point *C*, we have

$$M_C = \left[(0.1)^2 + (1.5)^2\right]^{1/2} = 1.503 \text{ kN} \cdot \text{m}$$

Similarly, at *D* and *E*,

$$M_D = 2.121 \text{ kN} \cdot \text{m} \quad M_E = 1.304 \text{ kN} \cdot \text{m}$$

The maximum bending moment is at *D*. Note from Figure 9.5d that the torque is also maximum at *D*, $T_D = 1$ kN·m. The exact location along the shaft where the maximum stress occurs, the critical section, is therefore at *D*. Hence, at point *D*,

$$M_m = 0 \qquad M_a = 2.121 \text{ kN} \cdot \text{m}$$

$$T_m = 1 \text{ kN} \cdot \text{m} \qquad T_a = 0.1(1) = 0.1 \text{ kN} \cdot \text{m}$$

Using Equation 7.1, the endurance limit of the material is

$$S'_e = 0.5(S_u) = 0.5(810) = 405 \text{ MPa}$$

By Equation 7.7 and Table 7.2, we determine that, for a ground surface,

$$C_f = AS_u^b = 1.58\left(810^{-0.085}\right) = 0.894$$

For reliability of 50%, we have C_r = 1 from Table 7.3. *Assuming* that the shaft diameter will be larger than 51 mm, C_s = 0.70 by Equation 7.9. The temperature factor is found applying Equation 7.11:

$$C_t = 1 - 0.0058(T - 450) = 1 - 0.0058(500 - 450) = 0.71$$

We can now determine the modified endurance limit by Equation 7.6:

$$S_e = C_f C_r C_s C_t \left(1/K_f\right) S_e' = (0.894)(1)(0.70)(0.71)(1/1.4)\left(405 \times 10^6\right) = 128.5 \text{ MPa}$$

Because the loading is smooth, $K_{sb} = K_{st} = 1$ from Table 9.1.

Substituting the S_y = 605 MPa for S_{tt} and the numerical values obtained into Equation 9.14, we have

$$\frac{605\left(10^6\right)}{2} = \frac{32}{\pi D^3}\left[(1)\left(0 + \frac{605 \times 2121}{128.5}\right)^2 + (1)\left(\frac{3}{4}\right)\left(1000 + \frac{605 \times 100}{128.5}\right)^2\right]^{1/2}$$

Solving,

$$D = 0.0697 \text{ m} = 69.7 \text{ mm}$$

Comments: Since this is larger than 51 mm, our assumptions are *correct*. A diameter of 70 mm is therefore quite satisfactory.

Example 9.4: Factor of Safety for a Stepped Shaft under Torsional Shock Loading

A stepped shaft of diameters D and d with a shoulder fillet radius r has been machined from AISI 1095 annealed steel and fixed at end A (Figure 9.6). Determine the factor of safety n, using the maximum shear stress theory incorporated with the Goodman fatigue relation.

Given: Free end C of the shaft is made to rotate back and forth between 1.0° and 1.5° under torsional minor shock loading. The shaft is at room temperature.

Data:

$$L = 300 \text{ mm}, \quad d = 30 \text{ mm}, \quad D = 60 \text{ mm}, \quad r = 2 \text{ mm},$$

$$K_{st} = 1.5 \quad \text{(by Table 9.1)}, \quad G = 79 \text{ GPa} \quad \text{(from Table B.1)}$$

$$S_u = 658 \quad \text{and} \quad H_B = 192 \quad \text{(by Table B.4)}$$

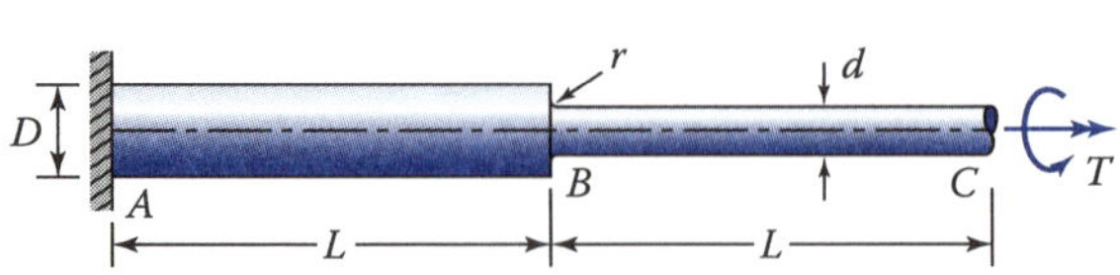

FIGURE 9.6
Example 9.4.

Design Assumption: A reliability of 95% is used.

Solution

From the geometry of the shaft, $D = 2d$ and $L_{AB} = L_{BC} = L$. The polar moment of inertia of the shaft segments are

$$J_{BC} = \frac{\pi d^4}{32} \quad J_{AB} = \frac{\pi D^4}{32} = 16 J_{BC}$$

in which

$$J_{BC} = \frac{\pi}{32}(0.030)^4 = 79.52\left(10^{-9}\right)\text{m}^4$$

The total angle of twist is

$$\phi = \frac{TL}{G}\left(\frac{1}{16J_{BC}} + \frac{1}{J_{BC}}\right)$$

or

$$T = \frac{16GJ_{BC}\phi}{17L}$$

Substituting the numerical data, this becomes $T = 19{,}708.5\phi$. Accordingly, for $\phi_{max} = 0.0262$ rad and $\phi_{min} = 0.0175$ rad, it follows that $T_{max} = 516.4$ N·m and $T_{min} = 344.9$ N·m. Hence,

$$T_m = 430.7\,\text{N}\cdot\text{m} \quad T_a = 85.8\,\text{N}\cdot\text{m}$$

The modified endurance limit, using Equation 7.6, is

$$S_e = C_f C_r C_s C_t \left(\frac{1}{K_f}\right) S_e'$$

where

$C_f = AS_u^b = 4.51(658^{-0.265}) = 0.808$ (by Equation 7.7 and Table 7.2)
$C_r = 0.87$ (from Table 7.3)
$C_s = 0.85$ (by Equation 7.9)
$C_t = 1$ (for normal temperature)
$S_e' = 0.29S_u = 190.8$ MPa (applying Equation 7.4)

and

$K_t = 1.6$ (from Figure C.8, for $D/d = 2$ and $r/d = 0.067$)
$q = 0.92$ (from Figure 7.9, for $r = 2$ mm and $H_B = 192$ annealed steel)
$K_f = 1 + 0.92(1.6 - 1) = 1.55$ (using Equation 7.13b)

Therefore,

$$S_e = (0.808)(0.87)(0.85)(1)(1/1.55)(190.8) = 73.55 \text{ MPa}$$

We now use Equation 9.13 with $M_m = M_a = 0$ to estimate the factor of safety:

$$\frac{S_u}{n} = \frac{32}{\pi d_{BC}^3}\left[K_{st}\left(T_m + \frac{S_u}{S_e}T_a\right)^2\right]^{1/2} \tag{9.17}$$

Introducing the numerical values,

$$\frac{658(10^6)}{n} = \frac{32}{\pi(0.03)^3}\left[1.5\left(430.2 + \frac{658}{73.55}86.2\right)^2\right]^{1/2}$$

This gives $n = 1.19$.

9.6 Interference Fits

Fits between parts, such as a shaft fitting in a hub, affect the accuracy of relative positioning of the components. *Press* or *shrink fit*, also termed *interference fit*, can sustain a load without relative motion between the two mating parts. A *clearance fit* provides ease with which the members can slide with respect to one another. *Tolerance* is the difference between the maximum and minimum size of a part. It affects both function and fabrication cost. Proper tolerancing of the elements is required for a successful design. Table 9.2 lists the eight classes of clearance and interference fits, together with their brief descriptions and applications.

The preferred limits and fits for cylindrical parts are given by the American National Standards Institute (ANSI) Standard B4.1-1967. This is widely used for establishing tolerances for various fits. The American Gear Manufacturers Association (AGMA) Standard 9003-A91, Flexible Couplings—Keyless Fits, contains formulas for the calculation of interference fits.

The interference fits are usually characterized by maintenance of constant pressures between two mating parts through the range of sizes. The amount of interference needed to create a tight joint varies directly with the diameter of the shaft. A simple rule of thumb

TABLE 9.2

Various Fits for Holes and Shafts

Type	Class	Some Common Applications
Clearance	1—loose fit	Road building and mining equipment, where accuracy is not essential
	2—free fit	Machines and automotive parts, where journal speeds of 600 rpm or higher
	3—medium fit	Precise machine tools and automotive part, where speeds under 600 rpm
	4—snug fit	Stationary parts, but can be freely assembled and disassembled
Interference	5—wringing	Parts requiring rigidity, but can be assembled by light tapping with a hammer
	6—tight fit	Semipermanent assemblies for shrink fits on light sections
	7—medium-force fit	Generator or motor armatures and car wheels press or shrink fits on medium sections
	8—high-force and shrink fit	Locomotive wheels and heavy-crankshaft disks, where shrink fit parts can be highly stressed

is to use 0.001 in., of interference for diameters up to 1 in., and 0.002 in., for diameters between 1 through 4 in. A detailed discussion of the state of stresses in shrink fits is found in Chapter 16, where we consider applications to various members.*

9.7 Critical Speed of Shafts

A rotating shaft becomes dynamically unstable at certain speeds, and large amplitudes of lateral vibration develop stresses to such a value that rupture may occur. The speed at which this phenomenon occurs is called a *critical speed*. Texts on vibration theory show that the frequency for free vibration when the shaft is *not* rotating is the same as its critical speed. That is, the critical speed of rotation numerically corresponds to the lateral natural frequency of vibration, which is induced when rotation is stopped and the shaft center is displaced laterally, then suddenly released. Hence, a natural frequency is also called a *critical frequency* or *critical speed* [6]. We shall here consider two simple approaches of obtaining the critical speed of shafts due to Rayleigh and Dunkerley.

9.7.1 Rayleigh Method

Equating the kinetic energy due to the rotation of the mounted shaft masses to the potential energy of the deflected shaft results in an expression, called the *Rayleigh equation*. This expression defines the critical speed of the shaft. A shaft has as many critical speeds as there are rotating masses. Unless otherwise specified, the term *critical speed* is used to refer to the *lowest* or fundamental critical speed. The critical speed n_{cr} (in cycles per second [cps]) for a shaft on two supports and carrying multiple masses is defined as follows:

$$n_{cr} = \frac{1}{2\pi}\left[\frac{g(W_1\delta_1 + W_2\delta_2 + \cdots + W_m\delta_m)}{W_1\delta_1^2 + W_2\delta_2^2 + \cdots + W_m\delta_m^2}\right]^{1/2} = \frac{1}{2\pi}\sqrt{\frac{g\sum W\delta}{\sum W\delta^2}} \tag{9.18}$$

The quantity W_m represents the concentrated weight (including load) of a rotating mass and δ_m is the respective static deflection of the weight, as shown in Figure 9.7. The acceleration of gravity is represented by g as 9.81 m/s² or 386 in./s².

Note that the Rayleigh equation only estimates the critical speed. It ignores the effects of the weight of the shaft, self-damping of the material, and the flexibility of the bearings or supports and assumes that all weights are concentrated. Tests have shown that the foregoing factors tend to lower the calculated critical speed. Thus, the approximate

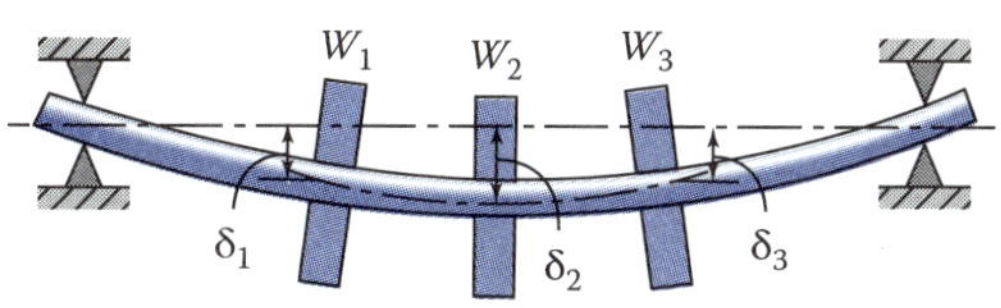

FIGURE 9.7
Simply supported shaft with concentrated loads (deflection greatly exaggerated).

* Some readers may prefer to study Sections 16.3 and 16.4 as a potential assembly method.

values of n_{cr} calculated from Equation 9.18 are always higher than the true fundamental frequency. More accurate approaches for determining the critical frequency, such as a modified Rayleigh's method (Rayleigh–Ritz) and Holzer's method, exist but are somewhat more complicated to implement [7,8].

9.7.2 Dunkerley's Method

The approach consists of reducing the actual multimass system into a number of simple subsystems, then calculating the critical speeds of each by a direct formula, and combining these critical speeds. Accordingly, the actual critical speed n_{cr} of the system is

$$\frac{1}{n_{cr}^2} = \frac{1}{n_1^2} + \frac{1}{n_2^2} + \cdots + \frac{1}{n_m^2} \tag{9.19}$$

Here, n_1, n_2, and n_m represent critical speeds if only mass 1, only mass 2, and only mass mth exists, respectively. Thus, referring to Equation 9.18, we have

$$n_1 = \frac{1}{2\pi}\sqrt{\frac{g}{\delta_i}} \tag{9.20}$$

The approximate values of n_{cr} calculated from Equation 9.19 are always lower than true fundamental frequency.

The main advantages of Dunkerley's and Rayleigh's approaches are that they use simple mechanics of materials formulas for beams. The exact range where the critical speed lies is well established by these two approaches taken together, and thus, they are very popular. Observe that the major difference between the Rayleigh and Dunkerley equations is in the deflections. In the former, the deflection at a specific mass location considers the deflections due to all the masses on the system; in the latter, the deflection is owing to the individual mass being evaluated.

Only rotations sufficiently below or above the critical speed result in dynamic stability of the shafts. In unusual situations, in very high-speed turbines, sometimes satisfactory operation is provided by quickly going through the critical speed and then running well above the critical speed. This practice is to be avoided if possible, as vibration may develop from other causes in the operation above critical speeds, even though the operation is stable. Interestingly, the critical speed of a shaft on three supports is also equal to the natural frequency of the shaft in lateral vibration [7].

Shaft critical speeds may readily be estimated by calculating static deflections at one or several points. The maximum allowable deflection of a shaft is usually determined by the critical speed and gear or bearing requirements. Critical speed requirements vary greatly with the specific application.

9.7.3 Shaft Whirl

Often shafts *cannot* be perfectly straight; also, when a rotor is mounted on a shaft, its center of mass does *not* often coincide with the center of the shaft. In such cases, during rotation, the shaft is subjected to a *centrifugal force* that tends to bend it in the direction of the *eccentricity* of the mass center. This further increases eccentricity and hence the centrifugal force.

Shaft whirl is a self-excited vibration caused by the speed of rotation acting on an eccentric mass. This will always occur when both rotation and eccentricity are present. *Whirling* or *whipping speed* is the speed at which the shaft tends to vibrate violently in transverse direction. When the rotation frequency is equal to one of the resonant frequencies of the shaft, whirling will take place. To prevent the shaft failure, operation at such whirling speeds must be avoided.

Example 9.5: Determining Critical Speed of a Hollow Shaft

A shaft with inner and outer diameters of d and D, respectively, is mounted between bearings and supporting two wheels as shown in Figure 9.8. Calculate the critical speed in rpm, applying (a) the Rayleigh method and (b) the Dunkerley method.

Given: $d = 30$ mm, $D = 50$ mm.

Assumptions: The shaft is made of $L = 1.5$ m long steel having $E = 210$ GPa. The weight of the shaft is ignored. Bearings act as simple supports.

Solution

The moment of inertia of the cross section is $I = \frac{\pi}{4}(25^4 - 15^4) = 267 \times 10^3\ \text{mm}^4$. The concentrated forces are $W_C = 20 \times 9.81 = 196.2$ N and $W_D = 30 \times 9.81 = 294.3$ N. Static deflections at C and D can be obtained by the equations for Case 6 of Table A.9:

$$\delta = \frac{Wbx}{6LEI}\left[L^2 - b^2 - x^2\right] \quad (0 \le x \le a) \tag{a}$$

$$\delta = \frac{Wa(L - x)}{6LEI}\left[2Lx - a^2 - x^2\right] \quad (a \le x \le L) \tag{b}$$

Deflection at C. Due to the load at C, [$L = 1.5$ m, $b = 1$ m, and $x = 0.5$ m, Equation (a)],

$$\delta'_C = \frac{196.2(1)(0.5)(1.5^2 - 1^2 - 0.5^2)}{6(1.5)(267 \times 10^{-9})(210 \times 10^9)} = 0.194 \text{ mm}$$

Owing to the load at D [$L = 1.5$ m, $b = 0.4$ m, and $x = 0.5$ m, Equation (a)],

$$\delta''_C = \frac{294.3(0.4)(0.5)(1.5^2 - 0.4^2 - 0.5^2)}{6(1.5)(267 \times 210)} = 0.215 \text{ mm}$$

The total deflection is then

$$\delta_C = 0.194 + 0.215 = 0.409 \text{ mm}$$

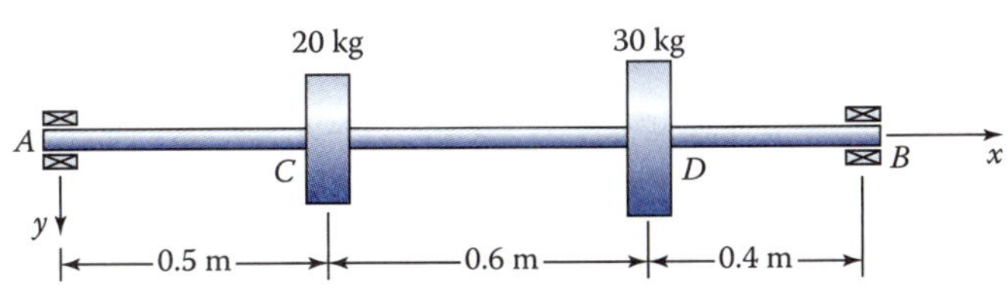

FIGURE 9.8
Example 9.5.

Deflection at D. Due to the load at C, [a = 0.5 m, x = 1.1 m, Equation (b)],

$$\delta'_D = \frac{196.2(0.5)(1.5-1.1)\left[2(1.5)(1.1)-0.5^2-1.1^2\right]}{6(1.5)(267\times 210)} = 0.143 \text{ mm}$$

Owing to the load at D [b = 0.4 m, x = 1.1 m, Equation (a)],

$$\delta''_D = \frac{294.3(0.4)(1.1)\left(1.5^2-0.4^2-1.1^2\right)}{6(1.5)(267\times 210)} = 0.226 \text{ mm}$$

and hence,

$$\delta_D = 0.143 + 0.226 = 0.369 \text{ mm}$$

a. Using Equation 9.18 with m = 2, we have

$$n_{cr} = \frac{1}{2\pi}\left[\frac{9.81\left(196.2\times 0.409\times 10^{-3} + 294.3\times 0.369\times 10^{-3}\right)}{196.2\left(0.409\times 10^{-3}\right)^2 + 294.3\left(0.369\times 10^{-3}\right)^2}\right]^{1/2}$$

$$= 25.37 \text{ cps} = 1522 \text{ rpm}$$

b. Equation 9.19 may be rewritten as

$$\frac{1}{n_{cr}^2} = \frac{1}{n_{cr,C}^2} + \frac{1}{n_{cr,D}^2} \tag{c}$$

Solving,

$$n_{cr} = \frac{n_{cr,C}\cdot n_{cr,D}}{\sqrt{n_{cr,C}^2 + n_{cr,D}^2}} \tag{9.21}$$

where

$$n_{cr,C} = \frac{1}{2\pi}\sqrt{\frac{g}{\delta'_C}} = \frac{1}{2\pi}\sqrt{\frac{9.81}{0.194\left(10^{-3}\right)}} = 35.79 \text{ cps} = 2147 \text{ rpm}$$

$$n_{cr,D} = \frac{1}{2\pi}\sqrt{\frac{g}{\delta''_D}} = \frac{1}{2\pi}\sqrt{\frac{9.81}{0.226\left(10^{-3}\right)}} = 33.16 \text{ cps} = 1990 \text{ rpm}$$

Equation 9.21 is therefore

$$n_{cr} = \frac{(2147)(1990)}{\sqrt{(2147)^2 + (1990)^2}} = 1459 \text{ rpm}$$

Comments: A comparison of the results obtained shows that, the Rayleigh's equation overestimates and the Dunkerley's equation underestimates the critical speed. It follows that the actual critical speed is between 1459 and 1522 rpm. Design of shaft should avoid this operation range.

Example 9.6: Critical Speed of a Stepped Shaft

Figure 9.9a shows a stepped round shaft supported by two bearings and carrying the flywheel weight *W*. Calculate the critical speed in rpm.

Given: The moment of inertia (2*I*) of the shaft in its central region is twice that of the moment of inertia (*I*) in the end parts and:

$$W = 400\ \text{N},\quad L = 1\ \text{m},\quad I = 0.3 \times 10^{-6}\ \text{m}^4.$$

Assumptions: The shaft is made of steel with $E = 200$ GPa. The shaft weight is ignored. Bearings act as simple supports.

Solution

The application of the moment-area method to obtain the static deflection at the midspan *C* is illustrated in Figure 9.9. The bending moment diagram is given in Figure 9.9b and the *M*/*EI* diagram in Figure 9.9c. Note that, in the latter figure, C_1 and C_2 denote the centroids of the triangular and trapezoidal areas, respectively.

The first moment of the various parts of the *M*/*EI* diagram are used to find the deflection. From the symmetry of the beam, the tangent to the deflection curve at *C* is horizontal. Hence, according to the second moment-area theorem defined by Equation 4.24, the deflection δ_C is obtained by taking the moment of the *M*/*EI* area diagram between *A* and *C* about point *A*. That is,

$$\delta_C = (\text{first moment of triangle}) + (\text{first moment of trapezoid})$$

$$= \left(\frac{L}{6}\right)\left(\frac{WL^2}{64EI}\right) + \left(\frac{L}{4} + \frac{5L}{36}\right)\left(\frac{3WL^2}{128EI}\right) = \frac{3WL^3}{256EI} \tag{d}$$

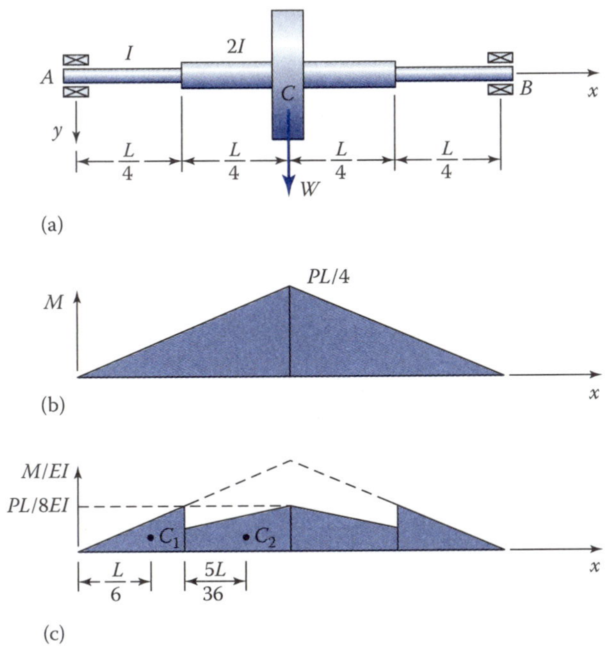

FIGURE 9.9
Example 9.6. Calculation of deflections of a stepped shaft by the moment-area method.

Substituting this, $m = 1$, and $\delta_1 = \delta_C$ into Equation 9.18, we have

$$n_{cr} = \frac{1}{2\pi}\sqrt{\frac{g}{\delta_C}}$$

$$= \frac{1}{2\pi}\left[\frac{9.81(256)(200\times 10^9)(0.3\times 10^{-6})}{3(400)(1)^3}\right]^{1/2}$$

$$= 56.40 \text{ cps} = 3384 \text{ rpm}$$

9.8 Mounting Parts

Mounting parts, such as keys, pins, screws, ring, collars, and splines, are usually used on shafts to attach the hub of rotating members such as gears, pulleys, sprockets, cams, and flywheels. Note that the portion of the mounted members in contact with the shaft is the *hub*. The hub is attached to the shaft in variety of ways, using one of the foregoing mounting elements. Each mounting configuration has its own advantages and disadvantages. Tables of dimensions for the mounting parts may be found in engineering handbooks and manufacturer's catalogs.

9.8.1 Keys

A *key* enables the transmission of torque from the shaft to the hub. Numerous kinds of keys are used to meet various design requirements. They are standardized as to size and shape in several styles. Figure 9.10 illustrates a variety of keys. The grooves in the shaft and hub into which the key fits form the *keyways* or key seats. The square, flat type of keys is most common in machine construction.

The *gib-head key* is tapered so that, when firmly driven, it prevents relative axial motion. Another advantage is that the hub position can be adjusted for the best location. A tapered key may have no head or a gib head (as in Figure 9.10d) to facilitate removal. The *Woodruff key* is semicircular in plan and of constant width (w). It is utilized widely in the automotive and machine tool industries. Woodruff keys yield better concentricity after assembly of the hub and shafts. They are self-aligning and accordingly preferred for tapered shafts.

9.8.2 Pins

A pin is employed for axial positioning and the transfer of relatively light torque or axial load (or both) to the hub. Some types of shaft pins are the straight round pin, tapered round pin, and roll pin (Figure 9.11). The so-called roll pin is a split-tubular spring pin. It has sufficient flexibility to accommodate itself to small amounts of misalignment and variations in hole diameters, so it does not come loose under vibrating loads.

9.8.3 Screws

Very wide keys can be held in place with countersunk flat head or *cap screws* if the shaft is not weakened. In addition to a key or pin, *setscrews* are often employed to keep the hub from shifting axially on the shaft. For light service, the rotation between shaft and hub also

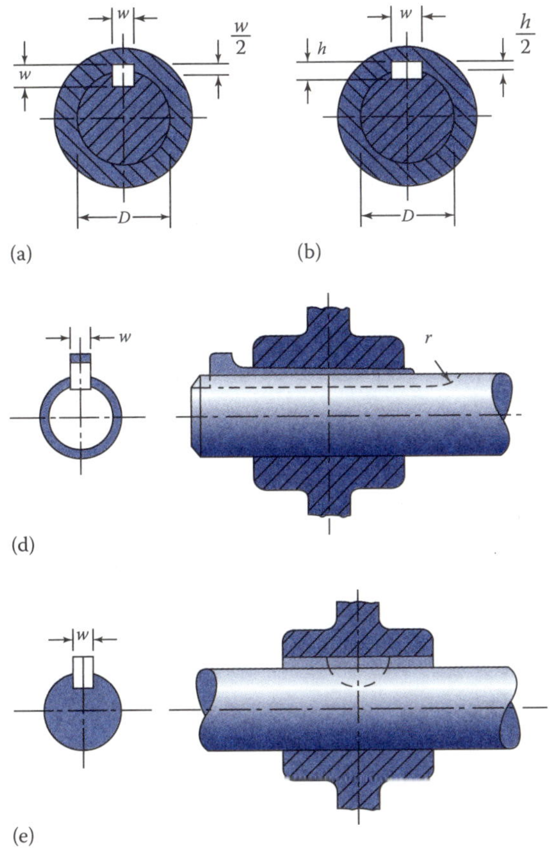

FIGURE 9.10
Common types of shaft keys: (a) square key ($w \approx D/4$), (b) flat key ($w \approx D/4$, $h \approx 3w/4$), (c) round key (often tapered), (d) gib-head key, and (e) Woodruff key.

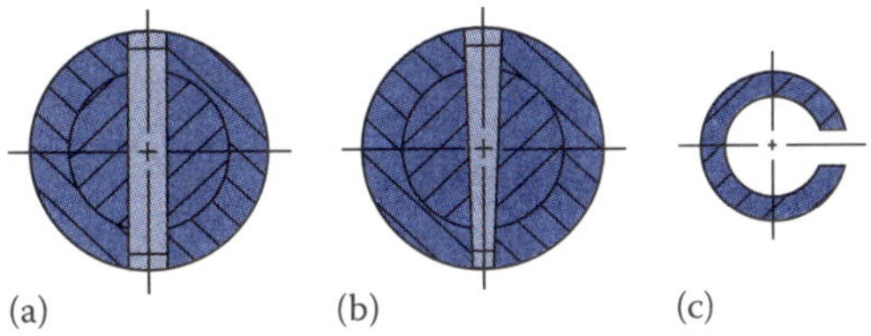

FIGURE 9.11
Some types of pins: (a) straight round pin, (b) tapered round pin, and (c) cross section of a split-tubular pin or so-called roll pin.

may be prevented by setscrews alone. Setscrews are sometimes used in combination with keys. Various types of screws and standardized screw threads are discussed in Chapter 15.

9.8.4 Rings and Collars

Retaining rings, commonly referred to as *snap rings*, are available in numerous varieties and require that a small groove of specific dimensions be machined in the shaft. Keys, pins, and snap rings can be avoided by the use of *clamp collars* that squeeze the outside

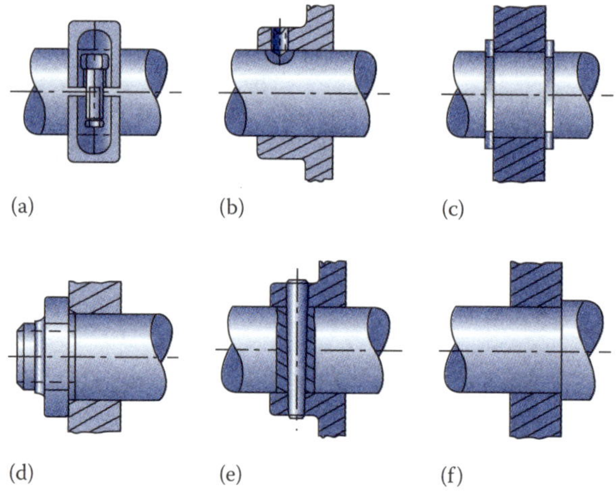

FIGURE 9.12
Various means of securing hubs for axial motion: (a) clamp collar, (b) setscrew, (c) snap rings, (d) nut, (e) tapered pin, and (f) interference fit.

diameter of the shaft with high pressure to clamp something to it. The hub bore and clamp collar have a matching slight taper. The clamp collar with axial slits is forced into the space between hub and shaft by tightening the bolts.

9.8.5 Methods of Axially Positioning of Hubs

Figure 9.12 shows common methods of axially positioning and retaining hubs into shafts. Axial loads acting on shafts or members mounted on the shaft are transmitted as follows: light loads by clamp joints, setscrews, snap rings, and tapered keys (Figure 9.10); medium loads by nuts, pins, and clamp joints; and heavy loads by press or shrink fits. Interference fits are also used to position and retain bearings into hubs.

9.9 Stresses in Keys

The distribution of the force on the surfaces of a key is very complicated. Obviously, it depends on the fit of the key in the grooves of the shaft and hub. The stress varies nonuniformly along the key length; it is highest near the ends.

Owing to many uncertainties, an exact stress analysis cannot be made. However, it is commonly assumed in practice that a key is fitted as depicted in Figure 9.13. This implies that the entire torque T is carried by a tangential force F located at the shaft surface and uniformly distributed along the full length of the key:

$$T = Fr \tag{9.22}$$

where r is the shaft radius.

Shear and compressive or bearing stresses are calculated for the keys from force F, using a sufficiently large factor of safety. For steady loads, a *factor of safety* of 2 is commonly applied. On the other hand, for minor to high shock loads, a factor of safety of 2.5–4.5 should be used.

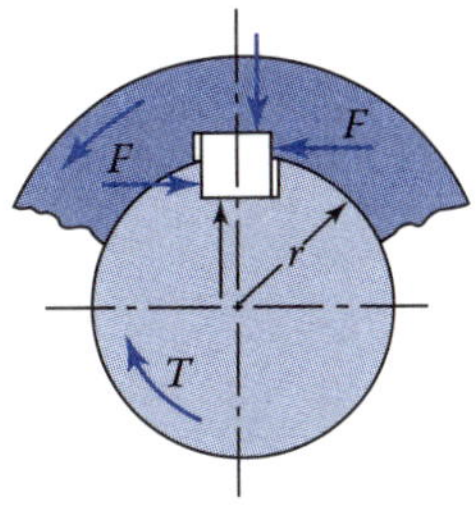

FIGURE 9.13
Forces on a key tightly fitted at the top and bottom.

For keyways, the concentration of stress depends on the values of the fillet radius at the ends and along the bottom of the keyways. For end-milled key seats in shafts under either bending or torsion loading, the theoretical *stress-concentration factors* range from 2 to about 4, contingent on the ratio r of r/D [9]. The quantity r represents the fillet radius (see Figure 9.10d) and D is the shaft diameter. The approximate values of the fatigue stress-concentration factor range between 1.3 and 2 [4].

Example 9.7: Design of a Shaft Key

A shaft of diameter D rotates at 600 rpm and transmits 100 hp through a gear. A square key of width w is to be used (Figure 9.10a) as a mounting part. Determine the required length of the key.

Given: $D = 50$ mm, $w = 12$ mm.

Design Decisions: The shaft and key will be made of AISI 1035 cold-drawn steel having the same tensile and compressive yield strength and that yield strength in shear is $S_{ys'} = S_y/2$. The transmitted power produces intermittent minor shocks and a factor of safety of $n = 2.5$ is used.

Solution

From Table B.3, for AISI 1035 CD steel, we find $S_y = 460$ MPa. Through the use of Equation 1.16,

$$T = \frac{7121(100)}{600} = 1.187 \text{ kN} \cdot \text{m}$$

The force F at the surface of the shaft (Figure 9.13) is

$$F = \frac{T}{r} = \frac{1.187}{0.025} = 47.48 \text{ kN}$$

On the basis of shear stress in the key,

$$\frac{S_y}{2n} = \frac{F}{wL} \quad \text{or} \quad L = \frac{2Fn}{S_y w} \tag{9.23}$$

Substitution of the given numerical values yields

$$L = \frac{2(47,480)(2.5)}{460(10^6)(0.012)} = 43 \text{ mm}$$

Based on compression or bearing on the key or shaft (Figure 9.10a),

$$\frac{S_{yc}}{n} = \frac{F}{(w/2)L} \quad \text{or} \quad L = \frac{2Fn}{S_{yc}w} \tag{9.24}$$

Inasmuch as $S_y = S_{yc}$, this also results in $L = 43$ mm.

9.10 Splines

When axial movement between the shaft and hub is required, relative rotation is prevented by means of splines machined on the shaft and into the hub. For example, splines are used to connect the transmission output shaft to the drive shaft in automobiles, where the suspension movement causes axial motion between the components. Splines are essentially *built-in keys*. They can transform more torque than can be handled by keys. There are two forms of splines (Figure 9.14): straight or square tooth splines and involute tooth splines. The former is relatively simple and employed in some machine tools, automatic equipment, and so on. The latter has an involute curve in its outline, which is in widespread use on gears. The involute tooth has less stress concentration than the square tooth and, hence, is stronger. Also easier to cut and fit, the involute splines are becoming the prominent spline form.

Formulas for the dimensions of splines are based on the nominal shaft diameter. Figure 9.14a shows the standard SAE 6 and 10 straight spline fittings. Note that the values of root diameter d, width w, and depth h of the internal spline are based on the nominal shaft diameter D or about the root diameter of the external spline. According to the SAE, the torque capacity (in lb·in.) of *straight-sided splines* with sliding is

$$T = pnr_mhL_c \tag{9.25}$$

where

T = the theoretical torque capacity
n = the number of splines
$r_m = (D + d)/4$, mean or pitch radius (see Figure 9.14)
h = the depth of the spline
L_c = the length of the spline contact
p = the spline pressure

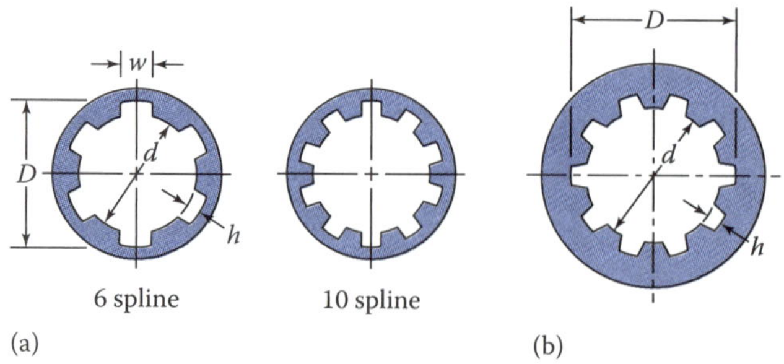

FIGURE 9.14
Some common types of splines: (a) straight-sided and (b) involute.

The SAE states that, in actual practice, owing to the inaccuracies in spacing and tooth form, the contact length L_c is about 25% of the spline length.

Involute splines (Figure 9.14b) have a general form of internal and external involute gear teeth, discussed in detail in Chapter 11, with modified dimensions. The length L_c of spline contact required to transmit a torque, as suggested by the SAE, is

$$L_c = \frac{D^2\left(1 - d_i^4 / D^4\right)}{d_m^3} \tag{9.26}$$

where

D = the nominal shaft diameter
d_m = the mean or pitch diameter
d_i = the internal diameter (if any) of a hollow shaft

The shear area at the mean diameter of the spline is $A_s = \pi d_m L_c/2$. By the SAE assumption, only one-quarter of the shear area is to be stressed. The shear stress is estimated as

$$\tau = \frac{T}{(d_m/2)(A_s/4)} = \frac{8T}{d_m A_s}$$

or

$$\tau = \frac{16T}{\pi d_m^2 L_c} \tag{9.27}$$

Here, T represents the torque on the shaft and L_c is given by Equation 9.25 or 9.26. If bending is present, the flexure stress in the spline must also be calculated.

9.11 Couplings

Couplings are used semipermanently to connect two shafts. They allow machines and shafts to be manufactured in separate units, followed by assembly. A wide variety of commercial shaft couplings are available. They may be grouped into two broad classes: rigid and flexible. A rigid coupling locks the two shafts together, allowing no relative motion between them, although some axial adjustment is possible at assembly. No provision is made for misalignment between the two shafts connected, nor does it reduce shock or vibration across it from one shaft to the other. However, shafts are often subject to some radial, angular, and axial misalignment. In these situations, flexible couplings must be used. Severe misalignment must be corrected; slight misalignment can be absorbed by flexible couplings. This prevents fatigue failure or destruction of bearings.

9.11.1 Clamped Rigid Couplings

Collinear shafts can be connected by clamp couplings that are made in several designs. The most common one-piece split coupling clamps around both shafts by means of bolts and

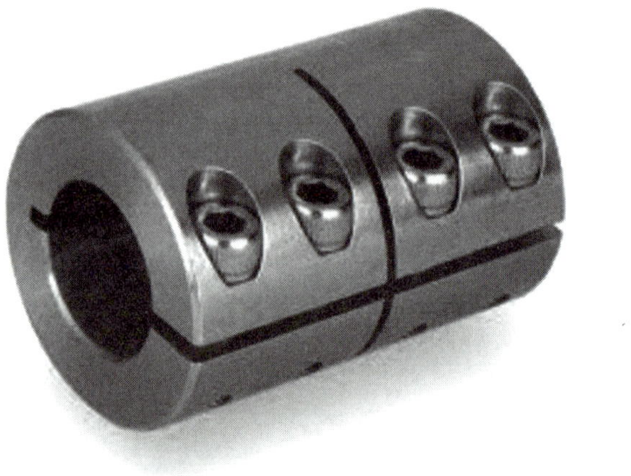

FIGURE 9.15
A rigid coupling: one-piece clamp with keyway.

transmits torque. It is necessary to key the shafts to the coupling (Figure 9.15). The torque is transmitted mainly by friction due to the clamping action and partially by the key. Clamp couplings are widely used in heavy-duty service.

9.11.2 Flanged Rigid Couplings

Collinear shafts can also be connected by flanged couplings, similar to those shown in Figure 9.16. The flanged portion at the outside diameter serves a safety function by shielding the bolt heads and nuts. The load is taken to be divided equally among the bolts. Rigid couplings are simple in design. They are generally restricted to relatively low-speed applications where good shaft alignment or shaft flexibility can be expected.

Keyed couplings are the most widely used rigid couplings (Figure 9.16a). They can transmit substantial torques. The coupling halves are attached to the shaft ends by keys. As can be seen in the figure, flange alignment is obtained by fitting a shallow machined projection on one flange face to a female recess cut in the face of the other flange. Another common way to obtain flange alignment is to permit one shaft to act as a pilot and enter the mating flange. Keyed couplings employ standard keys as discussed in Section 9.8.

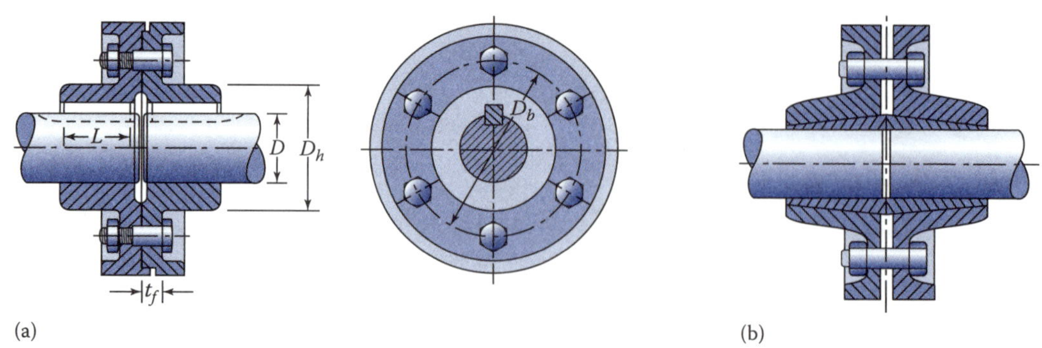

FIGURE 9.16
Flanged rigid couplings: (a) keyed type and (b) compression type.

Compression couplings have a split double cone that does not move axially but is squeezed against the shaft by the wedging of the flanges, as shown in Figure 9.16b. This kind of coupling transmits torque only by the frictional force between the shaft and the split double cone, eliminating the need for a key and keyway in the coupling.

In specifying a rigid coupling using ground and fitted flange bolts, the designer should check the strength of various parts. These include direct shear failure of the bolts, bearing of the projected area of the bolt in contact with the side of the hole, shear of the flange at the hub, and shear or crushing of the key. Note that, in contrast to fitted bolts, a flange coupling designed on the basis of friction-torque capacity requires a somewhat different analysis than that just described.

For flanged rigid couplings, it is usually assumed that shear stress in any one bolt is uniform and governed by the distance from its center to the center of the coupling. Friction between the flanges is disregarded. Then, if the shear stress in a bolt is multiplied by its cross-sectional area, the force in the bolt is ascertained. The moment of the forces developed by the bolts around the axis of a shaft estimates the torque capacity of a coupling.

Example 9.8: Torque Capacity of a Rigid Coupling

A flanged keyed coupling is keyed to a shaft (Figure 9.16a). Calculate the torque that can be transmitted.

Given: There are 6 bolts of 25 mm diameter. The bolt circle diameter is $D_b = 150$ mm.

Assumptions: The torque capacity is controlled by an allowable shear strength of 210 MPa in the bolts.

Solution

Area in shear for one bolt is

$$A = \left(\frac{1}{4}\right)\pi(25)^2 = 491\ \text{mm}^2$$

Allowable force for one bolt is

$$P_{\text{all}} = AS_{ys} = 491(210) = 103.1\ \text{kN}$$

Inasmuch as 6 bolts are available at a 75 mm distance from the central axis, we have

$$T_{\text{all}} = 103.1 \times 10^3 \times 0.075 \times 6 = 46.4\ \text{kN}\cdot\text{m}$$

9.11.3 Flexible Couplings

Flexible couplings are employed to connect shafts subject to some small amount of misalignment. One class of flexible couplings contains a flexing insert such as rubber or spring. The insert cushions the effect of shock and impact loads that could be transferred between shafts. A shear type of rubber-inserted coupling can be used for higher speeds and horsepowers. A chain coupling type consists of two identical sprockets coupled by a roller chain.

Figure 9.17 shows the two identical hubs of a *square-jawed coupling* with an elastomer (i.e., rubber) insert. In operation, the halves slide along the shafts on which they are

FIGURE 9.17
A jaw coupling showing jaws and an elastomer insert. (Courtesy of Magnaloy Coupling Co., Alpena, MI.)

mounted until they engage with the elastomer. The clearances permit some axial, angular, and parallel misalignment. Clearly, the jaws are subjected to bearing and shear stresses. The force acting on the jaw producing these stresses depends on the horsepower and speed that the coupling is to transmit.

Many other types of flexible couplings are available. Examples include helical and bellow couplings. Both are one-piece designs that use their elastic deflections to allow axial and parallel misalignments. Details, dimensions, and load ratings may be found in the catalogs of various manufacturers or mechanical engineering handbooks.

9.12 Universal Joints

A universal joint (U-joint) is a kinematic linkage used to connect two shafts that have permanent intersecting axes. U-joints permit substantial misalignment of shafts. They come in two common types: the *Hooke* or Cardan *coupling*, which does not have constant velocity across a single joint, and the Rzeppa, Bendix-Weiss, or Thompson coupling, which does. Both types can deal with very large angular misalignment. Shaft angles up to 30° may be used [10]. Typical applications for U-joints include automotive drive shafts, mechanical control mechanisms, rolling mill drives, and farm tractors.

Hooke's coupling is the simplest kind of U-joint. It consists of a yoke on each shaft connected by a central cross-link. Figure 9.18 depicts a double-Hooke joint, where plain bearings are used at the yoke-to-cross connections. These joints are employed mostly with equal yoke alignment angles (α) in the two joints, as shown in the figure. The use of equal angles provides uniform angular velocity in the driven shaft. A pair of Hooke's couplings is often used in a rear-drive automobile drive shaft. Note that a familiar application of the Rzeppa-type constant velocity (CV) ball joint is in front-wheel-drive automobiles, where the drive shaft is short and shaft angles can be large. For further information, see texts on mechanics of machinery [6] and manufacturers' product literature on U-joints.

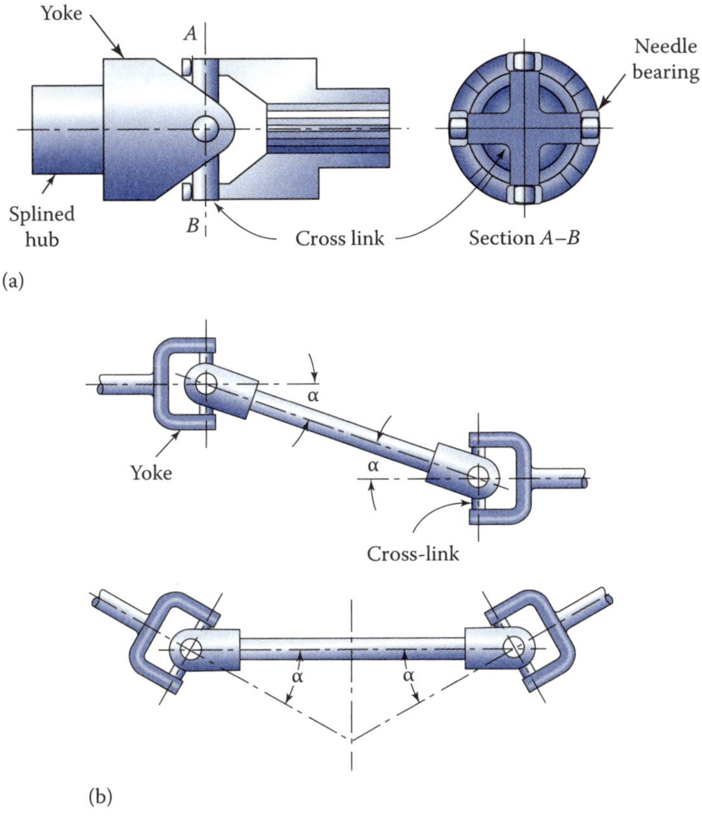

FIGURE 9.18
Simple U-joints: (a) Hook's coupling and (b) two arrangements of a pair of Hooke's couplings for achieving constant velocity ratio.

Problems

The bearings of the shafts described in the following problems act as simple supports.

Sections 9.1 through 9.4

9.1 Design a solid shaft for a 15 hp motor operating at a speed n.

Given: $G = 80$ GPa, $S_{ys} = 150$ MPa, $n = 2500$ rpm

The angle of twist is limited to 2° per meter length.

Design Assumptions: The shaft is made of steel. A factor of safety of 3 is used.

9.2 Repeat Example 9.1, assuming that the shaft is made of an ASTM-50 gray cast iron (see Table B.2) and applying the maximum principal stress theory of failure based on a safety factor of $n = 2.5$.

9.3 A 40 hp motor, through a set of gears, drives a shaft at a speed n, as shown in Figure P9.3.

a. Based on a safety factor of 2, design solid shafts AC and BC.

b. Determine the total angle of twist between A and B.

Given: $G = 82$ GPa, $S_{ys} = 210$ MPa, $n = 1200$ rpm.

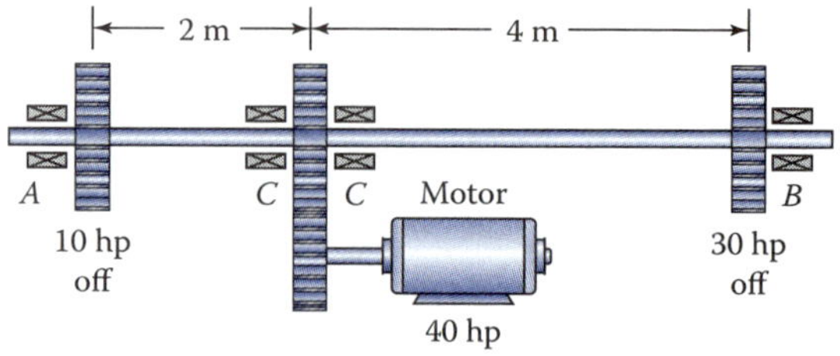

FIGURE P9.3

9.4 Because of atmospheric corrosion, an ASTM-A36 steel shaft of diameter D_s is to be replaced by an aluminum alloy 2014-T6 shaft. Determine the diameter of aluminum shaft D_a in terms of D_s. What is the weight ratio of the shafts?

Assumption: Both shafts have the same angular stiffness.

9.5 A solid steel shaft of diameter D carries end loads P, M, and T. Determine the factor of safety n, assuming that failure occurs according to the following criteria:

a. Maximum shear stress

b. Maximum energy of distortion

Given: $S_y = 260$ MPa, $D = 100$ mm, $P = 50$ kN, $M = 5$ kN·m, $T = 8$ kN·m.

9.6 An ASTM-A242 high-strength steel circular shaft AB having a diameter D, simply supported on bearings at its ends, carries a torque (T) and two loads (F_y, F_z) as illustrated in Figure P9.6. Find the safety factor of n on the basis of

a. Tensile yield strength

b. Shear yield strength

Given: $a = 0.6$ m, $D = 45$ mm, $F_y = 2.4$ kN, $F_z = 1.5$ kN, $T = 1.25$ kN·m. $S_y = 345$ MPa, $S_{ys} = 210$ MPa (Table B.1).

9.7 A solid steel shaft carries belt tensions (at an angle α from the y axis in the yz plane) at pulley C, as shown in Figure P9.7. For $\alpha = 0$ and a factor of safety of n, design the shaft according to the following failure criteria:

a. Maximum shear stress

b. Maximum energy of distortion

Given: $S_y = 250$ MPa, $n = 1.5$.

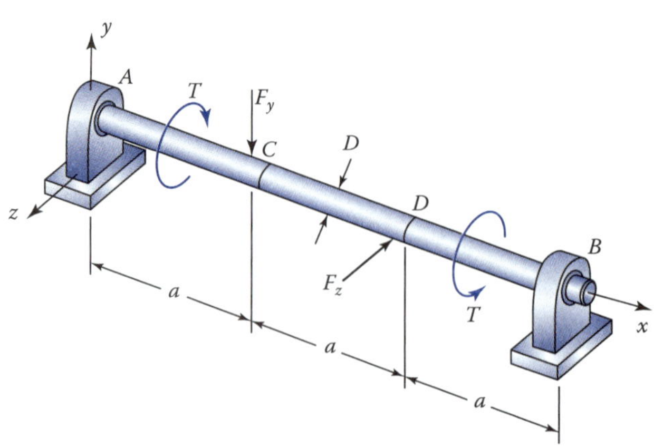

FIGURE P9.6

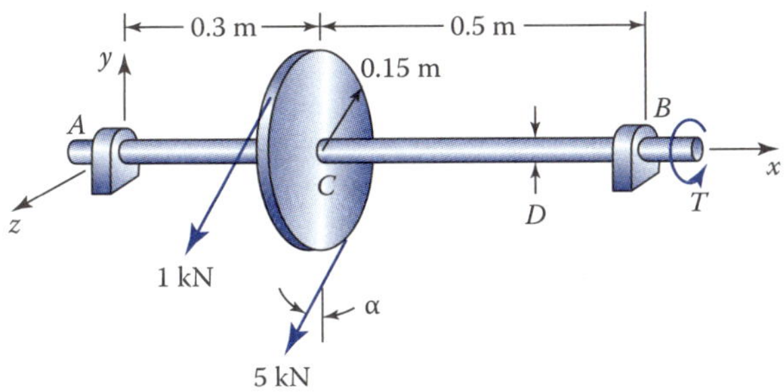

FIGURE P9.7

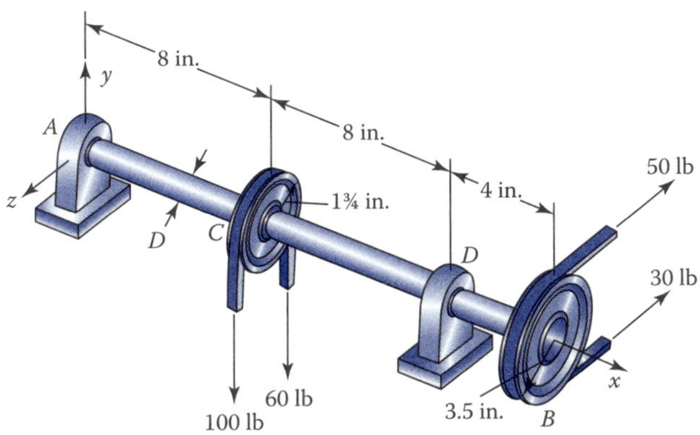

FIGURE P9.8

9.8 A circular shaft–pulley assembly is supported at A and D and subjected to tensile forces through belts at B and C as illustrated in Figure P9.8. Compute the required minimum shaft diameter D on the basis of the maximum shear stress theory of failure based on a safety factor of $n = 1.5$ against the yielding.

Assumption: The shaft is made of an AISI 1050HR steel (Table B.3).

9.9 Reconsider Problem 9.8, but applying the maximum energy of distortion theory of failure with a factor of safety $n = 1.2$ against the yielding.

9.10 A solid shaft is used to transmit 90 hp to a series of chemical mixing vats at a speed of n. Calculate the shaft diameter according to the following failure criteria:

a. Maximum shear stress

b. Maximum energy of distortion

Given: $n = 110$ rpm.

Design Decisions: The shaft is made of type 302 cold-rolled stainless steel. Since the atmosphere may be corrosive, a safety factor of 4 is used.

9.11 A shaft–pulley assembly is supported and loaded as shown in Figure P9.11. What is the diameter D of the shaft, through the use of the maximum principal stress theory of failure with a factor of safety $n = 1.4$?

Assumption: The shaft is made an AISI 1040 CD steel (see Table B.3).

9.12 Resolve Problem 9.11, for the case in which the tensions on pulley C are in the horizontal (z) direction and the factor of safety is $n = 1.5$.

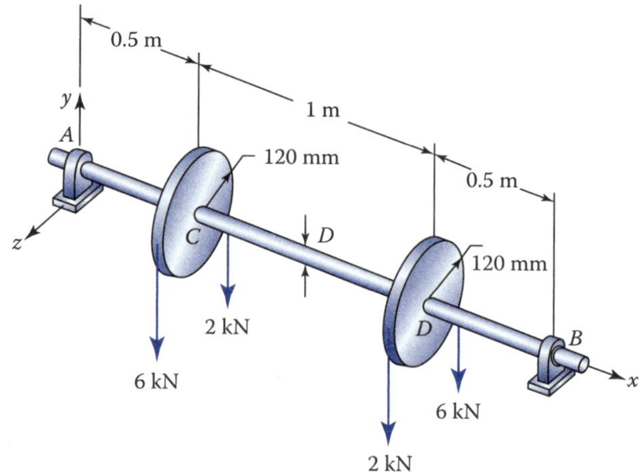

FIGURE P9.11

Section 9.5

9.13 A revolving solid steel shaft *AB* with a machined surface carries minor shock belt tensions (in the *yz* plane) for the case in which $\alpha = 30°$ (Figure P9.7). Design the shaft segment *BC* by using the maximum shear stress theory incorporated with the Goodman failure relation.

Given: $S_u = 520$ MPa, $n = 1.5$, and $K_f = 1.2$.

The operating temperature is 500°C maximum.

Assumptions: The survival rate is 90%.

9.14 A solid shaft of diameter *D* rotates and supports the loading depicted in Figure P9.14. Determine the factor of safety *n* for the shaft on the basis of maximum energy of distortion theory of failure combined with the Goodman criterion.

Given: $D = 2.5$ in. and $S_u = 180$ ksi. The torque fluctuates 10% each way from mean value, and the survival rate is 99%.

Design Decision: The shaft is ground from unnotched steel.

9.15 A solid shaft of diameter *D* rotates and supports the loading shown in Figure P9.15. Calculate the factor of safety *n* for the shaft using the maximum shear stress theory of failure incorporated with the Soderberg criterion.

Given: $D = 3.5$ in., $S_y = 130$ ksi, and $S_u = 200$ ksi. The torque involves heavy shocks and fluctuates 20% each way from the mean value, and the survival rate is 90%. The maximum operating temperature is 950°F.

Design Decision: The shaft is to be hot rolled from an unnotched steel.

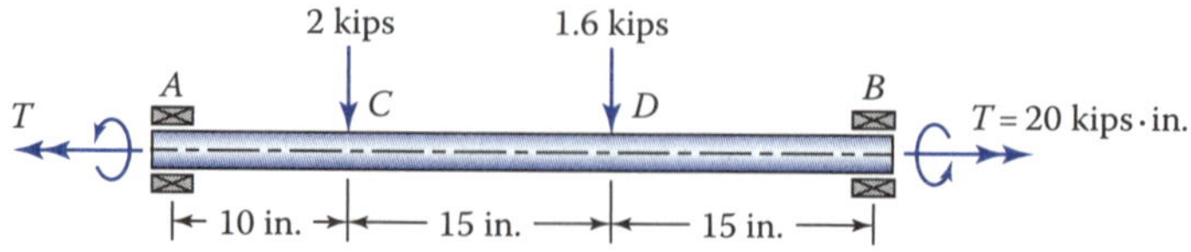

FIGURE P9.14

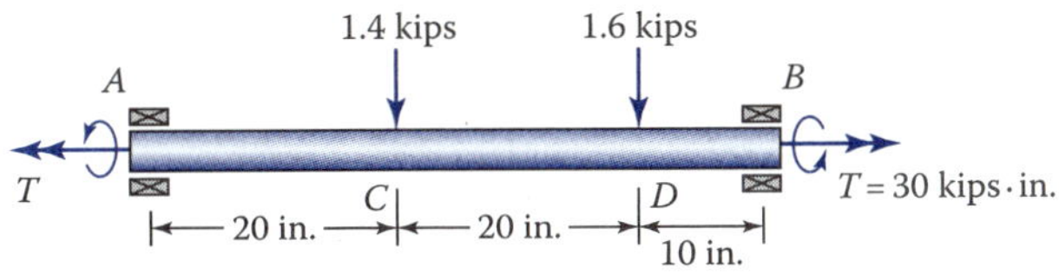

FIGURE P9.15

9.16 A 40 mm diameter shaft, made of AISI 1060 HR steel that has a 5 mm diameter hole drilled transversely through it, carries a steady torque of 600 N · m and in phase with a completely reversed bending moment of 120 N · m. What is, applying the maximum energy of distortion theory of failure combined with Goodman criteria, the factor of safety n?

Given: S_u= 520 MPa, S_y= 440 MPa, 149 Bhn (from Table B.3).

Assumption: The operating temperature will be 480°C maximum at a survival rate of 99%.

9.17 Repeat Problem 9.16, through the use of the Soderberg criterion.

9.18 A rotating solid shaft is acted on by repetitive steady moments M and heavy shock torques T at its ends. Calculate, on the basis of the maximum energy of distortion failure criterion associated with the Goodman theory, the required shaft diameter D.

Given: $M = 200$ N · m, $T = 500$ N · m, $S_u = 455$ MPa, $n = 1.5$, $K_f = 2.2$.

Design Decisions: The shaft is machined from 1020 HR steel. A survival rate of 95% is used.

9.19 Figure P9.19 shows a rotating stepped shaft supported in (frictionless) ball bearings at A and B and loaded by nonrotating force P and torque T. All dimensions are in millimeters. Determine the factor of safety n for the shaft, based on the maximum shear stress theory of failure incorporated with the Soderberg fatigue relation.

Given: $P = 5$ kN, $T = 600$ kN · m, $S_y = 600$ MPa, $S_u = 1000$ MPa, $K_t = 1.8$.

The torque fluctuates 15% each way from mean value, and the survival rate is 98%.

Assumption: The shaft is to be hot rolled from steel.

9.20 When accelerates through a bend at high speeds, an AISI 1050 CD steel drive shaft of a sports car is subjected with minor shocks to a mean moment M_m, alternating bending moment M_a, and a steady torque T_m. What is the shaft diameter D according to the maximum shear stress theory combined with Goodman fatigue criterion?

Given: M_m = 200 N · m, M_a = 600 N · m, T_m = 360 N · m, K_s = 1.5 (by Table 9.1), S_u = 690 MPa, S_y = 580 MPa (from Table B.3).

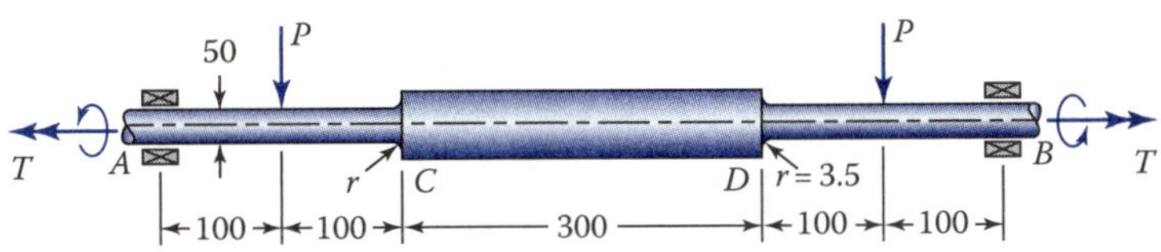

FIGURE P9.19

Assumptions: A factor of safety of $n = 3.5$ is used. The reliability will be 95% at an operating temperature of 800°F maximum. Fatigue stress-concentration factor for bending and torsion is $K_f = 1.2$.

9.21 Reconsider Problem 9.20, with the exceptions that the drive shaft of the car is under heavy shocks and use the maximum energy of distortion theory combined with the Soderberg criterion.

9.22 A solid shaft of diameter D rotates and carries the minor shock loading as shown in Figure P9.22. Calculate the factor of safety n for the shaft using the maximum energy of distortion theory of failure combined with the Goodman criterion.

Given: $D = 75$ mm, $S_y = 550$ MPa, $S_u = 660$ MPa. The torque fluctuates 5% each way from the mean value, and the shaft is to be machined from unnotched steel.

9.23 Redo Problem 9.22 using the maximum shear stress theory of failure incorporated with the Soderberg fatigue relation and a survival rate of 90%.

9.24 A revolving shaft, made of solid AISI 1040 cold-drawn steel, supports the loading depicted in Figure P9.24. The pulley weighs 300 lb and the gear weighs 100 lb. Design the shaft by using the maximum energy of distortion theory of failure incorporated with the Goodman fatigue criterion.

Given: $K_f = 1.8$, $n = 1.6$.

Assumption: A survival rate of 95% is used.

9.25 Redo Problem 9.24 using the maximum shear stress theory of failure incorporated with the Soderberg criterion, a survival rate of 99.9%, $K_f = 1.2$, $n = 2$, and neglecting the weights of pulley and gear.

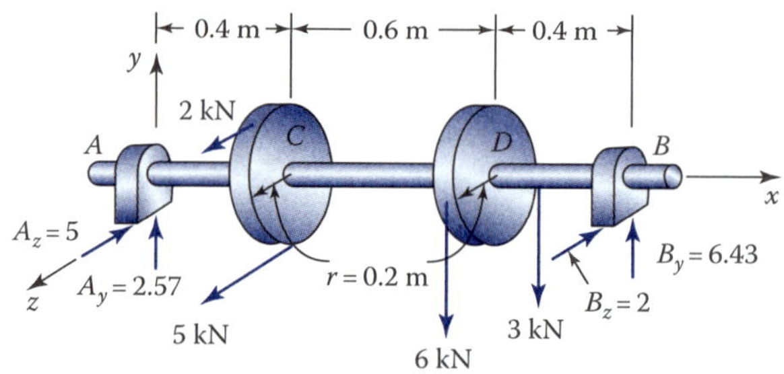

FIGURE P9.22

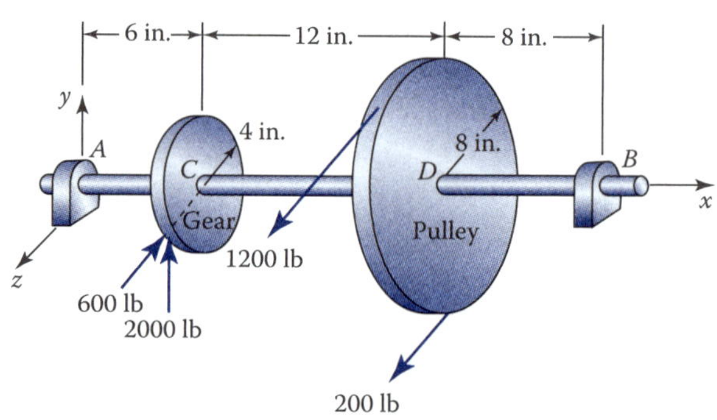

FIGURE P9.24

Sections 9.6 through 9.12

9.26 A solid steel shaft of diameter *D* is supported and loaded as shown in Figure P9.26. Determine the critical speed n_{cr} in rpm.

Given: $D = 25$ mm, $E = 210$ GPa.

9.27 Calculate the critical speed n_{cr} in rpm of the steel shaft of Figure P9.26, if the maximum allowable static deflection is 0.5 mm.

Given: $E = 200$ GPa.

9.28 A uniform steel shaft with an overhang is loaded as shown in Figure P9.28. Determine the critical speed n_{cr} in rpm.

Given: $E = 210$ GPa.

9.29 Determine the value of the critical speed of rotation for the simply supported shaft carrying two loads at *D* and *C* as illustrated in Figure P9.29, applying

a. The Rayleigh method
b. The Dunkerley method

Given: The shaft is made of $D = 2\,¼$ in. diameter a cold-rolled steel with $E = 29 \times 10^6$ psi.

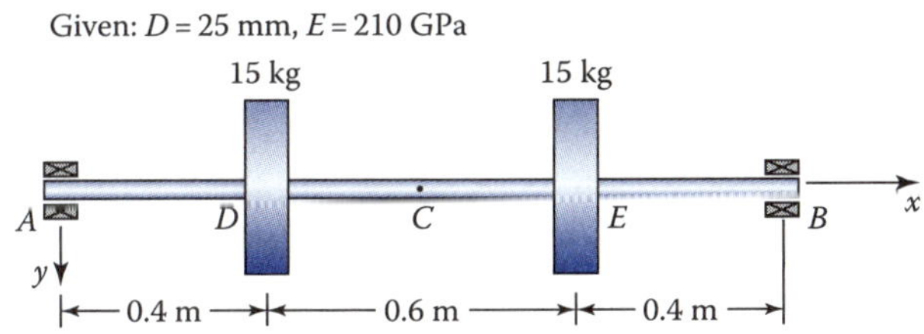

FIGURE P9.26

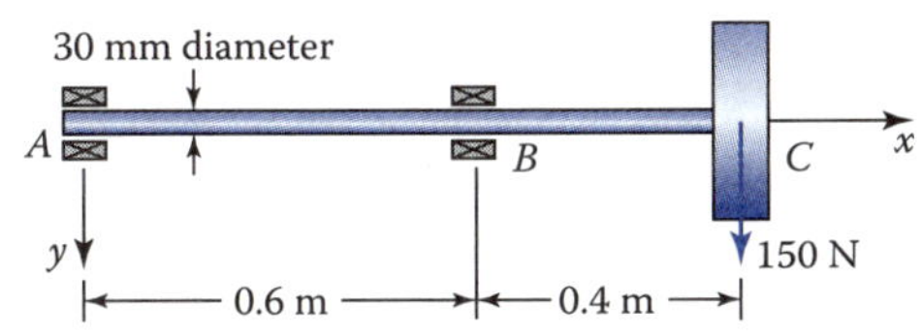

FIGURE P9.28

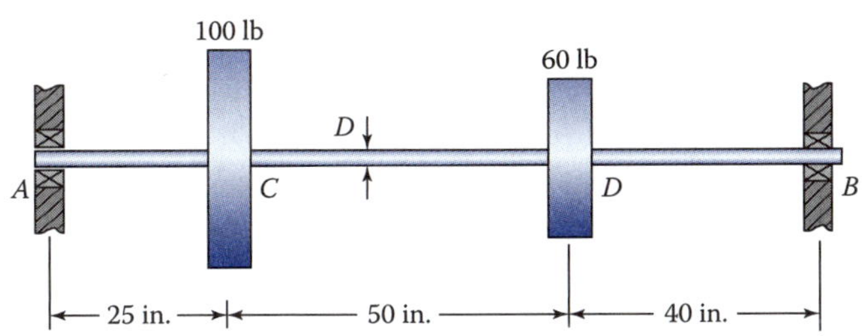

FIGURE P9.29

9.30 Compute the value of the critical speed of rotation for the simply supported and loaded shaft as illustrated in Figure P9.29, through the use of

a. The Rayleigh method

b. The Dunkerley method

Given: The shaft is constructed of $D = 3½$ in. diameter wrought steel having $E = 30 \times 10^6$ psi.

9.31 Determine the value of the critical speed of rotation for the outboard motor shaft loaded as illustrated in Figure P9.31, using

a. The Rayleigh method

b. The Dunkerley method

Given: The shaft is made of $D = 1⅞$ in. diameter cold-rolled bronze of $E = 16 \times 10^6$ psi.

9.32 A motor drive shaft of diameter D that is made of AISI 1050 CD steel transmits 120 hp at a speed of 900 rpm through a keyed coupling, similar to that shown in Figure 9.16, to the transmission input shaft of a driven machine (Figure P9.32). Find the length L of the key on the basis of

a. Bearing on shaft

b. Bearing on key

c. Shear in key

Given: *Shaft*: $D = 1½$ in, $S_y = 84.1$ ksi (see Table B.3).

Key: Square steel key of width $w = ⅜$ in. (Figure 9.10a) with $S_y = 44$ ksi, and $S_{ys} = S_y/2 = 22$ ksi is used.

Assumption: Safety factor of $n = 2.3$ will apply.

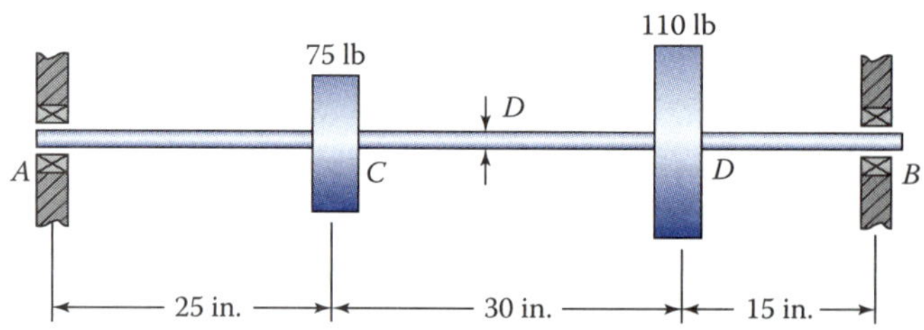

FIGURE P9.31

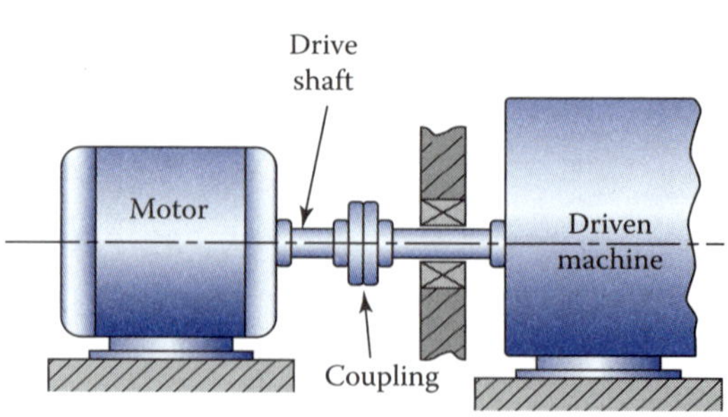

FIGURE P9.32

9.33 A solid shaft of diameter D has a ¾ × ⅜ in., flat key. Determine the required length L of key based on the maximum steady torque that can be transmitted by the shaft.

Given: $D = 3$ in., $S_{ys} = 0.58\ S_y$, $n = 2$.

Design Decision: The shaft and key are made of cold-drawn steels of AISI 1030 and AISI 1020, respectively.

9.34 A ⅜ × ⅜ × 3 in. key is used to hold a 3 in. long hub in a 1½ in. diameter shaft. What is the factor of safety against shear failure of the key if the torque transmitted is 3.5 kip·in.

Assumption: Key and shaft are of the same material with an allowable stress in shear of 10 ksi.

9.35 A 20 × 20 mm square key made of AISI 1050 HR steel is used on a 60 mm diameter shaft constructed of AISI 1095 HR steel to attach a hub of a rotating pulley as illustrated in Figure 9.10a. Compute the required key length on the basis of shock torque loading:

a. Bearing on the shaft
b. Bearing on key
c. Shear in key

Given: $S_y = 460$ MPa for shaft and $S_y = 340$ MPa for key (from Table B.3).

Assumption: A factor of safety $n = 4$ will be used. Yield strength in shear of both materials is taken as $S_{ys} = S_y/2$.

9.36 Figure P9.36a shows the free-body diagram of a shaft coupling. Observe that the flanges of this coupling are joined by $N = 6$ bolts with a bolt circle of radius $R = 80$ mm and shear force in each bolt is denoted by F (Figure P9.36b). What is the allowable bolt diameter d_b so that the bolts supply the same torque capacity $T = 5$ kN·m of the shaft?

Assumption: Bolts are made of AISI 1030 HR steel with $S_y = 260$ MPa (see Table B.3). A factor of safety $n = 1.2$ applies.

9.37 A $w \times w$ square key of length L will be used on a shaft of diameter D to attach a hub of a rotating coupling as illustrated in Figure 9.16a. The shaft is subjected to a steady load of 40 hp at $n_o = 120$ rpm. Determine the following stresses and the factor of safety n against yielding:

a. Bearing and shear in key
b. Shear in bolts
c. Bearing on bolts in flange

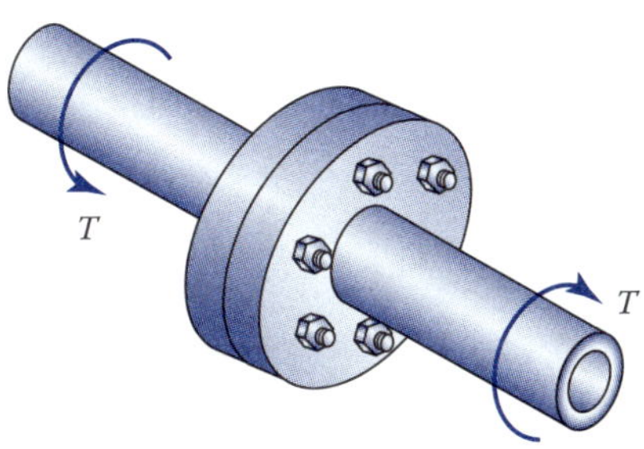

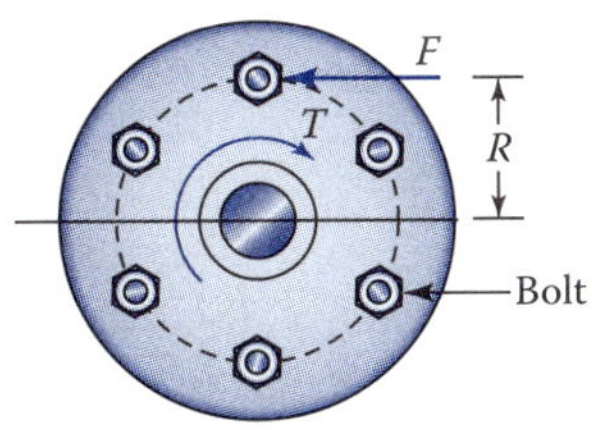

FIGURE P9.36

Given: $D = 60$ mm, $w = 10$ mm, $L = 75$ mm, bolt diameter $d_b = 15$ mm, $D_b = 144$ mm, flange thickness $t_f = 15$ mm.

Assumption: All parts are constructed of AISI 1080 HR steel with yield strength in tension $S_y = 420$ MPa (by Table B.3) and yield strength in shear $S_{ys} = 420/2 = 210$ MPa.

9.W Through the use of the website at www.grainger.com, conduct a search for flexible couplings, both rated for ½ hp at 1725 rpm:

a. ½ in. bore, 2½ in. long

b. ½ in. bore, 3½ in. long

List the manufacturer and description in each case.

9.38 For the coupling shown in Figure 9.16a, the key is $\frac{9}{16} \times \frac{9}{16} \times 3\frac{1}{2}$ in., bolt diameter $D_b = 6$ in., hub diameters $D_h = 4$ in., and $D = 2$ in. Six ⅜ in. bolts are used and flange thickness is $t_f = ⅞$ in. Determine

a. The shear and bearing stresses in the key

b. The shear stress in the bolts

c. The bearing stress on bolts in the flange

d. The shear stress in the flange at the hub or web

Given: The shaft carries a steady load of 60 hp at 200 rpm.

10

Bearings and Lubrication

10.1 Introduction

The goal of a bearing is to provide relative positioning and rotational freedom while transmitting a load between two parts, commonly a shaft and its housing. The object of lubrication is to reduce the friction, wear, and heating between two surfaces moving relative to each other. This is done by inserting a substance, called *lubricant*, between the moving surfaces. The study of lubrication and the design of bearings are concerned mainly with phenomena related to the oil film between the moving parts. Note that *tribology* may be defined as the study of the lubrication, friction, and wear of moving or stationary parts. The literature on this complex subject is voluminous. Much is collected in the *CRC Handbook of Lubrication*, sponsored by the American Society of Lubrication Engineers [1]. Also see [2–4]. The website www.machinedesign.com includes general information on bearings and lubrication.

There are two parts in this chapter. In Part A, the fundamentals of lubrication with particular emphasis on the design of journal (the so-called sleeve or sliding) bearings is discussed. The basic forms of journal bearings are simple. In Part B, concern is with rolling bearings, also known as *rolling-element bearings*, and antifriction bearings. We describe the most common types of rolling bearings, bearing dimensions, bearing load, and bearing life. There is also a brief discussion on materials, mounting, and lubricants of rolling bearings. Rolling-element bearings are employed to transfer the main load through elements in rolling contact, and they have been brought to their present state of perfection only after a long period of development. Either ball bearings or roller bearings, they are made by all major bearing manufacturers worldwide.

Part A: Lubrication and Journal Bearings

Journal bearings support loads perpendicular to the shaft axis by pressure developed in the liquid. A journal bearing is a typical sliding bearing requiring sliding of the load-carrying member on its support. Sleeve thrust bearings support loads in the direction of the shaft axis. We begin with a description of the lubrications and journal bearings. The general relationship between film velocity rate, viscosity, coefficient of friction, and load is then developed. This is followed by discussions of the hydrodynamic lubrication theory, design, and heat balance of bearings. Techniques for supplying oil to bearings and bearing materials are also considered.

10.2 Lubricants

As noted previously, the introduction of a lubricant to a sliding surface reduces the coefficient of friction. In addition, lubricants can act as contaminants to the metal surfaces and coat them with monolayers of molecules that inhibit adhesion between metals. Although usually in the liquid state, solids and gases are also used as lubricants. A brief description of the classification and characteristics of lubricants are as follows. Lubricant manufacturers should be consulted for particular application.

10.2.1 Liquid Lubricants

Liquid lubricants are largely petroleum-based or synthetic *oils.* They are characterized by their viscosity, but other properties are also important. Characteristics such as acidity, resistance to oxidation, antifoaming, pour, flash, and fire deterioration are related to the quality of oil needed for a particular operation. Many oils are marketed under the name of application, such as compressor or turbine oils. Oils for vehicle engines are classified by their viscosity as well as by the presence of additives for various service conditions (see Section 10.5).

Synthetic lubricants are mainly silicones. They have high-temperature stability, low-temperature fluidity, and high-internal resistance. Because of their higher cost, synthetic lubricants are used only when their special properties are needed, for instance, in the hydraulic control systems of aircraft. *Water* and *air* are used as lubricants where contamination by oil is prohibitive. In addition, often the lubricant is the water or air in which the machine is immersed. Air or an inert gas has very low internal resistance. It operates well from low to high temperatures. Gas lubricants are necessary at extremely high speeds involving low loading conditions.

Greases are liquid lubricants that have been thickened (by mixing with soaps) to provide properties not available in the liquid lubricant alone. Mineral oils are the most commonly used liquid for this purpose. Greases are often used where the lubricant is required to stay in position. Unlike oils, greases cannot circulate and thereby serve a cooling and cleaning function; however, they are expected to accomplish all functions of fluid lubricants. The many types of greases have properties suitable for a wide variety of operating conditions. Typical uses of greases include vehicle suspension and steering, for gears and bearings in lightly loaded and intermittent service, with infrequent lubrication by hand or grease gun.

10.2.2 Solid Lubricants

Solid lubricants are of two types: graphite and powdered metal. They are used for bearing operating at high temperatures (e.g., in electric motors). Other kinds include Teflon and some chemical coatings. Solid lubricants may be brushed or sprayed directly into the bearing surfaces. To improve retention, they are mixed with adhesives. Determination of composite bearing materials with low wear rates as well as frictional coefficients is an active area of contemporary design and research.

10.3 Types of Journal Bearings

The journal bearing or sleeve bearing supports a load in the radial direction. It has two main parts: a shaft called the *journal* and a hollow cylinder or sleeve that carries the shaft, called the *bearing* (Figure 10.1). When assembly operations do not require that a bearing be of two pieces, the bearing insert can be made as a one-piece cylindrical shell pressed into a hole in the housing. This insert is also called a *bushing*.

A *full-journal bearing*, or so-called 360° journal bearing, is made with the full bearing thickness around the whole circumference, as depicted in Figure 10.1a. Circumferential or any (usually axial or diagonal) grooving may be cut in the two-piece and one-piece bearings, respectively. Preferably, the oil is brought at the center of the bearing so that it will flow out both ends, thus increasing the flow and cooling action. In most applications, the journal rotates (with a speed n) within a stationary bearing and the relative motion is sliding. However, the journal may remain stationary and the bearing rotates or both the journal and bearing rotate. All situations require an oil film between the moving parts to minimize friction and wear. The pressure distribution around a bearing varies greatly. The coefficient of friction, f, is the ratio of the tangential friction force, discussed in Section 10.6, to the load carried by the bearing.

Sleeve bearings are employed in numerous fields. Two typical services a bearing is to perform are as follows. The crankshaft and connecting rod bearings of an automobile engine must operate for thousands of miles at high temperatures and under variable loading. The journal bearings used in the steam turbines and power generator sets must have very high reliability. *Gas bearings* using air or more inert gases as the lubricant and film find applications for lightly loaded, high-speed shafts, such as in gas-cycle machinery, gyros, and high-speed dental drills. Also, when the loads are light and the service relatively unimportant, a *nylon bearing* that requires no lubrication must be used.

A *partial bearing* is used when the radial load on a bearing always acts in one direction; hence, the bearing surface needs to extend only part way around the periphery. Often, an oil cap is placed around the remainder of the circumference. An angle (e.g., $\theta = 60°$) describes the angular length of a partial bearing (Figure 10.1b). Rail freight car axle bearings are

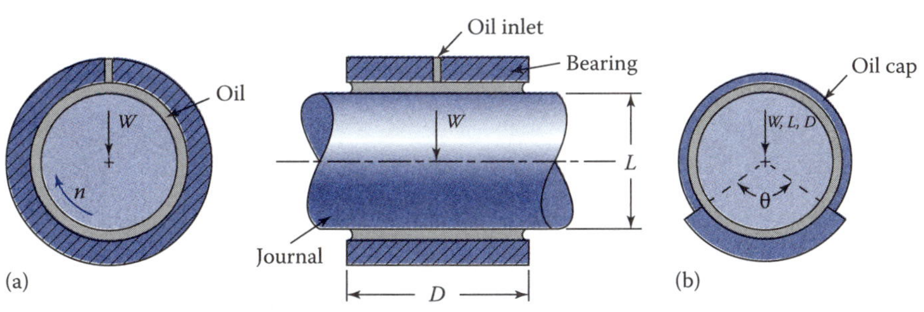

FIGURE 10.1
(a) Full-journal bearing and (b) partial-journal bearing. *Notes*: W, load; L, bearing length; D, journal diameter; n, journal rotational speed; θ, angle of partial bearing.

an example. A partial bearing having *zero* clearance is known as a *fitted bearing*. Zero clearance means that the radii of the journal and bearing are equal. We consider only the more common full bearing.

10.4 Forms of Lubrication

Lubrications commonly are classified according to the degree with which the lubricant separates the sliding surfaces. Five distinct forms or *types of lubrication* occur in bearings: hydrodynamic, mixed, boundary, elastohydrodynamic, and hydrostatic. The bearings are often designated according to the form of lubrication used.

Let us reconsider a journal bearing with a load W as depicted in Figure 10.1a. The bearing clearance space is filled with oil; however, when the journal is not rotating, the load squeezes out the oil film at the bottom. Slow clockwise rotation of the shaft causes it to roll to the right. As the rotating speed rises, oil adhering to the journal surface comes into the contact zone and pressure builds up just ahead of the contact zone to *float* the shaft. The high pressure of the oil flow to the right moves the shaft slightly to the left of center. Equilibrium is obtained with the full separation of the journal and bearing surfaces with an eccentricity (e) of the journal in the bearing.

To gain insight into the possible lubrication states, consider the experimentally determined curve between the shaft speed n and coefficient of friction f in a journal bearing (Figure 10.2). Clearly, the numerical values for the curve in the figure depend on features of the particular bearing design. Note that bearings operate under boundary conditions at start-up or shutdown. At slow speeds, the coefficient of friction remains about the same in region of boundary lubrication. As n is increased, a mixed lubrication situation is initiated (point A), and f drops rapidly until hydrodynamic lubrication is established (point B). At higher speeds, f rises slowly. For extremely large velocities (beyond point C), instability and turbulence may be established in the lubricant. Note that regions to the left and right of point B represent thin-film and thick-film lubrications, respectively. We now briefly discuss the conditions that induce the foregoing lubrication states.

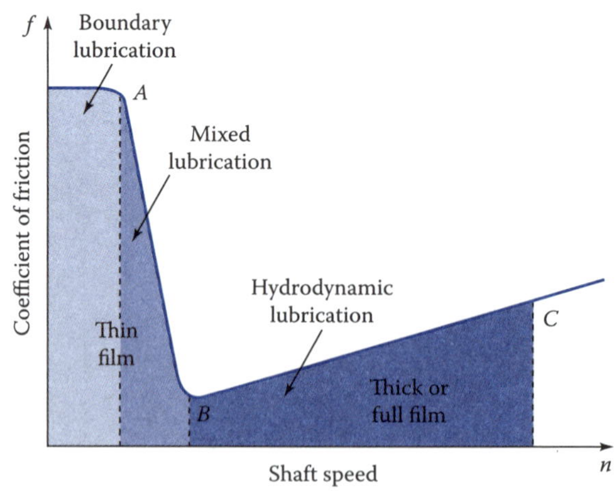

FIGURE 10.2
The change in the coefficient of friction f with shaft speed n in a journal bearing.

In addition to the shaft speed (*n*), two other parameters that influence the type of lubrication and resulting coefficient of friction are oil *viscosity* (η) and the *bearing unit load* (*P*). Viscosity is discussed in the next section. Bearing unit load is defined as follows:

$$P = \frac{W}{A_p} = \frac{W}{DL} \tag{a}$$

where W, A_p, D, and L denote the unit load and *projected area*, diameter, and length of the bearing, respectively. Interestingly, the higher the viscosity, the lower the rotating speed required to *float* the journal at a prescribed unit load. Also, the smaller the bearing unit load, the lower the rotating speed (and the viscosity) required to *float* the journal.

10.4.1 Hydrodynamic Lubrication

Hydrodynamic lubrication means that load-carrying surfaces of the bearing are separated by a (relatively thick) layer of fluid, called *fluid film*. For this condition to occur, a relative motion must exist between the two surfaces and a pressure must be developed. The pressure is created internally by the relative velocity, viscosity of the fluid, and the wedging action that results when the two surfaces are not parallel. This technique does not depend on the introduction of the lubricant under pressure. It does require, however, the existence of an adequate fluid supply at all times.

In a journal bearing at rest, the shaft sits in contact with the bottom of the bearing. As soon as the shaft rotates, its centerline shifts eccentrically within the bearing. Thus, a flow is set up within the small thickness of the oil film. When the rotating speed increases sufficiently, the shaft moves up on a wedge of pumped oil and ends its metal contact with the bearing; hydrodynamic lubrication is established. In a hydrodynamically lubricated sleeve-bearing surface, wear does not occur. Friction losses originate only within the lubricant film. Typical minimum film thickness (denoted h_o) ranges from 0.008 to 0.020 mm. Coefficients of friction *f* commonly range from 0.002 to 0.010. Hydrodynamic lubrication is also known *as fluid film* or *fluid lubrication*. The design of journal bearings is based on this most desirable type of lubrication.

10.4.2 Mixed Lubrication

Mixed lubrication describes a combination of partial lubricant film plus intermittent contact between the surfaces. Under this condition, the wear between the surfaces depends on the properties of the surfaces and the lubricant viscosity. Typical values of the coefficient of friction are 0.004–0.10.

For example, if the lubricant is supplied by hand oiling and by drop or mechanical feed (see Section 10.9), the bearing is operating under mixed oil-film conditions. This lubricating condition may also be present where lubrication is deficient, viscosity is too low, bearing is overloaded, clearance is too tight, bearing speed is too low, and bearing assembly is misaligned.

10.4.3 Boundary Lubrication

Boundary lubrication refers to the situations in which the fluid film gets thinner and partial metal-to-metal contact can occur. This depends on such factors as surface finish and wear-in and surface chemical reaction. Boundary lubrication occurs in journal bearings at low speeds and high loads, as when starting or stopping a rotating machinery. The properties of the sliding metallic surfaces and the lubricant are significant factors in limiting wear. The coefficient of friction is about 0.10. Boundary lubrication is less desirable than the other types, inasmuch as it allows the surface asperities to contact and wear rapidly. Design for this type of lubrication is largely empirical. Electric motor shaft bearings, office machinery bearings, power screw support bearings, and electric fan bearings represent some examples of boundary lubrication bearings.

Note that the initial boundary lubrication can be avoided by the introduction of pressurized oil on the loaded side of the journal, thereby hydraulically lifting it at start-up and again at shutdown. This is a common practice on large machines (e.g., power turbines), to provide sleeves and shaft a wear-free long life. The foregoing, called hydrostatic lubrication, is discussed later.

10.4.4 Elastohydrodynamic Lubrication

Elastohydrodynamic lubrication is concerned with the interrelation between the hydrodynamic action of full-fluid films and the elastic deformation of the supporting materials. It occurs when the lubricant is introduced between surfaces in rolling contact, such as mating gears and rolling bearings. Under loaded contact, balls and rollers, as well as cams and gear teeth, develop a small area of contact because of local elastic deformation owing to high stress (e.g., 700–3500 MPa). Factors that have major effect on creating elastohydrodynamic lubrication are increased relative velocity, increased oil viscosity, and increased radius of curvature at the contact. The mathematical explanation requires the Hertz contact stress analysis, as discussed in Chapter 3 and fluid mechanics [1].

10.4.5 Hydrostatic Lubrication

Hydrostatic lubrication refers to the continuous supply of flow of lubricant to the sliding interface at some elevated hydrostatic pressure. It does not require motion of the surfaces relative to another. This mechanism creates full-film lubrication. Some special applications involving hydrostatic lifts, thrust bearings, and oil lifts needed during the start-up of heavily loaded bearings are of the hydrostatic forms. Obviously, in hydrostatic lubrication, the pressure is developed externally by a pump, and the fluid (typically oil) enters the bearing opposite the load. The advantages of this technique include notably low friction and high load-carrying capacity at low speeds at all times. Disadvantages are the cost and the need for an external source of fluid pressurization.

Consider a simplified sketch of a vertical shaft *hydrostatic thrust bearing* shown in Figure 10.3. The rotating shaft supports a vertical load W. High-pressure oil at p is supplied into the recess of radius r_o at the center of the bearing from an external pump. Oil flows radially outward the annulus of depth h, finally escapes at the periphery of the shaft, and then finally returns through a system of piping to the reservoir at about atmospheric pressure. The oil film is present whether the shaft rotates or not. It can be shown that [5] the load-carrying capacity is given in the following form:

$$W = \frac{p\pi}{2}\left[\frac{r^2 - r_o^2}{ln(r/r_o)}\right] \tag{10.1}$$

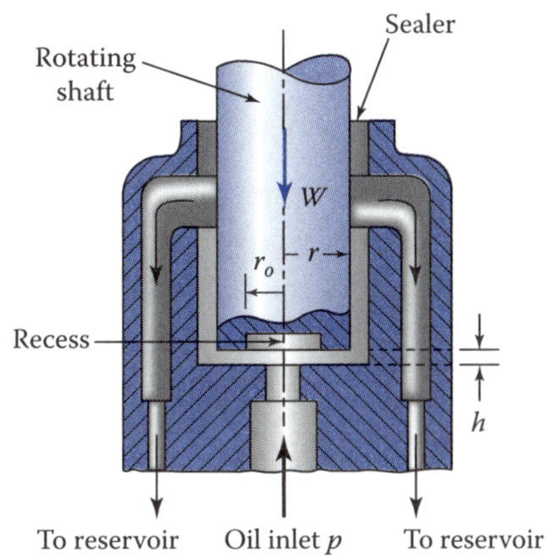

FIGURE 10.3
Schematic representation of a hydrostatic thrust bearing.

The preceding is applicable even if the recess is eliminated. In this case, r_o becomes the radius of the inlet oil-supply pipe.

Hydrostatic bearings are used in various special applications. Some examples are telescopes and radar tracking units subjected to heavy loads at very low speeds as well as the machine tools and gyroscopes under high speed but light loads. Further details may be found in [5].

Finally, we note that friction and wear (discussed in Sections 8.3 through 8.6) are significant considerations when boundary lubrication or metal-to-metal contact occurs. Recall that the *depth of wear* δ (by letting $A_\alpha = A_p$ and $P = W$ is given by

$$\delta = K\frac{Wl}{HA_p} \tag{8.3′}$$

The quantities *K*, *l*, and *H* represent the wear coefficient, length of sliding, and Brinell hardness of bearing material, respectively. As noted previously, *W* and A_p designate the load and projected area (*DL*) of the bearing. Practically, it is useful to include in this relationship *motion-related* and *environmental factors* depending upon motion type, load, and speed [10]. Observe that the properties of sliding surfaces of the lubricant are important factors in limiting wear under lubrication conditions. Table 10.3 of Section 10.11 furnishes designer limits of the unit bearing load *P*, sliding velocity *V*, as well as *PV* for various materials. Sliding velocity for continuous motion is $V = Dn$. Clearly, for an acceptable bearing design configuration, operating values of the preceding quantities must be less than the values listed in the table. An application of Equation 8.3 is illustrated in the following numerical problem.

Example 10.1: Preliminary Design of a Boundary-Lubricated Journal Bearing

A 1¼ in. steel shaft having 450 Bhn with an excellent lubrication rotates continuously at a load of 40 lb at 20 rpm for 3.5 years in a sleeve of bronze–lead having 170 Bhn (Figure 10.4). Estimate the largest length *L* of the sleeve.

Given: $D = 1¼$ in., $H = 450$ Bhn, $n = 30$ rpm, $W = 40$ lb, $t = 2$ years.

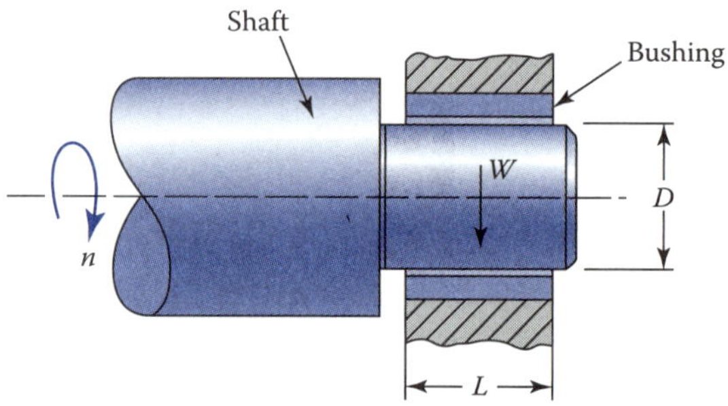

FIGURE 10.4
Example 10.1. Bushing of bronze–lead with steel shaft.

Assumptions: Maximum wear of the bearing is to be δ = 0.002 in. Bronze is partially compatible with steel, and lead is incompatible.

Solution

A conservative value of $K = 1 \times 10^{-7}$ will be taken for partially compatible materials and excellent lubrication from Table 8.3. The hardness of sleeve, softer material in bearing, must be used (see Section 8.5) and thus

$$H = 1.424(170) = 242 \text{ ksi.}$$

The length of sliding equals

$$l = n\pi Dt = 30\pi(1.25)(60 \times 24 \times 365 \times 2) = 123.8 \times 10^{6} \text{ in.}$$

The bearing length, from Equation 8.3, is given by

$$L = \frac{KWl}{HD\delta} \tag{b}$$

Substituting the data, we have

$$L = \frac{(1 \times 10^{-7})(40)(123.8 \times 10^{6})}{242{,}000(1.25)(0.002)} = 0.82 \text{ in.}$$

Comments: The next largest available standard length, probably L = 1.0 in., should be used. Note, as a check, that

$$P = \frac{W}{DL} = \frac{40}{(1.25)(1)} = 32 \text{ psi}$$

$$V = \frac{\pi Dn}{12} = \frac{\pi(1.25)(30)}{12} = 9.82 \text{ fpm}$$

and

$$PV = (32)(9.82) = 314.2 \text{ psi} \cdot \text{fpm}$$

The foregoing results are well below the maximum allowable values given in Table 10.1 for bronze–lead [10,11].

TABLE 10.1

Average Sleeve Bearing Pressures in Current Practice

	Average Pressure $P = W/DL$	
Application	**MPa**	**(psi)**
Relatively steady loads		
Centrifugal pumps	0.7–1.3	(100–180)
Gear reducer	0.8–1.7	(120–250)
Steam turbines	1.0–2.1	(150–300)
Electric motors	0.8–1.7	(120–250)
Rapidly fluctuating loads		
Automotive gasoline engines		
Main bearings	4–5	(600–750)
Connecting rod bearings	12–16	(1700–2300)
Diesel engines		
Main bearings	6–12	(900–1700)
Connecting rod bearings	8–16	(1150–2300)

10.5 Lubricant Viscosity

When two plates having relative motions are separated by a lubricant (e.g., oil), a flow takes place. In most lubrication problems, conditions are such that the flow is laminar. In *laminar flow*, the fluid is in layers that are maintained as the flow progresses. When this condition is not met, the flow is called *turbulent*. The laminar flow and internal resistance to shear of the fluid can be demonstrated by referring to the system depicted in Figure 10.5a. The figure shows that the lower plate is stationary, while the upper plate moves to the right with a velocity U under the action of the force F. Inasmuch as most fluids tend to *wet* and adhere to solid surfaces, it can be taken that, when the plate moves, it does not slide along on top of the film (Figure 10.5b). The plot of fluid velocity u against y across the film (shown in the figure) is known as the *velocity profile*.

Newton's law of viscous flow states that the shear stress in the fluid is proportional to the rate of change of velocity with respect to y (Figure 10.5a). That is,

$$\tau = \eta \frac{du}{dy} \tag{10.2}$$

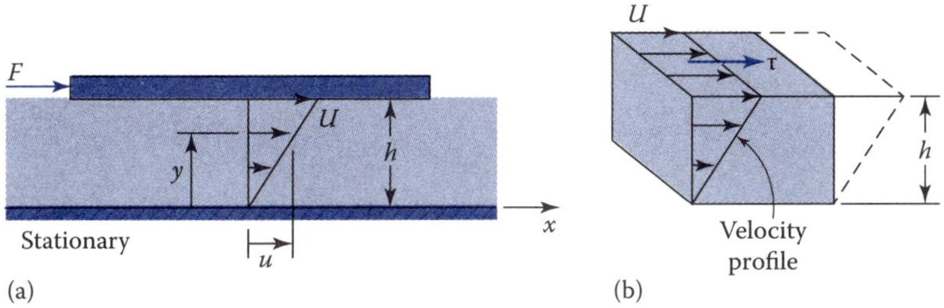

FIGURE 10.5
Laminar flow: (a) flat plate moving on fluid film and (b) a fluid element.

The factor of proportionality η is called the *absolute viscosity* or simply the *viscosity*. The viscosity is a measure of the ability of the fluid *to resist* shear stress. Newtonian fluids include air, water, and most oils. Those fluids to which Equation 10.2 does not apply are called non-Newtonian. Examples are lubricating greases and some oils with additives. Let the distance between the two plates, the film thickness, be denoted by h, as shown in Figure 10.5. Because the velocity varies linearly across the film, we have $du/dy = U/h$ and $\tau = F/A$. Substitution of these relations into Equation 10.2 results in

$$F = \eta \frac{AU}{h} \tag{10.3}$$

In the foregoing, A represents the area of the upper plate.

10.5.1 Units of Viscosity

In SI, viscosity is measured in newton-second per square meter ($N \cdot s/m^2$) or pascal-seconds. The U.S. customary unit of viscosity is the pound-force-second per square inch ($lb \cdot s/in.^2$), called the *reyn*. The conversion between the two units is the same as stress:

$$1 \text{ reyn} = 6890 \text{ Pa} \cdot \text{s}$$

The reyn and pascal-second are such large units that microreyn (μreyn) and millipascal-second (mPa·s) are more commonly used.

In the former metric system, centimeter–gram–second (cgs), the unit of viscosity, is poise (p), having dimensions of dyne-second per square centimeter ($dyne \cdot s/cm^2$). Note that 1 centipoise (cp) is equal to 1 millipascal-second (1 cp = 1 mPa·s). It has been customary to use the cp, which is 1/100 of a poise. The conversion from cgs units to U.S. customary units is as follows: 1 reyn = $6.89 (10^6)$ cp. To obtain viscosity in μreyn, multiply the cp value by 0.145.

10.5.2 Viscosity in terms of Saybolt Universal Seconds

The American Society for Testing and Materials (ASTM) standard method for determining viscosity employs an instrument known as the Saybolt universal viscometer. The approach consists of measuring the time in seconds needed for 60 cm^3 of oil at a specified temperature to flow through a capillary tube 17.6 mm diameter and 12.25 mm long (Figure 10.6). The time, measured in seconds, is known as *Saybolt universal seconds*, S.

Kinematic viscosity, also called *Saybolt universal viscosity* (SUV) in seconds, is defined by

$$\nu = \frac{\text{Absolute viscosity}}{\text{Mass density}} = \frac{\eta}{\rho} \tag{10.4}$$

The mass density ρ is in g/cm^3 of oil (which is numerically equal to the specific gravity). In SI, η and ρ have units $N \cdot s/m^2$ and $N \cdot s^4/m^4$, respectively. Thus, the kinematic viscosity ν has the unit of m^2/s. In the former metric system, a unit of cm^2/s was named a *stoke*, abbreviated *St*.

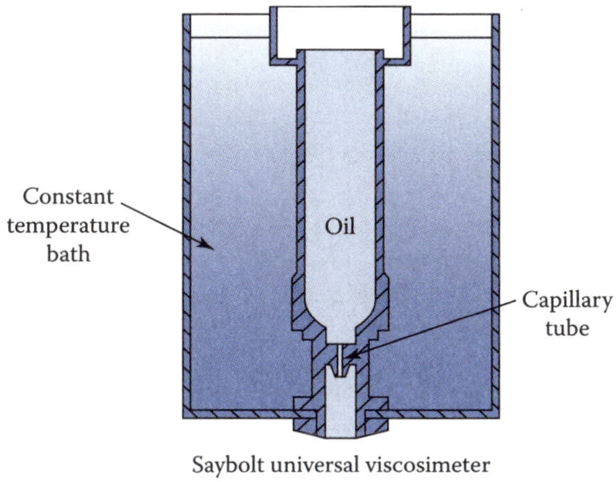

FIGURE 10.6
Saybolt universal viscometer.

Absolute viscosity is needed for calculation of oil pressure and flows within a bearing. It can be found from Saybolt viscometer measurements by the following formulas:

$$\eta = \left(0.22S - \frac{180}{S}\right)\rho \quad (\text{mPa}\cdot\text{s, or cp}) \tag{10.5a}$$

$$\eta = 0.145\left(0.22S - \frac{180}{S}\right)\rho \quad (\mu\text{reyn}) \tag{10.5b}$$

Here, Saybolt time S is in seconds. Interestingly, for petroleum oils, the mass density at 60°F (15.6°C) is approximately 0.89 g/cm^3. The mass density, at any other temperature, is given by

$$\rho = 0.89 - 0.00063(°\text{C} - 15.6) \tag{10.6a}$$

$$\rho = 0.89 - 0.00035(°\text{F} - 60) \tag{10.6b}$$

both in g/cm^3.

10.5.3 Effects of Temperature and Pressure

The viscosity of a liquid varies inversely with temperature and directly with pressure, both nonlinearly. In contrast, gases such as air have an increased viscosity with increased temperature. Figure 10.7 shows the absolute viscosity of various fluids and how they vary. The Society of Automotive Engineers (SAE) and the International Standards Organization (ISO) classify oils according to viscosity. Viscosity–temperature curves for typical SAE numbered oils are given in Figure 10.8. These oil types must exhibit particular viscosity behavior at 100°C. In addition, the SAE classifies identifications such as 10W, 20W, 30W, and 40W. Accordingly, for instance, a 20W-40 *multigrade*, also called *multiviscosity*, oil must

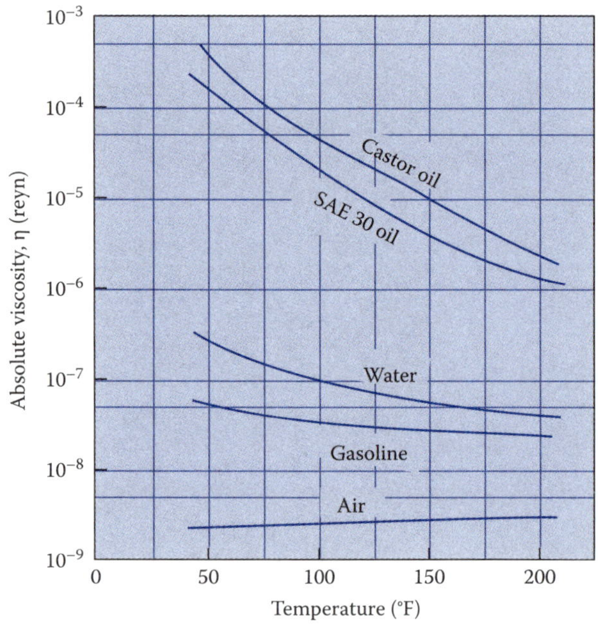

FIGURE 10.7
Variation in viscosity with temperature of several fluids.

satisfy the 20W behavior at –18°C and the SAE 40 viscosity behavior at 100°C. The viscosity of multigrade oils varies less with temperature than that of single-grade oils.

A widely used means for specifying the rate of change of viscosity with temperature is known as the *viscosity index*, abbreviated *VI*. It compares an oil to oils with very small and very large rates of change in viscosity. A contemporary basis for viscosity index rating is given in the American National Standards Institute (ANSI)/ASTM Specification D2270.

Example 10.2: Viscosity and SAE Number of an Oil

An engine oil has a kinematic viscosity at 80°C corresponding to 62 s as found from a Saybolt viscometer. Calculate the absolute viscosity in millipascal-second and microreyns. What is the corresponding SAE number?

Solution

Through the use of Equation 10.6a, we have

$$\rho = 0.89 - 0.00063(80 - 15.6) = 0.849\ \text{g/cm}^3$$

Then, applying Equation 10.5a,

$$\eta = 0.849\left[0.22(62) - \frac{180}{62}\right] = 9.116\ \text{mPa}\cdot\text{s}$$

Equation 10.5b gives

$$\eta = 0.145(9.116) = 1.322\ \mu\text{reyn}$$

Referring to Figure 10.8, the viscosity at 80°C is near to that of an SAE 20 oil.

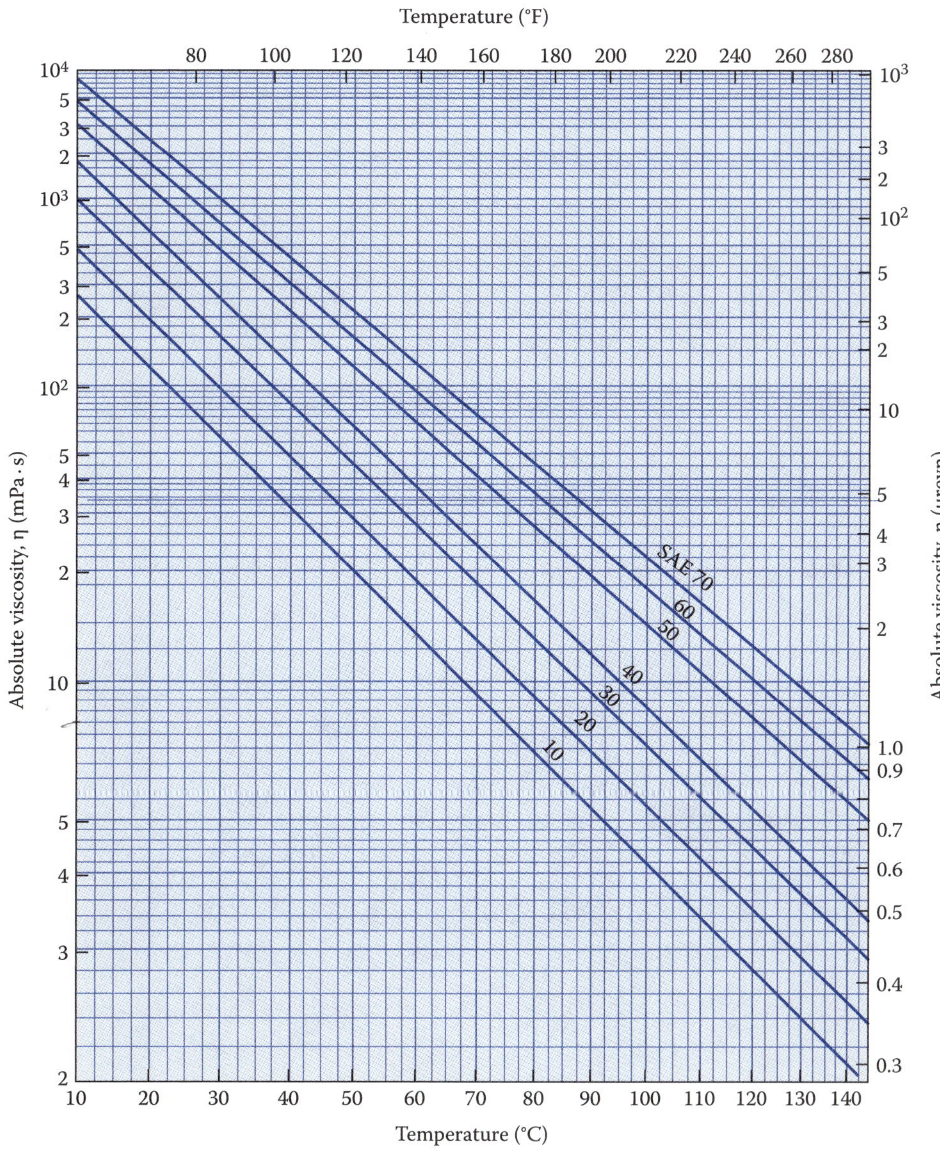

FIGURE 10.8
Viscosity versus temperature curve for typical SAE-graded oils.

10.6 Petroff's Bearing Equation

The phenomenon of bearing friction was first explained by N. Petroff in 1883. He analyzed a journal bearing based on the assumption that the shaft is concentric in bearing. Obviously, this operation condition could not occur in an actual journal bearing. However, by Petroff's approach, if the load applied is very low and the speed and viscosity are fairly high, approximate results are obtained. Usually, Petroff's equation is applied in preliminary design calculations.

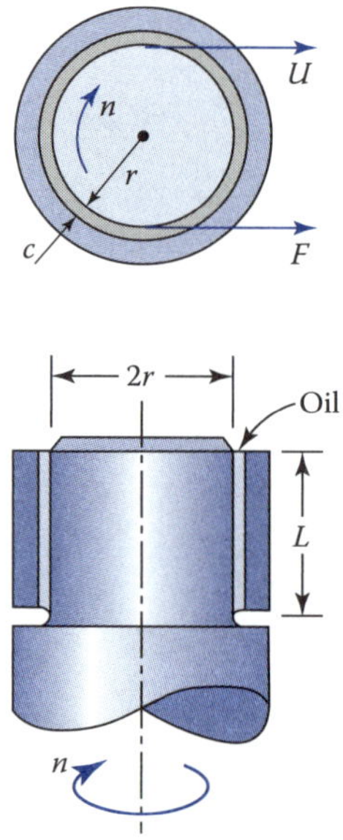

FIGURE 10.9
Journal-centered bearing.

10.6.1 Friction Torque

Let us assume that the moving flat plate shown in Figure 10.5a is wrapped into a cylindrical shaft (Figure 10.9). Now the thickness h becomes the radial clearance c that is taken to be completely filled with lubricant and from which leakage is negligible. Note that the radial clearance represents the difference in radii of the bearing and journal.

The developed journal area A is $2\pi rL$. Carrying A and h into Equation 10.3 yields the tangential friction force $F = 2\pi\eta UrL/c$, in which the tangential velocity U of the journal is $2\pi rn$. The frictional torque owing to resistance of fluid equals $T_f = Fr$. Equation for no-load torque is then

$$T_f = \frac{4\pi^2\eta Lr^3 n}{c} \tag{10.7}$$

where

T_f = the frictional torque
η = the absolute viscosity
L = the length of bearing
r = the journal radius
n = the journal speed, revolutions per second, rps
c = the radial clearance or film thickness

When a small load W is supported by the bearing, the pressure P of the projected area equals $P = W/2rL$ (Figure 10.10). The frictional force is fW, where f represents the coefficient of friction. Hence, the friction torque due to load is

$$T_f = fWr = 2r^2 fLP \tag{10.8}$$

Clearly, load W will cause the shaft to become somewhat eccentric in its bearing.

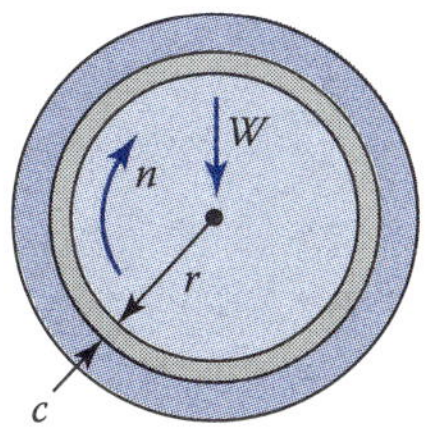

FIGURE 10.10
Lightly loaded journal bearing.

According to Petroff's approach, the effect of load W in Equation 10.7 can be considered negligible. Therefore, Equation 10.7 can be equated to Equation 10.8. In so doing, we obtain the coefficient of friction in the following form:

$$f = 2\pi^2 \frac{\eta n}{P} \frac{r}{c} \tag{10.9}$$

This is known as *Petroff's equation* or *Petroff's law*. Through the use of Equation 10.9, reasonable estimates of the coefficient of friction in lightly loaded bearings can be obtained. The two dimensionless quantities n/P and r/c are significant parameters in lubrication, as is observed in Section 10.8.

10.6.2 Friction Power

Having the expression for the friction torque available, friction power for the bearing may be obtained from the general relations given in Section 10.11. In the SI units, by Equations 1.15 and 1.16,

$$\text{kW} = \frac{T_f n}{159} \quad (T_f \text{ in N} \cdot \text{m}) \tag{10.10}$$

$$\text{hp} = \frac{T_f n}{119} \quad (T_f \text{ in N} \cdot \text{m}) \tag{10.11}$$

where n is in rps.

In U.S. customary units, using Equation 1.17, we have

$$\text{hp} = \frac{T_f n}{1050} \quad (T_f \text{ in lb} \cdot \text{in.}) \tag{10.12}$$

As before, journal speed n is in rps.

Example 10.3: Friction Power Using Petroff's Approach

An 80 mm diameter shaft is supported by a full-journal bearing of 120 mm length with a radial clearance of 0.05 mm. It is lubricated by SAE 10 oil at 70°C. The shaft rotates 1200 rpm and is under a radial load of 500 N. Apply Petroff's equation to determine

a. The bearing coefficient of friction
b. The friction torque and power loss

Solution

From Figure 10.8, $\eta = 9.2$ mPa·s. We have

$$P = \frac{500}{(0.08)(0.12)} = 52.08 \text{ kPa}$$

$$n = \frac{1200}{60} = 20 \text{ rps}$$

a. Substitution of the given data into Equation 10.9 gives

$$f = 2\pi^2 \frac{(0.0092)(20)}{52,080} \frac{40}{0.05} = 0.0558$$

b. Equations 10.8 and 10.10 are therefore

$$T_f = fWD/2 = (0.0558)(500)(0.04) = 1.116 \text{ N}\cdot\text{m}$$

$$\text{kW} = \frac{1.116(20)}{159} = 0.14$$

10.7 Hydrodynamic Lubrication Theory

Recall from Section 10.4 that, in hydrodynamic lubrication, oil is drawn into the wedge-shaped opening produced by two nonparallel surfaces having relative motion. The velocity profile of the lubricant is different at the wider and narrower sections. As a result, sufficient pressure is built up in the oil film to support the applied vertical load without causing metal-to-metal contact. This technique is utilized in the thrust bearings for hydraulic turbines and propeller shafts of ships as well as in the conventional journal bearings for piston engines and compressors.

10.7.1 Reynolds's Equation of Hydrodynamic Lubrication

Hydrodynamic lubrication theory is based on Osborne Reynolds's study of the laboratory investigation of railroad bearings by Beauchamp Tower in the early 1880s in England [6]. The initial Reynolds's differential equation for hydrodynamic lubrication was used by him to explain Tower's results. A simplifying assumption of Reynolds's analysis was that the oil films were so thin in comparison with the bearing radius that the curvature could be disregarded. This enabled him to replace the curved partial bearing with a flat bearing. Other presuppositions include those discussed in Section 10.5. The following is a brief outline of the development of Reynolds's fluid flow equation for two typical bearings.

10.7.1.1 Long Bearings

Consider a journal rotating in the clockwise direction supported by a lubricant film of variable thickness h on a fixed sleeve (Figure 10.11a). Assume that the lubricant velocity u

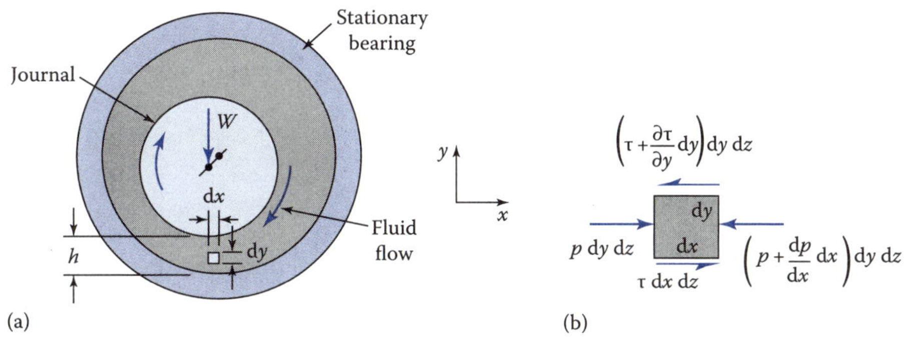

FIGURE 10.11
(a) An eccentric journal and (b) pressure and viscous forces acting on an oil fluid element of sides dx, dy, and dz, isolated from part a.

and shear stress τ vary in both the x and y directions, while pressure p depends on the x direction alone and bearing *side leakage* is *neglected*. The summation of the x-directed forces on the fluid film (Figure 10.11b) gives

$$p\mathrm{d}y\mathrm{d}z + \tau\mathrm{d}x\mathrm{d}z - \left(p + \frac{\mathrm{d}p}{\mathrm{d}x}\mathrm{d}x\right)\mathrm{d}y\mathrm{d}z - \left(\tau + \frac{\partial\tau}{\partial y}\mathrm{d}y\right)\mathrm{d}x\mathrm{d}z = 0$$

This reduces to

$$\frac{\mathrm{d}p}{\mathrm{d}x} = \frac{\partial\tau}{\partial y} \tag{a}$$

From Newton's law of flow,

$$\tau = -\eta\frac{\partial u}{\partial y} \tag{b}$$

in which the minus sign indicates a negative velocity ingredient.

Carrying Equation (b) into Equation (a) and rearranging, we obtain

$$\frac{\partial^2 u}{\partial y^2} = \frac{1}{\eta}\frac{\mathrm{d}p}{\mathrm{d}x}$$

Integrating twice with respect to y results in

$$u = \frac{1}{\eta}\left(\frac{\mathrm{d}p}{\mathrm{d}x}\frac{y^2}{2} + c_1 y + c_2\right) \tag{c}$$

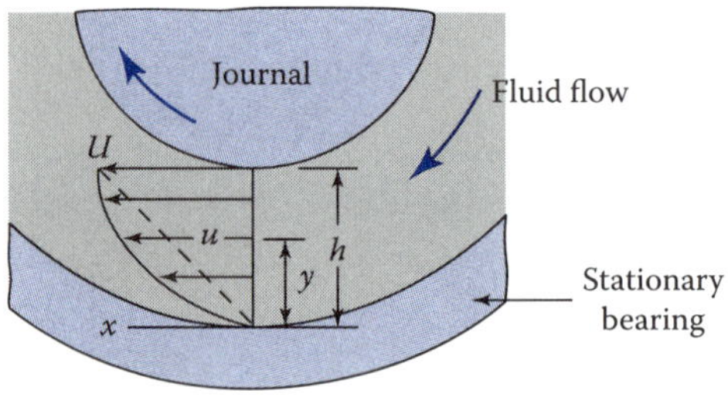

FIGURE 10.12
Velocity profile of the oil.

Assuming that no slip occurs between the lubricant and the boundary surfaces (Figure 10.12) leads to

$$u = 0 \quad (\text{at } y = 0), \quad u = U \quad (\text{at } y = h)$$

The quantity U represents the journal surface velocity. The constants c_1 and c_2 are evaluated by introducing these conditions into Equation (c):

$$c_1 = \frac{U\eta}{h} - \frac{h}{2}\frac{\mathrm{d}p}{\mathrm{d}x}, \quad c_2 = 0$$

Hence,

$$u = \frac{1}{2\eta}\frac{\mathrm{d}p}{\mathrm{d}x}\left(y^2 - hy\right) + \frac{U}{h}y \tag{10.13}$$

It is interesting to observe from this equation that the velocity distribution across the film is obtained by superimposing a parabolic distribution (the first term) onto a linear distribution (the second term). The former and the latter are indicated by the solid and dashed lines in Figure 10.12, respectively.

Let the volume of the lubricant per unit time flowing (in the x direction) across the section containing the element in Figure 10.11 be denoted by Q. For a width of unity in the z direction, using Equation 10.13, we have

$$Q = \int_0^h u\,\mathrm{d}y = \frac{Uh}{2} - \frac{h^3}{12\eta}\frac{\mathrm{d}p}{\mathrm{d}x} \tag{d}$$

Based on the assumptions of lubricant incompressibility and no side leakage, the flow rate must be identical for all sections: $\mathrm{d}Q/\mathrm{d}x = 0$. So, differentiating Equation (d) and setting the result equal to 0 yield

$$\frac{\mathrm{d}}{\mathrm{d}x}\left(\frac{h^3}{\eta}\frac{\mathrm{d}p}{\mathrm{d}x}\right) = 6U\frac{\mathrm{d}h}{\mathrm{d}x} \tag{10.14}$$

This is *Reynolds's equation for 1D flow.*

For the case in which the axial (z-directed) fluid flow *includes leakage*, the preceding expression may be generalized to obtain the *2D Reynolds's equation:*

$$\frac{\partial}{\partial x}\left(\frac{h^3}{\eta}\frac{\partial p}{\partial x}\right)+\frac{\partial}{\partial z}\left(\frac{h^3}{\eta}\frac{\partial p}{\partial z}\right)=6U\frac{\partial h}{\partial x} \tag{10.15}$$

The solutions of Equations 10.14 and 10.15 provide reasonable approximations for bearings of $L/D > 1.5$. Here, L and D represent the length and diameter of the bearing, respectively. Long bearings are sometimes used to restrain a shaft from vibration and position the shaft accurately in transmission shafts and machine tools, respectively.

10.7.1.2 Short Bearings

The circumferential flow of oil around the bearing may be taken to be negligible in comparison to the flow in the z direction for a short bearing. On the basis of this premise, F.W. Ocvirk and G.B. Dubois [7] proposed that the x term in Equation 10.15 may be omitted. In so doing, we obtain

$$\frac{\partial}{\partial z}\left(\frac{h^3}{\eta}\frac{\partial p}{\partial z}\right)=6U\frac{\partial h}{\partial x} \tag{10.16}$$

The foregoing equation can readily be integrated to give an expression for pressure in the oil film. Often, this procedure is referred to as *Ocvirk's short bearing approximation*. The solution of Equation 10.16 has moderate accuracy for bearings of L/D ratios up to about 0.75. In modern power machines, the trend is toward the use of short bearings.

We should mention that the exact solution of Reynolds's equation is a challenging problem that has interested many investigators ever since then, and it is still the starting point for lubrication studies. A mathematical treatment of the hydrodynamic lubrication is beyond the scope of this volume. Fortunately, it is possible to make design calculations from the graphs obtained by mathematical analysis, as will be observed in the next section.

10.8 Design of Journal Bearings

In actual bearings, a full continuous fluid film does not exist. The film ruptures, and bearing load W is supported by a partial film located beneath the journal. Petroff's law may be applied only to estimate the values of coefficient of friction. As noted previously, mathematical solutions to Reynolds's equations give reasonably good results for hydrodynamic or journal bearings of some commonly encountered proportions.

The design of journal bearings usually involves two suitable combinations of variables: *variables under control* (viscosity, load, radius and length of bearing, and clearance) and *dependent variables* or performance factors (coefficients of friction, temperature rise, oil flow, and minimum oil-film thickness). Essentially, in bearing design, limits for the latter group of variables are defined. Then, the former group is decided on so that these limitations are not exceeded. The following is a brief discussion of the quantities under control.

10.8.1 Lubricants

Recall that lubricants are characterized by their viscosity (η). Their choice is based on such factors as type of machine, method of lubrication, and load features.

10.8.2 Bearing Load

Usually, the load acting on a bearing is particularized. The value of the load per projected area, P, depends on the length and diameter of the bearing. Obviously, the smaller is P, the greater the bearing life.

10.8.3 Length–Diameter Ratio

Various factors are considered in choosing proper length-to-diameter ratios, or L/D values. Bearings with a length-to-diameter ratio less than 1 (short bearings) accommodate the shaft deflections and misalignments that are expected to be severe. Long bearings ($L/D > 1$) must be used in applications where shaft alignment is important.

10.8.4 Clearance

The effects of varying dimensions and clearance ratios are very significant in a bearing design. The radial clearance c (Figure 10.10) is contingent to some extent on the desired quality. Suitable values to be used for radial bearing clearance rely on factors that include materials, manufacturing accuracy, load-carrying capacity, minimum film thickness, and oil flow. Furthermore, the clearance may increase because of wear. The *clearance ratios* (c/r) typically vary from 0.001 to 0.002 and occasionally as high as 0.003. It would seem that large clearances increase the flow that reduces film temperature and hence increase bearing life. However, very large clearances result in a decrease in minimum film thickness. Therefore, some iteration ordinarily is needed to obtain a proper value for the clearance.

10.8.5 Design Charts

A.A. Raimondi and J. Boyd applied digital computer techniques toward the solution of Reynolds's equation and present the results in the form of design charts and tables [8]. These provide accurate results for bearings of all proportions. Most charts utilize the bearing characteristic number, or the *Sommerfeld number*:

$$S = \left(\frac{r}{c}\right)^2 \frac{\eta n}{P} \tag{10.17}$$

where

S = the bearing characteristic number, dimensionless
r = the journal radius
c = the radial clearance
η = the viscosity, reyns
n = the relative speed between journal and bearing, rps
P = the load per projected area

Notations used in the charts are illustrated in Figure 10.13. The center of the journal is shown at O and the center of the bearing is at O'. The minimum oil-film thickness h_0

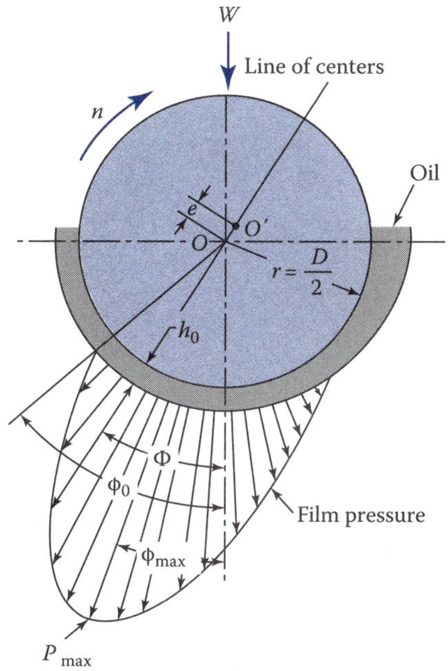

FIGURE 10.13
Radial pressure distribution in a journal bearing.

occurs at the line of centers. The distance between these centers represents the eccentricity, denoted by e. The eccentricity ratio ϵ is defined by

$$\epsilon = \frac{e}{c} \tag{10.18}$$

The minimum film thickness is then

$$h_0 = c - e = c(1-\epsilon) \tag{10.19}$$

The foregoing gives

$$\epsilon = 1 - \frac{h_0}{c} \tag{10.20}$$

As depicted in the figure, the angular location of the minimum oil-film thickness is designated by Φ. The terminating position and position of maximum film pressure p_{max} of the lubricant are denoted by ϕ_0 and ϕ_{max}, respectively.

Load per projected area, the average pressure or the so-called unit loading, is

$$P = \frac{W}{DL} \tag{10.21}$$

where,

W = the load
D = the journal diameter
L = the journal length

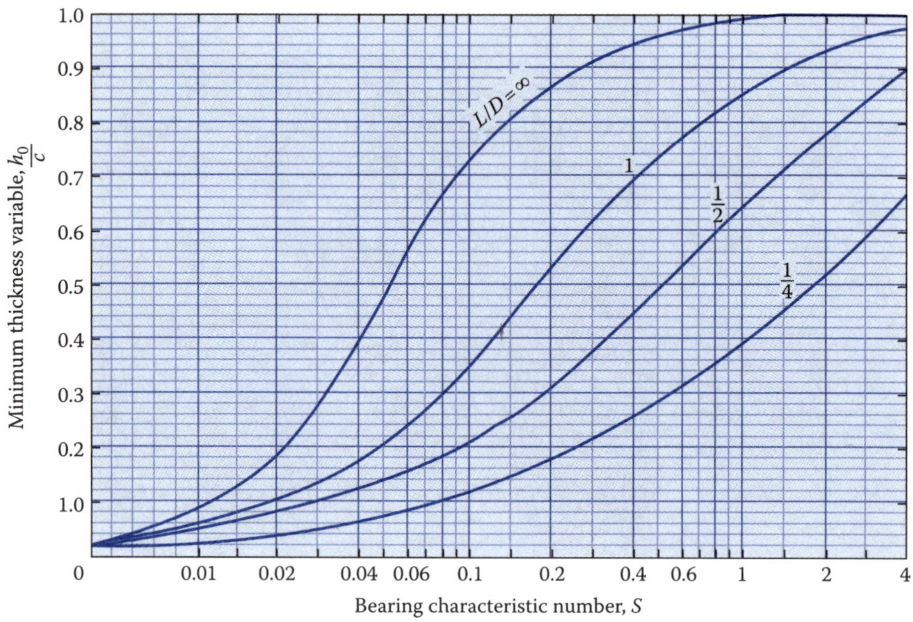

FIGURE 10.14
Chart for minimum film-thickness variable. (From Raimondi, A.A. and Boyd, J., *Trans. ASLE I*, 1, 159, 1958.)

Note that L and D are also referred to as the *bearing length* and *diameter*, respectively. Table 10.1 furnishes some representative values of P in common use.

Design charts by Raimondi and Boyd provide solutions for journal bearings having various length–diameter (L/D) ratios. Only portions of three selected charts are reproduced in Figures 10.14 through 10.16, for *full bearings*. All charts give the plots of *dimensionless* bearing parameters as functions of the dimensionless Sommerfeld variable, S. Note that the S scale on the charts is logarithmic except for a linear portion between 0 and 0.01. Space does not permit the inclusion of charts for partial bearings and thrust bearings. Those seeking more complete information can find it in the references cited.

The use of the design charts are illustrated in the solution of the following numerical problem.

Example 10.4: Performance Factors of Journal Bearings Using the Design Charts

A full-journal bearing of diameter D, length L, with a radial clearance c, carries a load of W at a speed of n. It is lubricated by SAE 30 oil, supplied at atmospheric pressure, and the average temperature of the oil film is t (Figure 10.17). Using the design charts, analyze the bearing.

Given: The numerical values are (see Figure 10.17)

$$D = 60 \text{ mm}, r = 30 \text{ mm}, L = 30 \text{ mm}, c = 0.05 \text{ mm}, n = 30 \text{ rps}, W = 3.6 \text{ kN}, t = 60°\text{C}$$

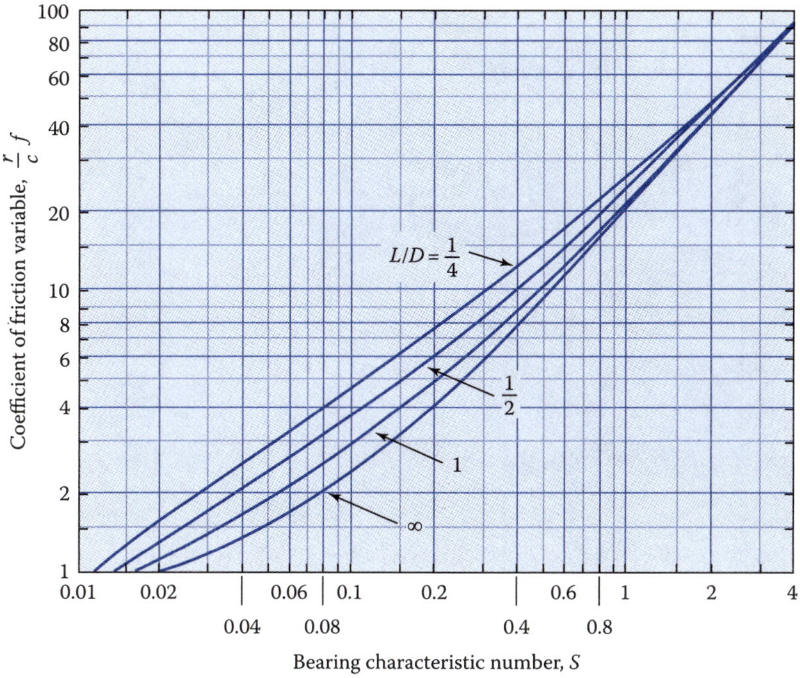

FIGURE 10.15
Chart for coefficient of friction variable. (From Raimondi, A.A. and Boyd, J., *Trans. ASLE I*, 1, 159, 1958.)

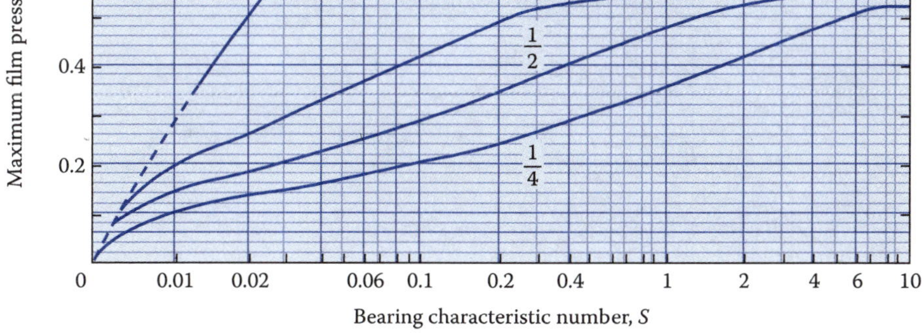

FIGURE 10.16
Chart for film maximum pressure. (From Raimondi, A.A. and Boyd, J., *Trans. ASLE I*, 1, 159, 1958.)

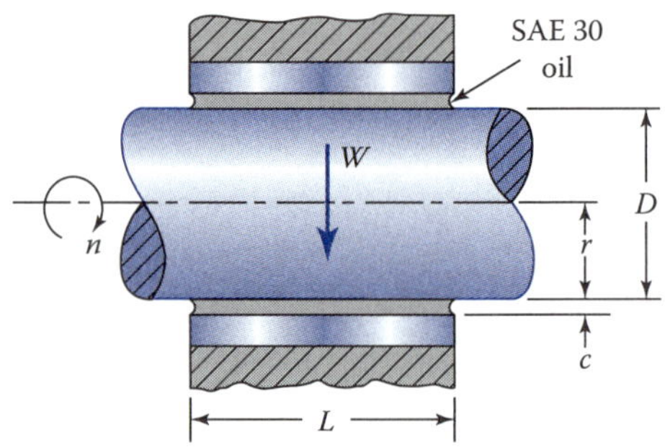

FIGURE 10.17
Example 10.4. A journal bearing.

Solution

The variables under control of the designer are

$$P = \frac{W}{DL} = \frac{3600}{(0.03)(0.06)} = 2 \text{ MPa}$$

$$\eta = 27 \text{ mPa} \cdot \text{s (from Figure 10.8)}$$

$$S = \left(\frac{r}{c}\right)^2 \left(\frac{\eta n}{P}\right) = \left(\frac{30}{0.05}\right)^2 \frac{(0.027)(30)}{2 \times 10^6} = 0.146$$

The determination of the dependent variables proceeds as described in detail in the following paragraphs. Note that the procedure can also be carried out conveniently in tabular form.

Minimum film thickness (Figure 10.14). Use $S = 0.146$ with $L/D = 1/2$ to enter the chart in this figure:

$$\frac{h_0}{c} = 0.25 \quad \text{or} \quad h_0 = 0.0125 \text{ mm}$$

Then, by Equation 10.19,

$$e = c - h = 0.05 - 0.0125 = 0.0375 \text{ mm}$$

The eccentricity ratio is then $\epsilon = e/c = 0.0375/0.05 = 0.75$.

Comments: The permissible oil-film thickness depends largely on the surface roughness of the journal and bearing. The surface finish should therefore be specified and closely controlled if the design calculations indicate that the bearing operates with a very thin oil film.

Coefficient of friction (Figure 10.15). Use $S = 0.146$ with $L/D = 1/2$. Hence, from the chart in this figure,

$$\frac{r}{c} f = 4.8 \quad \text{or} \quad f = 0.008$$

The friction torque is then, applying Equation 10.8,

$$T_f = fWr = 0.008(3600)(0.03) = 0.864 \text{ N} \cdot \text{m}$$

The frictional power lost in the bearing, from Equation 10.10, is

$$\text{kW} = \frac{T_f n}{159} = \frac{(0.864)(30)}{159} = 0.163$$

Film pressure (Figure 10.16). Use $S = 0.146$ with $L/D = 1/2$ to enter the chart in this figure:

$$\frac{P}{p_{\max}} = 0.32$$

The foregoing gives $p_{\max} = 2/0.32 = 6.25$ MPa.

Comments: A temperature rise Δt in the oil film owing to the fluid friction can be determined based on the assumption that the oil film carries away all the heat generated [10]. The Raimondi–Boyd papers also contain charts to obtain oil flow Q, side leakage Q_s, and conservative estimates of the Δt. In addition, they include charts to find the angular locations of the minimum film thickness, maximum pressure, and the terminating position of the oil. These charts are not presented in this book.

10.9 Lubricant Supply to Journal Bearings

The hydrodynamic analysis assumes that oil is available to flow into the journal bearing at least as fast as it leaks out at the ends. A variety of methods of lubrication are used for journal bearings. The system chosen for a specific problem depends to a large extent on the type of service the bearing is to perform. Some typical techniques for supplying oil to the bearing are briefly described as follows.

10.9.1 Splash Method

The splash system of lubrication is used effectively when a machine has a rotating part, such as crank or gear enclosed in a housing. The moving part runs through a reservoir of oil in the enclosed casing. This causes a spray of oil to soak the casing, lubricating the bearing. The term *oil bath* refers to a system where oil is supplied by partially submerging the journal into the oil reservoir, as in the railroad partial bearings.

10.9.2 Miscellaneous Methods

A number of simple methods of lubrication also are used. Bearings that are used in low-speed, light-load applications can be lubricated by hand oiling. A wick-feed oiler, as the name implies, depends on an absorbent material serving as a wick to supply oil to the bearing. A drop-feed oiler permits oil from a reservoir to flow through a needle valve to the bearing. A ring-oiled bearing uses a ring that is often located over the journal at the center of the bearing.

Self-contained bearings contain the lubricant in the bearing housing, which is sealed to prevent oil loss. Oil may be gravity fed from a reservoir or cup above the bearing. Obviously, a bearing of this type is economically more desirable because it requires no expensive cooling or lubricant-circulating system. Self-contained bearings are known as *pillow-block* or *pedestal bearings*.

10.9.3 Pressure-Fed Systems

In the pressure-fed lubrication systems, a continuous supply of oil is furnished to the bearing by a small pump. The oil is returned to a reservoir after circulating through the bearing. The pump may be located within the machine housing above the sump and driven by one of the shafts. This complete system is the commonly used method. An example is the pressure-fed lubrication system of a piston-type engine or compressor. Here, oil supplied by the pump fills grooves in the main bearings. Holes drilled in the crankshaft transfer oil from these grooves to the connection rod sleeve bearings. Note that, in most automotive engines, the piston pins are splash lubricated.

10.9.4 Methods for Oil Distribution

Figure 10.18a illustrates a bearing with a *circumferential groove* used to distribute oil in a tangential direction. The oil flows either by gravity or under pressure into the groove through an oil supply hole placed in the groove opposite the portion of the oil film supporting the load. The effect of the groove is to create two half bearings, each having a smaller L/D ratio than the original. As a result, the pressure distribution does not vary, as the smooth curve shown by the dashed line indicates; hydrodynamic pressure drops to nearly 0 at the groove. Although the oil film is broken in half, the efficient cooling obtained allows these bearings to carry larger loads without overheating.

An *axial groove* fed by the oil hole (Figure 10.18b) generally gives sufficient flow at low or ambient oil pressure. A wide variety of groove types give even better oil distribution. In all flow problems, it is assumed that provision has been made to keep the entrance full.

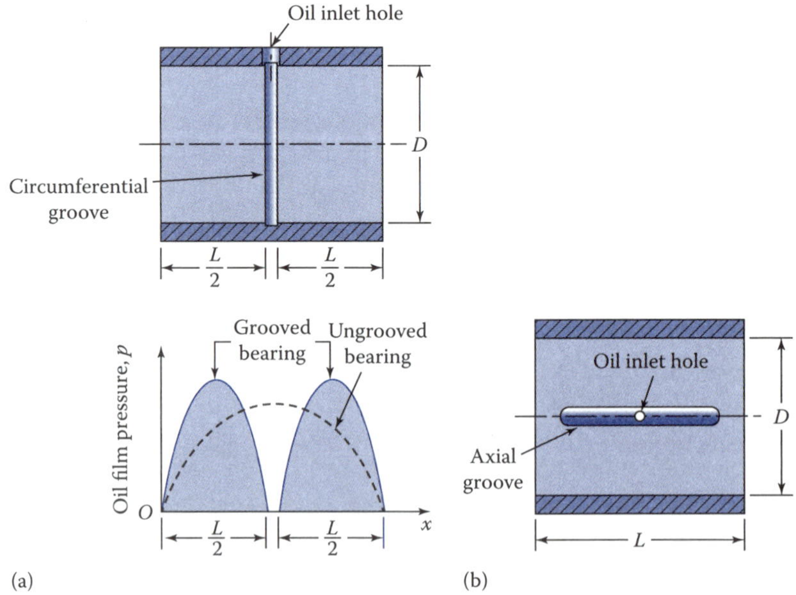

FIGURE 10.18
Common methods used for oil distribution: (a) bearing with circumferential groove and comparison of the axial pressure distribution with or without a groove and (b) bearing with axial groove.

10.10 Heat Balance of Journal Bearings

The frictional loss of energy in a bearing is transferred into heat, raising the temperature of the lubricant and the adjacent parts in a bearing. The *heat balance of a bearing* refers to the balance between the heat developed and dissipated in a bearing. The usual desired value for the average oil temperature is about 70°C for a satisfactory balance. If the average temperature rises above 105°C, deterioration of the lubricant as well as the bearing material can occur [11].

In a pressure-fed system, as the oil flows through the bearing, it absorbs heat from the bearing. The oil is then returned to a sump, where it is cooled before being recirculated. Based on this method, the lubricant carries most of the generated heat, and hence, design charts give a reasonably accurate value of a temperature rise in the oil.

10.10.1 Heat Dissipated

Here, we consider heat balance in self-contained bearings, where the lubricant is stored in the bearing housing itself. Bearings of this type dissipate heat to the surrounding atmosphere by conduction, convection, and radiation heat transfer. Practically, a precise value of the rate of heat flow cannot be calculated with any accuracy. The heat dissipated from the bearing housing may only be approximated by

$$H = CA\left(t_o - t_a\right) \tag{10.22}$$

The foregoing gives

$$t_o = t_a + \frac{H}{AC} \tag{10.23}$$

where

H = the time rate of heat lost, W
C = the overall heat transfer coefficient, $W/m^2 \cdot °C$
A = the surface area of housing, m^2
t_o = the average oil-film temperature, °C
t_a = the temperature of surrounding air, °C

Rough estimates of values for coefficient C are given in Table 10.2. For simple ring-oiled bearings, the bearing housing area may be estimated as 12.5 times the bearing projected area (i.e., 12.5*DL*). It is to be emphasized that Equation 10.23 should be used only when *ballpark* results are sufficient.

TABLE 10.2

Heat Transfer Coefficient C for Self-Contained Bearings

Lubrication System	Conditions	C
Oil ring	Still air	7.4
	Average air circulation	8.5
Oil bath	Still air	9.6
	Average air circulation	11.3

10.10.2 Heat Developed

Under equilibrium conditions, the rate at which heat develops within a bearing is equal to the rate at which heat dissipates:

$$fW(2\pi rn) = H \tag{10.24}$$

where

f = the coefficient of friction
W = the load
r = the journal radius
n = the journal speed (as defined in Section 10.5)
H = given by Equation 10.22

A heat-balance computation, involving finding average film temperature at the equilibrium, is a trial-and-error procedure [12].

10.11 Materials for Journal Bearings

The operating conditions for journal bearing materials are such that rather strict requirements must be placed on the material to be used. For instance, in thick-film lubrication, any material with sufficient compressive strength and a smooth surface is an adequate bearing material. Small bushings and thrust bearings are often expected to run with thin-film lubrication. Any foreign particles larger than the minimum film thickness present in the oil damage the shaft surface unless they can become imbedded in a relatively soft bearing material.

In this section, we discuss some of the types of bearing materials in widespread usage. Special uses are for many other materials, such as glass, silver, ceramics, and sapphires [13,14]. The pressure P, velocity V, or the PV product serves as an *index* to temperature at the sliding interface, and it is widely used as a design parameter for boundary-lubricated bearings. Table 10.3 lists design limit values of these quantities for a variety of journal bearing materials.

10.11.1 Alloys

Babbitt alloys are the most commonly used materials, usually having a tin or lead base. They possess low melting points, moduli of elasticity, yield strength, and good plastic flow. In a bearing, the foregoing gives good conformability and embeddability characteristics. *Conformability* measures the capability of the bearing to adapt to shaft misalignment and deflection. *Embedability* is the bearing's capability to ingest harder, foreign particles. Shafts for babbitt bearings should have a minimum hardness of 150–200 Bhn and a ground surface finish.

Compressive and fatigue strengths of babbitts are low, particularly above about 77°C. Babbitts can rarely be used above about 121°C. However, these shortcomings are improved by using a thin internal babbitt surface on a steel (or aluminum) backing.

TABLE 10.3

Design Limits of Boundary-Lubricated Sleeve Bearings Operating in Contact with Steel Shafts

	Unit Load *P*		Temperature *t*		Velocity *V*		*PV*	
Sleeve Material	**MPa**	**(ksi)**	**°C**	**(°F)**	**m/s**	**(fpm)**	**MPa·m/s**	**(ksi·fpm)**
Porous metals								
Bronze	14	(2)	232	(450)	6.1	(1200)	1.8	(50)
Lead–bronze	5.5	(0.8)	232	(450)	7.6	(1500)	2.1	(60)
Copper–iron	28	(4)	—	—	1.1	(225)	1.2	(35)
Iron	21	(3)	232	(450)	2.0	(400)	1.0	(30)
Bronze–iron	17	(2.5)	232	(450)	4.1	(800)	1.2	(35)
Lead–iron	7	(1)	232	(450)	4.1	(800)	1.8	(50)
Aluminum	14	(2)	121	(250)	6.1	(1200)	1.8	(50)
Nonmetals								
Phenolics	41	(6)	93	(200)	13	(2500)	0.53	(15)
Nylon	14	(2)	93	(200)	3.0	(600)	0.11	(3)
Teflon	3.5	(0.5)	260	(500)	0.25	(50)	0.035	(1)
Teflon fabric	414	(60)	260	(500)	0.76	(150)	0.88	(25)
Polycarbonate	7	(1)	104	(220)	5.1	(1000)	0.11	(3)
Acetal	14	(2)	93	(200)	3.0	(600)	0.11	(3)
Carbon graphite	4	(0.6)	400	(750)	13	(2500)	0.53	(15)
Rubber	0.35	(0.05)	66	(150)	20	(4000)	—	—
Wood	14	(2)	71	(160)	10	(2000)	0.42	(12)

Sources: Based on Budynas, R. and Nisbett, K. *Shigley's Mechanical Engineering Design*. 9th ed., New York, McGraw-Hill, 2010; Rothbart, H.A. and Brown, T.H. Jr., eds., *Mechanical Design Handbook*. 2nd ed., New York, McGraw-Hill, 2006.

For small and medium bearings under higher pressure (as in internal combustion engines), babbitt layers 0.025–2.5 mm thick are used: while in medium and large bearings under low pressure, the babbitt is often cast in thicknesses of 3–13 mm into a thicker steel shell (Figure 10.19).

Copper alloys are principally bronze and aluminum alloys. They are generally stronger and harder, have greater load capacity and fatigue strength, but less compatible (i.e., antiweld and antiscoring) than babbitt bearings. Owing to their thermal conductivity, corrosion resistance, and low cost, aluminum alloys are in widespread usage for bearings in internal combustion engines. A thin layer of babbitt is placed inside an aluminum bearing to improve its conformability and embedability.

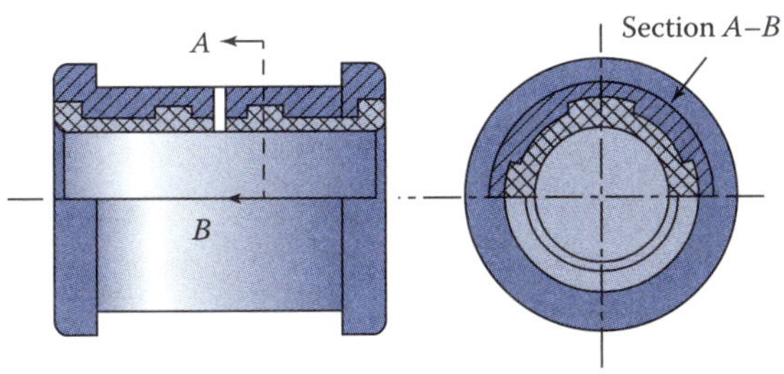

FIGURE 10.19
Babbitt metal bearing cast into a steel shell.

10.11.2 Sintered Materials

Sintered materials, porous metal bearings or insertable powder-metallurgy bushings, have found wide acceptance. These self-lubricated bearings have interconnected pores in which oil is stored in the factory. The pores act as a reservoir for oil, expelling it when heated by shaft rubbing, reabsorbing it when inactive. The low cost and lifetime use in a machine, without further lubrication, are their prime advantages.

10.11.3 Nonmetallic Materials

A variety of *plastics* are used as bearing materials. No corrosion, quiet operation, moldability, and excellent compatibility are their advantages. The last characteristic often implies that no lubrication is required. *Carbon-graphite* bearings can be used at high temperatures. They are chemically inert. These bearings are useful in ovens and in pumps for acids and fuel oils. *Rubber* and other elastomers are excellent bearing material for water pumps and propellers. They are generally placed inside a noncorrodible metal shell and can provide vibration isolation, compensate for misalignment, and have good conformability.

Part B: Rolling-Element Bearings

Recall from Section 10.1 that rolling-element bearings are also known as *rolling bearings* or *antifriction bearings.* The Anti-friction Bearing Manufacturing Association (AFBMA) and the ISO standardized bearing dimensions and the basis for their selection. The load, speed, and operating viscosity of the lubricant affect the friction characteristics of a rolling bearing. These bearings provide coefficients of friction between 0.001 and 0.002. The designer must deal with such matters as fatigue, friction, heat, lubrication, kinematic problems, material properties, machining tolerances, assembly, use, and cost. A complete history of the rolling-element bearings is given in [15]. The following is a comparision of rolling and sliding bearings.

Some *advantages* of rolling-element bearings over the sliding or journal bearings are

1. Low starting and good operating friction torque
2. Ease of lubrication
3. Requiring less axial space
4. Generally, taking both radial and axial loads
5. Rapid replacement
6. Warning of impending failure by increasing noisiness
7. Good low-temperature starting

The *disadvantages* of rolling-element bearings compared to sliding bearings include

1. Greater diametral space
2. More severe alignment requirements
3. Higher initial cost
4. Noisier normal operation
5. Finite life due to eventual failure by fatigue
6. Ease of damage by foreign matter
7. Poor damping ability

10.12 Types and Dimensions of Rolling Bearings

Rolling bearings can carry radial, thrust, or combinations of the two loads, depending on their design. Accordingly, most rolling bearings are categorized in one of the three groups: radial for carrying loads that are primarily radial, *thrust* or axial contact for supporting loads that are primarily axial, and *angular contact* for carrying combined axial and radial loads. As noted earlier, the rolling-element bearings are of two types: ball bearings and roller bearings. The former are capable of higher speeds, and the latter can take greater loads. The rolling bearings are precise, yet simple machine elements. They are made in a wide variety of types and sizes (Figure 10.20). Most bearing manufacturers provide engineering manuals and brochures containing descriptions of the various kinds available. Only some common types are considered here.

10.12.1 Ball Bearings

A ball bearing is employed in almost every type of machine or mechanism with rotating parts. Figure 10.21 illustrates the various parts, surfaces, and edges of a ball bearing. Observe that the basic bearing consists of an inner ring, an outer ring, the balls, and the *separator* (also known as cage or retainer). To increase the contact area and hence permit larger loads to be carried, the balls run in curvilinear grooves in the rings called *raceways*. The radius of the raceway is very little larger than the radius of the ball.

The *deep-groove* (Conrad-type) *bearing* (Figure 10.22a) can stand a radial load as well as some thrust load. The balls are inserted into grooves by moving the inner ring to an eccentric position. They are separated after loading, and then the retainers are inserted.

FIGURE 10.20
Various rolling-element bearings. (Courtesy of SKF, Lansdale, PA.)

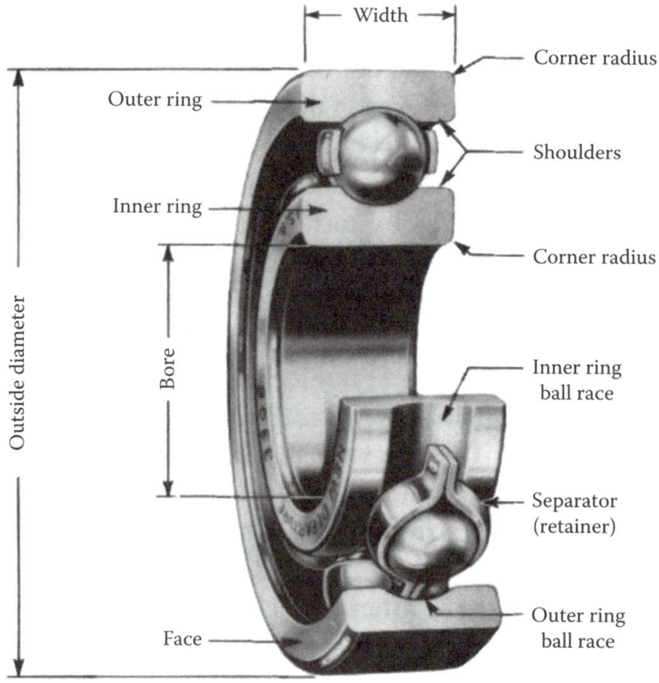

FIGURE 10.21
Ball bearing geometry and nomenclature. (Courtesy of New Departure-Hyatt Division, General Motors Corporation, Detroit, MI.)

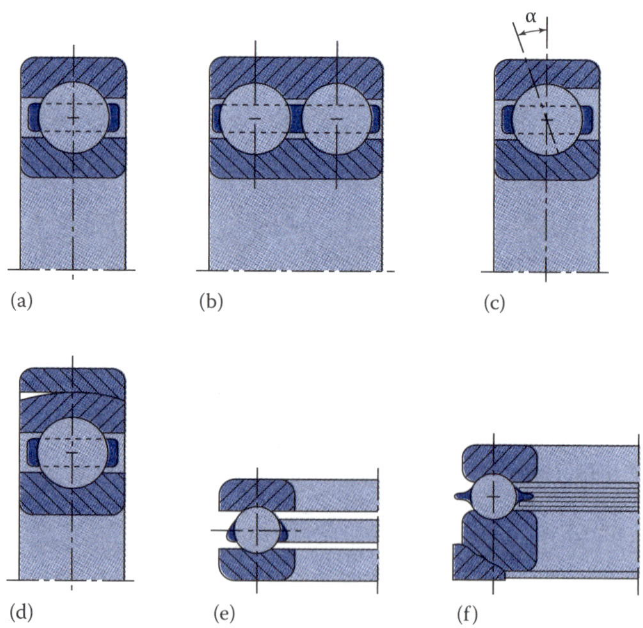

FIGURE 10.22
Some types of ball bearings: (a) deep groove (Conrad), (b) double row, (c) angular contact, (d) external self-aligning, (e) thrust, and (f) self-aligning thrust. (Courtesy of the Timken Company, Canton, OH.)

Obviously, an increase in radial load capacity may be obtained by using rings with deep grooves or by employing a double-row radial bearing (Figure 10.22b).

The *angular-contact bearing* (Figure 10.22c) has a two-shouldered ball groove in one ring and a single-shouldered ball groove in the other ring. It can support greater thrust capacity in one direction as well as radial loads. The cutaway shoulder allows bearing assembly and use of a one-piece machined cage. The contact angle α is defined in the figure. Typical values of α for angular ball bearings vary from 15° to 40°.

The *self-aligning bearing* has an outer raceway ball path ground in a spherical shape so it can accommodate large amounts of angular misalignments or shaft deflections. These bearings can support both radial and axial loads and are available in two types: self-aligning external (Figure 10.22d) and self-aligning internal. *Thrust bearings* are designed to carry a pure axial load only, as shown in Figure 10.22e and f. They are made exclusively for machinery with vertically oriented shafts and have modest speed capacity.

It should be noted that, although separators do not support load, they can exert an essential influence on bearing efficiency. Without a separator in a bearing, the rolling elements contact one another during operation and undergo rigorous sliding friction. The main role of a separator is to keep the proper distance between the rolling elements and secure proper load distribution and balance within the bearing. Obviously, the separator is also sustains control of rolling elements from falling out of the bearing during handling.

10.12.2 Roller Bearings

A roller bearing uses straight, tapered, or contoured cylindrical rollers. When shock and impact loads are present or when a large bearing is needed, these bearings usually are employed. Roller bearings can support much higher static and dynamic (shock) loads than comparably sized ball bearings, since they have line contact instead of point contact. A roller bearing generally consists of the same elements as a ball bearing. These bearings can be grouped into five basic types: cylindrical roller bearings, spherical roller bearings, tapered thrust roller bearings, needle roller bearings, and tapered roller bearings (Figure 10.23). Straight roller bearings provide purely radial load support in most applications; they cannot resist thrust loads. The spherical roller bearings have the advantage of

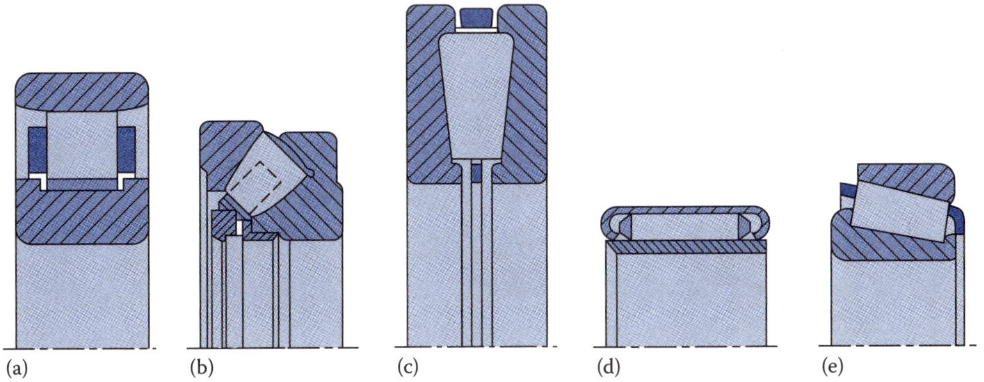

FIGURE 10.23
Some types of roller bearings: (a) straight cylindrical, (b) spherical, (c) tapered thrust, (d) needle, and (e) tapered. (Courtesy of the Timken Company, Canton, OH.)

accommodating some shaft misalignments in heavy-duty rolling mill and industrial gear drives. Needle bearings are in widespread usage where radial space is limited.

Tapered roller bearings combine the advantages of ball and straight roller bearings, as they can stand either radial or thrust loads or any combination of the two. The centerlines of the conical roller intersect at a common apex on the centerline of rotation. Tapered roller bearings have numerous features that make them complicated [10], and space does not permit their discussion in this text. Note that pairs of single-row roller bearings are usually employed for wheel bearings and some other applications. Double-row and four-row roller types are used to support heavier loads. Selection and analysis of most bearing types are identical to that presented in the following sections.

10.12.3 Special Bearings

Rolling-element bearings are available in many other types and arrangements. Detailed information is available in the literature published by the several manufacturers and other references. Two common samples are shown in Figure 10.24. Note that these bearings package standard ball or roller bearings in cast-iron housings. They can be readily attached to horizontal or vertical surfaces.

10.12.4 Standard Dimensions for Bearings

The AFBMA established standard boundary dimensions for the rolling-element bearings, shafts, and housing shoulders. These dimensions are illustrated in Figure 10.25: D is the bearing bore, D_o is the outside diameter (OD), w is the width, d_s is the shaft shoulder diameter, d_h is the housing diameter, and r is the fillet radius. For a given bore, there are various widths and ODs. Similarly, for a particular OD, we can find many bearings with different bores and widths.

In basic AFBMA plan, the bearings are identified by a two-digit number, called the *dimension series code.* The first and second digits represent the width series and the diameter series, respectively. This code does not disclose the dimensions directly, however; and it is required to resort to tabulations. Tables 10.4 and 10.5 furnish dimensions of some 02- and 03-series of ball and cylindrical roller bearings. The load ratings of these bearings,

(a) (b)

FIGURE 10.24
Special bearings: (a) pillow block and (b) flange. (Courtesy of Emerson Power Transmission, Sealmaster Bearings, Aurora, IL.)

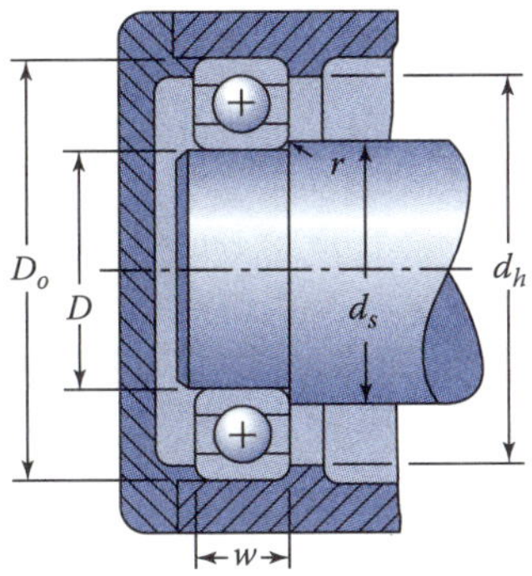

FIGURE 10.25
Dimensions of ball bearing, shaft, and housing.

TABLE 10.4

Dimensions and Basic Load Ratings for 02-Series Ball Bearings

				Load Ratings (kN)			
				Deep Groove		Angular Contact	
Bore, D (mm)	OD, D_o (mm)	Width, w (mm)	Fillet Radius, r (mm)	C	C_s	C	C_s
10	30	9	0.6	5.07	2.24	4.94	2.12
12	32	10	0.6	6.89	3.10	7.02	3.05
15	35	11	0.6	7.80	3.55	8.06	3.65
17	40	12	0.6	9.56	4.50	9.95	4.75
20	47	14	1.0	12.7	6.20	13.3	6.55
25	52	15	1.0	14.0	6.95	14.8	7.65
30	62	16	1.0	19.5	10.0	20.3	11.0
35	72	17	1.0	25.5	13.7	27.0	15.0
40	80	18	1.0	30.7	16.6	31.9	18.6
45	85	19	1.0	33.2	18.6	35.8	21.2
50	90	20	1.0	35.1	19.6	37.7	22.8
55	100	21	1.5	43.6	25.0	46.2	28.5
60	110	22	1.5	47.5	28.0	55.9	35.5
65	120	23	1.5	55.5	34.0	63.7	41.5
70	125	24	1.5	61.8	37.5	68.9	45.5
75	130	25	1.5	66.3	40.5	71.5	49.0
80	140	26	2.0	70.2	45.0	80.6	55.0
85	150	28	2.0	83.2	53.0	90.4	63.0
90	160	30	2.0	95.6	62.0	106	73.5
95	170	32	2.0	108	69.5	121	85.0

Source: Bamberger, E.N. et al., *Life Adjustment Factors for Ball and Roller Bearings: An Engineering Design Guide*, New York, ASME, 1971.

Note: Bearing life capacities, C, for 10^6 revolution life with 90% reliability. To convert from kN to kips, divide the given values by 4.448.

TABLE 10.5

Dimensions and Basic Load Ratings for Straight Cylindrical Bearings

	02-Series			03-Series		
Bore, D (mm)	OD, D_o (mm)	Width, w (mm)	Load Rating, C (kN)	OD, D_o (mm)	Width, w (mm)	Load Rating, C (kN)
25	52	15	16.8	62	17	28.6
30	62	16	22.4	72	19	36.9
35	72	17	31.9	80	21	44.6
40	80	18	41.8	90	23	56.1
45	85	19	44.0	100	25	72.1
50	90	20	45.7	110	27	88.0
55	100	21	56.1	120	29	102
60	110	22	64.4	130	31	123
65	120	23	76.5	140	33	138
70	125	24	79.2	150	35	151
75	130	25	91.3	160	37	183
80	140	26	106	170	39	190
85	150	28	119	180	41	212
90	160	30	142	190	43	242
95	170	32	165	200	45	264

Source: Bamberger, E.N. et al., *Life Adjustment Factors for Ball and Roller Bearings: An Engineering Design Guide*, New York, ASME, 1971.

Note: Bearing life capacities, C, for 10^6 revolution life with 90% reliability. To convert from kN to kips, divide the given values by 4.448.

discussed in the next section, are also included in the table. More detailed information is readily available in the latest AFBMA Standards [16], engineering handbooks, and manufacturers' catalogs and journals [17,18].

10.13 Rolling Bearing Life

When the ball or roller of an antifriction bearing rolls into a loading region, contact (i.e., Hertzian) stresses occur on the raceways and on the rolling element. Owing to these stresses, which are higher than the endurance limit of the material, the bearing has a limited life. If a bearing is well maintained and operating at moderate temperatures, metal fatigue is cause of failure alone. Failure consists of pitting, spalling, or chipping load-carrying surfaces, as discussed in Section 8.9.

Practically, the life of an individual bearing or any one group of identical bearings cannot be accurately predicted. Hence, the AFBMA established the following definitions associated with the life of a bearing. We note that *bearing life* is defined as the number of revolutions or hours at some uniform speed at which the bearing operates until fatigue failure.

Rating life L_{10} refers to the number of revolutions (or hours at a uniform speed) that 90% of a group of identical roller bearings will complete or exceed before the first evidence of fatigue develops. The term *minimum life* is also used to denote the rating life. *Median life* refers to the life that 50% of the group of bearings would complete or exceed. Test results show that the median life is about five times the L_{10} life.

Basic dynamic load rating C is the constant radial load that a group of apparently identical bearings can take for a rating life of 1 million (i.e., 10^6) revolutions of the inner ring in a stationary load (outer ring does not rotate).

Basic static load rating C_s refers to the maximum allowable static load that does not impair the running characteristics of the bearing. The basic load ratings for different types of bearings are listed in Tables 10.4 and 10.5. The value of C_s depends on the bearing material, number of rolling elements per row, the bearing contact angle, and the ball or roller diameter. Except for an additional parameter relating the load pattern, the value of C is based on the same factors that determine C_s.

10.14 Equivalent Radial Load

Catalog ratings are based only on the radial load. However, with the exception of thrust bearings, bearings are usually operated with some combined radial and axial loads. It is then necessary to define an equivalent radial load that has the same effect on bearing life as the applied loading. The AFBMA recommends, for rolling bearings, the maximum of the values of these two equations:

$$P = XVF_r + YF_a \tag{10.25}$$

$$P = VF_r \tag{10.26}$$

where

P = the equivalent radial load

F_r = the applied radial load

F_a = the applied axial load (thrust)

V = a rotational factor

$$= \begin{cases} 1.0 \ (\text{for inner-ring rotation}) \\ 1.2 \ (\text{for outer-ring rotation}) \end{cases}$$

X = a radial factor

Y = a thrust factor

The equivalent load factors X and Y depend on the geometry of the bearing, including the number of balls and the ball diameter. The AFBMA recommendations are based on the ratio of the axial load F_a to the basic static load rating C_s and a variable reference value e. For deep-groove (single-row and double-row) and angular-contact ball bearings, the values of X and Y are given in Tables 10.6 and 10.7. Straight cylindrical roller bearings are very limited in their thrust capacity because axial loads produce sliding friction at the roller ends. So, the equivalent load for these bearings can also be estimated using Equation 10.26.

10.14.1 Equivalent Shock Loading

Some applications have various degrees of shock loading, which has the effect of increasing the equivalent radial load. Therefore, a shock or service factor, K_s, can be substituted into Equations 10.25 and 10.26 to account for any shock and impact conditions to which the

TABLE 10.6

Factors for Deep-Groove Ball Bearings

		$F_a/VF_r \leq e$		$F_a/VF_r > e$	
F_a/C_s	e	X	Y	X	Y
0.014[a]	0.19				2.30
0.21	0.21				2.15
0.028	0.22				1.99
0.042	0.24				1.85
0.056	0.26				1.71
0.070	0.27	1.0	0	0.56	1.63
0.084	0.28				1.55
0.110	0.30				1.45
0.17	0.34				1.31
0.28	0.38				1.15
0.42	0.42				1.04
0.56	0.44				1.00

Source: Based on Bamberger, E.N. et al., *Life Adjustment Factors for Ball and Roller Bearings: An Engineering Design Guide*, New York, ASME, 1971.

[a] Use 0.014 if $F_a/C_s < 0.014$.

TABLE 10.7

Factors for Commonly Used Angular-Contact Ball Bearings

			Single-Row Bearing		Double-Row Bearing			
			$F_a/VF_r > e$		$F_a/VF_r \leq e$		$F_a/VF_r > e$	
Contact Angle (α)	e	$\frac{iF_a}{C_s}$[a]	X	Y	X	Y	X	Y
	0.38	0.015		1.47		1.65		2.39
	0.40	0.029		1.40		1.57		2.28
	0.43	0.058		1.30		1.46		2.11
	0.46	0.087		1.23		1.38		2.00
15°	0.47	0.12	0.44	1.19	1.0	1.34	0.72	1.93
	0.50	0.17		1.12		1.26		1.82
	0.55	0.29		1.02		1.14		1.66
	0.56	0.44		1.00		1.12		1.63
	0.56	0.58		1.00		1.12		1.63
25°	0.68		0.41	0.87	1.0	0.92	0.67	1.41
35°	0.95		0.37	0.66	1.0	0.66	0.60	1.07

Source: Adapted from Bamberger, E.N. et al., *Life Adjustment Factors for Ball and Roller Bearings: An Engineering Design Guide*, New York, ASME, 1971.

[a] Number of rows of balls.

TABLE 10.8
Shock or Service Factors K_s

Type of Load	Ball Bearing	Roller Bearing
Constant or steady	1.0	1.0
Light shocks	1.5	1.0
Moderate shocks	2.0	1.3
Heavy shocks	2.5	1.7
Extreme shocks	3.0	2.0

bearing may be subjected. In so doing, the equivalent radial load becomes the larger of the values given by the two equations:

$$P = K_s\left(XVF_r + YF_a\right) \tag{10.27}$$

$$P = K_s VF_r \tag{10.28}$$

Values to be used for K_s depend on the judgment and experience of the designer, but Table 10.8 may serve as a guide.

10.15 Selection of Rolling Bearings

Each group of seemingly identical bearings may differ slightly metallurgically, in surface finish, in roundness of rolling elements, and so on. Consequently, no two bearings within the same family may have the exact number of operating hours to fatigue failure after having been subjected to the identical speed and load condition. Therefore, the selection of rolling bearings is often made from tables of standard types and sizes containing data on their load and life ratings.

Usually, the basic static load rating C_s has little effect in the ball or roller bearing selection. However, if a bearing in a machine is stationary over an extended period of time with a load higher than C_s, local permanent deformation can occur. In general, the bearings cannot operate at very low speeds under loading that exceeds the basic static load rating.

The basic dynamic load rating C enters directly into the process of selecting a bearing, as is observed in the following formulation for the bearing life. Extensive testing of rolling bearings and subsequent statistical analysis has shown that load and life of a bearing are related statistically. This relationship can be expressed as

$$L_{10} = \left(\frac{C}{P}\right)^a \tag{10.29}$$

where

L_{10} = the rating life, in 10^6 revolution
C = the basic load rating (from Tables 10.4 and 10.5)
P = the equivalent radial load (from Section 10.14)

$$a = \begin{cases} 3\left(\text{for ball bearings}\right) \\ 10/3\left(\text{for roller bearings}\right) \end{cases}$$

We note that the load C is simply a reference value (see Section 10.13) that permits bearing life to be predicted at any level of actual load applied. Alternatively, the foregoing equation may be written in the following form:

$$L_{10} = \frac{10^6}{60n}\left(\frac{C}{P}\right)^a \tag{10.30}$$

where
L_{10} represents the rating life, in h
n is the rotational speed, in rpm

When *two groups* of identical bearings are run with different loads P_1 and P_2, the ratio of their rating lives L'_{10} and L''_{10}, by Equation 10.29, is

$$\frac{L'_{10}}{L''_{10}} = \left(\frac{P_2}{P_1}\right)^a \tag{10.31}$$

Good agreement between this relation and experimental data has been realized. Rearranging the foregoing, we have

$$L'_{10}P_1^a = L''_{10}P_2^a = 10^6C^a \tag{10.32}$$

Clearly, the terms in Equation 10.32 are constant 10^6C^a, as previously defined.

10.15.1 Reliability Requirement

Recall from Section 10.13 that the definition of rating life L_{10} is based on a 90% reliability (or 10% failure). In some applications, the foregoing survival rate cannot be tolerated (e.g., nuclear power plant controls, medical and hospital equipment). As mentioned in Section 6.14, the distribution of bearing failures at a constant load can be best approximated by the Weibull distribution.

Using the general Weibull equation [9,19,20] together with extensive experimental data, the AFBMA formulated recommended *life adjustment factors*, K_r, plotted in Figure 10.26. This curve can be applied to both ball and roller bearings but is restricted to reliabilities no greater than 99%. The expected bearing life is the product of the rating life and the adjustment factor. Combining this factor with Equation 10.29, we have

$$L_5 = K_r\left(\frac{C}{P}\right)^a \tag{10.33}$$

The quantity L_5 represents the rating life for any given reliability *greater* than 90%.

Most manufacturers' handbooks contain specific data on bearing design lives for many classes of machinery. For reference, Table 10.9 may be used when such information is unavailable.

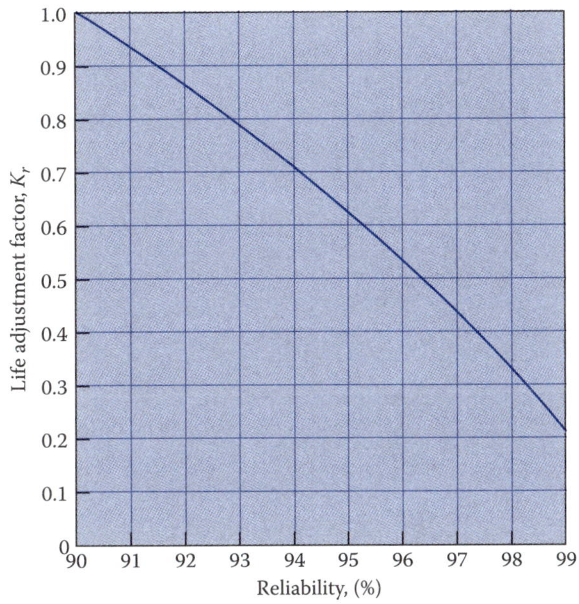

FIGURE 10.26
Reliability factor K_r.

TABLE 10.9
Representative Rolling Bearing Design Lives

Type of Application	Life (kh)
Instruments and apparatus for infrequent use	Up to 0.5
Aircraft engines	0.5–2
Machines used intermittently	
Service interruption is of minor importance	4–8
Reliability is of great importance	8–14
Machines used in an 8 h service working day	
Not always fully utilized	14–20
Fully utilized	20–30
Machines for continuous 24 h service	50–60
Reliability is of extreme importance	100–200

Example 10.5: Median Life of a Deep-Groove Ball Bearing

A 50 mm bore (02-series) deep-groove ball bearing, such as shown in Figure 10.22a, carries a combined load of 9 kN radially and 6 kN axially at 1200 rpm. Calculate

a. The equivalent radial load
b. The median life in hours

Assumptions: The inner ring rotates and the load is steady.

Solution

Referring to Table 10.4, we find that, for a 50 mm bore bearing, $C = 35.1$ kN and $C_s = 19.6$ kN.

a. To obtain the values of the radial load factors X and Y, it is necessary to obtain

$$\frac{F_a}{C_s} = \frac{6}{19.6} = 0.306, \quad \frac{F_a}{VF_r} = \frac{6}{1(9)} = 0.667.$$

We find from Table 10.6 that $F_a/VF_r > e$: $X = 0.56$ and $Y = 1.13$ by interpolation. Applying Equation 10.25,

$$P = XVF_r + YF_a = (0.56)(1)(9) + (1.13)(6) = 11.82 \text{ kN}$$

Through the use of Equation 10.26, $P = VF_r = 1(9) = 9$ kN.

b. Since 11.95 > 9 kN, the larger value is used for life calculation. The rating life, from Equation 10.29, is

$$L_{10} = \left(\frac{C}{P}\right)^a = \left(\frac{35.1}{11.82}\right)^3 = 26.19(10^6) \text{ rev}$$

By Equation 10.30,

$$L_{10} = \left(\frac{10^6}{60n}\right)\left(\frac{C}{P}\right)^a = \frac{10^6(26.19)}{60(1200)} = 364 \text{ h}$$

The median life is therefore $5L_{10} = 1820$ h.

Example 10.6: The Median Life of a Deep-Groove Ball Bearing under Moderate Shock

Redo Example 10.5, but the outer ring rotates, and the bearing is subjected to a moderate shock load.

Solution

We now have $V = 1.2$; hence,

$$\frac{F_a}{VF_r} = \frac{6}{1.2(9)} = 0.556$$

Table 10.6 shows that still $F_a/VF_r > e$; therefore, $X = 0.56$ and $Y = 1.13$, as before.

a. Applying Equation 10.27,

$$P = K_s(XVF_r + YF_a) = 2(0.56 \times 1.2 \times 9 + 1.13 \times 6) = 25.66 \text{ kN}$$

From Equation 10.28, $P = K_s VF_r = 2(1.2 \times 9) = 21.6$ kN.

b. Inasmuch as 25.66 > 21.6 kN, we use the larger value for calculating the rating life. Through the use of Equation 10.30,

$$L_{10} = \left(\frac{10^6}{60n}\right)\left(\frac{C}{P}\right)^a = \frac{10^6}{60(1200)}\left(\frac{35.1}{25.66}\right)^3 = 35.5 \text{ h}$$

and the median life is $5L_{10} = 177.5$ h.

Example 10.7: Extending a Ball Bearing's Expected Life

What change in the loading of a ball bearing will increase the expected life by 25%?

Solution

Let L'_{10} and P_1 be the initial life and load L''_{10} and P_2 be the new life and load. Then, Equation 10.32 with $a = 3$ and $L''_{10} = 1.25L'_{10}$ gives

$$P_2^3 = \frac{L'_{10}P_1^3}{1.25L'_{10}} = 0.8P_1^3$$

from which $P_2 = 0.928\,P_1$.

Comment: A reduction of the load to about 93% of its initial value causes a 25% increase in the expected life of a ball bearing.

Example 10.8: Expected Life of a Ball Bearing with a Low Rate of Failure

Determine the expected life of the bearing in Example 10.5, if only a 6% probability of failure can be permitted.

Solution

From Figure 10.26, for a reliability of 94%, $K_r = 0.7$. Using Equation 10.33, the expected rating life is

$$L_5 = K_r\frac{10^6}{60n}\left(\frac{C}{P}\right)^a = 0.7(364) = 254.8\text{ h}$$

Comment: To improve the reliability of the bearing in Example 10.5 from 90% to 94%, a reduction of median life from 1820 h to $5L_{10} = 1274$ h is required.

10.16 Materials and Lubricants of Rolling Bearings

Most *balls* and *rings* are made from high-carbon chromium steel (SAE 52100) and heat-treated to high strength and hardness, and the surfaces smoothly ground and polished. *Separators* are usually made of low-carbon steel and copper alloy, such as bronze. Unlike ball bearings, roller bearings are often fabricated of case-hardened steel alloys. Modern steel manufacturing processes have resulted in bearing steels with reduced level of impurities.

The most usual kind of separator is made from two strips of carbon steel that are pressed and riveted together. These, termed ribbon separators, are the least expensive to manufacture and are well suited for most applications. In addition, they are lightweight and often require small space. Angular-contact ball bearings permit the use of a one-piece separator. The simplicity and strength of one-piece separators allow their lubrication from various desirable materials. Reinforced phenolic and bronze represent the two most ordinarily employed materials. Bronze separators have considerable strength with low-friction characteristics and can be operated at temperatures to 230°C.

As pointed out in Section 10.4, elastohydrodynamic lubrication occurs in rolling bearings in which deformation of the parts must be taken into account as well as increased viscosity of the oil owing to the high pressure. This small elastic flattening of parts, together with the increase in viscosity, provides a film, although very thin, that is much thicker than would prevail with complete rigid parts. In addition to providing a film between the sliding and rolling parts, a lubricant may help distribute and dissipate heat, prevent corrosion of the bearing surfaces, and protect the parts from the entrance of foreign particles.

Depending on the load, speed, and temperature requirements, bearing lubricants are either *greases* or *oils.* Where bearing speeds are higher or loading is severe, oil is preferred. *Synthetic* and *dry lubricants* are also widely used for special applications. Greases are suitable for low-speed operation and permit bearings to be prepacked.

10.17 Mounting and Closure of Rolling Bearings

Rolling-element bearings are generally mounted with the rotating inner or outer ring with a press fit. Then the stationary ring is mounted with a push fit. Bearing manufacturers' literature contains extensive information and illustrations on mountings. Here, we discuss only the basic principle of mounting ball bearings properly.

Figure 10.27 shows a common method of mounting, where the inner rings are backed up against the shaft shoulders and held in position by round nuts threaded into the shaft. As noted, the outer ring of the left-hand bearing is backed up against a housing shoulder and retained in position, but the outer ring of the right-hand bearing floats in the housing. This allows the outer ring to slide both ways in its mounting to avoid thermal-expansion-induced axial forces on the bearings, which would seriously shorten their life. An alternative bearing mounting is illustrated in Figure 10.28. Here, the inner ring is backed up against the shaft shoulder, as before, however, no retaining device is needed; threads are eliminated. With this assembly, the outer rings of both

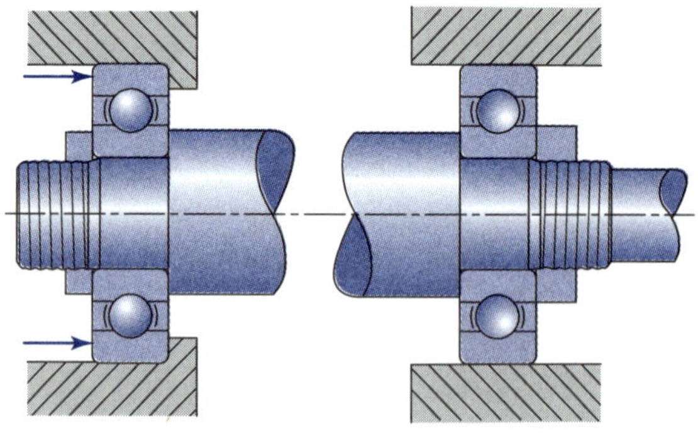

FIGURE 10.27
A common bearing mounting. (Courtesy of the Timken Company, Canton, OH.) *Note*: The outer ring of the left-hand bearing is held in position by a device (not shown).

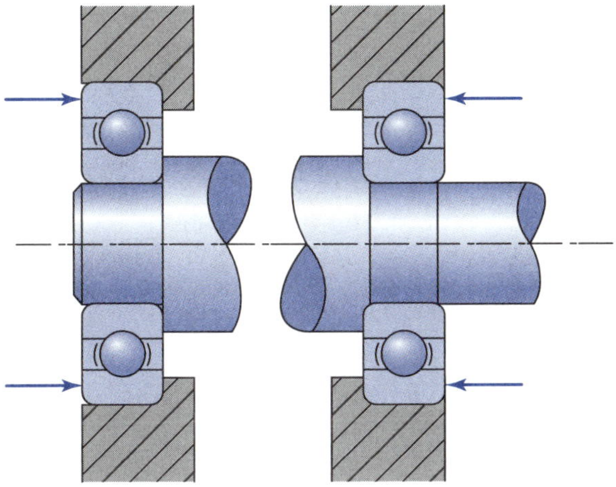

FIGURE 10.28
An alternative bearing mounting. (Courtesy of the Timken Company, Canton, OH.) *Note*: The outer rings of both bearings are held in position by devices (not shown).

bearings are completely retained. As a result, accurate dimensions in axial direction or the use of adjusting devices is required.

Duplexing of angular contact ball bearings arises when maximum stiffness and resistance to shaft misalignment is required, such as in machine tools and instruments. Bearings for *duplex mounting* have their rings ground with an offset, so that, when a pair of bearings is rigidly assembled, a controlled axial preload is automatically achieved. Figure 10.29a and b shows face-to-face (DF) and back-to-back (DB) mounting arrangements, respectively, which take heavy radial and thrust loads from either direction. The latter has greater mounting stiffness. Clearly, a tandem (DT) mounting arrangement is employed when the thrust is in the same direction (Figure 10.29c). Single-row ball bearings are often loaded by the axial load built in during assembly, as shown in Figure 10.27. Preloading helps to remove the internal clearance often found in bearings to increase the fatigue life and decrease the shaft slope at the bearings.

Note that the majority of bearings may be supplied with side shields. The shields are not complete closures, but they offer a measure of protection against dust or dirt. A sealed bearing is generally to be lubricated for life. The roller bearings are not often supplied in a sealed and self-lubricated form, as are most ball bearing types.

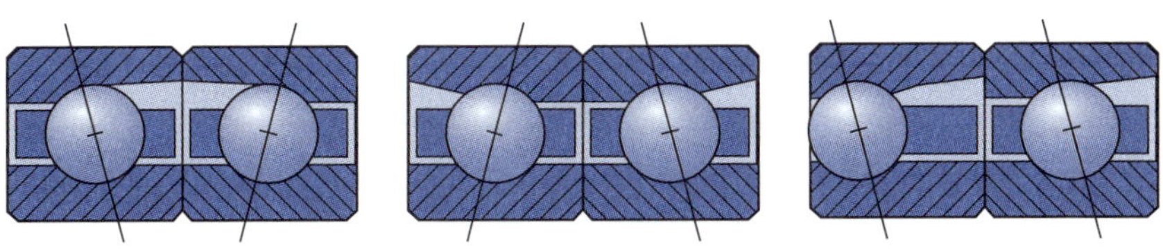

FIGURE 10.29
Mounting arrangements of angular ball bearings: (a) face to face, (b) back to back, and (c) tandem. (Courtesy of the Timken Company, Canton, OH.)

Problems

Sections 10.1 through 10.6

10.1 In a journal bearing, a 24 mm diameter steel shaft is to operate continuously for 1500 h inside of a bronze sleeve having a Brinell hardness of 65 (Figure P10.1). Bearing metals are taken to be partially compatible (Table 8.3). Estimate the depth of wear for two conditions:

a. Good boundary lubrication

b. Excellent boundary lubrication

Given: $D = 24$ mm, $L = 12$ mm, $W = 150$ N, $n = 18$ rpm, $t = 1200$ h, $H = 9.81 \times 65 = 638$ MPa.

10.2 Reconsider Problem 10.1, for the case in which a sleeve made of 2014-T4 wrought aluminum alloy (see Table B.6) with the following data: $D = 1$ in., $L = 1$ in., $W = 25$ lb, $n = 1500$ h.

10.3 The allowable depth of wear of a 1 in. diameter and 1 in. long brass bushing with a Brinell hardness of 60 is 0.006 in. The bearing is to operate 1.2 years with excellent boundary lubrication at a load of 100 lb and bearing metals are partially incompatible (Table 8.3). What is the number of revolutions of the shaft?

10.4 A lightly loaded journal bearing 220 mm in length and 160 mm in diameter consumes 2 hp in friction when running at 1200 rpm. Diametral clearance is 0.18 mm and SAE 30 oil is used. Find the temperature of the oil film.

10.5 A journal bearing has a 4 in. length, a 3 in. diameter, and a c/r ratio of 0.002, carries a 500 lb radial load at 24,000 rpm, and is supplied with an oil having a viscosity of 0.6 μreyns. Using the Petroff approach, estimate

a. The frictional torque developed

b. The frictional horsepower

c. The coefficient of friction

10.6 A Petroff bearing has a 120 mm length, a 120 mm diameter, a 0.05 mm radial clearance, a speed of 600 rpm, and a radial load of 8 kN. Assume that the coefficient of friction is 0.01 and the average oil-film temperature is 70°C. Determine

a. The viscosity of the oil

b. The approximate SAE grade of the oil

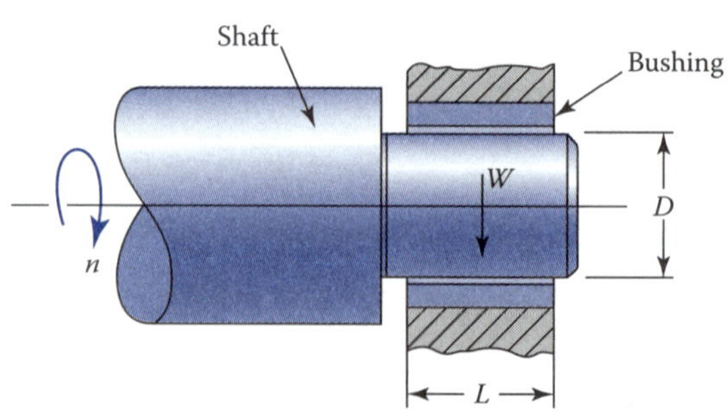

FIGURE P10.1

10.7 A journal bearing having a 125 mm diameter, a 125 mm length, and c/r ratio of 0.0004 carries a radial load of 12 kN. A frictional force of 80 N is developed at a speed of 240 rpm. What is the viscosity of the oil according to the Petroff approach?

10.8 A journal bearing 6 in. in diameter and 1.5 in. long carries a radial load of 500 lb at 1500 rpm; $c/r = 0.001$. It is lubricated by SAE 30 oil at 180°F. Estimate, using the Petroff approach,

a. The bearing coefficient of friction

b. The friction power loss

10.9 A 6 in. diameter and 8 in. long journal bearing under a 400 lb load consumes 18 hp in friction at 2100 rpm. Diametral clearance equals 0.007 in. and SAE 30 oil is used.

Find:

a. The temperature of the oil film

b. The coefficient of friction

Sections 10.7 through 10.11

10.10 A 4 in. diameter × 2 in. long bearing turns at 1800 rpm; $c/r = 0.001$; $h_0 = 0.001$ in. SAE 30 oil is used at 200°F. Through the use of the design charts, find the load W.

10.11 Redo Problem 10.7 employing the design charts.

10.12 Resolve Problem 10.8 using the design charts.

10.13 A shaft of diameter D is supported by a bearing of length L with a radial clearance c. The bearing is lubricated by SAE 60 oil of viscosity of 3 μreyn (Figure P10.13). Compute, using the design charts,

a. The eccentricity

b. The friction power loss

c. Maximum film thickness

Given: $D = 1.2$ in., $L = 1.2$ in., $c = 0.0012$ in., $n = 40$ rps, $W = 320$ lb.

10.14 A 120 mm diameter and 60 mm long journal bearing supports a weight W at a speed of n (Figure P10.13). It is lubricated by SAE 40 oil and the average temperature of the oil film equals 80°C. What is the minimum oil-film thickness?

Given: $n = 1500$ rpm, $W = 15$ kN, $S = 0.15$.

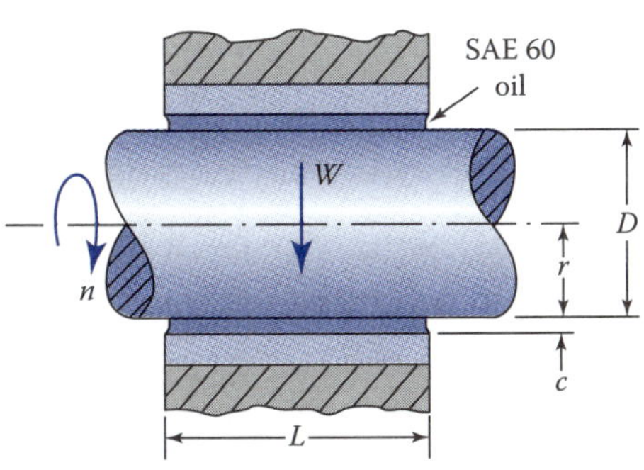

FIGURE P10.13

10.15 A 4 in. diameter shaft is supported by a bearing 4 in. long with a minimum oil-film thickness of 0.001 in. and radial clearance of 0.0025 in. It is lubricated by SAE 20 oil. The bearing carries a load of 100 psi of projected area at 900 rpm. Employing the design charts, determine

a. The temperature of the oil film

b. The coefficient of friction

c. The friction power

10.16 A 25 mm diameter × 25 mm long bearing carries a radial load of 1.5 kN at 1000 rpm; $c/r = 0.0008$; $\eta = 50$ mPa s. Using the design charts, determine

a. The minimum oil-film thickness

b. The friction power loss

10.17 An 80 mm diameter × 40 mm long bearing supports a radial load of 4 kN at 600 rpm; $c/r = 0.002$. SAE 40 oil is used at 65°C. Employing the design charts, determine

a. The minimum oil-film thickness

b. The maximum oil pressure

10.18 A 50 mm diameter × 50 mm long bearing having a c/r ratio of 0.001 consumes 0.16 hp in friction at an operating speed of 1630 rpm. It is lubricated by SAE 20 oil at 83°C. (*Hint*: Try $S = 0.03$.) Using the design charts, determine

a. The radial load for the bearing

b. The minimum oil-film thickness

c. The eccentricity ratio

10.19 A journal bearing having an L/D ratio of ½, a 100 mm diameter, a c/r ratio of 0.0015, and an operating speed of 900 rpm carries a radial load of 8 kN. The minimum oil-film thickness is to be 0.025 mm. Using the design charts, determine

a. The viscosity of the oil

b. The friction force and power developed

10.20 A 100 mm diameter × 50 mm long ring-oiled bearing supports a radial load 6 kN at 300 rpm in still air; $c/r = 0.001$, and $\eta = 20$ mPa·s. If the temperature of the surrounding air of the housing is 20°C, estimate the average film temperature.

10.21 Redo Problem 10.20 for an oil-bath lubrication system in an average air circulation condition when the temperature of the air surrounding air of the housing is $t_a = 30$°C.

Sections 10.12 through 10.17

10.W Use the website at www.grainger.com to conduct a search for roller bearings. Locate a thrust ball bearing ¼ in. bore, 9⁄16 in. OD, and 7⁄32 in. width. List the manufacturer and description.

10.22 A 25 mm (02-series) deep-groove ball bearing carries a combined load of 2 kN radially and 3 kN axially at 1500 rpm. The outer ring rotates and the load is steady. Determine the rating life in hours.

10.23 Resolve Problem 10.22, for a single-row, angular-contact ball bearing having 35° contact angle.

10.24 Redo Problem 10.22, if the inner ring rotates and the bearing is subjected to a light shock load.

10.25 A 2 in. bore (02-series) double-row angular-contact ball bearing supports a combined load of 1.5 kips axially and 5 kips radially. The contact angle is 25°, the outer ring stationary, and the load steady. What is the median life in hours at the speed of 700 rpm?

10.26 A 25 mm bore (02-series) deep-groove ball bearing carries 900 lb radially and 450 lb axially at 3500 rpm with internal ring rotating. Compute the rating life in hours with a survival rate of 95%.

10.27 A 30 mm Conrad-type deep-groove ball bearing is under a combined load of 4.5 kN radially and 1.7 kN axially at a speed of 600 rpm. If the outer ring is stationary, what is the rating life in hours?

10.28 Redo Problem 10.27, if the outer ring rotates and bearing carries a heavy shock load with a reliability a 94% survival rate.

10.29 What percentage change in loading of a ball bearing causes the expected life be doubled?

10.30 Resolve Problem 10.29 for a roller bearing.

10.31 A 60 mm bore (02-series) double-row, angular-contact ball bearing has a 15° contact angle. The outer ring rotates, and the bearing carries a combined steady load of 5 kN radially and 1.5 kN axially at 1000 rpm. Calculate the median life in hours.

10.32 Determine the expected rating lives in hours of a 35 mm bore (02- and 03-series) straight cylindrical bearings operating at 2400 rpm. Radial load is 5 kN, with heavy shock, and the outer rings rotate.

10.33 Calculate the median lives in hours of a 75 mm bore (02- and 03-series) straight cylindrical bearings operating at 2000 rpm. Radial load is 25 kN, with light shock, and inner rings rotate.

10.34 Select two (02- and 03-series) straight cylindrical bearings for an industrial machine intended for a rating life of 24 h operation at 2400 rpm. The radial load is 12.5 kN, with extreme shock, and the inner rings rotate.

10.35 Select a (02-series) deep-groove ball bearing for a machine intended for a median life of 40 h operation at 900 rpm. The bearing is subjected to a radial load of 8 kN, with heavy shock, and the outer ring rotates.

10.36 Determine the expected rating life of the deep-groove ball bearing in Problem 10.22, if only a 5% probability of failure can be permitted at 1200 rpm.

10.37 Calculate the expected median life of the straight cylindrical bearing in Problem 10.32, if only a 2% probability of failure can be permitted.

11

Spur Gears

11.1 Introduction

Gears are used to transmit torque, rotary motion, and power from one shaft to another. They have a long history. In about 2600 BC, the Chinese used primitive gearsets, most likely made of wood and their teeth merely pegs inserted in wheels. In the fifteenth century AD, Leonardo da Vinci showed many gear arrangements in his drawings. Presently, a wide variety of gear types have been developed that operate quietly and with very low friction losses. Smooth, vibrationless action is secured by giving the proper geometric form to the outline of the teeth.

Compared to various other means of power transmission (e.g., belts and chains), gears are the most rugged and durable. They have transmission *efficiency* as high as 98%. However, gears are generally more costly than belts and chains. As we shall see, two modes of failure affect gear teeth: fatigue fracture owing to fluctuating bending stress at the root of the tooth and fatigue (wear) of the tooth surface. Both must be checked when designing the gears. The shapes and sizes of the teeth are standardized by the *American Gear Manufacturers Association* (AGMA). The methods of AGMA are widely employed in design and analysis of gearing. Selection of the proper materials to obtain satisfactory strength, fatigue, and wear properties is important. The AGMA approach requires extensive use of charts and graphs accompanied by equations that facilitate application of computer-aided design. Gear design strength and life rating equations have been computer modeled and programmed by most gear suppliers. It is not necessary for designers to create their own computer programs [1–4].

There are four principal *types of gearing*: spur, helical, bevel, and worm gears (Figure 11.1). Note that spur and helical gears have teeth parallel and inclined to the axis of rotation, respectively. Bevel gears have teeth on conical surfaces. The geometry of a worm is similar to that of a screw. Of all types, the spur gear is the simplest. Here, we introduce the general gearing terminology, develop fundamental geometric relationships of the tooth form, and deal mainly with spur gears. A review of the nomenclature and kinematics is followed by a detailed discussion of the stresses and a number of factors influencing gear design. The basis of the AGMA method and its use is illustrated. Other gear types are dealt with in the next chapter. For general information on gear types, gear drives, and gearboxes, see the website at www.machinedesign.com. The site at www.powertransmission.com lists websites for numerous manufacturers of gears and gear drives.

FIGURE 11.1
A variety of gears, including spur gears, rack and pinion, helical gears, bevel gears, worm, and worm gear. (Courtesy of Quality Transmission Components; www.qtcgears.com.)

11.2 Geometry and Nomenclature

Consider two virtual friction cylinders (or disks) having no slip at the point of contact, represented by the circles in Figure 11.2a. A friction cylinder can be transformed into *spur gear* by placing teeth on it that run parallel to the axis of the cylinder. The surfaces of the rolling cylinders, shown by the dashed lines in the figures, then become the *pitch circles*. The diameters are the pitch diameters, and the cylinders represent the pitch cylinders. The teeth, which lie in axial paths on the cylinder, are arranged to extend both outside and inside the pitch circles (Figure 11.2b). All calculations are

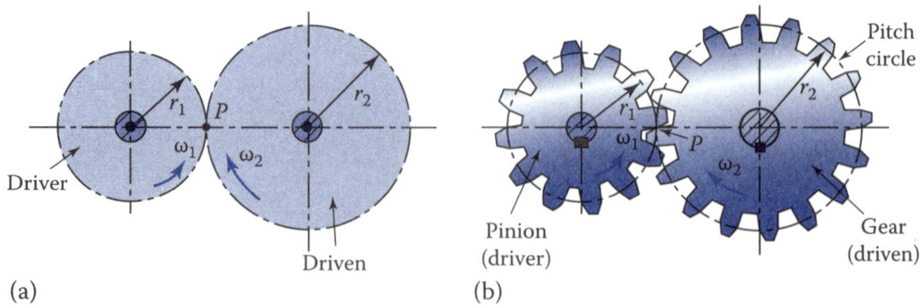

FIGURE 11.2
Spur gears are used to connect parallel shafts: (a) friction cylinders and (b) an external gearset.

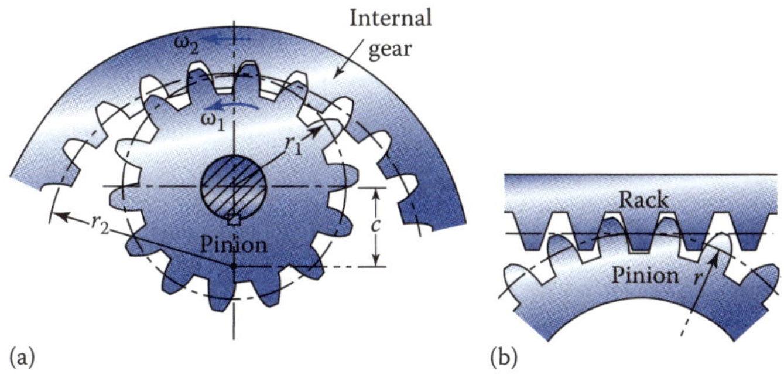

FIGURE 11.3
Gearsets: (a) internal gear and pinion and (b) rack and pinion.

based on the pitch circle. Note that spur gears are used to transmit rotary motion between parallel shafts.

A *pinion* is the smaller of the two mating gears, which is also referred to as a *pair of gears* or *gearset*. The larger is often called the *gear*. In most applications, the pinion is the driving element, whereas the gear is the driven element. This reduces speed, but increases torque, from the power source (engine, motor, turbine): Machinery being driven runs slower. In some cases, gears with teeth cut on the inside of the rim are needed. Such a gear is known as an *internal gear* or an *annulus* (Figure 11.3a). A *rack* (Figure 11.3b) can be thought of as a segment of an internal gear of infinite diameter.

11.2.1 Properties of Gear Tooth

The face and flank portion of the tooth surface are divided by the pitch cylinder. The *circular pitch p* is the distance, on the pitch circle, from a point on one tooth to a corresponding on the next. This leads to the definition

$$p = \frac{\pi d}{N} \tag{11.1}$$

where

p = the circular pitch, in.
d = the pitch diameter, in.
N = the number of teeth

The *diametral pitch P* is defined as the number of teeth in the gear per inch of pitch diameter. Therefore,

$$P = \frac{N}{d} \tag{11.2}$$

This measure is used in the U.S. specification of gears. The units of P are teeth/in. or in.$^{-1}$. Both circular and diametral pitches prescribe the tooth size. The latter is a more convenient definition. Combining Equations 11.1 and 11.2 yields the useful relationship

$$pP = \pi \tag{11.3}$$

For two gears to mesh, they must have the same pitch.

In SI units, the size of teeth is specified by the *module* (denoted by m) measured in *millimeters*. We have

$$m = \frac{d}{N} \tag{11.4}$$

where pitch diameter d and pitch radius r must be in millimeters and N is the number of teeth. Carrying the foregoing expression into Equation 11.1 results in the circular pitch in millimeters:

$$p = \pi m \tag{11.5a}$$

The diametral pitch, using Equation 11.3, is then

$$P = \frac{1}{m} \tag{11.5b}$$

It is measured in teeth/mm or mm^{-1}. Note that metric gears are not interchangeable with U.S. gears, as the standards for tooth size are different.

The addendum a is the radial distance between the top land and the pitch circle as shown in Figure 11.4. The dedendum b_d represents the radial distance from the bottom land to the pitch circle. The *face width b* of the tooth is measured along the axis of the gear. The whole depth h is the sum of the addendum and dedendum. The clearance circle represents a circle tangent to the addendum circle of the mating gear. The clearance f represents the amount by which the dedendum in a given gear exceeds the addendum of the mating gear. Clearance is required to prevent the end of the tooth of one gear from riding on the bottom of the mating gear. The difference between the whole depth and clearance represents the working depth h_k. The distance between the centers of the two gears in mesh is called the *center distance c*. Using Equation 11.2 with $d = 2r$,

$$c = r_1 + r_2 = \frac{N_1 + N_2}{2P} \tag{11.6}$$

Here, subscripts 1 and 2 refer to driver and driven gears, respectively.

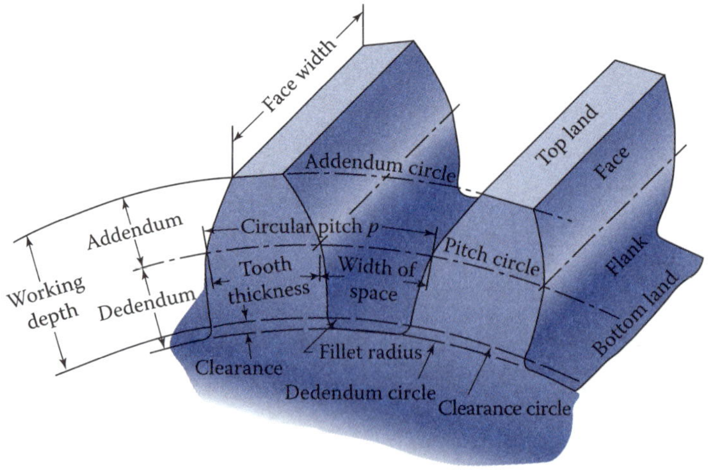

FIGURE 11.4
Nomenclature of the spur gear teeth.

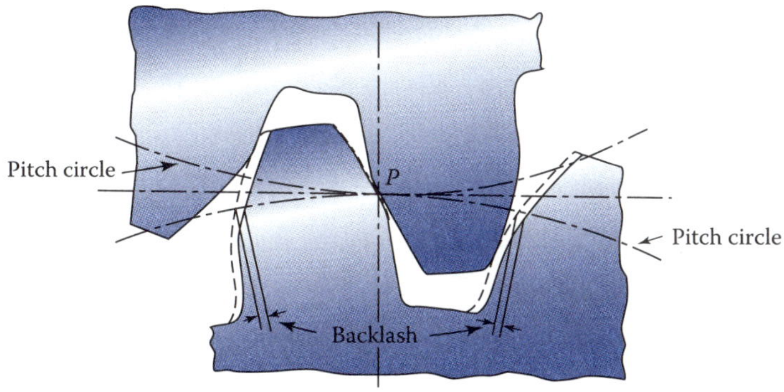

FIGURE 11.5
Depiction of backlash in meshing gears.

The width of space between teeth must be made slightly larger than the gear tooth thickness t, both measured on the pitch circle. Otherwise, the gears cannot mesh without jamming. The difference between the foregoing dimensions is known as *backlash*. That is, the backlash is the gap between mating teeth measured along the circumference of the pitch circle, as schematically shown in Figure 11.5. Manufacturing tolerances preclude a 0 backlash, since all teeth cannot be exactly the same dimensions and all must mesh without jamming. The amount of backlash must be limited to the minimum amount necessary to ensure satisfactory meshing of gears. Excessive backlash increases noise and impact loading whenever torque reversals occur.

Example 11.1: Geometric Properties of a Gearset

A diametral pitch P set of gears consists of an N_1 tooth pinion and N_2 tooth gear (Figure 11.2b).

Find: The pitch diameters, module, circular pith, and center distance

Given: $N_1 = 19$, $N_2 = 124$, $P = 16$ in.$^{-1}$.

Solution

Through the use of Equation 11.4, diameters of pinion and gear are

$$d_1 = \frac{N_1}{P} = \frac{19}{16} = 1.1875 \text{ in.} = 30.16 \text{ mm}$$

$$d_2 = \frac{N_2}{P} = \frac{124}{16} = 7.75 \text{ in.} = 196.85 \text{ mm}$$

Note that in SI units, from Equation 11.5b, the module is

$$m = \frac{1}{P}(25.4) = \frac{1}{16}(25.4) = 1.5875 \text{ mm}$$

and alternatively Equation 11.4 gives the preceding result for the diameters.

Applying Equation 11.3, the circular pitch equals

$$p = \frac{\pi}{P} = \frac{\pi}{16} = 0.1963 \text{ in.} = 4.99 \text{ mm}$$

The center distance, by Equation 11.6, is therefore

$$c = \frac{1}{2}(d_1 + d_2) = \frac{1}{2}(1.1875 + 7.75) = 4.4688 \text{ in.} = 113.51 \text{ mm}$$

11.3 Fundamentals

The main requirement of gear tooth geometry is the provision that angular ratios are exactly constant. We assume that the teeth are perfectly formed, perfectly smooth, and absolutely rigid. Although manufacturing inaccuracies and tooth deflections induce slight deviations in velocity ratio, acceptable tooth profiles are based on theoretical curves that meet this criterion.

11.3.1 Basic Law of Gearing

For quiet, vibrationless operation, the velocities of two mating gears must be the same at all times. This condition is satisfied when the pitch circle of the driver is moving with constant velocity and the velocity of the pitch circle of the driven gear neither increases nor decreases at any instant while the two teeth are touching. The basic law of gearing states that *as the gears rotate, the common normal at the point of contact between the teeth must always pass through a fixed point on the line of centers*. The fixed point is called the *pitch point* P (Figure 11.2). If two gears in mesh satisfy the basic law, the gears are said to produce *conjugate action*.

According to the fundamental law, when two gears are in mesh, their pitch circles roll on one another without slipping. Denoting the pitch radii by r_1 and r_2 and angular velocities as ω_1 and ω_2, respectively, the *pitch-line velocity* is then

$$V = r_1\omega_1 = r_2\omega_2 \tag{11.7}$$

Several useful relations for determining the *speed ratio* may be written as follows:

$$r_s = \frac{\omega_2}{\omega_1} = \frac{n_2}{n_1} = \frac{N_1}{N_2} = \frac{d_1}{d_2} \tag{11.8}$$

where

r_s = the speed or velocity ratio
ω = the angular velocity, rad/s
n = the speed, rpm
N = the number of teeth
d = the pitch circle diameter

Subscripts 1 and 2 refer to the driver and driven gears, respectively.

11.3.2 Involute Tooth Form

To obtain conjugate action, most gear profiles are cut to conform to an involute curve. Our discussions are limited to toothed wheel gearing of the involute form. The involute curve may be generated graphically by wrapping a string around a fixed cylinder, then tracing the path a point on the string (kept taut) makes as the string is unwrapped from the cylinder. When the involute is applied to gearing, the cylinder around which the string is wrapped is defined as the *base circle* (Figure 11.6).

Gear teeth are cut in the shape of an involute curve between the base and the addendum circles, while that part of the tooth between the base and dedendum circles is generally a radial line. Figure 11.7 shows two involutes, on separate cylinders in mesh, representing the gear teeth. Note especially that conjugate involute action can occur only outside both base circles.

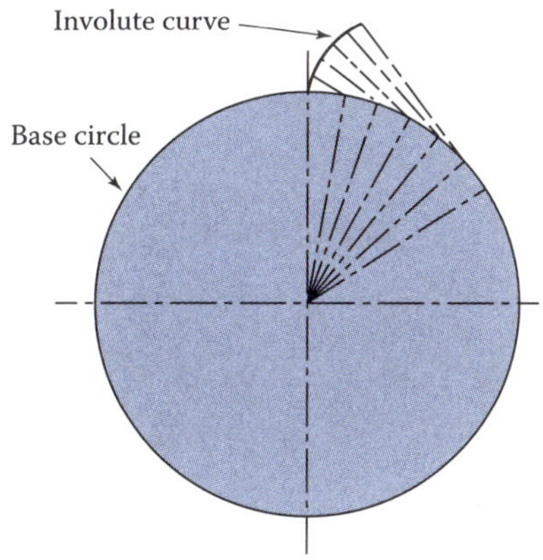

FIGURE 11.6
Development of the involute curve.

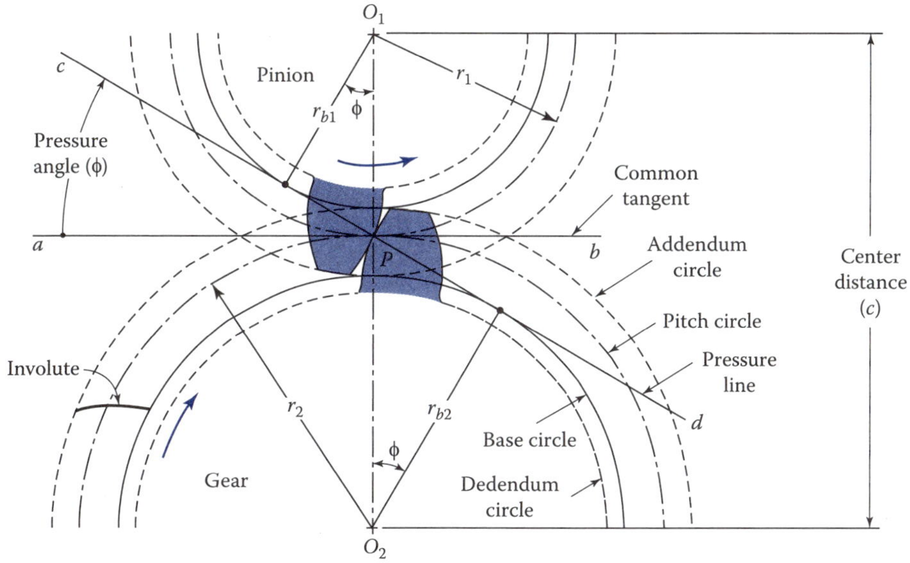

FIGURE 11.7
Involute gear teeth contact form and pressure angle.

11.4 Gear Tooth Action and Systems of Gearing

To illustrate the action occurring when two gears are in mesh, consider Figure 11.7. The pitch radii r_1 and r_2 are mutually tangent along the line of centers O_1O_2, at the pitch point *P*. Line *ab* is the common tangent through the pitch point. Note that line *cd* is normal to the teeth that are in contact and always passes through *P* at an angle ϕ to *ab*. Line *cd* is also tangent to both base circles. This line, called *line of action* or *pressure line*, represents (neglecting the sliding friction) the direction in which the resultant force acts between the gears.

The angle ϕ is known as the *pressure angle*, which is measured in a direction opposite to the direction of rotation of the driver. The involute is the only geometric profile satisfying the basic law of gearing that maintains a constant-pressure angle as the gears rotate. Gears to be run together must be cut to the same nominal pressure angle.

As pointed out, the base circle is tangent to the pressure line. Referring to Figure 11.7, the *radius of the base circle* is then

$$r_b = r \cos \phi \tag{11.9}$$

where *r* represents the pitch circle radius. The *base pitch* p_b refers to the distance measured on the base circle between corresponding adjacent teeth:

$$p_b = p \cos \phi \tag{11.10}$$

where *p* is the circular pitch.

Note that changing the center distance has no effect on the base circle, because this is used to generate the tooth profiles. That is, the base circle is basic to a gear. Increasing the center *c* distance increases the pressure angle ϕ, but the teeth are still conjugate; the requirement for uniform motion transmission is still satisfied. Therefore, with an involute tooth form, center distance errors do not affect the velocity ratio.

11.4.1 Standard Gear Teeth

Most gears are cut to operate with standard pressure angles of 20° or 25°. The tooth proportions for some involute, spur gear teeth are given in Table 11.1 in terms of the diametral pitch *P*. Full-depth involute is a commonly used system of gearing. The table shows that

TABLE 11.1

Commonly Used Standard Tooth Systems for Spur Gears

Item	20° Full Depth	20° Stub	25° Full Depth
Addendum a	$1/P$	$0.8/P$	$1/P$
Dedendum b_d	$1.25/P$	$1/P$	$1.25/P$
Clearance f	$0.25/P$	$0.2/P$	$0.25/P$
Working depth h_k	$2/P$	$1.6/P$	$2/P$
Whole depth h	$2.25/P$	$1.8/P$	$2.25/P$
Tooth thickness t	$1.571/P$	$1.571/P$	$1.571/P$

the 20° stub-tooth involute system has shorter addenda and dedenda than the full-depth systems. The short addendum reduces the duration of contact. Because of insufficient overlap of contact, vibration may occur, especially in gears with few stub teeth.

As a general rule, spur gears should be designed with a *face width b* greater than $9/P$ and less than $13/P$. Unless otherwise specified, we use the term pressure angle to refer to a pressure angle of *full-depth* teeth.

We observe from Table 11.1 the following relationship for all standard pressure angles:

$$f = b_d - a \tag{11.11}$$

Having the addendum a, dedendum b_d, and hence clearance f available, some other gear dimensions can readily be found. These include (Figures 11.4 and 11.7)

$$\begin{aligned} \text{Outside radius:} \quad & r_o = r + a \\ \text{Root radius:} \quad & r_r = r - b_d \\ \text{Total depth:} \quad & h_t = a + b_d \\ \text{Working depth:} \quad & h_w = h_t - f \end{aligned} \tag{11.12}$$

Clearly, the foregoing formulas may also be written in terms of diametral pitch and number of teeth using Equation 11.2.

Figure 11.8 depicts the actual sizes of 20° pressure angle, standard, full-depth teeth, for several standard pitches from $P = 4$ to $P = 80$. Note the inverse relationship between

FIGURE 11.8
Actual size gear teeth of various diametral pitches. (Courtesy of Bourn & Koch Machine Tool Co., Rockford, IL.)

P and tooth size. With SI units, the *standard* values of *metric module mm* are listed in the following:

0.3	0.4	0.5	0.8	1	1.25
1.5	2	3	4	5	6
8	10	12	16	20	25

The *conversion* from one standard to the other is $m = 25.4/P$. The most widely used pressure angle ϕ, in both U.S. customary and SI units, is 20°.

Example 11.2: Gear Tooth and Gear Mesh Parameters

Two parallel shafts *A* and *B* with center distance *c* are to be connected by 2 teeth/in. diametral pitch, 20° pressure angle, and spur gears 1 and 2 providing a velocity ratio of r_s (Figure 11.9). Determine, for each gear,

a. The number of teeth *N*
b. The radius of the base circle r_b and outside diameter d_o
c. Clearance *f*
d. The pitch-line velocity *V*, if gear 2 rotates at speed n_2

Given: $n_2 = 50$ rpm, $r_s = 1/3$, $c = 14$ in., $P = 2\ \text{in.}^{-1}$, $\phi = 20°$.

Design Decision: Common stock gear sizes are considered.

Solution

a. Using Equation 11.6, we have $r_1 + r_2 = c = 14$ in., $r_1/r_2 = 1/3$. Hence, $r_1 = 3.5$ in., $r_2 = 10.5$ in., or $d_1 = 7$ in., $d_2 = 21$ in. Equation 11.2 leads to

$$N_1 = 7(2) = 14, \quad N_2 = 21(2) = 42$$

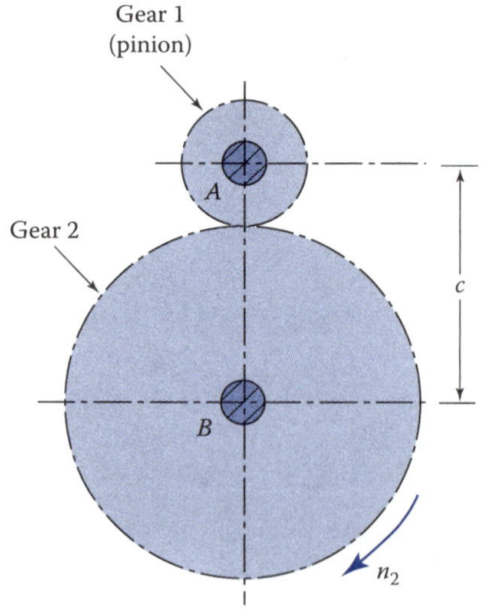

FIGURE 11.9
Example 11.2.

b. Base circle radii, applying Equation (11.9), are

$$r_{b1} = 3.5\cos 20° = 3.289 \text{ in.}$$

$$r_{b2} = 10.5\cos 20° = 9.867 \text{ in.}$$

From Table 11.1, the addendum $a = 1/2 = 0.5$ in. Then

$$d_{o1} = 7 + 2(0.5) = 8 \text{ in.}$$

$$d_{o2} = 21 + 2(0.5) = 22 \text{ in.}$$

c. We have $f = b_d - a$. Table 11.1 gives the dedendum $b_d = 1.25/2 = 0.625$ in., and hence,

$$f = 0.625 - 0.5 = 0.125 \text{ in.}$$

for the pinion and gear. Note as a check that from Table 11.1, $f = 0.25/2 = 0.125$ in.

d. Substituting the given data, Equation 11.7 results in

$$V = r_2\omega_2 = \frac{10.5}{12}\left(500 \times \frac{2\pi}{60}\right) = 45.81 \text{ fps}$$

11.5 Contact Ratio and Interference

Inasmuch as the tips of gear teeth lie on the addendum circle, *contact* between two gears *starts* when the addendum circle of the driven gear intersects the pressure line and *ends* when the addendum circle of the driver intersects the pressure line. The length of action or *length of contact* L_c can be derived from the mating gear and pinion geometry [5,6] in the form

$$L_c = \sqrt{(r_p + a_p)^2 - (r_p\cos\phi)^2} + \sqrt{(r_g + a_g)^2 - (r_g\cos\phi)^2} - c\sin\phi \tag{11.13}$$

where

r = the pitch radius
a = the addendum
c = the center distance
ϕ = the pressure angle

The subscripts p and g present pinion and gear, respectively. When two gears are in mesh, it is desirable to have at least one pair of teeth in contact at all times.

The method often used to show how many teeth are in contact at any time represents thus the *contact ratio* C_r, which is defined as the length of contact divided by the base pitch:

$$C_r = \frac{L_c}{p_b} \tag{a}$$

Thus, inserting Equations 11.13 and 11.10 into Equation (a) defines the contact ratio in terms of circular pitch:

$$C_r = \frac{1}{p\cos\phi}\left[\sqrt{(r_p + a_p)^2 - (r_p\cos\phi)^2} + \sqrt{(r_g + a_g)^2 - (r_g\cos\phi)^2}\right] - \frac{c\tan\phi}{p} \tag{11.14}$$

Obviously, the length of contact must be somewhat greater than a base pitch, so that a new pair of teeth comes into contact before the pair that had been carrying the load separates. Observe from Equation 11.14 that for smaller teeth (larger diametral pitch) and larger pressure angle, the contact ratio will be larger. If the contact ratio is 1, then one tooth is leaving contact just as the next is beginning contact. The minimum acceptable contact ratio for smooth operation equals 1.2. Most gears are designed with contact ratios between 1.4 and 2. For instance, a ratio of 1.5 indicates that one pair of teeth is always in contact and the second in contact 50% of the time. Contact ratio of 2 or more indicates that at least two pairs of teeth are theoretically in contact at all times. It is obvious that their *actual* contact relies upon the precision of manufacturing, tooth stiffness, and applied loading. Generally, the greater is the contact ratio or considerable overlap of gear actions, the smoother and quieter the operation of gears.

Since the part of a gear tooth below the base line is cut as a radial line and not an involute curve, if contact should take place below the base circle, nonconjugate action would result. Hence, the basic law of gearing would not hold. The contact of these portions of tooth profiles that are not conjugate is called *interference*. When interference occurs, the gears do not operate without modification. Removal of the portion of tooth below the base circle and cutting away the interfering material result in an *undercut tooth*. Undercutting causes early tooth failure. Interference and its attendant undercutting can be prevented as follows: remove a portion of the tips of the tooth, increase the pressure angle, or use minimum required tooth numbers [7]. The method to be used depends largely on the application and the designer's experience.

Example 11.3: Contact Ratio of Meshing Gear and Pinion

A gearset has N_1 tooth pinion, N_2 tooth gear, pressure angle ϕ, and diametral pitch P (Figure 11.7).

Find:

a. The contact ratio
b. The pressure angle and contact ratio, if the center distance is increased by 0.2 in.

Given: $N_1 = 15$, $N_2 = 45$, $\phi = 20°$, $P = 2.5$ in.$^{-1}$

$$a = \frac{1}{P} = \frac{1}{2.5} = 0.4 \text{ in. (by Table 11.1)}$$

Assumption: Standard gear sizes are considered.

Solution

Applying Equation 11.2, the pitch diameter for the pinion and gear are found to be

$$d_1 = \frac{15}{2.5} = 6 \text{ in.} = 152.4 \text{ mm} \quad \text{and} \quad d_2 = \frac{45}{2.5} = 18 \text{ in.} = 457.2 \text{ mm}$$

Hence, the gear pitch radii are

$$r_1 = 3 \text{ in.} = 76.2 \text{ mm} \quad \text{and} \quad r_2 = 9 \text{ in.} = 228.6 \text{ mm}$$

a. The center distance c is the sum of the pitch radii. So

$$c = 3 + 9 = 12 \text{ in.} = 304.8 \text{ mm}$$

The radii of the base circles, using Equation 11.9, are

$$r_{b1} = 3\cos 20° = 2.819 \text{ in.} = 71.6 \text{ mm}$$

$$r_{b2} = 9\cos 20° = 8.457 \text{ in.} = 214.8 \text{ mm}$$

Substitution of the numerical values into Equation 11.14 gives the contact ratio as

$$C_r = \frac{2.5}{\pi\cos 20°}\left[\sqrt{(3+0.4)^2-(2.819)^2}+\sqrt{(9+0.4)^2-(8.457)^2}\right]-\frac{12\tan 20°}{\pi/2.5}$$
$$= 1.61$$

Comment: The result, about 1.6, represents a suitable value.

b. For the case in which the center distance is increased by 0.2 in., we have $c = 12.2$ in. It follows that

$$c = \frac{1}{2}(d_1+d_2), \quad d_1+d_2 = 2(12.2) = 24.4 \text{ in.} \tag{b}$$

By Equation 11.2,

$$\frac{N_1}{d_1} = \frac{N_2}{d_2}, \quad \frac{15}{d_1} = \frac{45}{d_2} \tag{c}$$

Solving Equations (b) and (c), we have $d_1 = 6.1$ in. and $d_2 = 18.3$ in. or $r_1 = 3.05$ in. and $r_2 = 9.15$ in.

The diametral pitch becomes $P = N_1/d_1 = 15/6.1 = 2.459 \text{ in}\cdot^{-1}$. The addendum is therefore $a = a_1 = a_2 = 1/2.459 = 0.407$ in. Base radii of the gears will remain the same. The new pressure angle can now be obtained from Equation 11.9:

$$\phi_{\text{new}} = \cos^{-1}\left(\frac{r_{b1}}{r_1}\right) = \cos^{-1}\left(\frac{2.819}{3.05}\right) = 22.44°$$

Through the use of Equation 11.14, the new contact ratio is then

$$C_{r,\text{new}} = \frac{2.459}{\pi\cos 22.44°}\left[\sqrt{(3.05+0.407)^2-(2.819)^2}+\sqrt{(9.15+0.407)^2-(8.457)^2}\right]-\frac{12.2\tan 22.44°}{\pi/2.459}$$
$$= 1.52$$

Comment: Results show that increasing the center distance leads to an increase in pressure angle but decrease in the contact ratio.

11.6 Gear Trains

Up to now in our discussion of gears, we have been concerned with no more than a pair of gears in mesh. Various applications exist where many pairs of gears are in mesh. Such a system is generally called a *gear train*. Typical examples include the gear trains in

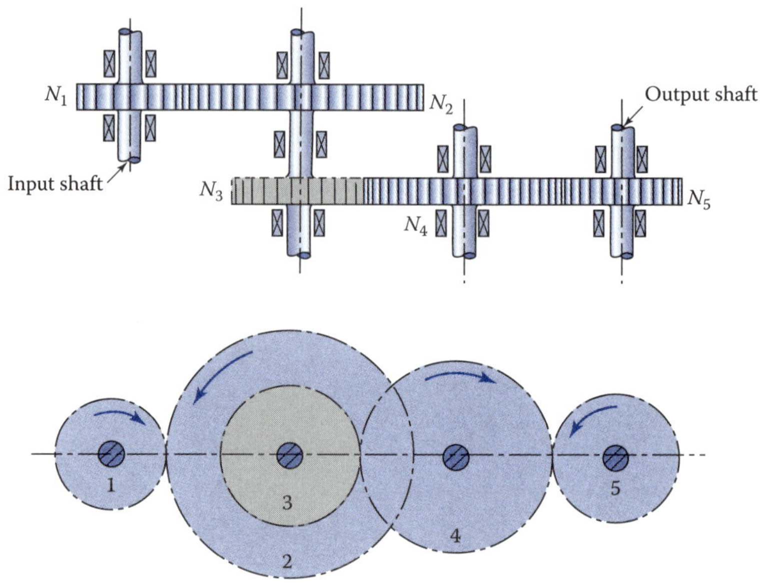

FIGURE 11.10
Gear train: A two-stage gear reducer.

odometers and mechanical watches or clocks. A gearset, the simplest form of gear train, is often limited to a ratio of 10:1. Gear trains are used to obtain a desired velocity or speed of an output shaft while the input shaft runs at different speed. The velocity ratio between the input and output gears is constant. Detailed kinematic relationships for gear trains may be found in [3]. AGMA suggests equations that can be used to determine thermal capacity for gear trains (see also Section 12.11).

The speed ratio of a conventional gear train can be readily obtained from an expanded version of Equation 11.8, if the number of teeth in each driver and driven gear is known. Consider, for example, a gear train made of five gears, with gears 2 and 3 mounted on the same shaft (Figure 11.10). The speed ratio between gears 5 and 1 is given by

$$e_{51} = \frac{n_5}{n_1} \tag{11.15}$$

The speed ratio is equal to the so-called gear value:

$$e_{51} = \left(-\frac{N_1}{N_2}\right)\left(-\frac{N_3}{N_4}\right)\left(-\frac{N_4}{N_5}\right) \tag{11.16}$$

In the foregoing expression, the *minus* signs indicate that the pinion and gear rotate in *opposite* directions, as depicted in the figure. The intermediate gears, called *idler gears*, do not influence the overall speed ratio. In this case, gear 4 is an idler (its tooth numbers cancel in the preceding equation); hence, it affects only the direction of rotation of gear 5.

Consider, for example, pinion 1 in a gear train is driven 800 rpm by a motor (Figure 11.10). Let $N_1 = 18$, $N_2 = 64$, $N_3 = 24$, $N_4 = 36$, and $N_5 = 20$. Then Equation 11.16 results in

$$e_{51} = -\frac{N_1 N_3}{N_2 N_5} = -\frac{(18)(24)}{(64)(20)} = -0.3375$$

The output speed, using Equation 11.15, is thus

$$n_5 = e_{51}n_1 = (-0.3375)(800) = -270 \text{ rpm}$$

The negative sign means that the direction is counterclockwise, as shown in the figure.

Additional ratios can be inserted into Equation 11.16 if the train consists of a larger number of gears. This equation can be generalized for any number of gearsets in the train to obtain the gear value:

$$e = \pm\frac{\text{Product of number of teeth on driver gears}}{\text{Product of number of teeth on driver gears}} \tag{11.17}$$

Clearly, to ascertain the correct algebraic sign for the overall train ratio, the signs of the ratios of the individual pairs must be indicated in this expression. Note also that pitch diameters can be used in Equation 11.17 as well. For spur gears, e is positive when the last gear rotates in the identical sense as the first; it is negative when the last rotates in the opposite sense. If the gear has internal teeth, its diameter is negative and the members rotate in the same direction.

11.6.1 Planetary Gear Trains

Also referred to as the *epicyclic* trains, planetary gear trains permit some of the gear axes to rotate about one another. Such trains always include a *sun gear*, an *arm*, and one or more *planet gears* (Figure 11.11). It is obvious that the maximum number of planets is limited by the space available and the teeth of each planet must align simultaneously with the teeth of the sun and the ring.

A planetary train must have two inputs: the motion of any two elements of the train; for example, the sun gear rotates at a speed of n_s (CW) and that the ring rotates at n_r (CCW) in Figure 11.11. The output would then be the motion of the arm. While power flow through a conventional gear train and the sense of motion for its members may be seen readily, it is often difficult to ascertain the behavior of a planetary train by observation. Planetary gear trains are thus more complicated to analyze than ordinary gear trains.

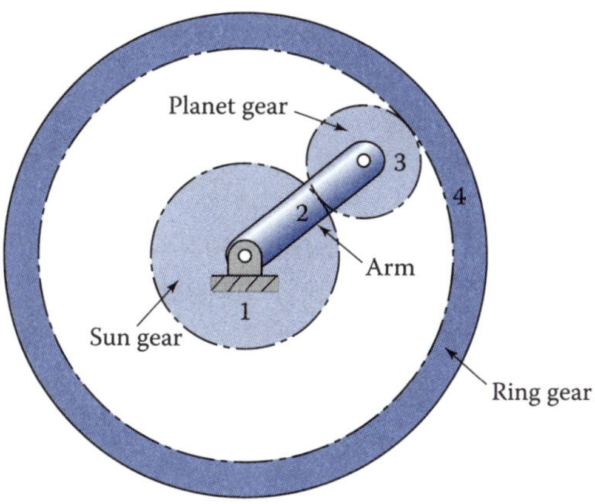

FIGURE 11.11
A planetary gear train.

However, planetary gear trains have several advantages over conventional trains. These include higher train ratios obtainable in smaller packages and bidirectional outputs available from a single unidirectional input. The foregoing features make planetary trains popular as automatic transmissions and drives in motor vehicles [4], where they provide desired forward gear reductions and a reverse motion. Manufacturing precision and the use of the helical gears contribute greatly to the quietness of planetary systems.

It can be shown that [8] the *gear value* of any *planetary train* is given in the following convenient form:

$$e = \frac{n_L - n_A}{n_F - n_A} \tag{11.18}$$

where

e = the gear value, defined by Equation 11.17
n_F = the speed of the first gear in the train
n_L = the speed of the last gear in the train
n_A = the speed of the arm

Note that both the first and last gears chosen must not be orbiting when two of the velocities are specified. Equation 11.18 can be used to compute the unknown velocity. That is, either the velocities of the arm and one gear or the velocities of the first and last gears must be known.

Example 11.4: Analysis of a Planetary Gear Train

In the epicyclic gear train illustrated in Figure 11.11, the sun gear is driven clockwise at 60 rpm and has N_1 teeth, the planet gear N_3 teeth, and the ring gear N_4 teeth. The sun gear is the input and the arm is the output. The ring gear is held stationary. What is the velocity of the arm?

Given: $N_1 = 30$, $N_3 = 20$, $N_4 = 80$.

Assumption: The sun gear is the first gear in the train and the ring gear is the last.

Solution

Refer to Figure 11.11.

The gear value, through the use of Equation 11.17, is

$$e = \left(-\frac{N_1}{N_3}\right)\left(+\frac{N_3}{N_4}\right) = \left(-\frac{30}{20}\right)\left(+\frac{20}{80}\right) = -0.375$$

Observe the signs on the gearset ratios: One is an external set (–) and one an internal set (+). Substitution of this equation together with $n_F = n_1 = 60$ rpm and $n_L = n_4 = 0$ into Equation 11.18 gives

$$-0.375 = \frac{n_4 - n_2}{n_1 - n_2} = \frac{0 - n_2}{60 - n_2}$$

from which $n_2 = 16.4$ rpm.

Comment: The sun gear rotates 3.66 times as fast and in the same direction as the arm.

11.7 Transmitted Load

With a pair of gears or gearset, power is transmitted by the load that the tooth of one gear exerts on the tooth of the other. As pointed out in Section 11.4, the transmitted load F_n is normal to the tooth surface; therefore, it acts along the pressure line or the line of action (Figure 11.12). This force between teeth can be resolved into *tangential force* and radial force components, respectively:

$$F_t = F_n \cos\phi$$
$$F_r = F_n \sin\phi = F_t \tan\phi \tag{11.19}$$

The quantity ϕ is the pressure angle in degrees. The tangential component F_t, when multiplied by the pitch-line velocity, accounts for the power transmitted, as is shown in Section 1.10. However, radial component F_r does no work but tends to push the gears apart.

The velocity along the pressure line is equal to the tangential velocity of the base circles. The tangential *velocity* of the *pitch circle* (in feet per minute, fpm) is given by

$$V = \frac{\pi d n}{12} \tag{11.20}$$

where
 d represents the pitch diameter in in.
 n is the speed in rpm

In design, we assume that the tangential force remains *constant* as the contact between two teeth moves from the top of the tooth to the bottom of the tooth. The applied torque and the transmitted load are related by

$$T = \frac{d}{2} F_n \cos\phi = \frac{d}{2} F_t \tag{11.21}$$

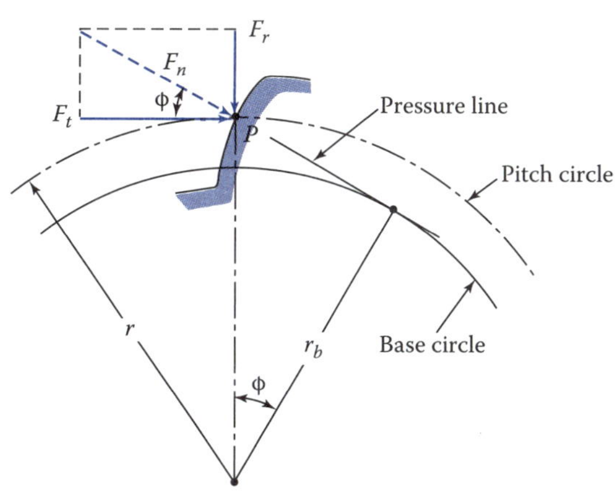

FIGURE 11.12
Gear tooth force F_n, shown resolved at pitch point P.

The horsepower is defined by

$$\text{hp} = \frac{Tn}{63,000} \tag{1.17}$$

in which the torque T is in pounds-inch and n is in rpm. Carrying Equations 11.20 and 11.21 into the preceding expression, we obtain the *tangential load transmitted:*

$$F_t = \frac{33,000\ \text{hp}}{V} \tag{11.22}$$

where V is given by Equation 11.20. Recall from Section 1.11 that 1 hp equals 0.7457 kW.

In SI units, the preceding equations are given by the relationships

$$\text{kW} = \frac{F_t V}{1000} = \frac{Tn}{9549} \tag{1.15}$$

$$\text{hp} = \frac{F_t V}{745.7} = \frac{Tn}{7121} \tag{1.16}$$

$$F_t = \frac{1000\ \text{kW}}{V} = \frac{745.7\ \text{hp}}{V} \tag{11.23}$$

In the foregoing, we have

F_t = the transmitted tangential load (N)
d = the gear pitch-diameter
n = the speed (rpm)
T = the torque (N·m)
$V = \pi dn/60$ = pitch-line velocity (in meters per second, m/s)

11.7.1 Dynamic Effects

The tangential force F_t is readily obtained by Equation 11.22. However, this is not the entire force that acts between the gear and teeth. Tooth inaccuracies and deflections, misalignments, and the like produce dynamic effects that also act on the teeth. The *dynamic load* F_d or total gear tooth load, in U.S. customary units, is estimated using one of the following formulas:

$$F_d = \frac{600 + V}{600} F_t \quad (\text{for } 0 < V \le 2000\ \text{fpm}) \tag{11.24a}$$

$$F_d = \frac{1200 + V}{1200} F_t \quad (\text{for } 2000 < V \le 4000\ \text{fpm}) \tag{11.24b}$$

$$F_d = \frac{78 + \sqrt{V}}{78} F_t \quad (\text{for } V > 4000\ \text{fpm}) \tag{11.24c}$$

where V is the pitch-line velocity in fpm. To convert to m/s, *divide* the given values in these equations by 196.8. Clearly, the dynamic load occurs in the time while a tooth goes through mesh. Note that the preceding relations form the basis of the AGMA dynamic factors, discussed in Section 11.9.

Example 11.5: Gear Force Analysis

The three meshing gears shown in Figure 11.13a have a module of 5 mm and a 20° pressure angle. Driving gear 1 transmits 40 kW at 2000 rpm to idler gear 2 on shaft *B*. Output gear 3 is mounted to shaft *C*, which drives a machine. Determine and show, on a free-body diagram,

a. The tangential and radial forces acting on gear 2
b. The reaction on shaft *B*

Assumptions: The idler gear and shaft transmit power from the input gear to the output gear. No idler shaft torque is applied to the idler gear. Friction losses in the bearings and gears are omitted.

Solution

The pitch diameters of gears 1 and 3, from Equation 11.4, are $d_1 = N_1 m = 20(5) = 100$ mm and $d_3 = N_3 m = 30(5) = 150$ mm.

a. Through the use of Equation 1.15,

$$T = \frac{9549\text{ kW}}{n} = \frac{9549(40)}{2000} = 191\text{ N}\cdot\text{m}$$

By Equations 11.21 and 11.19, the tangential and radial forces of gear 1 on gear 2 are then

$$F_{t,12} = \frac{T_1}{r_1} = \frac{191}{0.05} = 3.82\text{ kN}, \quad F_{r,12} = 3.82\tan 20° = 1.39\text{ kN}$$

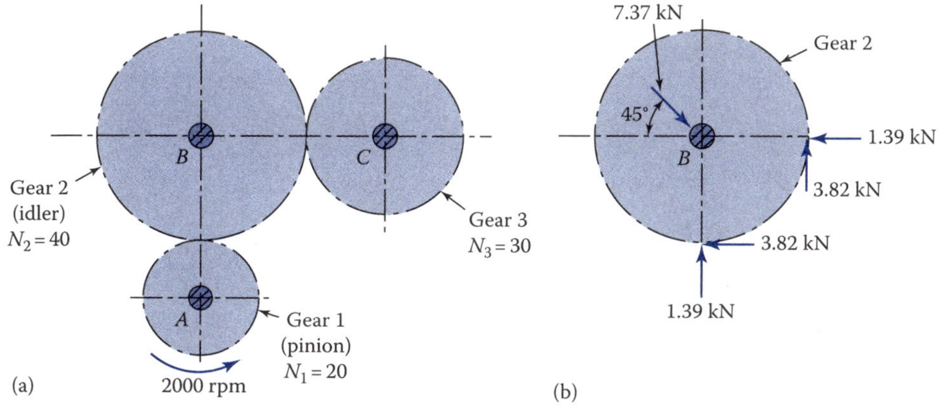

FIGURE 11.13
Example 11.5. (a) a gearset and (b) free-body diagram of the forces acting on gear 2 and reaction on shaft *B*.

Inasmuch as gear 2 is an idler, it carries no torque, so the tangential reaction of gear 3 on 2 is also equal to $F_{t,12}$. Accordingly, we have

$$F_{t,32} = 3.82 \text{ kN}, \quad F_{r,32} = 1.39 \text{ kN}$$

The forces are shown in proper directions in Figure 11.13b.

b. Equilibrium of x- and y-directed forces acting on the idler gear gives $R_{Bx} = R_{By} = 3.82 + 1.39 = 5.21$ kN. The reaction on the shaft B is then

$$R_B = \sqrt{5.21^2 + 5.21^2} = 7.37 \text{ kN}$$

acting as depicted in Figure 11.13b.

Comments: When a combination of numerous gears is used as in a gear train, usually the shafts supporting the gears lie in different planes and the problem becomes a little more involved. For this case, the tangential and radial force components of one gear must be further resolved into components in the same plane as the components of the meshing gear. Hence, forces along two mutually perpendicular directions may be added algebraically.

11.8 Bending Strength of a Gear Tooth: The Lewis Formula

Wilfred Lewis was the first to present the application of the bending equation to a gear tooth. The formula was announced in 1892, and it still serves the basis for gear tooth bending stress analysis. Simplifying *assumptions* in the Lewis approach are as follows [9]:

1. A full load is applied to the tip of a single tooth.
2. The radial load component is negligible.
3. The load is distributed uniformly across the full-face width.
4. The forces owing to tooth sliding friction are negligible.
5. The stress concentration in the tooth fillet is negligible.

To develop the basic Lewis equation, consider a cantilever subjected to a load F_t, uniformly distributed across its width b (Figure 11.14a). We have the section modulus $I/c = bt^2/6$. So the maximum bending stress is

$$\sigma = \frac{Mc}{I} = \frac{6F_t L}{bt^2} \tag{a}$$

This flexure formula yields results of acceptable accuracy at cross sections away from the point of load application (see Section 3.1).

We now treat the tooth as a cantilever fixed at BD (Figure 11.14b). It was noted already that the normal force F_n is considered as acting through the corner tip of the tooth along

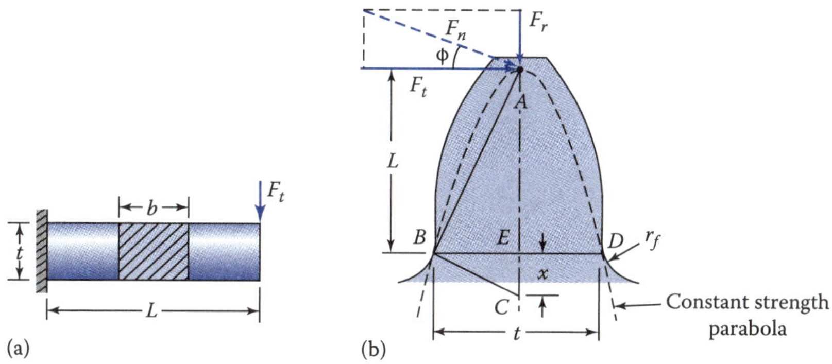

FIGURE 11.14
Beam strength of a gear tooth: (a) cantilever beam and (b) gear tooth as cantilever.

the pressure line. The radial component F_r causes a uniform compressive stress over the cross section. This compressive stress is small enough compared to the bending stress, due to the tangential load F_t, to be ignored in determining the strength of the tooth. Clearly, the compressive stress increases the bending stress on the compressive side of the tooth and decreases the resultant stress on the tensile side. Therefore, for many materials that are stronger in compression than in tension, the assumption made results in a stronger tooth design. Also note that because gear teeth are subjected to *fatigue failures* that start on the tension side of the tooth, the compressive stress reduces the resultant tensile stress and thus strengthens the tooth.

11.8.1 Uniform Strength Gear Tooth

In a gear tooth of *constant* strength, the stress is uniform; hence, $b/6F_t$ = constant = C, and Equation (a) then leads to $L = Ct^2$. The foregoing expression represents a parabola inscribed through point A, as shown by the dashed lines in Figure 11.14b. This parabola is tangent to the tooth profile at points B and D, where the maximum compressive and tensile stresses occur, respectively. The tensile stress is the cause of fatigue failure in a gear tooth. Referring to the figure, by similar triangles ABE and BCE, we write $(t/2)/x = L/(t/2)$ or $L = t^2/4x$. Carrying this into Equation (a) and multiplying the numerator and denominator by the circular pitch p, we have

$$\sigma = \frac{F_t p}{b(2/3)xp} \tag{b}$$

The Lewis form factor is defined as

$$y = \frac{2x}{3p} \tag{11.25}$$

Finally, substitution of the preceding into Equation (b) results in the original Lewis formula:

$$\sigma = \frac{F_t}{bpy} \tag{11.26}$$

Because the diametral pitch rather than circular pitch is often used to designate gears, the following substitution may be made: $p = \pi/P$ and $Y = \pi y$. Then the *Lewis form factor* is expressed as

$$Y = \frac{2xP}{3} \tag{11.27}$$

Similarly, the *Lewis formula* becomes

$$\sigma = \frac{F_t P}{bY} \tag{11.28}$$

When using SI units, in terms of module $m = 1/P$,

$$\sigma = \frac{F_t}{mbY} \tag{11.29}$$

Both Y and y are the functions of tooth shape (but not size) and thus vary with the number of teeth in the gear. Some values of Y determined from Equation 11.27 are listed in Table 11.2. For nonstandard gears, the factor Y (or y) can be obtained by a graphical layout of the gear or digital computation.

Let bending stress σ in Equation 11.28 be designated by the allowable static bending stress σ_o and so tangential load F_t by the *allowable bending load* F_b. Then this equation becomes

$$F_b = \sigma_o b \frac{Y}{P} \tag{11.30}$$

or, in SI units,

$$F_b = \sigma_o bYm \tag{11.31}$$

TABLE 11.2

Values of the Lewis Form Factor for Some Common Full-Depth Teeth

No. of Teeth	20° Y	25° Y	No. of Teeth	20° Y	25° Y
12	0.245	0.277	26	0.344	0.407
13	0.264	0.293	28	0.352	0.417
14	0.276	0.307	30	0.358	0.425
15	0.289	0.320	35	0.373	0.443
16	0.295	0.332	40	0.389	0.457
17	0.302	0.342	50	0.408	0.477
18	0.308	0.352	60	0.421	0.491
19	0.314	0.361	75	0.433	0.506
20	0.320	0.369	100	0.446	0.521
21	0.326	0.377	150	0.458	0.537
22	0.330	0.384	200	0.463	0.545
24	0.337	0.396	300	0.471	0.554
25	0.340	0.402	Rack	0.484	0.566

TABLE 11.3

Allowable Static Bending Stresses for Use in the Lewis Equation

Material	Treatment	σ_o ksi	σ_o (MPa)	Average Bhn
Cast iron				
ASTM 35		12	(82.7)	210
ASTM 50		15	(103)	220
Cast steel				
0.20% C		20	(138)	180
0.20% C	WQ&T	25	(172)	250
Forged steel				
SAE 1020	WQ&T	18	(124)	155
SAE 1030		20	(138)	180
SAE 1040		25	(172)	200
SAE 1045	WQ&T	32	(221)	205
SAE 1050	WQ&T	35	(241)	220
Alloy steels				
SAE 2345	OQ&T	50	(345)	475
SAE 4340	OQ&T	65	(448)	475
SAE 6145	OQ&T	67	(462)	475
SAE 65 (phosphor bronze)		12	(82.7)	100

Note: WQ&T, water-quenched and tempered; OQ&T, oil-quenched and tempered.

The values of σ_o for some materials of different hardness are listed in Table 11.3. Note that the tip-load condition assumed in the preceding derivation occurs when another pair of the teeth is still in contact. Actually, the heaviest loads occur near the middle of the tooth while a single pair of teeth is in contact. For this case, the derivation of the Lewis equation would follow exactly as in the previous case.

11.8.2 Effect of Stress Concentration

The stress in a gear tooth is greatly influenced by size of the fillet radius r_f (Figure 11.14b). It is very difficult to obtain the theoretical values of the stress-concentration factor K_t for the rather complex shape of the gear tooth. Experimental techniques and the finite element method are used for this purpose [9,10]. Since the gear tooth is subjected to fatigue loading, the factor K_t should be modified by the notch sensitivity factor q to obtain the fatigue stress-concentration factor K_f. The Lewis formula can be modified to include the effect of the stress concentration. In so doing, Equations 11.28 and 11.30 become, respectively,

$$\sigma = \frac{K_f F_t P}{bY} \tag{11.32}$$

$$F_b = \frac{\sigma_o b}{K_f} \frac{Y}{P} \tag{11.33}$$

As a reasonable approximation, $K_f = 1.5$ may be used in these equations.

11.8.3 Requirement for Satisfactory Gear Performance

The load capacity of a pair of gears is based on either the bending or wear (Section 11.10) capacity, whichever is smaller. For *satisfactory gear performance*, it is necessary that the dynamic load should not exceed the allowable load capacity. That is,

$$F_b \geq F_d \tag{11.34}$$

in which the dynamic load F_d is given by Equation 11.24. Note that this *dynamic load* approach can be used for all gear types [5].

The Lewis equation is important, since it serves as the basis for the AGMA approach to the bending strength of the gear tooth, discussed in the next section. Equations 11.33 and 11.34 are quite useful in estimating the capacity of gear drives when the life and reliability are not significant considerations. They are quite useful in preliminary gear design for a variety of applications. When a gearset is to be designed to transmit a load F_b, the gear material should be chosen so that the values of the product $\sigma_o Y$ are approximately the same for both gears.

Example 11.6: Power Transmitted by a Gear Based on Bending Strength and Using the Lewis Formula

A 25° pressure angle, 25-tooth spur gear having a module of 2 mm, and a 45 mm face width are to operate at 900 rpm. Determine

a. The allowable bending load applying the Lewis formula
b. The maximum tangential load and power that the gear can transmit

Design Decisions: The gear is made of SAE 1040 steel. A fatigue stress-concentration factor of 1.5 is used.

Solution

We have $Y = 0.402$ for 25 teeth (Table 11.2) and $\sigma_o = 172$ MPa (Table 11.3). The pitch diameter is $d = mN = 2(25) = 50$ mm and $V = \pi dn = \pi\,(0.05)(15) = 2.356$ m/s = 463.7 fpm.

a. Using Equation 11.33 with $1/P = m$, we have

$$F_b = \frac{\sigma_o bYm}{K_f} = \frac{1}{1.5}(172 \times 45 \times 0.402 \times 2) = 4.149 \text{ kN}$$

b. From Equation 11.24a, the dynamic load is

$$F_d = \frac{600 + 463.7}{600} F_t = 1.77 F_t$$

The limiting value of the transmitted load, applying Equation 11.34, is

$$4.149 = 1.77 F_t \quad \text{or} \quad F_t = 2.344 \text{ kN}$$

The corresponding gear power, by Equation 1.15, is

$$\text{kW} = \frac{F_t \pi dn}{60}$$

$$= \frac{(2.344)\pi(0.05)900}{60} = 5.52$$

11.9 Design for the Bending Strength of a Gear Tooth: The AGMA Method

The fundamental formula for the bending stress of a gear tooth is the AGMA modification of the Lewis equation. This formula applies to the original Lewis equation correction factors that compensate for some of the simplifying presuppositions made in the derivation as well as for important factors not initially considered. In the AGMA method to the design and analysis of gearing, the bending strength of a gear tooth is also modified by various factors to obtain the allowable bending stress.

In this section, we present selective AGMA bending strength equations for a gear tooth. They are based on certain assumptions about the tooth and gear tooth geometry. It should be mentioned that some definitions and symbols used are different than that given by the AGMA. Nevertheless, procedures introduced here and in Section 11.11 are representative of current practice [11,12]. For further information, see the latest AGMA standards [1] and the relevant literature.

Bending stress is defined by the formula

$$\sigma = F_t K_o K_\upsilon \frac{P}{b} \frac{K_s K_m}{J} \quad \text{(U.S. customary units)} \tag{11.35}$$

$$\sigma = F_t K_o K_\upsilon \frac{1.0}{bm} \frac{K_s K_m}{J} \quad \text{(SI units)} \tag{11.35'}$$

where

σ = the calculated bending stress at the root of the tooth
F_t = the transmitted tangential load
K_o = the overload factor
K_υ = the velocity or dynamic factor
P = the diametral pitch
b = the face width
m = the metric module
K_s = the size factor
K_m = the mounting factor
J = the geometry factor

Allowable bending stress, or the design stress value, is

$$\sigma_{\text{all}} = \frac{S_t K_L}{K_T K_R} \tag{11.36}$$

where

σ_{all} = the allowable bending stress
S_t = the bending strength
K_L = the life factor
K_T = the temperature factor
K_R = the reliability factor

TABLE 11.4
Overload Correction Factor K_o

	Load on Driven Machine		
Source of Power	**Uniform**	**Moderate Shock**	**Heavy Shock**
Uniform	1.00	1.25	1.75
Light shock	1.25	1.50	2.00
Medium shock	1.50	1.75	2.25

As a *design specification*, the bending stress must not exceed the design stress value:

$$\sigma \leq \sigma_{\text{all}} \tag{11.37}$$

Note that there are three groups of terms in Equation 11.35: The first refers to the loading characteristics, the second to the gear geometry, and the third to the tooth form. Obviously, the essence of this equation is the Lewis formula with the updated geometry factor J introduced for the form factor Y. The K factors are modifiers to account for various conditions. Equation 11.36 defines the allowable bending stress. The specification in the AGMA approach for designing for strength is given by Equation 11.37. That is, the calculated stress σ of Equation 11.35 must always be less than or equal to the allowable stress σ_{all} as determined by Equation 11.36. To facilitate the use of the Equations 11.35 through 11.37, the following concise description of the correction factors is given.

The *overload factor* K_o is used to compensate for situations in which the actual load exceeds the transmitted load F_t. Table 11.4 gives some suggested values for K_o. The velocity or dynamic factor K_r shows the severity of impact as successive pairs of teeth engage. This depends on pitch velocity and manufacturing accuracy. Figure 11.15 depicts some commonly employed approximate factors pertaining to representative gear manufacturing processes. It is seen from the figure that dynamic factors become higher, when hobs or milling cutters are used to form the teeth or inaccurate teeth are generated. For more detailed information, consult the appropriate AGMA standard.

The *size factor* K_s attempts to account for any nonuniformity of the material properties. It depends on the tooth size, diameter of parts, and other tooth and gear dimensions. For most standard steel gears, the size factor is usually taken as unity. A value of 1.25–1.5 would be a conservative assumption in cases of very large teeth. The *mounting factor* K_m reflects the accuracy of mating gear alignment. Table 11.5 is used as a basis for rough estimates. The *geometry factor* J relies on the tooth shape, the position at which the highest load is applied, and the contact ratio. The equation for J includes a modified value of the Lewis factor Y and a fatigue stress-concentration factor K_f. Figure 11.16 may be used to estimate the geometry factor for only 20° and 25° standard spur gears.

The bending strength S_t for standard gear materials varies with such factors as material quality, heat treatment, mechanical treatment, and material composition. Some selected values for AGMA fatigue strength for bending are found in Table 11.6. These values are based on a reliability of 99%, corresponding to 10^7 tooth load cycles. Note that in the table, Bhn and R_c denote the Brinell and Rockwell hardness numbers, respectively.

The *life factor* K_L rectifies the allowable stress for the required number of stress cycles other than 10^7. Values of this factor are furnished in Table 11.7. The *temperature factor* K_T

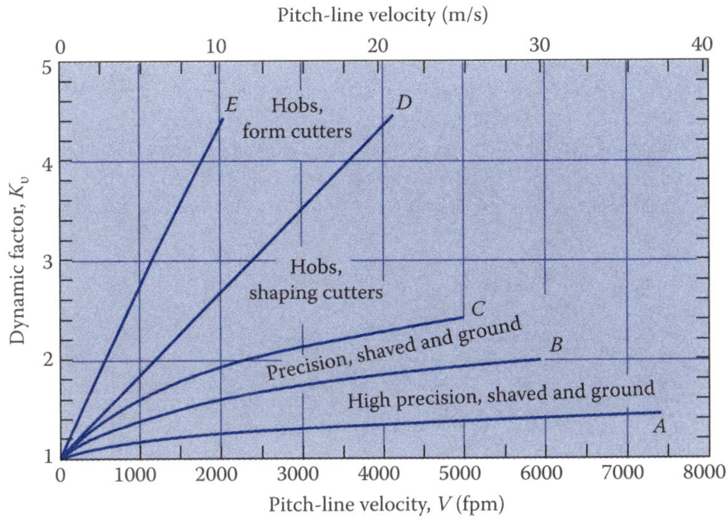

FIGURE 11.15
Dynamic factor K_v. (From AGMA, Standards of the American Gear Manufacturers Association, Alexandria, VA, ANSI/AGMA 2001–C95, revised AGMA 2001–C95.)
Notes:

$$\text{Curve } A,\quad K_v = \sqrt{\frac{78+\sqrt{V}}{78}};\quad \text{Curve } B,\quad K_v = \frac{78+\sqrt{V}}{78}$$

$$\text{Curve } C,\quad K_v = \frac{50+\sqrt{V}}{50};\quad \text{Curve } D,\quad K_v = \frac{1200+V}{1200}$$

$$\text{Curve } E,\quad K_v = \frac{600+V}{600},$$

where V is in feet per minute, fpm. To covert to meters per second (m/s), *divide* the given values by 196.8.

TABLE 11.5
Mounting Correction Factor K_m

	Face Width (in.)			
Condition of Support	**0–2**	**6**	**9**	**16 up**
Accurate mounting, low bearing clearances, maximum deflection, precision gears	1.3	1.4	1.5	1.8
Less rigid mountings, less accurate gears, contact across the full face	1.6	1.7	1.8	2.2
Accuracy and mounting such that less than full-face contact exists	Over 2.2			

is applied to adjust the allowable stress for the effect of operating temperature. Usually, for gear lubricant temperatures up to $T < 160°F$, $K_T = 1$ is used. For $T > 160°F$, use $K_T = (460 + T)/620$. The *reliability factor* K_R corrects the allowable stress for the reliabilities other than 99%. Table 11.8 lists some K_R values applied to the fatigue strength for bending of the material.

The use of the AGMA formulas and graphs is illustrated in the solution of the following numerical problem.

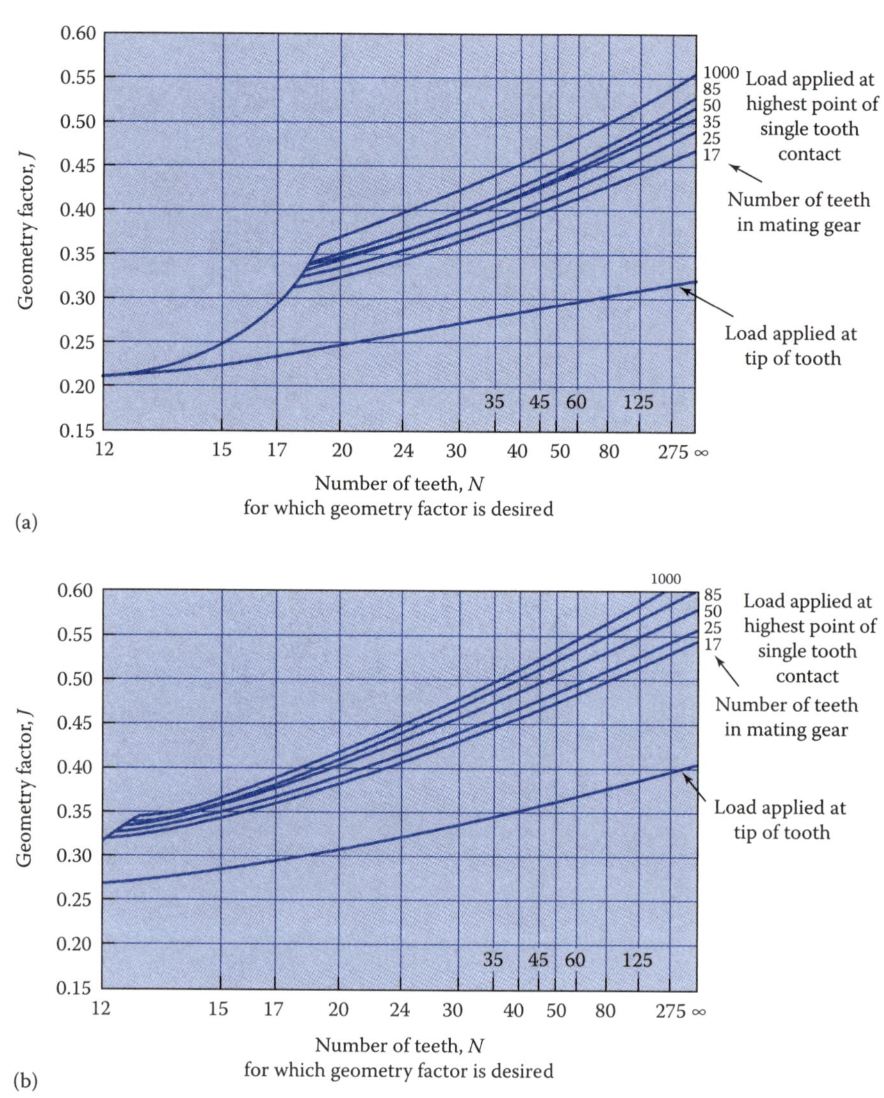

FIGURE 11.16
Geometry factors for spur gears (based on tooth fillet radius of 0.35/*P*): (a) 20° full-depth teeth and (b) 25° full-depth teeth. (From ANSI/AGMA Standard 218.01.)

Example 11.7: Design of a Speed Reducer for Bending Strength by the AGMA Method

A conveyor drive involving heavy shock torsional loading is to be operated by an electric motor turning at a speed of n, as shown schematically in Figure 11.17. The speed ratio of the spur gears connecting the motor and conveyor or speed reducer is to be r_s = 1:2. Determine the maximum horsepower that the gearset can transmit, based on bending strength and applying the AGMA formulas.

Given: Both gears are of the same 300 Bhn steel and have a face width of b = 1.5 in. Pinion rotates at n = 1600 rpm. P = 10 in.$^{-1}$ and N_p = 18.

Design Decisions: Rational values of the factors are chosen, as indicated in the parentheses in the solution.

TABLE 11.6

Bending Strength S_t of Spur, Helical, and Bevel Gear Teeth

Material	Heat Treatment	Minimum Hardness or Tensile Strength	S_t ksi	S_t (MPa)
Steel	Normalized	140 Bhn	19–25	(131–172)
	Q&T	180 Bhn	25–33	(172–223)
	Q&T	300 Bhn	36–47	(248–324)
	Q&T	400 Bhn	42–56	(290–386)
	Case carburized	55 R_C	55–65	(380–448)
		60 R_C	55–70	(379–483)
	Nitrided AISI-4140	48 R_C case 300 Bhn core	34–45	(234–310)
Cast iron				
AGMA grade 30		175 Bhn	8.5	(58.6)
AGMA grade 40		200 Bhn	13	(89.6)
Nodular iron ASTM grade				
60-40-18			15	(103)
80-55-06	Annealed		20	(138)
100-70-18	Normalized		26	(179)
120-90-02	Q&T		30	(207)
Bronze, AGMA 2C	Sand cast	40 ksi (276 MPa)	5.7	(39.3)

Source: ANSI/AGMA Standard 218.01.
Note: Q&T, Quenched and tempered.

TABLE 11.7

Life Factor K_L for Spur and Helical Steel Gears

Number of Cycles	160 Bhn	250 Bhn	450 Bhn	Case Carburized (55–63 R_C)
10^3	1.6	2.4	3.4	2.7–4.6
10^4	1.4	1.9	2.4	2.0–3.1
10^5	1.2	1.4	1.7	1.5–2.1
10^6	1.1	1.1	1.2	1.1–1.4
10^7	1.0	1.0	1.0	1.0

Source: ANSI/AGMA Standard 218.01.

TABLE 11.8

Reliability Factor K_R

Reliability (%)	50	90	99	99.9	99.99
Factor K_R	0.70	0.85	1.00	1.25	1.50

Source: From ANSI/AGMA Standard 2001–C95.

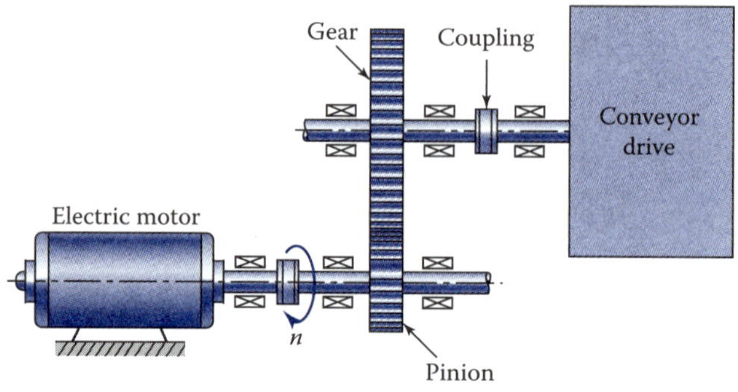

FIGURE 11.17
Example 11.7. Schematic arrangement of motor, gear, and conveyor drive.

Solution

The pinion pitch diameter and number of teeth of the gear are

$$d_p = \frac{N_p}{P} = \frac{18}{10} = 1.8 \text{ in.,} \quad N_g = N_p\left(\frac{1}{r_s}\right) = 18(2) = 36$$

The pitch-line velocity, using Equation 11.20, is

$$V = \frac{\pi d_p n_p}{12} = \frac{\pi(1.8)(1600)}{12} = 754 \text{ fpm}$$

The allowable bending stress is estimated from Equation 11.36:

$$\sigma_{\text{all}} = \frac{S_t K_L}{K_T K_R} \tag{a}$$

where

S_t = 41.5 ksi (from Table 11.6, for average strength)
K_L = 1.0 (from Table 11.7, for indefinite life)
K_T = 1 (oil temperature should be <160°F)
K_R = 1.25 (by Table 11.8, for 99.9% reliability)

Carrying the foregoing values into Equation (a) results in

$$\sigma_{\text{all}} = \frac{41.5(1)}{1(1.25)} = 33.2 \text{ ksi}$$

The maximum allowable transmitted load is now obtained, from Equation 11.35 by setting $\sigma_{\text{all}} = \sigma$, as

$$F_t = \frac{33{,}200}{K_o K_v}\frac{b}{P}\frac{J}{K_s K_m} \tag{b}$$

In the foregoing, we have

$P = 10$
$b = 1.5$ in.
$K_v = 1.55$ (from curve C of Figure 11.15)
$J = 0.235$ (from Figure 11.16a, the load acts at the tip of the tooth, $N_p = 18$)
$K_o = 1.75$ (by Table 11.4)
$K_s = 1.0$ (for standard gears)
$K_m = 1.6$ (from Table 11.5)

Equation (b) yields

$$F_t = \frac{33,200(1.5)(0.235)}{(1.75)(1.55)(10)(1.0)(1.6)} = 270 \text{ lb}$$

The allowable power is then, by Equation 11.22,

$$\text{hp} = \frac{F_t V}{33,000} = \frac{270(754)}{33,000} = 6.2$$

11.10 Wear Strength of a Gear Tooth: The Buckingham Formula

The failure of the surfaces of gear teeth is called *wear.* As noted in Section 8.9, wear is a broad term, which encompasses a number of kinds of surface failures. So it is evident that gear tooth surface durability is a more complex matter than the capacity to withstand gear tooth bending failure. Tests have shown that pitting, a surface fatigue failure due to repeated high contact stress, occurs on those portions of a gear tooth that have relatively little sliding compared with rolling. Clearly, spur gears and helical gears have pitting near the pitch line, where the motion is almost pure rolling. We therefore are concerned with only gear tooth surface fatigue failure or pitting and designate it by the term wear [13,14].

As was the case with the rolling bearings, gear teeth are subjected to Hertz contact stresses, and the lubrication is often elastohydrodynamic (Section 10.16). The gear teeth must be sufficiently strong to carry the wear load. The surface stresses in gear teeth were first investigated in a systematic way by Buckingham [5], who considered the teeth as two parallel cylinders in contact. The radii of the teeth as two parallel cylinders are taken as the radii of curvature for the involutes when the teeth make contact at pitch point *P* (Figure 11.18). Hence,

$$R_1 = \frac{d_p}{2}\sin\phi, \quad R_2 = \frac{d_g}{2}\sin\phi = \frac{N_g d_p}{2N_p}\sin\phi \tag{a}$$

where

ϕ represents the pressure angle

d_p and d_g are the pitch diameters of the pinion and gear, respectively

The last form in the second of these equations is from the relationship $d_p/N_p = d_g/N_g$.

Generally, good correlation has been observed between spur gear surface fatigue failure and the computed elastic contact stress (see Section 8.7). The maximum contact pressure p_o between the two cylinders may be computed from the equation given in Table 8.4, for $\nu = 0.3$:

$$p_o = 0.592\left[\frac{F_a E_p E_g}{E_p + E_g}\left(\frac{1}{R_1} + \frac{1}{R_2}\right)\right]^{1/2} \tag{b}$$

where

F_a is the load per axial length pressing the cylinders together

E_p and E_g are the moduli of elasticity for the materials composing the gears

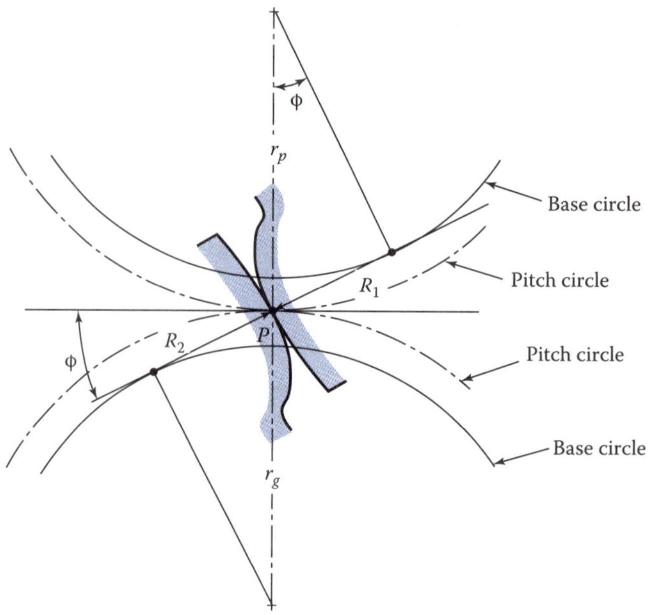

FIGURE 11.18
Radii of curvature R_1 and R_2 for tooth surfaces at pitch point P.

We observe from this equation that stress increases only as the square root of the load F_a. Likewise, stress decreases with decreased moduli of elasticity E_p and E_g. Moreover, larger gears have greater radii of curvature, hence lower stress.

We now let the maximum contact pressure p_o represent the *surface endurance limit* in compression, S_e, for the pinion and gear material. Substituting Equation (a) into Equation (b), we have

$$S_e^2 = 0.35 \frac{F_a E_p E_g}{E_p + E_g} \frac{2}{d_p \sin\phi} \left(1 + \frac{N_p}{N_g}\right) \tag{c}$$

Usually, both sides of this expression are multiplied by gear width b, and the total load $F_a b$ denoted by F_w. In so doing and solving Equation (c), we obtain the *allowable wear load:*

$$F_w = d_p b Q K \tag{11.38}$$

where

$$K = \frac{S_e^2 \sin\phi}{1.4} \left(\frac{1}{E_p} + \frac{1}{E_g}\right) \tag{11.39}$$

$$Q = \frac{2N_g}{N_p + N_g} \tag{11.40}$$

TABLE 11.9

Surface Endurance Limit S_e and Wear Load Factor K for Use in the Buckingham Equation

	S_e		K			
			$\phi = 20°$		$\phi = 25°$	
Materials in Pinion and Gear	**ksi**	**(MPa)**	**psi**	**(MPa)**	**psi**	**(MPa)**
Both steel gears, with average Bhn of pinion and gear						
150	50	(345)	41	(0.283)	51	(0.352)
200	70	(483)	79	(0.545)	98	(0.676)
250	90	(621)	131	(0.903)	162	(1.117)
300	110	(758)	196	(1.352)	242	(1.669)
350	130	(896)	270	(1.862)	333	(2.297)
400	150	(1034)	366	(2.524)	453	(3.124)
Steel (150 Bhn) and cast iron	50	(354)	60	(0.414)	74	(0.510)
Steel (200 Bhn) and cast iron	70	(483)	119	(0.821)	147	(1.014)
Steel (250 Bhn) and cast iron	90	(621)	196	(1.352)	242	(1.669)
Steel (150 Bhn) and phosphor bronze	59	(407)	62	(0.428)	77	(0.531)
Steel (200 Bhn) and phosphor bronze	65	(448)	100	(0.690)	123	(0.848)
Steel (250 Bhn) and phosphor bronze	85	(586)	184	(1.269)	228	(1.572)
Cast iron and cast iron	90	(621)	264	(1.821)	327	(2.555)
Cast iron and phosphor bronze	83	(572)	234	(1.614)	288	(1.986)

Equation 11.38 is known as the *Buckingham formula*. The typical values of S_e and *wear load factor* K for materials of different Brinell surface hardnesses Bhn, recommended by Buckingham, are shown in Table 11.9. Note that the values of K are obtained from Equation 11.39 for pressure angles of 20° and 25° full-depth teeth only. The allowable wear load serves as the basis for analyzing gear tooth surface durability.

For *satisfactory gear performance*, the usual requirement is that

$$F_w \geq F_d \tag{11.41}$$

Here, the dynamic load F_d is defined by Equations 11.24. To prevent too much pinion wear, particularly for high-speed gearing, a medium-hard pinion with low hardness gear is often used. This has the advantage of giving some increase in load capacity and slightly lower coefficient of friction on the teeth.

Example 11.8: Maximum Load Transmitted by a Gear Based on Wear Strength and Using the Buckingham Formula

The pinion of Example 11.6 is to be mated with a gear. Determine

a. The allowable load for wear for the gearset using the Buckingham formula
b. The maximum load that can be transmitted

Given: $N_p = 25$, $d_p = 50\,\text{mm}$, $b = 45\,\text{mm}$, $N_g = 60$.

Design Decision: The mating gear is made of 60-tooth cast iron.

Solution

a. Using Equation 11.40,

$$Q = \frac{2N_g}{N_p + N_g} = \frac{120}{85} = \frac{24}{17}$$

By Table 11.9,
$K = 1.014$ MPa for 200 Bhn
Applying Equation 11.38,

$$F_w = d_p bQK = (50)(45)\left(\frac{24}{17}\right)(1.014) = 3.221\ \text{kN}$$

b. The dynamic load, from Example 11.6, is $1.77F_t$. The wear-limiting value of the transmitted load, using Equation 11.41, is then

$$3.221 = 1.77F_t \quad \text{or} \quad F_t = 1.82\ \text{kN}$$

11.11 Design for the Wear Strength of a Gear Tooth: The AGMA Method

The Buckingham equation serves as the basis for analyzing only the contact stress on the gear tooth. The AGMA formula applies several factors, influencing the actual state of stress at the point of contact, not considered in the previous section. Similarly, in the AGMA method, the surface fatigue strength of the gear tooth is modified by a variety of factors to determine the allowable contact stress. The selective AGMA wear formulas for gears are as follows [1]:

Contact stress is defined by the formula

$$\sigma_c = C_p\left(F_t K_o K_\upsilon \frac{K_s}{bd}\frac{K_m C_f}{I}\right)^{1/2} \tag{11.42}$$

where

$$C_p = 0.564\left[\frac{1}{\dfrac{1-\nu_p^2}{E_p} + \dfrac{1-\nu_g^2}{E_g}}\right]^{1/2} \tag{11.43a}$$

$$I = \frac{\sin\phi\cos\phi}{2m_N}\frac{m_G}{m_G+1} \tag{11.43b}$$

In the foregoing,

σ_c = the calculated contact stress
C_p = the elastic coefficient
K_υ = the velocity or dynamic factor
K_s = the size factor
b = the face width
d = the pitch diameter
K_m = the load distribution factor
C_f = the surface condition factor
I = the geometry factor
m_G = the gear ratio = $d_g/d_p = N_g/N_p$ (for internal gears m_G is negative)
m_N = the load sharing ratio = 1 (for spur gears)
E = the modulus of elasticity
ν = the Poisson's ratio
ϕ = the pressure angle

Allowable contact stress or the design stress value is

$$\sigma_{c,\text{all}} = \frac{S_c C_L C_H}{K_T K_R} \tag{11.44}$$

where

$\sigma_{c,\text{all}}$ = the allowable contact stress
S_c = the surface fatigue strength
C_L = the life factor
C_H = the hardness ratio factor
K_T = the temperature factor
K_R = the reliability factor

As a *design specification*, the contact stress must not exceed the design stress value:

$$\sigma_c \le \sigma_{c,\text{all}} \tag{11.45}$$

A comparison of the foregoing fundamental equations with those given in Section 11.9 shows that some bending factors and wear factors are equal and so indicated by the same symbols. A brief description of the new wear factors follows.

The elastic coefficient C_p, defined by Equation 11.43a, accounts for differences in tooth material. In this expression, E_p and E_g are the moduli of elasticity for, respectively, pinion and gear, and ν_p and ν_g are their respective Poisson's ratios. The units of C_p are $\sqrt{\text{psi}}$ or $\sqrt{\text{MPa}}$, depending on the system of units used. For convenience, rounded values of C_p are given in Table 11.10, where $\nu = 0.3$ in all cases.

The *surface condition factor* C_f is used to account for such considerations as surface finish, residual stress, and plasticity effects. The C_f is usually taken as unity for a smooth

TABLE 11.10

AGMA Elastic Coefficients C_p for Spur Gears, in $\sqrt{\text{psi}}$ and ($\sqrt{\text{MPa}}$)

		Gear Material			
Pinion Material	***E*, ksi (GPa)**	**Steel**	**Cast Iron**	**Aluminum Bronze**	**Tin Bronze**
Steel	30,000	2300	2000	1950	1900
	(207)	(191)	(166)	(162)	(158)
Cast iron	19,000	2000	1800	1800	1750
	(131)	(166)	(149)	(149)	(145)
Aluminum bronze	17,500	1950	1800	1750	1700
	(121)	(162)	(149)	(145)	(141)
Tin bronze	16,000	1900	1750	1700	1650
	(110)	(158)	(145)	(141)	(137)

surface finish. When rough finishes are present or the possibility of high residual stress exists, a value of 1.25 is reasonable. If both rough finish and residual stress exist, 1.5 is the suggested value.

The surface fatigue strength S_c represents a function of such factors as the material of the pinion and gear, number of cycles of load application, size of the gears, type of heat treatment, mechanical treatment, and the presence of residual stresses. Table 11.11 may be used to estimate the values for S_c. The life factor C_L accounts for the expected life of the gear. Figure 11.19 can be used to obtain approximate values for C_L.

The hardness ratio factor C_H is used *only* for the gear. Its intent is to adjust the surface strengths for the effect of the hardness. The values of C_H are calculated from the expression

$$C_H = 1.0 + A\left(\frac{N_g}{N_p} - 1.0\right) \qquad \left(\text{for } \frac{H_{Bp}}{H_{Bg}} < 1.70\right) \tag{11.46}$$

where

$$A = 8.98\left(10^{-3}\right)\left(\frac{H_{Bp}}{H_{Bg}}\right) - 8.29(10^{-3})$$

The quantities H_{Bp} and H_{Bg} represent the Brinell hardness of the pinion and gear, respectively.

Example 11.9: Design of a Speed Reducer for Wear by the AGMA Method

Determine the maximum horsepower that the speed reducer gearset in Example 11.7 can transmit, based on wear strength and applying the AGMA method.

Given: Both gears are made of the same 300 Bhn steel of $E = 30 \times 10^6$ psi, $\nu = 0.3$, and have a face width of $b = 1.5$ in., $P = 10$ in.$^{-1}$, and $N_p = 18$.

Design Decisions: Rational values of the factors are chosen, as indicated in the parentheses in the solution.

TABLE 11.11

Surface Fatigue Strength or Allowable Contact Stress S_c

Material	Minimum Hardness or Tensile Strength	S_c ksi	S_c (MPa)
Steel	Through hardened		
	180 Bhn	85–95	(586–655)
	240 Bhn	105–115	(724–793)
	300 Bhn	120–135	(827–931)
	360 Bhn	145–160	(1000–1103)
	400 Bhn	155–170	(1069–1172)
	Case carburized		
	55 R_C	180–200	(1241–1379)
	60 R_C	200–225	(1379–1551)
	Flame or induction hardened		
	50 R_C	170–190	(1172–1310)
Cast iron			
AGMA grade 20		50–60	(345–414)
AGMA grade 30	175 Bhn	65–75	(448–517)
AGMA grade 40	200 Bhn	75–85	(517–586)
Nodular (ductile) iron		90%–100% of the S_c	
Annealed	165 Bhn	Value of steel with the same hardness	
Normalized	210 Bhn		
OQ&T	255 Bhn		
Tin bronze			
AGMA 2C (10%–12% tin)	40 ksi (276 MPa)	30	(207)
Aluminum bronze			
ASTM B 148–52 (alloy 9C-HT)	90 ksi (621 MPa)	65	(448)

Source: ANSI/AGMA Standard 218.01.
Note: OQ&T, Oil-quenched and tempered; HT, heat-treated.

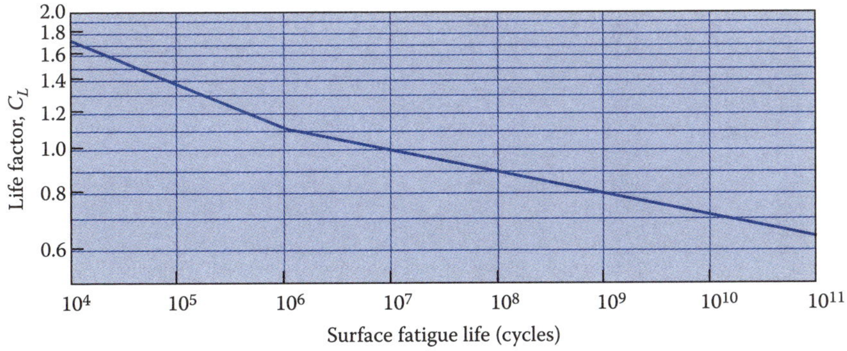

FIGURE 11.19
Life factor for steel gears. (From ANSI/AGMA Standard 218.01.)

Solution

Allowable contact stress is estimated from Equation 11.44 as

$$\sigma_{c,\text{all}} = \frac{S_c C_L C_H}{K_T K_R} \tag{a}$$

In the preceding,

$S_c = 127.5$ ksi (from Table 11.11, for average strength)
$C_L = 1.0$ (from Figure 11.19, for indefinite life)

$$C_H = 1.0 + A\left(\frac{N_g}{N_p} - 1.0\right) = 1.0 \quad \text{(by Equation 11.46)}$$

$K_T = 1.0$ and $K_R = 1.25$ (both from Example 11.7)

Hence,

$$\sigma_{c,\text{all}} = \frac{127{,}500(1.0)(1.0)}{(1.0)(1.25)} = 102 \text{ ksi}$$

The maximum allowable transmitted load is now determined, from Equation 11.42 setting $\sigma_{c,\text{all}} = \sigma_c$, as

$$F_t = \left(\frac{102{,}000}{C_p}\right)^2 \frac{1.0}{K_o K_v} \frac{bd}{K_s} \frac{I}{K_m C_f}$$

where

$C_p = 2300\sqrt{\text{psi}}$ (by Table 11.10)
$b = 1.5$ in., $d_p = 1.8$ in.
$K_v = 1.55$, $K_o = 1.75$, $K_s = 1$, $K_m = 1.6$ (all from Example 11.7)
$C_f = 1.0$ (for smooth surface finish)

$$I = \frac{\sin\phi\cos\phi}{2m_N} \frac{m_G}{m_G + 1} = 0.107 \quad \text{(using Equation 11.43b)}$$

Therefore,

$$F_t = \left(\frac{102{,}000}{2300}\right)^2 \frac{1.0}{(1.75)(1.55)} \frac{1.5(1.8)}{1.0} \frac{0.107}{1.6(1.0)} = 131 \text{ lb}$$

This value applies to both mating gear tooth surfaces. The corresponding power, using Equation 11.22 with $V = 754$ fpm (from Example 11.7), is

$$\text{hp} = \frac{F_t V}{33{,}000} = \frac{131(754)}{33{,}000} = 2.99$$

11.12 Materials for Gears

Gears are made from a wide variety of materials, both metallic and nonmetallic, covered in Chapter 2. The material used depends on which of several criteria is the most important to the problem at hand. When high strength is the prime consideration, steel should be chosen rather than cast iron or other materials. Test data for fatigue strengths of most materials can be found in contemporary technical literature and current publications of the AGMA (see Tables 11.6 and 11.11). For situations involving noise abatement, nonmetallic materials perform better than metallic ones. The characteristics of some common gear materials follow.

Cast irons have low cost, ease of casting, good machinability, high wear resistance, and good noise abatement, which make them one of the most commonly used gear materials. Cast iron gears typically have greater surface fatigue strength than bending fatigue strength. Nodular cast iron gears, containing a material such as magnesium or cerium, have higher bending strength and good surface durability. The combination of a steel pinion and cast iron gear represents a well-balanced design.

Steels usually require heat treatment to produce a high surface endurance capacity. Heat-treated steel gears must be designed to resist distortion; hence, alloy steels and oil quenching are often preferred. Through-hardened gears usually have 0.35%–0.6% carbon. Case-hardened gears are generally processed by flame hardening, induction hardening, carburizing, or nitriding. When gear accuracy is required, the gear must be ground.

Nonferrous metals such as the copper alloys (known as *bronzes*) are most often used for gears. They are useful in situations where corrosion resistance is important. Owing to their ability to reduce friction and wear, bronzes are generally employed for making worm wheel in a worm gearset. Aluminum, zinc, and titanium are also used to obtain alloys that are serviceable for gear materials.

Plastics such as acetal, polypropylene, nylon, and other nonmetallic materials have often been used to make gears. Teflon is sometimes added to these materials to lower the coefficient of friction. Plastic gears are generally quiet, durable, reasonably priced, and can operate under light loads without lubrication. However, they are limited in torque capacity by their low strength. Furthermore, plastics have low heat conductivity, resulting in heat distortion and weakening of the gear teeth.

Reinforced thermoplastics, formulated with fillers such as glass fiber, are desirable gear materials owing to their versatility. For best wear resistance, nonmetallic gears are often mated with cast iron or steel pinions having a hardness of at least 300 Bhn. The *composite gears* of thermosetting phenolic have been used in applications such as the camshaft drive gear driven by a steel pinion in some gasoline engines. Design procedures of gears made of plastic are identical to that of metal gears, but not yet as reliable. Prototype testing is therefore more significant than for gears made of metals.

11.13 Gear Manufacturing

Various methods are employed to manufacture gears. These can be divided into two classes: forming and finishing. Gear teeth are formed numerous ways by milling and generating processes. For high speed and heavy loads, a finishing operation may be required to bring the tooth outline to a sufficient degree of accuracy and surface finish. Finishing operations typically remove little or no material. Gear errors may be diminished somewhat by finishing the tooth profiles. A general discussion of some manufacturing processes is found in Chapter 2. This section can provide only a brief description of gear forming and finishing methods [8,11].

11.13.1 Forming Gear Teeth

Milling refers to removal of the material between the teeth from a blank on a milling machine that uses a formed circular cutter. The cutter must be made to the shape of the gear tooth for the tooth geometry and number of teeth of each particular gear. Gears having large-size teeth are often made by formed cutters. *Shaping* is the generation of teeth with a rack cutter or with a pinion cutter called a *shaper* (i.e., a small circular gear). The cutter works in a rapid reciprocating motion parallel to the axis of the gear blank while slowly translating with and into the blank, as depicted in Figure 11.20. When the blank and cutter have rolled a distance equal to the circular pitch, the cutter is returned to the starting point and the process is continued until all the teeth have been cut.

The process requires a special cutting machine that translates the tool and rotates the gear blank with the same velocity as though a rack were driving a gear. Internal gears can be cut with this method as well. *Hobbing* refers to a process that accounts for the major portion of gears made in high-quantity production. A hob is a cutting tool shaped like a worm or screw. Both the hob and blank must be rotated at the proper angular velocity ratio. The hob is then fed slowly across the face of the blank until all the teeth have been cut.

Other gear forming methods include die casting, drawing, extruding, sintering, stamping, and injection molding. These processes produce large volumes of gears that are low in cost but poor in quality. Gears are *die cast* by forcing molten metal under pressure into a form die. In a cold *drawing* process, the metal is drawn through several dies and emerges as a long piece of gear from which gears of smaller widths can be sliced. On the other hand, in an *extruding* process, the metal is pushed rather than pulled through the dies.

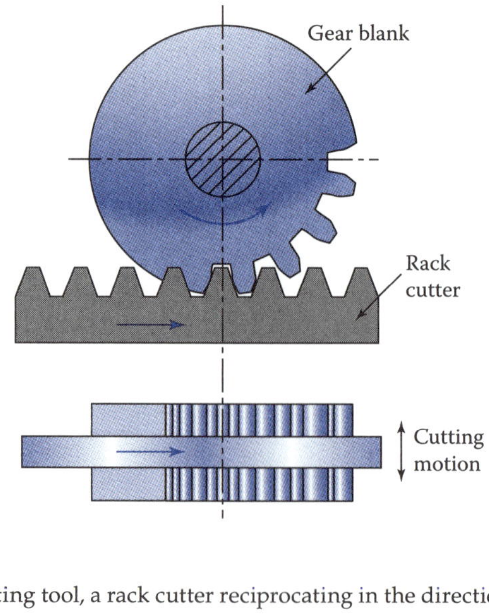

FIGURE 11.20
The generating action of a cutting tool, a rack cutter reciprocating in the direction of the axis of a gear blank.

Nonferrous materials such as copper and aluminum alloys are usually extruded. A *sintering* method consists of applying pressure and heat to a powdered metal (PM) to form the gear. A *stamping* process uses a press and a die to cut out the gear shape. An *injection molding* method is applied to produce nonmetallic gears in a variety of thermoplastics such as nylon and acetal.

11.13.2 Finishing Processes

Grinding is accomplished by the use of some form of abrasive grinding wheel. It can be used to give the final form to teeth after heat treatment. *Shaving* refers to a machine operation that removes small amounts of material. It is done prior to hardening and is widely used for gears made in large quantities. *Burnishing* runs the gear to be smoothed against a specially hardened gear, which flattens and spreads minute surface irregularities only. *Honing* employs a tool, known as a *hone*, to drive the gear to be finished. It makes minor tooth-form corrections and improves the smoothness of the surface of the hardened gear surface. *Lapping* runs a gear with another that has some abrasive material embedded in it. Sometimes two mating gears are similarly run.

Problems

Sections 11.1 through 11.7

11.W1 Use the website at www.grainger.com to conduct a search for spur gears. Select and list the manufacturer and description of spur gears with

a. A 16 pitch, 14½° pressure angle, and 48 teeth

b. A 24 pitch, 14½° pressure angle, and 12 teeth

11.W2 Search the website at www.powertransmission.com. List 10 websites for manufacturers of gears and gear drives.

11.1 A 20° pressure angle gear has 32 teeth and a diametral pitch of 4 teeth/in. Determine the whole depth, working depth, base circle radius, and outside radius.

11.2 Two modules of 3 mm gears are to be mounted on a center distance of 360 mm. The speed ratio is 1/3. Determine the number of teeth in each gear.

11.3 A 10 mm module gearset consists of an 18-tooth pinion and a 42-teeth gear. The gears are cut using a pressure angle of 20°. Find

a. The circular pitch, center distance, and radii of the circles

b. The new values of the pressure angle and the pitch circle diameters, if the center distance is increased by 8 mm

11.4 A 20° full-depth spur gear tooth has a diametral pitch of 6, measured along the pitch circle. What are the circular pitch and the tooth thickness?

11.5 Determine the dedendum, clearance, working depth, thickness, and circular pitch of a 20° full-depth spur gear tooth with a module of 12 mm.

11.6 A gearset with 1/4 speed ratio and 8 in. center distance has a diametral pitch of 5 (see Figure 11.9). Compute the number of teeth in each gear.

11.7 An 18-tooth pinion with a 3 mm module rotates 2400 rpm and drives a gear at 1200 rpm. What are the number of teeth in the gear, the center distance, and the pitch diameters of the gears?

11.8 Determine the approximate center distance for an external 25° pressure angle gear having a circular pitch of 0.5234 in. that drives an internal gear having 84 teeth, if the speed ratio is to be 1/4.

11.9 A gearset of 60-teeth driven gear and 24-tooth pinion has an 8 mm module and a pressure angle of 20°. Determine

a. The circular pitch, center distance, and base radius for the pinion gear and the gear

b. For the case in which the center distance is increased by 6 mm, the pitch diameters for the pinion and the gear

11.10 A gearset has a module of 4 mm and a speed ratio of 1/4. The pinion has 22 teeth. Determine the number of teeth of the driven gear, gear and pinion diameters, and the center distance.

11.11 A 3 mm module gearset of 25-tooth pinion that rotates at 3400 rpm and 48-tooth gear. The gears are cut using a pressure angle of 20°. What is the speed of the gear?

11.12 A 2 mm module gearset consists of 30-tooth pinion and 100-tooth gear. The gears are cut using a pressure angle of 20°. Determine the outside radii for the pinion and the gear.

11.13 What is the contact ratio of the gearset described in Problem 11.9?

11.14 The sun gear in Figure 11.11 is the input and driven clockwise at 120 rpm. If the ring gear is held stationary, determine the rotation speed and direction of the arm. **Given**: $N_1 = 24$ and $N_4 = 96$.

11.15 The gears shown in Figure P11.15 have a diametral pitch of 3 teeth/in. and a 25° pressure angle. Determine and show on a free-body diagram

a. The tangential and radial forces acting on each gear

b. The reactions on shaft C

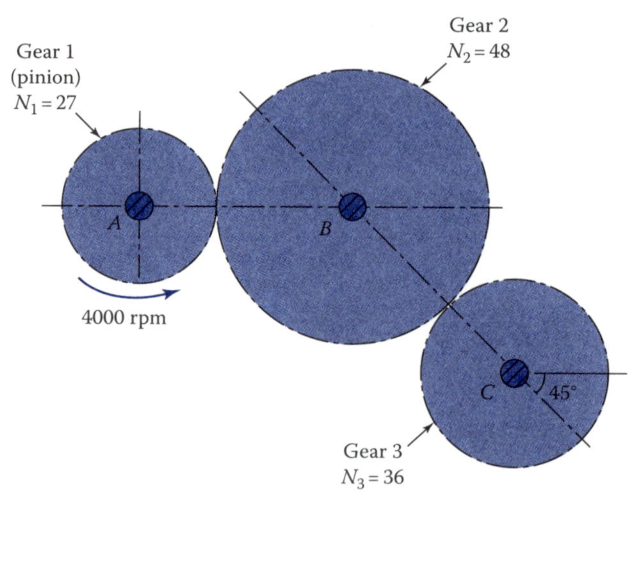

FIGURE P11.15

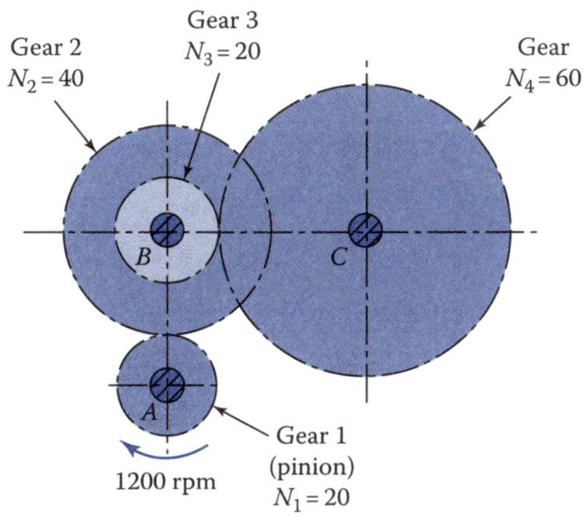

FIGURE P11.17

Given: Driving gear 1 transmits 30 hp at 4000 rpm through the idler to gear 3 on shaft *C*.

11.16 Redo Problem 11.15 for gears having a diametral pitch of 6 teeth/in., a 20° pressure angle, and a clockwise rotation of the driving gear 1.

11.17 The gears shown in Figure P11.17 have a diametral pitch of 4 teeth/in. and a 20° pressure angle. Determine and indicate on a free-body diagram

a. The tangential and radial forces on each gear

b. The reaction on shaft *B*

Design Decision: Driving gear 1 transmits 50 hp at 1200 rpm through the idler pair mounted on shaft *B* to gear 4.

11.18 The gears shown in Figure P11.18 have a module of 6 mm and a 20° pressure angle. Determine and show on a free-body diagram

a. The tangential and radial loads on each gear

b. The reactions on shaft *C*

Given: Driving gear 1 transmits 80 kW at 1600 rpm through the idler to gear 3 mounted on shaft *C*.

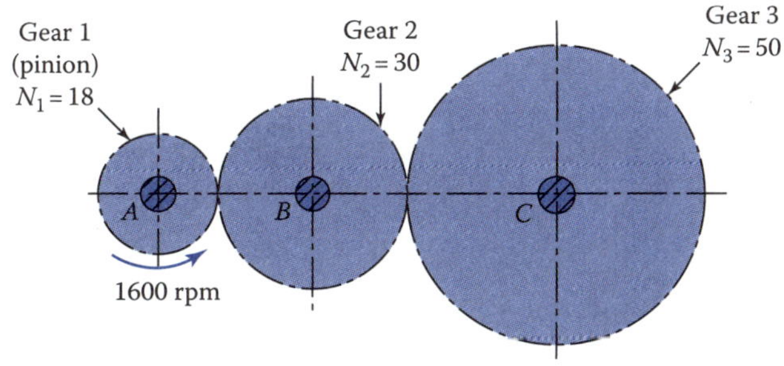

FIGURE P11.18

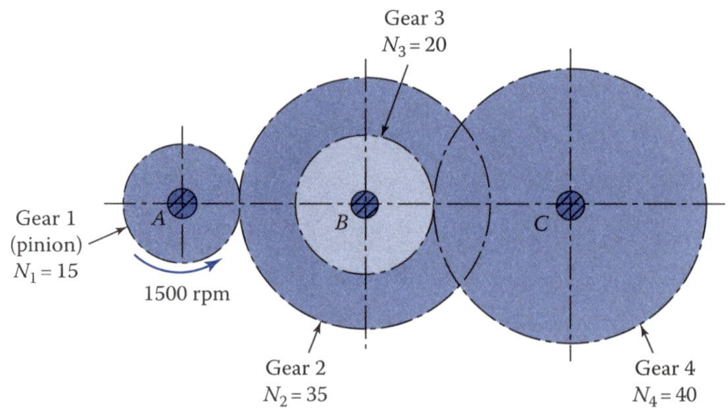

FIGURE P11.20

11.19 Resolve Problem 11.18 for gears having a module of 8 mm and a 25° pressure angle.

11.20 The gears shown in Figure P11.20 have a diametral pitch of 5 teeth/in. and 25° pressure angle. Determine and indicate on free-body diagrams

a. The tangential and radial forces on gears 2 and 3
b. The reactions on shaft *C*

Design Decision: Driving gear 1 transmits 10 hp at 1500 rpm through idler pair mounted on shaft *B* to gear 4.

11.21 The gears shown in Figure P11.21 have a diametral pitch of 6 teeth/in. and a 25° pressure angle. Determine and show on a free-body diagram

a. The tangential and radial forces on gears 2 and 3
b. The reactions on shaft *B*

Given: Driving gear 1 transmits 20 hp at 1800 rpm through an idler pair mounted on shaft *B* to gear 4.

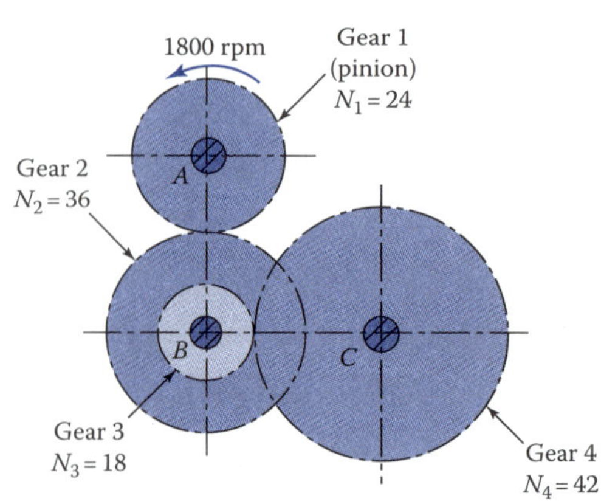

FIGURE P11.21

Sections 11.8 through 11.13

11.22 The gears shown in Figure P11.20 have a diametral pitch of 5 teeth/in., 25° pressure angle, and a tooth width of ½ in. Determine

a. The allowable bending load, using the Lewis equation and $K_f = 1.6$, for the tooth of gear 2
b. The allowable load for wear, applying the Buckingham equation, for gears 1 and 2
c. The maximum tangential load that gear 2 can transmit

Design Decisions: All gears are made of cast steel (0.20% C WQ&T, Table 11.3); gears 2 and 3 are mounted on shaft *B*.

11.23 The gears shown in Figure P11.20 have a module of 5 mm, tooth width of 15 mm, and a 20° pressure angle. Determine

a. The allowable bending load, applying the Lewis equation and $K_f = 1.5$, for the tooth of gear 3
b. The allowable load for wear, using the Buckingham equation, for gears 3 and 4
c. The maximum tangential load that gear 3 can transmit

Design Decisions: All gears are made of cast steel (0.20% C WQ&T, Table 11.3); gears 2 and 3 are mounted on shaft *B*.

11.24 Reconsider Problem 11.23, but to find the allowable bending load for gear 1, the allowable bending load for gears 1 and 2, and the maximum tangential load that the gear 2 can transmit.

11.25 A 20° pressure angle, 22-tooth spur gear with a diametral pitch of 10 teeth/in. and a 1¼ in. face width is to operate at 1500 rpm. Find

a. The allowable bending load by the Lewis formula
b. The maximum tangential load and power that the gear can transmit

Assumption: The gear is made of AISI 1045 WQ&T steel (see Table 11.3).

11.26 The gears shown in Figure P11.21 have a diametral pitch of 6 teeth/in., a 25° pressure angle, and a tooth width ½ in. Determine

a. The allowable bending load, applying the Lewis equation and $K_f = 1.4$, for the tooth of gear 2
b. The allowable load for wear, using the Buckingham equation, for gears 3 and 4
c. The maximum tangential load that gear 2 can transmit, based on bending strength

Design Decisions: All gears are made of steel hardened to 200 Bhn; gears 2 and 3 are mounted on shaft *B*.

11.27 The gears shown in Figure P11.21 have a module of 10 mm, a 20° pressure angle, and a tooth width of 15 mm. Determine

a. The allowable bending load, using the Lewis equation and $K_f = 1.5$, for the tooth of gear 4
b. The allowable load for wear, applying the Buckingham equation, for gears 1 and 2

Design Decisions: All gears are made of hardened steel (200 Bhn), and gears 2 and 3 are mounted on shaft *B*.

11.28 Reconsider Problem 11.27, except to determine bending load of gear 1 and the allowable load for wear for gears 3 and 4.

11.29 The 20° pressure angle and tooth width of ⅝ in. gears in a speed reducer are specified as follows:

Pinion: 1600 rpm, 24 teeth, 180 Bhn steel, hp = 1.2, $P = 12$ in.$^{-1}$,

$C_L = C_f = 1, \quad K_m = 1.6, \quad K_o = K_s = K_T = K_L = 1, \quad K_R = 1.25$

Gear: 60 teeth, AGMA 30 cast iron

Are the gears safe with regard to the AGMA bending strength?

Assumption: The manufacturing quality of the pinion and gear corresponds to curve D of Figure 11.15.

11.30 Determine whether the gears in Problem 11.29 are safe with regard to the AGMA wear strength.

11.31 The 20° pressure angle and tooth width of 15 mm gears in a gearbox are specified with the following data:

Pinion: 1200 rpm, 20 teeth, 300 Bhn, hp = 1.5, $m = 3$ mm,

$C_L = C_f = 1, \quad K_m = 1.7, \quad K_o = K_S = K_T = K_L = 1, \quad K_R = 1.3$

Gear: 50 teeth, AGMA 40 cast iron

Determine whether the gears are safe on the basis of the AGMA bending strength.

Assumption: The manufacturing quality of the pinion and the gear correspond to curve D of Figure 11.15.

11.32 Find whether the gears described Problem 11.31 are safe with regard to the AGMA wear strength.

11.33 A pair of cast iron (AGMA grade 40) gears has a diametral pitch of 5 teeth/in., a 20° pressure angle, and a width of 2 in. A 20-tooth pinion rotating at 90 rpm drives a 40-tooth gear. Determine the maximum horsepower that can be transmitted, based on wear strength and using the Buckingham equation.

11.34 Resolve Problem 11.33 using the AGMA method, if the life is to be no more than 10^6 cycles corresponding to a reliability of 99%.

Given: $E = 19 \times 10^6$ psi and $\nu = 0.3$.

Assumption: The gears are manufactured with precision.

11.35 A gearset of cast iron (AGMA grade 30) has a 5 mm module and a width of 40 mm. A 25-tooth pinion rotates at 120 rpm and drives a 50-tooth gear. The gears are cut using a pressure angle of 20°. What is the maximum horsepower that can be transmitted on the basis of wear strength by applying the Buckingham formula?

11.36 A pair of gears has a 20° pressure angle and a diametral pitch of 6 teeth/in. Determine the maximum horsepower that can be transmitted, based on bending strength and applying the Lewis equation and $K_f = 1.4$.

Design Decisions: The gear is made of phosphor bronze, has 60 teeth, and rotates at 240 rpm. The pinion is made of SAE 1040 steel and rotates at 600 rpm.

Given: Both gears have a width of 3.5 in.

11.37 Redo Problem 11.36 for two meshing gears that have widths of 80 mm and a module of 4 mm.

11.38 Resolve Problem 11.36, based on a reliability of 90% and moderate shock on the driven machine. Apply the AGMA method for $K_L = K_T = K_s = 1$.

Assumption: The manufacturing quality of the gearset corresponds to curve *C* of Figure 11.15.

11.39 Two meshing gears have face widths of 2 in. and diametral pitches of 5 teeth/in. The pinion is rotating at 600 rpm and has 28 teeth, and the velocity ratio is to be 4/5. Determine the horsepower that can be transmitted, based on wear strength and using the Buckingham equation.

Design Decisions: The gears are made both of steel hardened to a 350 Bhn and have a 20° pressure angle.

11.40 Resolve Problem 11.39 for a gearset that has a width of 60 mm and a module of 6 mm.

11.41 Redo Problem 11.39 based on a reliability of 99.9% and applying the AGMA method.

Given: $C_L = C_f = 1$, $K_o = K_s = K_T = 1$, $E = 30 \times 10^6$ psi, $\nu = 0.3$.

Assumption: The gearset is manufactured with high precision, shaved, and ground.

12

Helical, Bevel, and Worm Gears

12.1 Introduction

In Chapter 11, the kinematic relationships and factors that must be considered in designing straight-toothed or spur gears were presented. Several specialized forms of gears exist. We now deal with the three principal types of nonspur gearing: helical, worm, and bevel gears.* The geometry of these different types of gearing is considerably more complicated than for spur gears (see Figure 11.1). However, much of the discussion of the previous chapter applies equally well to the present chapter. Therefore, the treatment here of nonspur gears is relatively brief. As noted in Section 11.9, the reader should consult the American Gear Manufacturing Association (AGMA) standards for more information when faced with a real design problem involving gearing.

Helical, bevel, and worm gears can meet specific geometric or strength requirements that cannot be obtained from spur gears. Helical gears are very similar to spur gears. They have teeth that lie in helical paths on the cylinders instead of teeth parallel to the shaft axis. Bevel gears, with straight or spiral teeth cut on cones, can be employed to transmit motion between intersecting shafts. A worm gearset, consisting of screw meshing with a gear, can be used to obtain a large reduction in speed. The analysis of power screw force components, to be discussed in Section 15.3, also applies to worm gears. By noting that the thread angle of a screw corresponds to the pressure angle of the worm, expressions of the basic definitions and efficiency of power screws are directly used for a wormset in Sections 12.9 and 12.11.

12.2 Helical Gears

Like spur gears, helical gears are cut from a cylindrical gear blank and have involute teeth. The difference is that their teeth are at some helix angle to the shaft axis. These gears are used for transmitting power between parallel or nonparallel shafts. The former case is shown in Figure 12.1a. The helix can slope in either the upward or downward direction. The terms *right-hand* (RH) and *left-hand* (LH) *helical gears* are used to distinguish between the two types, as indicated in the figure. Note that the rule for

* For variations of the foregoing basic gear types and further details, see, for example, [1–10] and the websites wvvw.machinedesign.com and www.powertransmission.com.

(a) (b)

FIGURE 12.1
Helical gears: (a) opposite-hand pair meshed on parallel axes (most common type) and (b) same-hand pair meshed on crossed axes. (Courtesy of Boston Gear, Boston, MA.)

determining whether a helical gear is right or left handed is the same as that used for determining RH and LH screws.

Herringbone gear refers to a helical gear having half its face cut with teeth of one hand and the other half with the teeth of opposite hand (Figure 12.2). In nonparallel, nonintersecting shaft applications, gears with helical teeth are known as *crossed helical gears*, as shown in Figure 12.1b. Such gears have point contact, rather than the line contact of regular helical gears. This severely reduces their load-carrying capacity. Nevertheless, crossed helical gears are frequently used for the transmission of relatively small loads, such as distributor and speedometer drives of automobiles. We consider only conventional helical gears on parallel shafts.

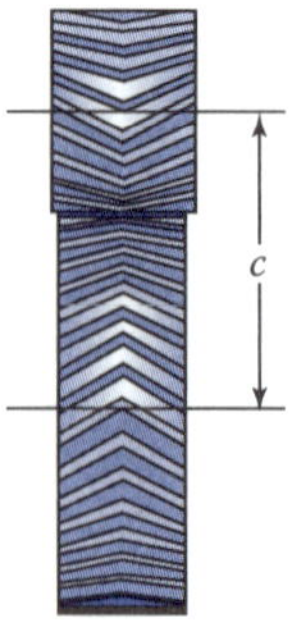

FIGURE 12.2
Atypical herringbone gearset.

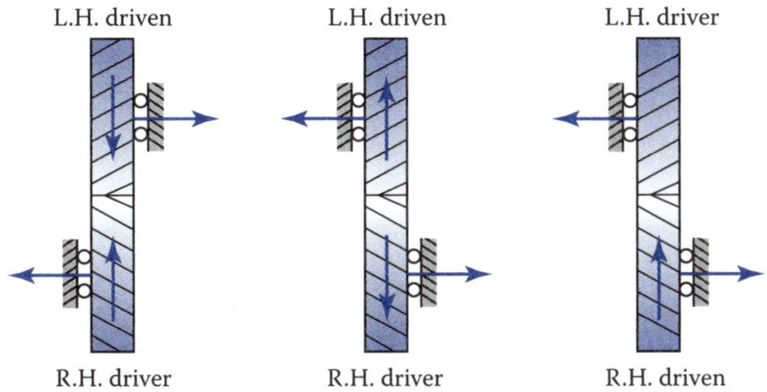

FIGURE 12.3
Direction, rotation, and thrust load for three helical gearsets with parallel shafts.

Figure 12.3 illustrates the thrust, rotation, and hand relations for some helical gearsets with parallel shafts. Note that the direction in which the thrust load acts is determined by applying the RH or LH rule to the driver. That is, for the RH driver, if the fingers of the RH are pointed in the direction of rotation of the gears, the thumb points in the direction of the thrust. The driven gear then has a thrust load acting in the direction opposite to that of the driver, as shown in the figure.

The *advantages* of helical gears over other basic gear types include more teeth in contact simultaneously, and the load transferred gradually and uniformly as successive teeth come into engagement. Gears with helical teeth operate more smoothly and carry larger loads at higher speeds than spur gears. The line of contact extends diagonally across the face of mating gears. When employed for the same applications as spur gears, these gears have quiet operation. The *disadvantages* of helical gears are greater cost than spur gears and the presence of an axial force component that requires thrust bearings on the shaft.

12.3 Helical Gear Geometry

Helical gear tooth proportions follow the same standards as those for spur gears. The teeth form the *helix* angle ψ with the gear axis, measured on an imaginary cylinder of pitch diameter d. The *usual* range of *values* of the helix angle is between 15° and 30°. Various relations may readily be developed from geometry of a basic rack. Figure 12.4 illustrated with dimensions in transverse plane (*At*), normal plane (*An*), and axial plane (*Ax*). The distances between similar pitch lines from tooth to tooth are the *circular pitch* p, the *normal circular pitch* p_n, and the *axial* (circular) *pitch* p_a.

Observe from the figure that the p and the pressure angle ϕ are measured in the transverse plane or the plane of rotation, as with spur gears. Hence, the p and the ϕ are also referred to as the *transverse circular pitch* and *transverse pressure angle*, respectively. The quantities p_n and normal pressure angle ϕ_n are measured in a normal plane. Referring to the triangles *ABC* and *ADC*, we write

$$p_n = p\cos\psi, \quad p_a = p\cot\psi = \frac{p_n}{\sin\psi} \tag{12.1}$$

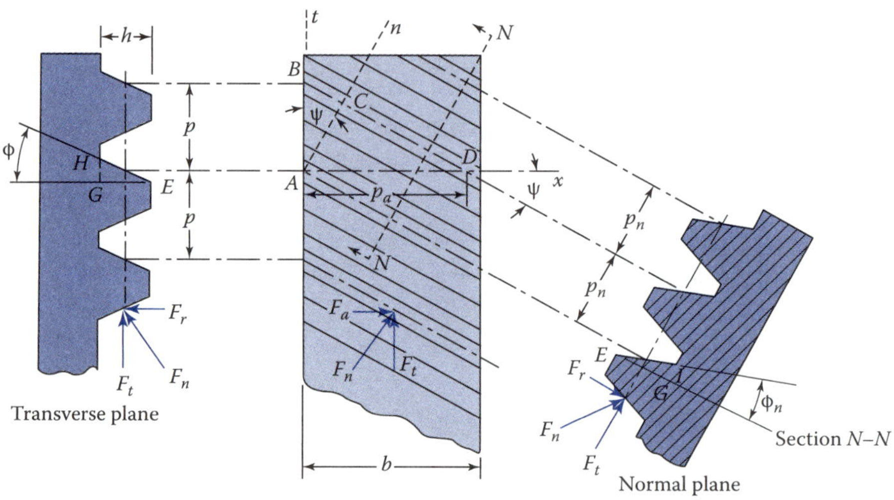

FIGURE 12.4
Portion of a helical rack displaying transverse and normal planes and resolution of forces.

Diametral pitch is more commonly employed than circular pitch to define tooth size. The product of circular and diametral pitch equals π for normal as well as transverse plane. Therefore,

$$Pp = \pi, \quad P_n p_n = \pi, \quad P_n = \frac{P}{\cos\psi}, \quad P = \frac{N}{d} \tag{12.2}$$

where
P = the diametral pitch
P_n = the normal diametral pitch
N = the number of teeth

In the triangles *EGH* and *EGI* (Figure 12.4), $EG = h$; however, $GH/GI = p/p_n = 1/\cos\psi$ from the first of Equations 12.1. Here, h is the whole depth of the gear teeth. We also have $\tan\phi = GH/h$, $\tan\phi_n = GI/h$, or $\tan\phi/\tan\phi_n = GH/GI$. Hence, pressure angles are related by

$$\tan\phi_n = \tan\phi\cos\psi \tag{12.3}$$

Other geometric quantities are expressed similarly to those for spur gears:

$$d = \frac{Np}{\pi} = \frac{Np_n}{\pi\cos\psi} = \frac{N}{P_n\cos\psi} \tag{12.4}$$

$$c = \frac{d_1 + d_2}{2} = \frac{p}{2\pi}(N_1 + N_2) = \frac{N_1 + N_2}{2P_n\cos\psi} \tag{12.5}$$

where c represents the center distance of mating gears (1 and 2).

12.3.1 Virtual Number of Teeth

Intersection of the normal plane N–N and the pitch cylinder of diameter d is an *ellipse* (Figure 12.4). The shape of the gear teeth generated in this plane, using the radius of curvature of the ellipse, would be a nearly virtual spur gear having the same properties as the actual helical gear. From analytic geometry, the radius of curvature r_c at the end of a semiminor axis of the ellipse is

$$r_c = \frac{d/2}{\cos^2 \psi} \tag{12.6}$$

The number of teeth of the equivalent spur gear in the normal plane, known as either the virtual or equivalent number of teeth, is then

$$N' = \frac{2\pi r_c}{p_n} = \frac{\pi d}{p_n \cos^2 \psi} \tag{12.7a}$$

Carrying the values for $\pi d/p_n$ from Equation 12.4 into Equation 12.7a, the *virtual number of teeth* N' may be expressed in the following convenient form:

$$N' = \frac{N}{\cos^3 \psi} \tag{12.7b}$$

As noted previously, N is the number of actual teeth.

It is necessary to know the virtual number of teeth in design: This is considered in finding the appropriate values of the geometric factors Y and J for helical gears as discussed in Section 12.5.

12.3.2 Contact Ratios

The total contact ratio C_{rt} of helical gears is the sum of transverse and axial contact ratios. That is,

$$C_{rt} = C_r + C_{ra} \tag{12.8}$$

The transverse contact ratio C_r is the same as defined for spur gears by Equation 11.14. Owing to the helix angle, the axial contact ratio C_{ra} is given by

$$C_{ra} = \frac{b}{p_a} = \frac{b \tan \psi}{p} \tag{a}$$

The quantities b and p_a represent the face width and axial pitch, respectively. The preceding ratio, indicating the degree of helical overlap in the mesh, should be higher than 1.15. Clearly, larger transverse contact ratio C_r and face width b will increase the overlapping of teeth and hence promote load sharing.

Example 12.1: Geometric Quantities for Helical Gears

Two helical gears have a center distance of c = 10 in., width b = 1.9 in., a pressure angle of ϕ = 25°, a helix angle of ψ = 30°, and a diametral pitch of P = 6 in.$^{-1}$ If the speed ratio is to be r_s = 1/3, calculate

a. The (transverse) circular, normal circular, and axial pitches
b. The number of teeth of each gear
c. The normal diametral pitch and normal pressure angle
d. The total contact ratio

Solution

a. Applying Equations 12.1 and 12.2,

$$p = \frac{\pi}{6} = 0.5236 \text{ in.}$$

$$p_n = 0.5236 \cos 30° = 0.453 \text{ in.}$$

$$p_a = 0.5236 \cot 30° = 0.907 \text{ in.}$$

b. Through the use of Equation 11.8,

$$r_s = \frac{1}{3} = \frac{N_1}{N_2} \quad \text{or} \quad N_2 = 3N_1$$

Equation 12.5 gives then

$$10 = \frac{0.5236}{2\pi}\left(N_1 + 3N_1\right) \quad \text{or} \quad N_1 = 30$$

and hence, N_2 = 90. Thus, d_1 = 30/6 = 5 in. = 127 mm and d_2 = 15 in. = 381 mm.

c. From Equation 12.2, we have

$$P_n = \frac{6}{\cos 30°} = 6.928 \text{ in.}^{-1}$$

By Equation 12.3,

$$\tan \phi_n = \tan 25° \cos 30°$$

from which

$$\phi_n = 22°$$

d. The addendum equals $a = a_1 = a_2 = 1/P$ = 1/6 in. Applying Equation 11.14, the contact ratio of spur gear is

$$C_r = \frac{1}{p\cos\phi}\left[\sqrt{\left(r_1 + a\right)^2 - \left(r_1 \cos\phi\right)^2} + \sqrt{\left(r_2 + a\right)^2 - \left(r_2 \cos\phi\right)^2}\right] - \frac{c\tan\phi}{p}$$

Introducing the data,

$$C_r = \frac{1}{0.5236\cos 25°}\left[\sqrt{(2.5+1/6)^2-(2.5\cos 25°)^2}+\sqrt{(7.5+1/6)^2-(7.5\cos 25°)^2}\right]-\frac{10\tan 25°}{0.5236}$$

$$= 1.53$$

The total contact ratio for the helical gear, by Equation 12.8, is

$$C_{rt} = C_r + C_{ra} = 1.53 + \frac{1.9\tan 30°}{0.5236} = 3.63$$

Comment: The result, about 3.63, is a reasonable value.

12.4 Helical Gear Tooth Loads

This section is concerned with the applied forces or loads acting on the tooth of a helical gear. As in the case of spur gears, the points of application of the force are in the pitch plane and in the center of the gear face. Again, the load is normal to the tooth surface and indicated by F_n. The transmitted load F_t is the same for spur or helical gears.

Figure 12.5 schematically shows the three components of the force acting against a helical gear tooth. Obviously, the inclined tooth develops the axial component, which is not present with spur gearing. We see from Figure 12.5 that the normal load F_n is at a compound angle defined by the normal pressure angle ϕ_n and the helix angle ψ in combination.

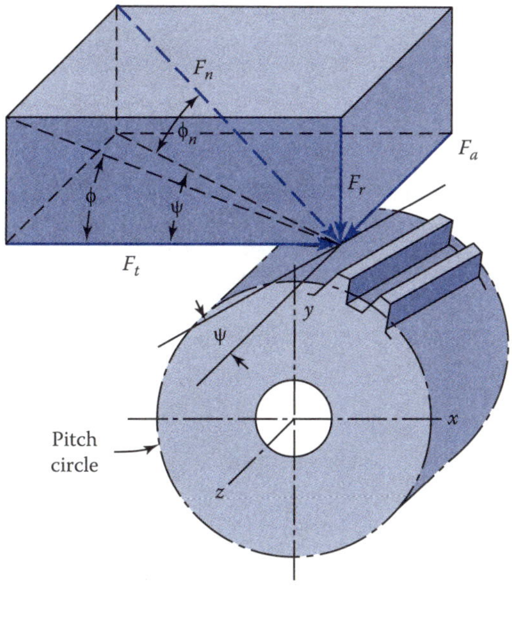

FIGURE 12.5
Components of tooth force acting on a helical gear.

The projection of F_n on the plane of rotation is inclined at an angle ϕ to the radial force component. Hence, the values of components of the tooth force are

$$F_r = F_n \sin\phi_n$$
$$F_t = F_n \cos\phi_n \cos\psi \quad (12.9)$$
$$F_a = F_n \cos\phi_n \sin\psi$$

where
F_n = the normal load or applied force
F_r = the radial component
F_t = the tangential component, also called the *transmitted load*
F_a = the axial component, also called the *thrust load*

Usually, the transmitted load F_t is obtained from Equation 11.22, and the other forces are desired. It is therefore convenient to rewrite Equations 12.9 as

$$F_r = F_t \tan\phi$$
$$F_a = F_t \tan\psi \quad (12.10)$$
$$F_n = \frac{F_t}{\cos\phi_n \cos\psi}$$

In the foregoing, the pressure angle ϕ is related to the helix angle ψ and normal pressure angle ϕ_n by Equation 12.3.

The thrust load requires the use of bearings that can resist axial forces as well as radial loads. Sometimes, the need for a thrust-resistant bearing can be eliminated by using a herringbone gear. Obviously, the axial thrust forces for each set of teeth in a herringbone gear cancel each other.

12.5 Helical Gear Tooth Bending and Wear Strengths

The equations for the bending and wear strengths of helical gear teeth are similar to those of spur gears. However, slight modifications must be made to take care of the effects of the helix angle ψ. The discussions of the factors presented in Chapter 11 on spur gears also apply to the helical gearsets. Adjusted forms of the fundamental equations are introduced in the following discussion.

12.5.1 Lewis Equation

The allowable bending load of the helical gear teeth is given by the following expression:

$$F_b = \frac{\sigma_o b}{K_f}\frac{Y}{P_n} \quad \text{(11.33, modified)}$$

The value of Y is obtained from Table 11.2 using the virtual teeth number N'.

12.5.2 Buckingham Equation

The limit load for wear for helical gears on parallel shafts may be written in the following form:

$$F_w = \frac{d_p b Q K}{\cos^2 \psi} \quad \text{(11.38, modified)}$$

where

$$Q = \frac{2N_g}{N_p + N_g} \quad \text{(11.40)}$$

The wear load factor K can be taken from Table 11.9 for the normal pressure angle ϕ_n.

For satisfactory helical gear performance, the usual *requirement* is that $F_b \geq F_d$ and $F_w \geq F_d$. The dynamic load F_d acting on helical gears can be estimated by the following formula:

$$F_d = \frac{78 + \sqrt{V}}{78} F_t \quad (\text{for } 0 < V > 4000 \text{ fpm}) \quad \text{(11.24c, modified)}$$

in which the pitch-line velocity, V in fpm, is defined by Equation 11.20. To convert to *m/s*, *divide* the given values in this expression by 196.8.

12.5.3 AGMA Equations

The formulas used for spur gears also apply to the helical gears. They were presented in Sections 11.9 and 11.11 with explanation of the terms. So the equation for *bending stress* is

$$\sigma = F_t K_o K_\upsilon \frac{P}{b} \frac{K_s K_m}{J} \quad \text{(11.35)}$$

$$\sigma = F_t K_o K_\upsilon \frac{1.0}{bm} \frac{K_s K_m}{J} \quad \text{(11.35, modified)}$$

Similarly, for *wear strength*, we have

$$\sigma_c = C_p \left(F_t K_o K_\upsilon \frac{K_s}{bd} \frac{K_m C_f}{I} \right)^{1/2} \quad \text{(11.42)}$$

where

$$I = \frac{\sin\phi \cos\phi}{2m_N} \frac{m_G}{m_G + 1} \quad \text{(11.43b)}$$

The charts and graphs previously given in Chapter 11 are valid equally well. The values of the geometry factor J for helical gears are obtained from Figure 12.6. The size factor $K_s = 1$ for helical gears.

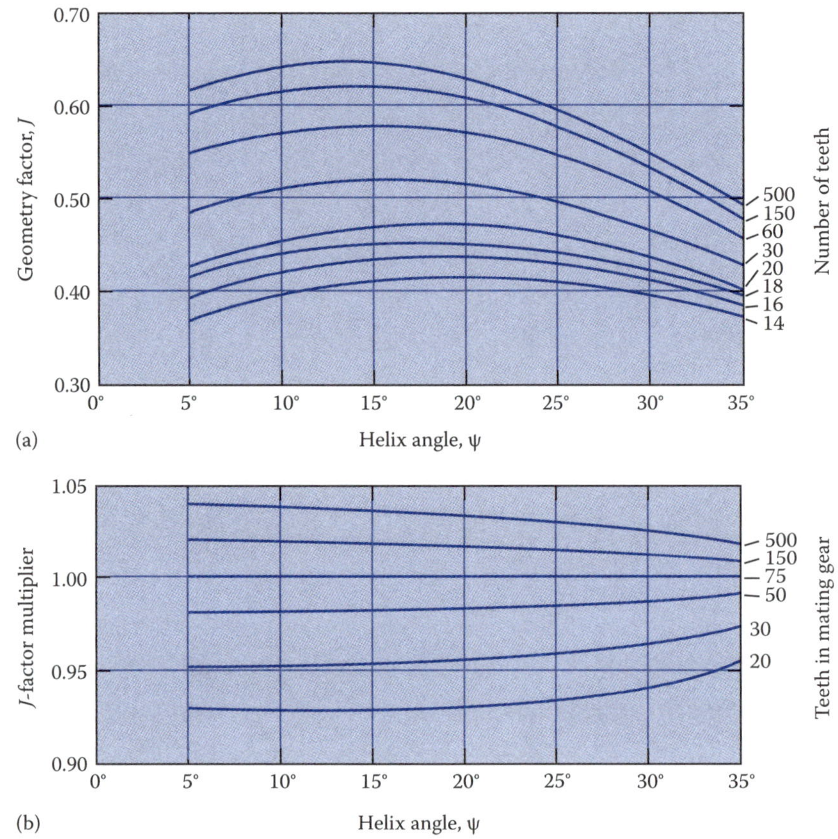

FIGURE 12.6
Helical gears with normal pressure angle $\phi_n = 20°$: (a) AGMA helical gear geometry factor *J*; (b) *J*-factor multipliers for use when the mating gear has other than 75 teeth (ANSI/AGMA Standard 218.01).

The calculation of the geometry factor I through the use of Equation 11.43b, for helical gears, requires the values of the load-sharing factor:

$$m_N = \frac{p_{nb}}{0.95Z} \tag{12.11}$$

Here

p_{nb} represents the normal base pitch = $p_n \cos \phi_n$
Z is the length of action in transverse plane

Equation 11.13 may be used to compute the value for L_0, in which the addendum equals $1/P_n$. Consult the appropriate AGMA standard for further details.

Allowable bending and *surface stresses* are calculated from equations given in Sections 11.9 and 11.11, repeated here, exactly as with spur gears:

$$\sigma_{\text{all}} = \frac{S_t K_L}{K_T K_R} \tag{11.36}$$

$$\sigma_{c,\text{all}} = \frac{S_c C_L C_H}{K_T K_R} \tag{11.44}$$

The design specifications and applications are the same as discussed in Chapter 11 for spur gears. Examples 12.2 and 12.3 further illustrate the analysis and design of helical gears.

Example 12.2: Electric Motor Geared to Drive a Machine

A motor at about $n = 2400$ rpm drives a machine by means of a helical gearset as shown in Figure 12.7. Calculate

a. The value of the helix angle
b. The allowable bending and wear loads using the Lewis and Buckingham formulas
c. The horsepower that can be transmitted by the gearset

Given: The gears have the following geometric quantities:

$$P_n = 5\text{ in.}^{-1},\quad \phi = 20°,\quad c = 9\text{ in.},\quad N_1 = 30,\quad N_2 = 42,\quad b = 2\text{ in.}$$

Design Assumptions: The gears are made of SAE 1045 steel, water-quenched and tempered (WQ&T), and hardened to 200 Bhn.

Solution

a. From Equations 12.1 through 12.5, we have

$$P = \frac{1}{2c}(N_1 + N_2) = \frac{72}{18}$$

$$d_1 = \frac{N_1}{P} = \frac{30}{72}(18) = 7.5\text{ in.}$$

$$d_2 = \frac{N_2}{P} = \frac{42}{72}(18) = 10.5\text{ in.}$$

$$\cos\psi_1 = \frac{N_1}{P_n d_1} = \frac{30}{5(7.5)} = 0.8 \quad \text{or} \quad \psi_1 = \psi_2 = 36.9°$$

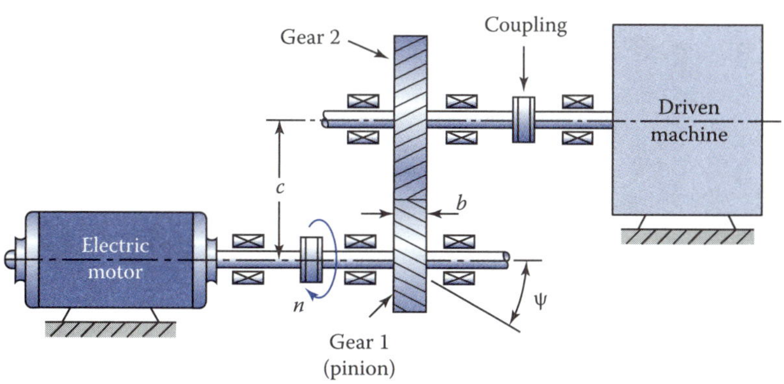

FIGURE 12.7
Example 12.20. Schematic arrangement of motor, gear, and driven machine.

b. The virtual number of teeth, using Equation 12.7b, is

$$N' = \frac{N}{\cos^3 \psi} = \frac{30}{(0.8)^3} = 58.6$$

Hence, interpolating in Table 11.2, $Y = 0.419$. By Table 11.3, $\sigma_o = 32$ ksi. Applying the Lewis equation, (11.33, modified) with $K_f = 1$,

$$F_b = \sigma_o b \frac{Y}{P_n} = 32(2)\frac{0.419}{5} = 5.363 \text{ kips}$$

By Table 11.9, $K = 79$ ksi. From Equation 11.40,

$$Q = \frac{2N_g}{N_p + N_g} = \frac{2(42)}{72} = \frac{7}{6}$$

The Buckingham formula, Equation 11.38′, yields

$$F_w = \frac{d_1 bQK}{\cos^2 \psi} = \frac{7.5(2)(7)(79)}{6(0.8)^2} = 2.16 \text{ kips}$$

c. The horsepower capacity is based on F_w since it is smaller than F_b. The pitch-line velocity equals

$$V = \frac{\pi d_1 n_1}{12} = \frac{\pi(7.5)(2400)}{12} = 4712 \text{ fpm}$$

The dynamic load, using Equation 11.24c′, is

$$F_d = \frac{78 + \sqrt{4712}}{78} F_t = 1.88 F_t$$

Equation 11.41, $F_w \geq F_d$, results in

$$2.16 = 1.88 F_t \quad \text{or} \quad F_t = 1.15 \text{ kips}$$

The corresponding gear power is therefore

$$\text{hp} = \frac{F_t V}{33{,}000} = \frac{1150(4712)}{33{,}000} = 164$$

Comments: Observe that the dynamic load is about twice the transmitted load, as expected for reliable operation.

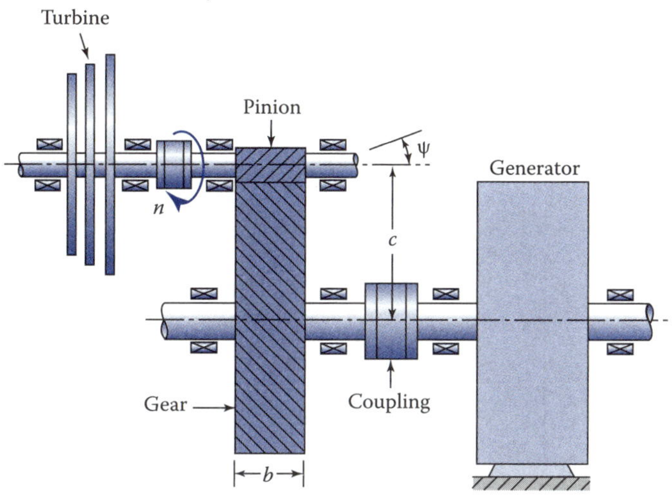

FIGURE 12.8
Example 12.3. Schematic arrangement of turbine, gear, and generator.

Example 12.3: High-Speed Turbine Geared to Drive a Generator

A turbine rotates at about n = 8000 rpm and drives, by means of a helical gearset, a 250 kW (335 hp) generator at 1000 rpm, as depicted in Figure 12.8. Determine

a. The gear dimensions and the gear tooth forces
b. The load capacity based on the bending strength and surface wear using the Lewis and Buckingham equations
c. The AGMA load capacity on the basis of strength only

Given: Gearset helix angle $\psi = 30°$

$$\text{Pinion: } N_p = 35, \quad \phi_n = 20°, \quad P_n = 10 \text{ in.}^{-1}$$

Design Assumptions:

1. Moderate shock load on the generator and a light shock on the turbine.
2. Mounting is accurate.
3. Reliability is 99.99%.
4. Both pinion and gear are through hardened, precision shaped, and ground to permit to run at high speeds.
5. The pinion is made of steel with 150 Bhn, and gear is cast iron.
6. The gearset goes on a maximum of $c = 18\frac{1}{2}$ in. center distance. However, to keep pitch-line velocity down, gears are designed with as small a center distance as possible.
7. To keep stresses down, a wide face width b = 8 in. large gearset is used.
8. The generator efficiency is 95%.

Solution
See Figures 12.6 and 12.8 and Tables 11.5 through 11.9.

a. The geometric quantities for the gearset are obtained by using Equations 12.1 through 12.5. Therefore,

$$\phi = \tan^{-1}\frac{\tan\phi_n}{\cos\psi} = \tan^{-1}\frac{\tan 20^\circ}{\cos 30^\circ} = 22.8^\circ$$

$$P = P_n\cos\psi = 10\cos 30^\circ = 8.66$$

$$d_p = \frac{N_p}{P} = \frac{35}{8.66} = 4.04 \text{ in.}$$

$$N_g = N_p\left(\frac{n_p}{n_g}\right) = 35\left(\frac{8000}{1000}\right) = 280$$

$$d_g = \frac{N_p}{P} = \frac{280}{8.66} = 32.3 \text{ in.}$$

It follows that

$$c = \frac{1}{2}(d_p + d_g)$$
$$= \frac{1}{2}(4.04 + 32.3) = 18.17 \text{ in.}$$

Comment: The condition that the center distance is not to exceed 18.5 is satisfied.

The pitch-line velocity equals

$$V = \frac{\pi dn}{12} = \frac{\pi(4.04)8000}{12} = 8461 \text{ fpm}$$

The power that the gear must transmit is about 335/0.95 = 353 hp. The transmitted load is then

$$F_t = \frac{33{,}000 \text{ hp}}{V} = \frac{33{,}000(353)}{8461} = 1.38 \text{ kips}$$

As a result, the radial, axial, and normal components of tooth force are, using Equation 12.10,

$$F_r = F_t\tan\phi = 1380\tan 22.8^\circ = 580 \text{ lb}$$

$$F_a = F_t\tan\psi = 1380\tan 30^\circ = 797 \text{ lb}$$

$$F_n = \frac{F_t}{\cos\phi_n\cos\psi} = \frac{1380}{\cos 20^\circ\cos 30^\circ} = 1696 \text{ lb}$$

b. From Equation 12.7b, we have

$$N' = \frac{N}{\cos^3\psi} = \frac{35}{\cos^3 30^\circ} = 53.9$$

Then, for 53.9 teeth and $\phi = 22.8°$ by interpolation from Table 11.2, $Y = 0.452$. Using Table 11.3, $\sigma_o \approx 18$ ksi. Applying Equation 11.33′ with $K_f = 1$,

$$F_b = \sigma_o b \frac{Y}{P_n} = 18(8)\frac{0.452}{10} = 6.51 \text{ kips}$$

Corresponding to $\phi = 22.8°$, interpolating in Table 11.9, $K \approx 68$ psi. Through the use of Equation 11.40,

$$Q = \frac{2N_g}{N_p + N_g} = \frac{2(280)}{35 + 280} = \frac{112}{63}$$

The limit load for wear, by Equation 11.38′, is

$$F_w = \frac{d_p bQK}{\cos^2 \psi} = \frac{4.04(8)(68)(112)}{63(\cos^2 30°)} = 5.21 \text{ kips}$$

Because the permissible load in wear is less than that allowable in bending, it is used as the dynamic load F_d. So from Equation 11.24c′,

$$F_d = \frac{78 + \sqrt{V}}{78} F_t$$

$$5.21 = \frac{78 + \sqrt{8461}}{78} F_t = 2.18 F_t$$

or

$$F_t = 2.39 \text{ kips}$$

c. Application of Equation 11.36 leads to

$$\sigma_{\text{all}} = \frac{S_t K_L}{K_T K_R}$$

where

$S_t = 20.5$ ksi (interpolating, Table 11.6 for a Bhn of 150)
$K_L = 1.0$ (indefinite life, Table 11.7)
$K_T = 1.0$ (from Section 11.9)
$K_R = 1.25$ (by Table 11.8)

The preceding equation is therefore

$$\sigma_{\text{all}} = \frac{20{,}500(1.0)}{(1.0)(1.25)} = 16.4 \text{ ksi}$$

The tangential force, by Equation 11.35 with $\sigma = \sigma_{all}$, is

$$F_t = \frac{\sigma_{all}}{K_o K_\upsilon} \frac{b}{P} \frac{J}{K_s K_m} \tag{a}$$

Here, we have

$K_o = 1.5$ (Table 11.4)
$K_\upsilon = 2.18$ (from curve *B* of Figure 11.15)
$K_s = 1.1$ (from section 12.5)
$K_m = 1.5$ (by Table 11.5)
$J = 0.47$ (for $N_p = 35$ and $\psi = 30$, Figure 12.6a)
J – multiplier = 1.02 (for $N_g = 280$ and $\psi = 30$, Figure 12.6b)
$J = 1.02 \times 0.47 = 0.48$

Equation (a) is then

$$F_t = \frac{16,400(8)(0.48)}{(1.5)(2.18)(8.66)(1.0)(1.5)} = 1.48 \text{ kips}$$

compared to the approximate result 2.39 kips for part b.

Comments: The tangential load capacity of the gearset, 1.48 kips, is larger than the force to be transmitted, 1.38 kips (allowance is made for 95% generator efficiency); the gears are safe. Since a wide face width is used, the design should be checked for combined bending and torsion at the pinion [1].

12.6 Bevel Gears

Bevel gears are cut on conical blanks to be used to transmit motion between intersecting shafts. The simplest bevel gear type is the straight-tooth bevel gear or *straight bevel gear* (Figure 12.9). As the name implies, the teeth are cut straight, parallel to the cone axis, like spur gears. Clearly, the teeth have a taper and, if extended inward, would intersect each other at the axis. Although bevel gears are usually made for a shaft angle of 90°, the only type we deal with here, they may be produced for almost any angle as well as with teeth lying in spiral paths on the pitch cones. When two straight bevel gears intersect at right angles and the gears have the same number of teeth, they are called *miter gears*.

Spiral teeth bevel gears, called just *spiral bevel gears* or *hypoid gears*, have shafts that do not intersect, as shown in Figure 11.1. We see that the teeth are cut at a spiral angle to the cone axis, analogous to helical gears. It is therefore possible to connect continuous nonintersecting shafts by such gears. Often spiral gears are most desirable for those applications involving large speed-reduction ratios and those requiring great smoothness and quiet operation. These gears are in widespread usage for automotive applications. *Zerol bevel gears* have curved teeth like spiral bevels; however, they have a 0 spiral angle similar to straight bevel gears. Bevel gears are noninterchangeable. Usually, they are made and replaced as matched pinion gearsets.

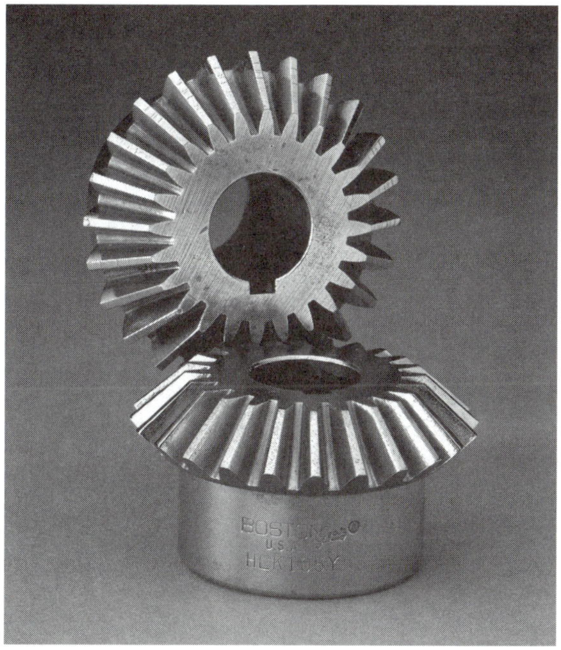

FIGURE 12.9
Bevel gears. (Courtesy of Boston Gear, Boston, MA.)

12.6.1 Straight Bevel Gears

Straight bevel gears are the most economical of the various bevel gear types. These gears are used primarily for relatively low-speed applications with pitch-line velocities up to 1000 fpm, where smoothness and quiet are not significant considerations. However, with the use of a finishing operation (e.g., grinding), higher speeds have been successfully handled by straight bevel gears.

12.6.1.1 Geometry

The geometry of bevel gears is shown in Figure 12.10. The size and shape of the teeth are defined at the large end on the back cones. They are similar to those of spur gear teeth. Note that the pitch cone and (developed) back cone elements are perpendicular. The *pitch angles* (also called pitch cone angles) are defined by the pitch cones joining at the apex. Standard straight bevel gears are cut by using a 20° pressure angle and full-depth teeth, which increase the contact ratio and the strength of the pinion.

The diametral pitch refers to the back cone of the gear. Therefore, the relationships between the geometric quantities and the speed for bevel gears are given as follows:

$$d_p = \frac{N_p}{P}, \quad d_g = \frac{N_g}{P} \tag{12.12a}$$

$$\tan\alpha_p = \frac{N_p}{N_g}, \quad \tan\alpha_g = \frac{N_g}{N_p} \tag{12.12b}$$

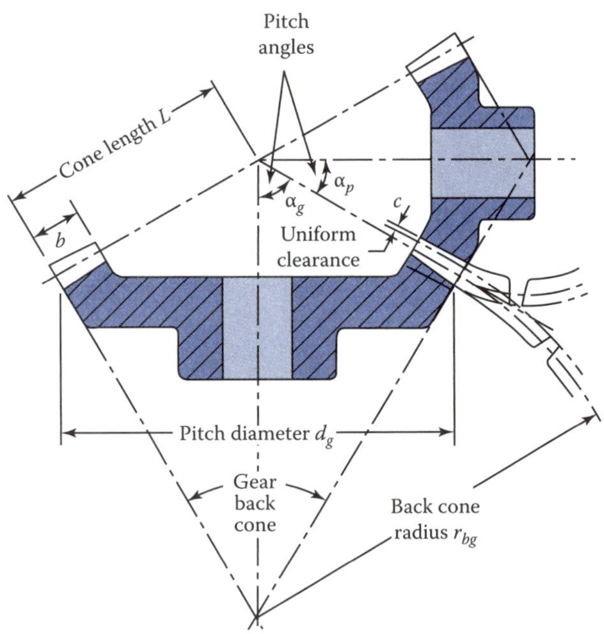

FIGURE 12.10
Notation for bevel gears.

$$r_s = \frac{\omega_g}{\omega_p} = \frac{N_p}{N_g} = \frac{d_p}{d_g} = \tan\alpha_p = \cot\alpha_g \tag{12.13}$$

where

d = the pitch diameter
P = the diametral pitch
N = the number of tooth
α = the pitch angle
ω = the angular speed
r_s = the speed ratio

In the preceding equations, the subscripts p and g refer to the pinion and gear, respectively.

It is to be noted that, for 20° pressure angle straight bevel gear teeth, the face width b should be made equal to

$$b = \frac{L}{3} \quad \text{or} \quad b = \frac{10}{P} \tag{a}$$

whichever is smaller. The uniform clearance is given by the following formula:

$$c = \frac{0.188}{P} + 0.002 \text{ in.} \tag{b}$$

The quantities L and c represent the pitch cone length and clearance, respectively (Figure 12.10).

12.6.2 Virtual Number of Teeth

In the discussion of helical gears, it was pointed out that the tooth profile in the normal plane is a spur gear having an ellipse as its radius of curvature. The result is that the form factors for spur gears apply, provided the equivalent or a virtual number of teeth N' are used in finding the tabular values. The identical situation exists with regard to bevel gears.

Figure 12.10 depicts the gear teeth profiles at the back cones. They relate to those of spur gears having radii of r_{bg} (gear) and r_{bp} (pinion). The preceding is referred to as the *Tredgold's approximation*. Accordingly, the *virtual number of teeth* N' in these imaginary spur gears

$$N'_p = 2r_{bp}P, \quad N'_g = 2r_{bg}P \tag{12.14}$$

This may be written in the following convenient form:

$$N'_p = \frac{N_p}{\cos\alpha_p}, \quad N'_g = \frac{N_g}{\cos\alpha_g} \tag{12.15}$$

in which r_{bg} is the back cone radius and N represents the actual number of teeth of bevel gear.

12.7 Tooth Loads of Straight Bevel Gears

In practice, the resultant tooth load is taken to be acting at the midpoint of the tooth face (Figure 12.11a). While the actual resultant occurs somewhere between the midpoint and the large end of the tooth, there is only a small error in making this simplifying assumption.

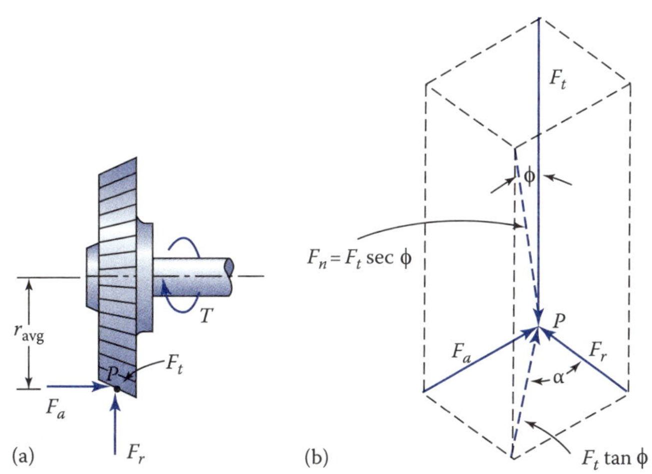

FIGURE 12.11
Forces at midpoint of bevel gear tooth: (a) three mutually perpendicular components; (b) the total (normal) force and its projections.

The transmitted *tangential load* or tangential component of the applied force, acting at the pitch point *P*, is then

$$F_t = \frac{T}{r_{avg}} \tag{12.16}$$

Here

T represents the torque applied
r_{avg} is the average pitch radius of the gear under consideration

The resultant force normal to the tooth surface at point *P* of the gear has value $F_n = F_t \sec\phi$ (Figure 12.11b). The projection of this force in the axial plane, $F_t \tan\phi$, is divided into the axial and radial components:

$$\begin{aligned} F_a &= F_t \tan\phi \sin\alpha \\ F_r &= F_t \tan\phi \cos\alpha \end{aligned} \tag{12.17}$$

where

F_t = the tangential force
F_a = the axial force
F_r = the radial force
ϕ = the pressure angle
α = the pitch angle

It is obvious that the three components F_t, F_a, and F_r are at right angles to each other. These forces can be used to ascertain the bearing reactions, by applying the equations of statics.

Example 12.4: Determining the Tooth Loads of a Bevel Gearset

A set of 20° pressure angle straight bevel gears is to be used to transmit 20 hp from a pinion operating at 500 rpm to a gear mounted on a shaft that intersects the shaft at an angle of 90° (Figure 12.12a). Calculate

a. The pitch angles and average radii for the gears
b. The forces on the gears
c. The torque produced about the gear shaft axis

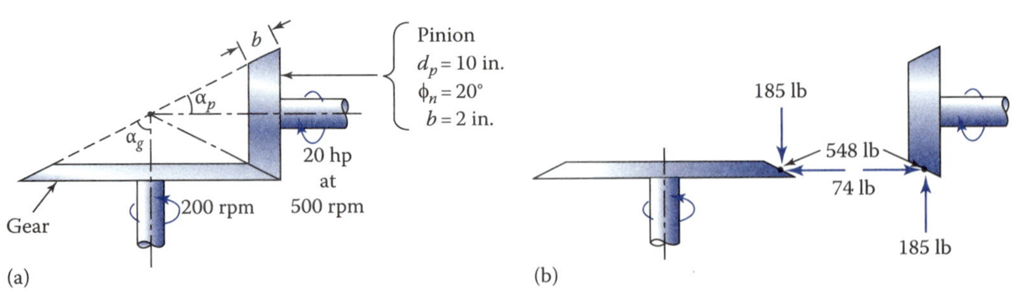

FIGURE 12.12
Example 12.4. Pitch cones of bevel gearset: (a) data; (b) axial and radial tooth forces.

Solution

a. Equation 12.13 gives

$$r_s = \frac{200}{500} = \frac{1}{2.5} \quad \text{or} \quad d_g = 2.5d_p = 25\text{ in.}$$

$$\alpha_p = \tan^{-1}\left(\frac{1}{2.5}\right) = 21.8° \quad \text{and} \quad \alpha_g = 90° - \alpha_p = 68.2°$$

Hence,

$$r_{g,\text{avg}} = r_g - \frac{b}{2}\sin\alpha_g = 12.5 - (1)\sin 68.2° = 11.6\text{ in.}$$

$$r_{p,\text{avg}} = r_p - \frac{b}{2}\sin\alpha_p = 5 - (1)\sin 21.8° = 4.6\text{ in.}$$

b. Through the use of Equation 11.22,

$$F_t = \frac{33{,}000\text{ hp}}{\pi d_{p.\text{avg}} n_p / 12} = \frac{33{,}000(20)(12)}{\pi(9.2)500} = 548\text{ lb}$$

From Equations 12.17, the pinion forces are

$$F_a = F_t \tan\phi\sin\alpha_p = 548(\tan 20°)(\sin 21.8°) = 74\text{ lb}$$

$$F_r = F_t \tan\phi\cos\alpha_p = 548(\tan 20°)(\cos 21.8°) = 185\text{ lb}$$

As shown in Figure 12.12b, the pinion thrust force equals the gear radial force, and the pinion radial force equals the gear thrust force.

c. Torque, $T = F_t(d_g/2) = 548(12.5) = 6.85$ kips·in.

12.8 Bevel Gear Tooth Bending and Wear Strengths

The expressions for bending and wear strengths are analogous to those for spur gears. However, slight modifications must be made to take care of the effects of the cone angle α. The adjusted forms of the basic formulas are introduced in the following paragraphs.

12.8.1 Lewis Equation

It is assumed that the bevel gear tooth is equivalent to a spur gear tooth whose cross section is the same as the cross section of the bevel tooth at the midpoint of the face *b*. The allowable *bending load* is given by

$$F_b = \frac{\sigma_0 b}{K_f}\frac{Y}{P} \tag{11.33}$$

The factor *Y* is read from Table 11.2, for a gear of *N′* virtual number of teeth.

12.8.2 Buckingham Equation

Due to the difficulty in achieving a bearing along the entire face width b, about three-quarters of b alone is considered as effective. So the allowable *wear load* can be expressed as

$$F_w = \frac{0.75 d_p b K Q'}{\cos \alpha_p} \quad \text{(11.38, re-modified)}$$

where

$$Q' = \frac{2N'_g}{N'_p + N'_g} \quad \text{(11.40, re-modified)}$$

In the preceding, we have

d_p = the diameter measured at the back of the tooth
N' = the virtual tooth number
α_p = the pitch angle
K = the wear load factor (from Table 11.9)

For the satisfactory operation of the bevel gearsets, the usual *requirement* is that $F_b \geq F_d$ and $F_w \geq F_d$ where the dynamic load F_d is given by Equation 11.24.

12.8.3 AGMA Equations

The formulas are the same as those presented in the discussions of the spur gears. But only some of the values of the correction factors are applicable to bevel gears [5]. For a complete treatment, consult the appropriate AGMA publications and the references listed [2]. We present only a brief summary of the method to bevel gear design as an introduction to the subject.

The equation for the *bending stress* at the root of bevel gear tooth is the same as for spur or helical gears. Therefore,

$$\sigma = F_t K_o K_\upsilon \frac{P}{b} \frac{K_s K_m}{J} \quad \text{(11.35)}$$

$$\sigma = F_t K_o K_\upsilon \frac{1}{bm} \frac{K_s K_m}{J} \quad \text{(11.35, modified)}$$

The tangential load F_t is obtained from Equation 12.16. Figure 12.13 gives the values for the foregoing geometry factor J for straight bevel gears. The AGMA standard also provides charts of the factors for zerol and spiral bevel gears. The remaining factors in Equation 11.35 can be taken to be the same as defined in Section 11.9. The allowable bending stress of bevel gear tooth is calculated from Equation 11.36, exactly as for spur and helical gears.

Surface stress for the wear of a bevel gear tooth is computed in a manner like that of spur or helical gears. Hence,

$$\sigma_c = C_p \left(F_t K_o K_\upsilon \frac{K_s}{bd} \frac{K_m C_f}{I} \right)^{1/2} \quad \text{(11.42)}$$

Only two modifications are required: Values of C_p are 1.23 times the values listed in Table 11.10, and values of I are taken from Figure 12.14. The *allowable* surface stress of a bevel gear tooth is obtained by Equation 11.44, exactly as for spur or helical gears.

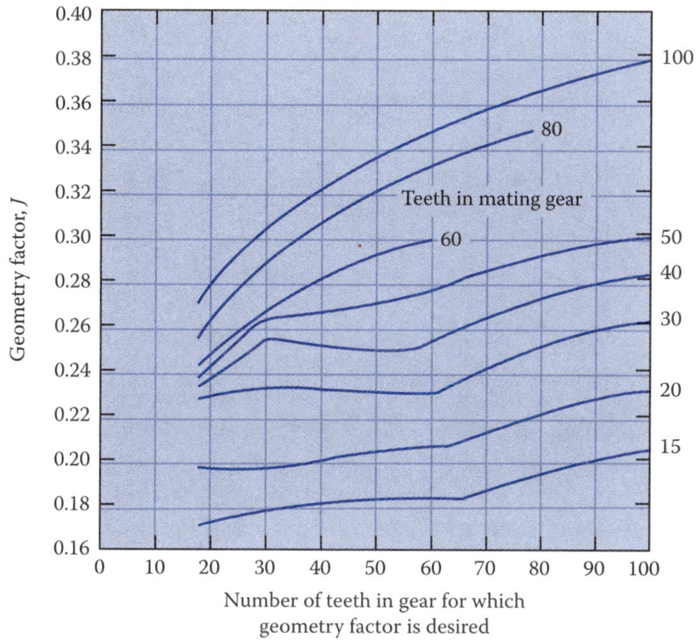

FIGURE 12.13
Geometry factors J for straight bevel gears. Pressure angle of 20° and shaft angle of 90° (AGMA Information Sheet 226.01).

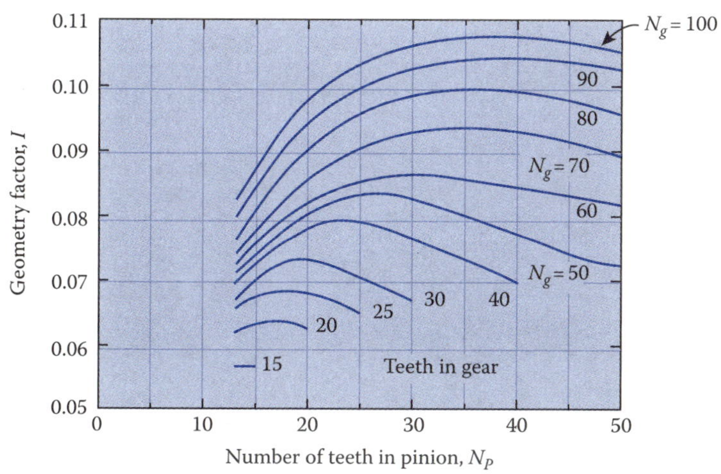

FIGURE 12.14
Geometry factors I for straight bevel gears. Pressure angle of 20° and shaft angle of 90° (AGMA Information Sheet 215.91).

12.9 Worm Gearsets

Worm gearing can be employed to transmit motion between nonparallel nonintersecting shafts, as shown in Figures 11.1, 12.15, and 12.18. A worm gearset, or simply wormset, consists of a *worm* (resembles a screw) and the *worm gear* (a special helical gear). The shafts on which the worm and gear are mounted are usually 90° apart. The meshing of two teeth takes place with a sliding action, without shock relevant to spur, helical, or bevel gears. This action, while occurring in quiet operation, may generate overheating, however. It is possible to obtain a large *speed reduction* (up to 360:1) and a high increase of torque by means of wormsets. Typical applications of worm gearsets include positioning devices that take advantage of their locking property (see Section 15.3).

Only a few materials are available for wormsets. The worms are highly stressed and usually made of case-hardened alloy steel. The gear is customarily made of one of the bronzes. The gear is hobbed, while the worm is ordinarily cut and ground or polished. The teeth of the worm must be properly shaped to provide conjugate surfaces. Tooth forms of worm gears are not involutes, and there are large sliding-velocity components in the mesh.

12.9.1 Worm Gear Geometry

Worms can be made with single, double, or more threads. Worm gearing may be either single enveloping (commonly used) or double enveloping. In a *single-enveloping* set (Figures 12.15 and 12.16b), the helical gear has its width cut into a concave surface, partially enclosing the worm when in mesh. To provide more contact, the worm may have an hourglass shape, in which case, the set is referred to as *double enveloping*. That is, with the helical gear cut concavely, the double-enveloping type also has the worm length cut concavely: Both the worm and the gear partially enclose each other. The geometry of the worm is very complicated, and reference should be made to the literature for details.

The *terminology* used to describe the worm (Figure 12.16a) and power screws (see Section 15.3) is very similar. In general, the worm is analogous to a screw thread and the worm gear is similar to its nut. The *axial pitch* of the worm gear p_w is the distance between corresponding points on adjacent teeth. The *lead L* is the axial distance the worm gear (nut) advances during one revolution of the worm. In a *multiple*-thread worm, the lead is found by multiplying the number of threads (or teeth) by the axial pitch.

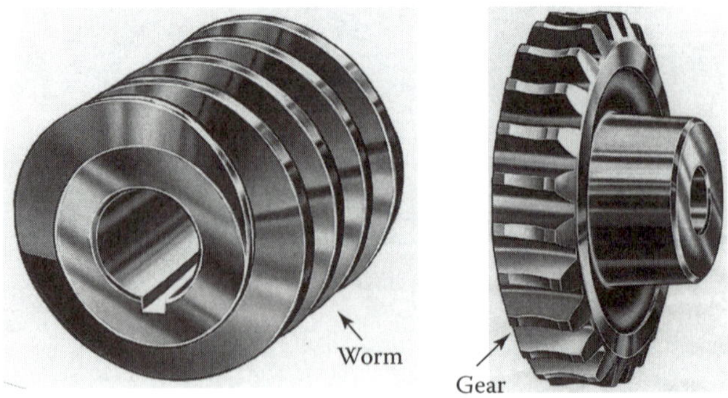

FIGURE 12.15
A single-enveloping wormset. (Courtesy of Martin Sprocket and Gear Co., Arlington, TX.)

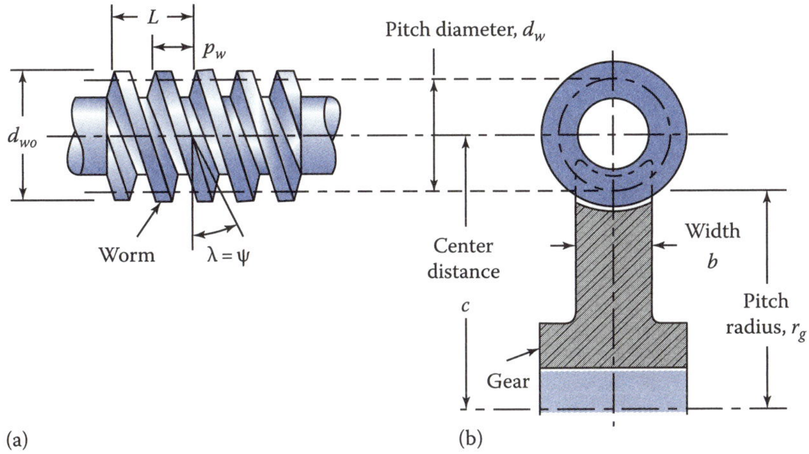

FIGURE 12.16
Notation for a worm gearset: (a) double-threaded worm; (b) worm gear (shown in a half-section view).

The pitch diameter of a worm d_w is not a function of its number of threads N_w. The speed ratio of a wormset is obtained by the ratio of gear teeth to worm threads:

$$r_s = \frac{\omega_g}{\omega_w} = \frac{N_w}{N_g} = \frac{L}{\pi d_g} \tag{12.18}$$

As in the case of a spur or helical gear, the pitch diameter of a worm gear is related to its circular pitch and number of teeth using Equation 12.1:

$$d_g = \frac{N_g p}{\pi} \tag{12.19}$$

The center distance between the two shafts equals $c = (d_w + d_g)/2$, as shown in the figure.

The *lead angle* of the worm (which corresponds to the screw lead angle) is the angle between a tangent to the helix angle and the plane of rotation. The lead and the lead angle of the worm have the following relationships:

$$L = p_w N_w \tag{12.20}$$

$$\tan\lambda = \frac{L}{\pi d_w} = \frac{V_g}{V_w} \tag{12.21}$$

where
L = the lead
P_w = the axial pitch
N_w = the number of threads
λ = the lead angle
d_w – the pitch diameter
V = the pitch line velocity

TABLE 12.1

Various Normal Pressure Angles for Wormsets

Pressure Angle, ϕ_n (°)	Maximum Lead Angle, λ (°)	Lewis Form Factor, Y
$14\frac{1}{2}$	15	0.314
20	25	0.392
25	35	0.470
30	45	0.550

For a 90° *shaft angle* (Figure 12.16), the lead angle of the worm and helix angle of the gear are equal:

$$\lambda = \psi$$

Note that λ and ψ are measured on the pitch surfaces.

We show in Section 15.3 that the normal pressure angle ϕ_n of the worm corresponds to the thread angle α_n of a screw. Normal pressure angles are related to the lead angle and the Lewis form factor Y, as shown in Table 12.1.

In conclusion, we point out that worm gears usually contain no less than 24 teeth, and the number of gear teeth plus worm threads should be more than 40. The face width b of the gear should not exceed half of the worm outside diameter d_{wo}. AGMA recommends the magnitude of the minimum and maximum values for *worm pitch diameter* d_W as follows:

$$\frac{c^{0.875}}{3} \leq d_W \leq \frac{c^{0.875}}{1.6} \tag{a}$$

Here, c represents the distance between the centers of the worm and the gear.

Example 12.5: Geometric Quantities of a Worm

A triple-threaded worm has a lead L of 75 mm. The gear has 48 teeth and is cut with a hob of modulus $m_n = 7$ mm perpendicular to the teeth. Calculate

a. The speed ratio r_s
b. The center distance c between the shafts if they are 90° apart

Solution

For a 90° shaft angle, we have $\lambda = \psi$.

a. The velocity ratio of the worm gearset is

$$r_s = \frac{N_w}{N_g} = \frac{3}{48} = \frac{1}{16}$$

b. Using Equation 12.20,

$$p_w = \frac{L}{N_w} = \frac{75}{3} = 25 \text{ mm}$$

From Equation 12.2 with $m_n = 1/p_n$, we obtain

$$p_n = \pi m_n = 7\pi = 21.99 \text{ mm}$$

Equation 12.1 results in

$$\cos\lambda = \frac{p_n}{p_w} = \frac{21.99}{25} = 0.88 \quad \text{or} \quad \lambda = 28.4°$$

Application of Equation 12.21 gives

$$d_w = \frac{L}{\pi \tan\lambda} = \frac{75}{\pi \tan 28.4°} = 44.15 \text{ mm}$$

Through the use of Equation 12.18,

$$d_g = \frac{L}{\pi r_s} = \frac{75}{\pi/16} = 381.97 \text{ mm}$$

We then have

$$c = \frac{1}{2}(d_w + d_g) = \frac{1}{2}(44.15 + 381.97) = 213.1 \text{ mm}$$

12.10 Worm Gear Bending and Wear Strengths

The approximate bending and wear strengths of worm gearsets can be obtained by equations analogous to those used for spur gears. Nevertheless, adjustments are made to account for the effects of the normal pressure angle ϕ_n and lead angle λ of the worm. The fundamental formulas for the allowable bending and wear loads for the gear teeth are as follows.

12.10.1 Lewis Equation

The bending stresses are much higher in the gear than in the worm. The following slightly *modified Lewis equation* is therefore applied to worm gear:

$$F_b = \frac{\sigma_o b}{K_f} \frac{Y}{P_n} \qquad \text{(11.33, modified)}$$

The value of Y can be taken from Table 12.1. It is to be noted that the normal diametral pitch P_n is defined by Equation 12.2.

TABLE 12.2

Worm Gear Factors K_w

Material		K_w (psi)		
Worm	**Gear**	$\lambda < 10°$	$\lambda < 25°$	$\lambda > 25°$
Steel, 250 Bhn	Bronze[a]	60	75	90
Steel, 500 Bhn	Bronze[a]	80	100	120
Hardened steel	Chilled bronze	120	150	180
Cast iron	Bronze[a]	150	185	225

[a] Sand cast.

12.10.2 Limit Load for Wear

The *wear equation* by Buckingham, frequently used for rough estimates, has the following form:

$$F_w = d_g b K_w \tag{12.22}$$

where

F_w = the allowable wear load do
d_g = the pitch diameter
b = the face width of the gear
K_w = a material and geometry factor, obtained from Table 12.2

For a satisfactory worm gearset, the usual *requirement* is that $F_b \geq F_d$ and $F_w \geq F_d$. The dynamic load F_d acting on worm gears can be approximated by

$$F_d = \frac{1200 + V}{1200} F_t \quad (\text{for } 0 < V > 4000 \text{ fpm}) \qquad \text{(11.24b, modified)}$$

As earlier, the pitch-line velocity V is in fpm in this formula.

12.10.3 AGMA Equations

The design of worm gearsets is more complicated and dissimilar to that of other gearing. The AGMA prescribes an input power rating formula for wormsets. This permits the worm gear dimensions to be obtained for a given power or torque–speed combination [5]. A wider variation in procedures is employed for estimating bending and surface strengths. Moreover, worm gear capacity is frequently limited not by fatigue strength but heat-dissipation or cooling capacity. The latter is discussed in the next section.

12.11 Thermal Capacity of Worm Gearsets

The power capacity of a wormset in continuous operation is often limited by the *heat-dissipation capacity* of the housing or casing. Lubricant temperature commonly should not exceed about 93°C (200°F). The basic relationship between temperature rise and heat dissipation can be expressed as follows:

$$H = CA\Delta t \tag{12.23}$$

where

H = the time rate of heat dissipation, lb · ft/min
C = the heat transfer, or cooling rate, coefficient (lb · ft per minute per square foot of housing surface area per °F)
A = the housing external surface area, ft^2
Δt = the temperature difference between oil and ambient air, °F

The values of A for a conventional housing, as recommended by AGMA, may be estimated [9] by the following formula:

$$A = 0.3c^{1.7} \tag{12.24}$$

Here, A is in square feet and c represents the distance between shafts (in inches). The approximate values of heat transfer rate C can be obtained from Figure 12.17. Note from the figure that C is greater at high velocities of the worm shaft, which causes a better circulation of the oil within the housing.

The manufacturer usually provides the means for cooling, such as external fins on housing and a fan installed on worm shaft, as shown in Figure 12.18. It is observed from the figure that the worm gear has spiral teeth and shaft at right angle to the worm shaft. Clearly, an extensive sump and corresponding large quantity of oil help increase heat

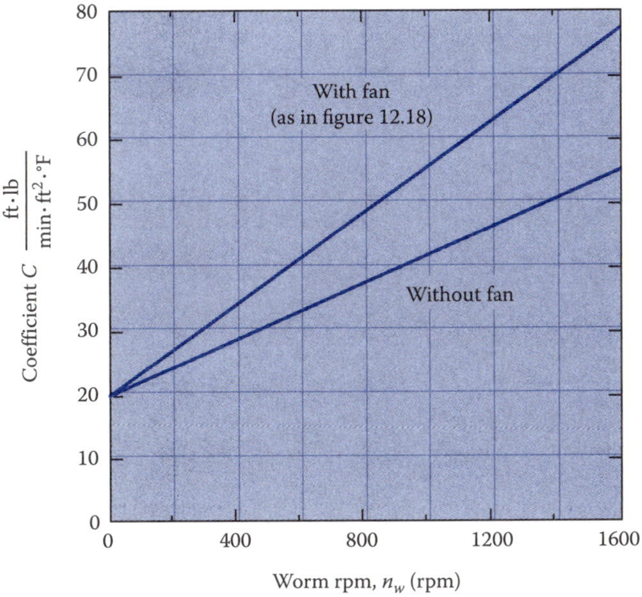

FIGURE 12.17
Heat transfer coefficient C for worm gear housing.

FIGURE 12.18
Worm gear speed reducer. (Courtesy of Cleveland Gear Company, Cleveland, OH.)

transfer, particularly during overloads. In some warm gear reduction units, oil in the sump may be externally circulated for cooling.

The heat-dissipation capacity H of the housing, as determined by Equation 12.23, in terms of horsepower is given in the following form:

$$(\mathrm{hp})_d = \frac{CA\Delta t}{33,000} \tag{12.25}$$

This loss of horsepower equals the difference between the input horsepower hp_i and output horsepower hp_o. Inasmuch as $e = \mathrm{hp}_o/\mathrm{hp}_i$, we have $\mathrm{hp}_d = \mathrm{hp}_i - e(\mathrm{hp}_o)$. The input horsepower capacity is therefore

$$\mathrm{hp}_i = \frac{\mathrm{hp}_d}{1-e} \tag{12.26}$$

The quantity e represents the efficiency.

12.11.1 Worm Gear Efficiency

The expression of the *efficiency* e for a worm gear reduction is the same as that used for a power screw and nut, developed in Section 15.4. In the notation of this chapter, Equation 15.13 is written as follows:

$$e = \frac{\cos\phi_n - f\tan\lambda}{\cos\phi_n + f\cot\lambda} \tag{12.27}$$

where

f is the coefficient of friction

ϕ_n represents the normal pressure angle

The value of f depends on the velocity of sliding V_s between the teeth:

$$V_s = \frac{V_w}{\cos\lambda} \quad (12.28)$$

The quantity V_s is the pitch-line velocity of the worm. Table 12.3 furnishes the values of the coefficient of friction.

Example 12.6: Design Analysis of a Worm Gear Speed Reducer

A worm gearset and its associated geometric quantities are schematically shown in Figure 12.19. Estimate

a. The heat-dissipation capacity
b. The efficiency
c. The input and output horsepower

Assumptions: The gearset is designed for continuous operation based on a limiting 100°F temperature rise of the housing without fan.

Solution

The speed ratio of the worm gearset is

$$r_s = \frac{N_w}{N_g} = \frac{4}{60} = \frac{1}{15}$$

a. Through the use of Equation 12.24,

$$A = 0.3c^{1.7} = 0.3(8)^{1.7} = 10.29\ \text{ft}^2$$

From Figure 12.17, we have

$$C = 42\ \text{lb}\cdot\text{ft}/(\text{min}\cdot\text{ft}^2\cdot{}^\circ\text{F})$$

Carrying the data into Equation 12.25,

$$\text{hp}_d = \frac{CA\Delta t}{33{,}000} = \frac{42(10.29)(100)}{33{,}000} = 1.31$$

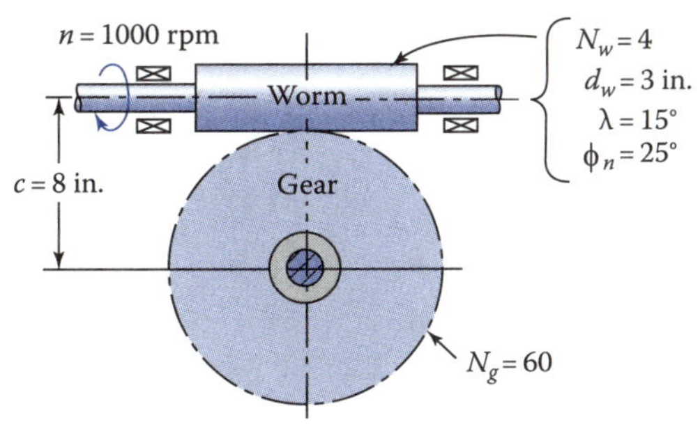

FIGURE 12.19
Example 12.6.

b. The pitch-line velocity of the worm is

$$V_w = \frac{\pi d_w n_w}{12} = \frac{\pi(3)(1000)}{12} = 785.4 \text{ fpm}$$

Applying Equation 12.28,

$$V_s = \frac{V_w}{\cos\lambda} = \frac{785.4}{\cos 15°} = 813 \text{ fpm}$$

By Table 12.3, $f = 0.0238$. Introducing the numerical values into Equation 12.27,

$$e = \frac{\cos 25° - 0.0238(\tan 15°)}{\cos 25° + 0.0238(\cot 15°)} = 0.904 \quad \text{or} \quad 90.4\%$$

c. Using Equation 12.26, the input horsepower is equal to

$$\text{hp}_i = \frac{\text{hp}_d}{1-e} = \frac{1.31}{1-0.904} = 13.65$$

The output horsepower is then

$$\text{hp}_o = 13.65 - 1.31 = 12.3$$

Comments: Because of the sliding friction inherent in the tooth action, usually worm gearsets have significantly lower efficiencies than those of spur gear drives. The latter can have efficiencies as high as 98% (Section 11.1).

TABLE 12.3

Worm Gear Coefficient of Friction f for Various Sliding Velocities V_s

V_s (fpm)	f	V_s (fpm)	f	V_s (fpm)	f
0	0.150	120	0.0519	1200	0.0200
1	0.115	140	0.0498	1400	0.0186
2	0.110	160	0.0477	1600	0.0175
5	0.099	180	0.0456	1800	0.0167
10	0.090	200	0.0435	2000	0.0160
20	0.080	250	0.0400	2200	0.0154
30	0.073	300	0.0365	2400	0.0149
40	0.0691	400	0.0327	2600	0.0146
50	0.0654	500	0.0295	2800	0.0143
60	0.0620	600	0.0274	3000	0.0140
70	0.0600	700	0.0255	4000	0.0131
80	0.0580	800	0.0240	5000	0.0126
90	0.0560	900	0.0227	6000	0.0122
100	0.0540	1000	0.0217	—	—

Source: From ANSI/AGMA Standard 6034-A87.

Problems

Sections 12.1 through 12.5

12.1 A helical gearset consists of a 20-tooth pinion rotating in a counterclockwise direction and driving a 40-tooth gear. Determine

a. The normal, transverse, and axial circular pitches

b. The diametral pitch and pressure angle

c. The pitch diameter of each gear

d. The directions of the thrusts and show these on a sketch of the gearset

Given: The pinion has a right-handed helix angle of 30°, a normal pressure angle of 25°, and a normal diametral pitch of 6 teeth/in.

12.2 Redo Problem 12.1 for a helical gearset that consists of an 18-tooth pinion having a ψ of 20° LH, a ϕ_n of 14½°, a P_n of 8 in.$^{-1}$, and driving a 55-tooth gear.

12.3 A helical gear has a 14½° normal pressure angle, a 40° helix angle, a circular diametral pitch of 8, and 30 teeth. Determine

a. The pitch diameter and the circular, the normal, and the axial pitches

b. The normal diametral pitch and the pressure angle

12.4 A helical gearset has a normal circular pitch of 0.625 in. The gear center distance is 10 in., the speed ratio equals ¼, and the pinion has 18 teeth. What is the required helix angle?

12.5 A 32-tooth gear has a pitch diameter of 260 mm, a normal module of m_n = 6 mm, and a normal pressure angle of 20°. Calculate the kW transmitted at 800 rpm.

Given: The force normal to the tooth surface is 10 N.

12.6 A 35-tooth helical gear has a helix angle of ψ = 30° and a pressure angle of ϕ = 20°. Determine

a. The pressure angle in the normal plane and the equivalent number of teeth

b. The pressure angle and teeth number on an equivalent strength spur gear

12.7 Determine the center distance for a helical gearset with a normal circular pitch of 14 mm. The helix angle is 15°, the speed ratio equals ⅓, and the pinion has 40 teeth.

12.8 A 35-tooth helical gear with ψ = 22° has a pressure angle of ϕ = 20°. Compute

a. The pressure angle in the normal plane and the equivalent number of teeth.

b. The pressure angle and teeth number on an equivalent strength spur gear.

12.9 A helical gearset has 1.5 mm module and a pressure angle of 20°. The width of gears is 40 mm, the pinion has 20 teeth, and gear has 120 teeth. Determine

a. The total contact ratio for the gears

b. The helix angle for the case in which the total contact ratio for gears equals 4.0

12.10 A left-handed helical pinion has a ϕ_n of 20°, a P_n of 10 in.$^{-1}$, a ψ of 45°, a N_p of 32, and an F_n of 90 lb and runs at 2400 rpm in the counterclockwise direction. The driven gear has 60 teeth. Determine and show on a sketch

a. The tangential, axial, and radial forces acting on each gear

b. The torque acting on the shaft of each gear

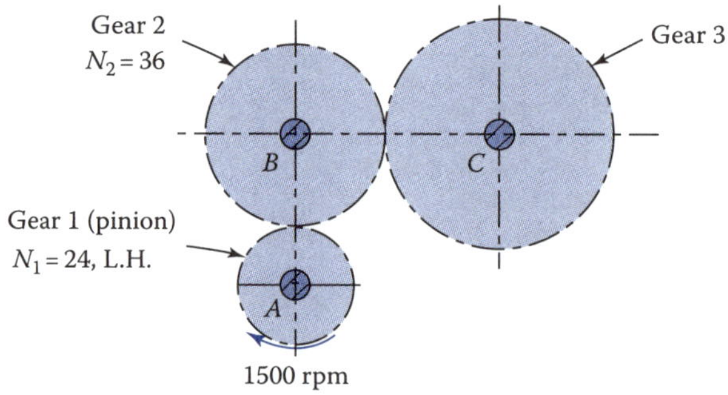

FIGURE P12.11

12.11 The helical gears depicted in Figure P12.11 have a normal diametral pitch of 4 teeth/in., a 25° pressure angle, and a helix angle of 20°. Calculate and show on a sketch

a. The tangential, radial, and axial forces acting on each gear

b. The torque acting on each shaft

Given: Gear 1 transmits 20 hp at 1500 rpm through the idler to gear 3 on shaft *C*; the speed ratio for gears 3–1 is to be ½.

12.12 A 22-tooth helical pinion has a normal pressure angle of 20°, a normal diametral pitch of 8 teeth/in., a face width of 2 in., and a helix angle of 25°; it rotates at 1800 rpm and transmits 30 hp to a 40-tooth gear. Determine the factor of safety based on bending strength, employing the Lewis equation.

Given: Fatigue stress-concentration factor $K_f = 1.5$.

Assumption: The pinion and gear are both steel, hardened to 250 Bhn.

12.13 A double-reduction helical gear train has in the normal plane 4 and 5.5 mm moduli for the high- and low-speed gears, respectively (Figure P12.13). The helix angles are different for these gearsets. Find

a. The total speed reduction produced by the four gears and the helix angle of the low-speed gears

b. The helix angle, for the case in which the low-speed gears are replaced by 20- and 32-tooth gears of the same modulus

Given: The helix angle of the high-speed gears is 29.8°, $N_1 = 30$, $N_2 = 75$, $N_3 = 25$, and $N_4 = 50$.

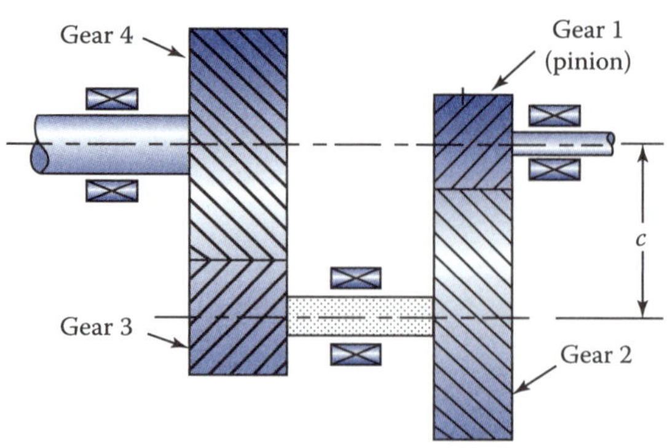

FIGURE P12.13

12.14 A 2 hp electric motor runs at 2400 rpm and drives a machine by means of a 25° normal pressure angle helical gearset with a normal module of $m_n = 4$ mm (see Figure 12.7). The helix angle is equal to 30°, and the pinion has 22 teeth. Find

a. The pressure angle and pitch diameter of the pinion

b. The pinion velocity V and transmitted load F_t from the motor

c. The radial, axial, normal, and normal loads

12.15 Resolve Problem 12.12 based on wear strength and using the Buckingham equation.

12.16 Two meshing helical gears are both made of SAE 1020 (WQ&T) steel, hardened to 150 Bhn, and are mounted on (parallel) shafts. Calculate the horsepower capacity of the gearset.

Requirement: Use the Lewis equation for bending strength and the Buckingham equation for wear strength.

Given: The number of teeth is 30 and 65, $\phi_n = 25°$, $\psi = 35°$, $P_n = 6$ in.$^{-1}$, and $b = 1.5$ in.; the pinion rotates at 2400 rpm.

12.17 Two meshing helical gears are made of SAE 1040 steel hardened to about 200 Bhn. and are mounted on parallel shafts 6 in. apart. Determine the horsepower capacity of the gearset

a. Applying the Lewis equation and $K_f = 1.4$ for bending strength and the Buckingham equation for wear strength

b. Applying the AGMA method on the basis of strength only

Given: The gears are to have a speed ratio of ⅓. A ϕ_n of 20°, a ψ of 30°, a P of 15 in.$^{-1}$, and a b of 2.5 in.; the pinion rotates at 900 rpm.

Design Assumptions:

1. The mounting is accurate.
2. Reliability is 90%.
3. The gearset has indefinite life.
4. Light shock loading acts on the pinion and uniform shock loading on the driven gear.
5. The pinion and gear are both high-precision ground.

12.W Review the website at www.bisonger.com and select a ¼ horsepower motor for

a. A single-speed reduction unit

b. A double-speed reduction unit

Sections 12.6 through 12.8

12.18 A 20° pressure angle straight bevel pinion having 20 teeth and a diametral pitch of 8 teeth/in. drives a 42-tooth gear. Determine

a. The pitch diameters

b. The pitch angles

c. The face width

d. The clearance

12.19 A pair of bevel gears is to transmit 15 hp at 500 rpm with a speed ratio of ½. The 20° pressure angle pinion has an 8 in. back cone pitch diameter, 2.5 in. face width, and a diametral pitch of 7 teeth/in. Calculate and show on a sketch the axial and radial forces acting on each gear.

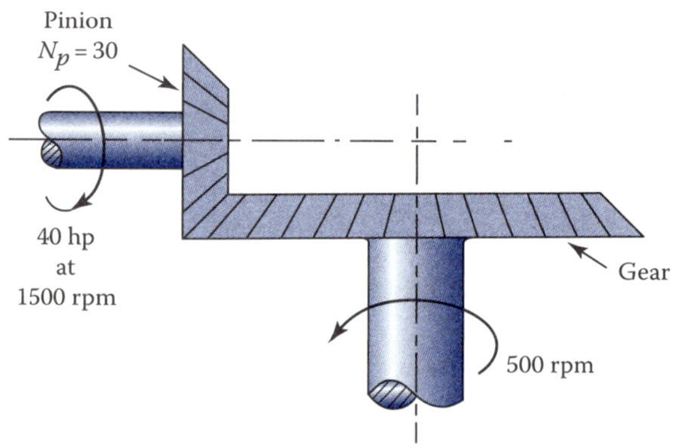

FIGURE P12.23

12.20 If the gears in Problem 12.19 are made of SAE 1020 steel (WQ&T), will they be satisfactory from a bending viewpoint? Employ the Lewis equation and $K_f = 1.4$.

12.21 If the pinion and gear in Problem 12.19 are made of steel (200 Bhn) and phospor bronze, respectively, will they be satisfactory from wear strength viewpoint? Use the Buckingham equation.

12.22 A pair of 20° pressure angle bevel gears of $N_1 = 30$ and $N_2 = 60$ has a diametral pitch P of 3 teeth/in. at the outside diameter. Calculate the horsepower capacity of the pair, based on the Lewis and Buckingham equations.

Given: Width of face b is 2.8 in., $K_f = 1.5$, and the pinion runs at 720 rpm.

Design Assumption: The gears are made of steel and hardened to about 200 Bhn.

12.23 A bevel gearset transmits 40 hp at 1500 rpm of 30-tooth pinion as shown in Figure P12.23. The gear speed is 500 rpm, face width equals 45 mm, $m = 4$ mm, and $\phi = 20°$. Determine

a. The pinion velocity V and transmitted load F_t

b. The axial and radial pinion forces and torques of pinion and gear shafts

12.24 A pair of 20° pressure angle bevel gears of $N_1 = 30$ and $N_2 = 60$ has a module m of 8.5 mm at the outside diameter. Determine the power capacity of the pair, using the Lewis and Buckingham equations.

Given: Face width b is 70 mm, $K_f = 1.5$, and the pinion rotates at 720 rpm.

Design Assumption: The gears are made of steel and hardened to about 200 Bhn.

Sections 12.9 through 12.11

12.25 Two shafts at right angles, with center distance of 6 in., are to be connected by worm gearset. Determine the pitch diameter, lead, and number of teeth of the worm.

Given: A speed ratio of 0.025, lead angle of worm of 35°, and a normal pitch of ⅜ in. for the worm gear

12.26 A double-threaded, 3 in. diameter worm has an input of 40 hp at 1800 rpm. The worm gear has 80 teeth and is 10 in. in diameter. Calculate the tangential force on the gear teeth if the efficiency is 90%.

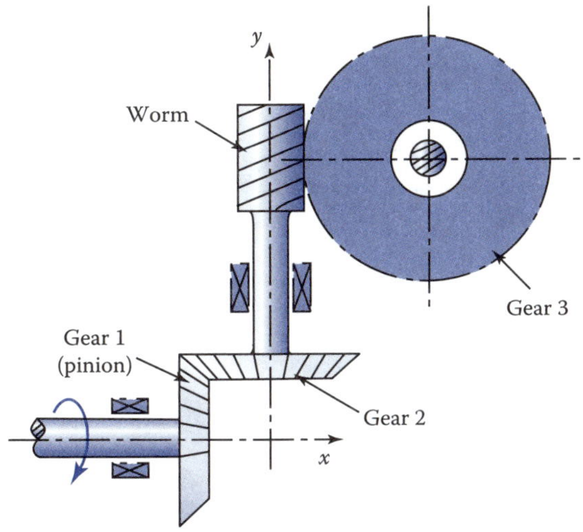

FIGURE P12.27

12.27 A bevel and a worm gearsets and their corresponding geometric properties are illustrated in Figure P12.27. Gear 1 rotates clockwise with a speed of k_1. Determine the speed and direction of rotation of the worm gear.

Given: $N_1 = 20$, $N_2 = 20$, $N_W = 5$, $N_3 = 50$, $n_1 = 250$ rpm

12.28 A worm gearset is schematically illustrated in Figure P12.28. Find

a. The gear ratio, gear diameter, and lead of worm
b. The helix angle and center distance of gears

Given: $N_W = 2$, $N_g = 40$, $P = 8$

Assumption: The worm diameter will be $d_W = 3.5p$, where p is the circular pitch of the gear.

12.29 A worm gearset consists of gear with 45 teeth and a double-threaded worm (Figure P12.28). Compute

a. The worm diameter d_W, gear pitch diameter d_g, and lead angle λ.
b. Limit load for wear through F_W, according to the Buckingham formula

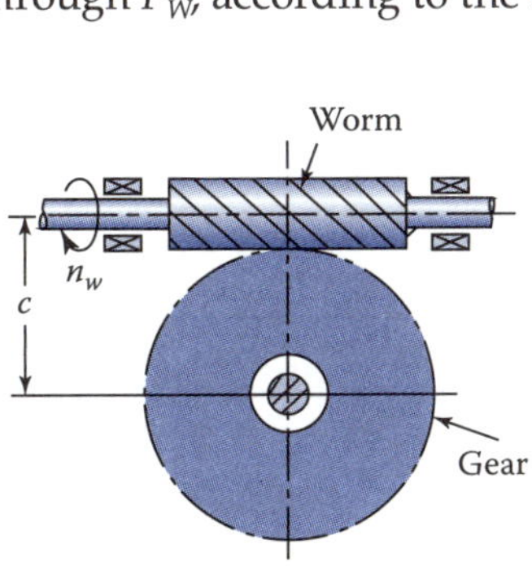

FIGURE P12.28

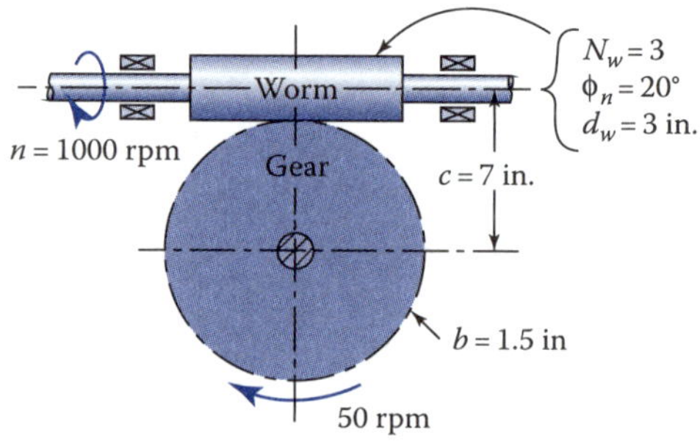

FIGURE P12.31

Given: $b = 25$ mm, $c = 210$ mm, $N_g = 65$, $N_W = 3$, $m = 6$ mm, $n_W = 1200$ rpm

Assumptions: The worm diameter will be $d_W = c^{0.875}/2$ (see Section 12.9). Gear and worm are made of bronze and cast iron, respectively.

12.30 A quadruple-threaded worm having 2.5 in. diameter meshes with a worm gear with a diametral pitch of 6 teeth/in. and 90 teeth. Find

a. The lead

b. The lead angle

c. The center distance

12.31 A 10 hp, 1000 rpm electric motor drives a 50 rpm machine through a worm gear reducer with a center distance of 7 in. (Figure P12.31). Determine

a. The value of the helix angle

b. The transmitted load

c. The power delivered to the driven machine

12.32 If the gears of Figure P12.31 are made of cast steel (WQ&T), will they be satisfactory from the bending strength viewpoint? Use the modified Lewis equation and $K_f = 1.4$.

12.33 If the worm and gear of Figure P12.31 are made of cast iron and bronze, respectively, will they be satisfactory from the wear viewpoint? Employ Equation 12.22 by Buckingham.

12.34 The worm gear reducer of Figure P12.31 is to be designed for continuous operation and a limiting 100°F temperature rise of the housing with a fan. Estimate the heat-dissipation capacity. Will overheating be a problem?

13

Belts, Chains, Clutches, and Brakes

13.1 Introduction

In contrast with bearings, friction is a useful and essential agent in belts, clutches, and brakes. Frictional forces are commonly developed on flat or cylindrical surfaces in contact with shorter pads or linkages or longer bands or belts. A number of these combinations are employed for brakes and clutches, and the band (chain) and wheel pair is used in belt (chain) drives as well. Hence, only a few different analyses are required, with surface forms affecting the equations more than the functions of the elements. Also, common operating problems relate to pressure distribution and wear, temperature rise and heat dissipation, and so on. The foregoing devices are thus effectively analyzed and studied together.

A *belt* or *chain drive* provides a convenient means for transferring motion from one shaft to another by means of a belt or chain connecting pulleys on the shafts. Part A of this chapter is devoted to the discussion of the flexible elements: belts and chains. In many cases, their use reduces vibration and shock transmission, simplifies the design of a machine substantially, and reduces the cost. Power is to be transferred between parallel or nonparallel shafts separated by a considerable distance. Thus, the designer is provided considerable flexibility in location of driver and driven machinery [1–11]. The websites www.machinedesign.com and www.powertransmission.com on mechanical systems include information on belts and chains as well as on clutches and brakes.

Brakes and clutches are essentially the same devices. Each is usually associated with rotation. The *brake* absorbs the kinetic energy of moving bodies and thus controls the speed. The function of the brake is to turn mechanical energy into heat. The *clutch* transmits power between two shafts or elements that must be often connected and disconnected. A brake acts likewise with the exception that one element is fixed. Clutches and brakes, treated in Part B of the chapter, are all of the friction type that relies on sliding between solid surfaces. Other kinds provide a magnetic, hydraulic, or mechanical connection between the two parts. The clutch is in common use to maintain constant torque on a shaft and serve as an emergency disconnection device to decouple the shaft from the motor in the event of a machine jam. In such cases, a brake also is fitted to bring the shaft (and machine) to a rapid stop in urgency. Brakes and clutches are used extensively in production machines of all types as well as in vehicle applications. They are *classified* as follows: disk or axial types, cone types, drum types with external shoes, drum types with internal shoes, and miscellaneous types [12–15].

Part A: Flexible Elements

In addition to gears (Chapters 11 and 12), belts and chains are in widespread use. Belts are frequently necessary to reduce the higher relative speeds of electric motors to the lower values required by mechanical equipment. Chains also can be employed for large reduction in speed if required. Belt drives are relatively quieter than chain drives, while the latter have greater life expectancy. However, neither belts nor chains have an infinite life and should be replaced at the first sign of wear or lack of elasticity.

13.2 Belts

There are four *main belt types*: flat, round, V, and timing. Flat and round belts may be used for long center distances between the pulleys in a belt drive. On the other hand, V and timing belts are employed for limited shorter center distances. Excluding timing belts, there is some slip and creep between the belt and the pulley, which is usually made of cast iron or formed steel. Characteristics of the principal belt types are furnished in Table 13.1. Catalogs of various manufacturers of the belts contain much practical information.

13.2.1 Flat and Round Belts

Flat belts and round belts are made of urethane or rubber-impregnated fabric reinforced with steel or nylon cords to take the tension load. One or both surfaces may have friction surface coating. Flat belts find considerable use in applications requiring small pulley diameters. Most often, both driver and driven pulleys lie in the same vertical plane. Flat belts are quiet and efficient at high speeds, and they can transmit large amounts of power. However, a flat belt must operate with higher tension to transmit the same torque as a V belt. *Crowned pulleys* are used for flat belts. Round belts run in grooved pulleys or *sheaves*. Deep-groove pulleys are employed for the drives that transmit power between horizontal and vertical shafts, or so-called quarter-turn drives, and for relatively long center distances.

TABLE 13.1

Characteristics of Some Common Belts

Belt Type	Geometric Form	Size Range	Center Distance between Pulleys
Flat	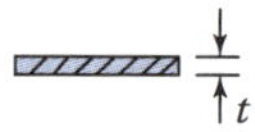	t = 0.75–5 mm	No upper limit
Round	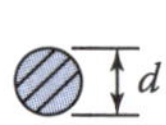	d = 3–19 mm	No upper limit
V	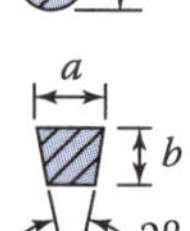	a = 13–38 mm b = 8–23 mm 2β = 34°–40°	Limited
Timing	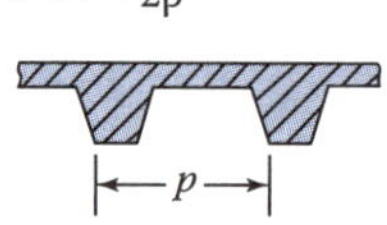	p = 2 mm and up	Limited

13.2.2 V Belts

A V belt is a rubber covered with impregnated fabric and reinforced with nylon, Dacron, rayon, glass, or steel tensile cords. V belts are the most common means of transmitting power between electric motors and driven machinery. They are also used in other household, automotive, and industrial applications. Usually, V-belt speed should be in the range of about 4000 fpm. These belts are produced in two series: the *standard* V belt, as shown in Figure 13.1, and the high-capacity V belt. Note that each standard section is designated by a letter for sizes in inch dimensions. Metric sizes are identified by numbers. V belts are slightly less efficient than flat belts. The *included angle* 2β for V belts, defined in the table, is usually from 34° to 40°.

Crowned pulleys and sheaves are also employed for V belts. The *wedging action* of the belt in the groove leads to a large increase in the tractive force produced by the belt, as discussed in Section 13.4. *Variable-pitch pulleys* permit an adjustment in the width of the groove. The effective pitch diameter of the pulley is thus varied. These pulleys are employed to change the input to output speed ratio of a V-belt drive. Some variable-pitch drives can change speed ratios when the belt is transmitting power. As for the number of V belts, as many as 12 or more can be used on a single sheave, making it a *multiple drive.* All belts in such a drive should stretch at the same rate to keep the load equally divided among them. A multiple V-belt drive (Figure 13.2) is used to satisfy high-power transmission requirements.

13.2.3 Timing Belts

A timing belt is made of rubberized fabric and steel wire and has evenly spaced teeth on the inside circumference. Also known as a *toothed* or *synchronous belt,* a timing belt does not stretch or slip and hence transmits power at a *constant* angular velocity ratio. This permits timing belts to be employed for many applications requiring precise speed ratio, such as driving an engine camshaft from the crankshaft. Toothed belts also allow the use of small pulleys and small arcs of contact. They are relatively lightweight and can efficiently operate at speeds up to at least 16,000 fpm. A timing belt fits into the grooves cut on the periphery of the wheels or *sprockets,* as shown in Figure 13.3. The sprockets come in sizes from 0.60 in. diameter to 35.8 in. and with teeth numbers ranging from 10 to 120. The *efficiency* of a toothed belt drive ranges from about 97% to 99%.

Figure 13.4 illustrates a portion of the timing-belt drive. The teeth are coated with nylon fabric. The tension member, usually steel wire, of a timing belt is positioned at the belt

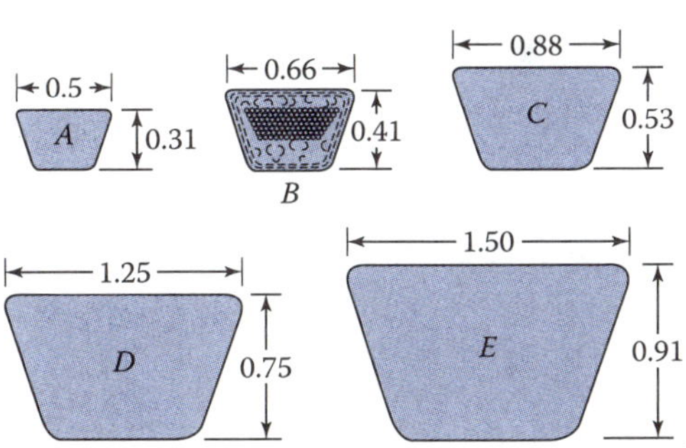

FIGURE 13.1
Standard cross sections of V belts. All dimensions are in inches.

FIGURE 13.2
Multiple V-belt drive. (Courtesy of T. B. Wood's Incorporated, Chambersburg, PA.)

FIGURE 13.3
Toothed or timing-belt drive for precise speed ratio. (Courtesy of T. B. Wood's Incorporated, Chambersburg, PA.)

pitch line. The pitch length therefore is the same regardless of the backing thickness. Note that, as in the case of gears, the circular pitch p is the distance, measured on the pitch circle, from a point on the tooth to a corresponding point on an adjacent tooth. Since timing belts are toothed, they provide *advantages* over ordinary belts; for example, for a timing belt, no initial tension is necessary and a fixed center drive may be used at any slow or fast speed. The *disadvantages* are the cost of the belt, the necessity of grooving the sprocket, and the dynamic fluctuations generated at the belt-tooth meshing frequency.

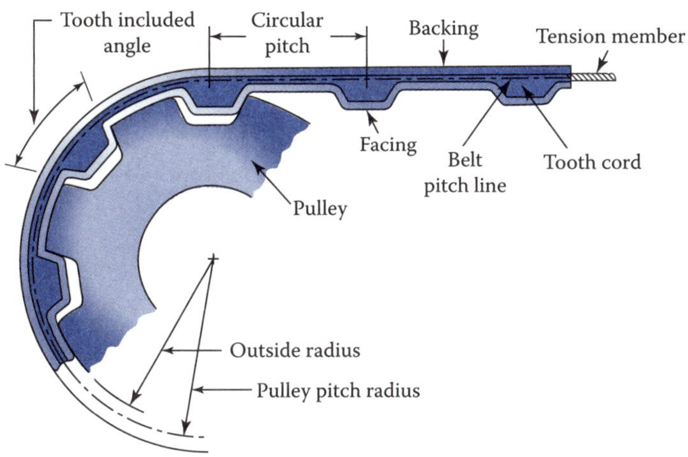

FIGURE 13.4
Timing-belt drive nomenclature.

13.3 Belt Drives

As noted previously, a belt drive transfers power from one shaft to another by using a belt and connecting pulleys on the shafts. Flat-belt drives produce very little noise and absorb more torsional vibration from the system than either V belt or other drives. A flat-belt drive has an efficiency of around 98%, which is nearly the same as for a gear drive. A V-belt drive can transmit more power than a flat-belt drive, as will be shown in the next section. However, the efficiency of a V-belt drive varies between 70% and 96% [1,2].

We present a conventional analysis that has long been used for the belt drives. Note that a number of theories describe the mechanics of the belt drives in more detailed mathematical form [3]. Figure 13.5a illustrates the usual belt drive, where the belt tension is such that the sag is visible when the belt is running. The friction force on the belt is taken to be uniform throughout the entire arc of contact. Due to friction of the rotation of the driver pulley, the *tight-side* tension is greater than the *slack-side* tension. Referring to the figure, the following basic relationships may be developed.

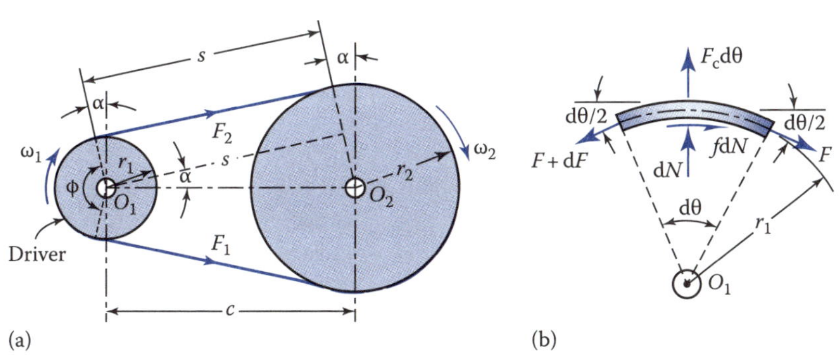

FIGURE 13.5
Belt drive: (a) forces in moving belt; (b) belt element on the verge of slipping on the small pulley.

13.3.1 Transmitted Power

For belt drives, the torque on a pulley is given as follows:

$$T = (F_1 - F_2)r \tag{13.1}$$

in which
- r = the pitch radius
- F_1 = the tension on the tight side
- F_2 = the tension on the slack side

We have $F_1 > F_2$. Note that the *pitch radius* is approximately measured from the center of the pulley to the neutral axis of the belt. The required *initial tension* F_i depends on the elastic characteristics of the belt. However, it is usually satisfactory to take

$$F_i = \frac{1}{2}(F_1 + F_2) \tag{13.2}$$

The transmitted power, through the use of Equation 1.17, is

$$\text{hp} = \frac{(F_1 - F_2)V}{33{,}000} = \frac{Tn}{63{,}000} \tag{13.3}$$

where

$$V = \frac{\pi dn}{12} \tag{13.4}$$

In the foregoing, we have
- T = the torque, lb · in.
- V = the belt velocity, fpm
- n = the speed of the pulley, rpm
- d = the pitch diameter of the pulley, in.

The speed ratio of the belt drive is given by

$$\frac{n_1}{n_2} = \frac{r_2}{r_1} \tag{13.5}$$

The numbers 1 and 2 refer to input and output, or small and large, pulleys, respectively.

13.3.2 Contact Angle

From the geometry of the drive (Figure 13.5a), angle a is found to be

$$\sin\alpha = \frac{r_2 - r_1}{c} \tag{13.6}$$

The *contact angle* on the small pulley ϕ or the so-called angle of wrap is therefore

$$\phi = \pi - 2\alpha \tag{13.7}$$

where

r_1 = the pitch radius of the small pulley
r_2 = the pitch radius of the large pulley
c = the center distance

The capacity of the belt drive is determined by the value of ϕ. This angle is particularly critical with pulleys of greatly differing size and shorter center distances.

13.3.3 Belt Length and Center Distance

The wrap angles on the small and large pulleys are $\pi - 2\alpha$ and $\pi + 2\alpha$, respectively. The distance between the beginning and end of contact, or span, $s = [c^2 - (r_2 - r_1)^2]^{1/2}$. The pitch length of the belt is obtained by the summation of the two arc lengths, $r_1(\pi - 2\alpha) + r_2(\pi + 2\alpha)$, with twice the span, $2s$. In so doing, we have

$$L = 2\left[c^2 - (r_2 - r_1)^2\right]^{1/2} + r_1(\pi - 2\alpha) + r_2(\pi + 2\alpha) \tag{13.8}$$

The span, the term in brackets on the right-hand side of this expression, can be approximated by two terms of a binomial expression and sin α from Equation 13.6 substituted for α. Then, we have the *belt pitch length* estimated by [4]

$$L = 2c + \pi(r_1 + r_2) + \frac{1}{c}(r_2 - r_1)^2 \tag{13.9}$$

This gives the approximate *center distance:*

$$c = \frac{1}{4}\left[b + \sqrt{b^2 - 8(r_2 - r_1)^2}\right] \tag{13.10}$$

in which

$$b = L - \pi(r_2 + r_1) \tag{13.11}$$

The values of *actual pitch lengths* of some standard V belts are listed in Table 13.2. These values are substituted into Equation 13.11 to obtain the actual center distances. Observe from the table that long center distances are not recommended for the V belt, since the excessive vibration of the slack side shortens the material life. In the case of flat belts, there is virtually no limit to the center distance, as noted in Table 13.1. Table 13.3 shows standard pitches with their letter identifications for toothed belts.

For nonstandard V belts, sometimes the center distance is given by the larger of

$$c = 3r_1 + r_2 \quad \text{or} \quad c = 2r_2 \tag{13.12}$$

This value of c may be used in Equation 13.9 to estimate the belt pitch length L. It is important to note that belt drives should be designed with provision for center distance adjustments, unless an idler pulley is employed, since belts tend to stretch in application.

TABLE 13.2

Pitch Lengths (in inches) of Standard V Belts

Cross Section							
A	**B**	**C**	**A**	**B**	**C**	**D**	**E**
27.3			113.3	113.8	114.9		
36.3	36.8		121.3	121.8	122.9	113.3	
43.3	43.8			145.8	146.9	147.3	
52.3	52.8	53.9		159.8	160.9	161.3	
61.3	61.8	62.9		174.8	175.9	173.3	
69.3	69.8	70.9		181.8	182.9	183.3	184.4
76.3	76.8	77.9		211.8	212.9	213.3	214.5
86.3	86.8	87.9		240.3	240.9	240.8	241.0
91.3	91.8	92.9		270.3	270.9	270.8	271.0
106.3	106.8	107.9		300.3	300.9	300.8	301.0

TABLE 13.3

Standard Pitches of Typical Timing Belts

Service	**Designation**	**Pitch, p (in.)**
Extra light	XL	$\frac{1}{5}$
Light	L	$\frac{3}{8}$
Heavy	H	$\frac{1}{2}$
Extra heavy	XH	$\frac{7}{8}$
Double extra heavy	XXH	$1\frac{1}{4}$

Example 13.1: Geometric Quantities of a V-Belt Drive

A V belt is to operate on sheaves of 8 and 12 in. pitch diameters (Figure 13.5a). Calculate

a. The center distance
b. The contact angle

Assumption: A B-section V belt is used, having the actual pitch length of 69.8 in. (Table 13.2).

Solution

a. Through the use of Equation 13.11,

$$b = L - \pi(r_1 + r_2) = 69.8 - \pi(6 + 4) = 38.38 \text{ in.}$$

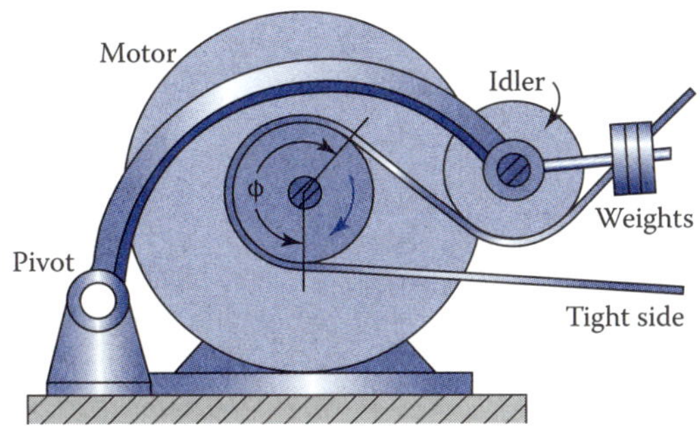

FIGURE 13.6
Weighed idler used to maintain slack-side tension.

Equation 13.10 is therefore

$$c = \frac{1}{4}\left[38.38 + \sqrt{38.38^2 - 8(6-4)^2}\right] = 19.09 \text{ in.}$$

b. Applying Equation 13.7,

$$\phi = \pi - 2\alpha = 180° - 2\sin^{-1}\left(\frac{6-4}{19.09}\right) = 168°$$

13.3.4 Maintaining the Initial Tension of the Belt

A flat belt stretches over a period of time, and some initial tension is lost. Of course, the simplest solution is to have excessive initial tension. However, this would overload the shafts and bearings and shorten the belt life. A *self-tightening drive* that automatically maintains the desired tension is illustrated in Figure 13.6. Note that a third pulley is forced against the slack side of the belt on top by weights (as in the figure) or by a spring. The extra pulley that rotates freely is termed an *idler pulley* or simply *idler.* The idler is positioned so that it increases the contact angle ϕ and thus the capacity of the drive.

There are various other approaches of maintaining necessary belt tension. These include using a pivoted-overhung motor drive, changing the belt and pulley materials to increase the coefficient of friction, and increasing the center distance during operation by employing a drive with an adjustable center distance. We note that, because of the resistance to stretch of their interior tension cords, timing and V belts do not require frequent adjustment of initial tension.

13.4 Belt Tension Relationships

The discussion of the preceding section pertains to belts that run slowly enough that centrifugal loading can be disregarded. We now develop a relationship between the tight- and slack-side tensions for the belt operating at maximum capacity. For this purpose, we first define the *centrifugal force* F_c, representing the inertia effect of the belt, in the following form:

$$F_c = \frac{w}{g} V^2 \tag{13.13}$$

where
w = the belt weight per unit length
V = the belt velocity
g = the acceleration of gravity

In SI units, F_c is expressed in N, w in N/m, V in m/s, and g in 9.81 m/s²; in U.S. customary units, F_c is measured in lb, w in lb/ft, V in fps, and g in 32.2 ft/s².

13.4.1 Flat or Round Belt Drives

Reconsider the belt drive of Figure 13.5a, running at its largest capacity. The free-body diagram of a belt element on the verge of slipping on small pulley is depicted in Figure 13.5b. The element is under normal force dN, tension F, centrifugal force F_c dθ, and friction force f dN, where f represents the coefficient of friction. Equilibrium of the forces in the horizontal direction is satisfied by

$$(F + \mathrm{d}F)\cos\frac{d\theta}{2} - f\mathrm{d}N - F\cos\frac{\mathrm{d}\theta}{2} = 0$$

Simplifying, and noting for small angles cos(dθ/2) = 1, we have

$$\mathrm{d}F = f\mathrm{d}N \tag{13.14}$$

Likewise, equilibrium of the vertical forces gives

$$\mathrm{d}N + F_c\mathrm{d}\theta - (F + \mathrm{d}F + F)\sin\frac{\mathrm{d}\theta}{2} = 0$$

We can take sin(dθ/2) = dθ/2, since dθ is a small angle, and neglect the higher-order term dF dθ. In so doing and introducing the value of dN from Equation 13.14, the preceding equation becomes

$$\frac{\mathrm{d}F}{F - F_c} = f\mathrm{d}\theta$$

The solution of this expression is obtained by integrating from minimum tension F_2 to maximum tension F_1 through the angle of contact ϕ of the belt (Figure 13.5a). Hence,

$$\int_{F_2}^{F_1} \frac{\mathrm{d}F}{F_1 F_c} = \ln\left(\frac{F_1 - F_c}{F_2 - F_c}\right) = f\phi \tag{13.15}$$

This may be written in the following convenient form:

$$\frac{F_1 - F_c}{F_2 - F_c} = e^{f\phi} \tag{13.16}$$

We see from this relation that centrifugal force tends to reduce the angles of contact ϕ.

Example 13.2: Maximum Tension of Flat-Belt Drive

A 12 hp, 2200 rpm electric motor drives a machine through the flat belt (Figure 13.7). The size of belt is 5 in. wide and 0.3 in thick and weighs 0.04 lb/in.[3] The center distance is equal to 6.5 ft. The pulley on motor shaft has a $r_1 = 2.5$ in. radius and the driven pulley is $r_2 = 7.5$ in. in radius.

Find: The belt tensions

Assumption: The coefficient of friction will be $f = 0.2$.

Solution

The cross-sectional area of the belt is 5(0.3) = 1.5 in.^{2} and its unit weight equals $w = 0.04(1.5) = 0.06$ lb/in. = 0.72 lb/ft. The belt velocity, using Equation 13.4,

$$V = \frac{\pi d_1 n_1}{12} = \frac{\pi(5)(2200)}{12} = 2880 \text{ fpm}$$

By Equation 13.3,

$$F_1 - F_2 = \frac{33{,}000 \text{ hp}}{V} = \frac{33{,}000(12)}{2880} = 137.5 \text{ lb} \tag{a}$$

Centrifugal force acting on the belt, applying Equation 13.13,

$$F_c = \frac{w}{g}V^2 = \frac{0.72}{32.2}\left(\frac{2880}{60}\right)^2 = 51.52 \text{ lb}$$

Through the use of Equation 13.6, we find

$$\alpha = \sin^{-1}\left[\frac{r_2 - r_2}{c}\right] = \sin^{-1}\left[\frac{7.5 - 2.5}{6.5(12)}\right] = 3.675°$$

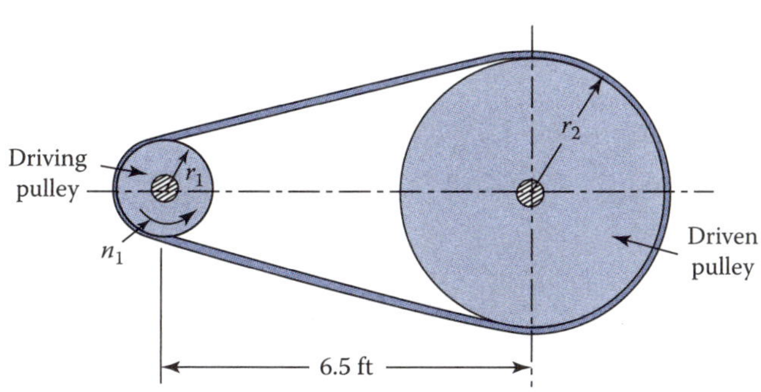

FIGURE 13.7
Example 13.2. A flat-belt drive.

Then the angle of wrap, $\phi = \pi - 2\alpha$, equals

$$\phi = 180° - 2(3.675°) = 172.65°$$

We have $e^{f\phi} = e^{(0.2)(172.65)(\pi/180)} = 1.827$. Substituting these into Equation 13.16 leads to

$$\frac{F_1 - 51.52}{F_2 - 51.52} = 1.827$$

or

$$F_1 = 1.827\,F_2 - 42.6 \qquad \text{(b)}$$

Solving Equations (a) and (b) results in

$$F_1 = 355.3 \text{ lb} \quad \text{and} \quad F_2 = 217.8 \text{ lb}$$

Comment: Equation 13.2 estimates the initial tension as (355.3 + 217.8)/2 = 286.6 lb and so the largest belt tension in the belt drive is F_1 = 355.3 lb.

13.4.2 V-Belt Drives

Figure 13.8 illustrates how a V belt rides in the sheave groove: with contact on the sides and clearance at the bottom. Obviously, this *wedging action* increases the normal force on the belt element from dN (Figure 13.5b) to dN/sin β. Following a procedure similar to that used in the preceding discussion, for a V-belt drive, we therefore obtain

$$\frac{F_1 - F_c}{F_2 - F_c} = e^{f\phi/\sin\beta} \qquad (13.17)$$

The quantity β is half the included angle of the V belt. It is interesting to observe that the (smaller) contact angle ϕ of the *driver pulley* leads to larger belt tension and hence is *critical.* Hence, design of belt drive is on the basis of small pulley geometry.

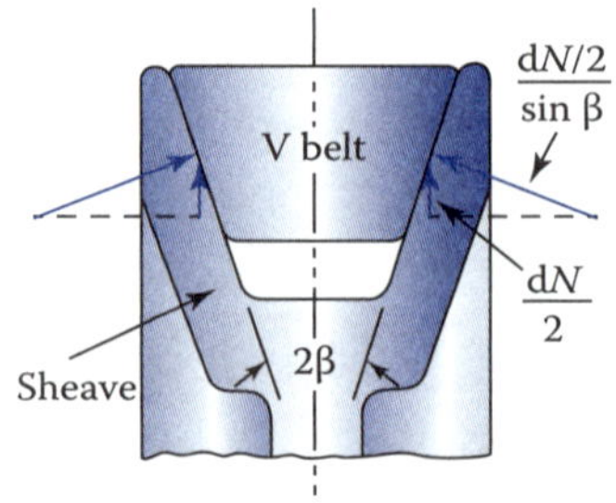

FIGURE 13.8
V belt in a sheave groove.

13.5 Design of V-Belt Drives

Of special concern is the design of belt drives for maximum tension and expected life or durability. In this section, attention is directed mainly to the former. Inasmuch as V-belt cross sections vary considerably, a rational design of belt drive usually relies on tables, charts, and guidelines given by the manufacturers. Design data are based on theory as well as results of extreme testing [5–7].

During circuit around sheaves, the force on the belt varies considerably, as depicted in Figure 13.9. Note the additional equivalent *tension* forces F_{b1} and F_{b2} in the cord due to *bending* around the pulleys. The peak or total force F_B in the belt at point B is the sum of the tight-side tension and the equivalent tension force owing to the bending around the small pulley. Therefore,

$$F_B = F_1 + F_{b1}$$

Likewise, at point E, the total force may be expressed as $F_E = F_2 + F_{b2}$. The tensions F_1 and F_2 are obtained from Equation 13.3. For standard V belts, the bending and centrifugal forces are given by empirical formulas. The peak forces F_B and F_E are key to the design of V-belt drives. Durability design is somewhat complicated by the induced flexure stresses in the belt. Expected V-belt life refers to a certain number of peak forces a belt can sustain before failure by fatigue. The fatigue performance of a V-belt drive is best obtained by experimental tests.

We now develop an *approximate design equation* for maximum tension in the belt. Let the speed and power of the belt drive be prescribed. Then, from Equation 13.3, the torque at the smaller pulley is

$$T = \frac{63,000 \text{ hp}}{n_1} \tag{13.18}$$

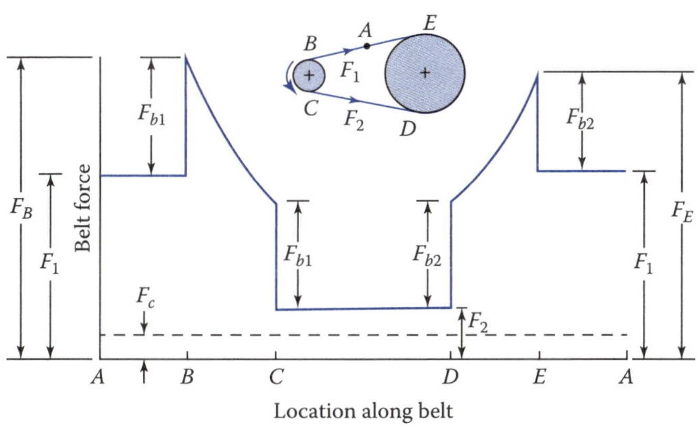

FIGURE 13.9
Forces in moving V belt.

in which n_1 represents the speed of the smaller pulley in revolutions per minute (rpm). The slack-side tension, as obtained from Equation 13.1, is

$$F_2 = F_1 - \frac{T_1}{r_1} \tag{13.19}$$

The quantity r_1 is the pitch radius of the smaller pulley. Substituting Equation 13.19 into Equation 13.17, after rearrangement, the tight-side tension may be expressed in the convenient form

$$F_1 = F_c + \left(\frac{\gamma}{\gamma - 1}\right)\frac{T_1}{r_1} \tag{13.20}$$

where

$$\gamma = e^{f\phi/\sin\beta} \tag{13.21}$$

Here the coefficient of friction f between rubber and dry steel is *usually* taken to be about 0.3.

In the case of a *flat belt* ($\beta = 90°$), $\sin\beta = 1$. We therefore see from Equation 13.20 that, for a given maximum tension F_1, a V belt can transmit more torque (and power). Consequently, V belts are usually preferred over flat belts. Table 13.4 may be used to estimate V-belt tensions. Note that, if the sheave diameters are equal, the contact angle is 180° and hence tight-side tension is five times as great as the slack-side tension F_2. Practically, the minimum contact angle is 90°, for which $F_1 = 2.24F_2$.

Since V belts are usually made from reinforced rubber, the required belt strength is governed mainly by the tension; that is, the effect of the additional force in the belt due to bending around the pulley may be neglected. However, the tight-side tension should be multiplied by a service factor K_s. The *maximum tension* is then

$$F_{max} = K_s F_1 \tag{13.22}$$

Service factors are listed by the manufacturers in great detail, usually based on the number of hours per day of overload, variations in loading the driving and driven shafts, starting overload, and variations in environmental conditions. Examples of the driven equipment in V-belt drives are blowers, pumps, compressors, fans, light-duty conveyors, dough mixers, generators, laundry machinery, machine tools, punches, presses, shears, printing machinery, bucket elevators, textile machinery, mills, and hoists. Typical service factors, relying on the characteristics of the driving and driven machinery, are given in Table 13.5. The horsepower should be multiplied by a service factor when selecting belt sizes. The design of a

TABLE 13.4

Ratios of V-Belt Tensions for Various Contact Angles

Contact Angle, ϕ	F_1/F_2	Contact Angle, ϕ	F_1/F_2	Contact Angle, ϕ	F_1/F_2
180°	5.00	150°	3.82	120°	2.92
175°	4.78	145°	3.66	115°	2.80
170°	4.57	140°	3.50	110°	2.67
165°	4.37	135°	3.34	105°	2.56
160°	4.18	130°	3.20	100°	2.44
155°	4.00	125°	3.06	90°	2.24

TABLE 13.5

Service Factors K_s for V-Belt Drives

	Driver (Motor or Engine)	
Driven Machine	**Normal Torque Characteristic**	**High or Nonuniform Torque**
Uniform	1.0–1.2	1.1–1.3
Light shock	1.1–1.3	1.2–1.4
Medium shock	1.2–1.2	1.4–1.6
Heavy shock	1.3–1.5	1.5–1.8

TABLE 13.6

Recommended Sheave Pitch Diameters (in.) for V-Belt Drives

	Motor Speed, rpm				
Motor hp	**575**	**695**	**870**	**1160**	**1750**
$\frac{1}{2}$	2.5	2.5	2.5	—	—
$\frac{3}{4}$	3.0	2.5	2.5	2.5	—
1.00	3.0	3.0	2.5	2.5	2.25

V-belt drive should use the largest possible pulleys. As the sheave sizes become smaller, the belt tension increases for a given horsepower output. Recommended pulley diameters for use with three electric motor sizes are given in Table 13.6.

Finally, we note that the shaft load at the pulley consists of torque T and force F_s. The latter is the vector sum of tensions F_1 and F_2. Referring to Figure 13.5a, the *shaft force* may therefore be expressed as

$$F_s = \left[(F_1 + F_2 \cos 2\alpha)^2 + (F_2 \sin 2\alpha)^2 \right]^{1/2} \tag{13.23}$$

The angle α is defined by Equation 13.6. Equation 13.23 yields results approximately equal to the scalar sum $F_1 + F_2$ in most cases. The designer will find it useful to have the shaft load for determining the reactions at the shaft bearings.

Example 13.3: Design Analysis of a V-Belt Drive

The capacity of a V-belt drive is to be 10 kW, based on a coefficient of friction of 0.3. Determine the required belt tensions and the maximum tension.

Given: A driver sheave has a radius of $r_1 = 100$ mm, a speed of $n_1 = 1800$ rpm, and a contact angle of $\phi = 153°$. The belt weighs 2.25 N/m and the included angle is 36°.

Assumptions: The driver is a normal torque motor and driven machine involves light shock load.

Solution

We have $\phi = 153° = 2.76$ rad and $\beta = 18°$. The tight-side tension is estimated from Equation 13.20 as

$$F_1 = F_c + \left(\frac{\gamma}{\gamma - 1} \right) \frac{T_1}{r_1} \tag{a}$$

where

$$F_c = \frac{w}{g}V^2 = \frac{2.25}{9.81}\left(\frac{\pi \times 0.2 \times 1800}{60}\right)^2 = 81.5 \text{ N}$$

$$\gamma = e^{f\phi/\sin\beta} = e^{0.3(2.67)/\sin 18^\circ} = 13.36$$

$$T_1 = \frac{9549 \text{ kW}}{n_1} = \frac{9549(10)}{1800} = 53.05 \text{ N}\cdot\text{m}$$

Carrying the preceding values into Equation (a), we have

$$F_1 = 81.5 + \left(\frac{13.36}{13.36 - 1}\right)\frac{53.05}{0.1} = 655 \text{ N}$$

Then, by Equation 13.19, the slack-side tension is

$$F_2 = 655 - \frac{53.05}{0.1} = 124.5 \text{ N}$$

Based upon a service factor of 1.2 (Table 13.5) to F_1, Equation 13.22 gives a maximum tensile force

$$F_{max} = 1.2(655) = 786 \text{ N}$$

applied to the belt.

The design of timing-belt drives is the same as that of flat- or V-belt drives. The manufacturers provide detailed information on sizes and strengths. The following case study illustrates an application.

13.6 Chain Drives

As pointed out in Section 13.1, chains are used for power transmission between parallel shafts. They can be employed for high loads and where precise speed ratios must be sustained. While precise location and alignment tolerances are not required, as with gear drives, the best performance can be expected when input and output sprockets lie in the same vertical plane [8,9]. Chain drives have shorter service lives than typical gear drives. They present no fire hazard and are unaffected by relatively high temperatures. Sometimes, an adjustable *idler* sprocket (toothed wheel) is placed on the outside of the chain near the driving sprocket to remove sag on the loose side and increase the number of teeth in contact. The only maintenance required after careful alignment of the elements is proper lubrication. Usually, a chain should have a sheet of metal casing for protection from atmospheric dust and to facilitate lubrication. Chains should be washed regularly in kerosene and then soaked in oil.

The *speed ratio* of a chain drive is expressed by the equation

$$\frac{n_2}{n_1} = \frac{N_1}{N_2}$$

In the preceding, we have

n_2 = the output speed
n_1 = the input speed
N_2 = the number of teeth in the output sprocket
N_1 = the number of teeth in the input sprocket

An odd number of teeth on the driving sprocket (17, 19, 21,…) is recommended, typically 17 and 25. Usually, an odd number of sprocket teeth causes each small sprocket tooth to contact many or all chain links, minimizing wear. The larger sprocket is ordinarily limited to about 120 teeth.

Center distance c should be greater than a value that just allows the sprockets to clear. It is $c = 2(r_1 + r_2)$; for smaller speed ratios, $n_1/n_2 < 3$. Here, r_1 and r_2 refer to pitch radii of the input and output sprockets, respectively. When longer chains are used, idlers may be required on the slack side of the chain. For the cases in which speed ratios $n_1/n_2 \geq 3$, the center distance should be $c = 2(r_2 - r_1)$. Having a tentative center distance c between shafts selected, *chain length L* may be estimated applying Equation 13.9. Finally, the center distance is recalculated through the use of Equation 13.10. The angle of contact for the chain drive is given by Equation 13.7. Note that, for a small sprocket, the angle of contact should not be less than 120°.

Chain pitch p represents the length of an individual link from pin center to pin center. The pitch radius of a sprocket with N teeth may be defined as

$$r = \frac{Np}{2\pi}$$

An even number of pitches in the chain is preferred to avoid a special link. *Chain velocity V* is defined as the number of feet coming off the sprocket in unit time. Therefore,

$$V = \frac{Npn}{12} \tag{13.24}$$

The *tensile force* that a chain transmits F_1 may be obtained from Equation 1.17 in the form

$$F_1 = \frac{33{,}000 \text{ hp}}{V} = \frac{396{,}000 \text{ hp}}{pNn} \tag{13.25}$$

In Equations 13.24 and 13.25, the chain pitch p is measured in inches and the sprocket speed n is in rpm. The total force or tension in the chain includes transmitted force F_1, a centrifugal force F_c, a small catenary tension, and a force from link action. There are also impact forces when link plates engage sprocket teeth, discussed in the next section.

13.7 Common Chain Types

There are various types of power transmission chains; however, roller chains are the most widely employed. Types of driven equipment with roller-chain drives include bakery machinery, blowers and fans, boat propellers, compressors, conveyors, clay-working machinery, crushers, elevators, feeders, food processors, dryers, machine tools, mills, pumps, pulp grinders, printing presses, carding machinery, and wood-working machinery. In some of these applications as well as in cranes, hoists, generators, ice machines, and a variety of laundry machinery, inverted chains are also used. Both common chain types are used on sprockets, well suited for heavy loads, and have high efficiency.

13.7.1 Roller Chains

Of its diverse applications, the most familiar is the roller-chain drive on a bicycle. A roller chain is generally made of hardened steel and sprockets of steel or cast iron. Nevertheless, stainless steel and bronze chains are obtainable where corrosion resistance is needed. The geometry of a roller chain is shown in Figure 13.10. The rollers rotate in bushings that are press fitted to the inner link plates. The pins are prevented from turning by the outer links' press-fit assembly. Roller chains have been standardized according to size by the American National Standards Institute (ANSI) [10]. The characteristics of representative standard sizes are listed in Table 13.7. These chains are manufactured in single (Figure 13.10a), double (Figure 13.10b), triple, and quadruple strands. Clearly, the use of multistrands increases the load capacity of a chain and sprocket system.

13.7.1.1 Chordal Action

The instantaneous chain velocity varies from the average velocity given by Equation 13.24. Consider a sprocket running at constant speed n and driving a roller chain in a counterclockwise direction, as illustrated in Figure 13.11. A chord representing the link between

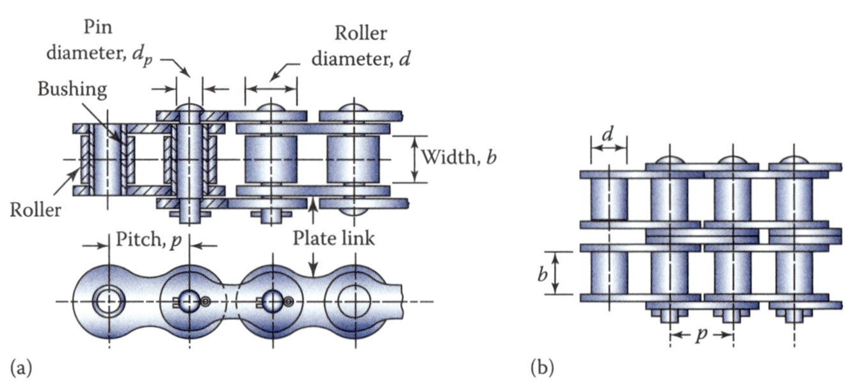

FIGURE 13.10
Portion of a roller chain: (a) single strand and (b) double strand.

TABLE 13.7
Sizes and Strengths of Standard Roller Chains

Chain No.	Pitch, p (in.)	Roller Diameter, d (in.)	Roller Width, b (in.)	Pin Diameter, d_p (in.)	Link Plate Thickness, t (in.)	Minimum Ultimate Strength (lb)
25	$\frac{1}{4}$	0.130	$\frac{1}{8}$	0.0905	0.030	780
35	$\frac{3}{8}$	0.200	$\frac{3}{16}$	0.141	0.050	1760
41	$\frac{1}{2}$	0.306	$\frac{1}{4}$	0.141	0.050	1500
40	$\frac{1}{2}$	$\frac{5}{16}$	$\frac{5}{16}$	0.156	0.060	3125
50	$\frac{5}{8}$	0.400	$\frac{3}{8}$	0.200	0.080	4480
60	$\frac{3}{4}$	$\frac{15}{32}$	$\frac{1}{2}$	0.234	0.094	7030
80	1	$\frac{5}{8}$	$\frac{5}{8}$	0.312	0.125	12.500
100	$1\frac{1}{4}$	$\frac{3}{4}$	$\frac{3}{4}$	0.375	0.156	19.530
120	$1\frac{1}{2}$	$\frac{7}{8}$	1	0.437	0.187	28.125
140	$1\frac{3}{4}$	1	1	0.500	0.219	38.280
160	2	$1\frac{1}{8}$	$1\frac{1}{4}$	0.562	0.250	50.000
180	$2\frac{1}{4}$	$1\frac{13}{32}$	$1\frac{13}{32}$	0.687	0.2811	63.280
200	$2\frac{1}{2}$	$2\frac{9}{16}$	$1\frac{1}{2}$	0.781	0.312	78.125
240	3	$1\frac{7}{8}$	$1\frac{7}{8}$	0.937	0.375	112.500

Source: ANSI/ASME Standard B29.1M-1993.

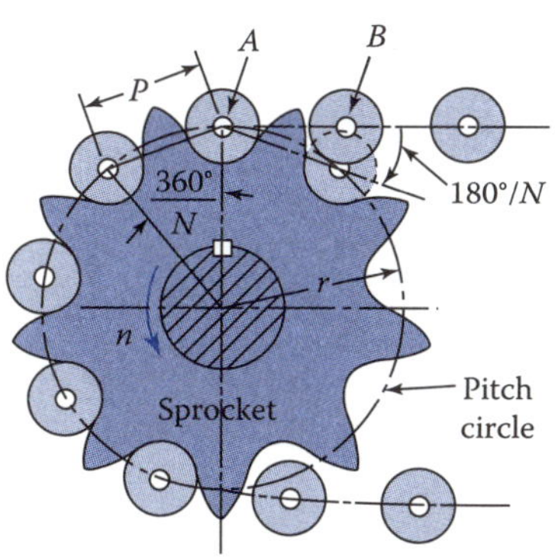

FIGURE 13.11
A roller chain and sprocket engagement.

centers has the pitch length of p. The chord subtends the pitch angle of the sprocket, 360°/N. From the geometry, sin (180°/2) = (p/2)/r. We therefore have

$$r = \frac{p}{2\sin(180°/N)}$$

where
 r = the radius of the sprocket
 p = the chain pitch
 N = the number of teeth in the sprocket

The angle 180°/N through which the link (AB) swings until the roller (B) seats on the sprocket is referred to as the *angle of articulation*. When the sprocket is in the position shown in the figure, the chain velocity is $2\pi rn$. This velocity changes to $2\pi r_1 n$, where $r_1 = r\cos(180°/N)$, after the sprocket is turned to the angle of articulation. The change of velocity ΔV is called the *chordal action:*

$$\Delta V = 2\pi rn\left[1 - \cos(180°/\mathrm{N})\right]$$

Rotation of the link through the angle of articulation causes impact among the rollers and the sprocket teeth as well as wear in the chain joint. The movement of the link up and down with rotation through the articulation angle develops an uneven chain exit velocity. Consequently, the driven shaft of a roller-chain drive may be given a pulsating motion, particularly at high-speed operations. The angle of articulation (hence the impact and chordal action) should be reduced as much as practicable, by increasing the number of sprocket teeth [1].

13.7.2 Power Capacity of Roller Chains

Equations and tables for roller-chain power capacity and selection were developed through the American Chain Association (ACA), as the result of many years of laboratory testing and field observation. Rated horsepower capacities are usually given in tabular form for each type of single-strand chain corresponding to a life expectancy of 15 kh for a variety of sprocket speeds. Table 13.8 is an example for ANSI No. 60 roller chains. Listed in Table 13.9 are the service factors that account for the abruptness associated with load application. Table 13.10 shows multiple-strand factors.

At lower speeds, the power capacity of roller chains is based upon the fatigue strength of the link plate. On the other hand, at higher speeds, the power relies on roller and bushing impact life. At extremely high speeds, the power capacity is on the basis of the galling or welding between pin and bushings. The *design power capacity* may be expressed as follows:

$$H_d = H_r K_1 K_2 \tag{13.26}$$

where
 H_r = the horsepower rating (from Table 13.8)
 K_1 = the service factor (from Table 13.9)
 K_2 = the multiple-strand factor (from Table 13.10)

TABLE 13.8

Rated Horsepower Capacity (H_r) for a Standard Single-Strand Roller Chain: ANSI No. 60, ¾ in. Pitch

No. of Teeth in Small Sprocket	rpm: Small Sprocket																							
	10	25	50	100	150	200	300	400	500	600	700	800	900	1000	1100	1200	1400	1600	1800	2000	2500	3000	3500	4000	4500
11	0.18	0.41	0.77	1.44	2.07	2.69	3.87	5.02	6.13	7.23	8.30	9.56	10.4	11.4	12.5	11.9	9.41	7.70	6.45	5.51	3.94	3.00	2.38	1.95	1.63
12	0.20	0.45	0.85	1.58	2.28	2.95	4.25	5.51	6.74	7.94	9.12	10.3	11.4	12.6	13.7	13.5	10.7	8.77	7.35	6.28	4.49	3.42	2.71	2.22	1.86
13	0.22	0.50	0.92	1.73	2.49	3.22	4.64	6.01	7.34	8.65	9.94	11.2	12.5	13.7	14.9	15.2	12.1	9.89	8.29	7.08	5.06	3.85	3.06	2.50	0
14	0.24	0.54	1.00	1.87	2.69	3.49	5.02	6.51	7.96	9.37	10.8	12.1	13.5	14.8	16.2	17.0	13.5	11.1	9.26	7.91	5.66	4.31	3.42	2.80	0
15	0.25	0.58	1.08	2.01	2.90	3.76	5.41	7.01	8.57	10.1	11.6	13.1	14.5	16.0	17.4	18.8	15.0	12.3	10.3	8.77	6.28	4.77	3.79	3.10	0
16	0.27	0.62	1.16	2.16	3.11	4.03	5.80	7.52	9.19	10.8	12.4	14.0	15.6	17.1	18.7	20.2	16.5	13.5	11.3	9.66	6.91	5.26	4.17	3.42	0
17	0.29	0.66	1.24	2.31	3.32	4.30	6.20	8.03	9.81	11.6	13.3	15.0	16.7	18.3	19.9	21.6	18.1	14.8	12.4	10.6	7.57	5.76	4.57	3.74	0
18	0.31	0.70	1.31	2.45	3.53	4.58	6.59	8.54	10.4	12.3	14.1	15.9	17.7	19.5	21.2	22.9	19.7	16.1	13.5	11.5	8.25	6.28	4.98	4.08	0
19	0.33	0.75	1.39	2.60	3.74	4.85	6.99	9.05	11.1	13.0	15.0	16.9	18.8	20.6	22.5	24.3	21.4	17.5	14.6	12.5	8.95	6.81	5.40	4.42	0
20	0.35	0.79	1.47	2.75	3.96	5.13	7.38	9.57	11.7	13.8	15.8	17.9	19.8	21.8	23.8	25.7	23.1	18.9	15.8	13.5	9.66	7.35	5.83	0	
21	0.36	0.83	1.55	2.90	4.17	5.40	7.78	10.1	12.3	14.5	16.7	18.8	20.9	23.0	25.1	27.1	24.8	20.3	17.0	14.5	10.4	7.91	6.28	0	
22	0.38	0.87	1.63	3.05	4.39	5.68	8.19	10.6	13.0	15.3	17.5	19.8	22.0	24.2	26.4	28.5	26.6	21.8	18.2	15.6	11.1	8.48	6.73	0	
23	0.40	0.92	1.71	3.19	4.60	5.96	8.59	11.1	13.6	16.0	18.4	20.8	23.1	25.4	27.7	29.9	28.4	23.3	19.5	16.7	11.9	9.07	7.19	0	
24	0.42	0.96	1.79	3.35	4.82	6.24	8.99	11.6	14.2	16.8	19.3	21.7	24.2	26.6	29.0	31.3	30.3	24.8	20.8	17.8	12.7	9.66	7.67	0	
25	044	1.00	1.87	3.50	5.04	8.52	9.40	12.2	14.9	17.5	20.1	22.7	25.3	27.8	30.3	32.7	32.2	26.4	22.1	18.9	13.5	10.3	8.15	0	
26	0.46	1.05	1.95	3.65	5.25	6.81	9.80	12.7	15.5	18.3	21.0	23.7	26.4	29.0	31.6	34.1	34.2	28.0	23.4	20.0	14.3	10.9	8.65	0	
28	0.50	1.13	2.12	3.95	5.69	7.37	10.6	13.8	16.8	19.8	22.8	25.7	28.5	31.4	34.2	37.0	38.2	31.3	26.2	22.4	16.0	12.2	0		
30	0.54	1.22	2.28	4.26	6.13	7.94	11.4	14.8	18.1	21.4	24.5	27.7	30.8	33.8	36.8	39.8	42.4	34.7	29.1	24.8	17.8	13.5	0		
32	0.57	1.31	2.45	4.56	6.57	8.52	12.3	15.9	19.4	22.9	26.3	29.7	33.0	36.3	39.5	42.7	46.7	38.2	32.0	27.3	19.6	14.9	0		
35	0.63	1.44	2.69	5.03	7.24	9.38	13.5	17.5	21.4	25.2	29.0	32.7	36.3	39.9	43.5	47.1	53.4	43.7	36.6	31.3	22.4	17.0	0		
40	0.73	1.67	3.11	5.81	8.37	10.8	15.6	20.2	24.7	29.1	33.5	37.7	42.0	46.1	50.3	54.4	62.5	53.4	44.7	38.2	27.3	0			
45	0.83	1.89	3.53	6.60	9.50	12.3	17.7	23.0	28.1	33.1	38.0	42.9	47.7	52.4	57.1	61.7	70.9	63.7	53.4	45.6	32.6	0			
		Type A					Type B											Type C							

Source: ANSI/ASME Standard B29.1M-1993.

TABLE 13.9

Service Factors (K_1) for Single-Strand Roller Chains

	Type of Input Power		
Type of Driven Load	**IC Engine Hydraulic Drive**	**Electric Motor or Turbine**	**IC Engine Mechanical Drive**
Smooth	1.0	1.0	1.2
Moderate shock	1.2	1.3	1.4
Heavy shock	1.4	1.5	1.7

Source: ANSI/ASME Standard B29.1M-1993.

TABLE 13.10

Multiple-Strand Factors (K_2) for Roller Chains

Number of Strands	Multiple-Strand Factor
2	1.7
3	2.5
4	3.3

Source: ANSI/ASME Standard B29.1M-1993.

Usually, a medium or light mineral oil is used as the lubricant. We observe from Table 13.8 that the proper lubrication of roller chains is essential to their performance. As speed increases, this requirement becomes more rigorous. The following *types of lubrication* systems are satisfactory:

Type A. Manual or drip lubrication; oil is applied periodically with brush or spout can.

Type B. Bath or disk lubrication; oil level is maintained in the casing at predetermined height.

Type C. Oil stream lubrication; oil is supplied by *circulating pump inside chain loop or lower span.*

Note that the limiting rpm for each lubrication type is read from the column to the left of the boundary line shown in the table. The chain manufacturer should be consulted for drives that exceed the speed and power requirements for the preceding lubrication kinds.

Example 13.4: Analysis of a Roller-Chain Drive

A three-strand ANSI No. 60, ¾ in. pitch roller chain transmits power from a N_1-tooth driver sprocket operating at n_1 rpm. Determine

a. The design power capacity
b. The tension in the chain
c. The factor of safety n of the chain on the basis of ultimate strength

Given: $N_1 = 19$, $p = \frac{3}{4}$ in., $n_1 = 1000$ rpm

Assumptions: The input power type is an internal combustion (IC) engine, mechanical drive. The type of driven load is moderate shock. With the exception of the tensile force, all forces are taken to be negligible.

Solution

See Tables 13.7 through 13.10.

a. For driver sprocket H_r = 20.6 hp, type B lubrication is required (Table 13.8). Service factor K_1 = 1.4 (Table 13.9). From Table 13.10 for three strands, K_2 = 2.5. Applying Equation 13.26, we have

$$H_d = 20.6(1.4)(2.5) = 72.1 \text{ hp}$$

b. The average chain velocity, by Equation 13.24, is

$$V_1 = \frac{19(0.75)(1000)}{12} = 1187.5 \text{ fpm}$$

Equation 13.25 results in

$$F_1 = \frac{33,000(72.1)}{1187.5} = 2.0 \text{ kips}$$

c. The ultimate strength, for a single-strand chain, is 7.03 kips (Table 13.7). The allowable load for a three-strand chain is then F_{all} = 7.03(3) = 21.09 kips. Hence, the factor of safety is

$$n = \frac{F_{all}}{F_1} = \frac{21.09}{2.0} = 10.5$$

Comment: The analysis is based on 15 kh of chain life, since other estimates are not available.

13.7.3 Inverted-Tooth Chains

The inverted-tooth chain, also referred to as the *silent chain*, is composed of a series of toothed link plates that are pin connected to allow articulation. The chain pitch is defined in Figure 13.12. An inverted-tooth chain ordinarily has guide links on the sides or in the

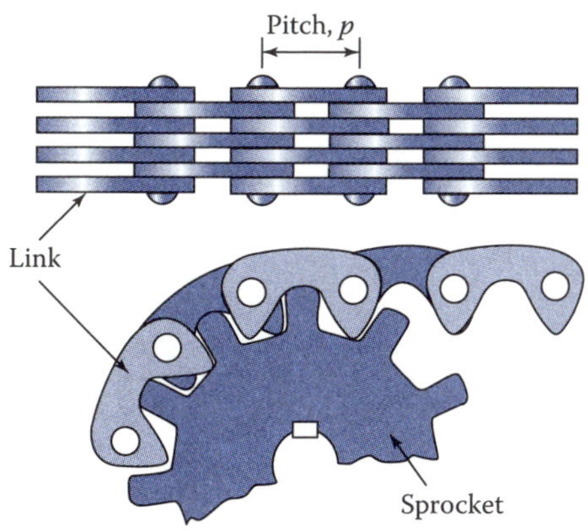

FIGURE 13.12
Portion of an inverted-tooth or *silent* chain.

center to keep it on the sprocket. To increase the chain life, different details of joint construction are used. Enclosures for the chain are customarily needed. Therefore, silent chains are more expensive than roller chains. Usually, when properly lubricated, at full load, drive efficiency is as high as 99%.

As the name suggests, these chains are quieter than roller chains. They may be run at higher speeds, because there is minor impact force when the chain link engages the sprocket. The inverted-tooth chain has a smooth flat surface, which can be conveniently used for conveying. Power capacities of silent chains are listed in tables analogous to those for roller chains. However, these chains reach maximum power at maximum speed, while roller chains reach highest power far below their maximum speed. Most of the remarks in the foregoing paragraphs relate as well to inverted-tooth chains and sprockets, which are also standardized by ANSI [11]. Regular pitches vary between ⅜ and 2 in. Sprockets may have 21–150 teeth. Center distance adjustment is periodically needed to compensate for wear.

Part B: High-Friction Devices

Our concern was with two flexible elements in the preceding sections. In Part B, we turn to clutches and brakes, which are high-friction devices. We consider the most commonly used types, having two or more surfaces pressed together with a normal force to produce a friction torque. *Performance analysis* of clutches and brakes involves the determination of the actuating force, torque transmitted, energy absorption, and temperature rise. The transmitted torque is associated with the actuating force, the coefficient of friction, and the geometry of the device. The temperature rise is related to energy absorbed in the form of friction heat during braking or clutching.

13.8 Materials for Brakes and Clutches

The materials used for clutches and brakes are of two types, those used for the disk or drums and those used for friction materials or linings. In the design of these devices, the selection of the friction materials is critical. Most linings are attached to the disks or drums by either riveting or bonding. The former has the advantage of low cost and relative ease in installation. The latter affords more friction area and greater effective thickness but is more expensive.

Drums are ordinarily made of *cast iron* with some alloying material added. Materials like stainless steel and Monel are used when good heat conduction is important. Many railroad brakes employ cast iron shoes, which are bearing on cast iron wheels or drums. Friction, thermal conductivity, resistance to wear, and thermal fatigue characteristics of drums are very important. They must have a sufficiently smooth surface finish to minimize wear of the lining.

Linings are often made of molded, woven, sintered, or solid materials. They are composed mainly of reinforcing fibers (to render strength and ability to resist high temperatures), metal particles (to obtain wear resistance and higher coefficient of friction), and bonding materials. The binder is ordinarily a thermosetting resin or rubber. In addition, a friction material should have good heat conductivity and impenetrability to moisture.

Because of health hazards associated with asbestos, it is banned on all current production, and alternative reinforcing materials are now in use.

A *woven* (cotton) lining is made as a fabric belt impregnated with metal particles and polymerized. These belts have flexibility, as required by band brakes. *Molded* linings typically use polymeric resin to bind a variety of powdered fillers or fibrous materials. Brass or zinc chips are sometimes added to improve heat conduction and wear resistance and reduce scoring of the drum and disks. These materials are the most commonly used in drum brakes and the least costly. *Sintered metal* pads are made of a mixture of copper and iron particles having friction modifiers molded, then heated to blend the material. They are the most costly but also the best suited for heavy-duty applications. Sintered metal-ceramic friction pads are similar, except that ceramic particles are added prior to sintering.

For sufficient performance of the brake or clutch, the requirements imposed on friction materials include the following: a high coefficient of friction having small variation on changes in pressure, velocity, and temperature; resistance to wear, seizing, and the tendency to grab; heat and thermal fatigue resistance. Tables 13.11 and 13.12 list approximate data related to allowable pressures and the coefficient of friction for a few linings [12]. For longer life, the lower values of the maximum pressure given should be used. As seen

TABLE 13.11

Properties of Common Brake and Clutch Friction Materials, Operating Dry

Material[a]	Dynamic Coefficient of Friction, f	Maximum Pressure p_{max} MPa	psi	Maximum Drum Temperature °C	°F
Molded	0.25–0.45	1.03–2.07	150–300	204–260	400–500
Woven	0.25–0.45	0.35–0.69	50–100	204–260	400–500
Sintered metal	0.15–0.45	1.03–2.07	150–300	232–677	450–1250
Cork	0.30–0.50	0.06–0.10	8–14	82	180
Wood	0.20–0.25	0.35–0.63	50–90	93	200
Cast iron, hard steel	0.15–0.25	0.70–0.17	100–250	260	500

[a] When rubbing against smooth cast iron or steel.

TABLE 13.12

Values of Friction Coefficients of Common Brake/Clutch Friction Materials, Operating in Oil

Material[a]	Dynamic Coefficient of Friction, f
Molded	0.06–0.09
Woven	0.08–0.10
Sintered metal	0.05–0.08
Cork	0.15–0.25
Wood	0.12–0.16
Cast iron, hard steel	0.03–0.06

[a] When rubbing against smooth cast iron or steel.

from the tables, the coefficients of friction are much smaller in oil than under dry friction, as expected. For more accurate information, consult the manufacturer or obtain test data [12–14].

13.9 Internal Expanding Drum Clutches and Brakes

Drum or rim clutches and brakes consist of three parts: the mating frictional surfaces, the means of transmitting the torque to and from the surfaces, and the actuating mechanism. Often, they are classified according to the operating mechanism. Figure 13.13 shows an internal expanding drum centrifugal clutch that engages automatically when the shaft speed exceeds a certain magnitude. The friction material is placed around the outer surface of the drum to engage the clutch.

The centrifugal clutch is in widespread use for automatic operation, such as to couple an engine to the drive train. When the engine speed increases, it automatically engages the clutch. This is particularly practical for electric motor drives, where during starting, the driven machine comes up to speed without shock. Used in chain saws for the same purpose, centrifugal clutches serve as an overload release that slips to allow the motor to continue running when the chain jams in the wood.

Magnetic, hydraulic, and pneumatic drum clutches are also useful in drives with complex loading cycles and in automatic machinery or robots. The expanding drum clutch is frequently used in textile machinery, excavators, and machine tools. Inasmuch as the analysis of drum clutch is similar to that for drum brakes, to be taken up in Sections 13.13 and 13.14, we do not discuss them at this time.

FIGURE 13.13
An internal expanding centrifugal-acting drum clutch. (Courtesy of Hillard Corporation, Elmira, New York.)

13.10 Disk Clutches and Brakes

Basic disk clutches and brakes are considered in this section. The former transmits torque from the input to the output shaft by the frictional force developed between the two disks or plates when they are pressed together. The latter is basically the same device, but one of the shafts is fixed. One of the friction surfaces of the clutch or brake is typically metal (cast iron or steel) and the other is usually a friction material or lining. Magnetic, hydraulic, and pneumatic operating mechanisms are also available in disk, cone, and multiple-disk clutches and brakes.

Uniform pressure and *uniform wear* are two basic conditions or *assumptions* that may occur at interface of the friction surfaces. The designer must decide on which assumption more closely approximates the particular clutch or brake being analyzed. The uniform wear assumption leads to lower calculated clutch or brake capacity than the assumption of uniform pressure, as observed in Example 13.5. Hence, disk clutches and brakes are ordinarily designed on the basis of uniform wear that gives conservative results. The following analysis can be used as a guide. Design considerations also include the characteristics of the machine whose brakes or clutches are to be a part and the environment in which the machine operates.

13.10.1 Disk Clutches

Friction clutches reduce shock by slipping during the engagement period. The single-plate or *disk clutch*, shown schematically in Figure 13.14, is employed in both automotive and industrial service. These devices are larger in diameter to give adequate torque capacity. Note that, in an automotive-type disk clutch, the input disk (flywheel) rotates with the crankshaft. The hub of the clutch output disk is spline-connected to the transmission shaft. Clearly, the device is disengaged by depressing the clutch pedal. The torque that can be transferred depends on the frictional force developed between the disks, coefficient of friction, and the geometry of the clutch. The axial force typically is quite large and can be applied mechanically (by spring, as in the figure), hydraulically, or electromagnetically. An advantage of the disk clutch over the drum clutch is the absence of centrifugal effects and efficient heat dissipation surfaces.

Multiple-disk clutches can have the friction lining on facing sides of a number (as many as 24) of alternative driving and driven disks or plates. The disks are usually thin

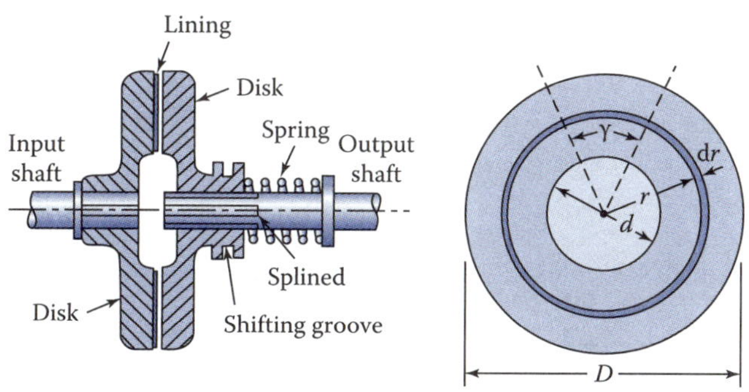

FIGURE 13.14
Basic disk clutch, shown in a disengaged position (for a brake, the *output* member is stationary).

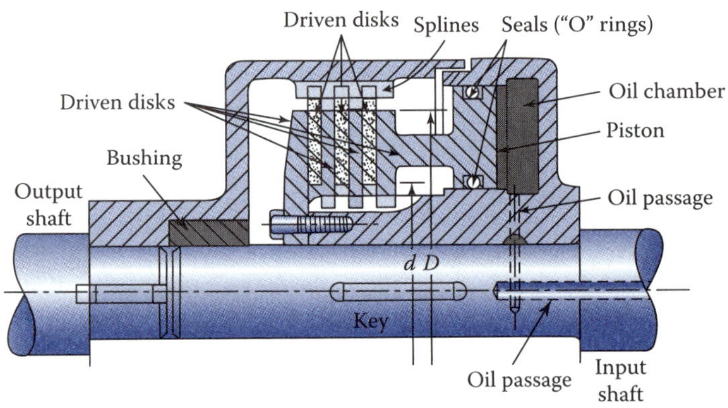

FIGURE 13.15
Half-section view of a multiple-disk clutch, hydraulically operated.

(about 1.5 mm) with small diameters. Thus, additional torque capacity with only a small increase in axial length is obtained. When the clutch is disengaged, the alternate disks are free to slide axially to disjoin. After the clutch is engaged, the disks are clamped tightly together to provide a *number of active friction surfaces N*. Disk clutches can be designed to operate either *dry* or *wet* with oil. The advantages of the latter are reduced wear, smoother action, and lower operating temperatures. As a result, most multiple-disk clutches operate either immersed in oil or in a spray. Multiple-disk clutches are compact and suitable for high-speed operations in various machineries. They are often operated automatically by either air or hydraulic cylinders (e.g., in automotive automatic transmissions).

Figure 13.15 shows a hydraulically operated clutch. In this device, the axial piston motion and force are produced by oil in an annular chamber, which is connected by an oil passage to an external pressure source. We see from the figure that, with the housing keyed to the input shaft, two disks and the piston are internally splined and an end plate is fastened. These are the driving disks. The three driven disks are externally splined to the housing keyed to the output shaft.

We develop the torque capacity equations for a single pair of friction surfaces, as in Figure 13.14. However, they can be modified for multiple disks by merely multiplying the values obtained by the number of active surfaces N. For example, $N = 6$ in the device depicted in Figure 13.15.

13.10.1.1 Uniform Wear

When the clutch disks are sufficiently rigid, it can be assumed that wear over the lining is uniform. This condition applies after an initial wearing-in has occurred. The uniform wear rate, which is taken to be proportional to the product of pressure and velocity pV, is constant. Note that the velocity at any point on the clutch surface varies with the radius and the angular velocity. Therefore, assuming a uniform angular velocity,

$$pr = C$$

where
p = the pressure
r = the radius
C = the constant

This equation indicates that the *maximum pressure* p_{max} takes place at the inside radius $r = d/2$ (Figure 13.14). Hence,

$$pr = C = p_{max}\frac{d}{2} \tag{13.27}$$

The total normal force that must be exerted by the actuating spring in Figure 13.14 is found by multiplying the area $2\pi r\,dr$ by the pressure $p = p_{max}d/2r$ and integrating over the friction surface. Hence, the *actuating force* F_a required equals

$$F_a = \int_{d/2}^{D/2} \pi p_{max} d\,dr = \frac{1}{2}\pi p_{max} d(D-d) \tag{13.28}$$

The friction torque or *torque capacity* is obtained by multiplying the force on the element by the coefficient of friction f and the radius and integrating over the area. It follows that

$$T = \int_{d/2}^{D/2} (\pi p_{max} d) fr\,dr = \frac{1}{8}\pi f p_{max} d(D^2 - d^2) \tag{13.29}$$

An expression relating the torque capacity to actuating force is obtained by solving Equation 13.28 for p_{max} and inserting its value into Equation 13.29. In so doing, we have

$$T = \frac{1}{4}F_a f(D+d) = F_a f r_{avg} \tag{13.30}$$

where r_{avg} is the average disk radius. Note the simple physical interpretation of this equation. As previously pointed out, for a multiple-disk clutch, Equations 13.29 and 13.30 must be multiplied by the number of active surfaces N.

In the design of clutches, the ratio of inside to outside diameters is an important parameter. It can be verified, applying Equation 13.29, that the maximum torque capacity for a prescribed outside diameter is attained when

$$d = \frac{D}{\sqrt{3}} = 0.577D \tag{13.31}$$

Usually employed proportions vary between $d = 0.45D$ and $d = 0.80D$.

13.10.1.2 Uniform Pressure

If the clutch disks are relatively flexible, the pressure p_{max} can approach a uniform distribution over the entire lining surface. For this condition, the wear is not constant. Referring to Figure 13.14, we readily obtain the actuating force and the torque capacity as follows:

$$F_a = \int_{d/2}^{D/2} 2\pi p_{max} r\,dr = \frac{1}{4}\pi p_{max}(D^2 - d^2) \tag{13.32}$$

$$T = \int_{d/2}^{D/2} (2\pi p_{max} r) fr\,dr = \frac{1}{12}\pi f p_{max}(D^3 - d^3) \tag{13.33}$$

These can be combined to yield the torque as a function of actuating force:

$$T = \frac{1}{3} F_a f \frac{D^3 - d^3}{D^2 - d^2} \tag{13.34}$$

The torque capacity for a multiple-disk clutch is obtained by multiplying Equations 13.33 and 13.34 by the number of active surfaces N.

Example 13.5: Design of a Disk Clutch

A disk clutch with a single friction surface has an outer diameter D and inner diameter d (Figure 13.14). Determine the torque that can be transmitted and the actuating force required of the spring, on the basis of

a. Uniform wear
b. Uniform pressure

Given: $D = 500$ mm, $d = 200$ mm

Design Decisions: Molded friction material and a steel disk are used, having $f = 0.35$ and $p_{max} = 1.5$ MPa (see Table 13.11).

Solution

a. Through the use of Equation 13.29, we have

$$T = \frac{1}{8}\pi(0.35)(1500)(0.2)(0.5^2 - 0.2^2) = 8.659 \text{ kN} \cdot \text{m}$$

From Equation 13.28,

$$F_a = \frac{1}{2}\pi(1500)(0.2)(0.5 - 0.2) = 141.4 \text{ kN}$$

b. Applying Equation 13.33,

$$T = \frac{1}{12}\pi(0.35)(1500)(0.5^3 - 0.2^3) = 16.08 \text{ kN} \cdot \text{m}$$

By Equation 13.32,

$$F_a = \frac{1}{4}\pi(1500)(0.5^2 - 0.2^2) = 247.4 \text{ kN}$$

Comment: The preceding results indicate that the uniform wear condition yielded a smaller torque and actuating force; it is therefore the more conservative of the two assumptions in terms of clutch capacity.

13.10.2 Disk Brakes

A disk brake is very similar to the disk clutch shown in Figure 13.14, with the exception that one of the shafts is replaced by a fixed member. Loads are balanced by locating friction linings or pads on both sides of the disk. Servo action can be obtained by addition of several machine parts. The torque capacity and actuating force requirements of disk brakes may readily be ascertained through the use of the foregoing procedures. The equations for the disk clutch can be adapted to the disk brake, if the brake pad is shaped like a sector of a circle and calculations are made accordingly, as illustrated in the next example.

13.10.2.1 Caliper-Type Disk Brakes

Usually, a caliper disk brake includes a disk-shaped rotor attached to the machine to be controlled and friction *pads*. The latter cover only a small portion of the disk surface, allowing the remainder exposed to dissipate heat. Figure 13.16 shows the geometry of contact area of an annular pad segment of a caliper-type disk brake. Observe that the lining pads (one each side) squeezed against both sides of the rotating disk by actuating force F_a. The expressions for this force and braking torque T may readily be obtained from Equations 13.28, 13.29, 13.32, or 13.33, by simply multiplying the selected equation by the ratio $\gamma/360°$, in which the angle γ represents the angle subtending brake pad sector. Also called included angle, the γ often lies in the range from 45° to 90°.

The linings are contained in a fixed-caliper assembly and forced against the disk by air pressure or hydraulically (Figure 13.17). Disk brakes have been employed in automotive

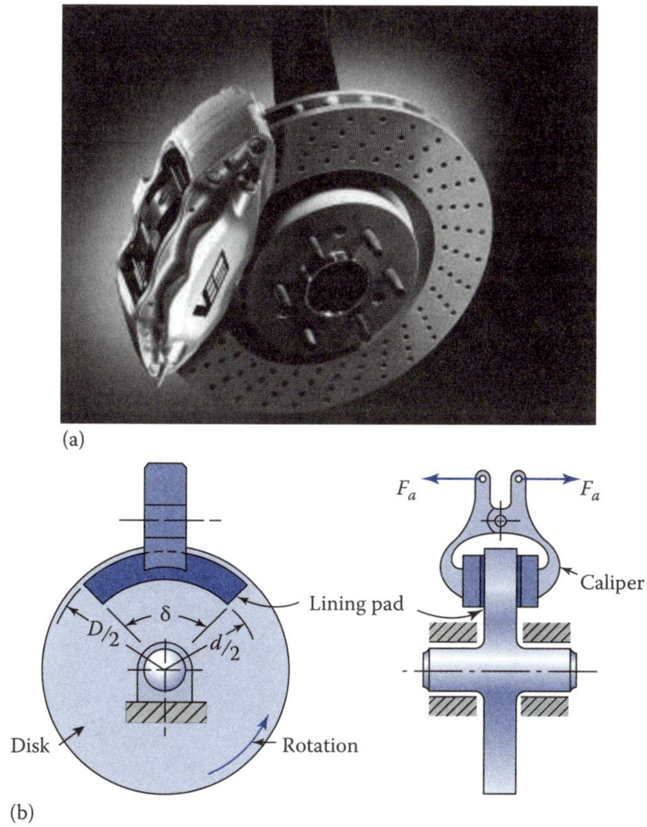

FIGURE 13.16
(a) A typical disk brake and (b) its schematic representation.

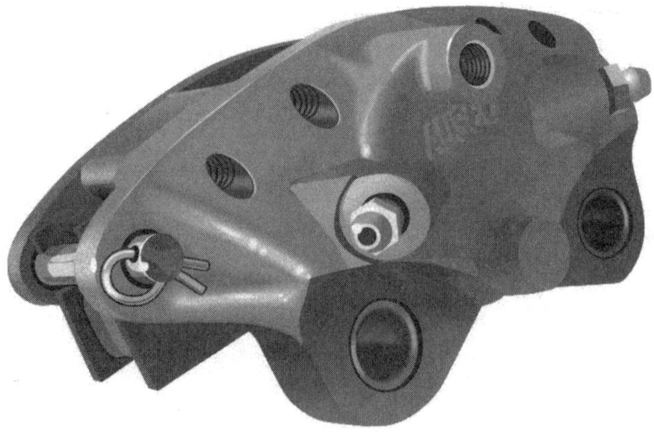

FIGURE 13.17
Caliper disk brake, hydraulically operated. (Courtesy of Ausco Products, Inc., Benton Harbor, MI.)

applications, due to their equal braking torque for either direction of rotation as well as greater cooling capacity than drum brakes (see Section 13.14). Most modern cars have disk brakes on the front wheels, and some have disk brakes on all four wheels. A disk brake of the brake system does the actual work of stopping the car. Disk brakes are also often preferred in heavy-duty industrial applications. Caliper disk brakes are widely used on the front wheel of most motorcycles. The common bicycle is another example, where the wheel rim forms the disk.

Example 13.6: Design of a Disk Brake

A disk brake has two pads of included angle $\gamma = 60°$ each, $D = 10$ in. and $d = 5$ in. (Figure 13.14). Determine

a. The actuating force required to apply one shoe
b. The torque capacity for both shoes

Design Decision: Sintered metal pads and cast iron disk are used with $f = 0.2$ and $p_{max} = 200$ psi.

Solution

a. Equation 13.28 may be written in the form

$$F_a = \frac{\gamma}{360}\left[\frac{1}{2}\pi p_{\max} d(D-d)\right] \tag{13.35}$$

Introducing the given numerical values,

$$F_a = \frac{60}{360}\left[\frac{1}{2}\pi(200)(5)(10-5)\right] = 1309 \text{ lb}$$

b. From Equation 13.30, we obtain

$$T = \frac{1}{4} F_a f (D + d)$$

$$= \frac{1}{4}(1309)(0.2)(10 + 5) = 982 \text{ lb} \cdot \text{in.}$$

13.11 Cone Clutches and Brakes

The cone clutch, Figure 13.18, can be considered as the general case of a disk clutch having a cone angle of 90°. Due to the increased frictional area and the wedging action of the parts, cone clutches convey a larger torque than disk clutches with the identical outside diameters and actuating forces. Practically, a cone clutch can have no more than one friction interface: hence, $N = 1$. Cone clutches are often used in low-speed applications. They could also be employed as cone brakes with some slight modifications.

13.11.1 Uniform Wear

The presupposition is made that the normal wear is proportional to the product of the normal pressure p and the radius. Let the radius r in Figure 13.18 locate the ring element running around the cone. The differential area is then equal to $dA = 2\pi r \, dr/\sin\alpha$. The normal force in the element equals $dF_n = p \, dA$, in which $p = p_{max}d/2r$. As before, p_{max} represents the maximum pressure. Hence, the total normal force is

$$F_n = \int_{d/2}^{D/2} \frac{p_{max} d}{2r} \frac{2\pi r dr}{\sin\alpha} = \frac{\pi d p_{max}}{\sin\alpha}(D - d) \tag{13.36}$$

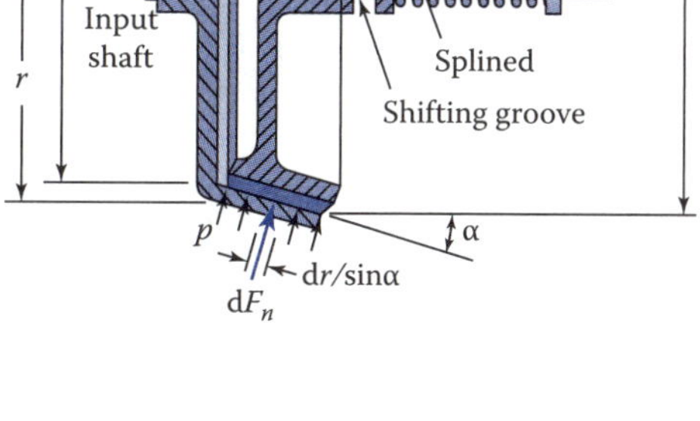

FIGURE 13.18
Cone clutch.

The corresponding axial force is $F_n \sin\alpha$. The actuating force is then

$$F_a = \frac{1}{2}\pi p_{\max} d(D-d) \tag{13.37}$$

which is the same as for a disk clutch, Equation 13.28.

The torque that can be transmitted by the ring element is equal to $dT = dF_n fr = 2\pi pfr^2\, dr/\sin\alpha$. The torque capacity of the clutch T is obtained by integrating the forgoing expression over the conical surface. In so doing, we obtain

$$T = \frac{\pi f p_{\max} d}{\sin\alpha}\int_{d/2}^{D/2} r\,dr = \frac{\pi f p_{\max} d}{8\sin\alpha}\left(D^2 - d^2\right) \tag{13.38}$$

In terms of the actuating force, we have

$$T = \frac{F_a f}{4\sin\alpha}(D+d) \tag{13.39}$$

13.11.2 Uniform Pressure

An analysis analogous to the uniform pressure made for disk clutches in Section 13.10 results in the following equations (Problem 13.28):

$$F_a = \frac{1}{4}\pi p_{\max}\left(D^2 - d^2\right) \tag{13.40}$$

$$T = \frac{\pi f p_{\max}}{12\sin\alpha}\left(D^3 - d^3\right) = \frac{F_a f}{3\sin\alpha}\frac{D^3 - d^3}{D^2 - d^2} \tag{13.41}$$

The cone angle α, cone diameter, and cone face width w are essential design parameters. The smaller the cone angle, the less actuating force is needed. This angle has a minimum value of 8°. It is because the clutch may bind or lock up if smaller angles are used. An angle of 12° is ordinarily regarded about optimum. The generally used values of α are in the range of 8°–15°.

Example 13.7: Pressure Capacity of a Cone Clutch

A cone clutch having an outside diameter D, inner diameter d, and width w (Figure 13.18) transmits a torque T.

Find: The contact pressure and the actuating force on the basis of

a. The uniform wear
b. The uniform pressure

Given: $D = 300$ mm, $d = 280$ mm, $w = 50$ mm, $T = 150$ N·m

Assumption: Coefficient of friction will be taken to be $f = 0.24$.

Solution

a. The half-cone angle of the clutch equals (Figure 13.18)

$$\sin\alpha = \frac{D-d}{2w} = \frac{300-280}{2(50)} = 0.2$$

or

$$\alpha = 11.54°$$

From Equation 13.38, the maximum pressure is found as

$$p_{\max} = \frac{8T\sin\alpha}{\pi f d\left(D^2 - d^2\right)} \tag{a}$$

Inserting the given data, we have

$$p_{\max} = \frac{8(150)(0.2)}{\pi(0.24)(0.28)\left(0.3^2 - 0.28^2\right)} = 98 \text{ kPa}$$

The actuating force, applying Equation 13.37, is then

$$F_a = \frac{1}{2}\pi p_{\max} d(D-d)$$
$$= \frac{1}{2}\pi\left(98\times10^3\right)(0.28)(0.3-0.28) = 862 \text{ Pa}$$

b. Making use of Equation 13.41, the maximum pressure is

$$p_{\max} = \frac{12T\sin\alpha}{\pi f\left(D^3 - d^3\right)}$$
$$= \frac{12(150)(0.2)}{\pi(0.24)\left(0.3^3 - 0.28^3\right)} = 94.6 \text{ kPa} \tag{b}$$

The actuating force, from Equation 13.40, is

$$F_a = \frac{1}{4}\pi\left(94.6\times10^3\right)\left(0.3^2 - 0.28^2\right) = 862 \text{ Pa}$$

Comment: The results show that uniform wear condition gives pressure capacity of about 3.5% larger than that of uniform rate.

13.12 Band Brakes

The *band brake*, the simplest of many braking devices, is employed in power excavators and hoisting and other machineries. Usually, the band is made of steel and lined with a woven friction material for flexibility. The braking action is secured by tightening the band wrapped around the drum that is to be slowed or halted. The difference in tensions at each end of the band ascertains the torque capacity.

Figure 13.19 shows a band brake with the drum rotating clockwise. For this case, friction forces acting on the band increase the tight-side tension F_1 and decrease the slack-side tension F_2. Consider the drum and band portion *above* the sectioning plane as a free body. Then, summation of the moments about the center of rotation of the drum gives the *torque capacity*, which is the same as for a belt drive:

$$T = (F_1 - F_2)r \tag{13.42}$$

The quantity r is the radius of the drum. Likewise, considering the lever and hand portion *below* the sectioning plane as free body, the *actuating force* is

$$F_a = F_2 \frac{c}{a} \tag{13.43}$$

The brake is actuated by the application of force F_a at the free end of the lever. It is obvious that a smaller force F_a is needed for operation when the tight side of the band is connected to the fixed support and the slack side attached to the lever, as shown in the figure.

An expression relating band tensions F_1 and F_2 is derived by following the same procedure used for flexible belts, with the exception that the centrifugal force acting on belts does not exist. Hence, referring to Section 13.3, the band tension relationship has the form

$$\frac{F_1}{F_2} = e^{f\phi} \tag{13.44}$$

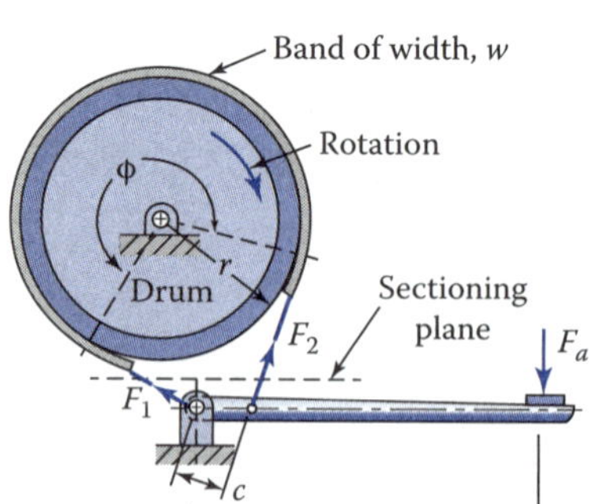

FIGURE 13.19
Simple band brake.

Here,

F_1 = the larger tensile force
F_2 = the smaller tensile force
f = the coefficient of friction
ϕ = the angle of contact between band and drum or the angle of wrap

Let the analysis of the belt shown in Figure 13.5b be applied to the band at the point of tangency for F_1. We now have $F_c = 0$ and $dN = p_{max}wr\,d\theta$. The inward components of the band forces are equal to $dN = F_1 d\theta$. These two forces are set equal to each other to yield

$$F_1 = wrp_{max} \tag{13.45}$$

The quantity p_{max} is the maximum pressure between the drum and lining, and w represents the width of band. An expression similar to this can also be written for the slack side.

The *differential band brake* is analogous to the simple band brake, with the exception that the tight-side tension helps the actuating force (Figure 13.20). A brake of this type is termed *self-energized*, since the friction force assists in applying the band. For a differential brake, Equation 13.43 becomes

$$F_a = \frac{1}{a}(cF_2 - sF_1) \tag{13.46}$$

In the case of a *self-locking* brake, the product sF_1 is greater than cF_2. Note that, when a brake is designed to be self-locking for one direction of rotation, it can be free to rotate in the opposite direction. A self-locking brake can then be employed when rotation is in one direction only.

Example 13.8: Design of a Differential Band Brake

A differential band brake similar to that in Figure 13.20 uses a woven lining having design values of $f = 0.3$ and $p_{max} = 375$ kPa. Determine

a. The torque capacity
b. The actuating force
c. The power capacity
d. The value of dimension s that would cause the brake to be self-locking

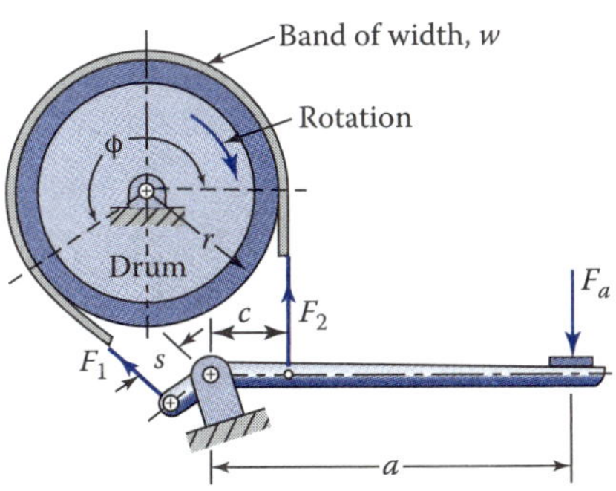

FIGURE 13.20
Differential brake.

Given: The speed is 250 rpm, $a = 500$ mm, $c = 150$ mm, $w = 60$ mm, $r = 200$ mm, $s = 25$ mm, and $\phi = 270°$.

Solution

a. Through the use of Equation 13.45, we obtain

$$F_1 = wrp_{max} = (0.06)(0.2)(375) = 4.5 \text{ kN}$$

Applying Equation 13.44,

$$F_2 = \frac{F_1}{e^{f\phi}} = \frac{4.5}{e^{0.3(1.5\pi)}} = 1.095 \text{ kN}$$

Then, Equation 13.42 gives $T = (4.5 - 1.095)(0.2) = 0.681$ kN m.

b. By Equation 13.46,

$$F_a = \frac{150(1.095) - 25(4.5)}{0.5} = 103.5 \text{ N}$$

c. From Equation 1.15,

$$\text{kW} = \frac{Tn}{9549} = \frac{681(250)}{9549} = 17.8$$

d. Using Equation 13.46, we have $F_a = 0$ for $s = 150(1.095)/4.5 = 36.5$ mm.

Comment: The brake is self-locking if $s \geq 36.5$ mm.

13.13 Short-Shoe Drum Brakes

A short-shoe drum brake consists of a short shoe pressed on the revolving drum by a lever. The schematic representation of a brake of this type is depicted in Figure 13.21. Inasmuch as the shoe is relatively short (i.e., the angle of contact is small, $\phi < 45°$), a *uniform pressure*

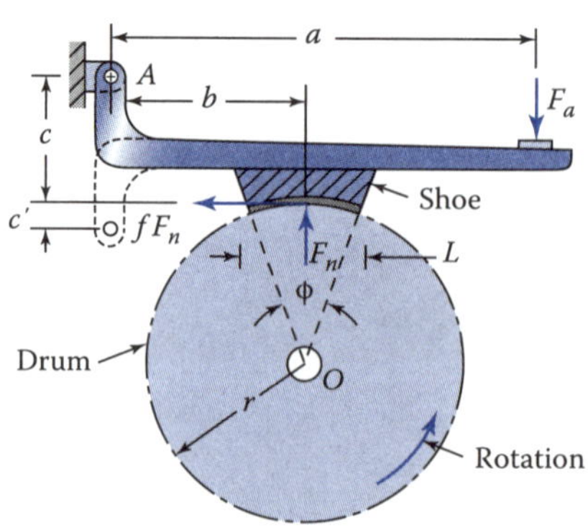

FIGURE 13.21
Short-shoe drum brake.

distribution may be taken between drum and shoe. Accordingly, the resultant normal force and the friction force act at the center of contact.

The *projected area A* of the shoe is the width multiplied by the chord length subtended by a ϕ^z arc of the radius of the drum. From the geometry of the figure, $A = wL = 2[r \sin(\phi/2)]w$. Hence, the normal force on the shoe is

$$F_n = p_{\max}\left[2\left(r\sin\frac{\phi}{2}\right)\right]w \tag{13.47}$$

In the foregoing, we have

F_n = the normal force

$p_{\max}$ = the maximum pressure between the drum and shoe

r = the radius of the drum

ϕ = the angle of contact

w = the width of the shoe

The value of the friction force is fF_n. The sum of moments about point O for the free-body diagram of the drum yields the *torque capacity* of the brake as

$$T = fF_n r \tag{13.48}$$

The quantity f represents the coefficient of friction.

We now consider the lever as the free body. Then taking moments about the pivot A, we have

$$F_a a + fF_n c - bF_n = 0$$

The preceding leads to the *actuating force*

$$F_a = \frac{F_n}{a}(b - fc) \tag{13.49}$$

in which a, b, and c represent the distances shown in Figure 13.21.

13.13.1 Self-Energizing and Self-Locking Brakes

For the brake with the *direction* of the rotation shown in the figure, the moment of the friction force assists in applying the shoe to the drum; this makes the brake *self-energizing*. If $b = fc$ or $b < fc$, the force F_a required to actuate the brake becomes 0 or negative, respectively. The brake is then said to be *self-locking* when

$$b \le fc \tag{13.50}$$

A self-locking brake requires only that the shoe be brought in contact with the drum (with $F_a = 0$) for the drum to be *loaded* against rotation in one direction. The self-energizing feature is useful, but the self-locking effect is generally undesirable. To secure proper utilization of the self-energizing effect while avoiding self-lock, the value of b must be at least 25%–50% greater than fc.

Note that, if the brake drum *rotation* is *reversed* from that indicated in Figure 13.21, the sign of fc in Equation 13.49 becomes negative and the brake is then *self-de-energized*. Also, if the pivot is located on the other side of the line of action of fF_n, as depicted by the dashed lines in the figure, the friction force tends to unseat the shoe. Then, the brake would not be self-energizing. Clearly, both pivot situations discussed are reversed if the direction of rotation is reversed.

Example 13.9: Design of a Short-Shoe Drum Brake

The brake shown in Figure 13.21 uses a sintered metal lining having design values of $f = 0.4$ and $p_{max} = 150$ psi. Determine

a. The torque capacity and actuating force
b. The reaction at pivot A

Given: $a = 12$ in., $w = 3$ in., $b = 5$ in., $c = 2$ in., $r = 4$ in., $\phi = 30°$

Solution

a. From Equation 13.47, we have

$$F_n = 150\left[2\left(4\sin\frac{30°}{2}\right)\right]3 = 931.7 \text{ lb}$$

Equation 13.48 yields

$$T = (0.4)(931.7)(4) = 1.491 \text{ kip} \cdot \text{in.}$$

Applying Equation 13.49,

$$F_a = \frac{931.7(5 - 0.4 \times 2)}{12} = 326.1 \text{ lb}$$

b. The conditions of equilibrium of the horizontal (x) and vertical (y) forces give

$$R_{Ax} = 931.7(0.4) = 372.7 \text{ lb}, \quad R_{Ay} = 605.6 \text{ lb}$$

The resultant radial reactional force is

$$R_A = \sqrt{372.7^2 + 605.6^2} = 711.1 \text{ lb}$$

13.14 Long-Shoe Drum Brakes

When the angle of contact between the shoe and the drum is about 45° or more, the short-shoe equations can lead to appreciable errors. Most shoe brakes have contact angles of 90° or greater, so a more accurate analysis is needed. The obvious problem relates the determination of the pressure distribution. The analysis that includes the effects of deflection is complicated and not warranted here. In the following development, we make the usual simplifying assumption: the pressure varies directly with the distance from the shoe pivot point. This is equivalent to the presupposition made earlier that the wear is proportional to the product of pressure and velocity.

13.14.1 External Long-Shoe Drum Brakes

Figure 13.22 illustrates an external long-shoe drum brake. The pressure p at some arbitrary angle θ is proportional to $c \sin\theta$. However, since c is a constant, p varies directly with $\sin\theta$.

As a result,

$$p = p_{max}\frac{\sin\theta}{(\sin\theta)_m} \tag{13.51}$$

Here

p_{max} is the maximum pressure between the lining and the drum

$(\sin\theta)_m$ is the maximum value of $\sin\theta$

Based on the geometry,

$$(\sin\theta)_m = \begin{cases} 1.0 & (\text{if } \theta_2 > 90°) \\ \sin\theta_2 & (\text{if } \theta_2 \le 90°) \end{cases} \tag{13.52}$$

Note from Equation 13.52 that the maximum pressure takes place at the location having the value of $(\sin\theta)_m$. External long-shoe brakes are customarily designed for $\theta_1 \ge 5°$, $\theta_2 < 120°$, and $\phi = 90°$, where ϕ is the angle of contact.

Let w represent the width of the lining. Then, the area of a small element, cut by two radii an angle $d\theta$ apart, is equal to $wr\,d\theta$. Multiplying by the pressure p and the arm $c\sin\theta$ and integrating over the entire shoe, the moment of normal forces, M_n, about pivot A results:

$$M_n = \int_{\theta_1}^{\theta_2} p(wr d\theta)c\sin\theta$$

$$= \frac{wrcp_{max}}{(\sin\theta)_m}\int_{\theta_1}^{\theta_2}\sin^2\theta d\theta$$

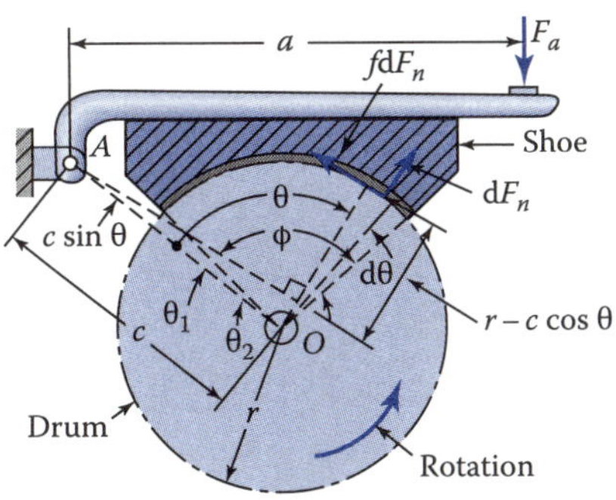

FIGURE 13.22
External long-shoe drum brake.

from which

$$M_n = \frac{wrcp_{\max}}{4(\sin\theta)_m}[2\phi - \sin 2\theta_2 + \sin 2\theta_1] \tag{13.53}$$

In a like manner, the moment of friction forces, M_f, about A is written in the form

$$M_f = \int_{\theta_1}^{\theta_2} fpwr\mathrm{d}\theta(r - c\cos\theta)$$

$$= \frac{fwrp_{\max}}{(\sin\theta)_m}\int_{\theta_1}^{\theta_2}\left(r\sin\theta - \frac{c}{2}\sin 2\theta\right)\mathrm{d}\theta$$

or

$$M_f = \frac{fwrp_{\max}}{4(\sin\theta)_m}\left[c(\cos 2\theta_2 - \cos 2\theta_1) - 4r(\cos\theta_2 - \cos\theta_1)\right] \tag{13.54}$$

Now, summation of the moments about the pivot point A results in the actuating force

$$F_a = \frac{1}{a}(M_n \mp M_f) \tag{13.55}$$

In this equation, the upper sign is for a self-energizing brake and the lower one for a self-de-energizing brake. Self-locking occurs when

$$M_f \geq M_n \tag{13.56}$$

As noted previously, it is often desirable to make a brake shoe self-energizing but not self-locking. This can be accomplished by designing the brake so that the ratio M_f/M_n is no greater than about 0.7.

The torque capacity of the brake is found by taking moments of the friction forces about the center of the drum O. In so doing, we have

$$T = \int_{\theta_1}^{\theta_2} fpwr\mathrm{d}\theta r$$

$$= \frac{fwrp_{\max}}{(\sin\theta)_m}\int_{\theta_1}^{\theta_2}\sin\theta\mathrm{d}\theta$$

from which

$$T = \frac{fwr^2p_{\max}}{(\sin\theta)_m}(\cos\theta_1 - \cos\theta_2) \tag{13.57}$$

Finally, pin reactions at A and O can readily be obtained from horizontal and vertical force equilibrium equations. Note that reversing the direction of the rotation changes the sign of the terms containing the coefficient of friction in the preceding equations.

Example 13.10: Design of a Long-Shoe Drum Brake

The long-shoe drum brake is actuated by a mechanism that exerts a force of $F_a = 4$ kN (Figure 13.23). Determine

a. The maximum pressure
b. The torque and power capacities

Design Decision: The lining is a molded material having a coefficient of friction $f = 0.35$ and a width $w = 75$ mm.

Solution

The angle of contact is $\phi = 90°$ or $\pi/2$ rad. From the geometry, $\alpha = \tan^{-1}(200/150) = 53.13°$. Hence,

$$\theta_1 = 8.13°, \quad \theta_2 = 98.13°$$

$$c = \sqrt{200^2 + 150^2} = 250 \text{ mm}$$

Inasmuch as $\theta_2 > 90°$, $(\sin\theta)_m = 1$.

a. Through the use of Equation 13.53,

$$M_n = \frac{(0.075)(0.15)(0.25)p_{\max}}{4(1)}\left[2\left(\frac{\pi}{2}\right) - \sin 196.26° + \sin 16.26°\right]$$

$$= 2.6\left(10^{-3}\right)p_{\max}$$

From Equation 13.54,

$$M_f = \frac{0.35(0.075)(0.15)p_{\max}}{4}\left[(0.25)(\cos 196.26° - \cos 16.26°) - 4(0.15)(\cos 98.13° - \cos 8.13°)\right]$$

$$= 0.196\left(10^{-3}\right)p_{\max}$$

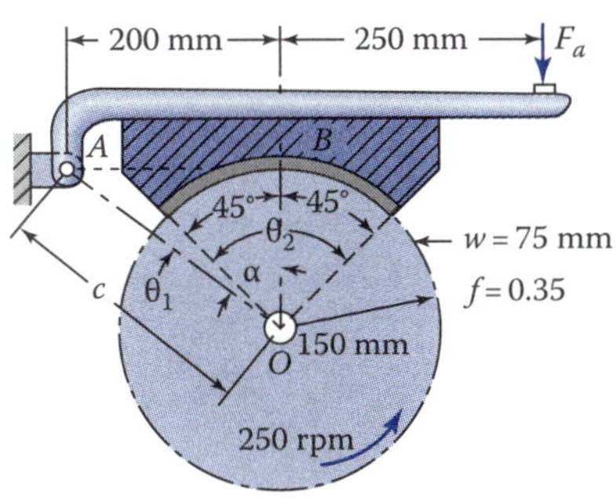

FIGURE 13.23
Example 13.10.

Applying Equation 13.55, we then have

$$4000(0.45) = (2.6 + 0.196)(10^{-3})p_{max}$$

or

$$P_{max} = 644 \text{ kPa}$$

b. Using Equation 13.57,

$$T = \frac{(0.35)(0.075)(0.15^2)(0.644 \times 10^6)}{1}(\cos 8.13° - \cos 98.13°)$$

$$= 430.3 \text{ N} \cdot \text{m}$$

By Equation 1.15, the corresponding power is

$$\text{kW} = \frac{Tn}{9549} = \frac{430.3(250)}{9549} = 11.28$$

13.14.1.1 Symmetrically Loaded Pivot-Shoe Brakes

A special case where the pivot is symmetrically located is illustrated in Figure 13.24. Observe from the figure that the magnitude of the friction forces with respect to the pivot A is zero. The largest pressure occurs at $\theta = 0°$ and the pressure variation can be expressed as

$$p = p_{max} \cos\theta \tag{a}$$

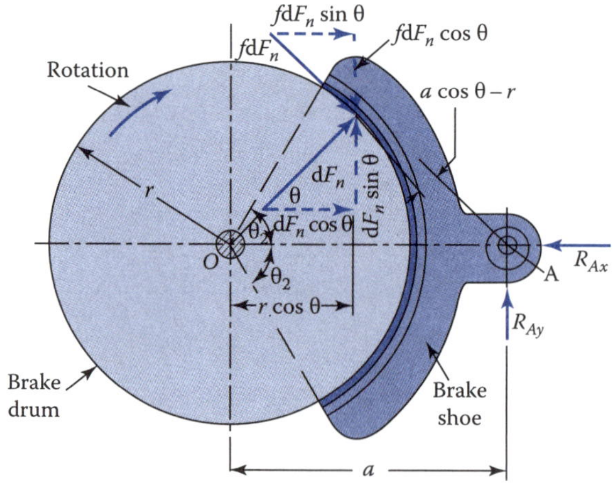

FIGURE 13.24
Brake with symmetrical pivoted shoe.

At any value of θ from the pivot, a differential normal force dF_n on the shoe is equal to

$$dF_n = pw(rd\theta) = p_{max}wr\cos\theta d\theta \tag{b}$$

in which w is the face width (perpendicular to the paper) of the friction material or the brake lining.

The distance a to the pivot is chosen such that the moment of friction forces M_f is zero. That is,

$$M_f = 2\int_0^{\theta_2} (f dF_n)(a\cos\theta - r) = 0$$

Carrying Equation (b) into this expression leads to

$$2fp_{max}wr\int_0^{\theta_2} \left(a\cos^2\theta - r\cos\theta\right) = 0$$

from which

$$a = \frac{4r\sin\theta_2}{2\theta_2 + \sin 2\theta_2} \tag{13.58}$$

Referring to Figure 13.24, the horizontal reaction may be expressed as

$$R_{Ax} = 2\int_0^{\theta_2} dF_n\cos\theta$$

This becomes, inserting Equations (b) and 13.58:

$$R_{Ax} = \frac{p_{max}}{2}(2\theta_2 + \sin 2\theta_2) = \frac{2wr^2}{a}p_{max}\sin\theta_2 \tag{13.59}$$

Here, owing to symmetry, $\int f dF_n \sin\theta = 0$. In a like manner, the vertical reaction has the form

$$R_{Ay} = 2\int_0^{\theta_2} f dF_n\cos\theta$$

We thus have

$$R_{Ay} = fR_x \tag{13.60}$$

in which, due to symmetry, $\int dF_n \sin\theta = 0$. The resultant pivot reaction is $R_A = \left[R_{Ax}^2 + R_{Ay}^2\right]^{1/2}$.

In addition, a final point is to be noted that

$$R_{Ax} = -F_n \quad \text{and} \quad R_{Ay} = -fF_n \tag{13.61a}$$

where F_n represents the resultant normal force on the shoe. The torque therefore is equal to

$$T = R_{Ay}a = afF_n \tag{13.61b}$$

The preceding is valid for the particular choice of the dimension a, defined by Equation 13.58.

13.14.2 Internal Long-Shoe Drum Brakes

Figure 13.25 shows an internal long-shoe drum brake. A brake of this type is widely used in automotive services. We see from the figure that both shoes pivot about anchor pins (A and B) and are forced against the inner surface of the drum by a piston in each end of the hydraulic wheel cylinder. The actuating forces are thus exerted hydraulically by pistons. The light return spring applies only enough force to take in the shoe against the adjusting cams. Each adjusting cam functions as a *stop* and is utilized to minimize the clearance between the shoe and drum.

The method of analysis and the resulting expressions for the internal brakes are identical with those of long-shoe external drum brakes just discussed. That is, Equations 13.51 through 13.54 apply as well to internal-shoe drum brakes. Note that, now, a positive result for M_n indicates clockwise moment about A of the left shoe or counterclockwise moment about B of the right-side shoe. A positive or negative result for friction moment M_f should be interpreted in the same manner as for a brake with external shoe.

Typically, in Figure 13.25, the *left shoe* is *self-energizing* and the right shoe is deenergizing. Should the direction of the rotation be reversed, the right shoe would be self-energizing and the left shoe would not. For a prescribed actuating force, the braking capacity with both shoes self-energizing is clearly higher than if only one were. Interestingly, automotive brakes are also made using two hydraulic wheel cylinders with both shoes self-energizing. Of course, this results in braking ability in reverse that is much less than for forward motion. Recently, the caliper-type disk brakes discussed

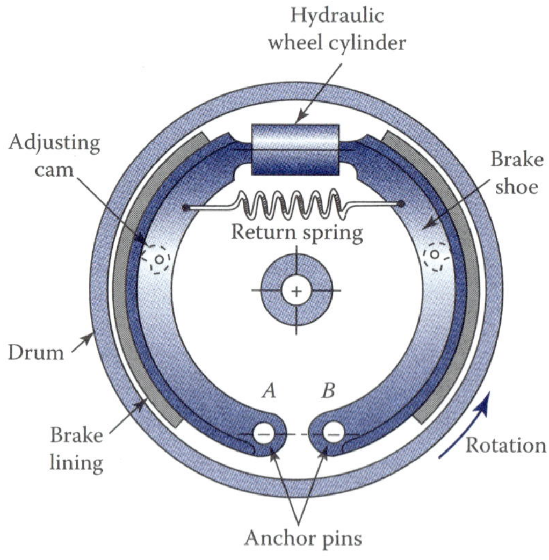

FIGURE 13.25
Brake with internal long shoe, automotive-type brake.

in Section 13.10 have replaced front drum brakes on most passenger cars due to their greater cooling capacity and other good qualities.

13.15 Energy Absorption and Cooling

The primary role of a brake is to absorb energy and dissipate the resulting heat without developing high temperatures. Clutches also absorb energy and dissipate heat but at a lower rate, since they connect two moving elements. The quality of heat dissipation depends on factors such as the size, shape, and condition of the surface of the various parts. Obviously, by increasing exposed surface areas (such as by fins and ribs) and the flow of the surrounding air, these devices can be cooled more conveniently. In addition, the length of time and the interval of brake application affect the temperature. With an increase of the temperature of the brake (or clutch), its coefficient of friction decreases. The result is *fading*; that is, the effectiveness of the device may be sharply deteriorated. The torque and power capacity of a brake or clutch is thus limited by the characteristics of the material and the ability of the device to dissipate heat. A satisfactory braking or clutching performance requires that the heat generation should not exceed the heat dissipation.

13.15.1 Energy Sources

The energy equation depends on the type of motion a body is going under. Let us consider a body of the weight W, mass m, and mass moment of inertia about its axis of rotation I. The sources of energy to be absorbed from the body by the clutch or brake are mainly as follows: Kinetic energy of translation is

$$E_k = \frac{1}{2} m\upsilon^2 \tag{13.62}$$

Kinetic energy of rotation is

$$E_k = \frac{1}{2} I\omega^2 \tag{13.63}$$

Potential energy is

$$E_p = Wh \tag{13.64}$$

In the foregoing

υ = the velocity
ω = the angular velocity
h = the vertical distance

To clarify the relevance to brakes of the kinetic and potential energies, refer to a winch crane (Figure 18.1). Suppose that the crane lowers a mass m of weight W translating at time t_1 with a velocity υ_1 at elevation h_1 and the gear shafts with mass moments of inertia I rotating at angular velocities ω_1. Shafts may be rotating at different speeds. If at time t_1 the internal brake (in the motor) is applied, then, at a time t_2, quantities will have reduced to υ_2, ω_2, and h_2. Therefore, during the time interval $t_2 - t_1$, we have [4] W_h as the work done by the brake; W_r as the work done by rolling friction, bearing friction, and air resistance; and W_m as the work done by drive motor. The conservation of the energy requires that the total work equals the change in energy:

$$W_b + W_r + W_m = \frac{1}{2}m\left(\upsilon_1^2 - \upsilon_2^2\right) + \Sigma\frac{1}{2}I\left(\omega_1^2 - \omega_2^2\right) + W\left(h_1 - h_2\right)$$

Here, the summation consists of multiplications made for different mass moments of inertia at their corresponding angular velocities.

The work required of the brake to stop, slow, or maintain speed is obtained by the solution of the preceding equation for W_b. This presents the mechanical energy transformed into heat at the brake and can be used to predict the temperature rise. Note that, in many machines such as slow speed hoists and winch cranes, W_r and W_m are negligible. Clearly, omitting these quantities results in a safer brake design.

13.15.2 Temperature Rise

When the motion of the body is halted or transmitted by a braking or clutching operation, the frictional energy E developed appears in the form of heat in the device. The temperature rise may be related to the energy absorbed in the form of friction heat by the familiar formula

$$\Delta t = \frac{E}{Cm} \tag{13.65}$$

where

Δt = the temperature rise,°C
E = the frictional energy the brake or clutch must absorb, J
C = the specific heat (use 500 J/kg°C for steel or cast iron)
m = the mass of brake or clutch parts, kg

The frictional energy E is ascertained as briefly discussed in the preceding. Then, through the use of Equation 13.65, the temperature rise of the brake or clutch assembly is obtained. The limiting temperatures for some commonly used brake and clutch linings are furnished in Table 13.11. These temperatures represent the largest values for steady operation.

Equations 13.62 through 13.65 illustrate what happens when a brake or clutch is operated. Many variables are involved, however, and such an analysis may only estimate experimental results. In practice, the rate of energy absorption and heat dissipated by a brake or clutch is of utmost importance. Brake and lining manufacturers include the effect of the rate of energy dissipation by assigning the appropriate limiting values of pV, a product of pressure and velocity, for specific kinds of brake design and service conditions [13–15]. Typical values of pV used in industry are given in Table 13.13.

TABLE 13.13

Representative pV Values for Shoe Brakes

Operation	Heat Dissipation	pV (MPa) (m/s)	pV (ksi) (ft/min)
Continuous	Poor	1.05	30
Occasional	Poor	2.10	60
Continuous	Good	3.00	85

Problems

Sections 13.1 through 13.7

13.1 A flat belt 4 in. wide and 3⁄16 in. thick operates on pulleys of diameters 5 in. and 15 in. and transmits 10 hp. Determine

a. The required belt tensions

b. The belt length

Given: Speed of the small pulley is 1500 rpm, the pulleys are 5 ft apart, the coefficient of friction is 0.30, and the weight of the belt material is 0.04 lb/in.^{3}.

13.2 A plastic flat belt 60 mm wide and 0.5 mm thick transmits 10 kW. Calculate

a. The torque at the small pulley

b. The contact angle

c. The maximum tension and stress in the belt

Given: The input pulley has a diameter of 300 mm, and it rotates at 2800 rpm, and the output pulley speed is 1600 rpm: the pulleys are 700 mm apart, the coefficient of friction is 0.2, and belt weight is 25 kN/m^3.

Assumptions: The driver is a high torque motor and the driven machine is under a medium shock load.

13.3 Rework Example 13.2 for the case in which the radius of the driven pulley equals $r_2 = 8$ in. and the coefficient of friction is changed to 0.25. Also find the length of the belt using Equation 13.9.

13.4 Figure P13.4 shows a flat-belt drive, where pulley B runs a machine tool and pulley A attached to the shaft of an electric motor. What is the largest torque that can be exerted by the belt on each pulley?

Given: $f = 0.15$, $F_1 = 2.5$ kN, $r_1 = 20$ mm, $r_2 = 150$ mm, $\phi = 120°$

Assumption: Belt runs slowly so that the centrifugal force may be neglected.

13.5 A 2 hp, 2500 rpm electric motor drives a machine through a flat belt. The driven shaft speed requirement equals 1000 rpm. Find

a. The pitch line velocity

b. The radius of the driven pulley r_2 and the angle ϕ wrap of the driving pulley

c. The values of tension forces F_1 and F_2

Given: $c = 25$ in., $r_1 = 1.5$ in., $f = 0.35$, $w = 0.01$ lb/in.

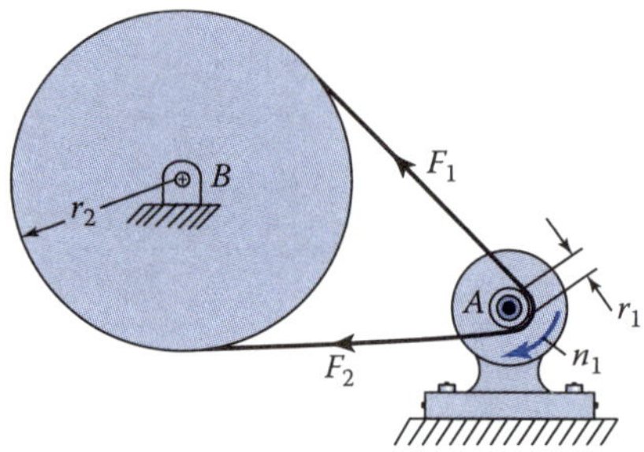

FIGURE P13.4

13.6 A timing belt similar to that shown in Figure P13.4 having 0.05 lb/in. weight is used to transfer power from an engine to a grinding wheel. The largest permissible force in the belt equals $F_1 = 500$ lb, and both engine and wheel rotate at the same speed of $n = 4200$ rpm. What is the optimum pulley pitch radius r for largest power transfer? Hint: maximum power, $P = V(F_1 - F_c)$, transfer occurs when $\partial P/\partial V = 0$, where F_c is the centrifugal force.

13.7 What is the largest power that can be transmitted by the pulley A of a V-belt drive illustrated in Figure P13.7?

Given: $n_1 = 3000$ rpm, $r_1 = 80$ mm, $\beta = 18°$, $\phi = 160°$, $f = 0.25$, $F_1 = 1100$ N, $w = 1.4$ N/m.

13.8 A V-belt drive has a 200 mm diameter small sheave with a 170° contact angle, 38° included angle, 0.15 coefficient of friction, 1600 rpm driver speed, belt weight of 8 N/m, and a tight-side tension of 3 kN. Determine the power capacity of the drive.

13.9 A V-belt drive has an included angle of 38°, belt weight of 3 N/m, belt cross-sectional area of 145 mm^2, coefficient of friction of 0.25, r_1, = 150 mm, $\phi = 160°$, $n_1 = 3000$ rpm, and $F_2 = 800$ N. Calculate

a. The maximum power transmitted

b. The maximum stress in the belt

13.10 A V-belt drive with an included angle of 34° is to have a capacity of 15 kW based on a coefficient of friction of 0.2 and a belt weight of 2.5 N/m. Determine the required maximum belt tension at full load.

Assumptions: The driver is a normal torque motor and the driven machine involves heavy shocks.

Design Decision: Speed is to be reduced from 2700 to 1800 rpm using a 200 mm diameter small sheave: shafts are 500 mm apart.

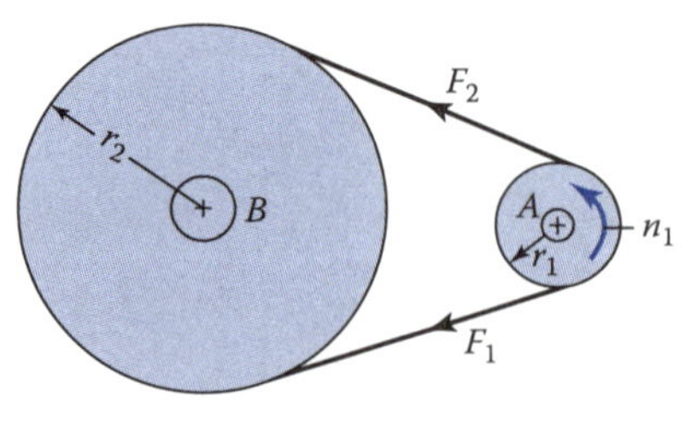

FIGURE P13.7

13.11 A two-strand ANSI No. 60, 19 mm (¾ in.) pitch roller-chain drive transmits power from a 22-tooth driver operating at 1400 rpm using an electric motor. The driven sprocket of a helicopter transmission rotates at 700 rpm under heavy shock. Find

a. The number of teeth on the driven sprocket

b. The design horsepower

c. The safety factor n of the chain based on ultimate strength

13.12 A two-strand ANSI No. 60, 19 mm (¾ in.) pitch roller-chain drive transmits power from an 18-tooth driver operating at 1600 rpm using an IC engine (hydraulic drive). The driven sprocket of a helicopter transmission rotates at 640 rpm under moderate shock. Calculate

a. The number of teeth on the driven sprocket

b. The design horsepower

c. The factor of safety n of the chain based on ultimate strength

13.13 A ⅝ in. pitch roller chain operates on a 22-tooth drive sprocket rotating at 4000 rpm and a driven sprocket rotating at 1000 rpm. Calculate the minimum center distance.

13.14 A 7/16 in. pitch inverted chain operates on a 14-tooth drive sprocket rotating at 4600 rpm and a driven sprocket at 2100 rpm. Determine the minimum center distance.

13.15 A four-strand ANSI No. 60, ¾ in. pitch roller-chain drive, under a moderate shock, transmits power from a 23-tooth driver operating at 1800 rpm using a turbine. The speed ratio is 3:1. Find

a. The number of teeth on the driven sprocket

b. Design power capacity

c. Tension in the chain

d. The safety factor n of the chain based on ultimate strength

13.16 Reconsider Problem 13.15, for the case in which a three-strand roller chain transmits power from a 32-tooth driving sprocket operating at 900 rpm.

Sections 13.8 through 13.11

13.W Search the website at www.sepac.com. List the selection (application procedure and application) factors to consider prior to choosing a brake or clutch.

13.17 A disk clutch has a single pair of friction surfaces of 250 mm outside diameter × 150 mm inside diameter. Determine the maximum pressure and the torque capacity, using the assumption of

a. Uniform wear

b. Uniform pressure

Given: The coefficient of friction is 0.3 and the actuating force equals 6 kN.

13.18 A disk clutch that has both sides effective, an outside diameter four times the inside diameter, used in an application where 40 hp is to be developed at 500 rpm. Determine, based on uniform pressure condition,

a. The inside and outside diameters

b. The actuating force required

Design Decisions: A friction material with $f = 0.25$ and $p_{max} = 20$ psi is used.

13.19 Resolve Problem 13.18 based on the assumption of uniform wear.

13.20 A disk clutch with a single pair of friction surfaces is to be used in a turbine with a maximum torque of 1.2 kip in. A sintered metal will contact steel in a dry environment. A safety factor of $n = 1.6$ is taken to consider for slippage at full turbine torque. Find, on the basis of uniform wear condition,

a. The outer diameter of the disk

b. The actuating force

Given: $d = 2$ in., $f = 0.3$, $p_{max} = 225$ psi (by Table 13.11)

13.21 A multiple-disk clutch having four active faces, 12 in. outer diameter, 6 in. inner diameter, and $f = 0.2$ is to carry 50 hp at 400 rpm. Determine, using the condition of uniform wear,

a. The actuating force required

b. The average pressure between the disks

13.22 A caliper brake has two annular pads, subtends an angle of 80° (Figure 13.16), and is actuated by a pair of hydraulic 40 mm cylinders. Determine, for the uniform wear condition,

a. The maximum pressure p_{max}

b. The actuating force F_a

c. The required hydraulic cylinder pressure p_{hyd}

Given: $f = 0.3$, $T = 1.8$ kN·m, $d = 200$ mm, $D = 280$ mm, $\gamma = 80°$

13.23 Redo Problem 13.22, on the basis of the uniform pressure assumption.

13.24 A 10 in. outside diameter cone clutch with 8° cone angle is to transmit 50 hp at 800 rpm. Calculate the face width w of the cone, on the basis of the uniform pressure assumption.

Design Decision: The maximum lining pressure will be 60 psi and the coefficient of friction $f = 0.3$.

13.25 Redo Problem 13.24 using the condition of uniform wear.

13.26 A cone clutch has a mean diameter of 500 mm, a cone angle of 10°, and a cone face width of $w = 80$ mm. Determine, using the uniform wear assumption,

a. The actuating force and torque capacity

b. The power capacity for a speed of 500 rpm

Design Decision: The lining has $f = 0.2$ and $p_{max} = 0.5$ MPa.

13.27 A cone clutch has an average diameter of 250 mm, a cone angle of 12°, and $f = 0.2$. Calculate the torque that the brake can transmit.

Assumptions: A uniform pressure of 400 kPa. Actuating force equals 5 kN.

13.28 Verify, based on the assumption of uniform pressure, that the actuating force and torque capacity for a cone clutch (Figure 13.18) are given by Equations 13.40 and 13.41.

Section 13.12

13.29 A band brake uses a 100 mm wide woven lining having design values of $f = 0.3$ and $p_{max} = 0.7$ MPa (Figure P13.29). Determine band tensions and power capacity at 150 rpm.

Given: $\phi = 240°$ and $r = 200$ mm

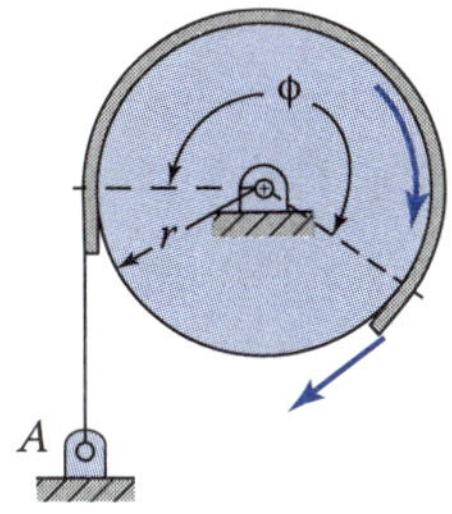

FIGURE P13.29

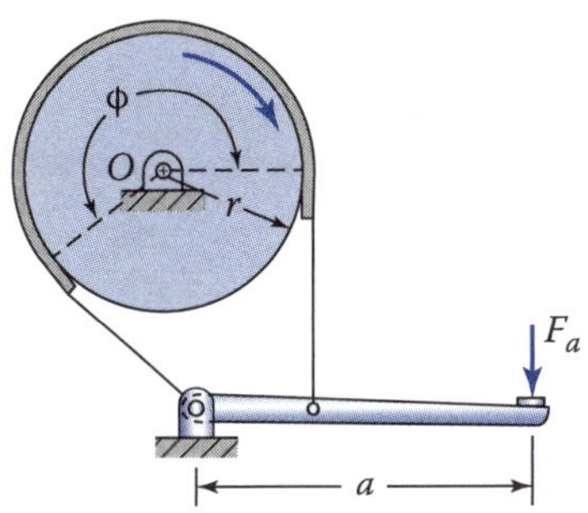

FIGURE P13.30

13.30 The drum of the band brake depicted in Figure P13.30 has a moment of inertia of $I = 20$ lb · in. · s^2 about point *O*. Calculate the actuating force F_a necessary to decelerate the drum at a rate of $\alpha = 200$ rad/s^2. Note that the torque is expressed by $T = I\alpha$.

Given: $\phi = 210°$, $a = 12$ in., $r = c = 5$ in., and $f = 0.3$.

13.31 The band brake shown in Figure P13.30 has a power capacity of 40 kW at 600 rpm. Determine the belt tensions.

Given: $\phi = 250°$, $r = 250$ mm, $a = 500$ mm, and $f = 0.4$.

13.32 The band brake depicted in Figure P13.30 uses a woven lining having design values of $p_{max} = 0.6$ MPa and $f = 0.4$. Calculate

a. The band tensions and the actuating force

b. The power capacity at 200 rpm

Given: The band width $w = 75$ mm, $\phi = 240°$, $r = 150$ mm, and $a = 400$ mm.

13.33 The differential brake depicted in Figure P13.33 is to absorb 10 kW at 220 rpm. Determine

a. The angle of wrap

b. The length of arm *s* from the geometry of the brake

Given: The maximum pressure between the lining and the drum is 0.8 MPa, $f = 0.14$, and $w = 60$ mm.

13.34 The differential brake depicted in Figure 13.20 has $a = 12$ in., $c = 2$ in., $s = 3.2$ in., $r = 4$ in., $n = 300$ rpm, $\phi = 210°$, $f = 0.12$, and $F_a = 300$ lb. If F_a is acting upward, determine the horsepower capacity.

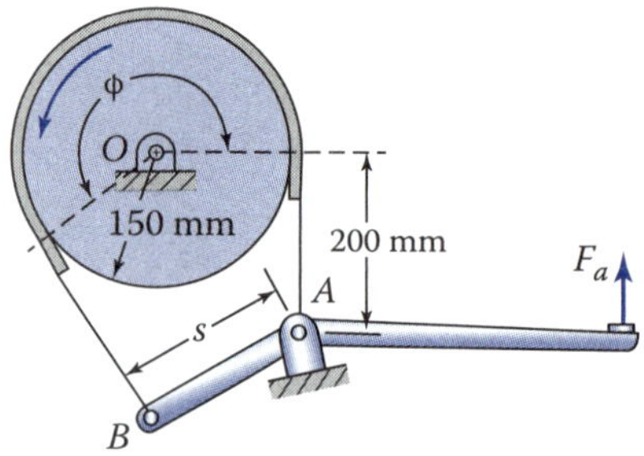

FIGURE P13.33

13.35 The differential band brake shown in Figure 13.20 has a = 250 mm, c = 100 mm, s = 50 mm, r = 200 mm, ϕ = 210°, and a woven lining material with f = 0.4. Determine the actuating force F_a required. Will the brake be self-locking?

Requirement: A power of 15 kW is to be developed at 900 rpm.

13.36 Redo Problem 13.35 for counterclockwise rotation of the drum.

13.37 The differential band brake shown in Figure 13.20 has the given dimensions r = 100 mm, w = 20 mm, a = 200 mm, c = 40 mm, s = 10 mm, wrap angle ϕ = 265°, lining coefficient of friction f = 0.25, and p_{max} = 500 kPa. Find

a. The torque capacity

b. Actuating force and value of distance s, if the brake force locks

13.38 Figure 13.20 illustrates a differential band brake that uses a lining with a design coefficient of friction f = 0.25 and maximum pressure p_{max} = 70 psi. Determine

a. The torque capacity

b. The actuating force

c. The value of the dimension s causing self-locking

Given: a = 25 in., c = 6 in., r = 10 in., s = 1.4 in., w = 3 in., ϕ = 260°.

Sections 13.13 through 13.15

13.39 A short-shoe drum brake having f = 0.25, a = 1 m, b = 0.4 m, c = 50 mm, and r = 0.3 m is to absorb 25 kW at 800 rpm (Figure 13.21). Determine

a. The actuating force and whether the brake is self-locking

b. The pin reaction at A

13.40 Resolve Problem 13.39 for clockwise rotation of the drum.

13.41 Redo Example 13.10 using short-shoe analysis, that is, assuming that the total normal and friction forces are concentrated at point B. Compare the results with the more exact results of Example 13.10.

13.42 Shown in Figure P13.42 is a short-shoe external drum brake. The material of the shoe and drum produces a coefficient of friction 0.25 and a maximum pressure of 800 kPa. Find

a. The limiting lever force F_a and the braking torque T. Is the brake self-energizing or de-energizing for the direction shown?

b. The radial force on the lever pivot A.

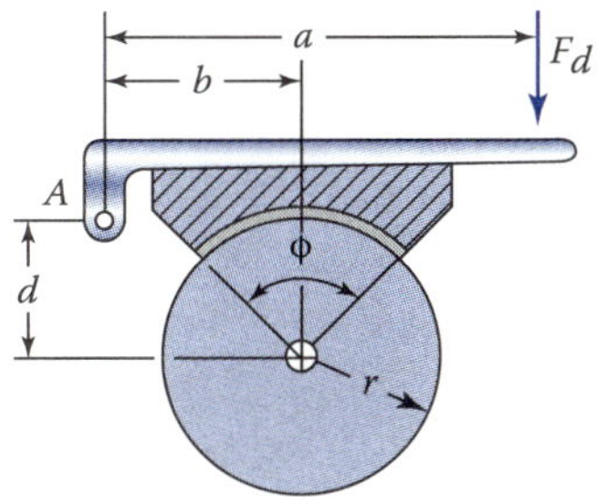

FIGURE P13.42

Given: a = 500 mm, b = 200 mm, r = 150 mm, d = 175 mm, w = 50 mm, ϕ = 44°, f = 0.25, p_{max} = 800 kPa.

13.43 A short-shoe brake in Figure 13.21 sustains 250 N·m of torque at a drum rotation 600 rpm. Find

a. The normal force F_n acting on the shoe

b. The actuating force F_a

Given: a = 900 mm, b = 320 mm, c = 34 mm, r = 350 mm, f = 0.4

13.44 A short-shoe drum brake illustrated in Figure 13.21 uses a lining material having p_{max} = 700 kN and the coefficient of friction of f = 0.2. What is the maximum value of the actuating force F_a?

Given: a = 200 mm, b = 120 mm, c = 30 mm, r = 100 mm, w = 40 mm, ϕ = 36°.

13.45 An external, long-shoe drum brake has a torque capacity T = 1.2 kip in. (Figure 13.22). The lining is a woven material with coefficient of friction f = 0.35. Determine the maximum pressure p_{max} between the lining and drum for two cases:

a. The contact angle is ϕ = 90° (θ_1 = 0°, θ_2 = 90°).

b. The contact angle equals ϕ = 45° (θ_1 = 20°, θ_2 = 65°).

Given: c = 4 in., r = 3 in., w = 1 in.

13.46 Figure P13.46 depicts a long-shoe drum brake. Determine the value of dimension b in terms of the radius r so that the friction forces neither assist nor resist in applying the shoe to the drum.

13.47 The long-shoe brake shown in Figure P13.47 has p_{max} = 900 kPa, f = 0.3, and w = 50 mm. Calculate

a. The actuating force

b. The power capacity at 600 rpm

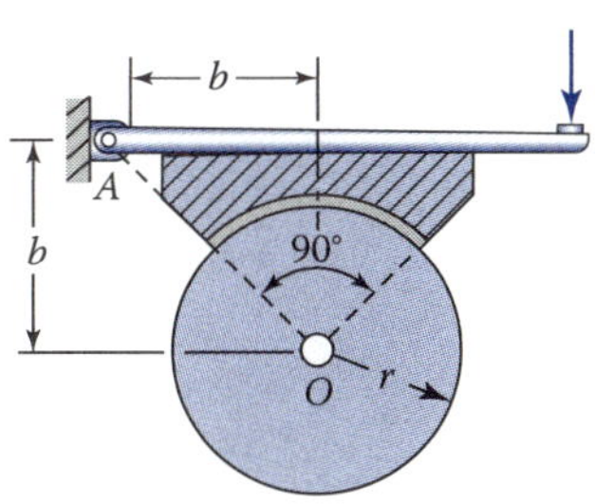

FIGURE P13.46

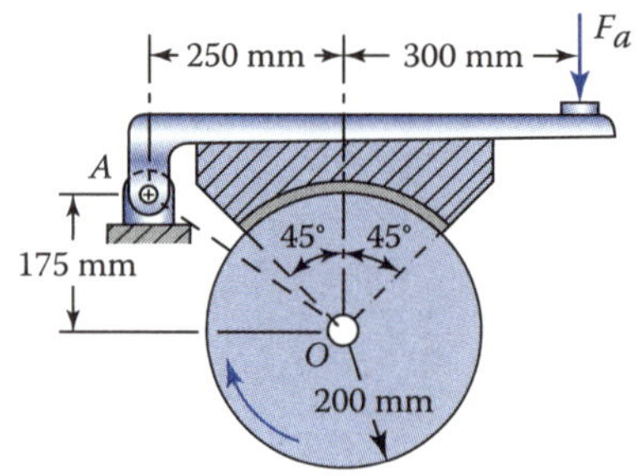

FIGURE P13.47

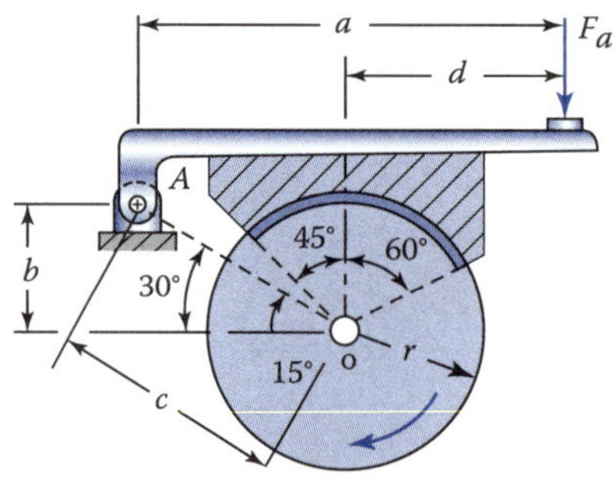

FIGURE P13.48

13.48 A long-shoe brake is shown in Figure P13.48. Determine

a. The actuating force

b. The power capacity at 500 rpm

Given: $b = 150$ mm, $d = 250$ mm, $r = 200$ mm, $w = 60$ mm, $f = 0.3$, and $p_{max} = 800$ kPa.

13.49 Consider a symmetrically loaded pivoted shoe brake similar to that illustrated in Figure 13.24 with a sintered metal friction material. Find

a. The distance a between the pivot A and the center of drum

b. Reactions at pivot A

c. The torque capacity of the brake

Given: $r = 125$ mm, $w = 50$ mm, $\phi = 90°$, $p_{max} = 1.55$ MPa, and $f = 0.3$ (from Table 13.11).

14

Mechanical Springs

14.1 Introduction

Springs are used to exert forces or torques in a mechanism or primarily to store the energy of impact loads. These flexible members often operate with high values for the ultimate stresses and with varying loads. Helical springs are round or rectangular wire, and flat springs (cantilever or simply supported beams) are in widespread usage. Springs come in a number of other kinds, such as disk, ring, spiral, and torsion bar springs. Numerous standard spring configurations are available as stock catalog items from spring manufacturers. Figure 14.1 shows various compression, tension, and torsion springs. The designer must understand and appropriately apply spring theory to specify or design a component.

Pneumatic springs of diverse types take advantage of the elastic compressibility of gases as compressed air in automotive air shock absorbers. For applications involving very large forces with small displacements, hydraulic springs have proven very effective. Our concern in this text is only with springs of common geometric form made of solid metal or rubber. For more information on others, see [1–5]. As discussed in Section 1.4, mechanical components are usually designed on the basis of strength. Generally, displacement is of minor significance. Often deflection is checked whether it is reasonable. However, in the design of springs, displacement is as important as strength. A notable deflection is essential to most spring applications.

14.2 Torsion Bars

A *torsion bar* is a straight hollow or solid bar fixed at one end and twisted at the other, where it is supported. This is the simplest of all spring forms, as shown by the portion *AB* in Figure 14.2a. Typical applications include counterbalancing for automobile hoods and trunk lids. A torsion bar with splined ends (Figure 14.2b) is used for a vehicle suspension spring or sway bar. Usually, one end fits into a socket on the chassis, and the other into the pivoted end of an arm. The arm is part of a linkage, permitting the wheel to rise and fall in approximately parallel motion. Note that in a passenger car, the bar may have about ¾ m length, 25 mm diameter, and twist 30°–45°.

FIGURE 14.1
A collection of wire springs. (Courtesy of Rockford Spring Co.)

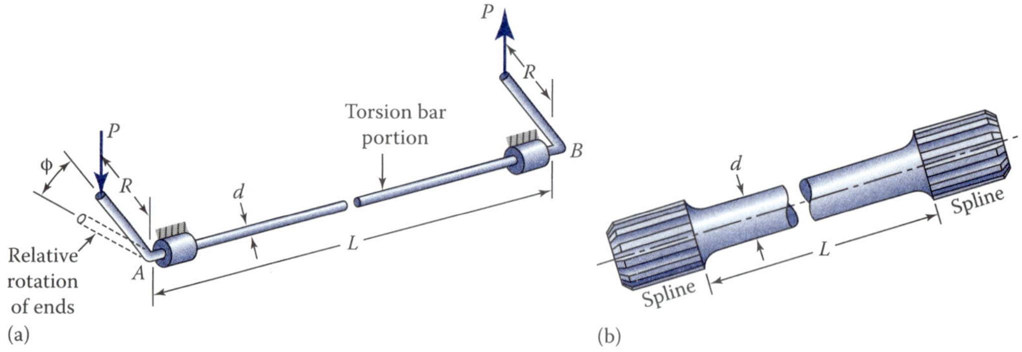

FIGURE 14.2
Torsion bar springs: (a) rod with bent ends and (b) rod with splined ends.

The stress in a torsion bar is mainly one of torsional shear. Hence, the equations for stress, angular displacement, and stiffness are given in Sections 3.5 and 4.3. Referring to Figure 14.2a, we can write

$$T = PR, \quad \delta = \phi R, \quad k = \frac{T}{\phi}$$

in which the angle of twist $\phi = TL/GJ$. For the solid *round torsion bar*, the moment of inertia is $J = \pi d^4/32$. We therefore have the formulas

$$\tau = \frac{16PR}{\pi d^3} \tag{14.1}$$

$$\delta = \frac{TLR}{GJ} = \frac{32PLR^2}{\pi d^4 G} \tag{14.2}$$

and

$$k = \frac{\pi d^4 G}{32L} \tag{14.3}$$

Here,

τ = the torsional shear stress
P = the load
δ = the relative displacement between ends
G = the modulus of rigidity
d = the bar diameter
R = the moment arm
L = the bar length
k = the spring rate

The foregoing basic equations are supplemented, in the case of torsion springs with noncircular cross sections, by Table 3.1. Note that at the end parts of the spring that do not lie between supports A and B, there is a shear load P and an associated direct shear stress acting on cross-sectional areas. Usually, the effects of the curvature and the effect of bending are neglected at these portions of the bar. When designing a torsion bar, the required diameter d is obtained by Equation 14.1. Then based on the allowable shear strength, Equation 14.2 gives the bar length L necessary to provide the required deflection δ.

14.3 Helical Tension and Compression Springs

In this section, attention is directed to closely coiled standard helical tension and compression springs. They provide a push or pull force and are capable of large deflection. The standard form has constant coil diameter, pitch (axial distance between coils), and spring rate

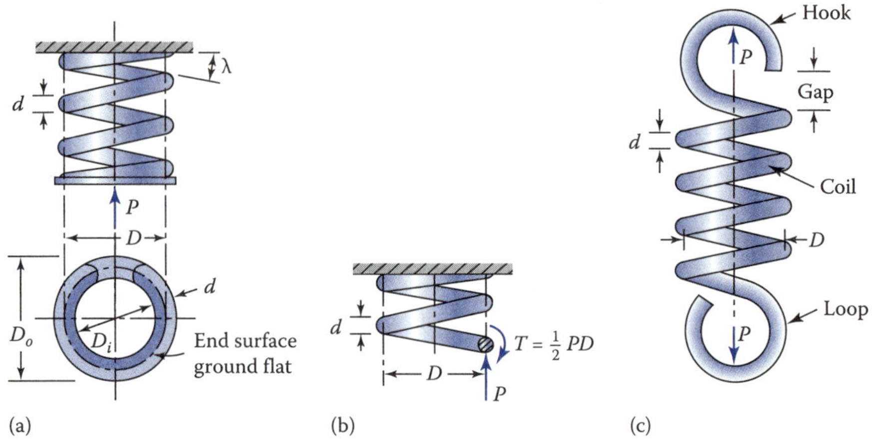

FIGURE 14.3
Helical springs: (a) compression spring, (b) free body of top portion of compression spring, and (c) tension spring.

(slope of its deflection curve). It is the most common spring configuration. Variable-pitch, barrel, and hourglass springs are employed to minimize resonant surging and vibration.

A helical spring of circular cross section is composed of a slender *wire of diameter d* wound into a helix of *mean coil diameter D, coil pitch p,* and *pitch angle* λ. The top portion, isolated from the compression spring of Figure 14.3a, is shown in Figure 14.3b. A section taken perpendicular to the axis of the spring wire can be considered nearly vertical. Hence, centric load P applied to the spring is resisted by a transverse shear force P and a torque $T = PD/2$ acting on the cross section of the coil, as depicted in the figure. Figure 14.3c shows a helical tension spring.

For a helical spring, the ratio of the mean coil diameter to wire diameter is termed the *spring index C*:

$$C = \frac{D}{d} \tag{14.4}$$

The springs of ordinary geometry have $C > 3$ and $\lambda < 12°$. In the majority of springs, C varies from about 6 to 12. At $C > 12$, the spring is likely to buckle and also tangles readily when handled in bulk. The outside diameter $D_o = D + d$ and the inside diameter $D_i = D - d$ are of interest primarily to define the smallest hole in which the spring will fit or the largest pin over which the spring can be placed. Usually, the minimum diametral clearance between the D_o and the hole or between D_i and a pin is about $0.1D$ for $D < 13$ mm or $0.05\ D$ for $D > 13$ mm.

14.3.1 Stresses

An exact analysis by the theory of elasticity shows that the transverse or direct shear stress acting on an element at the inside coil diameter has the value

$$\tau_d = 1.23\frac{P}{A} = 1.23\frac{P}{\pi d^2/4} \tag{14.5}$$

This expression may be rewritten in the form

$$\tau_d = \frac{8PD}{\pi d^3} \times \frac{0.615}{C}$$

The torsional shear stress, by neglecting the initial curvature of the wire, is

$$\tau_t = \frac{16T}{\pi d^3} = \frac{8PD}{\pi d^3}$$

The superposition of the preceding stresses gives the maximum or *total shear stress* in the wire on the inside of the coil:

$$\tau = K_s \frac{8PD}{\pi d^3} = K_s \frac{8PC}{\pi d^2} \tag{14.6}$$

In the foregoing,

$$K_s = 1 + \frac{0.615}{C} \tag{14.7}$$

is called the *direct shear factor*.

For a slender wire, the C has large values, and clearly, the maximum shear stress is caused primarily by torsion. In this case, a helical compression or tension spring can be thought of as a torsion bar wound into a helix. On the other hand, in a heavy spring, where C has small values, the effect of direct shear stress cannot be disregarded.

The intensity of the torsional stress increases on the inside of the spring because of the curvature. The following more accurate relationship, known as the *Wahl formula*, includes the curvature effect [2]:

$$\tau = K_w \frac{8PD}{\pi d^3} = K_w \frac{8PC}{\pi d^2} \tag{14.8}$$

The *Wahl factor* K_w is defined by

$$K_w = \frac{4C-1}{4C-4} + \frac{0.615}{C} \tag{14.9}$$

The first term in Equation 14.9, which accounts for the effect of curvature, is basically a stress-concentration factor. The second term gives a correction factor for direct shear only. The Wahl factor may be used for most calculations. A more exact theory shows that it is accurate within 2% for $C \geq 3$. Figure 14.4 illustrates the variation of the K_w as a function of C.

After some minor local yielding under static loading, typical spring materials relieve the local stress concentration due to the curvature (see Section 8.8). Therefore, we use

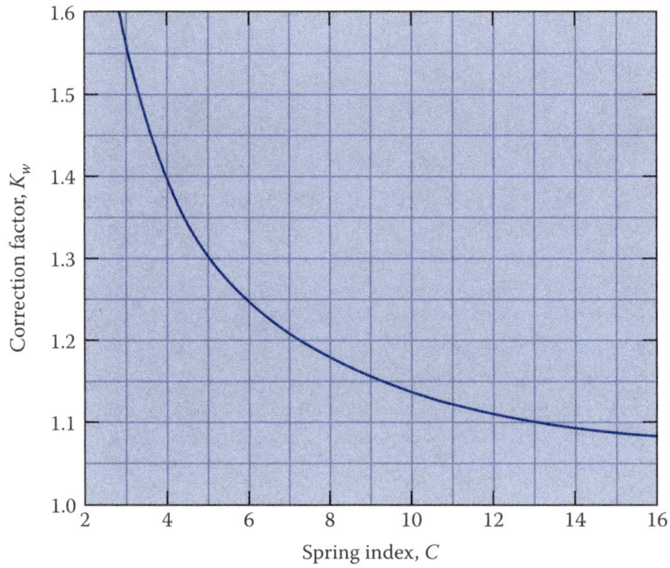

FIGURE 14.4
Stress correction factors for curvature and direct shear for helical springs.

Equations 14.8 and 14.6 for *alternating* loading and *static* or *mean* loading, respectively. We note that, occasionally,

$$K_s = 1 + \frac{0.5}{C} \tag{14.7'}$$

is used in static applications for compression springs, instead of Equation 14.7. This is based on the assumption that the transverse shear stresses are uniformly distributed subsequent to some yielding under static loading.

Interestingly, the free-body diagram of Figure 14.3b contains no bending loading for closely coiled springs. However, for springs with a pitch angle of λ greater than 15° and deflection of each coil greater than $D/4$, bending stresses should be taken into account [2]. In addition, it is rarely possible to have exactly centric axial loading P, and any eccentricity introduces bending and changes the torsional moment arm. This gives rise to stresses on one side of the spring higher than indicated by the foregoing equations. Also observe from Figure 14.3b that in addition to creating a transverse shear stress, a small component of force P produces axial compression of the spring wire. In critical spring designs involving relatively large values of λ, this factor should be considered.

14.3.2 Deflection

In determining the deflection of a closely coiled spring, it is common practice to ignore the effect of direct shear. Therefore, the twist causes one end of the wire segment to rotate an angle $d\phi$ relative to the other, where $\phi = TL/GJ$. This corresponds to a deflection $d\delta$, at the axis of the spring:

$$d\delta = \frac{D}{2} d\phi = \frac{D}{2} d\left(\frac{TL}{GJ}\right)$$

The total deflection δ, of spring of length $L = \pi D N_a$, is then

$$\delta = \frac{8PD^3N_a}{Gd^4} = \frac{8PC^3N_a}{Gd} \tag{14.10}$$

in which

N_a is the number of active coils
G represents the modulus of rigidity

An alternative derivation of this equation may readily be accomplished using Castigliano's theorem.

14.3.3 Spring Rate

The elastic behavior of a spring may be expressed conveniently by the slope of its force–deflection curve or *spring rate k*. Through the use of Equation 14.10, we have

$$k = \frac{P}{\delta} = \frac{Gd^4}{8D^3N_a} = \frac{dG}{8C^3N_a} \tag{14.11}$$

Also referred to as the spring constant or spring scale, the spring rate has units of N/m in SI and lb/in. in the U.S. customary system. The standard helical spring has a spring rate k that is basically linear over most of its operating range. The first and last few percent of its deflection have a nonlinear rate. Often, in spring design, the spring rate is defined between about 15% and 85% of its total and working deflections [1].

Occasionally, helical compression springs are wound in the form of a cone (Figure 14.5), where the coil radius and hence the torsional stresses vary throughout the length. The maximum stress in a *conical spring* is given by Equation 14.8. The deflection and spring rate can be estimated from Equations 14.10 and 14.11, using the average value of mean coil diameter for D.

Example 14.1: Finding the Spring Rate

A helical compression spring of an average coil diameter D, wire diameter d, and number of active coils N_a supports an axial load P (Figure 14.3a). Calculate the value of P that will cause a shear stress of τ_{all}, the corresponding deflection, and rate of the spring.

Given: $D = 48$ mm, $d = 6$ mm, $N_a = 5$, $\tau_{all} = 360$ MPa

Design Decisions: A steel wire of $G = 79$ GPa is used.

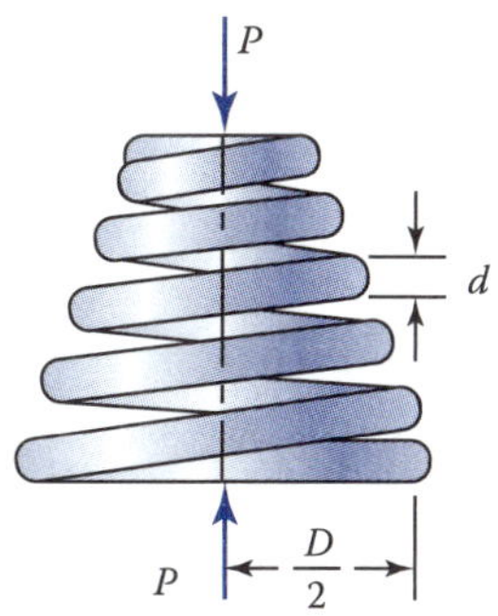

FIGURE 14.5
Conical–helical compression spring.

Solution

The mean diameter of the spring is $D = 48 - 6 = 42$ mm. The spring index is equal to $C = 42/6 = 7$. Applying Equation 14.6, we have

$$360 = \frac{8P(42)}{\pi(6)^3}\left(1 + \frac{0.615}{7}\right) = 0.539P$$

Solving

$P = 668$ N

From Equation 14.10,

$$\delta = \frac{8(668)(7)^3(5)}{(79 \times 10^3)(6)} = 19.34 \text{ mm}$$

The spring rate is therefore $k = 668/0.01934 = 34.54$ kN/m.

14.4 Spring Materials

Springs are manufactured either by hot- or cold-working processes, depending on the size and strength properties needed. Ordinarily, preheated wire should not be used if spring index $C < 4$ in. or if diameter $d > 1/4$ in.; that is, small sizes should be wound cold. Heavy-duty springs (e.g., vehicle suspension parts) are usually hot worked. Winding of the springs causes residual stresses owing to bending. Customarily, in spring forming, such stresses are relieved by heat treatment.

A limited number of materials are suitable for usage as springs. These include carbon steels, alloy steels, phosphor bronze, brass, beryllium copper, and a variety of nickel alloys. Plastics are used when loads are light. Blocks of rubber often form springs, as in bumpers and vibration isolation mountings of various machines such as electric motors and internal combustion engines. The UNS steels (see Section 2.12) listed in Table B.3 should be used in designing hot-rolled or forged heavy-duty coil springs, as well as flat springs, leaf springs, and torsion bars. The typical spring material has a high ultimate and yield strengths to provide maximum energy storage. For springs under dynamic loading, the fatigue strength properties of the material are of main significance. The website www.acxesspring.com includes information on commonly used spring materials.

Experiment shows that for common spring materials, the ultimate strength obtained from a torsion test can be estimated in terms of the ultimate strength determined from a simple tension test, as noted in Section 7.5. The ultimate strength in shear is then

$$S_{us} = 0.67S_u \tag{7.5a}$$

Similarly, yield strength in shear is

$$S_{ys} = 0.577S_y \tag{7.5b}$$

Here, the quantities S_u and S_y are the ultimate strength or tensile strength and the yield strength in tension, respectively.

TABLE 14.1
Some Common Spring-Wire Materials

Material	ASTM No.	Description
Hard-drawn wire 0.60–0.70C	A227	Least-expensive general-purpose spring steel. Suitable for static loading only and in temperature range 0°C–120°C. Available in diameters 0.7–16 mm.
Music wire 0.80–0.95C	A228	Toughest high-carbon steel wire widely used in the smaller coil diameters. It has the highest tensile and fatigue strengths of any spring material. The temperature restrictions are the same as for hard-drawn wire. Available from 0.1 to 6.5 mm in diameter.
Oil-tempered wire 0.60–0.70C	A229	Used for many types of coil springs and less expensive than music wire. Suitable for static loading only and in the temperature range 0°C–180°C. Available in diameters 0.5–16 mm.
Chrome-vanadium	A232	Suitable for severe service conditions and shock loads. Widely used for aircraft engine valve springs, where fatigue resistance and long endurance needed, and for temperatures to 220°C. Available in diameters from 0.8 to 11 mm.
Chrome-silicon	A401	Suitable for fatigue loading and in temperatures up to 250°C. Second highest in strength to music wire. Available from 1.6 to 9.5 mm in diameter.

14.4.1 Spring Wire

Round wire is the most often utilized spring material. It is readily available in a selection of alloys and wide range of sizes. Rectangular wire is also attainable but only in limited sizes. A brief description of commonly used high-carbon (C) and alloy spring steels is given in Table 14.1.

14.4.1.1 Ultimate Strength in Tension

Spring materials may be compared by examining their tensile strengths varying with the wire size. The material and its processing also have an effect on tensile strength. The strength properties for some common spring steels may be estimated by the formula

$$S_u = Ad^b \tag{14.12}$$

where
S_u = the ultimate tensile strength
A = a coefficient
b = an exponent
d = the wire diameter (in. or mm)

Values of coefficient A and exponent b pertaining to the materials presented in Table 14.1 are furnished in SI and U.S. customary units in Table 14.2. Likewise, the strengths of stainless steel wire and hard phosphor bronze wire are given in the form of graphs, tables, and formulas. The preceding equation provides a convenient means to calculate

TABLE 14.2

Coefficients and Exponents for Equation 14.12

Material	ASTM No.	b	A MPa	A ksi
Hard–drawn wire	A227	−0.201	1510	137
Music wire	A228	−0.163	2060	186
Oil–tempered wire	A229	−0.193	1610	146
Chrome–vanadium wire	A232	−0.155	1790	173
Chrome–silicon wire	A401	−0.091	1960	218

Source: Associated Spring-Barnes Group, Design Handbook, Associated Spring-Barnes Group, Bristol, CN, 1987.

steel wire tensile strength within a spring-design *computer program* and allows fast iterating to a proper design solution.

14.4.1.2 Yield Strength in Shear and Endurance Limit in Shear

Data of extensive testing [5] indicate that a semilogarithmic plot of torsional yield strength S_{ys} (and hence S_u) versus wire diameter is almost a straight line for some materials (Figure 14.6). Note from the figure that the strength increases with a reduction in diameter. There is also ample experimental evidence that the relationships between the ultimate strength in tension, the yield strength in shear, and the endurance limit in shear S'_{es} are as given in Table 14.3. Observe that the test data values of S'_{es} were developed with actual conditions of surface and size factors of the wire materials, to be discussed in Section 14.7. We use these values, assuming 50% reliability.

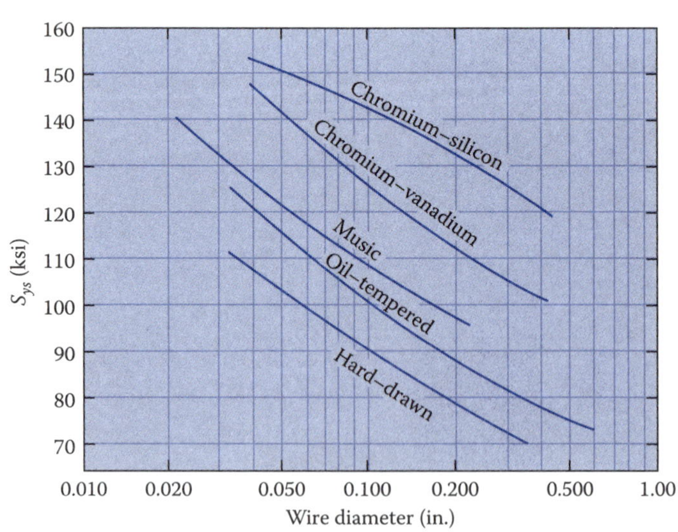

FIGURE 14.6
Yield strength in shear of spring wire. (From Carlson, H.C.R., *Mech. Eng.*, 78, 331, 1956; Carlson, H.C.R., *Spring Designer's Handbook*, Marcel Dekker, New York, 1978.)

TABLE 14.3

Approximate Strength Ratios of Some Common Spring Materials

Material	S_{ys}/S_u	S'_{es}/S_u
Hard–drawn wire	0.42	0.21
Music wire	0.40	0.23
Oil–tempered wire	0.45	0.22
Chrome–vanadium wire	0.52	0.20
Chrome–silicon wire	0.52	0.20

Source: Associated Spring-Barnes Group, Design Handbook, Associated Spring-Barnes Group, Bristol, CN, 1987.

Notes: S_{ys}, yield strength in shear; S_u, ultimate strength in tension; S'_{es}, endurance limit (or strength) in shear.

Example 14.2: Allowable Load of a Helical Compression Spring

A helical compression spring for mechanical device is subjected to an axial load *P*. Determine

a. The yield strength in the shear of the wire
b. The allowable load *P* corresponding to yielding

Design Decisions: Use a 0.0625 in. music wire. The mean diameter of the helix is $D = 0.5$ in. A safety factor of 1.5 is applied due to uncertainty about the yielding.

Solution

The spring index is $C = D/d = 0.5/0.0625 = 8$.

a. Through the use of Equation 14.12 and Table 14.2, we have

$$S_u = Ad^b = 186(0.0625^{-0.163}) = 292 \text{ ksi}$$

Then, by Table 14.3, $S_{ys} = 0.4(292) = 117$ ksi.

b. The allowable load is obtained by applying Equation 14.6 as

$$P_{\text{all}} = \frac{\tau_{\text{all}}\pi d^2}{8K_sC}$$

where

$$\tau_{\text{all}} = \frac{S_{ys}}{n} = \frac{117}{1.5} = 78 \text{ ksi}$$

$$K_s = 1 + \frac{0.615}{8} = 1.077 \quad \text{(from Equation 14.7)}$$

Hence,

$$P_{\text{all}} = \frac{\pi(78{,}000)(0.0625)^2}{8(1.077)(8)} - 13.9 \text{ lb}$$

14.5 Helical Compression Springs

End details are four *standard* types on helical compressive springs. They are plain, plain–ground, squared, and squared–ground, as shown in Figure 14.7. A spring with plain ends has ends that are the same as if a long spring had been cut into sections. A spring with plain ends that are squared, or closed, is obtained by deforming the ends to 0° helix angle. Springs should always be both *squared and ground* for significant applications, because a better transfer of load is obtained. A spring with squared and ground ends compressed between rigid plates can be considered to have fixed ends. This represents the most common end condition.

Figure 14.7 shows how the type of end used affects the number of active coils N_a and the solid height of the spring. Square ends effectively decrease the *number of total coils* N_t by approximately two; that is,

$$N_t = N_a + 2 \tag{14.13}$$

Grinding by itself removes one active coil. To obtain basically uniform contact pressure over the full end turns, special end members must be used (such as countered end plates) for all end conditions except squared and ground.

Working deflection corresponds to the working load P_w on a compression spring. Referring to Figure 14.8, the *solid deflection* δ_s is defined as follows:

$$\delta_s = h_f - h_s \tag{14.14}$$

where

h_f represents the free (no load) height

h_s is the solid height or shut height under solid load P_s

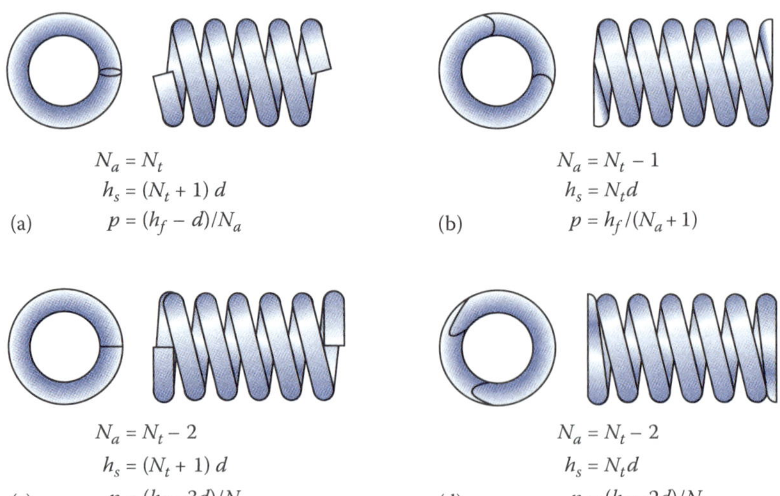

FIGURE 14.7
Common types of ends for helical compression springs and corresponding spring solid height equations: (a) plain ends, (b) plain–ground ends, (c) squared or closed ends, and (d) squared–ground ends. *Note*: p, pitch; h_f, free height (Figure 14.8).

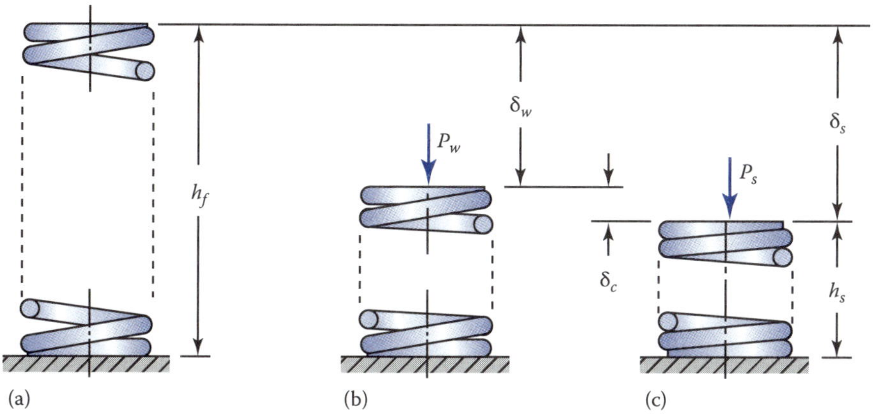

FIGURE 14.8
Deflections of a helical compression spring: (a) free height, (b) working deflection, and (c) solid deflection.

For *special* applications where space is limited, solid height of ground springs can be obtained by the expression

$$h_s = (N_t - 0.5)(1.01d) \tag{14.15}$$

Clash allowance (r_c) refers to difference in spring length between maximum load and spring solid position. It is defined as a ratio of margin of extra deflection or clash deflection δ_c to the working deflection δ_w:

$$r_c = \frac{\delta_c}{\delta_w} \tag{14.16}$$

Usually, a minimum clash allowance of 10%–15% is used to avoid reaching the solid height in service. A maximum clash allowance of 20% is satisfactory for most applications. Based on this value, an overload P_s of 20% deflects the spring to its maximum deflection δ_s, and higher overload has no effect on deflection or stress. Hence, with a sufficient safety factor, a compression spring is protected against failure after it reaches its solid deflection.

14.5.1 Design Procedure for Static Loading

The two *basic requirements* of a helical spring design are an allowable stress level and the desired spring rate. The stress requirement can be fulfilled by many combinations of D and d. Having D and d selected, N_a is determined on the basis of the required spring rate. Finally, the free height can be obtained for a prescribed clash allowance. Note that in some situations, the outside diameter, inside diameter, or working deflection may be limited. Clearly, when the spring comes out too large or too heavy, a stronger material must be used.

If the resulting design is likely to fail by buckling Section 14.6, the process would be repeated with another combination of D and d. In any case, spring design is essentially an iterative problem. Some assumptions must be made to prescribe the values of enough variables to calculate the stresses and deflections. Usually, charts, nomographs, and computer programs have been used to simplify the spring design problem [6–9].

Example 14.3: Design of a Hard-Drawn Wire Compression Spring

A helical compression coil spring made of hard-drawn round wire with squared and ground ends (Figure 14.7d) has spring rate *k*, diameter *d*, and spring index *C*. The allowable force associated with a solid length is P_{all}.

Find: The wire diameter and the mean coil diameter for the case in which the spring is compressed solid

Given: $C = 9$, $P_{all} = 45$ N

Assumptions: Static loading conditions will be considered. Factor of safety based on yielding is $n = 1.8$.

Solution

The direct shear factor, from Equation 14.7, is $K_s = 1 + (0.615/9) = 1.068$. The ultimate strength is estimated using Equation 14.12 and Table 14.2 as

$$S_u = Ad^b = 1.51\ (10^9)\ d^{-0.201}$$

in which *d* is in millimeters. Expressing *d* in meters, the foregoing becomes

$$S_u = 1.51\ (10^9)\ (1000)^{-0.201}\ d^{-0.201} = 376.7\ (10^6)\ d^{-0.201}$$

The yield strength in shear, referring to Table 14.3, is then

$$S_{ys} = 0.42\ S_u = 158.2(10^6)\ d^{-0.201} \tag{a}$$

Substitution of the given numerical values into Equation 14.6 together with τ_{all}/n, the maximum design shear stress is expressed as

$$\tau_{all} = \frac{8nK_sCP_{all}}{\pi d^2}$$

$$= \frac{8(1.8)(1.068\times 9)(45)}{\pi d^2} = \frac{1982.6}{d^2} \tag{b}$$

Finally, equating Equations (a) and (b) results in

$$158.2\ (10^6)\ d^{-0.201} = 1982.6\ d^{-2}$$

from which

$$d = 0.00188\ \text{m} = 1.88\ \text{mm}$$

Thus, the mean coil diameter equals

$$D = Cd = 9\ (1.88) = 16.92\ \text{mm}$$

Comment: A standard 1.9 mm diameter hard-drawn wire should be used.

14.6 Buckling of Helical Compression Springs

A compression spring is loaded as a column and can buckle if it is too slender. In this section, we examine the problem of buckling of springs by their resistance to bending. For this purpose, consider a spring of length L and coil radius $D/2$ subjected to bending moment M (Figure 14.9a). The effect is an angular rotation 6. The bending and

$$M_\alpha = M\sin\alpha, \quad T_\alpha = M\cos\alpha \tag{a}$$

Derivation of the equation for helical spring deflection is readily accomplished using Castigliano's theorem as follows. Application of Equation 5.35 gives twisting moments at any section are (Figure 14.9b)

$$\theta = \frac{1}{EI}\int_0^L M_\alpha \frac{\partial M_\alpha}{\partial C}\,dx + \frac{1}{GJ}\int_0^L T_\alpha \frac{\partial T_\alpha}{\partial C}\,dx \tag{b}$$

In which $C = M$. Introducing Equations (a) into (b) together with $dx = ds = (D/2)\, d\alpha$, we obtain

$$\theta = M\int_0^{2\pi Na}\left(\frac{\sin^2\alpha}{EI} + \frac{\cos^2\alpha}{GJ}\right)\left(\frac{D}{2}\,d\alpha\right)$$

Here, $G = E/2(1 + \nu)$, and for a round wire, $J = 2I = \pi d^4/32$. Hence, the angular rotation of the entire spring is, after integrating,

$$\theta = \frac{64MDN_a}{Ed^4}\left(1 + \frac{\nu}{2}\right) \tag{14.17}$$

By analogy to a simple beam in pure bending, we may write using Equations 4.14 and 4.15

$$\theta = \frac{ML}{EI_e} \tag{c}$$

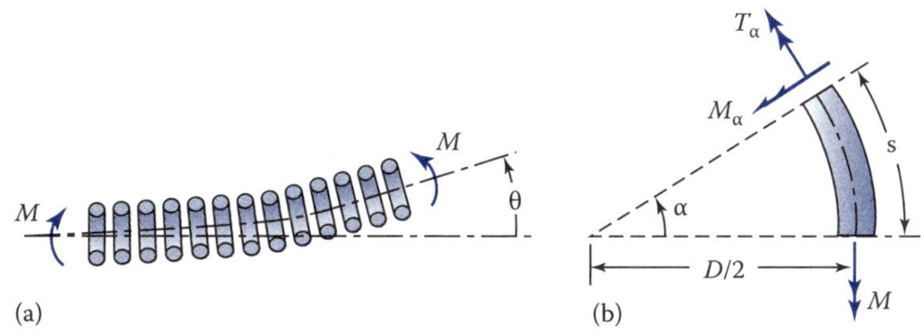

FIGURE 14.9
(a) Bending and (b) moment resultants at a cut section of a helical compression spring.

The *equivalent* moment of inertia of the spring coil I_e is obtained by eliminating θ from Equations 14.17 and (c). In so doing, we have

$$I_e = \frac{Ld^4}{64DN_a(1+\nu/2)} \quad (14.18)$$

The preceding result may be used directly in Equation 5.58 to ascertain the Euler buckling load of the spring in the form

$$P_{cr} = \frac{\pi^2 EI_e}{L_e^2} \quad (14.19)$$

The quantity L_e denotes the effective column length (see Figure 5.17). The allowable value of compressive load is then found from $P_{all} = P_{cr}/n$, in which n represents a factor of safety.

14.6.1 Aspect Ratio

It is important to point out that the measure of slenderness ratio for solid columns is not directly applicable to springs due to their much different form. An identical slenderness ratio is established as the *aspect ratio* of free length to mean coil diameter, h_f/D. In compression springs, it is important that the aspect ratios be not so great that buckling occurs. Figure 14.10 shows the results for the two end conditions given in Figure 5.17c and d [1,10].

Curve A in Figure 14.10 represents the springs supported between flat surfaces, a commonly used case. Observe from the figure that buckling occurs for conditions above and to the right of each curve. Clearly, as in the case of solid columns, the end conditions of the spring affect its tendency to buckle. *Curve B* renders the springs having one end free to tip. In these cases, the springs will buckle with smaller aspect ratios, as depicted in the figure.

We use Figure 14.10, rather than Equation 14.19, to check readily for possible buckling of the spring. Note that if buckling is indicated, the preferred solution is to redesign the spring. Otherwise, the buckling can be avoided by placing the spring either inside or outside a tube that provides a small clearance.

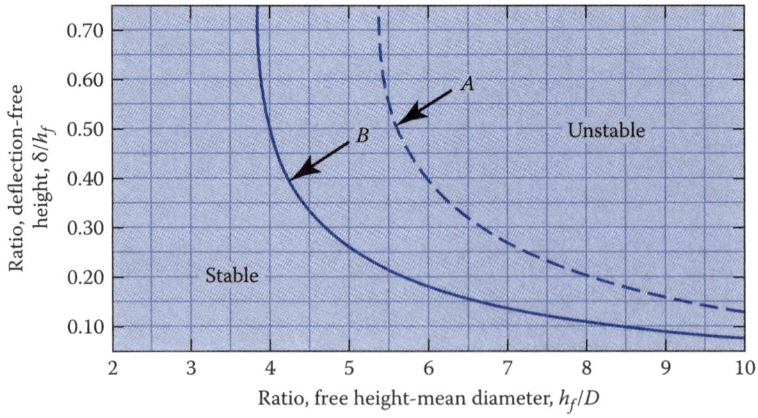

FIGURE 14.10
Buckling conditions for helical springs: (*A*) with parallel end plates (depicts the case of Figure 5.17d) and (*B*) one end plate is free to tip (depicts the case of Figure 5.17c).

Example 14.4: Stability of a Hard-Drawn Wire Compression Spring

Reconsider the hard-drawn compression spring discussed in Example 14.3.

Find:

a. The solid length
b. Whether the spring will buckle in service
c. The pitch of the body coil

Assumptions: The modulus of rigidity of the wire will be $G = 79$ GPa. The spring rate equals $k = 1.4$ kN/m.

Solution

Refer to the numerical values given in Example 14.3. The solid deflection is

$$\delta_s = \frac{P_{\text{all}}}{k} = \frac{45}{1400} = 0.03214 \text{ m} = 32.14 \text{ mm}$$

a. The number of active coils, by Equation 14.11, is

$$N_a = \frac{Gd}{8kC^3} = \frac{(79 \times 10^9)(1.88 \times 10^{-3})}{8(1400)(9^3)} = 18.19$$

For the squared and ground ends, observe from Figure 14.7d that

$$N_t = N_a + 2 = 20.19$$

and the solid length

$$h_s = N_t d = (20.19)\ (1.88) = 37.96 \text{ mm}$$

b. Applying Equation 14.14, the free length is equal to

$$h_f = h_s + \delta_s = 37.96 + 32.14 = 70.1 \text{ mm}$$

For the case under consideration, we have

$$\frac{\delta_s}{h_f} = \frac{32.14}{70.1} = 0.46 \quad \text{and} \quad \frac{h_f}{D} = \frac{70.1}{16.92} = 4.14$$

Case A in Figure 14.10 illustrates that the spring is far outside of the buckling zone and obviously safe.

c. From Figure 14.7d, the pitch is

$$p = \frac{1}{N_a}(h_f - 2d) = \frac{1}{18.19}[70.1 - 2(1.88)] = 3.647 \text{ mm}$$

Comments: With the values of D, N_t, and h_f obtained here and in the previous example, a compression spring can be drawn or made.

14.7 Fatigue of Springs

Spring failures under fatigue loads are typical of that in torsional shear. A crack initiates at the surface on the inside of the coil and acts at 45° to the radial shear plane in the direction perpendicular to the tensile stress (see Figure 3.27). We note that helical springs are never used as both compression and extension springs: They do not normally experience a stress reversal. Moreover, these springs are assembled with a preload in addition to the working stress. The stress is thus prevented from being 0. The extreme case occurs if the preload drops to 0; that is, minimum shear τ_{min} stress equals 0.

Inasmuch as most failures are caused by fatigue, a poor surface is the worst disadvantage of hot-formed springs. *Presetting* refers to a process used in the manufacture of compression springs to produce residual stresses (see Section 3.14). This is done by making the spring longer than required and then compressing it to its solid height. Shot peening, discussed in Section 2.11, and presetting are two operations that add to the strength and durability of steel springs. The former, done after cooling, introduces a layer of compressive residual stresses. In a like manner, the latter always initiates surface residual stresses opposite to those caused by subsequent load application in the same direction as the presetting load. Maximum fatigue strengthening can be acquired using both the foregoing operations. The *set* spring loses some free length but gains the benefits described in the preceding. On the other hand, shot peening is most effective against cyclic loading in fatigue, while it has little benefit for statically loaded springs.

Data on fatigue strengths of round-wire helical springs are voluminous. Note that Equation 7.1 defines the uncorrected endurance limit for fully reversed bending of steels as $S'_e = 700$ MPa for $S_u \geq 1400$ MPa. It can readily be verified by Equation 14.12 and Table 14.2 that most spring wires smaller than 10 mm diameter are in this strength category. We conclude therefore that the torsional endurance limit of these spring-wire materials may be regarded as independent of size or their particular alloy composition. On this basis, the best data for the torsional endurance limit of spring steel wire of $d < 10$ mm are by [1]

$$
\begin{aligned}
S'_{es} &= 45 \text{ ksi } (310 \text{ MPa}) \quad \text{for shot unpeened springs} \\
S'_{es} &= 67.5 \text{ ksi } (465 \text{ MPa}) \quad \text{for shot unpeened springs}
\end{aligned}
\tag{14.20}
$$

Equation 14.20 apply for infinite life with $\tau_{min} = 0$. As in the case of Table 14.2, the foregoing values were corrected for surface condition and size.

It should be mentioned that *corrosion*, even in a mild form, greatly reduces the fatigue strength. Also if the spring operates under conditions of elevated temperature, there is a danger of creep or permanent deformation unless very low fluctuating stress values are used. Such effects become noticeable above 350°C, and the ordinary spring steels cannot be used, as noted in Table 14.1.

14.8 Design of Helical Compression Springs for Fatigue Loading

Springs are almost always subject to fluctuating or fatigue loads. The design process for dynamic loading is analogous to that for static loading with some significant variations. It is still an iterative problem. The design of helical springs for both static and fatigue loading can be readily computerized [9].

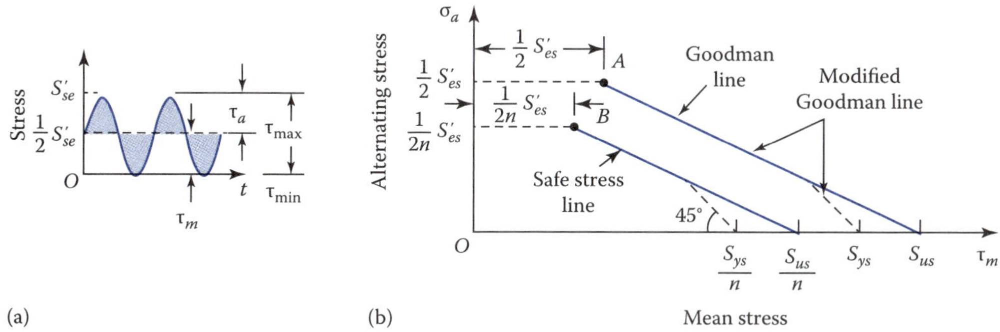

FIGURE 14.11
Fatigue loading: (a) endurance limit in pulsating shear test and (b) modified Goodman criteria for spring.

As pointed out in the preceding section, the stress–time diagram of Figure 14.11a expresses the worst condition that could occur for helical springs for pulsating shear when there is no preload; that is, when $\tau_{min} = 0$. We assume that the endurance limit in shear S'_{es} is the value of shear (see Table 14.3) for which a part on the verge of failure after an *infinite* number of cycles. In many cases, S'_{es} may be based on 1 million or 10 million cycles per shear loading.

A dynamically loaded spring operates between two force levels, P_{max} and P_{min}. Therefore, referring to Section 7.8, we define the *mean* and *alternating* axial *spring forces* as

$$P_m = \frac{1}{2}(P_{max} + P_{min}), \quad P_a = \frac{1}{2}(P_{max} - P_{min})$$

The most common spring-loading situation may involve both positive P_{max} and P_{min}. The direct shear factor K_s is used for the mean stress τ_m only (see Section 14.3). We apply the Wahl factor K_w to the alternating stress τ_a. Equations 14.6 and 14.8 become then

$$\tau_m = K_s \frac{8P_m C}{\pi d^2} \tag{14.21}$$

$$\tau_a = K_w \frac{8P_a C}{\pi d^2} \tag{14.22}$$

The factors K_s and K_w are given by Equations 14.7 and 14.9, respectively. The notch sensitivity of high-hardness steels is near unity, $q \approx 1$. Hence, for analyzing fatigue loading, there is no need to correct K_w to the fatigue stress-concentration factor K_f. Note that the *clash allowance* for fatigue design should be based on the *maximum load.*

14.8.1 Goodman Criteria Helical Springs

When the shear endurance limit S'_{es} of a spring wire is given, the Goodman or other fatigue failure criteria listed in Table 7.4 may be used. A torsional Goodman diagram can be constructed for any spring-loading situation. For $\tau_{min} = 0$, the alternating stress is equal to the mean stress or $S'_{es}/2$. Hence, the line of failure can be drawn from point A to ultimate strength in shear S_{us} on the τ_m axis shown in Figure 14.11b. The line representing the safe stress is parallel to the line of failure and can be drawn (from point *B*) after dividing the endurance limit in shear by the factor of safety *n*. Recall from Chapter 7 that the stress points (τ_m, τ_a) falling on or below the safe stress line constitute a satisfactory spring design. The modified Goodman criterion includes another line (shown dashed in the figure), drawn from the yield strength in shear S_{ys} on the τ_m axis with a slope upward and to the left at 45°.

The equation of the safe stress line is found by substituting the two stress points in the general equation of a line. In so doing, for the Goodman criterion,

$$\tau_a = \frac{\frac{1}{2}S'_{es}\left(S_{us}/n - \tau_m\right)}{S_{us} - \dfrac{1}{2}S'_{es}} \tag{14.23}$$

From Section 14.4, the ultimate strength in shear is given by $S_{us} = 0.67S_u$, in which S_u represents the ultimate tensile strength. When the ratio of range to mean stress is known, it may be convenient to rewrite Equation 14.23 as follows:

$$\tau_m = \frac{S_{us}/n}{\dfrac{\left(\tau_a/\tau_m\right)\left(2S_{us} - S'_{es}\right)}{S'_{es}} + 1} \tag{14.24}$$

An alternative form of Equation 14.23 gives the factor of safety guarding against a failure:

$$n = \frac{S_{us}S'_{es}}{\tau_a\left(2S_{us} - S'_{es}\right) + \tau_m S'_{es}} \tag{14.25}$$

We note that Equations 14.23 through 14.25 could also be written based on the Soderberg criterion by replacing S_{us} by S_{ys}. Having found the mean shear stress τ_m, we may use it and the mean load P_m to obtain wire diameter *d*. Hence, through the use of Equation 14.21,

$$d^3 = K_s \frac{8P_m D}{\pi\tau_m} \quad \text{or} \quad d^2 = K_s \frac{8P_m C}{\pi\tau_m} \tag{14.26}$$

When the safety factor is too low, the wire diameter, spring index, or the material can be altered to improve the result. The complete design includes consideration of the buckling discussed in Section 14.6 and surging of the springs, as is illustrated in the next example. Subsequent to several iterations, a reasonable combination of parameters can often be obtained.

14.8.2 Compression Spring Surge

A sudden compression of the end of a helical spring may form a compression wave that travels along the spring and is reproduced at the far end. This vibration wave, when it approaches resonance, is termed *surging*. It causes the coils to impact one another. The large forces due to both the excessive coil deflection and impacts fail the spring. To prevent this condition, the spring should not be cycled at a frequency close to its natural frequency.

Typically, the natural frequency of the spring should be greater than about 13 times that of any applied forcing frequency.

The *natural frequency* f_n of a helical compression spring depends on its end conditions. It can be shown that [2] for a spring with *fixed–fixed ends,*

$$f_n = \frac{d}{2\pi D^2 N_a}\sqrt{\frac{Gg}{2\gamma}} \quad \text{Hz} \tag{14.27}$$

This is twice that of a spring with *fixed–free ends.* Here, g is the acceleration of gravity and γ represents the weight per unit volume of the spring material. When d and D are in inches, we have $g \approx 386$ in./s². For steel spring, $G = 11.5 \times 10^6$ psi and $\gamma = 0.285$ lb/in.³ Equation 14.27 then becomes

$$f_n = \frac{14,040d}{D^2 N_a} \quad \text{Hz} \tag{14.28}$$

In SI units, when d and D are in millimeters,

$$f_n = \frac{356,620d}{D^2 N_a} \quad \text{Hz} \tag{14.29}$$

The surge of a spring decreases the ability of the spring to control the motion of the machine part involved. For example, the engine valve (shown in a closed position in Figure P14.25) might tend to oscillate rather than to operate properly. In addition, the spring material under a compression wave is subjected to higher stresses, which may cause early fatigue failure. It is obvious therefore that springs used in high-speed machinery must have natural frequencies of vibration considerably in excess of the natural frequency of the motion they control.

Example 14.5: Helical Compression Spring: Design for Cyclic Loading

A helical compression spring for a cam follower is subjected to the load that varies between P_{min} and P_{max}. Apply the Goodman criterion to determine

a. The wire diameter
b. The free height
c. The surge frequency
d. Whether the spring will buckle in service

Given: $P_{min} = 300$ N, $P_{max} = 600$ N

Design Decisions: We use a chrome-vanadium ASTM A232 wire of $G = 79$ GPa; $r_c = 20\%$, $N_a = 10$, and $C = 7$. Both ends of spring are squared and ground. A safety factor of 1.3 is used due to uncertainty about the load.

Solution

The mean and alternating loads are

$$P_m = \frac{1}{2}(600 + 300) = 450 \text{ N}, \quad P_a = \frac{1}{2}(600 - 300) = 150 \text{ N}$$

Equations 14.7 and 14.9 give

$$K_s = 1 + \frac{0.615}{7} = 1.088, \quad K_w = \frac{28-1}{28-4} + \frac{0.615}{7} = 1.213$$

So we have, using Equations 14.21 and 14.22, $\tau_a/\tau_m = K_w P_a / K_s P_m = 0.372$.

a. Tentatively *select* a 6 mm wire diameter. Then from Equation 14.12 and Table 14.2, we have

$$S_u = Ad^b = 1790(6^{-0.155}) = 1356 \text{ MPa}$$

By Equation 7.5 and Table 14.3, $S_{us} = 0.67(1356) = 908.5$ MPa and $S'_{es} = 0.2(1356) = 271$ MPa. Substitution of the numerical values into Equation 14.24 results in

$$\tau_m = \frac{908.5/1.3}{\frac{(0.372)(2\times 908.5 - 271)}{271} + 1} = 224 \text{ MPa}$$

Applying Equation 14.26,

$$d^2 = K_s \frac{8P_m C}{\pi\tau_m} = 1.088 \frac{8(450)(7)}{\pi(224\times 10^6)}$$

Hence, $D = 7(6.24) = 43.68$ mm. Inasmuch as $S_u = 1790(6.24^{-0.155}) = 1348 < 1356$ MPa, $d = 6.24$ mm is satisfactory.

b. From Figure 14.7d, $h_s = (N_a + 2)d = 74.88$ mm. Using Equation 14.11,

$$k = \frac{dG}{8C^3 N_a} = \frac{(6.24)(79{,}000)}{8(7)^3(10)} = 17.97 \text{ N/mm}$$

With a 20% clash allowance,

$$\delta_s = 1.2\frac{P_{\max}}{k} = 1.2(33.39) = 40.07 \text{ mm}$$

Thus,

$$h_f = 74.88 + 40.07 = 115 \text{ mm}$$

c. Through the use of Equation 14.29,

$$f_n = \frac{356{,}620d}{D^2 N_a} = \frac{356{,}620(6.24)}{(43.68)^2(10)}$$
$$= 116.6 \text{ cps} = 6996 \text{ cpm}$$

Comment: If this corresponds to operating speeds (for equipment mounted on this spring), it may be necessary to redesign the spring.

d. Check for buckling for extreme case of deflection ($\delta = \delta_s$):

$$\frac{\delta_s}{h_f} = \frac{40.07}{115} = 0.35, \quad \frac{h_f}{D} = \frac{115}{43.68} = 2.63$$

Since (2.63, 0.35) is inside of the stable region of curve *A* in Figure 14.10, the spring will not buckle.

14.9 Helical Extension Springs

Figure 14.3c illustrates a round-wire helical extension spring. Observe that a hook and loop are provided to permit a pull force to be applied. The significant dimensions of a standard end hook or loop are shown in Figure 14.12. Most of the preceding discussion of compression springs applies equally to helical extension springs. The natural frequency of a helical extension spring with both ends fixed against axial deflection is the same as that for a helical spring in compression.

In extension springs, however, the coils are usually close wound so that there is an *initial tension* or so-termed preload P_i. No deflection therefore occurs until the initial tension built into the spring is overcome; that is, the applied load *P* becomes larger than initial tension ($P > P_i$). It is recommended that [1], the preload be built so that the resulting *initial* torsional shear stress can be estimated as

$$\tau_i = 0.7\frac{S_u}{C} \tag{a}$$

Here, S_u and *C* present ultimate strength and spring index, respectively.

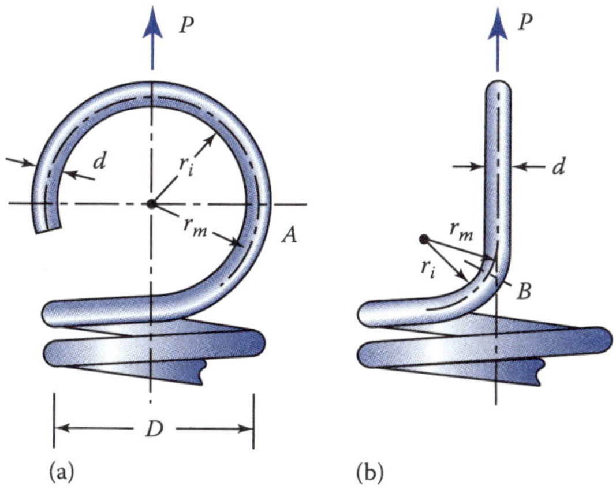

FIGURE 14.12
Front view and side view points of maximum stress in hook in conventional extension springs: (a) stress at the cross section through *A* is due to axial force and bending and (b) stress at the cross section through *B* is due primarily to torsion.

14.9.1 Coil Body

Coil deflection of helical extension springs, through the use of Equation 14.10 with $P = P - P_i$, is given as follows:

$$\delta = \frac{8N_aC^3(P - P_i)}{dG} \tag{14.30}$$

The reduced coil diameter results in a lower stress because of the shorter moment arm. Hence, hook stresses can be reduced by winding the last few coils with a decreasing diameter D. No stress-concentration factor is needed for the axial component of the load.

Active coils refer to all coils in the spring, not counting the end coils, which are bent to form a hook (Figure 14.3c). Depending on the details of the design, each end hook adds the equivalent of 0.1–0.5 helical coil. For an extension spring with two end hooks, the total number of coils is then

$$N_t = N_a + 2(0.1 \text{ to } 0.5) \tag{14.31}$$

As earlier, N_a represents the number of active coils.

The *spring rate* is expressed, by the application of Equation 14.30, in the form

$$k = \frac{P - P_i}{\delta} = \frac{dG}{8N_aC^3} \tag{14.32}$$

The spring load is therefore

$$P = P_i + k\delta \tag{14.33}$$

The quantities k and δ are given by Equations 14.30 and 14.32, respectively.

14.9.2 End Hook Bending and Shear

Critical stresses occur in the end hooks or end loops of extension springs. The hooks must be designed so that the stress-concentration effects produced by the presence of bends are decreased as much as possible. It is obvious that sharp bends should be avoided, since the stress-concentration factor is higher for sharp bends. *Maximum bending stress* at section A (Figure 14.12a) and *maximum torsional stress* at section B (Figure 14.12b) in the bend of the end coil may be approximated, respectively, by the formulas

$$\sigma_A = K\frac{16PD}{\pi d^3} \tag{14.34a}$$

$$\tau_B = K\frac{8PD}{\pi d^3} \tag{14.34b}$$

In each case, the stress-concentration factor K is given by

$$K = \frac{r_m}{r_i} \tag{b}$$

where
r_m is the mean radius
r_i represents the inside radius

The estimated permissible normal stress value in Equation 14.34a is the yield strength in tension, defined by Equation 14.11b. Recall that the allowable shear stress in Equation 14.34b is given by Table 14.3.

The *stresses in coils* are obtained from the same formulas as used in compression springs. In extension springs, a mechanical stop is desirable to limit deflection to an allowable value, while in compression springs, deflection is restricted by the solid deflection. Maximum stress values may be 70% of those used for extension springs of the identical compression springs.

Example 14.6: Load-Carrying Capacity of a Helical Extension Spring Hook

A helical extension spring with hook ends is made of a music wire of mean coil radius D, wire diameter d, mean hook radius r_m, and inner hook radius r_i (Figure 14.12). The preload is P_i and the free end is h_f.

Find:

a. The material properties and initial torsional stress in the wire using Equation (a)
b. Maximum load when yielding in tension impends at section A
c. Distance between the hook ends

Given: $d = 2.5$ mm, $D = 12.5$ mm, $(r_m)_A = 6.25$ mm, $(r_m)_B = 3.75$ mm

$$N_a = 150, \quad P = 50 \text{ N}, \quad h_f = 290 \text{ mm}$$

$$A = 2060 \quad \text{and} \quad b = -0.163 \text{ (from Table 14.2)}$$

Assumption: Modulus of rigidity will be $G = 79$ GPa.

Solution

a. Ultimate tensile strength, estimated from Equation 14.12, is

$$S_u = Ad^b = 2060\,(2.5)^{-0.163} = 1774 \text{ MPa}$$

By Equation 14.11b and Table 14.3, we obtain $S_y = S_{ys}/0.577 = (0.40/0.577)$ $S_y = 0.693S_u$. The yield strength is thus

$$S_y = 0.693\,(1774 \times 10^6) = 1229.4(10^6)$$

The spring index equals $C = D/d = 12.5/2.5 = 5$. Equation (a) results in then

$$\tau_i = 0.7\frac{S_u}{C} = 0.7\frac{1774}{5} = 248.4 \text{ MPa}$$

b. *Combined normal stress* at section A in the hook is obtained by superimposing bending and axial stresses. The former is defined by Equation 14.34a, and the latter equals $P/(\pi d^2/4)$. At the onset of yield, we therefore have

$$\sigma_A = K\frac{16PD}{\pi d^3} + \frac{4P}{\pi d^2} = S_y \tag{14.35}$$

where $K = r_m/r_i$ with $r_m = 6.25$ mm and $r_i = 6.25–2.5/2 = 5$ mm. Introducing the given data, Equation 14.34a leads to

$$\sigma_A = \left(\frac{6.25}{5}\right)\left[\frac{16P(12.5)}{\pi(2.5)^3(10^{-6})}\right] + \frac{4P}{\pi(2.5)^2(10^{-6})} = 1229.4(10^6)$$

or

$$(5.09296P + 0.2037P)10^6 = 1229.4(10^6)$$

Solving the maximum load when yielding begins in the hook gives

$$P = 232.1 \text{ N}$$

c. Inserting the given data into Equation 14.11, we obtain the spring rate as

$$k = \frac{dG}{8N_a C^3} = \frac{(2.5\times10^{-3})(79\times10^9)}{8(150)(5)^3} = 1317 \text{ N/m}$$

The deflection from Equation 14.32 is then

$$\delta = \frac{P - P_i}{k} = \frac{232.1 - 50}{1317} = 0.1383 \text{ m} = 138.3 \text{ mm}$$

The distance between hook ends equals

$$h_f + \delta = 290 + 138.3 = 428.3 \text{ mm}$$

Comments: Force P required to cause the torsional stress at section B in the hook may also readily be determined, using Equation 14.34b and Table 14.3. In so doing, a smaller load is obtained (see Problem 14.27), which shows that failure by yielding first takes place by shear stress in the hook.

14.10 Torsion Springs

Torsion springs are of two general types: helical and spiral. The primary stress in a torsion spring is bending, with a moment being applied to each end of the wire. The analysis of curved beams discussed in Sections 3.7 and 16.7 is applicable. Springs of this kind are employed in door hinges, automotive starters, and so on, where torque is needed.

The yield strength S_y for torsion springs can be estimated from Table 14.3. Based on the energy of distortion criterion, we divide the S_{ys} in each part in Table 14.3 by the quantity 0.577. The endurance limit S_e for torsion springs can be found in a like manner: the S'_{es} in each part in Table 14.3 is divided by 0.577. The process of designing of torsion springs is very similar to that of the helical compression springs.

14.10.1 Helical Torsion Springs

As depicted in Figure 14.13, helical torsion springs are wound in a way similar to extension or compression springs but with the ends shaped to transmit torque. These coil ends can have a variety of forms to suit the application. The coils are usually close wound like an extension spring but have no initial tension. We note that forces (P) should always be applied to arms of helical torsion springs to close the coil, as shown in the figure, rather than open it. The spring is usually placed over a supporting rod. The rod diameter is about 90% smaller than the inside diameter of the spring. Square or rectangular wire is in widespread use in torsion springs. However, round wire is often used in ordinary applications, since it costs less. The torque about the axis of the helix acts as a bending moment on each section of the wire. The material is therefore stressed in flexure. The bending stress can be obtained from curved beam theory. It is convenient to write the flexure formula in the form

$$\sigma = K\frac{Mc}{I} \tag{a}$$

where

σ = the maximum bending stress
M = the bending moment
c = the distance from the neutral axis to the extreme fiber
I = the moment of inertia about the neutral axis
K = the stress-concentration factor

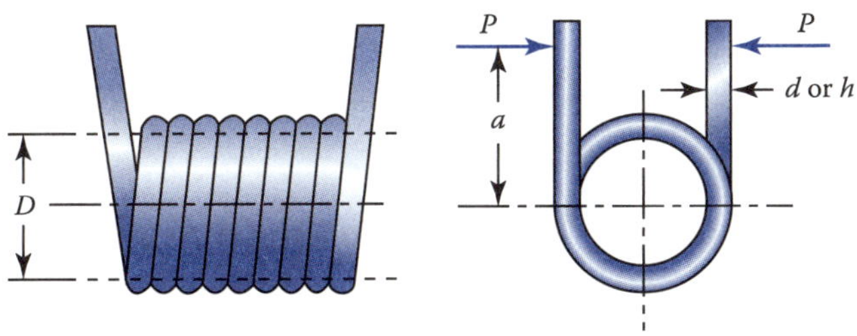

FIGURE 14.13
Helical torsion spring.

Wahl analytically determined the values for the stress-concentration factors [2]. For *round wire,*

$$K_i = \frac{4C^2 - C - 1}{4C(C-1)}$$
$$K_o = \frac{4C^2 + C - 1}{4C(C+1)} \tag{14.36}$$

In the foregoing, the spring index $C = D/d$: the subscripts i and o refer to the inner and outer fibers, respectively. For *rectangular wire,*

$$K_i = \frac{3C^2 - C - 0.8}{3C(C-1)}$$
$$K_o = \frac{3C^2 + C - 0.8}{3C(C+1)} \tag{14.37}$$

where $C = D/h$. The quantity h represents the depth of the rectangular cross section. We see from these expressions that $K_i > K_o$, as expected.

The *maximum compressive bending stress* at the inner fiber of the helical torsion spring is therefore

$$\sigma_i = K_i \frac{Mc}{I} \tag{14.38}$$

Carrying the bending moment $M = Pa$ and the section modulus I/c of round and rectangular wires into Equation 14.38 gives the bending stress. In so doing, stress on the inner fiber of the coil is

$$\sigma_i = \frac{32Pa}{\pi d^3} K_i \quad \text{(round wire)} \tag{14.39}$$

$$\sigma_i = \frac{6Pa}{bh^2} K_i \quad \text{(rectangular wire)} \tag{14.40}$$

The quantity b is the width of rectangular cross section.

For commonly employed values of the spring index, $k = M/\theta_{rev}$, the curvature has no effect on the *angular deflection*. Through the use of Equations 4.14 and 4.15, we have

$$\theta_{rev} = \frac{1}{2\pi}\theta_{rad} = \frac{1}{2\pi}\frac{ML_w}{EI} \tag{14.41}$$

where

θ = the angular deflection
L_w = the length of wire = $\pi D N_a$
EI = the flexural rigidity

For springs of round wire, to account for the friction between coils, based on experience [1], Equation 14.41 is multiplied by the factor of 1.06. Interestingly, angle θ in some cases can be many complete turns, as in Example 14.7.

14.10.2 Fatigue Loading

A dynamically loaded torsion spring operates between two moment levels M_{max} and M_{min}. The tensile stress components occurring at the outside coil diameter of a round-wire helical torsion are then

$$\sigma_{o,\max} = K_o \frac{32M_{\max}}{\pi d^3}, \quad \sigma_{o,\min} = K_o \frac{32M_{\min}}{\pi d^3}$$

Hence, the mean and alternating stresses are

$$\sigma_{o,m} = \frac{\sigma_{o,\max} + \sigma_{o,\min}}{2}, \quad \sigma_{o,a} = \frac{\sigma_{o,\max} - \sigma_{o,\min}}{2}$$

Having the mean and alternating stresses available, helical torsion springs are designed by following a procedure similar to that of helical compression springs.

14.10.3 Spiral Torsion Springs

A spiral torsion spring (Figure 14.14) can also be analyzed by the foregoing procedure. Therefore, the highest stress occurring on the inner edge of the wire is given by Equations 14.39 and 14.40. Likewise, Equation 14.41 can be applied directly to ascertain the angular deflection. Spiral springs are usually made of thin rectangular wire.

Example 14.7: Spiral Torsion Spring: Design for Static Loading

For a torsional window-shade spring (Figure 14.14), determine the maximum operating moment and corresponding angular deflection.

Design Decisions: We select a music wire of $E = 207$ GPa; $d = 1.625$ mm, $D = 25$ mm, and $N_a = 350$. A safety factor of 1.5 is used.

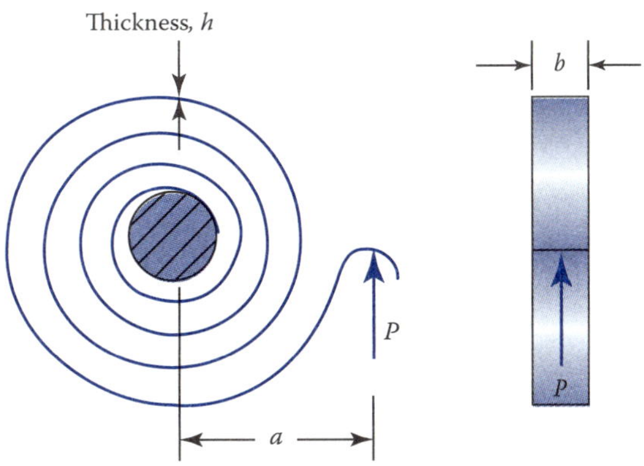

FIGURE 14.14
Spiral torsion spring.

Solution

By Equation 14.12 and Table 14.2,

$$S_u = Ad^b = 2060\left(1.625^{-0.163}\right) = 1903 \text{ MPa}$$

From Equation 7.5b and Table 14.3,

$$S_y = \frac{S_{ys}}{0.577} = 0.4\frac{1903}{0.577} = 1319 \text{ MPa}$$

Applying Equation 14.36 with $C = 25/1.625 = 15.38$,

$$K_i = \frac{4(15.38)^2 - 15.38 - 1}{4(15.38)(15.38 - 1)} = 1.051$$

Through the use of Equation 14.39, we have

$$M = Pa = \frac{\pi d^3 S_y/n}{32K_i} = \frac{\pi(1.625)^3(1319/1.5)}{32(1.051)}$$

$$= 352.5 \text{ N}\cdot\text{mm}$$

The geometric properties of the spring are $L_w = \pi DN_a = \pi(25)(350) = 27{,}489$ mm and $I = \pi(1.625)^4/64 = 0.342$ mm^4. Equation 14.41 results in

$$\theta_{\text{rad}} = \frac{ML_w}{EI} = \frac{352.5(27,489)}{(207\times 10^3)(0.342)} = 136.9 \text{ rad}$$

Comment: The maximum moment winds the spring $136.9/2\pi = 21.8$ turns.

14.11 Leaf Springs

A *leaf spring* is usually arranged as a cantilever or simply supported member. This thin beam or plate is also known as a flat spring, although it usually has some initial curvature. Springs in the form of a cantilever are often used as electrical contacts. For springs with uniform sections, we may use the results of Chapters 3 and 4. Recall from Sections 4.4 and 4.10 that when the width of the cross section is large compared with the depth, it is necessary to multiply the deflection as given by the formula for a narrow beam section by $(1 - \nu^2)$, where ν is the Poisson's ratio.

A *cantilever spring* of *uniform stress a* with a constant depth h and length L in a plan view looks like the triangle depicted in Figure 14.15 (see Section 3.8). However, near the free end, the wedge-shaped profile must be modified to have adequate strength to resist the

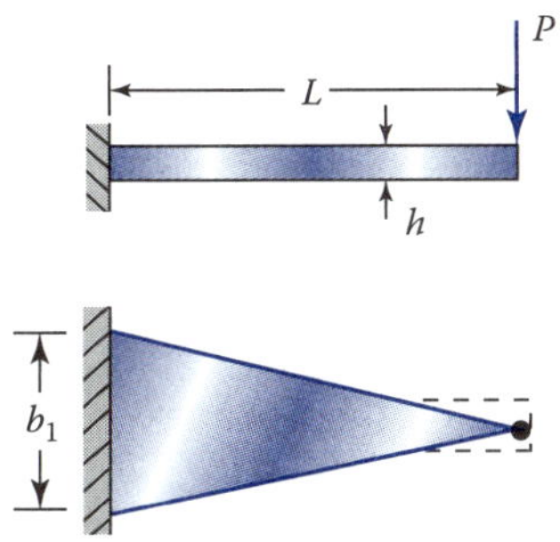

FIGURE 14.15
Cantilever spring of uniform stress.

shear force as depicted by the dashed lines in the figure. From the flexure formula, due to a concentrated load P applied at the free end, we have

$$\sigma = \frac{6PL}{b_1 h^2} \tag{a}$$

As the cross section varies, end deflection δ may conveniently be obtained using Castigliano's theorem (see Section 5.5). It can be shown that

$$\delta - \left(1 - \nu^2\right)\frac{6P}{Eb_1}\left(\frac{L}{h}\right)^3 \tag{b}$$

The corresponding spring rate is

$$k = \frac{P}{\delta} = \frac{Eb_1}{6\left(1 - \nu^2\right)}\left(\frac{h}{L}\right)^3 \tag{c}$$

The quantity E represents the modulus of elasticity.

14.11.1 Multileaf Springs

Springs of varying width present a space problem. Multileaf springs are in widespread usage, particularly in automotive and railway services. An exact analysis of these springs is mathematically complex. For small deflections, an approximate solution can be obtained by the usual equations of beams, as shown in the following brief discussion.

A multileaf spring, approximating a triangular spring of uniform strength, is shown in Figure 14.16. Note that each half of the spring acts as a cantilever of length L. We observe from the figure that a constant strength triangle is cut into a series of leaves of equal width and rearranged in the form of a multileaf spring. Therefore, letting $b_1 = nb$, the stress and deflection for the *ideal leaf spring* are

$$\sigma = \frac{6PL}{nbh^2} \tag{14.42}$$

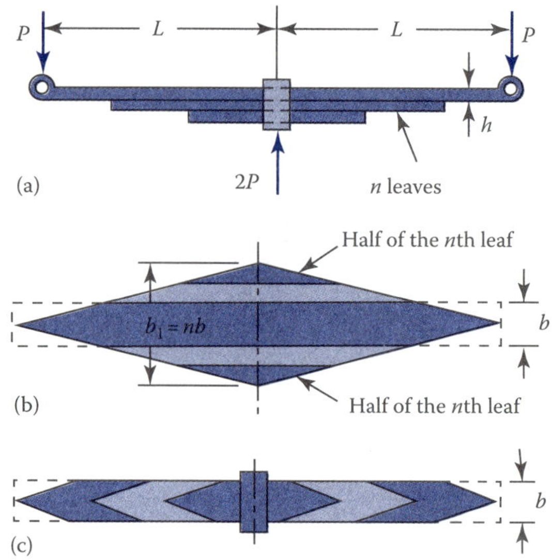

FIGURE 14.16
Multileaf spring: (a) front view of actual spring, (b) top view of approximation, and (c) top view of equivalent spring.

$$\delta = \left(1-\nu^2\right)\frac{6P}{Enb}\left(\frac{L}{h}\right)^3 \tag{14.43}$$

The spring rate is then

$$k = \frac{P}{\delta} = \frac{Enb}{6\left(1-\nu^2\right)}\left(\frac{h}{L}\right)^3 \tag{14.44}$$

In the preceding equations, the quantity n represents the number of leaves.

A central bolt or clamp, used to hold the leaves together, causes a stress concentration. The triangular spring and equivalent multileaf spring have the *identical* stress and deflection characteristics, with the exception that the interleaf friction provides damping in the multileaf spring. Also the multileaf spring can resist full load in only one direction; that is, leaves tend to separate when loaded in opposite direction. However, this is partially overcome by clips, as in vehicle suspension springs (Figure 14.17).

Example 14.8: Design of a Nine-Leaf Cantilever Spring

A steel 0.9 m long cantilever spring has 80 mm wide nine leaves. The spring is subjected to a concentrated load P at its free end.

Find: The depth of the leaves and the largest bending stress

Given: $b = 80$ mm, $L = 0.9$ m, $P = 2.5$ kN, $n = 9$
$E = 200$ GPa, $\nu = 0.3$

Assumption: Maximum vertical deflection caused by the load will be limited to 50 mm.

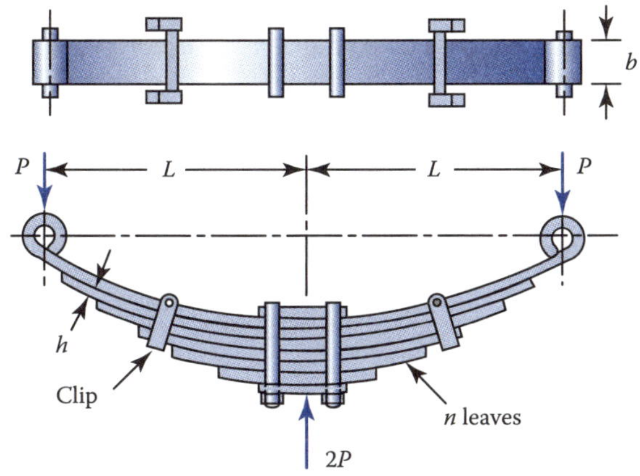

FIGURE 14.17
Example 14.8. Automotive-type leaf spring.

Solution

Equation 14.43 may be rearranged into the form

$$h^3 = (1-\nu^2)\frac{6PL^3}{Enb\delta} \tag{d}$$

Inserting the given data, we have

$$h^3 = (1-0.3^2)\frac{6(2500)(0.9)^3}{(200\times 10^9)(9)(0.08)(0.05)} = 1.382(10^{-6})$$

or

$$h = 0.0111 \text{ m} = 11.1 \text{ mm}$$

Equation 14.42 results in the maximum stress as

$$\sigma_{max} = \frac{6PL}{nbh^2} = \frac{6(2500)(0.9)}{9(0.08)(0.0111)^2} = 152.2 \text{ MPa}$$

The Goodman criterion may be used in the design of leaf springs subject to cyclic loading, as illustrated in the solution of the following numerical problem.

Example 14.9: Automotive-Type Multileaf Spring: Design for Fatigue Loading

A six-leaf spring is subjected to a load at the center that varies between P_{max} and P_{min} (Figure 14.17). Estimate the total length $2L$ and width of each leaf.

Given: $P_{min} = 160$ lb, $P_{max} = 800$ lb, $n = 6$

Assumptions: Stress concentration at the center is such that $K_f = 1.2$. Use a survival rate of 50% and $C_f = C_s = 1$.

Design Decisions: We use a steel alloy spring of $S_u = 200$ ksi, $S_e' = 78$ ksi, $E = 30 \times 10^6$ psi, $\nu = 0.3$, $h = 0.25$ in., and $k = 140$ lb/in. The material is shot peened. A safety factor of $n_s = 1.4$ is applied.

Solution

From Table 7.3, $C_r = 1$. The modified endurance limit, by Equation 7.6, $S_e = (1)(1)(1)(1/1.2)78 = 65$ ksi. Each half of a spring acts as a cantilever supporting half of the total load. The mean and the alternating loads are therefore

$$P_m = \frac{400+80}{2} = 240 \text{ lb}, \quad P_a = \frac{400-80}{2} = 160 \text{ lb}$$

Inasmuch as bending stress is directly proportional to the load, we have $\sigma_a/\sigma_m = P_a/P_m = 2/3$.

The mean stress, using Equation 14.42, is

$$\sigma_m = \frac{6P_mL}{nbh^2} = \frac{6(240)L}{6b(0.25)^2} = 3840\frac{L}{b} \tag{e}$$

Substituting the given numerical values into Equation 7.20, we have

$$\sigma_m = \frac{S_u/n_s}{\dfrac{\sigma_a}{\sigma_m}\dfrac{S_u}{S_e}+1} = \frac{200/1.4}{\dfrac{2}{3}\dfrac{200}{65}+1} = 46.82 \text{ ksi}$$

From Equation (e),

$$46{,}820 = 3840\frac{L}{b} \quad \text{or} \quad b = 0.082L \tag{f}$$

Because the spring is loaded at the center with $2P$, Equation 14.44 becomes $k = Enbh^3/3L^3(1-\nu^2)$. Introducing the given data results in

$$140 = \frac{(30\times10^6)(6)(0.082L)(0.25)^3}{3L^3(0.91)}$$

$$L = 24.56 \text{ in.}$$

Hence, the *overall length* is $2L = 49.12$ in. The *width* of *each* of the six *leaves* using Equation (f) equals $b = 0.082(24.56) = 2.014$ in.

14.12 Miscellaneous Springs

Many spring functions may also be acquired by the elastic bending of thin plates and shells of various shapes and by the blocks of rubber. Hence, there are spring washers, clips, constant-force springs, volute springs, rubber springs, and so on. A *volute spring* is a wide, thin strip of steel wound flat so that the coils fit inside one another, as shown in Figure 14.18. These springs have more lateral stability than helical compression springs,

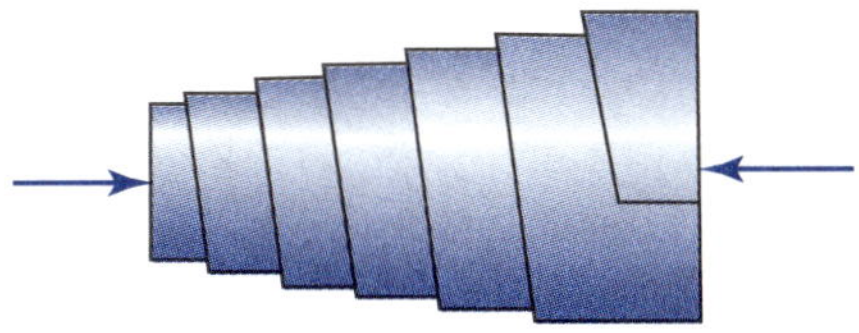

FIGURE 14.18
Volute spring.

and rubbing of adjacent turns provides high damping. Here, we briefly discuss three commonly encountered types of miscellaneous springs.

14.12.1 Constant-Force Springs

The constant-force (Neg'ator) spring is a prestressed strip of flat spring stock that coils around a bushing or successive layers of itself (Figure 14.19). Usually, the inner coil is fastened to a flanged drum. When the spring is deflected by pulling on the outer end of the coil, a nearly constant resisting force develops, and there is a tendency for the material to recoil around itself. A uniform-force spring is widely employed for counterbalancing loads (such as in window sash), cable retractors, returning typewriter carriages, and making constant-torque spring motors. It provides very large deflection at about a constant pull force [1].

14.12.2 Belleville Springs

Belleville springs or *washers*, also known as coned-disk springs (Figure 14.20), patented by Belleville in 1867, are often used for supporting very *large* loads with *small* deflections. Some applications include various bolted connections, clutch plate supports, and gun recoil mechanisms. On loading, the disk tends to flatten out, spring action being obtained thus. The load–deflection characteristics are changed by varying the ratio h/t between cone

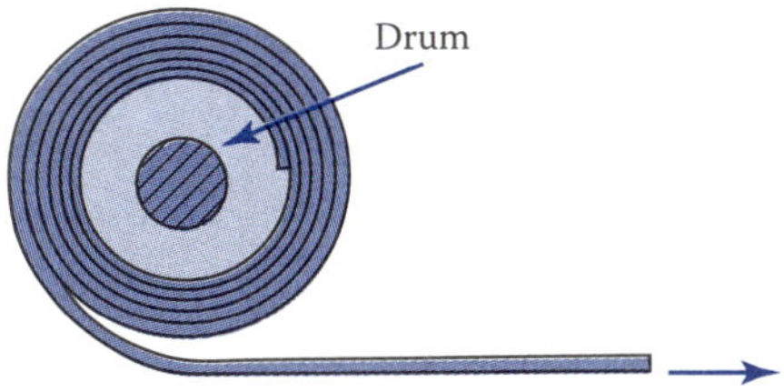

FIGURE 14.19
Constant-force spring.

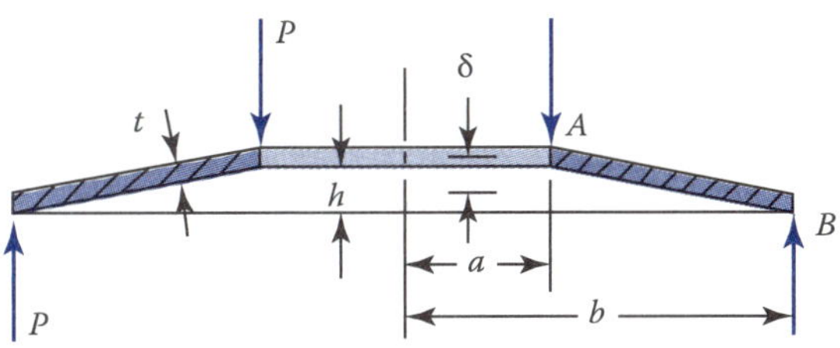

FIGURE 14.20
Cross section through a Belleville spring.

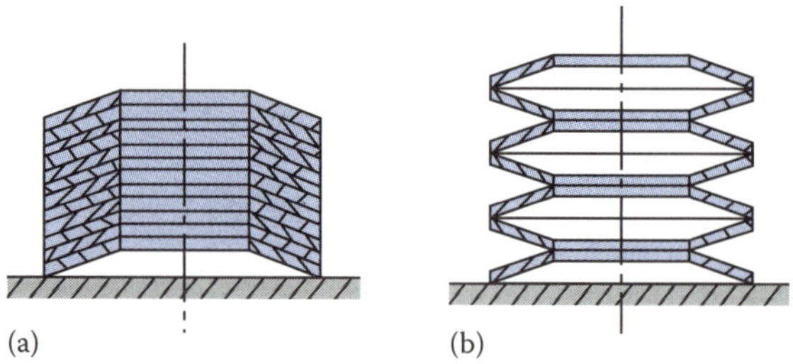

FIGURE 14.21
Belleville springs or washers: (a) in parallel stack and (b) in series stack.

height h and thickness t. Belleville springs are extremely compact and may be used singly or in combination of multiples of identical springs to meet needed characteristics. The forces associated with a coned-disk spring can be multiplied by stocking them in parallel (Figure 14.21a). On the other hand, the deflection corresponding to a given force can be increased by stacking the springs in series as shown in Figure 14.21b.

The theory of the Belleville springs is complicated. The following formulas are based on the simplifying assumption that radial cross sections of the spring do not distort during deflection. The results are in approximate agreement with available test data [1,2]. As is the case for a truncated cone shell, the upper edge of the spring is in compression, and the lower edge is in tension [11].

The *load–deflection* relationship can be expressed in the form

$$P = \frac{E\delta}{(1-\nu^2)Kb^2}\left[\left(h - \frac{\delta}{2}\right)(h-\delta)t + t^3\right] \tag{14.45a}$$

where

$$K = \frac{6}{\pi \ln\alpha}\left(\frac{\alpha - 1}{\alpha}\right)^2 \tag{14.45b}$$

The load at the *flat position* ($\delta = h$) is given by

$$P_{\text{flat}} = \frac{Eht^3}{(1-\nu^2)Kb^2} \tag{14.45c}$$

where
- P = the load
- δ = the deflection
- a = the inside radius
- b = the outside radius
- h = the cone height
- t = the thickness
- α = the radius ratio = b/a

Zero deflection and load ($\delta = 0$ and $P = 0$) are taken at *the free position* depicted in Figure 14.20.

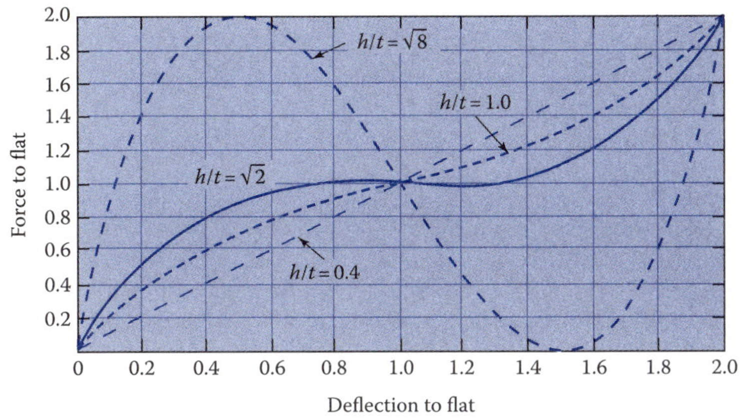

FIGURE 14.22
Force–deflection curves for Belleville springs.

Load–deflection characteristics are changed by varying the ratio between cone height and thickness, h/t. Figure 14.22 illustrates force–deflection curves for Belleville washers with four different h/t ratios. These curves are generated by applying Equation 14.45, where 1.0 deflection and 1.0 force refer to the deflection at the flat condition and the force at the flat condition, respectively [12]. We see from the figure that coned-disk springs have nonlinear P–δ properties. For low values ($h/t = 0.4$), the spring acts almost linearly, and large h/t values result in prominent nonlinear behavior. At $h/t = \sqrt{2}$, the central portion of the curve approximates a horizontal line: that is, the load is nearly constant over a considerable deflection range. In the range $\sqrt{2} < h/t \leq \sqrt{8}$, a prescribed force corresponds to more than one deflection. A phenomenon occurring at $h/t > \sqrt{2}$ is termed *snap-through buckling*, at which the spring deflection becomes unstable.

Interestingly, in snap-through buckling, the spring quickly deflects or snaps to the next stable position. It can be shown that if $h/t > \sqrt{8}$, the spring can snap into a deflection position for which the calculated force becomes negative. Then a load in the direction opposite to the initial load will be required to return the spring to its unloaded configuration.

Stress distribution in the washer is nonuniform. The *largest* stress σ_A occurs at the upper inner edge A (convex side) at deflection δ and is compressive. The outside lower edge B (concave side) has the largest tensile stress σ_B. The expressions for the foregoing stresses are

$$\sigma_A = -\frac{E\delta}{(1-\nu^2)Kb^2}\left[c_1\left(h-\frac{\delta}{2}\right)+c_2 t\right] \tag{14.46a}$$

$$\sigma_B = -\frac{E\delta}{(1-\nu^2)Kb^2}\left[c_3\left(h-\frac{\delta}{2}\right)+c_4 t\right] \tag{14.46b}$$

where

$$c_1 = \frac{6}{\pi \ln\alpha}\left(\frac{\alpha-1}{\ln\alpha}-1\right)$$

$$c_2 = \frac{6}{\pi \ln\alpha}\left(\frac{\alpha-1}{2}\right) \tag{14.46c}$$

$$c_3 = \left[\frac{\alpha\ln\alpha-(\alpha-1)}{\ln\alpha}\right]\left[\frac{\alpha}{(\alpha-1)^2}\right]$$

$$c_4 = \frac{\alpha}{2(\alpha-1)}$$

Stresses are highly concentrated at the edges of Belleville springs. When the yield strength is exceeded, a redistribution of the stresses occurs due to localized yielding. The compressive stress σ_A given by Equation 14.46a controls the design for static loading. If the spring is under *dynamic* loading, the alternating and mean stresses are determined from tensile stress σ_B defined by Equation 14.46b. The factor of safety, according to the Goodman criterion, is found from Equation 7.22.

14.12.3 Rubber Springs

A rubber spring and cushioning device is referred to as a *rubber mount* (Figure 14.23). Springs of this type are widely used due to their essentially shock and vibration damping qualities and low elastic moduli. The foregoing properties help dissipate energy and prevent sound transmission. Stresses and deformations in the rubber mounts for small deflections can be derived by the use of appropriate equations of mechanics of materials.

A cylindrical rubber spring with *direct shear loading* is shown in Figure 14.23a. The rubber is bonded to a steel ring on the outside and a steel shaft in the center. The shear stress τ at radius r is

$$\tau = \frac{P}{2\pi r h} \tag{14.47}$$

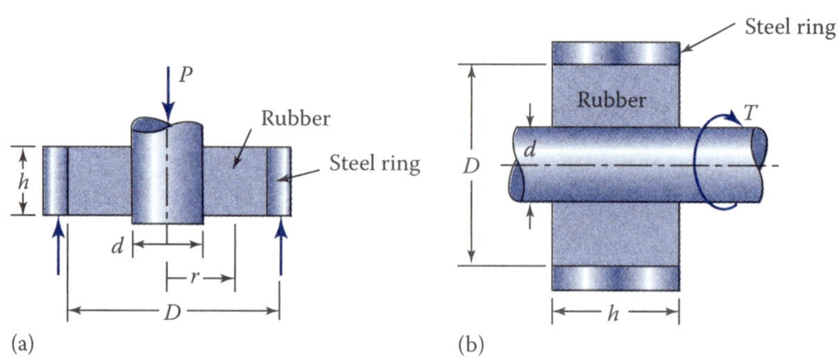

FIGURE 14.23
Cylindrical rubber mounts: (a) with shear loading, and (b) with torsion loading.

Maximum deflection δ occurs at inner edge ($r = d/2$):

$$\delta = \frac{P}{2\pi hG} \ln \frac{D}{a} \tag{14.48}$$

where
P = the load
d = the inner radius
h = the depth of mount
D = the outer radius

A cylindrical spring with *torsional shear loading* is depicted in Figure 14.23b. The maximum shear stress taking place at the inner edge ($r = d/2$) is given by

$$\tau_{\max} = \frac{2T}{\pi d^2 h} \tag{14.49}$$

The angular rotation of the shaft, or maximum angle of twist ϕ, is

$$\phi = \frac{T}{\pi hG}\left(\frac{1}{d^2} - \frac{1}{D^2}\right) \tag{14.50}$$

The quantity T represents the torque.

Note that rubber does not follow Hooke's law but becomes increasingly stiff as the deformation is increased. The modulus of elasticity is contingent on the durometer hardness number of the rubber chosen for the mount. The results of calculations must therefore be considered only approximate.

Problems

Sections 14.1 through 14.6

14.1 A steel torsion bar is used as a counterbalance spring for the trunk lid of an automobile (Figure 14.2a). Determine, when one end of the bar rotates 80° relative to the other end,

a. The change in torque
b. The change in shear stress

Given: L = 1.25 m, d = 8 mm, G = 79 GPa

14.2 A steel bar supports a load of 2 kN with a moment arm R = 150 mm (Figure 14.2a). Calculate

a. The wire diameter
b. The length for a deflection of 40 mm

Given: n = 1.5, S_{ys} = 350 MPa, G = 79 GPa

14.3 A high-strength ASTM A242 steel torsion bar ($G = 79$ GPa) with splined ends shown in Figure 14.2b has a length $L = 1.2$ m and diameter $d = 12$ mm. Find, if relative rotation between the ends is changed by 20°, the torque and the shear stress.

14.4 A helical spring must *exert* a force of 1 kN after being released 20 mm from its most highly compressed position. Determine the number of active coils.

Design Assumptions: The loading is static. $\tau_{all} = 450$ MPa, $G = 29$ GPa, $d = 7$ mm, and $C = 5$.

14.5 A helical coil spring of mean diameter $D = 2$ in. and wire diameter $d = 3/8$ in., wound with a coil pitch $p = 1/2$ in. (Figure 14.3a), is compressed solid. The material is ASTM A229 oil-tempered steel ($G = 11.5 \times 10^6$ psi). Determine the force required to compress spring to solid and corresponding shear stress. Will the spring return to its original free length after the force is removed?

14.6 Figure 14.5 illustrates a conical–helical compression spring of five active coils fabricated of ASTM A232 hard-drawn steel wire. Find which coil will deflect to zero pitch first and the corresponding force required. What is the total spring deflection?

Given: $D_{max} = 60$ mm, $D_{min} = 25$ mm, $d = 4$ mm, $p = 6$ mm

14.7 A pair of concentric helical compression springs made of structural steel supports weight $W = 2$ kN of an equipment (Figure P14.7). Both springs are made of a structural steel with the modulus of rigidity $G = 79$ GPa and have the same length. Find, in each spring,

a. The deflection

b. The largest stress

Given:

Outer spring: $D_o = 40$ mm, $d_o = 7$ mm, $N_o = 4$

Inner spring: $D_i = 22$ mm, $d_i = 4.5$ mm, $N_i = 8$

14.W1 Using the website at www.leesspring.com, rework Example 14.1.

14.W2 Check the site at www.acxesspring.com to review the common spring materials presented. List five commonly employed wire spring materials and their mechanical properties.

14.8 A helical compression spring used for static loading has $d = 3$ mm, $D = 15$ mm, $N_a = 10$, and squared ends. Determine

a. The spring rate and the solid height

b. The maximum load that can be applied without causing yielding

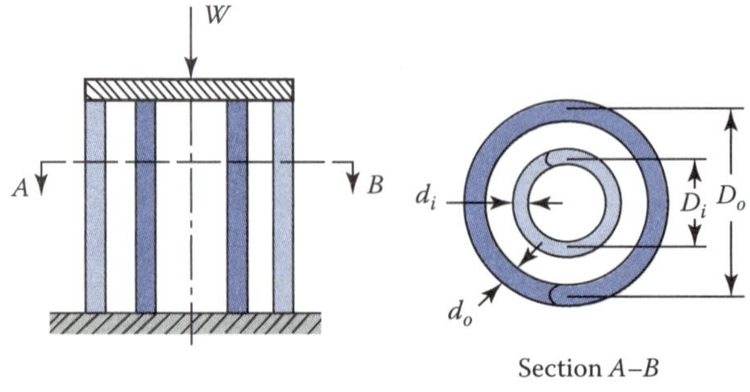

FIGURE P14.7

Design Decision: The spring is made of ASTM A227 hard-drawn steel wire of $G = 79$ GPa.

14.9 A helical compression spring is to support a 2 kN load. Determine

a. The wire diameter

b. The free height

c. Whether buckling will occur in service

Given: The spring has $r_c = 10\%$, $C = 5$, and $k = 90$ N/mm.

Assumptions: Both ends are squared and ground and constrained by parallel plates.

Design Decisions: The spring is made of steel of $S_{ys} = 500$ MPa, $S'_{es} = 280$ MPa, and $G = 79$ GPa. Use a safety factor of 1.3.

14.10 A helical compression spring with ends squared and ground has $d = 1.8$ mm, $D = 15$ mm, $r_c = 15\%$, and $h_s = 21.6$ mm. Determine, using a safety factor of 2,

a. The free height

b. Whether the spring will buckle in service, if one end is free to tip

Design Decision: The spring is made of steel having $S_{ys} = 900$ MPa and $G = 79$ GPa.

14.11 Design a helical compression spring with squared and ground ends for a static load of 40 lb. $C = 8$, $k = 50$ lb/in., $r_c = 20\%$, and $n = 2.5$. Also check for possible buckling.

Assumption: The ends are constrained by parallel plates.

Design Decision: The spring is made of steel of $S_{ys} = 60$ ksi and $G = 11.5 \times 10^6$ psi.

14.12 A machine that requires a helical compression spring of $k = 120$ lb/in., $\tau_{all} = 75$ ksi, $r_c = 10\%$, $D = 3$ in., and is to support a static load of 400 lb. Determine

a. The wire diameter

b. The free height

c. Whether the spring will buckle in service, if one end is free to tip

Assumption: The ends are squared.

Design Decision: The spring is made of steel having $G = 11.5 \times 10^6$ psi.

14.13 A helical compression spring has mean coil diameter $D = 20$ mm, wire diameter $d = 2.5$ mm, and the number of active coils $N_a = 11$ (Figure 14.8a). The material is ASTM A228 music wire ($G = 79$ GPa). Find

a. The largest static load and the spring rate

b. The free length, for which spring would become solid under the load found in (a)

c. Whether the buckling occurs, for the case in which one end plate is free to tilt

14.14 A helical compression spring is fabricated from ASTM A229 oil-tempered wire (Figure 14.3a) and has the mean coil diameter $D = 10$ mm, wire diameter $d = 0.9$ mm, total number of active coils $N_a = 14.5$, and the modulus of rigidity $G = 79$ GPa. Find

a. The torsional yield strength of the wire

b. The static load corresponding to the yield strength and spring constant

c. The solid and free heights and the pitch of the body coil of the spring

d. Whether buckling will be possible, if the ends are squared–ground

14.15 A helical compression spring has $N_a = 16$ active coils, a free length of $h_f = 35$ mm, mean coil diameter $D = 14$ mm, and wire diameter $d = 1.5$ mm. The spring is made of ASTM A229 oil-tempered steel wire of $G = 79$ GPa. Determine, for the *static* conditions,

a. The spring rate, the solid height, pitch, and the solid deflection

b. The force required to compress the coils to solid height, corresponding shear stress, and the safety factor against yielding

c. Whether buckling will occur in service, if the ends are constrained by parallel plates

Sections 14.7 and 14.8

14.16 Redo Problem 14.9 for a load that varies between 2 and 4 kN, using the Soderberg relation. Also determine the surge frequency.

14.17 A helical compression spring with squared ends operates under a fluctuating load between $P_{min} = 0$ N and $P_{max} = 400$ N with the deflection varying by 10 mm. A shot peened steel spring wire is used (see Equation 14.20). Compute the wire diameter d, the number of active coils N_a, and free height h_f.

Assumptions: A clash allowance of 15% of the maximum deflection will be used.

Given: $D = 40$ mm, $S'_{es} = 465$ MPa, $K_w = 1.3$, $G = 79$ GPa

14.18 Redo Problem 14.17, for a wire without shot peening (see Equation 14.20) based on a clash allowance of 8% of the maximum deflection.

14.19 Reconsider Problem 14.15, for *dynamic* condition, with minimum load $P_{min} = 4$ N and maximum load $P_{max} = 14$ N. Compute

a. The alternating and mean stresses

b. The factor of safety against torsional yielding

c. The factor of safety against torsional endurance limit fatigue

14.20 A helical compression spring for a cam follower supports a load that varies between 30 and 180 N. Determine

a. The factor of safety, according to the Goodman criterion

b. The free height

c. The surge frequency

d. Whether the spring will buckle in service

Design Decisions: The spring is made of music wire. Both ends are squared and ground; one end is free to tip.

Given: $d = 3$ mm, $D = 15$ mm, $N_a = 22$, $r_c = 10\%$, $G = 79$ GPa

14.21 A helical compression spring, made of 0.2 in. diameter music wire, carries a fluctuating load. The spring index is 8 and the factor of safety is 1.2. If the average load on the spring is 100 lb, determine the allowable values for the maximum and minimum loads. Employ the Goodman theory.

14.22 A helical compression spring made of a music wire has $d = 5$ mm, $D = 24$ mm, and $G = 79$ GPa. Determine

a. The factor of safety, according to the Goodman relation

b. The number of active coils

Requirements: The height of the spring varies between 65 and 72 mm with corresponding loads of 400 and 240 N.

14.23 A steel helical compression spring is to exert a force of 4 lb when its height is 3 in. and a maximum load of 18 lb when compressed to a height of 2.6 in. Determine, using the Soderberg criterion with a safety factor of 1.6,

a. The wire diameter

b. The solid deflection

c. The surge frequency

d. Whether the spring will buckle in service, if ends are constrained by parallel plates

Given: The spring has $C = 6$, $S_{ys} = 80$ ksi, $S'_{es} = 45$ ksi, $G = 11.5 \times 10^6$ psi, $r_c = 10\%$.

Design Assumption: Ends will be squared and ground.

14.24 Resolve Problem 14.23 for the case in which the helical spring is to exert a force of 2 lb at 5 in. height and a maximum load of 10 lb at 4.2 in. height.

14.25 An engine valve spring must exert a force of 300 N when the valve is closed (as shown in Figure P14.25) and 500 N when the valve is open. Apply the Goodman theory with a safety factor of 1.6 to calculate

a. The wire diameter

b. The number of active coils

Given: The lift is 8 mm.

Design Decisions: The spring is made of steel having $S_{us} = 720$ MPa, $S'_{es} = 330$ MPa, $G = 79$ GPa, and $C = 6$.

14.26 A helical spring, made of hard-drawn wire having $G = 29$ GPa, supports a continuous load. Determine

a. The factor of safety based on the Soderberg criterion

b. The free height

c. The surge frequency

d. Whether the spring will buckle in service

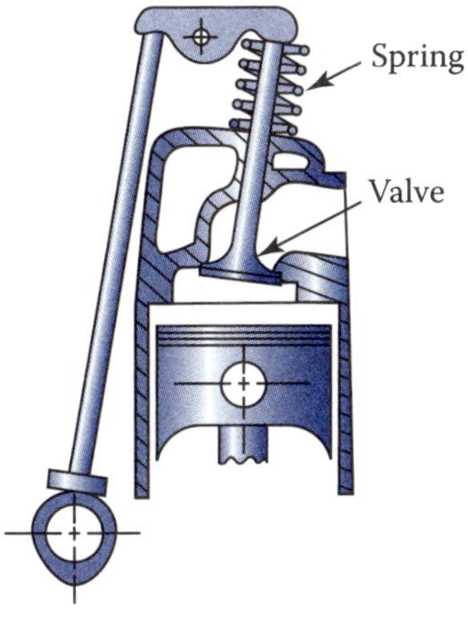

FIGURE P14.25

Given: $d = 6$ mm, $D = 30$ mm, $r_c = 20\%$

Design Requirements: The ends are squared and ground; one end is free to tip. In the most-compressed condition, the force is 600 N: after 13 mm of release, the minimum force is 340 N.

Sections 14.9 through 14.12

14.27 What is the value of the force required to cause the torsional stress (at point B) in the hook to reach shear yield strength in the hook of the extension spring discussed in Example 14.6?

14.28 A helical tension spring has $d_1 = 3$ mm and $D_1 = 30$ mm. If a second spring is made of the same material and the same number of coils with $D_2 = 240$ mm, find the wire diameter d_2 that would be required to give the same spring rate as the first spring.

14.29 An extension coil spring is made of 0.02 in. music wire and has a mean diameter of coil of 0.2 in. The spring is wound with a pretension of 0.2 lb, and the load fluctuates from this value up to 1.0 lb. Determine the factor of safety guarding against a fatigue failure. Use the Goodman criterion.

14.30 Consider a helical extension spring (Figure 14.12) of a shutter return of a small camera made of ASTM A228 music wire with the following given numerical values: $d = 0.6$ mm, $D = 2.4$ mm, $P = 6$ N, $G = 79$ GPa, $(r_m)_A = 1.2$ mm. Find

a. The maximum shear stress in the spring body away from the loop

b. The factor of safety with respect to the yielding (at point A) in the end loop

14.31 Design a window-shade spring similar to that depicted in Figure 14.14. Determine

a. The number of active coils, if a pull on shade of 15 N is exerted after being wound up to 16 revolutions

b. The maximum bending stress

Assumptions: The spring will be made of 1.2 mm square wire having $E = 207$ GPa, $D = 18$ mm, and a roller diameter of 32 mm.

14.32 Consider a torsion spring made of ASTM A227 hard-drawn steel wire (Figure 14.13) with the following given data: $a = 54$ mm, $d = 1.5$ mm, $D = 15$ mm, $N_a = 10$, and $E = 210$ GPa. Find the maximum operating moment and corresponding number of active coils.

Assumption: The largest angular deflection will be limited to 1.2 rad.

14.33 A torsion spring such as shown in Figure 14.14, made of ASTM 229 oil-tempered steel wire, has a diameter $d = 0.08$ in., mean coil diameter $D = 0.5$ in., the arm length $a = 0.4$ in., the number of active coils $N_a = 3.5$, and modulus of elasticity $E = 29 \times 10^6$ psi. Find

a. The maximum load P that can be applied, based on a safety factor of $n = 1.8$ against yielding

b. The corresponding angle of rotation in radians

14.34 A multileaf steel spring ($E = 200$ GPa, $\nu = 0.3$) for a truck wheel set having a maximum bending strength of 800 MPa supports a weight of 40 kN. Determine

a. The width b of the spring based on a safety factor of 2.5

b. The largest deflection of the spring

Given: $h = 22$ mm, $L = 0.7$ m, $P = 40$ kN, $n = 8$

14.35 Design a helical torsion spring similar to that shown in Figure 14.13. Calculate

a. The maximum operating moment

b. The maximum angular rotation

Assumptions: A safety factor of 1.4 is used. The spring is made of oil-tempered steel wire.

Given: $E = 30 \times 10^6$ psi, $d = 0.08$ in., $D = 0.6$ in., $N_a = 6$

14.36 A multileaf steel spring is to support a center load that varies between 300 and 1100 N (Figure 14.16). Estimate, using the Goodman criterion with a safety factor of 1.2,

a. The appropriate values of h and b for a spring of proportions $b = 40h$

b. The spring rate

Given: $S_u = 1400$ MPa, $S_e' = 500$ MPa, $E = 207$ GPa, and $\nu = 0.3$. The total length $2L$ is to be 800 mm.

Assumptions: Use $C_r = C_f = C_s = 1$. Stress concentration at the center is such that $K_f = 1.4$.

15

Power Screws, Fasteners, and Connections

15.1 Introduction

This chapter is devoted to the analysis and design of power screws, threaded fasteners, bolted joints in shear, and permanent connectors such as rivets and weldments. Adhesive bonding, brazing, and soldering are also discussed briefly. Power screws are threaded devices used mainly to move loads or accurately position objects. They are employed in machines for obtaining motion of translation and also for exerting forces. The kinematics of power screws is the same as that for nuts and screws, the only difference being the geometry of the threads. Power screws find applications as motion devices.

The success or failure of a design can depend on proper selection and use of its fasteners. A fastener is a device to connect or join two or more members. Many varieties of fasteners are available commercially. The threaded fasteners are used to fasten the various parts of an assembly together. We limit our consideration to detachable threaded fasteners such as bolts, nuts, and screws (Figure 15.1). General information for threaded fasteners as well as for other methods of joining is presented in some references listed at the end of this chapter and at the websites www.americanfastener.com and www.machinedesign.com. Listings of a variety of nuts, bolts, and washers are found at www.nutty.com. For bolted joint technology, see the website at www.boltscience.com.

Analysis of riveted, welded, and bonded connections cannot be made on as rigorous a basis as used for most structural and machine members. Their design is largely empirical and relies on available experimental results. As with the threaded fasteners, rivets exist in great variety. Note that while welding has replaced riveting and bonding to a considerable extent, rivets are customarily employed for certain types of joints. Often, rivets are used in joining smaller components in products associated with the automotive, business machines, appliances, and other fields. Welding speeds the manufacturing of parts and assembly of these components into structures and reduces the cost compared to casting and forging. Soldering, brazing, cementing, and adhesives are all means of bonding parts together. Other popular fastening and joining methods [20] include snap fasteners, which greatly simplify assembly of mechanical components.

15.2 Standard Thread Forms

Threads may be external on the screw or bolt and internal on the nut or threaded hole. The thread causes a screw to proceed into the nut when rotated. The basic arrangement of a helical thread cut around a cylinder or a hole, used as screw-type fasteners, power screws,

FIGURE 15.1
An assortment of threaded fasteners. (Courtesy of Clark Craft Fasteners.)

and worms, is as shown in Figure 15.2. Note that the length of unthreaded and threaded portions of shank is called the *shank* or bolt length. Also, observe the washer face, the fillet under the bolt head, and the start of the threads. Referring to the figure, some terms from geometry that relate to screw threads are defined as follows.

Pitch p is the axial distance measured from a point on one thread to the corresponding point on the adjacent thread. *Lead L* represents the axial distance that a nut moves, or advances, for one revolution of the screw. *Helix angle*, λ, also called the lead angle, may be

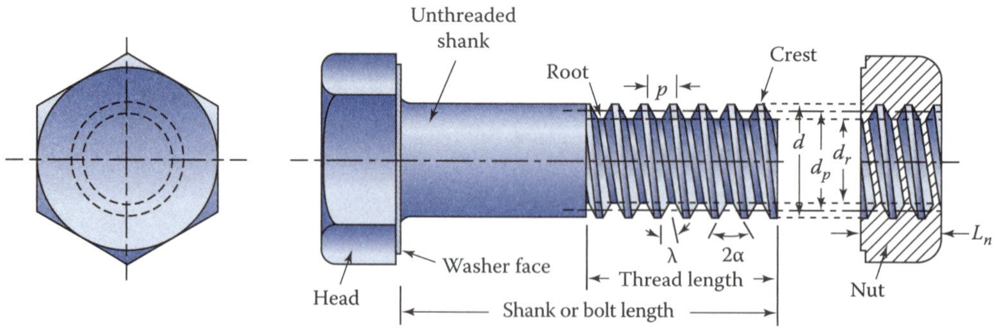

FIGURE 15.2
Hexagonal bolt and nut illustrate the terminology of threaded fasteners. *Notes*: *P*, the pitch; λ, the helix or lead angle; α, the thread angle; *d*, the major diameter; d_p, the pitch diameter; d_r, the root diameter; and L_n, the nut length.

cut either right handed (as in Figure 15.2) or left handed. All threads are assumed to be right handed, unless otherwise stated.

A single-threaded screw is made by cutting a single helical groove on the cylinder. For a *single thread*, the lead is the same as the pitch. Should a second thread be cut in the space between the grooves of the first (imagine two strings wound side by side around a pencil), a double-threaded screw would be formed. For a *multiple* (two or most)-*threaded screw*,

$$L = np \tag{15.1}$$

where

L = the lead

n = the number of threads

p = the pitch

We observe from this relationship that a multiple-threaded screw advances a nut more rapidly than a single-threaded screw of the same pitch. Most bolts and screws have a single thread, but worms and power screws sometimes have multiple threads. Some automotive power-steering screws occasionally use quintuple threads.

15.2.1 Unified and ISO Thread Form

For fasteners, the standard geometry of screw thread shown in Figure 15.3 is used. This is essentially the same for both the Unified National Standard (UNS), or so-called unified, and International Standards Organization (ISO) threads. The UNS (inch series) and ISO (metric series) threads are not interchangeable. In both systems, the *thread angle* is 60°, and the crests and roots of the thread may be either flat (as depicted in the figure) or rounded. The major diameter d and root (minor) diameter d_r refer to the largest and smallest diameters, respectively. The diameter of an imaginary cylinder, coaxial with the screw, intersecting the thread at the height that makes width of thread equal to the width of space, is called the *pitch diameter* d_p.

Tables 15.1 and 15.2 furnish a summary of the various sizes and pitches for the UNS and ISO systems. We see from these listings that the thread size is specified by giving the number of threads per inch N for the unified sizes and giving the pitch p for the metric sizes. The tensile *stress area* tabulated is on the basis of the average of the pitch and root

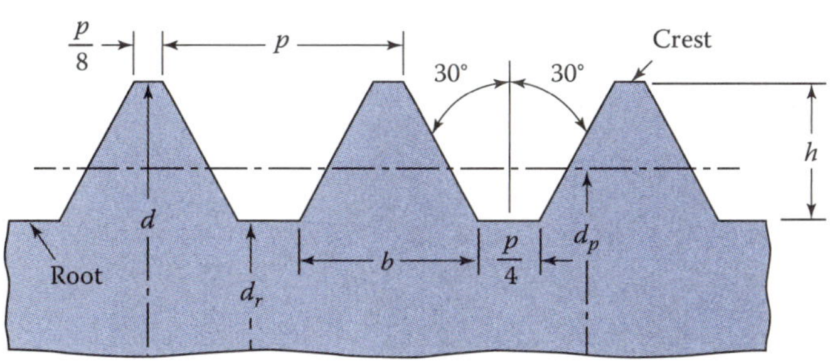

FIGURE 15.3
Unified and ISO thread forms. The portion of basic profile of the external thread is shown: h is the depth of thread and b is the thread thickness at the root.

TABLE 15.1

Dimensions of Unified Screw Threads

	Course Threads—UNC				Fine Threads—UNF		
Size	Major Diameter, d (in.)	Threads per Inch, $N = 1/p$	Minor Diameter d_r (in.)	Tensile Stress Area, A_t (in.^{2})	Threads per Inch, $N = 1/p$	Minor Diameter, d_r (in.)	Tensile Stress Area, A_t (in.^{2})
1	0.073	64	0.0538	0.00263	72	0.0560	0.00278
2	0.086	56	0.0641	0.00370	64	0.0668	0.00394
3	0.099	48	0.0734	0.00487	56	0.0771	0.00573
4	0.112	40	0.0813	0.00604	48	0.0864	0.00661
5	0.125	40	0.0943	0.00796	44	0.0971	0.00830
6	0.138	32	0.0997	0.00909	40	0.1073	0.01015
8	0.164	32	0.1257	0.0140	36	0.1299	0.01474
10	0.190	24	0.1389	0.0175	32	0.1517	0.0200
12	0.216	24	0.1649	0.0242	28	0.1722	0.0258
1/4	0.250	20	0.1887	0.0318	28	0.2062	0.0364
3/8	0.375	16	0.2983	0.0775	24	0.3239	0.0878
1/2	0.500	13	0.4056	0.1419	20	0.4387	0.1599
5/8	0.625	11	0.5135	0.226	18	0.5368	0.256
3/4	0.750	10	0.6273	0.334	16	0.6733	0.373
7/8	0.875	9	0.7387	0.462	14	0.7874	0.509
1	1.000	8	0.8466	0.606	12	0.8978	0.663

Source: *ANSI/ASME Standards*, B1.1–2014, B1.13–2005, New York, American Standards Institute, 2005.
Note: The pitch or mean diameter $d_m \approx d - 0.65p$.

diameters. This is the area used for calculation of axial stress (P/A). Extensive information for various inch-series threads may be found in ANSI Standards [1,2].

Coarse thread (designated as UNC) is most common and recommended for ordinary applications, where the screw is threaded into a softer material. It is used for general assembly work. *Fine thread* (denoted by UNF) is more resistant to loosening, because of its smaller helix angle. Fine threads are widely employed in automotive, aircraft, and other applications where vibrations are likely to occur. In identifying threads, the letter *A* is used for external threads, and *B* is used for internal threads. The UNS defines the threads according to *fit*. Class 1 fits have the widest tolerances and so are the loosest fits. Class 2 fits are most commonly used. Class 3 fit is the one having the least tolerance and is utilized for the highest precision applications. Clearly, cost increases with higher class of fit. An example of approved identification symbols is as follows:

1 in.-12 UNF-2A-LH

This defines 1 in. diameter × 12 threads per inch, unified fine-thread series, class 2 fit, external, and left-handed thread. Metric thread specification is given in Table 15.2.

15.2.2 Power Screw Thread Forms

Figure 15.4 depicts some thread forms used for power screws. The *Acme screw* is in widespread usage. They are sometimes modified to a stub form by making the thread shorter. This results in a larger minor diameter and a slightly stronger screw. A *square thread*

TABLE 15.2

Basic Dimensions of ISO (Metric) Screw Threads

	Coarse Threads		Fine Threads	
Nominal Diameter, d (mm)	**Pitch, p (mm)**	**Tensile Stress Area, A_t (mm²)**	**Pitch, p (mm)**	**Tensile Stress Area, A_t, (mm²)**
2	0.4	2.07		
3	0.5	5.03		
4	0.7	8.78		
5	0.8	14.2		
6	1	20.1		
7	1	28.9		
8	1.25	36.6	1.25	39.2
10	1.5	58.0	1.25	61.2
12	1.75	84.3	1.25	92.1
14	2	115	1.5	125
16	2	157	1.5	167
18	2.5	192	1.5	216
20	2.5	245	1.5	272
24	3	353	2	384
30	3.5	561	2	621
36	4	817	2	915
42	4.5	1120	9	1260
48	5	1470	2	1670
56	5.5	2030	2	2300
64	6	2680	2	3030

Source: *ANSI/ASME Standards*, B1.1–2014, B1.13–2005, New York, American Standards Institute, 2005.

Notes: Metric threads are specified by nominal diameter and pitch in millimeters, for example, $M10 \times 1.5$. The letter *M*, which proceeds the diameter, is the clue to the metric designation; root or minor diameter $d_r \approx d - 1.227p$.

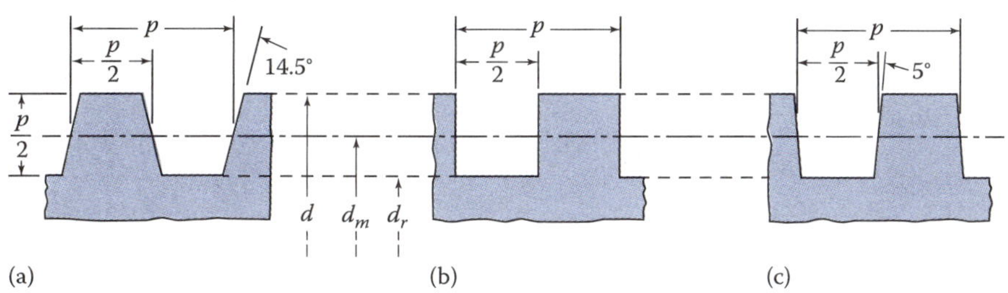

FIGURE 15.4

Typical power screw thread forms. All threads shown are external, $d_m = (d + d_r)/2$: (a) Acme, (b) square, and (c) modified square.

provides somewhat greater strength and efficiency but is rarely used, due to difficulties in manufacturing the 0° thread angle. The 5° thread angle of the *modified square thread* partially overcomes this and some other objections. Standard sizes for three power screw thread forms are listed in Table 15.3. The reader is referred to ANSI Standards for further details.

TABLE 15.3

Standard Sizes of Power Screw Threads

	Threads per Inch	
Major Diameter, *d* (in.)	**Acme, Acme Stub**	**Square and Modified Square**
$\frac{1}{4}$	16	10
$\frac{1}{2}$	10	$6\frac{1}{2}$
$\frac{5}{8}$	8	$5\frac{1}{2}$
$\frac{3}{4}$	6	5
$\frac{7}{8}$	6	$4\frac{1}{2}$
1	5	4
$1\frac{1}{4}$	5	$3\frac{1}{2}$
$1\frac{1}{2}$	4	2
$1\frac{3}{4}$	4	$2\frac{1}{2}$
2	4	$2\frac{1}{4}$
$2\frac{1}{4}$	3	$2\frac{1}{4}$
$2\frac{1}{2}$	3	2
$2\frac{3}{4}$	3	2
3	2	2
3	2	$1\frac{3}{4}$
4	2	$1\frac{1}{2}$
5	2	

Source: James. F.D. et al. eds., *Machinery's Handbook*, 23rd ed., Industrial Press, New York, 1974.

15.3 Mechanics of Power Screws

As noted previously, a power screw, sometimes called the *linear actuator* or *translation screw*, is in widespread usage in machinery to change angular motion into linear motion, to exert force, and to transmit power. Applications include the screws for vises, C-clamps, presses, micrometers, jacks (Figure 15.5), valve stems, and the lead screws for lathes and other equipment. In the usual configuration, the nut rotates in place, and the screw moves axially. In some designs, the screw rotates in place, and the nut moves axially. Forces may be large, but motion is usually slow and power is small. In all the foregoing cases, power screws operate on the same principle.

A simplified drawing of a screw jack having the Acme thread is shown in Figure 15.6. The load W can be lifted or lowered by the rotation of the nut that is supported by a washer, called a *thrust collar* (or a *thrust bearing*). It is, of course, assumed that the load and screw are prevented from turning when the nut rotates. Hence, there needs to be some friction at the

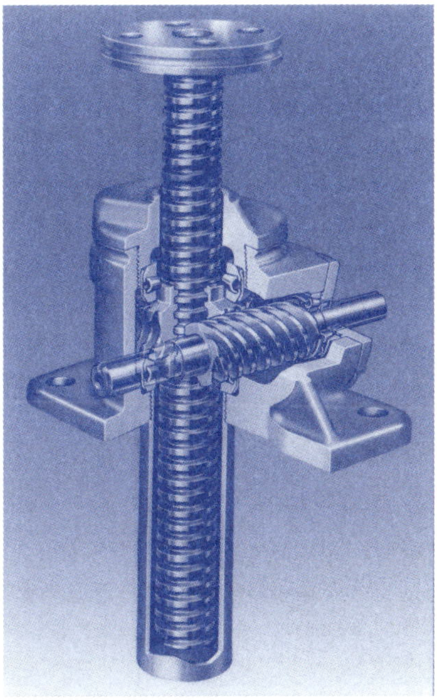

FIGURE 15.5
Worm gear screw jack. (Courtesy of Joyce/Dayton Corp.)

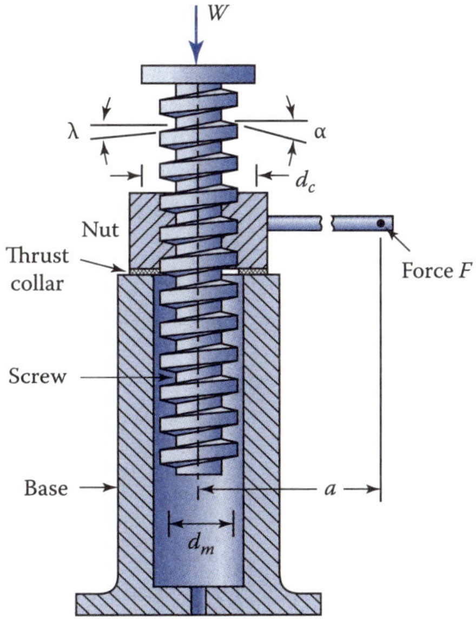

FIGURE 15.6
Schematic representation of power screw used as a screw jack. *Notes*: Only the nut rotates in this model: d_m represents the mean thread diameter and d_c is the mean collar diameter.

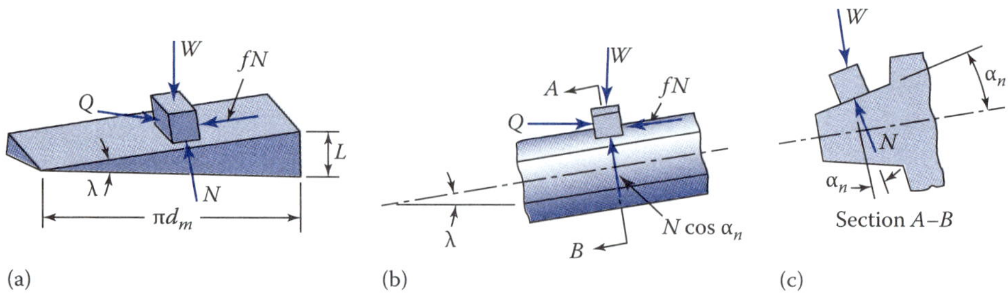

FIGURE 15.7
Forces acting on an Acme screw–nut interface when lifting load W: (a) a developed screw thread, (b) a segment of the thread, and (c) thread angle measured in the plane normal to thread, α_n.

load surface to prevent the screw from turning with the nut. Alternatively, the power screw could be turned against a nut that is prevented from turning to lift or lower the load. In either case, there is significant friction between the screw and nut as well as between the nut and the collar. Ordinarily, the screw is a hard steel, while the nut is made of a softer material (e.g., an alloy of aluminum, nickel, and bronze) to allow the parts to move smoothly.

In this section, we develop expressions for ascertaining the values of the torque needed to lift and lower the load using a jack. We see from Figure 15.6 that turning the nut forces each portion of the nut thread to climb an inclined plane. This plane is depicted by unwrapping or developing one revolution of the helix in Figure 15.7a, which includes a small block representing the nut being slid up the inclined plane of an Acme thread. The forces acting on the nut as a free-body diagram are also noted in the figure. Clearly, one edge of the thread forms the hypotenuse of the right triangle, having a base as the circumference of the mean-thread-diameter circle and as the lead. Therefore,

$$\tan\lambda = \frac{L}{\pi d_m} \tag{15.2}$$

where

λ = the helix or lead angle
L = the lead
d_m = the mean diameter of thread contact surface

The preceding notation is the same as for worms (see Section 12.9) except that unnecessary subscripts are omitted.

15.3.1 Torque to Lift the Load

The *sum* of all loads and normal forces acting on the entire thread surface in contact are denoted by W and N, respectively. To lift or raise the load, a tangential force Q acts to the right, and the friction force fN acts to oppose the motion (Figure 15.7). The quantity f represents the coefficient of sliding friction between the nut and screw or the coefficient of *thread friction*. The thread angle increases the frictional force by the wedging action of the threads. The conditions of equilibrium of the horizontal and vertical forces give

$$\begin{aligned} \sum F_h = 0: \quad & Q - N\left(f\cos\lambda + \cos\alpha_n \sin\lambda\right) = 0 \\ \sum F_v = 0: \quad & W + N\left(f\sin\lambda - \cos\alpha_n \cos\lambda\right) = 0 \end{aligned} \tag{a}$$

where α_n is the normal thread angle and the other variables are defined in the figure. Inasmuch as we are not interested in the normal force *N*, we eliminate it from the foregoing equations and solve the result for *Q*. In so doing, we have

$$Q = W = \frac{f\cos\lambda + \cos\alpha_n \sin\lambda}{\cos\alpha_n \cos\lambda - f\sin\lambda} \tag{15.3}$$

The screw torque required to move the load up the inclined plane, after dividing the numerator and denominator by cos λ, is then

$$T = \frac{1}{2}Qd_m = \frac{Wd_m}{2}\frac{f + \cos\alpha_n \tan\lambda}{\cos\alpha_n - f\tan\lambda} \tag{15.4}$$

But the thrust collar also contributes a friction force. That is, the normal reactive force acting on contact surface due to *W* results in an additional force f_cW. Here, f_c is the sliding coefficient of the *collar friction* between the thrust collar and the surface that supports the screw. It is assumed that this frictional force acts at the mean collar diameter d_c (Figure 15.6). The torque needed to overcome collar friction is

$$T = \frac{Wf_cd_c}{2} \tag{15.5}$$

The required total *torque* T_u *to lift the load* is found by addition of Equations 15.4 and 15.5:

$$T_u = \frac{Wd_m}{2}\frac{f + \cos\alpha_n \tan\lambda}{\cos\alpha_n - f\tan\lambda} + \frac{Wf_cd_c}{2} \tag{15.6}$$

15.3.2 Torque to Lower the Load

The analysis of lowering a load is exactly the same as that just described, with the exception that the directions of *Q* and *fN* (Figure 15.7b) are reversed. This leads to the equation for the total required *torque* T_d *to lower the load* as

$$T_d = \frac{Wd_m}{2}\frac{f - \cos\alpha_n \tan\lambda}{\cos\alpha_n + f\tan\lambda} + \frac{Wf_cd_c}{2} \tag{15.7}$$

15.3.3 Values of Friction Coefficients

When a plain *thrust collar* is used, as shown in Figure 15.6, values of *f* and f_c vary customarily between 0.08 and 0.20 under conditions of ordinary service, lubrication, and the common materials of steel and cast iron or bronze. The lowest value applies for good workmanship, the highest value for poor workmanship, and some in between value for other work quality. The preceding range includes both starting and running frictions. *Starting* friction can be about 4/3 times the *running* friction. Should a rolling *thrust bearing* be used, f_c would usually be low enough (about 0.008–0.02) that collar friction can be omitted. For this case, the second term in Equations 15.6 and 15.7 is eliminated.

15.3.4 Values of Thread Angle in the Normal Plane

A relationship between normal thread angle α_n, thread angle α, and helix angle λ can be obtained from a comparison of thread angles measured in axial plane and normal plane. Referring to Figures 15.6 and 15.7c, it can readily be verified that

$$\tan\alpha_n = \cos\lambda \tan\alpha \tag{15.8}$$

In most applications, λ is relatively small, and hence, $\cos\lambda \approx 1$. So, we can set $\alpha_n \approx \alpha$ and Equation 15.6 becomes

$$T_u = \frac{Wd_m}{2}\frac{f + \cos\alpha\tan\lambda}{\cos\alpha - f\tan\lambda} + \frac{Wf_c d_c}{2} \tag{15.9}$$

Obviously, for the case of the *square thread*, $\alpha = \alpha_n = 0$, and $\cos\alpha = 1$ in the preceding expressions.

15.4 Overhauling and Efficiency of Power Screws

A self-locking screw requires a positive torque to lower the load. This is a useful provision, particularly in screw jack applications. *Self-locking* refers to a condition in which the screw cannot be turned by applying an axial force of any magnitude to the nut. If collar friction is neglected, Equation 15.7 shows that the *condition* for self-locking is

$$f \geq \cos\alpha_n \tan\lambda \tag{15.10}$$

For a square thread, the foregoing equation reduces to

$$f \geq \tan\lambda \tag{15.10a}$$

In other words, self-locking is obtained when the coefficient of thread friction is equal to or greater than the tangent of the thread helix angle. Note that Equation 15.10 presumes a static situation and most power screws are self-locking.

Overhauling or back-driving screw is one that has low enough friction to enable the load to lower itself, by causing the screw to spin. In this situation, the inclined plane in Figure 15.7b moves to the right, and force Q must act to the left to preserve uniform motion. It can be shown that the *torque T_o of overhauling* screw is

$$T_o = \frac{Wd_m}{2}\frac{-f + \cos\alpha_n\tan\lambda}{\cos\alpha_n + f\tan\lambda} - \frac{Wf_c d_c}{2} \tag{15.11}$$

A negative external lowering torque must now be maintained to keep the load from lowering.

15.4.1 Screw Efficiency

Screw efficiency is the ratio of the torque required to raise a load without friction to the torque required with friction. Using Equation 15.6, *efficiency* is expressed in the form

$$e = \frac{d_m \tan\lambda}{d_m \dfrac{f + \cos\alpha_n \tan\lambda}{\cos\alpha_n - f \tan\lambda} + d_c f_c} \tag{15.12}$$

We observe from this equation that efficiency depends on only the screw geometry and the coefficient of friction. If the *collar friction* is *neglected,* the efficiency becomes

$$e = \frac{\cos\alpha_n - f \tan\lambda}{\cos\alpha_n + f \cot\lambda} \tag{15.13}$$

For a *square thread,* $\alpha_n = 0$ and Equation 15.13 simplifies to

$$e = \frac{1 - f \tan\lambda}{1 + f \cot\lambda} \tag{15.13a}$$

Equation 15.13 with α_n substituted from Equation 15.8 and $\alpha = 14.5°$ is plotted in Figure 15.8 for five values of f. We see from the curves that the power screws have very low mechanical efficiency when the helix angle is in the neighborhood of either 0° or 90°. They generally have an efficiency of 30%–90%, depending on the λ and f. We mention that values for square threads are higher by less than 1% over those for Acme screws in the figure.

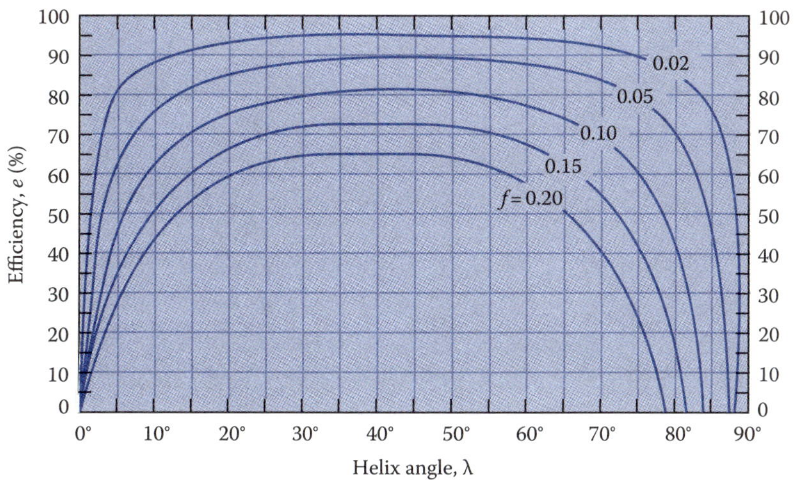

FIGURE 15.8
The efficiency of Acme screw threads (neglecting thrust collar friction).

Example 15.1: Quadruple-Threaded Power Screw

A screw jack with an Acme thread of diameter d, similar to that illustrated in Figure 15.6, is used to lift a load of W. Determine

a. The screw lead, mean diameter, and helix angle
b. The starting torque for lifting and for lowering the load
c. The efficiency of the jack when lifting the load, if collar friction is neglected
d. The length of a crank required, if $F = 150$ N is exerted by an operator

Design Assumptions: The screw and nut are lubricated with oil. Coefficients of friction are estimated as $f = 0.12$ and $f_c = 0.09$.

Given: $d = 30$ mm and $W = 6$ kN. The screw is quadruple threaded having a pitch of $p = 4$ mm. The mean diameter of the collar is $d_c = 40$ mm.

Solution

a. From Figure 15.4, $d_m = d - p/2 = 30 - 2 = 28$ mm. Through the use of Equations 15.1 and 15.2, we have

$$L = np = 4(4) = 16\,\text{mm}$$

$$\lambda = \tan^{-1}\frac{16}{\pi(28)} = 10.31°$$

b. The coefficients of friction for starting are $f = \frac{4}{3}(0.12) = 0.16$ and $f_c = \frac{4}{3}(0.09) = 0.12$. For an Acme thread, $\alpha = 14.5°$ (Figure 15.4a), by Equation 15.8.

$$\alpha_n = \tan^{-1}(\cos\lambda\tan\alpha)$$
$$= \tan^{-1}(\cos 10.31°\tan 14.5°) = 14.28°$$

Then, application of Equations 15.6 and 15.7 results in

$$T_u = \frac{6(28)}{2}\,\frac{0.16 + \cos 14.28°\tan 10.31°}{\cos 14.28° - (0.16)\tan 10.31°} + \frac{6(0.12)40}{2}$$
$$= 30.05 + 14.4 = 44.45\,\text{N}\cdot\text{m}$$

$$T_d = \frac{6(28)}{2}\,\frac{0.16 - \cos 14.28°\tan 10.31°}{\cos 14.28° + (0.16)\tan 10.31°} + 14.4$$
$$= -1.37 + 14.4 = 13.03\,\text{N}\cdot\text{m}$$

Comments: The minus sign in the first term of T_d means that the screw alone is not self-locking and would rotate under the action of the load, except that the collar friction must be overcome too. Since T_d is positive, the screw does not overhaul.

c. The running torque needed to lift the load is based on $f = 0.12$. Using Equation 15.13, we have

$$e = \frac{\cos 14.28° - (0.12)\tan 10.31°}{\cos 14.28° + (0.12)\cot 10.31°}$$
$$= 0.582 = 58.2\%$$

d. The length of the crank arm is

$$a = \frac{T_n}{F} = \frac{44.45}{150} = 0.296 \text{ m} = 296 \text{ mm}$$

Example 15.2: Single-Threaded Power Screw

Given: The screw jack (Figure 15.6) discussed in the previous example has a single-threaded Acme screw instead of a quadruple thread.

Find: The torque required for lifting the load and efficiency of the jack.

Solution

Refer to Example 15.1.
Now the lead is equal to the pitch, $L = p = 4$ mm. The helix angle is therefore

$$\lambda = \tan^{-1}\left(\frac{1}{\pi d_m}\right) = \tan^{-1}\left(\frac{4}{28\pi}\right) = 2.604°$$

Through the use of Equation 15.8, we have

$$\alpha_n = \tan^{-1}(\cos\lambda \tan\alpha) = \tan^{-1}(\cos 2.604° \tan 14.5°) = 14.49°$$

Then, applying Equation 15.6, the torque required to raise the load is equal to

$$T_u = \frac{6(28)}{2}\frac{0.16 + \cos 14.49° \tan 2.604°}{\cos 14.49° - (0.16)\tan 2.604°} + \frac{6(0.12)40}{2}$$
$$= 17.84 + 14.4 = 32.24 \,\text{N}\cdot\text{m}$$

Equation 15.13 results in the efficiency in lifting the load as follows:

$$e = \frac{\cos 14.49° - (0.12)\tan 2.604°}{\cos 14.49° + (0.12)\cot 2.604°}$$
$$= 0.267 = 26.7\%$$

Comments: A comparison of the results obtained here with those of Example 15.1 shows that to lift the load, the single-treaded screw requires lower torque than the quadruple. However, the former is less efficient than the latter by 54.1%.

Example 15.3: Self-Locking of Quadruple- and Single-Threaded Screws

Given: The quadruple-threaded and single-threaded screws discussed in the preceding two examples.

Find: The tread coefficient of friction necessary to ensure that self-locking takes place.

Assumption: Rolling element bearings have been installed at the collar, so that the collar friction can be disregarded.

Solution

Refer to Examples 15.1 and 15.2.
Coefficient of friction for self-locking is specified by Equation 15.10 as

$$f \geq \cos\alpha_n \tan\lambda$$

Therefore, for the quadruple-threaded screw self-locking does not occur since

$$0.12 < \cos 14.28° \tan 10.31° = 0.176$$

But, for the single-threaded screw, self-locking occurs since

$$0.12 > \cos 14.49° \tan 2.604° = 0.044$$

Comment: The foregoing results indicate that the quadruple-threaded screw requires four times the friction coefficient of friction of the single-threaded screw.

15.5 Ball Screws

A *ball screw*, or so-called ball-bearing screw, is a linear actuator that transmits force or motion with minimum friction. A cutaway illustration of a ball screw and two of its precision assemblies supported by ball bearings at the ends are shown in Figure 15.9. Note that

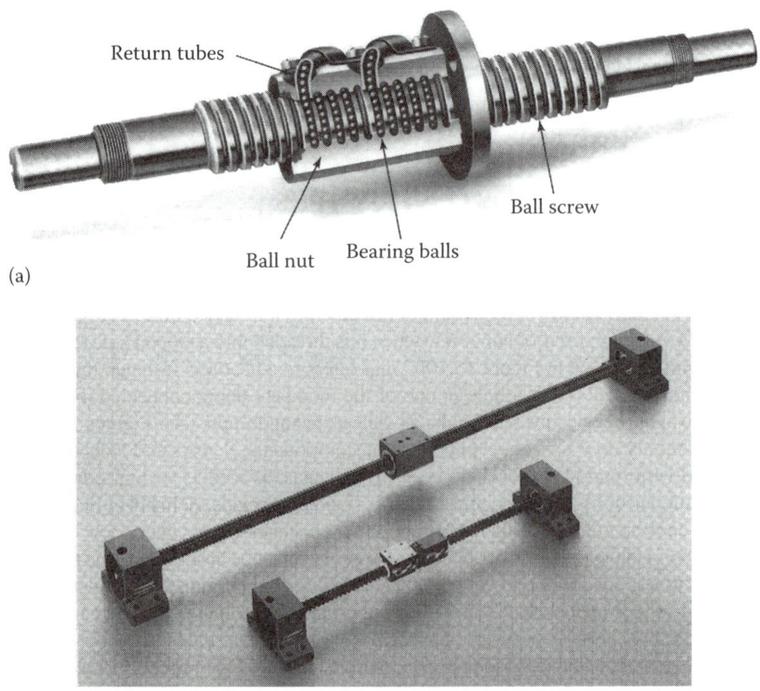

FIGURE 15.9
Ball screw used as a positioning device: (a) cutaway of a ball screw and (b) two assemblies. (Courtesy of Thomson Industries, Inc.)

a circular groove is cut to proper conformity with the balls. The groove has a thread helix angle matching the thread angle of the groove within the nut. The *balls* are contained within the *nut* to produce an approximate rolling contact with the screw threads. The rotation of the screw (or nut) is converted into a linear motion and force with very little friction torque. During the motion, the balls are diverted from one end and middle of the nut and carried by two ball-return tubes (or ball guides) located outside of the nut to the middle and opposite end of the nut. Such recirculation allows the nut to travel the full length of the screw.

A ball screw can support greater load than that of ordinary power screws of identical diameter. The smaller size and lighter weight are usually an advantage. A thin film lubricant is required for these screws. Certain dimensions of ball screws have been standardized by ANSI [3,4] but mainly for use in machine tools. Capacity ratings for ball screws are obtained by methods and equations identical to those for ball bearings, which can be found in manufacturers' catalogs.

Efficiencies of 90% or greater are possible with ball screws over a wide range of helix angles when converting rotary into axial motion. Ball screws may be preferred by the designers if higher screw efficiencies are required. As a positioning device, these screws are used in many applications. Examples include the steering mechanism of automobiles, hospital bed mechanisms, automatic door closers, antenna drives, the aircraft control (e.g., a ball or jack screw and gimbal nut assembly as an actuator on a linkage for extending and retracting the wing flaps) and landing-gear actuator, jet aircraft engine thrust reverser actuators, and machine tool controls. Because of the low friction of ball screws, they are not self-locked. An auxiliary brake is required to hold a load driven by a ball screw for some applications.

15.6 Threaded Fastener Types

The common element among screw fasteners used to connect or join two or more parts is their thread. Screws and *bolts* are the most familiar threaded fastener types. The only difference between a screw and a bolt is that the bolt needs a *nut* to be used as a fastener (Figure 15.10a). On the other hand, a screw fits into a threaded hole. The same fastener is termed a *machine screw* or *cap screw* when it is threaded into a tapped hole rather than used

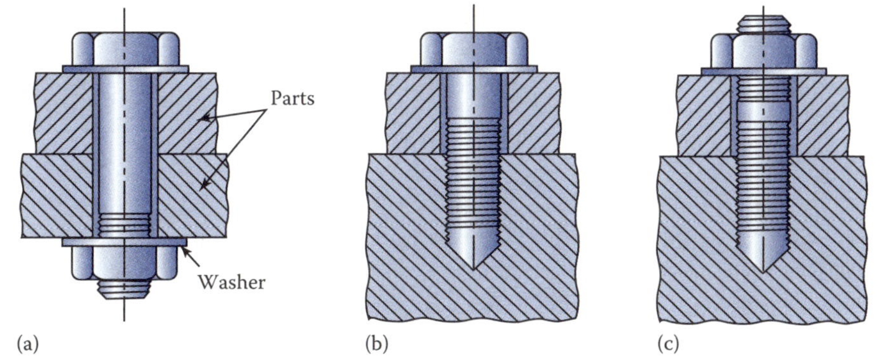

FIGURE 15.10
Typical threaded fasteners: (a) bolt and nut, (b) cap screw, and (c) stud.

with a nut, as shown in Figure 15.10b. *Stud* refers to a headless fastener, threaded on both ends, and screwed into the hole in one of the members being connected (Figure 15.10c).

Hexagon-head screws and bolts as well as hexagon nuts (see Figures 15.1 and 15.2) are commonly used for connecting machine components. Screws and bolts are also manufactured with round heads, square heads, oval heads, and various other head styles. Conventional bolts and nuts generally use standard threads, defined in Section 15.2. An almost endless number of threaded (and other) fasteners exist; many new types are constantly being developed [5–8]. Threaded fasteners must be designed so that they are lighter in weight, less susceptible to corrosion, and more resilient to loosening under vibration.

Flat or plain *washers* (Figure 15.1) are often used to increase the area of contact between the bolt head or nut and clamped part in a connection, as shown in Figure 15.10. They prevent stress concentration by the sharp edges of the bolt holes. Flat washer sizes are standardized to bolt size. A plain washer also forestalls marring of the clamped part surface by the nut when it is tightened. Belleville washers, discussed in Section 14.12, provide a controlled axial force over changes in bolt length.

Lock washers help prevent spontaneous loosening of standard nuts. The split lock washer acts as a spring under the nut. *Lock nuts* prevent too-spontaneous loosening of nuts due to vibration. Simply, two nuts jammed together on the bolt or a nut with a cotter pin serve for this purpose as well. The cotter pin is a wire that fits in diametrically opposite slots in the nut and passes through a drilled hole in the bolt. Lock nuts are considered to be more effective in preventing loosening than lock washers.

15.6.1 Fastener Materials and Strengths

A fastener is classified according to a grade or property class that defines its strength and material. Most fasteners are made from steel of specifications standardized by the SAE, ASTM, and ISO. The SAE grade (inch series) and SAE class (metric series) of steel-threaded members are numbered according to tensile strength. The *proof strength* S_p corresponds to the axial stress at which the bolt or screw begins to develop a permanent set. It is close to but lower than the material yield strength. The *proof load* F_p is defined by

$$F_p = S_p A_t \tag{15.14}$$

Here, the *tensile stress area, A*, represents the minimum radial plane area for fracture through the threaded part of a bolt or screw. Numerical values of A_t are listed in Tables 15.1 and 15.2. The proof strength is obtained from Tables 15.4 or 15.5. For other materials, an approximate value is about 10% less than for yield strength, that is, $S_p = 0.9S_y$ based on a 2% offset.

Threads are generally formed by *rolling* and *cutting* or grinding. The former is stronger than the latter in fatigue and impact because of cold working. Hence, high-strength screws and bolts have rolled threads. The rolling should be done subsequent to hardening the bolt. The material of the nut must be selected carefully to match that of the bolt. The washers should be of hardened steel, where the bolt or nut compression load needs to be distributed over a large area of clamped part.

A soft washer bends rather than uniformly distributes the load. Fasteners are also made of a variety of materials, including aluminum, brass, copper, nickel, Monel, stainless steel, titanium, beryllium, and plastics. Appropriate coatings may be used in special applications

TABLE 15.4

SAE Specifications and Strengths for Steel Bolts

SAE Grade	Size Range Diameter, *d* (in.)	Proof Strength,[a] S_p (ksi)	Yield Strength,[b] S_y (ksi)	Tensile Strength,[b] S_u (ksi)	Material Carbon Content
1	$\frac{1}{4}-1\frac{1}{2}$	33	36	60	Low or medium
2	$\frac{1}{4}-\frac{3}{4}$	55	57	74	Low or medium
2	$\frac{7}{8}-1\frac{1}{2}$	33	36	60	Low or medium
5	$\frac{1}{4}-1$	85	92	120	Medium, CD
5	$1\frac{1}{8}-1\frac{1}{2}$	74	81	105	Medium, CD
7	$\frac{1}{4}-1\frac{1}{2}$	105	115	133	Medium, alloy, Q&T
8	$\frac{1}{4}-1\frac{1}{2}$	120	130	150	Medium, alloy, Q&T

Source: Society of Automotive Engineers Standard J429k, 2011.

[a] Corresponds to permanent set not over 0.0001 in.

[b] Offset of 0.2%.

Q&T, quenched and tempered.

TABLE 15.5

Metric Specifications and Strengths for Steel Bolts

Class Number	Size Range Diameter, *d* (mm)	Proof Strength, S_p (MPa)	Yield Strength, S_y (MPa)	Tensile Strength, S_u (MPa)	Material Carbon Content
4.6	M5–M36	225	240	400	Low or medium
4.8	M1.6–M16	310	340	420	Low or medium
5.8	M5–M24	380	420	520	Low or medium
8.8	M3–M36	600	660	830	Medium, Q&T
9.8	M1,6–M 16	650	720	900	Medium, Q&T
10.9	M5–M36	830	940	1040	Low, martensite, Q&T
12.9	M1.6–M36	970	1100	1220	Alloy, Q&T

Source: Society of Automotive Engineers Standard J429k, 2011.

in place of a more expensive material, for corrosion protection and to reduce thread friction and wear. Obviously, a designer has many options in selecting the fastener's material to suit the particular application.

15.7 Stresses in Screws

Stress distribution of the thread engagement between the screw and the nut is nonuniform. In reality, inaccuracies in thread spacing cause virtually all the load to be taken by the first pair of contacting threads, and a large stress concentration is present here. While the stress concentration is to some extent relieved by the bending of the threads and the expansion of the nut, most bolt failures occur at this point. A concentration of stress also exists in the screw where the load is transferred through the nut to the adjoining member. Obviously, factors such as fillet radii at the thread roots and surface finish have significant effects on the actual stress values. For ordinary threads, the *stress concentration factor* K_t varies between 2 and 4 [9].

Note that the screws should always have enough ductility to permit local yielding at thread roots without damage. For *static loading*, it is commonly assumed that the load carried by a screw and nut is about *uniformly* distributed throughout thread engagement. The stress distribution for threads with steady loads is usually determined by photoelastic analysis. A variety of methods are used to obtain a more nearly equal distribution of loads among the threads, including increasing the flexibility of the nut (or bolt), making the nut from a softer material than the bolt, and cutting the thread of the nut on a very small taper. A rule of thumb for the length of full thread engagement is 1.0d in steel, 1.5d in gray cast iron, and 2.0d in aluminum castings, where d is the nominal thread size.

The following expressions for stresses in power screws and threaded fasteners are obtained through the use of the elementary formulas for stress. They enable the analyst to achieve a reasonable design for a static load. When bolts are subjected to fluctuating loads, stress concentration is very important.

15.7.1 Axial Stress

Power screws may be under tensile or compressive stress; threaded fasteners normally carry only tension. The axial stress σ is then

$$\sigma = \frac{P}{A} \tag{15.15}$$

where

P represents the tensile or compressive load

$$A \text{ is the } \begin{cases} A_t \text{ from Tables 15.1 and 15.2} & \text{(threaded fasteners)} \\ \pi d_r^2, \quad d_r \text{ is the root diameter} & \text{(power screws)} \end{cases}$$

15.7.2 Torsional Shear Stress

Power screws in operation and threaded fasteners during tightening are subject to torsion. The shear stress τ is given by

$$\tau = \frac{Tc}{J} = \frac{16T}{\pi d_r^3} \tag{15.16}$$

In the foregoing, we have

$$T = \begin{cases} \text{applied torque, for } f_c = 0 & \text{(power screws)} \\ \text{half the wrench torque} & \text{(threaded fastenrs)} \end{cases}$$

$$d = \begin{cases} \text{from Figure 15.4} & \text{(power screws)} \\ \text{from Tables 15.1 and 15.2} & \text{(threaded screws)} \end{cases}$$

15.7.3 Combined Torsion and Axial Stress

The combined stress of Equations 15.15 and 15.16 can be treated as in Section 6.7, with the energy of distortion theory employed as a criterion for yielding.

15.7.4 Bearing Stress

The direct compression or bearing stress σ_b is the pressure between the surface of the screw thread and the contacting surface of the nut:

$$\sigma_b = \frac{P}{\pi d_m h n_e} = \frac{Pp}{\pi d_m h L_n} \qquad (15.17)$$

where

P = the load
d_m = the pitch or mean screw thread diameter
h = the depth of thread (Figure 15.3)
n_e = the number of threads in engagement (= L_n/p)
L_n = the nut length
p = the pitch

Exact values of σ_b are given in ANSI B 1.1-1989 and various handbooks.

15.7.5 Direct Shear Stress

The screw thread is considered to be loaded as a cantilevered beam. The load is assumed to be uniformly distributed over the mean screw diameter. Hence, both the threads on the screw and the threads on the nut experience a transverse shear stress $\tau = 3P/2A$ at their roots. Here, A is the cross-sectional area of the built-in end of the beam: $A = \pi d_r b n_e$ for the screw and $A = \pi d b n_e$ for the nut. Therefore, shear stress, for the *screw*, is

$$\tau = \frac{3P}{2\pi d_r b n_e} \qquad (15.18)$$

and, for the *nut*, is

$$\tau = \frac{3P}{2\pi d b n_e} \qquad (15.19)$$

in which

d_r = the diameter of the screw
d = the major diameter of the screw
b = the thread thickness at the root (Figure 15.3)

The remaining terms are as defined earlier.

The *design formulas* for screw threads are obtained by incorporating K_t and replacing σ or σ_b by S_y/n and τ by S_{ys}/n in the preceding equations. For the nut, for example, Equations 15.17 and 15.19 with $n_e = L_n/p$ may be written as follows:

$$\frac{S_y}{n} = \frac{K_t P p}{\pi d_m h L_n} \qquad (15.17a)$$

and

$$\frac{S_{ys}}{n} = \frac{3K_t Pp}{2\pi dbL_n} \tag{15.19a}$$

Here, S_y, S_{ys}, and n represent the yield strength in tension, yield strength in shear, and safety factor, respectively. Application of such formulas is illustrated in Case Study 18.7.

15.7.6 Buckling Stress for Power Screws

For the case in which the unsupported screw length is equal to or larger than about eight times the root diameter, the screw must be treated as a column. So, critical stresses are obtained as discussed in Sections 5.10 and 5.13.

15.8 Bolt Tightening and Preload

Bolts are commonly used to hold parts together in opposing to forces likely to pull, or sometimes slide, them apart. Typical examples include connecting rod bolts and cylinder head bolts. *Bolt tightening* is prestressing at assembly. In general, bolted joints should be tightened to produce an *initial tensile force*, usually the so-called preload F_i. The advantages of an initial tension are especially noticeable in applications involving fluctuating loading, as demonstrated in Section 15.12, and in making a leakproof connection in pressure vessels. An increase of fatigue strength is obtained when initial tension is present in the bolt. The parts to be joined may or may not be separated by a gasket. In this section, we consider the situation when no gasket is used.

The *bolt strength* is the main factor in the design and analysis of bolted connections. Recall from Section 15.6 that the proof load F_p is the load that a bolt can carry without developing a permanent deformation. For both static and fatigue loading, the preload is often prescribed by

$$F_i = \begin{cases} 0.75F_p & \text{(reused connections)} \\ 0.9F_p & \text{(permanent connections)} \end{cases} \tag{15.20}$$

where the proof load $F_p = S_p A_t$ from Equation 15.14. The amount of initial tension is clearly a significant factor in bolt design. It is usually maintained fairly constant in value.

15.8.1 Torque Requirement

The most important factor determining the preload in a bolt is the *torque required* to *tighten* the *bolt*. The torque may be applied manually by means of a wrench that has a dial attachment indicating the magnitude of the torque being enforced. Pneumatic or air wrenches give more consistent results than a manual torque wrench and are employed extensively.

An expression relating applied torque to initial tension can be obtained using Equation 15.6 developed for power screws. Observe that load W of a screw jack is equivalent to F_i, for a bolt and that collar friction in the jack corresponds to friction on the flat surface of the nut or under the screwhead. It can readily be shown that [7], for standard screw threads, Equation 15.6 has the form

$$T = KdF_i \tag{15.21}$$

where

T = the tightening torque
d = the nominal bolt diameter
K = the torque coefficient
F_i = the initial tension or preload

For dry surface and *unlubricated* bolts or *average* condition of thread friction, taking $f = f_c = 0.15$, Equation 15.6 results in $K = 0.2$. It is suggested that, for *lubricated* bolts, a value of 0.15 be used for torque coefficient. For various-plated bolts, see [10].

Note that Equation 15.21 represents an approximate relationship between the induced initial tension and applied torque. Tests have shown that a typical joint loses about 5% or more of its preload owing to various relaxation effects. The exact tightening torque needed in a particular situation can likely be best ascertained experimentally through calibration. That is, a prototype can be built and accurate torque testing equipment used on it. Interestingly, bolts and washers are available with built-in sensors indicating a degree of tightness. Electronic assembly equipment is available [2,7].

15.9 Tension Joints under Static Loading

A principal utilization of bolts and nuts is clamping parts together in situations where the applied loads put the *bolts in tension*. Attention here is directed toward preloaded tension joints under static loading. We treat the case of two plates or parts fastened with a bolt and subjected to an external separating load P, as depicted in Figure 15.11a. The preload F_i, an initial tension, is applied to the bolt by tightening the nut prior to the load P. Clearly, the bolt axial load and the *clamping force* between the two parts F_p are both equal to F_i.

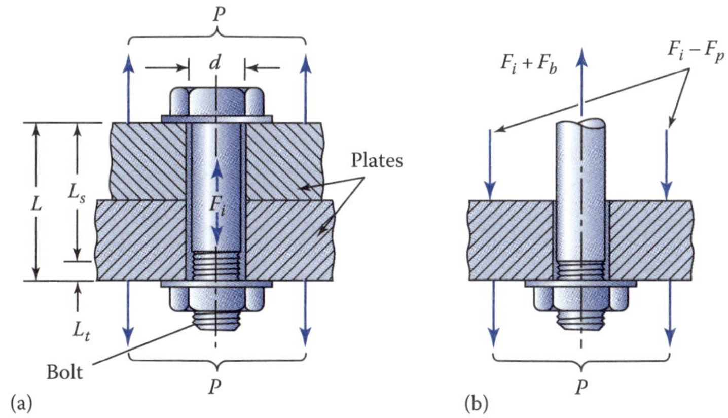

FIGURE 15.11
A bolted connection: (a) complete joint with preload F_i and external load P and (b) isolated portion depicting increased bolt force F_b and decreased force on parts or plates F_p.

To determine what portion of the externally applied load is carried by the bolt and what portion by the connected parts in the assembly, refer the free-body diagram shown in Figure 15.11b. The equilibrium condition of forces requires that

$$P = F_b + F_p \tag{a}$$

The quantity F_b is the *increased* bolt (tensile) force, and F_p represents the *decreased* clamping (compression) force between the parts. It is taken that the parts have not been separated by the application of the external load. The deformation of the bolt and the parts are defined by

$$\delta_b = \frac{F_b}{k_b}, \quad \delta_p = \frac{F_p}{k_p} \tag{b}$$

Here, k_b and k_p represent the *stiffness constants* for the bolt and parts, respectively.

Because of the setup of the members in Figure 15.11a, the deformations given by Equation (b) are equal. The *compatibility condition* is then

$$\frac{F_b}{k_b} = \frac{F_p}{k_p} \tag{c}$$

Combining Equations (a) and (c) yields

$$F_b = \frac{k_b}{k_b + k_p} P = CP, \quad F_p = \frac{k_p}{k_b + k_p} P = (1 - C)P \tag{d}$$

The term C, called the *joint's stiffness factor* or simply the *joint constant*, is defined in Equation (d) as

$$C = \frac{k_b}{k_b + k_p} \tag{15.22}$$

Note that, typically, k_b is small in comparison with k_p and C is a small fraction.

The total forces on the bolt and parts are, respectively,

$$F_b = CP + F_i \quad (\text{for } F_p < 0) \tag{15.23}$$

$$F_p = (1 - C)P - F_i \quad (\text{for } F_p < 0) \tag{15.24}$$

where

F_b = the bolt axial tensile force
F_p = the lamping force on the two parts
F_i = the initial tension or preload

A graphical representation of Equations 15.23 and 15.24 is given in Figure 15.12a. Clearly, if load P is sufficient to bring the clamping force F_p to zero (point A), we have bolt force $F_b = P$ (point B). As indicated in the expressions, the foregoing results are valid only as long as some clamping force prevails on the parts: with no preload (loosened joint),

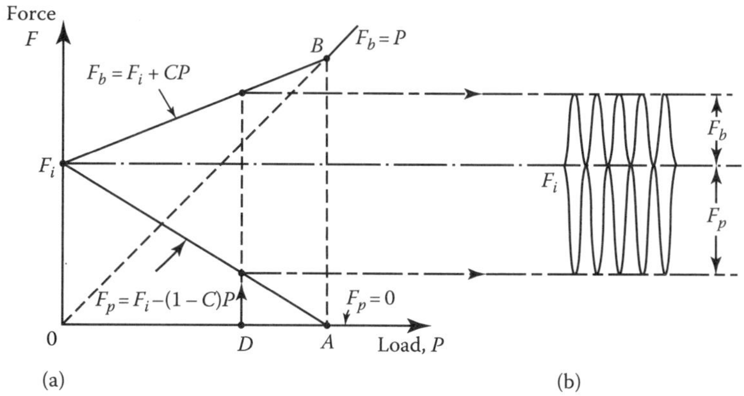

FIGURE 15.12
Preload in a bolted connection: (a) force relationship and (b) variations in F_b and F_p related to variations in P between 0 and D.

$C = 1$, $F_i = 0$. We see that the ratios C and $1 - C$ in Equations 15.23 and 15.24 describe the proportions of the external load carried by the bolt and the parts, respectively. In all situations, the *parts* take a *greater portion* of the external load. This is significant when *fluctuating* loading is present, where variations in F_b and F_p are readily found from Figure 15.12a and b, as indicated. We shall discuss this loading situation in detail in Section 15.12.

15.9.1 Deflections due to Preload

Figure 15.13 illustrates the load–deflection behavior of both bolt and parts on force (F)–deflection (δ) axes. Observe that the slope of the bolt line is positive, since its length increases with increasing force. On the contrary, the slope of the parts is negative, because its length decreases with the increasing force. As is often the case, the figure shows that $k_p > k_b$. It is obvious that the force in both bolt and parts is identical as long as they remain in contact. A preload force F_i is applied by tightening the bolt and $F_b = F_p = F_i$. The deflections of the bolt δ_b and parts δ_p are controlled by the spring rates of reach points A and B on their respective load–deflection lines.

For the case in which an external load P is applied to the joint, there will be an additional deflection added to both bolt and parts. Although the quantitative amount is the same,

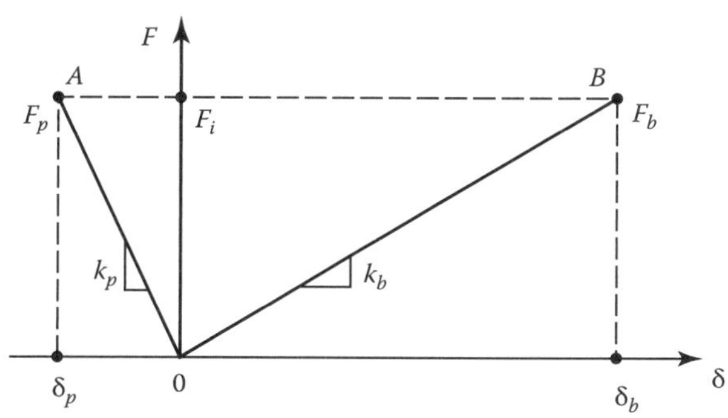

FIGURE 15.13
Preload versus initial deflections.

$\Delta\delta$, for the bolt, the deflection is an *increased elongation* while for the parts the *contraction is decreased*. The deflection $\Delta\delta$ causes a new load situation in both bolt and parts. As a result, the applied load is split into two components, one taken by the parts and one taken by the bolt. It will be seen in Section 15.12 that the preload effect is even greater for joints under dynamic loads than for statically loaded joints.

15.9.2 Factors of Safety for a Joint

The tensile stress σ_b in the bolt can be found by dividing both terms of Equation 15.24 by the tensile stress area A_t:

$$\sigma_b = \frac{CP}{A_t} + \frac{F_i}{A_t} \tag{15.25}$$

A means of ensuring a safe joint requires that the external load be smaller than that needed to cause the joint separate. Let nP be the value of external load that would cause bolt failure and the limiting value of σ_b be the proof strength S_p. Substituting these, Equation 15.25 becomes

$$\frac{CPn}{A_t} + \frac{F_i}{A_t} = S_p \tag{15.26}$$

It should be mentioned that the factor of safety is not applied to the preload. The foregoing can be rewritten to give the *bolt safety factor:*

$$n = \frac{S_p A_t - F_i}{CP} \tag{15.27}$$

As noted earlier, the tensile stress area A_t is furnished in Tables 15.1 and 15.2, and S_p is listed in Tables 15.4 and 15.5.

15.9.3 Joint-Separating Force

Equation 15.27 suggests that the safety factor n is maximized by having no preload on the bolt. We also note that for $n > 1$, the bolt stress is smaller than the proof strength. Separation occurs when in $F_p = 0$ in Equation 15.24:

$$P_s = \frac{F_i}{(1-C)} \tag{15.28a}$$

Therefore, the *load safety factor* guarding against joint separation is

$$n_s = \frac{P_s}{P} = \frac{F_i}{P(1-C)} \tag{15.28b}$$

Here, P is the *maximum* load applied to the joint.

Example 15.4: Load-Carrying Capacity of a Bolted Joint

Given: A ½ in.-13UNC grade 5 steel bolt clamps two steel plates and loaded as shown in Figure 15.11a.

Find: The maximum load based on a safety factor of 2.

Assumption: The connection will be permanent. Joint stiffness is taken as $C = 0.35$ (a detailed discussion about this is in Section 15.11).

Solution

For the ½-13UNC grade 5 steel bolt, we have

$A_t = 0.1419$ in^2. (by Table 15.1)

$S_p = 85$ ksi (from Table 15.4)

Applying Equation 15.27, the maximum load that the bolt can safely support is then

$$P_{\max,b} = \frac{S_p A_t - F_i}{nC} = \frac{(85)(0.1419) - 10.86}{2(0.35)} = 1.716 \text{ kips}$$

By Equation 15.28b, the maximum load before separation takes place equals

$$P_{\max,p} = \frac{F_i}{n(1-C)} = \frac{10.86}{2(1-0.35)} = 8.35 \text{ kips}$$

Comment: Failure owing to separation of art will not take place prior to bolt failure.

15.10 Gasketed Joints

Sometimes, a sealing or gasketing material must be placed between the parts connected. Gaskets are made of materials that are soft relative to other joint parts. Obviously, the stiffer and thinner is the gasket, the better. The stiffness factor of a gasketed joint can be defined as

$$C = \frac{k_b}{k_b + k_c} \tag{15.29a}$$

The quantity k_c represents the combined constant found from

$$\frac{1}{k_c} = \frac{1}{k_g} + \frac{1}{k_p} \tag{15.29b}$$

where k_g and k_p are the spring rates of the gasket and connected parts, respectively.

When a full gasket extends over the entire diameter of a joint, the gasket pressure is

$$p = \frac{F_p}{A_g} \tag{a}$$

in which A_g is the gasket area per bolt and F_p represents the clamping force on parts. For a load factor n_s, Equation 15.24 becomes

$$F_p = (1-C)n_s P - F_i \tag{b}$$

Carrying Equation (a) into (b), *gasket pressure* may be expressed in the form

$$p = \frac{1}{A_g}\left[F_i - n_s P(1-C)\right] \tag{15.30}$$

We point out that to maintain the uniformity of pressure, bolts should not be spaced more than *six* bolt diameters apart.

15.11 Determining the Joint Stiffness Constants

Application of the equations developed in Section 15.9 requires a determination of the spring rates of bolt and parts or at least a reasonable approximation of their relative values. Recall from Chapter 4 that the axial deflection is found from the equation $\delta = PL/AE$ and spring rate by $k = P/\delta$. Thus, we have for the bolt and parts, respectively,

$$k_b = \frac{A_b E_b}{L} \tag{15.31a}$$

$$k_p = \frac{A_p E_p}{L} \tag{15.31b}$$

where

k_b = the stiffness constant for bolt
k_p = the stiffness constant for parts
A_b = the cross-sectional area of bolt
A_p = the effective cross-sectional area of parts
E = the modulus of elasticity
L = the grip, which represents approximate length of clamped zone

15.11.1 Bolt Stiffness

When the thread stops immediately above the nut as shown in Figure 15.11, the gross cross-sectional area of the bolt must be used in approximating k_b, since the unthreaded portion is stretched by the load. Otherwise, a bolt is treated as a spring in series when considering the threaded and unthreaded portions of the shank. For a bolt of axially loaded *thread length* L_t and the unthreaded *shank length* L_s (Figure 15.11a), the spring constant is

$$\frac{1}{k_b} = \frac{L_t}{A_t E_b} + \frac{L_s}{A_b E_b} \tag{15.32}$$

in which A_b is the gross cross-sectional area and A_t represents the tensile stress area of the bolt. Note that, ordinarily, a bolt (or cap screw) has as little of its length threaded as practicable to maximize bolt stiffness. We then use Equation 15.31a in calculating the bolt spring rate k_b.

We note that a bolt or cap screw ordinarily has as little of its length threaded as practicable to increase bolt stiffness. Nevertheless, for standardized threads, the thread length is prescribed, as shown in the following expressions:

Metric threads (in mm)

$$L_t = \begin{cases} 2d + 6 & L \le 125 \\ 2d + 12 & 125 < L \le 200 \\ 2d + 25 & L > 200 \end{cases} \tag{15.33a}$$

Inch series

$$L_t = \begin{cases} 2d + 0.25\,\text{in.} & L \le 6\,\text{in.} \\ 2d + 0.50\,\text{in.} & L > 6\,\text{in.} \end{cases} \tag{15.33b}$$

Here (see Figure 15.11a), L_t is the threaded length, L represents the total bolt length, and d is the diameter.

15.11.2 Stiffness of Clamped Parts

The spring constant of clamped parts is seldom easy to ascertain and frequently approximated by employing an empirical procedure. Accordingly, the stress induced in the joint is assumed to be *uniform* throughout a region surrounding the bolt hole [11,12]. The region is often represented by a double-cone-shaped *barrel* geometry of a half-apex angle 30°, as depicted in Figure 15.14. The stress is taken to be 0 outside the region. The effective cross-sectional area A_p is equal to about the average area of the shaded section shown in the figure:

$$A_p = \frac{\pi}{4}\left[\left(\frac{d_w + d_2}{2}\right)^2 - d^2\right]$$

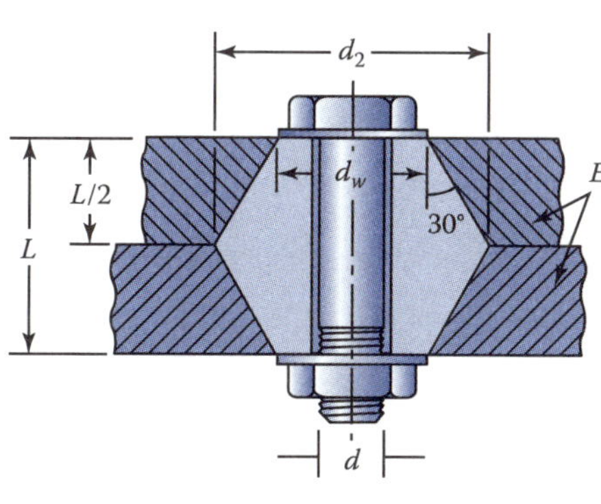

FIGURE 15.14
A method for estimating the effective cross-sectional area of clamped parts A_p.

The quantities $d_2 = d_w + L \tan 30°$ and d_w represent the washer (or washer face) diameter. Note that $d_w = 1.5d$ for *standard* hexagon-headed bolts and cap screws. The preceding expression of A_p is used for estimating k_p from Equation 15.31. It can be shown that [11], for connections using standard hexagon-headed bolts, the *stiffness constant for parts* is given by

$$k_p = \frac{0.58\pi E_p d}{2\ln\left(5\dfrac{0.58L + 0.5d}{0.58L + 2.5d}\right)} \tag{15.34}$$

where

d = the bolt diameter
L = the grip
E_p = the modulus of elasticity of the single or two identical parts

We should mention that the spring rate of clamped parts can be determined with good accuracy by experimentation or finite element analysis [13]. Various handbooks list rough estimates of the stiffness constant *ratio* k_p/k_b for typical gasketed and ungasketed joints. Sometimes, $k_p = 3k_b$ is used for ungasketed *ordinary* joints.

Example 15.5: Preloaded Bolt Connecting the Head and Cylinder of a Pressure Vessel

Figure 15.15 illustrates a portion of a cover plate bolted to the end of a thick-walled cylindrical pressure vessel. A total of N_b bolts are to be used to resist a separating force P. Determine

a. The joint constant
b. The number N_b for a permanent connection
c. The tightening torque for an average condition of thread friction

Given: The required joint dimensions and materials are shown in the figure. The applied load P = 55 kips.

Design Assumptions: The effects of the flanges on the joint stiffness are omitted. The connection is permanent. A bolt safety factor of n = 1.5 is used.

Solution

a. Referring to Figure 15.15, Equation 15.34 gives

$$k_p = \frac{0.58\pi(E_s/2)(0.75)}{2\ln\left[5\dfrac{0.58(2)+0.5(0.75)}{0.58(2)+2.5(0.75)}\right]} = 0.368E_s$$

Through the use of Equation 15.31a,

$$k_b = \frac{AE_s}{L} = \frac{\pi d^2 E_s}{4L} = \frac{\pi(0.75)^2 E_s}{4(2)} = 0.221E_s$$

Equation 15.22 is therefore

$$C = \frac{k_b}{k_b + k_p} = \frac{0.221}{0.221 + 0.368} = 0.375$$

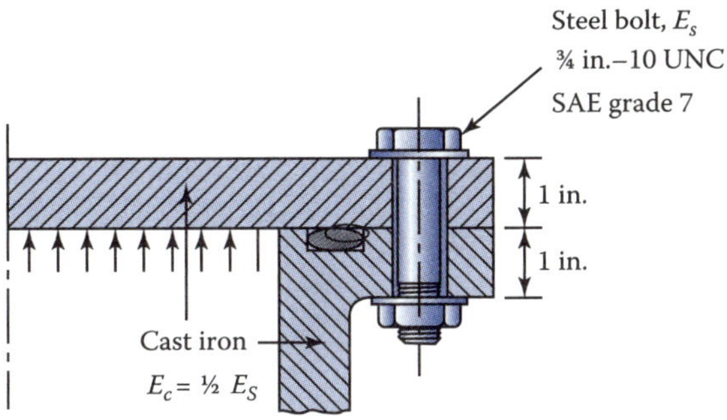

FIGURE 15.15
Example 15.5. Portion of a bolted connection subjected to pressure.

b. From Tables 15.1 and 15.4, we have $A_t = 0.334$ in.^{2} and $S_p = 105$ ksi. Applying Equation 15.20,

$$F_i = 0.9S_pA_t = 0.90(105)(0.334) = 31.6 \text{ kips}$$

For N_b bolts, Equation 15.26 can be written in the form

$$\frac{C(P/N_b)n}{A_t} + \frac{F_i}{A_t} = S_p$$

from which

$$N_b = \frac{CPn}{S_pA_t - F_i}$$

Substituting the numerical values, we have

$$N_b = \frac{(0.375)(55)(1.5)}{105(0.334) - 31.6} = 8.92$$

Comment: Nine bolts should be used.

c. By Equation 15.21,

$$T = 0.2dF_i = 0.2(0.75)(31.6) = 4.74 \text{ kips} \cdot \text{in.}$$

Example 15.6: Preloaded Bolt Clamping of a Cylinder under External Load

A steel bolt-and-nut clamps a steel cylinder of known cross section and length subjected to an external load P, as illustrated in Figure 15.16.

Given: $D = 20$ mm, $L = 65$ mm, $d = 10$ mm, $E = E_b = E_p = 200$ GPa

$$P = 8 \text{ kN} \qquad A_t = 58 \text{ mm}^2 \text{ (from Table 15.2)}$$

$$S_p = 380 \text{ MPa} \quad \text{and} \quad S_y = 420 \text{ MPa (by Table 15.5)}$$

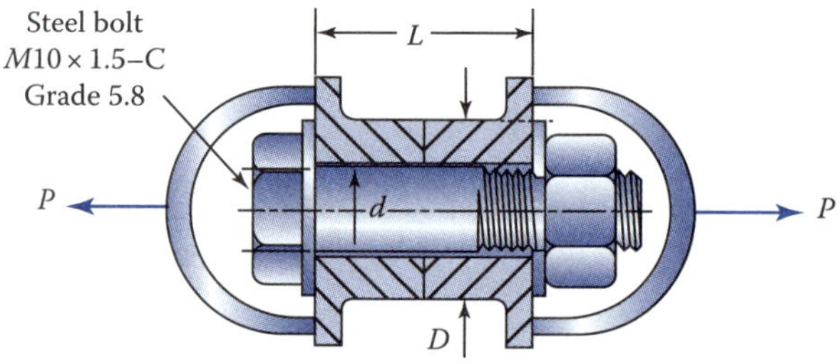

FIGURE 15.16
Example 15.6. A bolted connection carries an axial load.

Find:

a. Preload and bolt tightening torque
b. Joint stiffness factor
c. Maximum tensile stress in the bolt
d. Factors of safety against yielding and separation

Assumptions: Connection is reused. The effects of the flanges on the joint stiffness will be omitted.

Solution

See Figures 15.11 and 15.16.
The cross-sectional area of the parts is equal to $A_p = \pi(D^2 - d^2)/4 = \pi(20^2 - 10^2)/4 = 235.6\,\text{mm}^2$.

a. Through the use of Equation 15.20, the preload is

$$F_i = 0.75F_p = 0.75S_pA_t = 0.75(380)(58) = 16.53\ \text{kN}$$

This corresponds to an estimated bolt tightening torque (see Section 15.8) of

$$T = 0.2F_id = 0.2(16.53)(10) = 33.06\ \text{N}\cdot\text{m}$$

b. From Equation 15.33a, the lengths of thread L_t and shank L_s of the bolt (Figure 15.12) are

$$L_t = 2d + 6 = 2(10) + 6 = 26\ \text{mm}$$

$$L_s = L - L_t = 65 - 26 = 39\ \text{mm}$$

The stiffness constant for the bolt, by Equation 15.32, is

$$\frac{1}{k_b} = \frac{L_t}{A_tE} + \frac{L_s}{A_sE} = \frac{1}{200\left(10^6\right)}\left[\frac{26}{58} + \frac{39(4)}{\pi(10)^2}\right], \quad k_b = 2.117\left(10^8\right)\text{N/m}$$

By Equation 15.31b, the stiffness constant for the parts is

$$k_p = \frac{A_pE}{L} = \frac{235.6 \times 10^{-6}\left(200 \times 10^9\right)}{65 \times 10^{-3}} = 7.249\left(10^8\right)\text{N/m}$$

The joint stiffness factor, using Equation 15.22, is therefore

$$C = \frac{k_b}{k_p + k_b} = \frac{2.117}{7.249 + 2.117} = 0.226$$

Comment: The results indicate that $k_p \approx 3.4\ k_b$.

c. From Equations 15.23 and 15.24, the forces on the bolt and parts are

$$F_b = F_i + CP = 16.53 + 0.226(8) = 18.34 \text{ kN}$$

$$F_p = F_i - (1 - C)P = 16.53 - (1 - 0.226)(8) = 10.34 \text{ kN}$$

The largest tensile stress in the bolt equals

$$\sigma_b = \frac{F_b}{A_t} = \frac{18.34\left(10^3\right)}{58\left(10^{-6}\right)} = 316 \text{ MPa}$$

Comment: No stress-concentration factor applies for a statically loaded ductile material.

d. The factor of safety with respect to onset of yielding is equal to

$$n = \frac{S_y}{\sigma_b} = \frac{420}{316} = 1.33$$

Applying Equation 15.28, the load required to separate the joint and factor of safety against joint separation are

$$P_s = \frac{F_i}{(1 - C)} = \frac{16.53}{(1 - 0.226)} = 21.36 \text{ kN}$$

$$n_s = \frac{P_s}{P} = \frac{21.36}{8} = 2.67$$

Comment: Both safety factors found against yielding and separation are acceptable.

15.12 Tension Joints under Dynamic Loading

Bolted joints with preload and subjected to fatigue loading can be analyzed directly by the methods discussed in Chapter 7. Since failure owing to fluctuating loading is more apt to occur to the bolt, our attention is directed toward the bolt in this section. As previously noted, the use of initial tension is important in problems for which the *bolt carries cyclic loading*. The maximum and minimum loads on the bolt are higher because of the initial tension. Consequently, the mean load is greater, but the alternating load component is reduced. Therefore, the fatigue effects, which depend primarily on the variations of the stress, are likewise reduced.

Reconsider the joint shown in Figure 15.11a, but let the applied force P vary between some minimum and maximum values, both positive. The mean and alternating loads are given by

$$P_m = \frac{1}{2}(P_{max} + P_{min}), \quad P_a = \frac{1}{2}(P_{max} - P_{min})$$

Substituting P_m and P_a in place of P in Equation 15.23, the mean and alternating forces felt by the bolt are

$$F_{bm} = CP_m + F_i \tag{15.35a}$$

$$F_{ba} = CP_a \tag{15.35b}$$

The mean and range stresses in the bolt are then

$$\sigma_{bm} = \frac{CP_m}{A_t} + \frac{F_i}{A_t} \tag{15.36a}$$

$$\sigma_{ba} = \frac{CP_a}{A_t} \tag{15.36b}$$

in which C represents the joint constant and A_t is the tensile stress area. We observe from Equation 15.36 that as long as separation does not occur, the alternating stress experienced by the bolt is reduced by the joint stiffness rate C. The mean stress is increased by the bolt preload.

For the bolted joints, the Goodman criterion given by Equation 7.16 may be written as follows:

$$\frac{\sigma_{ba}}{S_e} + \frac{\sigma_{bm}}{S_u} = 1$$

As before, the safety factor is not applied to the initial tension. Hence, introducing Equation 15.36 into this equation, we have

$$\frac{CP_a n}{A_t S_e} + \frac{CP_m n + F_i}{A_t S_u} = 1$$

The preceding is solved to give the *factor of safety* guarding against *fatigue failure* of the *bolt:*

$$n = \frac{S_u A_t - F_i}{C\left[P_a\left(\frac{S_u}{S_e}\right) + P_m\right]} \tag{15.37}$$

TABLE 15.6

Fatigue Stress Concentration Factors K_f for Steel-Threaded Members

SAE Grade (Unified Thread)	Metric Grade (ISO Thread)	Rolled Threads	Cut Threads	Fillet
0–2	3.6–5.8	2.2	2.8	2.1
4–8	6.6–10.9	3.0	3.8	2.3

Alternatively,

$$n = \frac{S_u - \sigma_i}{C\left[\sigma_a\left(\dfrac{S_u}{S_e}\right) + \sigma_m\right]} \tag{15.38}$$

Here, $\sigma_a = P_a/A_t$, $\sigma_m = P_m/A_t$, and $\sigma_i = F_i/A_t$. Recall from Section 7.9 that this equation represents the Soderberg criterion when ultimate strength S_u is replaced by the yield strength S_y.

The modified endurance limit S_e is obtained from Equation 7.6. For threaded finishes having good quality, a surface factor of $C_f = 1$ may be applicable. The size factor $C_s = 1$ (see Section 7.7), and by Equation 7.3, we have $S'_e = 0.45S_u$ for reversed axial loading. As a result,

$$S_e = C_r C_t \left(\frac{1}{K_f}\right)(0.45S_u) \tag{15.39}$$

where C_r and C_t are the reliability and temperature factors. Table 15.6 gives average stress-concentration factors for the fillet under the bolt and also at the beginning of the threads on the shank [9]. Cutting is the simplest method of producing threads. Rolling the threads provides a smoother thread finish than cutting. The fillet between the head and the shank reduces the K_f, as shown in the table. Unless otherwise specified, the threads are usually assumed to be rolled.

A very *common case* is that the fatigue loading fluctuates between 0 and some maximum value, such as in a bolted pressure vessel cycled from 0 to a maximum pressure. In this situation, the minimum tensile loading $P_{\min} = 0$. The effect of initial tension with regard to fatigue loading is illustrated in the solution of the following sample problem.

Example 15.7: Preloaded Fasteners in Fatigue Loading

Figure 15.17a illustrates the connection of two steel parts with a single ⅝ in.–11UNC grade 5 bolt having rolled threads. Determine

a. Whether the bolt fails when no preload is present
b. If the bolt is safe with preload
c. The fatigue factor of safety n when preload is present
d. The static safety factors n and n_s

Design Assumptions: The bolt may be reused when the joint is taken apart. Survival rate is 90%. Operating temperature is normal.

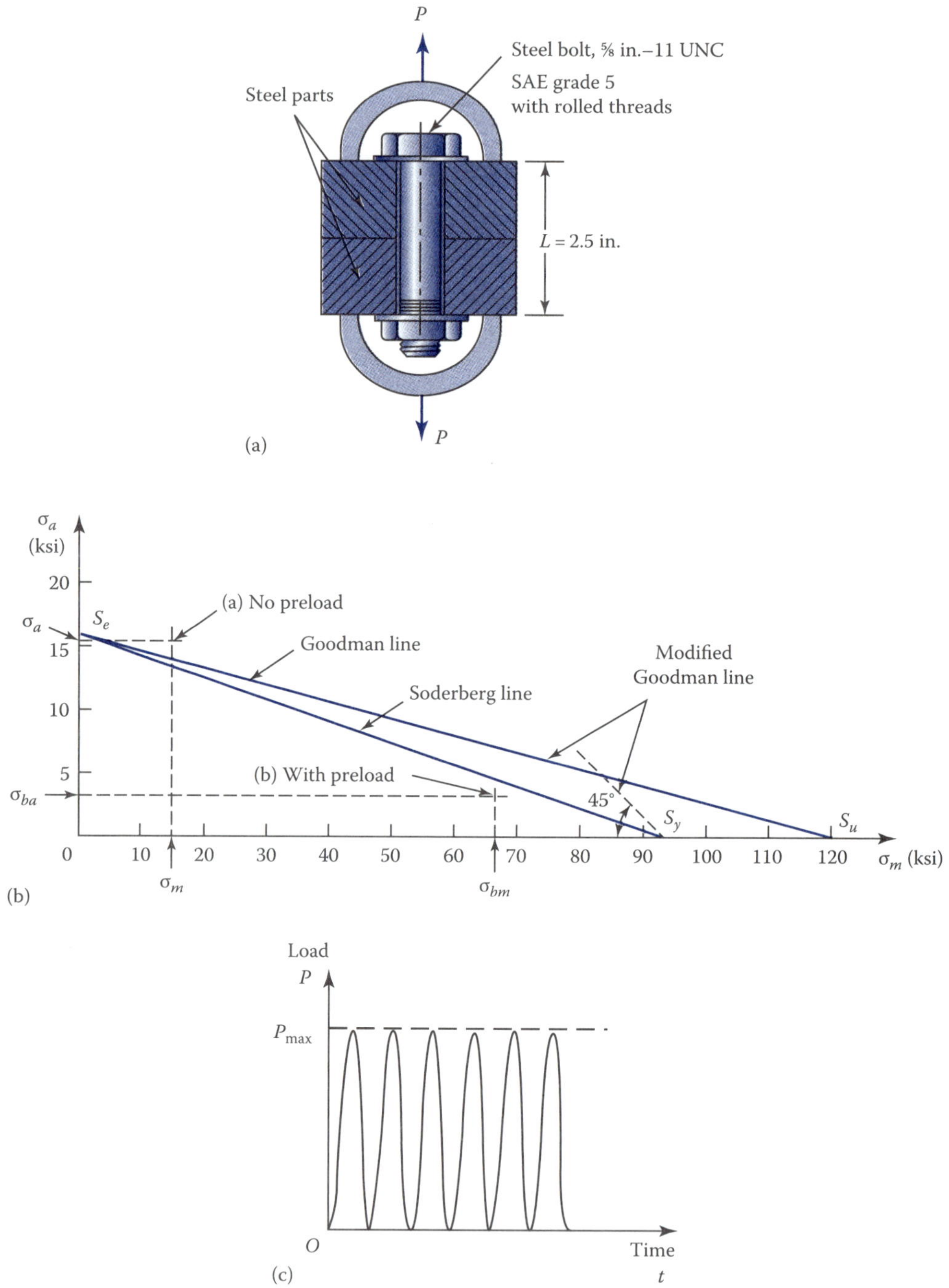

FIGURE 15.17
Example 15.7. (a) bolted parts carrying fluctuating loads, (b) alternating separating load as function of time, and (c) fatigue diagram for bolts.

Given: The joint is subjected to a load P that varies continuously between 0 and 7 kips (Figure 15.17b).

Solution

See Figure 15.17.

From Table 7.3, the reliability factor is $C_r = 0.89$. The temperature factor is $C_t = 1$ (Section 7.7). Also, $S_p = 85$ ksi, $S_y = 92$ ksi, $S_u = 120$ ksi (from Table 15.4), $K_f = 3$ (by Table 15.6), and

$$A_t = 0.226 \text{ in.}^2 \text{ (from Table 15.1)}$$

Equation 15.39 results in

$$S_e = (0.89)(1)\left(\frac{1}{3}\right)(0.45 \times 120) = 16 \text{ ksi}$$

The Soderberg and Goodman fatigue failure lines are shown in Figure 15.17c.

a. For loosely held parts, when $F_i = 0$, load on the bolt equals the load on parts:

$$P_m = \frac{1}{2}(7+0) = 3.5 \text{ kips}, \quad P_a = \frac{1}{2}(7-0) = 3.5 \text{ kips}$$

$$\sigma_a = \sigma_m = \frac{3.5}{0.226} = 15.5 \text{ ksi}$$

A plot of the stresses shown in Figure 15.17c indicates that failure will occur.

b. Through the use of Equation 15.20,

$$F_i = 0.75 S_p A_t = 0.75(85)(0.226) = 14.4 \text{ kips}$$

The grip is $L = 2.5$ in. By Equations 15.31a and 15.34 with $E_b = E_p = E$, we obtain

$$k_b = \frac{\pi d^2 E}{4L} = \frac{\pi(0.625)^2 E}{4(2.5)} = 0.123E$$

$$k_p = \frac{0.58\pi E(0.625)}{2\ln\left[5\dfrac{0.58(2.5)+0.5(0.625)}{0.58(2.5)+2.5(0.625)}\right]} = 0.53E$$

The joint constant is then

$$C = \frac{k_b}{k_b + k_p} = \frac{0.123}{0.123 + 0.53} = 0.188$$

Comment: The foregoing means that only about 20% of the external load fluctuation is felt by the bolt and hence about 80% goes to decrease clamping pressure.

Applying Equations 15.35 and 15.36,

$$F_{bm} = CP_m + F_i$$

$$= 0.188(3.5) + 14.4 = 15.1 \text{ kips}$$

$$\sigma_{bm} = \frac{15.1}{0.226} = 66.8 \text{ ksi}$$

$$F_{ba} = CP_a = 0.188(3.5) = 0.66 \text{ kips}$$

$$\sigma_{ba} = \frac{0.66}{0.226} = 2.92 \text{ ksi}$$

A plot on the fatigue diagram shows that *failure* will *not occur* (Figure 15.17c).

c. Equation 15.37 with $P_a = P_m$ becomes

$$n = \frac{S_u A_t - F_i}{CP_a\left[\left(\frac{S_u}{S_e}\right) + 1\right]} \tag{15.40}$$

Introducing the given numerical values,

$$n = \frac{(120)(0.226) - 14.4}{(0.188)(3.5)\left[\left(\frac{120}{16}\right) + 1\right]}$$

from which $n = 2.27$.

Comment: This is the factor of safety guarding against the fatigue failure. Observe from Figure 15.17c that the Goodman criteria led to a less conservative (higher) value for n.

d. Substitution of the given data into Equations 15.27 and 15.28 gives

$$n = \frac{85(0.226) - 14.4}{(0.188)(7)} = 3.66$$

$$n_s = \frac{14.4}{7(1 - 0.188)} = 2.53$$

Comments: The factor of 3.66 prevents the bolt stress from becoming equal to proof strength. On the other hand, the factor of 2.53 guards against joint separation and the bolt taking the entire load.

15.13 Riveted and Bolted Joints Loaded in Shear

A rivet consists of a cylindrical body, known as the *shank*, usually with a rounded end called the *head*. The purpose of the rivet is to join together two plates while securing proper strength and tightness. If the rivet is heated prior to being placed in the hole, it is referred to as a hot-driven rivet, while if it is not heated, it is referred to as a cold-driven rivet.

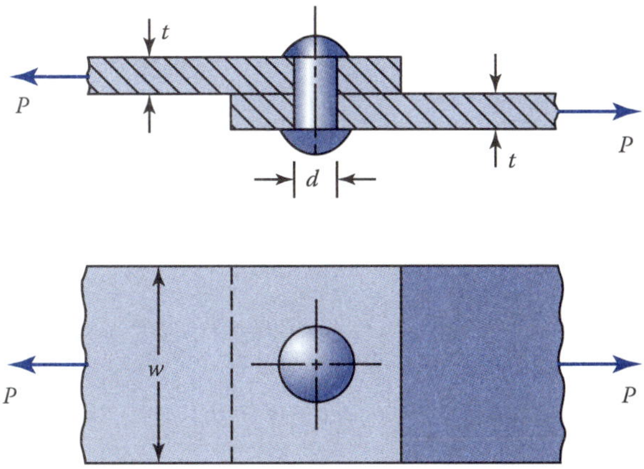

FIGURE 15.18
Riveted connection loaded in shear.

Rivets and bolts are ordinarily used in the construction of buildings, bridges, aircraft, and ships. The design of riveted and bolted connections is governed by construction codes formulated by such societies as the AISC [14] and the ASME.

Riveted and bolted joints loaded in shear are treated *exactly alike* in design and analysis. Figure 15.18 illustrates a simple riveted connection loaded in shear. It is obvious that the loading is eccentric and an unbalanced moment *Pt* exists. Hence, bending stress will be present. However, the usual procedure is to ignore the bending stress and compensate for its presence by a larger factor of safety. Table 15.7 lists various types of failure of the connection shown in the figure.

The *effective diameters* in a riveted joint are defined as follows. For a drilled hole, $d_e = d + \frac{1}{16}$ in (about 1.5 mm), and for a *punched hole*, $d_e = d + \frac{1}{8}$ in (about 3 mm). Here, *d* represents the diameter of the rivet. Unless specified otherwise, we *assume* that the holes have

TABLE 15.7

Types of Failure for Riveted Connections (Figure 15.18)

A. Shearing Failure of Rivet

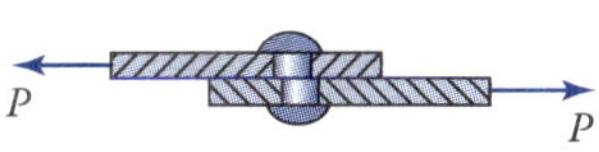

$\tau = 4P/\pi d^2$

B. Tensile Failure of Plate

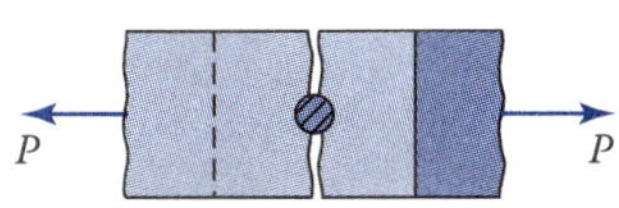

$\sigma_t = P/(w - d_e)t$

C. Bearing Failure of Plate or Rivet

$\sigma_b = P/dt$

D. Shearing Failure of Edge of Plate

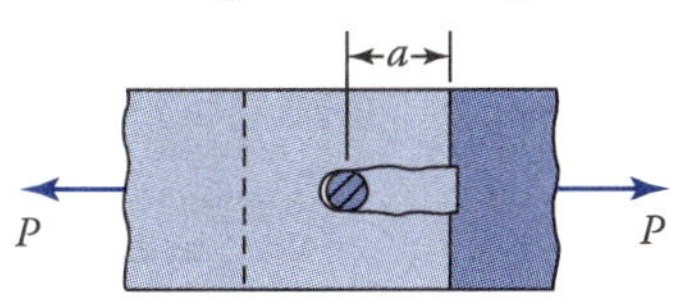

$\tau_t = P/2at$

Notes: *P*, applied shear load; *d*, diameter of rivet; *w*, width of plate; *t*, thickness of the thinnest plate; d_e, effective hole diameter; *a*, the closest distance from rivet to the edge of plate.

been *punched*. Usually, shearing, or tearing, failure is avoided by spacing the rivet at least 1.5d away from the plate edge. To sum up, essentially, three modes of failure must be considered in determining the capacity of a riveted or bolted connection: *shearing* failure of the rivet, *bearing* failure of the plate or rivet, and *tensile* failure of the plate. The associated normal and shear equations are given in the table.

Example 15.8: Capacity of a Riveted Connection

The standard AISC connection for the W310 × 52 beam consists of two 102 × 102 × 6.4 mm angles, each 215 mm long, 22 mm rivets spaced 75 mm apart are used in 24 mm holes (Figure 15.19). Calculate the maximum load that the connection can carry.

Design Decisions: The allowable stresses are 100 MPa in shear and 335 MPa in bearing of rivets. Tensile failure cannot occur in this connection; only shearing and bearing capacities need to be investigated.

Solution

The web thickness of the beam is $t_w = 7.6$ mm (from Table A.6), and the cross-sectional area of one rivet is $A_r = \pi(22)^2/4 = 380\ \text{mm}^2$.

Bearing on the web of the beam:

$$P_b = 3(7.6)(22)(335) = 168\ \text{kN} \quad \text{(governs)}$$

Shear of six rivets:

$$P_b = 6(380)(100) = 228\ \text{kN}$$

Bearing of six rivets on angles:

$$P_b = 6(22)(6.4)(335) = 283\ \text{kN}$$

Comment: The capacity of this connection, the smallest of the forces obtained in the foregoing, is 168 kN.

15.13.1 Joint Types and Efficiency

Most connections have many rivets or bolts in a variety of models. Riveted or bolted connections loaded in shear are of two types: *lap joints* and *butt joints*. In a lap joint, sometimes called a *single-shear joint*, the two plates to be jointed overlap each other (Figure 15.20a). On the other hand, in a butt, also termed a *double-shear joint*, the two plates to be connected (main plates) butt against one other (Figure 15.20b). *Pitch* is defined as the distance between

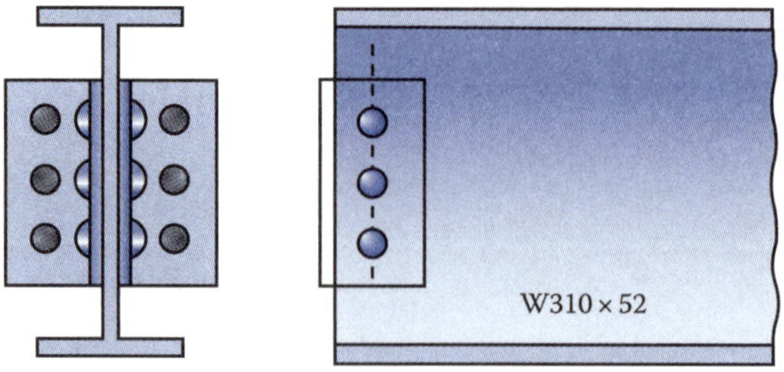

FIGURE 15.19
Example 15.8.

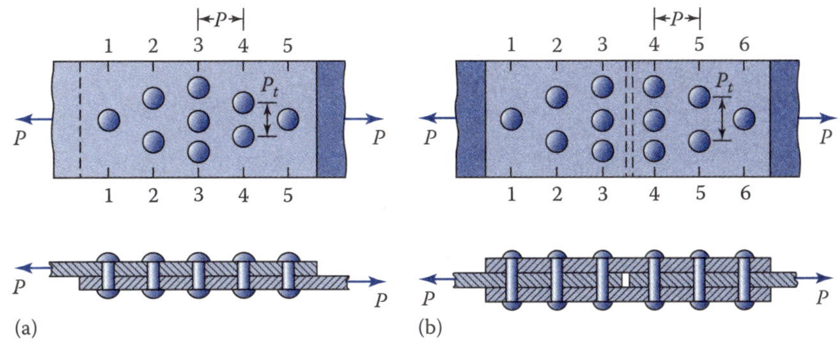

FIGURE 15.20
Types of riveted connections: (a) lap joint and (b) butt joint.

adjacent rivet centers. It represents a significant geometric property of a joint. The *axial* pitch p for rivets is measured along a line parallel to the edge of the plate, while the corresponding distance along a line perpendicular to the edge of the plate is known as the *transverse* pitch p_t. Both kinds of pitch are depicted in the figure. The smallest symmetric group of rivets that repeats itself along the length of a joint is called a *repeating section*. The strength analysis of a riveted connection is based on its repeated section (see Example 15.9).

The *efficiency of joints* is defined as follows:

$$e = \frac{P_{\text{all}}}{P_t} \tag{15.41}$$

In the foregoing equation, P_{all} is the smallest of the allowable loads in shear, bearing, and tension; P_t represents the static tensile yield load (strength) of plate with no hole. The most efficient joint would be as strong in tension, shear, and bearing as the original plate to be joined is in tension. This can never be realized, since there must be at least one rivet hole in the plate: The allowable load of joint in tension therefore always is less than the strength of the plate with no holes.

For centrally applied loads, it is often assumed that the rivets are about equally stressed. In many cases, this cannot be justified by elastic analysis; however, ductile deformations permit an equal redistribution of the applied force, before the ultimate capacity of connection is reached. Also, it is usually taken that the row of rivets immediately *adjacent* to the load carries the full load. Thus, the maximum load supported by such a row occurs when there is only one rivet in that row. The *actual load* carried by an interior row can be obtained from

$$P_i = \frac{n - n'}{n} P \tag{15.42}$$

where

P = the externally applied load
P_i = the actual load, or portion of P, acting on a particular row i
n = the total number of rivets in the joint
n' = the total number of rivets in the row between the row being checked and the external load

For instance, load on row 3 of the joint in Figure 15.20a equals $P_3 = (9 - 3)P/9 = 2P/3$. Likewise, load on row 2 or 5 of the joint in Figure 15.20b is $P_2 = P_5 = (12 - 1)P/12 = 11P/12$.

Example 15.9: Strength Analysis of a Multiple-Riveted Lap Joint

Figure 15.21a shows a multiple-riveted lap joint subjected to an axial load P. The dimensions are given in inches. Calculate the allowable load and efficiency of the joint.

Given: All rivets are ¾ in. in diameter.

Design Assumptions: The allowable stresses are 20 ksi in tension, 15 ksi in shear, and 30 ksi in bearing.

Solution

The analysis is on the basis of the repeating section, which has four rivets and $L = 6$ in. (Figure 15.21b).

The plate in tension, without holes:

$$P_t = 20\left(6\times\frac{1}{2}\right) = 60 \text{ kips}$$

The rivet shear:

$$P_s = 4\left[\frac{\pi}{4}\left(\frac{3}{4}\right)^2\right](15) = 26.51 \text{ kips} \quad \text{(governs)}$$

The plate bearing:

$$P_b = 4\left(\frac{1}{2}\times\frac{3}{4}\right)(30) = 45 \text{ kips}$$

The tension across sections 1–1 through 3–3 of the *bottom plate*, using Equation 15.42:

$$\frac{4-3}{4}P_1 = \frac{1}{2}\left[6-\left(\frac{3}{4}+\frac{1}{8}\right)\right](20); \quad P_1 = 205 \text{ kips}$$

$$\frac{4-1}{4}P_2 = \frac{1}{2}\left[6-2\left(\frac{3}{4}+\frac{1}{8}\right)\right](20); \quad P_2 = 56.7 \text{ kips}$$

$$P_3 = \frac{1}{2}\left[6-\left(\frac{3}{4}+\frac{1}{8}\right)\right](20); \quad P_3 = 51.25 \text{ kips}$$

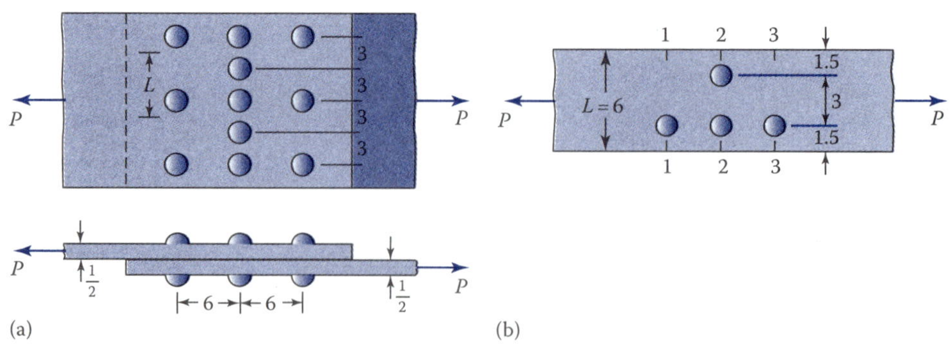

FIGURE 15.21
Example 15.9. (a) a riveted lap joint and (b) enlarged view of a repeating group of rivets.

The maximum allowable force that the joint can safely carry is the smallest of the force obtained in the preceding, $P_{all} = 26.51$ kips. The efficiency of this joint, from Equation 15.41, is then

$$e = \frac{26.51}{60} \times 100 = 44.2\%$$

15.14 Shear of Rivets or Bolts due to Eccentric Loading

For the case in which the load is applied eccentrically to a connection having a group of bolts or rivets, the effects of the torque or moment, as well as the direct force, must be considered. A typical structural problem is the situation that occurs when a horizontal beam is supported by a vertical column (Figure 15.22a). In this case, each bolt is subjected to a twisting moment $M = Pe$ and a direct shear force P. An enlarged view of bolt group with loading (P and M) acting at the centroid C of the group and the reactional shear forces acting at the cross section of each bolt are shown in Figure 15.22b.

Let us assume that the reactional *tangential force* due to moment, the so-called moment load or secondary shear, on a bolt varies directly with the distance from the centroid C of the group of bolts and is directed perpendicular to the centroid. As a result,

$$\frac{F_1}{r_1} = \frac{F_2}{r_2} = \frac{F_3}{r_3} = \frac{F_4}{r_4}$$

In the preceding, F_i and r_i ($i = 1,\ldots,4$) are the tangential force and radial distance from C to the center of each bolt, respectively. The externally applied moment and tangential forces are related as follows:

$$M = Pe = F_1r_1 + F_2r_2 + F_3r_3 + F_4r_4$$

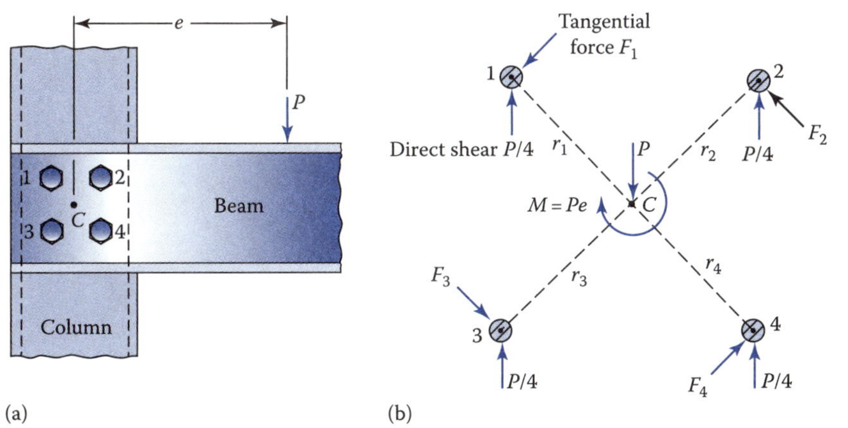

FIGURE 15.22
(a) Bolted joint with eccentric load. (b) Bolt group with loading and reactional shear forces (15.43).

Solving these equations simultaneously, we obtain

$$F_1 = \frac{Per_1}{r_1^2 + r_2^2 + r_3^2 + r_4^2}$$

This expression can be written in the following general form:

$$F_i = \frac{Mr_j}{\sum_{j=1}^{n} r_j^2} \tag{15.43}$$

where

F_i = the tangential force
$M = Pe$, externally applied moment
n = the number of bolts in the group
i = the particular bolt whose load is to be found

It is customary to assume that the reactional *direct force F/n* is the same for all bolts of the joint. The vectorial sum of the tangential force and direct force is the resultant shear force on the bolt (Figure 15.22b). Clearly, only the bolt having the maximum resultant shear force needs to be considered. An inspection of the vector force diagram is often enough to eliminate all but two or three bolts as candidates for the worst-loaded bolt.

Example 15.10: Bolt Shear Forces due to Eccentric Loading

A gusset plate is attached to a column by three identical bolts and vertically loaded, as shown in Figure 15.23a. The dimensions are in millimeters. Calculate the maximum bolt shear force and stress.

Assumption: The bolt tends to shear across its major diameter.

Solution

For the bolt group, point C corresponds to the centroid of the triangular pattern, as shown in Figure 15.23b. This free-body diagram illustrates the bolt reactions and the

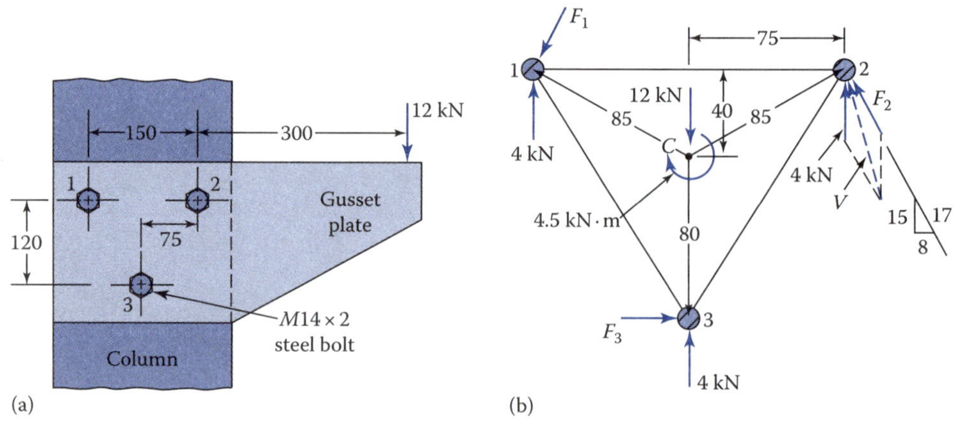

FIGURE 15.23
Example 15.10. (a) bolted connection and (b) bolt shear force and moment equilibrium.

external loading replaced at the centroid. Each bolt supports one-third of the vertical shear load, 4 kN, plus a tangential force F_i. The distances from the centroid to bolts are

$$r_1 = r_2 = \sqrt{(40)^2 + (75)^2} = 85.0 \text{ mm}, \quad r_3 = 80 \text{ mm}$$

Equation 15.43 then results in

$$F_1 = F_2 = \frac{Mr_1}{r_1^2 + r_2^2 + r_3^2} = \frac{4500(85)}{2(85)^2 + (80)^2}$$

$$= \frac{382,500}{20,850} = 18.35 \text{ kN}$$

$$F_3 = \frac{4,500(80)}{20,850} = 17.27 \text{ kN}$$

The vector sum of the two shear forces, obviously greatest for the bolt 2, can be obtained algebraically (or graphically):

$$V_2 = \left[\left(\frac{15}{17} \times 18.35 + 4\right)^2 + \left(\frac{8}{17} \times 18.35\right)^2\right]^{1/2} = 21.96 \text{ kN}$$

The bolt shear stress area is $A_s = \pi d^2/4 = \pi(14)^2/4 = 153.9 \text{ mm}^2$. Hence,

$$\tau = \frac{V_2}{A_s} = \frac{21,960}{153.9(10^{-6})} = 142.7 \text{ MPa}$$

Example 15.11: Shear Stress in Rivets Owing to Eccentric Loading

A riveted joint is under an inclined eccentric force P, as indicated in Figure 15.24a. Calculate the maximum shear stress in the rivets.

Given: The rivets are 1 in. in diameter. $P = 10$ kips.

Solution

For simplicity in computations, the applied load P is first resolved into horizontal and vertical components. Each rivet carries one-half of the load. The centroid of the

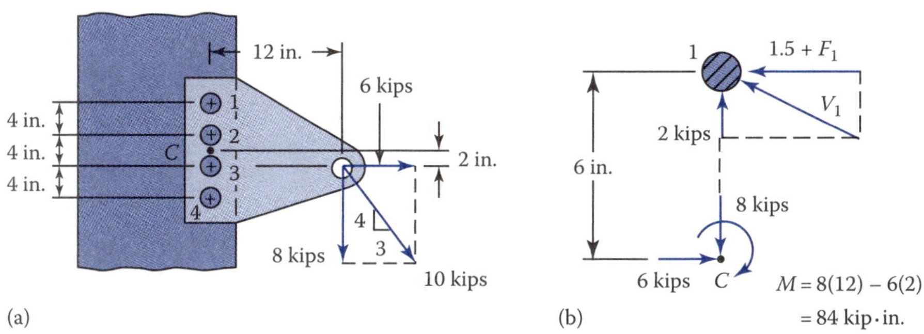

FIGURE 15.24
Example 15.11. (a) riveted connection and (b) enlarged view showing loading acting at the centroid and reactions on rivet 1.

rivet group is between the top and bottom rivets at C. An inspection of Figure 15.24a shows that the top rivet 1 is under the highest stress (Figure 15.24b). Through the use of Equation 15.43,

$$F_1 = \frac{Mr_1}{r_1^2 + r_2^2 + r_3^2 + r_4^2} = \frac{84(6)}{2(6)^2 + 2(2)^2} = 6.3 \text{ kip} \cdot \text{in}$$

Vector sum of the shear forces is

$$V_1 = \left[(1.5 + 6.3)^2 + 2^2\right]^{1/2} = 8.052 \text{ kips}$$

We then have

$$\tau = \frac{V_1}{\pi d^2/4} = \frac{4(8.052)}{\pi(1)^2} = 10.25 \text{ ksi}$$

15.15 Welding

A *weld* is a joint between two surfaces produced by the application of localized heat. Here, we briefly discuss only welding between metal surfaces: thermoplastics can be welded much like metals. A *weldment* is fabricated by welding together a variety of metal forms cut to particular configuration. Nearly all wielding is by fusion processes. Establishment of metallurgical bond between two parts by melting together the base metals with a filler metal is called the *fusion process*. Heat is brought about usually by an electric arc, electric current, or gas flame. Metals and alloys to arc and gas welding must be properly selected. Properties of welding filler material must be matched with those of base metal when possible. The joint strength would then be equal to the strength of the base metal, giving an efficiency of almost 100% for static loads.

15.15.1 Welding Processes and Properties

Metallic arc welding, the so-called shielded metal arc welding (SMAW), refers to a process where the heat is applied by an arc passing between an electrode and the work. The electrode is composed of suitable filler material with coating ordinarily similar to that of base metal. It is melted and fed into the joint as the weld is being formed. The coating is vaporized to provide a shielding gas-preventing oxidation at the weld as well as acts as a flux and directs the arc. Either direct or alternating current can be used with this process. A weld thickness greater than about ⅜ in. is often produced on successive layers. In *metal–inert gas arc welding* or gas–metal arc welding (GMAW), heat is applied by a gas flame. In this process, a bare or plated wire is continuously fed into the weld from a large spool. The wire serves as electrode and becomes the filler in the union. Uniform-quality welds are attainable with metal–gas welding.

Resistance welding uses electric-current-generated heat that passes through the parts to be welded while they are clamped together firmly. Filler material is not ordinarily

TABLE 15.8

Typical Weld-Metal Properties

	Ultimate Strength		Yield Strength		
AWS Electrode Number	**ksi**	**(MPa)**	**ksi**	**(MPa)**	**Percent Elongation**
E6010	62	(427)	50	(345)	22
E6012	67	(462)	55	(379)	17
E6020	62	(427)	50	(345)	25
E7014	72	(496)	60	(414)	17
E7028	72	(496)	60	(414)	22

Source: American Welding Society Code AWSD.1.77, American Welding Society, Miami, FL.

employed. Usually, thin metal parts may be connected by spot or continuous resistance welding. A spot weld is made by a pair of electrodes that apply pressure to either side of a lap joint and devise a complete circuit. Laser beam welding, plasma arc welding, and electron beam welding are utilized for special applications. The suitability of several metals and alloys to arc and gas welding is very important.

Materials and symbols for welding have been standardized by the ASTM and the American Welding Society (AWS). Numerous different kinds of electrodes have been standardized to fit a variety of conditions encountered in the welding of machinery and structures. Table 15.8 presents the characteristics for some E60 and E70 electrode classes. Note that the AWS *numbering system* is based on the use of an *E* prefix followed by four digits. The first two numbers on the left identify the approximate strength in ksi. The last digit denotes a group of welding technique variables, such as current supply. The next to last digit refers to a welding position number (1 for all and 2 for horizontal positions, respectively). Welding electrodes are available in diameters from $\frac{1}{16}$ to $\frac{5}{16}$ in. It should be mentioned that the *electrode* material is often the strongest material present in a joint [15,16].

15.15.2 Strength of Welded Joints

Among numerous configurations of welds, we consider only two common butt and fillet types. The geometry of a typical *butt weld* loaded in tension and shear is shown in Figure 15.25. The equations for the stresses due to the loading are also given in the figure. Note that the height h for a butt weld does not include the bulge or reinforcement used to compensate for voids or inclusions in the weld metal. Plates of ¼ in. and heavier should be beveled before welding as indicated.

Figure 15.26 illustrates two *fillet welds* loaded in shear and transverse tension. The corresponding average stress formula is written under each figure. Now the size of the fillet weld is defined as the leg length h. Normally, the two legs are of the same length h. In welding design, stresses are calculated for the *throat section:* minimum cross-sectional area A_t located at 45° to the legs. We have $A_t = tL = 0.707hL$, where t and L represent the throat length and length of weld, respectively (Figure 15.26a). We note that actual stress distribution in a weld is somewhat complicated and design depends on the stiffness of the base material and other factors that have been neglected. Particularly, for stress situation on the throat area in Figure 15.26b, no exact solutions are available.

The foregoing average results are valid for design, however, because weld strengths are on the basis of tests on joints of these types. Having the material strengths available for a

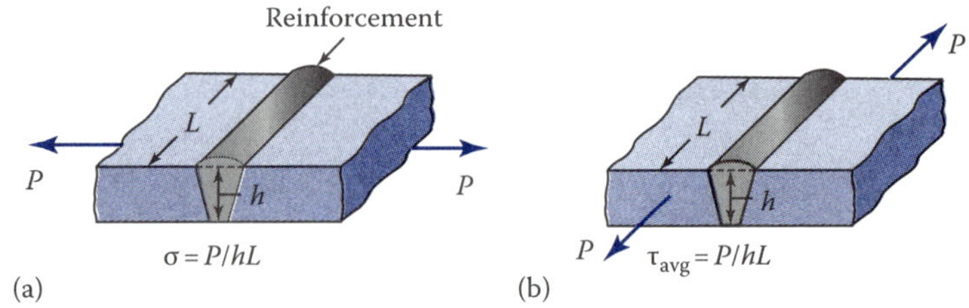

FIGURE 15.25
Butt weld: (a) tension loading and (b) shear loading.

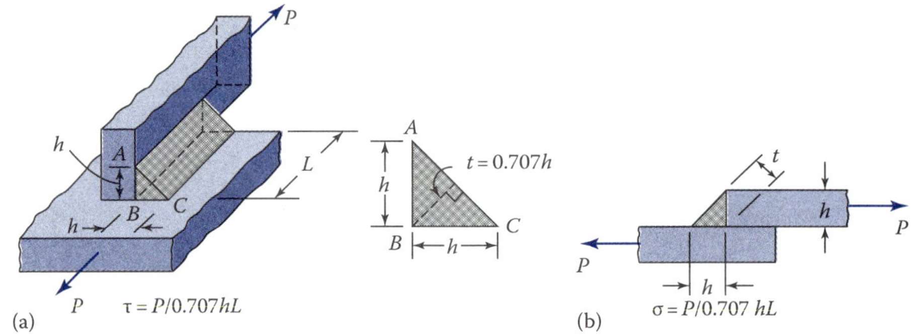

FIGURE 15.26
Fillet weld: (a) shear loading and (b) transverse tension loading. *Notes*: h is the length of weld leg, t is the throat length, and L is the weld length.

welded joint, the required weld size h can be obtained for a prescribed safety factor. The usual equation of the *factor of safety n* applies for *static loads:*

$$n = \frac{S_{ys}}{\tau} = \frac{0.5S_y}{\tau} \tag{15.44}$$

The quantities S_y and S_{ys} represent tensile yield and shear yield strengths of weld material, respectively.

15.15.3 Stress Concentration and Fatigue in Welds

Abrupt changes in geometry take place in welds, and hence, stress concentrations are present. The weld and the plates at the base and reinforcement should be thoroughly blended together (Figure 15.25). The stresses are highest in the immediate vicinity of the weld. Sharp corners at the *toe* and *heel*, points A and B in Figure 15.26, should be rounded. Since welds are ductile materials, stress concentration effects are ignored for static loads. As has always been the case, when the loading fluctuates, a stress concentration factor is applied to alternating component. Approximate values for fatigue strength reduction factors are listed in Table 15.9.

Under *cyclic loading*, the welds fail long before the welded members. The fatigue factor of safety and working stresses in welds are defined by the AISC as well as AWS codes for buildings and bridges [2,17]. The codes allow the use of a variety of ASTM structural steels. For ASTM steels, tensile yield strength is one-half of the ultimate strength in tension, $S_y = 0.5S_u$ for static or fatigue loads. Unless otherwise specified, an *as-forged surface* should always be used for weldments. Also, prudent design would suggest taking the size factor

TABLE 15.9

Fatigue Stress Concentration Factors K_f for Welds

Type of Weld	K_f
Reinforced butt weld	1.2
Toe of transverse fillet weld	1.5
End of parallel fillet weld	2.7
T-butt joint with sharp corners	2.0

Source: American Welding Society Code AWSD.1.77, American Welding Society, Miami, FL.

$C_s = 0.7$. Design calculations for fatigue loading can be made by the methods described in Section 7.11, as illustrated in the following sample problem.

Example 15.12: Design of a Butt Welding for Fatigue Loading

The tensile load P on a butt weld (Figure 15.25a) fluctuates continuously between 20 and 100 kN. Plates are 20 mm thick. Determine the required length L of the weld, applying the Goodman criterion.

Assumptions: Use an E6010 welding rod with a factor of safety of 2.5.

Solution

By Table 15.8 for E6010, S_u = 427 MPa. The endurance limit of the weld metal, from Equation 7.6, is

$$S_e = C_f C_r C_s C_t (1/K_f) S_e'$$

Referring to Section 7.7, we have

$C_r = 1$ (based on 50% reliability)
$C_s = 0.7$ (lacking information)
$C_f = AS_u^b = 272(427)^{-0.995} = 0.657$ (by Equation [7.7])
$C_t = 1$ (normal room temperature)
$K_f = 1.2$ (from Table 15.9)
$S_e' = 0.5S_u = 0.5(427) = 213.5$ MPa

Hence,

$$S_e = (1)(0.7)(0.657)(1)(1/1.2)(213.5) = 81.82 \text{ MPa}$$

The mean and alternating loads are given by

$$P_m = \frac{100+20}{2} = 60 \text{ kN}, \quad P_a \frac{100-20}{2} = 40 \text{ kN}$$

Corresponding stresses are

$$\sigma_m = \frac{60{,}000}{20L} = \frac{3000}{L}$$

$$\sigma_a = \frac{40{,}000}{20L} = \frac{2000}{L}$$

Through the use of Equation 7.16, we have

$$\frac{S_u}{n} = \sigma_m + \frac{S_u}{S_e}\sigma_a; \qquad \frac{427}{2.5} = \frac{3000}{L} + \frac{427}{81.82}\left(\frac{2000}{L}\right)$$

Solving,

$$L = 78.67 \text{ mm}$$

Comment: A 79 mm long weld should be used.

15.16 Welded Joints Subjected to Eccentric Loading

When a welded joint is under eccentrically applied loading, the effect of torque or moment must be taken into account as well as the direct load. The exact stress distribution in such a joint is complicated. A detailed study of both the rigidity of the parts being joined and the geometry of the weld is required. The following procedure, which is based on simplifying assumptions, leads to reasonably accurate results for most applications.

15.16.1 Torsion in Welded Joints

Figure 15.27 illustrates an eccentrically loaded joint, with the centroid of all the weld areas or weld group at point C. The load P is applied at a distance e from C, in the plane of the group.

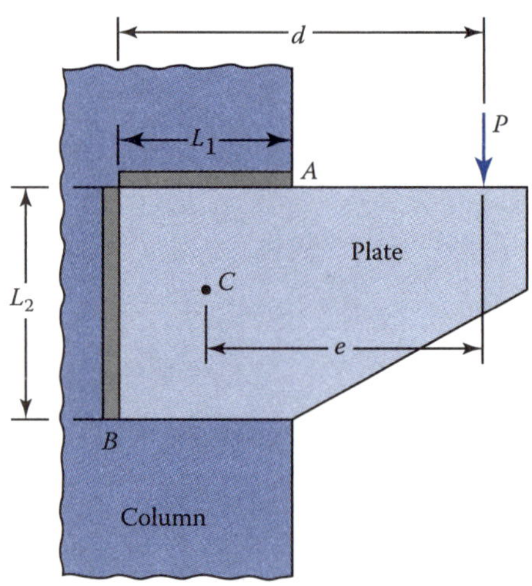

FIGURE 15.27
Welded joint in plane eccentric loading. *Notes*: C is the centroid of the weld group.

As a result, the welded connection is under torsion $T = Pe$ and the direct load P. The latter force causes a direct shear stress in the welds:

$$\tau_d = \frac{P}{A} \tag{15.45}$$

in which P is the applied load and A represents the throat area of all the welds. The preceding stress is taken to be uniformly distributed over the length of all welds. The torque causes the following torsional shear stress in the welds:

$$\tau_t = \frac{Tr}{J} \tag{15.46}$$

where

T = the torque

r = the distance from C to the point in the weld of interest

J = the polar moment of inertia of the weld group about C (based on the throat area)

Resultant shear stress in the weld at radius r is given by the vector sum of the direct shear stress and torsional stress:

$$\tau = \left(\tau_d^2 + \tau_t^2\right)^{1/2} \tag{15.47}$$

Note that r usually represents the *farthest* distance from the centroid of the weld group.

15.16.2 Bending in Welded Joints

Consider an angle welded to a column, as depicted in Figure 15.28. Load P acts at a distance e, out of plane of the weld group, producing bending in addition to direct shear. We again take a linear distribution of shear stress due to moment $M = Pe$ and a uniform distribution of direct shear stress. The latter stress τ_d is given by Equation 15.45. The moment causes the shear stress:

$$\tau_m = \frac{Mc}{I} \tag{15.48}$$

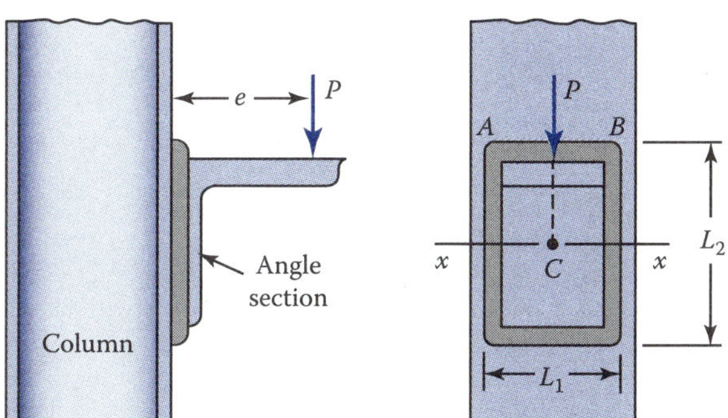

FIGURE 15.28
Welded joint under out-of-plane loading.

Here, the distance c is measured from C to the farthest point on the weld. As in the previous case, the resultant shear stress τ in the weld is estimated by the vector sum of the direct shear stress and the moment-induced stress:

$$\tau = \left(\tau_d^2 + \tau_m^2\right)^{1/2} \tag{15.49}$$

On the basis of the geometry and loading of Figure 15.28, we note that τ_d is downward and τ_m along edge AB is outward.

15.16.2.1 Centroid of the Weld Group

Let A_i denote the weld segment area and x_i and y_i the coordinates to the centroid of any (straight line) segment of the weld group. Then, the centroid C of the weld group is located at

$$\bar{x} = \frac{\sum A_i x_i}{\sum A_i}, \quad \bar{y} = \frac{\sum A_i y_i}{\sum A_i} \tag{15.50}$$

in which $i = 1. 2,\ldots, n$ for n welds. In the case of symmetric weld group, the location of the centroid is obvious.

15.16.2.2 Moments of Inertia of a Weld (Figure 15.29)

For simplicity, we assume that the effective weld width in the plane of the paper is the same as throat length $t = 0.707h$, shown in Figure 15.26a. The parallel axis theorem can be applied to find the moments of inertia about x and y axes through the centroid of the weld group:

$$\begin{aligned} I_x &= I_{x'} + Ay_1^2 = \frac{tL^3}{12} + Lty_1^2 \\ I_y &= I_{y'} + Ax_1^2 = \frac{Lt^3}{12} + Ltx_1^2 = Ltx_1^2 \end{aligned} \tag{15.51}$$

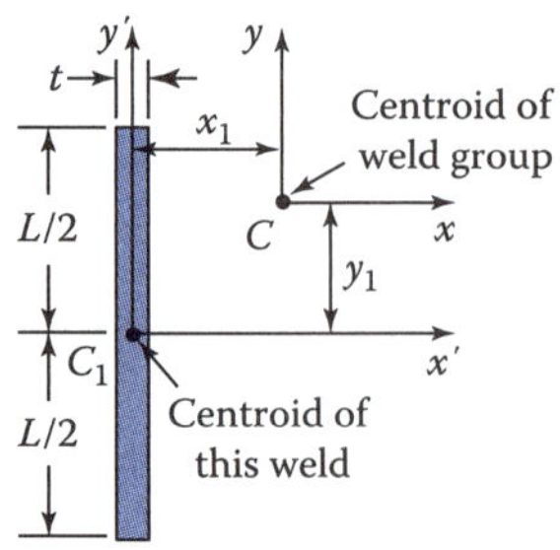

FIGURE 15.29
Moments of inertia of a weld parallel to the y axis.

Note that t is assumed to be *very small* in comparison with the other dimensions and hence $I_{y'} = Lt^3/12 = 0$ in the second of the preceding equations. The polar moment of inertia about an axis through C perpendicular to the plane of the weld is then

$$J = I_x + I_y = \frac{tL^3}{12} + Lt\left(x_1^2 + y_1^2\right) \tag{15.52}$$

The values of I and J for each weld about C should be calculated by using Equations 15.51 and 15.52: the results are added to obtain the moment and product of inertia of the entire joint. It should be mentioned that the moment and polar moment of inertias for the most common fillet welds encountered are listed in some publications [11]. The detailed procedure is illustrated in Case Study 18.9 and in the following sample problem.

Example 15.13: Design of a Welded Joint under Out-of-Plane Eccentric Loading

A welded joint is subjected to out-of-plane eccentric force P (Figure 15.28). What weld size is required?

Given: $L_1 = 60$ mm, $L_2 = 90$ mm, $e = 50$ mm, $P = 15$ kN

Assumption: An E6010 welding rod with factor of safety $n = 3$ is used.

Solution

By Table 15.8, for E6010, $S_y = 345$ MPa. The centroid lies at the intersection of the two axes of symmetry of the area enclosed by the weld group. The moment of inertia is

$$I_x = 2(60)t(45)^2 + \frac{2(90)^3 t}{12} = 364{,}500t \text{ mm}^4$$

The total weld area equals $A = 2(60t + 90t) = 300t$ mm². Moment is $M = 15(50) = 750$ kN mm. The maximum shear stress, using Equation 15.49, is

$$\tau = \left[\left(\frac{15{,}000}{300t}\right)^2 + \left(\frac{750{,}000 \times 45}{364{,}500t}\right)^2\right]^{1/2} = \frac{105.2}{t} \text{ N/mm}^2$$

Applying Equation 15.44, we have

$$n\tau = 0.5S_y; \quad 3\left(\frac{105.2}{t}\right) = 0.5(345) \quad \text{or} \quad t = 1.83 \text{ mm}$$

Hence,

$$h = \frac{t}{0.707} = \frac{1.83}{0.707} = 2.59 \text{ mm}$$

Comment: A nominal size of 3 mm fillet welds should be used throughout.

15.17 Brazing and Soldering

Brazing and soldering differ from welding essentially in that the temperatures are always below the melting point of the parts to be united, but the parts are heated above the melting point of the solder. It is important that the surfaces initially be clean. Soldering or brazing filler material acts somewhat similar to a molten metal glue or cement, which sets directly on cooling. Brazing or soldering can thus be categorized as bonding.

15.17.1 Brazing Process

Brazing starts with heating the workpieces to a temperature above 450°C. On contact with the parts to be united, the filler material melts and flows into the space between the workpieces. The filler materials are customarily alloys of copper, silver, or nickel. These may be handheld and fed into the joint (free of feeding) or preplaced as washers, shims, rings, slugs, and the like. Dissimilar metals, cast, and wrought metals, as well as nonmetals and metals, can be brazed together. Brazing is ordinarily accomplished by heating parts with a torch or in a furnace. Sometimes, other brazing methods are used. A brief description of some processes of brazing follows. Note that, in all metals, either flux or an inert gas atmosphere is required.

Torch brazing utilizes acetylene, propane, and other fuel gas, burned with oxygen or air. It may be manual or mechanized. On the other hand, *furnace* brazing uses the heat of a gas-fired, electric, or other kind of furnace to raise the parts to brazing temperature. A technique that utilizes a high-frequency current to generate the required heat is referred to as *induction* brazing. As the name suggests, *dip* brazing involves the immersion of the parts in a molten bath. A method that utilizes resistance-welding machines to supply the heat is called *resistance* brazing. As currents are large, water cooling of electrodes is essential.

15.17.2 Soldering Process

The procedure of soldering is identical to that of brazing. However, in soldering, the filler metal has a melting temperature below 450°C and relatively low strength. Heating can be done with a torch or a high-frequency induction heating coil. Surfaces must be clean and covered with flux that is liquid at the soldering temperature. The flux is drawn into the joint and dissolves any oxidation present at the joint. When the soldering temperature is reached, the solder replaces the flux at the joint.

Cast iron, wrought iron, and carbon steels can be soldered to each other or to brass, copper, nickel, silver, Monel, and other nonferrous alloys. Nearly all solders are tin–lead alloys, but alloys including antimony, zinc, and aluminum are also employed. The strength of a soldered union depends on numerous factors, such as the quality of the solder, thickness of the joint, smoothness of the surfaces, kind of materials soldered, and soldering temperature. Some common soldering applications involve electrical and electronic parts, sealing seams in radiators and in thin cans.

15.18 Adhesive Bonding

Adhesives are substances able to hold materials together by surface attachment. Nearly all structural adhesives are thermosetting as opposed to thermoplastic or heat-softening types, such as rubber cement and hot metals. Epoxies and urethanes are versatile and in widespread use as the structural adhesives [6]. Numerous other adhesive materials are used for various applications. Some remain liquid in the presence of oxygen, but they harden in restricted spaces, such as on bolt threads or in the spaces between a shaft and hub. *Adhesive bonding* is extensively utilized in the automotive and aircraft industries. Retaining compounds of adhesives can be employed to assemble cylindrical parts formerly needing press or shrink fits. In such cases, they eliminate press-fit stresses and reduce machining costs. Ordinary engineering adhesives have shear strengths varying from 25 to 40 MPa. The website at www.3m.com/bonding includes information and data on adhesives.

The *advantages* of adhesive bonding over mechanical fastening include the capacity to bond alike and dissimilar materials of different thickness, economic and rapid assembly, insulating characteristics, weight reduction, vibration dumping, and uniform stress distribution. On the other hand, examples of the *disadvantages* of the adhesive bonding are the preparation of surfaces to be connected, long cure times, possible need for heat and pressure for curing, service temperature sensitivity, service deterioration, tendency to creep under prolonged loading, and questionable long-term durability. The upper service temperature of most ordinarily employed adhesives is restricted to about 400°F. However, simpler, cheaper, stronger, and more easy-to-apply adhesives can be expected in the future.

15.18.1 Design of Bonded Joints

A design technique of rapidly growing significance is metal-to-metal adhesive bonding. Organic materials can be bonded as well. In cementing together metals, specific adhesion becomes important, inasmuch as the penetration of adhesive into the surface is insignificant. A number of metal-to-metal adhesives have been refined, but their use has been confined mainly to lap or spot joints of relatively limited area. Metal-to-metal adhesives, as employed in making plymetal, have practical applications.

Three common methods of applying adhesive bonding are illustrated in Table 15.10. Here, based on an approximate analysis of joints, stresses are assumed to be uniform over the bonded surfaces. The actual stress distribution varies over the area with *aspect ratio* b/L. The highest and lowest stresses occur at the edges and in the center, respectively. Adhesive joints should be properly designed to support only shear or compression and very small tension. Connection geometry is most significant when relatively high-strength materials are united. Large bond areas are recommended, such as in a lap joint (case A of the table), particularly connecting the metals. Nevertheless, this shear joint has noteworthy stress concentration of about 2 at the ends for an aspect ratio of 1.

It should be pointed out that the lap joints may be inexpensive because no preparation is required except, possibly, surface cleaning, while the machining of a scarf joint is impractical. The exact stress distribution depends on the thickness and elasticity of the joined members and adhesives. Stress concentration can arise because of the abrupt angles and changes in material properties. Load eccentricity is an important aspect in the state of stress of a single lap joint. In addition, often, the residual stresses associated with the mismatch in coefficient of thermal expansion between the adhesive and adherents may be significant [18,19].

TABLE 15.10

Some Common Types of Adhesive Joints

Configuration	Average Stress
A. Lap 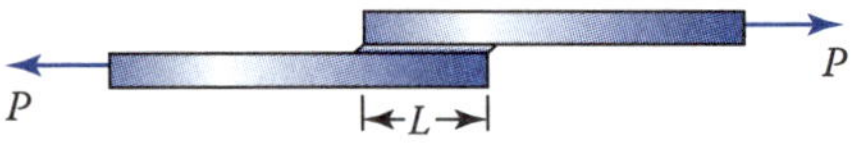	$\tau = \frac{P}{bL}$
B. Double lap 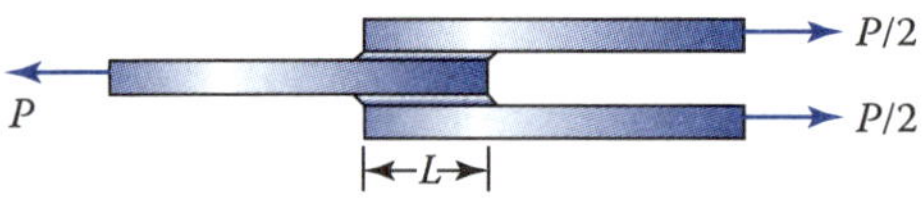	$\tau = \frac{P}{2bL}$
C. Scarf 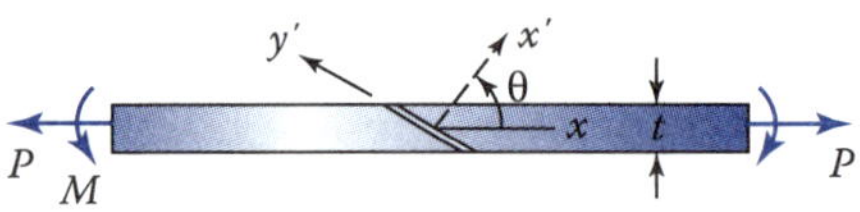	Axial Loading $\sigma_{x'} = \frac{P}{bt}\cos^2\theta, \quad \tau_{x'y'} = -\frac{P}{2bt}\sin 2\theta$ Bending $\sigma_{x'} = \frac{6M}{bt^2}\cos^2\theta, \quad \tau_{x'y'} = -\frac{3M}{bt^2}\sin 2\theta$

Notes: P, centric load; M, moment; b, width of plate; t, thickness of thinnest plate; and L, length of lap.

Problems

Sections 15.1 through 15.7

15.1 A power screw is 75 mm in diameter and has a thread pitch of 15 mm. Determine the thread depth, the thread width or the width at pitch line, the mean and root diameters, and the lead, for the case in which

a. Square threads are used

b. Acme threads are used

15.2 A 1½ in. diameter, double-thread Acme screw is to be used in an application similar to that of Figure 15.6. Determine

a. The screw lead, mean diameter, and helix angle

b. The starting torques for lifting and lowering the load

c. The efficiency, if collar friction is negligible

d. The force F to be exerted by an operator, for $a = 15$ in.

Given: $f = 0.1, \quad f_c = 0.08, \quad d_c = 2$ in., $\quad W = 1.5$ kips

15.3 What helix angle would be required so that the screw of Problem 15.2 would just begin to overhaul? What would be the efficiency of a screw with this helix angle, for the case in which the collar friction is negligible?

15.4 A 32 mm diameter power screw has a double-square thread with a pitch of 4 mm. Determine the power required to drive the screw.

Design Requirement: The nut is to move at a velocity of 40 mm/s and lift a load of $W = 6$ kN.

Given: The mean diameter of the collar is 50 mm. Coefficients of friction are estimated as $f = 0.1$ and $f_c = 0.15$.

15.5 A square-thread screw has a mean diameter of 1¾ in. and a lead of $L = 1$ in. Determine the coefficient of thread friction.

Given: The screw consumes 5 hp when lifting a 2 kips weight at the rate of 25 fpm.

Design Assumption: The collar friction is negligible.

15.6 A 2¾ in. diameter square-thread screw is used to lift or lower a load of $W = 50$ kips at a rate of 2 fpm. Determine

a. The revolutions per minute of the screw

b. The motor horsepower required to lift the weight, if the screw efficiency is $e = 85\%$ and $f = 0.15$

Design Assumption: Because the screw is supported by a thrust ball bearing, the collar friction can be neglected.

15.7 A square-threaded power screw (Figure 15.4b) with single thread lifts a load. The given numerical values are as follows: the mean screw diameter is $d = 24$ mm, the pitch is $p = 6$ mm, the collar diameter is $d_c = 36$ mm, the coefficient of friction for thread and collar is $f_c = f = 0.11$, and the load is $W = 100$ kN. Find

a. The major diameter of the screw and the value of the screw torque needed to lift the load

b. For the case in which $f_c = 0$, the minimum value of the f to prevent screw from overhauling

15.8 A triple-threaded Acme screw of the major diameter $d = 50$ mm and pitch $p = 8$ mm is used in a jack with a plain thrust collar of mean diameter $d_c = 68$ mm (Figure 15.6). Find

a. The lead, thread depth, mean pitch diameter, and helix angle of the screw

b. The starting torque for lowering a load of $W = 15$ kN

Assumption: Coefficients of running friction are $f_c = 0.12$ and $f = 0.13$.

15.9 Reconsider Problem 15.8, knowing that the screw is lifting a load of $W = 15$ kN at a rate of 0.02 mps. Find

a. The efficiency of the jack in this situation

b. Weather the screw overhauls when a ball thrust bearing with $f_c = 0$ is used in place of the plain thrust bearing

15.10 A square-thread screw has an efficiency of 70% when lifting a weight. Determine the torque that a brake mounted on the screw must exert when lowering the load at a uniform rate.

Given: The coefficient of thread friction is estimated as $f = 0.12$ with collar friction negligible; the load is 50 kN and the mean diameter is 30 mm.

15.11 Determine the pitch that must be provided on a square-thread screw to lift a 2.5 kips weight at 40 fpm with power consumption of 5 hp.

Given: The mean diameter is 1.875 in. and $f = 0.15$.

Design Assumption: The collar friction is negligible.

15.12 A 1 in.–8 UNC screw supports a tensile of 12 kips. Determine

a. The axial stress in the screw

b. The minimum length of nut engagement, if the allowable bearing stress is not to exceed 10 ksi

c. The shear stresses in the nut and screw

15.13 A 50 mm diameter square-thread screw having a pitch of 8 mm carries a tensile load of 15 kN. Determine

a. The axial stress in the screw

b. The minimum length of nut engagement needed, if the allowable bearing stress is not to exceed 10 MPa

c. The shear stresses in the nut and screw

Sections 15.8 through 15.12

15.W1 Search the website at www.nutty.com. Perform a product search for various types of nuts, bolts, and washers. Review and list 15 commonly used configurations and descriptions of each of these elements.

15.W2 Use the site at www.boltscience.com to review the current information related to bolted joint technology. List three usual causes of relative motion of threads.

15.14 The joint shown in Figure P15.14 has a 15 mm diameter bolt and a grip length of L = 50 mm. Calculate the maximum load that can be carried by the part without losing all the initial compression in the part.

Given: The tightening torque of the nut for average condition of thread friction is 72 N m by Equation 15.21.

15.15 The bolt of the joint shown in Figure P15.14 is ⅞ in. –9 UNC, SAE grade 5, with a rolled thread. Apply the Goodman criterion to determine

a. The permissible value of preload F_i if the bolt is to be safe for continuous operation with $n = 2$

b. The tightening torque for an average condition of thread friction

Given: The value of load P on the part ranges continuously from 8 to 16 kips; the grip is L = 2 in. The survival rate is 95%.

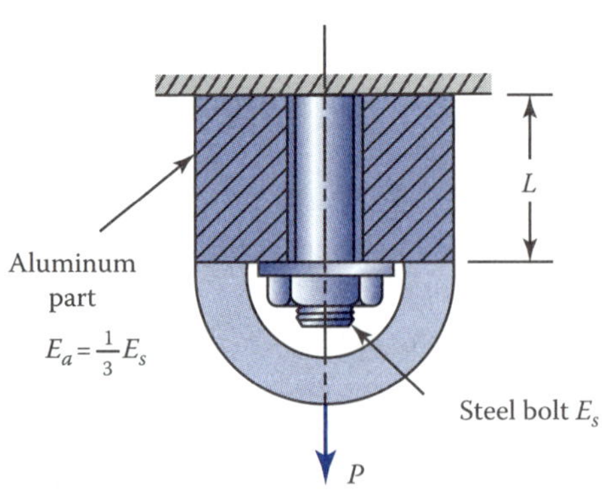

FIGURE P15.14

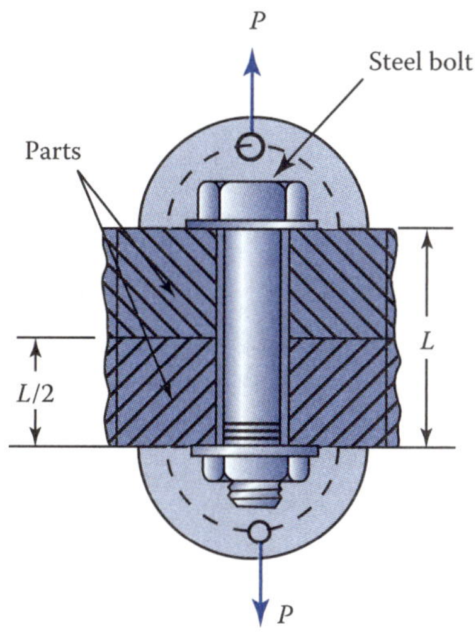

FIGURE P15.17

15.16 The bolt of the joint depicted in Figure P15.14 is $M20 \times 2.5$-C, grade 10.9, with cut thread, $S_y = 620$ MPa, and $S_u = 750$ MPa. Calculate

a. The maximum and minimum values of the fluctuating load P on the part, on the basis of the Soderberg theory

b. The tightening torque, if bolt is lubricated

Given: The grip is $L = 50$ mm; the preload equals $F_i = 25$ kN; the average stress in the root of the screw is 160 MPa; the survival rate equals 90%.

Design Requirement: The safety factor is 2.2. The operating temperature is not elevated.

15.17 A bolted connection has been tightened by applying torque T to the nut to produce an initial preload force $F = 4.2$ kN in the $M5 \times 0.8$-C, ISO grade 4.6 steel bolt, and resists an external load of $P = 5$ kN (Figure P15.17). The clamped parts each is $L/2 = 20$ mm thick. Stiffness of parts k_p is four times the bolt stiffness k_b.

Determine:

a. The tension in the bolt and compression in the parts when the load is applied

b. Whether the parts will separate or remain in contact under the load

15.18 A bolted joint with two class number 4.8 steel bolts is to support an external load of $P = 11.6$ kN, and the stiffness constant ratio is to be $k_p/k_b = 3$ (Figure P15.18). Determine

a. The required bolt preload

b. The thread size of the bolt

15.19 Figure P15.19 depicts a partial section from a permanent connection. Determine

a. The total force and stress in each bolt

b. The tightening torque for an average condition of thread friction

Given: A total of six bolts are used to resist an external load of $P = 18$ kips.

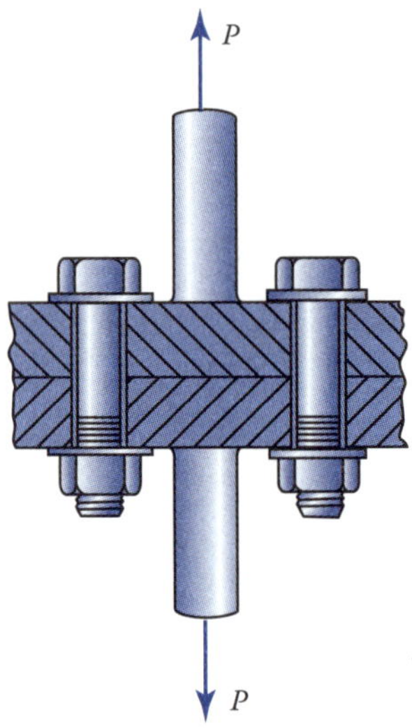

FIGURE P15.18

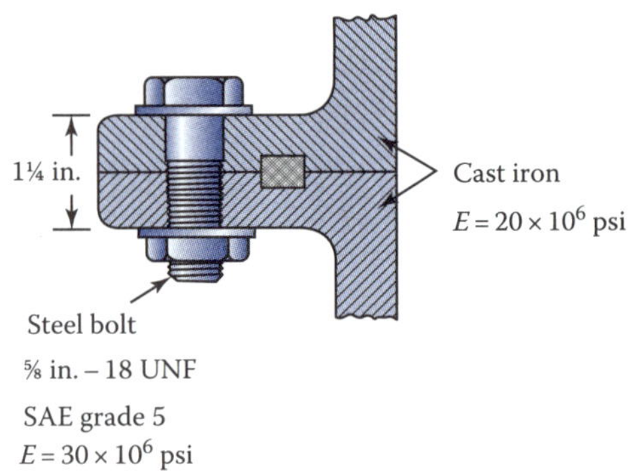

FIGURE P15.19

15.20 A section of the connection illustrated in Figure P15.19 carries an external load that fluctuates between 0 and 4 kips. Using the Goodman criterion, determine the factor of safety n guarding against the fatigue failure of the bolt.

15.21 A joint of two steel parts with a single cold-rolled steel bolt supports an external load P that alternates continuously between 0 and 35 kN (Figure P15.17). Find

a. The minimum required preload when compression of the two parts is lost

b. The minimum force in the parts for the alternating load, knowing that the preload equals $F_i = 38$ kN

Assumption: Clamped parts have stiffness k_p four times the bolt stiffness k_b.

15.22 Redo Problem 15.21 for the case in which the clamped parts have stiffness k_p three times the bolt stiffness k_b.

15.23 A connection of two steel plates and a steel bolt has an initial compression force of F_i = 2600 lb (Figure P15.17). Clamping plates have stiffness k_p three times the bolt stiffness k_b. Determine

a. The external force P that would reduce the clamping force to 500 lb

b. Knowing that P is repeatedly applied and removed, the mean force P_m and alternating force P_a applied on the bolt

15.24 A joint of two steel plates and a steel bolt has an initial compression force F_i = 8000 N (Figure P15.17). The clamped plates have stiffness k_p two times the bolt stiffness k_b. Find

a. The external separating force that would reduce the clamping force to 600 N

b. When the force P is repeatedly applied and removed, the mean force P_m and alternating force P_a on the bolt

15.25 Figure P15.17 shows a connection of two plates and a steel bolt with initial compression force of F_i = 6000 N. The clamped plates have stiffness k_p four times the bolt stiffness k_b. Find

a. The external separating force P that would reduce the clamping force to 1800 N

b. For the case in which the force P is repeatedly applied and removed, the mean force P_m and alternating force P_a applied to the bolt

15.26 The bolt of connection shown in Figure P15.26 is $M20 \times 2.5$. ISO course thread having S_y = 630 MPa. Determine

a. The total force on the bolt, if the joint is reusable

b. The tightening torque, if the bolts are lubricated

Given: The grip is L = 60 mm; the joint carries an external load of P = 40 kN.

Design Assumption: The bolt will be made of steel of modulus of elasticity E_s, and the parts are cast iron with modulus of elasticity $E_c = E_s/2$.

15.27 The connection shown in Figure P15.26 carries an external loading P that value ranges from 1 to 5 kips. Determine

a. If the bolt fails without preload

b. Whether the bolt is safe when preload present

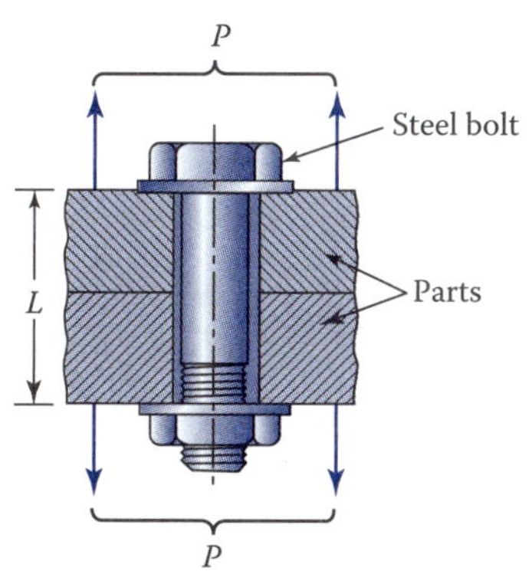

FIGURE P15.26

c. The fatigue factor of safety n when preload is present

d. The load factor n_s guarding against joint separation

Design Assumptions: The bolt is made of steel (E_s), and the parts are cast iron with modulus of elasticity $E_c = E_s/2$. The operating temperature is normal. The bolt may be reused when the joint is taken apart. The survival rate is 90%.

Given: The steel bolt is ½ in.–13 UNC, SAE grade 2, with rolled threads; the grip is $L = 2$ in.

15.28 The assembly shown in Figure P15.26 uses an $M14 \times 2$, ISO grade 8.8 course cut threads. Apply the Goodman criterion to determine the fatigue safety factor n of the bolt with and without initial tension.

Given: The joint constant is $C = 0.31$. The joint carries a load P varying from 0 to 10 kN. The operating temperature is 490°C maximum.

Design Assumptions: The bolt may be reused when the joint is taken apart. Survival probability is 95%.

15.29 Determine the maximum load P the joint described in Problem 15.28 can carry based on a static safety factor of 2.

Design Assumptions: The joint is reusable.

15.30 Figure P15.30 shows a portion of a high-pressure boiler accumulator having flat heads. The end plates are affixed using a number of bolts of $M16 \times 2$-C, grade 5.8, with rolled threads. Determine

a. The factor of safety n of the bolt against fatigue failure with and without preload

b. The load factor n_s, against joint separation

Given: The fully modified endurance limit is $S_e = 100$ MPa. External load P varies from 0 to 12 kN/bolt.

Design Assumptions: Clamped parts have a stiffness k_p, five times the bolt stiffness k_b. The connection is permanent.

Sections 15.13 and 15.14

15.31 A double-riveted lap joint with plates of thickness t is to support a load P as shown in Figure P15.31. The rivets are 19 mm in diameter and spaced 50 mm apart in each row. Determine the shear, bearing, and tensile stresses.

Given: $P = 32\,\text{kN}$, $t = 10\,\text{mm}$

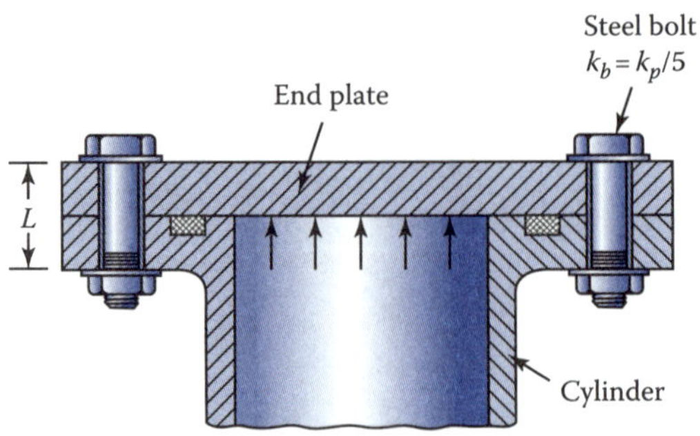

FIGURE P15.30

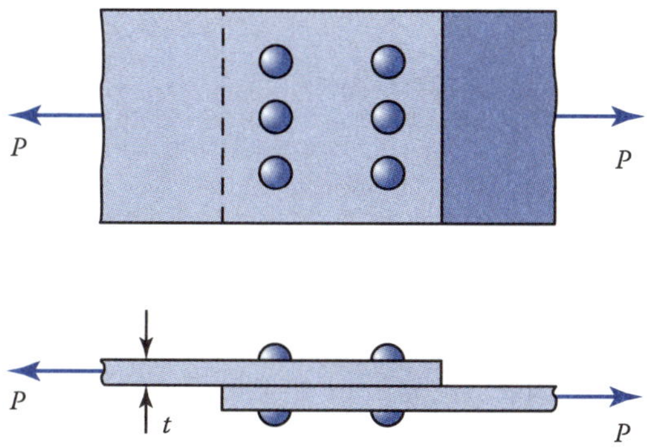

FIGURE P15.31

15.32 A double-riveted longitudinal lap joint (Figure P15.31) is made of plates of thickness t. Determine the efficiency of the joint.

Given: The ¾ in. diameter rivets have been drilled 2½ in. apart in each row and t = ⅜ in.

Design Assumptions: The allowable stresses are 22 ksi in tension, 15 ksi in shear, and 48 ksi in bearing.

15.33 Figure P15.33 shows a bolted lap joint that uses ⅝ in.–11 UNC, SAE grade 8 bolts. Determine the allowable value of the load P, for the following safety factors: 2, shear on bolts; 3, bearing of bolts; 2.5, bearing on members; and 3.5, tension of members.

Design Assumption: The members are made of cold-drawn AISI 1035 steel with $S_{ys} = 0.577S_y$.

15.34 The bolted connection shown in Figure P15.34 uses $M14 \times 2$ course pitch thread bolts having S_y = 640 MPa and S_{ys} = 370 MPa. A tensile load P = 20 kN is applied to the connection. The dimensions are in millimeters. Determine the factor of safety n for all possible modes of failure.

Design Assumption: Members are made of hot-rolled 1020 steel.

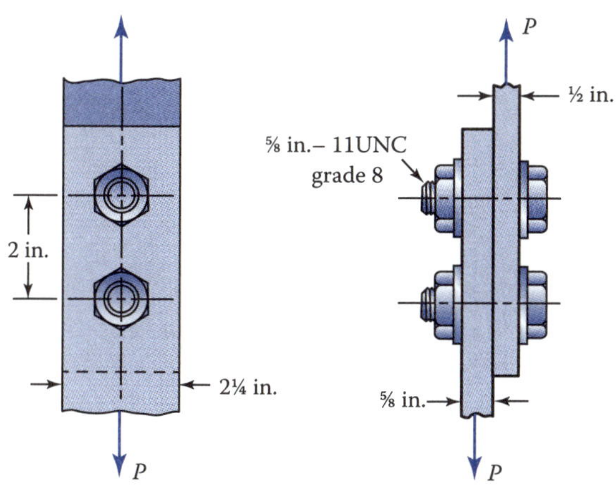

FIGURE P15.33

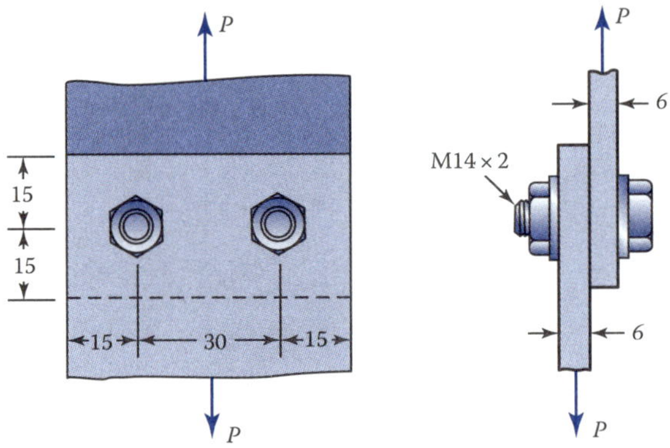

FIGURE P15.34

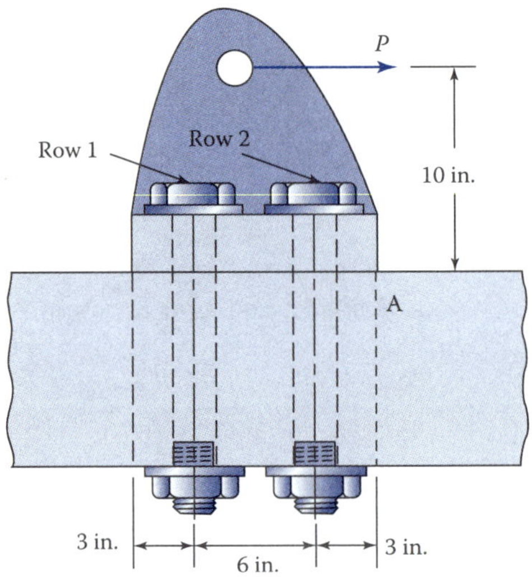

FIGURE P15.35

15.35 A machine part is fastened to a frame by means of ½ in.–13 UNC (Table 15.1) two rows of steel bolts, as shown in Figure P15.35. Each row also has two bolts. Determine the maximum allowable value of P.

Design Decisions: The allowable stresses for the bolts are 20 ksi in tension and 12 ksi in shear.

15.36 A narrow walkway bracket, bolted to a steel bridge as depicted in Figure P15.36, supports a maximum load of $P = 5$ kips with a safety factor on the basis of proof strength of $n = 2.3$. Find the required thread size of the bolt.

Given: Three SAE grade 5 steel bolts with proof strength of $S_p = 85$ ksi (from Table 15.4).

15.37 A machine bracket is attached with bolts that each must support a static load of $P = 500$ lb. Find, based on a safety factor of $n = 3$,

a. The size SAE grade 2 UNC thread bolt needed

b. The nut length (apply Equation 15.19a of Case Study 18.7)

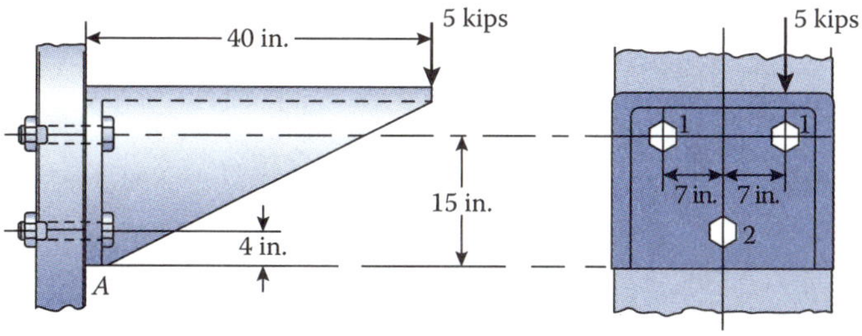

FIGURE P15.36

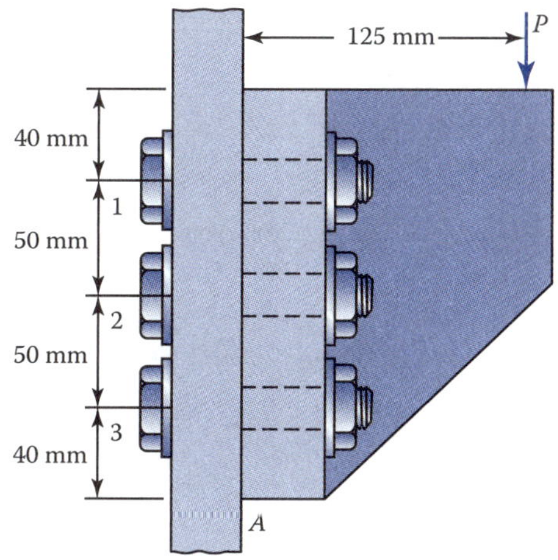

FIGURE P15.39

Assumption: The loads are equally distributed between the threads, and a stress concentration factor of $K_t = 4$ will be used.

15.38 Repeat Problem 15.37, based on a safety factor of $n = 5$ on the proof strength and using a load of $P = 800$ lb.

15.39 Three $M20 \times 2.5$ coarse-thread steel bolts (Table 15.2) are used to connect a part to a vertical column, as shown in Figure P15.39. Calculate the maximum allowable value of P.

Design Decisions: The allowable stresses for the bolt are 145 MPa in tension and 80 MPa in shear.

15.40 A riveted structural connection supports a load of 10 kN, as shown in Figure P15.40. What is the value of the force on the most heavily loaded rivet in the bracket? Determine the values of the shear stress for 20 mm rivets and the bearing stress if the Gusset plate is 15 mm thick.

Given: The applied loading is $P = 10$ kN.

15.41 The riveted connection shown in Figure P15.41 supports a load P. Determine the distance d.

Design Decision: The maximum shear stress on the most heavily loaded rivet is 100 MPa.

Given: The applied loading equals $P = 50$ kN.

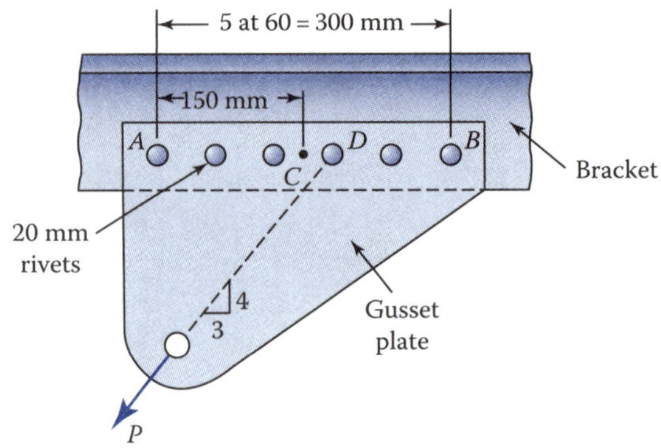

FIGURE P15.40

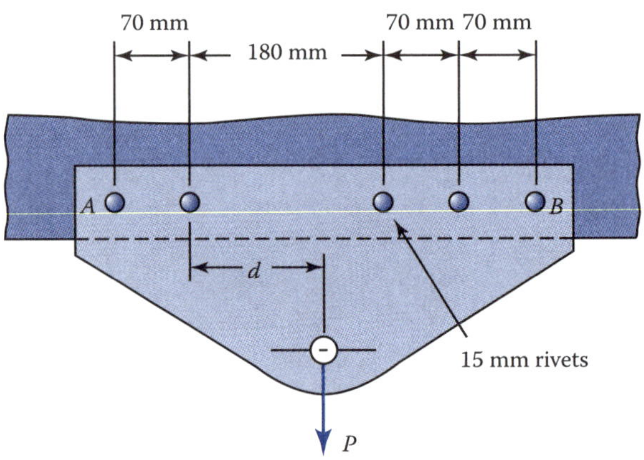

FIGURE P15.41

15.42 Determine the value of the load P for the riveted joint shown in Figure P15.41.

Design Assumption: The allowable rivet stress in shear is 100 MPa.

Given: d = 90 mm.

Sections 15.15 through 15.18

15.43 The plates in Figure 15.26a are 10 mm thick × 40 mm wide and made of steel having S_y = 250 MPa. They are welded together by a fillet weld with h = 1 mm leg. L = 60 mm long, S_y = 350 MPa, and S_{ys} = 200 MPa. Using a safety factor of 2.5 based on yield strength, determine the load P that can be carried by the joint.

15.44 Two ⅝ in. thick AISI 1050 normalized steel plates are butt welded using AWS, number E6020, welding rods (Figure 15.25a). The weld length equals L = 3.5 in. Compute the maximum tensile load that can be applied to the connection with a factor of safety of n = 4.

15.45 Resolve Problem 15.44, knowing that the plate is in shear loading (as shown in Figure 15.25b) and the factor of safety is n = 3.

15.46 Two AISI 1035 CD steel plates are double-fillet welded using AWS E6012 steel rods (as shown in Figure P15.46). For the weld dimensions of h = 8 mm and L = 70 mm, what is the largest tensile load P that can be applied on the basis of a safety factor of n = 3.5?

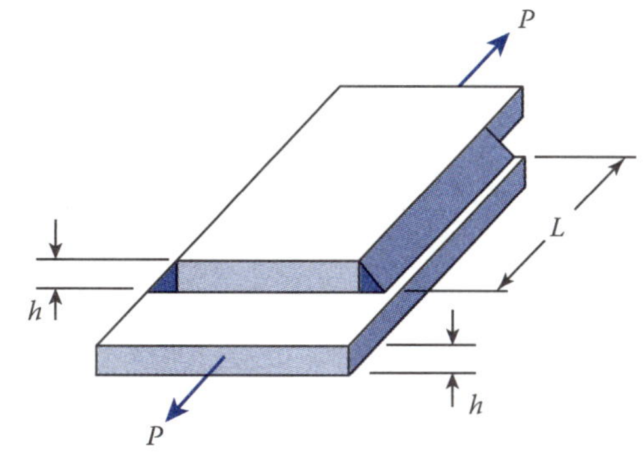

FIGURE P15.46

15.47 Two AISI 1050 HR steel plates are welded with a double fillet using AWS E6014 steel rods (Figure P15.46). The weld dimensions are $h = 5$ mm and $L = 60$ mm. Find the largest tensile load P that can be applied with a factor of safety of $n = 4$.

15.48 Determine the lengths L_1 and L_2 of welds for the connection of a 75 × 10 mm steel plate ($\sigma_{all} = 140$ MPa) to a machine frame (Figure P15.48).

Given: 12 mm fillet welds having a strength of 1.2 kN per linear millimeter.

15.49 Calculate the required weld size for the bracket in Figure 15.28 if a load $P = 3$ kips is applied with eccentricity $e = 10$ in.

Design Assumptions: 8 ksi is allowed in shear: $L_1 = 4$ in. and $L_2 = 5$ in.

15.50 Resolve Problem 15.49 if the load P varies continuously from 2 to 4 kips. Apply the Goodman criterion.

Given: $S_u = 60$ ksi, $n = 2.5$.

15.51 Determine the required length of weld L in Figure P15.51 if an E7014 electrode is used with a safety factor $n = 2.5$.

Given: $P = 100$ kN, $a = 60$ mm, $h = 12$ mm.

15.52 Resolve Problem 15.51 if the load P varies continuously between 80 and 120 kN.

Design Decision: Use the Soderberg criterion.

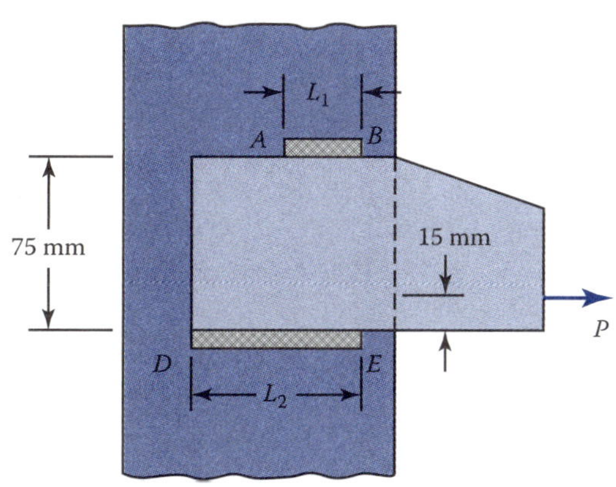

FIGURE P15.48

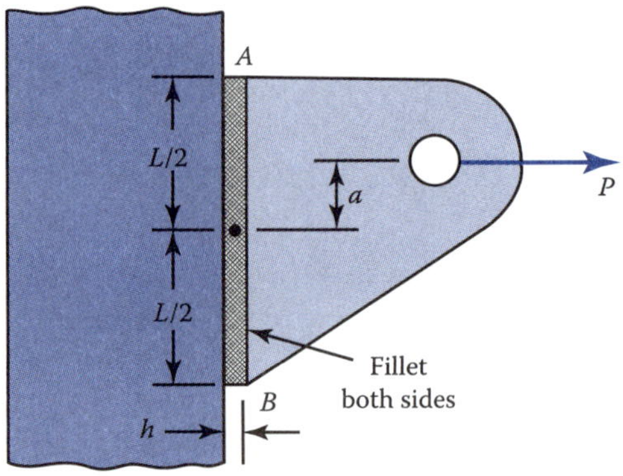

FIGURE P15.51

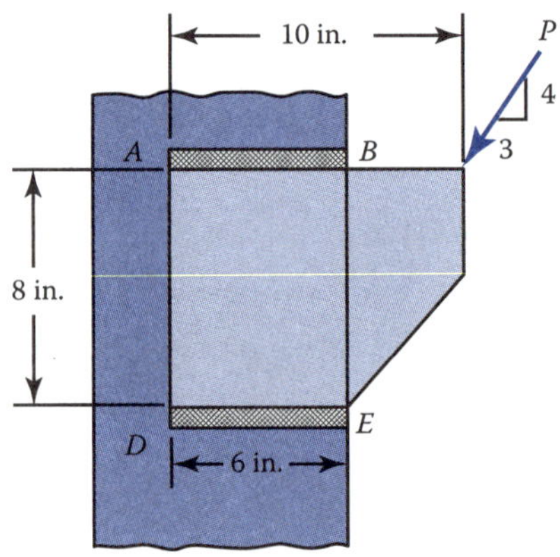

FIGURE P15.54

15.53 Load P in Figure P15.51 varies continuously from 0 to P_{max}. Determine the value of P_{max} if an E6010 electrode is used, with a safety factor of 2. Apply the Goodman theory.

Given: $a = 3$ in., $L = 10$ in., $h = ¼$ in.

15.54 Calculate the size h of the two welds required to attach a plate to a frame as shown in Figure P15.54 if the plate supports an inclined force $P = 10$ kips.

Design Decisions: Use $n = 3$ and $S_y = 50$ ksi for the weld material.

15.55 The value of load P in Figure P15.54 ranges continuously between 2 and 10 kips. Using $S_u = 60$ ksi and $n = 1.5$, determine the required weld size. Employ the Goodman criterion.

16

Miscellaneous Mechanical Components

16.1 Introduction

In the class of axisymmetrically loaded members, the basic problem may be defined in terms of the radial coordinate. Typical examples are thick-walled cylinders, flywheels, press and shrink fits, curved beams subjected to pure bending, and thin-walled cylinders. This chapter concerns mainly *exact* stress distribution in this group of machine and structural members. The methods of the mechanics of materials and the theory of elasticity are applied. Consideration is given to thermal and plastic stresses, the material strength, and an appropriate theory of failure to obtain a safe and reliable design in Sections 16.6 and 16.7. Following this, bending of curved frames is taken up. We also discuss briefly symmetric bending of circular plates, axisymmetrically loaded shells, and filament-wound cylinders in Sections 16.9 through 16.13.

The buckling of thin-walled cylinders under axial compression and critical pressures in vessels is treated in the concluding section. There are several other problems of practical interest dealing with axisymmetric stress and deformation in a member. Among these are various situations involving rings reinforcing a juncture, hoses, semicircular barrel vaults, torsion of circular shafts of variable diameter, local stresses around a spherical cavity, and pressure between two spheres in contact (discussed in Section 8.7). For more detailed treatment of the members with axisymmetric loading, see, for example, [1–4].

16.2 Basic Relations

In the cases of axially loaded members, torsion of circular bars, and pure bending of beams, simplifying assumptions associated with deformation patterns are made so that strain (and stress) distribution for a cross section of each member can be ascertained. A basic hypothesis has been that plane sections remain plane subsequent to the loading. However, in axisymmetric and more complex problems, it is usually impossible to make similar assumptions regarding deformation. So analysis begins with consideration of a general infinitesimal element. Hooke's law is stated, and the solution is found after stresses acting on any element and its displacements are known. At the boundaries of a member, the equilibrium of known forces (or prescribed displacement) must be satisfied by the corresponding infinitesimal elements.

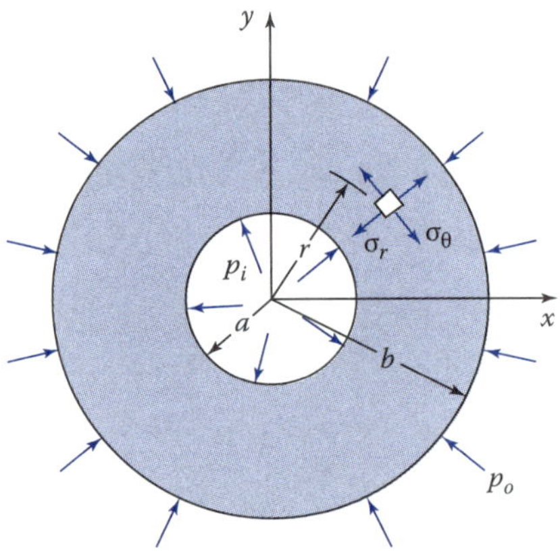

FIGURE 16.1
Thick-walled cylinder.

Here, we present the basic relations of an axially symmetric 2D problem referring to the geometry and notation of the thick-walled cylinder (Figure 16.1). The inside radius of the cylinder is a and the outside radius is b. The tangential stresses σ_θ and the radial stresses σ_r in the wall at a distance r from the center of the cylinder are caused by pressure. A typical infinitesimal element of unit thickness isolated from the cylinder is defined by two radii, r and $r + \mathrm{d}r$, and an angle $\mathrm{d}\theta$, as shown in Figure 16.2. The quantity F_r represents the radial body force per unit volume. The conditions of symmetry dictate that the stresses and deformations are independent of the angle θ and that the shear stresses must be 0. Note that the radial stresses acting on the parallel faces of the element differ by $\mathrm{d}\sigma_r$, but the tangential stresses do not vary among the faces of the element. There can be no tangential displacement in an axisymmetrically loaded member of revolution; that is, $\upsilon = 0$. A point represented in the element has *radial* displacement u as a consequence of loading.

It can be demonstrated that [2] Equation 3.53 (in the absence of body forces), 3.55, and 2.6 can be written in polar coordinates as given in the following outline.

Equation of Equilibrium

$$\frac{\mathrm{d}\sigma_r}{\mathrm{d}r} + \frac{\sigma_r - \sigma_\theta}{r} = 0 \tag{16.1}$$

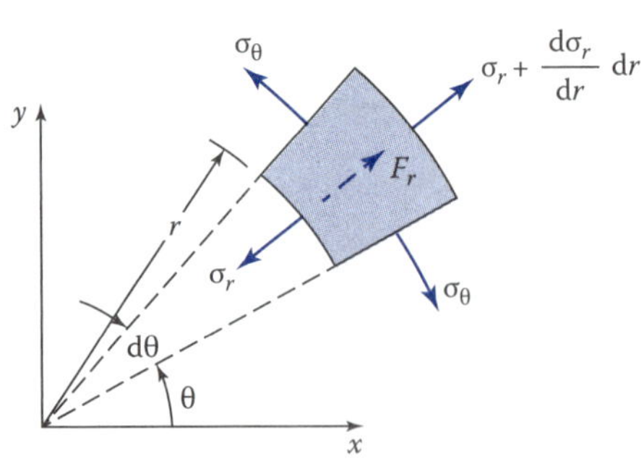

FIGURE 16.2
Stress element of unit thickness.

Strain–Displacement Relations

$$\varepsilon_r = \frac{du}{dr}, \quad \varepsilon_\theta = \frac{u}{r} \tag{16.2}$$

and the shear strain $\gamma_{r\theta} = 0$. Here, ε_θ and ε_r are the tangential strain and radial strain, respectively. Substitution of the second into the first of Equation 16.2 gives a simple *compatibility condition* among the strains. This ensures the geometrically possible form of variation of strains from point to point within the member.

Hooke's Law

$$\varepsilon_r = \frac{1}{E}(\sigma_r - \nu\sigma_\theta), \quad \varepsilon_\theta = \frac{1}{E}(\sigma_\theta - \nu\sigma_r) \tag{16.3}$$

The quantity E represents the modulus of elasticity, and ν is Poisson's ratio. The foregoing governing equations are sufficient to obtain a unique solution to a 2D axisymmetric problem with specific *boundary conditions*. Applications to thick-walled cylinders, rotating disks, and pure bending of curved beams are illustrated in sections that follow.

16.3 Thick-Walled Cylinders under Pressure

The circular cylinder is usually divided into thin-walled and thick-walled classifications. In a thin-walled cylinder, the tangential stress may be regarded as constant with thickness, as discussed in Section 3.4. When the wall thickness exceeds the inner radius by more than 10%, the cylinder is usually considered *thick walled*. For this case, the variation of stress with radius can no longer be neglected. Thick-walled cylinders, which we deal with here, are used extensively in industry as pressure vessels, storage tanks, hydraulic and pneumatic tubes, rolling-element bearings, or gears pressed into shafts, pipes, gun barrels, and the like.

16.3.1 Solution of the Basic Relations

In a thick-walled cylinder subjected to uniform internal or external pressure, the deformation is symmetrical about the axial (z) axis. The equilibrium condition and strain–displacement relations, Equations 16.1 and 16.2, apply to any point on a ring of unit length cut from the cylinder (Figure 16.1). When ends of the cylinder are open and unconstrained, so that $\sigma_z = 0$, the cylinder is in a condition of plane stress. Then, by Hooke's law (Equation 16.3), the strains are

$$\begin{aligned} \frac{du}{dr} &= \frac{1}{E}(\sigma_r - \nu\sigma_\theta) \\ \frac{u}{r} &= \frac{1}{E}(\sigma_\theta - \nu\sigma_r) \end{aligned} \tag{16.4}$$

The preceding equations give the radial and tangential stresses, in terms of the radial displacement:

$$\sigma_r = \frac{E}{1-\nu^2}(\varepsilon_r + \nu\varepsilon_\theta) = \frac{E}{1-\nu^2}\left(\frac{du}{dr} + \nu\frac{u}{r}\right)$$
$$\sigma_\theta = \frac{E}{1-\nu^2}(\varepsilon_\theta + \nu\varepsilon_r) = \frac{E}{1-\nu^2}\left(\frac{u}{r} + \nu\frac{du}{dr}\right) \tag{16.5}$$

Introducing this into Equation 16.1 results in the desired differential equation:

$$\frac{d^2u}{dr^2} + \frac{1}{r}\frac{du}{dr} - \frac{u}{r^2} = 0 \tag{16.6}$$

The solution of this *equidimensional equation* is

$$u = c_1 r + \frac{c_2}{r} \tag{16.7}$$

The stresses may now be expressed in terms of the constants of integration c_1 and c_2 by inserting Equation 16.7 into 16.5 as

$$\sigma_r = \frac{E}{1-\nu^2}\left[c_1(1+\nu) - c_2\left(\frac{1-\nu}{r^2}\right)\right] \tag{a}$$

$$\sigma_\theta = \frac{E}{1-\nu^2}\left[c_1(1+\nu) - c_2\left(\frac{1-\nu}{r^2}\right)\right] \tag{b}$$

16.3.2 Stress and Radial Displacement for Cylinder

For a cylinder under internal and external pressures p_i and p_o, respectively, the boundary conditions are

$$(\sigma_r)_{r=a} = -p_i, \quad (\sigma_r)_{r=b} = -p_o \tag{16.8}$$

In the foregoing, the negative signs are used to indicate compressive stress. The constants are ascertained by introducing Equation 16.8 into (a); the resulting expressions are carried into Equations 16.7, (a), and (b). In so doing, the *radial* and *tangential stresses* and *radial displacement* are obtained in the following forms:

$$\sigma_r = \frac{a^2p_i - b^2p_o}{b^2-a^2} - \frac{(p_i-p_o)a^2b^2}{(b^2-a^2)r^2} \tag{16.9}$$

$$\sigma_\theta = \frac{a^2p_i - b^2p_o}{b^2-a^2} - \frac{(p_i-p_o)a^2b^2}{(b^2-a^2)r^2} \tag{16.10}$$

$$u = \frac{1-\nu}{E}\frac{\left(a^2 p_i - b^2 p_o\right)}{b^2 - a^2} + \frac{1+\nu}{E}\frac{\left(p_i - p_o\right)a^2 b^2}{\left(b^2 - a^2\right)r^2} \tag{16.11}$$

These equations were first derived by French engineer G. Lame in 1833, for whom they are named. The maximum numerical value of σ_r occurs at $r = a$ to be p_i, provided that p_i exceeds p_o. When $p_o > p_i$, the maximum σ_r is found at $r = b$ and equals p_o. On the other hand, the maximum σ_θ occurs at either the inner or outer edge depending on the pressure ratio [1].

The *maximum shear stress* at any point in the cylinder, through the use of Equations 16.9 and 16.10, is found as

$$\tau_{max} = \frac{1}{2}\left(\sigma_\theta - \sigma_r\right) = \frac{\left(p_i - p_o\right)a^2 b^2}{\left(b^2 - a^2\right)r^2} \tag{16.12}$$

The largest value of this stress corresponds to $p_o = 0$ and $r = a$:

$$\tau_{max} = \frac{p_i b^2}{b^2 - a^2} \tag{16.13}$$

that occurs on the planes making an angle of 45° with the planes of the principal stresses (σ_r and σ_θ). The pressure p_y that initiates *yielding* at the inner surface, by setting $\tau_{max} = S_y/2$ in Equation 16.13, is

$$p_y = \frac{b^2 - a^2}{2b^2} S_y \tag{16.14}$$

where S_y is the tensile yield strength.

In the case of a pressurized *closed-ended cylinder*, the longitudinal stresses are in addition to σ_r and σ_θ. For a transverse section some distance from the ends, σ_z may be taken uniformly distributed over the wall thickness. The magnitude of the *longitudinal stress* is obtained by equating the net force acting on an end attributable to pressure loading to the internal *z*-directed force in the cylinder wall:

$$\sigma_z = \frac{p_i a^2 - p_o b^2}{b^2 - a^2} \tag{16.15}$$

where it is again assumed that the ends of the cylinder are not constrained. Also note that Equations 16.9 through 16.15 are applicable only away from the ends. The difficult problem of determining deformations and stresses near the junction of the thick-walled caps and the thick-walled cylinder lies outside of the scope of our analysis. This usually is treated by experimental approaches or by the finite element method, since its analytical solution depends on a general 3D study in the theory of elasticity. For thin-walled cylinders, stress in the vicinity of the end cap junctions is presented in Section 16.12.

16.3.3 Special Cases

16.3.3.1 Internal Pressure Only

In this case, $p_o = 0$, and Equations 16.9 through 16.11 become

$$\sigma_r = \frac{a^2 p_i}{b^2 - a^2}\left(1 - \frac{b^2}{r^2}\right) \tag{16.16a}$$

$$\sigma_\theta = \frac{a^2 p_i}{b^2 - a^2}\left(1 + \frac{b^2}{r^2}\right) \tag{16.16b}$$

$$u = \frac{a^2 p_i r}{E(b^2 - r^2)}\left[(1-\nu) + (1+\nu)\frac{b^2}{r^2}\right] \tag{16.16c}$$

Since $b/r \geq 1$, σ_r is always compressive stress and is maximum at $r = a$. As for σ_θ, it is always a tensile stress and also has a maximum at $r = a$:

$$\sigma_{\theta,\max} = p_i \frac{b^2 + a^2}{b^2 - a^2} \tag{16.17}$$

To illustrate the variation of stress and radial distance for the case of no external pressure, dimensionless stress and displacement are plotted against dimensionless radius in Figure 16.3 for $b/a = 4$.

16.3.3.2 External Pressure Only

For this case, $p_i = 0$, and Equations 16.9 through 16.11 simplify to

$$\sigma_r = -\frac{b^2 p_o}{b^2 - a^2}\left(1 - \frac{a^2}{r^2}\right) \tag{16.18a}$$

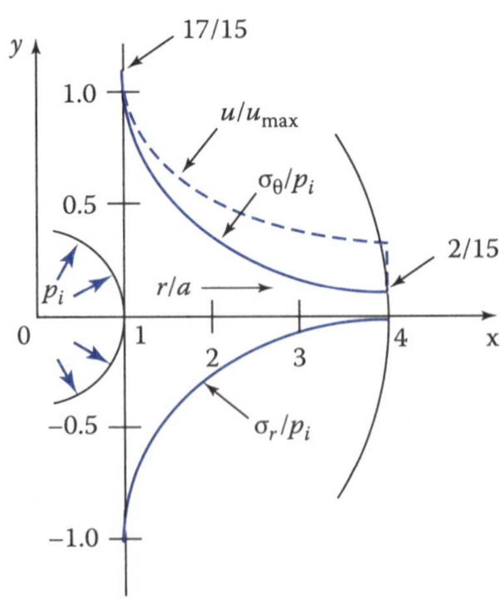

FIGURE 16.3
Distribution of stress and displacement in a thick-walled cylinder with $b/a = 4$ under internal pressure.

$$\sigma_\theta = -\frac{b^2 p_o}{b^2 - a^2}\left(1 + \frac{a^2}{r^2}\right) \tag{16.18b}$$

$$u = -\frac{b^2 p_o r}{E\left(b^2 - a^2\right)}\left[(1-\nu) + (1+\nu)\right]\frac{a^2}{r^2} \tag{16.18c}$$

Inasmuch as $a^2/r^2 \geq 1$, the maximum σ_r occurs at $r = b$ and is always compressive. The maximum σ_θ is found at $r = a$ and is likewise always compressive:

$$\sigma_{\theta,\max} = -2p_o \frac{b^2}{b^2 - a^2} \tag{16.19}$$

16.3.3.3 Cylinder with an Eccentric Bore

The problem corresponding to cylinders having eccentric bore was solved by G.B. Jeffrey [1,3]. For the case of $p_o = 0$ and the eccentricity $e < a/2$ (Figure 16.4), the maximum tangential stress takes place at the internal surface at the thinnest part (point A). The result is as follows:

$$\sigma_{\theta,\max} = p_i\left[\frac{2b^2\left(b^2 + a^2 - 2ae - e^2\right)}{\left(a^2 + b^2\right)\left(b^2 - a^2 - 2ae - e^2\right)} - 1\right] \tag{16.20}$$

When $e = 0$, this coincides with Equation 16.17.

16.3.3.4 Thick-Walled Spheres

Equations for thick-walled spheres may be derived following a procedure similar to that employed for thick-walled cylinders. Clearly, the notation of Figure 16.1 applies, with the sketch now representing a diametral cross section of a sphere. It can be shown that [1] the radial and tangential stresses are

$$\sigma_r = \frac{p_i a^3}{b^3 - a^3}\left(1 - \frac{b^3}{r^3}\right) - \frac{p_o b^3}{b^3 - a^3}\left(1 - \frac{a^3}{r^3}\right) \tag{16.21}$$

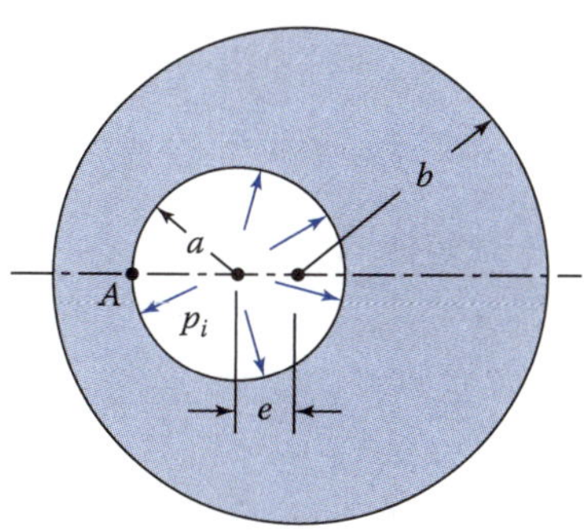

FIGURE 16.4
Thick-walled cylinder with eccentric bore (with $e < a/2$) under internal pressure.

$$\sigma_\theta = \frac{p_i a^3}{b^3 - a^3}\left(1 + \frac{b^3}{2r^3}\right) - \frac{p_o b^3}{b^3 - a^3}\left(1 + \frac{a^3}{2r^3}\right) \quad (16.22)$$

Thick-walled spheres are used as vessels in high-pressure applications (e.g., in deep-sea vehicles). They yield lower stresses than other shapes and, under external pressure, the greatest resistance to buckling.

16.4 Compound Cylinders: Press or Shrink Fits

A composite or *compound cylinder* is made by *shrink or press fitting* an outer cylinder on an inner cylinder. Recall from Section 9.6 that a press or shrink fit is also called *interference fit*. Contact pressure is caused by interference of metal between the two cylinders. Examples of compound cylinders are seen in various machine (Figure 16.5) and structural members, compressors, extrusion presses, conduits, and the like. A fit is obtained by machining the hub hole to a slightly smaller diameter than that of the shaft. Figure 16.6 depicts a shaft and hub assembled by shrink fit; after the hub is heated, the contact comes through contraction on cooling. Alternatively, the two parts are forced slowly in press to form a press fit. The stresses and displacements resulting from the contact pressure p may readily be obtained from the equations of the preceding section.

Note from Figure 16.3 that most material is underutilized (i.e., only the innermost layer carries high stress) in a thick-walled cylinder subjected to internal pressure. A similar conclusion applies to a cylinder under external pressure alone. The cylinders may be strengthened and the material used more effectively by shrink or press fits or by plastic flow [2]. Both cases are used in high-pressure technology. The technical literature contains an abundance of specialized information on multilayered cylinders in the form of graphs and formulas [3].

FIGURE 16.5
A bushing press fit into a gear.

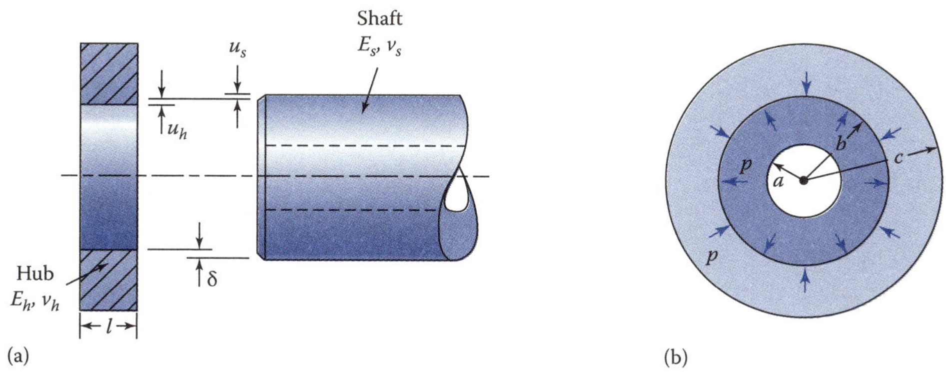

FIGURE 16.6
Notation for shrink and press fits: (a) unassembled parts and (b) after assembly.

In the unassembled stage (Figure 16.6a), the external radius of the shaft is larger than the internal radius of the hub by the amount δ. The increase u_h in the radius of the hub, using Equation 16.16c, is

$$u_h = \frac{bp}{E_h}\left(\frac{b^2+c^2}{c^2-b^2}+\nu_h\right) \tag{16.23}$$

The decrease u_s in the radius of the shaft, by Equation 16.18c, is

$$u_s = -\frac{bp}{E_s}\left(\frac{a^2+b^2}{b^2-a^2}-\nu_s\right) \tag{16.24}$$

In the preceding, the subscripts *h* and *s* refer to the hub and shaft, respectively.

Radial interference or so-called shrinking allowance δ is equal to the sum of the absolute values of the expansion $|u_h|$ and of shaft contraction $|u_s|$:

$$\delta = \frac{bp}{E_h}\left(\frac{b^2+c^2}{c^2-b^2}+\nu_h\right)+\frac{bp}{E_s}\left(\frac{a^2+b^2}{b^2-a^2}-\nu_s\right) \tag{16.25}$$

When the hub and shaft are composed of the same material ($E_h = E_s = E$, $\nu_h = \nu_s$), the contact pressure from Equation 16.25 may be obtained as (Figure 16.6b)

$$p = \frac{E\delta}{b}\frac{\left(b^2-a^2\right)\left(c^2-b^2\right)}{2b^2\left(c^2-a^2\right)} \tag{16.26}$$

The stresses and displacements in the hub are then determined using Equation 16.16 by treating the contact pressure as p_i. Likewise, by regarding the contact pressure as p_o, the stresses and deformations in the shaft are calculated, applying Equation 16.18.

An interference fit creates stress concentration in the shaft and hub at each end of the hub, owing to the abrupt change from uncompressed to compressed material. Some design modifications are often made in the faces of the hub close to the shaft diameter to reduce the stress concentrations at each sharp corner. Usually, for a press or shrink fit, a *stress concentration factor* K_t is used. The value of K_t depending on the contact pressure, the design of the hub, and the maximum bending stress in the shaft rarely exceeds 2 [5]. Note that an approximation of the torque capacity of the assembly may be made on the basis of a *coefficient of friction* of about $f = 0.15$ between shaft and hub. The AGMA standard suggests a value of $0.15 < f < 0.20$ for shrink or press hubs, based on a ground finish on both surfaces.

Example 16.1: Designing a Press Fit

A steel shaft of inner radius a and outer radius b is to be press fit in a cast iron disk having outer radius c and axial thickness or length of hub engagement of l (Figure 16.6). Determine

a. The radial interference
b. The force required to press together the parts and the torque capacity of the joint

Given: $a = 25$ mm, $b = 50$ mm, $c = 125$ mm, and $l = 100$ mm. The material properties are $E_s = 210$ GPa, $\nu_s = 0.3$, $E_h = 70$ GPa, and $\nu_h = 0.25$.

Assumptions: The maximum tangential stress in the disk is not to exceed 30 MPa; the contact pressure is uniform; and $f = 0.15$.

Solution

a. Through the use of Equation 16.17, with $p_i = p$, $a = b$, and $b = c$, we have

$$p = \sigma_{\theta,\max}\frac{c^2 - b^2}{b^2 + c^2} = 30\frac{125^2 - 50^2}{50^2 + 125^2} = 21.72 \text{ MPa}$$

From Equation 16.25,

$$\delta = \frac{0.05(21.72)}{70 \times 10^3}\left(\frac{50^2 + 125^2}{125^2 - 50^2} + 0.25\right) + \frac{0.05(21.72)}{210 \times 10^3}\left(\frac{25^2 + 50^2}{50^2 - 25^2} - 0.3\right)$$

$$= 0.0253 + 0.0071 = 0.0324 \text{ mm}$$

b. The *force* (axial or tangential) required for the assembly:

$$F = 2\pi bpfl \tag{16.27a}$$

Introducing the required numerical values,

$$F = 2\pi(50)(21.72)(0.15)(100) = 102.4 \text{ kN}$$

The *torque capacity* or torque carried by the press fit is then

$$T = Fb = 2\pi b^2 fpl \tag{16.27b}$$

Inserting the given data, we obtain

$$T = 102.4(0.05) = 5.12 \text{ kN} \cdot \text{m}$$

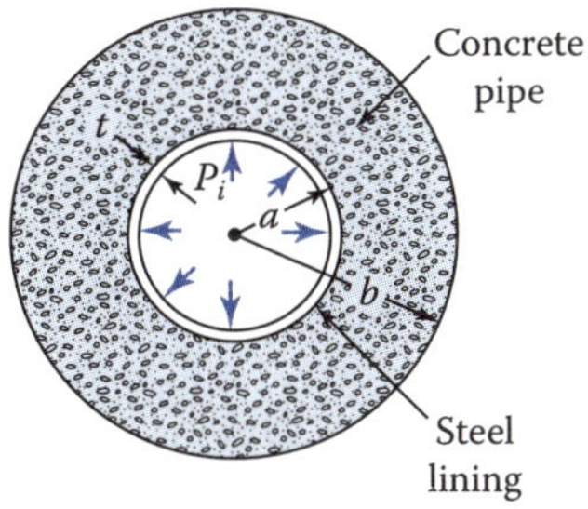

FIGURE 16.7
Example 16.2.

Example 16.2: Design of a Duplex Hydraulic Conduit

A thick-walled concrete pipe (E_c, ν_c) with a thin-walled steel cylindrical liner or sleeve (E_s) of outer radius a is under internal pressure p_i as shown in Figure 16.7. Develop an expression for the pressure p transmitted to the concrete pipe.

Design Decision: For practical purposes, we take

$$\frac{E_s}{E_c} = 15, \quad \nu_c = 0.2, \quad \frac{t}{a-t} = \frac{t}{a} \tag{a}$$

and $a \pm t = a$, since $a/t > 10$ for a thin-walled cylinder.

Solution

The sleeve is under internal pressure p_i and external pressure p:

$$\sigma_\theta = \frac{p_i - p}{t}(a-t) = (p_i - p)\left(\frac{a}{t} - 1\right) \tag{b}$$

Also from Hooke's law and the second of Equation 16.2 with $r = a$,

$$\sigma_\theta = E_s \varepsilon_\theta = E_s \frac{u}{a} \tag{c}$$

The radial displacement at the bore ($r = a$) of pipe, using Equation 16.16c, is

$$u = \frac{pa}{E_c}\left(\frac{a^2 + b^2}{b^2 - a^2} + \nu_c\right) \tag{d}$$

Evaluating u from Equations (b) and (c) and carrying into Equation (d) lead to an expression from which the interface pressure can be obtained. In so doing, we obtain

$$p = \frac{p_i}{1 + \left(\frac{E_s}{E_o}\right)\left(\frac{t}{a-t}\right)\left(\frac{a^2 + b^2}{b^2 - a^2} + \nu_c\right)} \tag{16.28}$$

A design formula for the interface pressure is obtained upon substitution of Equation (a) into the preceding equation:

$$p = \frac{p_i}{1 + 15\left(\frac{t}{a}\right)\left(\frac{R^2+1}{R^2-1} + 0.2\right)} \tag{16.29}$$

where the pipe radius ratio $R = b/a$. This formula can be used to prepare design curves for steel-lined concrete conduits [3].

Comments: It is interesting to observe from Equation 16.29 that as the sleeve thickness t increases, the pressure p transmitted to the concrete decreases. But for any given t/a ratio, the p increases as the R increases.

16.5 Disk Flywheels

A *flywheel* is often used to smooth out changes in the speed of a shaft caused by torque fluctuations. Flywheels are therefore found in small and large machinery, such as compressors, punch presses (see Example 1.5), rock crushers, and internal combustion engines. Considerable stress may be induced in these members at high speed. Analysis of this effect is important, since failure of rotating disks is particularly hazardous. Designing of energy-storing flywheels for hybrid electric cars is an active area of contemporary research. *Disk flywheels,* rotating annular disks of constant thickness, are often made of various materials, such as ceramics, composites, high-strength steel, aluminum and titanium alloys, inexpensive lead alloys (in children toys), and cast iron [6]. In this section, attention is directed to the design analysis of these flywheels using both equilibrium and energy approaches.

16.5.1 Stress and Displacement

Figure 16.8 illustrates a flywheel of axial thickness or length of hub engagement l, with inner radius a and outer radius b, shrunk onto a shaft. Let the contact pressure between the two parts be designated by p. An element of the disk is loaded by an outwardly directed *centrifugal force* $F_r = \rho\omega^2 r$ (Figure 16.2). Here, ρ is the mass density ($N \cdot s^2/m^4$ or

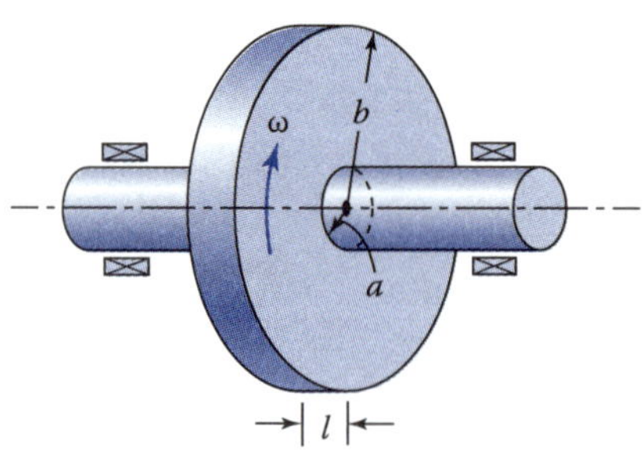

FIGURE 16.8
A flywheel shrunk onto a shaft.

lb s²/in.⁴), and ω represents the angular velocity or speed (rad/s). The condition of equilibrium, Equation 16.1, becomes

$$\frac{d\sigma_r}{dr} + \frac{\sigma_r - \sigma_\theta}{r} + \rho\omega^2 r = 0 \tag{16.30}$$

The boundary conditions are $\sigma_r = -p$ at the inner surface ($r = a$) and $\sigma = 0$ at the outer surface ($r = b$).

The solution of Equation 16.30 is obtained by following a procedure similar to that used in Section 16.2. It can be shown that [2] the combined radial stress (σ_r), tangential stress (σ_θ), and displacement (u) of a disk due to contact pressure p and angular speed ω are

$$\sigma_r = (\sigma_r)_p + \frac{3+\nu}{8}\left(a^2 + b^2 - \frac{a^2b^2}{r^2} - r^2\right)\rho\omega^2 \tag{16.31a}$$

$$\sigma_\theta = (\sigma_\theta)_p + \frac{3+\nu}{8}\left(a^2 + b^2 - \frac{1+3\nu}{3+\nu}r^2 + \frac{a^2b^2}{r^2}\right)\rho\omega^2 \tag{16.31b}$$

$$u = (u)_p + \frac{(3+\nu)(1-\nu)}{8E}\left(a^2 + b^2 - \frac{1+\nu}{3+\nu}r^2 + \frac{1+\nu}{1-\nu}\frac{a^2b^2}{r^2}\right)\rho\omega^2 r \tag{16.31c}$$

Here, $(\sigma_r)_p$, $(\sigma_\theta)_p$, and $(u)_p$ are given by Equation 16.16 with $p_i = p$. The quantity ν is Poisson's ratio.

In most cases, *tangential stress controls the design*. This stress is a maximum at the inner boundary ($r = a$) and is equal to

$$\sigma_{\theta,\max} = p\frac{a^2 + b^2}{b^2 - a^2} + \frac{\rho\omega^2}{4}\left[(1-\nu)a^2 + (3+\nu)b^2\right] \tag{16.32}$$

Clearly, the preceding problem in which pressure and rotation appear simultaneously could also be solved by superposition. Note that, due to *rotation* only, maximum radial stress occurs at $r = \sqrt{ab}$ and is given by

$$\sigma_{r,\max} = \frac{3+\nu}{8}(b-a)^2\rho\omega^2 \tag{16.33}$$

Owing to the internal *pressure* alone, the largest radial stress is at the inner boundary and equals $\sigma_{r,\max} = -p$.

Customarily, inertial stress and displacement of a shaft are neglected. Therefore, for a shaft, we have approximately

$$\sigma_r = \sigma_\theta = -p$$
$$u = -\frac{1-\nu}{E_s}pr \tag{16.34}$$

Note, however, that the contact pressure p depends on angular speed ω. For a given contact pressure p at angular speed ω, the required initial radial interference δ may be obtained using Equations 16.31c and 16.34 for u. Hence, with $r = a$, we have

$$\delta = \frac{ap}{Ed}\left(\frac{a^2+b^2}{b^2-a^2}+\nu\right)+\frac{ap}{E_s}(1-\nu)+\frac{a\rho\omega^2}{4E_d}\left[(1-\nu)a^2+(3+\nu)b^2\right] \tag{16.35}$$

in which E_d and E_s represent the moduli of elasticity of the disk and shaft, respectively. The preceding equation is valid as long as a positive contact pressure is maintained.

Example 16.3: Rotating Blade Design Analysis

A disk of uniform thickness is used at 12,000 rpm as a rotating blade for cutting blocks of paper or thin plywood. The disk is mounted on a shaft of 1 in. radius and clamped, as shown in Figure 16.9. Determine

a. The factor of safety n according to the maximum shear stress criterion
b. The values of the maximum radial stress and displacement at outer edge

Assumptions: The cutting forces are relatively small, and speed is steady: loading is considered static. The disk outside radius is taken as 15 in. The stress concentrations due to clamping and sharpening at the periphery are disregarded.

Design Decision: The disk material is a high-strength ASTM A242 steel.

Solution

The material properties are (Table B.1)

$$\rho = \frac{0.284}{386} = 7.358\times 10^{-4}\ \text{lb}\cdot\text{s}^2/\text{in.}^4, \quad \nu = 0.3$$

$$E = 29\times 10^6\ \text{psi}, \quad S_{ys} = 30\ \text{ksi}$$

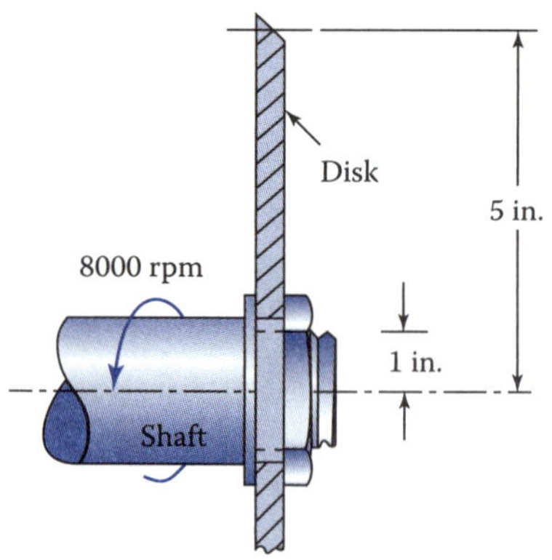

FIGURE 16.9
Example 16.3. A rotating blade (only a partial view shown) and shaft assembly.

We have

$$\rho\omega^2 = 7.358\times 10^{-4}\left(\frac{12{,}000\times 2\pi}{60}\right)^2 = 1161.929$$

a. The tangential stress, expressed by Equation 16.31b with $p = 0$, has the form

$$\sigma_\theta = \frac{3+\nu}{8}\left(a^2 + b^2 - \frac{1+3\nu}{3+\nu}r^2 + \frac{a^2b^2}{r^2}\right)\rho\omega^2$$

The stresses in the inner and outer edges of the blade are, from the preceding equation,

$$(\sigma_\theta)_{r=1} = \frac{3.3}{8}\left(1^2 + 5^2 - \frac{1.9\times 1^2}{3.3} + \frac{1^2\times 5^2}{1^2}\right)(1161.929) = 24.17 \text{ ksi}$$

$$(\sigma_\theta)_{r=5} = \frac{3.3}{8}\left(1^2 + 5^2 - \frac{1.9\times 5^2}{3.3} + 1^2\right)(1161.929) = 6.04 \text{ ksi}$$

The maximum shear stress occurs at the inner surface ($r = 1$ in.), where $\sigma_r = 0$:

$$\tau_{\max} = \frac{\sigma_\theta}{2} = \frac{24.17}{2} = 12.09 \text{ ksi}$$

The factor of safety, based on the maximum shear stress theory, is then

$$n = \frac{S_{ys}}{\tau_{\max}} = \frac{30}{12.09} = 2.48$$

Comment: Should there be starts and stops, the condition is one of fatigue failure, and a lower value of n would be obtained by the techniques of Section 7.11.

b. The largest radial stress in the disk, from Equation 16.33, is given by

$$\sigma_{r,\max} = \frac{3+\nu}{8}(b-a)^2\rho\omega^2$$

$$= \frac{3.3}{8}(5-1)^2(1161.929) = 7.67 \text{ ksi}$$

The radial displacement of the disk is expressed by Equation 16.31c with $p = 0$. Hence,

$$(u)_{r=5} = \frac{(3.3)(0.7)}{8\times 29\times 10^6}\left(1^2 + 5^2 - \frac{1.3\times 1^2}{3.3} + \frac{1.3\times 1^2}{0.7}\right)(1161.929)(5)$$

$$= 1.589\times 10^{-3} \text{ in.}$$

is the radial displacement at the outer periphery.

Example 16.4: Design of a Flywheel–Shaft Assembly

A 400 mm diameter flywheel is to be shrunk onto a 50 mm diameter shaft. Determine

a. The required radial interference
b. The maximum tangential stress in the assembly
c. The speed at which the contact pressure becomes 0

Requirement: At a maximum speed of $n = 5000$ rpm, a contact pressure of $p = 8$ MPa is to be maintained.

Design Decisions: Both the flywheel and shaft are made of steel having $\rho = 7.8$ kN s^2/m^4, $E = 200$ GPa, and $\nu = 0.3$.

Solution

a. Applying Equation 16.35, we have

$$\delta = \frac{25(10^{-3})p}{200(10^9)}\left(\frac{25^2+200^2}{200^2-25^2}+1\right)+\frac{25(7.8)\omega^2}{4(200\times 10^9)}\left[0.7(0.025)^2+3.3(0.2)^2\right]$$

$$=(0.254p+32.282\omega^2)10^{-12} \tag{a}$$

For $p = 8$ MPa and $\omega = 5000(2\pi/60) = 523.6$ rad/s, Equation (a) leads to $\delta = 0.011$ mm.

b. Using Equation 16.32,

$$\sigma_{\theta,\max} = 8\frac{25^2+200^2}{200^2-25^2}+\frac{7800(523.6)^2}{4}\left[0.7(0.025)^2+3.3(0.2)^2\right]$$

$$= 8.254+70.80 = 79.05 \text{ MPa}$$

c. Inserting $\delta = 0.011 \times 10^{-3}$ m and $p = 0$ into Equation (a) results in

$$\omega = \left(\frac{0.011\times 10^9}{32.282}\right)^{1/2} = 583.7 \text{ rad/s}$$

Therefore,

$$n = 583.7\frac{60}{2\pi} = 5574 \text{ rpm}$$

Comment: At this speed, the shrink fit becomes completely ineffective.

The constant-thickness disks discussed in the foregoing do not make optimum use of the material. Often disks are not flat, being thicker at the center than at the rim. Other types of rotating disks, offering many advantages over flat disks, are variable-thickness and uniform-stress disks. For these cases, the procedure outlined here must be modified. A number of problems of this type are discussed in Reference 1.

16.5.2 Energy Stored

Heavy disks often serve as flywheels designed to store energy to maintain reasonably constant speed in a machine in spite of variations in input and output power. A flywheel absorbs and stores energy when speeded up and releases energy to the system when needed by slowing its rotational speed. The change in kinetic energy ΔE_k stored in a flywheel by a change in speed from ω_{max} to ω_{min} by Equation 1.10 is

$$\Delta E_k = \frac{1}{2} I \left(\omega_{max}^2 - \omega_{min}^2 \right) \tag{16.36}$$

The mass moment of inertia I about the axis of an annular disk flywheel of outer radius b and inner radius a (Figure 16.8) is given by

$$I = \frac{\pi}{2g} \left(b^4 - a^4 \right) l\gamma \tag{b}$$

The weight of the disks is

$$W = \pi \left(b^2 - a^2 \right) l\gamma \tag{c}$$

where
- l = the length of the hub engagement
- $\gamma = \rho g$ is the specific weight
- g = the acceleration of gravity

Substitution of Equation (c) into (b) gives

$$I = \frac{W}{2g} \left(b^2 + a^2 \right) \tag{16.37}$$

For a conservative system, the change in the kinetic energy is available as work output:

$$\Delta E_k = T \Delta \phi \tag{16.38}$$

in which
- T represents the torque
- $\Delta\phi$ is the change in the angular rotation of the disk in radians

Ordinarily, there are two stages to the flywheel design [6]. First, the amount of energy needed for the required degree of smoothing must be found from Equation 16.38 and the moment of inertia needed to absorb that energy calculated by Equation 16.36. Then, a flywheel geometry must be defined by Equation 16.37.

Example 16.5: Flywheel Braking-Torque Requirement

A flywheel of outer diameter D, inner diameter d, and weight W rotates at speed n. Determine the average braking torque required to stop the wheel in one-third revolution.

Given: $D = 10$ in., $d = 2$ in., $W = 30$ lb, $n = 3600$ rpm

Solution

From Equations 16.36 and 16.38,

$$T\phi = \frac{1}{2} I\omega^2 \tag{16.39}$$

where

$\phi = 2\pi/3$ rad

$\omega = 3600(2\pi/60) = 377$ rad/s

$I = \frac{30}{2(386)}(5^2 + 1^2) = 1.01 \text{ lb} \cdot \text{s}^2 \cdot \text{in.}^4$

Hence,

$$T = \frac{1.01(377)^2}{2(2\pi/3)} = 34.27 \text{ kips} \cdot \text{in.}$$

16.6 Thermal Stresses in Cylinders

Here, we are concerned with the stress and displacement associated with an axisymmetric temperature *T*(*r*) dependent on the radial dimension alone. Examples include heat exchanger tubes, chemical reaction vessels, clad reactor elements, nozzle sections of rockets, annular fins, and turbine disks. The deformation also is symmetrical about the axis, and we may use the method developed in Section 16.3. The results of this section are restricted to the static, steady-state problem.

Consider a cylinder with ends assumed to be restrained so that the axial displacement $w = 0$. This exemplifies the case of plane strain (i.e., $\varepsilon_z = 0$). The applicable strain–stress relationships, from Equation 16.5 with reference to Equation 2.16, are assumed as follows:

$$\begin{aligned}
\varepsilon_r &= \frac{1}{E}\left[\sigma_r - \nu(\sigma_\theta + \sigma_z)\right] + \alpha T \\
\varepsilon_\theta &= \frac{1}{E}\left[\sigma_\theta - \nu(\sigma_r + \sigma_z)\right] + \alpha T \\
\varepsilon_z &= \frac{1}{E}\left[\sigma_z - \nu(\sigma_z + \sigma_\theta)\right] + \alpha T
\end{aligned} \tag{16.40}$$

The third of the foregoing equations, setting $\varepsilon_z = 0$, leads to

$$\sigma_z = \nu(\sigma_r + \sigma_\theta) - \alpha E T \tag{a}$$

Introducing this expression into the first two of Equation 16.40 results in

$$\begin{aligned}
\varepsilon_r &= \frac{1+\nu}{E}\left[(1-\nu)\sigma_r - \nu\sigma_\theta + \alpha E T\right] \\
\varepsilon_\theta &= \frac{1+\nu}{E}\left[(1-\nu)\sigma_\theta - \nu\sigma_r + \alpha E T\right]
\end{aligned} \tag{16.41}$$

Equations 16.2 and 16.3 are unchanged for the case under discussion. Following a procedure similar to that outlined in Section 16.3, we obtain

$$u = \frac{(1+\nu)\alpha}{(1-\nu)r}\int_a^r Tr\mathrm{d}r + c_1 r + \frac{c_2}{r} \tag{b}$$

$$\sigma_r = \frac{E}{1+\nu}\left[\frac{(1+\nu)\alpha}{(1-\nu)r^2}\int_a^r Tr\,\mathrm{d}r + \frac{c_1}{1-2\nu} - \frac{c_2}{r^2}\right] \tag{c}$$

$$\sigma_\theta = \sigma_r + r\frac{\mathrm{d}\sigma_r}{\mathrm{d}r} \tag{d}$$

Equation (a) can then be written as

$$\sigma_z = -\frac{\alpha E T}{1-\nu} + \frac{2\nu E c_1}{(1+\nu)(1-2\nu)} \tag{e}$$

Here, c_1 and c_2 are constants of integration, to be determined from known boundary conditions at the edges of the cylinder.

16.6.1 Cylinder with a Central Hole

We now determine the expressions for stress components for a hollow cylinder subjected to a radial temperature change (Figure 16.10). The inner and outer edges are free of mechanical loads:

$$(\sigma_r)_{r=a} = 0, \quad (\sigma_r)_{r=b} = 0 \tag{f}$$

Substituting these boundary conditions into Equation (c), we have

$$\begin{aligned} c_1 &= \frac{(1+\nu)(1-2\nu)\alpha}{(1-\nu)(b^2-a^2)}\int_a^b Tr\ \mathrm{d}r \\ c_2 &= \frac{(1+\nu)a^2\alpha}{(1-\nu)(b^2-a^2)}\int_a^b Tr\ \mathrm{d}r \end{aligned} \tag{g}$$

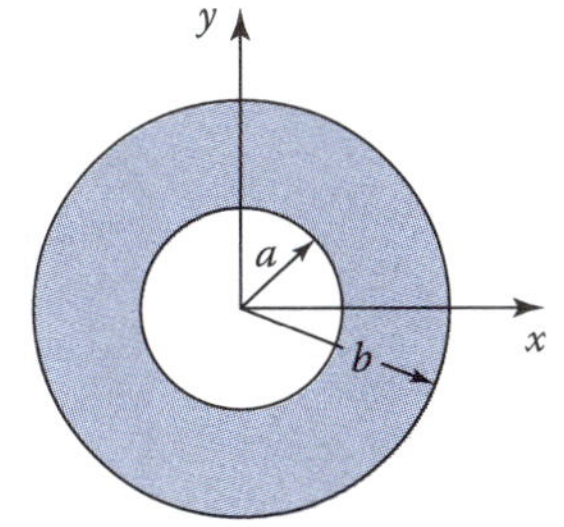

FIGURE 16.10
Cross section of a long circular cylinder under thermal loading.

Equations (c) through (e) provide the stresses as follows:

$$\sigma_r = \frac{E\alpha}{(1-\nu)r^2}\left(-\int_a^r Tr\,dr + \frac{r^2-a^2}{b^2-a^2}\int_a^b Tr\,dr\right)$$

$$\sigma_\theta = \frac{E\alpha}{(1-\nu)r^2}\left(-Tr^2 + \int_a^r Tr\,dr + \frac{r^2+a^2}{b^2-a^2}\int_a^b Tr\,dr\right) \tag{16.42}$$

$$\sigma_z = \frac{\alpha E}{1-\nu}\left(\frac{2\nu}{b^2-a^2}\int_a^b Tr\,dr - T\right) \tag{16.43a}$$

If the *ends* are *free*, it can be shown that the axial stress is given by

$$\sigma_z = \frac{\alpha E}{1-\nu}\left(\frac{2}{b^2-a^2}\int_a^b Tr\,dr - T\right) \tag{16.43b}$$

Similarly, an expression for the displacement u may also be obtained by carrying Equation (g) into Equation (b). For a given temperature distribution $T(r)$ over the cylinder, we can find the stresses using Equations 16.42 and 16.43 for each particular case.

16.6.2 Steady-Flow Temperature Change *T*(*r*)

When the walls of the cylinder are at temperatures T_a and T_b, at the inner ($r = a$) and outer ($r = b$) surfaces, respectively, the temperature distribution may be represented in the form [2]

$$T = \frac{T_a - T_b}{\ln(b/a)}\ln\frac{b}{r} \tag{16.44}$$

This expression can be used with Equations 16.42 and 16.43 to determine stress components for steady-state temperature distribution in a thick-walled cylinder. The results are

$$\sigma_r = \frac{\alpha E(T_a - T_b)}{2(1-\nu)\ln(b/a)}\left[-\ln\frac{b}{r} - \frac{a^2(r^2-b^2)}{r^2(b^2-a^2)}\ln\frac{b}{a}\right]$$

$$\sigma_\theta = \frac{\alpha E(T_a - T_b)}{2(1-\nu)\ln(b/a)}\left[1-\ln\frac{b}{r} - \frac{a^2(r^2+b^2)}{r^2(b^2-a^2)}\ln\frac{b}{a}\right] \tag{16.45}$$

$$\sigma_z = \frac{\alpha E(T_a - T_b)}{2(1-\nu)\ln(b/a)}\left[1-2\ln\frac{b}{r} - \frac{2a^2}{(b^2-a^2)}\ln\frac{b}{a}\right]$$

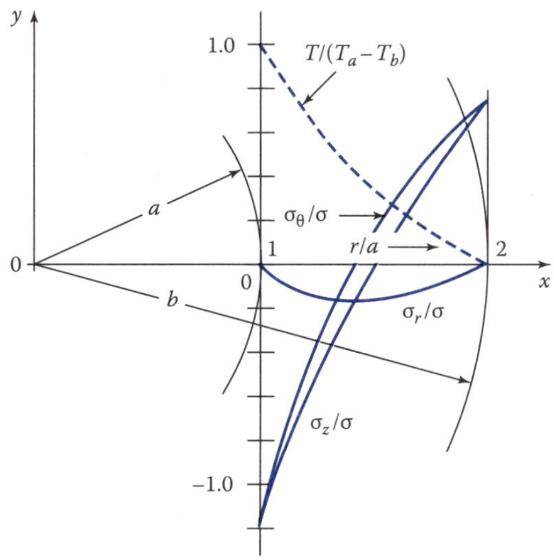

FIGURE 16.11
Thermal stress distribution in a thick-walled cylinder with $b/a = 2$ and $T_a > T_b$. *Note:* $\sigma = \alpha E(T_a - T_b)/2(1 - \nu)$.

The dimensionless distribution of the temperature and stress over the cylinder wall for the particular case when $b/a = 2$ is shown in Figure 16.11. We see from the figure that the tangential σ_θ and axial σ_z stresses at the outer surface are equal and *tensile*. This is why internal heating may cause external cracks in materials weak in tension, such as in the chimneys and conduits of concrete masonry. On the contrary, the radial stress is compressive at all points and becomes 0 at the inner and outer edges of the cylinder.

Note that, in practice, a pressure loading is usually superimposed on the thermal stresses, as in chemical reaction pressure vessels. In this case, the internal pressure gives a tangential stress (Figure 16.3), causing a partial cancellation of compressive stress due to temperature. Also, when a cylinder (or disk) is rotating, stresses owing to the inertia may be superimposed over those due to temperature change and pressure.

16.6.3 Special Case

In a *thin-walled cylinder*, as in the cylinder liner of an engine or compressor, we can simplify Equation 16.45. In this situation, it can be readily verified that the temperature distribution is nearly linear and the stresses have the values

$$(\sigma_\theta)_{r=a} = -\frac{E\alpha(T_a - T_b)}{2(1-\nu)}$$
$$(\sigma_\theta)_{r=b} = \frac{E\alpha(T_a - T_b)}{2(1-\nu)} \tag{16.46}$$

The preceding equations coincide with the stress expressions of an annular plate that is heated on sides and its edges are clamped [4]. Equations 16.46 can also be used with sufficient accuracy in the case of a *thin-walled spherical shell.*

*16.7 Exact Stresses in Curved Beams

Curved beams or bars in the form of hooks, C-clamps, press frames, chain links, and brackets are often used as machine or structural elements. Stresses in curved beams of rectangular cross sections already were discussed briefly in Section 3.7. Here, we are concerned with applications of the theory of elasticity. The mechanics of materials approaches to initially curved bars or frames is taken up in the next section. In both cases, only elastic cases are treated.

Figure 16.12a shows a beam of narrow rectangular cross section and circular axis subjected to equal end couples M such that pure bending occurs in the plane of the curvature. Since the bending moment is constant throughout the length of the bar, stress distribution is the same in all radial cross sections. This is the case of a plane stress problem with *axial symmetry about* θ. But, unlike the axisymmetrically loaded members of revolution treated in the preceding sections, there is a θ-dependent tangential displacement [1]. The condition of equilibrium is given by Equation 16.1 as

$$\frac{d\sigma_r}{dr} + \frac{\sigma_r - \sigma_\theta}{r} = 0 \tag{a}$$

The conditions at the curved boundaries are

$$(\sigma_r)_{r=a} = (\sigma_r)_{r=b} = 0 \tag{b}$$

The conditions at the straight edges or ends are expressed as

$$t\int_a^b \sigma_\theta \, dr = 0, \quad t\int_a^b r\sigma_\theta \, dr = M \tag{c}$$

Shear stress is also taken to be 0 throughout the beam.

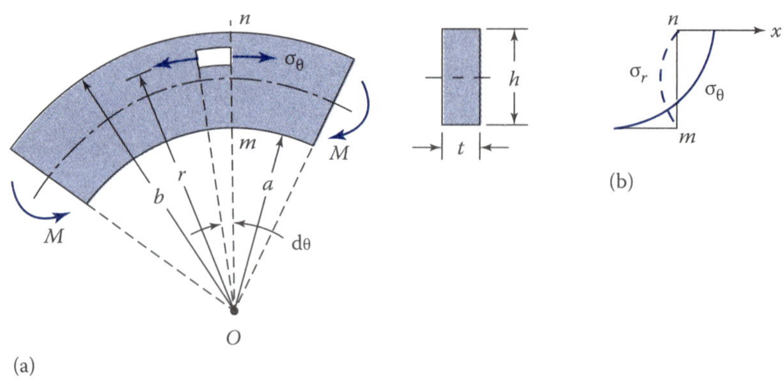

FIGURE 16.12
(a) A thin curved beam in pure bending. (b) Distribution of stresses.

Solution of Equation (a) is determined by following a procedure somewhat similar to that outlined in Section 16.2. It can be shown that the tangential and radial stress distributions in the beam are expressed as

$$\sigma_r = \frac{4M}{tb^2N}\left[\left(1-\frac{a^2}{b^2}\right)\ln\frac{r}{a}-\left(1-\frac{a^2}{r^2}\right)\ln\frac{b}{a}\right] \tag{16.47}$$

$$\sigma_\theta = \frac{4M}{tb^2N}\left[\left(1-\frac{a^2}{b^2}\right)\left(1+\ln\frac{r}{a}\right)-\left(1+\frac{a^2}{r^2}\right)\ln\frac{b}{a}\right] \tag{16.48}$$

where

$$N = \left(1-\frac{a^2}{b^2}\right)^2 - 4\frac{a^2}{b^2}\ln^2\frac{b}{a} \tag{16.49}$$

The bending moment is taken as *positive* when it tends to *decrease* the radius of curvature of the beam, as in Figure 16.12a. Using this sign convention, σ_r as determined from Equation 16.47 is always negative, meaning that it is compressive. Similarly, when σ_θ is found to be positive, it is tensile; otherwise, it is compressive. A sketch of the stresses at section *mn* is presented in Figure 16.12b. Observe that the maximum stress magnitude is at the extreme fiber of the inner (concave) side.

16.8 Curved Beam Formula

The approximate approach to the curved beams by E. Winkler (1835–1888) is now explored. The fundamental assumptions of the elementary theory of straight beams are also valid for *Winkler's theory*. Only elastic bending is treated, with the usual condition that the modulus of elasticity is identical in tension and compression. Consider the pure bending of a curved beam of uniform cross section having a vertical (y) axis of symmetry (Figure 16.13a). An expression for the tangential stress is derived by applying the three principles of analysis based on the familiar hypothesis: *Plane sections* perpendicular to the axis of the beam *remain plane* after bending. This is depicted by the line *ef* in relation to a beam segment abed subtended by the central angle θ.

Figure 16.13a shows that the deformation pattern of curved beams is the same as for straight beams. The initial length of a beam fiber such as *gh* depends on the distance r from the center of the curvature O. The total deformation of beam fibers as the beam rotates through a small angle $d\theta$ follows a linear law. The *tangential strain* on the fiber *gh* may be expressed as

$$\varepsilon_\theta = \frac{(R-r)d\theta}{r\theta} \tag{d}$$

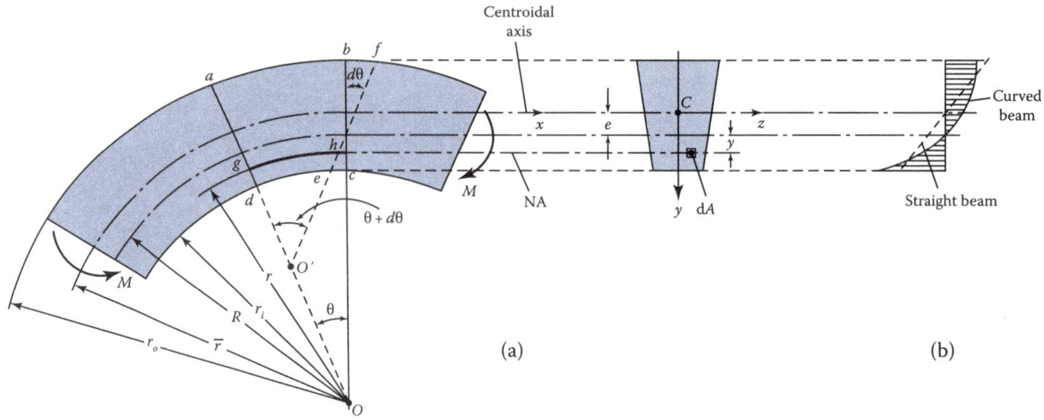

FIGURE 16.13
(a) Pure bending of a beam with a cross-sectional axis of symmetry. (b) Stress distribution.

We see from this expression that ε_θ does *not* vary linearly over the depth of the beam as it does for straight beams. The tangential stress σ_θ on an element dA of the cross-sectional area is, using *Hooke's law,*

$$\sigma_\theta = E\varepsilon_\theta \tag{e}$$

Equations of equilibrium, $\Sigma F_x = 0$ and $\Sigma M_z = 0$, respectively, are

$$\int \sigma_\theta \mathrm{d}A = 0, \quad -\int \sigma_\theta y \mathrm{d}A = M \tag{f}$$

Integration in these equations extends over the entire cross-sectional area A.

We now substitute Equation (e) together with (d) into Equation (f). After rearrangement, we find *radius of the neutral axis* R as follows:

$$R = \frac{A}{\int \frac{\mathrm{d}A}{r}} \tag{16.50}$$

Here, A represents the cross-sectional area of the beam. The integral in this expression may be evaluated for various cross-sectional forms (see Example 16.6, Case Study 18.8 and Problems 16.28–16.30). For the purposes of reference, Table 16.1 furnishes some commonly used cases [7,8].

The distance e between the *centroid and the neutral axis* of the cross section (Figure 16.13) is equal to

$$e = \bar{r} - R \tag{16.51}$$

Hence, in a curved member, the *neutral axis does not coincide with the centroidal axis*. Clearly, this conclusion differs from the situation found to be true for straight elastic beams.

TABLE 16.1

Properties for a Variety of Cross-Sectional Shapes

Cross Section	Radius of Neutral Surface
A. Rectangle 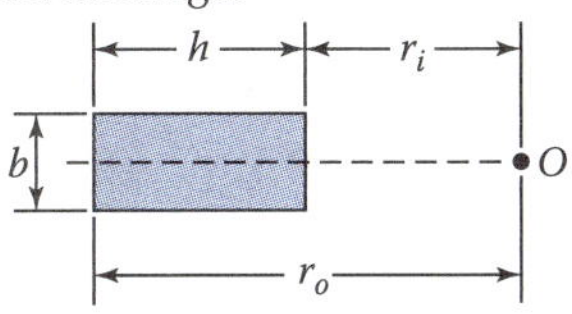	$R = \dfrac{h}{\ln \dfrac{r_o}{r_i}}$ $A = bh$
B. Circle 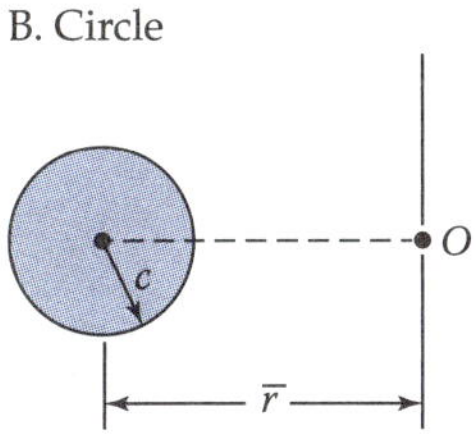	$R = \dfrac{A}{2\pi\left(\bar{r} - \sqrt{\bar{r}^2 - c^2}\right)}$ $A = \pi c^2$
C. Ellipse 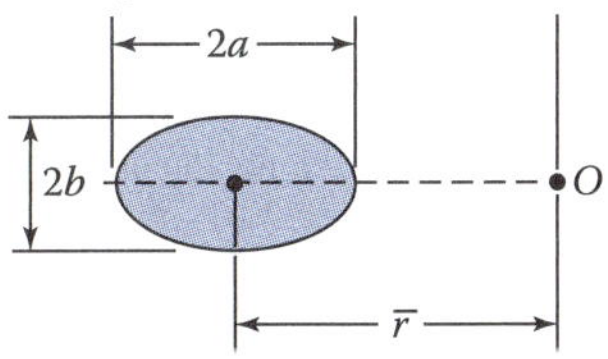	$R = \dfrac{A}{\dfrac{2\pi b}{a}\left(\bar{r} - \sqrt{\bar{r}^2 - a^2}\right)}$ $A = \pi ab$
D. Triangle 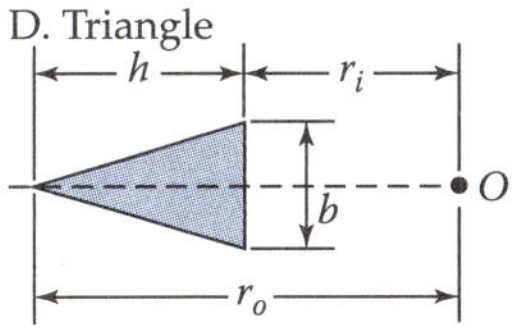	$R = \dfrac{A}{\dfrac{br_o}{h}\left(\ln \dfrac{r_o}{r_i}\right) - b}$ $A = \dfrac{1}{2} bh$
E. Trapezoid 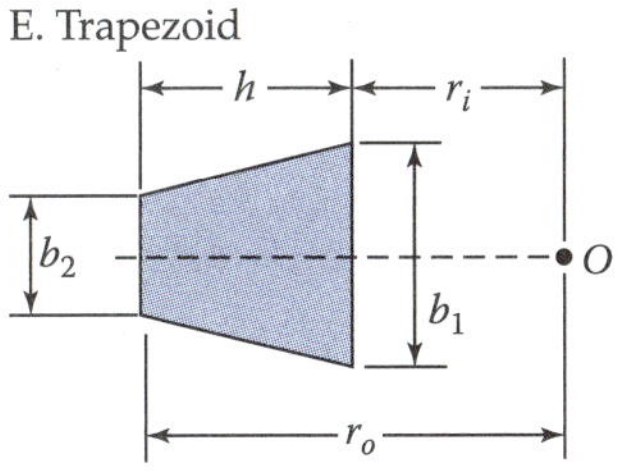	$R = \dfrac{1}{\dfrac{1}{h}\left[(b_1 r_o - b_2 r_i) \cdot \ln \dfrac{r_o}{r_i} - h(b_1 - b_2)\right]}$ $A = \dfrac{1}{2}(b_1 + b_2)h$

It can be verified that the normal stress acting on a curved beam at a distance r from the center of curvature is

$$\sigma_\theta = -\frac{M(R - r)}{Aer} \tag{16.52}$$

in which e is given by Equation 16.51. Equation 16.52 is called Winkler's formula or *curved-beam formula*. It shows that the stress distribution in a curved beam follows a *hyperbolic* pattern. The *sign convention* applied to bending moment is that it is positive when directed toward the concave side of the beam, as indicated in Figure 16.13. A positive value found using Equation 16.52 means a tensile stress.

A *comparison* of this result with one that follows from the formula for straight beams is illustrated in Figure 7.13b. It can be shown that [see 1], the linear and hyperbolic stress distributions are about the same for $r_o/r_i = 1.1$. That is, for beams of only slight curvature, the flexure formula provides acceptable results, while requiring simple computation. When beam curvature increases ($r_o/r_i > 1.3$), the stress on the concave side rapidly increases over the one given by the flexure formula.

The tangential stress given by Equation 16.52 may be superimposed to the stress produced by a centric normal load *P*. Hence, the *combined stress* in a curved beam is

$$\sigma_\theta = \frac{P}{A} - \frac{M(R-r)}{Aer} \tag{16.53}$$

As usual, a negative sign is associated with a compressive load. It is obvious that the conditions of axisymmetry do not apply for a beam subjected to combined loading.

Example 16.6: Determining Stresses in Curved Frame Using Various Methods

A circular frame of rectangular cross section and mean radius $\bar{r}$ is subjected to a load *P* as shown in Figure 16.14a. Compute the tangential stresses at points *A* and *B*, using

a. Winkler's curved-beam theory
b. The elementary theory
c. The elasticity theory

Given: $P = 200$ kN, $b = 100$ mm, $h = 200$ mm, $\bar{r} = 300$ mm

Solution:

a. With reference to Figure 16.14, we first derive the expression for the radius *R* of the neutral axis. In this case, $A = bh$ and $dA = bdr$. Integrating Equation 16.51 between the limits r_i and r_o, readily gives

$$R = \frac{A}{\int_A \frac{dA}{r}} = \frac{bh}{\int_{r_i}^{r_o} \frac{bdr}{r}} = \frac{h}{\int_{r_i}^{r_o} \frac{dr}{r}}$$

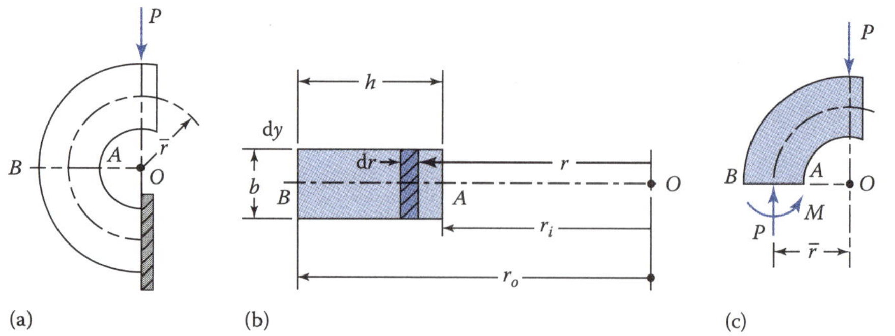

FIGURE 16.14
Example 16.6. (a) Curved frame with a vertical load at free end, (b) rectangular cross section, and (c) stress resultants at a cross section *A-B*.

Therefore,

$$R = \frac{h}{\ln\frac{r_o}{r_i}} \tag{16.54}$$

The given data lead to

$$A = bh = (100)(200) = 20(10)^3 \text{ mm}^2$$

$$r_i = \bar{r} - \frac{1}{2}h = 300 - 100 = 200 \text{ mm}$$

$$r_o = \bar{r} + \frac{1}{2}h = 300 + 100 = 400 \text{ mm}$$

Equations 16.54 and 16.51 result in yield, respectively,

$$R = \frac{h}{\ln\frac{r_o}{r_i}} = \frac{200}{\ln 2} = 288.5390 \text{ mm}$$

$$e = \bar{r} - R = 300 - 288.5390 = 11.4610 \text{ mm}$$

Comment: Observe that the radius of the neutral axis R must be calculated with five significant figures.

The tangential stresses are due to the compressive normal load $-P$ and the moment $M = P\bar{r}$ acting at the centroid C of the cross section (Figure 16.14c). The maximum compression and tension values of σ_θ occur at points A and B, respectively. Substituting the given numerical values, Equation 16.53 results in then

$$(\sigma_\theta)_A = -\frac{P}{A} - \frac{P\bar{r}(R - r_i)}{Aer_i} = -\frac{P}{A}\left[1 + \frac{\bar{r}(R - r_i)}{er_i}\right] \tag{16.55a}$$

$$= -\frac{200 \times 10^3}{20(10^{-3})}\left[1 + \frac{300(288.539 - 200)}{11.461(200)}\right] = -125.9 \text{ MPa}$$

$$(\sigma_\theta)_B = -\frac{P}{A} - \frac{P\bar{r}(R - r_o)}{Aer_o} = -\frac{P}{A}\left[1 + \frac{\bar{r}(R - r_o)}{er_o}\right] \tag{16.55b}$$

$$= -\frac{200 \times 10^3}{20(10^{-3})}\left[1 + \frac{300(288.539 - 400)}{11.461(400)}\right] = 62.9 \text{ MPa}$$

The negative sign of $(\sigma_\theta)_A$ means a compressive stress at A. The largest tensile stress is at B.

Comment: The stress caused by the axial force, $P/A = 200(10^3)/(20 \times 10^{-3}) =$ 10 MPa, is negligibly small when compared to the combined stresses at points A and B of the cross section.

b. Through the use of the flexure formula, with $M = P\bar{r} = 200(300) = 60\ \text{kN}\cdot\text{m}$,

$$(\sigma_\theta)_B = -(\sigma_\theta)_A = \frac{Mc}{I} = \frac{60{,}000(0.1)}{(0.1)(0.2)^3/12} = 90\ \text{MPa}$$

c. From Equation 16.49, with $a = 300 - 100 = 200$ mm and $b = 300 + 100 = 400$ mm, we have

$$N = \left[1-(0.5)^2\right]^2 - 4(0.5)^2 \ln^2 2 = 0.082$$

Superposition of $-P/A$ and Equation 16.48 at $r = a$ gives

$$(\sigma_\theta)_A = -\frac{200{,}000}{0.02} + \frac{4(60{,}000)}{(0.1)(0.4)^2(0.082)}\left[(1-0.25)(1+0)-(1+1)\ln^2\right]$$

$$= (-10 - 116.4)(10^6) = -126.4\ \text{MPa}$$

Likewise, at $r = b$, we obtain $(\sigma_\theta)_B = -10 + 73.8 = 63.8$ MPa.

Comments: The foregoing shows that the results of the Winkler and elasticity theories are in good agreement. However, the usual flexure formula provides a result of unacceptable accuracy for the tangential stress in this nonslender curved beam.

*16.9 Circular Plates

The deflection w of a plate of circular shape depends on its radial position alone if the applied load and conditions of end support are independent of the angle θ. For this case, radial and tangential moments M_r and M_θ per unit length and force Q_r per unit length act on the circular plate element (Figure 16.15b). To develop the basic equations of a circular plate, we need only to transform the appropriate formulations of Sections 4.10 and 4.11 from the Cartesian to polar coordinates.

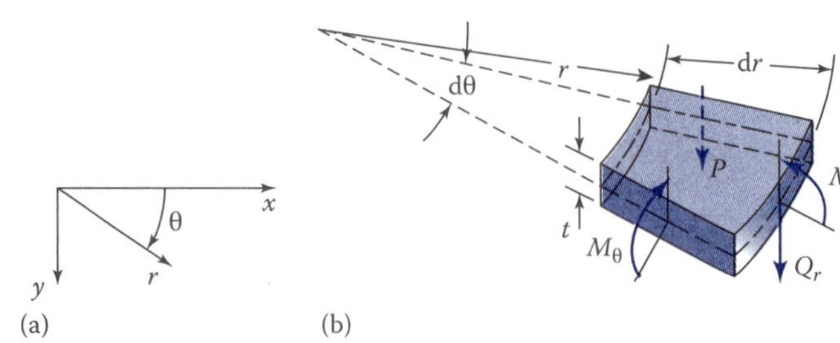

FIGURE 16.15
(a) Polar coordinates. (b) Axisymmetrically loaded circular plate element. (Only shear force and moments on the positive faces are shown.)

The polar coordinate set (r, θ) and the Cartesian set (x, y) are related by the following equations (Figure 16.15a):

$$x = r\cos\theta, \quad r^2 = x^2 + y^2$$

$$y = r\sin\theta, \quad \theta = \tan^{-1}\frac{y}{x}$$

Applying these relationships, the radial and tangential bending moments are found from Equation 4.48 as

$$M_r = -D\left(\frac{d^2w}{dr^2} + \frac{\nu}{r}\frac{dw}{dr}\right) \tag{16.56a}$$

$$M_\theta = -D\left(\frac{1}{r}\frac{dw}{dr} + \nu\frac{d^2w}{dr^2}\right) \tag{16.56b}$$

and the twisting moment $M_{r\theta} = 0$. The flexural rigidity, defined by Equation 4.49, is $D = Et^3/12(1 - \nu^2)$. To obtain the maximum stresses in polar coordinates, it is necessary to replace subscripts x by r and y by θ in Equation 4.50:

$$\sigma_{r,\max} = \frac{6M_r}{t^2}, \quad \sigma_{\theta,\max} = \frac{6M_\theta}{t^2}, \tag{16.57}$$

in which t is the thickness of the plate.

The differential equation describing the surface deflection, as determined from Equation 4.51 in a similar fashion [4], has the form

$$\nabla^4 w = \frac{1}{r}\frac{d}{dr}\left\{r\frac{d}{dr}\left[\frac{1}{r}\frac{d}{dr}\left(r\frac{dw}{dr}\right)\right]\right\} = \frac{p}{D} \tag{16.58}$$

The deflection w is obtained by successive integration when $p(r)$ is given:

$$w = \int\frac{1}{r}\int r\int\frac{1}{r}\int\frac{pr}{D}\,dr\,dr\,dr\,dr \tag{16.59}$$

When the plate is under a *uniform loading* $p = p_o$, the general solution of Equation 16.58 is

$$w = c_1\ln r + c_2 r^2\ln r + c_3 r^2 + c_4 + \frac{p_o r^4}{64D} \tag{16.60}$$

Here, the c's are constants of integration.

Example 16.7: Pump Diaphragm Stress Analysis

Determine deflection and stress for a clamped circular plate of radius a, representing a pump diaphragm, subjected to uniformly distributed load p_o (Figure 16.16).

Assumptions: Presuppositions of the thin-plate bending theory given in Section 4.10 applies.

Design Decision: The boundary conditions are

$$w = 0, \quad \frac{dw}{dr} = 0, \quad (r = a)$$

Solution

The terms involving logarithms in Equation 16.60 give an infinite displacement at $r = 0$ for all values of c_1 and c_2 except 0, so $c_1 = c_2 = 0$. Satisfying the preceding conditions, we have

$$c_3 = \frac{p_o a^2}{32D}, \quad c_4 = \frac{p_o a^4}{64D}$$

The deflection is then

$$w = \frac{p_o}{64D}\left(a^2 - r^2\right)^2 \tag{16.61}$$

The maximum displacement, occurring at the center of the plate, is

$$w_{\max} = \frac{p_o a^4}{64D} \tag{16.62}$$

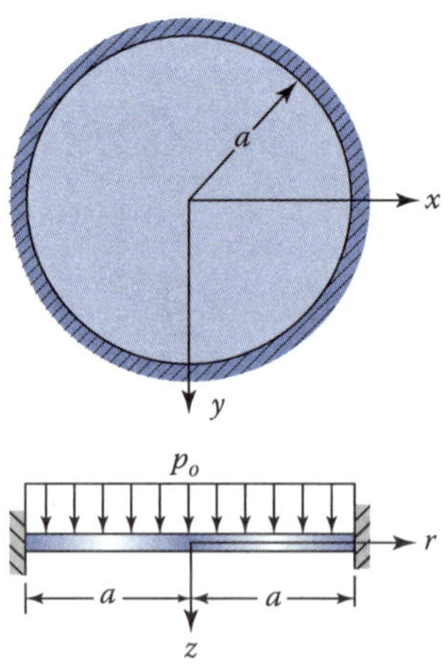

FIGURE 16.16
Example 16.7.

Expressions for the bending moments may be determined by Equation 16.61 in the form

$$M_r = \frac{p_o}{16}\left[(1+\nu)a^2 - (3+\nu)r^2\right]$$
$$M_\theta = \frac{p_o}{16}\left[(1+\nu)a^2 - (1+3\nu)r^2\right] \tag{16.63}$$

Algebraically extreme values of the moments are found at the center and at the edge. At the edge ($r = a$), Equations 16.63 result in

$$M_r = -\frac{p_o a^2}{8}, \quad M_\theta = -\frac{\nu p_o a^2}{8}$$

while at $r = 0$, $M_r = M_\theta = (1 + \nu)p_o a^2/16$. We observe that the maximum moment occurs at the edge. Hence, we have the value of the maximum stress as

$$\sigma_{r,\max} = \frac{6M_r}{t^2} = -\frac{3p_o}{4}\left(\frac{a}{t}\right)^2$$

Comment: The minus sign means a compressive stress in the bottom half of the plate.

A similar procedure may be applied to symmetrically loaded circular plates subjected to different end conditions and loading. For reference purposes, Table 16.2 presents the maximum principal stress and maximum deflection for some commonly encountered cases. Design calculations are facilitated by this type of compilation.

TABLE 16.2

Equations for Maximum Stresses $\sigma_{\max}$ and Maximum Deflections $w_{\max}$ That Occur at the Center of Circular Plates

Supporting and Loading	$\sigma_{\max}$	$w_{\max}$
Edge simply supported, load uniform	$\frac{3}{8}(3+\nu)p_o\frac{a}{t^2}$	$\frac{3}{16}(1-\nu)(5+\nu)\frac{p_o a^2}{Et^3}$
Edge simply supported, load at center, $P = \pi r_c^2 p_o$ $r_c \to 0$, but $r_c > 0$	$\frac{3(1+\nu)P}{2\pi t^2}\left(\frac{1}{1+\nu} + \ln\frac{a}{r_c} - \frac{1-\nu}{1+\nu}\frac{r_c^2}{4a^2}\right)$	$\frac{3(1-\nu)(1+\nu)Pa^2}{4\pi Et^3}$
Fixed edge, load at center, $P = \pi r_c^2 p_o$ $r_c \to 0$, but $r_c > 0$	$\frac{3(1+\nu)P}{2\pi t^2}\left(\ln\frac{a}{r_c} + \frac{r_c^2}{4a^2}\right)$ $(\text{for } a > 1.7r_c)$	$\frac{3(1-\nu^2)Pa^2}{4\pi Et^2}$

Notes: p_o, uniform load per unit area; a, radius of plate; r_c, radius of central loaded area; t, thickness of plate; ν, Poisson's ratio; E, modulus of elasticity.

*16.10 Thin Shells of Revolution

Structural members resembling curved plates are known as *shells*. Examples include aircraft fuselages, pressure vessels, and a variety of containers. As was the case of plates, we limit our consideration to isotropic, homogeneous, elastic shells having a constant thickness that is small relative to the remaining dimensions. The surface bisecting the shell thickness is called the *midsurface*. To specify the geometry of a shell, we need to know only the form of the midsurface and the thickness of the shell at each point. In practice, if the ratio of thickness t to radius of curvature r equals or is less than 1/20, the shell is classified as *thin*.

16.10.1 Shell Theories

The stress analysis of shells normally embraces two distinct theories. The *membrane theory* is limited to moment-free membranes, which usually applies to a rather large proportion of the entire shell. The *bending theory* or general theory includes the effects of bending and enables us to treat discontinuities in the stress distribution occurring in a limited region in the vicinity of a load application, an abrupt change in the geometry, or in regions near boundaries. This method usually involves a membrane solution, corrected in those areas in which discontinuity effects are pronounced [4]. The *following assumptions* are generally made in the *small-deflection analysis* of thin shells:

1. The ratio of the shell thickness to the radius of curvature of the midsurface is small compared to unity.
2. Displacements are very small compared with the shell thickness.
3. Sections perpendicular to the midsurface remain so to the deformed midsurface after bending.
4. The component of stress normal to the midsurface, σ_z, is negligible.

16.10.2 Geometry of Shell of Revolution

Consider a particular type of shell having a surface of revolution, as shown in Figure 16.17. The midsurface of this shell is generated by rotation of a so-called meridian curve about an axis lying in the plane of the curve. The figure shows that a point on the shell is conveniently located by coordinates θ, ϕ, r, and elemental surface $ABCD$ is defined by two meridians and two parallel circles or parallels.

The planes associated with the principal radii of curvature r_ϕ and r_θ at any point on the midsurface of the shell are the *meridian plane* and the *parallel plane* at the point in question, respectively. The radii of curvature r_ϕ and r_θ are then related to sides CD and AC. The radius r_θ generates the shell surface in the direction perpendicular to the direction of the tangent to the meridian curve. The two radii r_ϕ and r_θ are related since $r_0 = r_\theta \sin\phi$ (Figure 16.17). Hence, lengths AC and CD of the curvilinear shell element are $r_0 d\theta = r_\theta \sin\phi d\theta$ and $r_\phi d\phi$, respectively.

16.10.3 Symmetrically Loaded Shells of Revolution

In axisymmetric problems involving shells of revolution, there are only two unknown *membrane forces per unit length*, the so-called *hoop force* per unit length N_θ and *meridional force* per unit length N_ϕ. The governing equations for these forces are developed from two conditions

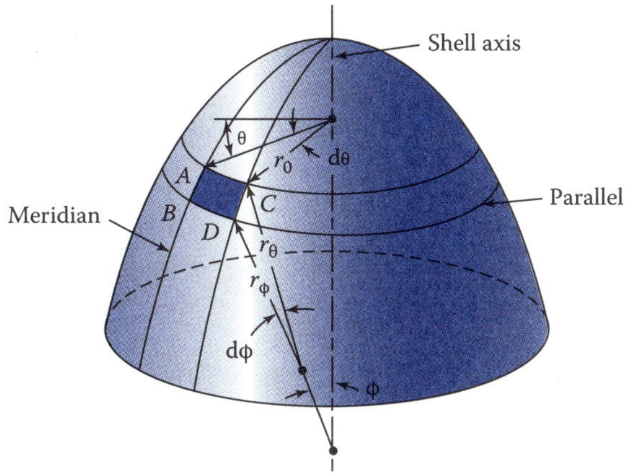

FIGURE 16.17
Geometry of shell of revolution.

of equilibrium. Figure 16.18a shows the element *ABCD* cut from the shell of Figure 16.17. Due to the condition of symmetry, the membrane forces and the loading display no variation with θ. We take the coordinate axes *x* and *y* tangent at *A* to the lines of principal curvature and the axis *z* normal to the midsurface (Figure 16.18a). Externally applied forces per unit area are represented by the components p_y and p_z in the *y* and *z* directions, respectively.

16.10.3.1 Equations of Equilibrium

It can be verified that [4] the equilibrium of *z*-directed forces (Figure 16.18a) results in a basic relation for the axisymmetrically loaded shell of revolution:

$$\frac{N_\phi}{r_\phi} + \frac{N_\theta}{r_\theta} = -p_z \tag{16.64a}$$

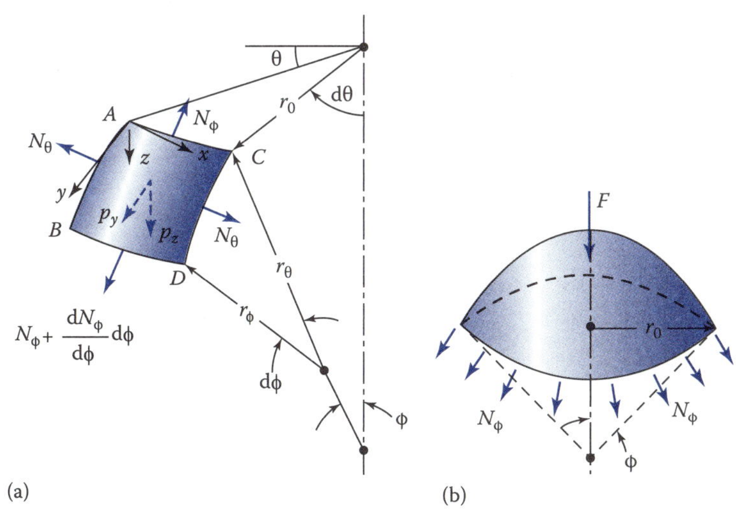

FIGURE 16.18
(a) Shell element with membrane forces per unit length and loading. (b) Truncated shell.

This is often referred to as the *membrane equation* of the shell. An equation for the equilibrium of y-directed forces of the shell element may be derived similarly. But instead of solving the z and y equilibrium equations simultaneously, it is more convenient to determine N_ϕ from the equilibrium of the vertical forces acting on the portion of the shell subtended by the angle ϕ (Figure 16.18b):

$$N_\phi = -\frac{F}{2\pi r_0 \sin\phi} \tag{16.64b}$$

Then calculate N_θ from Equation 16.64a. Here, F represents the *resultant* of all *external loading* acting on the portion of the shell.

16.10.3.2 Compatibility of Deformations

In axisymmetrically loaded shells of revolution, due to their freedom of motion in the z direction, strains are produced to assure consistency with the stress distribution. These strains are compatible with one another. When a shell is under a concentrated surface loading or is constrained at its boundaries, membrane theory cannot satisfy the conditions on deformation everywhere. In such cases, however, the departure from membrane behavior is limited to a narrow region in the vicinity of the loading or the boundary. Membrane theory remains valid for the major portion of the shell, but the complete solution can be determined only by applying the bending theory [4].

16.11 Special Cases of Shells of Revolution

Many types of shells are used as machine and structural components. The membrane stresses in any particular axisymmetrically loaded shell in the form of a surface of revolution may readily be found from Equation 16.64. Presented in this section is a brief discussion of three commonly encountered members. Note that the following groups do not relate only to the special types of axisymmetric loading but are quite general.

16.11.1 Spherical Shell

Figure 16.19a shows a spherical shell. For this case, we need only set the mean radius $r = r_\theta = r_\phi$ and hence $r_0 = r\sin\phi$. Equations 16.64 then reduce to

$$\begin{aligned} N_\phi + N_\theta &= -p_z r \\ N_\phi &= -\frac{F}{2\pi r \sin^2\phi} \end{aligned} \tag{16.65}$$

Example 16.8 illustrates the application.

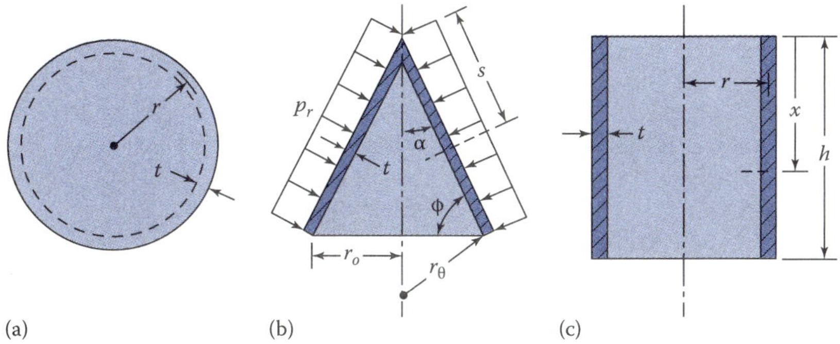

FIGURE 16.19
Common types of shells of revolution: (a) spherical, (b) conical, and (c) vertical cylindrical.

The simplest case is that of a spherical shell under internal pressure. We have $p = -p_z$, $\phi = 90°$, and $F = -\pi r^2 p$. Owing to the symmetry of the spherical shell, $N_\phi = N_\theta = N$. Carrying these quantities into Equation 16.65 gives the membrane stress in a *spherical pressure vessel*:

$$\sigma = \frac{N}{t} = \frac{pr}{2t} \tag{16.66}$$

The quantity t is the thickness of the shell. Equation 16.66 represents uniform stresses in all directions of a pressurized sphere. As pointed out in Section 3.4, a sphere is the most favorably stressed shape for a vessel requiring the minimum wall thickness. It is used for extremely high-pressure operations and employed in space vehicles and missiles for the storage of liquefied gases at lower pressures but with lightweight thin walls.

16.11.2 Conical Shells

Figure 16.19b illustrates a conical shell under a radial loading p_r. We need only to set $r_\phi = \infty$ in Equation 16.64a. This, together with Equation 16.64b, gives the following pair of equations for obtaining the membrane forces per unit length under uniform pressure $p_z = p_r$:

$$\begin{aligned} N_\theta &= p_r r_\theta = -\frac{p_r r_0}{\sin\theta} \\ N_\phi &= N_s = -\frac{F}{2\pi r_0 \sin\phi} \end{aligned} \tag{16.67}$$

in which s represents the direction of the generator. The tangential or hoop and meridian stresses in a conical shell are found by dividing N_θ and N_s by the thickness t of the shell.

16.11.3 Circular Cylindrical Shells

To obtain the membrane forces in a circular shell (Figure 16.19c), we set $\phi = \pi/2$, $p_z = p_r$, and mean radius $r = r_0$ = constant in the cone expressions. Equations 16.67 are then

$$N_x = -\frac{F}{2\pi r}, \quad N_\theta = -p_r r \tag{16.68}$$

where x represents the axial direction of the cylinder.

For a closed-ended *cylindrical vessel* subjected to internal pressure $p = -p_r$ and $F = -\pi r^2 p$, the preceding equations then give the tangential or hoop and the axial stresses as

$$\sigma_\theta = \frac{pr}{t}, \quad \sigma_x = \frac{pr}{2t} \tag{16.69}$$

The quantity t is the thickness of the shell. We note that for thin-walled cylindrical pressure vessels, $\sigma_\theta = 2\sigma_x$.

Example 16.8: Force Analysis of a Spherical Dome

Determine expressions for the meridional and the hoop stresses in a spherical dome of radius a and thickness t, carrying only its own weight p per unit area (Figure 16.20).

Assumptions: The simple support shown in the figure is free to move as the shell deforms under loading; no bending is produced in the neighborhood of the edge.

Solution

The weight of that portion of the dome corresponding to the angle ϕ is obtained as

$$F = \int_0^\phi p(2\pi a \sin\phi \cdot a\, d\phi) = 2\pi a^2 p(1-\cos\phi)$$

Also,

$$p_z = p\cos\phi$$

Introducing into Equation 16.65 the preceding expressions for p_z and F, we have

$$\begin{aligned} N_\phi &= -\frac{ap(1-\cos\phi)}{\sin^2\phi} = -\frac{ap}{1+\cos\phi} \\ N_\theta &= -ap\left(\cos\phi - \frac{1}{1+\cos\phi}\right) \end{aligned} \tag{16.70}$$

Division of the results by t gives the expressions for the membrane stresses.

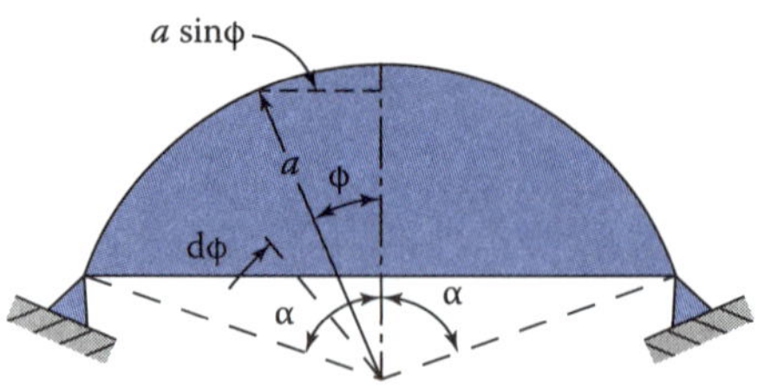

FIGURE 16.20
Example 16.8.

Comments: In Equation 16.70, the negative signs mean compression. Obviously, N_ϕ is always compressive. The sign of N_ϕ, on the other hand, depends on ϕ. By the second equation, when $N_\theta = 0$, $\phi = 51°50'$. For ϕ smaller than this value, N_θ is compressive. If $\phi > 51°50'$, N_θ is tensile.

Example 16.9: Design of a Parabolic Pressure Vessel

A parabolic shell is closed at the top by a thick plate and subjected to internal pressure p (Figure 16.21a). At level *A–A*, calculate the minimum permissible thickness t_{all} of the shell.

Given: $p = 200$ psi

Design Decision: The parabola is $y = x^2/4$, in which x and y are in inches.

Assumptions: At level *A–A*, the allowable membrane stress is $\sigma_{all} = 16.5$ ksi. Section *A–A* is away from the top.

Solution

Let the load resultant for the portion of the shell *below* plane *A–A* be *F* (Figure 16.21b). At level *A–A*, $y = 25$ in.:

$$x = r_0 = \sqrt{4(25)} = 10 \text{ in.}$$

and hence, $dy/dx = x/2 = 5$. From geometry,

$$r_\theta = 10\left(\frac{\sqrt{26}}{5}\right) = 10.2 \text{ in.}$$

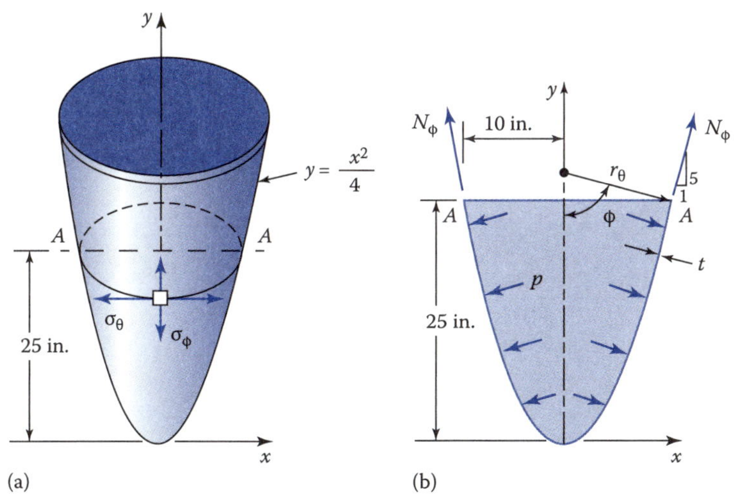

FIGURE 16.21
Example 16.9.

The familiar expression for the curvature then gives

$$r_\phi = \frac{\left[1+(dy/dx)^2\right]^{1.5}}{d^2y/dx^2} = \frac{(1+5^2)^{1.5}}{1/2} = 265.15 \text{ in.}$$

The membrane forces per unit length at *A–A* can now be obtained by applying Equation 16.64:

$$N_\phi = -\frac{F}{2\pi r_0 \sin\phi} = -\frac{-p\pi r_0^2}{2\pi r_0\left(5/\sqrt{26}\right)} = \frac{\sqrt{26}pr_0}{10}$$

$$= \frac{\sqrt{26}(200)(10)}{10} = 1.02 \text{ kips}$$

$$\frac{N_\phi}{r_\phi} + \frac{N_\theta}{r_\theta} = -p_z; \quad \frac{1.02}{265.15} + \frac{N_\theta}{10.2} = 0.2$$

Solving, $N_\theta = 2.0$ kips. Inasmuch as $N_\theta > N_\phi$, we have

$$t_{\text{all}} = \frac{N_\theta}{\sigma_{\text{all}}} = \frac{2.0}{16.5} = 0.121 \text{ in.}$$

Comment: The required shell thickness should be ⅛ in. at level *A–A*.

Example 16.10: Design of a Conical Storage Tank

A thin-walled container of conical shape supported from the top and filled with a heavy-liquid metal of specific weight γ is shown in Figure 16.22 (see Example 6.6). Determine

a. The expressions for the tangential stress σ_θ and meridional stress σ_ϕ
b. The factor of safety n against yielding, using the maximum shear stress criterion

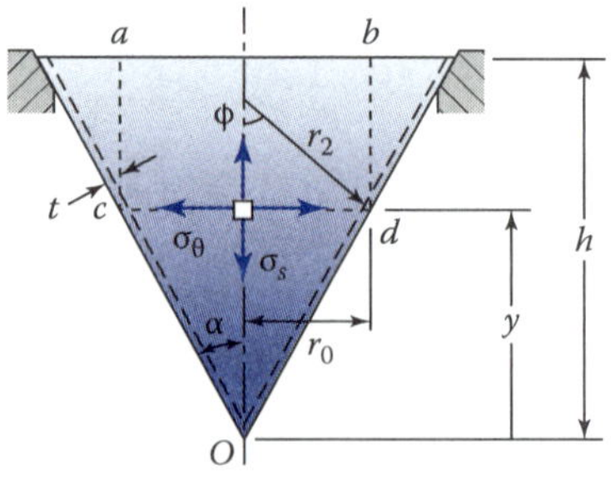

FIGURE 16.22
Conical tank.

Given: The container is made of ASTM A36 structural steel.

$$h = 3 \text{ m}, \quad \alpha = 45^\circ$$

$$t = 3 \text{ mm}, \quad \gamma = 80 \text{ kN/m}^3$$

$$S_y = 250 \text{ MPa} \quad \text{(from Table B.1)}$$

Assumption: The simple support depicted in the figure, free to move as the shell deforms under loading, ensures that no bending is produced in the neighborhood of the edge.

Solution

Referring to Figure 16.22, we write

$$\phi = \frac{\pi}{2} - \alpha, \quad r_0 = y \tan\alpha$$

where α is the half-cone angle. At any arbitrary level y, the pressure is

$$p = p_z = \gamma(h - y)$$

a. Substitution of the foregoing expressions into the first of Equation 16.67, after division by t, results in the hoop stress:

$$\sigma_\theta = \frac{\gamma(h-y)y}{t} \frac{\tan\alpha}{\cos\alpha} \tag{16.71a}$$

The load equals the weight of the liquid of volume *acOdb*. Then,

$$F = -\pi\gamma y^2\left(h - y + \frac{1}{3}y\right)\tan^2\alpha$$

Carrying this value into the second of Equation 16.67, and dividing the resulting expression by t, gives the meridional stress:

$$\sigma_\phi = \sigma_s = \frac{\gamma(h - 2y/3)y}{2t} \frac{\tan\alpha}{\cos\alpha} \tag{16.72a}$$

Comments: We see from these expressions that at the bottom ($y = 0$), the pressure and the membrane stresses vanish. The exact stress distribution at the apex and edge is obtained by application of the bending theory [4].

b. Differentiating with respect to y and equating to 0, it can be verified that the maximum values of the preceding membrane stresses and their location are as follows:

$$\sigma_{\theta,\max} = \frac{\gamma h^2}{4t} \frac{\tan\alpha}{\cos\alpha} \quad (at\ y = h/2) \tag{16.71b}$$

$$\sigma_{s,\max} = \frac{3\gamma h^2}{16t} \frac{\tan\alpha}{\cos\alpha} \quad (\text{at } y = 3h/4) \tag{16.72b}$$

Note that the largest stress occurs at midheight and is given by Equation 16.71b. The design equation is given by Equation 6.18 in Example 6.6 as

$$\frac{\gamma h^2}{4t}\frac{\tan\alpha}{\cos\alpha} = \frac{S_y}{n} \tag{16.73}$$

Introducing the required values, we have

$$\frac{80(3)^2}{4(3)}\frac{\tan 45^\circ}{\cos 45^\circ} = \frac{250}{n}$$

from which $n = 2.95$.

16.12 Pressure Vessels and Piping

The discussions of the last two sections are limited to the membrane stresses occurring over the entire wall thickness of thin shells. As noted previously, the membrane theory must be augmented by bending theory in the vicinity of a discontinuity in load or geometry and in regions near boundaries. The latter approach considers membrane forces as well as shear forces and moments acting on the shell structure. Complete bending theory is mathematically intricate, and the first solutions involving shell-bending stresses date back to only 1920. Since then, great many problems in shell-like structures have been worked out.

Ever-broadening use of vessels for storage, industrial processing, and power generation under unique conditions of temperature, pressure, and environment has given special emphasis to analytical, numerical, and experimental techniques for determining the appropriate working stresses. The finite element method has gained considerable favor in the design of vessels over other methods. A discontinuity of the membrane action in a vessel occurs at all points of external constraint or at the junction of the cylindrical shell and its head or end, possessing different stiffness characteristics. Any incompatibility of deformation at the joint produces bending moments and shear forces. The stresses due to this bending and shear are called *discontinuity stresses*. Most of the situations involving pressure vessels and piping are covered in codes.

Since the bending is of a local character, the discontinuity stresses become negligibly small within a short distance. The narrow region at the edge of *spherical*, *elliptical*, and *conical* vessel heads can be assumed as *nearly cylindrical* in shape. Cylindrical shell (see Figure 16.23) equations can be used to obtain an approximate solution applicable at the juncture of vessels having spherical, elliptical, or conical ends. In the case of *flat-end* vessels, expressions for circular plates and cylindrical shells are utilized. The derivation of the discontinuity stresses in pressure vessels, using the bending and membrane theories of shells and the method of superposition, is well known [3] and is not given here. As in thick-walled cylinders, the tangential or *hoop stress* is usually the *largest* and most critical in the design of the thin-walled vessels and piping.

16.12.1 Short Catalog of Results

Results of some common thin-walled vessel and pipe problems encountered in routine design are listed in the following paragraphs. Thin shell equations under uniform

FIGURE 16.23
A cylindrical pressure vessel.

pressure apply to internal pressure p. They also pertain to cases of external pressure if the sign of p is changed. However, stresses so obtained are valid only if the pressure is not significant relative to that which causes failure by elastic instability. A degree of caution is necessary when applying the formulas for which there is uncertainty as to applicability and restriction of use. Particular emphasis should be given to the fact that high loading, extreme temperature, and rigorous performance requirements present difficult design challenges.

For a thin shell ($a/t > 10$) of radius a and thickness t, the geometric parameters are expressed by the following:

Cylinder

$$\beta^4 = \frac{Et}{4a^2D} \tag{16.74a}$$

It has the dimensions of L^{-1}, the reciprocal of length.

Sphere

$$\lambda = \beta a \tag{16.74b}$$

The flexural rigidity of the shell is

$$D = \frac{Et^3}{12(1-\nu^2)} \tag{16.75}$$

where
- E represents the modulus of elasticity
- ν is the Poisson's ratio

Referring to the preceding two sections,

σ_x = the axial bending stress in cylinder
σ_θ = the circumferential bending stress
σ_ϕ = the meridional bending stress in sphere
w = the radial displacement or deflection of cylinder

Moments M and forces V and P per unit length have units (in SI) N · m/m, or simply N, and N/m, respectively.

The circular cylindrical shell or thin-walled pipe, of special importance in engineering, is often classified into long and short. A *long cylindrical shell* loaded at one end or at the midspan is defined as of length

$$L > \frac{2\pi}{\beta} \tag{16.76}$$

Conditions at one end do not affect to the other end if the ends are far enough apart. Most cylindrical shells are mathematically long even when physically short. Cases 1 and 2 apply to *long* thin-walled pipes.

CASE 16.1 Long Pipe, Radial Center Load *P* per Unit Length

$$w = \frac{Pe^{-\beta x}}{8D\beta^3}\left[\sqrt{2}\sin\left(\beta x + \frac{\pi}{4}\right)\right] \tag{a}$$

$$w_{\max} = \frac{Pa^2\beta}{2Et} \quad (\text{at } x = 0)$$

$$\sigma_{x,\max} = \frac{3P}{2\beta t^2}, \quad \sigma_{\theta,\max} = \frac{P\beta}{2}\left(-\frac{a}{t} + \frac{3\nu}{\beta^2 t^2}\right) \quad (at\ x = 0)$$

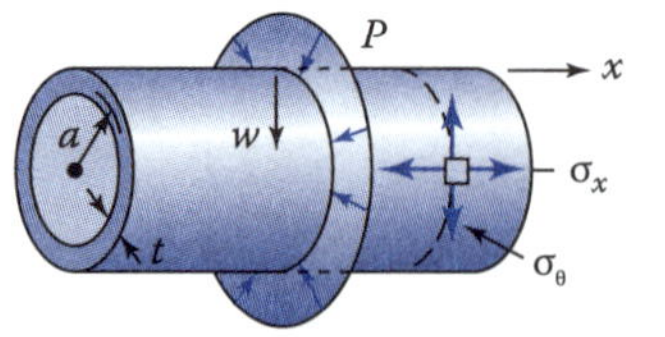

Equation (a) shows that the deflection (and moment) attenuates with distance as an exponentially damped sine wave of wave length, for $\nu = 0.3$, $2\pi/\beta \approx 4.89\sqrt{at}$; bending is of local character.

CASE 16.2 Long Pipe, End Moment *M* per Unit Length

$$w_{\max} = -\frac{M}{2D\beta^2} \quad (at\ x = 0)$$

$$\sigma_{x,\max} = \frac{6M}{t^2}, \quad \sigma_{\theta,\max} = \nu\sigma_{x,\max} \quad (\text{at } x = 0)$$

CASE 16.3 Cylindrical Vessel with Hemispherical Heads, Pressure *p*

$$\sigma_{x,\max} = 1.293\frac{pa}{2t} \quad (\text{at } x = \pi/4\beta)$$

$$\sigma_{\theta,\max} = 1.032\frac{pa}{t} \quad (\text{at } x = 1.85/\beta)$$

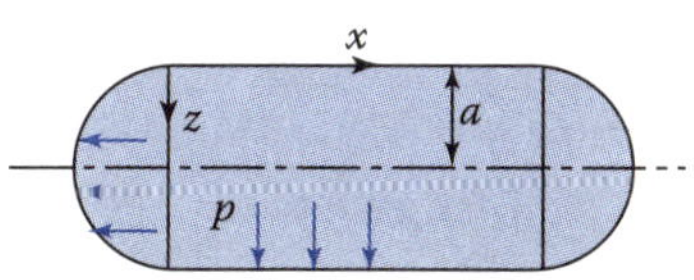

The thickness t is uniform and $\nu = 0.3$. The foregoing maximum discontinuity stresses occur at the outer surface of the cylinder.

CASE 16.4 Cylindrical Vessel with Ellipsoidal Heads, Pressure *p*

$$\sigma_{x,\max} = 2.172\frac{pa}{2t}$$

$$\sigma_{\theta,\max} = 1.128\frac{pa}{t}$$

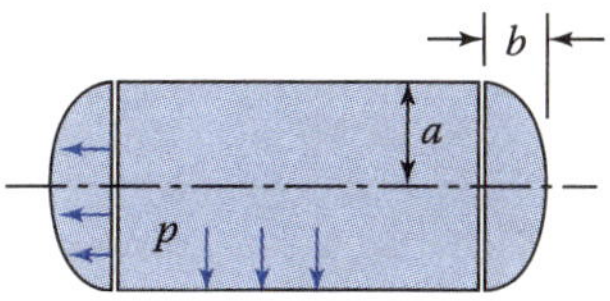

The thickness t is uniform, $\nu = 0.3$, and $b = a/2$. The preceding maximum discontinuity stresses take place at the same locations of the cylinder as in Case 16.3. Interestingly, the membrane stress in the spherical heads ($pa/2t$) is lower than in the cylinder (pa/t). But for $a = 2b$, the largest membrane stress in the elliptical heads equals that in the cylinder.

16.12.2 ASME Code for Pressure Vessels

With the advent of nuclear plants and outer space and undersea explorations, much attention is being given to the analysis and design of pressure vessels. We observe in the preceding section that the discontinuity stress in the vicinity of the joint of a cylindrical shell and its head is quite a bit larger than the membrane stress in either portion. When a vessel is properly designed and constructed, however, these stresses are reduced greatly, and it becomes unnecessary to consider them.

The ASME *code for pressure vessels* [9,10] lists formulas for calculating the required minimum thickness of the shell and the ends. The following factors and a host of others contributing to an ideal vessel design are described by the code: approved techniques for joining the head to the shell, formulas for computing the thickness of shell and end, materials used in combination, temperature ranges, maximum allowable stress values, corrosion, types of closure, and so on. Table 16.3 gives the minimum required vessel thicknesses.

The required wall thickness for tubes and pipes under internal pressure is obtained according to the rules for a shell in the code. In its present status, the code applies when the pressure does not exceed 3000 psi. Pressures in excess of this amount may require special attention in the design and construction of the vessels, closures, and branch connections of piping systems. Also, the preceding is only a partial description of the code specifications that, if complied with, provide assurance that discontinuity stresses may be neglected. For the complete requirements, reference should be made to the current edition of the code. The ASME publishes relevant books, conference papers, and a quarterly *Journal of Pressure Vessel Technology*.

TABLE 16.3

ASME Design Code for Pressure Vessels

Geometry	Required Thickness (in.)
Cylindrical shell	$t = \dfrac{pr}{\sigma e - 0.6p}$
Hemispherical head	$t = \dfrac{pr}{2\sigma e - 0.2p}$
Conical head	$t = \dfrac{pD}{2\cos\alpha(\sigma e - 0.6p)}$
Flat head	$t = d\sqrt{\dfrac{cp}{\sigma}}$
Ellipsoidal head	$t = \dfrac{pD}{2\sigma e - 0.2p}$
Wholly spherical shell	$t = \dfrac{pr}{2\sigma e - 0.2p}$

Notes: p, internal pressure (psi); r, inside radius of shell or hemispherical head (in.); σ, maximum allowable stress (psi); e, lowest joint efficiency ($e < 1.0$); D, inside diameter of conical head or inside length of the major axis of an ellipsoidal head (in.); α, half the apex angle of the conical head; d, diameter of flat head (in.); c, a numerical coefficient depending on the method of attachment of head (such as $c = 0.5$ for circular plates welded to the end of the shell).

16.13 Filament-Wound Pressure Vessels

A unique class of composites, formed by wrapping of high-strength filaments over a mandrel, followed by impregnation of the windings with a plastic binder and removal of the mandrel in pieces, is called *filament-wound cylinders*. A common system is the glass filament/epoxy resin combination. Filament structures of this type have an exceptional strength/weight ratio and reliability. They are in widespread use as lightweight vessels and thrust chambers in spacecraft, rockets, and airborne vehicles. Basic filament vessels contain longitudinal, circumferential, or helical windings. A combination of these windings is used if necessary.

Consider a *filament-wound vessel* with closed ends subjected to an internal pressure p (Figure 16.24). The tangential and axial stresses due to p are, from Equation 16.69,

$$\sigma_\theta = \frac{pa}{t}, \quad \sigma_x = \frac{pa}{2t}$$

Hence,

$$\frac{\sigma_\theta}{\sigma_x} = 2 \tag{16.77}$$

The quantities a and t represent the average radius and wall thickness, respectively, of the vessel composed entirely of filament of tensile strength S_u and the binder.

The maximum tensile force carried by the filament may be expressed by

$$F = S_u w t$$

where w is the filament width wrapped at angle ψ. The corresponding circumferential force is

$$F_\theta = F \sin \psi$$

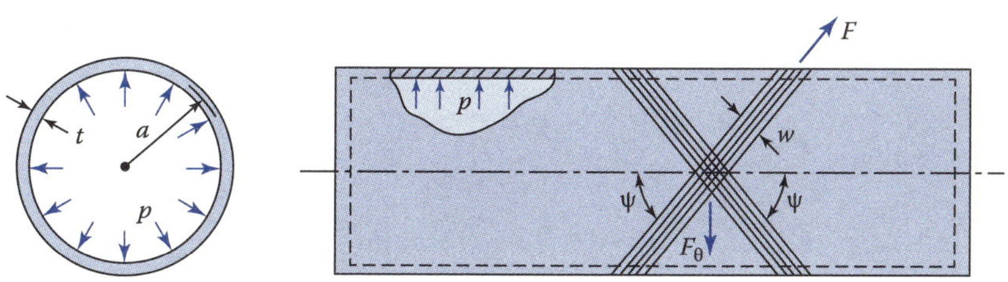

FIGURE 16.24
Filament-wrapped cylindrical pressure vessel. *Notes*: F, tension force in the filament; ψ, helix angle; w, width of filaments; F_0, tangential component of filament force; p, pressure.

The filament cross-sectional area $A = wt/\sin\psi$. The tangential stress filament then can carry

$$\sigma_\theta = \frac{F_\theta}{A} = S_u \sin^2 \psi \tag{16.78}$$

In a like manner, the axial stress may readily be ascertained in the form

$$\sigma_x = S_u \cos^2 \psi \tag{16.79}$$

The preceding expressions give

$$\frac{\sigma_\theta}{\sigma_x} = \tan^2 \psi \tag{16.80}$$

The *optimum helix angle of filament*, by Equation 16.77 and 16.80, is

$$\frac{\sigma_\theta}{\sigma_x} = 2 = \tan^2 \psi \tag{16.81}$$

which yields $\psi = 54.7°$. This represents the condition of helical wrapping to support an internal pressure.

Note that by additional use of circumferential filaments, the helix angle may be decreased for convenience in wrapping. Clearly, the preceding analysis applies only to the cylindrical portion of the vessel, away from the ends. Filament winding is also accomplished by laying down a pattern over a base material and forming a so-called filament-overlay composite [3,11], for example, thin-walled (polyethylene) pipe overlaid with (nylon) cord or a wire of the same material of the shell.

16.14 Buckling of Cylindrical and Spherical Shells

The procedures of analysis and design of thin-walled curved structures or shells in a state of static equilibrium have been taken up in some detail in Sections 16.10 through 16.12. But, thin shell structures subjected to compression may fail by instability of the walls. The failure by *buckling* is similar to that which occurs in the compression of wing and fuselage structures, missile casings, long pipes, pressure vessels, and so forth. This section provides only a brief introduction to the buckling of a thin-walled homogeneous shell. Formulas for critical stresses for pressurized vessels are also furnished. Shell stability is an important area of engineering design [12].

16.14.1 Cylindrical Shells under Uniaxial Compression

Consider a simply supported circular cylindrical shell of radius a, length L, and thickness t, subjected to uniform axial pressure (Figure 16.25). We note that the instability is particularly significant under this type of loading, because axial buckling is often synonymous with collapse of the shells. To determine the critical loading, the methods described in Chapter 6 may be applied.

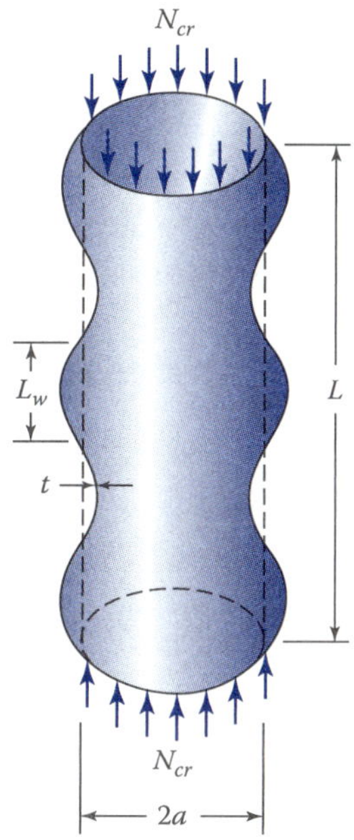

FIGURE 16.25
Buckled shape of a cylindrical shell under uniaxial pressure. The initial form is depicted by the dashed lines.

It can be shown that [4] for $\nu = 0.3$, the length of half-sine waves L_w into which the shell buckles may be expressed in the form

$$L_w = \pi \sqrt[4]{\frac{a^2t^2}{12(1-\nu^2)}} \approx 1.72\sqrt{at} \tag{16.82}$$

The corresponding minimum *buckling load* is

$$N_{cr} = \frac{Et^2}{a\sqrt{3(1-\nu^2)}} = 0.605\frac{Et^2}{a} \tag{16.83}$$

The *critical stress* is then $\sigma_{cr} = N_{cr}/t = 0.605E(t/a)$. Observe that the σ_{cr} depends on material properties, thickness, and radius and is independent of cylinder length. A thin-walled tube can wrinkle like tissue paper when subjected to axial compression. The value of the critical load defined by Equation 16.83 often does not agree with experimental data. To relate the theoretical values introduced to actual test data, it is necessary to incorporate an empirical factor.

16.14.2 Cylindrical and Spherical Pressure Vessels

The determination of critical shell stresses is a very laborious process. The results are listed in Table 16.4 for four practical applications [3,9]. With them comes the off-repeated warning that they are valid only for pressure vessels of $a/t > 10$.

We point out that there are situations in which vessels under internal pressure can fail by buckling, such as hydraulic tubing loaded by rams. Various types of pressure vessel

TABLE 16.4

Critical Pressure p_{cr} Formulas for Buckling of Thin-Walled Vessels

A. Long cylindrical shells under external pressure

$$p_{cr} = \frac{Et^3}{4a^3(1-\nu^2)}$$

B. Long cylindrical shells under internal pressure from ram action

$$p_{cr} = \pi^2 \frac{Eta}{L^2}$$

C. Spherical shells under external pressure

$$p_{cr} = \frac{2Et^2}{a^2\sqrt{3(1-\nu^2)}}$$

D. Spherical end caps under external pressure

$$p_{cr} = \frac{2Et}{a\sqrt{3(1-\nu^2)}}$$

Notes: t, wall thickness; a, mean radius; L, length; E, modulus of elasticity; ν, Poisson's ratio.

heads may also collapse when subjected to internal pressure owing to the existence of large tangential compressive stresses in the knuckle region. The problems involving buckling by external pressure have been of considerable interest, particularly for vacuum systems and structures, like submarines. Interestingly, *spheres*, in addition to the desirable features noted in Section 16.11, have the *greatest* buckling resistance. In single or tandem configuration, they form the pressure-carrying structure and living space in most deep-submergence oceanographic vehicles.

Problems

Sections 16.1 through 16.6

16.1 A cylinder of inner radius a and the outer radius $3a$ is subjected to an internal pressure p_i. Determine the limiting values of the p_i applying

a. The maximum shear stress theory

b. The maximum energy of distortion theory

Design Decision: The cylinder is made of steel of $S_y = 260$ MPa.

16.2 A solid steel shaft of radius b is pressed into a steel disk of outer radius $2b$ and the length of hub engagement $l = 3b$ (Figure 16.6). Determine the value of the radial interference in terms of b.

Given: The shearing stress in the shaft caused by the torque that the joint is to carry equals 100 MPa; $E = 210$ GPa, $f = 0.15$.

16.3 For an ASTM A36 structural steel cylinder of inner radius $a = 120$ mm and outer radius $b = 180$ mm, find

a. When $p_o = 0$, the largest internal pressure and the maximum displacement

b. When $p_i = 0$, the largest external pressure

Assumption: Maximum tangential stress is not to exceed the yield strength of the material.

16.4 A cylinder of inner radius a and outer radius $2a$ is under internal pressure p_i. Calculate the allowable value of p_i using

a. The maximum principal stress theory

b. The Coulomb–Mohr theory

Design Decisions: The cylinder is made of aluminum of S_u = 350 MPa and S_{uc} = 650 MPa.

16.5 A cast iron disk is to be shrunk on a 125 mm diameter steel shaft. Determine

a. The contact pressure

b. The minimum allowable outside diameter of the disk

Requirement: The tangential stress in the disk is not to exceed 60 MPa.

Given: The radial interference is 0.05 mm, E_h = 100 GPa, E_s = 200 GPa, and $\nu = 0.3$.

16.6 A cast iron pinion with 100 mm dedendum diameter and l = 50 mm hub engagement length is to transmit a maximum torque of 150 N·m at low speeds. Calculate

a. The required radial interference on a 25 mm diameter steel shaft

b. The maximum stress in the gear due to a press fit

Given: E_c = 100 GPa, E_s = 200 GPa, $\nu = 0.3$, $f = 0.15$

16.7 A bronze bushing 50 mm in outer diameter and 30 mm in inner diameter is to be pressed into a hollow steel cylinder of 100 mm outer diameter. Determine the tangential stresses for the steel and bronze at the boundary between the two parts.

Given: E_b = 105 GPa, E_s = 210 GPa, $\nu = 0.3$

Design Requirement: The radial interference is $\delta = 0.025$ mm.

16.8 A cast iron cylinder of outer radius 150 mm is to be shrink fitted over a 50 mm radius steel shaft. Calculate the maximum tangential and radial stresses in both parts.

Given: E_c = 120 GPa, $\nu_c = 0.25$, E_s = 210 GPa, $\nu_s = 0.3$

Design Requirement: The radial interference is $\delta = 0.03$ mm.

16.9 When a steel disk of external diameter $4b$ is shrunk onto a steel shaft of diameter of $2b$, the internal diameter of the disk is increased by an amount λ. What reduction occurs in the diameter of the shaft?

Given: $\nu = 0.3$

16.10 A brass tube of inner radius a and outer radius b is shrink fitted at p = 90 MPa into a brass collar of outer radius c (Figure 16.6). Determine the speed at which the contact pressure becomes zero.

Given: a = 20 mm, b = 30 mm, c = 40 mm, $\rho = 8.5\ \text{kN}\cdot\text{s}^2/\text{m}^4$, $\nu = 0.34$ (Table B.1)

16.11 A thick-walled disk flywheel has inner and outer radii of b and $4b$, respectively. Determine

a. The radius b

b. The kinetic energy delivered for a 5% drop in speed

Given: The maximum speed is 3600 rpm with a maximum stress from rotation equals to 75 MPa and l = 50 mm.

Design Decisions: The disk is made of steel having $\rho = 7.8\ \text{kN}\cdot\text{s}^2/\text{m}^4$ and $\nu = 0.3$.

16.12 A 200 mm diameter disk is shrunk onto a 40 mm diameter shaft. Determine

a. The initial contact pressure required if the contact pressure is to be 3.2 MPa at 2400 rpm
b. The maximum stress when not rotating

Design Decision: Both members are made of steel having $E = 210$ GPa, $\nu = 0.3$, and $\rho = 7.8\ \text{kN}\cdot\text{s}^2/\text{m}^4$.

16.13 A flywheel of 600 mm outer diameter and 100 mm inner diameter is to be press fit on a solid shaft with a radial interference of 0.02 mm. Calculate

a. The maximum stress in the assembly at standstill
b. The speed n in rpm at which the press fit loosens as a result of rotation

Design Decision: Both members are made of steel with $E = 200$ GPa, $\nu = 0.3$, and $\rho = 7.8\ \text{kN}\cdot\text{s}^2/\text{m}^4$.

16.14 A solid steel shaft of radius b is to be press fit into a wrought iron hub of outer radius c and length l. Find

a. The interface pressure
b. The force needed for the press fit
c. The torque capacity of the assembly

Given: $b = 60$ mm, $c = 120$ mm, $l = 200$ mm, $E_s = 200$ GPa, $E_i = 190$ GPa, $\nu = 0.3$ (Table B.1).

Assumptions: $f = 0.18$, and the maximum tangential stress will be 30 MPa.

16.15 A 60 mm thick steel flywheel has inner and outer radii, a and b, respectively. Determine the average braking torque required.

Given: $a = 50$ mm, $b = 200$ mm, $\rho = 7.8\ \text{kN}\cdot\text{s}^2/\text{m}^4$

Design Requirement: The flywheel speed must be reduced from 2400 to 1200 rpm in 2 rev.

16.16 A rolled steel disk flywheel has inner radius a, outer radius b, and length of hub engagement of l. It rotates on a shaft at a normal speed of 3000 rpm with a 10% drop during working cycle. Determine

a. The maximum stress
b. The energy delivered per cycle

Given: $a = 25$ mm, $b = 250$ mm, $l = 60$ mm, $\rho = 7.8\ \text{kN}\cdot\text{s}^2/\text{m}^4$, $\nu = 0.3$

Sections 16.7 and 16.8

16.17 The cross section of the circular cast iron frame of Figure 16.14a has a channel form, as shown in Figure P16.17. The dimensions are in millimeters. Determine the maximum load P.

Given: $r_A = r_i = 215$ mm

Design Decisions: Stress does not exceed 100 MPa on the critical section. Winkler's formula is used.

16.18 A curved wrought iron frame with a rectangular cross section is acted upon by the bending moment as illustrated in Figure P16.18. Find

a. The tangential stresses σ_i and σ_o of the inside and outside fibers, respectively, applying the curved-beam formula
b. Redo item (a) by the flexure formula

Given: $M = 900\ \text{N}\cdot\text{m}$, $b = 20$ mm, $h = 40$ mm, $r_i = 50$ mm

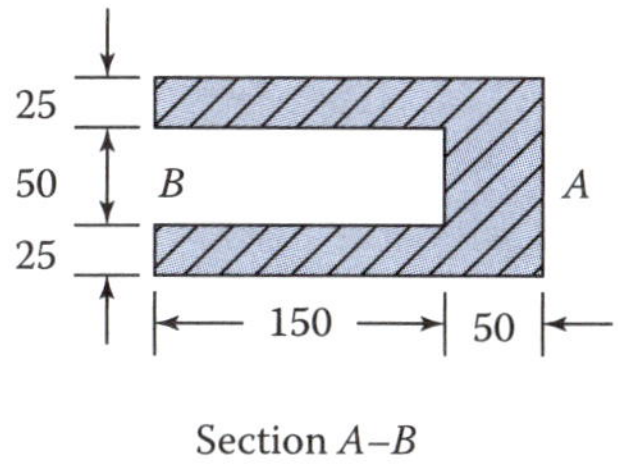

FIGURE P16.17

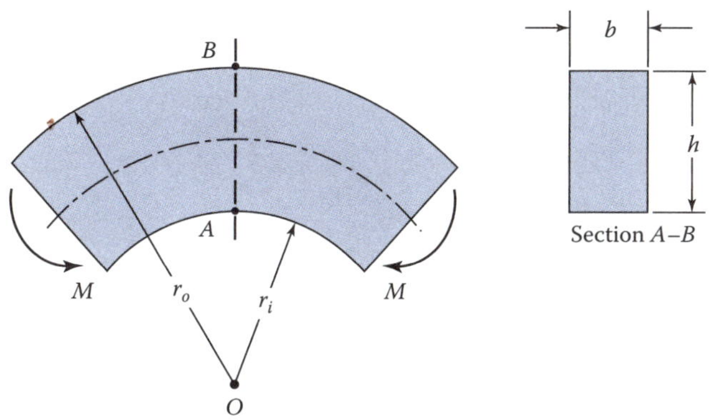

FIGURE P16.18

16.19 A rectangular aluminum machine frame is curved to a radius $\bar{r}$ along the centroidal axis and carries end moments M as shown in Figure P16.18. Compute the circumferential stresses σ_i and σ_o of the inner and outer fibers, respectively.

Given: $M = 12$ kips · in., $\bar{r} = 8$ in., $b = 1\frac{3}{4}$ in., $h = 2\frac{1}{2}$ in.

16.20 The allowable stress in compression for the clamp body shown in Figure P16.20 is 120 MPa. Calculate, applying Winkler's formula, the maximum permissible load the member can carry. The dimensions are in millimeters.

16.21 A steel frame with a square cross section is curved to a radius $\bar{r}$ along the centroidal axis and subjected to end moments M as illustrated in Figure P16.21. Find the largest allowable value of the bending moment M, knowing that the permissible stress is σ_{all}.

Given: $r_i = 220$ mm, $r_o = 280$ mm, $b = h = 60$ mm, $\sigma_{all} = 150$ MPa.

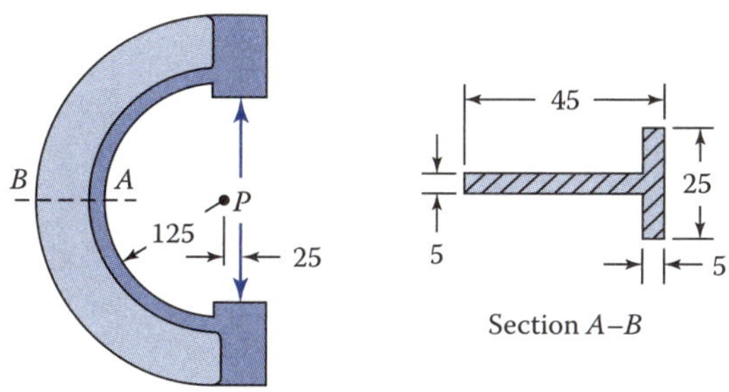

FIGURE P16.20

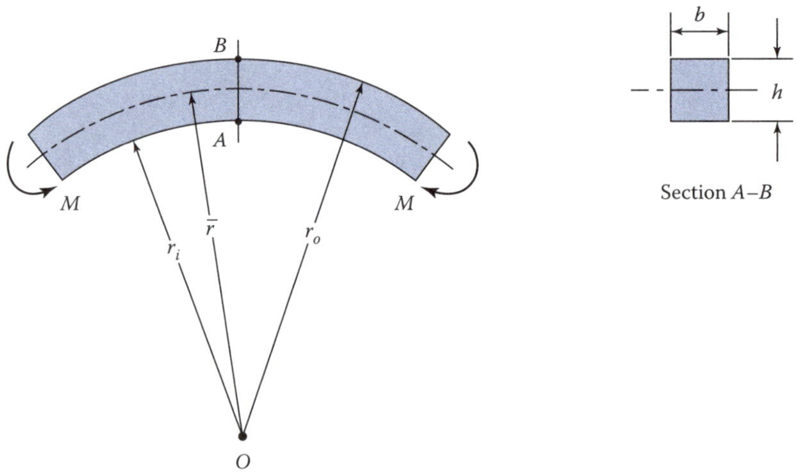

FIGURE P16.21

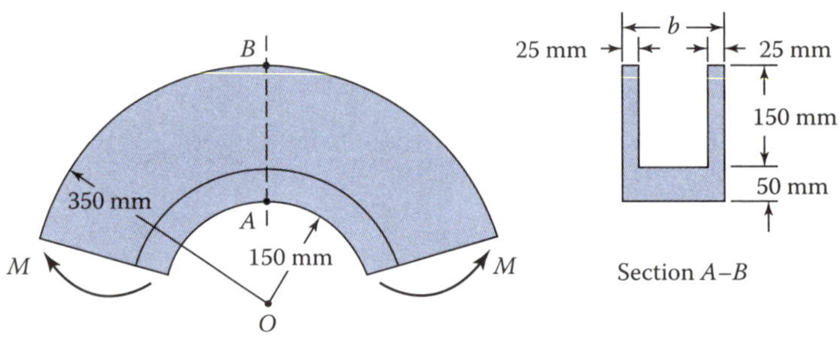

FIGURE P16.22

16.22 Figure P16.22 shows a beam of channel-shaped cross section subjected to end moments M. What is the dimension b required in order that the tangential stresses at points A and B of the beam are equal in magnitude?

16.23 Calculate, using Winkler's formula, the maximum distance d for which tangential stress does not exceed 80 MPa on the cross section A-B of the frame shown in Figure P16.23.

Given: $P = 25$ kN

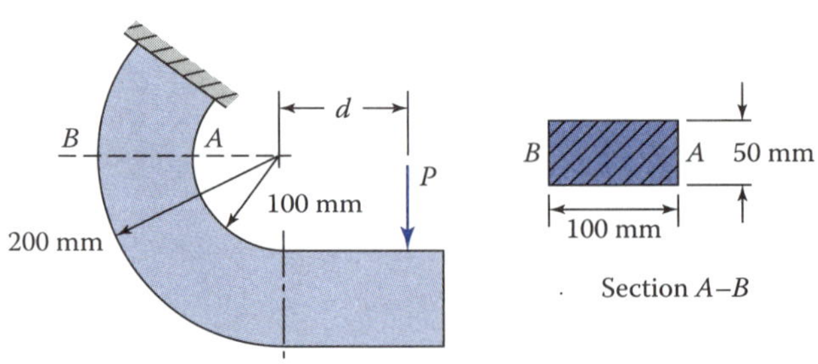

FIGURE P16.23

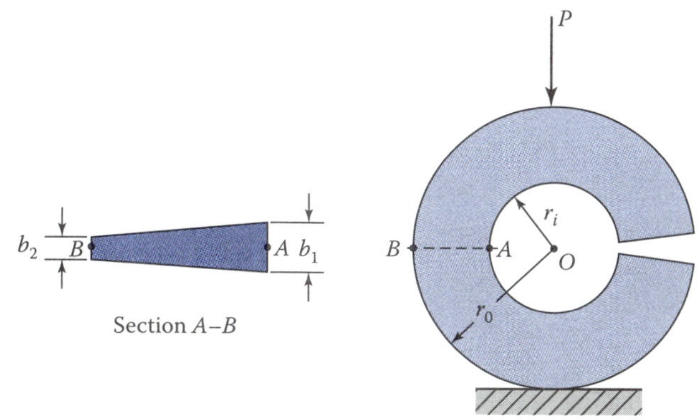

FIGURE P16.24

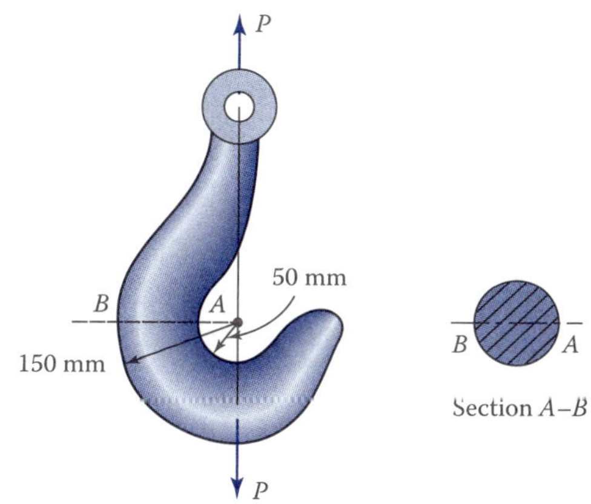

FIGURE P16.25

16.24 Figure P16.24 illustrates a split-ring frame with an inner radius r_i, outer radius r_o, and a trapezoidal cross-sectional area. What are the values of the circumferential stresses at points A and B?

Given: $r_i = 3.2$ in., $r_o = 8$ in., $b_1 = 3$ in., $b_2 = 2$ in., $P = 15$ kips

16.25 For the crane hook of circular cross section in Figure P16.25, determine

a. The maximum load P that may be supported without exceeding a stress of 150 MPa at point A

b. The tangential stress at point B for the load found in part A

Design Decision: Use Winkler's formula.

16.26 A steel machine frame of an elliptical cross section is fixed at one end and acted upon a concentrated load P at the free end as shown in Figure P16.26. Find the tangential stresses at points A and B.

Given: $r_i = 5$ in., $r_o = 9$ in., $a = 4$ in., $b = 2$ in., $P = 25$ kips

16.27 The allowable stress for the cast iron frame with an elliptical cross section illustrated in Figure P16.26 is σ_{all}. What is the maximum load P that can be applied in the frame?

Given: $r_i = 130$ mm, $r_o = 200$ mm, $a = 70$ mm, $b = 35$ mm, $\sigma_{all} = 90$ MPa

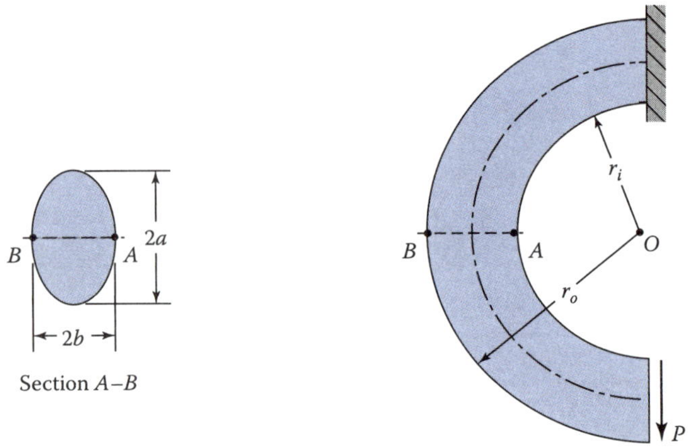

FIGURE P16.26

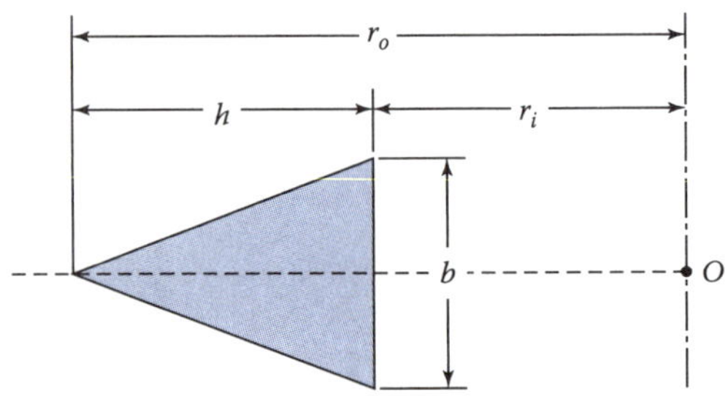

FIGURE P16.28

16.28 Figure P16.28 illustrates the triangular cross section of a machine frame. Derive the expression for the radius R along the neutral axis and compare the result with that listed in Table 16.1.

16.29 Consider the circular cross section of a machine frame shown in Figure P16.29. Develop the expression for the radius R along the neutral axis and compare the result with that furnished in Table 16.1.

16.30 The trapezoidal cross section of a structural frame is illustrated in Figure P16.30. Determine the expression for the radius R along the neutral axis and compare the result with that given in Table 16.1.

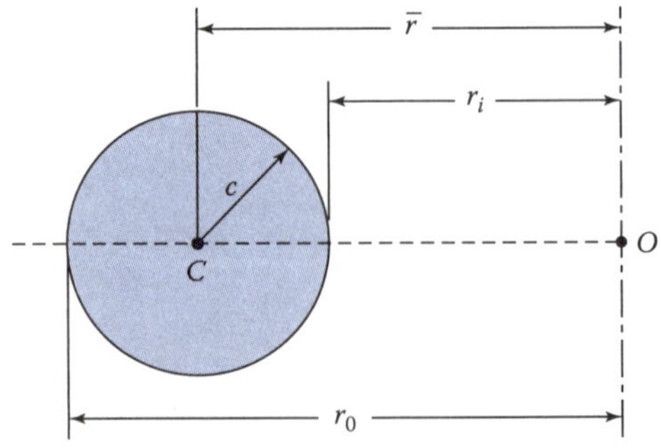

FIGURE P16.29

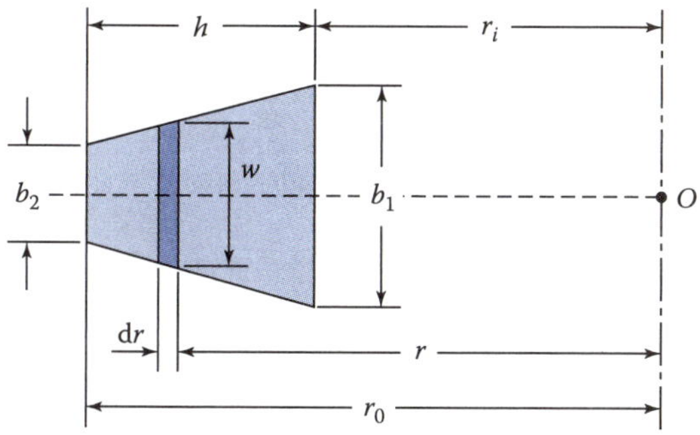

FIGURE P16.30

Sections 16.9 through 16.14

16.31 Determine maximum deflection and maximum tangential stress for a simply supported circular plate of radius a under uniformly distributed load p_o.

16.32 A circular clamped-edge window of an aircraft is under uniform pressure differential p_o between the cabin and the outside. Calculate the maximum value of the safety factor n, based on the following criteria:

a. The maximum shear stress

b. The maximum energy of distortion

Given: $\nu = 1/3$

Design Decisions: The plate is made of a material having yield strength S_y, thickness t, and radius a.

16.33 A pressure vessel control system includes a thin plate that is to close an electrical circuit by deflecting 0.05 in. at the center when the pressure reaches a value of 500 psi. Design the plate.

Given: $E = 30 \times 10^6$ psi, $\nu = 0.3$, $S_y = 80$ ksi; the plate has a radius of $a = 1.5$ in.

Assumption: The plate is simply supported at the edge.

16.34 An aluminum alloy flat-clamped disk valve ($S_y = 150$ MPa, $\nu = 1/3$) of 10 mm thickness is under a liquid pressure of 200 kPa. Design the plate according to the maximum shear stress failure criterion.

Requirement: The factor of safety is 2.

16.35 A football of uniform skin thickness t is subjected to an internal pressure of p (Figure P16.35). Calculate the maximum stress at point A.

Given: $t = 2$ mm, $p = 100$ kPa

16.36 In the toroidal shell of Figure P16.36 under internal pressure p, the maximum membrane forces per unit length are

$$N_\theta = \frac{pa}{2}, \quad N_\phi = \frac{pa}{b + a\sin\phi}\left(\frac{a}{2}\sin\phi + b\right)$$

Calculate the required minimum thickness t of the vessel.

Given: $p = 2.2$ MPa, $a = 50$mm, $b = 250$ mm, $S_y = 240$ MPa, $n = 1.2$

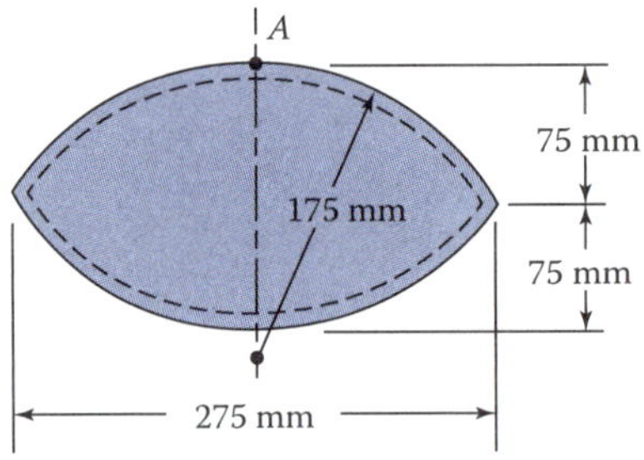

FIGURE P16.35

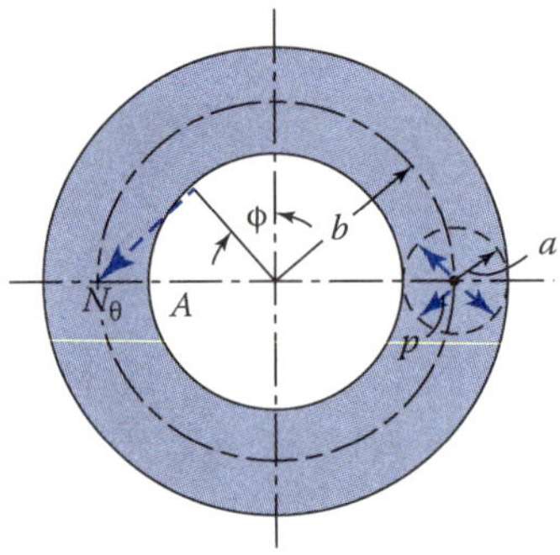

FIGURE P16.36

16.37 A circular cylindrical pressure vessel of radius a is under an internal pressure p. Determine the required wall thickness t, using a factor of safety of n and the following failure criteria:

a. The maximum shear

b. The maximum energy of distortion

c. The Coulomb–Mohr theory

Design Decision: The cylinder is made of a material with yield strength S_y and ultimate strength in compression equals twice the ultimate strength in tension. $S_{uc} = 2S_u$.

16.38 A closed-ended vertical cylindrical tank of radius r and height h (Figure 16.19c) is completely filled with a liquid of density γ and subjected to an additional gas pressure of p. Determine the wall thickness t required

a. At the bottom

b. At $x = 3h/4$

c. At the ends, approximating these as clamped thin circular plates

Given: $r = 5$ m, $h = 16$ m, $\gamma = 15$ kN/m^3, $p = 200$ kPa, and allowable stress is $\sigma_{all} = 150$ MPa.

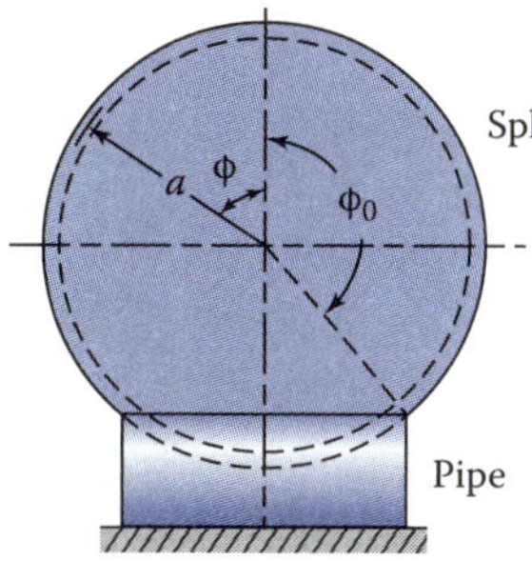

FIGURE P16.39

16.39 A spherical tank is filled with a liquid of density γ and supported on a cylindrical pipe, as shown in Figure P16.39. Verify that the membrane force per unit length is given by the expressions:

$$N_\phi = \frac{\gamma a^2}{6\sin^2\phi}\left[1-\cos^2\phi(3-2\cos\phi)\right] = \frac{\gamma a^2}{6}\left(1-\frac{2\cos^2\phi}{1+\cos\phi}\right)$$

$$N_\theta = \frac{\gamma a^2}{6}\left(5-6\cos\phi+\frac{2\cos^2\phi}{1+\cos\phi}\right) \tag{P16.39}$$

The preceding equations are valid for $\phi > \phi_0$.

16.40 During a stage of firing, a long cylindrical missile casing of diameter *d* and thickness t is subjected to axial compression. Calculate

a. The half-length of the sine waves into which the shell buckles

b. The limiting stress for failure

Given: $d = 1.5$ m, $t = 12$ mm

Design Decisions: The cylinder is made of steel of $S_y = 350$ MPa and $E = 210$ GPa.

16.41 A long pipe of diameter *d* and thickness *t* is to be used in a structure as a column. Determine the maximum axial load that can be applied without causing the shell to buckle.

Given: $d = 4$ ft, $t = \frac{1}{2}$ in.

Design Decision: The pipe is made of steel having $E = 30 \times 10^6$ psi.

17

*Finite Element Analysis in Design**

17.1 Introduction

In real design problems, generally, structures are composed of a large assemblage of various members. In addition, the built-up structures or machines and their components involve complicated geometries, loadings, and material properties. Given these factors, it becomes apparent that the classical methods can no longer be used. For complex structures, the designer has to resort to more general approaches of analysis. The most widely used of these techniques is the finite element stiffness or displacement method. Unless otherwise specified, we refer to it as the *finite element method* (FEM).

Finite element analysis (FEA) is a numerical approach and well suited to digital computers. The method is based on the formulations of a simultaneous set of algebraic equations relating forces to corresponding displacements at discrete preselected points (called *nodes*) on the structure. These governing algebraic equations, also referred to as force–displacement relations, are expressed in matrix notation. With the advent of high-speed, large-storage capacity digital computers, the FEM gained great prominence throughout the industries in the solution of practical analysis and design problems of high complexity. The literature related to the FEA is extensive (e.g., [1–10]). Numerous commercial FEA software programs are available, including some directed at the learning process. Most of the developments have now been coded into commercial programs. The FEM offers numerous advantages, including

1. Structural shape of components that can readily be described
2. Ability to deal with discontinuities
3. Ability to handle composite and anisotropic materials
4. Ease of dealing with dynamic and thermal loadings
5. Ability to treat combined load conditions
6. Ability to handle nonlinear structural problems
7. Capacity for complete automation

The basic concept of the finite element approach is that the real structure can be discretized by a finite number of elements, connected not only at their nodes but along the interelement boundaries as well. Usually, triangular or rectangular shapes of elements

* The material presented in this chapter is *optional* and the entire chapter can be omitted without destroying the continuity of the text.

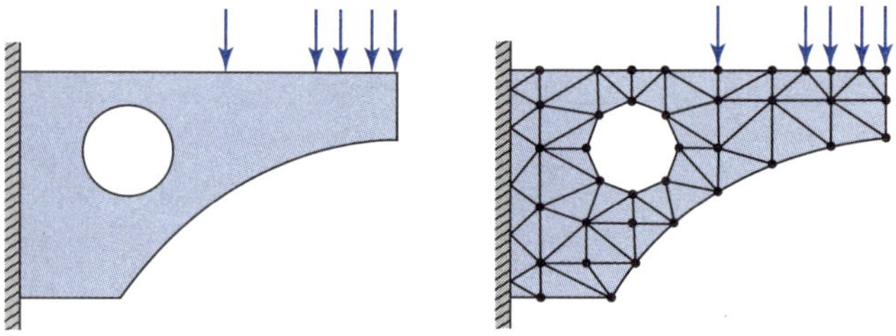

FIGURE 17.1
Tapered plate bracket and its triangular finite element model.

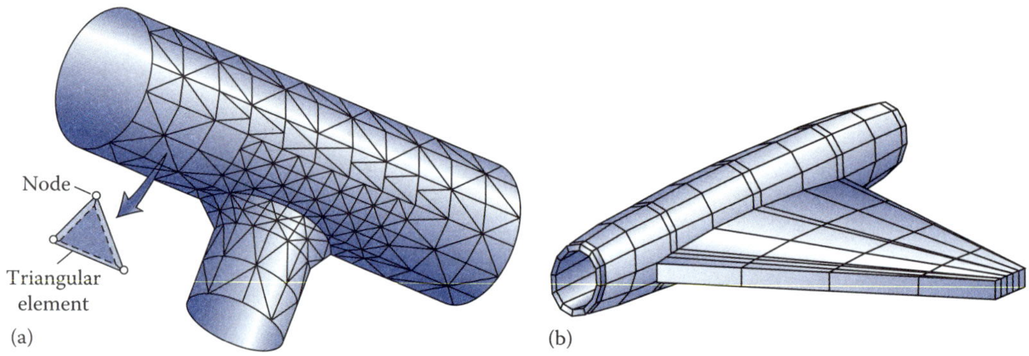

FIGURE 17.2
Finite element models of some components: (a) pipe connection and (b) fuselage and a wing.

are used in the FEM. Figure 17.1 depicts how a real structure is modeled using triangular element shapes. The types of elements commonly employed in structural idealization are the truss, beam, 2D elements, shell and plate bending, and 3D elements. The models of a pipe joint and an aircraft structure [1] created using triangular, beam, plate, and shell elements are shown in Figure 17.2.

Note that the network of elements and nodes that discretize the region is termed *mesh*. The mesh density increases as more elements are placed within a given region. *Mesh refinement* is when the mesh is modified from one analysis of a model to the next analysis to give improved solutions. Results usually improve when the mesh density is increased in areas of high stress concentrations and when geometric transition zones are meshed smoothly. Generally, but not always, the FEA results converge toward the exact solutions as mesh is continuously refined.

To adequately treat the subject of the FEA would require a far more lengthy presentation than could be justified here. Nevertheless, the subject is so important that any engineer concerned with the analysis and design of members should have at least an understanding of FEA. The fundamentals presented can clearly indicate the potential of the FEA as well as its complexities. It can be covered as an option, used as a *teaser* for a student's advance study of the topic, or as a professional reference. For simplicity, only four basic structural elements are discussed here: the 1D axial element or truss element, the beam element or plane frame element, the 2D element, and the axisymmetric element. Sections 17.3, 17.5, and 17.8 present the formulation and general procedure for treating typical problems by the FEM. Solutions of axial stress, plane stress, and axisymmetric problems are demonstrated in various examples and case studies.

17.2 Bar Element

An axial element, also called a *truss bar* or simply *bar element*, can be considered as the simplest form of structural finite element. An element of this type with length L, modulus of elasticity E, and cross-sectional area A is denoted by e (Figure 17.3). The two ends or joints or nodes are numbered 1 and 2, respectively. It is necessary to develop a set of two equations in matrix form to relate the *joint forces* ($\bar{F}_1$ and $\bar{F}_2$) to the *joint displacements* ($\bar{u}_1$ and $\bar{u}_2$).

17.2.1 Direct Equilibrium Method

The following derivation by the direct equilibrium approach is simple and clear. However, this method is practically applicable only for truss and frame elements. The equilibrium of the x-directed forces requires that $\bar{F}_1 = -\bar{F}_2$ (Figure 17.3). Because AE/L is the spring rate of the element, we have

$$\bar{F}_1 = \frac{AE}{L}(\bar{u}_1 - \bar{u}_2), \quad \bar{F}_2 = \frac{AE}{L}(\bar{u}_2 - \bar{u}_1)$$

This may be written in the following matrix form:

$$\begin{Bmatrix} \bar{F}_1 \\ \bar{F}_2 \end{Bmatrix}_e = \frac{AE}{L}\begin{bmatrix} 1 & -1 \\ -1 & 1 \end{bmatrix}\begin{Bmatrix} \bar{u}_1 \\ \bar{u}_2 \end{Bmatrix}_e \tag{17.1a}$$

or symbolically

$$\{\bar{F}\}_e = [\bar{k}]_e\{\bar{u}\}_e \tag{17.1b}$$

The quantity $[\bar{k}]_e$ is called the *stiffness matrix* of the element. Clearly, it relates the joint displacement to the joint forces on the element.

17.2.2 Energy Method

The energy technique is more general, easier to apply, and powerful than the direct approach just discussed, especially for sophisticated types of finite elements. To employ this method, it is necessary to first define a displacement function for the element (Figure 17.3):

$$\bar{u} = a_1 + a_2\bar{x} \tag{17.2}$$

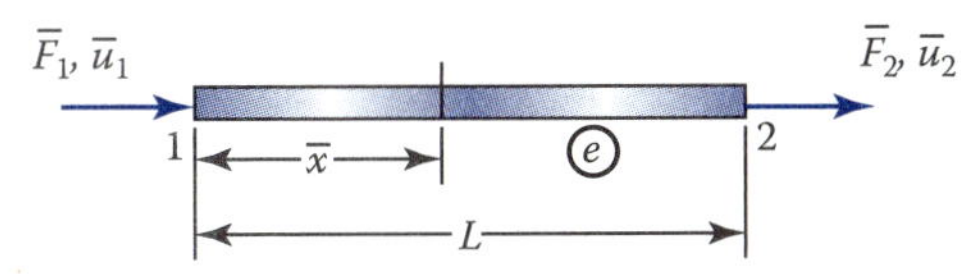

FIGURE 17.3
Axial (truss bar or bar) element.

in which a_1 and a_2 are constants. Clearly, Equation 17.2 represents a linear continuous displacement variation along the x-axis of the element. The axial displacements of joints 1 (at $\bar{x} = 0$) and 2 (at $\bar{x} = L$), respectively, are therefore

$$\bar{u}_1 = a_1, \qquad \bar{u}_2 = a_1 + a_2 L$$

Solving the preceding expressions, $a_1 = \bar{u}_1$ and $a_2 = -(\bar{u}_1 - \bar{u}_2)/L$. Carrying these into Equation 17.2, we have

$$\bar{u} = \left(1 - \frac{\bar{x}}{L}\right)\bar{u}_1 + \frac{\bar{x}}{L}\bar{u}_2 \tag{17.3}$$

Then, by Equation 3.54, the strain is

$$\varepsilon_x = \frac{d\bar{u}}{d\bar{x}} = \frac{1}{L}(-\bar{u}_1 + \bar{u}_2) \tag{17.4}$$

So the element axial force is

$$\bar{F} = (E\varepsilon_x)A = \frac{AE}{L}(-\bar{u}_1 + \bar{u}_2) \tag{17.5}$$

The strain energy in the element is obtained by substituting Equation 17.5 into Equation 5.10 in the following form:

$$U = \int_0^L \frac{F^2 dx}{2AE} = \frac{AE}{2L}\left(\bar{u}_1^2 - 2\bar{u}_1\bar{u}_2 + \bar{u}_2^2\right) \tag{17.6}$$

Applying Castigliano's first theorem, Equation 5.46, we obtain

$$\bar{F}_1 = \frac{\partial U}{\partial \bar{u}_1} = \frac{AE}{L}(\bar{u}_1 + \bar{u}_2)$$

$$\bar{F}_2 = \frac{\partial U}{\partial \bar{u}_2} = \frac{AE}{L}(-\bar{u}_1 + \bar{u}_2)$$

The matrix forms of the preceding equations are the same as those given by Equation 17.1.

17.2.3 Global Stiffness Matrix

We now develop the global stiffness matrix for an element oriented arbitrarily in a 2D plane. The *local coordinates* are chosen to conveniently represent the individual element, whereas the *global* or reference *coordinates* are chosen to be convenient for the whole structure. We designate the local and global coordinate systems for an axial element by $\bar{x}, \bar{y}$ and x, y, respectively (Figure 17.4).

The figure depicts a typical axial element e lying along the $\bar{x}$ axis, which is oriented at an angle θ, measured *counterclockwise*, from the reference axis x. In the local coordinate

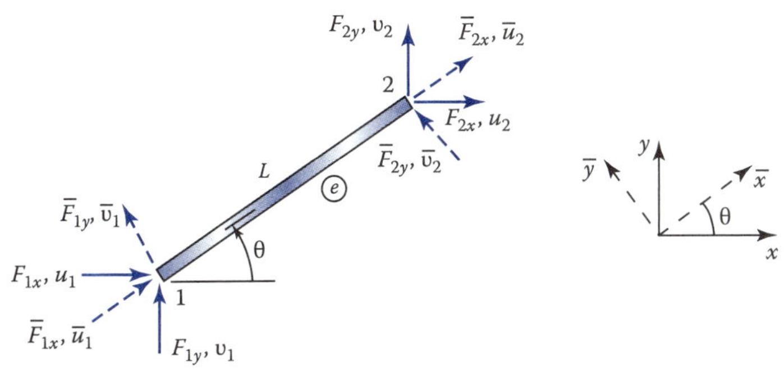

FIGURE 17.4
Local ($\bar{x}$, $\bar{y}$) and global (x, y) coordinates for a typical axial element e. All forces and displacements have a positive sense.

system, each joint has an axial force $\bar{F}_x$, a transverse force $\bar{F}_y$, an axial displacement $\bar{u}$, and a transverse displacement $\bar{\upsilon}$. Referring to Figure 17.4, Equation 17.1a is expanded as

$$\begin{Bmatrix} \bar{F}_{1x} \\ \bar{F}_{1y} \\ \bar{F}_{2x} \\ \bar{F}_{2y} \end{Bmatrix}_e = \frac{AE}{L}\begin{bmatrix} 1 & 0 & -1 & 0 \\ 0 & 0 & 0 & 0 \\ -1 & 0 & 1 & 0 \\ 0 & 0 & 0 & 0 \end{bmatrix}\begin{Bmatrix} \bar{u}_1 \\ \bar{\upsilon}_1 \\ \bar{u}_2 \\ \bar{\upsilon}_2 \end{Bmatrix}_e \tag{17.7a}$$

or

$$\{\bar{F}\}_e = [\bar{k}]_e \{\bar{\delta}\}_e \tag{17.7b}$$

Clearly, $\{\bar{\delta}\}_e$ represents the nodal displacements in the local coordinate system.

We see from Figure 17.4 that the two local and global forces at joint 1 may be related by the following expressions:

$$\bar{F}_{1x} = F_{1x}\cos\theta + F_{1y}\sin\theta$$

$$\bar{F}_{1y} = -F_{1x}\sin\theta + F_{1y}\cos\theta$$

Similar expressions apply at joint 2. For brevity, we designate

$$c = \cos\theta \quad \text{and} \quad s = \sin\theta$$

Thus, the local and global forces are related in the following matrix form:

$$\begin{Bmatrix} \bar{F}_{1x} \\ \bar{F}_{1y} \\ \bar{F}_{2x} \\ \bar{F}_{2y} \end{Bmatrix}_e = \begin{bmatrix} c & s & 0 & 0 \\ -s & c & 0 & 0 \\ 0 & 0 & c & s \\ 0 & 0 & -s & c \end{bmatrix}\begin{Bmatrix} F_{1x} \\ F_{1y} \\ F_{2x} \\ F_{2y} \end{Bmatrix}_e \tag{17.8a}$$

or symbolically

$$\{\bar{F}\}_e = [T]\{F\}_e \tag{17.8b}$$

In the foregoing, $[T]$ is the *coordinate transformation matrix:*

$$[T] = \begin{bmatrix} c & s & 0 & 0 \\ -s & c & 0 & 0 \\ 0 & 0 & c & s \\ 0 & 0 & -s & c \end{bmatrix} \tag{17.9}$$

and $\{F\}_e$ represents the *global nodal force matrix:*

$$\{F\}_e = \begin{Bmatrix} F_{1x} \\ F_{1y} \\ F_{2x} \\ F_{2y} \end{Bmatrix}_e \tag{17.10}$$

In as much as the displacement transforms in the same manner as forces, we have

$$\begin{Bmatrix} \bar{u}_1 \\ \bar{\upsilon}_1 \\ \bar{u}_2 \\ \bar{\upsilon}_2 \end{Bmatrix}_e = [T] \begin{Bmatrix} u_1 \\ \upsilon_1 \\ u_2 \\ \upsilon_2 \end{Bmatrix}_e \tag{17.11a}$$

or

$$\{\bar{\delta}\}_e = [T]\{\delta\}_e \tag{17.11b}$$

Here, $\{\delta\}_e$ is the global nodal displacements. Carrying Equations 17.11b and 17.8b into 17.7b leads to

$$[T]\{F\}_e = [\bar{k}]_e [T]\{\delta\}_e$$

or

$$\{F\}_e = [T]^{-1}[\bar{k}]_e [T]\{\delta\}_e$$

Note that the transformation matrix $[T]$ is an orthogonal matrix; that is, its inverse is the same as its transpose: $[T]^{-1} = [T]^T$, where the superscript T denotes the transpose. The transpose of a matrix is obtained by interchanging the rows and columns. The global *force–displacement relations* for an element e are

$$\{F\}_e = [k]_e\{\delta\}_e \tag{17.12}$$

where

$$[k]_e = [T]^T\left[\bar{k}\right]_e[T] \tag{17.13}$$

Finally, to evaluate the *global stiffness matrix* for the element, we substitute Equation 17.9 and $[k]_e$ from Equation 17.7a into Equation 17.13:

$$[k]_e = \frac{AE}{L}\begin{bmatrix} c^2 & cs & -c^2 & -cs \\ cs & s^2 & -cs & -s^2 \\ -c^2 & -cs & c^2 & cs \\ -cs & -s^2 & cs & s^2 \end{bmatrix} = \frac{AE}{L}\begin{bmatrix} c^2 & cs & -c^2 & -cs \\ & s^2 & -cs & -s^2 \\ & & c^2 & cs \\ \text{Symmetric} & & & s^2 \end{bmatrix} \tag{17.14}$$

This relationship shows that the element stiffness matrix depends on its dimensions, orientation, and material property.

17.2.4 Axial Force in an Element

Reconsider the general case of an axial element oriented arbitrarily in a 2D plane, depicted in Figure 17.4. It can be shown that equation for the axial force is expressed in the following matrix form:

$$F_{12} = \frac{AE}{L}\begin{bmatrix} c & s \end{bmatrix}\begin{Bmatrix} u_2 - u_1 \\ \upsilon_2 - \upsilon_1 \end{Bmatrix} \tag{17.15}$$

This may be written for an element with nodes ij as follows:

$$F_{ij} = \left(\frac{AE}{L}\right)_{ij}\begin{bmatrix} c & s \end{bmatrix}_{ij}\begin{Bmatrix} u_j - u_i \\ \upsilon_j - \upsilon_i \end{Bmatrix} \tag{17.16}$$

A positive (negative) value obtained for F_{ij} indicates that the element is in tension (compression). The axial stress in the element is given by $\sigma_{ij} = F_{ij}/A$.

17.3 Formulation of the Finite Element Method

Development of the governing equations appropriate to a truss demonstrates the formulation of the structural stiffness method, or the FEM. As noted previously, a truss is an assemblage of axial elements that may be differently oriented. To derive truss equations, the global element relations given by Equation 17.12 must be assembled. The preceding leads to the following force–displacement relations for the entire truss, the *system equations*:

$$\{F\}=[K]\{\delta\} \tag{17.17}$$

The *global nodal matrix* $\{F\}$ and the *global stiffness matrix* $[K]$ are

$$\{F\}=\sum_{1}^{n}\{F\}_e \tag{17.18a}$$

$$[K]=\sum_{1}^{n}[k]_e \tag{17.18b}$$

Here, e designates an element and n is the number of elements making up the truss. It is noted that $[K]$ relates the global nodal force $\{F\}$ to the global displacement $\{\delta\}$ for the entire truss.

17.3.1 Method of Assemblage of the Values of $[k]_e$

The element stiffness matrices in Equation 17.18 must be properly added together or *superimposed*. To carry out proper summation, a convenient method is to label the columns and rows of each element stiffness matrix according to the displacement components associated with it. In so doing, the truss stiffness matrix $[K]$ is obtained simply by adding terms from the individual element stiffness matrix into their corresponding locations in $[K]$. This approach of assemblage of the element stiffness matrix is given in Case Study 17.1. An alternative way is to expand the $[k]_e$ for each element to the order of the truss stiffness matrix by adding rows and columns of zeros. However, for the problem involving a large number of elements, it becomes tedious to apply this approach.

17.3.2 Procedure for Solving a Problem

We now illustrate the use of the equations developed in the preceding paragraphs. The general procedure for solving a structural problem by application of the finite element method may be summarized as shown in Figure 17.5. This outline is better understood when applied to planar structures, as shown in the solution of the following sample problem.

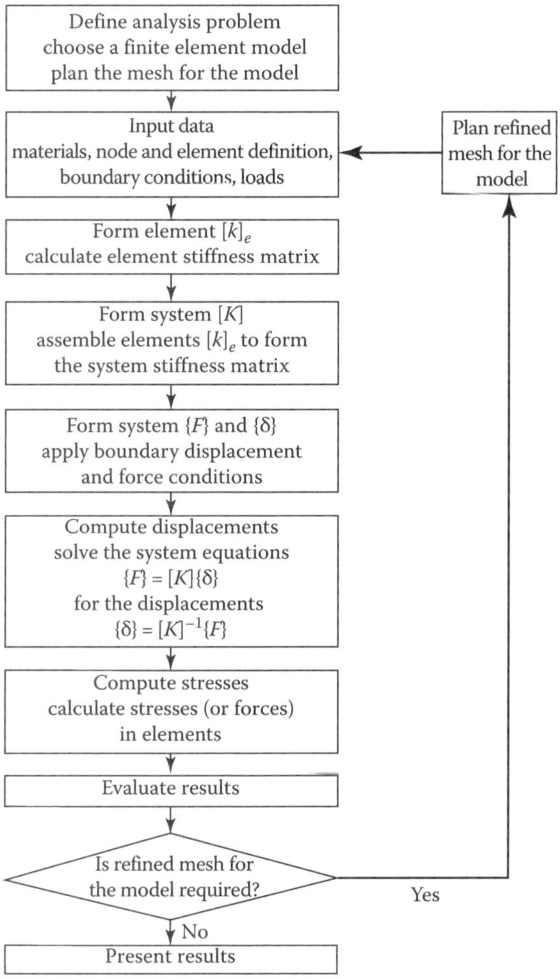

FIGURE 17.5
Finite element block diagram. (From Ugural, A.C., *Stresses in Beams, Plates and Shells*, 3rd ed., CRC Press, Boca Raton, FL, 2009.)

CASE STUDY 17.1 Analysis and Design of a Truss

A three-bar truss 123 (Figure 17.6a) is subjected to a horizontal force P acting at joint 2. Analyze the truss and calculate the required cross-sectional area of each member.

Assumptions: All members will have the same yield strength S_y, length L, and axial rigidity AE. Use a factor of safety of $n = 1.5$ on yielding.

Given: $S_y = 240$ MPa, $P = 200$ kN.

Solution: The reactions are noted in Figure 17.6a. The node numbering is arbitrary for each element.

Input data. At each node, there are two displacements and two nodal force components (Figure 17.6b). Recall that θ is measured counterclockwise from the positive x-axis to each element (Table 17.1). Inasmuch as the terms in $[k]_e$ involve c^2, s^2, and cs, a change in angle from θ to $\theta + \pi$, causing both c and s to change sign, does not affect

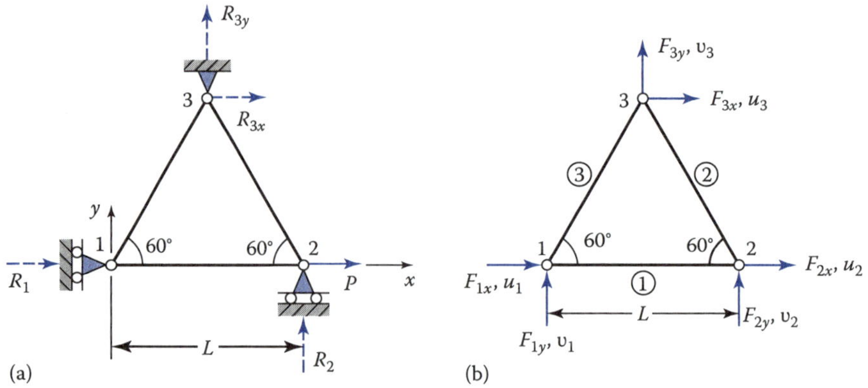

FIGURE 17.6
(a) Basic plane truss and (b) finite element model.

TABLE 17.1
Data for the Truss of Figure 17.6

Element	θ	c	s	c²	cs	s²
1	0°	1	0	1	0	0
2	120°	−1/2	$\sqrt{3}/2$	1/4	$-\sqrt{3}/4$	3/4
3	60°	1/2	$\sqrt{3}/2$	1/4	$\sqrt{3}/4$	3/4

the signs of the terms in the stiffness matrix. For example, in the case of member 3, θ = 60° if measured counterclockwise at node 1 or 240° if measured counterclockwise at node 3. However, by substituting into Equation 17.14, $[k]_e$ remains unchanged.

Element stiffness matrix. Using Equation 17.14 and Table 17.1, we have for the elements 1, 2, and 3, respectively,

$$[k]_1 = \frac{AE}{L}\begin{bmatrix} 1 & 0 & -1 & 0 \\ 0 & 0 & 0 & 0 \\ -1 & 0 & 1 & 0 \\ 0 & 0 & 0 & 0 \end{bmatrix}\begin{matrix} u_1 \\ \upsilon_1 \\ u_2 \\ \upsilon_2 \end{matrix} \quad (\text{columns: } u_1,\ \upsilon_1,\ u_2,\ \upsilon_2)$$

$$[k]_2 = \frac{AE}{4L}\begin{bmatrix} 1 & -\sqrt{3} & -1 & \sqrt{3} \\ -\sqrt{3} & 3 & \sqrt{3} & -3 \\ -1 & \sqrt{3} & 1 & -\sqrt{3} \\ \sqrt{3} & -3 & -\sqrt{3} & 3 \end{bmatrix}\begin{matrix} u_2 \\ \upsilon_2 \\ u_3 \\ \upsilon_3 \end{matrix} \quad (\text{columns: } u_2,\ \upsilon_2,\ u_3,\ \upsilon_3)$$

$$[k]_3 = \frac{AE}{4L}\begin{bmatrix} 1 & \sqrt{3} & -1 & \sqrt{3} \\ \sqrt{3} & 3 & -\sqrt{3} & -3 \\ -1 & -\sqrt{3} & 1 & \sqrt{3} \\ -\sqrt{3} & -3 & \sqrt{3} & 3 \end{bmatrix}\begin{matrix} u_1 \\ \upsilon_1 \\ u_3 \\ \upsilon_3 \end{matrix} \quad (\text{columns: } u_1,\ \upsilon_1,\ u_3,\ \upsilon_3)$$

Note that the column and row of each stiffness matrix are labeled according to the nodal displacements associated with them.

System stiffness matrix. There are a total of six components of displacement for the truss before boundary constraints are imposed. Therefore, the *order* of the truss stiffness matrix must be 6 × 6. Subsequent to addition of the terms from each element stiffness matrices into their corresponding locations in [*K*], we readily obtain the global stiffness matrix for the truss:

$$[K]=\frac{AE}{4L}\begin{array}{c} \begin{matrix} u_1 & \upsilon_1 & u_2 & \upsilon_2 & u_3 & \upsilon_3 \end{matrix} \\ \begin{bmatrix} 4+1 & 0+\sqrt{3} & -4 & 0 & -1 & -\sqrt{3} \\ 0+\sqrt{3} & 0+3 & 0 & 0 & -\sqrt{3} & -3 \\ -4 & 0 & 4+1 & 0-\sqrt{3} & -1 & \sqrt{3} \\ 0 & 0 & 0-\sqrt{3} & 0+3 & \sqrt{3} & -3 \\ -1 & -\sqrt{3} & -1 & \sqrt{3} & 1+1 & \sqrt{3}-\sqrt{3} \\ -\sqrt{3} & -3 & \sqrt{3} & -3 & \sqrt{3}-\sqrt{3} & 3+3 \end{bmatrix} \end{array} \begin{matrix} u_1 \\ \upsilon_1 \\ u_2 \\ \upsilon_2 \\ u_3 \\ \upsilon_3 \end{matrix} \tag{a}$$

System force and displacement matrices. Accounting for the applied load and support constraints, with reference to Figure 17.6, the truss nodal force matrix is

$$\{F\}-\begin{Bmatrix} F_{1x} \\ F_{1y} \\ F_{2x} \\ F_{2y} \\ F_{3x} \\ F_{3y} \end{Bmatrix}-\begin{Bmatrix} R_1 \\ 0 \\ P \\ R_2 \\ R_{3x} \\ R_{3y} \end{Bmatrix} \tag{b}$$

Similarly, accounting for the support conditions, the truss nodal displacement matrix is

$$\{\delta\}=\begin{Bmatrix} u_1 \\ \upsilon_1 \\ u_2 \\ \upsilon_2 \\ u_3 \\ \upsilon_3 \end{Bmatrix}=\begin{Bmatrix} 0 \\ \upsilon_1 \\ u_2 \\ 0 \\ 0 \\ 0 \end{Bmatrix} \tag{c}$$

Displacements. Substituting Equations (a), (b), and (c) into Equation 17.18, the truss force–displacement relations are given by

$$\begin{Bmatrix} R_1 \\ 0 \\ P \\ R_2 \\ R_{3x} \\ R_{3y} \end{Bmatrix}=\frac{AE}{4L}\begin{bmatrix} 5 & \sqrt{3} & -4 & 0 & -1 & -\sqrt{3} \\ \sqrt{3} & 3 & 0 & 0 & -\sqrt{3} & -3 \\ -4 & 0 & 5 & -\sqrt{3} & -1 & \sqrt{3} \\ 0 & 0 & -\sqrt{3} & 3 & \sqrt{3} & -3 \\ -1 & -\sqrt{3} & -1 & \sqrt{3} & 2 & 0 \\ -\sqrt{3} & -3 & \sqrt{3} & -3 & 0 & 6 \end{bmatrix}\begin{Bmatrix} 0 \\ \upsilon_1 \\ u_2 \\ 0 \\ 0 \\ 0 \end{Bmatrix} \tag{d}$$

To determine υ_1 and u_2, only the part of Equation (d) relating to these displacements is considered. We then have

$$\begin{Bmatrix} 0 \\ P \end{Bmatrix} = \frac{AE}{4L}\begin{bmatrix} 3 & 0 \\ 0 & 5 \end{bmatrix}\begin{Bmatrix} \upsilon_1 \\ u_2 \end{Bmatrix}$$

Solving preceding equations simultaneously or by matrix inversion, the nodal displacements are obtained:

$$\begin{Bmatrix} \upsilon_1 \\ u_2 \end{Bmatrix} = \frac{4L}{15AE}\begin{bmatrix} 5 & 0 \\ 0 & 3 \end{bmatrix}\begin{Bmatrix} 0 \\ P \end{Bmatrix} = \frac{4PL}{5AE}\begin{Bmatrix} 0 \\ 1 \end{Bmatrix} \quad \text{(e)}$$

Reactions. The values of υ_1 and u_2 are used to determine reaction forces from Equation (d) as follows:

$$\begin{Bmatrix} R_1 \\ R_2 \\ R_{3x} \\ R_{3y} \end{Bmatrix} = \frac{AE}{4L}\begin{bmatrix} \sqrt{3} & -4 \\ 0 & -\sqrt{3} \\ -\sqrt{3} & -1 \\ -3 & \sqrt{3} \end{bmatrix}\begin{Bmatrix} \upsilon_1 \\ u_2 \end{Bmatrix} = \frac{P}{5}\begin{Bmatrix} -4 \\ -\sqrt{3} \\ -1 \\ \sqrt{3} \end{Bmatrix}$$

The results may be verified by applying the equations of equilibrium to the free-body diagram of the entire truss (Figure 17.6a).

Axial forces in elements. Using Equations 17.16 and (e) and Table 17.1, we obtain

$$F_{12} = \frac{AE}{L}\begin{bmatrix} 1 & 0 \end{bmatrix}\begin{Bmatrix} \dfrac{4PL}{5AE} \\ 0 \end{Bmatrix} = \frac{4}{5}P$$

$$F_{23} = \frac{AE}{L}\begin{bmatrix} -\dfrac{1}{2} & \dfrac{\sqrt{3}}{2} \end{bmatrix}\begin{Bmatrix} -\dfrac{4PL}{5AE} \\ 0 \end{Bmatrix} = \frac{2}{5}P$$

$$F_{13} = \frac{AE}{L}\begin{bmatrix} \dfrac{1}{2} & \dfrac{\sqrt{3}}{2} \end{bmatrix}\begin{Bmatrix} 0 \\ 0 \end{Bmatrix} = 0$$

Stresses in elements. Dividing the foregoing element forces by the cross-sectional area, we have $\sigma_{12} = 4P/5A_1$, $\sigma_{23} = 2P/5A_2$, and $\sigma_{13} = 0$.

Required cross-sectional areas of elements. The allowable stress is $\sigma_{\text{all}} = 240/1.5 = 160$ MPa. We then have $A_1 = 0.8(200 \times 10^3)/160 = 1000$ mm^2, $A_2 = 500$ mm^2, and A_3 = any area.

17.4 Beam and Frame Elements

Here, we formulate stiffness matrices for flexural or beam elements and axial–flexural or plane frame elements. Consider first an initially straight beam element of constant flexural rigidity *EI* and length *L*, as depicted in Figure 17.7. Such an element has a transverse

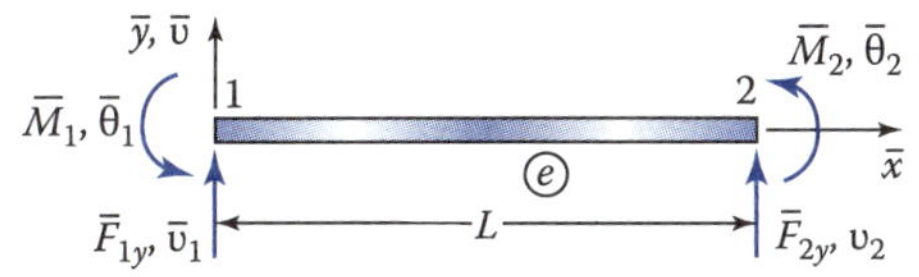

FIGURE 17.7
Beam element; all forces and displacements have a positive sense.

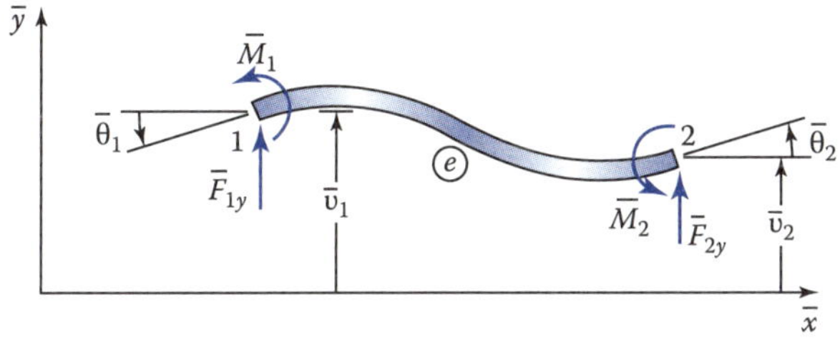

FIGURE 17.8
Deformed beam element.

deflection $\bar{\upsilon}$ and a slope $\bar{\theta}$ at each end or node. Corresponding to these displacements, a transverse shear force $\bar{F}_y$ and a bending moment $\bar{M}$ act at each node. The deflected configuration of the beam element is shown in Figure 17.8.

The linearly elastic behavior of a beam element is governed according to Equation 4.16c as $d^4\upsilon/dx^4 = 0$. The right-hand side of this equation is 0 because in the formulation of the stiffness matrix equations, we assume no loading between nodes. In the elements where there is a distributed load, the equivalent nodal load components are used. The solution is taken to be a cubic polynomial function of x:

$$\bar{\upsilon} = a_1 + a_2\bar{x} + a_3\bar{x}^2 + a_4\bar{x}^3 \tag{a}$$

The constant values of a are obtained by using the conditions at both ends. The stiffness matrix can again be obtained by the procedure discussed in Section 17.2. It can be verified that [8] the nodal force–displacement relations in the matrix form are

$$\begin{Bmatrix} \bar{F}_{1y} \\ \bar{M}_1 \\ \bar{F}_{2y} \\ \bar{M}_2 \end{Bmatrix}_e = \frac{EI}{L_3} \begin{bmatrix} 12 & 6L & -12 & 6L \\ & 4L^2 & -6L & 2L^2 \\ & & 12 & -6L \\ \text{Symmetric} & & & 4L^2 \end{bmatrix} \begin{Bmatrix} \bar{\upsilon}_1 \\ \bar{\theta}_1 \\ \bar{\upsilon}_2 \\ \bar{\theta}_2 \end{Bmatrix}_e \tag{17.19a}$$

or symbolically

$$\{\bar{F}\}_e = [\bar{k}]_e \{\delta\}_e \tag{17.19b}$$

The matrix $\{\bar{F}\}_e$ represents the force and moment components. Equation 17.19b defines the stiffness matrix $\{\bar{k}\}_e$ for a beam element lying along a local coordinate axis $\bar{x}$. Having developed the stiffness matrix, formulation and solution of problems involving beam elements proceed as discussed in Section 17.3.

Example 17.1: Displacements and Forces in a Statically Indeterminate Beam

A propped cantilevered beam of flexural rigidity EI is subjected to end load P as shown in Figure 17.9a. Using the FEM, find

a. The nodal displacements
b. The nodal forces and moments

Solution

We discretize the beam into elements with nodes 1, 2, and 3, as shown in Figure 17.9a. By Equation 17.19,

$$[k]_1 = \frac{EI}{L^3}\begin{bmatrix} 12 & 6L & -12 & 6L \\ & 4L^2 & -6L & 2L^2 \\ & & 12 & -6L \\ \text{Symmetric} & & & 4L^2 \end{bmatrix} \begin{matrix} \upsilon_1 \\ \theta_1 \\ \upsilon_2 \\ \theta_2 \end{matrix} \quad (\text{columns: } \upsilon_1, \theta_1, \upsilon_2, \theta_2)$$

$$[k]_2 = \frac{EI}{L^3}\begin{bmatrix} 12 & 6L & -12 & 6L \\ & 4L^2 & -6L & 2L^2 \\ & & 12 & -6L \\ \text{Symmetric} & & & 4L^2 \end{bmatrix} \begin{matrix} \upsilon_2 \\ \theta_2 \\ \upsilon_3 \\ \theta_3 \end{matrix} \quad (\text{columns: } \upsilon_2, \theta_2, \upsilon_3, \theta_3)$$

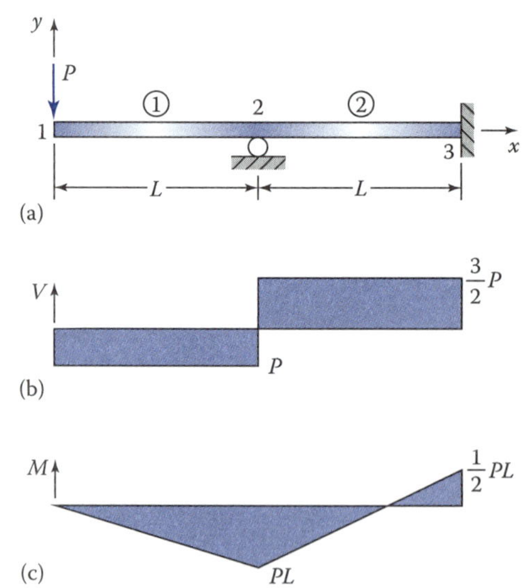

FIGURE 17.9
Example 17.1. (a) load diagram, (b) shear diagram, and (c) moment diagram.

a. The global stiffness matrix of the beam can now be assembled: $[K] = [k]_1 + [k]_2$. The governing equations for the beam are then

$$\begin{Bmatrix} F_{1y} \\ M_1 \\ F_{2y} \\ M_2 \\ F_{3y} \\ M_3 \end{Bmatrix} = \frac{EI}{L^3} \begin{bmatrix} 12 & 6L & -12 & 6L & 0 & 0 \\ & 4L^2 & -6L & 2L^2 & 0 & 0 \\ & & 24 & 0 & -12 & 6L \\ & & & 8L^2 & -6L & 2L^2 \\ & & & & 12 & -6L \\ \text{Symmetric} & & & & & 4L^2 \end{bmatrix} \begin{Bmatrix} \upsilon_1 \\ \theta_1 \\ \upsilon_2 \\ \theta_2 \\ \upsilon_3 \\ \theta_3 \end{Bmatrix} \quad (17.20a)$$

or

$$\{F\} = [K]\{\delta\} \quad (17.20b)$$

The boundary conditions are $\upsilon_2 = 0$, $\theta_3 = 0$, and $\upsilon_3 = 0$. Partitioning the first, second, and fourth of these equations associated with the unknown displacements,

$$\begin{Bmatrix} -P \\ 0 \\ 0 \end{Bmatrix} = \frac{EI}{L^3} \begin{bmatrix} 12 & 6L & 6L \\ 6L & 4L^2 & 2L^2 \\ 6L & 2L^2 & 8L^2 \end{bmatrix} \begin{Bmatrix} \upsilon_1 \\ \theta_1 \\ \theta_2 \end{Bmatrix}$$

Solving for nodal displacements, we obtain

$$\upsilon_1 = -\frac{7PL^3}{12EI}, \quad \theta_1 = \frac{3PL^2}{4EI}, \quad \theta_2 = \frac{PL^2}{4EI}$$

b. Introducing these equations into Equation 17.20a, after multiplying, the nodal forces and moments are found as

$$F_{1y} = -P, \quad M_1 = 0, \quad F_{2y} = \frac{5}{2}P,$$
$$M_2 = 0, \quad F_{3y} = -\frac{3}{2}P, \quad M_3 = \frac{1}{2}PL$$

Note that M_1 and M_2 are 0, since no reactive moments are present on the beam at nodes 1 and 2.

Comments: In general, it is necessary to determine the local nodal forces and moments associated with each element to analyze the entire structure. For the case under consideration, it may readily be observed from a free-body diagram of element 1 that $(M_2)_1 = -PL$. Hence, we obtain the shear and moment diagrams for the beam as shown in Figures 17.9b and c, respectively.

17.4.1 Arbitrarily Oriented Beam Element

In plane frame structures, the beam elements are no longer horizontal. They can be oriented in a 2D plane as shown in Figure 17.10. So it is necessary to expand $[k]_e$ to allow for the displacements transforming into u and υ displacements in the global system. The moments are unaffected. Referring to the figure, the global force and displacement matrices are, respectively,

$$\{F\}_e = \begin{Bmatrix} F_{1x} \\ F_{1y} \\ M_1 \\ F_{2x} \\ F_{2y} \\ M_2 \end{Bmatrix}_e \tag{17.21a}$$

$$\{\delta\}_e = \begin{Bmatrix} u_1 \\ \upsilon_1 \\ \theta_1 \\ u_2 \\ \upsilon_2 \\ \theta_2 \end{Bmatrix}_e \tag{17.21b}$$

Following a procedure similar to that described in Section 17.2, the coordinate transformation matrix now becomes

$$[T] = \begin{bmatrix} c & s & 0 & 0 & 0 & 0 \\ -s & c & 0 & 0 & 0 & 0 \\ 0 & 0 & 1 & 0 & 0 & 0 \\ 0 & 0 & 0 & c & s & 0 \\ 0 & 0 & 0 & -s & c & 0 \\ 0 & 0 & 0 & 0 & 0 & 1 \end{bmatrix} \tag{17.22}$$

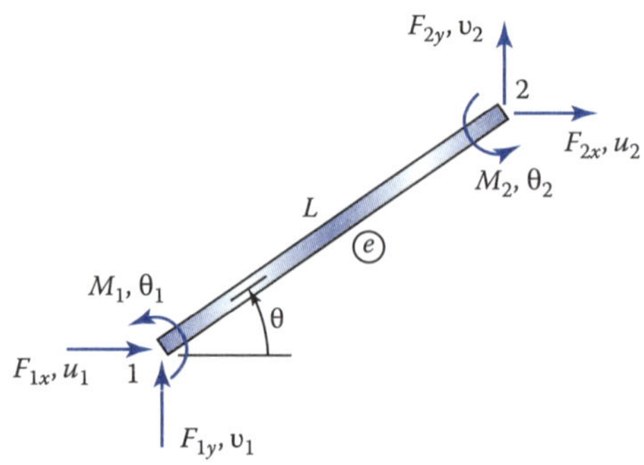

FIGURE 17.10
Global forces and displacements acting on an arbitrarily oriented beam element.

where, as before, $c = \cos\theta$ and $s = \sin\theta$. Substituting $[T]$ from Equation 17.22 and $[k]_e$ from Equation 17.19 into Equation 17.13, the *global stiffness matrix* is formed:

$$[k]_e = \frac{EI}{L^3}\begin{bmatrix} 12s^2 & -12cs & -6Ls & -12s^2 & 12cs & -6Ls \\ & 12c^2 & 6Lc & 12cs & -12c^2 & 6Lc \\ & & 4L^2 & 6Ls & -6Lc & 2L^2 \\ & & & 12s^2 & -12cs & 6Ls \\ & & & & 12c^2 & -6Lc \\ \text{Symmetric} & & & & & 4L^2 \end{bmatrix} \quad (17.23)$$

17.4.2 Arbitrarily Oriented Axial–Flexural Beam or Frame Element

When a horizontal axial element (Figure 17.3) and a horizontal beam element (Figure 17.7) are combined, we obtain the axial–flexural beam element. In this case, the solution for the axial displacements and the transverse deflections and rotations can be carried out separately and independently. Local nodal forces acting on an axially flexural beam or frame element oriented in the 2D plane with an angle θ with the x-axis are shown in Figure 17.11. For this element, the stiffness matrix must undergo the routine coordinate transformation procedure described previously. In so doing, we obtain the *global stiffness matrix* for the element that contains the axial force, shear force, and bending moment effects [7]:

$$[k]_e = \frac{E}{L}\begin{bmatrix} Ac^2 + \frac{12I}{L^2}s^2 & \left(A - \frac{12I}{L^2}\right)cs & -\frac{6I}{L}s & -\left(Ac^2 + \frac{12I}{L^2}s^2\right) & -\left(A - \frac{12I}{L^2}\right)cs & -\frac{6I}{L}s \\ & As^2 + \frac{12I}{L^2}c^2 & \frac{6I}{L}c & -\left(A - \frac{12I}{L^2}\right)cs & -\left(As^2 + \frac{12I}{L^2}c^2\right) & \frac{6I}{L}c \\ & & 4I & \frac{6I}{L}s & -\frac{6I}{L}c & 2I \\ & & & Ac^2 + \frac{12I}{L^2}s^2 & \left(A - \frac{12I}{L^2}\right)cs & \frac{6I}{L}s \\ & & & & As^2 + \frac{12I}{L^2}c^2 & -\frac{6I}{L}c \\ \text{Symmetric} & & & & & 4I \end{bmatrix}$$

(17.24)

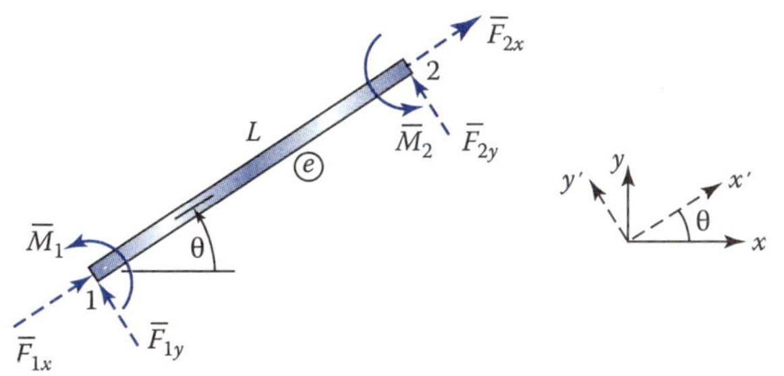

FIGURE 17.11
Local forces acting on arbitrarily oriented axial–flexural elements.

The global force and displacements are again given by Equation 17.21. Analysis and design of rigid-jointed frameworks can be undertaken by applying Equation 17.23 or 17.24. From the latter equation, we observe that the element stiffness matrix of a frame in general is a function of E, A, L, I and the angle of orientation θ of the element with respect to the global coordinate axes. With the element stiffness matrix developed, formulation and solution of a frame problem proceed as discussed in Section 17.3. The following example illustrates the procedure.

Example 17.2: Displacements in a Frame

A planar rectangular frame 1234 is fixed at both supports 1 and 4 (Figure 17.12). The load on the frame consists of a horizontal force P acting at joint 2 and a moment M applied at joint 3. By the FEA, calculate the nodal displacements.

Given: $P = 4$ kips, $M = 2$ kips · in., $L = 5$ ft, $E = 30 \times 10^6$ psi, and $A = 5$ in.^{2} for all elements; $I = 120$ in.^{4} for elements 1 and 3 and $I = 60$ in.^{4} for element 2.

Solution

The global coordinate axes xy are indicated in Figure 17.12. Through the use of Equation 17.24 and Table 17.2, the element stiffness matrices are obtained as

$$
\begin{array}{rcl}
 & \begin{array}{cccccc} u_1 & \upsilon_1 & \theta_1 & u_2 & \upsilon_2 & \theta_2 \end{array} & \\
[k]_1 = 5\left(10^5\right) & \begin{bmatrix} 0.4 & 0 & -12 & -0.4 & 0 & -12 \\ 0 & 5 & 0 & 0 & -5 & 0 \\ -12 & 0 & 480 & 12 & 0 & 240 \\ -0.4 & 0 & 12 & 0.4 & 0 & 12 \\ 0 & -5 & 0 & 0 & 5 & 0 \\ -12 & 0 & 240 & 12 & 0 & 480 \end{bmatrix} & \begin{matrix} u_1 \\ \upsilon_1 \\ \theta_1 \\ u_2 \\ \upsilon_2 \\ \theta_2 \end{matrix}
\end{array}
$$

$$
\begin{array}{rcl}
 & \begin{array}{cccccc} u_2 & \upsilon_2 & \theta_2 & u_3 & \upsilon_3 & \theta_3 \end{array} & \\
[k]_2 = 5\left(10^5\right) & \begin{bmatrix} 5 & 0 & 0 & -5 & 0 & 0 \\ 0 & 0.2 & 6 & 0 & -0.2 & 6 \\ 0 & 6 & 240 & 0 & -6 & 120 \\ -5 & 0 & 0 & 5 & 0 & 0 \\ 0 & -0.2 & -6 & 0 & 0.2 & -6 \\ 0 & 6 & 120 & 0 & -6 & 240 \end{bmatrix} & \begin{matrix} u_2 \\ \upsilon_2 \\ \theta_2 \\ u_3 \\ \upsilon_3 \\ \theta_3 \end{matrix}
\end{array}
$$

$$
\begin{array}{rcl}
 & \begin{array}{cccccc} u_3 & \upsilon_3 & \theta_3 & u_4 & \upsilon_4 & \theta_4 \end{array} & \\
[k]_3 = 5\left(10^5\right) & \begin{bmatrix} 0.4 & 0 & 12 & -0.4 & 0 & 12 \\ 0 & 5 & 0 & 0 & -5 & 0 \\ 12 & 0 & 480 & -12 & 0 & 240 \\ -0.4 & 0 & -12 & 0.4 & 0 & -12 \\ 0 & -5 & 0 & 0 & 5 & 0 \\ 12 & 0 & 240 & -12 & 0 & 480 \end{bmatrix} & \begin{matrix} u_3 \\ \upsilon_3 \\ \theta_3 \\ u_4 \\ \upsilon_4 \\ \theta_4 \end{matrix}
\end{array}
$$

We superpose the element stiffness matrices and apply the boundary conditions:

$$u_1 = \upsilon_1 = \theta_1 = 0, \qquad u_4 = \upsilon_4 = \theta_4 = 0$$

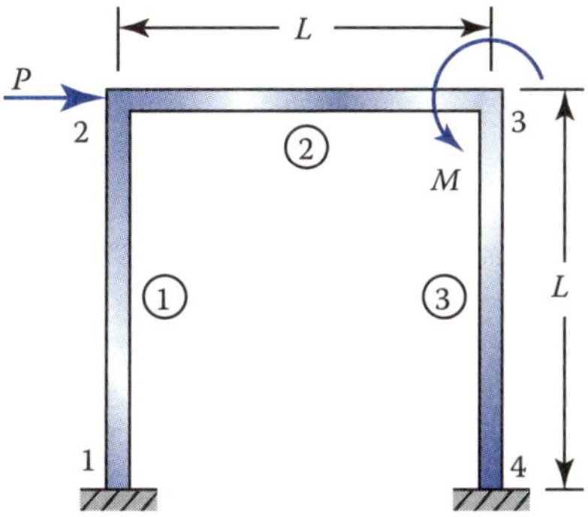

FIGURE 17.12
Example 17.3. Plane frame.

TABLE 17.2

Data for the Frame of Figure 17.12

Element	**1**	**2**	**3**
θ	90°	0°	270°
c	0	1	0
s	1	0	−1
$12I/L^2$	0.4	0.2	0.4
$6I/L$	12	6	12
E/L	5×10^5	5×10^5	5×10^5

at nodes 1 and 4. This leads to the following reduced set of equations:

$$\begin{Bmatrix} 4000 \\ 0 \\ 0 \\ 0 \\ 0 \\ 2000 \end{Bmatrix} = 5\left(10^5\right) \begin{array}{c} \begin{array}{cccccc} u_2 & v_2 & \theta_2 & u_3 & v_3 & \theta_3 \end{array} \\ \begin{bmatrix} 5.4 & 0 & 12 & -5 & 0 & 0 \\ & 5.2 & 6 & 0 & -0.2 & 6 \\ & & 720 & 0 & -6 & 120 \\ & & & 5.4 & 0 & 12 \\ & & & & 5.2 & -6 \\ \text{Symmetric} & & & & & 720 \end{bmatrix} \end{array} \begin{Bmatrix} u_2 \\ v_2 \\ \theta_2 \\ u_3 \\ v_3 \\ \theta_3 \end{Bmatrix}$$

Solving, the nodal deflections and rotations are

$$\begin{Bmatrix} u_2 \\ v_2 \\ \theta_2 \\ u_3 \\ v_3 \\ \theta_3 \end{Bmatrix} = \begin{Bmatrix} 18.208 \\ 0.582 \\ -0.271 \\ 17.408 \\ -0.582 \\ 0.248 \end{Bmatrix} \left(10^{-3}\right) \quad \begin{array}{l} \text{in.} \\ \text{in.} \\ \text{rad} \\ \text{in.} \\ \text{in.} \\ \text{rad} \end{array}$$

The negative sign indicates a downward displacement or clockwise rotation.

17.5 Two-Dimensional Elements

So far, we have dealt with only line elements connected at common nodes, forming trusses and frames. In this section, attention is directed toward the properties of 2D finite elements of an isotropic elastic structure and general formulation of the FEM for plane structures. To begin with, the plate shown in Figure 17.13a is discretized, as depicted in Figure 17.13b. The finite elements are connected not only at their nodes but also along the interelement boundaries. All formulations are based on a counterclockwise labeling of the nodes *i*, *j*, and *m*. The simplest constant strain triangular (CST) finite element is used to clearly demonstrate the basic formulative method. The nodal displacements, represented by *u* and υ in the *x* and *y* directions, respectively, are the primary unknowns.

17.5.1 Displacement Functions

Consider a typical finite element *e* with nodes *i*, *j*, and *m* (Figure 17.13b). The nodal displacements are expressed in the following convenient matrix form:

$$\{\delta\}_e = \begin{Bmatrix} u_i \\ \upsilon_i \\ u_j \\ \upsilon_j \\ u_m \\ \upsilon_m \end{Bmatrix} \tag{17.25}$$

The displacement functions, describing the displacements at any point within the element, $\{f\}_e$, are represented by

$$\{f\}_e = \begin{Bmatrix} u(x,y) \\ \upsilon(x,y) \end{Bmatrix} \tag{17.26a}$$

$$\{f\}_e = [N]\{\delta\}_e \tag{17.26b}$$

In the foregoing, the matrix $[N]$ is a function of position, to be obtained in the next section.

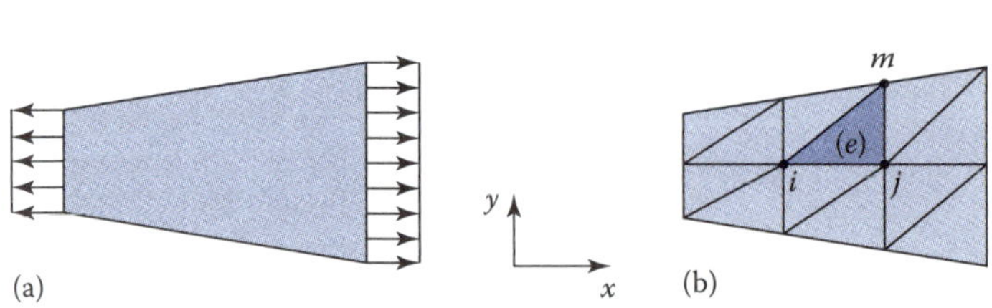

FIGURE 17.13
Plate in tension: (a) before and (b) after division into finite elements.

17.5.2 Strain, Stress, and Displacement Matrices

The strain and stress are defined in terms of displacement functions. The strain matrix may be written as follows:

$$\{\varepsilon\}_e = \begin{Bmatrix} \varepsilon_x \\ \varepsilon_y \\ \gamma_{xy} \end{Bmatrix}_e = \begin{Bmatrix} \dfrac{\partial u}{\partial x} \\ \dfrac{\partial \upsilon}{\partial y} \\ \dfrac{\partial \upsilon}{\partial x} + \dfrac{\partial u}{\partial y} \end{Bmatrix} \tag{17.27a}$$

or

$$\{\varepsilon\}_e = [B]\{\delta\}_e \tag{17.27b}$$

in which $[B]$ is also obtained in the next section.

In a like manner, the stresses throughout the element are, by Hooke's law,

$$\{\sigma\}_e = \frac{E}{1-\nu^2}\begin{bmatrix} 1 & \nu & 0 \\ \nu & 1 & 0 \\ 0 & 0 & (1-\nu)/2 \end{bmatrix}\begin{Bmatrix} \varepsilon_x \\ \varepsilon_y \\ \gamma_{xy} \end{Bmatrix}_e \tag{17.28a}$$

or

$$\{\sigma\}_e = [D]\{\varepsilon\}_e \tag{17.28b}$$

Clearly, the *elasticity matrix* is

$$[D] = \frac{E}{1-\nu^2}\begin{bmatrix} 1 & \nu & 0 \\ \nu & 1 & 0 \\ 0 & 0 & (1-\nu)/2 \end{bmatrix} \tag{17.29a}$$

In general, we write

$$[D] = \lambda\begin{bmatrix} 1 & D_{12} & 0 \\ D_{12} & 1 & 0 \\ 0 & 0 & D_{33} \end{bmatrix} \tag{17.29b}$$

Recall from Sections 3.9 and 3.11 that 2D problems are of two classes: plane stress and plane strain. The constants λ, D_{12}, and D_{33} for a plane problem are defined in Table 17.3 [4].

TABLE 17.3

Elastic Constants for 2D Problems

Quantity	Plane Strain	Plane Stress
λ	$\frac{E}{1-\nu^2}$	$\frac{E(1-\nu)}{(1+\nu)(1-2\nu)}$
D_{12}	ν	$\frac{\nu}{1-\nu}$
D_{33}	$\frac{1-\nu}{2}$	$\frac{1-2\nu}{2(1-\nu)}$

17.5.3 Governing Equations for 2D Problems

Through the use of the principle of minimum potential energy, we can develop the expressions for a *plane stress* and *plane strain* element. For this purpose, the total potential energy $\Pi = U - W$ (see Section 5.7) is expressed in terms of 2D element properties. Then, the minimizing condition, $\partial\Pi/\partial\{\delta\}_e = 0$, results in [4]

$$\{F\}_e = [k]_e\{\delta\}_e \tag{17.12}$$

This is of the same form as obtained in Section 17.2 and $\{\delta\}_e$ represents the element nodal displacement matrix. However, the element *stiffness matrix* $[k]_e$ and element *nodal force* matrix $\{F\}_e$ are now given by

$$[k]_e = \int_V [B]^T[D][B]\,\mathrm{d}V \tag{17.30}$$

$$\{F\}_e = \int_s [N]^T\{p\}\,\mathrm{d}s \tag{17.31}$$

where

p = the boundary surface force per unit area
s = the boundary surface over which the forces p act
V = the volume of the element
T = the transpose of a matrix

We next assemble the element stiffness and nodal force matrices. This gives the following global governing equations for the entire member, the system equations:

$$\{F\} = [K]\{\delta\} \tag{17.17}$$

where

$$\{F\} = \sum_{1}^{n}\{F\}_e \quad [K] = \sum_{1}^{n}[K]_e \tag{17.18}$$

as before. Now, n represents the number of finite elements making up the member. Note that, in the preceding formulations, the finite element stiffness matrix has been derived for a general orientation of global coordinates (x, y). Equation 17.17 is therefore applicable to all elements. Hence, no transformation from local to global equations is necessary. The general procedure for solving a problem by the FEM is already shown in Figure 17.5. This outline is better understood when applied to a triangular element in the section to follow.

17.6 Triangular Element

We now develop the basic *CST* plane stress and strain element. Boundaries of irregularly shaped members can be closely approximated and the expressions related to the triangular elements are simple. The treatment given here is brief. Various types of 2D finite elements yield better solutions. Examples include *linear strain triangular* (LST) elements, triangular elements with additional side and interior nodes, rectangular elements with corner nodes, and rectangular elements with additional side modes [1,3,7]. The LST element has six nodes: usual corner nodes and three additional nodes conveniently located at the midpoints of the sides. Hence, the element has 12 unknown displacements. The procedures for development of the equations for the LST element follow the same steps as that of the CST element.

17.6.1 Displacement Function

Consider the triangular finite element i, j, m shown in Figure 17.14. The nodal displacement matrix $\{\delta\}_e$ is given by Equation 17.25. The displacements u and υ throughout the element can be assumed in the following linear form:

$$\{f\}_e = \begin{Bmatrix} u(x,y) \\ \upsilon(x,y) \end{Bmatrix} = \begin{Bmatrix} \alpha_1 + \alpha_2 x + \alpha_3 y \\ \alpha_4 + \alpha_5 x + \alpha_6 y \end{Bmatrix} \tag{17.32}$$

where the α represents constants. The foregoing expressions ensure that the compatibility of displacements on the boundaries of adjacent elements is satisfied.

The nodal displacements of the element are

$$\begin{aligned} u_i &= \alpha_1 + \alpha_2 x_i + \alpha_3 y_i & \upsilon_i &= \alpha_4 + \alpha_5 x_i + \alpha_6 y_i \\ u_j &= \alpha_1 + \alpha_2 x_j + \alpha_3 y_j & \upsilon_j &= \alpha_4 + \alpha_5 x_j + \alpha_6 y_j \\ u_m &= \alpha_1 + \alpha_2 x_m + \alpha_3 y_m & \upsilon_m &= \alpha_4 + \alpha_5 x_m + \alpha_6 y_m \end{aligned}$$

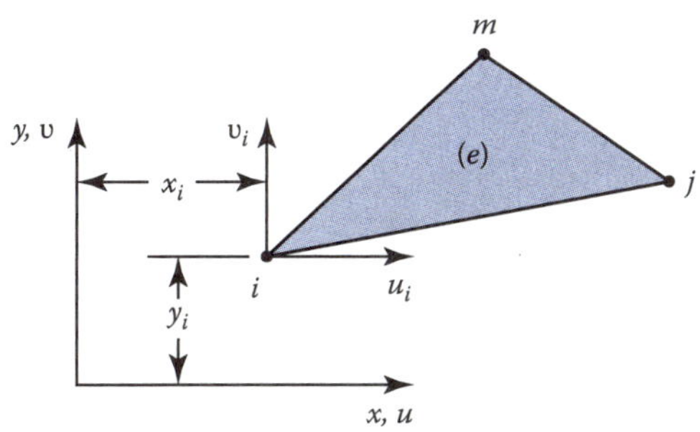

FIGURE 17.14
Basic triangular element.

Solving these equations gives [3]

$$\begin{Bmatrix} \alpha_1 \\ \alpha_2 \\ \alpha_3 \end{Bmatrix} = \frac{1}{2A} \begin{bmatrix} a_i & a_j & a_m \\ b_i & b_j & b_m \\ c_i & c_j & c_m \end{bmatrix} \begin{Bmatrix} u_i \\ u_j \\ u_m \end{Bmatrix}$$
$$\begin{Bmatrix} \alpha_4 \\ \alpha_5 \\ \alpha_6 \end{Bmatrix} = \frac{1}{2A} \begin{bmatrix} a_i & a_j & a_m \\ b_i & b_j & b_m \\ c_i & c_j & c_m \end{bmatrix} \begin{Bmatrix} \upsilon_i \\ \upsilon_j \\ \upsilon_m \end{Bmatrix} \tag{a}$$

The quantity A represents the area of the triangle:

$$A = \frac{1}{2}\left[x_i\left(y_j - y_m\right) + x_j\left(y_m - y_i\right) + x_m\left(y_i - y_j\right)\right] \tag{17.33}$$

and

$$\begin{array}{lll} a_i = x_j y_m - y_j x_m & a_j = y_i x_m - x_i y_m & a_m = x_i y_j - y_i x_j \\ b_i = y_j - y_m & b_j = y_m - y_i & b_m = y_i - y_j \\ c_i = x_m - x_j & c_j = x_i - x_m & c_m = x_j - x_i \end{array} \tag{17.34}$$

Substituting Equation (a) into Equation 17.32, the displacement function is provided by

$$\{f\}_e = \begin{bmatrix} N_i & 0 & N_j & 0 & N_m & 0 \\ 0 & N_i & 0 & N_j & 0 & N_m \end{bmatrix} \begin{Bmatrix} u_i \\ \upsilon_i \\ u_j \\ \upsilon_j \\ u_m \\ \upsilon_m \end{Bmatrix} = [N]\{\delta\}_e \tag{17.35}$$

in which

$$\begin{aligned} N_i &= \frac{1}{2A}\left(a_i + b_i x + c_i y\right) \\ N_j &= \frac{1}{2A}\left(a_j + b_j x + c_j y\right) \\ N_m &= \frac{1}{2A}\left(a_m + b_m x + c_m y\right) \end{aligned} \tag{17.36}$$

The *strain matrix* is obtained by carrying Equation 17.35 into Equation 17.27a:

$$\begin{Bmatrix} \varepsilon_x \\ \varepsilon_y \\ \gamma_{xy} \end{Bmatrix}_e = \frac{1}{2A}\begin{bmatrix} b_i & 0 & b_j & 0 & b_m & 0 \\ 0 & c_i & 0 & c_j & 0 & c_m \\ c_i & b_i & c_j & b_j & c_m & b_m \end{bmatrix}\{\delta\}_e \tag{17.37}$$

Introducing Equation 17.37 into Equation 17.27b, we have

$$[B] = \frac{1}{2A}\begin{bmatrix} b_i & 0 & \vdots & b_j & 0 & \vdots & b_m & 0 \\ 0 & c_i & \vdots & 0 & c_j & \vdots & 0 & c_m \\ c_i & b_i & \vdots & c_j & b_j & \vdots & c_m & b_m \end{bmatrix} \tag{17.38a}$$

or

$$[B] = [B_i][B_j][B_m] \tag{17.38b}$$

where

$$[B_i] = \frac{1}{2A}\begin{bmatrix} b_i & 0 \\ 0 & c_i \\ c_i & b_i \end{bmatrix}, \quad [B_j] = \frac{1}{2A}\begin{bmatrix} b_j & 0 \\ 0 & c_j \\ c_j & b_j \end{bmatrix}, \quad [B_m] = \frac{1}{2A}\begin{bmatrix} b_m & 0 \\ 0 & c_m \\ c_m & b_m \end{bmatrix} \tag{17.39}$$

Clearly, matrix [*B*] depends only on the element nodal coordinates, as seen from Equation 17.34. Hence, the strain (and stress) is observed to be *constant*, and as already noted, the element of Figure 17.14 is called a constant strain triangle.

17.6.2 Stiffness Matrix

For an element of constant thickness *f*, the stiffness matrix can be obtained from Equation (17.30) as follows:

$$[K]_e = [B]^T[D][B]tA \tag{17.40}$$

This equation is assembled together with the elasticity matrix [*D*] and [*B*] given by Equations 17.29 and 17.38. Expanding the resulting expression, the stiffness matrix is usually written in a partitioned form of order 6 × 6. We point out that the element stiffness matrix is generally developed in most computer programs by performing the matrix triple products shown by Equation 17.40. The explicit form of the stiffness matrix is rather lengthy and given in the specific publications on the subject (see, e.g., [2]).

17.6.3 Element Nodal Forces due to Surface Loading

The nodal force attributable to applied external loading may be obtained either by evaluating the static resultants or applying Equation 17.31. Some examples, found by the former

TABLE 17.4

Nodal Forces F_j and F_m of a CST Element of Thickness t due to Some Common Loadings

A. Linear load $p(y)$ per unit area

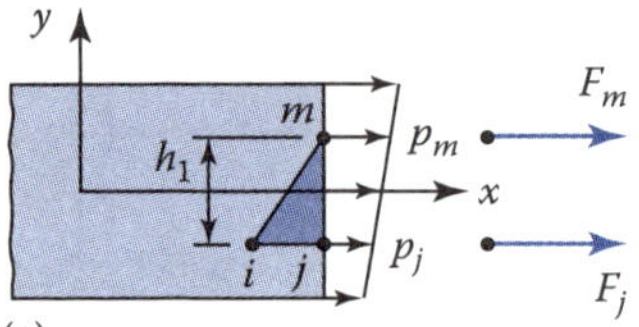

(a)

$$F_j = \frac{h_1 t}{6}(2p_j + p_m)$$

$$F_m = \frac{h_1 t}{6}(2p_m + p_j)$$

B. Uniform load P

$$F_j = F_m = \frac{1}{2} p h_1 t$$

C. End shear load P

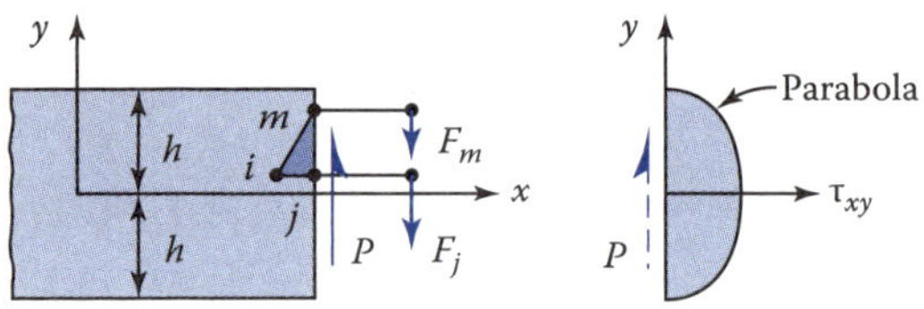

(c)

$$F_m = \frac{3P}{4h}\left\{\frac{1}{2}(y_m - y_j) + \frac{1}{3h^2}\left[\frac{y_j}{4}(y_m^2 + y_j y_m + y_j^2) - \frac{3}{4}y_m^3\right]\right\}$$

$$F_j = \frac{3P}{4h}\left\{\frac{1}{2}(y_m - y_j) - \frac{1}{3h^2}\left[\left(y_m - \frac{3}{4}y_j\right)(y_m^2 + y_j y_m + y_j^2) - \frac{3}{4}y_m^3\right]\right\}$$

approach, are given in Table 17.4. An expanded form of Equation 17.40, together with those expressions given for the nodal forces, characterizes the CST element. The unknown displacements, strains, and stresses may now be determined applying the general outline given in Figure 17.5. The basic procedure employed in the finite method using CST or any other element is illustrated in the next section.

17.7 Plane Stress Case Studies

Here, we present four case studies limited to plane stress situations and CST finite elements. A plate under tension, a deep beam or plate in pure bending, a plate with a hole subjected to an axial loading, and a disk carrying concentrated diametral compression are the members analyzed. There are very few elasticity or *exact* solutions to 2D problems, especially for any but the simplest forms. As will become evident from the following discussion, the designer and stress analyst can reach a very accurate solution by applying proper techniques and modeling. Accuracy is usually limited by the willingness to model all the significant features of the problem and pursue the analysis until convergence is reached.

It should be mentioned that an *exact* solution is unattainable using the FEM, and we seek instead an acceptable solution. The goal is then the establishment of a finite element that ensures convergence to the exact solution. The literature contains many comparisons among the various elements. The efficiency of a finite element solution can, in certain situations, be enhanced using a *mix* of elements. A denser mesh, for instance, within a region of severely changing or localized stress may save much time and effort.

CASE STUDY 17.2 Steel Plate in Tension

A cantilever plate of depth h, length L, and thickness t supports a uniaxial tension load p as shown in Figure 17.15a. Outline the determination of deflections, strains, and stresses.

Given: $p = 4$ ksi, $E = 30 \times 10^6$ psi, $\nu = 0.3$, $t = ½$ in., $L = 20$ in., $h = 10$ in.

Assumption: The plate is divided into two CST elements.

Solution

The discretized plate is depicted in Figure 17.15b. The origin of coordinates is placed at node 1, for convenience: however, it may be located at any point in the x, y plane. The area of each element is

$$A = \frac{1}{2}hL = \frac{1}{2}(10)(20) = 100 \text{ in.}^2$$

The statically equivalent forces at nodes 2 and 3, [4(10 × ½)/2 = 10 kips], are shown in the figure. For plane stress, elasticity matrix [D] is given by Equation 17.29a.

Stiffness matrix. For element a, on assigning $i = 1$, $j = 3$, and $m = 4$, Equation 17.34 gives

$$\begin{aligned} b_1 &= y_3 - y_4 = 10 - 10 = 0 \\ b_3 &= y_4 - y_1 = 10 - 0 = 10 \\ b_4 &= y_1 - y_3 = 0 - 10 = -10 \\ \\ c_1 &= x_4 - x_3 = 0 - 20 = -20 \\ c_3 &= x_1 - x_4 = 0 - 0 = 0 \\ c_4 &= x_3 - x_1 = 20 - 0 = 20 \end{aligned} \tag{a}$$

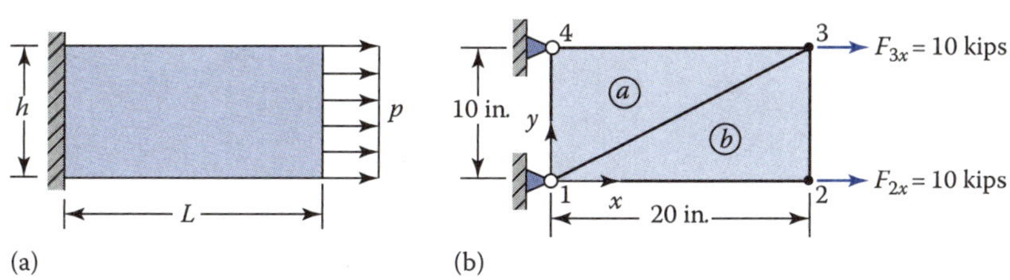

FIGURE 17.15
Cantilever plate: (a) before and (b) after being discretized.

Substitution of these and the given data into Equation 17.40, after performing the matrix multiplications, results in stiffness matrix $[k]_a$. Similarly, for element b, assignment of $i = 1, j = 2$, and $m = 3$ into Equation 17.34 leads to

$$b_1 = y_2 - y_3 = 0 - 10 = -10$$
$$b_2 = y_3 - y_1 = 10 - 0 = 10$$
$$b_3 = y_1 - y_2 = 0 - 0 = 0$$

$$c_1 = x_3 - x_2 = 20 - 20 = 0$$
$$c_2 = x_1 - x_3 = 0 - 20 = -20$$
$$c_3 = x_2 - x_1 = 20 - 0 = 20$$

and $[k]_b$ is determined. The displacements u_2, υ_2 and u_4, υ_4 are not involved in elements a and b, respectively. So, before summing $[k]_a$ and $[k]_b$ to form the system matrix, rows and columns of zeros must be added to each element matrix to account for the absence of these displacements, as mentioned in Section 17.3. Finally, superimposition of the resulting matrices gives the system matrix $[K]$.

Nodal displacements. The boundary conditions are $u_1 = \upsilon_1 = u_4 = \upsilon_4 = 0$. The force–displacement relationship of the system is

$$\begin{Bmatrix} R_{1x} \\ R_{1y} \\ 10 \\ 0 \\ 10 \\ 0 \\ R_{4x} \\ R_{4y} \end{Bmatrix} = [K] \begin{Bmatrix} 0 \\ 0 \\ u_2 \\ \upsilon_2 \\ u_3 \\ \upsilon_3 \\ 0 \\ 0 \end{Bmatrix}$$

Next, to compare the quantities involved, we introduce the results without going through the computation of the $[K]$. It can be verified [3] that the preceding derivations yield

$$\begin{Bmatrix} 10 \\ 0 \\ 10 \\ 0 \end{Bmatrix} = \frac{187.5}{0.91} \begin{bmatrix} 48 & 0 & -28 & 14 \\ 0 & 87 & 12 & -80 \\ -28 & 12 & 48 & -26 \\ 14 & -80 & -26 & 87 \end{bmatrix} \begin{Bmatrix} u_2 \\ \upsilon_2 \\ u_3 \\ \upsilon_3 \end{Bmatrix} \tag{a}$$

Solving,

$$\begin{Bmatrix} u_2 \\ \upsilon_2 \\ u_3 \\ \upsilon_3 \end{Bmatrix} = \begin{Bmatrix} 2.4383 \\ 0.0163 \\ 2.6548 \\ 0.4261 \end{Bmatrix} \left(10^{-3}\right) \text{in.} \tag{b}$$

Stresses. For element *a*, carrying Equations (a) and (b) into 17.37, we obtain the strain matrix $\{\varepsilon\}_a$. Equation 17.28, $[D]\,\{\varepsilon\}_a$, then results in

$$\begin{Bmatrix} \sigma_x \\ \sigma_y \\ \tau_{xy} \end{Bmatrix}_a = \begin{Bmatrix} 4020 \\ 1204 \\ 9.6 \end{Bmatrix} \text{psi}$$

Element *b* is treated in a like manner.

Comments: Due to constant *x*-directed stress of 4000 psi applied on the edge of the plate, the normal stress is expected to be about 4000 psi in the element *a* (or *b*). The foregoing result for σ_x is therefore quite good. Interestingly, the support of the element *a* at nodes 1 and 4 causes a relatively high stress of $\sigma_y = 1204$ psi. Also note that the value of shear stress τ_{xy} is negligibly small, as anticipated. The effect of element size on solution accuracy will be demonstrated in the next example.

CASE STUDY 17.3 Displacements of a Deep Beam by the Exact and Finite Element Methods

A deep beam (or plate) is loaded by couples *M*, acting about its axis at the ends as shown in Figure 17.16a. Determine the stresses and displacements, using the methods of theory of elasticity and finite element.

Given: $L = 76.2$ mm, $h = 50.8$ mm, thickness $t = 25.4$ mm, $p = 6895$ kPa, $E = 207$ GPa, $\nu = 0.15$.

Assumption: The weight of the member is negligible.

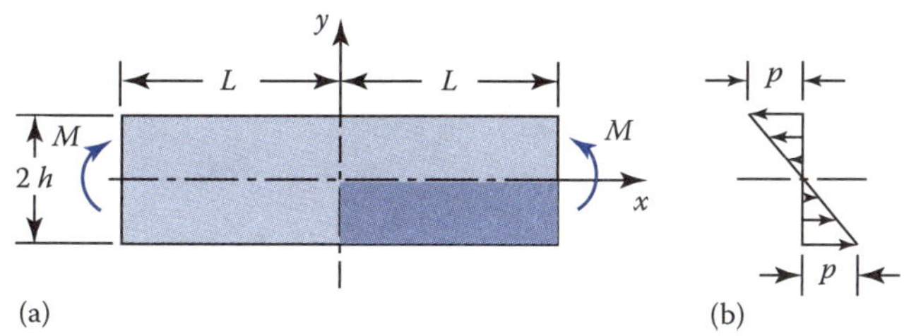

FIGURE 17.16
Beam in pure bending.

Elasticity or *Exact* Solution

To begin with, we replace the end moments with the statically equivalent load per unit area $p = Mh/I$, as depicted in Figure 17.16b. The stresses are then

$$\sigma_x = -\frac{y}{h}p, \quad \sigma_y = \tau_{xy} = 0 \tag{17.41}$$

By Equations 2.6 and 3.55,

$$\frac{\partial u}{\partial x} = -\frac{yp}{Eh}, \quad \frac{\partial \upsilon}{\partial y} = \frac{\nu yp}{Eh}, \quad \frac{\partial \upsilon}{\partial x} + \frac{\partial u}{\partial y} = 0$$

It can be shown that, satisfying the conditions $u(0, 0) = 0$ and $u(L, 0) = 0$, we have

$$u = -\frac{p}{Eh}xy, \quad \upsilon = \frac{p}{2Eh}\left(x^2 + \nu y^2\right) \tag{17.42}$$

Introducing the data into Equations 17.41 and 17.42 gives

$$\begin{gathered}\sigma_x = -\frac{1}{0.0508}yp \\ u(0.0762, -0.0508) = 2.54 \times 10^{-6} \\ \sigma_{x,\max} = 6895 \text{ kPa} \\ \upsilon(0.0762, 0) = 1.904 \times 10^{-6} \text{ m}\end{gathered} \tag{c}$$

Finite Element Solution

Based on symmetry and antisymmetry, only a quadrant (indicated by the shaded part in the figure) of the beam needs to be analyzed.

Boundary conditions. The discretized quarter plate is composed of 12 triangular elements (Figure 17.17a). Since no deformation takes place along the x and y axes, the boundary conditions are

$$u_1 = u_2 = u_3 = u_6 = u_9 = u_{12} = 0, \quad \upsilon_3 = 0$$

Nodal forces. Applying the equation in Case A of Table 17.4 to Figure 17.16b and inserting data given, we have

$$\begin{aligned}F_{10x} &= \frac{0.0254 \times 0.0254}{6}(2 \times 6895 + 3447.5)(10^3) \\ &= 1853.5 \text{ N} \\ F_{11x} &= \frac{0.0254 \times 0.0254}{6}(2 \times 3447.5 + 6895)(10^3) + \frac{0.0254 \times 0.0254}{6}(2 \times 3447.5 + 0)(10^3) \\ &= 2224 \text{ N} \\ F_{12x} &= \frac{0.0254 \times 0.0254}{6}(0 + 3447.5)(10^3) = 370.7 \text{ N}\end{aligned}$$

The other nodal forces are 0.

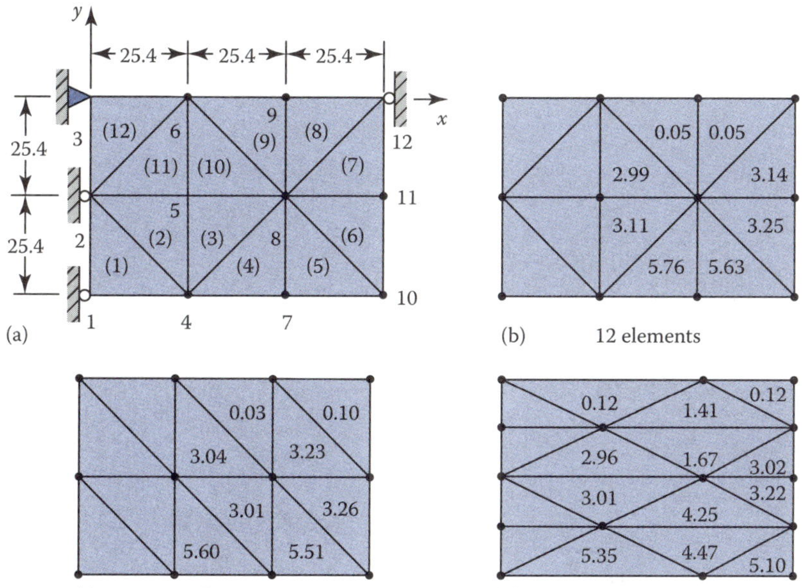

FIGURE 17.17
Effect of element size and orientation on stress in beam shown in Figure 17.16. *Note*: Stress (megapascal) obtained for an element is assigned to the centroid.

TABLE 17.5
Displacement

Case	Number of Nodes	Displacement (10^6 m)	
		v_{12}	u_{10}
Figure 17.17b	12	1.547	2.133
Figure 17.17c	12	1.745	2.062
Figure 17.17d	15	1.572	1.976
Exact solution	—	1.905	2.540

Results [4]. The values of the stresses, obtained by a procedure identical to that of the preceding example, are shown in Figure 17.17b. Note the considerable difference between the exact solution, Equation (c), and that resulting from the coarse mesh arrangement used. To illustrate the effect of element size and orientation, the results corresponding to different grid configurations are given in Figure 17.17c and d. The displacements corresponding to Figure 17.17b through d and the exact solution are listed in Table 17.5.

Comments: Figure 17.17b and c shows the effect of element orientation for the same number of elements and node locations. Observe that elements having large differences between their sides, or so-called weak elements, such as Figure 17.17d, yield less accurate results: even the number of nodes is larger than those of Figure 17.17b and c. Using *equilateral* or nearly equilateral, well-formed elements of finer mesh gives solutions approaching the exact values.

CASE STUDY 17.4 Stress Concentration in a Plate with a Hole in Uniaxial Tension

A thin plate containing a small circular hole of radius a is subjected to uniform tensile load of intensity σ_0 at its edges, as shown in Figure 17.18a. Apply the FEA to determine the theoretical stress-concentration factor.

Given: $L = 24$ in., $a = 2$ in., $h = 20$ in., $\sigma_0 = 6$ ksi, $E = 10\times10^6$ psi, $\nu = 0.3$.

Solution

Owing to the symmetry, only any one-quarter of the plate needs to be analyzed (Figure 17.18b). The solution for the case in which the quarter plate discretized to contain 202 CST elements is given in [1]. The roller boundary conditions are also indicated in the figure. The values of the normal edge stress σ_x obtained by the FEM and the theory of elasticity are plotted in Figure 17.18c for comparison. We see from the figure that the agreement is reasonably good. The stress-concentration factor for σ_x is $K_t \approx 3\sigma_0/\sigma_0 = 3$.

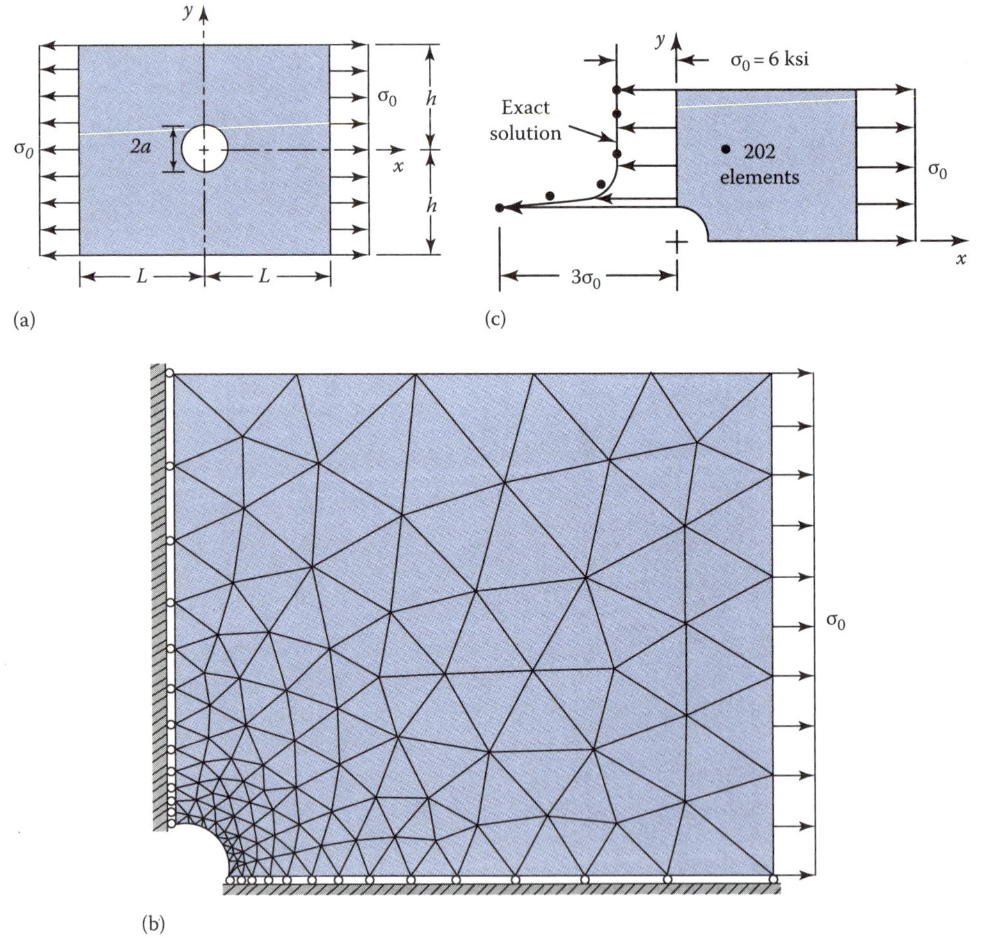

FIGURE 17.18
(a) Circular hole in a plate under uniaxial tension, (b) one-quadrant plate model, and (c) uniaxial stress (σ_x) distribution.

CASE STUDY 17.5 Circular Disk with a Diametral Load

A circular disk of constant thickness t is subjected to a pair of compression forces, as shown in Figure 17.19a. Demonstrate the accuracy of the finite element solution using well-formed (i.e., nearly equilateral) CST elements.

Given: The radius $a = 8t$ and $\nu = 0.3$.

Assumption: The disk is in a state of plane stress.

Solution

Because of symmetry, only a quarter of the disk needs to be considered (Figure 17.19b). The exact analytical as well as 30 and 109 CST element solutions of the normal stress σ_y are available [5]. A plot of these results is shown in Figure 17.20. Note the considerable difference between the exact solution and the results for 30 elements but the particularly good agreement between the results for 109 elements and the exact solution.

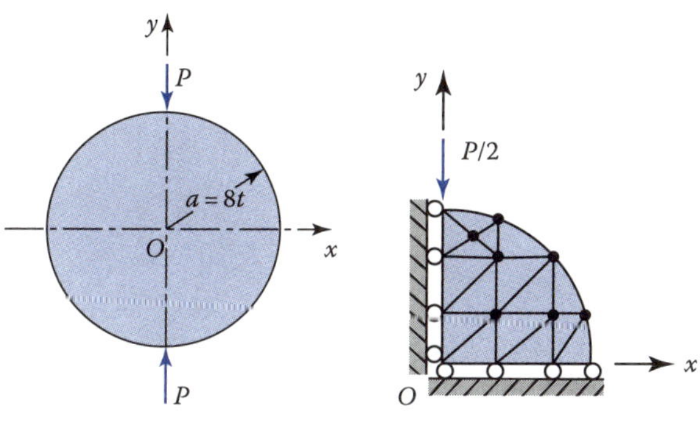

FIGURE 17.19
(a) Circular thin disk subjected to a diametral load and (b) one-quarter circular disk model with a coarse mesh.

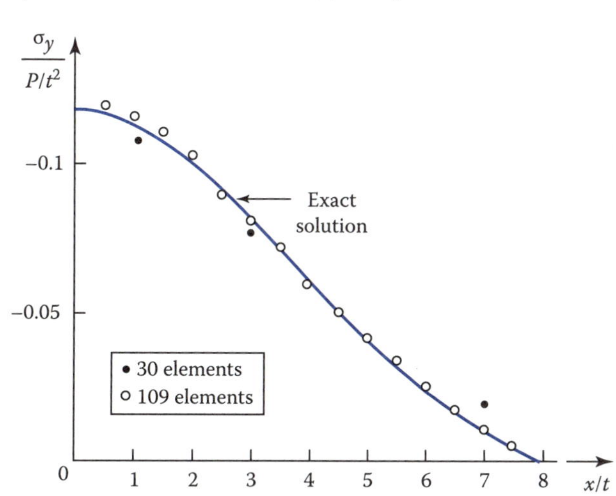

FIGURE 17.20
Normal stress (σ_y) distribution along the x-axis of the circular disk.

17.8 Axisymmetric Element

This section can provide only a brief introduction to the FEM for determination of strain and stress in an axisymmetrically loaded member, formed as a solid of revolution with material properties, support conditions, and loading, all of which are symmetrical about the z-axis. The finite elements of the body of revolution or rings are used to discretize the axisymmetric member. We use a simple element e of triangular cross section and cylindrical coordinates (r, θ, z), shown in Figure 17.21. Note that a node now is a circle. Therefore, the elemental volume dV appearing in Equation 17.30 is the volume of the ring element (i.e., $2\pi r dr dz$). Although the element lies in 3D space, any of its arbitrarily selected vertical cross sections is a *plane triangle*. As observed earlier, no tangential displacement can exist in the axisymmetric member, $\upsilon = 0$. Inasmuch as only the radial displacement u and axial displacement ir in a (rz) plane are present, the expressions for displacements for plane strain may readily be extended to the axisymmetric analysis.

Consider the triangular finite element i, j, m depicted in Figure 17.21. Properties belonging to the element are denoted by e. Referring to Equations 17.27 and 16.3, we define the strain–displacement matrix of the following form:

$$\{\varepsilon\}_e = \begin{Bmatrix} \varepsilon_r \\ \varepsilon_z \\ \varepsilon_\theta \\ \gamma_{rz} \end{Bmatrix} = \begin{Bmatrix} \dfrac{\partial u}{\partial r} \\ \dfrac{\partial w}{\partial z} \\ \dfrac{u}{r} \\ \dfrac{\partial u}{\partial z} + \dfrac{\partial w}{\partial r} \end{Bmatrix}_e \tag{17.43}$$

The nodal displacement matrix for the element is represented by

$$\{\delta\}_e = \begin{Bmatrix} u_i \\ u_j \\ u_m \\ w_i \\ w_j \\ w_m \end{Bmatrix}_e \tag{17.44}$$

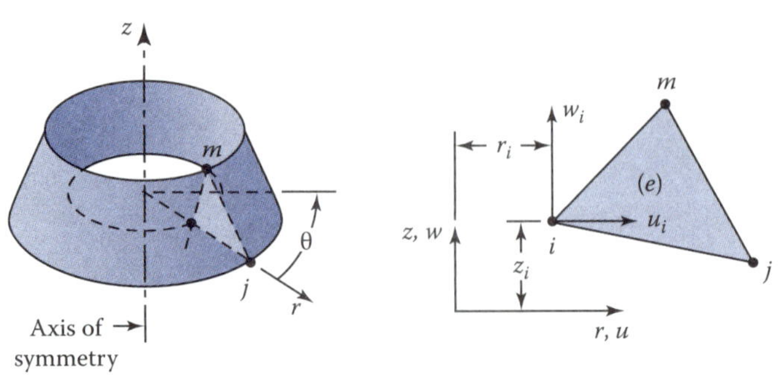

FIGURE 17.21
Symmetric solid finite element with triangular cross section.

The element strain–nodal displacement matrix may be expressed as

$$\{\varepsilon\}_e = [B]\{\delta\}_e \tag{17.45}$$

In the preceding, $[B]$ is yet to be determined. The stress–strain relationship is as follows:

$$\{\sigma\}_e = \begin{Bmatrix} \sigma_r \\ \sigma_z \\ \sigma_\theta \\ \tau_{rz} \end{Bmatrix}_e = [D]\{\varepsilon\}_e \tag{17.46}$$

The *elasticity matrix* is given by

$$[D] = \frac{E}{(1+\nu)(1-2\nu)} \begin{bmatrix} 1-\nu & \nu & \nu & 0 \\ & 1-\nu & \nu & 0 \\ & & 1-\nu & 0 \\ \text{Symmetric} & & & (1-2\nu)/2 \end{bmatrix} \tag{17.47}$$

Following a procedure basically identical to that described in Section 17.6, it can be shown that [10]

$$[B] = \frac{1}{2A} \begin{bmatrix} b_i & b_j & b_m & 0 & 0 & 0 \\ 0 & 0 & 0 & c_i & c_j & c_m \\ d_i & d_j & d_m & 0 & 0 & 0 \\ c_i & c_j & c_m & b_j & b_j & b_m \end{bmatrix} \tag{17.48}$$

The quantity A represents the area of the triangle defined by Equation 17.33 and

$$\begin{aligned} b_i &= z_j - z_m & b_j &= z_m - z_i & b_m &= z_i - z_j \\ c_i &= r_m - r_j & c_j &= r_i - r_m & c_m &= r_j - r_i \\ d_n &= \frac{a_n}{r} + b_n + \frac{c_n z}{r} & & (n = i, j, m) \end{aligned} \tag{17.49}$$

Note that the matrix $[B]$ includes the coordinates r and z. Therefore, the strains are *not constant*, as in the case of plane strain and plane stress. The *stiffness matrix* can now be obtained through the use of Equation 17.30:

$$[k]_e = \int_V [B]^T [D][B] \mathrm{d}V \tag{17.30}$$

This may be rewritten, after integrating along the circumferential or ring boundary, as follows:

$$[k]_e = 2\pi \int r[B]^T [D][B] \mathrm{d}r\mathrm{d}z \tag{17.50}$$

where the matrices $[D]$ and $[B]$ are defined by Equations 17.47 and 17.48, respectively.

Let the radial and axial components of force per unit length be denoted by f_r and f_z, respectively, of the circumferential boundary of a node or a radius r. The total radial and axial *nodal forces* are then

$$F_r = 2\pi r f_r, \quad F_z = 2\pi r f_z \tag{17.51}$$

Equations 17.50 and 17.51 characterize the triangular element for an axisymmetric problem. These are introduced into Equation 17.18 and subsequently into Equation 17.17 to evaluate the nodal displacements by satisfying the prescribed boundary conditions. The strains and stresses are then found using Equations 17.45 and 17.46. Most of what was learned in studying the 2D problems in Sections 17.5 through 17.7 also applies to axisymmetric problems.

CASE STUDY 17.6 Thick-Walled High-Pressure Steel Cylinder

A thick-walled pipe with the inner and outer radii a and b, respectively, is subjected to external pressure p_o (Figure 17.22a). Sketch the stress distribution, as obtained by the finite element and exact approaches of analysis, across the wall of the cylinder.

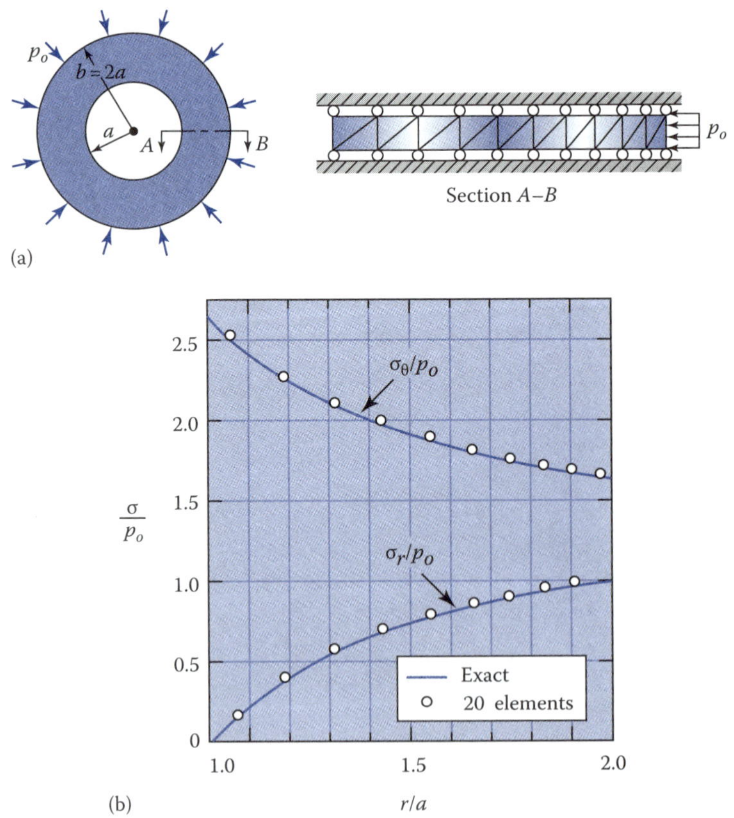

FIGURE 17.22
(a) Thick-walled cylinder under external pressure and modeling of its section *A-B* and (b) distribution of tangential and radial stresses in the cylinder for $b = 2a$.

Given: $b = 2a$, $\nu = 0.3$.

Solution

Only a slice of the cylinder needs to be analyzed (Figure 17.22a). A total of 20 triangular elements, with gradually decreasing size toward the loaded region, are used. The boundary conditions for the mesh system may be represented by discrete rollers at the nodes along the top and bottom faces of the slice; that is, at these nodes, the w nodal displacements are to be constrained to vanish.

Using a general-purpose finite element program [1,7,10], the tangential and radial stresses, σ_θ and σ_r, are obtained. The nondimensional results are plotted in Figure 17.22b. Also shown are the exact results given by Equations 16.18a and 16.18b. We see and excellent agreement between the solutions determined by the two methods.

Problems

Sections 17.1 through 17.3

17.1 A fixed-end composite rod is acted upon a concentrated load P at node 2 as illustrated in Figure P17.1. The aluminum rod 1–3 has cross-sectional area A and modulus of elasticity E. The copper rod 3–4 is with cross-sectional area $2A$ and elastic modulus $E/2$. Find

a. The system stiffness matrix
b. The displacements of nodes 2 and 3
c. The nodal forces and reactions at the supports

17.2 Consider a stepped *steel* bar 1–4 held between rigid supports and that carries a concentrated load P at node 3 as illustrated in Figure P17.2. Determine

a. The system stiffness matrix
b. The displacements of nodes 2 and 3
c. The nodal forces and reactions at supports

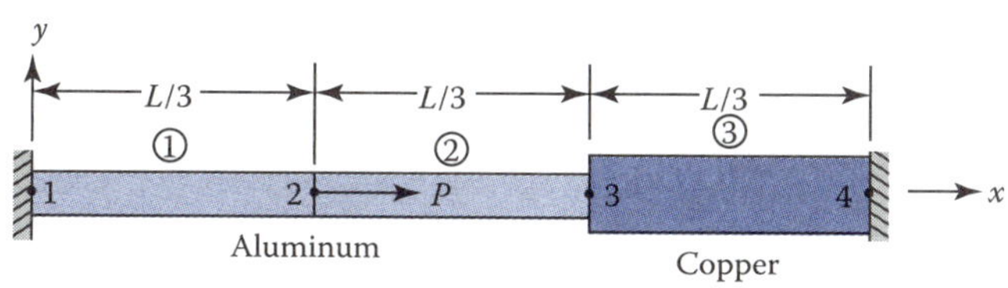

FIGURE P17.1

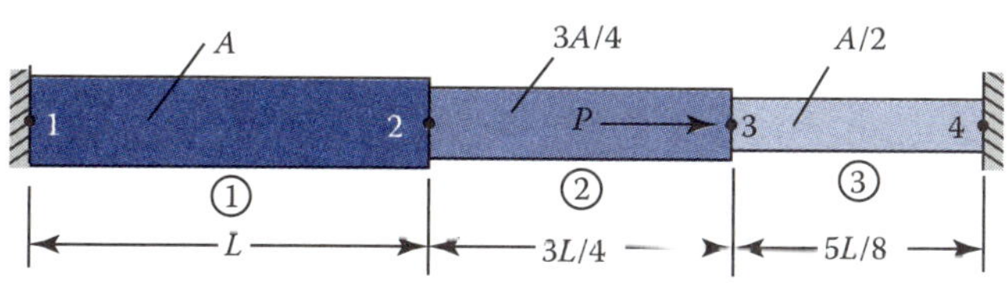

FIGURE P17.2

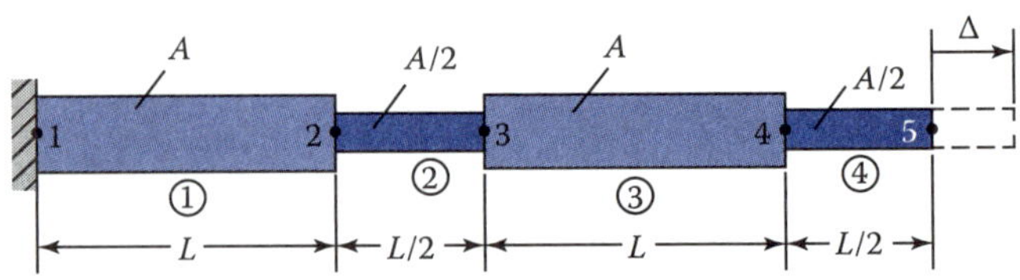

FIGURE P17.3

17.3 A stepped *brass* rod 1–5 is built in at the left end and given a displacement Δ at the right end as depicted by the dashed lines in Figure P17.3. Find

a. The system stiffness matrix
b. The displacements of nodes 2, 3, and 4
c. The nodal forces and reactions at support

17.4 The bar element 1–3 of a plane linkage mechanism shown in Figure P17.4 with length L, cross-sectional area A, and modulus of elasticity E is oriented at an angle θ counterclockwise from the x-axis. Find

a. The global stiffness matrix of the element
b. The local displacements $\bar{u}_1$, $\bar{v}_1$, $\bar{u}_3$, and $\bar{v}_3$ of the element
c. The axial stress in the element

Given: $u_1 = 1.5$ mm, $v_1 = 1.2$ mm, $u_3 = -2.2$ mm, $v_3 = 0$, $\theta = 30°$, $A = 1200$ mm^2, $L = 1.6$ m, $E = 72$ GPa.

Assumption: The bar is made of a 2014-T6 aluminum alloy.

17.5 Resolve Problem 17.4, for the case in which an ASTM-A36 structural steel bar element is oriented at an angle $\theta = 60°$ counterclockwise from the x-axis.

17.6 Figure P17.6 shows a plane truss containing five members each having axial rigidity AE supported at joints 2, 3, and 4. What is the global stiffness matrix for each element?

17.7 A planar truss consisting of five members is supported at joints 1 and 4 as shown in Figure P17.7. Determine the global stiffness matrix for each element.

Assumption: All bars have the same axial rigidity AE.

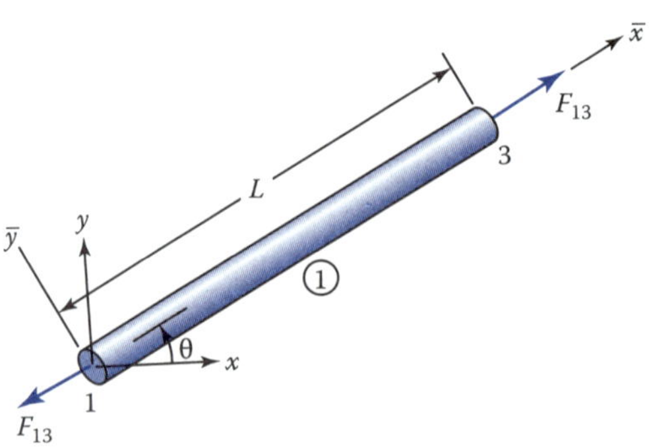

FIGURE P17.4

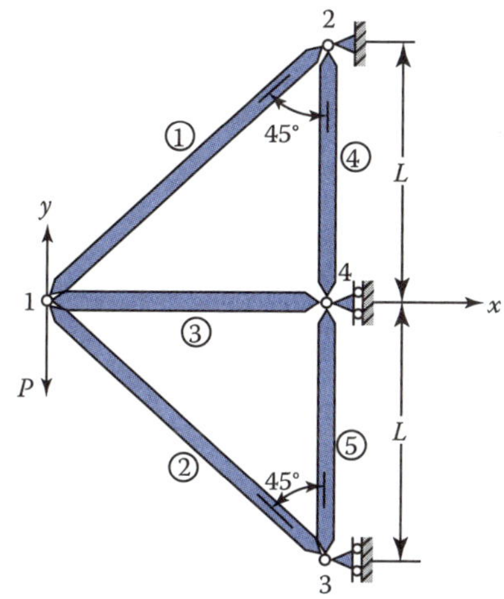

FIGURE P17.6

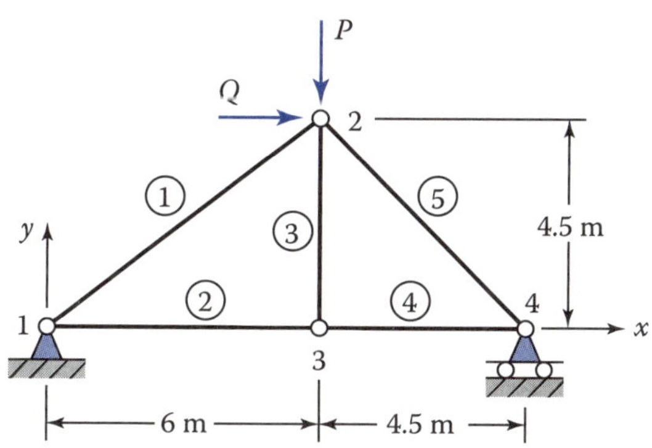

FIGURE P17.7

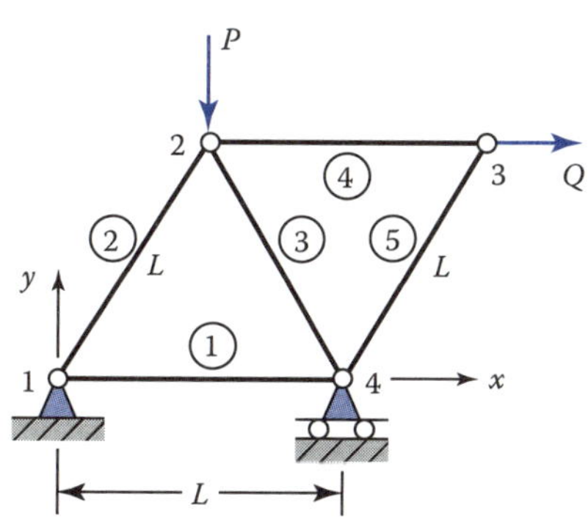

FIGURE P17.8

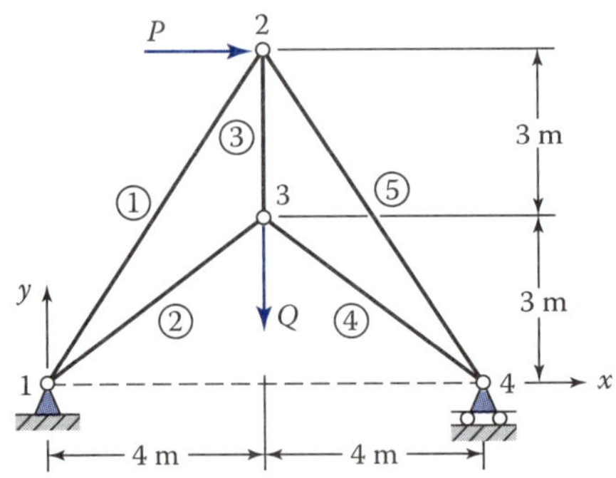

FIGURE P17.9

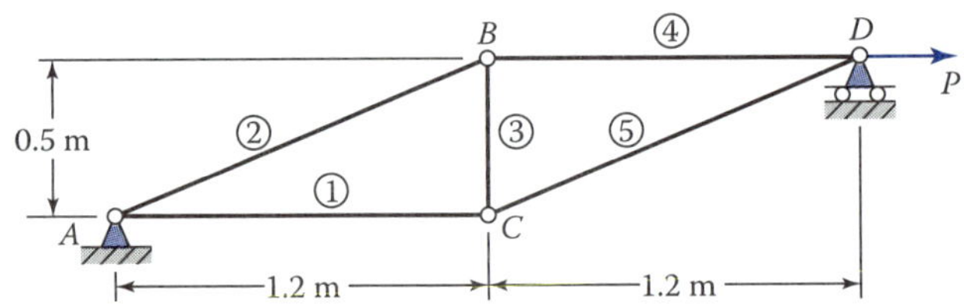

FIGURE P17.10

17.8 through 17.10 The plane truss is loaded and supported as shown in Figures P17.8 through P17.10. Determine

a. The global stiffness matrix for each element

b. The system matrix and the system force–displacement equations

Assumption: The axial rigidity *AE* is the same for each element.

17.11 A vertical load 10 kN acts at joint 2 of the two-bar truss 123 shown in Figure P17.11. Determine

a. The global stiffness matrix for each member

b. The system stiffness matrix

c. The nodal displacements

d. The reactions

e. The axial forces in each member and show the results on a sketch of each member

Assumption: The axial rigidity *AE* = 30 MN is the same for each bar.

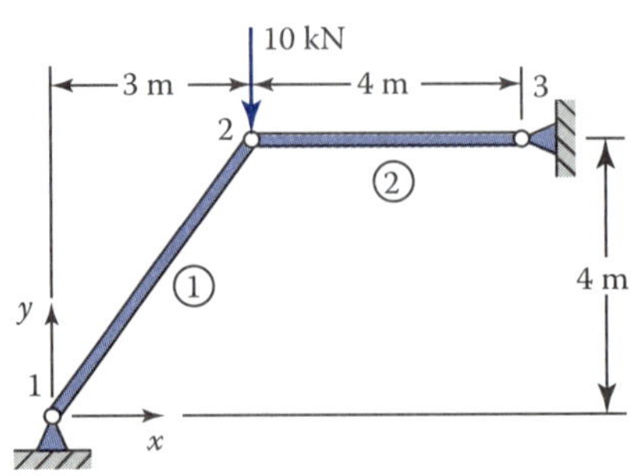

FIGURE P17.11

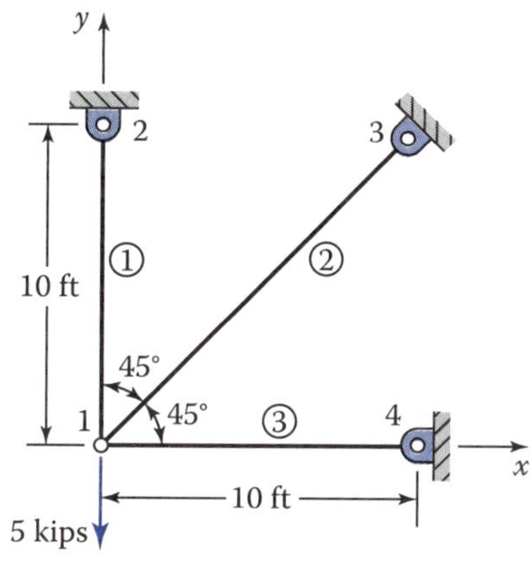

FIGURE P17.12

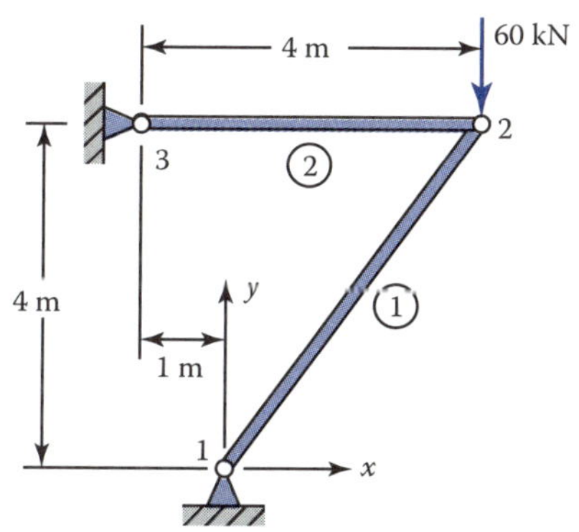

FIGURE P17.13

17.12 Redo Problem 17.11 for the structure shown in Figure P17.12, with $A = 1.8$ in.^{2} and $E = 30 \times 10^6$ psi.

17.13 Solve Problem 17.11 for the structure shown in Figure P17.13, with $AE = 10$ MN for each bar.

17.14 Resolve Problem 17.11 for the truss shown in Figure P17.14, with $AE = 125$ MN for each member.

17.15 The two-bar plane structure shown in Figure P17.15, due to loading $P = 100$ kN, settles an amount of $u_1 = 25$ mm downward at support 1. Determine

a. The global stiffness matrix for each member

b. The system matrix

c. The nodal displacements

d. The reactions

e. The axial forces in each member

Given: $E = 210$ GPa, $A = 5 \times 10^{-4}$ m^2 for each bar.

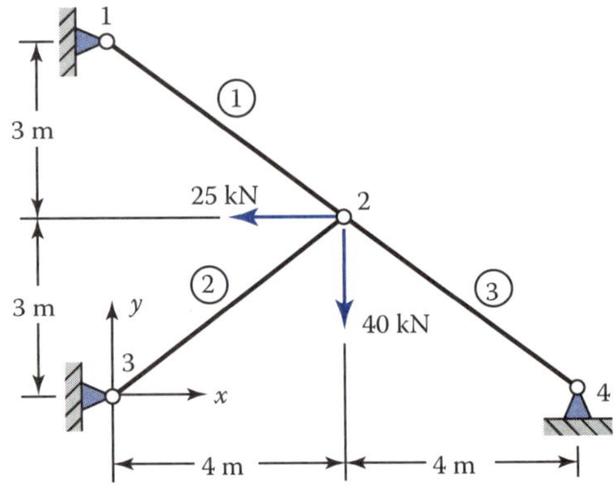

FIGURE P17.14

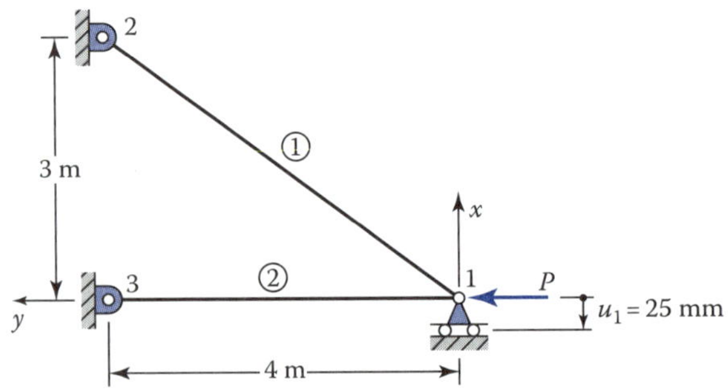

FIGURE P17.15

17.16 A plane truss is loaded and supported as shown in Figure P17.16. Determine

a. The global stiffness matrix for each member

b. The system stiffness matrix

c. The nodal displacements

d. The reactions

e. The axial forces in each member

Assumption: The axial rigidity $AE = 20$ MN is the same for each bar.

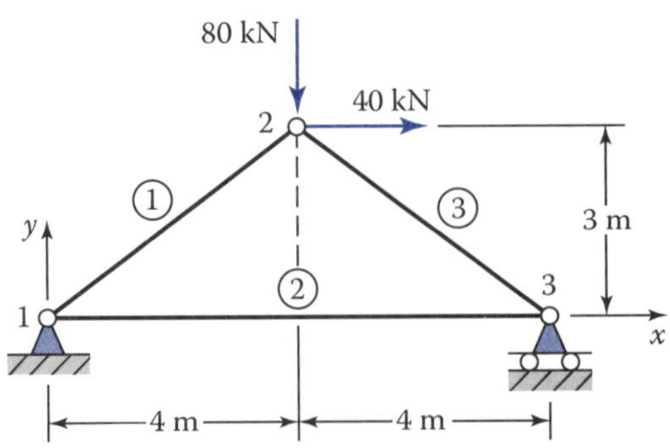

FIGURE P17.16

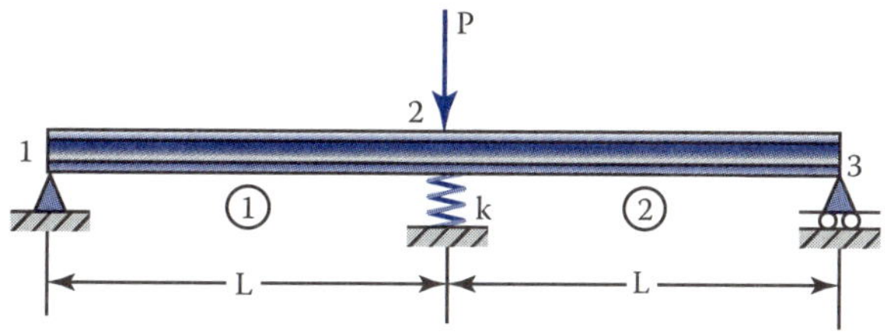

FIGURE P17.17

Sections 17.4 through 17.8

17.17 A steel beam supported by a pin, a spring of stiffness k, and a roller at points 1, 2, and 3, respectively, is acted upon by a concentrated load P at point 2 as shown in Figure P17.17. Calculate

a. The nodal displacements

b. The nodal forces and spring force

Given: $L = 4$ m, $P = 20$ kN, $EI = 12$ MN·m^2, $k = 180$ kN/m.

17.18 A cantilever aluminum beam is supported at its free end by a spring of stiffness k and carries a concentrated load P as shown in Figure P17.18. Calculate

a. The nodal displacements

b. The nodal forces and spring force

Given: $L = 22$ ft, $P = 1.8$ kips, $EI = 60\ (10^8)$ lb·in.^{2}, $k = 1.2$ kips/in.

17.19 A propped cantilever beam of constant flexural rigidity EI with a vertical load of 10 kips at its midspan is shown in Figure P17.19. Determine

a. The stiffness matrix for each element

b. The system stiffness matrix and nodal displacements

c. The member end forces and moments

d. Sketch the shear and moment diagrams

Given: $EI = 216 \times 10^6$ lb·in.^{2}

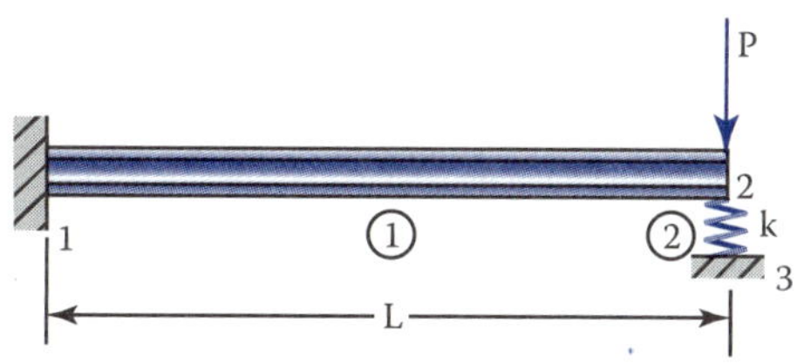

FIGURE P17.18

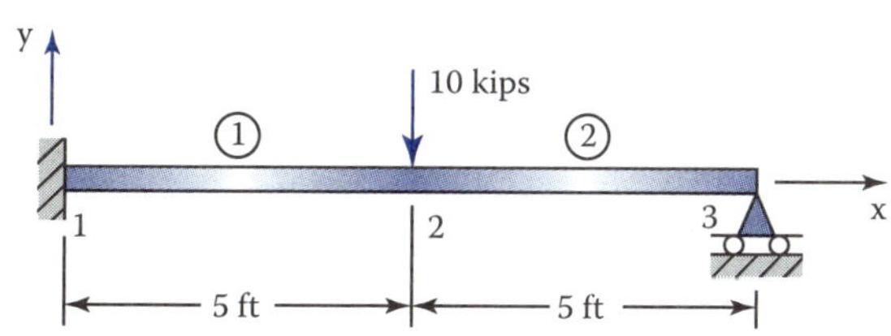

FIGURE P17.19

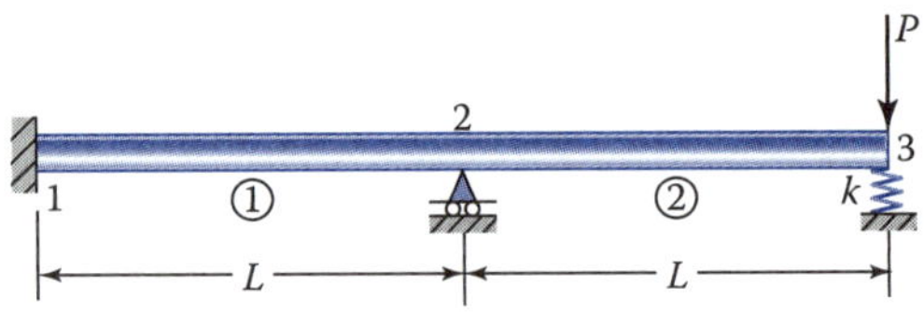

FIGURE P17.20

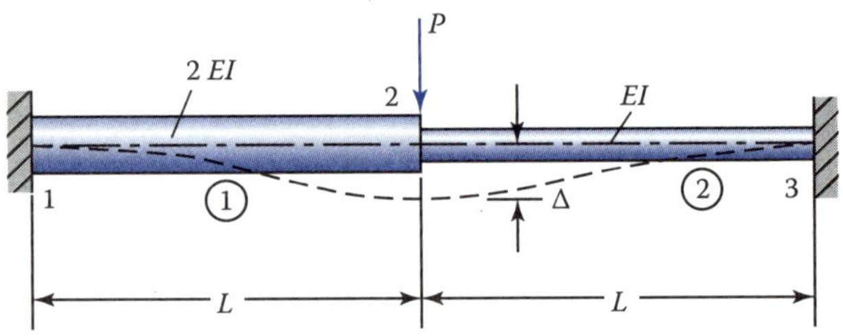

FIGURE P17.21

17.20 A prismatic steel beam fixed at end 1, simply supported at point 2, carries a load P at its free end 3 where supported on a spring of stiffness k (Figure P17.20). Find

a. The stiffness matrix for each element

b. The system matrix

c. The nodal displacements v_3, θ_3, and θ_3

17.21 A fixed-end stepped steel beam is acted upon by a concentrated center load P that causes a vertical deflection at the midpoint 2 as shown by the dashed lines in Figure P17.21. Find, in terms of EI, L, and Δ, as required

a. The load P

b. The slope at point 2

17.22 A plane frame 123 with hinged supports at joints 1 and 3 is subjected to a horizontal load of 30 kN (Figure P17.22). Determine

a. The global stiffness matrix for each member

b. The system stiffness matrix

c. The displacements u_2, v_2, and θ_2

Design Assumptions: Members 12 and 23 are identical with a square cross-sectional area of $A = b \times h = 900$ mm² and $E = 70$ GPa.

17.23 A frame 123 is fixed at supports 1 and 2 as shown in Figure P17.23. A horizontal load of 40 kN acts at joint 2. Determine

a. The global stiffness matrix of each member

b. The system stiffness matrix

c. Displacements u_2, v_2, and θ_2

Given: $E = 200$ GPa, $I_1 = 5\sqrt{2}\times10^6\ \text{mm}^4$, $I_2 = 5\times10^6\ \text{mm}^4$, $A_1 = 2.5\sqrt{2}\times10^3\ \text{mm}^2$, $A_2 = 2.5\times10^3\ \text{mm}^2$

17.24 Redo Case Study 17.2, by discretizing the plate into four elements, as shown in Figure P17.24. Use a computer program with CST (or LST) elements.

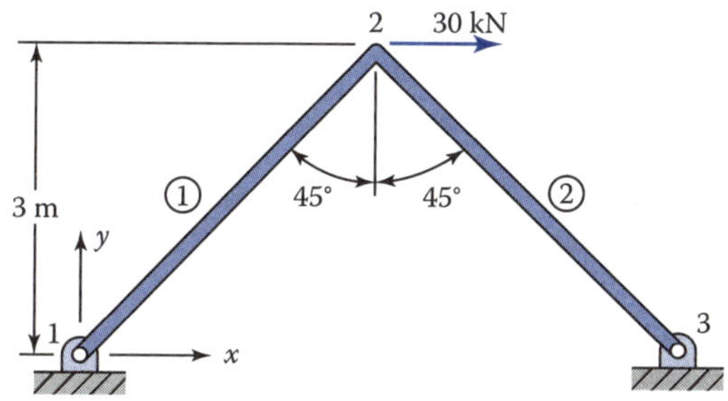

FIGURE P17.22

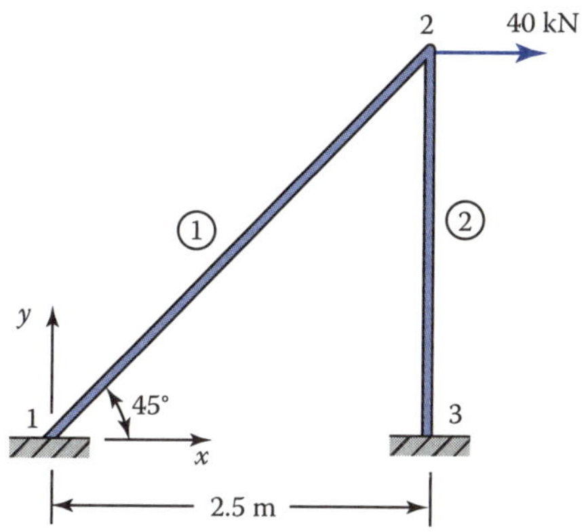

FIGURE P17.23

17.25 Verify the results introduced in Case Study 17.4 using a computer program with CST (or LST) elements.

17.26 A steel plate with a hole is under a uniform axial tension loading P (Figure P17.26). The dimensions are in millimeters.

a. Analyze the stresses using a computer program with the CST (or LST) elements.

b. Compare the stress-concentration factor K_t, obtained in part with that found from Figure C.5.

Given: $P = 4$ kN and plate thickness $t = 10$ mm.

17.27 Redo Problem 17.26 for the plate shown in Figure P17.27.

17.28 A simply supported beam is subjected to a uniform loading of intensity w (Figure P17.28). Take $L = 10\,h$, $t = 1$, and $\nu = 0.3$. Refine the meshes to calculate the

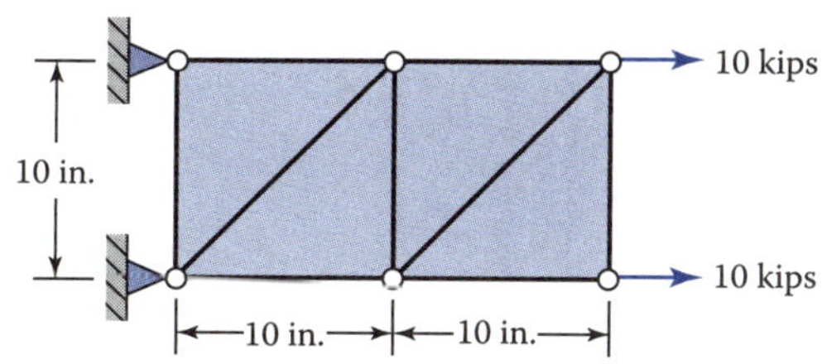

FIGURE P17.24

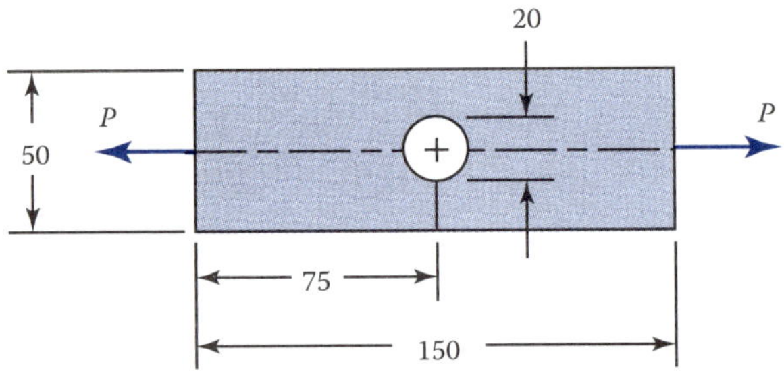

FIGURE P17.26

stress and deflection within 5% accuracy, by using a computer program with the CST (or LST) elements.

Given: The *exact* solution [9] is in which t represents the thickness.

$$\sigma_{x,\max} = \frac{3wL^2}{4th^2}, \quad v_{\max} = -\frac{5wL^4}{16Eth^3} - \frac{3wL^2(1+\nu)}{5Eht}$$

17.29 Redo Problem 17.28 for the case in which a cantilevered beam is under a uniform loading of intensity w (Figure P17.29).

Given: The *exact* solution has the following form:

$$\sigma_{x,\max} = \frac{3wL^2}{4th^2}, \quad \upsilon_{\max} = -\frac{3wL^4}{16Eth^3} - \frac{3wL^2(1+\nu)}{5Eht}$$

where t is the thickness.

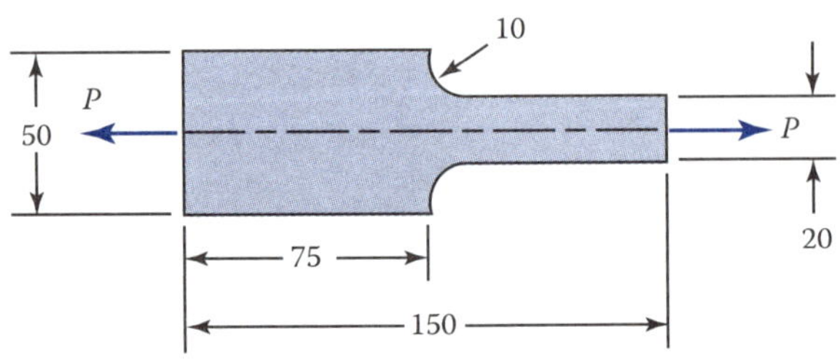

FIGURE P17.27

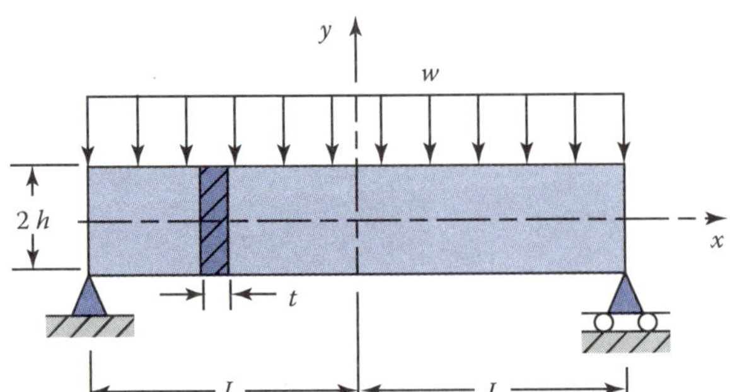

FIGURE P17.28

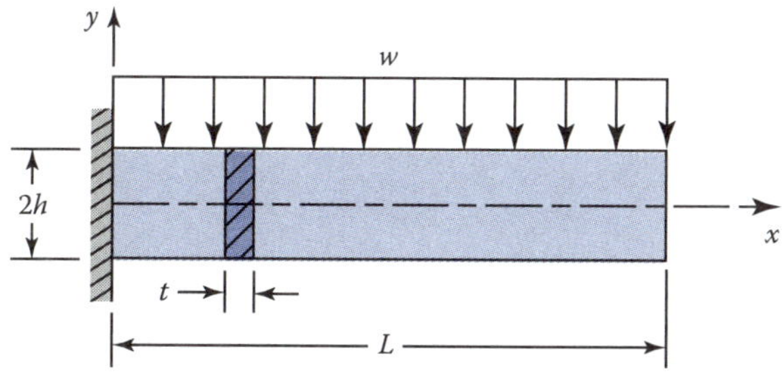

FIGURE P17.29

17.30 Verify the results introduced in Case Study 17.5, using a finite element computer program with CST (or LST) elements.

17.31 A cast iron cylinder ($\nu = 0.2$) of a hydraulic device having inner radius a and outer radius $4a$ is subjected to an external pressure p_o. Using a finite element computer program with CST (or LST) elements, determine the distribution of tangential and radial stresses.

17.32 Redo Problem 17.31, if the cylinder is under an internal pressure p_i. Compare the results with the exact solution shown in Figure 16.3.

17.33 Resolve Problem 17.31 for the case in which the cylinder is subjected to internal pressure p_i and external pressure $p_o = 0.5p_i$.

18

Case Studies in Machine Design

18.1 Introduction

As observed earlier, design is an iteration process. When presented with a design problem statement, some simplifying assumptions are necessary from the start. Engineering software should be used for effective component design. A general case study in *component* or *machine design* may include the following: a step-by-step proposal of the product, the relevant trade study, the configuration development, the detailed design and construction process, and the prototype tests. Thus, a case study presents a product in action. It covers *lessons learned* during the development of a device, such as product goals, market needs, and engineering/manufacturing relationships. All of these can contribute to a continuous improvement program, resulting in a superior product [1,2].

We shall here present two case studies in *preliminary design* that are larger scale than those introduced in preceding chapters. These case studies show how the design of any one component may be affected by the design of related parts. Because of space limitations, only certain important aspects of these studies are discussed. A *floor crane with electric winch* and a *high-speed cutter* are the systems analyzed. Further information on types, pricing, maintenance, and lifespan of these machines can be found on manufacturer websites. Clearly, advancing from the design of individual parts to the design of a complete machine is a major step. The objective of this chapter is to help prepare the reader for attempting this step.

18.2 Floor Crane with Electric Winch

A *crane* is a type of machine that is generally equipped with a hoist, winding drum, cable or chain, and sheaves. Coming in many forms, cranes can be employed both to lift or lower materials and to move them horizontally. A crane creates mechanical advantage and hence moves loads beyond the normal capability of a human. Such machines are often employed in the transport industry for the loading/unloading of freight, in the construction industry for the movement of materials, and in the manufacturing industry for the assembly of heavy equipment.

The earliest cranes were constructed from wood, with cast iron and steel taking over during the Industrial Revolution. They were powered by men or animals and employed for the construction of tall buildings. Larger cranes were developed using treadwheels that permit the lifting of heavier weights. In the Middle Ages, harbor cranes were introduced to load and unload ships and assist with their construction—some were built into stone towers for extra strength and stability. For many centuries, hoists in watermills and windmills were driven by the harnessed natural power. The first *mechanical* power was provided by steam engines, which lead to the earliest steam crane in the early nineteenth century. Many remained in use well into the late twentieth century.

Modern cranes, which commonly include an *electric winch*, use internal combustion engines or electric motors with hydraulic systems in order to provide a much greater lifting capability than was previously possible. Cranes exist in various forms, each tailored to a specific use. The following photographs represent some examples of modern cranes. Sizes range from a small overhead crane, used in workshops, to a huge tower crane, used for constructing tall structures. Minicranes are also employed to facilitate construction of high buildings by reaching tight spaces. Larger floating cranes are often used to build oil rigs and salvage sunken ships. They contain a stationary frame with an I-beam that suspends from a trolley, which is designed for easy moving in a straight or curvilinear direction.

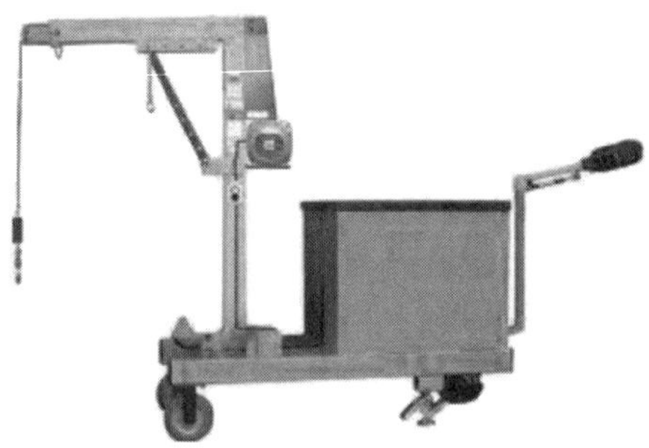

Typical counterbalance floor cranes with electric winches.

This section is concerned with a *floor crane with electric winch* supported by solid plastic wheels and used for lifting and moving loads in the laboratory or machine shop (see Figure 18.1). It has electric power capacity to lift a load (P). The concrete or sand counterbalance weight (W) on the base prevents the crane from tipping forward when the crane is pushed by a horizontal force (F) acting at a height (H) from the ground. For safety purposes, the drive system includes a torque limiter coupled to a drum and allows the crane to lift no more than the working load (P).

Other key features include the following:

- All welded robust construction
- Manufactured from hollow steel box sections for combined strength and lightness
- Heavy-duty swivel hook
- Fitted with hard wearing polyurethane front wheels and swivel casters on the rear (with brakes) for easy movement of the crane
- Offers added productivity and ergonomic advantages over manual models
- In addition to precision, allows the operator freedom to work close to machines or over obstructions

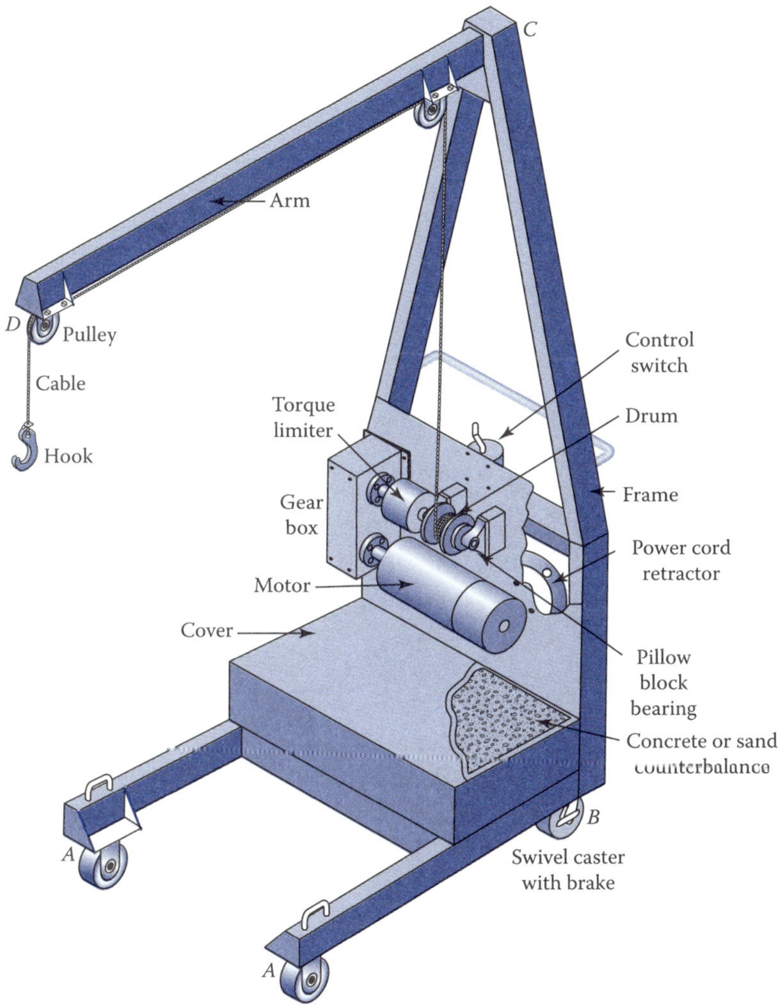

FIGURE 18.1
Schematic drawing of floor crane with electric winch.

CASE STUDY 18.1 Entire Frame Load Analysis

Consider the crane winch depicted in Figure 18.1. The entire frame of this machine is illustrated in Figure 18.2. Determine

a. The design load on the front and rear wheels
b. The factor of safety n_t for the crane tipping forward from the loading

Given: The geometry of each element is known. The cable and hook are rated at 15 kN, which gives a safety factor of 5. The 85 mm diameter drum is about 20 times the cable diameter. The crane frame carries the load P, counterweight W_C, weights of parts W_i (i = 1, 2, 3, 4, 5), and the push force F, as shown in Figure 18.2. The frame is made of b = 50 mm and h = 100 mm structural steel tubing of t = 6 mm thickness with weight w Newton per meter (Table A.4).

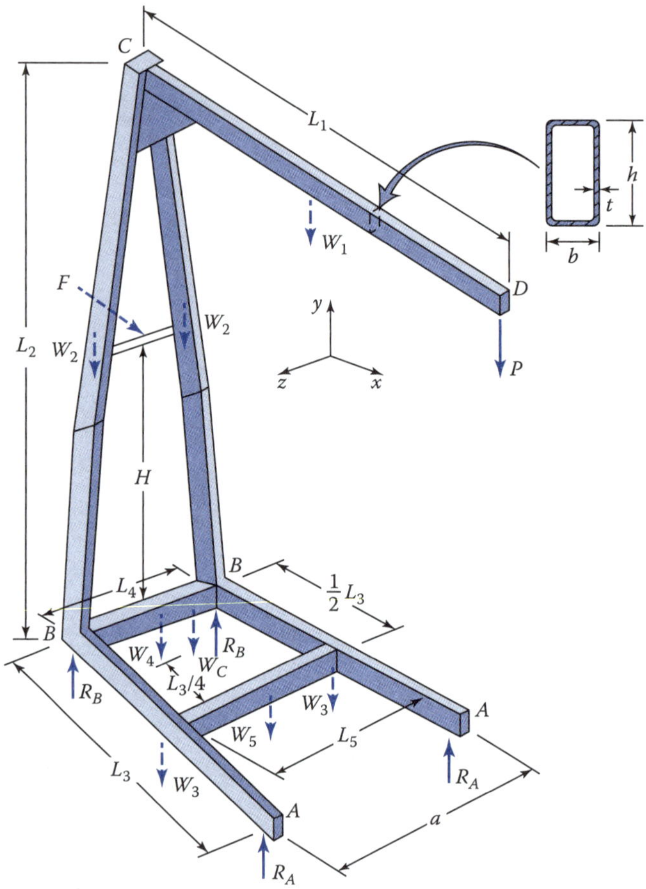

FIGURE 18.2
Simple sketch of the crane winch frame shown in Figure 18.1.

Data:

$$P = 3\text{ kN} \qquad F = 100\text{ N} \qquad W_C = 2.7\text{ kN}$$
$$w \approx 130\text{ N/m} \qquad a = 0.8\text{ m} \qquad H = 1\text{ m}$$
$$L_1 = 1.5\text{ m} \qquad L_2 = 2\text{ m} \qquad L_3 = 1\text{ m}$$
$$L_4 = 0.5\text{ m} \qquad L_5 = 0.65\text{ m}$$

and

$$W_1 = wL_1 = 130(1.5) = 195\text{ N} \quad W_2 \approx wL_2 = 260\text{ N}$$
$$W_3 = 130\text{ N} \quad W_4 = 65\text{ N} \quad W_5 = 84.5\text{ N}$$

For dimensions and properties of a selected range of frequently used crane members, refer to manufacturers' catalogs.

Assumptions:

1. A line speed of 0.12 m/s is used, as suggested by several catalogs for lifting. The efficiency of the speed reduction unit or gearbox is 95%. The electric motor has 0.5 hp capacity to lift 3 kN load for the preceding line speed and efficiency and includes an internal brake to hold the load when it is inoperative. The gear ratios (see Case Study 18.4) satisfy the drive system requirements.
2. Only the weights of concrete counterbalance and main frame parts are considered. All frame parts are weld connected to one another.
3. Compression forces caused by the cable running along the members are ignored. All forces are static; F is x directed (horizontal) and remaining forces are parallel to the xy plane. Note that the horizontal component of the reaction at B equals $F/2$, not indicated in Figure 18.2.

Solution

See Figure 18.2; Section 1.9.

a. Reactional forces R_A and R_B acting on the wheels are determined by applying conditions of equilibrium, $\sum M_z = 0$ at B and $\sum F_y = 0$, to the free-body diagram shown in the figure with $F = 0$. Therefore,

$$R_A = \frac{1}{2}\left(P\frac{L_1}{L_3} + \frac{1}{2}W_1\frac{L_1}{L_3} + \frac{1}{4}W_C + W_3 + \frac{1}{2}W_5\right)$$

$$R_B = -R_A + \frac{1}{2}P + \frac{1}{2}W_1 + W_2 + W_3 + \frac{1}{2}W_4 + \frac{1}{2}W_C + \frac{1}{2}W_5 \qquad (18.1)$$

Substitution of the given data into the foregoing results in

$$R_A = \frac{1}{2}\left[3000(1.5) + \frac{1}{2}(195)(1.5) + \frac{1}{4}(2700) + 130 + \frac{1}{2}(84.5)\right] = 2747 \text{ N}$$

$$R_B = -2747 + \frac{1}{2}(3000) + \frac{1}{2}(195) + 260 + 130 + \frac{1}{2}(65) + \frac{1}{2}(2700) + \frac{1}{2}(84.5) = 665 \text{ N}$$

Note that, when the crane is unloaded ($P = 0$), Equations 18.1 give

$$R_A = 497 \text{ N} \quad R_B = 1415 \text{ N}$$

Comment: Design loads on front and rear wheels are 2747 N and 1415 N, respectively.

b. The factor of safety n_t is applied to tipping loads. The condition $\sum M_z = 0$ at point A is

$$n_t\left[P(L_1 - L_3) + FH\right] = W_1\left(L_3 - \frac{1}{2}L_1\right) + 2W_2L_3 + W_4L_3 + \frac{3}{4}W_CL_3 + \frac{1}{2}(2W_3 + W_5)L_3 \qquad (18.2)$$

Introducing the given numerical values,

$$n_t\left[3000(0.5)+100(1)\right]=195(0.25)+2(260)(1)+65(1)+\frac{3}{4}(2700)(1)$$
$$+\frac{1}{2}\left[2(130)+84.5\right](1)$$

from which $n_t = 1.77$.

Comments: For the preceding forward tipping analysis, the rear wheels are assumed to be locked and the friction is taken to be sufficiently high to prevent sliding. Side-to-side tipping may be checked similarly.

CASE STUDY 18.2 Design Analysis of Arm *CD*

The arm *CD* of a winch crane is represented schematically in Figure 18.2. Determine the maximum stress and the factor of safety against yielding. What is the deflection under the load using the method of superposition?

Given: The geometry and loading are known from Case Study 18.1. The frame is made of ASTM-A36 structural steel tubing. From Table B.1,

$$S_y = 250 \text{ MPa} \quad E = 200 \text{ GPa}$$

Assumptions: The loading is static. The displacements of welded joint *C* are negligibly small; hence, part *CD* of the frame is considered a cantilever beam.

Solution

See Figures 18.2 and 18.3; and Table B.1, Section 3.7.

We observe from Figure 18.2 that the maximum bending moment occurs at points *B* and *C* and $M_B = M_C = M$. Since two vertical beams resist moment at *B*, the critical

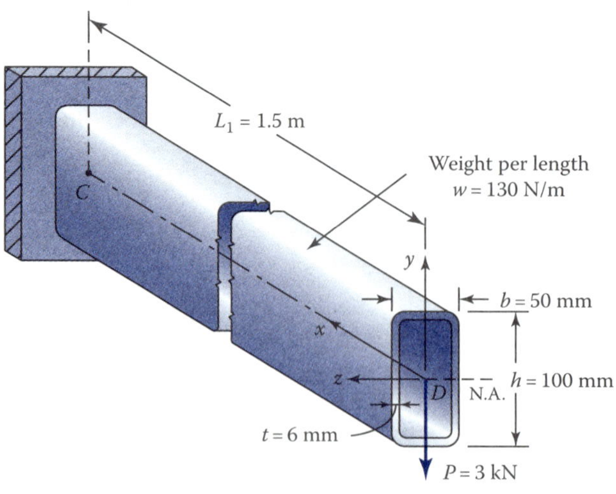

FIGURE 18.3
Part *CD* of the crane arm shown in Figure 18.1.

section is at *C* of cantilever *CD* carrying its own weight per unit length *w* and concentrated load *P* at the free end (Figure 18.3).

The bending moment *M* and shear force *V* at the cross section through the point *C*, from static equilibrium, have the following values:

$$M = PL_1 + \frac{1}{2}wL_1^2$$

$$= 3000(1.5) + \frac{1}{2}(130)(1.5)^2 = 4646\ \text{N}\cdot\text{m}$$

$$V = 3\ \text{kN}$$

The cross-sectional area properties of the tubular beam are

$$A = bh - (b-2t)(h-2t)$$

$$= 50\times 100 - 38\times 88 = 1.66\left(10^{-3}\right)\text{m}^2$$

$$I = \frac{1}{12}bh^3 - \frac{1}{12}(b-2t)(h-2t)^3$$

$$= \frac{1}{12}\left[\left(50\times 100^3\right) - (38)(88)^3\right] = 2.01\left(10^{-6}\right)\text{m}^4$$

where *I* represents the moment of inertia about the neutral axis.

Therefore, the maximum bending stress at the top of outer fiber of section through *C* equals

$$\sigma_{\text{max}} = \frac{Mc}{I} = \frac{4646(0.05)}{2.01\left(10^{-6}\right)} = 115.6\ \text{MPa}$$

where the shear strain is zero. The highest value of the shear stress occurs at the neutral axis. Referring to Figure 18.3 and Equation 3.21, the first moment of the area about the NA is

$$Q_{\text{max}} = b\left(\frac{h}{2}\right)\left(\frac{h}{4}\right) - (b-2t)\left(\frac{h}{2}-t\right)\left(\frac{h/2-t}{2}\right) \tag{18.3}$$

$$= 50(50)(25) - (38)(44)(22) = 25.716\left(10^{-6}\right)\text{m}^3$$

Hence,

$$\tau_{\text{max}} = \frac{VQ_{\text{max}}}{I(2t)}$$

$$= \frac{3000(25.716)}{2.01(2\times 0.006)} = 3.199\ \text{MPa}$$

The factor of safety against yielding is then equal to

$$n = \frac{S_y}{\sigma_{max}} = \frac{250}{115.6} = 2.16$$

This is satisfactory because the frame is made of average material operated in ordinary environment and subjected to known loads.

Comments: At joint *C*, as well as at *B*, a thin (about 6 mm) steel gusset should be added at each side (not shown in the figure). These enlarge the weld area of the joints and help reduce stress in the welds. Case Study 18.9 illustrates the design analysis of the welded joint at *C*.

When the load *P* and the weight *w* of the cantilever depicted in the figure act alone, displacements at *D* (from cases 1 and 3 of Table A.9) are $PL_1^3 / 3EI$ and $wL_1^4 / 8EI$, respectively. It follows that the deflection υ_D at the free end owing to the combined loading is

$$\upsilon_D = -\frac{PL_1^3}{3EI} - \frac{wL_1^4}{8EI}$$

Substituting the given numerical values into the preceding expression, we have

$$\upsilon_D = -\frac{1}{200(10^3)(2.01)}\left[\frac{3000(1.5)^3}{3} + \frac{130(1.5)^4}{8}\right]$$

$$= -8.6 \text{ mm}$$

Here, the minus sign means a downward displacement.

Comment: Since $\upsilon_D \ll h/2$, the magnitude of the deflection obtained is well within the acceptable range (see Section 3.7).

CASE STUDY 18.3 Deflection of Arm *CD* due to Bending and Shear

A schematic representation of a winch crane arm CD is shown in Figure 18.3. Find the deflection at the free end *D* applying the energy method.

Given: The dimensions and loading of the frame are known from Case Study 18.1.

Data:

$$E = 200 \text{ GPa}, \quad G = 76.9 \text{ GPa},$$

$$I = 2.01(10^{-6}) \text{ m}^4$$

Assumption: The loading is static.

Requirement: Deflections owing to the bending and transverse shear are considered.

Solution

See Figure 18.3; Table 5.1, Section 5.5.

The form factor for shear for the rectangular box section, from Table 5.1, is

$$\alpha = \frac{A}{A_{\text{web}}} = \frac{A}{ht}$$

Here h is the beam depth and t represents the wall thickness. The moment and shear force, at an arbitrary section x distance from the free end of the beam, are expressed as follows:

$$M = -Px - \frac{1}{2}wx^2, \quad V = P + wx$$

Therefore,

$$\frac{\partial M}{\partial P} = -x, \quad \frac{\partial V}{\partial P} = 1$$

After substitution of all the preceding equations, Equation 5.36 becomes

$$\begin{aligned} \upsilon_D &= \frac{1}{EI}\int_0^{L_1} M\frac{\partial M}{\partial P}\,\mathrm{d}x + \frac{1}{AG}\int_0^{L_1} \alpha V\frac{\partial V}{\partial P}\,\mathrm{d}x \\ &= \frac{1}{EI}\int_0^{L_1}\left(Px + \frac{1}{2}wx^2\right)(x)\,\mathrm{d}x \\ &\quad + \frac{1}{htG}\int_0^{L_1}(P + wx)(1)\,\mathrm{d}x \end{aligned}$$

Integrating,

$$\upsilon_D = \frac{1}{EI}\left(\frac{PL_1^3}{3} + \frac{wL_1^4}{8}\right) + \frac{1}{htG}\left(PL_1 + \frac{wL_1^2}{2}\right) \qquad (18.4)$$

Substituting the given numerical values into this equation, we obtain

$$\begin{aligned} \upsilon_D &= \frac{1}{200\times 10^3(2.01)}\left[\frac{3000(1.5)^3}{3} + \frac{130(1.5)^4}{8}\right] \\ &\quad + \frac{1}{100(6)(76.9\times 10^3)}\left[3000(1.5) + \frac{130(1.5)^2}{2}\right] \\ &= (8.6 + 0.1)10^{-3} = 8.7\ \text{mm} \end{aligned}$$

Since P is vertical and directed downward, υ_D represents a vertical displacement and is positive downward.

Comments: If the effect of shear force is omitted, $\upsilon_D = 8.6$ mm; the resultant error in deflection is about 1.2%. The contribution of shear force to the displacement of the frame can therefore be neglected.

CASE STUDY 18.4 Design of the Spur Gear Train

The spur gearbox of the crane winch (Figure 18.1) is illustrated in Figure 18.4. Analyze the design of each gearset using the AGMA method.

Given: The geometry and properties of each element are known. A 0.5 hp 1725 rpm electric motor at 95% efficiency delivers 0.475 hp to the 85 mm diameter drum. The maximum capacity of the crane is $P = 3$ kN. All gears have $\phi = 20°$ pressure angle. Shafts 1 or 2, 3, and 4 are supported by 12, 19, and 25 mm bore flanged bearings, respectively.

TABLE 18.4

	Module m (mm)	Number of Teeth N	Pitch Diameter d (mm)	Face Width b (mm)
Gear 1 (pinion)	1.3	15	20	14
Gear 2	1.3	60	80	14
Gear 3 (pinion)	1.6	18	28.8	20
Gear 4	1.6	72	115.2	20
Gear 5 (pinion)	2.5	15	37.5	32
Gear 6	2.5	60	150	32

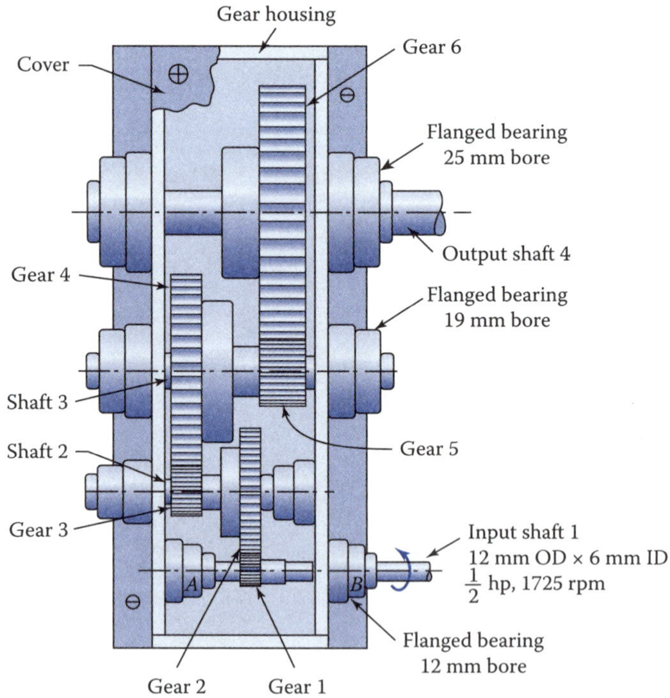

FIGURE 18.4
Gearbox of the winch crane shown in Figure 18.1.

Assumptions:

1. The pinions are made of carburized 55 R_c steel. Gears are Q&T, 180 Bhn steel. Hence, by Tables 11.6 and 11.11, we have

$$\text{Pinions:} \quad S_t = 414 \text{ MPa}, \quad S_c = 1310 \text{ MPa}$$

$$\text{Gears:} \quad S_t = 198 \text{ MPa}, \quad S_c = 620 \text{ MPa}$$

2. All gears and pinions are high-precision shaved and ground: manufacturing quality corresponds to curve *A* in Figure 11.15.
3. Loads are applied at the highest point of single-tooth contact.

Design decisions: The following reasonable values of the bending and wear strength factors for pinions and gears are chosen (from Tables 11.4, 11.5, 11.7, 11.8, 11.10):

$$K_o = 1.5, \quad K_s = 1.0 \left(\text{from Section 11.9}\right)$$

$$K_m = 1.6, \quad K_L = 1.1$$

$$K_T = 1.0 \left(\text{from Section 11.9}\right), \quad K_R = 1.25,$$

$$C_H = 1.0 \left(\text{by Equation 11.40}\right), \quad C_p = 191\sqrt{\text{MPa}}$$

$$C_f = 1.25 \left(\text{from Section 11.11}\right),$$

$$C_L = 1.1 \left(\text{from Figure 11.19}\right)$$

Solution

See Figures 18.1 and 18.4; and Sections 11.9 and 11.11.
The operating line velocity of the hoist at the maximum load is, by Equation 1.16,

$$V = \frac{745.7 \text{ hp}}{P} = \frac{745.7(0.475)}{3000} = 0.12 \text{ m/s}$$

The operating speed of the drum shaft is

$$n = \frac{V}{\pi d} = \frac{(0.12)(60)}{\pi(0.085)} \approx 27 \text{ rpm}$$

This agrees with the values suggested by several catalogs for light lifting. The gear train in Figure 18.4 fits all the parameters in the lifting system.

Gearset I: (Pair of gears 1 and 2. Figure 18.4)
The input torque and the transmitted load on gear 1 are

$$T_1 = \frac{7121\,\text{hp}}{n_1} = \frac{7121(0.5)}{1725} = 2.06\ \text{N}\cdot\text{m}$$

$$F_{t1} = \frac{T_1(2)}{d_1} = \frac{2.06(2)}{0.02} = 206\ \text{N}$$

The radial load is then

$$F_{r1} = F_{t1}\tan\phi = 206\tan 20^\circ = 75\ \text{N}$$

The pitch-line velocity is determined as

$$V_1 = \pi d_1 n_1 = \pi(0.02)\left(\frac{1725}{60}\right) = 1.81\ \text{m/s} = 356\ \text{fpm}$$

Then, from curve *A* of Figure 11.15, the dynamic factor is

$$K_\upsilon = \sqrt{\frac{78+\sqrt{356}}{78}} = 1.11$$

Equation 11.37b with $m_g = 4$ gives

$$I = \frac{\sin 20^\circ \cos 20^\circ}{2}\frac{4}{4+1} = 0.129$$

By Figure 11.16a, we have

$$J = 0.25 \quad (\text{for pinion, } N_p = 15 \text{ and } N_g = 60)$$

$$J = 0.42 \quad (\text{for gear, } N_g = 60 \text{ and } N_p = 15)$$

Gear 1 (pinion). Substituting the numerical values into Equations 11.35 and 11.36

$$\begin{aligned}\sigma &= F_{t1}K_oK_\upsilon\frac{1.0}{bm}\frac{K_sK_m}{J}\\ &= 206(1.5)(1.11)\frac{1.0}{14(1.3)10^{-6}}\frac{1.0(1.6)}{0.25}\\ &= 120.6\ \text{MPa}\end{aligned}$$

$$\sigma_{\text{all}} = \frac{S_tK_L}{K_TK_R} = \frac{414(1.1)}{(1.0)(1.25)} = 364.3\ \text{MPa}$$

Similarly, Equations 11.42 and 11.44 lead to

$$\sigma_c = C_p\left(F_{t1}K_oK_\upsilon \frac{K_s}{bd}\frac{K_mC_f}{I}\right)^{1/2}$$

$$= 191(10^3)\left[206(1.5)(1.11)\frac{1.0}{14(20)10^{-6}}\frac{1.6(1.25)}{0.129}\right]^{1/2}$$

$$= 832.4 \text{ MPa}$$

$$\sigma_{c,\text{all}} = \frac{S_eC_LC_H}{K_TK_R} = \frac{1310(1.1)(1.0)}{(1.0)(1.25)} = 1153 \text{ MPa}$$

Gear 2. We have $F_{t2} = F_{t1} = 206$ N. Substitution of the data into Equations 11.35′ and 11.36 gives

$$\sigma = F_{t2}K_oK_\upsilon \frac{1.0}{bm}\frac{K_sK_m}{J}$$

$$= 206(1.5)(1.11)\frac{1.0}{14(1.3)10^{-6}}\frac{1.0(1.6)}{0.42}$$

$$= 71.79 \text{ MPa}$$

$$\sigma_{\text{all}} = \frac{S_tK_L}{K_TK_R} = \frac{198(1.1)}{(1.0)(1.25)} = 174.2 \text{ MPa}$$

In a like manner, through the use of Equations 11.42 and 11.44,

$$\sigma_c = C_p\left(F_{t2}K_oK_\upsilon \frac{K_s}{bd}\frac{K_mC_f}{I}\right)^{1/2}$$

$$= 191(10^3)\left[206(1.5)(1.11)\frac{1.0}{14(80)10^{-6}}\frac{1.6(1.25)}{0.129}\right]^{1/2}$$

$$= 416.2 \text{ MPa}$$

$$\sigma_{c,\text{all}} = \frac{S_cC_LC_H}{K_TK_R}$$

$$= \frac{620(1.1)(1.0)}{(1.0)(1.25)} = 545.6 \text{ Pa}$$

Comments: Inasmuch as $\sigma < \sigma_{\text{all}}$ and $\sigma_c < \sigma_{c,\text{all}}$, the pair of gears 1 and 2 is *safe* with regard to the AGMA bending and wear strengths, respectively.

Gearsets II and III: (Pairs of gears 3–4 and 5–6. Figure 18.4)
Shaft 2 rotates at the speed

$$n_2 = n_1 \frac{N_1}{N_2} = 1725\left(\frac{15}{60}\right) \approx 431 \text{ rpm}$$

Hence, for gear 3 (pinion), we have

$$T_3 = \frac{7121 \text{ hp}}{n_2} = \frac{7121(0.5)}{431} = 8.261 \text{ N}\cdot\text{m}$$

$$F_{t3} = \frac{T_3(2)}{d_3} = \frac{8.261(2)}{0.0288} = 573.7 \text{ N}$$

$$F_{r3} = 573.7 \tan 20° = 208.8 \text{ N}$$

$$V_3 = \pi d_3 n_3 = \pi(0.0288)\left(\frac{431}{60}\right) = 0.65 \text{ m/s}$$

Shaft 3 runs at

$$n_3 = n_2 \frac{N_3}{N_4} = 431\left(\frac{18}{72}\right) \approx 108$$

It follows for gears 5 (pinion) and 6 that

$$T_5 = \frac{7121(0.5)}{108} = 32.97 \text{ N}\cdot\text{m}$$

$$F_{t5} = \frac{32.97(2)}{0.0375} = 1758 \text{ N} = F_{t6}$$

$$F_{r5} = 1758 \tan 20° = 639.9 \text{ N} = F_{r6}$$

$$V_5 = \pi d_5 n_3 = \pi(0.0375)\left(\frac{108}{60}\right) = 0.21 \text{ m/s} = V_6$$

The output shaft rotates at

$$n_4 = n_3 \frac{N_5}{N_6} = 108\left(\frac{15}{60}\right) \approx 27 \text{ rpm}$$

The speed ratio between the output and input shafts (or gears 6 and 1) of the spur gear train can now be obtained as

$$r_s = \frac{n_4}{n_1} = \left(-\frac{N_1}{N_2}\right)\left(-\frac{N_3}{N_4}\right)\left(-\frac{N_5}{N_6}\right)$$

$$= \left(-\frac{15}{60}\right)\left(-\frac{18}{72}\right)\left(-\frac{15}{60}\right) = -\frac{1}{64}$$

Here, the minus sign means that the pinion and gear rotate in opposite directions.

Comments: Having the tangential forces and pitch-line velocities available, the design analysis for gearsets 2 and 3 can readily be made by following a procedure identical to that described for gearset 1.

CASE STUDY 18.5 Gearbox Shafting Design

Figure 18.5 shows the input shaft of the crane gearbox, supported in the gearbox by bearings *A* and *B* and driven by electric motor. Determine

a. The factor of safety *n* for the shaft using the maximum energy of distortion theory incorporated with the Goodman criterion
b. The rotational displacements or slopes at the bearings
c. The stresses in the shaft key

Given: The geometry and dimensions of the hollow shaft and square shaft key are known.

Data: Refer to Figure 18.5a and Case Study 18.4.

$$F_t = 206\text{ N},\quad F_r = 75\text{ N},\quad T = 2.06\text{ N}\cdot\text{m},$$

$$a = 66\text{ mm},\quad b = 84\text{ mm},\quad L = 150\text{ N}\cdot\text{m},$$

$$d = 6\text{ mm},\quad d_p = 20\text{ mm},\quad D = 12\text{ mm},$$

$$w = 2.4\text{ mm},\quad L_k = 25\text{ mm},$$

$$I = \frac{\pi}{64}\left(D^4 - d^4\right) = 954.3\text{ mm}^4$$

The operating environment is room air at a maximum temperature of 50°C.

Assumption: Bearings act as simple supports.

Design Decisions:

1. The shaft and shaft key are made of 1030 *CD* steel with machined surfaces:

$$S_u = 520\text{ MPa},\quad S_y = 440\text{ MPa (from Table B.3)}$$

$$E = 210\text{ GPa}$$

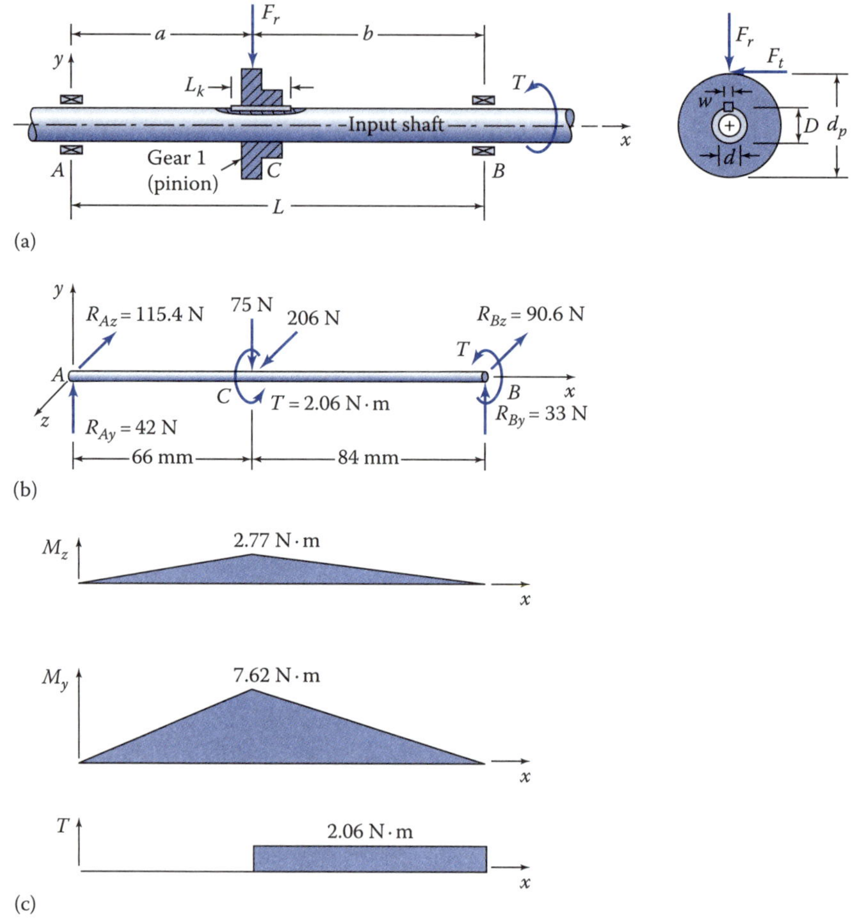

FIGURE 18.5
Drive shaft supported in the gearbox of the winch crane shown in Figure 18.1: (a) shaft layout, (b) loading diagram, and (c) moment and torque diagrams.

2. At the keyway, $K_f = 2$.
3. The shaft rotates and carries steady loading at normal temperature.
4. The factor of safety is $n = 3$ against shear of shaft key.
5. A survival rate of 99.9% is used.

Solution

See Figure 18.5; Table A.9, Section 9.5.

a. The reactions at *A* and *B*, as determined by the conditions of equilibrium, are indicated in Figure 18.5b. The moment and torque diagrams are obtained in the usual manner and drawn in Figure 18.5c. Observe that the critical section is at point *C*. We have

$$M_C = \left[(2.77)^2 + (7.62)^2\right]^{1/2} = 8.11\ \text{N}\cdot\text{m}$$

$$T_C = 2.06\ \text{N}\cdot\text{m}$$

The mean and alternating moments and torques, by Equation 9.10, are then

$$M_m = 0 \quad M_a = 8.11 \text{ N}\cdot\text{m}$$

$$T_m = 2.06 \text{ N}\cdot\text{m} \quad T_a = 0$$

The modified endurance limit, through the use of Equations 7.1 and 7.6 and referring to Section 7.7, is

$$S_e = C_f C_r C_s C_t \left(\frac{1}{K_f}\right) S'_e$$

where

$C_f = 4.51(520^{-0.265}) = 0.86$
$C_r = 0.76$ (by Table 7.3)
$C_s \approx 0.85$ (using Equation 7.9)
$C_1 = 1$
$K_f = 2$
$S'_e = 0.5(520) = 260$ MPa (from Equation 7.1)

Hence,

$$S_e = (0.86)(0.75)(0.85)(1)\left(\frac{1}{2}\right)(260) = 71.27 \text{ MPa}$$

Since the loading is steady, the shock factors $K_{sb} = K_{st} = 1$ by Table 9.1.

Substituting the numerical values into Equation 9.12 and replacing D^3 with $D^3[1 - (d/D)^4]$, we obtain

$$\frac{520(10^6)}{n} = \frac{32}{\pi(0.012)^3\left[1-(6/12)^4\right]} \times \left[\left(0+\frac{520\times 8.11}{71.27}\right)^2 + \frac{3}{4}(2.06)^2\right]^{1/2}$$

from which

$$n = 1.4$$

b. The slopes at ends A and B (Figure 18.5b) are given by Case 6 of Table A.9. Note that $L^2 - b^2 = (L + b)(L - b) = (L + b)a$ and similarity $L^2 - a^2 = (L + a)b$. Introducing the given data, the results are

$$\theta_A = -\frac{F_t ab(L+b)}{6EIL}$$

$$= -\frac{206(66)(84)(150+84)}{6(210\times10^3)(954.3)(150)}$$

$$= -1.482(10^{-3})\text{ rad} = -0.085°$$

$$\theta_B = \frac{F_t ab(L+a)}{6EIL}$$

$$= \frac{206(66)(84)(150+66)}{6(210\times10^3)(954.3)(150)}$$

$$= 1.368(10^{-3})\text{ rad} = 0.078°$$

where a minus sign means a clockwise rotation.

Comments: Inasmuch as the bearing and gear stiffnesses are ignored, the negligibly small values of θ_A and θ_B estimated by the preceding equations represent higher angles than the true slopes. Therefore, self-aligning bearings are not necessary.

c. The compressive forces acting on the sides of the shaft key equal F_t = 206 N (Figure 18.5). The shear stress in the shaft key is

$$\tau = \frac{F_t}{wL_k} = \frac{206}{(0.0024)(0.025)} = 3.433\text{ MPa}$$

We have, from Equation 6.20,

$$S_{ys} = 0.577S_y = (0.577)(440) = 253.9\text{ MPa}$$

The allowable shear stress in the shaft key is

$$\tau_{\text{all}} = \frac{S_{ys}}{n} = \frac{253.9}{3} = 84.63\text{ MPa}$$

Since $\tau_{\text{all}} \gg \tau$, shear should not occur at shaft key. We obtain the same result on the basis of compression or bearing on key (see Section 9.9).

Comments: On following a procedure similar to that in the preceding solution, the design of the remaining three shafts and the associated keys in the gearbox of the winch crane can be analyzed in a like manner.

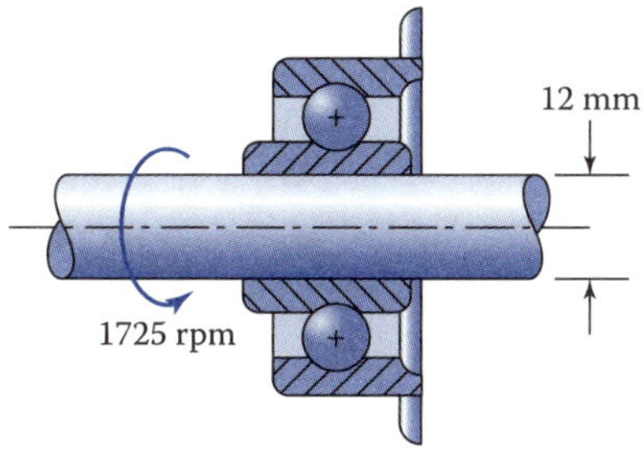

FIGURE 18.6
Flanged ball bearing at the left end of the input shaft in the gearbox of a winch crane shown in Figure 18.1.

CASE STUDY 18.6 Selection of Gearbox Shaft and Bearing

Figure 18.6 shows a flanged ball bearing of the input shaft in the gearbox (see Case Study 18.4) of a winch crane. Analyze the load-carrying capacity of the bearing.

Given: The shaft has a 12 mm diameter and operates at 1725 rpm. Rating life is 30 kh.

Assumptions: Thrust loads are negligible. Bearings at both ends of the shaft are taken to be identical 02-series deep groove and subjected to light-shock loading. The inner ring rotates.

Solution

See Figures 18.5, 18.6; and 18.4; Tables 10.4, 10.6, and 10.8; Section 10.13.

Referring to Figure 18.5b, the forces acting on bearings at the shaft end are

$$R_A = \left[(42)^2 + (115.4)^2\right]^{1/2} = 122.8\,\text{N}$$

$$R_B = \left[(33)^2 + (90.6)^2\right]^{1/2} = 96.42\,\text{N}$$

Since $R_A > R_B$, we analyze the bearing at the left end A of the shaft. Through the use of Equation 10.27 with axial thrust $F_a = 0$, the equivalent radial load is

$$P = K_s XVF_r$$

Here,

$$F_r = 122.8\,\text{N}$$

$$K_s = 1.5 \quad (\text{from Table 10.8})$$

$$X = 1.0, \quad Y = 0 \quad (\text{by Table 10.6})$$

$$V = 1 \quad (\text{from Section 10.14})$$

Therefore, we have

$$P = 1.5(1.0)(1)(122.8) = 184.2\,\text{N}$$

The basic dynamic load rating C, applying Equation 10.30, is given by

$$C = P\left(\frac{60nL_{10}}{10^6}\right)^{1/a}$$

where $a = 3$ for ball bearings. Introducing the given data into this equation,

$$C = 184.2\left[\frac{60(1725)(30{,}000)}{10^6}\right]^{1/3} = 2.69 \text{ kN}$$

We see from Table 10.4 that a 02-series deep-groove ball bearing with a bore of 12 mm has a load rating of $C = 6.89$ kN. This is well above the estimated value of 2.69 kN, and the bearing is quite satisfactory. Following this procedure, other bearings for the shafts in the gearbox may be analyzed in a like manner.

Comment: The final selection of the bearings would be made on the basis of standard shaft and housing dimensions.

CASE STUDY 18.7 Screw Design for Swivel Hook

The steel crane hook supported by a trunnion or crosspiece as shown in Figure 18.7 is rated at $P = 3$ kN. Determine the necessary nut length L_n. Observe that a ball-thrust bearing permits rotation of the hook for positioning the load. The lower race of the bearing and a third (bottom) ring have matching spherical surfaces to allow self-alignment of the hook with the bearing load. Usually, bearing size selected for a given load and service has internationally standardized dimensions.

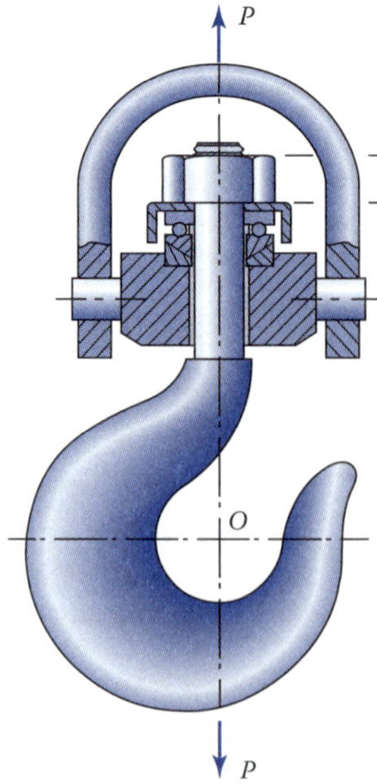

FIGURE 18.7
Swivel hook for the winch crane (Figure 18.1) showing the section of the trunnion with a thrust-ball bearing.

Assumption: Both the threaded portion of the shank or bolt and the nut are made of $M12 \times 1.75$ class 5.8 rolled coarse threads. A stress concentration factor of $K_t = 3.5$ and a safety factor of $n = 5$ are used for threads.

Given: From Table 15.2,

$$p = 1.75 \text{ mm}, \quad d = 12 \text{ mm},$$

$$d_m \approx 10.925 \text{ mm}, \quad d_r \approx 9.85 \text{ mm}.$$

$$h = \frac{1}{2}(d - d_r) = 1.075 \text{ mm},$$

$$S_y = 420 \text{ MPa (from Table 15.5)}$$

Solution

See Figures 18.1 and 18.7; Section 15.7.

Bearing strength. For the nut, apply the following design formula:

$$\frac{K_t P p}{\pi d_m h L_n} = \frac{S_y}{n} \tag{15.17a}$$

Substituting the given numerical values, we have

$$\frac{(3.5)(3000)(1.75)}{\pi(10.925)(1.075)L_n} = \frac{420}{5}$$

Solving,

$$L_n = 5.9 \text{ mm}$$

Shear strength. Based on the energy of distortion theory of failure,

$$S_{ys} = 0.577 \quad S_y = 0.577 \times 420 = 242.3 \text{ MPa}$$

From Figure 15.3, the thread thickness at the root is

$$b = \frac{p}{8} + 2h \tan 30° = 1.46 \text{ mm}$$

The design formula is

$$\frac{3K_t P p}{2\pi d b L_n} = \frac{S_{ys}}{n} \tag{15.19a}$$

Inserting the data given,

$$\frac{3(3.5)(3000)(1.75)}{2\pi(12)(1.46)L_n} = \frac{242.3}{5}$$

from which

$$L_n = 10.3 \text{ mm}$$

Comment: A standard nut length of 10 mm should be used.

CASE STUDY 18.8 Swivel Hook Design Analysis

A crane hook for the winch crane, shown in Figure 18.8a, is rated at $P = 3$ kN. Determine the tangential stresses at points A and B using Winkler's formula. Note that, for a large number of manufactured crane hooks, the critical section AB can be closely approximated by a trapezoidal area with half an ellipse at the inner radius and an arc of a circle at the outer radius, as shown in Figure 18.8b. The solution for standardized crane hooks is expedited by readily available computer programs.

Assumptions: The critical section AB is taken to be trapezoidal. The hook is made of A1SI 1020-HR steel with a safety factor of n against yielding.

Given:

$$r_i = 20 \text{ mm}, \quad b_1 = 30 \text{ mm}, \quad b_2 = 10 \text{ mm}$$

$$h = 42 \text{ mm}, \quad n = 5,$$

$$S_y = 210 \text{ MPa}, \quad (\text{from Table B.3})$$

Solution

See Figures 18.7 and 18.8; Section 16.8.

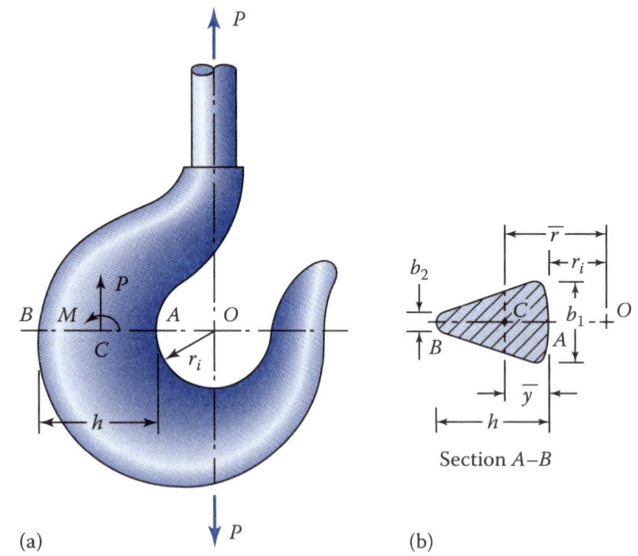

FIGURE 18.8
Part of the hook for a winch crane (Figure 18.1): (a) stress resultants at cross section A–B and (b) the critically stressed, modified trapezoidal section.

Referring to Figure 18.8b, we obtain the following quantities. The cross-sectional area is

$$A = \frac{1}{2}(b_1 + b_2)h = \frac{1}{2}(30 + 10)(42) = 840 \text{ mm}^2$$

The distance to the centroid *C* from the inner edge is

$$\bar{y} = \frac{h(b_1 + 2b_2)}{3(b_1 + b_2)} = \frac{42(30 + 2 \times 10)}{3(30 + 10)} = 17.5 \text{ mm}$$

Hence,

$$\bar{r} = r_i + \bar{y} = 20 + 17.5 = 37.5 \text{ mm}$$

By case *E* of Table 16.1, the radius of the neutral axis, with $r_o = r_i + h = 62$ mm, is then

$$\begin{aligned} R &= \frac{A}{\frac{1}{h}\left[(b_1 r_o - b_2 r_i)\ln\frac{r_o}{r_i} - h(b_1 - b_2)\right]} \\ &= \frac{840}{\frac{1}{42}\left[(30 \times 62 - 10 \times 20)\ln\frac{62}{20} - 42(30 - 10)\right]} \\ &= 33.9843 \text{ mm} \end{aligned}$$

Equation 16.51 leads to

$$e = \bar{r} - R = 37.5 - 33.9843 = 3.5157 \text{ mm}$$

The circumferential stresses are determined through the use of Equations 16.55 with a tensile normal load *P* and bending moment *M* = –*PR*. Therefore,

$$(\sigma_\theta)_A = \frac{P}{A}\left[1 + \frac{\bar{r}(R - r_i)}{er_i}\right]$$

$$(\sigma_\theta)_B = \frac{P}{A}\left[1 + \frac{\bar{r}(R - r_o)}{er_0}\right]$$

Introducing the required values into the preceding expression, we have

$$(\sigma_\theta)_A = \frac{3000}{840(10^{-6})}\left[1+\frac{37.5(33.9843-20)}{3.5157(20)}\right]$$

$$= 30.21 \text{ MPa}$$

$$(\sigma_\theta)_B = \frac{3000}{840(10^{-6})}\left[1+\frac{37.5(33.9843-62)}{3.5157(62)}\right]$$

$$= -13.64 \text{ MPa}$$

where a minus sign means compression.

Comment: The allowable stress $\sigma_{all} = 210/5 = 42$ MPa is larger than the maximum stress of 30.21 MPa. That is, the crane hook can support a load of 3 kN with a factor of safety of 5 without yielding.

CASE STUDY 18.9 Design of Welded Joint *C*

The welded joint *C*, with identical fillets on both sides of the vertical frame of the winch crane frame, is under in-joint, plane eccentric loading, as shown in Figure 18.9a. Determine the weld size *h* at the joint.

Given: $L_1 = 100$ mm $L_2 = 150$ mm

$e = 1.5$ m $P = 3$ kN

Assumptions: An E6010 welding rod with a factor of safety $n = 2.2$ is used. The vertical frame of the crane is taken to be a rigid column.

Solution

See Figures 18.1 and 18.9; Section 15.16.

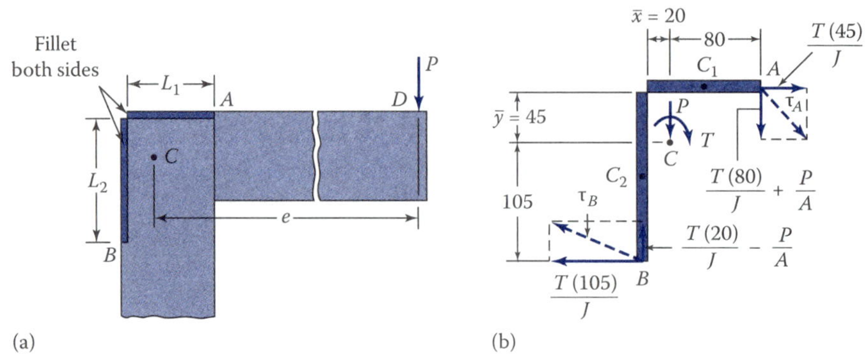

FIGURE 18.9
(a) Welded joint *C* of the winch crane shown in Figure 18.1 and (b) enlarged view of the weld group. Loading acts at the centroid *C* of the group and shear stresses at weld ends *A* and *B*.

Area properties. The centroid of the weld group (Figure 15.29) is given by

$$\bar{x} = \frac{A_1x_1 + A_2x_2}{A_1 + A_2} = \frac{(100t)(50) + (150t)(0)}{100t + 150t}$$

$$= 20 \text{ mm}$$

$$\bar{y} = \frac{A_1y_1 + A_2y_2}{A_1 + A_2} = \frac{(100t)(0) + (150t)(75)}{100t + 150t} = 45 \text{ mm}$$

The torque equals

$$T = Pe = (3000)(1500) = 4.5 \text{ MN} \cdot \text{mm}$$

The centroidal moments of inertia are tL^3

$$I_x = \Sigma \frac{tL^3}{12} + Lty^2$$

$$= \frac{t(100)^3}{12} + (100)t(45)^2 + 0 + (150)t\ (75 - 45)^2 = 420{,}833\,t$$

$$I_y = \Sigma \frac{tL^3}{12} + Ltx^2$$

$$= 0 + (100)t(50 - 20)^2$$

$$+ \frac{t(150)^3}{12} + (150)t(20)^2 = 431{,}250t$$

$$J = 852{,}083t \text{ mm}^4$$

Since there are fillets at both sides of the column, the area properties are multiplied by 2.

Stresses. From Table 15.8, we have S_y = 345 MPa. By inspection of Figure 18.9b, either at point A or B, the combined torsional and direct shear stresses are greatest. At point A,

$$\tau_\upsilon = \frac{P}{A} + \frac{Tr_{i1}}{J} = \frac{3(10)^3}{2(250t)} + \frac{4.5(10)^6(80)}{2(852{,}083t)}$$

$$= \frac{6}{t} + \frac{211.2}{t}$$

$$\tau_h = \frac{Tr_x}{J} = \frac{4.5(10^6)(45)}{2(852{,}083t)} = \frac{118.8}{t}$$

$$\tau_A = \left(\tau_\upsilon^2 + \tau_h^2\right)^{1/2} = \frac{247.6}{t}$$

Similarly, at point B,

$$\tau_v = -\frac{6}{t} + \frac{4.5(10^6)(20)}{2(852{,}083t)} = -\frac{6}{t} + \frac{52.8}{t}$$

$$\tau_h = \frac{4.5(10^6)(105)}{2(852{,}083t)} = \frac{277.3}{t}$$

$$\tau_B = \frac{281.2}{t}\ \text{N/mm} \quad \text{(governs)}$$

Weld size. Therefore, by Equation 15.44,

$$n\tau_B = 0.5S_y, \quad 2.2\left(\frac{281.2}{t}\right) = 0.5(345)$$

from which t = 3.59 mm. Referring to Figure 15.26a, we obtain

$$h = \frac{t}{0.707} = \frac{3.59}{0.707} = 5.08\ \text{mm}$$

Comment: A nominal size of 5 mm fillet welds should be used in the joint.

18.3 High-Speed Cutter

Cutting machines, commonly known as saws, may be movable or stationary. The working member of a mechanically powered cutter is a thin steel blade or disk with sharp teeth. Traditional *high-speed cutting machines* have three main types as illustrated in the following photos. A *circular blade* saw uses a metal disk that rotates to cut the material and can create narrow slots. While these saws are equipped with a blade for cutting wood, masonry, plastic, or metal, there are also purpose-made circular saws specially designed for particular materials. A *reciprocating blade* saw uses a push-and-pull motion of the blade. It often has a mechanism to lift up the saw blade on the return stroke. A *continuous band saw*, which usually rides on two wheels circulating in the same plane, produces a uniform cutting action as a result of an evenly distributed tooth load. This machine also provides better cutting quality, and its output averages twice that of a straight-knife cutting machine. Band saws are particularly useful for cutting irregular or curved shapes. In addition to these traditional cutters, modern laser cutting and water cutting machines are increasing in use.

A circular-blade table saw

A reciprocating-blade hacksaw

A continuous-band saw

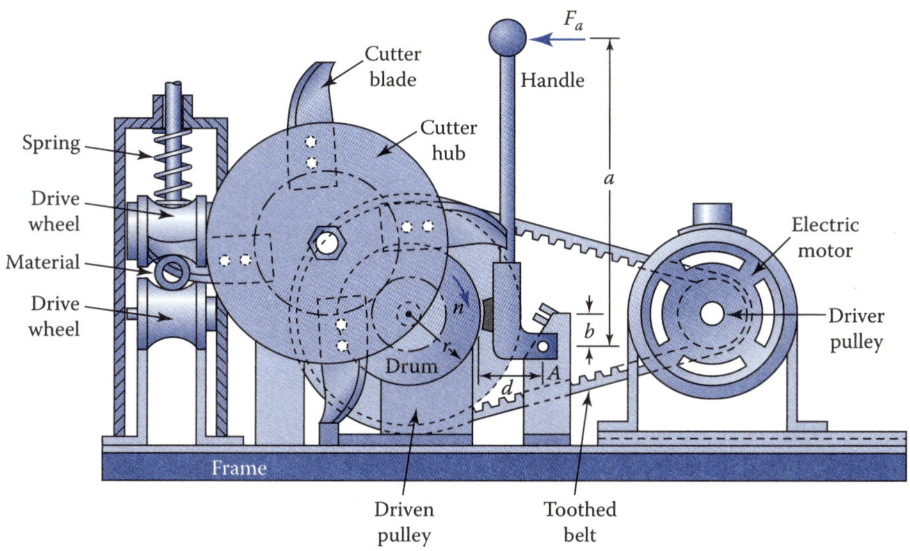

FIGURE 18.10
Schematic drawing of high-speed cutting machine.

In this section, attention is directed to a simple high-speed blade cutter assembly portrayed in Figure 18.10, which is used for flexible materials including PVC and other plastics. The unit is compact and designed for benchtop installation. The drive wheel is a part of the automatic feeding mechanism (not shown in the figure). These wheels drive the material through the feed tube to the cutting wheel. The feed mechanism includes a compression spring for smooth operation. The variable cut length and rate of output (as high as 1000 per minute) are accomplished by changing the number of blades in the rotary cutting wheel and changing the reduction drive ratio between the motor and the shaft on which the cutter is keyed.

CASE STUDY 18.10 Belt Design

Consider the toothed belt of a high-speed cutter shown in Figure 18.10. Determine

a. The belt length
b. The maximum center distance
c. The maximum belt tension

Design Requirements: The center distance between the motor (driver) pulley and driven pulley should not exceed $c = 17$ in. A belt coefficient of friction of $f = 1.0$ is used.

Given: A 2 hp, $n_1 = 1800$ rpm, AC motor is used. The belt weighs $w = 0.007$ lb/in. The driver pulley radius $r_1 = 1\frac{1}{4}$ in. Driven pulley radius $r_2 = 2\frac{1}{4}$ in.

Assumptions: The driver is a normal torque motor. The cutter, and hence the driven shaft, resists heavy shock loads. The machine cuts uniform lengths of flexible materials of cross sections up to 2 in. in diameter. Operation is fully automatic, requiring minimal operator involvement.

Solution
See Figure 18.10 and Table 13.5; Sections 13.3, 13.4, and 13.5.

a. The appropriate belt pitch length is determined using Equation 13.9:

$$L = 2c + \pi(r_1 + r_2) + \frac{1}{c}(r_2 - r_1)^2 \quad \text{(a)}$$

Substitution of the given data yields

$$\begin{aligned} L &= 2(17) + \pi(1.25 + 2.25) + \frac{1}{17}(2.25 - 1.25)^2 \\ &= 45.05 \text{ in.} \end{aligned}$$

Comment: A standard toothed or timing belt with maximum length of 45 in. is selected.

b. An estimate of the center distance is given by Equation 13.10:

$$c = \frac{1}{4}\left[b + \sqrt{b^2 - 8(r_2 - r_1)^2}\right] \quad \text{(b)}$$

Here, from Equation 13.11,

$$b = L - \pi(r_2 + r_1) \quad \text{(c)}$$

Carrying the given numerical values into Equation (c), we have

$$b = 45.05 - \pi(2.25 + 1.25) = 34 \text{ in.}$$

Equation (b) is then

$$c = \frac{1}{4}\left[34 + \sqrt{(34)^2 - 8(2.25 - 1.25)^2}\right] = 16.97 \text{ in.}$$

Comment: The requirement that $c < 17$ in. is satisfied.

c. The contact angle ϕ, from Equations 13.7 and 13.6, equals

$$\begin{aligned} \phi &= \pi - 2\sin^{-1}\left(\frac{r_2 - r_1}{c}\right) \\ &= \pi - 2\sin^{-1}\left(\frac{2.25 - 1.25}{16.97}\right) = 3.024 \text{ rad} \end{aligned}$$

The tight-side tension in the belt is calculated, through the use of Equation 13.20 in which, for a toothed (or flat) belt, sin β = 1 and

$$F_c = \frac{w}{g}V^2$$

$$= \frac{0.007}{386.4}\left[\frac{\pi(2.25)1800}{60}\right]^2 = 1.006 \text{ lb}$$

$$\gamma = e^{f\phi} = e^{(1)(3.024)} = 20.57$$

$$T_1 = \frac{33{,}000 \text{ hp}}{n_1}$$

$$= \frac{33{,}000(2)}{1800} = 36.67 \text{ lb}\cdot\text{in.}$$

Therefore,

$$F_1 = 1.006 + \left(\frac{20.57}{20.57-1}\right)\frac{36.67}{1.25} = 31.84 \text{ lb}$$

It follows that

$$F_2 = F_1 - \frac{T_1}{r_1}$$

$$= 31.84 - \frac{36.67}{1.25} = 2.50 \text{ lb}$$

The service factor, from Table 13.5, K_s = 1.4. The maximum belt tensile load is then obtained by Equation 13.22 as

$$F_{\max} = K_s F_1$$

$$= 1.4(31.84) = 44.58 \text{ lb}$$

Comment: Recall from Section 13.2 that toothed belts can provide safe operation at speeds up to at least 16,000 fpm. This is well above the belt velocity. $V = \pi(2.5)(1800)/12 =$ 1178 fpm, of the cutting machine.

CASE STUDY 18.11 Brake Design Analysis

A short-shoe brake is used on the drum, which is keyed to the center shaft of the high-speed cutter as shown in Figure 18.10. The driven pulley is also keyed to that shaft. Determine the actuating force F_a.

Assumptions: The brake shoe material is molded material. The drum is made of iron. The lining rubs against the smooth drum surface, operating dry.

Given: The dram radius r = 3 in., torque T = 270 lb·in. (CW), a = 12 in., b = 1.2 in., d = 2.5 in., and the width of shoe w = 1.5 in. (Figure 18.10). By Table 13.11, p_{max} = 200 psi and f = 0.35.

Requirement: The shoe must be self-actuating.

Solution

The normal force, through the use of Equation 13.48, is

$$F_n = \frac{T}{fr}$$

$$= \frac{270}{(0.35)3} = 257.1 \text{ lb}$$

The angle of contact, applying Equation 13.47, is then

$$\phi = 2\sin^{-1}\frac{F_n}{2p_{max}rw}$$

$$= 2\sin^{-1}\frac{257.1}{2(200)(3)(1.5)} = 16.4°$$

The actuating force is obtained from Equation 13.49 with $d = c$ as follows:

$$F_a = \frac{F_n}{a}(b - fc)$$

$$= \frac{257.1}{12}(1.2 - 0.35 \times 2.5)$$

$$= 6.964 \text{ lb}$$

Comments: Since $\phi < 45°$, the short-shoe drum brake approximations apply. A positive value of F_a means that the brake is not self-locking.

CASE STUDY 18.12 Spring Design of Feed Mechanism

A helical compression spring of feed mechanism for a high-speed cutter is shown in Figure 18.11. The spring is to support a load P without exceeding a deflection δ. Determine a satisfactory design. Will the spring buckle in service?

Given: P = 60 N, δ = 15 mm.

Assumptions: Clash allowance r_c = 20%, spring index C = 6, and safety factor n = 2.2. Loading is applied steadily. Ends are squared-ground and supported between flat surfaces.

Design Decision: Hard-drawn ASTM A227 wire of G = 79 GPa is used.

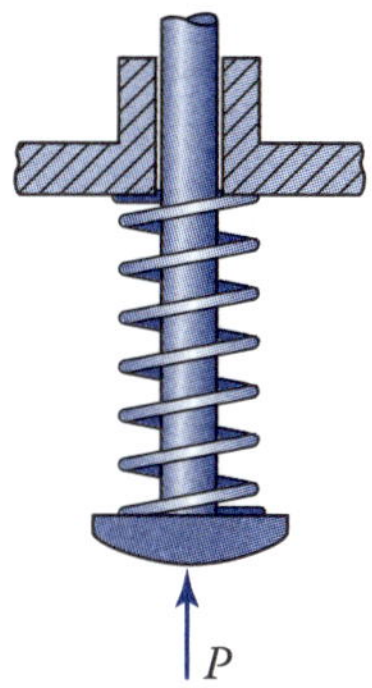

FIGURE 18.11
Compression spring for feed mechanism of high-speed cutter shown in Figure 18.10.

Solution

See Figures 18.10 and 18.11; Sections 14.4, 14.5, and 14.6.

Arbitrarily *select* a 2 mm diameter wire. Then, using Equation 14.12 and Table 14.2,

$$S_u = Ad^b = 1510\left(2^{-0.201}\right) = 1314 \text{ MPa}$$

Corresponding yield strength in shear is $S_{ys} = 0.42(1314) = 552$ MPa (from Table 14.3).

Stress requirement. Rearranging Equation 14.6 and setting $\tau = S_{ys}/n$,

$$d^2 = \frac{8PCn}{\pi S_{ys}}\left(1 + \frac{0.615}{C}\right)$$

$$= \frac{8(60)(6)(2.2)}{\pi\left(552 \times 10^6\right)}\left(1 + \frac{0.615}{6}\right)$$

$$d = 2.01 \text{ mm}$$

We then have $D = 6(2.01) = 12.06$ mm. Since $S_u = 1510(2.01^{-0.201}) = 1312 < 1314$ MPa, $d = 2.01$ mm is satisfactory.

Spring rate requirement. By Equation 14.11,

$$k = \frac{P}{\delta} = \frac{dG}{8C^3N_a}, \quad \frac{60}{15} = \frac{2.01(79,000)}{8(6)^3 N_a}$$

Solving, $N_a = 22.97$. From Figure 14.7d, $h_s = (N_a + 2)d = 50.2$ mm. With a 20% c clash allowance, the solid deflection is 120% of the working deflection. Hence,

$$\delta_s = 1.2\frac{P}{k} = 1.2\frac{60}{4} = 18 \text{ mm}$$

The free height is given by $h_f = h_s + \delta_s = 68.2$ mm.

Check for buckling. For the extreme case of deflection ($\delta = \delta_s$),

$$\frac{\delta_s}{h_f} = \frac{18}{68.2} = 0.26, \quad \frac{h_f}{D} = \frac{68.2}{12.06} = 5.66$$

Curve A in Figure 14.10 shows that the spring is far outside the buckling region and clearly safe.

Comment: Having the foregoing values of D, N_a, and h_f available, a technician can draw or make the compression spring for the high-speed cutter.

Problems

Sections 18.1 through 18.2

18.1 A *loader*, also known as a front loader, bucket loader, scoop loader, or shovel, is a type of tractor. A typical mobile front-end loader truck is used for hydraulically raising or lowering a pipe, log, or lumber (Figure P18.1). There are two identical pin-connected arm–linkage–hydraulic cylinder systems, one set on each side of a central vertical plane in the fore-and-aft direction. The two systems, one of which is shown in the figure, share the load equally. The arm ABD is attached to linkages AE and BG at A and B and is controlled by the hydraulic cylinder CF. The action of the clamp that grabs, holds firmly, or releases the pipe is also controlled by two hydraulic cylinders (not shown). A *falling object protective structure* (FOPS), which consists of welded side supports, is used to keep the operator safe within the cab of the loader.

The pipe weight exerts a vertical force P at the right end D of the arm, as illustrated in Figure P18.1. Each pin is made of steel with a yield shear strength of S_{ys}. Factor of safety against shear by yielding of pins is n. Find, for the position shown,

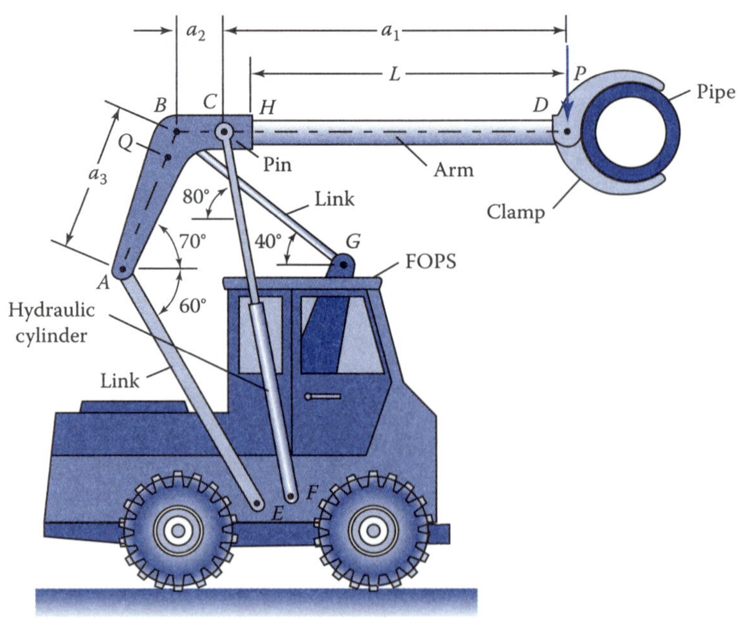

FIGURE P18.1
Schematic of loader truck.

a. The forces exerted by the hydraulic cylinder (*CF*) and the links (*AE*, *BG*) on the arm

b. The required pin diameter at *A*, *B*, and *C*

Given: $P = 15$ kN, $a_1 = 2.6$ m, $a_2 = 0.16$ m, $a_3 = 1.0$ m, $L = 2.5$ m, $S_{ys} = 150$ MPa, $n = 2.4$.

Assumptions:

1. Friction in the joints is omitted. The accelerations are insignificant. All forces are coplanar, 2D, and static.
2. Each connection is made with a pin in double shear.
3. Weights of the members are disregarded compared to the forces they support and so can be omitted. For the particular position shown in the figure, part *BD* of the arm is horizontal. Forward or side-to-side tipping of the unit will not occur.

18.2 The portion of a hydraulically controlled loader arm of Figure P18.1 shown in Figure P18.2 carries a concentrated load of *P* at its free end. The arm is made of ASTM-A242 high-strength steel tubing with the ultimate strengths in tension and shear that are S_u and S_{us}, respectively. What are the values of maximum normal stress, maximum shear stress, and the factor of safety?

Given: $c_2 = 75$ mm, $c_1 = 50$ mm, $L = 2.5$ m, $P = 15$ kN, $S_u = 480$ MPa, $S_{us} = 280$ MPa (by Table B.1).

Assumptions: Part *HD* of the arm will be modeled as cantilever with the more massive portion at its end serving as *ground frame*. The critical point *K* is at the fixed end through *H*. The effect of shear in the stress distribution is neglected.

18.3 At a critical point *Q* in a loader arm (Figure P18.1), the material is under the state of plane stress.

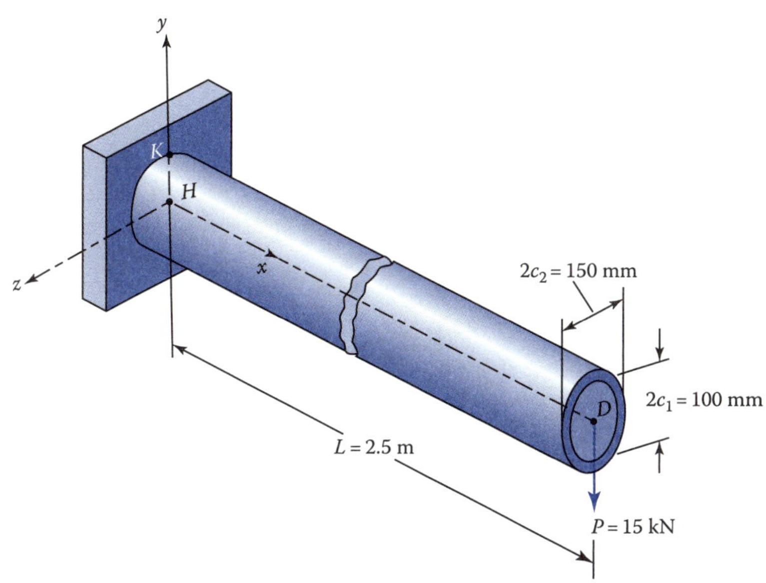

FIGURE P18.2
Part of loader arm as a cantilever beam.

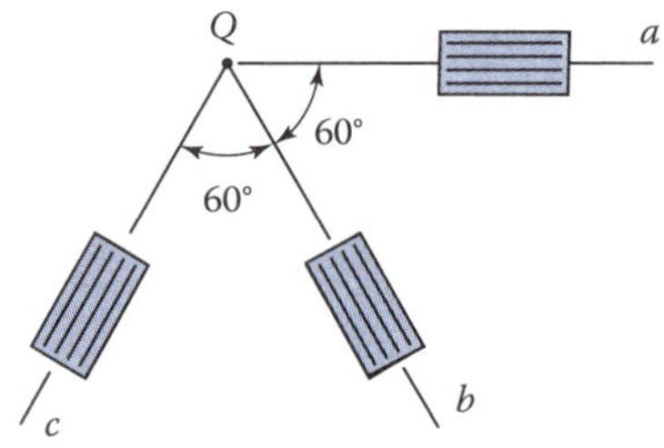

FIGURE P18.5
A 60° strain rosette.

Given:

$$\varepsilon_x = 1000\mu, \quad \varepsilon_y = -200\mu, \quad \gamma_{xy} = 700\mu.$$

Using Mohr's circle, find the principal strains and the maximum shear strains.

18.4 Reconsider the state of strain of the loader arm discussed in Problem 18.3, which is made of a steel with modulus of elasticity $E = 210$ GPa and Poisson's ratio $\nu = 0.28$. Find at point Q the principal stresses, maximum shear stress, and their orientations. What is the value of the normal stresses that occur on the planes of maximum shear stress?

18.5 At a critical point Q on the surface of the loader arm illustrated in Figure P18.1, the 60 rosette readings show the normal strains during a static test:

$$\varepsilon_a = 1104\mu, \quad \varepsilon_b = 432\mu, \quad \varepsilon_c = -96\mu$$

The forgoing correspond to $\theta_a = 0°$, $\theta_b = -60°$, and $\theta_c = -120°$ (Figure P18.5). Find the strain components ε_x, ε_y, and γ_{xy}.

18.6 A standard 1.0 in. nominal diameter steel pipe link BG of length $L_{BG} = 1.6$ m with pinned ends is subjected to compression load of $F_{BG} = 11.34$ kN (Figure P18.1). What is the allowable stress for the link, using AISC formulas?

Given:

$$A = 0.494\,\text{in.}^2 = 318\,\text{mm}^2, \quad r = 0.421\ \text{in.} = 10.69\ \text{mm}$$

$$E = 200\ \text{GPa}, \quad S_y = 250\ \text{MPa} \quad (\text{by Table B.1})$$

18.7 Find the largest length L_m for which the steel pipe link BG discussed in Problem 18.6 can safely support the loading of $F_{BG} = 11.34$ kN.

18.8 An idealized FOPS frame of the loader truck discussed in Problem 18.1 and depicted in Figure P18.8 carries a load of W at the center C of the assembly. Drive the expressions for

a. The total strain energy U due to bending and shear of the members in the form

$$U_t = W^2\left[\frac{1}{EI}\left(\frac{c^3}{96} + \frac{b^3}{384}\right) + \frac{3}{20GA}\left(c + \frac{b}{2}\right)\right] \tag{P18.8}$$

b. The corresponding maximum static deflection at C, using Castigliano's theorem

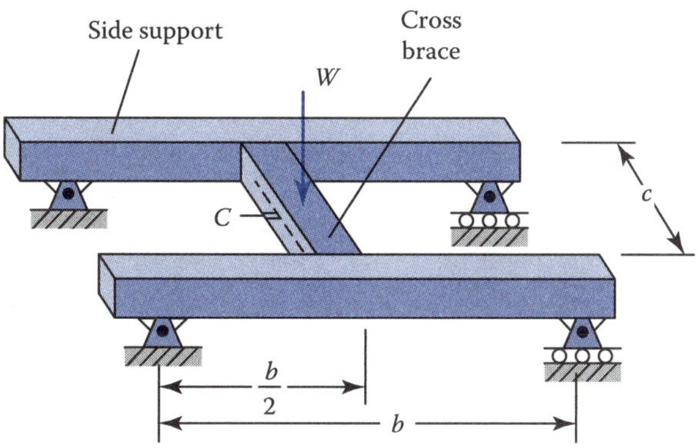

FIGURE P18.8
Idealized FOPS frame of the loader truck shown in Figure P18.1.

Assumptions: Each member is made of a square steel tube with the same modulus of elasticity E, modulus of rigidity G, cross-sectional area A, and moment of inertia I. Side support beams of lengths b are taken to be simply supported and a cross brace of length c is placed and welded at the middle of the side supports. Note that, although a typical FOPS frame will usually have a number of cross braces, for simplicity, here only one *equivalent* cross brace is considered (Figure P18.8).

18.9 Reconsider Problem 18.8, with the exception that an object of weight W *drops* from a *height* h striking at the midspan C of the FOPS (see Figure P18.8). Compute the values of

a. The largest static deflection

b. The maximum dynamic deflection

Given:

$$b = 1.2\text{ m},\quad c = 0.8\text{ m},\quad h = 250\text{ mm},\quad W = 15\text{ kN},$$

$$E = 200\text{ GPa},\quad G = 79\text{ GPa}\quad (\text{by Table B.1})$$

$$A = 1.27\text{ in.}^2 = 819\times10^{-6}\text{ m}^2,\quad I = 0.668\text{ in.}^4 = 0.278\times10^{-6}\text{ m}^4\quad (\text{from Table A.4})$$

Requirement: A 2 in. × 2 in. nominal size steel tube of thickness $\frac{3}{16}$ in. (see Table A.4) will be used for both the side supports and the cross brace.

18.10 *Hydraulic cylinders* that used *extrusion* presses are subject to extremely high internal pressures, such as p = 3.5 ksi. *Extrusion* is a process by which pressure is applied to a material (in soft state) in a cylinder causing it to flow through a restricted tapered hole or die as depicted in Figure 18.10a. Usually, cylinder and die are made of heat-treated steel. Extrusion is used more commonly with materials that melt at low temperatures, such as aluminum, copper, magnesium, lead, tin, and zinc. Using different die patterns, extrusion of long tubes, rods, and various shapes (such as channels, I-beams, and angles) is often performed hot in hydraulic presses (Figure P18.10b). The extruded metal then passes through a water-cooling station. Experiments indicate that the die shape and die length have considerable effect on the extrusion force P required [4,5].

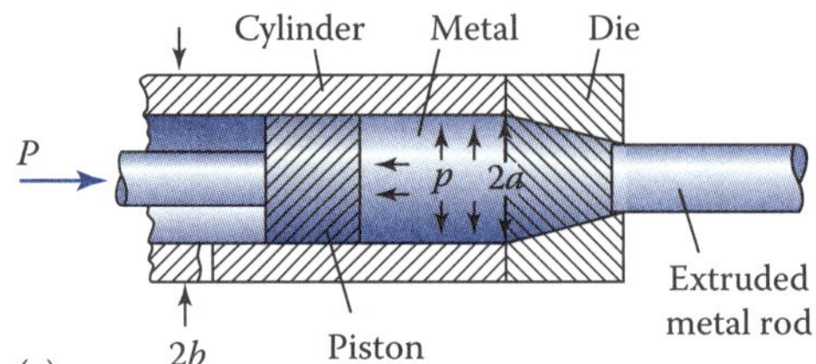

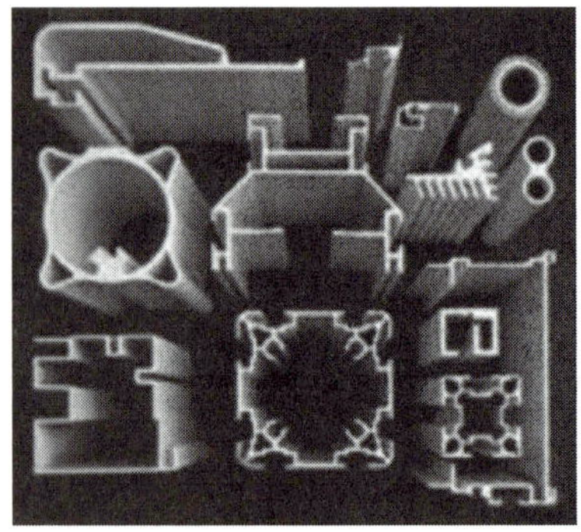

FIGURE P18.10
(a) Schematic portion of extrusion process. (b) Typical extruded shapes.

Given: The hydraulic cylinder for an extrusion press made of AISI 1040 OQ&T steel (at 425°C) is subjected to a high internal pressure $p = 3$ ksi. The inner and outer radii of the cylinder are $a = 14$ in. and b, respectively (Figure P18.10a). From Table B.4, $S_y =$ 80 ksi (552 MPa). Factor of safety against yielding is $n = 5$. Find, applying *thin-walled* pressure vessel equations,

a. The outer radius b of the cylinder based on the maximum distortion theory of failure
b. Whether the thin-walled analysis applies

18.11 Repeat Problem 18.10, through the use of the *thick-walled* pressurized cylinder equations.

Given: $a = 14$ in., $p = 3$ ksi, $S_y = 80$ ksi, and $n = 5$.

18.12 A crane boom or basic plane truss with all members have the same axial rigidity AE supports a horizontal force P and a load W acting at joint 2 as illustrated in Figure P18.12. Through the use of finite element method,

a. Drive the stiffness matrix of each element, system stiffness matrix, and force–displacement relationships
b. Compute the nodal displacements, reactions, stresses, and safety factors against yielding of each member

Assumptions: All members are made of ASTM A36 structural steel. Friction in the pin joints will be neglected. Recall that θ is measured *counterclockwise* from the positive x-axis to each element (see Figure P18.12 and Table P18.12).

Input data:

$$P = 24 \text{ kN}, \quad W = 36 \text{ kN}, \quad S_y = 250 \text{ MPa}, \quad E = 200 \text{ GPa}$$

$$L_1 = L_2 = L = 2.4 \text{ m}, \quad L_3 = 2.4\sqrt{2} \text{ m}, \quad A = 480 \text{ mm}^2$$

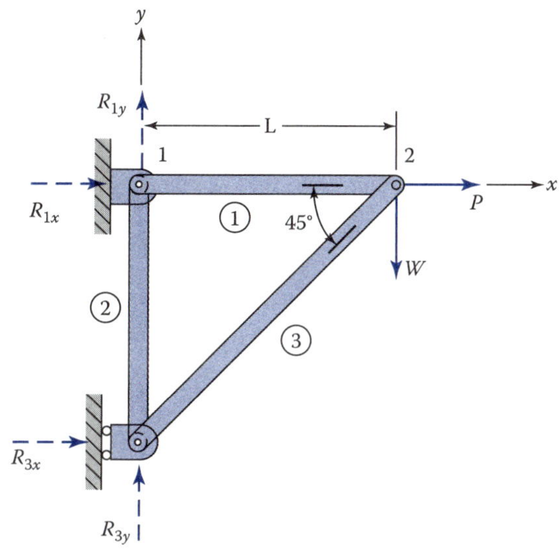

FIGURE P18.12
A three-bar plane truss.

TABLE P18.12

Data for the Truss of Figure P18.12

Element	θ	c	s	c^2	cs	s^2	AE/L
1	0°	1	0	1	0	0	$4(10^7)$
2	270°	0	−1	0	0	1	$4(10^7)$
3	225°	$-1/\sqrt{2}$	$-1/\sqrt{2}$	0.5	0.5	0.5	$2\sqrt{2}(10^7)$

Appendix A: Units, Properties of Shapes, and Beam Deflections

Knowledge of the units of typical quantities and the characteristics of common areas and masses is essential in mechanical analysis and design. Quantities given in SI units can be converted to U.S. customary units by multiplying with the conversion factors furnished in Table A.1. To reverse the process, the number in customary units is divided by the factor. Prefixes can be attached to SI units to form multiples and submultiples (see Table A.2). Properties of most standard shapes encountered in practice are given in various handbooks. Tables A.3 through A.5 present several typical cases. Data for Tables A.4 and A.6 through A.8 were compiled from the listings found in the *AISC* Manual of Steel Construction (Chicago, American Institute of Steel Construction, 2011).

Representative expressions for deflection and slope for selected beams are given in Tables A.9 and A.10. Restrictions on the application of these equations include constancy of the flexural rigidity EI, symmetry of the cross section about the vertical y axis, and the magnitude of displacement υ of the beam. In addition, equations apply to beams long in proportion to their depth and not disproportionally wide. Displacements are restricted to linearly elastic region, as shown by the presence of the elastic modulus E in the formulas.

TABLE A.1

Conversion Factors: SI Units to U.S. Customary Units

Quantity	SI Unit	U.S. Equivalent
Acceleration	m/s^2 (meter per square second)	3.2808 ft/s^2
Area	m^2 (square meter)	10.76 ft^2
Force	N (newton)	0.2248 lb
Intensity of force	N/m (newton per meter)	0.0685 lb/ft
Length	m (meter)	3.2808 ft
Mass	kg (kilogram)	2.2051 lb
Moment of a force	N·m (newton meter)	0.7376 lb·ft
Moment of inertia of a plane area	m^4 (meter to fourth power)	2.4025×10^6 $in.^4$
Moment of inertia of a mass	$kg \cdot m^2$ (kilogram meter squared)	0.7376 $ft \cdot s^2$
Power	W (watt)	0.7376 ft·lb/s
	kW (kilowatt)	1.3410 hp
Pressure or stress	Pa (pascal)	0.145×10^{-3} psi
Specific weight	kN/m^3 (kilonewton per cubic meter)	3.684×10^{-3} $lb/in.^3$
Velocity	m/s (meter per second)	3.2808 ft/s
Volume	m^3 (cubic meter)	35.3147 ft^3
Work or energy	J (joule, newton meter)	0.7376 ft·lb

Notes: 1 mile, mi = 5280 ft = 1609 m; 1 kilogram, kg = 2.20946 lb = 9.807 N; 1 joule, J = 1 N·m; 1 inch, in. = 25.4 mm; 1 foot, ft = 12 in. = 304.6 mm; 1 acceleration of gravity, $g = 9.8066\ m/s^2 = 32.174\ ft/s^2$.

TABLE A.2

SI Prefixes

Prefix	Symbol	Factor
Tera	T	10^{12} = 1 000 000 000 000
Giga	G	10^9 = 1 000 000 000
Mega	M	10^6 = 1 000 000
Kilo	k	10^3 = 1 000
Hecto	h	10^2 = 100
Deka	da	10^1 = 10
Deci	d	10^{-1} = 0.1
Centi	c	10^{-2} = 0.01
Milli	m	10^{-3} = 0.001
Micro	μ	10^{-6} = 0.000 001
Nano	n	10^{-9} = 0.000 000 001
Pico	P	10^{-12} = 0.000 000 000 001

Note: The use of the prefixes hecto, deka, and centi is not recommended. However, they are sometimes encountered in practice.

TABLE A.3

Properties of Areas

(1) Rectangle

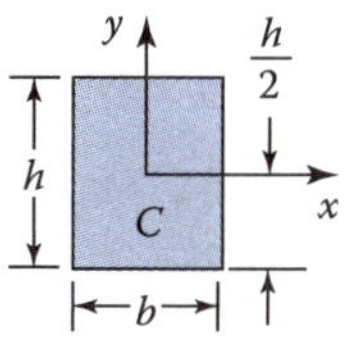

$$A = bh$$

$$I_x = \frac{bh^3}{12}$$

$$J_c = \frac{bh\left(b^2 + h^2\right)}{12}$$

(2) Circle

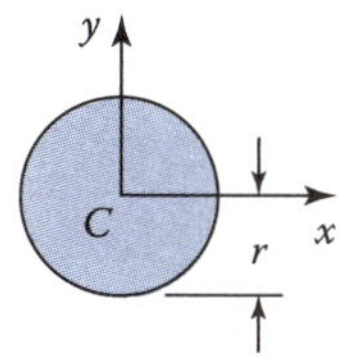

$$A = \pi r^2$$

$$I_x = \frac{\pi r^4}{4}$$

$$J_c = \frac{\pi r^4}{2}$$

(3) Right triangle

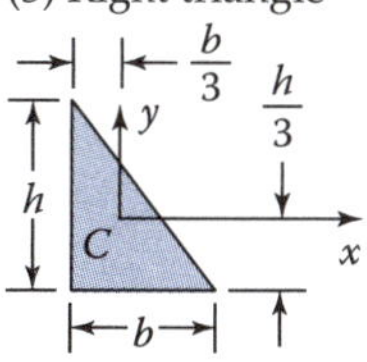

$$A = \frac{bh}{2}$$

$$I_x = \frac{bh^3}{36} \qquad I_{xy} = -\frac{b^2h^2}{72}$$

$$J_c = \frac{bh\left(b^2 + h^2\right)}{36}$$

(4) Semicircle

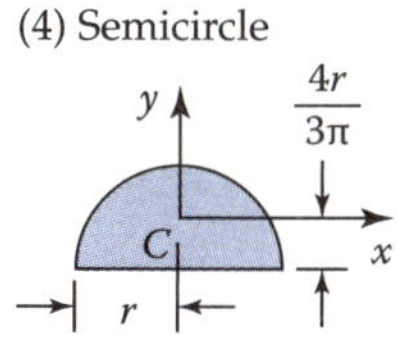

$$A = \frac{\pi r^2}{4}$$

$$I_x = 0.110r^2$$

$$I_y = \frac{\pi r^4}{8}$$

(5) Ellipse

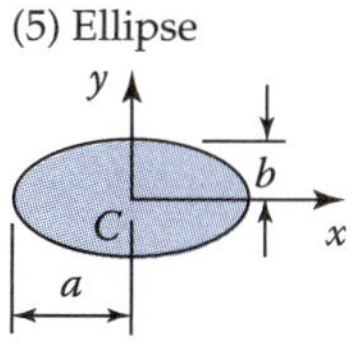

$$A = \pi ab$$

$$I_x = \frac{\pi ab^3}{4}$$

$$J_c = \frac{\pi ab\left(a^2 + b^2\right)}{4}$$

(6) Thin tube

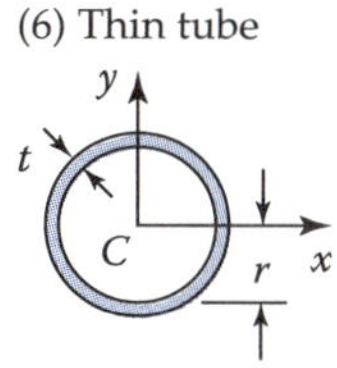

$$A = 2\pi rt$$

$$I_x = \pi r^3 t$$

$$J_c = 2\pi r^3 t$$

(Continued)

TABLE A.3 (*Continued*)

Properties of Areas

(7) Isosceles triangle

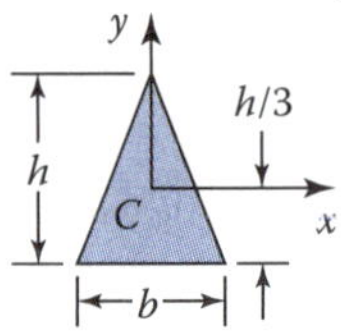

$$A = \frac{bh}{2}$$

$$I_x = \frac{bh^3}{36} \qquad I_y = \frac{hb^3}{48}$$

$$J_c \frac{bh}{144}\left(4h^2 + 3b^2\right)$$

(8) Half of thin tube

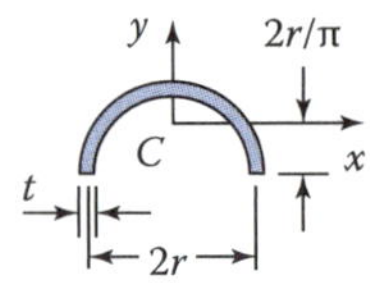

$$A = \pi r t$$

$$I_x \approx 0.095\pi r^3 t$$

$$I_y = 0.5\pi r^3 t$$

(9) Triangle

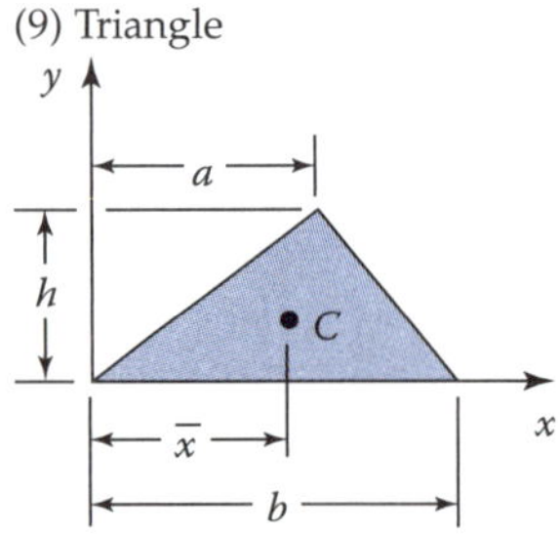

$$A = \frac{bh}{2}$$

$$\bar{x} = \frac{(a+b)}{3}$$

(10) Parabolic spandrel ($y = kx^2$)

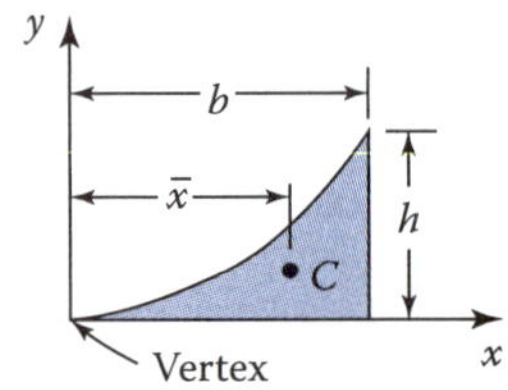

$$A = \frac{bh}{3}$$

$$\bar{x} = \frac{3b}{4}$$

(11) Parabola ($y = kx^2$)

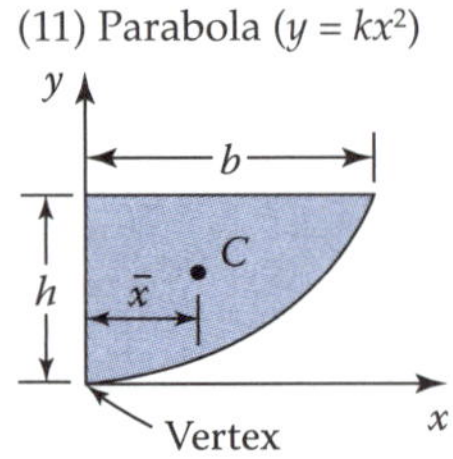

$$A = \frac{2bh}{3}$$

$$\bar{x} = \frac{3b}{8}$$

(12) General spandrel ($y = kx^n$)

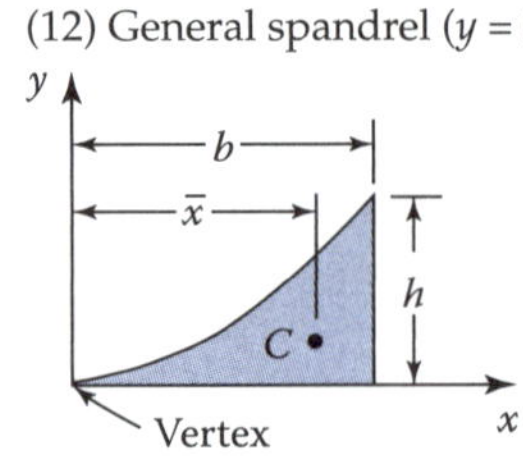

$$A = \frac{bh}{n+1}$$

$$\bar{x} = \frac{n+1}{n+2}b$$

Notes: A, area; I, moment of inertia; J, polar moment of inertia.

TABLE A.4

Properties of Some Steel Pipe and Tubing

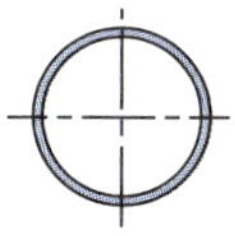

Standard Weight Pipe Dimensions and Properties

Dimensions					**Properties**			
Nominal Diameter (in.)	**Outside Diameter (in.)**	**Inside Diameter (in.)**	**Wall Thickness (in.)**	**Weight per Foot (lb/ft) Plain Ends**	A (in.^{2})	I (in.^{4})	S (in.^{3})	r (in.)
$\frac{1}{2}$	0.840	0.622	0.109	0.85	0.250	0.017	0.041	0.261
$\frac{3}{4}$	1.050	0.824	0.113	1.13	0.333	0.037	0.071	0.334
1	1.315	1.049	0.133	1.68	0.494	0.087	0.133	0.421
$1\frac{1}{4}$	1.660	1.380	0.140	2.27	0.669	0.195	0.235	0.540
$1\frac{1}{2}$	1.900	1.610	0.145	2.72	0.799	0.310	0.326	0.623
2	2.375	2.067	0.154	3.65	1.07	0.666	0.561	0.787
$2\frac{1}{2}$	2.875	2.469	0.203	5.79	1.70	1.53	1.06	0.947
3	3.500	3.068	0.216	7.58	2.23	3.02	1.72	1.16
4	4.500	4.026	0.237	10.79	3.17	7.23	3.21	1.51

(*Continued*)

TABLE A.4 (*Continued*)

Properties of Some Steel Pipe and Tubing

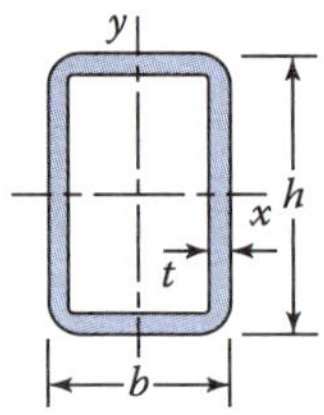

Square and Rectangular Structural Tubing Dimensions and Properties

Dimensions			Properties[a]						
Nominal[b] Size (in.)	Wall Thickness (in.)	Weight per Foot (lb/ft)	A (in.^{2})	I_x (in.^{3})	S_x (in.^{3})	r_x (in.)	I_y (in.^{4})	S_y (in.^{3})	r_y (in.)
2 × 2	$\frac{3}{16}$	4.32	1.27	0.668	0.668	0.726			
	$\frac{1}{4}$	5.41	1.59	0.766	0.766	0.694			
2.5 × 2.5	$\frac{3}{16}$	5.59	1.64	1.42	1.14	0.930			
	$\frac{1}{4}$	7.11	2.09	1.69	1.35	0.899			
3 × 2	$\frac{3}{16}$	5.99	1.64	1.86	1.24	1.06	0.977	0.977	0.771
	$\frac{1}{4}$	7.11	2.09	2.21	1.47	1.03	1.15	1.15	0.742
3 × 3	$\frac{3}{16}$	6.87	2.02	2.60	1.73	1.13			
	$\frac{1}{4}$	8.81	2.59	3.16	2.10	1.10			
4 × 2	$\frac{3}{16}$	6.87	2.02	3.87	1.93	1.38	1.29	1.29	0.798
	$\frac{1}{4}$	8.81	2.59	4.69	2.35	1.35	1.54	1.54	0.770
4 × 4	$\frac{3}{16}$	9.42	2.77	6.59	3.30	1.54			
	$\frac{1}{4}$	12.21	3.59	8.22	4.11	1.51			

Notes: A, area; S, section modulus; I, moment of inertia; r, radius of gyration.

[a] Properties are based on a nominal outside corner radius equal to two times the wall thickness (t).

[b] Outside dimensions across flat sides ($h \times b$).

TABLE A.5

Mass and Mass Moments of Inertia of Solids

(1) Slender rod

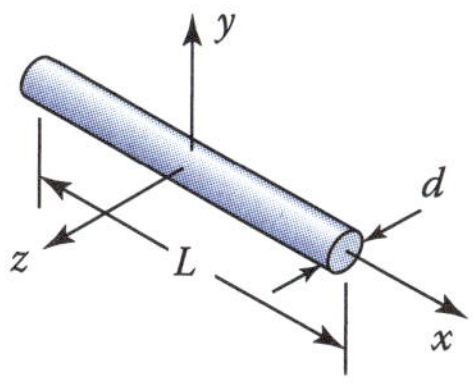

$$m = \frac{\pi d^2 L\rho}{4}$$

$$I_y = I_z = \frac{mL^2}{12}$$

(2) Thin disk

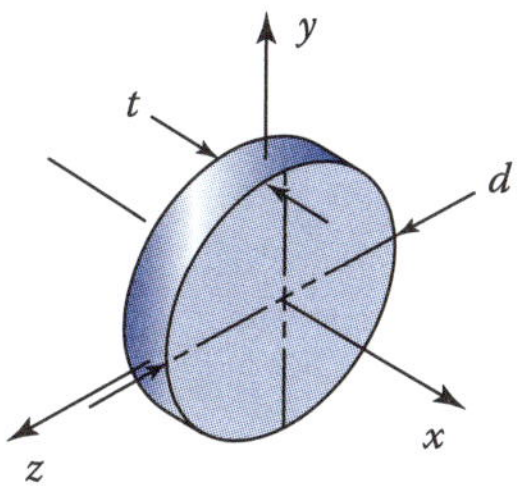

$$m = \frac{\pi d^2 t\rho}{4}$$

$$I_x = \frac{md^2}{8}$$

$$I_y = I_z = \frac{md^2}{16}$$

(3) Rectangular prism

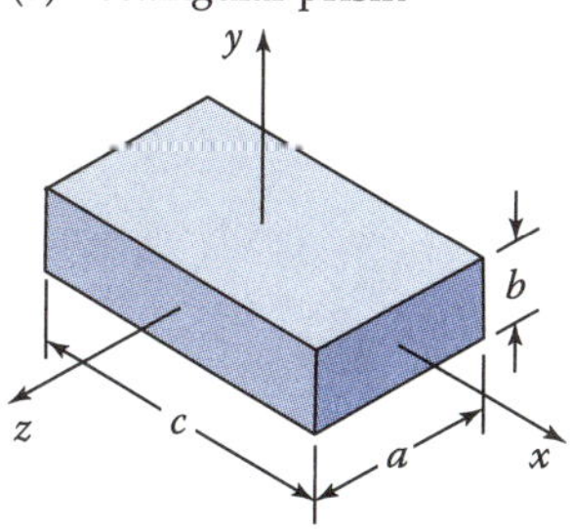

$$m = abc\rho$$

$$I_x = \frac{m}{12}\left(a^2 + b^2\right)$$

$$I_y = \frac{m}{12}\left(a^2 + c^2\right)$$

$$I_z = \frac{m}{12}\left(b^2 + a^2\right)$$

(4) Cylinder

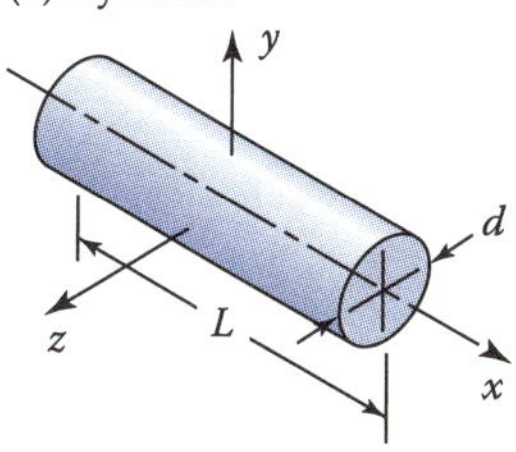

$$m = \frac{\pi d^2 L\rho}{4}$$

$$I_x = \frac{md^2}{8}$$

$$I_y = I_z = \frac{m}{48}\left(3d^2 + 4L^2\right)$$

(5) Hollow cylinder

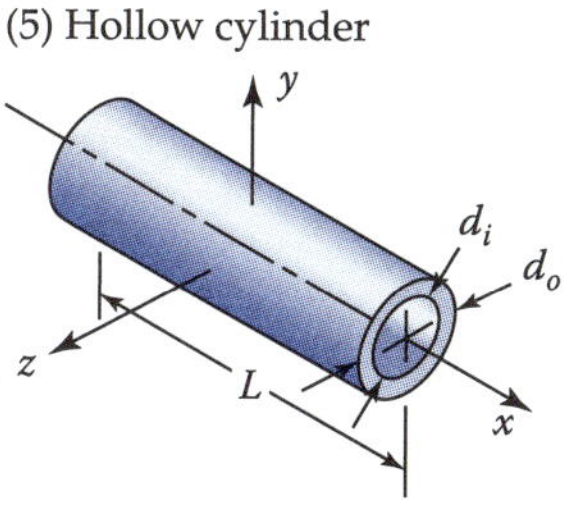

$$m = \frac{\pi L\rho}{4}\left(d_o^2 - d_i^2\right)$$

$$I_x = \frac{m}{8}\left(d_o^2 + d_i^2\right)$$

$$I_y = I_z = \frac{m}{48}\left(3d_o^2 + 3d_i^2 + 4L^2\right)$$

Notes: ρ, mass density; m, mass; I, mass moment of inertia.

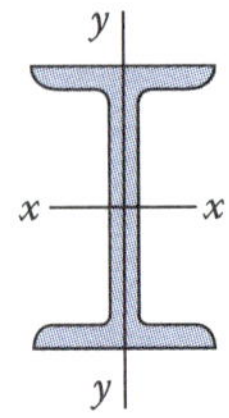

TABLE A.6

Properties of Rolled-Steel (*W*) Shapes, Wide-Flange Sections

			Flange			Axis x-x			Axis y-y	
Designation[a]	Area (10^3 mm^2)	Depth (mm)	Width (mm)	Thickness (mm)	Web Thickness (mm)	I (10^6 mm^4)	r (mm)	S (10^3 mm^3)	I (10^6 mm^4)	r (mm)
W 610 × 155	19.7	611	324	19.0	12.7	1290	256	4220	108	73.9
× 125	15.9	612	229	19.6	11.9	985	249	3220	39.3	49.7
W 460 × 158	20.1	476	284	23.9	15.0	795	199	3340	91.6	67.6
× 74	9.48	457	190	14.5	9.0	333	188	1457	16.7	41.9
× 52	6.65	450	152	10.8	7.6	212	179	942	6.4	31.0
W 410 × 114	14.6	420	261	19.3	11.6	462	178	2200	57.4	62.7
× 85	10.8	417	181	18.2	10.9	316	171	1516	17.9	40.6
× 60	7.61	407	178	12.8	7.7	216	168	1061	12	39.9
W 360 × 216	27.5	375	394	27.7	17.3	712	161	3800	282	101.1
× 122	15.5	363	257	21.7	13.0	367	154	2020	61.6	63.0
× 79	10.1	354	205	16.8	9.4	225	150	1271	24.0	48.8
W 310 × 107	13.6	311	306	17.0	10.9	248	135	1595	81.2	77.2
× 74	9.48	310	205	16.3	9.4	164	132	1058	23.4	49.8
× 52	6.65	317	167	13.2	7.6	119	133	748	10.2	39.1
W 250 × 80	10.2	256	255	15.6	9.4	126	111	985	42.8	65
× 67	8.58	257	204	15.7	8.9	103	110	803	22.2	51.1
× 49	6.26	247	202	11.0	7.4	70.8	106	573	15.2	49.3
W 200 × 71	9.11	216	206	17.4	10.2	76.6	91.7	709	25.3	52.8
× 59	7.55	210	205	14.2	9.1	60.8	89.7	579	20.4	51.8
× 52	6.65	206	204	12.6	7.9	52.9	89.2	514	17.7	51.6
W 150 × 37	4.47	162	154	11.6	8.1	22.2	69	274	7.12	38.6
× 30	3.79	157	153	9.3	6.6	17.2	67.6	219	5.54	38.1
× 24	3.06	160	102	10.3	6.6	13.2	66	167	1.84	24.6
× 18	2.29	153	102	7.1	5.8	9.2	63.2	120	1.25	23.3

Notes: t, moment of inertia; S, section modulus; r, radius of gyration.

[a] A wide-flange shape is designated by letter *W* followed by the nominal depth in millimeters and the mass in kilogram per meter.

(*Continued*)

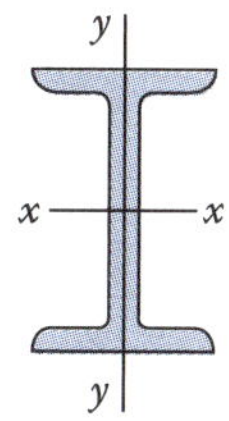

TABLE A.6 ***(Continued)***

Properties of Rolled-Steel (*W*) Shapes, Wide-Flange Sections

U.S. Customary Units

			Flange			Axis *x-x*			Axis *y-y*	
Designation[a]	**Area 10^2 (in.^{2})**	**Depth (in.)**	**Width (in.)**	**Thickness (in.)**	**Web Thickness (in.)**	***I* (in.^{4})**	***r* (in.)**	***S* (in.^{3})**	***I* (in.^{4})**	***r* (in.)**
W 24 × 104	30.6	24.06	12.750	0.750	0.500	3100	10.1	258	259	2.91
× 84	24.7	24.10	9.020	0.770	0.470	2370	9.79	196	94.4	1.95
W 18 × 106	31.1	18.73	11.200	0.940	0.590	1910	7.84	204	220	2.66
× 50	14.7	17.99	7.495	0.570	0.355	800	7.38	88.9	40.1	1.65
× 35	10.3	17.70	6.000	0.425	0.300	510	7.04	57.6	15.3	1.22
W 16 × 77	22.6	16.52	10.295	0.760	0.455	1110	7.00	134	138	2.47
× 57	16.8	16.43	7.120	0.715	0.430	758	6.72	92.2	43.1	1.60
× 40	11.8	16.01	6.995	0.505	0.305	518	6.63	64.7	28.9	1.57
W 14 × 145	42.7	14.78	15.500	1.090	0.680	1710	6.33	232	677	3.98
× 82	24.1	14.31	10.130	0.855	0.510	882	6.05	123	148	2.48
× 53	15.6	13.92	8.060	0.660	0.370	541	5.89	77.8	57.7	1.92
W 12 × 72	21.1	12.25	8.080	0.670	0.430	597	5.31	97.4	195	3.04
× 50	14.7	12.19	8.080	0.640	0.370	394	5.18	64.7	56.3	1.96
× 35	10.3	12.50	6.560	0.520	0.300	285	5.25	45.6	24.5	1.54
W 10 × 54	15.8	10.09	10.030	0.615	0.370	303	4.37	60.0	103	2.56
× 45	13.3	10.10	8.020	0.620	0.350	248	4.33	49.1	53.4	2.01
× 33	9.71	9.73	7.960	0.435	0.290	170	4.19	35.0	36.6	1.94
W 8 × 48	14.1	8.50	8.110	0.685	0.400	184	3.61	43.3	60.9	2.08
× 40	11.7	8.25	8.070	0.560	0.360	146	3.53	35.5	49.1	2.04
× 35	10.3	8.12	8.020	0.495	0.310	127	3.51	31.2	42.6	2.03
W 6 × 25	7.34	6.38	6.080	0.455	0.320	53.4	2.70	16.7	17.1	1.52
× 20	5.88	6.20	6.020	0.365	0.260	41.4	2.66	13.4	13.3	1.50
× 16	4.47	6.28	4.030	0.405	0.260	32.1	2.60	10.2	4.43	0.967
× 12	3.35	6.03	4.000	0.280	0.230	22.1	2.49	7.31	2.99	0.918

Source: The American Institute of Steel Construction, Chicago, IL.

[a] A wide-flange shape is designated by letter *W* followed by the nominal depth in inches and the weight in pounds per foot.

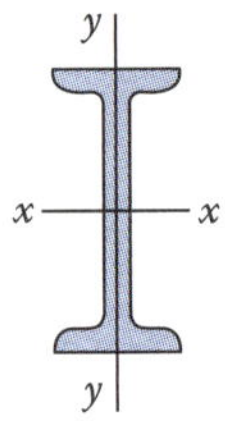

TABLE A.7

Properties of Rolled-Steel (*S*) Shapes, American Standard I Beams

SI Units

			Flange			Axis *x-x*			Axis *y-y*	
Designation[a]	Area (10^3 mm^2)	Depth (mm)	Width (mm)	Thickness (mm)	Web Thickness (mm)	*I* (10^6 mm^4)	*r* (mm)	*S* (10^3 mm^3)	*I* (10^6 mm^4)	*r* (mm)
S 610 × 149	19.0	610	184	22.1	19.0	995	229	3260	19.9	32.3
× 119	15.2	610	178	22.1	12.7	878	241	2880	17.6	34.0
S 510 × 141	18.0	508	183	23.3	20.3	670	193	2640	20.7	33.8
× 112	14.3	508	162	20.1	16.3	533	193	2100	12.3	29.5
S 460 × 104	13.3	457	159	17.6	18.1	385	170	1685	10.0	27.4
× 81	10.4	457	152	17.6	11.7	335	180	1466	8.66	29.0
S 380 × 74	9.5	381	143	15.8	14.0	202	146	1060	6.53	26.2
× 64	8.13	381	140	15.8	10.4	186	151	977	5.99	27.2
S 310 × 74	9.48	305	139	16.8	17.4	127	116	833	6.53	256.2
× 52	6.64	305	129	13.8	10.9	95.3	120	625	4.11	24.9
S 250 × 52	6.64	254	126	12.5	15.1	61.2	96	482	3.48	22.9
× 38	4.81	254	118	12.5	7.9	51.6	103	406	2.83	24.2
S 200 × 34	4.37	203	106	10.8	11.2	27	78.7	266	1.79	20.3
× 27	3.5	203	102	10.8	6.9	24	82.8	236	1.55	21.1
S 150 × 26	3.27	152	90	9.1	11.8	11.0	57.9	144	0.96	17.2
× 19	2.36	152	84	9.1	5.8	9.20	62.2	121	0.76	17.9
S 100 × 14	1.80	102	70	7.4	8.3	2.83	39.6	55.5	0.38	14.5
× 11	1.45	102	67	7.4	4.8	2.53	41.6	49.6	0.32	14.8

[a] An American standard beam is designated by letter *S* followed by the nominal depth in millimeters and the mass in kilograms per meter.

(*Continued*)

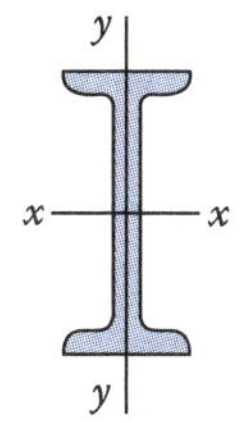

TABLE A.7 (*Continued*)

Properties of Rolled-Steel (*S*) Shapes, American Standard I Beams

U.S. Customary Units										
			Flange			Axis x-x			Axis y-y	
Designation[a]	Area (in.^{2})	Death (in.)	Width (in.)	Thickness (in.)	Web Thickness (in.)	I (in.^{4})	r (in.)	S (in.^{3})	I (in.^{4})	r (in.)
S 24 × 100	29.4	24.00	7.247	0.871	0.747	2390	9.01	199	47.8	1.27
× 79.9	23.5	24.00	7.001	0.871	0.501	2110	9.47	175	42.3	1.34
S 20 × 95	27.9	20.00	7.200	0.916	0.800	1610	7.60	161	49.1	1.33
× 75	22.1	20.00	6.391	0.789	0.641	1280	7.60	128	29.6	1.16
S 18 × 70	20.6	18.00	6.251	0.691	0.711	926	6.71	103	24.1	1.08
× 54.7	16.1	18.00	6.001	0.691	0.461	804	7.07	89.4	20.8	1.14
S 15 × 50	14.7	15.00	5.501	0.622	0.550	486	5.75	64.8	15.7	1.03
× 42.9	12.6	15.00	5.501	0.622	0.411	447	5.95	59.6	14.4	1.07
S 12 × 50	14.7	12.00	5.477	0.659	0.687	305	4.55	50.8	15.7	1.03
× 35	10.3	12.00	5.078	0.544	0.428	229	4.72	38.2	9.87	0.980
S 10 × 35	10.3	10.00	4.944	0.491	0.594	147	3.78	29.4	8.36	0.901
× 25.4	7.46	10.00	4.661	0.491	0.311	124	4.07	24.7	6.79	0.954
S 8.23	6.77	8.00	4.171	0.425	0.441	64.9	3.10	16.2	4.31	0.798
× 18.4	5.41	8.00	4.001	0.425	0.271	57.6	3.26	14.4	3.73	0.831
S 6 × 17.25	5.07	6.00	3.565	0.359	0.465	26.3	2.28	8.77	2.31	0.675
× 12.5	3.67	6.00	3.332	0.359	0.232	22.1	2.45	7.37	1.82	0.705
S 4 × 9.5	2.79	4.00	2.796	0.293	0.326	6.79	1.56	3.39	0.903	0.569
× 7.7	2.26	4.00	2.663	0.293	0.193	6.08	1.64	3.04	0.764	0.581

Source: The American Institute of Steel Construction, Chicago, IL.

[a] An American standard beam is designated by letter *S* followed by the nominal depth in inches and the weight in pounds per foot.

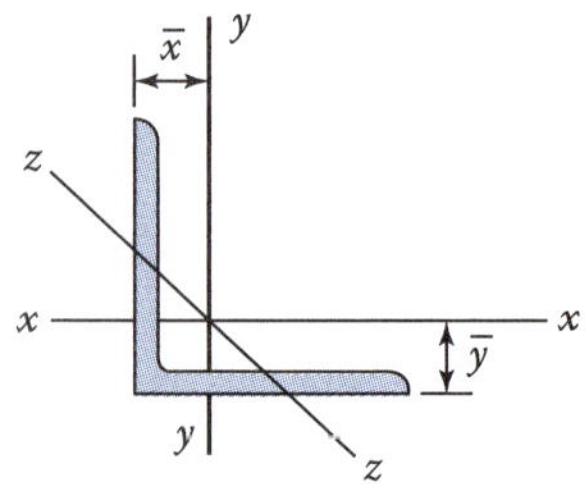

TABLE A.8

Properties of Rolled-Steel (*L*) Shapes, Angles with Equal Legs

SI Units							
			Axis *x-x* or *y-y*				Axis *z-z*,
Size and Thickness (mm)	**Mass (kg/m)**	**Area (10^3 mm^2)**	***I* (10^6 mm^4)**	***r* (mm)**	***S* (10^3 mm^3)**	**$\bar{x}$ or $\bar{y}$ (mm)**	***r* (mm)**
L 203 × 203 × 25.4	75.9	9.68	37	61.8	259	60.2	39.6
× 19	57.9	7.36	29	62.8	200	57.9	40.1
× 12.7	39.3	5.0	20.2	63.6	137	55.6	40.4
L 152 × 152 × 25.4	55.7	7.1	14.8	45.6	140.4	47.2	29.7
× 15.9	36	4.59	10.1	46.8	92.8	43.9	30.0
× 9.5	22.2	2.8	6.41	47.8	57.8	41.7	30.2
L 127 × 127 × 19	35.1	4.48	6.53	38.2	74.2	38.6	24.8
× 12.7	24.1	3.07	4.70	39.2	51.8	36.3	25.0
× 9.5	18.3	2.33	3.64	39.5	39.7	35.3	25.1
L 102 × 102 × 19	27.5	3.51	3.19	30.1	46.0	32.3	19.8
× 12.7	19	2.42	2.31	30.9	32.3	30.0	19.9
× 6.4	9.8	1.25	1.27	31.8	17.2	27.7	20.2
L 89 × 89 × 9.5	12.6	1.6	1.20	27.3	18.9	25.7	17.5
× 6.4	8.6	1.09	0.84	27.7	13.0	24.6	17.6
L 76 × 76 × 12.7	14	1.77	0.92	22.8	17.5	23.7	14.8
× 6.4	7.3	0.93	0.52	23.6	9.46	21.4	15.0

(*Continued*)

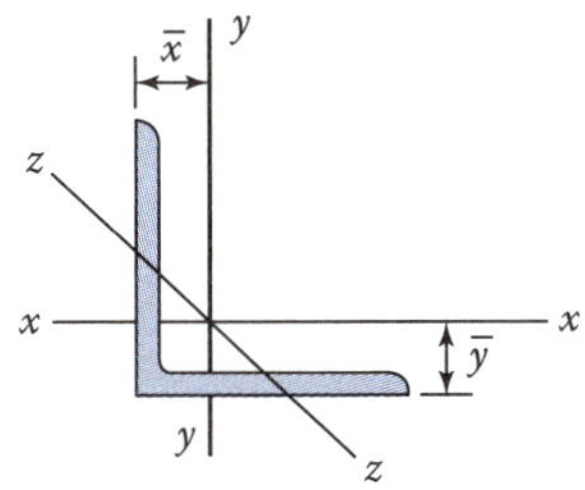

TABLE A.8 (*Continued*)

Properties of Rolled-Steel (*L*) Shapes, Angles with Equal Legs

			Axis x-x or y-y				Axis z-z
Size and Thickness (in.)	**Weight (lb/ft)**	**Area (in.^{2})**	I **(in.^{4})**	r **(in.)**	S **(in.^{3})**	$\bar{x}$ **or** $\bar{y}$ **(in.)**	r **(in.)**
$L\,8\times 8\times 1$	51.0	15.0	89.0	2.44	15.8	2.37	1.56
$\times\frac{3}{4}$	38.9	11.4	69.7	2.47	12.2	2.28	1.58
$\times\frac{1}{2}$	26.4	7.75	48.6	2.50	8.36	2.19	1.59
$L\,6\times 6\times 1$	37.4	11.0	35.5	1.80	8.57	1.86	1.17
$\times\frac{5}{8}$	24.2	7.11	24.2	1.84	5.66	1.73	1.18
$\times\frac{3}{8}$	14.9	4.36	15.4	1.88	3.53	1.64	1.19
$L\,5\times 5\times\frac{3}{4}$	23.6	6.94	15.7	1.51	4.53	1.52	0.975
$\times\frac{1}{2}$	16.2	4.75	11.3	1.54	3.16	1.43	0.983
$\times\frac{3}{8}$	12.3	3.61	8.74	1.56	2.42	1.39	0.990
$L\,4\times 4\times\frac{3}{4}$	18.5	5.44	7.67	1.19	2.81	1.27	0.778
$\times\frac{1}{2}$	12.8	3.75	5.56	1.22	1.97	1.18	0.782
$\times\frac{1}{4}$	6.6	1.94	3.04	1.25	1.05	1.09	0.795
$L\,3\frac{1}{2}\times 3\frac{1}{2}\times\frac{3}{8}$	8.5	2.48	2.87	1.07	1.15	1.01	
$\times\frac{1}{4}$	5.8	1.69	2.01	1.09	0.794	0.968	0.694
$L\,3\times 3\times\frac{1}{2}$	9.4	2.75	2.22	0.898	1.07	0.932	0.584
$\times\frac{1}{4}$	4.9	1.44	1.24	0.930	0.577	0.842	0.592

Source: The American Institute of Steel Construction, Chicago, IL.

TABLE A.9

Deflections and Slopes of Variously Loaded Beams

Load and Support	Maximum Deflection	Slope at End	Equation of Elastic Curve
(1) 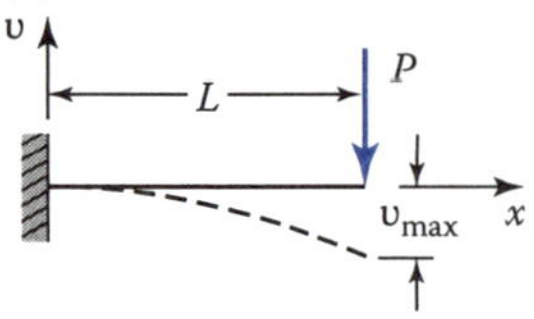	$-\frac{PL^3}{3EI}$	$-\frac{PL^2}{2EI}$	$\upsilon = \frac{Px^2}{6EI}(x-3L)$
(2) 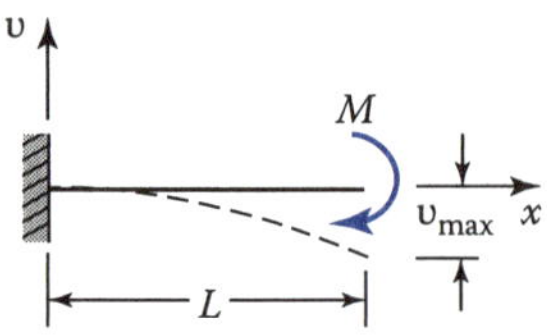	$-\frac{ML^2}{2EI}$	$-\frac{ML}{EI}$	$\upsilon = \frac{Mx^2}{2EI}$
(3) 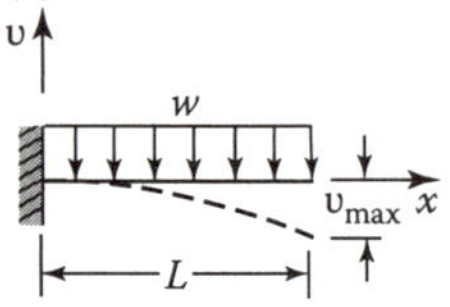	$-\frac{wL^4}{8EI}$	$-\frac{wL^3}{6EI}$	$\upsilon = \frac{wx^2}{24EI}(x^2-4Lx+6L^2)$
(4) 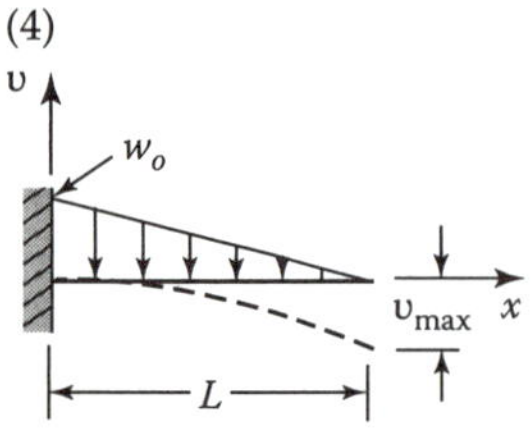	$-\frac{w_0L^4}{30EI}$	$-\frac{w_0L^3}{24EI}$	$\upsilon = \frac{w_0x^3}{120EIL}(x^3-5Lx^2+10L^2x-10L^3)$
(5) 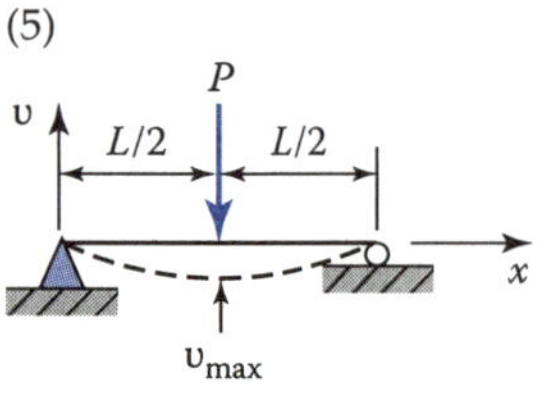	$-\frac{PL^3}{48EI}$	$\pm\frac{PL^2}{16EI}$	$\upsilon = \frac{Px}{48EI}(4x^2-3L^2) \quad (x \le L/2)$
(6) 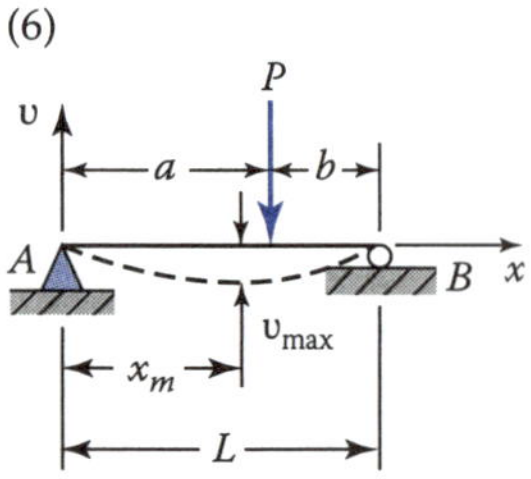	For $a > b$: $-\frac{Pb(L^2-b^2)^{3/2}}{9\sqrt{3}EIL}$ $x_m = \sqrt{\frac{L^2-b^2}{3}}$	$\theta_A = -\frac{Pb(L^2-b^2)}{6EIL}$ $\theta_B = -\frac{Pa(L^2-a^2)}{6EIL}$	$\upsilon = \frac{Pbx}{6EIL}(x^2-L^2+b^2) \quad (x \le a)$ $\upsilon = \frac{Pa(L-x)}{6EIL}(x^2+a^2-2Lx) \quad (a \le x \le L)$

(Continued)

TABLE A.9 (*Continued*)

Deflections and Slopes of Variously Loaded Beams

Load and Support	Maximum Deflection	Slope at End	Equation of Elastic Curve
(7) 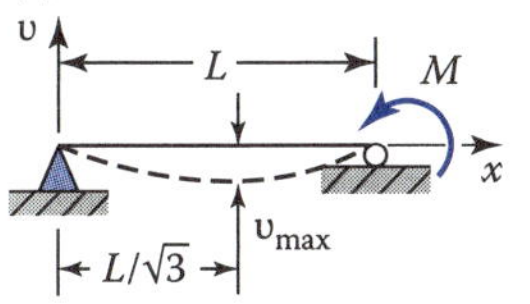	$-\dfrac{ML^2}{9\sqrt{3}EI}$	$\theta_A = -\dfrac{ML}{6EI}$ $\theta_B = -\dfrac{ML}{3EI}$	$\upsilon = \dfrac{Mx}{6EIL}\left(x^2 - L^2\right)$
(8) 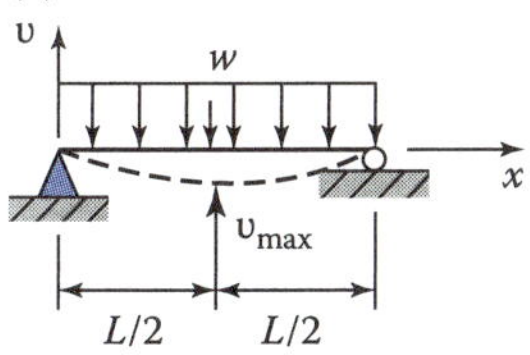	$-\dfrac{5wL^4}{384EI}$	$\pm\dfrac{wL^3}{24EI}$	$\upsilon = -\dfrac{wx}{24EI}\left(x^3 - 2Lx^2 + L^3\right)$
(9) 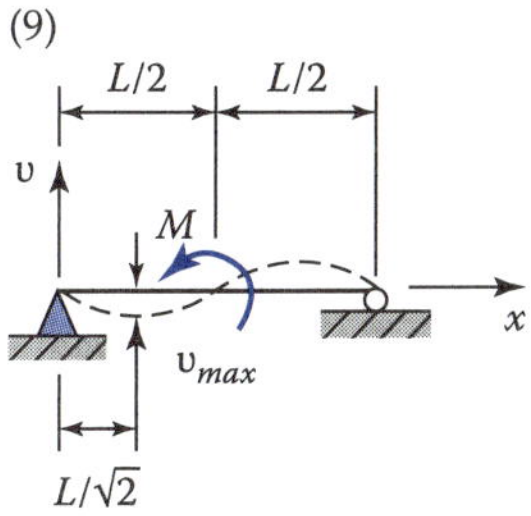	$\pm\dfrac{ML^2}{36\sqrt{12}EI}$	$\pm\dfrac{ML}{24EI}$	$\upsilon = \dfrac{Mx}{24EIL}\left(4x^2 - L^2\right) \quad \left(x \ge L/2\right)$
(10) 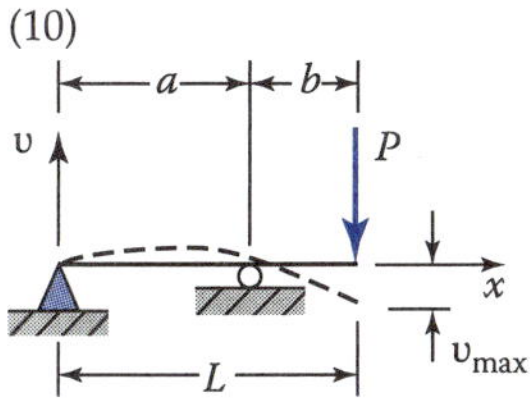	$-\dfrac{Pb^2L}{3EI}$	$\theta_A = -\dfrac{PAb}{6EI}$ $\theta_B = -\dfrac{Pb}{6EI}(2L + b)$	$\upsilon = \dfrac{Pbx}{6aEI}\left(a^2 - x^2\right) \quad \left(0 \le x \le a\right)$

TABLE A.10

Reactions and Deflections of Statically Indeterminate Beams

Load and Support	Reactions[a]	Deflections
(1) 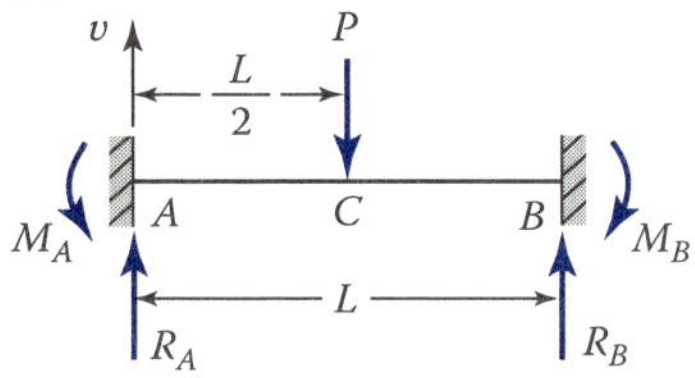	$R_A = R_B = \frac{P}{2}$ $M_A = M_B = \frac{PL}{8}$	$\upsilon_{max} = \upsilon_C = -\frac{PL^3}{192EI}$
(2) 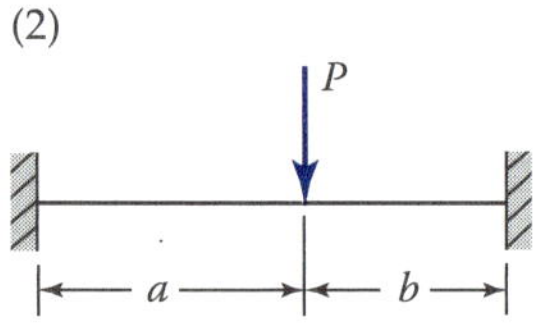	$R_A = \frac{Pb^2}{L^3}(3a+b), \quad R_B = \frac{Pb^2}{L^3}(a+3b)$ $M_A = \frac{Pab^2}{L^2} \quad M_B = \frac{Pa^2b}{L^2}$	For $a > b$: $\upsilon_C = -\frac{Pb^2}{48EI}(3L-4b)$
(3) 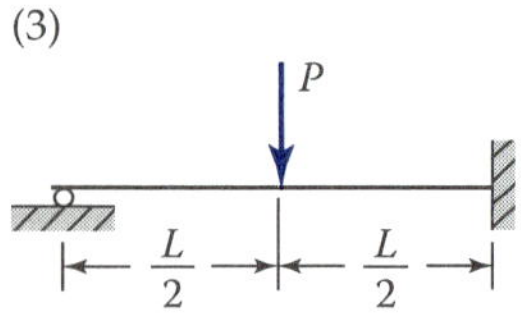	$R_A = \frac{5}{16}P, \quad R_B = \frac{11}{16}P$ $M_B = \frac{13}{16}PL$	$\upsilon_C = -\frac{7PL^3}{768EL}$
(4) 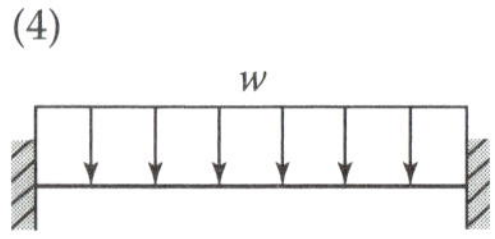	$R_A = R_B = \frac{wL}{2}$ $M_A = M_B = \frac{wL^2}{12}$	$\upsilon_{max} = \upsilon_C = -\frac{wL^4}{384EI}$
(5) 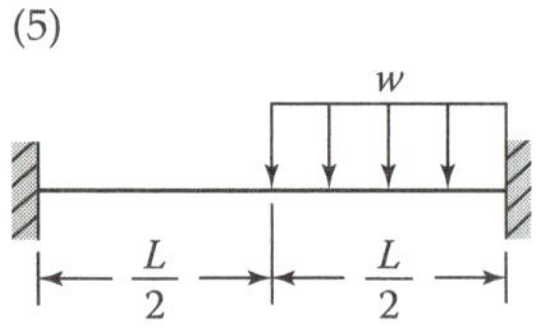	$R_A = \frac{3}{32}wL, \quad R_B = \frac{13}{32}wL$ $M_A = \frac{5}{192}wL^2, \quad M_B = \frac{11}{192}wL^2$	$\upsilon_C = -\frac{wL^4}{768EI}$
(6) 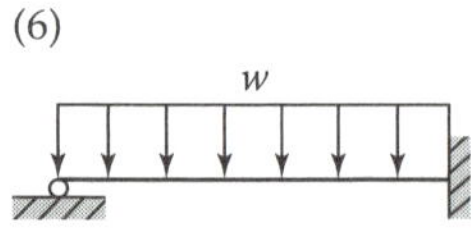	$R_A = \frac{3}{8}wL, \quad R_B = \frac{5}{8}wL$ $M_B = \frac{1}{8}wL^2$	$\upsilon_C = -\frac{wL^4}{192EI}$

[a] For all the cases tabulated, the senses of the reactions and the notations are the same as those shown in case 1.

Appendix B: Material Properties

The properties of materials vary widely, depending on numerous factors, including chemical composition, manufacturing processes, internal defects, heat treatment, temperature, and dimensions of test specimens. Hence, the values furnished in Tables B.1 through B.10 are representative but are not necessarily suitable for a specific application. In some cases, a range of values given in the listings show the possible variations in characteristics.

Unless otherwise indicated, the modulus of elasticity E and other properties are for materials in tension. The specific data were compiled from broad tabulations listed in the references cited. For details, see, for example, [1,2] of Chapter 2. Note that the reference issues of *Machine Design Materials* (Cleveland: Penton/IPC) also constitute an excellent source of data on a great variety of materials.

TABLE B.1

Average Properties of Common Engineering Materials[a]

	SI Units										
Material	Density (Mg/m³)	Ultimate Strength (MPa) Tension	Ultimate Strength (MPa) Compression[c]	Ultimate Strength (MPa) Shear	Yield Strength[b] (MPa) Tension	Yield Strength[b] (MPa) Shear	Modulus of Elasticity (GPa)	Modulus of Rigidity (GPa)	Coefficient of Thermal Expansion (10^{-6}/°C)	Elongation in 50 mm (%)	Poisson's Ratio
Steel											0.27–0.3
Structural, ASTM-A36	7.86	400	—	—	250	145	200	79	11.7	30	
High strength, ASTM-A242	7.86	480	—	—	345	210	200	79	11.7	21	
Stainless (302), cold rolled	7.92	860	—	—	520	—	190	73	17.3	12	
Cast iron											0.2–0.3
Cray, ASTM A-48	7.2	170	650	240	—	—	70	28	12.1	0.5	
Malleable, ASTM A-47	7.3	340	620	330	230	—	165	64	12.1	10	
Wrought iron	7.7	350	—	240	210	130	190	70	12.1	35	0.3
Aluminum											0.33
Alloy 2014-T6	2.8	480	—	290	410	220	72	28	23	13	
Alloy 6061-T6	2.71	300	—	185	260	140	70	26	23.6	17	
Brass, yellow											0.34
Cold rolled	8.47	540	—	300	435	250	105	39	20	8	
Annealed	8.47	330	—	220	105	65	105	39	20	60	
Bronze, cold rolled (510)	8.86	560	—	—	520	275	110	41	17.8	10	0.34
Copper, hard drawn	8.86	380	—	—	260	160	120	40	16.8	4	0.33

(*Continued*)

TABLE B.1 (*Continued*)

Average Properties of Common Engineering Materials[a]

	SI Units									
		Ultimate Strength (MPa)			Yield Strength[b] (MPa)						
Material	Density (Mg/m^3)	Tension	Compression[c]	Shear	Tension	Shear	Modulus of Elasticity (GPa)	Modulus of Rigidity (GPa)	Coefficient of Thermal Expansion ($10^{-6}/°C$)	Elongation in 50 mm (%)	Poisson's Ratio
Magnesium alloys	1.8	140–340	—	165	80–280	—	45	17	27	2–20	0.35
Nickel	8.08	310–760	—	—	140–620	—	210	80	13	2–50	0.31
Titanium alloys	4.4	900–970	—	—	760–900	—	100–120	39–44	8–10	10	0.33
Zinc alloys	6.6	280–390	—	—	210–320	—	83	31	27	1–10	0.33
Concrete											0.1–0.2
Medium strength	2.32	—	28	—	—	—	24	—	10	—	
High strength	2.32	—	40	—	—	—	30	—	10	—	
Timber[d] (air dry)											
Douglas fir	0.54	—	55	7.6	—	—	12	—	4	—	
Southern pine	0.58	—	60	10	—	—	11	—	4	—	
Glass, 98% silica	2.19	—	50	—	—	—	65	28	80	—	0.2–0.27
Graphite	0.77	20	240	35	—	—	70	—	7	—	
Rubber	0.91	14	—	—	—	—	—	—	162	600	0.45–0.5

[a] Properties may vary widely with changes in composition, heat treatment, and method of manufacture.
[b] Offset of 0.7%.
[c] For ductile metals, the compression strength is assumed to be the same as that in tension.
[d] Loaded parallel to the grain.

(*Continued*)

TABLE B.1 (*Continued*)

Average Properties of Common Engineering Materials[a]

Material	Specific Weight (lb/in.^{3})	Ultimate Strength (ksi) Tension	Ultimate Strength (ksi) Compression[c]	Ultimate Strength (ksi) Shear	Yield Strength[b] (ksi) Tension	Yield Strength[b] (ksi) Shear	Modulus of Elasticity (10^6 psi)	Modulus of Rigidity (10^6 psi)	Coefficient of Thermal Expansion (10^{-6}/°F)	Elongation in 2 in. (%)	Poisson's Ratio
U.S. Customary Units											
Steel											0.27–0.3
Structural, ASTM-A36	0.284	58	—	—	36	21	29	11.5	6.5	30	
High strength, ASTM-A242	0.284	70	—	—	50	30	29	11.5	6.5	21	
Stainless (302), cold rolled	0.286	125	—	—	75	—	28	10.6	9.6	12	
Cast iron											0.2–0.3
Gray, ASTM A-48	0.260	25	95	35	—	—	10	4.1	6.7	0.5	
Malleable, ASTM A-47	0.264	50	90	48	33	—	24	9.3	6.7	10	
Wrought iron	0.278	50		35	30	18	27	10	6.7	35	0.3
Aluminum											0.33
Alloy 2014-T6	0.101	70	—	42	60	32	10.6	4.1	12.8	13	
Alloy 6061-T6	0.098	43	—	27	38	20	10.0	3.8	13.1	17	
Brass, yellow											0.34
Cold rolled	0.306	78	—	43	63	36	15	5.6	11.3	8	
Annealed	0.306	48	—	32	15	9	15	5.6	11.3	60	
Bronze, cold rolled (510)	0.320	81	—	—	75	40	16	5.9	9.9	10	0.34
Magnesium alloys	0.065	20–49	—	24	11–40	—	6.5	2.4	15	2–20	0.35
Copper, hard drawn	0.320	55	—	—	38	23	17	6	9.3	4	0.33

(*Continued*)

TABLE B.1 ***(Continued)***

Average Properties of Common Engineering Materials[a]

	U.S. Customary Units										
	Specific Weight	Ultimate Strength (ksi)			Yield Strength[b] (ksi)		Modulus of Elasticity	Modulus of Rigidity	Coefficient of Thermal	Elongation	Poisson's
Material	**(lb/in.^{3})**	**Tension**	**Compression[c]**	**Shear**	**Tension**	**Shear**	**(10^6 psi)**	**(10^6 psi)**	**Expansion (10^{-6}/°F)**	**in 2 in. (%)**	**Ratio**
Nickel	0.320	45–110	—	—	20–90	—	30	11.4	7.2	2–50	0.31
Titanium alloys	0.160	130–140	—	—	110–130	—	15–17	5.6–6.4	4.5–5.5	10	0.33
Zinc alloys	0.240	40–57	—	—	30–46	—	12	4.5	15	1–10	0.33
Concrete											0.1–0.2
Medium strength	0.084	—	4	—	—	—	3.5	—	5.5	—	
High strength	0.084	—	6	—	—	—	4.3	—	5.5	—	
Timber[d] (air dry)											
Douglas fir	0.020		7.9	1.1			1.7		2.2		
Southern pine	0.021		8.6	1.4			1.6		2.2		
Class, 98% silica	0.079		7				9.6	4.1	44		0.2–0.27
Graphite	0.028	3	35	5		—	10	—	3.9		
Rubber	0.033	2	—	—	—	—	—	—	90	600	0.45–0.5

[a] Properties may vary widely with changes in composition, heat treatment, and method of manufacture.
[b] Offset of 0.7%.
[c] For ductile metals, the compression strength is assumed to be the same as that in tension.
[d] Loaded parallel to the grain.

TABLE B.2

Typical Mechanical Properties of Gray Cast Iron

ASTM Class[a]	Ultimate Strength S_u (MPa)	Compressive Strength S_{uc} (MPa)	Modulus of Elasticity (GPa) Tension	Modulus of Elasticity (GPa) Torsion	Brinell Hardness H_B	Fatigue Stress Concentration Factor K_f
20	150	575	66–97	27–39	156	1.00
25	180	670	79–102	32–41	174	1.05
30	215	755	90–113	36–45	201	1.10
35	250	860	100–120	40–48	212	1.15
40	295	970	110–138	44–54	235	1.25
50	365	1135	130–157	50–54	262	1.35
60	435	1295	141–162	54–59	302	1.50

Note: To convert from MPa to ksi, divide given values by 6.895.
[a] Minimum values of S_u (in ksi) are given by the class number.

TABLE B.3

Mechanical Properties of Some Hot-Rolled (HR) and Cold-Drawn (CD) Steels

UNS Number	AISI/SAE Number	Processing	Ultimate Strength[a] S_u (MPa)	Yield Strength[a] S_y (MPa)	Elongation in 50 mm (%)	Reduction in Area (%)	Brinell Hardness (H_B)
G10060	1006	HR	300	170	30	55	86
		CD	330	280	20	45	95
G10100	1010	HR	320	180	28	50	95
		CD	370	300	20	40	105
G10150	1015	HR	340	190	28	50	101
		CD	390	320	18	40	111
G10200	1020	HR	380	210	25	50	111
		CD	470	390	15	40	131
G10300	1030	HR	470	260	20	42	137
		CD	520	440	12	35	149
G10350	1035	HR	500	270	18	40	143
		CD	550	460	12	35	163
G10400	1040	HR	520	290	18	40	149
		CD	590	490	12	35	170
G10450	1045	HR	570	310	16	40	163
		CD	630	530	12	35	179
G10500	1050	HR	620	340	15	35	179
		CD	690	580	10	30	197
G10600	1060	HR	680	370	12	30	201
G10800	1080	HR	770	420	10	25	229
G10950	1095	HR	830	460	10	25	248

Source: *ASM Handbook*, vol. 1, ASM International, Materials Park, OH, 1991.
Note: To convert from MPa to ksi, divide given values by 6.895.
[a] Values listed are estimated ASTM minimum values in the size range 18–32 mm.

TABLE B.4

Mechanical Properties of Selected Heat-Treated Steels

AISI Number	Treatment	Temperature (°C)	Ultimate Strength S_u (MPa)	Yield Strength S_y (MPa)	Elongation in 50 mm (%)	Reduction in Area (%)	Brinell Hardness (H_B)
1030	WQ&T	205	848	648	17	47	495
	WQ&T	425	731	579	23	60	302
	WQ&T	650	586	441	32	70	207
	Normalized	925	521	345	32	61	149
	Annealed	870	430	317	35	64	137
1040	OQ&T	205	779	593	19	48	262
	OQ&T	425	758	552	21	54	241
	OQ&T	650	634	434	29	65	192
	Normalized	900	590	374	28	55	170
	Annealed	790	519	353	30	57	149
1050	WQ&T	205	1120	807	9	27	514
	WQ&T	425	1090	793	13	36	444
	WQ&T	650	717	538	28	65	235
	Normalized	900	748	427	20	39	217
	Annealed	790	636	365	24	40	187
1060	OQ&T	425	1080	765	14	41	311
	OQ&T	540	965	669	17	45	277
	OQ&T	650	800	524	23	54	229
	Normalized	900	776	421	18	37	229
	Annealed	790	626	372	11	38	179
1095	OQ&T	315	1260	813	10	30	375
	OQ&T	425	1210	772	12	32	363
	OQ&T	650	896	552	21	47	269
	Normalized	900	1010	500	9	13	293
	Annealed	790	658	380	13	21	192
4130	WQ&T	205	1630	1460	10	41	467
	WQ&T	425	1280	1190	13	49	380
	WQ&T	650	814	703	22	64	245
	Normalized	870	670	436	25	59	197
	Annealed	865	560	361	28	56	156
4140	OQ&T	205	1770	1640	8	38	510
	OQ&T	425	1250	1140	13	49	370
	OQ&T	650	758	655	22	63	230
	Normalized	870	870	1020	18	47	302
	Annealed	815	655	417	26	57	197

Source: *ASM Metals Reference Book*, 2nd ed. Materials Park, OH, American Society for Metals, 1983.

Notes: To convert from MPa to ksi, divide given values by 6.895.

Values tabulated for 25 mm round sections and of gage length 50 mm. The properties for quenched and tempered steel are from a single heat: OQ&T, oil-quenched and tempered; WQ&T, water-quenched and tempered.

TABLE B.5

Mechanical Properties of Some Annealed (An.) and Cold-Worked (CW) Wrought Stainless Steels

	Ultimate Strength S_u (MPa)		Yield Strength S_y (MPa)		Elongation in 50 mm (%)		Izod Impact J (N m)	
AISI Type	**An.**	**CW**	**An.**	**CW**	**An.**	**CW**	**An.**	**CW**
Austenitic								
302	586	758	241	517	60	35	149	122
303	620	758	241	552	50	22	115	47
304	586	758	241	517	60	55	149	122
347, 348	620	758	241	448	50	40	149	—
Martensitic								
410	517	724	276	586	35	17	122	102
414	793	896[a]	620[a]	862	20	15[a]	68	—
431	862	896[a]	655[a]	862[a]	20	15[a]	68	—
440 A, B, C	724	796[a]	414	620[a]	14	7[a]	3	3[a]
Ferritic								
430, 430F	517	572	296	434	27	20	—	—
446	572	586	365	483	23	20	3	—

Sources: *Metals Handbook*, ASM International, Materials Park, OH, 1985.
Note: To convert from MPa to ksi, divide given values by 6.895.
[a] Annealed and cold drawn.

TABLE B.6

Mechanical Properties of Some Aluminum Alloys

	Ultimate Strength S_u		Yield Strength S_y			
Alloy	**(MPa)**	**(ksi)**	**(MPa)**	**(ksi)**	**Elongation in 50 mm (%)**	**Brinell Hardness (H_B)**
Wrought:						
1100-H14	125	(18)	115	(17)	20	32
2011-T3	380	(55)	295	(43)	15	95
2014-T4	425	(62)	290	(42)	20	105
2024-T4	470	(68)	325	(47)	19	120
6061-T6	310	(45)	275	(40)	17	95
6063-T6	240	(35)	215	(31)	12	73
7075-T6	570	(83)	505	(73)	11	150
Cast						
201-T4[a]	365	(53)	215	(31)	20	—
295-T6[a]	250	(36)	165	(24)	5	—
355-T6[a]	240	(35)	175	(25)	3	—
-T6[b]	290	(42)	190	(27)	4	—
356-T6[a]	230	(33)	165	(24)	2	—
-T6[b]	265	(38)	185	(27)	5	—
520-T4[a]	330	(48)	180	(26)	16	—

Sources: *Materials Engineering, Materials Selector,* Penton Publication, Cleveland, OH, 1991.
[a] Sand casting.
[b] Permanent-mold casting.

TABLE B.7

Mechanical Properties of Some Copper Alloys

Alloy	UNS Number	Ultimate Strength S_u (MPa)	Yield Strength S_y (MPa)	Elongation in 50 mm (%)
Wrought				
Leaded				
Beryllium copper	C17300	469–1379	172–1227	43–3
Phos bronze	C54400	469–517	393–434	20–15
Aluminum				
Silicon bronze	C64200	517–703	241–469	32–22
Silicon bronze	C65500	400–745	152–414	60–13
Manganese bronze	C67500	448–579	207–414	33–19
Cast				
Leaded				
Red brass	C83600	255	117	30
Yellow brass	C85200	262	90	35
Manganese bronze	C86200	655	331	20
Bearing bronze	C93200	241	124	20
Aluminum bronze	C95400	586–724	241–372	18–8
Copper nickel	C96200	310	172	20

Source: *Machine Design, Materials Reference Issue*, Penton Publication, Cleveland, OH, 1991.
Note: To convert from MPa to ksi, divide given values by 6.895.

TABLE B.8

Selected Mechanical Properties of Some Common Plastics

Plastic	Ultimate Strength S_u (MPa)	Ultimate Strength S_u (ksi)	Elongation in 50 mm (%)	Izod Impact Strength J	Izod Impact Strength (ft·lb)
Acrylic	72	(10.5)	6	0.5	(0.4)
Cellulose acetate	14–18	(2–7)	—	1.4–9.5	(1–7)
Epoxy (glass filled)	69–138	(10–20)	4	2.7–41	(2–30)
Fluorocarbon	23	(3.4)	300	4.1	(3)
Nylon (6/6)	83	(12)	60	1.4	(1)
Phenolic (wood-flour filled)	48	(7)	0.4–0.8	0.4	(0.3)
Polycarbonate	62–72	(9–10.5)	110–125	16–22	(12–16)
Polyester (25% glass filled)	110–160	(16–23)	1–3	1.4–2.6	(1.0–1.9)
Polypropylene	34	(5)	10–20	0.7–3.0	(0.5–2.2)

Sources: *Materials Engineering, Materials Selector*, Penton Publication, Cleveland, OH, 1991.

TABLE B.9

Materials and Selected Members of Each Class

Class	Members	Abbreviation
Engineering alloys	Aluminum alloys	Al alloys
(the metals and alloys of engineering)	Copper alloys	Cu alloys
	Lead alloys	Lead alloys
	Magnesium alloys	Mg alloys
	Molybdenum alloys	Mo alloys
	Nickel alloys	Ni alloys
	Steels	Steels
	Tin alloys	Tin alloys
	Titanium alloys	Ti alloys
	Tungsten alloys	W alloys
	Zinc alloys	Zn alloys
Engineering polymers	Epoxies	EP
(the thermoplastics and thermosets of engineering)	Melamines	MEL
	Polycarbonate	PC
	Polyesters	PEST
	Polyethylene, high density	HDPE
	Polyethylene, low density	LDPE
	Polyformaldehyde	PF
	Polymethylmethacrylate	PMMA
	Polypropylene	PP
	Polytetrafluoroethylene	PTFE
	Polyvinyl chloride	PVC
Engineering ceramics	Alumina	Al_2O_3
(fine ceramics capable of load-bearing applications)	Diamond	C
	Sialons	Sialons
	Silicon carbide	SiC
	Silicon nitride	Si_3N_4
	Zirconia	ZrO_2
Engineering composites	Carbon-fiber-reinforced polymer	CFRP
(the composites of engineering practice)	Glass-fiber-reinforced polymer	GFRP
A distinction is drawn between the properties of a ply (uniply) and of a laminate (laminates).	Kevlar-fiber-reinforced polymer	KFRP
Porous ceramics	Brick	Brick
(traditional ceramics, cements, rocks, and minerals)	Cement	Cement
	Common rocks	Rocks
	Concrete	Concrete
	Porcelain	Pcln
	Pottery	Pot
Glasses	Borosilicate glass	B-glass
(ordinary silicate glass)	Soda glass	Na-glass
	Silica	SiO_2

(Continued)

TABLE B.9 (*Continued*)

Materials and Selected Members of Each Class

Class	Members	Abbreviation
Woods	Ash	Ash
Separate envelopes[a] describe properties: parallel to the grain, normal to it, and wood products	Balsa	Balsa
	Fir	Fir
	Oak	Oak
	Pine	Pine
	Wood products (ply, etc.)	Wood products
Elastomers	Natural rubber	Rubber
(natural and artificial rubbers)	Hard butyl rubber	Hard butyl
	Polyurethanes	PU
	Silicone rubber	Silicone
	Soft butyl rubber	Soft butyl
Polymer foams	Cork	Cork
(foamed polymers of engineering)	Polyester	PEST
	Polystyrene	PS
	Polyurethane	PU

Source: Ashby, M.F., *Material Selection in Mechanical Design*, 4th ed. Butterworth Heinemann, U.K., 2011.

[a] Data for members of a particular class of material *cluster* together and are enclosed by *envelopes* in Ashby's charts (see Figure 2.21).

TABLE B.10

Properties of Some Natural Rubbers

Name/Repeat Unit	Density (kg/m³)	Ultimate Strength ksi (MPa)	Maximum Elongation (%)	Modulus of Elasticity at 100% Elongation psi (MPa)	Minimum Service Temperature °C (°F)	Maximum Service Temperature °C (°F)	Abrasion Resistance	Tear Resistance	Oxidation Resistance
Natural polyisoprene (natural rubber, NR)	920–1037	3.5–4.6 (24–32)	500–760	480–850 (3.3–5.9)	–60 (–75)	120 (250)	Excellent	Excellent	Good
Styrene-butadiene (SBR, GRS)	940	1.8–3.0 (12–21)	450–500	300–1500 (2.1–10.3)	–60 (–75)	120 (250)	Excellent	Fair	Good
Acrylonitrile butadiene (nitrile, Buna A, NBR)	980	1.0–3.5 (7–24)	400–600	490 (3.4)	–50 (–60)	150 (300)	Excellent	Good	Fair–Good
Chloroprene (neoprene, CR)	1230–1250	0.5–3.5 (3.5–24)	100–800	100–3000 (0.7–20)	–50 (–60)	105 (225)	Excellent	Good	Very good
Polybutadiene (BR)	910	2.0–2.5 (14–17)	450	300–1500 (2.1–10.3)	–100 (–150)	90 (200)	Excellent	Good	Good
Polyurethane	1020–1250	0.8–8.0 (5.5–55)	250–800	25–5000 (0.17–34.5)	–55 (–65)	120 (250)	Excellent	Outstanding	Excellent
Polydimethylsiloxane (silicone)	1100–1600	1.5 (10)	100–800	—	–115 (–175)	315 (600)	Poor	Fair	Excellent

Source: Adapted from Materials Engineering, a Penton Publication.

Appendix C: Stress-Concentration Factors

In the following charts, the theoretical or geometric stress-concentration factors K_t for some common cases are presented as an aid to the reader in the solution of practical problems. These graphs were selected from the extensive charts found in [8,9] of Chapter 3. Equations to estimate most of these curves have been included to allow automatic generation of the K_t during calculations. Figures C.1 through C.6 are for flat bars and Figures C.7 through C.13 relate to cylindrical members. Note that the results pertain to an isotropic material and for use in Equation 3.42.

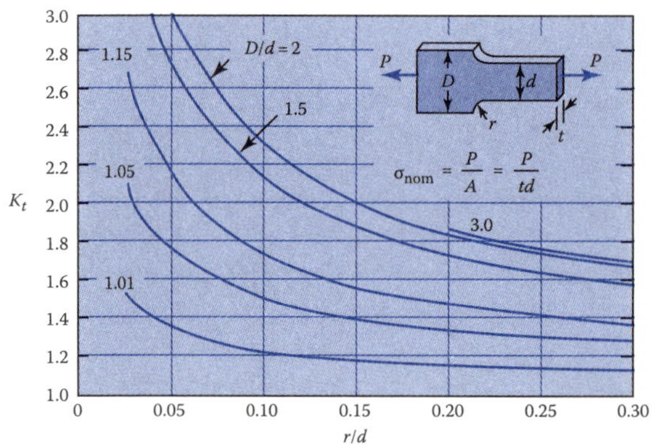

Approximate formula $K_t \approx B\left(\frac{r}{d}\right)^a$, where

D/d	B	a
2.00	1.100	−0.321
1.50	1.077	−0.296
1.15	1.014	−0.239
1.05	0.998	−0.138
1.01	0.977	−0.107

FIGURE C.1
Theoretical stress-concentration factor K_t for a filleted bar in axial tension [8 and 9, Chapter 3].

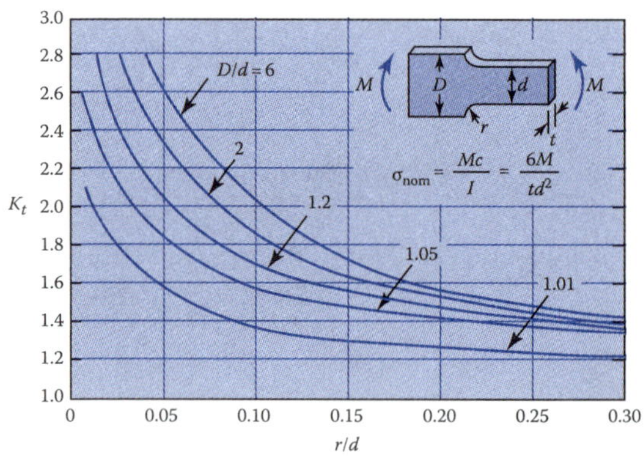

Approximate formula $K_t \approx B\left(\frac{r}{d}\right)^a$, where

D/d	B	a
6.00	0.896	−0.358
2.00	0.932	−0.303
1.20	0.996	−0.238
1.05	1.023	−0.192
1.01	0.967	−0.154

FIGURE C.2
Theoretical stress-concentration factor K_t for a filleted bar in bending [8 and 9, Chapter 3].

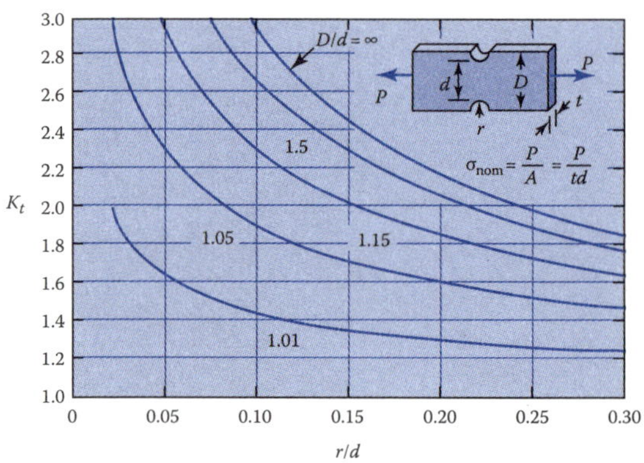

Approximate formula $K_t \approx B\left(\frac{r}{d}\right)^a$, where

D/d	B	a
∞	1.110	−0.417
1.50	1.133	−0.366
1.15	1.095	−0.325
1.05	1.091	−0.242
1.01	1.043	−0.142

FIGURE C.3
Theoretical stress-concentration factor K_t for a notched bar in axial tension [8 and 9, Chapter 3].

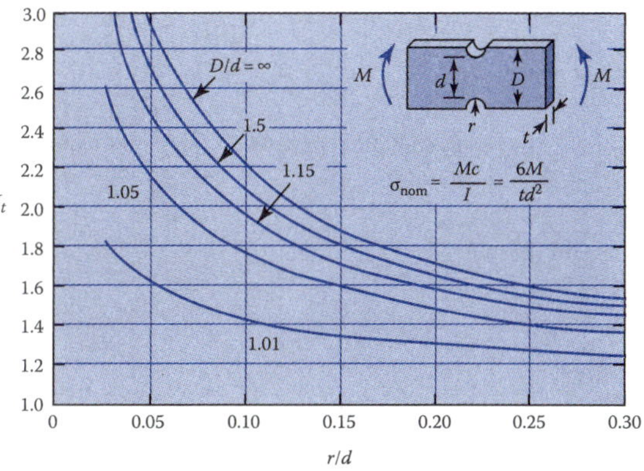

Approximate formula $K_t \approx B\left(\frac{r}{d}\right)^a$, where

D/d	***B***	***a***
∞	0.971	−0.357
1.50	0.983	−0.334
1.15	0.993	−0.303
1.05	1.025	−0.240
1.01	1.061	−0.134

FIGURE C.4
Theoretical stress-concentration factor K_t for a notched bar in bending [8 and 9, Chapter 3].

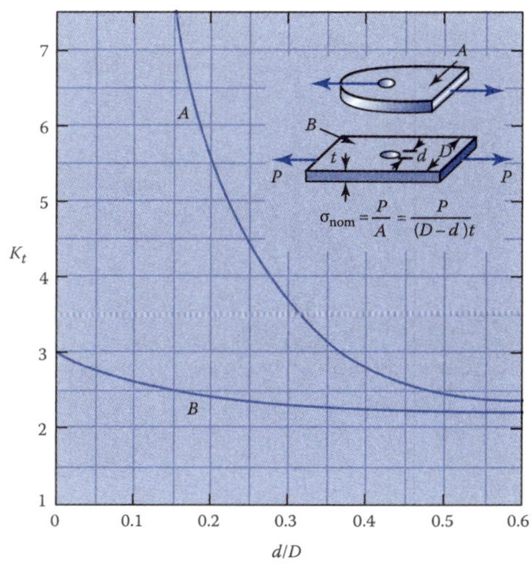

FIGURE C.5
Theoretical stress-concentration factor K_t: *A*, for a flat bar loaded in tension by a pin through the transverse hole; *B*, for a flat bar with a transverse hole in axial tension [8, Chapter 3].

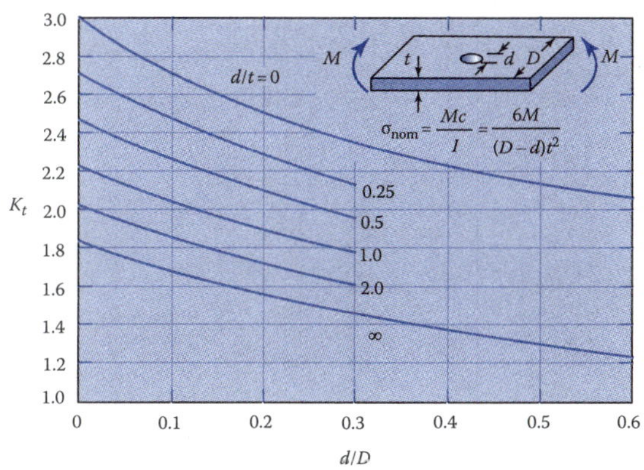

FIGURE C.6
Theoretical stress-concentration factor K_t for a flat bar with a transverse hole in bending [8, Chapter 3].

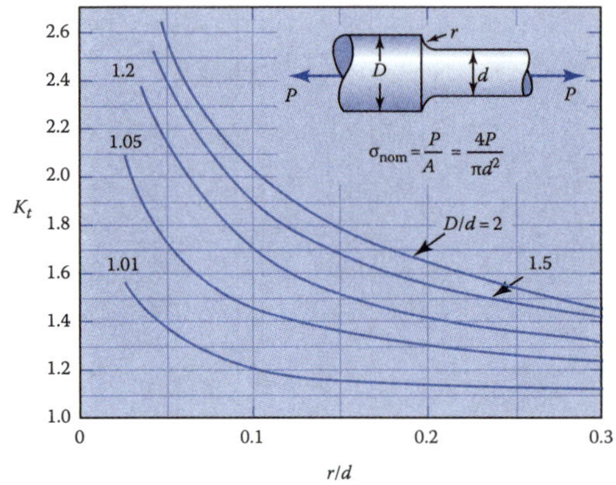

Approximate formula $K_t \approx B\left(\frac{r}{d}\right)^a$, where

D/d	B	a
2.00	1.015	−0.300
1.50	1.000	−0.282
1.20	0.963	−0.255
1.05	1.005	−0.171
1.01	0.984	−0.105

FIGURE C.7
Theoretical stress-concentration factor K_t for a shaft with a shoulder fillet in axial tension [8 and 9, Chapter 3].

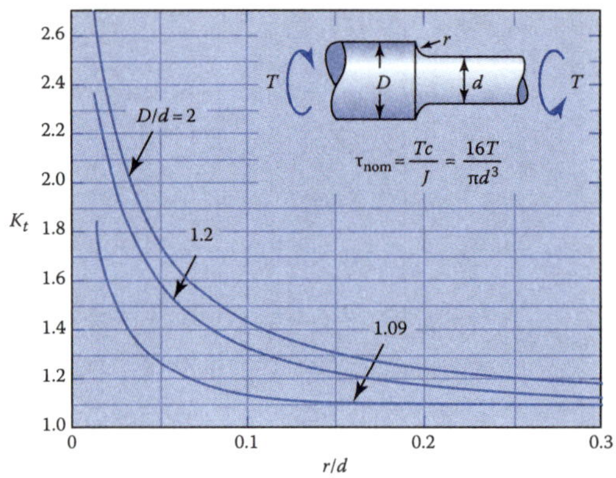

Approximate formula $K_t \approx B\left(\frac{r}{d}\right)^a$, where

D/d	B	a
2.00	0.863	−0.239
1.20	0.833	−0.216
1.09	0.903	−0.127

FIGURE C.8
Theoretical stress-concentration factor K_t for a shaft with a shoulder fillet in torsion [8 and 9, Chapter 3].

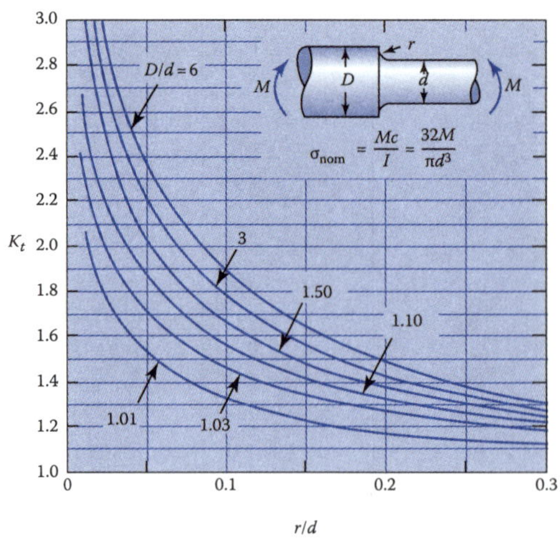

Approximate formula $K_t \approx B\left(\frac{r}{d}\right)^a$, where

D/d	B	a
6.00	0.879	−0.332
3.00	0.893	−0.309
1.50	0.938	−0.258
1.10	0.951	−0.238
1.03	0.981	−0.184
1.01	0.919	−0.170

FIGURE C.9
Theoretical stress-concentration factor K_t for a shaft with a shoulder fillet in bending [8 and 9, Chapter 3].

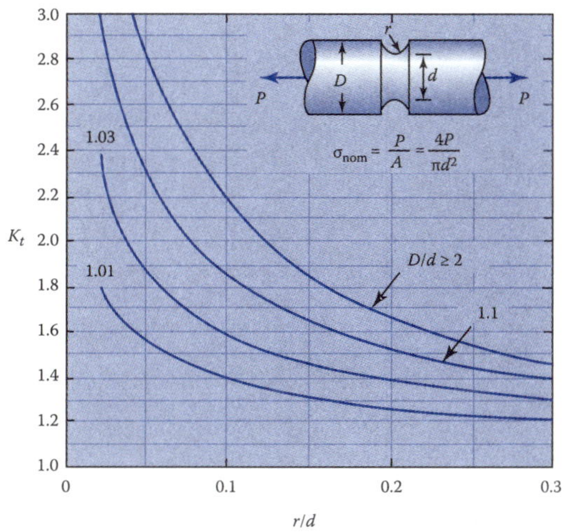

FIGURE C.10
Theoretical stress-concentration factor K_t for a grooved shaft in axial tension [8, Chapter 3].

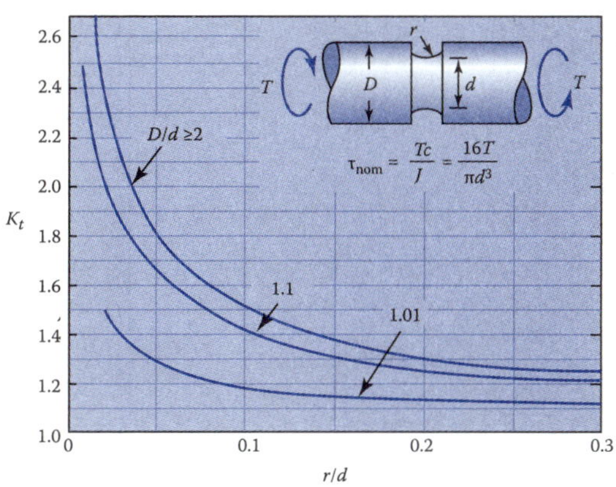

Approximate formula $K_t \approx B\left(\frac{r}{d}\right)^a$, where

D/d	*B*	*a*
2.00	0.890	−0.241
1.10	0.923	−0.197
1.01	0.972	−0.102

FIGURE C.11
Theoretical stress-concentration factor K_t for a grooved shaft in torsion [8 and 9, Chapter 3].

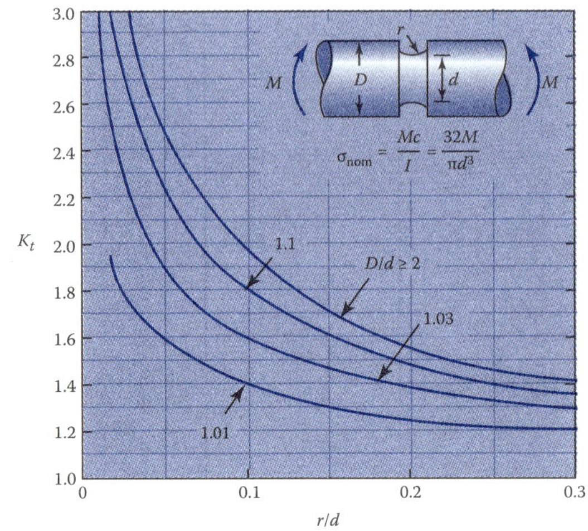

Approximate formula $K_t \approx B\left(\frac{r}{d}\right)^a$, where

D/d	*B*	*a*
2.00	0.936	−0.331
1.10	0.955	−0.283
1.03	0.990	−0.215
1.01	0.994	−0.152

FIGURE C.12
Theoretical stress-concentration factor K_t for a grooved shaft in bending [8 and 9, Chapter 3].

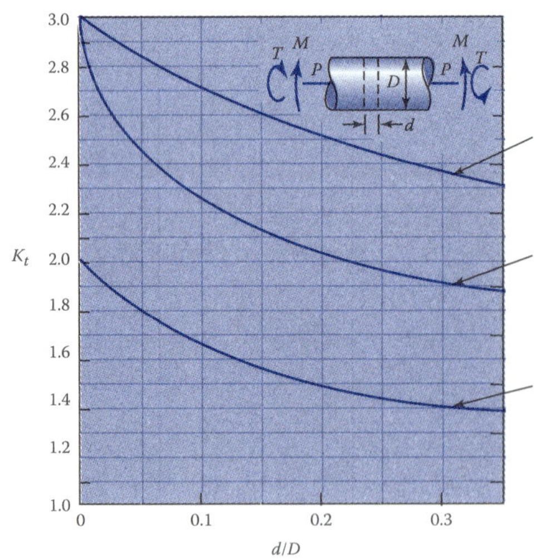

Axial load:

$$\sigma_{nom} = \frac{P}{A} = \frac{P}{(\pi D^2/4) - Dd}$$

Bending, in this plane:

$$\sigma_{nom} = \frac{Mc}{I} = \frac{M}{(\pi D^3/32) - (dD^2/6)}$$

Torsion:

$$\tau_{nom} = \frac{Tc}{J} = \frac{T}{(\pi D^3/16) - (dD^2/6)}$$

FIGURE C.13
Theoretical stress-concentration factor K_t for a shaft with a transverse hole in axial tension, bending, and torsion [8, Chapter 3].

Appendix D: Solution of the Stress Cubic Equation

Principal Stresses

Numerous methods for solving a cubic equation are in common use. The following is a practical approach for calculating the roots of stress cubic equation (see Section 3.15):

$$\sigma_i^3 - I_1\sigma_i^2 + I_2\sigma_i - I_3 = 0 \quad (i = 1, 2, 3) \tag{3.48}$$

where

$$\begin{aligned} I_1 &= \sigma_x + \sigma_y + \sigma_z \\ I_2 &= \sigma_x\sigma_y + \sigma_x\sigma_z + \sigma_y\sigma_z - \tau_{xy}^2 - \tau_{yz}^2 - \tau_{xz}^2 \\ I_3 &= \sigma_x\sigma_y\sigma_z + 2\tau_{xy}\tau_{yz}\tau_{xz} - \sigma_x\tau_{yz}^2 - \sigma_y\tau_{xz}^2 - \sigma_z\tau_{xy}^2 \end{aligned} \tag{3.49}$$

In accordance with the method, expressions that provide direct means for solving both 2D and 3D stress problems are (see [3] of Chapter 3)

$$\begin{aligned} \sigma_a &= 2S\left[\cos(\alpha/3)\right] + \frac{1}{3}I_1 \\ \sigma_b &= 2S\left\{\cos\left[(\alpha/3) + 120^\circ\right]\right\} + \frac{1}{3}I_1 \\ \sigma_c &= 2S\left\{\cos\left[(\alpha/3) + 240^\circ\right]\right\} + \frac{1}{3}I_1 \end{aligned} \tag{D.1}$$

Here, the constants are expressed by

$$
\begin{aligned}
S &= \left(\frac{1}{3}R\right)^{1/2} \\
\alpha &= \cos^{-1}\left(-\frac{Q}{2T}\right) \\
R &= \frac{1}{3}I_1^2 - I_2 \\
Q &= \frac{1}{3}I_1I_2 - I_3 - \frac{2}{27}I_1^3 \\
T &= \left(\frac{1}{27}R^3\right)^{1/2}
\end{aligned}
\tag{D.2}
$$

The invariants I_1, I_2, and I_3 are represented in terms of the given stress components by Equation 3.49. The principal stresses found from Equation D.1 are *redesignated* using numerical subscripts so that algebraically $\sigma_1 > \sigma_2 > \sigma_3$.

Direction Cosines

The values of the direction cosines of a principal stress are determined using Equations 3.46 and 3.44, as already discussed in Section 3.15. However, the following simpler method is preferred:

$$
\begin{bmatrix}
(\sigma_x - \sigma_i) & \tau_{xy} & \tau_{xz} \\
\tau_{xy} & (\sigma_y - \sigma_i) & \tau_{yz} \\
\tau_{xz} & \tau_{yz} & (\sigma_z - \sigma_i)
\end{bmatrix}
\begin{Bmatrix} l_i \\ m_i \\ n_i \end{Bmatrix} = 0
\tag{3.46}
$$

The *cofactors* of the determinant of the preceding matrix on the elements of the first row are given by

$$
\begin{aligned}
a_i &= \begin{vmatrix} (\sigma_y - \sigma_i) & \tau_{yz} \\ \tau_{yz} & (\sigma_z - \sigma_i) \end{vmatrix} \\
b_i &= -\begin{vmatrix} \tau_{xy} & \tau_{yz} \\ \tau_{xz} & (\sigma_z - \sigma_i) \end{vmatrix} \\
c_i &= \begin{vmatrix} \tau_{xy} & (\sigma_y - \sigma_i) \\ \tau_{xz} & \tau_{yz} \end{vmatrix}
\end{aligned}
\tag{D.3}
$$

Let us introduce the notation

$$k_i = \frac{1}{\left(a_i^2 + b_i^2 + c_i^2\right)^{1/2}} \tag{D.4}$$

The *direction cosines* are then expressed in the form

$$l_i = a_i k_i \qquad m_i = b_i k_i \qquad n_i = c_i k_i \tag{D.5}$$

Clearly, Equation D.5 gives $l_i^2 + m_i^2 + n_i^2 = 1$.

The foregoing procedures are well adopted to a quality scientific calculator or digital computer.

Appendix E: Introduction to MATLAB®

This appendix represents a brief introduction to MATLAB (*mat*rix *lab*oratory). MATLAB is a programming medium for plotting of functions, data analysis, matrix manipulation, and visualization, and numerical computation. Through the use of MATLAB, technical computing problems can be solved more quickly than with traditional programming tools. Application of MATLAB involves a wide variety of situations such as image processing, communications, design, test and measurement, and financial modeling and analysis. For more than a million engineers and scientists in industry and academia, MATLAB is the language of technical computing. Details on the subject are available from MathWorks at www.mathworks.com/products/matlab/. There are also numerous printed publications on MATLAB for engineers.

The MATLAB application is built around the MATLAB language, also known as *M*-code. The simplest way to execute *M*-code is to type it in at the prompt (≫) in the Command Window or in the Edit Window (recommended), which allows you to type all commends without executing them. The MATLAB teaching codes consist of 37 short text files containing MATLAB commands for performing basic linear algebra computations. Some familiar *M*-codes are *determ.m* (matrix determinant), *cofactor.m* (matrix of cofactors), *cramer.m* (solve the system of equations $Ax = C$), *inverse.m* (matrix inverse by Gauss–Jordon elimination), *eigen2.m* (characteristic polynomial, eigenvalues, eigen vectors), and *plot2d.m* (two-dimensional plot). In MathWorks, new users can select *Help* on the toolbar at the top of the MATLAB command window, then select *MATLAB Help* and *Getting Started*.

MATLAB (*professional or student version*) is now used in most universities. Among other features, it has the ability to treat engineering design problems, allowing the iteration process to proceed in an easy and rapid manner. Students are able to complete homework problems without use of a calculator. To learn the basics as well as to gain proficiency in MATLAB, many illustrations are available with the *Help* feature. The functions generally used in MATLAB can also be found at *Help > Function Browser > Mathematics > Elementary Math.* When editing a MATLAB program (as shown next), each step must be checked since any wrong result along the way will lead to an incorrect final solution.

A sample MATLAB® solution of Example 16.5

Code

%A flywheel of outer diameter *D*, inner diameter *d*, and weight *W*, rotates at speed *n*.

%**Find**: The average braking torque required to stop the wheel in one-third revolution.

%**Given**:

```
  D = 10; d = 2; W = 30; n = 3600; g = 386
%We have
  phi = 2*pi/3,   omega = n*(2*pi/60),   I = (W/(2*g))*((D/2)^2 + (d/2)^2)
% Equation 16.39 is thus,
  T = (I*omega^2)/(2*phi)
```

Solution:

>> phi = 2.0944   omega = 376.9911   *I* = 1.0104   *T* = 3.4281e + 04 >>

For further demonstration of this tool, a variety of text examples and case studies (listed in Table E.1) are re-solved using MATLAB on the CRC Website: http://www.crcpress.com/product/isbn/9781439887806. Note the differences in numerical precision between textbook and MATLAB answers. Recall from Section 1.5 that significant digits used in textbook calculations are based on a common engineering rule. MATLAB uses

TABLE E.1

MATLAB® Solution Contents

Example 1.1	Member forces in a pin-connected frame
Example 1.6	Strains in a plate
Example 2.3	Volume change of a cylinder under biaxial loads
Example 2.4	Material resilience on an axially loaded rod
Example 3.2	Design of a monoplane wing rod
Example 3.6	Maximum stresses in a simply supported beam
Example 4.11	Impact loading on a rod
Example 4.14	Impact loading on a shaft
Example 5.19	Stress in a strut of a clamping assembly
Example 5.21	Steel connecting rod buckling analysis
Example 6.2	Design of a wide plate with a central crack
Example 6.4	Failure of a rod under combined torsion and axial loading
Example 7.1	Endurance limit of a torsion bar
Example 7.7	Fatigue life of instrument panel with a crack
Example 8.2	Maximum contact pressure between a cylindrical rod and a beam
Example 8.3	Ball bearing load capacity
Example 9.3	Shaft design for repeated torsion and bending
Example 9.5	Determining critical speed of a hollow shaft
Example 10.1	Preliminary design of a boundary-lubricated journal bearing
Example 10.5	Median life of a deep-groove ball bearing
Example 11.2	Gear tooth and gear mesh parameters
Example 11.3	Contact ratio of meshing gear and pinion
Example 12.2	Electric motor geared to drive a machine
Example 12.5	Geometric quantities of a worm
Example 13.3	Design analysis of a V-belt drive
Example 13.6	Design of a disk brake
Example 14.2	Allowable load of a helical compression spring
Example 14.8	Design of a nine-leaf cantilever spring
Example 15.1	Quadruple-threaded power screw
Example 15.5	Preloaded bolt connecting the head and cylinder of a pressure vessel
Example 16.1	Designing a press fit
Example 16.9	Design of a parabolic pressure vessel
Example 17.2	Displacements in a frame
Case study 1.1	Bolt cutter loading analysis
Case study 3.1	Bolt cutter stress analysis
Case study 4.1	Bolt cutter deflection analysis
Case study 7.1	Camshaft fatigue design of intermittent-motion mechanism
Case study 8.1	Cam and follower stress analysis of an intermittent-motion mechanism
Case study 9.1	Motor-belt-drive shaft design for steady loading
Case study 18.1	Entire frame load analysis
Case study 18.2	Design analysis of Arm *CD*
Case study 18.7	Screw design for swivel hook
Case study 18.11	Brake design analysis

double-precision floating-point numbers in its calculations that result in 16 decimal places. Changing the display format will not change the accuracy of the solution. Values with decimal fractions are printed in the default short format that shows four digits after the decimal point. In order to quickly view MATLAB calculations for a problem solution, type in only the data and equations seen in bold, without the comment lines (these do not affect results).

Answers to Selected Problems

Chapter 1

1.3 $V_D = 4$ kips, $M_D = 4$ kip · ft,
$V_E = 2.5$ kips, $M_E = 6$ kip · ft.

1.5 (a) $R_{Cx} = 75$ kN←, $R_{Cy} = 50$ kN↑,
$R_{Bx} = 75$ kN←, $R_{By} = 10$ kN↑.
(b) $F_D = 85$ kN, $V_D = 30$ kN,
$M_D = 37.5$ kN · m.

1.7 (a) $T = 200$ N · m, $R_A = R_B = 2$ kN.

1.9 $R_B = 12.58$ kips.

1.10 (a) $T = 0.6$ kN · m.
(b) $R_A = 4.427$ kN.
$R_E = 8.854$ kN.

1.12 $R_y = 200$ N, $T = 30$ N m, $M_z = 96$ N · m.

1.15 $V_y = 64$ N, $M_z = 31.2$ N · m.

1.17 $T_d = 0.3$ kN · m.

1.23 $T = 5.348$ kN, $T_{DE} = 401.1$ N · m.

1.24 $e = 90\%$.

1.28 (a) $\varepsilon_{c,\max} = 2000\mu$.
(b) $\varepsilon_r = 1000\mu$.

1.31 $\Delta L_{BD} = 0.232$ mm.

1.32 $\varepsilon_x = \varepsilon_y = -363\mu$.
$\gamma_{xy} = 1651\mu$.

Chapter 2

2.2 $E = 30 \times 10^6$ psi.

2.5 (a) $E = 53$ GPa.
(b) $\nu = 0.25$.
(c) $G = 21.2$ GPa.

2.7 (a) $\nu = 0.25$.
(b) $E = 8.335 \times 10^6$ psi.
(c) $a' = 2.9885$ in.
(d) $G = 3.334 \times 10^6$ psi.

2.8 $\varepsilon_x = 1327\mu$.

2.12 $L' = 99.96$ mm, a′ = 49.98 mm,
b′ = 9.996 mm.

2.15 $n = 3.16$.

2.20 $d = 0.667$ in.

2.25 $S_u = 78$ ksi, $S_y = 51.9$ ksi.

Chapter 3

3.2 $a_{\text{min}} = 73$ mm.

3.4 (a) $\sigma_{BD} = -1.018$ MPa.

(b) $\tau_A = 9.697$ MPa.

3.5 $\alpha = 54.7°$.

3.10 (a) $d = 26.1$ mm.

(b) $d = 18.43$ mm.

3.13 $b = \dfrac{d}{\sqrt{3}}, h = d\sqrt{\dfrac{2}{3}}$.

3.16 $b = 56.5$ mm.

3.18 $h = h_1\left(\dfrac{x}{L}\right)^{3/2}$.

3.21 $t = 0.399$ in.

3.24 $M = 38.45$ kip · in.

3.26 (a) $\sigma_x = -37.1$ kPa, $\sigma_y = -2.9$ kPa,

$\tau_{xy} = 47$ kPa.

(b) $\tau_{\text{max}} = 50$ kPa.

3.29 (a) $\sigma_x = 25$ ksi, $\sigma_y = -5$ ksi, $\tau_{xy} = -8.66$ ksi.

(b) $\sigma_1 = 27.32$ ksi, $\sigma_2 = -7.32$ ksi, $\theta_p' = 15°$.

3.31 $\sigma_{x'} = 140.3$ MPa, $\sigma_{y'} = 1.07$ MPa,

$\tau_{x'y'} = 12.28$ MPa.

3.34 Point A: (a) $\sigma_1 = 864$ psi, $\sigma_2 = 234$ psi.

(b) $\tau_{\text{max}} = 315$ psi.

3.36 $\sigma_{x'} = 19.6$ ksi, $\tau_{x'y'} = 2.87$ ksi.

3.43 (a) $\gamma_{\text{max}} = 566\mu$.

(b) $\Delta L_{AC} = 1.42 \times 10^{-4}$ in.

3.45 $p_{\text{all}} = 1.281$ MPa.

3.51 $p_{\text{all}} = 29.3$ kN.

3.55 (a) $\sigma_1 = 26.49$ ksi, $\sigma_2 = 9.512$ ksi, $\sigma_3 = -3$ ksi.

(b) $\tau_{\text{max}} = 14.75$ ksi.

3.61 $\sigma = 24.98$ MPa

$\tau = 21.27$ MPa.

Chapter 4

4.1 (a) $d = 8.74$ mm.

(b) $k = 2000$ kN/m.

4.4 $R_A = 7.041$ kips, $R_B = 0.959$ kips.

4.6 (a) $P = 178.2$ kN.
(b) $\delta_a = 0.0653$ mm.

4.11 (a) $\phi_D = 6.82°$.
(b) $\tau_{AB} = 41.92$ MPa.

4.16 $h = 197$ mm.

4.20 $\upsilon_1 = \dfrac{M_o x}{6EIL}\left(-6aL + 3a^2 + 2L^2 + x^2\right)$.

4.24 $R = \dfrac{5P\left(E_2 I_2\right)}{2\left(E_1 I_1 + E_2 I_2\right)}$.

4.29 $R_A = R_B = \dfrac{P}{2},\ M_A = -M_B = \dfrac{PL}{8}$

$\upsilon = -\dfrac{Px^2}{48EI}(3L - 4x)$.

4.33 $R_A = \dfrac{3}{4}P\downarrow,\ R_B = \dfrac{7}{4}P\uparrow,\ M_A = \dfrac{1}{2}Pa$

4.36 (a) $\delta_{\max} = 7.37$ mm.
(b) $\sigma_{\max} = 198$ MPa.

4.38 $d = 1.943$ in.

4.40 (a) $\upsilon_{\max} = 7$ mm.
(b) $\sigma_{\max} = 53.1$ MPa.

4.42 (a) $\phi_{\max} = 4.24°$.
(b) $\tau_{\max} = 243.3$ MPa.

4.44 $M_{\max} = 46.875$ lb·in.

4.46 (b) $\sigma_{\max} = 0.75 p_o \left(\dfrac{b}{t}\right)^2$.

(c) $w_{\max} = 0.071$ in.,
$\sigma_{\max} = 9.375$ ksi.

Chapter 5

5.1 $d/a = \sqrt{8/\pi}$

5.8 $U_s = \dfrac{1}{20}\dfrac{w^2 L^3}{AG}$.

5.11 $\upsilon_A = 2Pa(a + L)\left[\dfrac{a}{6EI} + \dfrac{3}{5AGL}\right]$.

5.19 $\delta_B = \dfrac{P}{12EI}\left(4L^3 + 6\pi RL^3 + 24R^2 L + 3\pi R^3\right)$

5.21 $R_A = \dfrac{2}{3}\dfrac{M_o}{L}\uparrow,\ R_B = 2\dfrac{M_o}{L}\downarrow,$

$R_c = \dfrac{4}{3}\dfrac{M_o}{L}\uparrow$.

5.23 $\delta_A = 60\dfrac{wa^4}{EI} \rightarrow$.

5.25 $\delta_B = \dfrac{PR^3}{2EI}$.

5.27 (b) $\phi_B = \dfrac{1}{GJ}\left(\dfrac{T_oL}{2} + PaL\right)$.

5.30 $(\delta_C)_v = 2.828\dfrac{PL}{AE} \downarrow$,

$(\delta_C)_h = 3.828\dfrac{PL}{AE} \rightarrow$.

5.34 $F = \dfrac{4P}{\pi} \uparrow$.

5.36 $\delta_C = \dfrac{PL}{3E}\left(\dfrac{11}{A} + \dfrac{8L^2}{I}\right) \downarrow$.

5.41 $\upsilon = \dfrac{w_o}{EI}\left(\dfrac{L}{\pi}\right)^4 \sin\dfrac{\pi x}{L}$.

5.43 (a) $\upsilon = \dfrac{Px^2}{6EI}(3L - x)$.

(b) $\upsilon_{\max} = \dfrac{PL^3}{3EI}$, $\theta_{\max} = \dfrac{PL^2}{2EI}$.

5.44 $\upsilon_A = \dfrac{Pc^2(L-c)^2}{4EIL}$.

5.45 $d = 29$ mm.

5.46 $d = 25.7$ mm.

5.49 $Q = 295$ N.

5.52 $P_{\text{all}} = 4.5$ kN.

5.54 $F_{\text{all}} = 129.2$ kN.

5.55 $P_{\text{all}} = 13.83$ kips.

5.72 $d = 109$ mm.

5.73 $P_{\text{all}} = 637.5$ kips.

5.76 $a = 48$ mm.

Chapter 6

6.1 $\sigma = 231.6$ MPa.

6.3 P = 256 kN.

$\sigma = 97.5$ MPa.

6.8 $M = 2.76$ kN·m.

6.10 $P = 490$ lb.

6.12 (a) $n = 2.86$.

(b) $n = 2.61$.

6.14 $F = 1.163$ kN.
6.19 (a) $t = 0.208$ in.
(b) $t = 0.18$ in.
6.21 (a) $n = 1.94$.
(b) $n = 1.82$.
6.26 (a) $p = 7.44$ MPa.
(b) $p = 10.02$ MPa.
6.29 $T = 18.81$ kips·in.
6.32 $\tau = 111.1$ MPa.
6.38 $R \approx 99.94\%$.
6.41 (a) $\sigma = 7.645$ ksi.
(b) $R \approx 76\%$.
6.44 10%.

Chapter 7

7.2 (a) $D = 48.4$ mm.
(b) $D = 45.2$ mm.
7.7 $S_e = 38.5$ MPa.
7.15 (a) $n = 4.64$.
(b) $n = 1.57$.
7.19 (a) T = 624.9 N·m.
7.22 $n = 1.98$.
7.26 $t = 0.736$ in.
7.28 $h = 1.45$ mm.
7.30 $h = 0.021$ in.
7.32 $P_o = 30.23$ N.
7.34 $n = 1.63$.
7.38 $n = 1.4$.
7.40 $n = 1.81$.

Chapter 8

8.3 $V_s = 284 \times 10^{-6}$ in.^{3}, $V_f = 101 \times 10^{-6}$ in.^{3}
8.7 (a) $a = 2.2162$ mm.
(b) $\delta = 0.0677$ mm.
(c) $\tau_{max} = 68.48$ MPa.
8.9 (a) $a = 0.135$ mm.
(b) $p_o = 943.1$ MPa.
8.12 (a) $p_o = 97.4$ ksi.
(b) $\delta = 1.84 \times 10^{-4}$ in.
8.18 $p_o = 414.9$ MPa.

8.21 (a) $a = 0.2033$ mm.
(b) $p_o = 939.4$ MPa.
(c) $\tau_{yz,\,max} = 282$ MPa.

8.22 $p_o = 1083$ MPa.

Chapter 9

9.3 (a) $D_{AC} = 14.22$ mm, $D_{BC} = 20.52$ mm.
(b) $\phi_{AB} = 7.91°$.

9.4 $W_a/W_s = 0.598$.

9.7 (a) $D = 42.71$ mm.
(b) $D = 42.3$ mm.

9.13 $D = 63.5$ mm.

9.15 $n = 2.07$.

9.19 $n = 1.93$.

9.24 $D = 2.24$ in.

9.26 $n_{cr} = 594$ rpm.

9.28 $n_{cr} = 967$ rpm.

9.34 $n = 1.205$.

9.37 (a) $n = 1.99$.
(b) $n = 6.75$.
(c) $n = 17.2$.

Chapter 10

10.3 $n = 25.9$ rpm.

10.5 (a) $T_f = 42.64$ lb · in.
(b) hp = 16.24.
(c) $f = 0.057$.

10.7 $\eta = 25.95$ mPa · s.

10.10 $W = 563$ lb.

10.12 (a) $f = 0.02$.
(b) hp = 0.714.

10.16 (a) $h_0 = 0.008$ mm.
(b) kW = 0.017.

10.17 (a) $h_0 = 0.013$ mm.
(b) $p_{max} = 4.808$ MPa.

10.19 (a) $\eta = 52.8$ mPa · s.
(b) kW = 0.377.

10.20 $t = 85.2$°C.

10.22 $L_{10} = 344.8$ h.

10.24 $L_{10} = 119.5$ h.

10.30 18.8%.

10.32 $L_{10} = 949.3$ h (for 03 series).

10.36 $L_5 = 267$ h.

Chapter 11

11.1 $h = 0.563$ in., $h_k = 0.5$ in.,
$r_b = 3.759$ in., $r_o = 4.25$ in.

11.2 $N_1 = 60$, $N_2 = 180$.

11.10 $N_g = 88$, $d_p = 88$ mm, $c = 220$ mm.

11.16 (a) $F_{t1} = 210$ lb, $F_{r1} = 76.43$ lb.
(b) $R_C = 223.5$ lb, $T_C = 630$ lb · in.

11.18 (a) $F_{t1} = 8.843$ kN, $F_{r1} = 3.219$ kN.
(b) $R_C = 9.411$ kN, $T_C = 1.326$ kN · m.

11.20 (a) $F_{t2} = 280$ lb, $F_{r2} = 130.6$ lb,
$F_{t3} = 490$ lb, $F_{r3} = 228.5$ lb.
(b) $R_C = 540.7$ lb,
$T_C = 1960$ lb · in.

11.22 (a) $F_b = 692.2$ lb.
(b) $F_w = 340.2$ lb.
(c) $F_t = 114.8$ lb.

11.23 (a) $F_b = 2.75$ kN.
(b) $F_w = 1.81$ kN.
(c) $F_t = 617.5$ kN.

11.27 (a) $F_b = 6.76$ kN.
(b) $F_w = 2.35$ kN.

11.29 No.

11.33 hp = 6.95.

11.36 hp = 19.57.

11.39 hp = 36.33.

Chapter 12

12.1 (a) $p_n = 0.524$ in., $p = 0.605$ in.,
$p_a = 1.048$ in.
(b) $P = 5.196$, $\phi = 28.3°$.
(c) $d_p = 3.849$ in., $d_g = 7.698$ in.

12.5 kW = 75.58.

12.7 $c = 369.1$ mm.

12.11 (a) $F_{t1} = F_{t2} = F_{t3} = 263.1$ lb.
(b) $T_1 = 840$ lb · in., $T_2 = 0$, $T_3 = 1680$ lb · in.

12.15 $n = 1.21$.

12.17 (a) hp = 17.63.
(b) hp = 31.1.

12.18 (a) $d_p = 2.5$ in., $d_g = 5.25$ in.
(b) $\alpha_p = 25.46°$, $\alpha_g = 64.54°$.
(c) $b = 0.969$ in.
(d) $c = 0.026$ in.

12.20 The gears are safe.

12.24 kW = 29.25.

12.26 $F_t = 10.08$ kips.

12.31 (a) $\lambda = 10.39°$.
(b) $F_{wt} = F_{ga} = 420.2$ lb.
(c) $(\text{hp})_m = 8.74$.

12.34 $(\text{hp})_d = 1.392$. No.

Chapter 13

13.1 (a) $F_1 = 296.6$ lb, $F_2 = 128.5$ lb.
(b) $L = 151.8$ in.

13.4 $T_A = 13$ N·m, $T_B = 101$ N·m.

13.8 kW = 34.7.

13.10 $F_{\text{max}} = 1.143$ kN.

13.13 $c = 13.128$ in.

13.17 (a) $p_{\text{max}} = 254.6$ kPa, $T = 180$ N·m.
(b) $p_{\text{max}} = 191$ kPa, $T = 183.8$ N·m.

13.19 (a) $D = 17.64$ in.
(b) $F_a = 1.833$ kips.

13.21 (a) $F_a = 2.187$ kips.
(b) $p_{\text{avg}} = 25.78$ kips.

13.24 $w = 1.451$ in.

13.27 $T = 602$ N·m.

13.29 $F_1 = 14$ kN, $F_2 = 3.983$ kN,
kW = 31.46.

13.31 $F_1 = 3{,}085$ N, $F_2 = 538.6$ N.

13.34 hp = 12.76.

13.36 $F_a = 366.04$ N. No.

13.39 (a) $F_a = 1.542$ kN. No.
(b) $R_A = 2.632$ kN.

13.46 $b = 1.414r$.

Chapter 14

14.1 (a) $T = 35.48$ N·m.
(b) $\tau = 353$ MPa.

14.4 $N_a = 7.49$.

14.8 (a) $h_s = 39$ mm.
(b) $P_{\max} = 320.4$ N.

14.10 (a) $h_f = 45.53$ mm.
(b) The spring is safe.

14.16 (a) $d = 14.94$ mm.
(b) $h_f = 274.6$ mm.
(c) The spring is safe.
$f_n = 4370$ cpm.

14.21 $P_{\min} = 72.4$ lb,
$P_{\max} = 127.6$ lb.

14.22 (a) $n = 2.49$.
(b) $N_a = 17.3$.

14.23 (a) $d = 0.103$ in.
(c) $f_n = 9270$ cpm.
(d) The spring is safe.

14.25 (a) $d = 5.41$ mm.
(b) $N_a = 9.89$.

14.29 $n = 1.30$.

14.35 (a) $M = 6.016$ lb · in.
(b) $\theta = 64.6°$.

Chapter 15

15.4 kW = 1.23.

15.6 (a) $n = 48$ rpm.
(b) $(\text{hp})_{\text{req}} = 12.1$.

15.10 $T_o = 145.3$ N · m.

15.13 (a) $\sigma = 10.8$ MPa.
(b) $L_{ne} = 20.8$ mm.
(c) Nut: $\tau = 13.8$ MPa,
screw: $\tau = 16.4$ MPa.

15.14 $P = 54.67$ kN.

15.16 (a) $P_{\max} = 37.27$ kN, $P_{\min} = 21.53$ kN.
(b) $T = 75$ N · m.

15.20 $n = 2.77$.

15.26 (a) $P_b = 118.5$ kN.
(b) $T = 312.6$ N · m.

15.28 $n = 2.07$ (with preload).
$n = 1.40$ (without preload).

15.30 (a) $n = 4.5$ (with preload),
$n = 2.19$ (no preload).
(b) $n_s = 5.37$.

15.32 $e = 64.3\%$.

15.35 $P_{all} = 4.57$ kips.

15.40 $V_B = 2.15$ kN, $\tau_B = 6.844$ MPa,
$\sigma_B = 7.167$ MPa.

15.41 $d = 54.3$ mm.

15.43 $P = 23.76$ kN.

15.49 $h = 0.19$ in.

15.51 $L = 199.6$ mm.

15.54 $h = 0.22$ in.

Chapter 16

16.5 (a) $p = 30.71$ MPa.
(b) $2c = 220$ mm.

16.7 Steel: $\sigma_{\theta,\max} = 62.2$ MPa,
bronze: $\sigma_{\theta,\max} = -116.8$ MPa.

16.9 $\Delta d_s = 0.356\lambda$.

16.12 (a) $p = 5.167$ MPa.
(b) $\sigma_\theta = 5.596$ MPa.

16.13 (a) $\sigma_{\theta.\max} = 41.11$ MPa.
(b) $n = 3539$ rpm.

16.14 (a) $p = 18$ MPa.
(b) $F = 244.3$ kN.
(c) $\tau = 14.66$ kN.

16.20 $P = 1.922$ kN.

16.22 $b = 156$ mm.

16.25 (a) $P = 84.58$ kN.
(b) $(\sigma_\theta)_B = -50$ MPa.

16.33 $t = 0.108$ in.

16.34 $a = 238.1$ mm.

16.35 $\sigma_\theta = 2.93$ MPa.

16.38 (a) $t = 14.67$ mm.
(b) $t = 12.67$ mm.
(c) Top end, $t = 158$ mm;
bottom end, $t = 235$ mm.

Chapter 17

17.11 (c) $\begin{Bmatrix} u_2 \\ \upsilon_2 \end{Bmatrix} = \begin{Bmatrix} 1.0 \\ -3.4 \end{Bmatrix}$ mm.

(d) $\begin{Bmatrix} F_{1x} \\ F_{1y} \\ F_{3x} \\ F_{3y} \end{Bmatrix} = \begin{Bmatrix} 7,632 \\ 10,176 \\ -7,500 \\ 0 \end{Bmatrix}$ N.

(e) $F_{12} = -12.72$ kN (T),
$F_{23} = -7.5$ kN (C).

17.12 (c) $\begin{Bmatrix} u_1 \\ \upsilon_1 \end{Bmatrix} = \begin{Bmatrix} 2.30 \\ -8.81 \end{Bmatrix} (10^{-3})$ in.

(d) $\begin{Bmatrix} R_{2y} \\ R_{3x} \\ R_{3y} \\ R_{4x} \end{Bmatrix} = \begin{Bmatrix} 3964.5 \\ 1037 \\ 1037 \\ -1035 \end{Bmatrix}$ lb.

(e) $F_{12} = 3964.5$ lb (T),
$F_{13} = 1464.5$ lb (T),
$F_{14} = -1035$ lb (C).

17.13 (c) $\begin{Bmatrix} u_2 \\ \upsilon_2 \end{Bmatrix} = \begin{Bmatrix} 18 \\ -60.4 \end{Bmatrix}$ mm.

(d) $\begin{Bmatrix} R_{1x} \\ R_{1y} \\ R_{3x} \end{Bmatrix} = \begin{Bmatrix} 45,024 \\ 60,032 \\ -45 \end{Bmatrix}$ kN.

(e) $F_{12} = -375.2$ kN (C),
$F_{23} = 180$ kN (T).

17.15 (c) $\upsilon_1 = 8.87$ mm.

(d) $\begin{Bmatrix} F_{2x} \\ F_{2y} \\ F_{3x} \\ F_{3y} \end{Bmatrix} = \begin{Bmatrix} 99.59 \\ 132.79 \\ 0 \\ -232.84 \end{Bmatrix}$ kN.

(e) $F_{12} = 166$ kN (T),
$F_{13} = -232.8$ kN (C).

17.19 (b) $\begin{Bmatrix} \upsilon_2 \\ \theta_2 \\ \theta_3 \end{Bmatrix} = \begin{Bmatrix} -0.7292 \text{ in.} \\ -0.00521 \text{ rad} \\ 0.02083 \text{ rad} \end{Bmatrix}$

(c) $\begin{Bmatrix} F_{1y} \\ M_1 \\ F_{3y} \end{Bmatrix} = \begin{Bmatrix} 6.876 \text{ kips} \\ 225 \text{ kip} \cdot \text{in} \\ 3.125 \text{ kips} \end{Bmatrix}$

Chapter 18

18.1 (a) $F_{-CF} = -42$ kN (C).
$F_{AE} = 45.35$ kN (T).
$F_{BG} = -20.08$ kN (C).

18.3 $\varepsilon_1 = 1094.6\ \mu$, $\varepsilon_2 = 294.6\ \mu$, $\gamma_{max} = 1389\ \mu$.

18.5 $\varepsilon_x = 1104\ \mu$, $\varepsilon_y = -144\ \mu$, $\gamma_{xy} = -610\ \mu$.

18.7 $L_m = 1.814$ m.

18.10 (a) $b = 16.27$ mm.
(b) No.

References

Chapter 1

1. American Society of Mechanical Engineers. *Code of Ethics for Engineers*. New York: ASME, 2009.
2. Ullman, D.G. *The Mechanical Design Process*, 4th ed. New York: McGraw-Hill, 2009.
3. Nevins, J.L. and D.E. Whitney, eds. *Concurrent Design Products and Processes*. New York: McGraw-Hill, 1989.
4. Budynas, R. and K. Nisbett. *Shigley's Mechanical Engineering Design*, 9th ed. New York: McGraw-Hill, 2010.
5. Juvinall, R.C. and K.M. Marshek. *Fundamentals of Machine Component Design*, 5th ed. Hoboken, NJ: Wiley, 2011.
6. Norton, R.L. *Machine Design: An Integrated Approach*, 5th ed. Upper Saddle River, NJ: Prentice Hall, 2013.
7. Zeid, I. *CAD/CAM Theory and Practice*, 2nd ed. New York: McGraw-Hill, 2010.
8. Ugural, A.C. and S.K. Fenster. *Advanced Mechanics of Materials and Applied Elasticity*, 5th ed. Upper Saddle River, NJ: Prentice Hall, 2012.
9. Ugural, A.C. *Mechanics of Materials*. Hoboken, NJ: Wiley, 2008.
10. Dimarogonas, A.D. *Machine Design: A CAD Approach*. New York: Wiley, 2001.
11. Ugural, A.C. *Living Better: A Guide to Health, Happiness, and Managing Stress*. New York: Eloquent Books, 2009. www.eloquentbooks.com or www.amazon.com.
12. Vidosic, J.P. *Machine Design Projects*. New York: Ronald Press, 1957.
13. *ASTM International Standard for Metric Practice*. Publication E 380.86. West Conshohocken, PA: American Society for Testing and Materials, 2014.
14. *Minimum Design Loads for Buildings and Other Structures, ASCE-7*. Reston, VA: American Society of Civil Engineers, 2010.

Chapter 2

1. Avallone, E.A., T. Baumeister III, and A. Sadegh. eds. *Mark's Standard Handbook for Mechanical Engineers*, 11th ed. New York: McGraw-Hill, 2006.
2. *Annual Book of ASTM*. Philadelphia, PA: American Society for Testing and Materials, 2014.
3. Ashby, M.J. *Material Selection in Mechanical Design*, 4th ed. Oxford, U.K.: Butterworth Heinemann, 2011.
4. Lewis, G. *Selection of Engineering Materials*. Upper Saddle River, NJ: Prentice Hall. 1990.
5. Marin, J. *Mechanical Behavior of Engineering Materials.* Upper Saddle River, NJ: Prentice Hall, 1962.
6. Lakes, R.S. Advances in negative Poisson's ratio materials. *Advanced Materials*, 5(4) (1993), 293–296.
7. Mechanical Testing. *Metals Handbook*. 9th ed., Vol. 8. Metals Park, OH: The American Society for Metals (ASM), 1989.
8. *Mechanical Engineering*. 120(8) (August 1995), 54–58.
9. *Manual of Steel Construction*. 14th ed. Chicago, IL: The American Institute of Steel Construction (AISC), 2010.
10. Hyer, M.W. *Stress Analysis of Fiber-Reinforced Composite Materials*, New York: McGraw-Hill, 1998.

Chapter 3

1. Timoshenko, S.P. and J.N. Goodier. *Theory of Elasticity*, 3rd ed. New York: McGraw-Hill, 1970.
2. Ugural, A.C. *Mechanics of Materials*. Hoboken, NJ: Wiley, 2008.
3. Ugural, A.C. and S.K. Fenster. *Advanced Mechanics of Materials and Applied Elasticity*, 5th ed. Upper Saddle River, NJ: Prentice Hall, 2012.
4. Young, W.C., R.C. Budynas, and A.H. Sadegh. *Roark's Formulas for Stress and Strain*, 8th ed. New York: McGraw-Hill, 2011.
5. Ugural, A.C. *Stresses in Beams, Plates and Shells*, 3rd ed. Boca Raton, FL: CRC Press, 2009.
6. Chen, F.Y. *Mohr's Circle and Its Application in Engineering Design*. New York: ASME Paper 76-DET-99. 1976.
7. Dally, J.W. and W.F. Riley. *Experimental Stress Analysis*, 4th ed. Knoxville, TN: Colloge House Enterprises, 2005.
8. Pilkey, D.F. and W.D. Pilkey. *Peterson's Stress Concentration Factors*, 3rd ed. Hoboken, NJ: Willey, 2008.
9. Norton, R.L. *Machine Design: An Integrated Approach*, 5th ed. Upper Saddle River, NJ: Prentice Hall, 2013.

Chapter 4

1. Ugural, A.C. and S.K. Fenster. *Advanced Mechanics of Materials and Applied Elasticity*, 5th ed. Upper Saddle River, NJ: Prentice Hall, 2012.
2. Young, W.C., R.C. Budynas, and A. Sadegh. *Roark's Formulas for Stress and Strain*, 8th ed. New York: McGraw-Hill, 2011.
3. Ugural, A.C. *Mechanics of Materials*. Hoboken, NJ: Wiley, 2008.
4. West, H.H and L.F. Geschwindner, *Fundamentals of Structural Analysis*, 2nd ed. New York: Wiley. 2002.
5. Timoshenko, S.P. and J.N. Goodier. *Theory of Elasticity*, 3rd ed. New York: McGraw-Hill, 1970.
6. Piersol, A.G. and T.L. Paez. *Harris' Shock and Vibration Handbook*, 6th ed. New York: McGraw-Hill, 2009.
7. Burr, A.H. and J.B. Cheatham. *Mechanical Analysis and Design*. 2nd ed. Upper Saddle River, NJ: Prentice Hall, 1995.
8. Kolsky, H. *Stress Waves in Solids*, 2nd ed. New York: Dover, 2012.
9. Ugural, A.C. *Stresses in Beams, Plates and Shells*. 3rd ed. Boca Raton, FL: CRC Press, 2009.

Chapter 5

1. Langhaar, H.L. *Energy Methods in Applied Mechanics*. Malabar, FL: Krieger, 1989.
2. Oden, J.T. and E.A. Ripperger. *Mechanics of Elastic Structures*. 2nd ed. New York: McGraw-Hill, 1981.
3. Sokolnikoff, I.S. *Mathematical Theory of Elasticity*. 2nd ed. Malabar, FL: Krieger, 1986.
4. Ugural, A.C. and S.K. Fenster. *Advanced Mechanics of Materials and Applied Elasticity*. 5th ed. Upper Saddle River, NJ: Prentice-Hall, 2012.
5. Faupel, J.H. and F.E. Fisher. *Engineering Design*, 2nd ed. New York: Wiley, 1981.
6. Burr, A.H. and J.B. Cheatham. *Mechanical Analysis and Design*, 2nd ed. Upper Saddle River, NJ: Prentice Hall, 1995.
7. Timoshenko, S.P. and J.M. Gere. *Theory of Elastic Stability*, 2nd ed. New York: McGraw-Hill, 1961.
8. Young, W.C., R.C. Budynas, and A.M. Sedegh. *Roark's Formulas for Stress and Strain*. 7th ed. New York: McGraw-Hill, 2011.
9. Shigley, J.E., C.R. Mischke, and Brown, T.H., eds. *Standard Handbook of Machine Design*. New York: McGraw-Hill, 1986.

10. *Manual of Steel Construction.* 14th ed. New York: American Institute of Steel Construction. 2011.
11. *Specifications and Guidelines for Aluminum Structures.* Washington, DC: Aluminum Association, 2000.
12. *National Design Specifications for Wood Construction and Design Values for Wood Construction, NDS Supplements.* Washington, DC: American Wood Council, 2012.
13. Ugural, A.C. *Stresses in Beams, Plates and Shells.* 3rd ed. Boca Raton, FL: CRC Press, 2009.

Chapter 6

1. Nadai, A. *Theory of Flow and Fracture of Solids,* vol .1, 2nd ed. New York: McGraw-Hill. 1950.
2. Marin, J. *Mechanical Behavior of Engineering Materials.* Upper Saddle River, NJ: Prentice Hall. 1962.
3. Dowling, N.E. *Mechanical Behavior of Materials,* 4th ed. Upper Saddle River, NJ: Prentice Hall. 2012.
4. American Society of Metals. *Metals Handbook.* Boca Raton, FL: CRC Press, 1998.
5. Irwin, G.R. Fracture Mechanics. In *Proceedings, First Symposium on Naval Structural Mechanics,* Pergamon, New York, 1960, p. 557.
6. Boresi, A.P. and R.J. Schmidt. *Advanced Mechanics of Materials,* 6th ed. New York: Wiley. 2003.
7. Meguid, S.A. *Engineering Fracture Mechanics.* London, U.K.: Elsevier, 1989.
8. Irwin, G.R. Mechanical Testing. *Metals Handbook,* 9th ed., vol. 8. Metals Park, OH: American Society for Metals (ASM), 1989, pp. 437–493.
9. *ASM International Guide in Selecting Engineering Materials.* Metals Park, OH: American Society for Metals, 1989.
10. Fauppel, J.H. and F.E. Fisher. *Engineering Design,* 2nd ed. New York: Wiley. 1981.
11. Ugural, A.C. and S.K. Fenster. *Advanced Mechanics of Materials and Applied Elasticity,* 5th ed. Upper Saddle River, NJ: Prentice Hall, 2012.
12. Juvinall, R.C. and K.M. Marshek. *Fundamentals of Machine Component Design,* 5th ed. Hoboken, NJ: Wiley, 2011.
13. Kennedy, J.B. and A.M. Neville. *Basic Statistical Methods for Engineers and Scientists,* 3rd ed. New York: Harper and Row, 1986.
14. Burr, A.H. and J.B. Cheatham. *Mechanical Analysis and Design,* 5th ed. Upper Saddle River, NJ: Prentice Hall, 1995.

Chapter 7

1. American Society of Metals. Failure analysis and prevention. *Metals Handbook,* 9th ed. Metals Park, OH: ASM International, 1986.
2. Engel, L. and H. Klingele. *An Atlas of Metal Damage.* Munich, Germany: Hanser Verlag, 1981.
3. Marin, J. *Mechanical Behavior of Engineering Materials.* Upper Saddle River, NJ: Prentice Hall, 1962.
4. Dowling, N.E. *Mechanical Behavior of Materials,* 4th ed. Upper Saddle River, NJ: Prentice Hall, 2012.
5. Rice, R.C., ed. *Fatigue Design Handbook,* 3rd ed. Warrendale, PA: Society of Automotive Engineers, 1997.
6. Stevens, R.I. and H.O. Fuchs et al. *Metal Fatigue in Engineering.* 2nd ed. Hoboken, NJ: Wiley, 2000.
7. Bannantine, J.A., J.J. Comer, and J.L. Handrock. *Fundamentals of Metal Fatigue Analysis.* Upper Saddle River, NJ: Prentice Hall, 1990.
8. Burr, A.H. and J.B. Cheatham. *Mechanical Analysis and Design,* 2nd ed. Upper Saddle River, NJ: Prentice Hall, 1995.

9. Norton, R.L. *Machine Design: An Integrated Approach,* 5th ed. Upper Saddle River, NJ: Prentice Hall, 2013.
10. Budynas, R. and K. Nisbett, *Shigley's Mechanical Engineering Design,* 9th ed. New York: McGraw-Hill, 2010.
11. Juvinall, R.C. and K.M. Marshek. *Fundamentals of Machine Component Design,* 5th ed. Hoboken, NJ: Wiley, 2011.
12. Sines, G. and J.L. Waisman, eds. *Metal Fatigue.* New York: McGraw-Hill. 1959, pp. 296–298.
13. Juvinall, R.C. *Engineering Consideration of Stress, Strain, and Strength.* New York: McGraw-Hill, 1967.
14. Ugural, A.C. and S.K. Fenster. *Advanced Mechanics of Materials and Applied Elasticity,* 5th ed. Upper Saddle River, NJ: Prentice Hall, 2012.
15. Sullivan, J.L. Fatigue life under combined stress. *Machine Design.* January 25, 1979.
16. Paris, P.C. and F. Erdogan. A critical analysis of crack propagation laws. *Transactions of the ASME. Journal of Basic Engineering,* 85(1963), 528.
17. Barsom, J.M. and S.T. Rolfe. *Fracture and Fatigue Control in Structures,* 3rd ed. Oxford, U.K.: Butterworth Heinemann, 2010.
18. Stulen, F.B., H.N. Cummings and W.C. Schulte. Preventing fatigue failures—Part 5. *Machine Design,* 33, 116–120. November 13, 1961.
19. Suresh, S. *Fatigue of Materials.* Cambridge, 2nd ed. U.K.: Cambridge University Press, 1998.

Chapter 8

1. Davis, J.R. ed. *Surface Engineering for Corrosion and Wear Resistance*. Materials Park, OH: ASM International, 2001.
2. Cotell, C.M., J.A. Sprague, and F.A. Smidt. *Surface Engineering,* Vol. 5, *ASM Handbook.* Materials Park, OH: ASM International, 2007.
3. Ashby, M.F. *Materials Selection in Mechanical Design,* 4th ed. Oxford, U.K.: Butterworth Heinemann. 2011.
4. Horger, O.J. ed. *ASME Handbook: Metals Engineering-Design.* 2nd ed. New York: McGraw-Hill, 1965.
5. Fontana, M.G. and N.D. Greene, *Corrosion Engineering,* 3rd ed. New York: McGraw-Hill, 1990.
6. Lipson, C. *Wear Considerations in Design.* Upper Saddle River, NJ: Prentice-Hall, 1967.
7. Avallone, E., T. Beaumeister, and A.M. Sadegh, eds. *Mark's Mechanical Engineers Handbook,* 11th ed. New York: McGraw-Hill, 2006.
8. Peterson, M.B. and W.O. Winer, eds. *Wear Control Handbook.* New York: The American Society of Mechanical Engineers, 1980.
9. Robinowicz, E. *Wear Coefficients-Metals.* New York: American Society of Mechanical Engineers, 1980, pp. 475–506.
10. Juvinall, R.C. and K.M. Marshek. *Fundamentals of Machine Component Design,* 5th ed. Hoboken, NJ: Wiley, 2011.
11. Ugural, A.C. and S.K. Fenster. *Advanced Mechanics of Materials and Applied Elasticity.* 5th ed. Upper Saddle River, NJ: Prentice Hall, 2012.
12. Timoshenko, S.P. and J.N. Goodier. *Theory of Elasticity,* 3rd ed. New York: McGraw-Hill, 1970.
13. Young, W.C., R.C. Budynas, and A.M. Sadegh. *Roark's Formulas for Stress and Strain,* 8th ed. New York: McGraw-Hill, 2011.
14. Budynas, R.G. and J.K. Nisbett. *Shigley's Mechanical Engineering Design,* 9th ed. New York: McGraw-Hill, 2010.
15. Burr, A.H. and J.B. Cheatham. *Mechanical Analysis and Design.* 2nd ed. Upper Saddle River, NJ: Prentice Hall, 1995.

Chapter 9

1. Borchardt, H.A. A comprehensive method for designing shafts to insure adequate fatigue life. *Machine Design*, April 25, 1963.
2. Design of Transmission Shafting. ANSI/ASME B106.1M-1985.
3. Loewenthal, S.H. Proposed design procedure for transmission shafting under fatigue loading. Technical Note TM-78927, NASA, Cleveland, OH, 1978.
4. Juvinall, R.C. and K.M. Marshek. *Fundamentals of Machine Component Design*. 5th ed. Hoboken, NJ: Wiley, 2011.
5. Burr, A.H. and J.B. Cheatham. *Mechanical Analysis and Design*. 2nd ed. Upper Saddle River, NJ: Prentice Hall, 1995.
6. Norton, R.L. *Design of Machinery*. 3rd ed. New York: McGraw-Hill, 2003.
7. Timoshenko, S., D.H. Young, and W. Weaver, Jr. *Vibration Problems in Engineering*. 5th ed. Hoboken, NJ: Wiley, 1990.
8. Huston, R. and J. Harold. *Practical Stress Analysis in Engineering Design*, 3rd ed. Boca Raton, FL: CRC Press, 2004.
9. Pilkey, D.F. and W.D. Pilkey. *Peterson's Stress Concentration Factors*, 3rd ed. Hoboken, NJ: Wiley, 2005.
10. Avollone, E.A., T. Baumeister III, and A.M. Sadegh, eds. *Mark's Standard Handbook for Mechanical Engineers*. 11th ed. New York: McGraw-Hill, 2006.

Chapter 10

1. Booser, E.R., ed. *CRC Handbook of Lubrication and Tribology*, 2nd ed. Vols. I and II. Boca Raton, FL: CRC Press, 2006, 1983 and 1988.
2. Lansdown, A.R. *Lubrication and Lubrication Selection*, 3rd ed. New York: ASME, 2004.
3. Hamrock, B.J. et al. *Fundamentals of Fluid Film Lubrication*, 2nd ed. Boca Raton, FL: CRC Press, 2004.
4. Szeri, A.Z. *Tribology: Friction, Lubrication, and Wear*. New York: McGraw-Hill, 1980.
5. Avallone, A.E., T. Beaumeister III, and A.M. Sadegh, eds. *Mark's Standard Handbook for Mechanical Engineers*. 11th ed. New York: McGraw-Hill, 2006.
6. Reynolds, O. On the theory of lubrication and its application to Mr. Beauchamp Tower's experiments. *Philosophical Transactions of the Royal Society (London)*, 177(1886), 157–234.
7. Ocvirk, F.W. Short bearing approximation for full journal bearings. Technical Note 2208. Washington, DC: NACA. 1952; Also see Dubois, G.B. and F.W. Ocvirk. The short bearing approximation for plain journal bearings. *Transactions of the ASME*, 77(1955), 1173–1178; Ocvirk, F.W. and G.B. Dubois. Surface finish and clearance effects on journal-bearing load capacity and friction. *Transactions of the ASME Journal of Basic Engineering*, 81(1959), 245.
8. Raimondi, A.A. and J. Boyd. A solution for finite journal bearings and its application to analysis and design. Parts I, II, and III. *Transactions of the ASLE* I, 1(1958), 159–209; Reprinted in *Lubrication Science and Technology*. New York: Pergamon Press. 1958.
9. Juvinall, R.E. and M. Marshek. *Fundamentals of Machine Component Design*. 5th ed. Hoboken, NJ: Wiley, 2011.
10. Budynas, R. and K. Nisbett. *Shigley's Mechanical Engineering Design*. 9th ed. New York: McGraw-Hill, 2010.
11. Rothbart, H.A. and T.H. Brown, Jr., eds. *Mechanical Design Handbook*. 2nd ed. New York: McGraw-Hill, 2006.
12. Burr, A.H. and J.B. Cheatham. *Mechanical Analysis and Design*. 2nd ed. Upper Saddle River, NJ: Prentice Hall, 1995.
13. Fuller, D.D. *Theory and Practice of Lubrication for Engineers*. 2nd ed. Hoboken, NJ: Wiley, 1984.
14. O'Conner, J.J. and J. Boyd. *Standard Handbook of Lubrication Engineering*. New York: McGraw-Hill, 1968.

15. Hamrock, J. and D. Dowson. *Ball Bearing Lubrication: The Elastohydrodynamics of Elliptical Contacts*. Hoboken, NJ: Wiley, 1981.
16. Standards of the Anti-friction Bearing Manufacturing Association (AFBMA). New York, 1990.
17. Ball bearing general catalog. Sandusky, OH: New Departure-Hyatt Bearing Division, General Motors Corporation.
18. *Timken Engineering Journal*. Canton, OH: The Timken Roller Bearing Company.
19. Bamberger, E.N. et al. *Life Adjustment Factors for Ball and Roller Bearings: An Engineering Design Guide*. New York: ASME, 1971.
20. Harris, T.A. and M.N. Kotzalas. *Rolling Bearing Analysis*. 5th ed. Boca Raton, FL: CRC Press, 2006.

Chapter 11

1. AGMA. *Standards of the American Gear Manufacturers Association*, Alexandria, VA: ANSI/AGMA 2001–C95 (revised AGMA 2001–C95).
2. Townsend, D.P., ed. *Dudley's Gear Handbook*. 2nd ed. New York: McGraw-Hill, 1992.
3. Drago, R.J. *Fundamentals of Gear Design*. Boston, MA: Butterworth-Heinemann Publishers, 1988.
4. Avallone, E.A., T. Baumeister III, and A.M. Sadegh, eds. *Marks' Standard Handbook for Mechanical Engineers*. 11th ed. New York: McGraw-Hill, 2006.
5. Buckingham, E. *Analytical Mechanics of Gears*. Mineola, New York: Dover, 2014.
6. Juvinall, R.C. and K.M. Marshek. *Fundamentals of Machine Component Design*. 5th ed. Hoboken, NJ: Wiley, 2011.
7. Mabie, H.H. and C.F. Reinholtz. *Mechanisms and Dynamics of Machinery*. 4th ed. Hoboken, NJ: Wiley, 1987.
8. Norton, R.L. *Machine Design: An Integrated Approach*. 5th ed. Upper Saddle River, NJ: Prentice Hall, 2013.
9. Lewis, W. *Investigation of the Strength of Gear Teeth*. Philadelphia, PA: Proceedings of the Engineers Club, 1893, pp. 16–23; Reprinted in *Gear Technology* 9(6) (November–December 1992), 19.
10. Pilkey, D.F. and W.D. Pilkey. *Peterson's Stress Concentration Factors*, 9th ed. Hoboken, NJ: Wiley, 2008.
11. Shigley, E.J. and C.E. Mischke, eds. *Standard Handbook of Machine Design*, 3rd ed. New York: McGraw-Hill, 2004.
12. Burr, A.H. and J.B. Cheatham, *Mechanical Analysis and Design*. 2nd ed. Upper Saddle River, NJ: Prentice Hall, 1995.
13. Dudley, D.W. Gear wear. In M.B. Peterson and W.O. Winer, eds., *Wear Control Handbook*. New York: ASME, 1980, p. 764.
14. Stephen, P.R. and D.W. Dudley. *Dudley's Handbook of Practical Gear Design and Manufacture*, 2nd ed. Boca Raton: CRC Press, 1994.

Chapter 12

1. Stephen, P.R. and D.W. Dudley. *Dudley's Handbook of Practical Gear Design and Manufacture*, 2nd ed. Boca Raton, FL: CRC Press, 1994.
2. Townsend, D.P. ed. *Dudley's Gear Handbook*. 2nd ed. New York: McGraw-Hill. 1992.
3. Buckingham, E. and H.H. Ryffel. *Design of Worm and Spiral Gears*. New York: Industrial Press, 1999.
4. 1979 Mechanical drives reference issue. *Machine Design*. June 29, 1979.
5. American Manufacturers Association. *Standards of the American Manufacturers Association*. Arlington, VA: AMA. 1993.
6. Avallone, E.A., T. Baumeister III, and A.M. Sadegh, eds. *Mark's Standard Handbook for Mechanical Engineers*. 11th ed. New York: McGraw-Hill, 2006.
7. Budynas, R. and K. Nisbett. *Shigtey's Mechanical Engineering Design*, 9th ed. New York: McGraw- Hill, 2011.

8. Norton, R.L. *Machine Design: An Integrated Approach.* 5th ed. Upper Saddle River, NJ: Prentice Hall, 2013.
9. Juvinall, R.C. and K.M. Marshek. *Fundamentals of Machine Component Design.* 5th ed. Hoboken, NJ: Wiley, 2011.
10. Rothbart, H.A. and T.H. Brown, Jr., eds. *Mechanical Design and Systems Handbook.* 2nd ed. New York: McGraw-Hill, 1985.

Chapter 13

1. Budynas, R. and K. Nisbett. *Shigley's Mechanical Engineering Design*, 9th ed. New York: McGraw-Hill, 2010.
2. Wallin, A.W. Efficiency of synchronous belts and V-belts. In *Proceedings of the National Conference Power Transmission*, vol. 5, Illinois Institute of Technology, Chicago, IL, November 7–9, 1978, pp. 265–271.
3. Alciatore, D.G. and A.E. Traver. Multiple belt drive mechanics: Creep theory vs. shear theory. *Transactions of the ASME. Journal of Mechanisms, Transmission, and Automation in Design*, 112(1990), 65–70.
4. Burr, A.H. and J.B. Cheatham. *Mechanical Analysis and Design.* 2nd ed. Upper Saddle River, NJ: Prentice Hall, 1995.
5. Gates Rubber Co. *V-Belt Drive Design Manual.* Denver, CO: Gates Rubber Co., 1999.
6 Rubber Manufacturers Association. *Specifications for Drives Using Classical Multiple V Belts.* American National Standard, IP-20. Washington, DC: Rubber Manufacturers Association, 2010.
7. Erickson, W.D. *Belt Selection and Application for Engineers.* New York: Marcel Dekker, 1987.
8. American Chain Association (ACA), *Standard Handbook of Chains: Chains for Power Transmission and Material Handling*, 2nd ed. Boca Raton, FL: Taylor & Francis, 2006.
9. Binder, R.C. *Mechanics of Roller Chain Drive.* Upper Saddle River, NJ: Prentice Hall, 1956.
10. ANSI/ASME. *Precision Power Transmission Roller Chains, Attachments, and Sprockets.* ANSI/ASME Standard B29.1M. New York: ASME. 1993.
11. ANSI/ASME. *Inverted-Tooth (Silent) Chains and Sprockets.* ANSI/ASME Standard B29.2M-82. New York: American Society of Mechanical Engineers. 1993.
12. Orthwein, W.C. *Clutches and Brakes: Design and Selection.* Boca Raton, FL: CRC Press, 2004.
13. Remling, J. *Brakes.* Hoboken, NJ: Wiley, 1978.
14. Neale, M.J. ed. *Tribology Handbook*, 2nd ed. Cambridge, UK: Elsevier, 1996.
15. Crouse, W.H. Automotive brakes. In *Automotive Chasis and Body.* 5th ed. New York: McGraw-Hill, 1976.

Chapter 14

1. Associated Spring-Barnes Group. *Design Handbook.* Bristol, CN: Associated Spring-Barnes Group. 1987.
2. Wahl, A.M. *Mechanical Springs.* 2nd ed. New York: McGraw-Hill, 1991.
3. Shigley, J.E. and T.H. Brown. *Standard Handbook of Machine Design.* 2nd ed. New York: McGraw-Hill, 1996.
4. Carlson, H.C.R. Selection and application of spring materials. *Mechanical Engineering* 78 (1956), 331–334.
5. Carlson, H.C.R. *Spring Designer's Handbook.* New York: Marcel Dekker, 1978.
6. Rothbart, H.A. and T.H. Brown, Jr., eds. *Mechanical Design and Systems Handbook.* 2nd ed. New York: McGraw-Hill, 2006.
7. Avallone, E.A., T. Baumeister III, and A.H. Sadegh, eds. *Mark's Standard Handbook for Mechanical Engineers.* 11th ed. New York: McGraw-Hill, 2006.
8. Samonov, C. Computer-aided design of helical compression springs. ASME Paper No. 80-DET-69, 1980.

9. Dietrich, A. Home computers aid spring design. *Design Engineering* (June 1981), 31–35.
10. Juvinall, R.C. and K.M. Marshak. *Fundamentals of Machine Component Design*, 5th ed. Hoboken, NJ: Wiley, 2011.
11. Ugural, A.C. *Stresses in Beams, Plates and Shells*. 3rd ed. Boca Raton, FL: CRC Press, 2009.
12. Norton, R.L. *Machine Design: An Integrated Approach*, 5th ed. Upper Saddle River, NJ: Prentice Hall, 2013.

Chapter 15

1. *ANSI/ASME Standards*, B1.1–2014, B1.13–2005. New York: American Standards Institute, 2005.
2. James, F.D. et al., eds. *Machinery's Handbook*. 23rd ed. New York: Industrial Press, 1974.
3. Burr, A.H. and J.B. Cheatham. *Mechanical Analysis and Design*. 2nd ed. Upper Saddle River, NJ: Prentice Hall, 1995.
4. ANSI B5.48. *Ball Screws*. New York: ASME, 1987.
5. Parmley, R.O. *Standard Handbook of Fastening and Joining*, 3rd ed. New York: McGraw-Hill, 1996.
6. Avallone, E.A., T. Baumeister III, and A.M. Sadegh, eds. *Mark's Standard Handbook for Mechanical Engineers*. 11th ed. New York: McGraw-Hill, 2006.
7. Kulak, G.I., J.W. Fisher, and H.A. Struik. *Guide to Design Criteria for Bolted and Riveted Joints*. 2nd ed. New York: Wiley, 1987.
8. Bickford, J.H. *An Introduction to the Design and Behavior of Bolted Joints*, 3rd ed. New York: Marcel Dekker, 1995.
9. Pilkey, D.F. and W.D. Pilkey. *Peterson's Stress Concentration Factors*, 8th ed. Hoboken, NJ: Wiley, 2008.
10. Fauppel, J.H. and F.E. Fisher. *Engineering Design*. 2nd ed. Hoboken, NJ: Wiley, 1981.
11. Budynas, R. and K. Nisbett. *Shigley's Mechanical Engineering Design*, 9th ed. New York: McGraw-Hill, 2011.
12. Juvinall, R.C. and K.M. Marshak. *Fundamentals of Machine Component Design*. 5th ed. Hoboken, NJ: Wiley, 2011.
13. Wileman, J., M. Choundury, and I. Green. Computational stiffness in bolted connections. *Transactions of the ASME, Journal of Mechanical Design* 113 (December 1991), 432–437.
14. American Institute of Steel Construction. *Manual of Steel Construction*. New York: AISC, 2011.
15. Johnston, B.G. and F.J. Lin. *Basic Steel Design*. Upper Saddle River, NJ: Prentice Hall, 1995.
16. *American Welding Society Code AWSD.1.77*. Miami, FL: American Welding Society.
17. Osgood, C.C. *Fatigue Design*, 2nd ed. Hoboken, NJ: Wiley, 1982.
18. Pocius, A.V. *Adhesion and Adhesives Technology:* An *Introduction*, 2nd ed. New York: Hanser, 2004.
19. Brinson, H.F. ed. *Engineering Materials Handbook*. vol. 3. *Adhesives and Sealants*. Metals Park, OH: ASM International. 1990.
20. Parmley, R.O. ed. *Standard Handbook for Fastening and Joining*. 3rd ed. New York: McGraw-Hill, 1997.
21. Society of Automotive Engineers Standard J1199, 2011.

Chapter 16

1. Timoshenko, S.P. and J.N. Goodier. *Theory of Elasticity*. 3rd ed. New York: McGraw-Hill, 1970.
2. Ugural, A.C. and S.K. Fenster. *Advanced Mechanics of Materials and Applied Elasticity*. 5th ed. Upper saddle River, NJ: Prentice Hall, 2012.
3. Faupel, J.H. and F.E. Fisher. *Engineering Design*. 2nd ed. New York: Wiley, 1981.
4. Ugural, A.C. *Stresses in Beams, Plates and Shells*. 3rd ed. Boca Raton, FL: CRC Press, 2009.
5. Norton, R.L. *Machine Design: An Integrated Approach*. 5th ed. Upper Saddle River, NJ: Prentice Hall, 2013.
6. Ashby, M.F. *Materials Selection in Mechanical Design*. Oxford, U.K.: Butterworth Heinemann, 2011.

7. Ugural, A.C. *Mechanics of Materials*. Hoboken, NJ: Wiley, 2008.
8. Young, W.C., R.C. Budynas, and A.M. Sadegh. *Roark's Formulas for Stress and Strain*, 8th ed. New York: McGraw-Hill, 2011.
9. *Boiler and Pressure Vessel Code*. New York: ASME, 2013.
10. Chuse, R. and S.M. Eber. *Pressure Vessels, the ASME Code Simplified*. 6th ed. New York: McGraw-Hill, 1984.
11. Fiberglass-reinforced Plastic Pressure Vessels. Section X. *ASME Boiler and Pressure Vessel Code*. New York: ASME, 2013.
12. Timoshenko, S.P. and J.M. Gere. *Theory of Elastic Stability*, 2nd ed. New York: McGraw-Hill, 1961.

Chapter 17

1. Yang, T.Y. *Finite Element Structural Analysis*. Upper Saddle River, NJ: Prentice Hall, 1986.
2. Gallagher, R.H. *Finite Element Analysis—Fundamentals*. Upper Saddle River, NJ: Prentice Hall, 1975.
3. Logan, D.L. *A First Course in the Finite Element Method*, 5th ed. Stanford, CT: Cengage Learning, 2011.
4. Ugural, A.C. and S.K. Fenster. *Advanced Mechanics of Materials and Applied Elasticity*. 5th ed. Upper Saddle River, NJ: Prentice Hall, 2012.
5. Boresi, A.P. and R.J. Schmidth. *Advanced Mechanics of Material*. 6th ed. Hoboken, NJ: Wiley, 2002.
6. Ugural, A.C. *Stresses in Beams, Plates and Shells*. 3rd ed. Boca Raton, FL: CRC Press, 2009.
7. Zienkiewitcz, O.C. and R.I. Taylor. *The Finite Element Method*, 6th ed. Oxford, U.K.: Elsevier, 2005.
8. Cook, R.D. et al. *Concepts and Applications of Finite Element Analysis*, 4th ed. Hoboken, NJ: Wiley, 2002.
9. Segerlind, L.J. *Applied Finite Element Analysis*. 2nd ed. Hoboken, NJ: Wiley, 1984.
10. Bathe, K.I. *Finite Element Procedures in Engineering Analysis*. Upper Saddle River, NJ: Prentice Hall, 2007.

Chapter 18

1. *A Case Study in Aircraft Design: The Boeing* 727. Reston, VA: AIAA, 1998.
2. *Products in Action*. Canonsburg, PA: ANSYS, Inc. Available at: www.ansvs.com.
3. Burr, A.H. and J.B. Cheatham, *Mechanical Analysis and Design*, 2nd ed. Upper Saddle River, NJ: Prentice Hall, 1995.
4. Ugural, A.C. *Mechanics of Materials*. Hoboken, NJ: Wiley, 2008.
5. Faupel, J.H. and F.E. Fisher. *Engineering Design*, 2nd ed. Hoboken, NJ: Wiley, 1986.

Index

C

Y

Z